# Biophysical Chemistry

# Biophysical Chemistry

Dagmar Klostermeier • Markus G. Rudolph

CRC Press
Taylor & Francis Group
Boca Raton  London  New York

CRC Press is an imprint of the
Taylor & Francis Group, an **informa** business

CRC Press
Taylor & Francis Group
6000 Broken Sound Parkway NW, Suite 300
Boca Raton, FL 33487-2742

First issued in paperback 2020

Version Date: 20160831

ISBN-13: 978-0-367-57238-9 (pbk)
ISBN-13: 978-1-4822-5223-1 (hbk)

**Library of Congress Cataloging-in-Publication Data**

Names: Klostermeier, Dagmar, author. | Rudolph, Markus G., author. Title: Biophysical chemistry / Dagmar Klostermeier and Markus G. Rudolph. Description: Boca Raton, FL : CRC Press, Taylor & Francis Group, [2017] | Includes bibliographical references and index. Identifiers: LCCN 2016035573| ISBN 9781482252231 (hardback ; alk. paper) | ISBN 1482252236 (hardback ; alk. paper) | ISBN 9781482252255 (e-book) | ISBN 1482252252 (e-book) | ISBN 9781482252262 (e-book) | ISBN 1482252260 (e-book) | ISBN 9781482252248 (e-book) | ISBN 1482252244 (e-book) Subjects: LCSH: Physical biochemistry. | Thermodynamics. | Molecular structure. Classification: LCC QD476.2 .K56 2017 | DDC 572/.43--dc23 LC record available at https://lccn.loc.gov/2016035573

Visit the Taylor & Francis Web site at
http://www.taylorandfrancis.com

and the CRC Press Web site at
http://www.crcpress.com

Printed and bound by CPI Group (UK) Ltd, Croydon, CR0 4YY

# Contents

## PART I — Thermodynamics

## PART II — Kinetics

# PART III — Molecular Structure and Stability

## PART IV — Methods

# Appendix

# Preface

Writing this book, we wanted to present the concepts of physical chemistry and molecular structure that underlie biochemical processes from a biochemist's perspective. The content of the book is equally well accessible to students and scientists with backgrounds in physics, chemistry or biology, but is particularly intended for bachelor and master students of life science curricula. The book will also be useful as an introductory text to more advanced students from other fields, and should serve as a reference for Ph.D. students and young researchers.

The book is organized in four parts: thermodynamics, kinetics, molecular structure and stability, and biophysical methods. To limit theoretical treatments to biologically relevant systems, we left out quantum mechanics and statistical thermodynamics. Each individual chapter is followed by a set of problems that should illustrate the principles treated in the chapter and put them into a biochemical context. The commented solutions to the problems are collected in a separate Solutions Manual. References for the work cited in the text are summarized at the end of each chapter, together with advanced literature that provides additional information and may serve as a starting point for delving into a particular subject.

The first part provides background on thermodynamic principles. Early on in the text, we have tried to illustrate the relevance of these thermodynamic concepts for biochemical systems. Electrochemistry is also treated in this part because it is presented from a thermodynamic point of view.

The second part presents kinetics in a somewhat non-traditional way. We have chosen to start with pre-steady state kinetics. This approach allows us to introduce many basic principles, which we can then build upon in the subsequent treatment of the more traditional steady-state enzymology. We also included a rather complete derivation of integrated rate laws and velocity equations.

In the third part, molecular interactions relevant for biological molecules are reviewed in light of the thermodynamic principles from the first part of the book. The principles of protein and nucleic acid structures are presented, striving for a balance between comprehensiveness and a focus on aspects that are important for life sciences. This part assumes a basic knowledge of the concepts of chemical bonds, orbitals and hybridization.

The fourth part deals with biophysical techniques and their applications to the study of biological macromolecules and their interactions. The selection naturally reflects our own preferences and areas of expertise. We included widely used techniques that should be accessible to most researchers when trying to answer a particular biochemical question. Standard biochemical methods, such as gel electrophoresis and chromatography, have been omitted. For the techniques most frequently used in a biophysical laboratory, a section on "potential pitfalls" may help to avoid common mistakes. Many well-established methods, among them fluorescence microscopy and mass spectrometry, have been developed much further in recent years. We incorporated these novel developments and include specific application examples.

Cross-references within and between all parts of the book serve to emphasize common themes and to highlight recurrent principles. Boxes throughout the main text illustrate the relevance of a particular topic, provide additional information on applications, or show derivations of formulas, but may be skipped without breaking the flow of reading.

The appendix is specifically tailored to the requirements of the four main Parts. We only list constants and units discussed in the text, and restrict the treatment of mathematical concepts to operations that are used in the book. The appendix is meant to be a convenient reference, not a substitute for a thorough training in mathematics.

We hope that this book is an enjoyable read, provides inspiration, and motivates students and researchers alike to dive deeper into biophysical chemistry.

**Dagmar Klostermeier**
**Markus G. Rudolph**
*Muenster*

# Acknowledgments

We are grateful to our colleagues and friends who have taken the time to carefully read large parts of the drafted manuscript: David Banner, Garry Crosson, Hans-Joachim Galla, Walter Huber, Jürgen Köhler, John Ladbury, Erwin Peterman, Jochen Reinstein, Alison Rodger, Paul Rösch, Franz-Xaver Schmid, Bernhard Schmidt, Joachim Seelig, Heinz-Jürgen Steinhoff, Dmitri Svergun, and Ian Wilson. Your constructive comments on how to improve the text, both scientifically and educationally, are much appreciated. A special "Thank You" goes to David Banner for proof-reading of the entire manuscript. Of course, any remaining errors in the book remain the sole responsibility of the authors.

We also thank our editors at CRC Press, Francesca McGowan and Emily Wells, for their commitment and constructive support, and for swiftly answering any queries that came up during the writing of this book.

# Authors

**Dagmar Klostermeier** studied Biochemistry and holds a PhD in Biochemistry. After her postdoctoral training with David Millar at The Scripps Research Institute in La Jolla, USA, she was a junior research group leader at the University of Bayreuth, Germany, and a Professor for Biophysical Chemistry at the University of Basel, Switzerland. In 2011, she became Professor for Biophysical Chemistry at the University of Muenster, Germany, and is teaching Physical and Biophysical Chemistry. Her research aims to decipher the mechanisms of ATP-driven molecular machines by biophysical techniques, including single-molecule fluorescence microscopy.

**Markus G. Rudolph** studied Biochemistry and holds a PhD in Biochemistry. He completed his postdoctoral training in Structural Biology in the laboratory of Ian A. Wilson at The Scripps Research Institute in La Jolla, USA, before leading his own research group at the University of Goettingen, Germany. In 2006, he moved to Hoffmann-La Roche Ltd. in Basel, Switzerland, as a researcher in Structural Biology and Biophysics. His main research interests are the structural biology and catalytic mechanisms of enzymes, and structure-based drug design.

# Part I
# Thermodynamics

The field of thermodynamics (gr. *thermos*: heat, *dynamis*: force) is centered around energy in all forms. It refers to the analysis of the available energy, the distribution of energy, and the conversion of energy from one form into the other. Thermodynamic considerations are invaluable for the understanding of the importance of energy in biological systems, its storage and availability, and the energetic costs involved in maintaining cellular functions and stability. The thermodynamic framework is based on four laws, simple rules that have been derived empirically from observing the real world. These laws and concepts allow a quantitative description of complex biological systems, and enable us to understand and, more importantly, to predict their behavior in specific circumstances. Before we begin to construct the building of thermodynamics, however, we will have to start with a set of definitions.

# Systems and Their Surroundings

A *system* is the part of the universe we are currently focusing on, and that we are attempting to describe quantitatively. Systems are separated from the rest of the universe by walls that allow them to interact with their *surroundings*. By surroundings we refer to that part of the rest of the universe that interacts with the system (Figure 1.1). An *isolated system* is separated from its surroundings by walls that do not allow exchange of heat or matter. A *closed system* exchanges heat with its surroundings, but does not exchange matter, and an *open system* can exchange heat and matter with its surroundings. Walls that allow heat exchange are called *diathermic*, whereas walls that do not allow the exchange of heat are called *adiabatic*.

These definitions imply that all living organisms are open systems: they exchange heat and matter with their surroundings. Open systems are not so easy to describe quantitatively, however. We will therefore start with isolated and closed systems in order to illustrate the underlying concepts, before we move to open systems (Section 2.6).

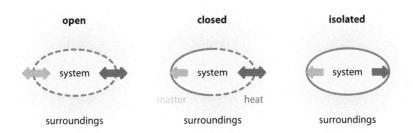

FIGURE 1.1: **Systems and surroundings: open, closed, and isolated systems.** A system is the part of the universe we want to understand (pale yellow). It is in contact with its surroundings (blue). System and surroundings are separated by walls. Open systems exchange heat (red) and matter (green) with their surroundings; closed systems exchange heat, but not matter; and isolated systems exchange neither heat nor matter with their surroundings. Open and closed systems are separated from their surroundings by diathermic (heat-permeable) walls, isolated systems are surrounded by adiabatic (thermally insulated) walls.

Energy exchanged between a system and its surroundings is always counted from the perspective of the system. A negative value means energy flows from the system into the surroundings: the energy of the system is reduced. A positive value means that energy flows into the system, and the energy of the system is increased (Section 2.2.2).

## QUESTIONS

1.1 What are open, closed or isolated systems? (1) a protein in solution in a sealed test tube, (2) the lungs, (3) a bird, (4) a lake, (5) a biochemical reaction in a reaction tube, (6) coffee in a thermos flask, (7) the sample chamber of a differential scanning calorimeter, (8) the sample chamber of an isothermal titration calorimeter, (9) a cuvette in a photometer (no lid).

1.2 What is the sign of the transferred energy from the perspective of the system and the surroundings for
(1) a swimmer in a pool with cold water?
(2) a heated house?
(3) a coffee cup?
(4) an ice cold drink in a warm bar?
(5) a cold hand shaking a warm hand?

1.3 Is the earth an isolated, closed, or open system?

# State Functions and the Laws of Thermodynamics

N ow that we have defined different types of systems, we can start to dive into the laws of thermodynamics. Firstly, we have to familiarize ourselves with the properties of state functions. *State functions*, as their name suggests, describe the state of a system as a function of individual *state variables* (sometimes referred to as state parameters or also state function). We will see in the following that state functions are important and very convenient tools to describe changes in the state of a system.

## 2.1 GENERAL CONSIDERATIONS: STATE VARIABLES AND STATE FUNCTIONS

The physical state of matter is described by a number of parameters, termed *state variables*. For example, the state of a gas is characterized by its volume, temperature, and pressure. The gas volume is delimited by the volume of the container in which the gas is kept. The temperature defines the thermal energy of the gas, and the velocity with which the gas particles move randomly within the container. The pressure is caused by the moving gas particles hitting the container walls. The *ideal gas law* describes the connection between the state variables volume $V$, temperature $T$ and pressure $p$ for an ideal gas and connects them to the amount of substance $n$ (eq. 2.1):

$$pV = nRT \text{ or } p = \frac{nRT}{V} \qquad \text{eq. 2.1}$$

with the pressure $p$ in Pa, the volume $V$ in m³, the amount of substance $n$ in mol, and the absolute temperature $T$ in K. $R$, the general gas constant, is the proportionality constant between the products $pV$ and $nT$, and has a value of $R = 8.3143$ J mol⁻¹ K⁻¹. Note that throughout this book, we use the standard international units (SI units; see Section 28.2). The ideal gas law defines the state of an ideal gas unambiguously: if three out of the four state variables $p$, $V$, $T$, and $n$ are known, the fourth is determined automatically by eq. 2.1. Similarly, any thermodynamic *state function* defines the state of a system unambiguously, without leaving room for interpretation. If n–1 of its n parameters are known, the remaining value can be directly calculated from the state function.

State functions derive their name from the property that they only depend on the current state of a system. As an important consequence of this property, state functions are independent of the pathway by which this state has been reached. This formal definition has very intriguing implications, and we will see later on that this property of state functions simplifies the description of systems tremendously. State functions are called *intensive state functions* when they are independent of the system size (e.g. temperature $T$ or pressure $p$). *Extensive state functions*, on the other hand, depend on the system size (e.g. mass $m$, volume $V$, internal energy $U$, enthalpy $H$, entropy $S$, free energy $G$). An extensive state function can be normalized to the size of the system, e.g. by dividing through the amount of substance, the volume or the mass. The resulting number is independent of the system size, and thus falls under the definition of an intensive state function. Normalized state variables are called *specific* (normalized by mass $m$), *molar* (normalized by amount of substance $n$) or *density* (normalized by volume $V$). Examples are the specific heat capacity $C$, the molar enthalpy $H_m$, molar entropy $S_m$, molar free energy $G_m$ and chemical potential $\mu$, or the density $\rho$.

We have noted before that state functions are independent of the pathway by which a certain state is reached. The highly convenient consequence is that changes in state functions during certain processes can be expressed simply by the differences between their final value and their initial value (Figure 2.1). This concept can be illustrated using the state variable temperature $T$ as an example. When energy is transferred to a system, such that the starting temperature $T_{initial}$ of the system increases to the final value $T_{final}$, the change in temperature, $\Delta T$, can be calculated as the difference between final and initial values of $T$ (eq. 2.2):

$$\Delta T = T_{final} - T_{initial} \qquad \text{eq. 2.2}$$

Here, $\Delta T$ refers to a large, measurable change of the temperature. In general, we refer to large, measureable changes in a state function $X$ by $\Delta X$. For state functions, this change can always be calculated from the final and initial values (eq. 2.3):

$$\Delta X = X_{final} - X_{initial} \qquad \text{eq. 2.3}$$

With d$X$, we refer to a small, infinitesimal change in the state function $X$. In contrast, with $\delta Y$ we refer to a small, infinitesimal change in a function $Y$ that is NOT a state function.

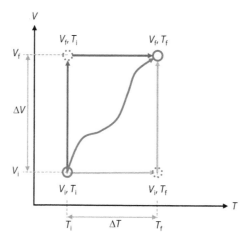

**FIGURE 2.1:**   **Properties of state functions.** A system changes from an initial state with temperature $T_i$ and a volume $V_i$ to a final state with temperature $T_f$ and volume $V_f$. Such a change in state can follow different paths: along the red path, a change from initial to final volume at constant (initial) temperature is followed by a change from the initial to the final temperature at constant (final) volume. Along the yellow path, the temperature changes first from its initial to the final value at constant (initial) volume, and the volume changes from its initial to its final value at constant (final) temperature. The blue path shows a pathway where temperature and volume change continuously throughout the transition. Temperature and volume are state functions, and their change in the transition can be calculated from the final and initial values: $\Delta V = V_f - V_i$ and $\Delta T = T_f - T_i$. $\Delta V$ and $\Delta T$ are therefore identical for all three pathways, and are generally independent of the pathway.

A system is in *equilibrium* when its state and the parameters that describe this state do not further change with time. A system is in *thermal equilibrium*, if its temperature remains constant with time (Figure 2.2). It is important to note that this does not necessarily imply that there is no heat exchange, it just indicates that the heat entering and exiting the system is balanced, and the net temperature change is zero. We will revisit this principle of *dynamic equilibria* when we discuss the kinetics of reversible reactions (Sections 8.1, 11.6.1).

Two systems are in thermal equilibrium when they have the same temperature (that is constant over time). If two systems, A and B, are in thermal equilibrium, and at the same time, the system B is in thermal equilibrium with system C, we can conclude that systems A and C must also be in thermal equilibrium. This statement is the $0^{th}$ law of thermodynamics: if out of three systems A, B, and C, systems A and B, and systems B and C are in pairwise equilibrium, then systems A and C are also in thermal equilibrium (Figure 2.2). This is a recurrent principle that also holds for chemical equilibria, and has important implications for coupled chemical reactions in biology (Section 3.4).

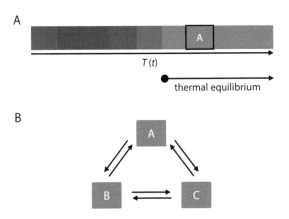

**FIGURE 2.2:**  **Thermal equilibrium and the $0^{th}$ law of thermodynamics**. A: A system is in thermal equilibrium when its temperature remains constant over time. If we let a hot system (red) equilibrate with its cooler surroundings, the system will eventually reach a constant temperature (blue) and thermal equilibrium (arrow). B: When system A is in thermal equilibrium with systems B and C, then systems B and C are also in thermal equilibrium. This relationship is the $0^{th}$ law of thermodynamics.

## 2.2 THE INTERNAL ENERGY *U* AND THE FIRST LAW OF THERMODYNAMICS

### 2.2.1 INTERNAL ENERGY, HEAT, AND WORK

Any system contains a certain amount of energy, which is called its *internal energy*. The internal energy $U$ is the sum of all forms of energy that are stored in the system. It can be regarded as the sum of the kinetic energy of all particles in a system plus the sum of their potential energy. Internal energy consists of two major components, *work w* and *heat q*. Heat is the form of energy associated with random movement of particles. In contrast, work is energy associated with ordered movement. Work can be subdivided into mechanical work, electrochemical work, expansion work, and other forms (Figure 2.3). In the following, we will introduce the laws of thermodynamics and define state functions by focusing on expansion work performed during state changes of ideal gases.

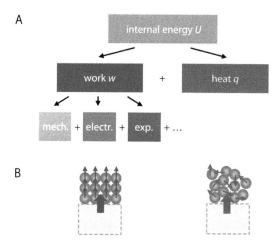

FIGURE 2.3:   **Internal energy, heat, work, and different types of work.** A: The internal energy of a system is divided into heat $q$ and work $w$. Work can be further subdivided into mechanical (mech.), electrical (electr.), expansion (exp.) work and other forms (...). B: Heat is associated with random movement of particles, work is associated with ordered movement.

### 2.2.2 THE FIRST LAW OF THERMODYNAMICS

Changes in the internal energy of a system, $dU$, can be brought about by adding or withdrawing heat ($\delta q$), or by performing work on the system or having the system perform work ($\delta w$) on the surroundings:

$$dU = \delta w + \delta q \qquad \text{eq. 2.4}$$

The internal energy of a system increases when energy is added, either by performing work on the system or by adding heat (or both). The internal energy of a system decreases if the system performs work and/or transfers heat to its surroundings. The notation in eq. 2.4, using $dU$, but $\delta w$ and $\delta q$, already implies that, in contrast to the state function $U$, $w$ and $q$ are not state functions. Indeed, despite $U$ being a state function, the distribution of changes in internal energy to work and heat depends on the pathway of the respective process, and hence $w$ and $q$ are not state functions. Work is a type of energy associated with directional movement, whereas heat is associated with random movement. We will see later that work and heat, although both are forms of energy, are not equivalent. Work can be converted into heat, but heat cannot be completely converted into work (Section 2.4.2).

Changes in work and heat are always formulated from the perspective of the system (Figure 2.4): $\delta w > 0$ means work is performed on the system, leading to an increase in its internal energy; $\delta w < 0$ means that the system performs work on the surroundings, leading to a decrease in internal energy of the system. Correspondingly, a change in heat $\delta q > 0$ means that energy in form of heat is added to a system, whereas $\delta q < 0$ means that heat is removed from the system, leading to an increase or decrease of the internal energy, respectively (eq. 2.4).

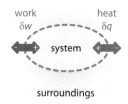

FIGURE 2.4:   **System-centered sign convention of heat and work.** Accounting of energy is always done from the perspective of the system. Heat transferred to the system enters the balance with a positive sign ($+\delta q$), heat transferred from the system to its surroundings is accounted for with a negative sign ($-\delta q$). Work performed on the system has a positive sign ($+\delta w$), and work performed by the system on its surroundings has a negative sign ($-\delta w$).

### 2.2.3 THE IDEAL GAS: A CONVENIENT SYSTEM TO UNDERSTAND THERMODYNAMIC PRINCIPLES

We have introduced before the ideal gas law which describes the relation between the state variables pressure, volume, temperature, and amount of substance (eq. 2.1). The ideal gas law tells us that the product of pressure $p$ and volume $V$ increases with the temperature $T$ of the gas (Figure 2.5). At a constant temperature (*isothermal* conditions), the product $pV$ will be constant. This relationship was found experimentally, and is called the *Boyle-Mariotte law*. Curves that describe $p$ as a function of $V$ for a constant temperature $T$ are called *isotherms* (Figure 2.5). If we keep the pressure constant (*isobaric* conditions), then the volume $V$ is proportional to the temperature $T$, a relationship known as *Charles' law* or the *first law of Gay-Lussac*. Plotting $V$ as a function of $T$ for different pressures gives a set of lines that are termed *isobars*. At constant volume (*isochoric* conditions), the pressure $p$ will increase proportionally with the temperature $T$. This relationship is called the *second law of Gay-Lussac*. The curves that describe $p$ as a function of $T$ for a constant volume are called *isochors*. Finally, the proportional relationship of the volume $V$ to the amount of substance $n$ at constant temperature and pressure (isothermal and isobaric conditions) is *Avogadro's law*. The ideal gas law combines the individual pairwise relationships between temperature, pressure, and volume, and the amount of substance, and introduces the proportionality constant between $pV$ and $nT$, the general gas constant $R$.

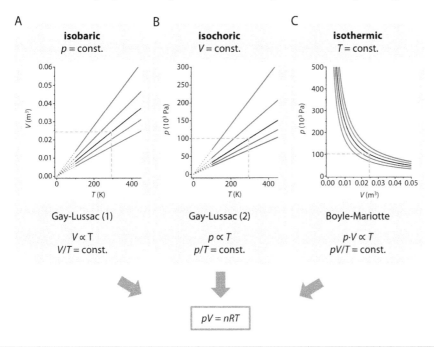

**FIGURE 2.5:** **The ideal gas law and its partial laws**. A: The first law of Gay-Lussac states the proportionality of the volume $V$ and the temperature $T$ of an ideal gas under isobaric conditions ($p$ = const., increasing pressure from red through black to blue). The broken lines indicate the extrapolation to the absolute temperature minimum. B: The second law of Gay-Lussac states the proportionality of the volume $V$ and the temperature $T$ of an ideal gas under isochoric conditions ($V$ = const.; decreasing volume from red through black to blue). The broken lines indicate the extrapolation to the absolute temperature minimum. C: The law of Boyle and Mariotte states the proportionality of the product of the pressure $p$ and the volume $V$ and the temperature $T$ of an ideal gas under isothermal conditions ($T$ = const.). All three individual laws are integrated into the ideal gas equation. The ideal gas law also contains Avogadro's law: the same number of particles (i.e. amount of substance $n$) of a gas takes up equal volumes at constant pressure $p$ and temperature $T$.

The proportionality of $V$ and $T$ has an important consequence because it defines a lower limit for the temperature scale. The volume $V$ reaches zero at a certain temperature, where all isobars intersect (Figure 2.5). The volume is always positive ($V > 0$), and lower temperatures are therefore not possible. This temperature is thus the absolute zero of the thermodynamic temperature scale ($T = 0$ K).

Because of the finite volume of atoms and molecules, a volume of zero cannot be reached, and the absolute minimum in temperature also cannot be reached. We will revisit this statement in the context of the third law of thermodynamics (Section 2.4.4).

The ideal gas law, and particularly the Boyle-Mariotte law, has important implications for scuba divers (Box 2.1). Many fish exploit the properties of gases and use gas reservoirs to control their buoyancy (Box 2.2).

---

### BOX 2.1: UNDER PRESSURE: IMPLICATIONS OF THE IDEAL GAS LAW FOR DIVING.

The mean air pressure at sea level fluctuates around $1.013 \cdot 10^5$ Pa (1.013 bar). The air pressure at sea level is caused by the air layers above. Water has a higher density than air, and under water we therefore experience a higher pressure than above the surface. At a depth of 10 m, the water column above us leads to an increase in pressure by another $1.013 \cdot 10^5$ Pa, and every additional ten meters will add approximately $1.013 \cdot 10^5$ Pa. Thus, we experience twice the normal pressure at 10 m (once from the air above us, once from the water), three times the normal pressure at 20 m, four times the normal pressure at 30 m, and so on. For physiological processes, the increased pressure is not a problem: most of our body is water, and water is almost incompressible. The most critical organ is our lung, a large air-filled reservoir. Following the Boyle-Mariotte law ($pV$ = const.), the air in our lungs is compressed while we descend during our dive and the external pressure increases, and expands upon ascent with decreasing pressure. As long as we take a deep breath to fill our lungs, then dive while holding our breath, and ascend again, the compression of air in our lungs is not problematic: we start our dive with full lungs, the air becomes compressed during the dive, and expands to its original volume when we reach the surface. The problems start when we extend our under-water time by breathing compressed air. Scuba divers carry a supply of compressed air to be able to breath under water. The air in their tanks is pressurized to typically $2 \cdot 10^7$ Pa (200 bar), and is down-regulated to environmental pressure for breathing. With every breath under water, divers fill their lungs with air at the pressure experienced at that particular depth. Breathing 20 meters below the surface means breathing air of three times the pressure (and a third of the volume) it would have at the surface. When we ascend to finish our dive, the surrounding pressure decreases with every meter of ascent, and the air in our lungs expands. If we have filled our lungs at 20 m and ascend without breathing, the volume will increase three times, which is far beyond the capacity of our lungs. The lung tissue is not able to accommodate this volume, and will eventually rupture, which leads to a life-threatening pulmonary barotrauma. To prevent rapid expansion of air in our lungs, we have to ascend slowly to ensure that the air expands slowly, and we have to breathe constantly. Every exhalation removes high-pressure air from the lungs, and every inhalation fills the lungs with air of lower pressure. The pressure of the air in the lungs and the external pressure are thereby equilibrated during the ascent, avoiding pressure differences, air expansion, and lung rupture. Failure to ascend slowly and to constantly release air from the lungs during the ascent can cause fatal lung injuries, and the first lesson scuba divers learn is: never hold your breath – a life-saving lesson from the Boyle-Mariotte law.

## BOX 2.2: WHY FISH DO NOT SINK: THE IDEAL GAS LAW AND BUOYANCY.

Many fish species contain a swim bladder, a gas-filled compartment. These fish can adjust the amount of gas in the swim bladder to control their buoyancy (Figure 2.6). Adding gas to the swim bladder from the bloodstream (or the gut) causes the fish to ascend, removing gas causes the fish to sink. When a fish with neutral buoyancy swims upward, the gas in the swim bladder expands because of the decreasing pressure. To regain neutral buoyancy, gas needs to be absorbed from the bladder. When the fish swims to deeper waters, the gas in the swim bladder is compressed by the increasing pressure, and gas needs to be added to prevent the fish from sinking further. Fish with a swim bladder can remain at constant depth without effort. In contrast, fish without a swim bladder, such as sharks, need to swim constantly in order not to sink.

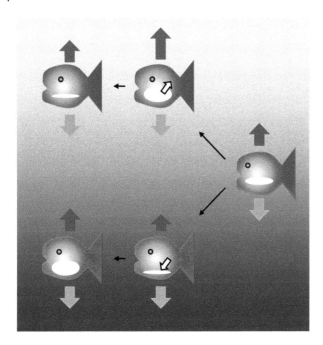

**FIGURE 2.6:** **Buoyancy control with a swim bladder**. The swim bladder (white) is a gas-filled reservoir that allows fish to control their buoyancy. When a fish with neutral buoyancy ascends (top), the air in the swim bladder expands, leading to positive buoyancy. To prevent further ascent and regain neutral buoyancy, the fish needs to decrease the amount of air in the bladder (white arrow). When a fish with neutral buoyancy descends (bottom), the air in the swim bladder is compressed, leading to negative buoyancy. To prevent further descent and regain neutral buoyancy, the fish increases the amount of air in the bladder (white arrow).

The same principle is used by divers who wear an air-filled vest as a buoyancy control device. The device is connected to the compressed air supply of the diver. Increasing the amount of air in the vest causes a diver with neutral buoyancy to ascend or prevents a diver with negative buoyancy from further sinking upon compression of the air at increasing depth. Conversely, releasing air from the vest through a valve makes a diver with neutral buoyancy descend, and prevents an ascending diver from uncontrolled ascent due to expansion of the air with decreasing pressure.

### 2.2.4 CHANGES IN THE STATE OF AN IDEAL GAS

We will now look at state changes of an ideal gas under different conditions, and calculate the changes in work and heat that are associated with these processes.

#### 2.2.4.1 Irreversible Isothermal Expansion and Compression

First, we start with an ideal gas in an initial state characterized by pressure and volume $p_1$, $V_1$ and a certain temperature $T$ (Figure 2.7). The gas is in a cylinder, closed off with a moveable piston. From the starting conditions ($p_1$, $V_1$, $T$), we now perform an *irreversible expansion* to a state with $p_2$, $V_2$, $T$: we <u>instantaneously</u> move the piston upwards, such that the volume the gas can fill is increased. At the external pressure $p_2$, the gas will then expand from the initial volume $V_1$ into the volume $V_2$. We perform this expansion at constant temperature $T$ by coupling the gas to a large heat reservoir. According to the Boyle-Mariotte law, the product $pV$ for the initial state ($p_1$, $V_1$) and the final state ($p_2$, $V_2$) is constant under these isothermal conditions. From the increase in volume, $V_2 > V_1$, we can therefore conclude that the pressure must decrease during the change in state, with $p_2 < p_1$.

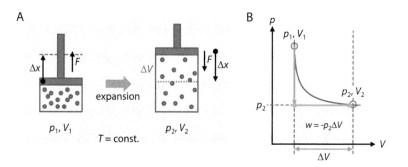

**FIGURE 2.7:    Irreversible expansion.** A: An ideal gas is kept in a container with a moveable piston at pressure $p_1$ and volume $V_1$. The temperature $T$ is maintained constant by coupling the gas to a large heat reservoir (blue). We now rapidly move the piston by $\Delta x$, providing a larger volume for the ideal gas to occupy. The ideal gas will expand into the volume ($V_2 > V_1$). The product of $p$ and $V$ is constant under isothermal conditions, hence $p_2 < p_1$. B: The states $p_1$, $V_1$ and $p_2$, $V_2$ are on the isotherm (red line). The irreversible expansion (orange arrow) does not follow the isotherm, however: the pressure is reduced from $p_1$ to $p_2$ instantaneously (gray arrow), and the gas experiences the external pressure $p_2$ throughout the expansion. The work performed during the expansion is $w = -p_2\Delta V$ (blue area, see text).

We can calculate the associated work with this expansion from the mechanical work $w_{mech}$ that is performed to move the piston to its new position. This work is the product of the force $F$ with which the piston is pushed out and the distance $\Delta x$ by which it moves:

$$w_{mech} = F \cdot \Delta x \qquad \text{eq. 2.5}$$

The force can be expressed as the product of the pressure $p$ on the piston and its area $A$. We have said above that we move the piston instantaneously. As a consequence, the complete change in state is performed under the final pressure $p_2$, and this is the pressure that acts on the piston throughout the expansion of the gas (Figure 2.7). We obtain for the work $w_{mech}$:

$$w_{mech} = p_2 \cdot A \cdot \Delta x \qquad \text{eq. 2.6}$$

By summarizing the product of the area $A$ and $\Delta x$ as the change in volume of the gas, $\Delta V$, the work for moving the piston becomes

$$w_{mech} = p_2 \cdot \Delta V \qquad \text{eq. 2.7}$$

The work $w$ associated with the irreversible expansion of the gas is of the same amount, but has the opposite sign:

$$w = -p_2 \Delta V \qquad \text{eq. 2.8}$$

The negative sign of the work $w$ reflects the system-centered perspective: the system performs work on the surroundings by moving the piston upward, therefore, the work $w$ has to be negative. Conversely, we can calculate the work associated with an irreversible compression from $p_2$, $V_2$ to $p_1$, $V_1$ from the product of the final pressure $p_1$, which is experienced throughout the compression, and the change in volume $\Delta V = V_1 - V_2$. To compress a gas, the piston needs to be pushed down, and we need to perform work on the system. The compression work therefore has a positive sign, and is stored in the system. The positive sign is directly obtained from eq. 2.8 because the volume change $\Delta V$ for the compression is negative, making $w$ positive.

During an irreversible expansion, the gas performs work on the surroundings, which would lead to a decrease in the internal energy of the system. However, if we perform the expansion of the gas under isothermal conditions, maintaining a constant temperature by coupling the system to a (large) heat reservoir such as a giant water bath, heat will be taken up by the gas upon expansion to compensate for the loss in internal energy due to the work performed on the surroundings. Conversely, heat will be released (i.e. taken up by the water bath) upon compression to compensate for the increase in internal energy because of the work performed <u>on</u> the system <u>by</u> the surroundings. The transferred heat $q$ during an irreversible isothermal expansion or compression is thus

$$q = -w = p_2 \, \Delta V \qquad \text{eq. 2.9}$$

Overall, the change in internal energy of the gas is then

$$\mathrm{d}U = q + w = p_2 \, \Delta V - p_2 \, \Delta V = 0 \qquad \text{eq. 2.10}$$

Thus, during an isothermal, irreversible expansion, the internal energy of the gas does not change. This is in agreement with the independence of the internal energy of an ideal gas on $p$ and $V$ (see eq. 2.20).

### 2.2.4.2 Reversible Isothermal Expansion and Compression

We now consider the case that instead of instantaneously moving the piston to the final position in one go, we perform the same isothermal expansion under *reversible* conditions by slowly moving the piston by infinitesimally small increments $\mathrm{d}x$. The volume then also changes in very small, infinitesimal steps $\mathrm{d}V = A \, \mathrm{d}x$. Under these conditions, the gas pressure can equilibrate to the pressure according to the ideal gas law at each point of the expansion (Figure 2.8). We can now calculate the mechanical work on the piston by integrating the associated force with the change in position, $\mathrm{d}x$.

$$w_{\mathrm{mech}} = \int F \, \mathrm{d}x \qquad \text{eq. 2.11}$$

By analogy to the considerations for the irreversible expansion, the expansion work performed by the gas during the reversible expansion is now also described by an integral

$$w = \int_{V_1}^{V_2} -p \, \mathrm{d}V \qquad \text{eq. 2.12}$$

Here, $p$ is the current pressure, which we can substitute for $nRT/V$ from the ideal gas law (eq. 2.1). The integral can then be evaluated as

$$w = \int_{V_1}^{V_2} -p\,dV = -nRT \int_{V_1}^{V_2} \frac{1}{V} dV = -nRT \ln \frac{V_2}{V_1} \qquad \text{eq. 2.13}$$

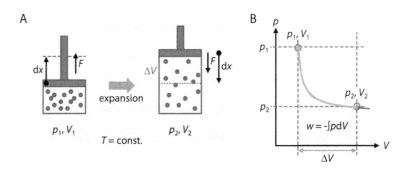

**FIGURE 2.8:  Isothermal expansion under reversible conditions.** To expand an ideal gas reversibly, we move the piston up slowly in small, infinitesimal steps d$x$. At each point of the expansion process, the pressure $p$ of the gas will equilibrate under the new conditions, and the volume will adjust according to the ideal gas law (orange arrow). We can calculate the work from the integral of $-p\,dV$, which corresponds to the blue area under the isotherm (red curve) between initial and final conditions. Note that the work performed by a reversibly expanding gas is larger than the work performed by an irreversibly expanding gas (Figure 2.7).

According to eq. 2.13, the work associated with a reversible isothermal expansion ($V_2 > V_1$, $\ln(V_2/V_1) > 0$) is negative, the system performs work on the surroundings. However, for a reversible isothermal compression from $p_2$, $V_2$ to $p_1$, $V_1$ ($V_1 < V_2$, $\ln(V_1/V_2) < 0$), the work according to eq. 2.13 is positive, work is performed on the system.

The work performed by the system on the surroundings during the reversible expansion would lead to a decrease in internal energy of the system. Again, under isothermal conditions heat is exchanged that is of the same magnitude as the work performed, but of opposite sign:

$$q = -w = nRT \ln \frac{V_2}{V_1} \qquad \text{eq. 2.14}$$

and the internal energy of the gas will remain constant during the expansion:

$$dU = q + w = nRT \ln \frac{V_2}{V_1} - nRT \ln \frac{V_2}{V_1} = 0 \qquad \text{eq. 2.15}$$

### 2.2.4.3  Comparison of Reversible and Irreversible Changes of State

If we now compare the work associated with the irreversible (eq. 2.8) and reversible (eq. 2.13) expansions, we see immediately that the work performed by the system during an irreversible expansion is smaller than the work performed during a reversible expansion (Figure 2.9). This is caused by the constant pressure $p$ during an irreversible expansion where the complete expansion occurs under the lower final pressure $p_2$. An irreversible compression, on the other hand, requires more work than the same compression under reversible conditions because the entire process occurs at the final higher pressure $p_1$ (Figure 2.9).

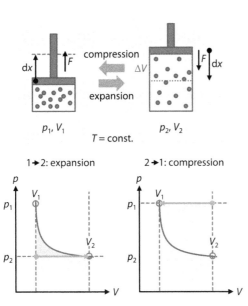

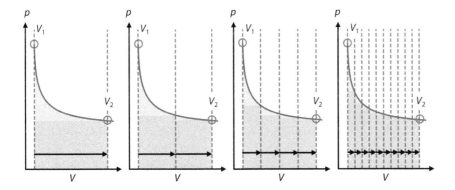

**FIGURE 2.9:** **Work for irreversible and reversible expansion and compression**. A gas expanding under irreversible conditions performs less work (blue area) than a gas that is expanding under reversible conditions (blue and light red area). Compressing a gas under irreversible conditions requires more work (blue area) than under reversible conditions (area under the red curve). The path for irreversible expansion and compression is indicated by the orange arrows. Reversible expansion and compression follow the isotherm (red curve).

Instead of expanding or compressing an ideal gas in one step from its initial to its final volume, we can perform the expansion or compression in several irreversible steps (Figure 2.10). If the numbers of steps is sufficiently large, the overall work for n irreversible steps will become equal to the work under reversible conditions. The reason for this is the smaller and smaller difference between the final pressure for each irreversible step and the pressure *p* under reversible conditions.

**FIGURE 2.10:** **From irreversible to reversible expansion of an ideal gas**. The red curve represents the *p–V* relationship according to the ideal gas law. We can expand an ideal gas from an initial volume $V_1$ to a final volume $V_2$ under irreversible or reversible conditions. The work performed by a reversibly expanding gas (blue and light red areas under the curve) is larger than the work performed by an irreversibly expanding gas (blue area). The excess of work performed by the reversibly expanding gas is highlighted in light red. We can now perform the same change of state from $V_1$ to $V_2$ in several irreversible steps (e.g. 2, 4 or 10 steps). For each individual irreversible expansion step, the work performed by the gas is indicated by a blue rectangle. The excess work performed by a reversibly expanding gas is indicated in light red. With more and more irreversible sub-steps, the difference between work performed under reversible and irreversible conditions becomes smaller and smaller. For an infinite number of steps, each step will be an expansion by an infinitesimal volume increment d*V*, which is the limiting case of an expansion under reversible conditions. The total work performed in an infinite number of irreversible expansion steps (sum of blue areas) will be identical to the work performed under reversible conditions (total area under the curve; blue and light red).

From Figure 2.10, we see that the work performed during a reversible expansion constitutes an upper limit for the work performed <u>by</u> the system:

$$\left|w_{\text{rev}}\right| \geq \left|w_{\text{irrev}}\right|$$

eq. 2.16

For the compression, on the other hand, the work performed under reversible conditions is a lower boundary for the work performed <u>on</u> the system:

$$\left|w_{\text{rev}}\right| \leq \left|w_{\text{irrev}}\right|$$

eq. 2.17

In other words, for a real process that is not entirely reversible, the work $w_{\text{irrev}}$ the system performs is most likely smaller than $w_{\text{rev}}$, whereas the work we have to perform on a system is most likely larger. This relation does not only apply to state changes of gases but to all processes: in general, we can have a system perform more work on the way from state 1 to state 2 if changes are reversible. We will come back to this in Section 2.4.2.

### 2.2.4.4 Adiabatic Expansion and Compression

So far, we have considered volume changes of an ideal gas under isothermal conditions, where the temperature of the gas remains constant because the gas is coupled to a large heat reservoir, such that heat released by the gas is taken up by the reservoir, and heat taken up by the gas is removed from the reservoir. We can also change the state of an ideal gas without allowing heat exchange with the surroundings. A change in state without heat exchange is called an *adiabatic process*. In this case, the temperature of the system will change. Upon expansion, the system cannot take up heat from the surroundings, and the temperature will decrease. Conversely, upon compression, no heat can be released, and the temperature of the system will increase. Adiabatic compression of air masses crossing a mountain range is the reason for warm winds in the valleys (Box 2.3).

---

**BOX 2.3: ADIABATIC COMPRESSION OF AIR MASSES GENERATES DRY, WARM WINDS.**

When winds move air towards high mountains, the air masses ascend to higher levels. They thereby reach regions with lower air pressure, and expand. The air does not exchange heat with its surroundings during ascent, and the expansion process is nearly adiabatic, resulting in cooling of the air. The cooler air cannot hold as much humidity as the warm air, which leads to cloud formation and relief rainfall. The now dry air rises further and crosses the highest ridge. On the other side of the mountains, the air descends rapidly, and again this process is adiabatic. The increase in pressure leads to compression of the air masses, and to an increase in temperature. This warmer air can take up more humidity, and will dry the area along its path (Figure 2.11).

*(Continued)*

---

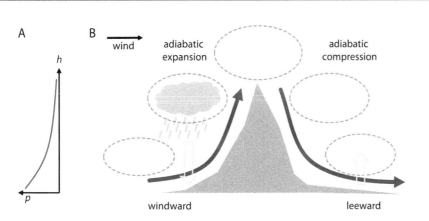

**FIGURE 2.11:    Adiabatic expansion and compression and dry fall winds.** A: The air pressure $p$ decreases exponentially with increasing height $h$ above sea level. B: Air masses (gray) are blown towards a mountain range by winds, and ascend on the windward side. During adiabatic expansion, the air becomes cooler (red-to-blue arrow) and can hold less humidity, leading to cloud formation and rainfall. On the wind-distant (leeward) side, the dry air is compressed adiabatically during descent. The air (blue-to-red arrow) and takes up humidity from the ground (blue arrow).

Dry, warm fall winds frequently occur on the northern side of the Alps (Föhn), and on the wind-distant side of other mountain ranges, such as the Sierra Nevada (Santa Ana winds), the Rocky Mountains (Chinook) or the Andes (Chanduy, Puelche, Zonda). Because of their dryness, these winds cause health and environmental problems, and promote wildfires.

To quantify the work performed during an adiabatic expansion, we have to take into account that now two state variables are changing: the temperature $T$ and the volume $V$. In this case, the property of a state function comes in handy: the internal energy of the system only depends on the initial and final states, not on the pathway on which the system changes from one state to the other. We can therefore describe the adiabatic expansion as a two-step process: an isothermal expansion from $V_1$ to $V_2$ at constant temperature $T_1$, followed by an isochoric change in temperature from $T_1$ to $T_2$ at a constant volume $V_2$ (Figure 2.12).

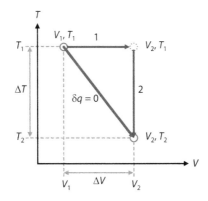

**FIGURE 2.12:    An adiabatic expansion is an expansion at constant temperature, followed by a change in temperature at constant volume.** In an adiabatic expansion, the volume of the gas increases from $V_1$ to $V_2$. No heat is exchanged, and the temperature of the gas will therefore decrease from $T_1$ to $T_2$ during the expansion. To quantitatively describe the work associated with this adiabatic change in state, we can make use of the fact that changes in state functions are independent of the pathway. This independence of the pathway allows us to dissect the adiabatic expansion (red arrow) into two steps (blue arrows): an isothermal expansion at a constant temperature $T_1$ (1), followed by an isochoric decrease in temperature from $T_1$ to $T_2$ at a constant volume $V_2$ (2).

The change in internal energy $dU$ during an adiabatic expansion is thus the sum of the changes in internal energy in the two steps, the isothermal expansion, and the isochoric change in temperature:

$$dU = dU_{T_1,V_1 \to T_1,V_2} + dU_{T_1,V_2 \to T_2,V_2} \qquad \text{eq. 2.18}$$

We can express the change in internal energy dU in the two steps as

$$dU = \left( \frac{\partial U}{\partial V} \right)_T dV + \left( \frac{\partial U}{\partial T} \right)_V dT \qquad \text{eq. 2.19}$$

where the first term describes the change in $U$ with the change in volume at constant temperature (isothermal expansion), and the second term corresponds to the temperature change at constant volume (isochoric process). The terms in brackets are termed *partial derivatives* (see Section 29.6), because the differentiation is only performed after one variable. These partial derivatives describe the dependence of $U$ on the volume $V$ for constant temperature, and its dependence on temperature $T$ at constant volume, respectively.

The internal energy of an ideal gas does not depend on volume or pressure, but only on the temperature $T$ of the gas:

$$U \propto nRT \qquad \text{eq. 2.20}$$

The relationship in eq. 2.20 is also called *Joule's law*. As a consequence of Joule's law, the internal energy of the gas remains constant during the isothermal expansion in the first step:

$$dU_{T_1,V_1 \to T_1,V_2} = 0 \qquad \text{eq. 2.21}$$

and

$$\left( \frac{\partial U}{\partial V} \right)_T = 0 \qquad \text{eq. 2.22}$$

We now define the temperature dependence of the internal energy $U$ at constant volume as

$$\left( \frac{\partial U}{\partial T} \right)_V = C_V \qquad \text{eq. 2.23}$$

$C_V$ is the *heat capacity* at constant volume (in J K$^{-1}$). For the second term in eq. 2.19, the change in internal energy upon the isochoric temperature change, we can then write

$$dU_{T_1,V_2 \to T_2,V_2} = C_V dT \qquad \text{eq. 2.24}$$

We can now combine the changes in internal energy in the two steps (eq. 2.21 and eq. 2.24) to calculate the total change in internal energy of the gas during the adiabatic expansion. As there is no heat exchange during an adiabatic expansion, the change in internal energy equals the expansion work:

$$dU = w = \int_{T_1}^{T_2} C_V (T_2 - T_1) \qquad \text{eq. 2.25}$$

Eq. 2.25 is in agreement with the system-centered sign convention: work upon expansion ($T_2 < T_1$) will be negative, meaning the system (gas) performs work on its surroundings. During a compression ($T_2 > T_1$), the work will be positive, the surroundings perform work on the system.

Table 2.1 summarizes the work associated with the state changes that we have discussed so far.

**TABLE 2.1:**
**Work associated with reversible, irreversible, and adiabatic expansion and compression.**

|  | irreversible | reversible | adiabatic |
|---|---|---|---|
| work (*w*) | $w = -p_{ex}\Delta V$ | $w = -nRT \ln \dfrac{V_2}{V_1}$ | $w = C_V (T_2 - T_1)$ |
| expansion | $\Delta V > 0 \rightarrow w < 0$ | $V_2 > V_1 \rightarrow w < 0$ | $T_2 < T_1 \rightarrow w < 0$ |
| compression | $\Delta V < 0 \rightarrow w > 0$ | $V_2 < V_1 \rightarrow w > 0$ | $T_2 > T_1 \rightarrow w > 0$ |

$p_{ex}$: external (final) pressure during an irreversible process, $V_1/T_1$: initial volume/temperature, $V_2/T_2$: final volume/temperature, $C_V$: heat capacity at constant volume.

We learnt earlier that the product of pressure and volume is constant for an ideal gas during isothermal processes (law of Boyle-Mariotte). If we know the initial pressure and volume of the gas before an isothermal state change and either the pressure or the volume after the change, we can therefore calculate the unknown volume or pressure from eq. 2.26:

$$p_1 V_1 = p_2 V_2 \qquad \text{eq. 2.26}$$

During an adiabatic process, the temperature changes, and so the product of pressure and volume before and after the process will be different. To derive a relationship between pressure and volume before and after an adiabatic process, we start from the change in internal energy d*U*. In the absence of other types of work and of heat exchange, the change in internal energy will equal the expansion work:

$$dU = w = -p\,dV \qquad \text{eq. 2.27}$$

We have derived an expression for d*U* during an adiabatic process in terms of the heat capacity $C_V$ at constant volume (eq. 2.24), and we can now equate these two expressions:

$$-p\,dV = C_V\,dT \qquad \text{eq. 2.28}$$

Substituting for *p* the expression $nRT/V$ from the ideal gas equation and separating the variables then yields

$$-\frac{nR}{V}\,dV = \frac{C_V}{T}\,dT \qquad \text{eq. 2.29}$$

which we can integrate from $T_1$ to $T_2$ and $V_1$ to $V_2$ to

$$-nR \ln \frac{V_2}{V_1} = C_V \ln \frac{T_2}{T_1} \qquad \text{eq. 2.30}$$

Eq. 2.30 can be rearranged to

$$\ln \frac{T_2}{T_1} = -\frac{nR}{C_V} \ln \frac{V_2}{V_1} \qquad \text{eq. 2.31}$$

which equals

$$\frac{T_2}{T_1} = \left( \frac{V_2}{V_1} \right)^{-\frac{nR}{C_V}} \qquad \text{eq. 2.32}$$

For an ideal gas, $nR$ is the difference between the heat capacity at constant pressure, $C_p$ (Section 2.3), and the heat capacity at constant volume $C_V$. Using this relationship, we can re-write eq. 2.32 as

$$\frac{T_2}{T_1} = \left(\frac{V_2}{V_1}\right)^{-\frac{C_p - C_V}{C_V}}$$

eq. 2.33

Now we define the ratio of $C_p/C_V$ as $\gamma$, and obtain

$$\frac{T_2}{T_1} = \left(\frac{V_2}{V_1}\right)^{1-\gamma}$$

eq. 2.34

The gas follows the ideal gas law before and after the change in state, and we can therefore express the temperatures $T_1$ and $T_2$ by the product of $p$ and $V$ as

$$\frac{p_2 V_2}{p_1 V_1} = \left(\frac{V_2}{V_1}\right)^{1-\gamma}$$

eq. 2.35

Rearrangement of eq. 2.35 finally leads us to the relationship

$$p_1 V_1^{\gamma} = p_2 V_2^{\gamma}$$

eq. 2.36

In contrast to isothermal processes, where the product $pV$ is constant, we see from eq. 2.36 that during adiabatic processes the product $pV^{\gamma}$ is constant. Note that this does not mean that the gas does not follow the ideal gas law during the adiabatic process! In fact, we have extensively used the ideal gas law in deriving eq. 2.36. The relationship in eq. 2.36 is a consequence of the change in temperature during the adiabatic process. In analogy to the definition of an isotherm, i.e. the $p$–$V$ curve during an isothermal process, the corresponding $p$–$V$ curve for adiabatic processes is called the *adiabate*. Because of the exponent $\gamma$, the adiabate is steeper than the isotherm (Figure 2.13).

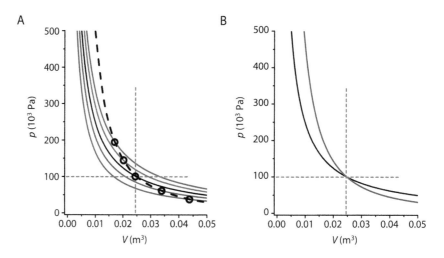

**FIGURE 2.13:   Comparison of isothermal and adiabatic processes**. A: For isothermal processes, the product $pV$ is constant, giving rise to a set of isotherms as described by the Boyle-Mariotte law (blue to red for increasing temperatures $T$). During adiabatic processes, no heat is exchanged. Therefore, adiabatic expansion leads to a decrease in temperature, and an adiabatic compression leads to an increase in temperature. This results in a crossing of the isotherms and gives rise to a steeper $p$–$V$ curve, the adiabate (black dotted line). For adiabatic processes, the product $pV^{\gamma}$ is constant. B: Direct comparison of isotherm (black) and adiabate (red).

### 2.2.5  Thermodynamic Cycles: Back and Forth or Round and Round

We will now take a look at *thermodynamic cycles*. Thermodynamic cycles consist of a sequence of processes, 1→2→…→n→1, starting from an initial state (1), and at the end returning to this initial state (1). In biochemistry we frequently find cyclic processes in enzyme catalysis where an enzyme populates a sequence of different states and returns to its original state at the end of the reaction sequence (see Chapters 11 and 12; Box 2.4).

The simplest form of a thermodynamic cycle is a coupling of two opposite changes, back and forth, such as the expansion of an ideal gas, followed by the compression to its original state. We first look at an irreversible expansion of an ideal gas at $p_1$, $V_1$ to $p_2$, $V_2$, followed by an irreversible compression back to the starting state with $p_1$, $V_1$. The volume change $\Delta V$ is $V_2 - V_1$ for the expansion and $V_1 - V_2$ for the compression, and thus equal in size, but opposite in sign. The work $w_{tot}$ over the complete process is then simply the sum of the work involved in the two steps:

$$w_{tot} = w_{exp} + w_{comp} = -p_2 \Delta V + p_1 \Delta V \qquad \text{eq. 2.37}$$

As $p_1 > p_2$, more work has to be performed on the system during compression at $p_1$ compared to the work the system has performed on its surroundings upon expansion at $p_2$. As a consequence, we obtain

$$w_{tot} > 0 \qquad \text{eq. 2.38}$$

If we now perform the sequence of expansion and compression to the initial state under reversible conditions, the pressure will be equal to the gas pressure according to the ideal gas law throughout the process. Under reversible conditions, we therefore obtain for $w_{tot}$:

$$w_{tot} = w_{exp} + w_{comp} = -nRT \ln \frac{V_2}{V_1} - nRT \ln \frac{V_1}{V_2} \qquad \text{eq. 2.39}$$

By inverting the argument of the logarithm for the compression, we change the sign of the second term, and see that now the work performed by the system during expansion equals the work that has to be performed on the system during compression of the gas. As a consequence, $w_{tot}$ is zero.

### 2.2.5.1 The Carnot Process

We can now think of a cyclic process of the sequence $1 \to 2 \to 3 \to 4 \to 1$, i.e. from an initial state 1 through intermediate states 2, 3, and 4 back to state 1. This cycle consists of four steps: (1) an isothermal expansion at the temperature $T_h$ ($1 \to 2$), (2) an adiabatic expansion ($2 \to 3$), (3) an isothermal compression at the temperature $T_c$ ($3 \to 4$), and (4) an adiabatic compression ($4 \to 1$). The temperatures $T_h$ and $T_c$ are maintained constant by coupling the system to two heat reservoirs, a hot reservoir at $T_h$ during the isothermal expansion, and a cold reservoir at $T_c$ during the isothermal compression. Such a cyclic process is called a *Carnot cycle* (Figure 2.14).

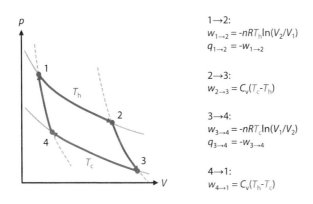

$$1 \to 2:$$
$$w_{1 \to 2} = -nRT_h \ln(V_2/V_1)$$
$$q_{1 \to 2} = -w_{1 \to 2}$$

$$2 \to 3:$$
$$w_{2 \to 3} = C_v(T_c - T_h)$$

$$3 \to 4:$$
$$w_{3 \to 4} = -nRT_c \ln(V_1/V_2)$$
$$q_{3 \to 4} = -w_{3 \to 4}$$

$$4 \to 1:$$
$$w_{4 \to 1} = C_v(T_h - T_c)$$

**FIGURE 2.14:    The Carnot cycle**. The Carnot process is a cyclic process in which an ideal gas is first expanded under isothermal conditions at the temperature $T_h$, then further expanded under adiabatic conditions, followed by an isothermal compression at the temperature $T_c$, and an adiabatic compression back to the starting state. The exchanged heat and work in each step is summarized on the right (see text).

In the Carnot process, the initial state of the system is equal to its final state after completion of one cycle. As a consequence, the internal energy $U$ (and other state functions) is constant over the complete cycle. The associated work with the isothermal processes $1\rightarrow2$ (expansion at $T_h$) and $3\rightarrow4$ (compression at $T_c$) is

$$w_{\text{isothermal}} = w_{1\rightarrow2} + w_{3\rightarrow4} = -nRT_h \ln\frac{V_2}{V_1} - nRT_c \ln\frac{V_4}{V_3} \qquad \text{eq. 2.40}$$

From eq. 2.34 for adiabatic changes, we can derive a relationship between $T_h$ and $T_c$ and $V_2$ and $V_3$ for the adiabatic expansion, or between $T_c$ and $T_h$ and $V_4$ and $V_1$ for the adiabatic compression. By solving both relationships for $T_h/T_c$ and equating them, we arrive at

$$\frac{V_1}{V_2} = \frac{V_4}{V_3} \qquad \text{eq. 2.41}$$

Using eq. 2.41, we can eliminate $V_3$ and $V_4$ from eq. 2.40, and obtain for the work associated with the two isothermal processes

$$w_{\text{isothermal}} = w_{1\rightarrow2} + w_{3\rightarrow4} = -nRT_h \ln\frac{V_2}{V_1} + nRT_c \ln\frac{V_2}{V_1} = -nR(T_h - T_c)\ln\frac{V_2}{V_1} \qquad \text{eq. 2.42}$$

The work for the two adiabatic processes $2\rightarrow3$ (expansion) and $4\rightarrow1$ (compression) is

$$w_{\text{adiabatic}} = w_{2\rightarrow3} + w_{4\rightarrow1} = C_V\left(T_c - T_h\right) + C_V\left(T_h - T_c\right) = 0 \qquad \text{eq. 2.43}$$

Finally, the work over the complete cycle is the sum of the work for all individual steps:

$$w_{\text{tot}} = w_{\text{isothermal}} + w_{\text{adiabatic}} = -nR(T_h - T_c)\ln\frac{V_2}{V_1} \qquad \text{eq. 2.44}$$

Because $T_h > T_c$ and $V_2 > V_1$, $w_{\text{tot}}$ is negative, and over the complete cycle the system thus performs the work $w_{\text{tot}}$ on the surroundings!

From the work involved in each step of the Carnot cycle, we can now calculate the heat exchanged. Heat is only exchanged during the isothermal steps, and the exchanged heat again has the same magnitude as the work performed, but the opposite sign:

$$q_{\text{isothermal}} = -\left(w_{1\rightarrow2} + w_{3\rightarrow4}\right) = nR(T_h - T_c)\ln\frac{V_2}{V_1} \qquad \text{eq. 2.45}$$

Again, because $T_h > T_c$ and $V_2 > V_1$, the overall heat exchange is positive, and the system absorbs heat during one cycle. In sum, during each complete Carnot cycle, the system takes up the heat $q_h$ from the hot reservoir at the temperature $T_h$, performs the work $w_{\text{tot}}$ on the surroundings, and transfers the heat $q_c$ to the cold reservoir at the temperature $T_c$ (Figure 2.15). Thus, the heat taken up by the system is partially converted into work.

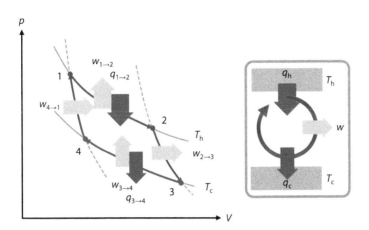

**FIGURE 2.15:**    **Heat and work during the Carnot process.** In the isothermal expansion (1→2), the system performs work on the surroundings ($w < 0$), and takes up heat ($q > 0$). During the adiabatic expansion (2→3), work is performed on the surroundings ($w < 0$), but no heat is exchanged. In the isothermal compression (3→4), the surroundings perform work on the system ($w > 0$), and heat is transferred from the system to the surroundings ($q < 0$). In the adiabatic compression (4→1), work is performed on the system ($w > 0$) without heat exchange. The work associated with the two adiabatic processes is equal, but of different sign, and cancels. The work performed on the surroundings upon isothermal expansion at $T_h$ is larger than the work performed on the system during the isothermal compression at $T_c$. Thus, over the complete process the system performs work on the surroundings. The heat taken up by the system from the hot reservoir during the isothermal expansion at $T_h$ is larger than the heat transferred from the system to the cold reservoir during the isothermal compression at $T_c$. Overall, the system thus takes up heat from the hot reservoir, converts part of it into work that is performed on the surroundings, and releases the remaining heat to the cold reservoir.

The *degree of efficiency* η of the Carnot cycle in Figures 2.14 and 2.15 can be described as the performed work $w_{tot}$ relative to the heat absorbed at $T_h$, $q_h$ (eq. 2.46). Note that by taking the negative work $-w_{tot}$, we now change the perspective from the system to a perspective centered on the surroundings: $w_{tot}$ is the work from the system-centered perspective, and negative values of $w_{tot}$ mean the system performs work on the surroundings. By inverting the sign to $-w_{tot}$, we convert the work into a positive number, the work that can be used for processes in the surroundings.

$$\eta_{\text{Carnot}} = \frac{-w_{tot}}{q_h} = \frac{nR(T_h - T_c)\ln\dfrac{V_2}{V_1}}{nRT_h\ln\dfrac{V_2}{V_1}} = \frac{T_h - T_c}{T_h} \qquad \text{eq. 2.46}$$

From eq. 2.46, we see that the degree of efficiency increases with increasing differences of $T_h$ and $T_c$. It would reach a limiting value of unity (i.e. 100% efficiency) only for $T_c = 0$ K, the absolute zero point. We have seen before (Section 2.2.3) that the absolute zero point cannot be reached. As a consequence, η will always remain $< 1$. Thus, from eq. 2.46 we can conclude that, for thermodynamic reasons, heat cannot be completely converted into work! This statement is one form of the 2nd law of thermodynamics. We will arrive at an alternative form of the 2nd law, the Clausius inequality, in Section 2.4.2.

The Carnot cycle in clockwise direction, following the sequence 1→2→3→4→1 (Figures 2.14, 2.15) that partially converts heat into work (Figures 2.14 and 2.15) is called a *forward Carnot cycle*. Such a process is the basis of a power heat pump such as a steam engine. We can also reverse the sequence, and run the process counter-clockwise, starting with an adiabatic expansion (1→4), followed by an isothermal expansion (4→3), an adiabatic compression (3→2) and an isothermal compression (2→1) (Figure 2.16). In this case, the Carnot process will transport heat from the cold reservoir ($T_c$) to the hot reservoir ($T_h$). We will see in Section 2.4.1 (see Figure 2.18) that the transfer of heat from cold to hot is against the spontaneous dissipation of heat, and thus an energetically unfavorable process. To drive this unfavorable process, work has to be performed on the system (Figure 2.16). Such a *reverse Carnot cycle* is the basic principle behind refrigerators, which need electric power to transport heat from their colder interior to the warmer room, or for heat pumps that take up heat from the cooler air or from deep in the ground, and pump it into the warmer living room, again at the expense of electric power. We will re-encounter this principle when we discuss transport across biological membranes (Section 4.3; Box 2.4).

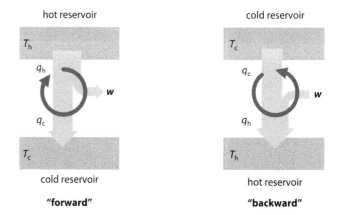

**FIGURE 2.16:   A Carnot cycle in forward and reverse mode.** In forward mode, following the sequence 1→2→3→4→1 (see Figure 2.14), the Carnot process converts part of the heat $q$ it takes up from the hot reservoir into work $w$ performed on the surroundings. The rest of the heat taken up is released to the cold reservoir. In reverse mode, following the sequence 1→4→3→2→1, the Carnot process takes up energy from a cold reservoir, and, together with work taken up from the surroundings, transfers a larger amount of heat to a hot reservoir. This reverse mode is the basis for the function of refrigerators and of heat pumps.

---

**BOX 2.4: CYCLIC PROCESSES IN BIOCHEMISTRY.**

In biochemistry, reactions typically occur at isothermal and isobaric conditions, nevertheless we find many reaction cycles that bear resemblance to the Carnot cycle. A number of enzymes convert one form of energy into a different form. An example that we will treat in Section 4.3 is the class of ATP-driven transporters that use the energy of ATP hydrolysis (chemical energy) to establish concentration gradients (osmotic energy). The transport cycle is similar to a reverse Carnot cycle: the transporter takes up cargo from a low concentration side, which corresponds to the process of taking up heat from a cold reservoir. The cargo is transported across a membrane and released on the distant side at a high concentration, which corresponds to the release of heat into a warm reservoir. Other transporters can use the energy of light for active transport. Further examples include motor proteins that couple the energy of ATP hydrolysis (chemical energy) to the generation of movement (mechanical energy), and enzymes that use the energy of ATP hydrolysis to alter the structure of their substrate, such as topoisomerases that can introduce torsional stress into DNA in an ATP-dependent reaction. During their catalytic cycles, these enzymes undergo conformational changes that contribute to coupling ATP hydrolysis to movement or to the catalyzed reaction. At the end of the catalytic cycle, the chemical energy has been converted into a different type of energy, and the enzyme is regenerated in its original state and can undergo further catalytic cycles.

Cyclic processes can also be exploited to generate heat. Bumble bees uses a substrate cycle to heat their bodies: they phosphorylate fructose-6-phosphate to fructose-1,6-bisphosphate, using ATP as a phosphoryl group donor. Fructose-1,6-bisphosphate is then dephosphorylated, leading to regeneration of fructose-6-phosphate. During each cycle, one ATP molecule is hydrolyzed. The energy of ATP hydrolysis is converted

*(Continued)*

into heat in this process. The resulting increase in body temperature warms up flight muscles to a temperature sufficient to generate force. This process is called *thermogenesis*. The heat generation through the substrate cycle enables bumble bees to fly at low temperatures, whereas honey bees have to wait until the outside temperatures warms them up.

## 2.2.6 The Temperature Dependence of the Internal Energy *U*

If we consider an ideal gas and allow only expansion work, the change d*U* in the internal energy *U* is described by

$$dU = -pdV + \delta q \qquad \text{eq. 2.47}$$

For isochoric processes ($V$ = const., d$V$ = 0), eq. 2.47 simplifies to

$$dU = \delta q_V \qquad \text{eq. 2.48}$$

where the index $V$ is an abbreviation that denotes the isochoric conditions. Thus, by limiting our considerations to changes with constant volume, we have converted a non-state function, the heat $q$, into a state function, the internal energy $U$ (under certain conditions, i.e. for isochoric processes). For processes at constant volume, the change in internal energy thus equals the amount of heat exchanged. This heat will affect the temperature of the system. We have defined the dependence of $U$ on the temperature at constant volume as the heat capacity $C_V$ at constant volume (eq. 2.23). Now we can relate the heat capacity $C_V$ to the heat exchanged in an isochoric process:

$$\left( \frac{\partial U}{\partial T} \right)_V = C_V = \frac{\delta q_V}{dT} \qquad \text{eq. 2.49}$$

We have derived the relationship between d$U$ and d$T$ before (eq. 2.24), and can now directly derive the temperature dependence of $U$ (at constant volume) by integrating eq. 2.24 from $T_0$ to $T$:

$$\int_{U(T_0)}^{U(T)} dU = \int_{T_0}^{T} C_V dT \qquad \text{eq. 2.50}$$

We assume that $C_V$ is independent of the temperature in the temperature interval from $T_0$ to $T$, and obtain the temperature dependence of the internal energy $U$ by evaluating the integrals as

$$U(T) - U(T_0) = C_V (T - T_0) \qquad \text{eq. 2.51}$$

or

$$U(T) = U(T_0) + C_V (T - T_0) \qquad \text{eq. 2.52}$$

Thus, if we know the internal energy of our system at some reference temperature $T_0$, and we know the heat capacity $C_V$ (at constant volume), we can calculate the internal energy $U$ of the system at any other temperature $T$, provided that $C_V$ is temperature-independent.

$U$ is a state function, hence changes in $U$ that are associated with any process can be described as

$$\Delta U = U_{\text{final}} - U_{\text{initial}} \qquad\qquad \text{eq. 2.53}$$

If we now consider the temperature dependence of $\Delta U$, $\mathrm{d}/\mathrm{d}T\,(\Delta U)$, we obtain

$$\frac{\mathrm{d}\left(\Delta U\right)}{\mathrm{d}T} = \frac{\mathrm{d}\left(U_{\text{final}}\right)}{\mathrm{d}T} - \frac{\mathrm{d}\left(U_{\text{initial}}\right)}{\mathrm{d}T} = C_{V,\,\text{final}} - C_{V,\,\text{initial}} \qquad\qquad \text{eq. 2.54}$$

Eq. 2.54 relates the change in internal energy during an isochoric process to the change in heat capacitiy $\Delta C_V$ between final and initial states:

$$\frac{\mathrm{d}\left(\Delta U\right)}{\mathrm{d}T} = \Delta C_V \qquad\qquad \text{eq. 2.55}$$

By integrating eq. 2.55, we obtain $\Delta U$, the change in the internal energy during this process, as a function of temperature $T$ and the change in heat capacity $\Delta C_V$ associated with the process:

$$\Delta U\left(T\right) = \Delta U\left(T_0\right) + \Delta C_V\left(T - T_0\right) \qquad\qquad \text{eq. 2.56}$$

The internal energy $U$ is a useful state function to describe processes at constant volume, where it corresponds to the exchanged heat. Biological processes typically take place at constant pressure, not at constant volume, but we will see in the following chapter that we can apply the principles we have learnt from considerations of the internal energy directly to isobaric processes.

## 2.3  THE ENTHALPY *H*

We have seen in Section 2.2 that the change in internal energy $U$ corresponds to the heat exchanged during an isochoric process (eq. 2.48). By analogy, we can also calculate the heat exchanged under isobaric conditions. First, we rearrange eq. 2.47 to

$$\delta q = \mathrm{d}U + p\mathrm{d}V \qquad\qquad \text{eq. 2.57}$$

We now define a new state function (we will see in a moment why this state function is useful), the enthalpy $H$, as

$$H = U + pV \qquad\qquad \text{eq. 2.58}$$

Complete differentiation of $H$ yields

$$\mathrm{d}H = \mathrm{d}U + p\mathrm{d}V + V\mathrm{d}p \qquad\qquad \text{eq. 2.59}$$

For isobaric processes ($p$ = const., $\mathrm{d}p$ = 0), eq. 2.59 simplifies to

$$\mathrm{d}H = \mathrm{d}U + p\mathrm{d}V \qquad\qquad \text{eq. 2.60}$$

If we now compare eq. 2.60 and eq. 2.57, we see that the right-hand sides are equal. Consequently, $\mathrm{d}H$ equals the heat exchanged at constant pressure:

$$dH = \delta q_p \qquad \text{eq. 2.61}$$

Changes $dH$ in the enthalpy $H$ equal the amount of heat transferred during isobaric processes. Thus, the enthalpy $H$ at constant pressure is an equivalent state function to the internal energy $U$ at constant volume.

We can extend the analogy of $dH$ (for isobaric processes) and $dU$ (for isochoric processes), and define a heat capacity $C_p$ at constant pressure as

$$\left(\frac{\partial H}{\partial T}\right)_p = C_p \qquad \text{eq. 2.62}$$

From eq. 2.62, we obtain

$$dH = C_p\, dT \qquad \text{eq. 2.63}$$

Separation of the variables and integration from $T_0$ to $T$ yields

$$H(T) = H(T_0) + C_p(T - T_0) \qquad \text{eq. 2.64}$$

which allows us to calculate the enthalpy $H$ for a system at any temperature $T$, if we know $H$ at a reference temperature $T_0$, $H(T_0)$, and $C_p$. Again, because $H$ is a state function, we can calculate changes in enthalpy associated with any given process simply by subtracting the enthalpies of the initial and final states:

$$\Delta H = H_{\text{final}} - H_{\text{initial}} \qquad \text{eq. 2.65}$$

For a chemical reaction, $H_{\text{initial}}$ is the enthalpy of the reactants, $H_{\text{final}}$ is the enthalpy of the products, and $\Delta H_r$ is the change in enthalpy of the reaction, i.e. the heat exchanged during the reaction at constant pressure:

$$\Delta H_r = H_{\text{products}} - H_{\text{reactants}} \qquad \text{eq. 2.66}$$

If we now consider the temperature dependence of $\Delta H$, $d/dT\,(\Delta H)$, eq. 2.65 becomes

$$\frac{d(\Delta H)}{dT} = \frac{d(H_{\text{final}})}{dT} - \frac{d(H_{\text{initial}})}{dT} = C_{p,\text{final}} - C_{p,\text{initial}} \qquad \text{eq. 2.67}$$

or

$$\frac{d(\Delta H)}{dT} = \Delta C_p \qquad \text{eq. 2.68}$$

The relationship between changes in enthalpy and heat capacity according to eq. 2.68 is called *Kirchhoff's law*. By integrating this equation, the temperature dependence of $\Delta H$ as a function of the change in heat capacity is obtained:

$$\Delta H(T) = \Delta H(T_0) + \Delta C_p(T - T_0) \qquad \text{eq. 2.69}$$

Thus, the change in enthalpy $\Delta H$ is temperature-dependent. The temperature dependence of $\Delta H$ is a consequence of the temperature dependence of $H$ for the final and initial states involved in the process. For chemical reactions, the enthalpy change $\Delta H_r$ depends on the temperature because of the different temperature dependence of the enthalpies of the reactants and the products ($C_{p,\text{reactant}} \neq C_{p,\text{products}}$, $\Delta C_p \neq 0$; Figure 2.17).

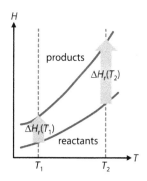

**FIGURE 2.17: Temperature dependence of enthalpy and enthalpy changes.** The enthalpies of the reactants and of the products depend on the temperature. In general, reactants and products will have different heat capacities, and thus a different dependence of their enthalpies on temperature. As a consequence, the change in enthalpy during the reaction, $\Delta H_r$, the difference between $H_{\text{products}}$ and $H_{\text{reactants}}$, also depends on temperature (arrows).

We can exploit the property of the enthalpy $H$ as a state function to calculate the enthalpy change for a process that is difficult to examine experimentally. To do so, we have to relate this process to others whose enthalpy changes are known, and use the *Hess law* of heat summation (Box 2.5).

---

**BOX 2.5: HESS LAW OF HEAT SUMMATION.**

Because $H$ is a state function, we can always calculate changes in $H$ from the difference $\Delta H = H_{\text{final}} - H_{\text{initial}}$. For a chemical reaction, this enthalpy change is $\Delta H_r = H_{\text{products}} - H_{\text{reactants}}$. We can also use the property of $H$ as a state function to calculate the enthalpy change for a process that is difficult to examine experimentally by relating this process to others whose enthalpy changes are known. In the simplest case, we can calculate the overall change in $\Delta H$ for a sequence of reactions (Scheme 2.1) by adding up the $\Delta H$ values for each individual step,

$$A \xrightarrow{\Delta H_1} B \xrightarrow{\Delta H_2} C \xrightarrow{\Delta H_3} \ldots \xrightarrow{\Delta H_{25}} Z \qquad \text{scheme 2.1}$$

and

$$\Delta H_{\text{tot}} = \sum_{i=1}^{25} \Delta H_i \qquad \text{eq. 2.70}$$

We can also construct cyclic schemes, e.g. with reactions from A to B to C and back to A (Scheme 2.2):

$$\text{scheme 2.2}$$

*(Continued)*

The change in enthalpy in one cycle (from A to A) has to be zero, because the initial and final states are identical. If we know all $\Delta H$ values except one, e.g. $\Delta H_2$ and $\Delta H_3$, we can calculate the missing $\Delta H_1$ value as

$$\Delta H_1 = \Delta H_{\text{tot}} - \Delta H_2 - \Delta H_3 = -\Delta H_2 - \Delta H_3 \qquad \text{eq. 2.71}$$

The additivity of heat and changes in $\Delta H$ is called the *Hess law* of heat summation. We can transfer this same principle to any other state function, e.g. to the changes in entropy $\Delta S$ (Section 2.4) or free energy $\Delta G$ (Section 2.5).

Isobaric processes with $\Delta H > 0$ are associated with an uptake of heat ($\delta q_p > 0$), and are called *endothermic*. Processes with $\Delta H < 0$ are accompanied by a release of heat ($\delta q_p < 0$), and are called *exothermic*. Most organisms live at constant pressure, and $\Delta H$ and $\Delta C_p$ (rather than $\Delta U$ and $\Delta C_v$) are therefore important parameters of life.

## 2.4 THE ENTROPY S AND THE SECOND LAW OF THERMODYNAMICS

### 2.4.1 PREDICTING SPONTANEITY OF PROCESSES: DISSIPATION OF HEAT AND MATTER

Until now we have considered the overall energy balance associated with expansion and compression of gases under different conditions, and have calculated the work and heat exchanged during these processes using the first law of thermodynamics. However, the first law of thermodynamics does not allow us to predict the spontaneous direction of a process. For example, both sodium nitrate ($NaNO_3$) and sodium hydroxide ($NaOH$) dissolve spontaneously in water. The dissolution of $NaOH$ is an exothermic reaction with $\Delta H < 0$: the solution becomes warm and heat is transferred to the surroundings. In contrast, the dissolution of $NaNO_3$ is an endothermic process, with $\Delta H > 0$. As a consequence, the solution becomes cold, as heat is drawn from the surroundings. Despite the opposite sign of $\Delta H$, both reactions are spontaneous. We can conclude that the sign of $\Delta H$ does therefore not have any predictive value for the spontaneity of reactions. The internal energy $U$ likewise does not indicate the direction of spontaneous processes. For example, we know from our everyday experience that we can leave a kettle with hot water on the stove, and it will slowly transfer heat to the room and cool down to room temperature. The reverse process, in which heat is spontaneously withdrawn from the room to bring water in the kettle to a boil, is never observed. More generally, bringing together a cold and a hot object will lead to heat exchange in one direction only, from the hot to the cold object, until both have the same temperature and are in thermal equilibrium. This process towards an even distribution of heat is called *dissipation* (Figure 2.18). However, the internal energy of the system, defined as the room including the kettle, or the two objects of different temperatures, does not change because heat is just moved from one element to the other within the system.

Dissipation is not only observed for heat, but also for matter: if we allow a gas, or matter in general, to occupy a larger volume, it will spontaneously do so. The reverse process, the spontaneous retraction of the gas, or matter in general, to only part of the available volume, is never observed (Figure 2.18). The expansion of a gas under isothermal conditions does not alter the internal energy of the gas. In general, the internal energy of a system does not change during the dissipation of matter.

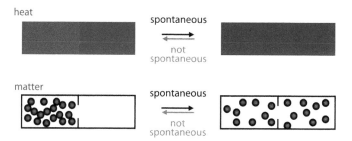

**FIGURE 2.18:** **Dissipation of heat and matter**. When we bring a hot (red) and a cold (blue) object in contact, heat will spontaneously flow from the hot to the cold object until both objects have the same temperature (thermal equilibrium, violet). The reverse process does not occur spontaneously. Similarly, when we provide a larger volume for a gas (red particles), the particles will spontaneously occupy the total available volume. The reverse process, in which particles accumulate in a small area of the available volume, does not occur spontaneously. Both heat and matter have a tendency to distribute evenly, a process called dissipation.

We have been able to intuitively predict the direction of these dissipative processes. If we compare initial and final states, we see that the final state, with heat or matter distributed throughout the system, is a state of lower order than the initial state. How can we formulate the underlying principle, a decrease in order, in a more objective and quantitative manner? What we need is a state function that only changes in one direction, such that the sign of its change predicts if a reaction is spontaneous. In the following, we will derive this state function from heat considerations (Section 2.4.2). Later, we will see that it is connected to the degree of order in the system (Section 2.4.5).

## 2.4.2 ENTROPY AND HEAT

To define a state function that predicts the spontaneous direction of a process, we return to the Carnot cycle we have considered in Section 2.2.5. We have calculated the heat exchanged over the complete cycle (eq. 2.45),

$$q_{\text{isothermal}} = q_{1\rightarrow2} + q_{3\rightarrow4} = nR\left(T_h - T_c\right)\ln\frac{V_2}{V_1} \qquad \text{eq. 2.45}$$

and we have seen that the overall heat exchanged is positive. This is in agreement with heat not being a state function. We have seen before that we can nevertheless convert heat into a state function under certain conditions, i.e. into the enthalpy $H$ under isobaric conditions, or into the internal energy $U$ under isochoric conditions. We will now illustrate how we can convert heat into a state function if we normalize it by the temperature. Instead of the heat we now sum up the "relative heat" for one Carnot cycle, i.e. the heat divided by the absolute temperature at which the process takes place. We obtain

$$\frac{q_{1\rightarrow2}}{T_h} + \frac{q_{3\rightarrow4}}{T_c} = -nR\ln\frac{V_2}{V_1} + nR\ln\frac{V_2}{V_1} = 0 \qquad \text{eq. 2.72}$$

While the heat does not qualify as a state function, the sum of the "relative heat" now equals zero over the cyclic process, and thus has the property of a state function! We therefore define this new state function as the *entropy S* with

$$S = \frac{q_{\text{rev}}}{T} \qquad \text{eq. 2.73}$$

The index "rev" denotes that this definition is derived for reversible processes. Comparing with eq. 2.72, we see that for a complete Carnot cycle consisting only of reversible processes, the change in entropy is

$$\mathrm{d}S = \frac{\delta q_{\mathrm{rev}}}{T} = 0 \qquad\qquad \text{eq. 2.74}$$

The equality (eq. 2.74) only holds true for reversible processes, i.e. for systems in equilibrium. For irreversible processes, eq. 2.74 becomes an inequality (see eq. 2.82).

We can derive a more general form of eq. 2.74 by including irreversible processes in the Carnot cycle. We know that irreversible processes, such as friction, will lead to the conversion of energy to heat, and will reduce the overall degree of efficiency $\eta$:

$$\eta_{\mathrm{irrev}} < \eta_{\mathrm{Carnot,rev}} \qquad\qquad \text{eq. 2.75}$$

We have defined the degree of efficiency for a Carnot process in general as the ratio of work performed by the system on the surroundings, from the perspective of the surroundings ($-w_{\mathrm{tot}}$), divided by the heat $q_h$ taken up from the high temperature reservoir (eq. 2.46). This general definition also holds for irreversible conditions. For reversible processes, we can calculate the individual heat terms, and express the degree of efficiency in terms of the temperatures $T_h$ and $T_c$. We can therefore relate the degrees of efficiency under irreversible and reversible conditions by:

$$\eta_{\mathrm{irrev}} = \frac{-w_{\mathrm{tot}}}{q_{1\to2}} = \frac{q_{1\to2} + q_{3\to4}}{q_{1\to2}} < \eta_{\mathrm{Carnot,rev}} = \frac{T_h - T_c}{T_h} \qquad\qquad \text{eq. 2.76}$$

leading to the relationship

$$\frac{q_{1\to2} + q_{3\to4}}{q_{1\to2}} < \frac{T_h - T_c}{T_h} \qquad\qquad \text{eq. 2.77}$$

We can rearrange this inequality to

$$1 + \frac{q_{3\to4}}{q_{1\to2}} < 1 - \frac{T_c}{T_h} \qquad\qquad \text{eq. 2.78}$$

and

$$\frac{q_{3\to4}}{q_{1\to2}} < -\frac{T_c}{T_h} \qquad\qquad \text{eq. 2.79}$$

and arrive at

$$\frac{q_{1\to2}}{T_h} + \frac{q_{3\to4}}{T_c} < 0 \qquad\qquad \text{eq. 2.80}$$

The left-hand side of eq. 2.80 is the sum of all $q/T$ terms, or the circular integral (see Section 29.6) over the Carnot cycle. The zero on the right-hand side is the overall entropy change over one Carnot cycle, which as a state function equals zero over a circular process. Eq. 2.80 can thus be rewritten as

$$\oint \frac{\delta q}{T} < \oint \mathrm{d}S \qquad\qquad \text{eq. 2.81}$$

The circle in eq. 2.81 denotes integration over a circular process (see Section 29.6). By comparing the arguments of the integrals only, we can conclude that for the entire cycle:

$$\frac{\delta q}{T} < \mathrm{d}S \qquad\qquad \text{eq. 2.82}$$

During a Carnot cycle with irreversible elements, the entropy increase is thus larger than the transferred heat divided by the temperature. For a reversible Carnot cycle, the inequality in eq. 2.82 becomes an equality.

For isolated systems, $\delta q$ is zero. The resulting form of eq. 2.82 is called the *Clausius inequality*:

$$\mathrm{d}S \geq 0 \qquad \text{eq. 2.83}$$

Thus, the entropy is the state function with the predictive value for spontaneous processes that we were looking for: in isolated systems, a process is spontaneous when the entropy increases. Processes with constant entropy are in equilibrium, and processes with a decrease in entropy do not occur spontaneously. Eq. 2.82 is an alternative formulation of the *second law of thermodynamics*. It is a simple statement with important consequences: the reason why heat cannot be converted completely into work (Box 2.6).

---

**BOX 2.6: REVERSIBILITY AND THE DEGREE OF EFFICIENCY.**

We have learnt in Section 2.2.4.3 that we can have a system perform most work under reversible conditions. In a car engine, the ignition and combustion of fuel causes a rapid and irreversible adiabatic expansion. This irreversibility is inevitable if we want to convert the energy of combustibles into work, yet it is the reason for the rather poor degrees of efficiency of these machines. Biological machines, on the other hand, achieve very high degrees of efficiencies because they perform their work under reversible conditions. Many of these machines use ATP as an energy source, and hydrolyze ATP to ADP and phosphate. However, these machines are true enzymes that also catalyze the reverse reaction (Section 14.1), the condensation of ADP and phosphate to ATP. The reversibility of reactions catalyzed by biological machines and motors is the key to their impressively high degrees of efficiencies.

---

### 2.4.3 TEMPERATURE DEPENDENCE OF THE ENTROPY

For processes at constant pressure, we can express changes in entropy as

$$\mathrm{d}S = \frac{\delta q_{\mathrm{rev}}}{T} = \frac{C_p \mathrm{d}T}{T} \qquad \text{eq. 2.84}$$

Integration of eq. 2.84 from $T_0$ to $T$ yields

$$S(T) - S(T_0) = C_p \ln\left(\frac{T}{T_0}\right) \qquad \text{eq. 2.85}$$

or

$$S(T) = S(T_0) + C_p \ln\left(\frac{T}{T_0}\right) \qquad \text{eq. 2.86}$$

By analogy, we obtain for the change in entropy during a reaction, $\Delta S(T)$

$$\Delta S(T) = \Delta S(T_0) + \Delta C_p \ln\left(\frac{T}{T_0}\right) \qquad \text{eq. 2.87}$$

where $\Delta C_p$ is the change in heat capacity during the reaction. The entropy change is temperature-dependent because of the different temperature dependencies of the entropies of reactants and products (Figure 2.19). Eq. 2.87 allows us to calculate $\Delta S$ at the temperature $T$ from $\Delta S$ at a certain reference temperature $T_0$, and from $\Delta C_p$. An analogous expression can be derived for isochoric processes, using $C_V$ as a reference value. For most biological systems that usually operate at constant pressure, these considerations are less relevant, however.

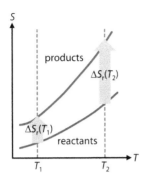

FIGURE 2.19:  **Temperature dependence of entropy and entropy changes**. The entropies of the reactants and of the products depend on temperature. In general, reactants and products will have different heat capacities, and thus a different dependence of their entropies on temperature. As a consequence, the change in entropy during the reaction, $\Delta S_r$, the difference between $S_{products}$ and $S_{reactants}$, also depends on temperature (arrows).

### 2.4.4 THE THIRD LAW OF THERMODYNAMICS AND ABSOLUTE ENTROPY

We have discussed before that there is an absolute temperature minimum that cannot be reached (Section 2.2.3). At this absolute minimum in temperature, all matter would be perfectly ordered, which corresponds to an entropy of $S = 0$. Although $T = 0$ K cannot be reached, we can still formulate the limit of the entropy towards the absolute temperature minimum:

$$\lim_{T \to 0} S(T) = 0$$

eq. 2.88

This statement is the *third law of thermodynamics*. As a consequence of the third law, the *absolute entropy* of a compound as a function of temperature can be calculated by summing up all entropy it has gained starting from $T = 0$ K (Figure 2.20; eq. 2.89). To do this, we have to integrate over the heat needed to increase the temperature of the compound in its solid state by one Kelvin, which is the heat capacity of the solid state $C_{p,s}$, integrated over the temperature range. At the melting temperature $T_{melt}$, heat will be absorbed during the phase transition to the liquid state without increase in temperature. Heat that is exchanged without changing the temperature of the system is termed *latent heat*. To calculate the associated entropy change, we have to divide the melting enthalpy $\Delta H_{melt}$ by the temperature $T_{melt}$ according to the heat definition of the entropy. Similarly, at the boiling temperature $T_{boil}$, heat $\Delta H_{boil}$ will be absorbed during the transition to the gas phase, again without increasing the temperature until the phase transition is completed, accounting for an entropy change of $\Delta H_{boil}/T_{boil}$. In the liquid state, between $T_{melt}$ and $T_{boil}$, the heat required for a temperature increase of one Kelvin is the heat capacity of the liquid state, $C_{p,l}$ or of the gaseous state, $C_{p,g}$, integrated over the temperature range. Note that the heat capacities for the different phases are different. By adding up all heat terms, divided by the temperature, we obtain the absolute entropy $S(T)$ as

$$S(T) = S(0) + \int_{0K}^{T_{melt}} \frac{C_{p,s}\,dT}{T} + \frac{\Delta H_{melt}}{T_{melt}} + \int_{T_{melt}}^{T_{boil}} \frac{C_{p,l}\,dT}{T} + \frac{\Delta H_{boil}}{T_{boil}} + \int_{T_{boil}}^{T} \left( \frac{C_{p,g}\,dT}{T} \right)$$

eq. 2.89

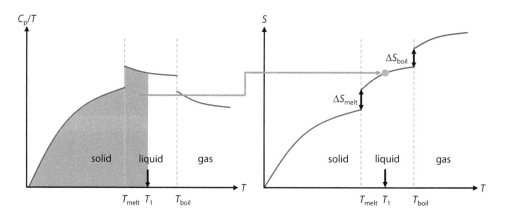

**FIGURE 2.20: Absolute entropy**. The absolute entropy of a compound can be calculated by summing up all entropy it has gained from $T = 0$ K. Following the heat definition of entropy, the heat required to increase the temperature by one Kelvin, i.e. the heat capacity $C_p$ ($C_{p,s}$, $C_{p,l}$, and $C_{p,g}$ for the solid, liquid, and gaseous states) has to be integrated over the relevant temperature ranges (left). Note that the heat capacities for the different phases are different. By integration over the temperature (orange area for $T_1$), the entropy as a function of temperature can be calculated (right; gray arrow, orange circle). At the melting temperature $T_{melt}$, heat will be absorbed during the phase transition to the liquid state without increase in temperature. The associated entropy increase $\Delta S_{melt}$ equals the melting enthalpy $\Delta H_{melt}$ divided by the temperature $T_{melt}$. Similarly, at the boiling temperature $T_{boil}$, heat $\Delta H_{boil}$ will be absorbed during the transition to the gas phase, again without increasing the temperature until the phase transition is completed, and the entropy increases by $\Delta S_{boil} = \Delta H_{boil}/T_{boil}$.

### 2.4.5 ENTROPY AND ORDER: THE STATISTIC INTERPRETATION

We introduced the state function entropy from the phenomenon of dissipation of matter and heat (Figure 2.18), and related entropy to order/disorder. In the following, we will see that the order or disorder of a particular state, and hence its entropy, can also be quantified by looking at the different ways by which a state can be generated. To illustrate this concept, we consider a system of four particles that can randomly distribute over two compartments (Figure 2.21). This system can exist in five different states: (1) all particles in the left compartment, (2) three particles left, one right, (3) two left, two right, (4) one left, three right, or (5) all four particles in the right compartment. The states (1) and (5), with all particles in one compartment, would be an "ordered" state, the state with two particles left, two right is what we usually perceive as random, or less ordered.

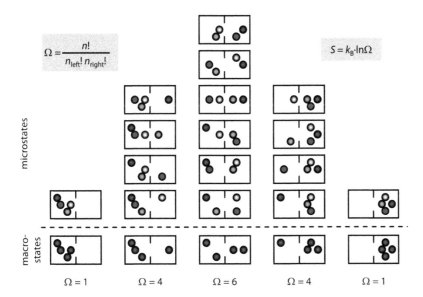

$$\Omega = \frac{n!}{n_{\text{left}}!\, n_{\text{right}}!}$$

$$S = k_\text{B} \cdot \ln\Omega$$

| $\Omega = 1$ | $\Omega = 4$ | $\Omega = 6$ | $\Omega = 4$ | $\Omega = 1$ |

**FIGURE 2.21:** **Macrostates, microstates, entropy, and statistics.** Four particles (red, bottom) can be distributed over two compartments, left and right, such that all four particles are in the left compartment, three are left and one is in the right compartment, two each are left and right, one is left and three are right, or all four are in the right compartment. These states are called macrostates. In the case that we can distinguish the four particles (red, blue, green, yellow), there are different sub-states for each macrostate, the microstates. The macrostates with all four particles in the left or in the right compartment can only be generated in one defined way. The macrostates with the 3–1 distribution can be generated in four different ways, with either the red, the blue, the green, or the yellow particle separated from the others. The macrostate with two particles on each side can be generated in six different ways. The number of microstates depends on the total number of particles *n* and the number of particles in the left and right compartment, $n_{\text{left}}$ and $n_{\text{right}}$. From the number of microstates, the entropy *S* can be calculated according to the Boltzmann equation (eq. 2.91).

How can order or randomness be expressed in quantitative terms? If we consider the same four-particle system, but now the four particles can be distinguished, e.g. by their different colors, we can tabulate the different arrangements to generate states (2) to (4) (Figure 2.21). For example, state (2) with one particle on the right can be generated in four different ways, with the red, yellow, green, or blue particle on the right-hand side. The same considerations lead to six possible ways to generate state (3), and again four different ways to generate state (4). There is still only one possibility to generate states (1) and (5). States 1–5 are called *macrostates*, and the different sub-states that lead to the net arrangement of particles in these states are called *microstates* (Figure 2.21). We can calculate the number of microstates $\Omega$ from the total number of particles, *n*, and the number of particles in the left and right compartment, $n_{\text{left}}$ and $n_{\text{right}}$, as

$$\Omega = \frac{n!}{n_{\text{left}}!\, n_{\text{right}}!} \qquad \text{eq. 2.90}$$

The number of microstates follows Pascal's triangle (coefficients for binomial distributions; see Section 29.3). We will revisit the concept of the number of ways to generate a macrostate when we look at binding of substrate molecules to enzymes with multiple active sites (Sections 11.4 and 11.5). According to the statistical definition of entropy, the number of possible microstates (eq. 2.90) of each macrostate defines the entropy *S* of this state as

$$S = k_B \ln\Omega \qquad \text{eq. 2.91}$$

where $k_\text{B}$ is the Boltzmann constant. Eq. 2.91 directly relates the randomness of a state, or the number of possibilities by which it can be generated, to its entropy. In this respect, the entropy is a measure of the order or randomness of a system: high entropy is related to a high degree of randomness or low order, and low entropy corresponds to a low degree of randomness or a highly ordered state.

Are the definitions of entropy by heat and the statistical definition by order equivalent? We will apply both definitions to the reversible isothermal expansion of 1 mol of an ideal gas from $V_1$ to $V_2$ to confirm the equivalency. First, we consider the heat definition. The heat exchanged during the isothermal expansion is

$$\delta q = -\delta w = p\,dV = \frac{RT}{V}\,dV$$

eq. 2.92

According to the heat definition, the entropy change $\Delta S$ associated with this process is thus

$$\Delta S = \frac{q^{\mathrm{rev}}}{T} = \int \frac{\delta q^{\mathrm{rev}}}{T} = \int_{V_1}^{V_2} \frac{R}{V}\,dV = R\ln\frac{V_2}{V_1}$$

eq. 2.93

Now we consider the statistical interpretation of the entropy. In the initial state, the gas occupies the volume $V_1$. We divide the volume $V_1$ into $n_1$ individual cells with a volume $V$ (Figure 2.22). One particle then has $n_1$ cells it can occupy, giving $n_1$ possibilities to generate the macrostate with one particle in the volume $V_1$. If we consider two particles, the possibilities are $(n_1)^2$, and for $n$ particles, they are $(n_1)^n$. In 1 mol of an ideal gas, we have $N_A$ particles (where $N_A$ is the Avogadro constant, $N_A = 6.022\cdot10^{23}$ mol$^{-1}$), that can be arranged in $(n_1)^{N_A}$ different ways, giving rise to $(n_1)^{N_A}$ microstates for the initial state. For the final state, we can divide the volume $V_2$ into $n_2$ cells of the volume $V$. The same number of $N_A$ particles can now be arranged in $(n_2)^{N_A}$ different ways; there are $(n_2)^{N_A}$ microstates for the final state. We can now relate the number of possibilities to generate the initial and final states to the entropy using the Boltzmann equation, and obtain

$$\Delta S = S_{\mathrm{final}} - S_{\mathrm{initial}} = k_B \ln\left(n_2^{N_A}\right) - k_B \ln\left(n_1^{N_A}\right) = k_B \ln\left(\frac{n_2}{n_1}\right)^{N_A} = N_A k_B \ln\left(\frac{n_2}{n_1}\right)$$

eq. 2.94

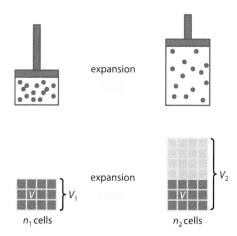

**FIGURE 2.22:   Statistical definition of entropy.** We consider the reversible isothermal expansion of one mol of an ideal gas from volume $V_1$ (left) to volume $V_2$ (right). The volume $V_1$ can be divided into $n_1$ cells with a volume $V$, and the volume $V_2$ can be described as $n_2$ cells of the same volume $V$. The number of microstates for the initial state depends on the number of possibilities for $N_A$ particles ($N_A$ = Avogadro constant, $N_A = 6.022\cdot10^{23}$ mol$^{-1}$) to distribute themselves over the $n_1$ cells of volume $V_1$, which is $n_1^{N_A}$. The number of microstates for the final states is $n_2^{N_A}$. From the number of microstates, we can calculate the entropy of the initial and final states, and $\Delta S$ as the difference. $\Delta S$ calculated from the statistical interpretation of the entropy equals the entropy change calculated from the heat definition (see text).

From our definitions above, we can substitute $n_2$ and $n_1$ for $V_2/V$ and $V_1/V$. Also, the product of $N_A$ and $k_B$ is the general gas constant $R$. We can therefore rearrange eq. 2.94 to

$$\Delta S = R\ln\left(\frac{V_2/V}{V_1/V}\right) = R\ln\frac{V_2}{V_1} \qquad\qquad \text{eq. 2.95}$$

Thus, we obtain the same result for the entropy change $\Delta S$ using the statistical (eq. 2.95) or the heat definition (eq. 2.93) of the entropy $S$, which illustrates that both concepts are equivalent.

## 2.5 THE FREE ENERGY *G*: COMBINING SYSTEM AND SURROUNDINGS

In Section 2.4.2, we derived the Clausius inequality (eq. 2.83), which tells us that spontaneous processes are associated with an increase in entropy, and that the entropy of an isolated system can only increase. By definition, isolated systems do not exchange heat with the surroundings, and entropy considerations can be limited to the system itself. In contrast, a closed system exchanges heat with its surroundings. To describe entropy changes associated with processes in closed systems, we therefore have to include the surroundings in our entropy considerations: we have to take into account the entropy change of the system itself, $\Delta S_{system}$, and of the surroundings, $\Delta S_{surroundings}$. The total change in entropy $\Delta S_{total}$ then becomes

$$\Delta S_{total} = \Delta S_{system} + \Delta S_{surroundings} \qquad\qquad \text{eq. 2.96}$$

For a reversible process, an entropy reduction within the system will be compensated by an increase in entropy of the surroundings of the same magnitude:

$$\Delta S_{surroundings} = -\Delta S_{system} \qquad\qquad \text{eq. 2.97}$$

For an irreversible process, however, more entropy will be generated in the surroundings:

$$\Delta S_{surroundings} > -\Delta S_{system} \qquad\qquad \text{eq. 2.98}$$

Instead of calculating entropy changes of the system and the surroundings separately, it would be convenient to combine these individual entropy changes for the system and the surroundings in one state function. Starting from the total change in entropy (eq. 2.96), we can express the change in entropy of the surroundings as the ratio of the heat transferred to or from the system, and the temperature at which this heat transfer takes place:

$$\Delta S_{total} = \Delta S_{system} - \frac{q_{system}}{T} \qquad\qquad \text{eq. 2.99}$$

For $p$ = const., the heat exchanged in this process equals $\Delta H$. At constant temperature $T$, we can therefore write

$$\Delta S_{total} = \Delta S_{system} - \frac{\Delta H_{system}}{T} \qquad\qquad \text{eq. 2.100}$$

The negative sign for $q$ and $\Delta H$ results from the system-centered perspective: the heat taken up by the surroundings is released from the system ($\delta q < 0$), and heat absorbed from the surroundings is taken up by the system ($\delta q > 0$).

By rearranging eq. 2.100 we obtain

$$-T\Delta S_{total} = \Delta H_{system} - T\Delta S_{system} \qquad\qquad \text{eq. 2.101}$$

On the right-hand side of eq. 2.101 we now only find changes in thermodynamic parameters of the system. Why is this relation useful? A closed system and its surroundings together constitute an isolated system, and $\Delta S_{total}$ represents the change in entropy in this isolated system. For spontaneous processes in isolated systems, $\Delta S$ is always positive, meaning $\Delta S_{total}$ must be positive for any spontaneous process that occurs in a closed system. The temperature $T$, the absolute temperature in K, is always positive. For a positive $\Delta S_{total}$, the term $-T\Delta S_{total}$ on the left-hand side in eq. 2.101 becomes negative. We name $-T\Delta S_{total}$ the change in free energy $\Delta G$ associated with the spontaneous process. $\Delta G$ summarizes entropy changes in system _and_ surroundings. Our criterion for spontaneous processes, an increase in total entropy $S_{total}$ or a negative $-T\Delta S_{total}$, can therefore now be expressed in terms of $\Delta G$: spontaneous processes are characterized by a decrease in the free energy ($\Delta G < 0$). Processes in equilibrium do not involve a change in free energy ($\Delta G = 0$), and processes associated with an increase in free energy ($\Delta G > 0$) do not occur spontaneously.

Processes with $\Delta G < 0$ are termed _exergonic_, processes with $\Delta G > 0$ are _endergonic_.

With $\Delta G$, we have implicitly introduced a new state function, the _free energy G_. The general definition of the free energy is

$$G = H - TS$$

eq. 2.102

$G$ is also called the _Gibbs free energy_. Complete differentiation of $G$ gives

$$dG = dH - TdS - SdT$$

eq. 2.103

For isothermal processes ($dT = 0$), eq. 2.103 simplifies to

$$dG = dH - TdS$$

eq. 2.104

For measureable changes of $\Delta G$ during a reaction, we can formulate the analogous relationship

$$\Delta G = \Delta H - T\Delta S$$

eq. 2.105

Eq. 2.105 is called the _Gibbs-Helmholtz equation_. Comparison of eq. 2.105 and eq. 2.101 shows that the right-hand sides are identical, and refer to the enthalpic and entropic changes within the closed system.

Changes in state functions can be calculated from the difference between the values for the final and initial states. For a chemical reaction, the changes in free energy $\Delta G_r$ therefore correspond to the difference in the free energies of the products (final state) and the reactants (initial state):

$$\Delta G_r = G_{\text{products}} - G_{\text{reactants}}$$

eq. 2.106

To calculate the values of the respective state function for reactants and products, we can use the elements as a reference state. This way, we can define the value of the state function for a molecule as the change in this state function when the molecule is generated from the elements (Figure 2.23). For example, the free energy $G$ of a compound can be defined relative to the free energy of the elements from which this compound can be generated. This value is called the _standard free energy of formation_, $\Delta G_f$. From the standard free energy of formation, the change in free energy associated with a chemical reaction can be determined as

$$\Delta G_r = \Sigma G_{\text{f,products}} - \Sigma G_{\text{f,reactants}}$$

eq. 2.107

Analogous to the standard free energy of formation $\Delta G_f$, standard enthalpies and entropies of formation, $\Delta H_f$ and $\Delta S_f$, can be defined, and their values are tabulated in reference books.

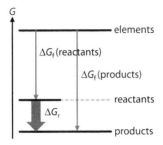

FIGURE 2.23:   **The standard free energy of formation and the change in free energy during a reaction**. The free energy $G$ is a state function, and its change during a chemical reaction can be calculated from the difference of the free energies of products and reactants (red arrow). The free energies of products and reactants can be defined relative to the elements. The free energy of a compound relative to its constituent elements is the free energy of formation, $\Delta G_f$ (blue arrows).

It is important to note the relationship between forward and reverse reactions in terms of changes in state functions: if a reaction is associated with a certain $\Delta G_{forward}$, the reverse reaction will be accompanied by $\Delta G_{reverse}$, the opposite change in $\Delta G$:

$$\Delta G_{forward} = -\Delta G_{reverse} \qquad\qquad \text{eq. 2.108}$$

Hence, if a forward reaction occurs spontaneously ($\Delta G_{forward} < 0$), the reverse reaction will not be spontaneous ($\Delta G_{reverse} > 0$), and *vice versa*.

### 2.5.1  ENTROPY- AND ENTHALPY-DRIVEN REACTIONS

We have now derived a decrease in $\Delta G$ as the criterion for a spontaneous reaction. According to the Gibbs-Helmholtz equation (eq. 2.105), $\Delta G$ is calculated as the difference between the change in enthalpy, $\Delta H$, and the change in entropy, $\Delta S$, converted into an energy value by multiplication with $T$ (eq. 2.105). It is the balance of the $\Delta H$ and the $-T\Delta S$ terms that determines whether a reaction is spontaneous. A large negative change in enthalpy and a large increase in entropy are both favorable and lead to a negative value for $\Delta G$. Likewise, an increase in enthalpy and a decrease in entropy are both unfavorable and $\Delta G$ becomes positive. However, $\Delta H$ and $\Delta S$ can also have opposite effects. If the unfavorable term (positive $\Delta H$ or negative $\Delta S$ and positive $-T\Delta S$) overcompensates the favorable term (positive $\Delta S$ and negative $-T\Delta S$ or negative $\Delta H$), $\Delta G$ is positive, and the reaction does not occur spontaneously. If, however, the favorable term overcompensates the unfavorable term, $\Delta G$ becomes negative, and the reaction will occur spontaneously. Reactions that are energetically favored because of a large negative enthalpy change, but associated with a decrease in entropy, are referred to as *enthalpy-driven reactions*. In contrast, spontaneous reactions associated with an increase in enthalpy that are favored by a large increase in entropy, are called *entropy-driven reactions*. As the entropy change in the Gibbs-Helmholtz equation (eq. 2.105) is weighted by temperature, an unfavorable reaction may become entropy-driven at a higher temperature, or enthalpy-driven at lower temperature (Figure 2.24).

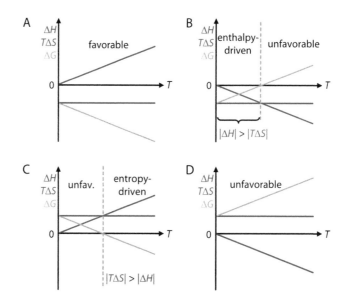

**FIGURE 2.24:   Temperature dependence of enthalpy and entropy changes and effects on the spontaneity of the reaction.** A: $\Delta H < 0$, $\Delta S > 0$. The reaction is favorable ($\Delta G < 0$) at all temperatures. B: $\Delta H < 0$, $\Delta S < 0$. The reaction is unfavorable at higher temperatures, but becomes favorable at low temperatures (enthalpy-driven). C: $\Delta H > 0$, $\Delta S > 0$. The reaction is unfavorable at low temperatures, but becomes favorable (entropy-driven) at higher temperatures. D: $\Delta H > 0$, $\Delta S < 0$. The reaction is unfavorable ($\Delta G > 0$) at all temperatures.

The free energy change $\Delta G$ expresses the balance between enthalpic and entropic changes. Many processes are associated with changes in enthalpy ($\Delta H$) and entropy ($T\Delta S$) that are substantial, but of opposite sign, and the resulting $\Delta G$ can be very modest. This effect is called *enthalpy-entropy compensation*. It is often encountered in biochemical processes, including protein folding (Box 2.7), hybridization of nucleic acids, protein-protein interactions, or the association of lipids into bilayers and membranes.

---

**BOX 2.7: ENTROPY-ENTHALPY COMPENSATION IN PROTEIN FOLDING.**

The three-dimensional structure of proteins is stabilized by numerous individual interactions involving side-chain and backbone atoms (Section 16.2), which constitutes a favorable (negative) enthalpic contribution to protein folding ($\Delta H \ll 0$). On the other hand, protein folding converts a flexible, disordered polypeptide chain into a highly ordered three-dimensionally structured protein; the entropic contribution for folding is unfavorable (negative contribution; $\Delta S \ll 0$). Entropic and enthalpic changes upon folding are substantial (hundreds of kJ mol$^{-1}$), but the resulting thermodynamic stability $\Delta G$, the difference between $\Delta H$ and $T\Delta S$ (eq. 2.105), is often very moderate (a few to tens of kJ mol$^{-1}$) because of *enthalpy-entropy compensation* (see Figure 2.25). We will see in the following Section (Section 2.5.2) that the temperature dependence of $\Delta H$ and $\Delta S$ (and thus of $\Delta G$) leads to enthalpy-driven protein unfolding at low temperatures, and entropy-driven unfolding at high temperatures (Figure 2.25).

### 2.5.2  PRESSURE AND TEMPERATURE DEPENDENCE OF THE FREE ENERGY

We will now derive the pressure- and temperature dependence of the free energy *G*. Starting from the definition of the free energy (eq. 2.102), we can differentiate *G* to

$$dG = dH - TdS - SdT \qquad\qquad \text{eq. 2.109}$$

From the definition of the enthalpy, d*H* can be calculated as

$$dH = dU + pdV + Vdp \qquad\qquad \text{eq. 2.59}$$

and by substitution we obtain

$$dG = dU + pdV + Vdp - TdS - SdT \qquad\qquad \text{eq. 2.110}$$

We can further substitute d*U* by

$$dU = \delta q + \delta w \qquad\qquad \text{eq. 2.4}$$

giving

$$dG = \delta q + \delta w + pdV + Vdp - TdS - SdT \qquad\qquad \text{eq. 2.111}$$

Now we include the (heat) definition of the entropy (eq. 2.74), and substitute *TdS* for δ*q*. Note that this restricts our considerations to reversible processes, because otherwise the equality *TdS* = δq does not hold. We further restrict our considerations to expansion work, and can substitute –*pdV* for δw. Altogether, we obtain

$$dG = TdS - pdV + pdV + Vdp - TdS - SdT \qquad\qquad \text{eq. 2.112}$$

The *TdS* and *pdV* terms in eq. 2.112 cancel and we arrive at

$$dG = Vdp - SdT \qquad\qquad \text{eq. 2.113}$$

For *T* = const. (d*T* = 0), eq. 2.113 further reduces to

$$dG = Vdp \qquad\qquad \text{eq. 2.114}$$

From eq. 2.114, we can derive the pressure dependence of *G* by integration from an initial pressure $p^0$ to the actual pressure *p*:

$$G = \int_{G(p^0)}^{G(p)} dG = \int_{p^0}^{p} Vdp \qquad\qquad \text{eq. 2.115}$$

Substitution of *V* according to the ideal gas equation (eq. 2.1; $p = nRT/V$) and evaluation of the integrals gives the pressure dependence of *G* as

$$G(p) = G(p^0) + nRT \ln\left(\frac{p}{p^0}\right)$$

eq. 2.116

Eq. 2.116 allows us to calculate $G$ for any given pressure $p$ if we know $G$ at the standard pressure $p^0$. By analogy, we can express the pressure dependence of changes in the free energy, $\Delta G$, as

$$\Delta G(p) = \Delta G(p^0) + nRT \ln\left(\frac{p}{p^0}\right)$$

eq. 2.117

The temperature dependence of $G$ and $\Delta G$ can in principle also be derived from eq. 2.113. At constant pressure ($dp = 0$), eq. 2.113 simplifies to

$$dG = -SdT$$

eq. 2.118

If $S$ is independent of temperature, we can directly integrate eq. 2.118 to obtain $G(T)$. However, we have seen in Section 2.4.3 that the entropy $S$ is also temperature-dependent. The temperature dependence of $G$ can be derived directly from $H(T)$ and $S(T)$ at any given temperature $T$ using the Gibbs-Helmholtz equation (eq. 2.105):

$$G(T) = H(T) - T \cdot S(T)$$

eq. 2.119

We have related the temperature dependence of $H$ and $S$ at constant pressure to the heat capacity $C_p$ in eq. 2.64 and eq. 2.86. By substituting these equations into eq. 2.119, we obtain

$$G(T) = H(T_0) + C_p(T - T_0) - T\left(S(T_0) + C_p \ln\frac{T}{T_0}\right)$$

eq. 2.120

Rearranging eq. 2.120 and summarizing $H(T_0) - T \cdot S(T_0)$ as $G(T_0)$ gives

$$G(T) = G(T_0) + C_p\left(T - T_0 - T\ln\frac{T}{T_0}\right)$$

eq. 2.121

Again, we can write the analogous expression for the change of $G$ during a reaction, $\Delta G(T)$:

$$\Delta G(T) = \Delta G(T_0) + \Delta C_p\left(T - T_0 - T\ln\frac{T}{T_0}\right)$$

eq. 2.122

Thus, if $\Delta H$, $\Delta S$, or $\Delta G$ are known at a reference temperature $T_0$, their values $\Delta S(T)$, $\Delta H(T)$, and $\Delta G(T)$ at the temperature $T$ can be calculated using the heat capacity $\Delta C_p$. In other words: the heat capacity $\Delta C_p$ determines the temperature dependence of $\Delta G$, $\Delta H$, and $\Delta S$. We will revisit the temperature and pressure dependence of $G$ and $\Delta G$ later when we discuss the effect of temperature and pressure on equilibrium constants (Section 3.1). Eq. 2.122 gives a bell-shaped profile for $\Delta G(T)$ as a function of temperature (Figure 2.25). The curvature of this curve is determined by $\Delta C_p$. $\Delta C_p$ thus determines the range where $\Delta G$ is negative, and where it becomes positive. A prominent example of the importance of changes in heat capacity is protein folding. $\Delta C_p$ determines the temperature range in which $\Delta G$ of unfolding is positive, and thus defines the temperature range where the protein is stably folded. The curve of $\Delta G_{unfold}$ as a function of temperature is also called the protein stability curve (Figure 2.25; Section 16.3.4). As a consequence of the curvature of the stability curve, proteins do not only unfold at high temperatures (heat denaturation) but also at low temperatures (cold denaturation). Heat denaturation is an entropy-driven reaction, cold denaturation is enthalpy-driven (Section 2.5.1; Box 2.7).

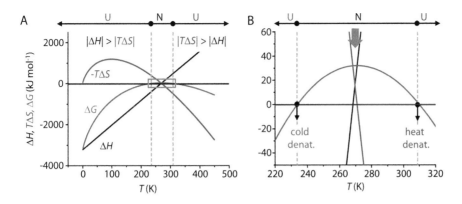

**FIGURE 2.25:    Stability curve of a protein**. A: Temperature dependence of $\Delta H$ (black), $-T\Delta S$ (red) and $\Delta G$ (blue) for unfolding of a protein. $\Delta H(T_m)$ = 500 kJ mol$^{-1}$, $\Delta C_p$ = 12 kJ mol$^{-1}$ K$^{-1}$, $T_m$ = 310 K. At low temperatures, $\Delta G$ for unfolding is negative because of a large negative $\Delta H$ term, the protein unfolds in an enthalpy-driven reaction (U, unfolded). At high temperatures, $\Delta G$ is negative because of a large negative $-T\Delta S$ term; the protein unfolds in an entropy-driven reaction (U). In the temperature range in-between, $\Delta G$ is positive, and the folded protein (N, native) is stable. The strong enthalpy-entropy compensation leads to moderate values for $\Delta G$ in the physiologically relevant temperature range (orange rectangle, magnified in B). B: Physiologically relevant range from A. At the maximum of the bell-shaped $\Delta G$ curve, the thermodynamic stability of the folded protein is maximal (orange arrow). As a consequence of the curvature, $\Delta G$ of unfolding becomes negative at high and low temperatures, i.e. unfolding becomes a spontaneous reaction, and the unfolded state is the more stable state at high and at low temperatures (heat and cold denaturation).

### 2.5.3  STANDARD STATES

We have seen in the previous sections that we can calculate the values of different state functions relative to certain reference states. We can calculate *H*, *S*, and *G* at a temperature *T* relative to the reference temperature $T_0$ (eq. 2.64, eq. 2.86 and eq. 2.121), and we can calculate *G* at a pressure *p* relative to the standard pressure $p^0$ (eq. 2.116). It therefore makes sense to define common reference states, which are called standard states. The pressure under standard conditions, $p^0$, is $10^5$ Pa (1 bar), and the standard temperature $T_0$ is 298.15 K (25°C). Values of the state functions and their changes at standard conditions are marked by a superscript "0", as in $\Delta H^0$, $\Delta S^0$, $\Delta G^0$. We can now use the equations we derived for the temperature and pressure dependence of the state functions, and calculate their values at any pressure *p* or temperature *T*, relative to the standard state.

We will see in Sections 2.6.5 and 2.6.6 that we can also calculate the concentration dependence of state functions relative to a standard state, which is required to describe the thermodynamics of mixtures. The standard state for components of mixtures is the pure component: pure solid, pure liquid, or pure gas, depending on the aggregation state of the pure compound at $T_0$ and $p^0$. For solutions, the standard state of the solvent is the pure solvent. In contrast, the standard state for the dissolved components, the *solutes*, is a concentration of 1 M.

The definition of the standard states for solutes has certain limitations when we look at the thermodynamics of biochemical reactions. Biochemical reactions at physiological conditions usually occur at a neutral pH value of around 7. Yet, the definition of the standard state refers to 1 M concentration of the solute. For $H^+$ ions, this corresponds to a solution with pH 0, a condition that is of little physiological relevance. Therefore, the biochemical standard state uses a different definition for protons: the standard state is pH 7 ($10^{-7}$ M protons), whereas for all other solutes, the standard state of 1 M concentration is maintained. The corresponding values for the state functions under biochemical standard conditions are denoted by a prime, and written as $\Delta H^{0\prime}$, $\Delta S^{0\prime}$, and $\Delta G^{0\prime}$. It is important to note that it does not matter which standard state we choose to calculate the value of the state function at a certain concentration, as long as we consistently stay in one of the reference systems (see Section 2.6.5; Box 2.10).

### 2.5.4 RELATION OF FREE ENERGY, ENTHALPY, AND ENTROPY TO MOLECULAR PROPERTIES

How do molecular properties contribute to enthalpy and entropy, and how can changes in these state functions during chemical reactions be related to molecular properties of the compounds involved? Enthalpy is the sum of all energies of each molecule in the system (intramolecular energy) and of the energy from interactions between these molecules (intermolecular energy; Figure 2.26). The intramolecular energy contains contributions from translational, rotational, and vibrational motions of each individual molecule (also referred to as the thermal energy), and the energy is stored in its chemical bonds. In macromolecules, intramolecular interactions also contribute to the intramolecular energy. The intermolecular energy contains contributions from van der Waals and ionic interactions and intermolecular hydrogen bonds (Section 15.3). Entropy, which we have introduced as randomness or ways of realization of a certain state in Section 2.4.5, cannot be defined on the level of an individual molecule. Instead, entropy is a measure of the ways how to distribute the energy in the entire system among the individual molecules. Energies associated with translation, rotation, and vibration can be distributed over each molecule (vibration) or among molecules (translation, rotation). In contrast, the energy of chemical bonds is fixed within each molecule and cannot be distributed within or among molecules. Bonds therefore do not contribute to entropy. Box 2.8 illustrates how changes in entropy and enthalpy during a chemical reaction can be related to molecular properties.

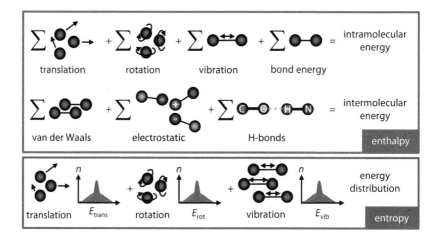

**FIGURE 2.26:** **Molecular components of entropy and enthalpy.** Enthalpy is the sum of all intramolecular and intermolecular energies, such as translation, rotation, and vibration of individual molecules and the energy of covalent bonds, or van der Waals and hydrogen bonding and ionic interactions between molecules. Entropy can only be defined on the basis of an ensemble of molecules and is related to the distribution of energy among molecules. While translational, rotational, and vibrational energies can be distributed among molecules, and contribute to the entropy of the system, energy in chemical bonds is located and cannot be distributed, and therefore does not contribute to entropy.

## BOX 2.8: MOLECULAR CONTRIBUTIONS TO ENTHALPY AND ENTROPY CHANGES IN ATP HYDROLYSIS.

ATP is hydrolyzed to ADP and inorganic phosphate. The reaction is associated with a free energy change at pH 7 of $\Delta G^{0\prime}$ of −31 kJ mol$^{-1}$ (310 K). The large negative $\Delta G^{0\prime}$ results from a negative enthalpy change ($\Delta H^{0\prime}$ = −20.1 kJ mol$^{-1}$) and a large increase in entropy ($\Delta S^{0\prime}$ = +33.5 J mol$^{-1}$ K$^{-1}$), i.e. both enthalpy and entropy change are favorable. What are the molecular reasons for these changes? We first take a look at the enthalpy. The magnitude of the enthalpy change is dominated by the disruption and formation of bonds. During ATP hydrolysis, one P-O bond is broken (between γ- and β-phosphate; Figure 2.27), which is associated with a positive $\Delta H^{0\prime}$. However, a new P-O bond is formed between the water molecule that performs the nucleophilic attack and the γ-phosphate, which is associated with a negative $\Delta H^{0\prime}$. The energies involved in bond breaking and bond formation are large, but because they are equal in magnitude but of opposite sign, they cancel and do not contribute to the overall $\Delta H^{0\prime}$. Apart from bonds being broken and formed, ATP hydrolysis leads to a distribution of the negative charges of the three phosphate groups (ATP$^{4-}$ at pH 7) onto two separate molecules, ADP (ADP$^{2-}$) and phosphate (PO$_4^{2-}$). The separation reduces the electrostatic repulsion, which is energetically favorable and associated with a negative $\Delta H^{0\prime}$. Finally, we have to consider hydration because the reaction occurs in aqueous solution. The hydration of ADP and phosphate is energetically more favorable compared to the hydration of ATP, and hydration also contributes to the negative $\Delta H^{0\prime}$. The negative change in enthalpy during ATP hydrolysis is thus a combined effect of the reduced electrostatic repulsion and the increased hydration of the products.

The entropic term depends on the possibilities to distribute energy among the molecules. ATP hydrolysis converts one reactant, ATP, into two products. The overall energy can now be distributed over both molecules, which causes an increase in entropy. On the other hand, the two product molecules are hydrated. Although hydration contributes favorably to the enthalpy change, it also leads to a higher order of water molecules, associated with a decrease in water entropy. The overall increase in entropy for the hydrolysis reaction tells us that the favorable entropy term, because of splitting one molecule into two, overcompensates the unfavorable term due to water ordering. The decrease in enthalpy and increase in entropy then add up to an overall decrease in free energy, and ATP hydrolysis is a spontaneous reaction (under standard conditions).

Note that the term "high-energy bond" often used to describe the large change in free energy associated with ATP hydrolysis is very misleading. On a molecular level, bond breaking and formation is isoenergetic, and does not contribute to the overall $\Delta H^{0\prime}$ and $\Delta G^{0\prime}$. It is also important to note that the large, negative $\Delta G^{0\prime}$ does not tell us anything about the velocity of ATP hydrolysis. In fact, ATP hydrolysis is a slow reaction because of a high activation barrier (Section 13.2). In the cell, hydrolysis of ATP is faster because it is catalyzed by enzymes. Synthesis and hydrolysis of ATP, and thus the energy level of the cell, can be regulated thereby.

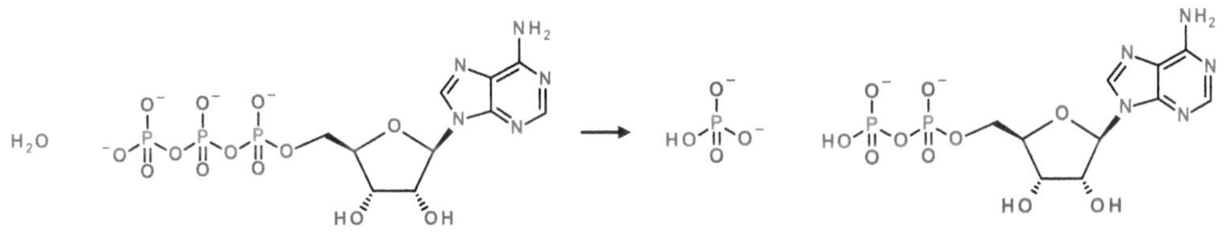

**FIGURE 2.27:    Hydrolysis of ATP to ADP and phosphate**. ATP hydrolysis to ADP and phosphate is associated with a decrease in enthalpy ($\Delta H^{0\prime}$ = −20.1 kJ mol$^{-1}$), and an increase in entropy ($\Delta S^{0\prime}$ = 33.5 J mol$^{-1}$ K$^{-1}$), leading to an overall decrease in free energy ($\Delta G^{0\prime}$ = −31 kJ mol$^{-1}$) at 310 K.

## 2.6 THE CHEMICAL POTENTIAL μ

So far, we have considered processes in isolated or closed systems that do not exchange matter with their surroundings. In other words, in the processes we discussed, the amount of substance $n$ (in mol) has remained constant. However, we often need to treat processes where $n$ is changing, e.g. chemical reactions or open systems. Open systems exchange matter with their surroundings, and their composition is not constant. Chemical reactions convert reactants into products. Over the course of the reaction, the reactants are consumed, and their amount $n$ is reduced. Conversely, products are formed, and their amount $n$ keeps increasing as long as the reaction proceeds. We have seen that $U$, $H$, $S$, and $G$ are extensive state functions: their values depend on the size of the system, and thus on $n$. The free energy $G$ is therefore not necessarily a useful state function when we consider chemical reactions and open systems where $n$ is not constant.

We can introduce a new state function that is independent of the amount $n$ by the following consideration: if we consider a pure compound, an infinitesimal change in the amount $n$ of this compound by dn leads to an infinitesimal change d$G$ in the free energy of the system (at constant pressure and temperature). We define the proportionality constant as the *chemical potential* μ of this compound:

$$dG = \mu dn \qquad \text{eq. 2.123}$$

or

$$\mu = \left(\frac{\partial G}{\partial n}\right)_{p,T} \qquad \text{eq. 2.124}$$

Eq. 2.124 is a partial derivative (see Section 29.6): $G$ is a function of $p$, $T$, and $n$, but with the partial derivative we only consider changes in $n$. For a pure compound, the chemical potential is simply equal to the free energy $G$ per mol, or molar free energy $G_m$:

$$\mu = \frac{G}{n} = G_m \qquad \text{eq. 2.125}$$

In analogy to eq. 2.125, we can also define the molar enthalpy $H_m$, the molar entropy $S_m$, or the molar volume $V_m$ of a pure substance by dividing the corresponding extensive properties by the amount of substance $n$. By the normalization to one mol, the extensive properties $G$, $H$, $S$, and $V$ become intrinsic properties $G_m$, $H_m$, $S_m$, or $V_m$.

### 2.6.1 THE CHEMICAL POTENTIAL AS A DRIVING FORCE FOR CHEMICAL REACTIONS

The general considerations on the chemical potential apply to pure substances, but it becomes much more useful for the description of mixtures. Most chemical reactions occur in mixtures of reactants and generate a mixture of products. $\Delta H_r$, $\Delta S_r$, and $\Delta G_r$ describe the changes in thermodynamic state functions for the reaction, but summarize the entire system. To understand the behavior of mixtures, it is often useful to describe the state from the perspective of one individual component. The chemical potential provides a convenient way to quantify the contribution of each component of the mixture to the free energy. For a mixture of $i$ components, we can sum up the changes in free energy d$G$ of the entire mixture caused by the increase in the amount $n_i$ of each component as

$$dG = \sum_i \mu_i dn_i \qquad \text{eq. 126}$$

where $\mu_i$ is the chemical potential of compound $i$. From eq. 2.126, we can infer that the free energy $G$ of a mixture can be calculated from the chemical potential of the individual components $i$ of the system and their amount $n_i$ (eq. 2.127):

$$G = \sum_i \mu_i n_i$$

eq. 2.127

Changes in free energy during a reaction are defined as the difference between $G$ of the final state (products) and $G$ of the initial state (reactants), leading to

$$\Delta G = G_{\text{products}} - G_{\text{reactants}} = \sum_{\text{products}} \mu_i n_i - \sum_{\text{reactants}} \mu_i n_i$$

eq. 2.128

or

$$\Delta G = \sum_i \mu_i \Delta n_i$$

eq. 2.129

$\Delta n_i$ denotes changes in the amount of substance for each component, with a positive sign for products that are being formed, and a negative sign for reactants being consumed. This convention is again a reflection of the system-centered perspective: reactants are disappearing from the system, and products are appearing. The value for $\Delta G$ in eq. 2.128 becomes negative if the sum of the chemical potential of the reactants is larger than the chemical potential of the products. For open systems and for chemical reactions, we can therefore formulate the condition $\Delta G < 0$ for spontaneous processes in terms of the chemical potentials of the reactants and the products as

$$\sum_{\text{reactants}} \mu_i n_i > \sum_{\text{products}} \mu_i n_i$$

eq. 2.130

A reaction with

$$\sum_{\text{reactants}} \mu_i n_i = \sum_{\text{products}} \mu_i n_i$$

eq. 2.131

is in equilibrium, and a reaction with

$$\sum_{\text{reactants}} \mu_i n_i < \sum_{\text{products}} \mu_i n_i$$

eq. 2.132

does not occur spontaneously (Table 2.2). Eq. 2.130 explains the term "chemical potential" for $\mu$. A potential is generally defined as the capability to perform work. In accordance with this definition, the chemical potential is a measure of the capability to drive a chemical reaction from reactants to products (Figure 2.28).

| **TABLE 2.2:** Criteria for spontaneous processes, equilibria and non-spontaneous processes in isolated, closed, and open systems. | | |
|---|---|---|
| | **spontaneous** | **equilibrium** | **non-spontaneous** |
| isolated | $\Delta S > 0$ | $\Delta S = 0$ | $\Delta S < 0$ |
| closed | $\Delta G < 0$ | $\Delta G = 0$ | $\Delta G > 0$ |
| open | $\sum n\mu_{reactants} > \sum n\mu_{products}$ | $\sum n\mu_{reactants} = \sum n\mu_{products}$ | $\sum n\mu_{reactants} < \sum n\mu_{products}$ |

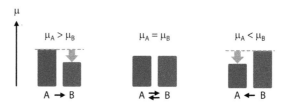

**FIGURE 2.28:   The chemical potential as a driving force for reactions**. The difference in chemical potential between reactant A (blue) and product B (red) is the driving force for a reaction from A to B (left, gray arrow). When the chemical potentials of A and B are identical (middle), there is no driving force for the reaction, and the reaction is in equilibrium. A higher chemical potential of B is the driving force for the reverse reaction from B to A (right, gray arrow).

## 2.6.2 THE CHEMICAL POTENTIAL AND STABLE STATES: PHASE DIAGRAMS

The chemical potential is a driving force for processes and chemical reactions. We have derived the chemical potential from the free energy, and have seen that the chemical potential of a pure substance is equivalent to the molar free energy. For pure substances, we can therefore describe the chemical potential according to the Gibbs-Helmholtz-equation (eq. 2.102), using the molar enthalpy $H_m$ and the molar entropy $S_m$ as parameters:

$$\mu = G_m = H_m - TS_m \qquad \text{eq. 2.133}$$

We now assume that $H_m$ and $S_m$ do not depend on temperature. Although this is a simplification (we have learnt that the heat capacity $C_p$ describes the temperature dependence of both $H$ and $S$; eq. 2.64 and eq. 2.86), it helps us to understand why compounds exist in solid, liquid, or gaseous states, why these states are stable in certain temperature ranges, and why they inter-convert at defined temperatures. For temperature-independent $H_m$ and $S_m$, eq. 2.133 describes a linear dependence of $\mu$ on the temperature, with a slope $-S_m$ and a $y$-axis intercept $H_m$. Thus, when we plot $\mu$ as a function of $T$ (Figure 2.29), the chemical potentials for the solid, liquid, and gaseous states of a pure substance are straight lines with different slopes and intercepts. To assign differences in slopes and intercepts, we have to compare the molar enthalpies and entropies for the three states. The solid phase has the highest order, and the order decreases from solid (s) to liquid (l) to gaseous state (g), which is equivalent to an increase in randomness from solid to gas. Therefore we can write for the molar entropies

$$S_m(g) > S_m(l) > S_m(s) \qquad \text{eq. 2.134}$$

Melting a solid requires energy to overcome interactions between particles in the solid phase. Likewise, evaporation of a liquid to a gas requires energy to overcome interactions between particles in the liquid. This energy remains in the system, which means that

$$H_m(g) > H_m(l) > H_m(s) \qquad \text{eq. 2.135}$$

Thus, the slopes of the $\mu(T)$-plot will increase from solid to liquid to gas, and so will the $y$-axis intercepts (Figure 2.29).

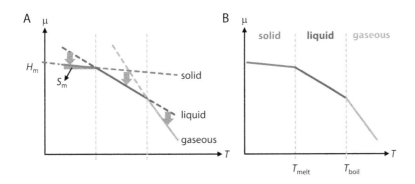

**FIGURE 2.29:    The chemical potential of solid, liquid, and gaseous states as a function of temperature.** A: The chemical potential depends linearly on temperature. The slope of the line is the (negative) molar entropy – $S_m$ the intercept with the y-axis is the molar enthalpy $H_m$. The molar entropy $S_m$ of a compound is low in the solid state, higher in the liquid state, and highest in the gaseous state. The molar enthalpy $H_m$ increases in the same order. Both slopes and y-axis intercepts therefore increase from solid state (blue) to liquid state (red) to gaseous state (green). Spontaneous processes are driven by a decrease in chemical potential (gray arrows). B: Phase diagram. The most stable state is always the one with the lowest chemical potential: the solid state at low temperatures, the liquid state at intermediate temperatures, and the gas at high temperatures. At the melting temperature $T_{melt}$, the chemical potentials of solid and liquid are equal, and solid and liquid are in equilibrium. Similarly, liquid and gas are in equilibrium at the boiling temperature $T_{boil}$.

Spontaneous processes are driven by a decrease in the chemical potential. If we now look at the straight lines in Figure 2.29 that represent the chemical potential of the different states, we see that they divide the temperature range into three different regions. At low temperatures, the chemical potential of the solid is lower than the chemical potential of the liquid state. As a consequence, the transition from solid to liquid is associated with an increase in the chemical potential, and does not occur spontaneously. The reverse process, the transition from liquid to solid, however, leads to a decrease in chemical potential, and is spontaneous. In this temperature range, the solid is thus the stable phase. The lines for solid and liquid states intersect at a certain temperature. At this temperature, the chemical potentials of the solid and liquid state are equal, and the two are in equilibrium. This temperature is the melting point of the solid (or the freezing point of the liquid, depending on the starting point). The part of the system that is in the liquid state is called the liquid phase, the solid part is the solid phase. A *phase* is defined as a continuous part of a system that is homogeneous in composition and physical state. At temperatures above the melting point, the chemical potential of the liquid state is lower than the chemical potential of the solid, and any solid present will be converted into liquid in a spontaneous reaction (with a decrease in chemical potential). At a higher temperature, the lines for the chemical potential of the liquid and gaseous states intersect, meaning the chemical potentials of these phases are equal, and they are in equilibrium. This temperature is the boiling or evaporation point of the liquid, or the condensation temperature of the gas. At temperatures above the evaporation temperature, the gas will be the phase with the lowest chemical potential, and all phase transitions towards the gas phase lead to a decrease in the chemical potential, and are therefore spontaneous. Thus, we can generalize that the state with the lowest chemical potential is the most stable state under the prevalent conditions. A diagram that shows the most stable state for a certain parameter, such as the temperature, is called a *phase diagram*. Phase diagrams are also used to describe phase transitions of biomolecules, such as protein folding and unfolding, the hybridization or dissociation of DNA strands, or phase transitions in lipid bilayers (Box 2.9).

## BOX 2.9: PHASE DIAGRAMS OF BIOMOLECULES.

Instead of considering solid, liquid, and gaseous state, we can also use the chemical potential to rationalize protein folding, unfolding, and stability (Figure 2.30). Eq. 2.133 also holds for the chemical potential of the folded state of a protein, and the chemical potential of the folded state will decrease linearly with temperature (again under the assumption that $H_m$ and $S_m$ are temperature-independent. The folded state is more ordered than the unfolded state of a protein, and the molar entropy of the folded state, $S_{m,folded}$, will be lower than the molar entropy of the unfolded state, $S_{m,unfolded}$. To unfold a protein, interactions between side chains and main-chain atoms need to be overcome, which requires energy. The molar enthalpy of the unfolded state, $H_{m,unfolded}$, will therefore be higher than the molar enthalpy of the folded state, $H_{m,folded}$. Thus, the curve for the chemical potential of the unfolded state has a higher slope and larger intercept than the curve for the folded state. The intersection point of the two curves defines the unfolding temperature of the respective protein. The chemical potentials of folded and unfolded states are equal, and both states are in equilibrium. At temperatures below the unfolding temperature, the folded state has a higher chemical potential, and folding is the spontaneous reaction. At higher temperatures, unfolding is spontaneous. The same scheme applies to phase changes of nucleic acids (melting of double-stranded DNA to single strands) or to phase transitions in lipid membranes (transition from solid phase (gel) to liquid phase).

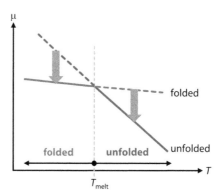

FIGURE 2.30: **Phase diagram for a protein that can be heat-denatured**. The chemical potential of the folded and unfolded state depends linearly on temperature, with the slope determined by the molar entropy $S_m$, and the $y$-axis intercept defined by the molar enthalpy $H_m$. The lines representing the chemical potential intersect at the melting temperature $T_{melt}$ ($\Delta\mu$ = zero, equilibrium). At temperatures below $T_{melt}$, the chemical potential of the unfolded state is higher, providing a driving force for folding (arrow). At temperatures above $T_{melt}$, the chemical potential of the folded state is higher, providing a driving force for unfolding. Hence, at lower temperatures, the folded state is more stable, at higher temperatures the unfolded state is more stable. Note that the chemical potential is only linear with temperature if $H_m$ and $S_m$ are temperature independent. When $H_m$ and $S_m$ are temperature-dependent, plots of the chemical potential $\mu$ as a function of temperature are non-linear (see also Figure 2.25).

### 2.6.3 Pressure and Temperature Dependence of the Chemical Potential

In analogy to the standard free energy $G^0$, the standard chemical potential $\mu^0$ of gases, liquids, and solids is defined as the chemical potential of the pure compounds under standard pressure ($10^5$ Pa) and at standard temperature (298.15 K). The standard chemical potential of elements is zero by definition. Pure elements hence serve as a reference state, as we have seen in the definition of enthalpies, entropies, and free energies of formation (Section 2.5; Figure 2.23).

The pressure- and temperature dependence of the chemical potential directly follows from the pressure- and temperature dependence of the free energy (eq. 2.116, eq. 2.119). The chemical potential $\mu$ at a pressure $p$ is

$$\mu(p,T) = \mu^0(T) + RT \ln\left(\frac{p}{p^0}\right)$$

eq. 2.136

$\mu^0(T)$ in eq. 2.136 refers to standard pressure $p^0$ ($10^5$ Pa), but does not have to be the value at standard temperature, as long as the temperature is constant. The second term corrects the chemical potential for deviations of the actual pressure $p$ from the standard pressure $p^0$. The temperature dependence of the chemical potential at constant pressure is described by the Gibbs-Helmholtz equation with the molar enthalpy $H_m$ and molar entropy $S_m$ at the actual temperature $T$:

$$\mu(T) = G_m(T) = H_m(T) - T \cdot S_m(T)$$

eq. 2.133

### 2.6.4 The Chemical Potential as a Partial Molar Property

In Section 2.6 (eq. 2.123 and 2.124), we introduced the chemical potential as the change in free energy caused by a change in amount of substance. In the next section (Section 2.6.5), we will see that the chemical potential is a very useful property to describe individual components in mixtures. The chemical potential $\mu_i$ of a component $i$ of a mixture is defined as the change in free energy of the entire system at constant pressure and temperature when the amount of substance of the component $i$ is changed, while the amount of all other components remains constant. We can express this definition of the chemical potential $\mu_i$ as the partial derivative of $G$ according to eq. 2.137:

$$\mu_i = \left(\frac{\partial G}{\partial n_i}\right)_{p,T,n_j \neq n_i}$$

eq. 2.137

The index $p,T,n_j \neq n_i$ indicates that pressure $p$ and temperature $T$ as well as the amounts $n_j$ of all other components of the mixture remain constant; the only changing parameter is $n_i$. The chemical potential $\mu_i$ is also called the *partial molar free energy* of the component $i$. For a pure compound, eq. 2.137 reduces to eq. 2.125.

By analogy to the definition of the chemical potential as the partial molar free energy, we can define a number of partial molar properties for individual components of mixtures, such as the partial molar enthalpy $H_{m,i}$,

$$H_{m,i} = \left(\frac{\partial H}{\partial n_i}\right)_{p,T,n_j \neq n_i}$$

eq. 2.138

the partial molar entropy $S_{m,i}$,

$$S_{m,i} = \left(\frac{\partial S}{\partial n_i}\right)_{p,T,n_j \neq n_i}$$

eq. 2.139

or the partial molar volume $V_{m,i}$:

$$V_{m,i} = \left(\frac{\partial V}{\partial n_i}\right)_{p,T,n_j \neq n_i}$$

eq. 2.140

Partial molar properties describe how the particular state function or state variable changes with the amount of one of the components of the mixture. It is important to note that the partial molar volume for components of a mixture is not identical to the molar volume of the pure compounds. Instead, all partial molar properties depend on the composition of the mixture. As a consequence, the partial molar properties of the components of the mixture are not independent: altering the partial molar property for one component of a mixture leads to a change in the partial molar property for the other components.

We can describe the volume, entropy, enthalpy, or free energy of the entire mixture by summing up the partial molar properties for each component, weighted by the amount of substance $n_i$ of each constituent:

$$V = \sum_i n_i V_{m,i}$$

$$S = \sum_i n_i S_{m,i}$$

$$H = \sum_i n_i H_{m,i}$$

$$G = \sum_i n_i G_{m,i} = \sum_i n_i \mu_i$$

eq. 2.141

The relation of molar volume, partial molar volume, and its dependence on the composition of the mixture can be illustrated considering a water-ethanol mixture as an example. The partial molar volume of pure ethanol, $V_{m,i}$, equals its molar volume $V_m$. The partial molar volume of pure water is also identical to its molar volume. In mixtures of ethanol and water, their partial molar volumes deviate from the molar volume of the pure compounds. The exact value for the partial molar volumes depends on the composition of the mixture, that is on the molar fractions of ethanol and water in the mixture. Because of favorable interactions between water and ethanol, the partial molar volume of water and ethanol in the mixture is smaller than the partial molar volume of the pure substances. As a consequence, when we mix 50 mL of ethanol with 50 mL of water, we do not obtain 100 mL of an ethanol/water mixture, but a smaller volume.

### 2.6.5 THE CHEMICAL POTENTIAL OF COMPOUNDS IN MIXTURES

To calculate the chemical potential for mixtures, we have to take into account the fraction of each component $i$ of the mixture. We again start our consideration from ideal gases and their mixtures. According to the ideal gas equation (eq. 2.1), the pressure $p$ of a gas is proportional to the ratio $n/V$ of amount of substance $n$ and the volume $V$, which is equal to the concentration $c$. Thus, the pressure of a gas is a measure of the concentration. For a mixture of gases, we can express the fraction of each component $i$ by its *partial pressure* $p_i$. The sum of the partial pressure values $p_i$ for all components is the pressure $p$ of the gas mixture:

$$p = \sum_i p_i$$

eq. 2.142

The additivity of partial pressures to the total pressure of a gas mixture is called *Dalton's law*. The chemical potential for the component $i$ in this gas mixture is

$$\mu_i\left(p^0, T\right) = \mu_i^0\left(p^0, T\right) + RT \ln \frac{p_i}{p^0}$$

eq. 2.143

$\mu_i^0$ refers to the chemical potential of the component $i$ in its standard state (pure gas) and at standard pressure. $p_i$ is the partial pressure of component $i$, and $p^0$ is the standard pressure.

Instead of using partial pressures, we can also relate the chemical potential of mixtures to the *mole fraction* $x_i$ of component $i$. The mole fraction of component $i$ in a mixture of $i$ different components is defined as

$$x_i = \frac{n_i}{\sum_i n_i} = \frac{n_i}{n}$$

eq. 2.144

$n_i$ is the amount of component $i$, $n$ is the total amount of substance in the mixture. Thus, $x_i$ is in the range of zero to one. The sum of $x_i$ for all components is one, which is very convenient in the treatment of binary mixtures (Section 2.6.7). From the ideal gas equation (eq. 2.1), we can derive

$$x_i = \frac{n_i}{n} = \frac{p_i V}{RT} \cdot \frac{RT}{pV} = \frac{p_i}{p}$$

eq. 2.145

By rearranging eq. 2.145 we obtain *Raoult's law*: in a mixture of (ideal) gases, the partial pressure $p_i$ of each component is the total pressure $p$ of the mixture multiplied by the mole fraction $x_i$ of component $i$:

$$p_i = x_i \cdot p$$

eq. 2.146

Combining eq. 2.143 and eq. 2.146, we can now write the concentration dependence of the chemical potential at standard pressure $p^0$ as

$$\mu_i = \mu_i^0\left(p^0, T\right) + RT \ln x_i$$

eq. 2.147

For $x_i \rightarrow 1$, i.e. pure component $i$, the logarithmic term approaches zero, and the chemical potential of component $i$ becomes the standard chemical potential $\mu_i^0$, which is just the chemical potential of the pure compound $i$. The mole fraction $x_i$ expresses the fraction of components not only in gas mixtures, but also in liquid mixtures and solutions:

$$x_i = \frac{n_i}{n} = \frac{c_i V}{cV} = \frac{c_i}{c}$$

eq. 2.148

where $n$ and $c$ denote the total amount and concentration of the mixture. Eq. 2.147 is therefore useful for the description of gaseous and liquid mixtures as well as solutions (Section 2.6.6).

We will see in the following that we can transfer our knowledge about the chemical potential of components of gas mixtures directly to liquids and liquid mixtures (Figure 2.31). First, we consider a pure liquid A with a chemical potential $\mu_A^*$. The star indicates that we refer to a pure compound. The liquid phase (l) is in equilibrium with the gas phase (g). As a consequence, the gas phase contains vapor of compound A with a partial pressure $p_A^*$. The pressure $p_A^*$ is called the *vapor pressure* of A.

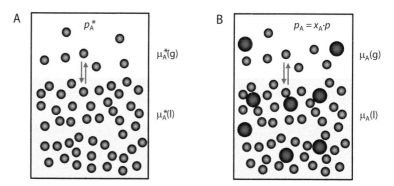

FIGURE 2.31:    **Vapor pressure of liquids and liquid mixtures**. A: A pure liquid (blue) is in equilibrium with the gas phase (white) above. The pressure of the gas phase is defined as the vapor pressure of the liquid $p_A^*$. B: In a liquid mixture, all components are in equilibrium in liquid and gas phases. Their partial vapor pressure in the gas phase is proportional to the mole fraction $x_i$ in the mixture.

Because liquid and gas phase (vapor) are in equilibrium, the chemical potential of the compound A in both phases must be equal:

$$\mu_A^*(l) = \mu_A^*(g) \qquad\qquad \text{eq. 2.149}$$

The stars in eq. 2.149 again mark properties of the pure compound. The chemical potential of A in the vapor is described by eq. 2.143. Together, we can therefore write

$$\mu_A^*(l) = \mu_A^0\left(p^0, T\right) + RT \ln \frac{p_A^*}{p^0} \qquad\qquad \text{eq. 2.150}$$

Eq.2.150 describes the chemical potential of the pure compound A in the liquid phase. To describe the chemical potential of compound A as part of a liquid mixture, we need to express the chemical potential of the gas phase as a function of the partial pressure $p_A$ (eq. 2.143). The chemical potential $\mu_A(l)$ of A in the liquid mixture then becomes

$$\mu_A(l) = \mu_A^0\left(p^0, T\right) + RT \ln \frac{p_A}{p^0} \qquad\qquad \text{eq. 2.151}$$

Subtraction of eq. 2.150 from eq. 2.151 gives

$$\mu_A(l) - \mu_A^*(l) = RT \ln \frac{p_A}{p^0} - RT \ln \frac{p_A^*}{p^0} \qquad\qquad \text{eq. 2.152}$$

which can be rearranged to

$$\mu_A(l) = \mu_A^*(l) + RT \ln \frac{p_A}{p_A^*} \qquad\qquad \text{eq. 2.153}$$

The chemical potential of compound A in a liquid mixture thus depends on the partial vapor pressure of A in the mixture relative to the vapor pressure of pure A. Following Raoult's law (eq. 2.146), the partial vapor pressure $p_A$ can be expressed in terms of the mole fraction $x_A$ and the vapor pressure of pure A, $p_A^*$, as

$$p_A = x_A \cdot p_A^* \qquad\qquad \text{eq. 2.154}$$

We can therefore convert eq. 2.153 into

$$\mu_A(l) = \mu_A^*(l) + RT \ln \frac{x_A \cdot p_A^*}{p_A^*} = \mu_A^*(l) + RT \ln x_A \qquad \text{eq. 2.155}$$

Eq. 2.155 corresponds to the dependence of the chemical potential on the mole fraction described in eq. 2.147. Thus, the chemical potential of components in gaseous and liquid mixtures depends on their mole fraction $x_i$.

Instead of using the mole fraction $x_i$, we can also describe the chemical potential of liquid mixtures in terms of molar concentration $c$ by

$$\mu_i = \mu_i^0 + RT \ln \frac{c_i}{c_i^0} \qquad \text{eq. 2.156}$$

In eq. 2.156, $\mu_i^0$ refers to the chemical potential of the compound in its standard state (pure liquid), $c_i$ is the actual concentration of component $i$, and $c_i^0$ is the concentration of $i$ in the standard state (pure liquid).

For solids, the chemical potential is concentration-independent. The chemical potential of each component in the solid mixture is therefore equal to the standard potential of the pure compound:

$$\mu_i = \mu_i^0 \qquad \text{eq. 2.157}$$

We will briefly revisit the chemical potential of solids in the context of electrochemistry (Chapter 5). The majority of biochemical reactions occur in solutions, however, and in the following sections, we therefore focus on the chemical potential and properties of solutions.

### 2.6.6 THE CHEMICAL POTENTIAL OF SOLUTIONS

Solutions are a special case of liquid mixtures in which one component, the *solvent*, is in large excess over the other components, the *solutes*. We can express the chemical potential of solvent and solutes by eq. 2.156:

$$\mu_i = \mu_i^0 + RT \ln \frac{c_i}{c_i^0} \qquad \text{eq. 2.156}$$

Although the overall form of the chemical potential is identical for solvent and solute(s), the relevant concentrations $c_i^0$ in their standard states are different. By definition, the solvent is present in large excess, and the solute is the minor component. Therefore, the standard state for the solvent is the pure liquid, whereas for solutes, the concentration in the standard state is defined as 1 M (Section 2.5.3). For aqueous solutions, we have to use a standard-state concentration for the solvent water (55.5 M), and 1 M for the solutes. We have seen before that the biochemical standard state defines a different reference state for protons in solutions (pH 7, $c^0 = 10^{-7}$ M). It is important to use a consistent system of standard states, otherwise the calculated values will not match (Box 2.10)!

---

**BOX 2.10: THE CHEMICAL POTENTIAL OF
PROTONS IN AQUEOUS SOLUTION.**

The chemical potential for H+ in a 0.1 M solution of HCl can be calculated by using the chemical standard state (1 M H+, corresponding to a pH of 0) and the corresponding standard state chemical potential:

$$\mu_{H^+} = \mu_{H^+}^0 + RT \ln \frac{0.1M}{1M}$$

*(Continued)*

Alternatively, we can use the biochemical standard state (pH 7 = $10^{-7}$ M for protons), and calculate the chemical potential as

$$\mu_{H^+} = \mu_{H^+}^{0'} + RT \ln \frac{0.1M}{10^{-7}M}$$

using the standard chemical potential at pH 7 (indicated by the prime) as a reference. Because the chemical potential is a state function, the result of both calculations must be identical, but the reference values $\mu^0$ and $\mu^{0'}$ differ.

We can also describe the chemical potential of solvent and solutes in terms of the mole fraction $x_i$ using eq. 2.147.

$$\mu_i = \mu_i^0 + RT \ln x_i \qquad \text{eq. 2.147}$$

For a dilute solution, the mole fraction is near unity for the solvent, and very small for the solutes. The reference chemical potential $\mu_i^0$ $(p,T)$ now refers to the chemical potential of the pure compound $(x_i = 1)$. We will see in the following section (Section 2.6.7) that the description of the chemical potential of solutions in terms of the mole fraction is very practical, and enables us to quantitatively describe important properties of solutions.

The general equation for the mole fraction dependence of the chemical potential of a solution only holds for *ideal solutions*. In an ideal solution, the molecules of the solute and the solvent do not interact among themselves or with each other. This is the exception, however, and in most solutions, interactions between solvent and solute do occur. Nevertheless, many dilute solutions ($x_{solute} \ll 1$) behave as ideal solutions, and their behavior can be described by eq. 2.147. Solutions in biological systems, however, frequently deviate from ideality because the solutes interact with the solvent, or the solutes interact with each other, and they are highly concentrated. For such *non-ideal solutions*, the chemical potential does not follow eq. 2.147:

$$\mu_i \neq \mu_i^0 + RT \ln x_i \qquad \text{eq. 2.158}$$

We can convert the inequality into an equation, however, by correcting the mole fraction to a value that corresponds to the effective mole fraction $a_i$, i.e. a mole fraction that under ideal conditions would lead to the real chemical potential of the solution (Figure 2.32):

$$\mu_i = \mu_i^0 + RT \ln a_i \qquad \text{eq. 2.159}$$

The effective mole fraction $a_i$ is called the *activity* of the component $i$. The activity can be expressed as the actual mole fraction $x_i$, multiplied by an *activity coefficient* $\gamma_i$:

$$a_i = x_i \cdot \gamma_i \qquad \text{eq. 2.160}$$

By combining eq. 2.159 and eq. 2.160, we obtain

$$\mu_i = \mu_i^0 + RT \ln x_i \gamma_i \qquad \text{eq. 2.161}$$

for the chemical potential of a non-ideal solution. Eq. 2.161 can also be written as

$$\mu_i = \mu_i^0 + RT \ln x_i + RT \ln \gamma_i \qquad \text{eq. 2.162}$$

The first two terms in eq. 2.162 correspond to the chemical potential of an ideal solution with a mole fraction $x_i$, while the last term corrects the chemical potential for non-ideal effects. The chemical potential of the non-ideal solution is thus just the chemical potential of an ideal solution, corrected by $RT \ln \gamma_i$ to account for non-ideality:

$$\mu_i = \mu_{i,\text{ideal}} + RT \ln \gamma_i \qquad \text{eq. 2.163}$$

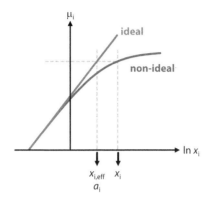

FIGURE 2.32:   **Dependence of the chemical potential on the mole fraction for ideal and non-ideal solutions**. In an ideal solution, the chemical potential of the solvent depends linearly on the logarithm of the mole fraction of the solute, $\ln x_i$ (blue). The chemical potential of a non-ideal solution deviates from this linear dependence (red). The effective mole fraction $x_{i,\text{eff}}$ (activity $a_i$) of a non-ideal solution corresponds to the mole fraction of an ideal solution with the same chemical potential.

### 2.6.7 COLLIGATIVE PROPERTIES

When we dissolve a solute in a pure solvent A, the mole fraction $x_A$ of the solvent decreases, and its chemical potential $\mu_A$ changes (eq. 2.147). The chemical potential $\mu_A$ of the solvent A is

$$\mu_A = \mu_A^0 + RT \ln x_A \qquad \text{eq. 2.164}$$

For a pure solvent ($x_A \rightarrow 1$), the chemical potential is $\mu_A^0$, the chemical potential at standard conditions, i.e. the pure liquid. If we now consider a solution of the solute B in solvent A, we can express the mole fraction $x_A$ of the solvent by the mole fraction $x_B$ of the solute as $(1-x_B)$. Eq. 2.164 then becomes

$$\mu_A = \mu_A^0 + RT \ln(1 - x_B) \qquad \text{eq. 2.165}$$

For a very dilute solution ($x_B \ll 1$), we can approximate the logarithm as

$$\ln(1 - x_B) \approx -x_B \qquad \text{eq. 2.166}$$

and obtain

$$\mu_A \approx \mu_A^0 - x_B RT \qquad \text{eq. 2.167}$$

The solute thus reduces the chemical potential of the solvent. Strikingly, this effect only depends on the mole fraction of the solute, but is independent of its chemical identity. Properties that only depend on the numbers of particles involved, but not on their identity, are called *colligative properties.*

   If we now consider the effect of the solute on the chemical potential of the solvent as a function of temperature, we see that colligative properties have an important consequence: at any given temperature, the chemical potential of the solvent in a solution will be lower than that of the pure solvent (Figure 2.33). The temperature at which the chemical potentials of solid and liquid states are equal, the melting point of the solid or freezing point of the solution, therefore shifts to lower temperatures (*freezing point depression*). Conversely, the temperature at which the chemical potentials of liquid and gaseous states are equal, the evaporation temperature, shifts to higher temperatures (*boiling point elevation*).

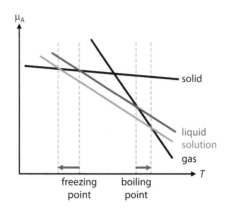

**FIGURE 2.33:   Effect of solute on the freezing and boiling points of the solvent.** The solute decreases the chemical potential of the solvent relative to the pure solvent at all temperatures (blue: liquid, cyan: solution). The chemical potential of solid or gaseous phases (black) is not affected. As a consequence, the intersection point for the chemical potential of the solid phase and the solution is shifted to lower temperatures: the freezing point (melting point of the solid phase) is decreased compared to the pure liquid. The intersection point for the chemical potential of the solution and of the gas phase is higher, the boiling point is increased compared to the pure solvent.

The change in boiling point caused by the solute can be calculated using the chemical potential of the solvent in liquid and gaseous phases. At the evaporation temperature, the liquid and gaseous phases of the solvent are in equilibrium, and we can write:

$$\mu_A\left(g\right) = \mu_A^0\left(l\right) + RT \ln x_A \qquad \text{eq. 2.168}$$

By rearranging to

$$\frac{\mu_A\left(g\right) - \mu_A^0\left(l\right)}{RT} = \ln x_A \qquad \text{eq. 2.169}$$

and by substitution of the chemical potential by the change in (molar) free energy associated with the phase transition, $\Delta G_{tr}$, and for $\Delta H_{tr}$ and $\Delta S_{tr}$ using the Gibbs-Helmholtz equation (eq. 2.105), we obtain

$$\frac{\Delta G_{tr}}{RT} = \frac{\Delta H_{tr} - T\Delta S_{tr}}{RT} = \ln x_A \qquad \text{eq. 2.170}$$

In equilibrium, $\Delta G_{tr}$ becomes zero. From the Gibbs-Helmholtz equation it follows that $\Delta S_{tr}$ can be substituted by $\Delta H_{tr}/T_{tr}$, which gives

$$\frac{\Delta H_{tr}}{RT} - \frac{T\Delta H_{tr}}{T_{tr} \cdot RT} = \ln x_A \qquad \text{eq. 2.171}$$

Eq. 2.171 can be rearranged to

$$\frac{\Delta H_{tr}}{R}\left(\frac{1}{T} - \frac{1}{T_{tr}}\right) = \ln x_A \qquad \text{eq. 2.172}$$

The term in brackets can be expanded to

$$\frac{T_{tr} - T}{T \cdot T_{tr}}$$

For small changes in the transition temperature ($T \approx T_{tr}$), this term can be approximated as

$$\frac{T_{tr} - T}{T \cdot T_{tr}} \approx -\frac{\Delta T}{T_{tr}^2} \qquad \text{eg. 2.173}$$

$\Delta T$ refers to the shift in the transition temperature ($T$–$T_{tr}$) caused by the solute. On the right-hand side of eq. 2.172, the mole fraction of the solvent A can be expressed as ($1$–$x_B$):

$$-\frac{\Delta H_{tr}}{R} \cdot \frac{\Delta T}{T_{tr}^2} = \ln x_A = \ln\left(1 - x_B\right) \qquad \text{eq. 2.174}$$

For small $x_B$, i.e. dilute solutions of B in A, we can again approximate ln ($1$–$x_B$) by $-x_B$ and obtain

$$-\frac{\Delta H_{tr}}{R} \cdot \frac{\Delta T}{T_{tr}^2} \approx -x_B \qquad \text{eq. 2.175}$$

Rearranging eq. 2.175 yields the shift $\Delta T$ in the transition temperature:

$$\Delta T \approx \frac{x_B \cdot T_{tr}^2 \cdot R}{\Delta H_{tr}} \qquad \text{eq. 2.176}$$

The shift in evaporation temperature by the solute thus depends on the mole fraction of the solute, $x_B$, and not on its identity: the boiling point increase by solutes is a colligative property. We can perform the same calculation to determine the shift in freezing point/melting point, which has an opposite sign (melting point reduction; see Figure 2.33). Due to the stronger dependence of the gas chemical potential on the temperature compared to the solid, and the higher slope of the corresponding $\mu(T)$ curves, the melting point reduction is always more pronounced than the boiling point increase (Figure 2.33).

As a consequence of the relationship between $\Delta T$ and $x_B$ (eq. 2.176), we can measure the shift in the transition temperature, $\Delta T$, and determine the mole fraction $x_B$, provided that the associated transition enthalpy $\Delta H_{tr}$ is known. An instrument that measures the boiling point elevation is called an *ebullioscope*. A *cryoscope* measures the depression of the freezing point. Ebullioscopy can be used to determine the alcohol content of liquids from the boiling point of the mixture. Measuring the melting temperature $T_m$ of a compound is a criterion for purity. If the melting point of a supposedly pure compound is lower than the reference value for the pure compound, impurities are present.

*Osmosis*, a phenomenon caused by diffusion through semipermeable membranes, is a biologically important example of a colligative property. Biological membranes act as selective barriers because they are not permeable for most solutes but can be readily passed by water. Membranes that are permeable for the solvent, in biological systems water, but not for the solute, are called *semipermeable* (Figure 2.34). If the solute concentration is high inside a membrane-limited compartment, but low outside, the chemical potential of water (the solvent) is higher on the outside than on the inside of the membrane. The higher chemical potential is a driving force for flow, and water will flow into the compartment to equilibrate the chemical potential. However, as the solute cannot pass the membrane, its concentration will always remain higher inside. No matter how much water flows into the cell, the equilibrium situation cannot be reached. Unlimited influx will eventually lead to bursting of the membrane. This is the reason why heavy rains in fall are dreaded by winemakers: the grapes have reached a high concentration of sugar, and the rain is almost pure water. Therefore, water will penetrate the skin of the grapes, causing them to burst and destroying the harvest.

We can also imagine the reverse situation, with pure water (or a dilute solution) inside the membrane-limited compartment, and a solution (or a solution of higher concentration) on the outside. In this case, the higher chemical potential of water in the inside will lead to uncontrolled efflux of water, called *plasmolysis*. We have all experienced plasmolysis when dressing lettuce: the higher salt concentration of the dressing compared to the interior of the leaves´ cells leads to an efflux of water, and the previously crisp leaves turn limp if we add the dressing too early.

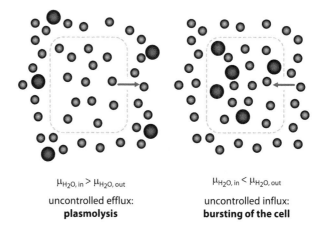

$\mu_{H_2O,\,in} > \mu_{H_2O,\,out}$

uncontrolled efflux:
**plasmolysis**

$\mu_{H_2O,\,in} < \mu_{H_2O,\,out}$

uncontrolled influx:
**bursting of the cell**

**FIGURE 2.34:    Diffusion through a semipermeable membrane.** Left: If the salt concentration (red) is higher outside the membrane-limited compartment, the chemical potential $\mu$ of water (blue) is higher on the inside than on the outside. Water spontaneously flows from the inside to the outside (blue arrow) to reduce the difference in chemical potential (plasmolysis). Right: If the salt concentration is higher inside the compartment, the higher chemical potential of water outside will drive influx of water (blue arrow), eventually leading to a bursting of the membrane.

Due to the limited elasticity of cellular membranes, the influx of water into the cell will cause a pressure $\pi$ to build up (Figure 2.35). This *osmotic pressure* will prevent further influx of water, and an equilibrium is reached.

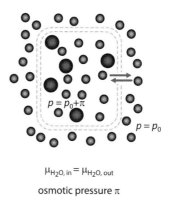

$p = p_0 + \pi$

$p = p_0$

$\mu_{H_2O,\,in} = \mu_{H_2O,\,out}$

osmotic pressure $\pi$

**FIGURE 2.35:    Osmosis.** A cell is separated from the exterior by a biological membrane, a lipid bilayer (gray broken lines). Inside, the solute (red) concentration is higher than outside. The higher chemical potential $\mu$ of water (blue) outside acts as a driving force for influx of water into the cell (blue arrow). Because of the limited elasticity of the membrane, the pressure inside will increase with the influx of water, and this counteracts further influx of water. The pressure difference between inside and outside in equilibrium is called the osmotic pressure $\pi$.

We can express the pressure- and mole fraction-dependent chemical potential of the solvent A in the solution inside the membrane-limited compartment as

$$\mu_A = \mu_A^0 + \int_{p^0}^{p^0 + \pi} V_{m,A}\,\mathrm{d}p + RT \ln x_A \qquad \text{eq. 2.177}$$

The chemical potential of the pure solvent on the outside is simply the standard potential $\mu_A^0$. In equilibrium, the chemical potentials outside and inside are equal:

$$\mu_A^0 + \int_{p^0}^{p^0+\pi} V_{m,A}\, dp + RT \ln x_A = \mu_A^0 \qquad \text{eq. 2.178}$$

We can now evaluate the integral

$$\int_{p^0}^{p^0+\pi} V_{m,A}\, dp = \pi \cdot V_{m,A} \qquad \text{eq. 2.179}$$

and obtain

$$\pi \cdot V_{m,A} = -RT \ln x_A \qquad \text{eq. 2.180}$$

Substituting $1-x_B$ for $x_A$ yields

$$\pi \cdot V_{m,A} = -RT \ln\left(1-x_B\right) \qquad \text{eq. 2.181}$$

Again, for dilute solutions and $x_B \ll 1$ we can approximate the logarithm as $-x_B$ and obtain

$$\pi \cdot V_{m,A} \approx RT x_B \qquad \text{eq. 2.182}$$

Rearrangement and substitution of $x_B$ by $n_B/n$ and $V_{m,A}$ by $V/n$ yields

$$\pi = \frac{x_B}{V_{m,A}} \cdot RT = \frac{n_B}{n} \cdot \frac{n}{V} \cdot RT \qquad \text{eq. 2.183}$$

and

$$\pi = \frac{n_B}{V} RT = c_B RT \qquad \text{eq. 2.184}$$

According to eq. 2.184, the osmotic pressure $\pi$ that builds up depends on the concentration or mole fraction of the solute B, but not on its identity. The effect of osmosis thus also is a colligative property. The osmotic effect can be used to determine concentrations. It is also exploited to exchange buffers by dialysis (Box 2.11), and is the principle behind dialysis for patients with kidney problems. Equilibrium dialysis can be used to determine equilibrium constants (Chapter 3; Box 2.11).

---

### BOX 2.11: DIALYSIS.

Dialysis is used for buffer exchange of protein solutions (Figure 2.36). The protein solution is filled into a dialysis tube that is a permeable for water and small ions, but not for the protein. The dialysis tube is placed into a large reservoir of the target buffer. Osmosis leads to equilibration of the ion and water chemical potentials across the membrane. Because of the large difference in volume of the solutions inside the dialysis tube and outside, this will lead to a near-equilibration of ion concentrations inside and outside, and effectively to an "exchange" of the buffer in which the protein is dissolved. Note that the osmotic pressure should be considered when setting up the dialysis: if the salt concentration is much higher inside than outside, the water flow into the dialysis bag can cause it to burst! Bursting can be avoided by leaving some void volume within the bag.

*(Continued)*

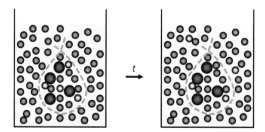

FIGURE 2.36: **Dialysis for buffer exchange**. A macromolecule in starting buffer (water, blue spheres, and salt 1, yellow spheres) is filled into a dialysis tube (gray broken line), and placed into a large reservoir of the desired buffer (not to scale, water and salt 2, green spheres). Water and ions can freely diffuse between the inside and outside of the tube, and concentrations will equilibrate. The reservoir is much larger than depicted, leading to (almost) equal salt concentrations in equilibrium and thus "exchange" of buffers. At the end of the dialysis, the solution will effectively be dissolved in the desired buffer.

Equilibrium dialysis can be used to determine dissociation constants of macromolecule-ligand complexes (Section 3.1). A macromolecule solution of known concentration is filled into one chamber, and the ligand into the second chamber, separated by a semipermeable membrane (Figure 2.37). The ligand can pass the membrane, but the macromolecule cannot and is confined to chamber one. The equilibrium is reached when the chemical potential of the ligand on both sides of the membrane is equal, or more specifically, when the chemical potential of the free ligand in both chambers is equal. The bound molecules do not contribute to the chemical potential of the ligand, and the equilibrium condition is

$$\mu_{\text{free ligand}}^{\text{left}} = \mu_{\text{ligand}}^{\text{right}} \qquad \text{eq. 2.185}$$

Eq. 2.185 implies that the concentration of the free ligand in the left chamber is equal to the ligand concentration in the right chamber. Thus, if we determine the total concentration of ligand in both chambers, $c_{\text{L,right}}$ (= $c_{\text{L,left,free}}$), and $c_{\text{L,left,tot}}$, we can calculate the concentration of the macromolecule-ligand complex, $c_{\text{L,left,bound}}$, from the difference $c_{\text{L,left,tot}} - c_{\text{L,right}}$ and obtain $K_d$ as

$$K_d = \frac{[M][L]}{[ML]} = \frac{\left([M]_{\text{left,tot}} - \left(c_{L,\text{left,tot}} - c_{L,\text{right}}\right)\right) \cdot c_{L,\text{right}}}{\left(c_{L,\text{left,tot}} - c_{L,\text{right}}\right)} \qquad \text{eq. 2.186}$$

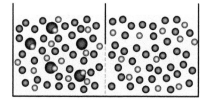

FIGURE 2.37: **Equilibrium dialysis.** The solution of a ligand (yellow, blue: water) for a macromolecule (red) is filled into two chambers separated by a semipermeable membrane (gray). Equilibrium is reached when the concentration of free ligand is equal in both compartments. By determining the total concentrations of the ligand in the left chamber and the ligand concentration in the right chamber, we can calculate the concentration of macromolecule-bound ligand, and the equilibrium constant $K_d$.

In *reverse osmosis* (Figure 2.38), external pressure is applied to concentrate a solute, or to desalinate water. The applied pressure provides the energy for the transport of water against a concentration gradient towards higher chemical potential (Section 4.3).

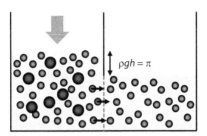

FIGURE 2.38:   **Reverse osmosis.** A solution (left) is separated from pure water (right) by a semipermeable membrane (broken gray line) that is permeable for water (blue), but impermeable for the solute (red). The osmotic effect causes a water flow from right to left until the hydrostatic pressure $\rho g h$ of the water column equals the osmotic pressure $\pi$. Applying external pressure to the side of the solution (gray arrow) provides energy for water to pass the membrane (black arrows) against the concentration gradient, leading to a concentration of the solution, and an increase in pure solvent on the right.

All living organisms consist of cells that are delimited by membranes. It is important for organisms to control salt concentrations in the cytoplasm and the water flow across membranes to be able to counteract osmotic effects (osmoregulation, Box 2.12).

---

### BOX 2.12: OSMOREGULATION.

Some organisms live in high-salt environments. Bacteria that thrive under high ionic strength conditions, termed halophiles, take up salts from their environment. Their complete metabolism is adapted to high ionic strength, and these organisms cannot survive when the ionic strength is reduced. Halotolerant bacteria produce osmotically active organic compounds such as amino acids, polyalcohols, or betaines when the external salt concentration increases. These low molecular mass compounds, also called *osmoprotectants*, help maintain a high osmotic value of the cytoplasm, and prevent constant efflux of water and plasmolysis.

Osmoregulation is also important for fresh-water and salt-water fish. The bodies of fresh-water fish have a higher salt concentration than the surroundings, they are *hypertonic*. Therefore, they continually take up water through their gills, and ions diffuse to the outside along the concentration gradient. Fresh-water fish have to take up ions actively, and rid themselves of water by producing highly diluted urine. Salt-water fish, in contrast, have a lower salt concentration in their body fluids than the surrounding ocean, they are *hypotonic*. To avoid constant loss of water and influx of ions, they take up salt water, and actively excrete ions.

Higher organisms excrete salts by specialized glands. The Galapagos lizards live in a high-salt environment: on the islands, there is virtually no fresh water available, and the lizards quench their thirst with amply available salt water. To eliminate the excess of salt from their bodies, they use nasal glands from which they expel highly concentrated salt drops at short intervals in a movement that very much looks like sneezing. Mangrove plants also secrete salt from glands. In vertebrates, osmoregulation is achieved by control of water excretion through the kidneys.

## QUESTIONS

2.1 A triathlete performs 1000 kJ of work and loses 400 kJ of heat while swimming in cold water. How does the internal energy of the triathlete change? How much does the pool (50 m length, 20 m width, 2 m depth) warm up? $C_p$ ($H_2O$) = 4.18 kJ $kg^{-1}$ $K^{-1}$, $\rho$ ($H_2O$) = 1 g $cm^{-3}$.

2.2 Isothermal expansion of an ideal gas does not change its internal energy. What is the associated change in enthalpy?

2.3 During normal breathing, humans exchange about 0.5 L of air in their lungs. Calculate the amount of worked that is performed during one exhalation against atmospheric pressure. How many grams of ATP need to be hydrolyzed for breathing during a day (30 breath cycles per minute, $\Delta G^{0'}$ (ATP hydrolysis) = −31 kJ $mol^{-1}$, M = 507.18 g $mol^{-1}$). A 100 g bar of chocolate provides about 500 kJ of energy. How long does the energy last just for breathing?

2.4 A diver grabs a bottle of compressed air. The pressure gauge states $p$ = 200·$10^5$ Pa (200 bar). Just after the descent into the dive, the pressure is reduced to 190·$10^5$ Pa. Why? The diver has directly descended to 30 m depth. What are the pressure and the temperature of the surrounding water? $T_{air,surface}$ = 30°C.

2.5 The diver breathes enriched air that contains 32% ($m/m_{tot}$) oxygen and 68% ($m/m_{tot}$) nitrogen (at $p$ = $p^0$ = $10^5$ Pa and $T$ = $T_0$ = 25°C). A partial oxygen pressure of >1.6 $10^5$ Pa is lethal for humans. Is it safe for the diver to dive down to the sea bed at 40 m? The diver starts with a 15 L tank of enriched air at $p$ = $p^0$. The air needed during descent and ascent can be neglected. How long can the diver stay at 45 m with his air supply for 0.5 L breathing volume and 30 breathing cycles $min^{-1}$?

2.6 On a winter day ($T$ = −10°C) you adjust the pressure of your car tires to 1.8·$10^5$ Pa. What is the pressure in summer ($T$ = 30°C)?

2.7 The enthalpy change for the reaction of glucose ($C_6H_{12}O_6$) to $CO_2$ and $H_2O$ is −2800 kJ $mol^{-1}$, for the reaction of ethanol to water it is −1370 kJ $mol^{-1}$. What is the enthalpy change during fermentation of glucose to ethanol? Is fermentation a useful metabolic pathway?

2.8 The heat capacity of water ($C_p$ = 75 J $mol^{-1}$ $K^{-1}$) is much higher than the heat capacity of air ($C_p$ = 20 J $mol^{-1}$ $K^{-1}$). Calculate the temperature change when 10 kJ of heat is transferred to 1 $m^3$ of water or air. $\rho(H_2O)$ = 1 g $cm^{-3}$; $\rho(air)$ = 1.2 mg $cm^{-3}$.

2.9 Calculate the change in entropy for the conversion of 1 mol ice ($H_2O(s)$, 0°C) into vapor ($H_2O(g)$, 100°C) at constant pressure. Sublimation enthalpy $\Delta H_{subl}$ = 47 kJ $mol^{-1}$, vaporization enthalpy $\Delta H_{vap}$ = 41 kJ $mol^{-1}$, $C_p$ = 75 J $mol^{-1}$ $K^{-1}$.

2.10 Calculate the entropy change when 200 kJ of heat are transferred to water at 0°C and 100°C (isothermal conditions). Explain the difference from the statistical interpretation of entropy.

2.11 Our metabolism generates heat of about 100 J $s^{-1}$. Calculate the change in entropy of the surrounding per hour at 25°C.

2.12 0.1 mol of a reactant react to products in an isobaric reaction at 25°C and $10^5$ Pa. During the reaction, 20 kJ of heat are transferred to the surroundings. Calculate $\Delta G^0$, $\Delta H^0$, and $\Delta S^0$ for this process.

2.13 Calculate the reaction enthalpy, entropy, and free energy, $\Delta H_r^0$, $\Delta S_r^0$, and $\Delta G_r^0$, for the amidation of glutamate to glutamine (1) at standard temperature, (2) at 37°C, and (3) at 75°C (a) taking into account the temperature dependence of $\Delta H$ $\Delta S$, and (b) assuming that $\Delta H$ and $\Delta S$ are temperature-independent. What can you conclude about the importance of $C_p$? What does the result mean for organisms that live at moderate temperature (37°C; mesophilic organisms) and thermophilic organisms that live at 75°C?

Glutamine: $\Delta H_f^0$ = −826 kJ $mol^{-1}$, $C_{p,m}$ = 184 J $mol^{-1}$ $K^{-1}$, $S_m$ = 195 J $mol^{-1}$ $K^{-1}$

Glutamate: $\Delta H_f^0$ = −1010 kJ $mol^{-1}$, $C_{p,m}$ = 175 J $mol^{-1}$ $K^{-1}$, $S_m$ = 188 J $mol^{-1}$ $K^{-1}$

$H_2O$: $\Delta H_f^0 = -290$ kJ mol$^{-1}$, $C_{p,m} = 75$ J mol$^{-1}$ K$^{-1}$, $S_m = 70$ J mol$^{-1}$ K$^{-1}$

$NH_4^+$: $\Delta H_f^0 = -133$ kJ mol$^{-1}$, $C_{p,m} = 80$ J mol$^{-1}$ K$^{-1}$, $S_m = 113$ J mol$^{-1}$ K$^{-1}$

2.14 The energy of one photon is E = h·$\nu$ = hc/$\lambda$. How many photons of light with $\lambda$ = 680 *nm* have to be absorbed to synthesize an ATP molecule (degree of efficiency 100%)? In what wavelength range is one photon sufficient for the synthesis of two ATP molecules? $\Delta G^{0'}$ of ATP synthesis: 31 kJ mol$^{-1}$.

2.15 Protein-protein interactions reduce the number of particles from two to one, which corresponds to a decrease in entropy. Why can the interaction nevertheless be entropically favorable and under what conditions?

2.16 How much does an ionic interaction between two oppositely charged side chains contribute to protein stability at the surface and in the interior of the protein?

2.17 You mix 10 mL glycerol and 90 mL water to obtain a 10% glycerol solution. The density of the mixture is $\rho_{mix} = 1.02567$ g cm$^{-3}$. What are the mole fraction of glycerol and the volume of the mixture? What is the reason for the volume change? What can you conclude for the necessity to take volume changes into account when stabilizing proteins by using 10% glycerol buffers? $M_m$(glycerol) = 92.09 g mol$^{-1}$, $M_m$($H_2O$) = 18 g mol$^{-1}$, $\rho$(glycerol) = 1.25802 g cm$^{-3}$, $\rho$($H_2O$) = 0.99708 g cm$^{-3}$.

2.18 Calculate the difference between and $\mu^0$ and $\mu^{0'}$ for protons in aqueous solutions.

2.19 Is mixing of two liquids to an ideal solution a spontaneous process?

2.20 A polysaccharide solution (c(PS) = 10 g L$^{-1}$ in $H_2O$) has an osmotic pressure of $5 \cdot 10^3$ N m$^{-2}$. What is the molar mass of the polysaccharide? What is the vapor pressure of the solution compared to pure water?

2.21 Giant sequoia trees reach a height of more than 100 m. They have to transport water into the top of their crown. Can this be explained by the osmotic pressure due to the solutes in the cytoplasm? $\rho$($H_2O$) = 1 g cm$^{-3}$.

2.22 What concentration of NaCl is required to prevent ice formation at $T = -1°C$ and at $T = -5°C$? $\Delta H_{melt}$($H_2O$) = 6 kJ mol$^{-1}$. The solubility of NaCl in $H_2O$ is 359 g L$^{-1}$, the molar mass $M_m$ is 58.44 g mol$^{-1}$. What is the maximum reduction in freezing point that can be achieved by a saturated NaCl solution?

## REFERENCES

Minton, A. P. (2006) Macromolecular crowding. *Curr Biol* **16**(8): R269–271.
· introductory note on molecular crowding in the cell

Luby-Phelps, K. (2013) The physical chemistry of cytoplasm and its influence on cell function: an update. *Mol Biol Cell* **24**(17): 2593–2596.
· review about molecular crowding and its impact on cellular processes

Ellis, R. J. (2001) Macromolecular crowding: obvious but underappreciated. *Trends Biochem Sci* **26**(10): 597–604.
· review about thermodynamic consequences of molecular crowding for cellular processes

# Energetics and Chemical Equilibria

## 3.1  THE FREE ENERGY CHANGE AND THE EQUILIBRIUM CONSTANT

We have seen in Section 2.6.1 (eq. 2.127) that the change in free enthalpy, $\Delta G$, during a reaction according to Scheme 3.1

$$a\text{A} + b\text{B} \rightleftharpoons c\text{C} + d\text{D} \qquad\qquad \text{scheme 3.1}$$

can be written as

$$\Delta G = \sum_i \mu_i n_i = \mu_C n_C + \mu_D n_D - \mu_A n_A - \mu_B n_B \qquad\qquad \text{eq. 3.1}$$

where $\mu_i$ is the chemical potential of compound $i$ and $n_i$ is the amount of substance.

The relationship in eq. 3.1 does not only hold for irreversible reactions that proceed completely from A and B to C and D, but also for reversible reactions in which C and D can react to A and B in a reverse reaction, and the final state is characterized by equilibrium concentrations of A, B, C, and D. We can substitute the chemical potential for each compound in eq. 3.1 by the standard chemical potential $\mu^0$ at standard concentrations of all compounds at pressure $p$ and temperature $T$, plus the concentration term:

$$\Delta G = c\,\mu_C^0 + d\,\mu_D^0 - a\,\mu_A^0 - b\,\mu_B^0 + RT\ln \frac{\left(\dfrac{c_C}{c_C^0}\right)^c \left(\dfrac{c_D}{c_D^0}\right)^d}{\left(\dfrac{c_A}{c_A^0}\right)^a \left(\dfrac{c_B}{c_B^0}\right)^b} \qquad\qquad \text{eq. 3.2}$$

In eq. 3.2, the standard terms are summarized at the beginning, and the concentration-dependent terms are summarized in the logarithmic term at the end. Each term in the argument of the logarithm contains the actual concentration ($c_A$, $c_B$, $c_C$, $c_D$), relative to the standard concentration $c^0_A$,

$c^0{}_B$, $c^0{}_C$, and $c^0{}_D$ of the respective compound. The ratio of the actual concentration and the standard concentration is dimensionless, and the argument of the logarithm is also a dimensionless number. Biochemical reactions typically take place in solution, and the standard concentration of the dissolved reactants is 1 M. Division of the actual concentration by 1 M does not change the numerical value, and eq. 3.2 is therefore often written in a simplified form as

$$\Delta G = c\,\mu_C^0 + d\,\mu_D^0 - a\,\mu_A^0 - b\,\mu_B^0 + RT \ln \frac{[C]^c [D]^d}{[A]^a [B]^b}$$

eq. 3.3

It is important to note that the values [A], [B], [C], and [D] in eq. 3.3 now denote the concentration <u>relative to the standard concentration</u>, and are dimensionless.

We can summarize all terms containing standard chemical potentials as $\Delta G^0$, and obtain

$$\Delta G = \Delta G^0 + RT \ln \frac{[C]^c [D]^d}{[A]^a [B]^b}$$

eq. 3.4

Eq. 3.4 allows calculating the free energy change $\Delta G$ involved in a reaction for any given concentration of reaction partners. When using this relationship, we always have to keep in mind that the concentrations are relative to the standard concentration.

In eq. 3.4, the activity coefficients $\gamma$ of the compounds A, B, C, and D have been neglected, and the equation therefore only holds for ideal solutions. For a non-ideal solution with $\gamma \neq 1$, we have to multiply the actual concentrations of A, B, C, and D by their activity coefficients $\gamma_A$, $\gamma_B$, $\gamma_C$, and $\gamma_D$, which gives

$$\Delta G = \Delta G^0 + RT \ln \frac{\left([C]\gamma_C\right)^c \left([D]\gamma_D\right)^d}{\left([A]\gamma_A\right)^a \left([B]\gamma_B\right)^b}$$

eq. 3.5

or

$$\Delta G = \Delta G^0 + RT \ln \frac{[C]^c [D]^d}{[A]^a [B]^b} + RT \ln \frac{(\gamma_C)^c (\gamma_D)^d}{(\gamma_A)^a (\gamma_B)^b}$$

eq. 3.6

The first two terms in eq. 3.6 are the standard free energy change for an ideal solution ($\gamma = 1$), and the last term corrects for non-ideal behavior.

In the following considerations, we will return to ideal solutions with activity coefficients of $\gamma = 1$. The corresponding equations for non-ideal solutions can easily be obtained by multiplication of each concentration with the activity coefficient.

In equilibrium, $\Delta G$ becomes zero, and we can write

$$\Delta G = \Delta G^0 + RT \ln \left( \frac{[C]^c [D]^d}{[A]^a [B]^b} \right)_{eq} = 0$$

eq. 3.7

The index "eq" in eq. 3.7 indicates concentrations at equilibrium. We now define the *equilibrium constant* $K_{eq}$ as

$$K_{eq} = \left( \frac{[C]^c [D]^d}{[A]^a [B]^b} \right)_{eq}$$

eq. 3.8

Note that $K_{eq}$ is always a dimensionless number as every concentration term is a concentration relative to the concentration in the standard state! We will see in Section 3.2 how we can convert this true equilibrium constant into a constant with dimensions related to concentrations. By combining eq. 3.7 and eq. 3.8, we then obtain

$$\Delta G^0 = -RT \ln K_{eq} \qquad \text{eq. 3.9}$$

or

$$K_{eq} = \exp\left(-\frac{\Delta G^0}{RT}\right) \qquad \text{eq. 3.10}$$

Thus, we can calculate the free energy change for the underlying reaction under standard conditions from the concentrations of the reaction partners under equilibrium conditions (relative to their respective standard states), and *vice versa*! It is important to note that the equilibrium concentrations are linked to the change in free energy under standard conditions (Box 3.1). The actual $\Delta G$ in equilibrium is always zero.

---

**BOX 3.1: THERMODYNAMIC INFORMATION FROM EQUILIBRIUM CONSTANTS: QUANTIFYING THE ENERGETIC CONTRIBUTION OF AN AMINO ACID SIDE CHAIN TO LIGAND BINDING.**

From eq. 3.9, it is evident that we can calculate free energies associated with equilibria. Specific binding of a ligand to a protein is often mediated by interactions between amino acid side chains and the ligands. We might suspect a certain amino acid, e.g. a tyrosine to contribute to ligand binding by forming a hydrogen bond. We can determine the dissociation equilibrium constant of the protein-ligand complex (see Section 19.5.5.1), and can calculate the $\Delta G^0$ of dissociation (and hence also the $\Delta G^0$ of binding) from the equilibrium constant. Then we create a variant in which we replace the tyrosine by phenylalanine, and determine the dissociation constant of the protein-ligand complex for this variant, preferentially by the same method. If the tyrosine indeed engages in a hydrogen bond with the ligand, then removal of the hydroxyl group by the exchange to phenylalanine will lead to a decrease in $\Delta G^0$. From the difference of the two $\Delta G^0$ values, we can calculate a $\Delta\Delta G^0$ value that reflects the thermodynamic contribution of the tyrosine to ligand binding (under standard conditions). If the tyrosine does not contribute to ligand binding, $\Delta G^0$ of the variant will be identical to the first value we have determined (provided that the amino acid exchange has not led to other side effects such as a conformational change of the protein).

---

### 3.1.1 TEMPERATURE DEPENDENCE OF THE EQUILIBRIUM CONSTANT

Eq. 3.9 shows that the equilibrium constant $K_{eq}$ is temperature-dependent. Expression of $\Delta G^0$ in eq. 3.9 by the Gibbs-Helmholtz equation (eq. 2.105) and division by $-RT$ yields

$$\ln K_{eq} = -\frac{\Delta H^0}{RT} + \frac{\Delta S^0}{R} \qquad \text{eq. 3.11}$$

This equation is called the *van't Hoff equation*. If we plot $\ln K_{eq}$ as a function of the inverse temperature $1/T$, we thus obtain a linear plot, the *van't Hoff plot*, with the slope related to $\Delta H^0$, and the intercept related to $\Delta S^0$ (Figure 3.1). The plot is only linear if $\Delta H$ and $\Delta S$ are independent of temperature over the temperature range considered, i.e. for $\Delta C_p = 0$. If $\Delta C_p$ does not equal zero, van't Hoff plots are curved. To determine thermodynamic data from curved van't Hoff plots, eq. 3.11 has to be derived taking the temperature dependence of $\Delta H$ and $\Delta S$ into account. The resulting equation can then be used to describe the non-linear van't Hoff plot and to extract $\Delta C_p$, $\Delta H^0$, and $\Delta S^0$. It is important to note that, particularly for biochemical reactions, the temperature range in which the equilibrium constant is determined is typically rather narrow, and the experimental data points in the van't Hoff plot are clustered. Nevertheless, the slope of the line through these points is typically well-determined, and $\Delta H^0$ can be determined reliably. In contrast, determination of the $y$-axis intercept and of $\Delta S^0$ require extrapolation from the cluster of data points over a large distance to the $y$-axis (Figure 3.1). Small uncertainties in the slope will therefore lead to large errors in the $y$-axis intercept and in $\Delta S^0$.

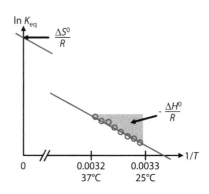

FIGURE 3.1:   **The van't Hoff plot**. The van't Hoff equation describes the temperature dependence of equilibrium constants. A plot of $\ln K_{eq}$ as a function of the inverse temperature $1/T$ is linear if $\Delta H$ and $\Delta S$ are temperature-independent over the temperature range considered, i.e. if $\Delta C_p^0$ is zero. $\Delta H^0$ can be determined from the slope, and $\Delta S^0$ from the $y$-axis intercept. The intercept has to be determined by extrapolation over a substantial inverse temperature range, leading to large errors in $\Delta S^0$. $\Delta C_p^0$ values $\neq 0$ lead to curved van't Hoff plots.

The van't Hoff equation is sometimes written in its differential form as

$$\frac{\mathrm{d}\ln K_{eq}}{\mathrm{d}(1/T)} = -\frac{\Delta H^0}{R}$$

eq. 3.12

or

$$\frac{\mathrm{d}\ln K_{eq}}{\mathrm{d}T} = -\frac{\Delta H^0}{RT^2}$$

eq. 3.13

Integration from $T_1$ to $T_2$ gives the relationship

$$\ln\left(\frac{K_{eq}(T_2)}{K_{eq}(T_1)}\right) = -\frac{\Delta H^0}{R}\left(\frac{1}{T_2} - \frac{1}{T_1}\right)$$

eq. 3.14

which allows to calculate the equilibrium constant at the temperature $T_2$ if the value at $T_1$ and the associated change in enthalpy $\Delta H^0$ are known.

### 3.1.2 THE PRINCIPLE OF LE CHATELIER

In Section 2.5.2, we have discussed the pressure- and temperature dependence of $G$ for closed systems, and have derived the equation

$$\mathrm{d}G = V\mathrm{d}p - S\mathrm{d}T$$

eq. 2.113

For open systems and chemical reactions, we now have to expand this equation by the changes in chemical potential for the components to

$$dG = Vdp - SdT + \sum_i \mu_i dn_i \qquad \text{eq. 3.15}$$

We can infer from eq. 3.15 that $G$ (and $\Delta G$) depend on pressure, temperature and amount of substance, and we have previously derived the pressure dependence of $G$ for $n$, $T$ = const., the temperature dependence for $p$, $n$ = const., and the dependence on amount of substance or concentration for $p$, $T$ = const. (Sections 2.5.2, 2.6.1). When equilibria are perturbed by changing one of these parameters, they will respond to this perturbation by favoring the reaction that counteracts the perturbation, and will reach a new equilibrium position. For example, an increase in temperature by transferring heat leads to a shift of the equilibrium in the endothermic direction of the chemical reaction, taking up the heat. Conversely, a decrease in temperature favors the exothermic direction, producing more heat. Likewise, an increase in pressure favors the direction of the reaction that involves a decrease in volume, and the equilibrium is shifted towards the side of the compounds with lower volume (fewer molecules, less gases), and *vice versa*. Changing the amount of substance by withdrawing a compound from the reaction favors the direction of the reaction in which this compound is produced, whereas adding the same compound in large excess favors the reaction in which it is consumed. This principle of "evading" changes in the reaction conditions is called the *principle of Le Chatelier*.

The principle of Le Chatelier is a recurrent theme in biochemical reactions. Removal of a product by a subsequent (irreversible) reaction draws an equilibrium to the side of products. The equilibrium between dihydroxyacetone phosphate and glyceraldehyde-3-phosphate, for example, is on the side of dihydroxyacetone phosphate ($K < 1$). However, in the glycolysis pathway, glyceraldehyde-3-phosphate is further metabolized *via* 1,3-bisphosphoglycerate, 3-phosphoglycerate, 2-phosphoglycerate, and phosphoenol pyruvate to pyruvate. The phosphate transfer from phosphoenol pyruvate to ADP in the last reaction is effectively irreversible, leading to a depletion of phosphoenol pyruvate. Following the principle of Le Chatelier, this removal of phosphoenol pyruvate from the equilibrium prior to the irreversible step, between 2-phosphoglycerate and phosphoenol pyruvate, causes the production of more phosphoenol pyruvate, which in turn leads to a depletion of 2-phosphoglycerate, and so on. Effectively, the irreversible step at the end affects all equilibria before that, eventually leading to further isomerization of dihydroxyacetone phosphate to glyceraldehyde-3-phosphate.

## 3.2 BINDING AND DISSOCIATION EQUILIBRIA AND AFFINITY

We have seen before that $\Delta G$ for a reverse reaction is of the same magnitude as $\Delta G$ for the forward reaction, but of opposite sign. From eq. 3.8, it follows for the associated equilibrium constants that

$$K_{\text{forward}} = \frac{1}{K_{\text{reverse}}} \qquad \text{eq. 3.16}$$

For a simple binding equilibrium of an enzyme E and its substrate S

$$\text{E} + \text{S} \rightleftharpoons \text{ES} \qquad \text{scheme 3.2}$$

we can write the equilibrium constant $K_{\text{association}}$ for the association reaction of E and S to ES

$$K_{\text{association}} = \frac{[ES]}{[E][S]} \qquad \text{eq. 3.17}$$

following the definition of concentration of products (ES) divided by the concentration of reactants (E, S). A high association constant is equivalent to a high concentration of ES complex in equilibrium, and we say the enzyme E has a high affinity for the substrate S. A low association constant is equivalent to a low affinity.

For the reverse reaction, the dissociation of ES to E and S, the equilibrium constant $K_{\text{dissociation}}$ is

$$K_{\text{dissociation}} = \frac{[E][S]}{[ES]} \qquad \text{eq. 3.18}$$

which is the inverse of the association constant, in accordance with eq. 3.16. A high value for the dissociation constant means a high concentration of E and S in equilibrium, and corresponds to a low substrate affinity. A low dissociation constant, in contrast, corresponds to a low concentration of free enzyme and substrate in equilibrium and a high affinity.

As discussed before (Section 3.1), all concentrations in eq. 3.17 and eq. 3.18 are relative to the concentration in the standard state, and thus dimensionless numbers. $K_{\text{eq}}$ is therefore also a dimensionless number. For convenience, the dimensionless equilibrium constant is generally converted into a constant with units related to concentrations (Box 3.2). The unit of the association constant depends on the numbers of molecules involved in the equilibrium. For the binary binding event (Scheme 3.2), the association constant has the unit $M^{-1}$ and the dissociation constant has the unit M. These are precisely the units we would obtain if we just entered the absolute concentrations into eq. 3.17 and eq. 3.18. For biochemical reactions in solution, we can in fact generally calculate equilibrium constants with units from the actual concentrations in equilibrium. This is because the standard state for a solute is 1 M, and division by 1 M does not alter the numerical value. From now on, we will always calculate equilibrium constants with dimensions for reactions in solution from absolute concentrations. The calculated value then automatically has a meaningful dimension (Box 3.2). Why is the unit meaningful? The unit M makes the dissociation constant a practical entity for argumentation and for comparison with actual concentrations. At a large excess of the substrate over enzyme ($S_0 \gg E_0$, where the index zero indicates total concentrations), we can estimate the degree of saturation of enzyme with substrate directly from the substrate concentration. At a substrate concentration that equals the $K_{\text{d}}$ value, the saturation of enzyme with substrate is 50%. At much lower concentrations "below the $K_{\text{d}}$", the enzyme occurs mostly in the free form. At higher concentrations "above the $K_{\text{d}}$", the enzyme is mostly encountered as the enzyme-substrate complex. Despite the inverse relationship of the dissociation constant and the binding strength or affinity, $K_{\text{dissociation}}$ or short $K_{\text{d}}$ is therefore typically used to characterize binding equilibria.

---

### BOX 3.2: FROM A DIMENSIONLESS EQUILIBRIUM CONSTANT TO AN EQUILIBRIUM CONSTANT WITH UNITS OF CONCENTRATION.

The definition of equilibrium constants according to eq. 3.8 and the association and dissociation constants defined in eq. 3.17 and eq. 3.18 are calculated from concentrations relative to the standard state (Section 2.5.3), and therefore lead to dimensionless numbers. The dimensionless equilibrium "number" can be converted to an equilibrium constant with dimensions of (inverse) concentration, depending on the particular equilibrium. For a general reaction according to Scheme 3.1

$$a\text{A} + b\text{B} \rightleftharpoons c\text{C} + d\text{D} \qquad \text{scheme 3.1}$$

the dimension of the equilibrium dissociation constant $K_{\text{d}}$ is

$$(-a - b + c + d)\text{M} \qquad \text{eq. 3.19}$$

This is precisely the unit we would obtain if we just entered all absolute concentrations with the unit M into the definition of the equilibrium constant (eq. 3.8). It is general practice in biochemistry to calculate equilibrium constants with dimensions from absolute concentrations. This is possible because the standard state of each compound for equilibria

*(Continued)*

in solution is 1 M, and the division by $c^0$ does not change the numerical value of the relative concentration. Nevertheless it is important to be aware of this shortcut: often, $K_d$ values are listed with the units µM, nM, or even pM. All these values are related to the change in free energy of the reaction under standard conditions, $\Delta G^0$. To calculate $\Delta G^0$ from the equilibrium constant, the value in M has to be used, otherwise the conversion is incorrect. Using concentrations relative to the standard state avoids this complication. Also, the use of equilibrium constants with units can be confusing when water is not only the solvent but also a reactant. For the solvent, the standard concentration is that of the pure solvent. As the solvent by definition is in large excess, the change in concentration due to the reaction is typically very small. In other words, the actual concentration almost equals the concentration of pure water. The ratio $c(H_2O)/c^0(H_2O)$ can therefore be approximated as unity, and the concentration term for water can be ignored, which is not so obvious when absolute concentrations are used.

From the definition of the dissociation constant $K_d$, we can calculate the conentration of the ES complex in equilibrium as a function of total substrate and enzyme concentrations, $E_0$ and $S_0$. First, we state the conservation of mass for $[E]$ and $[S]$:

$$E_0 = [E] + [ES] \qquad\qquad \text{eq. 3.20}$$

and

$$S_0 = [S] + [ES] \qquad\qquad \text{eq. 3.21}$$

We can then express $K_d$ as

$$K_d = \frac{(E_0 - [ES])(S_0 - [ES])}{[ES]} \qquad\qquad \text{eq. 3.22}$$

and obtain the quadratic equation

$$0 = [ES]^2 + (E_0 + S_0 + K_d) \cdot [ES] + E_0 S_0 \qquad\qquad \text{eq. 3.23}$$

This quadratic equation has two solutions (see Section 29.2)

$$[ES] = \frac{E_0 + S_0 + K_d}{2} \pm \sqrt{\left(\frac{E_0 + S_0 + K_d}{2}\right)^2 - E_0 S_0} \qquad\qquad \text{eq. 3.24}$$

of which only the solution with the negative sign in front of the square root is physically meaningful (the other solution gives $[ES] > E_0$ or $[ES] > S_0$, which is impossible). The concentration of the ES complex at any given total concentration of $E_0$ and $S_0$ as a function of the dissociation constant $K_d$ is

$$[ES] = \frac{E_0 + S_0 + K_d}{2} - \sqrt{\left(\frac{E_0 + S_0 + K_d}{2}\right)^2 - E_0 S_0} \qquad\qquad \text{eq. 3.25}$$

We will return to eq. 3.25 in the context of binding kinetics (Chapter 10). Eq. 3.25 also becomes important when we discuss how to determine $K_d$ values experimentally using spectroscopic probes (Section 19.5.5.1).

## 3.3 PROTOLYSIS EQUILIBRIA: THE DISSOCIATION OF ACIDS AND BASES IN WATER

A specific binding/dissociation event with important implications on biochemical reactions is the dissociation of an acid HA or a base BOH in water. These equilibria are described by the equilibrium constants $K_A$ and $K_B$:

$$K_A = \frac{[A^-][H^+]}{[HA]}$$ 

eq. 3.26

for the dissociation of the acid HA and

$$K_B = \frac{[B^+][OH^-]}{[BOH]}$$ 

eq. 3.27

for the dissociation of the base BOH in water. Often, the negative decadic logarithms of the $K_A$ and $K_B$ values are tabulated (see also Tables 16.1 and 27.1). In analogy to the pH value, these values are called $pK_A$ and $pK_B$. The $pK_A$ and $pK_B$ values are very convenient constants (see eq. 3.32): at a pH that equals the $pK_A$, half of the acid is deprotonated, the other half is protonated. At pH values well above the $pK_A$, the acid is mostly dissociated, and at pH values well below the $pK_A$, it is mostly protonated. Correspondingly, half of the base is dissociated at a pH that equals the $pK_B$, the other half is not. At pH values above the $pK_B$, the dissociation is less, and at pH values below the $pK_B$, the base is more and more dissociated.

The dissociated form of an acid HA, the anion $A^-$, can accept a proton from water and act as a base. Similarly, the pronated form of a base B, $BH^+$, can transfer a proton to water and act as an acid. Such pairs of HA and $A^-$ are called acid and *conjugate base*. The $pK_A$ and $pK_B$ values for acids and their conjugate bases are related: the $K_B$ for the reaction of $A^-$ with $H_2O$ is

$$K_B = \frac{[HA][OH^-]}{[A^-]}$$ 

eq. 3.28

By multiplying $K_A$ (eq. 3.26) and $K_B$, we obtain

$$K_A K_B = \frac{[A^-][H^+]}{[HA]} \frac{[HA][OH^-]}{[A^-]} = [H^+][OH^-] = K_W$$ 

eq. 3.29

The product of $K_A$ and $K_B$ of a conjugate acid and base thus equals the ionic product $K_W$ of water. The ionic product is $K_W = 10^{-14}$. The small value indicates that the equilibrium of water dissociation into $H^+$ and $OH^-$ lies very much on the side of undissociated water. By taking the negative decadic logarithm of eq. 3.29, we obtain

$$pK_A + pK_B = -\log K_W = 14$$ 

eq. 3.30

Hence, we can directly calculate the $pK_B$ for a conjugate base if we know the $pK_A$ for dissociation of the acid.

Taking the negative decadic logarithm of eq. 3.26 gives

$$-\log K_A = -\log \frac{[A^-][H^+]}{[HA]} = -\log \frac{[A^-]}{[HA]} - \log[H^+]$$ 

eq. 3.31

or

$$pH = pK_A + \log \frac{[A^-]}{[HA]}$$ 

eq. 3.32

Eq. 3.32 is the *Hendersen-Hasselbalch equation* that describes the pH of a solution as a function of the concentration ratio of acid and conjugate base. For $[A^-] = [HA]$, $pH = pK_A$. Under these conditions, the solution of acid and conjugate base acts as a pH buffer: addition of acid or base leads to very small changes in pH. The *buffer capacity* β is defined as the amount of acid (or base) added, divided by the change in pH:

$$\beta = \frac{\left[H^+\right]_{added}}{\Delta pH}$$
eq. 3.33

Many acids such as oxaloacetic acid, phosphoric acid or amino acids undergo multiple protolysis steps with different $pK_A$ values, and buffer the pH of a solution at different pH values.

# 3.4 THERMODYNAMIC CYCLES, LINKED FUNCTIONS AND APPARENT EQUILIBRIUM CONSTANTS

In biochemistry, equilibria are usually not isolated, but linked by compounds that are involved in several equilibria. Often, one species A can be converted into a different species C by alternative pathways, e.g. either *via* B or D (Figure 3.2). Such a pattern is called a *thermodynamic cycle*. Similar to a Carnot cycle (Section 2.2.5.1), we can start from A, convert it to B and C and D and back to A. The associated change in free energy $\Delta G$ is zero because initial and final states are identical and have the same free energy $G$. Similarly, the free energy change from A through B to C has to be identical to the free energy change for conversion of A through D to C because initial and final states are identical, and the pathway we chose to move from A to C is irrelevant for the change in the state function $G$. Under standard conditions, we can therefore write

$$\Delta G^0_{B-A} + \Delta G^0_{C-B} = \Delta G^0_{D-A} + \Delta G^0_{C-D}$$
eq. 3.34

Note that $\Delta G^0_{Y-X}$ corresponds to the $\Delta G^0$ value associated with the reaction from X to Y. In terms of the individual equilibrium constants, this corresponds to

$$K_{D/A} \cdot K_{C/D} = K_{B/A} \cdot K_{C/B}$$
eq. 3.35

$K_{Y/X}$ is the equilibrium constant for the equilibrium between X and Y. Eq. 3.34 and eq. 3.35 show that the four equilibria are *thermodynamically linked*. If we know $\Delta G^0$ or $K$ for three of the four equilibria, we can therefore directly calculate the missing fourth value.

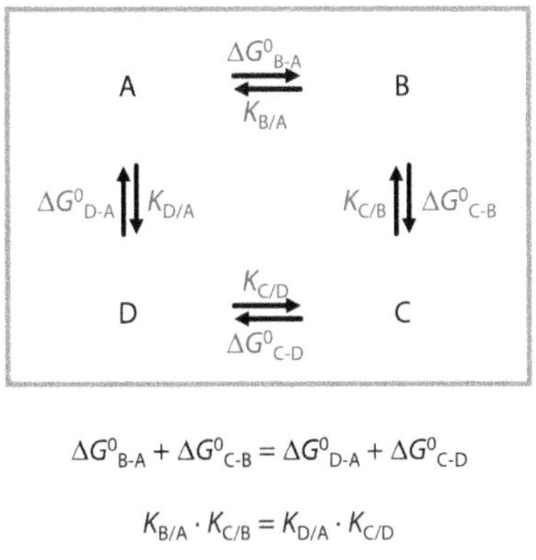

FIGURE 3.2:    **A thermodynamic cycle**. Compound A can be converted into compound C following two pathways, A→B→C or A→D→C. The four equilibria form a thermodynamic cycle. If we move through all four reactions from A to B to C to D and back to A, the initial and final states are identical, and the change in free energy must be zero. An equivalent statement to formulate the thermodynamic balance is that the free energy change for both pathways from A to C must be identical, i.e. $\Delta G^0_{B/A} + \Delta G^0_{C/B} = \Delta G^0_{D/A} + \Delta G^0_{C/D}$. As a consequence, the four equilibrium constants are also linked: $K_{B/A} \cdot K_{C/B} = K_{D/A} \cdot K_{C/D}$.

From eq. 3.35, we can also conclude that the ratios of the equilibrium constants for the two vertical processes (A→D, B→C) and for the two horizontal processes (A→B, D→C) must be equal:

$$\frac{K_{D/A}}{K_{C/B}} = \frac{K_{B/A}}{K_{C/D}} \qquad \text{eq. 3.36}$$

This insight is useful if we look at a common example of a biochemical thermodynamic cycle, the binding of two substrates to an enzyme (Figure 3.3). The enzyme can either bind substrate A first, and the EA complex then binds substrate B, or substrate B binds first, and the EB complex then binds substrate A. In both cases, the final product is the EAB complex. The binding equilibria form a thermodynamic cycle, and we know that

$$K_A \cdot K_B' = K_A' \cdot K_B \qquad \text{eq. 3.37}$$

where $K_A$ is the equilibrium constant for the dissociation of the EA complex, $K_A'$ the dissociation constant of A from EAB, $K_B$ the dissociation constant of the EB complex, and $K_B'$ the dissociation constant for B from the EAB complex. Binding of A and B can be independent, and in this case, $K_A$ equals $K_A'$, and $K_B$ equals $K_B'$, It is also possible that binding of A promotes binding of B. In this case, B will be bound more tightly by the EA complex than by E alone. In terms of equilibrium dissociation constants, this means $K_B'$ is decreased compared to $K_B$ (Section 3.2; eq. 3.18). From eq. 3.37, it follows that in this case B must also favor binding of A by the same factor α, i.e.

$$\alpha = \frac{K_A}{K_A'} = \frac{K_B}{K_B'} \qquad \text{eq. 3.38}$$

Thus, if binding of A favors binding of B, then for thermodynamic reasons binding of B must favor binding of A to the same extent! Similarly, if binding of A disfavors binding of B ($K_B' > K_B$), then binding of B must in turn disfavor binding of A by the same factor. This mutual effect because of thermodynamically coupled equilibria is referred to as *thermodynamic linkage*.

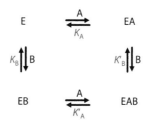

---

**FIGURE 3.3:** **Thermodynamic linkage in substrate binding**. An enzyme E binds two substrates, A and B, to form the EAB complex. $K_A$ is the dissociation constant of the EA complex, $K_B$ the dissociation constant of the EB complex. $K_A'$ and $K_B'$ are the corresponding constants in the presence of the second ligand. The four equilibria form a thermodynamic cycle, and the equilibrium constants are linked. If binding of A favors binding of B ($K_B' < K_B$), then binding of B also favors the binding of A to the same extent ($K_A' < K_A$) due to the thermodynamic balance within the cycle.

---

If the binding of A and B is interdependent, we can calculate the *interaction energy* $\Delta\Delta G^0_{\text{int}}$ for A and B binding by comparing the free energy changes upon formation of the EAB complex when binding of A and B is independent with the free energy change when binding is coupled. Note that the $\Delta G^0$ values in eq. 3.34 have been calculated from the equilibrium constants of dissociation, and are changes in free energy upon dissociation. The $\Delta G^0$ for dissociation of EAB in the absence of coupling is $\Delta G^0_A + \Delta G^0_B$, for coupling it is $\Delta G^0_A + \Delta G^{0'}_B$ for EAB→EA→E or $\Delta G^{0'}_A + \Delta G^0_B$ for EAB→EB→E. The interaction energy $\Delta\Delta G_{int}$ is the difference between these values:

$$\Delta\Delta G^0_{int} = \Delta G^0_A + \Delta G^0_B - \left(\Delta G^0_A + \Delta G^{0'}_B\right) = \Delta G^0_A + \Delta G^0_B - \left(\Delta G^{0'}_A + \Delta G^0_B\right) \qquad \text{eq. 3.39}$$

$\Delta\Delta G^0_{int}$ can be positive or negative. $\Delta\Delta G^0_{int} < 0$ means binding of one substrate increases the affinity for the second substrate, we say that binding of A and B shows *positive thermodynamic linkage*. In cases where $\Delta\Delta G^0_{int} > 0$, binding of one substrate decreases the affinity of the enzyme for the second substrate, the two substrates bind with *negative thermodynamic linkage*. Sometimes this phenomenon is also referred to as positive or negative cooperativity, although the term cooperativity describes a different effect (Section 11.5). Note that we can also calculate $\Delta\Delta G^0_{int}$ from the free energies of binding using eq. 3.39. In this case, $\Delta\Delta G^0_{int}$ is positive for positive thermodynamic linkage, and negative for negative thermodynamic linkage. Interaction energies can be determined in double mutant cycles (Box 3.3).

---

**BOX 3.3: INTERACTION ENERGY: DOUBLE MUTANT CYCLES.**

Thermodynamic cycles can be used to identify interactions between amino acids in a protein. In such an approach, the two amino acids of interest are exchanged to different amino acids that cannot engage in this putative interaction, most commonly alanine. The two amino acids are replaced individually and in combination. For all three modified (variant) proteins and the unmodified (wild-type) protein, the thermodynamic stability is determined experimentally (Section 16.3.4). The contribution of the amino acid to the stability is determined as the difference in $\Delta G^0_{unfold}$ for the variant compared to the wild-type ($\Delta\Delta G^0 = \Delta G^0_{unfold,wild-type} - \Delta G^0_{unfold,variant}$). If the two amino acids, e.g. a serine and a tyrosine, engage in an interaction that contributes to the thermodynamic stability of the protein, then exchanging either of them (e.g. serine to alanine or tyrosine to phenylalanine, Figure 3.4) will reduce the stability of the protein by this contribution. For variant 1 we obtain a destabilization by $\Delta\Delta G^0_1$, for variant 2 by $\Delta\Delta G^0_2$. If the two amino acids do not interact with each other, but with other partners, we will obtain a $\Delta\Delta G^0_3$ for the double variant that is the sum of $\Delta\Delta G^0_1$ and $\Delta\Delta G^0_2$ because we now remove both interactions. If the two amino acids interact with each other, however, then substitution of the second amino acid will not have any further destabilizing effect, and $\Delta\Delta G^0_3$ will equal $\Delta\Delta G^0_1$ and $\Delta\Delta G^0_2$. A similar approach can be used to identify cooperative interactions of two side chains in substrate binding, for example. Here, $\Delta\Delta G^0$ is the difference in free energy change of dissociation of the enzyme-substrate complex between wild-type and variant ($\Delta\Delta G^0 = \Delta G^0_{diss,wild-type} - \Delta G^0_{diss,variant}$). In this case, the energies can be additive ($\Delta\Delta G^0_3 = \Delta\Delta G^0_1 + \Delta\Delta G^0_2$; independent interactions of the two side chains with the substrates) or non-additive ($\Delta\Delta G^0_3 \neq \Delta\Delta G^0_1 + \Delta\Delta G^0_2$; cooperative interactions with the substrate). For $\Delta\Delta G^0_3 > (\Delta\Delta G^0_1 + \Delta\Delta G^0_2)$ the two amino acids show positive cooperativity in substrate binding, for $\Delta\Delta G^0_3 > (\Delta\Delta G^0_1 + \Delta\Delta G^0_2)$ they show negative cooperativity. The difference $(\Delta\Delta G^0_1 + \Delta\Delta G^0_2) - \Delta\Delta G^0_3$ is what we have defined as the interaction energy $\Delta\Delta G^0_{int}$ in a thermodynamic cycle (see text). If we calculate $\Delta\Delta G^0_{int}$ from dissociation free energies, a positive value corresponds to negative, a negative value to positive thermodynamic linkage. For binding free energies, a positive or negative $\Delta\Delta G^0_{int}$ corresponds to positive or negative thermodynamic linkage.

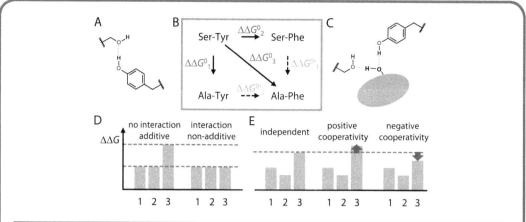

**FIGURE 3.4:    Double mutant cycles to identify interactions between side chains or cooperativity in substrate binding.** Interacting side chains (A) or cooperativity of two side chains in binding of substrate (blue; C) can be identified in double mutant cycles (B). In a double mutant cycle, the amino acids of interest (here Ser and Tyr) are individually and simultaneously replaced by other side chains that cannot engage in the interaction. The wild-type protein and the three variants form a thermodynamic cycle. The $\Delta\Delta G^0$ values are either differences in thermodynamic stability between wild-type and protein (bottom left) or differences in substrate binding energies (bottom right). $\Delta\Delta G_1^0$, $\Delta\Delta G_2^0$, and $\Delta\Delta G_3^0$ are measured experimentally. D: If the two side chains interact with each other, this interaction stabilizes the protein. Removal of the interaction by either side chain exchange leads to a decrease in thermodynamic stability by $\Delta\Delta G_1^0$ or $\Delta\Delta G_2^0$, respectively. Exchanging the second side chain does not have an additional destabilizing effect ($\Delta\Delta G_1^0 = \Delta\Delta G_2^0 = \Delta\Delta G_3^0$). If the two side chains interact with different partners, however, exchange of both side chains will remove both of these interactions, and $\Delta\Delta G_3^0$ is the sum of $\Delta\Delta G_1^0$ and $\Delta\Delta G_2^0$. E: If two side chains interact with a ligand, exchange of either of these amino acids leads to reduced binding energy ($\Delta\Delta G_1^0$, $\Delta\Delta G_2^0$). If the interactions are independent, simultaneous removal of both interactions leads to a change in binding energy that is the sum of the individual contributions (left). The $\Delta\Delta G$ values are not additive if there is positive (right) or negative cooperativity (center) between these interactions in substrate binding. The arrows indicate the interaction energy (see text).

A second important example of a thermodynamic cycle is represented by the following situation: an enzyme exists in two different conformations, E and E'. The enzyme contains a basic side chain that can be protonated to $EH^+$ or $E'H^+$ (Figure 3.5). The equilibrium constants for the conformational equilibrium are $K$ (deprotonated state) and $K'$ (protonated state), the equilibrium constants for the protonation equilibrium are $K_A$ (E) and $K'_A$ (E'). The four species E, E', $EH^+$, and $E'H^+$ are linked by four equilibria that form a thermodynamic cycle (Figure 3.5).

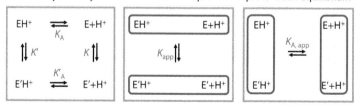

**FIGURE 3.5:    Thermodynamic cycles and linked functions.** The thermodynamic cycle (left) shows coupled equilibria of protonation and conformational equilibria of an enzyme E. The enzyme can exist in two states E and E'. Both E and E' can be protonated. The equilibrium constants for the individual equilibria are indicated. We can now focus on the overall conformational equilibrium by summing all species of the enzyme in conformational state E (E and $EH^+$) as one pseudo-species, and all species in state E' (E' and $E'H^+$) as a second pseudo-species, and defining an apparent equilibrium constant between these species (center). The apparent constant for the conformational equilibrium will then depend on the pH. Alternatively, we can focus on the protonation equilibrium, and sum together all deprotonated species (E and E') and all protonated species ($EH^+$ and $E'H^+$) as pseudo-species, and define an apparent equilibrium constant between those (right). The apparent protolysis constant will depend on the conformational equilibria: protonation and conformational change are linked functions.

If $K_A$ and $K_A'$ are identical, then $K$ and $K'$ are also identical: the conformational equilibrium is independent of the protonation state of the enzyme. Often, the two types of equilibria are coupled, and $K_A \neq K_A'$ and $K \neq K'$. In this case, we can define an *apparent equilibrium constant* for the overall conformational equilibrium, $K_{app}$, by collecting all species of E in the denominator, and all species of E' in the numerator (Figure 3.5). The apparent equilibrium constant is then

$$K_{app} = \frac{[E'] + [E'H^+]}{[E] + [EH^+]}$$

eq. 3.40

The protonation equilibria of E and E' are described by the equilibrium constants $K_A$ and $K_A'$. Protonation equilibria are typically formulated as dissociation reactions (protolysis):

$$K_A = \frac{[E][H^+]}{[EH^+]}$$

eq. 3.41

and

$$K_A' = \frac{[E'][H^+]}{[E'H^+]}$$

eq. 3.42

We can now solve eq. 3.41 and eq. 3.42 for EH⁺ and E'H⁺, respectively, and substitute these expressions into eq. 3.40 to obtain

$$K_{app} = \frac{[E'] + \dfrac{[E'][H^+]}{K_A'}}{[E] + \dfrac{[E][H^+]}{K_A}}$$

eq. 3.43

By factoring out $[E']$ in the numerator and $[E]$ in the denominator, eq. 3.43 simplifies to

$$K_{app} = \frac{[E']\left(1 + \dfrac{[H^+]}{K_A'}\right)}{[E]\left(1 + \dfrac{[H^+]}{K_A}\right)}$$

eq. 3.44

Substitution of $[E']/[E]$ by $K$ gives

$$K_{app} = K\frac{\left(1 + \dfrac{[H^+]}{K_A'}\right)}{\left(1 + \dfrac{[H^+]}{K_A}\right)}$$

eq. 3.45

We see that as a result of the coupling of the conformational and protonation equilibria, the apparent equilibrium constant for the conformational equilibrium is dependent on the H⁺ concentration and thus on the pH. We can distinguish two limiting cases: at high pH, the proton concentration is small, and the terms H⁺/$K_A$ and H⁺/$K_A'$ in the sums in the numerator and denominator of eq. 3.45 are small and can be neglected, giving

$$K_{app,high\ pH} = K \qquad \text{eq. 3.46}$$

Under these conditions, EH$^+$ or E'H$^+$ will not be populated. The conformational equilibrium will be established between E and E', and $K_{app}$ equals $K$. At low pH, i.e. high proton concentrations, the terms H$^+$/$K_A$ and H$^+$/$K'_A$ in eq. 3.45 are large and now dominate the sums in numerator and denominator. Eq. 3.45 then simplifies to

$$K_{app,low\ pH} = K\frac{K_A}{K'_A} \qquad \text{eq. 3.47}$$

Note that, according to the thermodynamic balance in the cycle, the right-hand side of eq. 3.47 equals $K'$! At low pH, both E and E' will be protonated, and the conformational equilibrium will be dominated by the equilibrium between EH$^+$ and E'H$^+$. The behavior of $K_{app}$ as a function of the pH between these two limiting cases is illustrated in Figure 3.6. The conformational equilibrium becomes pH-dependent because the protonation equilibria are different for the two conformations.

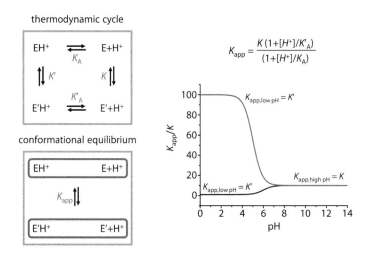

**FIGURE 3.6:    Linked functions and apparent equilibrium constants. $K_{app}$ as a function of pH.** The apparent equilibrium constant for the conformational equilibrium, $K_{app}$, depends on the pH. $K_A$ and $K'_A$ are the protolysis constants for the enzyme in the two conformations. Blue curve: $K_A = 1\ \mu M$ (p$K_A = 6$), $K'_A = 10\ \mu M$ (p$K'_A = 5$), $K = 10$; black curve: $K_A = 10\ \mu M$ (p$K_A = 5$), $K'_A = 1\ \mu M$ (p$K'_A = 6$), $K = 10$. At low pH, the enzyme is protonated, and $K_{app}$ approaches $K'$. At high pH, the enzyme is deprotonated, and $K_{app}$ approaches $K$. The inflection point of the pH dependence depends on the p$K_A$ value of the predominant enzyme conformation (E' in both cases).

We can also define an apparent protonation constant by collecting all protonated and deprotonated species (Figure 3.5). By convention, we formulate the equilibrium constant $K_{A,app}$ as a dissociation constant:

$$K_{A,app} = \frac{\left([E'] + [E]\right)\left[H^+\right]}{\left[E'H^+\right] + \left[EH^+\right]} \qquad \text{eq. 3.48}$$

We can now substitute the concentrations for E' by $K[E]$ and E'H$^+$ by $K'[EH^+]$ and obtain

$$K_{A,app} = \frac{\left(K[E] + [E]\right)\left[H^+\right]}{K'\left[EH^+\right] + \left[EH^+\right]} \qquad \text{eq. 3.49}$$

Now $[E]$ and $[EH^+]$ can be factored out, giving

$$K_{A,app} = \frac{(K+1)[E]\left[H^+\right]}{(K'+1)\left[EH^+\right]} \qquad \text{eq. 3.50}$$

The ratio of the concentrations in eq. 3.50 is $K_A$, and we can simplify to

$$K_{A,app} = K_A \frac{K+1}{K'+1}$$

eq. 3.51

The true protolysis constant $K_A$ of the protonated basic group in the enzyme is modified by the factor $(K + 1)/(K' + 1)$ because of the coupling to the conformational equilibrium. This factor is significantly different from unity when $K \neq K'$ and either $K$ or $K'$ are larger than unity (Figure 3.7).

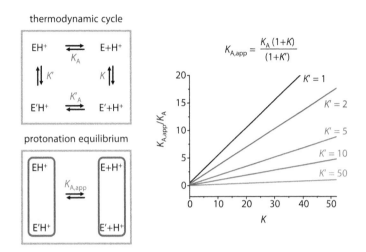

**FIGURE 3.7:    Linked functions and apparent equilibrium constants: $K_{A,app}$ for different values of $K$ and $K'$.** The apparent protolysis equilibrium constant, $K_{A,app}$, is the intrinsic protolysis constant $K_A$, multiplied by $(K + 1)/(K' + 1)$. $K$ and $K'$ are the equilibrium constants for the conformational equilibria of the deprotonated and protonated enzyme. The ratio of $K_{A,app}/K_A$ is shown as a function of $K$ for several values of $K'$. A coupled conformational equilibrium shifts the apparent $pK_A$ of an ionizable group when $K$ and $K'$ are different, and one of them is large.

An example of such behavior is chymotrypsin. α-chymotrypsin is a serine protease that consists of three peptide chains, A, B, and C, linked by disulfide bonds. The protonation state of the N-terminus of chain B strongly influences the conformational equilibrium of α-chymotrypsin: while the enzyme is predominantly in the inactive E state when it is deprotonated ($K < 1$), it is predominantly in the active E' state in the protonated form ($K' \gg 1$). From the thermodynamic balance, we know that the $K_A$ value of the ionizable group must also strongly differ in E and E'. According to eq. 3.51, $K_{A,app}$ is much smaller than $K_A$. Thus, a change in pH not only shifts the protonation equilibrium, but also dramatically alters the conformational equilibrium of chymotrypsin. At low pH, the enzyme is in the protonated state, and E'H+ is the predominant form. Raising the pH leads to a dissociation of E'H+ to E' and H+. In the deprotonated state, the conformational equilibrium is on the side of E, and E' is converted to E. The overall equilibrium is established between E'H+ and E (Figure 3.8).

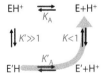

**FIGURE 3.8:    Linked ionization and conformational equilibria in α-chymotrypsin.** α-chymotrypsin exists in an active conformation (E') in which the N-terminus of the polypeptide chain B is protonated (E'H+). The conformational equilibrium between E and E' is characterized by an equilibrium constant $K < 1$ in the deprotonated state, and $K' \gg 1$ in the protonated state. At low pH (high H+ concentration), the enzyme is protonated, and because $K'$ is much larger than unity, E'H+ is the predominant species. With increasing pH, E'H+ dissociates, and because $K$ is less than unity, E' is converted to E (blue arrow).

By deriving the apparent equilibrium constants $K_{app}$ and $K_{A,app}$, we have seen that the conformational and protonation equilibria are coupled, or linked. Thermodynamically coupled properties are also called *linked functions*. Apparent equilibrium constants are often also called group or *macroscopic equilibrium constants*. In contrast, $K_A$, $K'_A$, $K$ and $K'$ are *microscopic equilibrium constants*. Note that apparent equilibrium constants are the <u>only</u> case where we have sums of concentrations in the numerator and the denominator! Microscopic equilibrium constants always have products of concentrations in the numerator and denominator.

## QUESTIONS

3.1   A reaction A + B $\rightleftarrows$ C + D leads to equilibrium concentrations of $[A]_{eq} = 1\,\mu M$, $[B]_{eq} = 2\,\mu M$, $[C]_{eq} = 4\,\mu M$, $[D]_{eq} = 8\,\mu M$. Calculate $K_{eq}$ and $\Delta G^0$. What is the maximum work the system can perform when 5 µM A and 10 µM B or 1 µM C and 1 µM D are mixed? $T = 298.15$ K.

3.2   A reaction A + B $\rightleftarrows$ C has an equilibrium constant

$$K_{eq} = \frac{[C]_{eq}}{[A]_{eq}[B]_{eq}} = 5\,\text{mM}^{-1}$$

What are the equilibrium concentrations of A and B when you start the reaction from 2 mM C? What are the equilibrium concentrations if you mix 1 mM A and 2 mM B?

3.3   $\Delta G^{0'}$ for ATP hydrolysis under standard conditions (pH 7, 25°C) is −31 kJ mol$^{-1}$. Calculate $\Delta G$ for steady-state conditions in the living cell of [ATP] = $10^{-3}$ M, [ADP] = $10^{-4}$ M, and [phosphate] = $10^{-2}$ M.

3.4   A reaction A $\rightleftarrows$ B has a free energy change of $\Delta G^{0'} = 25$ kJ mol$^{-1}$ under standard conditions. The $\Delta G^{0'}$ for ATP hydrolysis is −31 kJ mol$^{-1}$. What is the minimum ATP/ADP ratio required to drive the reaction towards B for $[P_i] = 10$ mM, $[A] = 5$ mM, $[B] = 1$ mM?

3.5   Adenylate cyclase catalyzes the formation of cyclic AMP (cAMP) from ATP:

$$ATP \rightarrow cAMP + PP_i$$

The equilibrium constant for this reaction at standard conditions is $K_{eq} = 0.065$. $\Delta G^{0'}$ for ATP hydrolysis to AMP is −33.5 kJ mol$^{-1}$. What is $\Delta G^{0'}$ for the hydrolysis of cAMP to AMP?

3.6   Consider the following scheme:
A $\rightleftarrows$ B            $K_{eq,1} = 2.4 \cdot 10^{-3}$
B + X $\rightleftarrows$ C + Y    $K_{eq,2} = 7.5 \cdot 10^{-2}$

What is the overall free energy change for the conversion of A to C?

3.7   The $K_d$ of two enzyme-substrate complexes differs by a factor of ten: $K_{d,1} = 10\,K_{d,2}$. What is the energetic difference $\Delta\Delta G$ in free energies of binding?

3.8   An enzyme binds its substrate with 1:1 stoichiometry. The equilibrium association constant is $K_{a,S} = 10^6$ M$^{-1}$. What is the equilibrium dissociation constant? Calculate the degree of binding, $\nu = [ES]/E_0$ at a concentration of 1 µM enzyme and 1 µM, 10 µM, and 100 µM substrate.

3.9   The equilibrium between the folded and denatured state of a protein is characterized by an equilibrium constant of folding, $K_{fold}$. Calculate the fraction $\theta$ of the folded protein.

3.10  Phosphoglyerate kinase couples the endergonic phosphorylation of 3-phosphoglycerate (3-PG) to 1,3-bisphosphoglycerate with the exergonic reaction of ATP hydrolysis. $\Delta G^{0'}$ for ATP hydrolysis is −31 kJ mol$^{-1}$, $\Delta G^{0'}$ for the phosphorylation of 3-PG to 1,3-BPG is +50 kJ mol$^{-1}$.

What is $\Delta G^{0'}$ for the coupled reaction? In yeast, the ATP/ADP ratio is 10 under optimal conditions, 2 under starving conditions. For what values of 3-PG/1,3-BPG does the overall reaction become exergonic?

3.11 A membrane transporter/cargo complex has a $K_d$ value of 10 μM at 25°C, at 37°C $K_d$ increases to 25 μM. What is $\Delta S$ and $\Delta H$ of binding at 25°C?

3.12 An acid HA has a dissociation constant $K_A$ = 0.1 mM. What are the pH of a 1 M and a 0.1 mM aqueous solution of the acid, a 1 M solution of the sodium salt of the acid, and a mixture of the sodium salt and the acid, both at a concentration of 1 M? $K_w = [OH^-][H^+] = 10^{-14}$.

3.13 The Tris base has a p$K_A$ of 8.3. What is the pH change for 1 L of a 50 mM Tris buffer at pH 8.3 upon addition of 10 mmol $H^+$? How much would the pH change when the same amount of protons is added to unbuffered water of pH 8.3? What is the buffer capacity $\beta$ of 50 mM Tris, pH 8.3, and water at pH 8.3? How much would the buffer capacity change if the Tris buffer had a concentration of 200 mM?

3.14 The protonation enthalpy for Tris is −4.2 kJ mol⁻¹, for hydrogen phosphate ($HPO_4^{2-}$) it is 46 kJ mol⁻¹. What is the change in p$K_A$ with temperature for both buffers? Which buffer would you use to investigate thermal unfolding of a protein?

3.15 A 2 mM solution of ATP is hydrolyzed at pH 8.0 in water and in 50 mM Tris buffer. What is the change in pH? p$K_A$(Tris) = 8.3.

3.16 The isoelectric point pI of amino acids is the pH at which the net charge is zero. Calculate the pI as a function of the p$K_A$ values for amino acids with acidic side chains, with basic side chains, and with side chains that are neither basic nor acidic.

3.17 Calculate the fractions of all histidine species at pH 4, 6, and 8. p$K_{A1}$ = 1.7 (carboxylic group) p$K_{A2}$ = 6.0 (side chain), p$K_{A3}$ = 9.1 (amino group).

3.18 An enzyme E binds a substrate S and a cofactor C. The equilibrium dissociation constant $K_{d,S}$ of the enzyme-substrate complex ES is 1 μM, for EC it is 10 μM. When the cofactor C is present, $K'_{d,S}$ is decreased to 0.1 μM. What is the value for the dissociation constant $K'_{d,C}$ of the enzyme-cofactor complex in the presence of substrate S? Calculate the interaction energy $\Delta\Delta G_{int}$ for cofactor and substrate binding.

3.19 A protein exists in two alternative conformations $N_1$ and $N_2$ in the native state and unfolds to U from $N_2$:

$$N_1 \overset{K_1}{\rightleftharpoons} N_2 \overset{K_2}{\rightleftharpoons} U$$

Calculate the apparent equilibrium constant for folding.

3.20 A ligand binds more tightly to the folded state (N) of a protein than to the unfolded state (U). Show that the ligand stabilizes the protein and calculate by how much ($\Delta\Delta G_{fold}$ = ?).

3.21 One effect of denaturants is a preferred interaction with the unfolded state of proteins. Show how this leads to destabilization and unfolding.

# REFERENCES

Horovitz, A. (1996) Double-mutant cycles: a powerful tool for analyzing protein structure and function. *Fold Des* **1**(6): R121–126.
· review about the application of double mutant cycles to proteins

Horovitz, A. and Fersht, A. R. (1990) Strategy for analysing the co-operativity of intramolecular interactions in peptides and proteins. *J Mol Biol* **214**(3): 613–617.
· theoretical treatment of interaction energies and cooperativity of interactions

Klostermeier, D. and Millar, D. P. (2002) Energetics of hydrogen bond networks in RNA: hydrogen bonds surrounding G+1 and U42 are the major determinants for the tertiary structure stability of the hairpin ribozyme. *Biochemistry* **41**(48): 14095–14102.
· application of double-mutant cycles to probe interactions in RNA

# Thermodynamics of Transport Processes

## 4.1 DIFFUSION

We encountered the spontaneous dissipation of matter in Section 2.4.1. A tendency of particles to achieve a uniform spatial distribution is equivalent to the tendency of molecules in solution to achieve uniform concentrations that are independent of position throughout the solution. Local differences in concentrations are usually equilibrated by diffusion. *Diffusion* is a spontaneous process in three dimensions. If a concentration gradient is present in one dimension, we can define the *flux J* as the number of particles that pass a certain cross-sectional area per time. The magnitude of the flux depends on the concentration gradient. For one-dimensional diffusion in the $x$-direction, the flux $J$ is described by *Fick's first law*:

$$J = -D\frac{\mathrm{d}c}{\mathrm{d}x}$$

eq. 4.1

$D$ is the diffusion coefficient (in the $x$-direction), and $\mathrm{d}c/\mathrm{d}x$ is the concentration gradient in the $x$-direction. The unit of the diffusion coefficient is area per time ($m^2\ s^{-1}$). For concentrations in particles per volume, the flux is also expressed in particles. For molar concentrations, the flux is expressed in mol. The derivation of Fick's first law is illustrated in Box 4.1. For three-dimensional diffusion, the components of the flux in each dimension $x$, $y$ and $z$ are described by eq. 4.1. The flux in three dimensions depends on the three-dimensional concentration gradient:

$$\vec{J} = -D\nabla c$$

eq. 4.2

The inverted triangle in eq. 4.2 is the gradient operator that symbolizes the concentration gradients in $x$-, $y$- and $z$-directions. The negative signs in eq. 4.1 and eq. 4.2 express that the flux is positive for a negative concentration gradient (from high to low concentrations), and negative for a positive concentration gradient (from low to high concentration). Fick's first law thus tells us that macroscopic diffusion in the presence of concentration gradients has a preferred direction: the molecules diffuse down the steepest gradient. This is in accordance with the concentration dependence of the chemical potential (Section 2.6.5): the higher chemical potential in regions with high concentrations drives the diffusion of molecules towards regions with lower concentration. Directed diffusion stops when the concentration is equilibrated over the entire volume, and the driving force for further directed movement is zero. Even when the macroscopic movement stops in equilibrium, the individual molecules do not come to a halt. Due to their thermal energy, molecules can also diffuse in the absence of concentration gradients. This microscopic movement is random, without a preferred direction. Such a random movement is called *Brownian diffusion* or *Brownian motion.*

*Fick's second law* relates the change in concentration with time, d$c$/d$t$, to the local concentration gradient d$c$/d$x$ and has the form

$$\frac{\mathrm{d}c}{\mathrm{d}t} = D\frac{\mathrm{d}^2 c}{\mathrm{d}x^2}$$

eq. 4.3

The derivation of Fick's second law is given in Box 4.1. We will revisit Fick's second law in the context of analytical ultracentrifugation (Section 26.2; eq. 26.26).

---

**BOX 4.1: DERIVATION OF FICK'S LAWS OF DIFFUSION.**

We consider a tube with a concentration gradient from one end to the other. The concentration of particles at position $x$ along the tube is $c(x)$ (Figure 4.1). The flux through the cross-section A of the tube at position $x$ can be calculated by accounting for molecules passing through the window from left to right and from right to left in a time window $\Delta t$. The average concentration of particles (in particles per volume) in a volume element of length 2b around $x - b$ and $x + b$ is $c(x - b)$ and $c(x + b)$. The number of particles $n_{\text{left}}$ in this volume element around $x - b$ is

$$n_{\text{left}} = c(x - b) \cdot 2b \cdot A$$

eq. 4.4

and the number of particles $n_{\text{right}}$ in the volume element around $x + b$ is

$$n_{\text{right}} = c(x + b) \cdot 2b \cdot A$$

eq. 4.5

The number of particles passing through the window at position $x$ in both directions in the time window $\Delta t$ is proportional to the number of particles in the volume elements left and right and to the time window $\Delta t$, giving the flux in number of particles as

$$n_{\text{left} \rightarrow \text{right}} = \text{const.} \cdot c(x - b) \cdot 2b \cdot A \cdot \Delta t$$

eq. 4.6

and

$$n_{\text{right} \rightarrow \text{left}} = \text{const.} \cdot c(x + b) \cdot 2b \cdot A \cdot \Delta t$$

eq. 4.7

The net flux $J$ through the cross-section at position $x$ is proportional to the difference of these numbers, divided by the area $A$ and the time window $\Delta t$:

*(Continued)*

$$J = \text{const.} \cdot \frac{c(x-b) \cdot 2b \cdot A \cdot \Delta t - c(x+b) \cdot 2b \cdot A \cdot \Delta t}{A \cdot \Delta t} \qquad \text{eq. 4.8}$$

We can express the concentrations at $x - b$ and $x + b$ as

$$c(x-b) = c(x) - b \cdot \frac{dc}{dx} \qquad \text{eq. 4.9}$$

and

$$c(x+b) = c(x) + b \cdot \frac{dc}{dx} \qquad \text{eq. 4.10}$$

Substituting these expressions into eq. 4.8 gives the flux $J$ as

$$J = -\text{const.} \cdot 4b^2 \frac{dc}{dx} \qquad \text{eq. 4.11}$$

Eq. 4.11 corresponds to Fick's first law (eq. 4.1) for $4b^2 \cdot \text{const.} = D$.

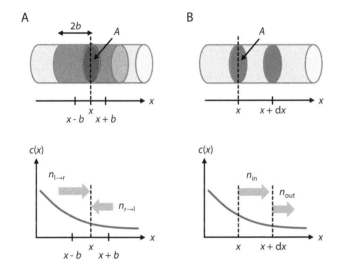

**FIGURE 4.1:    Derivation of Fick's laws of diffusion.** A: To derive Fick's first law, the flux of particles from left to right and from right to left through a window of area $A$ (purple) at position $x$ in a tube with a concentration gradient d$c$/d$x$ (bottom) is considered. The flux from left to right is proportional to the number of particles in the blue volume element around $x - b$, the flux from right to left is proportional to the number of particles in the red volume element around $x + b$. B: The derivation of Fick's second law considers the flux of particles into and out of a certain volume element between $x$ and $x + dx$. A net flux occurs only if the concentration gradient is different at $x$ and $x + dx$.

Fick's second law can be derived by similar considerations, starting from a tube with a concentration gradient d$c$/d$x$ along the $x$-axis (Figure 4.1). The concentrations at positions $x$ and $x + dx$ along the tubes are $c(x)$ and $c(x + dx)$, with $c(x) > c(x + dx)$. If the concentration gradient at $x$ and x+dx is identical, the flux of particles into the volume element from the left will be identical to the flux of particles from the volume element to the right, and the total flux through this element is zero. If the concentration gradients at $x$ and $x + dx$ are different, the flux in and out of the element differs, and a net flux is observed. For a flux $J(x)$ at position $x$, we can calculate the number of particles that pass a window of area $A$ at position $x$ during an infinitesimal time interval d$t$ as

*(Continued)*

$$n_{\text{in}} = J(x) A \mathrm{d}t \qquad \text{eq. 4.12}$$

and the number of particles that pass this window of area $A$ at position $x + \mathrm{d}x$ is

$$n_{\text{out}} = J(x + \mathrm{d}x) A \mathrm{d}t \qquad \text{eq. 4.13}$$

The net flux $n_{\text{net}}$ of particles is the difference between these two numbers, giving

$$n_{\text{net}} = J(x) A \mathrm{d}t - J(x + \mathrm{d}x) A \mathrm{d}t = \big(J(x) - J(x + \mathrm{d}x)\big) A \mathrm{d}t \qquad \text{eq. 4.14}$$

The flux $J(x+\mathrm{d}x)$ can be expressed in terms of the flux $J(x)$ and the gradient of the flux, $\mathrm{d}J/\mathrm{d}x$:

$$J(x + \mathrm{d}x) = J(x) + \frac{\mathrm{d}J}{\mathrm{d}x} \cdot \mathrm{d}x \qquad \text{eq. 4.15}$$

Combination of eq. 4.14 and eq. 4.15 gives

$$n_{\text{net}} = -\frac{\mathrm{d}J}{\mathrm{d}x} \cdot \mathrm{d}x \cdot A \mathrm{d}t \qquad \text{eq. 4.16}$$

We now have to divide by the volume of the element between $x$ and $x + \mathrm{d}x$, $\mathrm{d}V = A\mathrm{d}x$, to convert the change in particle numbers into a change in concentration. The net rate of this change in concentration, $\mathrm{d}c/\mathrm{d}t$, is obtained by dividing by $\mathrm{d}t$. Together, this gives

$$\frac{\mathrm{d}c}{\mathrm{d}t} = -\frac{\mathrm{d}J}{\mathrm{d}x} \qquad \text{eq. 4.17}$$

By expression of the flux $J$ by Fick's first law (eq. 4.1), we obtain

$$\frac{\mathrm{d}c}{\mathrm{d}t} = -\frac{\mathrm{d}\left(-D\dfrac{\mathrm{d}c}{\mathrm{d}x}\right)}{\mathrm{d}x} = D\frac{\mathrm{d}^2 c}{\mathrm{d}x^2} \qquad \text{eq. 4.18}$$

which is Fick's second law (eq. 4.3).

Membrane proteins can diffuse in biological membranes because of the membrane fluidity. This diffusion is restricted to the plane of the membrane, and is therefore only two-dimensional (lateral diffusion). The random path that the protein covers during lateral diffusion is called a *trajectory* (Figure 4.2). The mean square displacement $\langle r^2 \rangle$ of the protein at time $t$ is defined as

$$\langle r^2 \rangle(t) = \frac{r_{t_1}^2 + r_{t_2}^2 + \ldots + r_t^2}{n} \qquad \text{eq. 4.19}$$

where $r_{t_1}$ is the distance traveled in time interval 1, $r_{t_2}$ the distance traveled in time interval $t_2$, and $n$ is the number of time intervals. The distance $\langle r^2 \rangle$ is thus the cumulative mean of the squares of the distances traveled at time $t$ since leaving the starting point. For two-dimensional diffusion, $\langle r^2 \rangle$ follows eq. 4.20:

$$\langle r^2 \rangle = 4Dt \qquad \text{eq. 4.20}$$

where $D$ is the diffusion coefficient. The diffusion coefficient can be determined from the slope of a plot of the square displacement $\langle r^2 \rangle$ as a function of time $t$. The trajectory of proteins can be followed by imaging techniques with single-molecule resolution (Section 23.1.7.3).

Under certain conditions, diffusion of the membrane protein is limited to a defined area of the membrane (Box 4.2). This phenomenon is called anomalous, restricted, or *corralled diffusion* (Figure 4.2). For restricted diffusion, the mean square displacement $\langle r^2 \rangle$ is described by

$$\langle r^2 \rangle = \frac{d^2}{3}\left(1 - \exp\left(\frac{-12D_0 t}{d^2}\right)\right)$$

eq. 4.21

where $d$ is a measure of the dimension of the area within which the protein can undergo free diffusion, and $D_0$ is the diffusion coefficient at time $t = 0$.

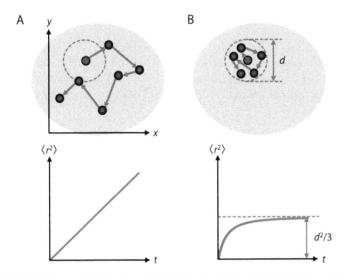

**FIGURE 4.2: Free and restricted diffusion.** A: A protein (red) diffuses freely within a membrane (gray). The position of the protein at different time points is indicated. The square of the distance the protein has traveled from its starting point increases linearly as a function of time. B: A protein (red) undergoes restricted (corralled) diffusion within a delimited area of the membrane (gray broken line). The mean square displacement now depends exponentially on time. The amplitude of the exponential function is related to the size of the membrane area within which the protein diffuses freely.

---

### BOX 4.2: MEMBRANE RAFTS AND CORRALLED DIFFUSION.

The *fluid mosaic model* of lipid membranes describes the membrane as a fluidic phase in which lipids and embedded proteins undergo free lateral diffusion. Many membrane studies are performed on reconstituted membranes with a well-defined lipid composition that differs from the composition and complexity of natural cell membranes. Diffusion coefficients of lipids and membrane proteins in reconstituted membranes are often one order of magnitude higher than diffusion coefficients within the cell membrane. The slower diffusion in biomembranes may be due to the large protein content and the molecular crowding. Furthermore, cell membranes interact with the actin cytoskeleton, and the actin filament network also affects the lateral diffusion of lipids and proteins within the membrane. In model membranes, the formation of lipid domains has been observed. These *lipid rafts* have a lipid composition that is distinct from the rest of the membrane, and are often enriched in cholesterol. The higher local fluidity within rafts may lead to

*(Continued)*

corralled diffusion of transmembrane proteins, and thus to their enrichment within these membrane domains. It is still a matter of debate if these rafts also exist in natural membranes. The membrane domains forming in biological membranes are typically small (tens of nanometers) and highly dynamic. Nevertheless, rafts are thought to play an important role in regulating the local concentration of transmembrane proteins including transporters and receptors. By spatial regulation of receptor concentration, dimerization, and activation, they might act as platforms for membrane signaling and trafficking.

## 4.2 THE CHEMIOSMOTIC HYPOTHESIS

The chemical potential is a driving force for chemical reactions. We have seen in Section 2.6.5 that the chemical potential depends on concentration. We now consider a system that is separated from the surroundings by a barrier that is impermeable for protons, such as a biological membrane. The chemical potential of protons inside and outside depends on the local proton concentration, and is described by

$$\mu_{H^+, \text{outside}} = \mu^0_{H^+} + RT \ln c_{H^+, \text{outside}}$$

eq. 4.22

and

$$\mu_{H^+, \text{inside}} = \mu^0_{H^+} + RT \ln c_{H^+, \text{inside}}$$

eq. 4.23

If the proton concentrations inside and outside are different, then

$$\mu_{H^+, \text{inside}} \neq \mu_{H^+, \text{outside}}$$

eq. 4.24

The difference of the chemical potential of protons inside and outside constitutes a driving force for the equilibration of the concentrations: in the absence of the barrier, protons would spontaneously move from the compartment with higher concentration (and higher chemical potential) to the compartment with lower concentration (and lower chemical potential). The flow would continue until the concentrations in both compartments are equal, the chemical potentials are equal, and there is no driving force for further proton flux. However, the impermeable membrane prevents such an equilibration of proton concentrations, and energy can be stored in form of a concentration gradient between protons (or other ions and molecules that cannot pass the membrane) on both sides of the membrane. The concept of storing energy in form of concentration gradients is called the *chemiosmotic hypothesis*. The energy that is stored in a concentration gradient can drive energetically unfavorable processes. Proton gradients are a central element in the intermediate storage of energy during oxidative phosphorylation and photosynthesis (Figure 4.3). In mitochondria, energetically favorable electron transport (Section 5.6) is coupled to the transport of protons out of the mitochondria into the intermembrane space between the inner and outer membrane. In a separate reaction, protons flow back into the mitochondria through a membrane protein called $F_1$–$F_o$ ATPase. The $F_1$–$F_o$ ATPase couples the endergonic synthesis of ATP from ADP and phosphate to the exergonic flow of protons from outside to inside. The energy is now stored in form of ATP, and ATP hydrolysis then provides the energy for numerous endergonic processes in the cell. Similarly, in the light reaction of photosynthesis the energy from electron transport chains in the thylakoid membrane of chloroplasts is converted into a proton gradient. The proton gradient is used to drive ATP synthesis by a transmembrane ATPase (Figure 4.3). The energy stored in a proton gradient is also called *proton-motive force* (PMF).

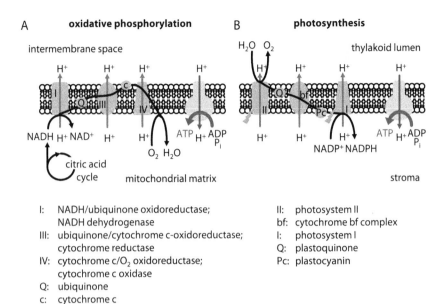

I: NADH/ubiquinone oxidoreductase;
   NADH dehydrogenase
III: ubiquinone/cytochrome c-oxidoreductase;
   cytochrome reductase
IV: cytochrome c/$O_2$ oxidoreductase;
   cytochrome c oxidase
Q: ubiquinone
c: cytochrome c

II: photosystem II
bf: cytochrome bf complex
I: photosystem I
Q: plastoquinone
Pc: plastocyanin

**FIGURE 4.3:** **Proton gradients for intermediate energy storage in oxidative phosphorylation and photosynthesis.** A: Oxidative phosphorylation takes place in the inner membrane of mitochondria. The citric acid cycle in the mitochondrial matrix generates reduction equivalents (NADH) that are oxidized to $NAD^+$ at the inner membrane by complex I. The electrons from NADH are transported through a transport chain (black line) *via* ubiquinone (Q) to complex III and *via* cytochrome c to complex IV that reduces the terminal electron acceptor oxygen to $H_2O$. The energetically favorable electron transport by complex I, III, and IV is coupled to transport of protons out of the mitochondrion into the intermembrane space. The proton gradient provides the energy for ATP synthesis at the ATP synthase (blue) that is coupled to influx of protons into the mitochondrion. B: The light reaction of photosynthesis takes place in the thylakoid membrane in chloroplasts. Photosystem II oxidizes $H_2O$ to oxygen in a light-induced reaction. The electrons are transported to photosystem I *via* plastoquinone (Q), the cytochrome bf complex and plastoquinone (Pc). Photosystem I transfers the electrons to the terminal electron acceptor in photosynthesis, $NADP^+$, and reduces it to NADPH. The energetically favorable electron transport by the cytochrome bf complex is coupled to proton transport into the thylakoid lumen. Photosystem I contributes to the increase in proton concentration in the lumen by generating two protons per oxidized $H_2O$ molecule. The proton-motive force then drives ATP synthesis at the ATP synthase (blue) that is coupled to an efflux of protons from the thylakoid into the chloroplast stroma.

Biological membranes are not only barriers for protons, but also for other ions and polar molecules, which allows the establishment of ion gradients and electromotive forces. Ion gradients serve as an energy source for transport (Section 4.3). The free energy change associated with the transport of protons or other ions across the membrane into the cell is

$$\Delta G = -RT \ln \frac{c_{in}}{c_{out}} + nF\Delta\phi \qquad \text{eq. 4.25}$$

with the concentrations inside and outside of the cell, $c_{in}$ and $c_{out}$, the number of charges $n$ of the ion, the Faraday constant $F$, and the potential difference across the membrane $\Delta\phi$. We will discuss the membrane potential $\Delta\phi$ in Section 5.7. Note that the concentrations $c_{in}$ and $c_{out}$ in eq. 4.25 are absolute concentrations: although we have to consider concentrations relative to the standard state, the standard-state concentration $c^0$ is identical inside and outside (1 M), and cancels.

Flux of ions along a concentration gradient is an exergonic, spontaneous process. If the ion flux is controlled, the energy can be coupled to the endergonic synthesis of ATP (Figure 4.4). Proton gradients are exploited to drive the energetically unfavorable synthesis of ATP from ADP and phosphate. Transport of ions or molecules against a concentration gradient, and thus building up concentration gradients, is an endergonic process, and requires energy input. Concentration gradients can be established by coupling transport to exergonic reactions, such as ATP hydrolysis, or redox reactions (oxidative phosphorylation, photosynthesis). Uncoupling flux along concentration gradients from ATP synthesis can be used to generate heat without muscle activity (Box 4.3). The reaction can also be reversed, and ATP hydrolysis can provide the energy for establishing proton (or other) gradients (Figure 4.4). The ATP-dependent establishment of an ion concentration gradient is important for the generation and maintenance of membrane potentials and in nerve impulse transfer (Section 5.7).

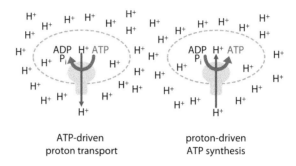

ATP-driven
proton transport

proton-driven
ATP synthesis

**FIGURE 4.4:** **Concentration gradients and energy storage**. The energy of ATP hydrolysis can be used to drive proton transport against a concentration gradient, converting the energy into a proton gradient (left). Conversely, the energy stored in a proton gradient can drive the synthesis of ATP from ADP and phosphate, coupled to the flux of protons along the concentration gradient (right).

---

**BOX 4.3: UNCOUPLING THE PROTON-MOTIVE
FORCE FROM ATP SYNTHESIS.**

Instead of coupling the proton-motive force to ATP synthesis, the energy can also be converted into heat by uncoupled influx of protons. This uncoupling is possible in the mitochondria of brown fat tissue. Instead of passing through the proton channel that is coupled to the ATP synthase, protons re-enter the cell through the ligand-gated proton channel thermogenin (see Section 4.3). Thermogenin is closed in the ligand-free state, but opens in response to binding of fatty acids. The fatty acids are generated in a signaling cascade. Upon binding of noradrenalin to receptors on the cell membrane of brown fat cells, a cAMP signaling cascade leads to activation of a lipase that hydrolyzes triglycerides to glycerol and fatty acids. The fatty acids are the activating ligands for thermogenin and lead to opening of the channel. Protons can now directly enter the mitochondria, and the energy of the proton gradient is not stored in form of ATP, but directly released as heat. This process is used by babies, and also by hibernating mammals. Uncoupling allows the generation of heat without muscle activity in form of trembling. Membrane-permeable hydrophobic molecules that are weak acids, such as 2,4-dinitrophenol, can also uncouple the proton-motive force from ATP synthesis by diffusing through the lipid membrane and transporting protons. While these substances are not very healthy, they have been useful tools in biochemical studies of the oxidative phosphorylation pathway.

## 4.3 ACTIVE AND PASSIVE TRANSPORT

We have seen before that the flux of ions or molecules from high concentration to low concentration along a concentration gradient is exergonic and thus in principle spontaneous. However, biological membranes act as barriers and impede spontaneous flow. Pore-forming proteins spanning the membrane can provide a channel for molecules to pass the membrane (Figure 4.5). Flow of molecules through these pores along a concentration gradient is called *passive transport*. Passive transport is typically regulated, and opening of the channel is triggered by ligand binding (*ligand-gated channel*) or by a change in membrane potential (*voltage-gated channel*) (Figure 4.5).

Transport of ions or molecules against a concentration gradient, from low to high concentrations, is an endergonic process: it does not occur spontaneously, but requires energy. Active transporters use the energy of ATP hydrolysis (*primary active transport*) or the energy of concentration gradients of other molecules (*secondary active transport*) to drive the transport of their cargo. In secondary active transport, molecules co-transported with the cargo can flow in the same (symport) or the opposite direction (antiport) (Figure 4.5).

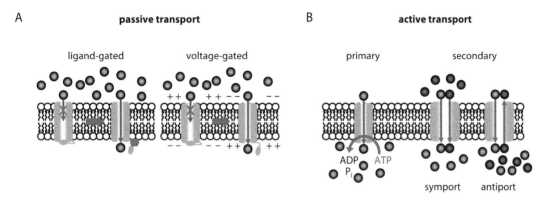

**FIGURE 4.5:** **Transport across biological membranes**. A: Passive transport along concentration gradients is mediated by channels. Channel opening is typically regulated, either by binding of ligands (ligand-gated channel) or by changes in the membrane potential (voltage-gated channel). B: Active transport against concentration gradients requires energy. In primary active transport, this energy stems from ATP hydrolysis. In secondary active transport, the energy is provided by the electromotive force of a second particle (red) that is co-transported along its own concentration gradient. Co-transport of this particle can occur in the same direction as the cargo (symport) or in the opposite direction (antiport).

The transport cycle of a primary active transporter needs to couple ATP hydrolysis to transport of the cargo. This coupling can be achieved by cycling of the transporter through four different states (Figure 4.6). At the beginning of the transport cycle, the cargo binding site of the transporter is exposed to the left-hand side of the membrane where the cargo will be taken up. The energy barrier for binding is low, and the cargo is rapidly bound (see Section 13.2 for the relation between energy barriers and reaction velocities). The bound state is a global minimum in the energy profile, and the cargo is stably bound. The energy barrier to the other side of the membrane is high, however, making release of the cargo to the right very slow. The transporter then switches into a second state with its binding site exposed towards the other side of the membrane. This transition can be induced e.g. by ATP-dependent phosphorylation. In this second state, the energy profile of the transporter-cargo interaction is altered: the bound state is now only a local minimum in the energy profile. The energetic barrier towards the left is high, and release towards the left is slow. Release to the right, however, is now favored both thermodynamically (energetically "downhill", exergonic) and kinetically (low energy barrier), and the cargo will be rapidly released to the right. At this stage one ATP has been hydrolyzed, and one cargo molecule has been transported from left to right. For further transport cycles, the transporter needs to be regenerated. Cleavage of the phosphate group will convert the transporter into its original conformation. The high-affinity cargo binding site is exposed towards the left, and the transporter is ready for another round of ATP-driven transport. Some transporters use the energy of light instead of ATP to transport ions against concentration gradient. In these cases, absorption of light and relaxation from the excited state (Sections 19.1 and 19.2) triggers conformational switches of the transporter. We will re-visit these light-driven ion pumps when we discuss the absorption of light and optical spectroscopy (Section 19.2).

The ATP-driven transport of protons against a concentration gradient is not too different from the reverse Carnot cycle (Section 2.2.5.1): ATP hydrolysis provides the energy (w) to take up material from the low concentration side (heat $q$ from the colder reservoir) and transport it to the high concentration side (heat $q$ released to the warmer reservoir).

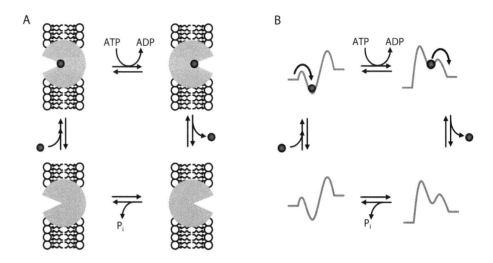

---

**FIGURE 4.6:** **Four-state model of a transporter: a thermodynamic cycle with energy input.** A: A transporter exists in two conformational states, one state with the high-affinity binding site for the cargo facing the exterior of the cell (left), and a second state with the low-affinity binding site facing towards the interior (right). Switching between these two conformations is achieved by phosphorylation and dephosphorylation. In both states, the transporter can be empty, or bound to cargo. For 100% coupling of ATP hydrolysis (phosphorylation) to transport of cargo, the transporter has to spend sufficient time in the outward facing state to bind cargo before the conformation is switched by phosphorylation. Similarly, it has to spend sufficient time in the inward facing state to ensure that the cargo is released before dephosphorylation switches it back to the outward facing state. B: Energy diagrams for the four states.

---

If the conformational switch of the transporter only occurs upon phosphorylation and dephosphorylation, and binding only occurs from the left and release only occurs to the right, then ATP hydrolysis and cargo transport are perfectly coupled; the coupling efficiency is 100%. To ensure high coupling efficiencies, timing has to be optimized: firstly, the transporter must stay long enough in the state with the high-affinity binding site towards the left, leaving enough time for the cargo to bind. If the transporter is phosphorylated before the cargo has bound, the transporter will undergo a futile cycle: one ATP will be hydrolyzed without a transport event. Secondly, the reporter needs to stay in the phosphorylated state, with the low-affinity binding site exposed to the right, long enough for the cargo to dissociate. Dephosphorylation of the transporter before the cargo has dissociated will bring the cargo back to the left, without net transport. The ATP has already been hydrolyzed at this stage, and premature switching back thus also leads to a futile cycle. The rates of cargo binding and release therefore have to be matched with the conformational cycle that is regulated by phosphorylation/dephosphorylation. To ensure that no cargo is transported in the absence of ATP hydrolysis, the conformational switch needs to be coupled strictly to phosphorylation and dephosphorylation. Spontaneous switching of a cargo-bound transporter without phosphorylation will lead to transport of one cargo molecule without ATP hydrolysis, which is referred to as *leakage*.

## 4.4 DIRECTED MOVEMENT BY THE BROWNIAN RATCHET

We have seen in Section 4.1 that thermal motion of molecules is a random process, and that molecules diffuse in no preferred direction in the absence of concentration gradients. However, it is possible to bias Brownian motion under certain conditions, and to enforce directional movement. We start with a thought experiment, and imagine a number of electric dipoles that are arranged head-to-tail to form a track of alternating charges (Figure 4.7). A charged particle above this track experiences repulsive forces near a charge of the same sign, and attractive electrostatic forces when it is located next to an opposite charge. We can translate this behavior into an energy landscape. For a negatively charged particle, the potential energy is low in the vicinity of positive charges on the track (attraction), and high close to negative charges (repulsion). Overall the particle therefore experiences a *saw-tooth-shaped energy potential* with periodic minima and maxima along the track (Figure 4.7). The distance between minima and maxima corresponds to the distances of the charges on the track. In thermal equilibrium, the preferred position of the particle along the track is determined by the energetic minima or wells in the saw-tooth potential. The thermal energy of the particles enables small excursions left and right, up to a potential energy level that equals their thermal energy. If the thermal energy is sufficient, the particle can cross the energy barrier to the left or to the right, and reach the next energetic minimum to the left or right along the track. The probability for such transitions depends on the height of the energy barrier (see Section 13.2), which is identical for movement in either direction. In other words, crossing to the next well left or right is of equal probability, and the particle cannot undergo directed movement under these conditions.

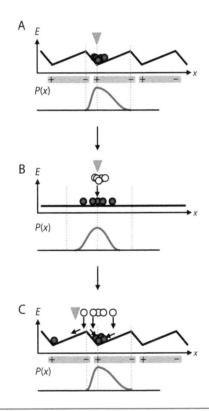

**FIGURE 4.7:    Principle of a Brownian ratchet.** A: Negatively charged particles above dipoles aligned head to tail experience a saw-tooth energy profile: their potential energy is low at the positive end of the dipoles (energetic minimum) and high energy at the negative end (energy maximum). In the presence of the saw-tooth potential, the particles distribute around the energetic minimum. Their probability distribution $P(x)$ (red) is a function of the position $x$ and is asymmetric, reflecting the asymmetry of the potential. As the barrier is equally high to either side, the probability for a particle to cross over to the left or right well is equal. B: If the saw-tooth profile is switched off (e.g. by neutralizing the charge of the particle in a chemical reaction), the energy profile is flat, and the potential energy is position-independent. The particles can now freely diffuse to left and right. The probability to move by a certain distance to left and right by diffusion is equal, the probability distribution of the particles at the end (red) is symmetric to the starting position. C: When the asymmetric saw-tooth profile is switched back on, the particles again experience a potential that depends on their actual position and will equilibrate to the position of the nearest energetic minimum. Because of the asymmetry of the potential, the position of the next barrier to the left is more likely to be crossed than that of the next barrier to the right. In our example, one particle is beyond the next barrier to the left at the end of the off-period, and equilibrates into the neighboring well on the left, while the remaining particles end up in the original well. The gray triangle indicates the mean position of the particles. The mean position has moved to the left in this one cycle.

Now imagine we switch off the electric potential. One way to achieve this would be by neutralizing the charge of the particle, e.g. by protonation of a negatively charged particle. The uncharged particle does not interact with the charges on the track. Therefore, the potential energy of the particle does not anymore depend on its position (Figure 4.7). The uncharged particle undergoes Brownian motion starting from the equilibrium position it occupied while the saw-tooth potential was still present. With a flat potential, the probabilities for excursions to left and right are equal, or more specifically, the probability for an excursion by $\Delta x$ to the left is equal to the probability for the same excursion by $\Delta x$ to the right. As a consequence, the probability to find the particle at a certain position along the track now symmetrically decreases from its starting position. Up to this point, there is still no directionality in the system.

What happens if we switch the potential back on by again conferring a negative charge to the particle, e.g. by deprotonation? The particle experiences the same saw-tooth potential as in the beginning. Its location is again determined by the energy landscape, and it relaxes to the next energetic minimum that it can reach in an energetically favorable reaction (energetically "downhill"). At this point of the thought experiment it is important to take note of the asymmetry of the saw-tooth energy potential (Figure 4.7): the change in potential along the $x$-axis is much steeper to the left of the minimum than to the right. As a consequence, the particle has to diffuse only a short distance $\Delta x$ during the time the potential was "off" to end up in a place corresponding to the next left well. To reach the next right well requires that the particle travels a much longer way. Traveling a short distance in a fixed time window is more probable than traveling a longer distance. Correspondingly, the chances that we catch the particle in a position that leads it into the next left well when we switch on the potential are higher than the chances that the particle has traveled far enough to be in the position of the next right well. Hence, the probability for a movement of the particle to the left is higher than for a movement to the right (Figure 4.7). By switching the saw-tooth energy potential on and off, we have biased the Brownian motion of the particle to directed movement to the left! Using the same principle, we can force a set of particles to undergo an overall movement to the left by switching the potential on and off, and generate directed motion. Such a rudimentary motor is called a *Brownian motor* or *Brownian ratchet*.

To achieve directed movement, the timing of the changes in potential is important. The potential needs to be off long enough for the particles to cover sufficient distance to leave the realm of one energy well, and to end up in the left neighboring well once the potential is switched on again. Similarly, the potential needs to be on long enough for the particles to equilibrate in the position of the energetic minimum. In our example, the timing is determined by the time point of deprotonation (to generate the negatively charged particle and establish the saw-tooth potential) and the time point of protonation (to generate an uncharged particle and switch off the saw-tooth potential). More generally speaking, the kinetics of the chemical reaction that modulates the interaction of the particle with the track is crucial for efficient directional motion. Kinetics and energetics need to be balanced for efficient movement, however. The asymmetry of the saw-tooth potential determines how strongly movement is biased into one direction (Figure 4.8). The distance between the minimum and the next maximum in the potential determines the distance the particle has to travel to cross over to the next well. The shorter this distance, the shorter the time the particle needs to cover this distance, and the more probable is reaching this well. A Brownian ratchet with a high spatial repetition frequency of the periodic potential can be driven by a fast chemical reaction, whereas a ratchet with a lower repetition frequency must be driven by a slower reaction. The shape of the potential and the distances between minima and maxima are determined by the structure of the units that assemble linearly into the track.

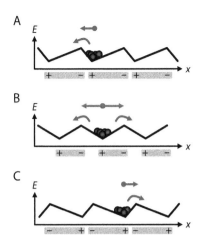

**FIGURE 4.8:    Shape of the potential and direction of movement.** Energy potential for negatively charged particles above a track made of electric dipoles. The traveling distance to reach the neighboring well is indicated by the blue arrow, the preferred direction of movement by the red arrow. A: Asymmetric saw-tooth potential that biases motion to the left. B: Symmetric saw-tooth potential. Movement to the left and right is equally probable. C: Asymmetric saw-tooth potential that biases motion to the right.

The remaining question from a thermodynamic point of view is "where does the energy for the directed movement come from?" In our model, switching the potential on and off requires energy, and this is the energy that is converted into directional movement. The energy potential was switched on and off because of the protonation/deprotonation reaction that neutralized or generated the negative charge of the particle. In general terms, the energy is thus provided by the chemical reaction that modulates the interaction between particle and track. It is important to note that a Brownian ratchet can only be driven by a reaction that is not in equilibrium. This is quite clear, because $\Delta G$ for a reaction in equilibrium is zero (Section 3.1; eq. 3.7), and this reaction cannot perform work.

Directional movement in biology is often achieved by biasing Brownian motion. The energy for this movement is typically provided by ATP hydrolysis. The proton transporter we discussed in Section 4.3 is an example of a Brownian ratchet: the transporter is switched between two conformations in an ATP-dependent process (phosphorylation/dephosphorylation). The phosphorylation/dephosphorylation reaction alters the interaction of the transporter with the cargo, and modulates the interaction energy landscape. As a result, one conformation of the transporter and the resulting energy potential enable rapid cargo binding from the outside with high affinity, and the other conformation is characterized by a low cargo affinity and rapid cargo dissociation to the inside. Conceptually the transporter therefore corresponds to the particle in our model Brownian ratchet, and the cargo corresponds to the track. The timing of the chemical reaction is key for the movement of the cargo across the membrane.

A motor example is the movement of myosin along actin filaments. The periodic structure of actin filaments (the track) generates a periodic energy potential that governs the interaction of myosin (the particle) with actin. Myosin is an ATPase: it binds and hydrolyzes ATP, and thereby alternates between an ATP and an ADP state that interact differently with the actin filament. Depending on its nucleotide state, myosin either binds tightly to the actin filament (potential on), or it binds with low affinity and becomes detached from the filament (potential off). The ATP- and ADP-bound states of myosin thus represent the charged and uncharged state of the particle in our conceptual model. The rate constants for ATP binding, hydrolysis, and dissociation of the products control how long each nucleotide state of myosin persists, and at which point in time the potential is switched on and off. The structure of actin and the periodicity of the actin filament determine the shape and periodicity of the interaction potential. Of course, the interactions between myosin and actin are much more complex than simple on- and off-switching of electrostatic interactions. Nevertheless, the principle of the movement of myosin relative to actin and of other biological motors relative to their track can be described by the conceptually simple model for a Brownian ratchet.

In summary, the Brownian ratchet describes the behavior of molecular motors on the basis of interactions between a moving unit and its track, modulated by a chemical reaction that provides energy input and the timing of changes in the interaction potential. We have restricted our considerations to electrostatic interactions, but biological motors can realize much more complex interactions with the track. Conformational dynamics subtly modulate these interactions, and influence the timing of changes, leading to higher levels of complexity and offering possibilities for regulation.

## QUESTIONS

4.1     Explain diffusion along concentration gradients based on the chemical potential and on entropy considerations.

4.2     A protein has a diffusion coefficient in the cell of $D = 10^{-11}$ m$^2$ s$^{-1}$. How long does it take the molecule to traverse the cell (diameter $r = 1$ µm)?

4.3     The concentration of chloride ions is higher in blood (0.1 M) than in brain cells (0.05 M). What is $\Delta G$ for the transport of Cl$^-$ from the blood into the brain and from the brain into the blood at 37°C?

4.4     The interior of our stomach has a pH of 2. The cells in the lining mucosa have a neutral pH (pH 7.2). Calculate the energy that is required for the secretion of H$^+$ from the mucosa cells into the gastric lumen.

4.5     The change in free energy for ATP hydrolysis is $\Delta G^{0\prime} = -31$ kJ mol$^{-1}$. What is the minimal proton gradient that is required for ATP synthesis under standard conditions and at intracellular concentrations of [phosphate] = $10^{-2}$ M, [ADP] = $10^{-4}$ M, and [ATP] = $10^{-3}$ M? Calculate the required pH outside the cell assuming a neutral pH in the cytoplasm. $T = 25$°C.

4.6     How can the movement of polymerases along DNA be described by the Brownian ratchet model?

## REFERENCES

Singer, S. J. and Nicolson, G. L. (1972) The fluid mosaic model of the structure of cell membranes. *Science* **175**(4023): 720–731.
· original paper that introduces the fluid mosaic model of membranes

Engelman, D. M. (2005) Membranes are more mosaic than fluid. *Nature* **438**(7068): 578–580.
· introductory review article on the mosaicity of membranes

Simons, K. and Gerl, M. J. (2010) Revitalizing membrane rafts: new tools and insights. *Nat Rev Mol Cell Biol* **11**(10): 688–699.
· review article on the role of membrane domains in signaling and trafficking and insight from modern fluorescent microscopy methods

Lingwood, D. and Simons, K. (2010) Lipid rafts as a membrane-organizing principle. *Science* **327**(5961): 46–50.
· review on rafts as platforms and their dynamics in cellular membranes

Sanderson, J. M. (2012) Resolving the kinetics of lipid, protein and peptide diffusion in membranes. *Mol Membr Biol* **29**(5): 118–143.
· review on diffusion of lipids and proteins in membranes

Kusumi, A., Nakada, C., Ritchie, K., Murase, K., Suzuki, K., Murakoshi, H., Kasai, R. S., Kondo, J. and Fujiwara, T. (2005) Paradigm shift of the plasma membrane concept from the two-dimensional continuum fluid to the partitioned fluid: high-speed single-molecule tracking of membrane molecules. *Annu Rev Biophys Biomol Struct* **34**: 351–378.
· review on current models for the diffusion of proteins in membranes

Heinemann, F., Vogel, S. K. and Schwille, P. (2013) Lateral membrane diffusion modulated by a minimal actin cortex. *Biophys J* **104**(7): 1465–1475.
· original paper on the effect of the actin cytoskeleton on protein diffusion in membranes

Mitchell, P. (2011) Chemiosmotic coupling in oxidative and photosynthetic phosphorylation. 1966. *Biochim Biophys Acta* **1807**(12): 1507–1538.
· reprint of the original paper from Peter Mitchell on the chemiosmotic hypothesis

Astumian, R. D. and Derenyi, I. (1998) Fluctuation driven transport and models of molecular motors and pumps. *Eur Biophys J* **27**(5): 474–489.
· review on the principle of a Brownian ratchet

Ait-Haddou, R. and Herzog, W. (2003) Brownian ratchet models of molecular motors. *Cell Biochem Biophys* **38**(2): 191–214.
· review on the Brownian ratchet and its application to biological motors

Astumian, R. D. (2005) Biasing the random walk of a molecular motor. *Journal of Physics: Condensed Matter* **17**(47): S3753.
· Brownian ratchet model and application to ATP-dependent enzymes and the movement of kinesin along microtubules

# Electrochemistry

Electrochemistry comprises reactions that involve charged particles, and the transfer of electrons. For their thermodynamic description, we have to expand our definition of the chemical potential, and have to include a term that represents the electrochemical work associated with the transport of charges in an electric field. We define the *electrochemical potential* $\tilde{\mu}_j$ of ions in the presence of electric fields as (eq. 5.1).

$$\tilde{\mu}_j = \mu_j + z_j F \phi \qquad \text{eq. 5.1}$$

$\mu_j$ is the chemical potential, $z$ is the number of charges of the ion, $F$ is the Faraday constant (96500 C mol$^{-1}$, i.e. the charge of one mol of elementary charges: the Faraday constant is the product of the elementary charge of one electron and the Avogadro constant), and $\phi$ is the electrostatic potential. The electrochemical potential is the driving force for electrochemical reactions, in the same way that the chemical potential is the driving force for a chemical reaction in the absence of electrochemical work. For uncharged particles ($z = 0$) or in the absence of an electrostatic potential, eq. 5.1 simplifies to the general definition of the chemical potential.

## 5.1 REDOX REACTIONS AND ELECTROCHEMICAL CELLS

Reactions that involve the transfer of electrons are called *redox reactions*. A *reduction* reaction is the uptake of one (or more) electrons. The reverse reaction, the release of one or more electrons is called an *oxidation*. As electrons are normally not stable in solution, reduction and oxidation are usually coupled, leading to the term redox reaction. A physiologically important oxidation is the reaction of NADH/H$^+$ to NAD$^+$:

$$\text{NADH/H}^+ \longrightarrow \text{NAD}^+ + 2\,\text{e}^- + 2\,\text{H}^+ \qquad \text{scheme 5.1}$$

A reduction that is central for our lives is the conversion of molecular oxygen to water:

$$O_2 + 4\,e^- + 4\,H^+ \longrightarrow 2\,H_2O \qquad\qquad \text{scheme 5.2}$$

The overall redox reaction is

$$2\,NADH/H^+ + O_2 \longrightarrow 2\,NAD^+ + 2\,H_2O \qquad\qquad \text{scheme 5.3}$$

We have encountered these reactions in the context of oxidative phosphorylation (Figure 4.3). To understand the quantitative concepts governing redox reactions, we will start with a very simple inorganic reaction in solution. If we add elemental zinc to a solution of a copper(II) salt, i.e. containing $Cu^{2+}$ ions, the following reaction will take place:

$$Zn + Cu^{2+} \longrightarrow Zn^{2+} + Cu \qquad\qquad \text{scheme 5.4}$$

The zinc metal will dissolve and metallic copper will appear. Conceptionally and physically, we can separate this reaction into two half-reactions that occur in separate *half-cells* (Figure 5.1). Imagine we have a zinc electrode in a solution of zinc ions. At the electrode, zinc and zinc ions are in equilibrium. Zinc can release two electrons, and will be converted into a zinc ion. Inversely, zinc ions can take up two electrons, and will be deposited on the electrode as elemental zinc. This is also a redox reaction, in which the oxidant and the reducing agent are the same element.

$$Zn^{2+} + 2\,e^- \rightleftharpoons Zn \qquad\qquad \text{scheme 5.5}$$

On the other hand, we can immerse a copper electrode in a solution containing copper ions, and will have the corresponding equilibrium between copper and copper ions.

$$Cu^{2+} + 2\,e^- \rightleftharpoons Cu \qquad\qquad \text{scheme 5.6}$$

For zinc, the equilibrium is on the side of the ion, meaning the release of two electrons is an exergonic reaction, and thus the dissolution of zinc metal is the spontaneous direction: the zinc from the electrode is oxidized. For copper, on the other hand, the equilibrium is on the side of elemental copper. Copper ions from the solution take up electrons, and are reduced to copper that is deposited on the electrode.

We can now combine the two half-cells to an *electrochemical cell* (Figure 5.1). Due to the different positions of the two ionization equilibria, the electrochemical potential of the two electrodes will be different, and this difference of potential leads to a (measureable) voltage. If we connect the half-cells with a conducting material, electrons released by zinc oxidation at the zinc electrode can then flow to the copper electrode, and can be used to reduce copper ions in the other half-cell: reduction and oxidation are coupled in a *redox reaction*.

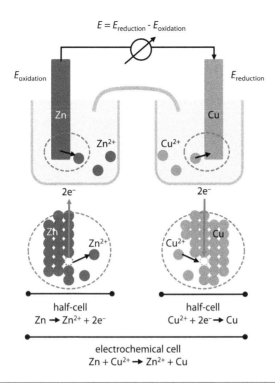

$$E = E_{reduction} - E_{oxidation}$$

$E_{oxidation}$

$E_{reduction}$

Zn

Cu

$Zn^{2+}$

$Cu^{2+}$

$2e^-$

$2e^-$

Zn

$Zn^{2+}$

$Cu^{2+}$

Cu

half-cell
$Zn \rightarrow Zn^{2+} + 2e^-$

half-cell
$Cu^{2+} + 2e^- \rightarrow Cu$

electrochemical cell
$Zn + Cu^{2+} \rightarrow Zn^{2+} + Cu$

**FIGURE 5.1:** **Electrochemical half-cells and cell.** When we place zinc metal (electrode) into a solution of a zinc salt, an equilibrium between zinc ions ($Zn^{2+}$) and elemental zinc (Zn) is established, leading to an electrochemical potential of the zinc half-cell (left). Similarly, when we place copper metal (Cu) into a solution of a copper salt, an equilibrium between Cu and $Cu^{2+}$ ions is established, causing an electrode potential. For $Zn/Zn^{2+}$, the equilibrium is on the side of the ions. Zn will be oxidized to $Zn^{2+}$, and the surplus of electrons causes a negative electrode potential. For $Cu/Cu^{2+}$, the equilibrium is on the side of Cu, and the $Cu^{2+}$ ions will be reduced at the electrode, leading to the deposition of Cu and a depletion of electrons (positive electrode potential, right). We can combine the two half-cells to an electrochemical cell, in which the two solutions are connected by an ion bridge. The difference in electrode potentials $E_{oxidation}$ and $E_{reduction}$ leads to a measureable voltage between the two half-cells, the electromotive force (see eq. 5.17).

The free energy change associated with the redox reaction (Scheme 5.4) can be calculated from the electrochemical potentials according to eq. 5.2:

$$\Delta G = \tilde{\mu}_{Zn^{2+}} + \tilde{\mu}_{Cu} - \tilde{\mu}_{Cu^{2+}} - \tilde{\mu}_{Zn} \qquad \text{eq. 5.2}$$

The electrochemical potentials of the (solid) electrode materials equal the standard chemical potentials (Section 2.6.5). If the electrochemical cell operates at standard pressure of $10^5$ Pa (1 bar), then

$$\tilde{\mu}_{Cu} = \tilde{\mu}_{Zn} = 0 \qquad \text{eq. 5.3}$$

(Section 2.6.3). With the help of eq. 2.156 that expresses the concentration dependence of the chemical potential we can then simplify eq. 5.2 to

$$\Delta G = \tilde{\mu}^0_{Zn^{2+}} - \tilde{\mu}^0_{Cu^{2+}} + RT \ln \frac{[Zn^{2+}]}{[Cu^{2+}]} = \Delta G^0 + RT \ln \frac{[Zn^{2+}]}{[Cu^{2+}]} \qquad \text{eq. 5.4}$$

The free energy change of the redox reaction equals the electrical work, which is the product of the transported charge (per mol) and the potential difference between the electrodes, $\Delta\phi$, in $V$. The transported charge in turn can be calculated from the Faraday constant $F$ (96500 C mol$^{-1}$) and the number of electrons $n$ that are exchanged:

$$\Delta G = -w_{el} = -n \cdot F \cdot \Delta\phi \qquad \text{eq. 5.5}$$

The negative sign in eq. 5.5 takes into account that the spontaneous reaction of the electrochemical cell with a negative $\Delta G$ allows the cell to perform work on the surroundings, and is in accordance with the system-centered perspective. We name the potential difference of an electrochemical cell working under reversible conditions the *electromotive force E* (or EMF, measured in V), and obtain eq. 5.6:

$$\Delta G = -n \cdot F \cdot E \qquad \text{eq. 5.6}$$

Under standard conditions, eq. 5.6 becomes

$$\Delta G^0 = -n \cdot F \cdot E^0 \qquad \text{eq. 5.7}$$

with the *standard electromotive force* $E^0$ of the electrochemical cell. By combining eq. 5.6 and eq. 5.7 with eq. 5.4, we can write

$$E = E^0 - \frac{RT}{nF} \ln \frac{\left[Zn^{2+}\right]}{\left[Cu^{2+}\right]} \qquad \text{eq. 5.8}$$

Eq. 5.8 allows us to calculate the electromotive force of an electrochemical cell from the standard electromotive force and the actual concentrations of the ions involved.

## 5.2 TYPES OF HALF-CELLS

There are different types of half-cells. So far we have considered half-cells consisting of a metal electrode immersed in a solution of a salt of this metal. Half-cells can also consist of a gas and an ionic solution. A prominent example we will come back to when we discuss reference measurements is the hydrogen electrode, where molecular hydrogen ($H_2$) dissociates into H atoms at a platinum electrode, and the H atoms are in equilibrium with hydronium ions (hydrated $H^+$):

$$2\,H^+ + 2\,e^- \rightleftharpoons H_2 \qquad \text{scheme 5.7}$$

A half-cell is also obtained when the same species is present in different oxidation states, such as

$$Fe^{3+} + e^- \rightleftharpoons Fe^{2+} \qquad \text{scheme 5.8}$$

or

$$Cu^{2+} + e^- \rightleftharpoons Cu^+ \qquad \text{scheme 5.9}$$

$Fe^{2+}/Fe^{3+}$ and $Cu^{2+}/Cu^+$ are redox pairs in components of the electron transport chains in oxidative phosphorylation and photosynthesis (see Section 5.6).

The last type of half-cells consists of a metal electrode, covered with a layer of an insoluble salt of the same metal, and in contact with a solution that contains the anion of this salt. Examples are the silver electrode:

$$AgCl\,(s) + e^- \rightleftharpoons Ag\,(s) + Cl^- \qquad \text{scheme 5.10}$$

or the calomel electrode

$$Hg_2Cl_2\,(s) + 2\,e^- \rightleftharpoons 2\,Hg\,(s) + 2\,Cl^- \qquad \text{scheme 5.11}$$

Silver and the calomel electrodes are frequently used as reference half-cells in pH measurements (Section 5.5).

# 5.3  STANDARD ELECTRODE POTENTIALS

It is not possible to measure the small potential differences between the solution and the electrode of a half-cell directly. Therefore, only relative half-cell electrode potentials against a second half-cell can be measured. The hydrogen electrode is used as a reference half-cell in these measurements, and its standard electrode potential has been defined as $E^0 = 0$ V. According to the standard electrode potentials, measured against the reference hydrogen electrode, different half-cells are ordered in the *electrochemical series* (Table 5.1), from highest positive to the highest negative electrode potential. In a half-cell with a negative standard potential, the release of electrons, the oxidation, is favored. In a half-cell with a positive standard potential, the uptake of electrons, the reduction, is favored.

**TABLE 5.1**
**The electrochemical series (I). Selection of standard electrode potentials in electrochemical order.**

| reduction (half-)reaction | $E^0$ (V) |
|---|---|
| $AgCl + e^- \rightarrow Ag + Cl^-$ | +1.98 |
| $Cl_2 + 2\,e^- \rightarrow 2\,Cl^-$ | +1.36 |
| $Ag^+ + e^- \rightarrow Ag$ | +0.80 |
| $Cu^{2+} + 2\,e^- \rightarrow Cu$ | +0.34 |
| $2\,H^+ + 2\,e^- \rightarrow H_2$ | 0.00 |
| $Zn^{2+} + 2\,e^- \rightarrow Zn$ | −0.76 |
| $Na^+ + e^- \rightarrow Na$ | −2.71 |

Standard potentials for biochemical redox systems are often given in the reference system of the biochemical standard state at pH 7, and are marked as $E^{0'}$. Biochemical standard potentials for biologically relevant reduction reactions are summarized in Table 5.2.

**TABLE 5.2**
**The electrochemical series (II). Biochemical standard potentials of biologically relevant reduction reactions in electrochemical order.**

| reduction (half-)reaction | $E^{0'}$ (V) |
|---|---|
| $O_2 + 4\,H^+ + 4\,e^- \rightarrow 2\,H_2O$ | +0.81 |
| $NO_3^- + 2\,H^+ + 2\,e^- \rightarrow NO_2^- + H_2O$ | +0.42 |
| $Fe^{3+} + e^- \rightarrow Fe^{2+}$ (cytochrome f) | +0.36 |
| $Cu^{2+} + e^- \rightarrow Cu^+$ (plastocyanin) | +0.35 |
| $Cu^{2+} + e^- \rightarrow Cu^+$ (azurin) | +0.30 |
| $O_2 + 2\,H^+ + 2\,e^- \rightarrow H_2O_2$ | +0.30 |
| $Fe^{3+} + e^- \rightarrow Fe^{2+}$ (cytochrome c) | +0.25 |
| $Fe^{3+} + e^- \rightarrow Fe^{2+}$ (cytochrome b) | +0.08 |
| dehydroascorbic acid $+ 2\,H^+ + 2\,e^- \rightarrow$ ascorbic acid | +0.08 |
| coenzyme Q $+ 2\,H^+ + 2\,e^- \rightarrow$ coenzyme QH$_2$ | +0.04 |
| fumarate $+ 2\,H^+ + 2\,e^- \rightarrow$ succinate | +0.03 |
| vitamin K$_1$ (ox) $+ 2\,H^+ + 2\,e^- \rightarrow$ vitamin K$_1$ (red) | −0.05 |
| oxaloacetate $+ 2\,H^+ + 2\,e^- \rightarrow$ malate | −0.17 |
| pyruvate $+ 2\,H^+ + 2\,e^- \rightarrow$ lactate | −0.18 |
| acetaldehyde $+ 2\,H^+ + 2\,e^- \rightarrow$ ethanol | −0.20 |
| riboflavin (ox) $+ 2\,H^+ + 2\,e^- \rightarrow$ riboflavin (red) | −0.21 |
| FAD $+ 2\,H^+ + 2\,e^- \rightarrow$ FADH$_2$ | −0.22 |
| glutathione (ox) $+ 2\,H^+ + 2\,e^- \rightarrow$ glutathione (red) | −0.23 |
| lipoic acid (ox) $+ 2\,H^+ + 2\,e^- \rightarrow$ lipoic acid (red) | −0.29 |
| $NAD^+ + H^+ + 2\,e^- \rightarrow$ NADH | −0.32 |
| cystine $+ 2\,H^+ + 2\,e^- \rightarrow 2$ cysteine | −0.34 |
| acetyl–CoA $+ 2\,H^+ + 2\,e^- \rightarrow$ acetaldehyde $+$ CoA | −0.41 |
| $2\,H_2O + 2\,e^- \rightarrow H_2 + 2\,OH^-$ | −0.42 |
| ferredoxin (ox) $+ e^- \rightarrow$ ferredoxin (red) | −0.43 |
| $O_2 + e^- \rightarrow O_2^-$ | −0.45 |

Note that a negative value for $E^0$ corresponds to a positive $\Delta G^0$, meaning the reduction reaction is thermodynamically disfavored, and the oxidation reaction is thermodynamically favored. In contrast, a positive value for $E^0$ corresponds to a negative $\Delta G^0$ for the reduction reaction, i.e. the reduction is thermodynamically favored. It is important to note that tabulated values for standard electrode potentials always refer to the reduction reaction. If the reaction proceeds in the opposite direction, towards oxidation, the sign of the standard electrode potential has to be inverted for calculations of the associated free energy change. We will come back to this sign inversion in Section 5.4.

## 5.4 THE NERNST EQUATION

For the simple electrochemical cell consisting of zinc and copper half-cells, we can calculate the free energy change $\Delta G$ associated with the redox reaction according to eq. 5.9:

$$\Delta G = \Delta G^0 + RT \ln \frac{[Zn^{2+}][Cu]}{[Zn][Cu^{2+}]} \qquad \text{eq. 5.9}$$

By combining eq. 5.9 with the relationships of $\Delta G$ and E or $\Delta G^0$ and $E^0$ (eq. 5.6 and eq. 5.7), we obtain for the electromotive force $E$ of the cell

$$E = E^0 - \frac{RT}{nF} \ln \frac{[Zn^{2+}][Cu]}{[Zn][Cu^{2+}]} \qquad \text{eq. 5.10}$$

At this point it is again important to remember that the concentrations in eq. 5.9 and eq. 5.10 are concentrations relative to the standard state. The solid electrodes are in their standard concentration, the concentration of the pure metals. Therefore, the terms for the electrodes in the argument of the logarithm have a value of unity, and can be neglected. For ions in solution, the standard concentration is 1 M. $\Delta G$ and the electromotive force $E$ therefore depend only on the ratio of the $Zn^{2+}$ and $Cu^{2+}$ concentrations in the half-cells: the standard concentration cancels. Eq. 5.9 thus leads to the same expression for the electromotive force that we have derived from considerations of the electrochemical potential (eq. 5.8).

Scheme 5.12 is a general scheme for the reactions in any electrochemical cell that contains a half-cell in which species 1 is oxidized and a half-cell where species 2 is reduced

$$
\begin{array}{r}
red_1 \rightleftharpoons ox_1 + n\,e^- \\
\underline{ox_2 + n\,e^- \rightleftharpoons red_2} \\
red_1 + ox_2 \rightleftharpoons ox_1 + red_2
\end{array}
\qquad \text{scheme 5.12}
$$

In analogy to the eq. 5.9, we can express $\Delta G$ for the cell reaction as

$$\Delta G = \Delta G^0 + RT \ln \frac{[ox_1][red_2]}{[red_1][ox_2]} \qquad \text{eq. 5.11}$$

and obtain the general form of the electromotive force as

$$E = E^0 - \frac{RT}{nF} \ln \frac{[ox_1][red_2]}{[red_1][ox_2]} \qquad \text{eq. 5.12}$$

This equation is equivalent to

$$E = E^0 + \frac{RT}{nF} \ln \frac{[red_1][ox_2]}{[ox_1][red_2]} \qquad \text{eq. 5.13}$$

and is called the *Nernst equation*.

For half-cell reactions (Scheme 5.13)

$$\text{ox} + n\,\text{e}^- \rightleftharpoons \text{red} \qquad \text{scheme 5.13}$$

we can write by analogy

$$\Delta G = \Delta G^0 + RT \ln \frac{[red]}{[ox]} \qquad \text{eq. 5.14}$$

and

$$E = E^0 - \frac{RT}{nF} \ln \frac{[red]}{[ox]} \qquad \text{eq. 5.15}$$

or

$$E = E^0 + \frac{RT}{nF} \ln \frac{[ox]}{[red]} \qquad \text{eq. 5.16}$$

We can calculate the potential difference $E$ for the $Zn/Zn^{2+}//Cu^{2+}/Cu$ electrochemical cell (Figure 5.1) directly from the individual electrode potentials of the half-cells. From the electrode potential of the half-cell in which the reduction reaction occurs, $E_{reduction}$, and the electrode potential from the half-cell in which the oxidation reaction occurs, $E_{oxidation}$, we obtain $E$ as

$$E = E_{reduction} - E_{oxidation} \qquad \text{eq. 5.17}$$

It is important to note that the electrode potentials $E_{reduction}$ and $E_{oxidation}$ in eq. 5.17 refer to the tabulated values for the reduction reactions. For the $Zn/Zn^{2+}//Cu^{2+}/Cu$ electrochemical cell, the standard electrode potential $E^0$ for the Zn half-cell is $-0.76$ V, for the Cu half-cell $E^0$ is 0.34 V (Table 5.1). According to eq. 5.17, the potential difference $E^0$ of the electrochemical cell under standard conditions is 0.34 V$-(-0.76$ V) = 1.1 V. This positive value corresponds to a negative value for $\Delta G^0$ (eq. 5.7), consistent with a spontaneous reaction.

Eq. 5.17 is also valid when the number of electrons that are taken up or released in the oxidation and reduction reactions differs. To write the overall reaction scheme and to calculate $\Delta G$ for these types of reactions, we have to multiply the stoichiometry coefficients of the reactions by the appropriate integer to arrive at the same number of electrons released in the oxidation and taken up in the reduction. However, the electrode potentials are proportional to the ratio of $\Delta G$ and $n$ ($E = -\Delta G/nF$; see eq. 5.6), and remain unchanged. The potential difference E can thus again be calculated directly from the tabulated values according to eq. 5.17.

We can also calculate the $\Delta G^{(0)}$ values for the complete cell reaction from the $\Delta G^{(0)}$ values of the individual reactions in the half-cells. Here we need to take into account what reaction is taking place in the half-cells. In our $Zn/Zn^{2+}//Cu^{2+}/Cu$ example, Zn is oxidized to $Zn^{2+}$ in one half-cell. The standard electrode potential is $-0.76$ V for the reduction, corresponding to a positive $\Delta G^0$ and a non-spontaneous reaction under standard conditions. For the reverse reaction, the oxidation of Zn, the electrode potential has the opposite sign, and $E^0 = +0.76$ V, which we can convert into a $\Delta G^0$ of $-176.7$ kJ mol$^{-1}$ according to eq. 5.7. In the Cu half-cell, the reduction reaction is taking place, and the electrode potential corresponds to the tabulated value of $E^0 = 0.34$ V, which can be converted to $\Delta G^0$ = –65.6 kJ mol$^{-1}$. The $\Delta G^0$ of the overall cell reaction is the sum of the $\Delta G^0$ values for the half-cell reactions, –212.3 kJ mol$^{-1}$. This value corresponds to an $E^0$ of 1.1 V for the electrochemical cell, the same value we have calculated from the electrode potentials of the half-cells according to eq. 5.17. However, the electrode potential is <u>not</u> a state function, and electrode potentials are not additive! Whenever we want to calculate the standard electrode potential for an overall reaction from the standard electrode potentials of the individual reactions, we have to take a detour over the $\Delta G$ values that are additive, and then convert $\Delta G$ for the overall reaction into $E$ (Box 5.1).

---

**BOX 5.1: NON-ADDITIVITY OF $E$.**

We consider a reduction reaction that occurs in two steps, the reduction of nitrate ions to nitrite, and the subsequent reduction of nitrite to ammonium (Scheme 5.14). The standard electrode potentials for the individual reactions are $E^0_1$ = 0.42 V and $E^0_2$ = 0.48 V.

$$NO_3^- + 2\,H^+ + 2\,e^- \longrightarrow NO_2^- + H_2O \qquad E^0_1 = 0.42\ V$$
$$NO_2^- + 8\,H^+ + 6\,e^- \longrightarrow NH_4^+ + 2\,H_2O \qquad E^0_2 = 0.48\ V$$
$$\overline{NO_3^- + 10\,H^+ + 8\,e^- \longrightarrow NH_4^+ + 3\,H_2O \qquad E^0_{tot} = ?}$$

scheme 5.14

To calculate the overall standard electrode potential, we convert $E^0_1$ and $E^0_2$ into $\Delta G^0$ values using eq. 5.7, and obtain $\Delta G^0_1$ = –81.06 kJ mol$^{-1}$ (with $n$ = 2) and $\Delta G^0_2$ = –277.92 kJ mol$^{-1}$ (with $n$ = 6). The change in free energy for the overall reaction is the sum of these values, $\Delta G^0_{tot}$ = –359.0 kJ mol$^{-1}$. From $\Delta G^0_{tot}$, we can calculate $E^0_{tot}$ = 0.465 V.

From eq. 5.7 and the additivity of $\Delta G$, we can derive a general expression for the electrode potential $E_{tot}$ of the overall reaction as a function of the electrode potentials $E_1$ and $E_2$ for the individual reactions:

$$E_{tot} = \frac{n_1 E_1 + n_2 E_2}{n_{tot}}$$

eq. 5.18

---

## 5.5 MEASURING pH VALUES

Using electrochemical cells, pH values of solutions can be measured. As a reference cell, either the hydrogen electrode (scheme 5.7) or the glass electrode (see Section 5.2) is used. The electrode potential of these electrodes depends directly on the concentration of H$^+$ ions, and thus of the pH of the test solution. According to the Nernst equation (eq. 5.16), the electrode potential of the H$^+$/H$_2$ half-cell is described by:

$$E = E^0_{H^+/H_2} + \frac{RT}{2F} \ln \frac{[H^+]^2}{p_{H_2}}$$

eq. 5.19

The standard potential for the H$^+$/H$_2$ half-cell is zero. The pressure $p_{H_2}$ is the standard pressure, and can be neglected. (Remember that $p_{H_2}$ is short for the actual pressure relative to the standard pressure $p^0$). The term $\ln[H^+]^2$ equals $2\cdot\ln[H^+]$, and the factor 2 in numerator and denominator cancels, leaving

$$E = \frac{RT}{F} \ln([H^+])$$

eq. 5.20

Converting the natural into the decadic logarithm by multiplying with $\ln 10 = 2.303$ (see Section 29.5) gives

$$E = \frac{2.303RT}{F} \log([H^+])$$

eq. 5.21

With the definition of the pH as the negative decadic logarithm of the proton concentration, we finally arrive at

$$E = -\frac{2.303RT}{F} pH$$

eq. 5.22

Eq. 5.22 describes the pH dependence of the electrode potential of the $H^+/H_2$ half-cell. To determine the pH value of a test solution, the hydrogen electrode is used in combination with a reference electrode with pH-independent electrode potential, and the pH can be calculated from the electromotive force. Hydrogen electrodes are not very convenient to work with, and glass electrodes are more commonly used for pH determination. Here, the potential at the measurement electrode is generated by the difference in proton concentrations between a reference solution ($[H^+]_{ref}$, typically pH 7) and the test solution of unknown pH ($[H^+]_{test}$), separated by a glass membrane. The potential $E$ at the glass membrane is

$$E = \frac{RT}{F} \ln\left(\frac{[H^+]_{test}}{[H^+]_{ref}}\right)$$

eq. 5.23

or

$$E = \frac{2.303RT}{F}(pH_{ref} - pH_{test}) = \frac{2.303RT}{F}\Delta pH$$

eq. 5.24

The potential is measured relative to a silver or a calomel electrode (Section 5.2) that serves as an internal reference electrode.

## 5.6 REDOX REACTIONS IN BIOLOGY

### 5.6.1 THE RESPIRATORY CHAIN

The oxidation of metabolites provides reduction equivalents in form of $NADH/H^+$ and $FADH_2$. These reduction equivalents are oxidized in the respiratory chain in the inner mitochondrial membrane (Figure 5.2). Molecular oxygen is the final electron acceptor, and is reduced to $H_2O$. The electrons from $NADH/H^+$ and $FADH_2$ are not directly transferred to oxygen, however, but pass a chain of electron acceptor/donor pairs. In a first reaction, the NADH/ubiquinone oxidoreductase, a transmembrane protein, catalyzes the oxidation of NADH, and the electron transfer to ubiquinone *via* flavin mononucleotide and an iron-sulfur cluster. In the second reaction, the ubiquinone/cytochrome c oxidoreductase catalyzes the electron transfer from ubiquinone to cytochrome c *via* cytochrome b, an iron-sulfur cluster, and cytochrome $c_1$. Finally, the cytochrome $c/O_2$ oxidoreductase catalyzes the electron transfer to molecular oxygen *via* cytochromes a and $a_3$. $NADH,H^+/NAD^+$ has the most negative electrode potential (–0.32 V), the potential of $O_2/H_2O$ is the most positive in the chain (0.81 V; Table 5.2). Each oxidation step in the respiratory chain goes from negative to more positive electrode potential, and is exergonic. All three

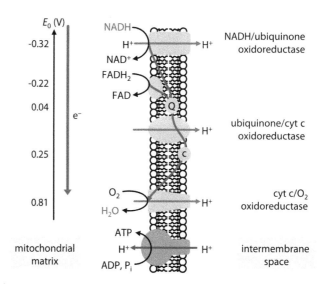

**FIGURE 5.2:** **Electron transport in the respiratory chain.** Electrons from oxidation of NADH to $NAD^+$ are transferred through a chain of electron acceptors and donors to oxygen that is reduced to $H_2O$. Electron transport is mediated by the transmembrane complexes NADH/ubiquinone reductase, ubiquinone/cytochrome c reductase amd cytochrome c/$O_2$ reductase (light blue). The electron flux is from negative to positive electrochemical potential (red arrow) through a series of exergonic reactions. Electron transfer is coupled to the endergonic transport of protons from the mitochondrial stroma to the intermembrane space, mediated by the transmembrane complexes. The proton transport generates a concentration gradient. Influx of protons through the $F_1–F_o$ ATP synthase (dark gray) leads to ATP synthesis. Electrons from $FADH_2$ are channeled into the transport chain after NADH/ubiquinone reductase, hence these electrons contribute less to the proton gradient and ATP synthesis. Flavin mononucleotide, cytochromes and FeS clusters that serve as electron acceptors and donors within the proton-pumping complexes are not depicted. The electrode potential $E^0$ is not to scale. Q: ubiquinone, cyt c: cytochrome c.

oxidoreductases couple the energy from electron transport to the endergonic transport of protons into the inter-membrane space, establishing a proton gradient (Section 4.2; Figure 4.3). The electrode potential of $FADH_2/FAD$ (–0.22 V; Table 5.2) is less negative than the NADH potential, and oxidation of $FADH_2$ provides less energy. Electrons from $FADH_2$ are channeled into the respiratory chain in the second reaction catalyzed by the ubiquinone/cytochrome c reductase. The oxidation of $FADH_2$ is thus only coupled to two proton pumps, and it contributes less to the establishment of the proton-motive force.

### 5.6.2 THE LIGHT REACTION IN PHOTOSYNTHESIS

During photosynthesis, absorption of light by chlorophyll molecules in the reaction centers of photosystems I and II (PS I, PS II) induces electronic transitions (see Section 19.1), converting these chlorophylls into a highly reducing form with a negative electrode potential. The reaction center of photosystem I, $P_{700}$, transfers its electrons in an electron transport chain *via* ferredoxin to the terminal electron acceptor $NADP^+$, whose reduction to NADPH is catalyzed by the ferredoxin/$NADP^+$ oxidoreductase. The oxidized $P_{700}$ has to be converted back into the reduced form, which is achieved by an electron transport chain from the oxidized reaction center of photosystem II, $P_{680}$. $P_{680}$ transfers its electrons to $P_{700}$ *via* plastoquinone, the cytochrome bf complex and plastocyanin. $P_{680}$ is reduced to its original state by electron transfer from water. The exergonic electron transport from $H_2O$ to $NADP^+$ is coupled to the endergonic establishment of a proton-motive force that is used for ATP synthesis (Section 4.2; Figure 4.3; Figure 5.3).

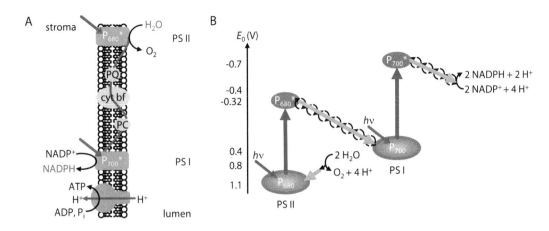

**FIGURE 5.3: Electron transport in the light reaction of photosynthesis**. A: Arrangement of the photo-systems I and II and the electron carriers in the chloroplast membrane. Electron transport from $H_2O$ to $NADP^+$ is coupled to proton transport, and the proton-motive force drives ATP synthesis. PQ: plastoquinone, PC: plastocyanin. B: Electron transport chain. Excitation of $P_{700}$ in PS I by light changes the redox potential, and $P_{700}^*$ can deliver electrons to $NADP^+$ through a transport chain of electron acceptors and donors. PS II ($P_{680}$) is excited by light and transports electrons to PS I through a transport chain. PS II ($P_{680}$) is reduced to the starting state by electrons from $H_2O$. The exergonic electron flux from negative to positive potential (orange arrows) is coupled to the transport of protons into the lumen. The generated proton-motive force is used for ATP synthesis.

To maintain metabolic processes, the $NAD^+/NADH$ ratio has to be balanced. A large portion of hangover symptoms after excessive alcohol consumption has been ascribed to the imbalance of the $NAD^+/NADH$ ratio because of the $NAD^+$ consumption in the oxidation of alcohol.

## 5.7 THE ELECTROCHEMICAL POTENTIAL AND MEMBRANE POTENTIALS

Biological membranes are impermeable for ions, and ion transport across the membrane is regulated (Section 4.3). We have discussed before that concentration gradients are useful to store energy (Section 4.2). To calculate the energy stored in an ion gradient, we have to take into account the energy stored in the concentration difference, plus the energy associated with the charge separation. We have defined the electrochemical potential of ions in the presence of electric fields, by correcting their chemical potential $\mu$ for the electrochemical work (see eq. 5.1 at the very beginning of this chapter). In the same way that the chemical potential is the driving force for a chemical reaction, the electrochemical potential is the driving force for an electrochemical reaction. If the electrochemical potential on the inside of a membrane is larger than the electrochemical potential outside, then the efflux of ions is the thermodynamically favored reaction:

$$\tilde{\mu}_{j,\text{in}} > \tilde{\mu}_{j,\text{out}}$$ eq. 5.25

Thus, ions would spontaneously pass the membrane from inside to outside if the membrane were ion-permeable. Conversely, if

$$\tilde{\mu}_{j,\text{in}} < \tilde{\mu}_{j,\text{out}}$$ eq. 5.26

then the spontaneous flow would be from the outside to the inside, and influx of ions is thermodynamically favored. Equilibrium is reached when the electrochemical potential on both sides of the membrane is equal:

$$\tilde{\mu}_{j,\text{in}} = \tilde{\mu}_{j,\text{out}}$$ eq. 5.27

The impermeability of biological membranes for ions allows the establishment and maintenance of ion gradients. By regulating opening and closing of ion channels in the membrane, the membrane potential can be modified. The transmission of neuronal signals is achieved by changes in the membrane potential of neurons. We will first derive the equation describing a membrane potential considering only sodium ions. In equilibrium, their electrochemical potential on both sides of the membrane must be equal:

$$\tilde{\mu}_{Na^+,\text{in}} = \tilde{\mu}_{Na^+,\text{out}}$$

eq. 5.28

From the definition of the electrochemical potential (eq. 5.1), we obtain

$$\mu^0_{Na^+} + RT\ln c_{Na^+,\text{in}} + z_{Na^+}F\phi_{Na^+,\text{in}} = \mu^0_{Na^+} + RT\ln c_{Na^+,\text{out}} + z_{Na^+}F\phi_{Na^+,\text{out}}$$

eq. 5.29

The charge of the sodium ions is one, and $z_{Na^+}$ can be neglected. We can rearrange eq. 5.29, and can calculate the potential difference $\Delta\phi$ as

$$\Delta\phi = \phi_{\text{in}} - \phi_{\text{out}} = \frac{RT}{F}\ln\frac{c_{Na^+,\text{out}}}{c_{Na^+,\text{in}}}$$

eq. 5.30

If more ions contribute to a membrane potential, eq. 5.30 becomes more complex because all relevant ions have to be taken into account. For the membrane potential of a neuron, Na$^+$ and K$^+$ are the most relevant ions. In the resting state of a neuron, the concentration of Na$^+$ ions is 150 mM on the outside and 15 mM on the inside. The K$^+$ concentration is 5 mM on the outside, and 140 mM on the inside. These concentrations are maintained by the activity of the Na$^+$/K$^+$ transporter, an antiporter that transports 3 Na$^+$ ions out of and 2 K$^+$ ions into the cell at the expense of ATP hydrolysis. From eq. 5.30, we can calculate the equilibrium potential at 37°C from the Na$^+$ gradient as +62 mV (positively charged on the outside, negative on the inside), for the K$^+$ gradient it is –90 mV (negatively charged on the outside, positive on the inside). The membrane potential of a neuron in the resting state is typically –60 to –80 mV (Figure 5.4), and thus much closer to the K$^+$ potential. This is caused by potassium channels that are open and permit flux of K$^+$ from the inside to the outside, along the concentration gradient. In contrast, Na$^+$ channels are closed in the resting state. The overall membrane potential is therefore predominantly determined by the K$^+$ equilibrium potential.

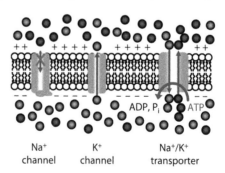

Na$^+$ channel     K$^+$ channel     Na$^+$/K$^+$ transporter

**FIGURE 5.4:** **Membrane potential of a neuron in the resting state.** The Na$^+$/K$^+$ transporter maintains the concentration gradient of Na$^+$ and K$^+$ ions across the neuronal membrane by actively transporting Na$^+$ ions out of and K$^+$ ions into the cell against the concentration gradient. Some K$^+$ channels are open and permit efflux of K$^+$. The resulting membrane potential is closer to the K$^+$ than to the Na$^+$ equilibrium potential.

For more than two ions, the membrane potential can be calculated from the concentrations of all cations $M_i^+$ and anions $X_i^-$ on the outside, multiplied with the (relative) permeability $P_i$, in the numerator of the logarithmic term, and the concentrations of all cations $M_i^+$ and anions $X_i^-$ on the inside, multiplied with the permeability $P_i$ in the denominator. The membrane potential is

$$\Delta\phi = \frac{RT}{F}\ln\frac{\sum P_i\left[M_i^+\right]_{\text{outside}} + \sum P_i\left[X_i^-\right]_{\text{outside}}}{\sum P_i\left[M_i^+\right]_{\text{inside}} + \sum P_i\left[X_i^-\right]_{\text{inside}}}$$

eq. 5.31

Eq. 5.31 implies that the membrane potential changes if the relative permeabilities of one (or more) ions change. This is indeed the mechanism of neuron activation, and the generation and transmission of electric signals in the neural system. The permeability is altered by the opening and closing of gated channels in the membrane in response to a certain signal. Opening of $K^+$ channels allows increased efflux of $K^+$ from the cell, and the membrane potential becomes more negative. This effect is called *hyperpolarization*. On the other hand, opening of $Na^+$ channels enables increased influx of $Na^+$, and reduces the negative charge on the inside (*depolarization*). Many of the gated channels involved in these processes are voltage-dependent, meaning they open and close in response to changes in membrane potential. Voltage-dependent $Na^+$ channels open in response to a slight depolarization of the membrane. Influx of $Na^+$ ions leads to further depolarization and opening of more $Na^+$ channels in a positive feedback loop, until a threshold value of the potential is reached. Quickly, all voltage-gated $Na^+$ channels will be open, and the membrane potential approaches the $Na^+$ equilibrium potential: an action potential has been generated (Figure 5.5). This action potential is localized, because the original depolarization was localized to a defined region of the neuron. The $Na^+$ channels only open for a short period of time, however, and are inactivated before the $Na^+$ equilibrium potential has been reached. At the same time, voltage-gated $K^+$ ion channels open, and $K^+$ ion flux out of the cell shifts the membrane potential back towards the $K^+$ equilibrium potential: the membrane becomes repolarized. With the voltage-gated $K^+$ channels open, the permeability for $K^+$ ions is larger than in the resting state. The membrane potential will therefore briefly overshoot and temporarily become more negative, before the equilibrium membrane potential of the resting state is reached. As soon as the $Na^+$ channels can be activated again, a new action potential can be generated. The transient inactivation of $Na^+$ channels ensures that the action potential moves along the axon in one direction, away from the area where it was generated (Figure 5.5).

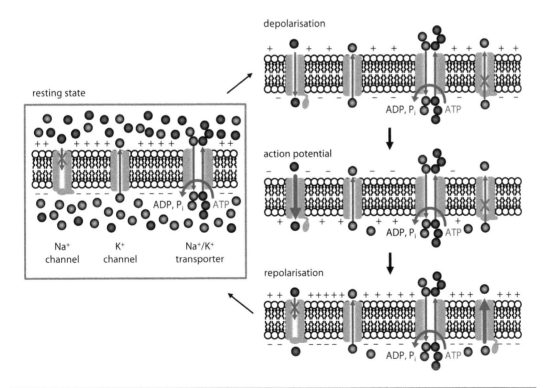

**FIGURE 5.5:**    **Generation and decline of an action potential**. When a signal leads to a slight depolarization, voltage-gated $Na^+$ channels open, and influx of sodium ions leads to a further depolarization of the cell. Before the $Na^+$ equilibrium potential has been reached, the $Na^+$ channels become inactivated, and voltage-gated $K^+$ channels open. The efflux of $K^+$ leads to a repolarization of the membrane. The membrane permeability for $K^+$ is higher than in the resting state, leading to a transient hyperpolarization, before the voltage-dependent $K^+$ channels close and the resting potential is re-established.

## 5.8 ELECTROPHYSIOLOGY: PATCH-CLAMP METHODS TO MEASURE ION FLUX THROUGH ION CHANNELS

In the previous section (Section 5.7) we have seen that controlled opening and closing of ion channels alters the ion permeability and thus the electrical conductivity of biological membranes. The current through these ion channels can be measured by the *patch-clamp* technique (Figure 5.6). In a patch-clamp experiment, a glass pipette with a micrometer opening is manually brought into contact with the membrane of a cell under visual control in a microscope. Suction is applied, which leads to a tight sealing of the glass walls at the opening with the membrane, while the cell remains intact (*cell-attached mode*). The glass-membrane contact has a high resistance and low conductivity, and prevents leakage of ions between the inside and outside of the micropipette. The part of the membrane that closes off the opening of the micropipette is called the *patch*. A flux of ions between inside and outside is possible if one or more ion channels are located within this patch, and can be measured as a current. This approach allows one to measure picoampere currents through single ion channels in their natural environment. For voltage-gated ion channels, different voltages can be applied, and the current-voltage response curve of a single channel can be measured. By patch-clamping a ligand-gated ion channel in the cell-attached mode, the response of the channel to different ligands can be tested. By measuring the current through the channel at different ligand concentrations, a dose-response curve for a particular ligand may be determined. In the cell-attached mode, the cytoplasmic solution cannot be manipulated because the cell remains intact. Therefore, the ligand has to be supplied with the solution inside the micropipette, and so cannot be changed after patching. Hence, the dose-response curve has to be measured point by point by clamping a new patch for each ligand concentration. By gently pulling on the membrane-sealed micropipette, the patch can be separated from the rest of the membrane. The cytosolic side of the membrane now faces to the outside of the micropipette, hence the name *inside-out configuration.* The solution outside the micropipette can easily be changed, and the inside-out configuration is therefore suitable to study the effect of intracellular, cytoplasmic ligands on ligand-gated ion channels. To study the effect of extracellular ligands such as neurotransmitters and drugs on ligand-gated channels, the *outside-out configuration* is used. To achieve this, more suction is applied in the cell-attached mode, which leads to opening of the membrane patch. The seal between pipette and membrane remains intact, and the solutions inside and outside the cell remain separated, but the solution in the micro-pipette is now in contact with the cytoplasm. In this *whole-cell mode*, the current through all channels in the entire cell membrane is measured. By pulling on the micropipette, the cell membrane ruptures, and membrane fractions sealed to the glass can form patches where the extracellular face of the membrane faces outside, hence the name outside-out configuration. In this mode, the response of single ion channels to extracellular ligands and their dose-response curves can be measured.

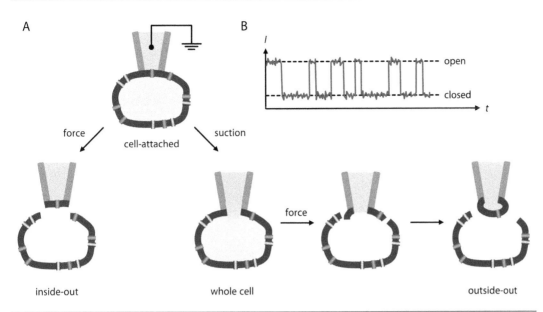

**FIGURE 5.6:** **Measurement of ion flux through ion channels by patch-clamp techniques.** A: Different patch-clamp configurations. A micropipette (gray/light red) is moved onto the surface of a cell and tightly attaches to the membrane upon suction. By applying force, the patch attached to the pipette can be extracted from the rest of the membrane. The intracellular (cytosolic) side of the membrane faces towards the outside of the pipette (inside-out). By increasing the suction instead of pulling, the patch ruptures, and the micropipette is sealed to the cell (whole cell). By force, the membrane can be ruptured. Membrane fragments sealed to the micropipette form a new patch. The extracellular side of the membrane faces the outside of the micropipette (outside-out). B: Schematic measurement: current I (in pA) as a function of time (in s). When the ion channel is open, the current is high. When the channel closes, no current is measured.

The patch-clamp method has been developed by Neher and Sakman in 1976 (Nobel Prize in Physiology or Medicine 1991). More recently, it has been combined with fluorescence microscopy to directly monitor the binding of fluorescently labeled ligands, conformational changes, or subunit stoichiometry in parallel to measuring the current.

## QUESTIONS

5.1 A pH meter provides a readout of 0 V for a solution of pH 7 at 25°C. What range of voltages is measured for pH 1–14?

5.2 An electrochemical cell consists of Zn in a $ZnSO_4$ solution and Cu in a $CuSO_4$ solution. The $CuSO_4$ solution has a concentration of 100 mM. What would be the minimal concentration of a $ZnSO_4$ solution to drive the reaction in the $Cu^{2+}/Cu$ half-cell towards $Cu^{2+}$? $E^0$ ($Zn^{2+}/Zn$) = –0.76 V, $E^0$ ($Cu^{2+}/Cu$) = 0.34 V.

5.3 The equilibrium constant $K = Cu^{2+}/(Cu^+)^2$ in aqueous solution is $10^6$ $M^{-1}$. The standard potential of a $Cu/Cu^{2+}$ electrode is $E^0$ = –0.34 V. What is the standard potential for the $Cu^+/Cu$ electrode and for the $Cu^{2+}/Cu^+$ electrode?

5.4 The standard potential for the reaction $NAD^+ + H^+ + 2 e^- \rightarrow NADH$ is 0.099 V at 20°C and 0.105 V at 30°C. Calculate the change in standard free energy, entropy, and enthalpy for 25°C, assuming a linear dependence of $E^0$ on temperature.

5.5 What is $E^{0\prime}$ for an $NAD^+/H^+/NADH$-Pt half-cell at pH 7? $E^0$ = –0.11 V.

5.6 A solution contains pyruvate, lactate, $NAD^+$, and NADH. The concentration ratios are pyruvate/lactate = 1, $NAD^+/NADH$ = 1. What reactions do occur, what is $E^0$ and what is $\Delta G^0$? How do these values change for a 500-fold excess of lactate over pyruvate and of $NAD^+$ over

NADH? At what ratio of lactate/pyruvate = $NAD^+$/NADH are the reactions in equilibrium? $E^{0\prime}(NAD^+/NADH) = -0.32$ V; $E^{0\prime}$(pyruvate/lactate) = $-0.19$ V.

5.7    Calculate the membrane potential of a cell for a concentration of $K^+$ ions that is 20-fold higher inside than outside. What membrane potential results from a $Na^+$ concentration that is 10-fold higher outside than inside? $T = 37°C$.

5.8    *E. coli* maintains a $\Delta$pH of 1 (inside - outside) and a $\Delta\phi$ of $-120$ mV across its cell membrane. What is the energy derived from this concentration gradient, and what is the maximum concentration of a metabolite M inside *versus* outside that can be achieved by its symport into the cell with $H^+$? What is the maximum concentration if M is negatively charged ($M^-$)?

5.9    Mitochondria maintain a $\Delta$pH of 1.4 (inside - outside) and a $\Delta\phi$ of $-0.14$ V (inside - outside). How many protons need to pass the membrane for synthesis of one ATP molecule? $\Delta G^{0\prime}$ (ATP synthesis) = 31 kJ mol$^{-1}$.

## REFERENCES

Neher, E. and Sakmann, B. (1976) Single-channel currents recorded from membrane of denervated frog muscle fibres. *Nature* **260**(5554): 799–802.
· original paper on the patch-clamp method

Kusch, J. and Zifarelli, G. (2014) Patch-clamp fluorometry: electrophysiology meets fluorescence. *Biophys J* **106**(6): 1250–1257.
· review about the combination of patch-clamp methods with fluorimetry to study ion channel conformational changes and ligand binding

Talwar, S. and Lynch, J. W. (2015) Investigating ion channel conformational changes using voltage clamp fluorometry. *Neuropharmacology* **98**: 3–12.
· review about the combination of voltage-clamp methods with fluorimetry to study conformational dynamics of ion channels

# Part II
# Kinetics

In Part I, we discussed energy changes associated with changes in the state of a system, transport processes, and chemical reactions. So far, we have not considered the velocity with which these reactions occur. In the following section, we will now derive concepts to describe the velocities of increasingly complex reactions, and to understand how enzymes catalyze biochemical reactions.

# Reaction Velocities and Rate Laws

uring a chemical reaction, the concentration of reactants decreases, and the concentration of products increases. The *reaction velocity v* is defined as the change in concentration of the reactants or products with time (Figure 6.1):

$$v = \frac{dc}{dt} = -\frac{d[reactants]}{dt} = \frac{d[products]}{dt}$$

eq. 6.1

The velocity of a reaction is always positive. For the definition of the reaction velocity in terms of the decrease of reactant concentration we therefore need a negative sign to convert it into a positive value. From eq. 6.1, we can derive the unit of the reaction velocity $v$ as M s$^{-1}$.

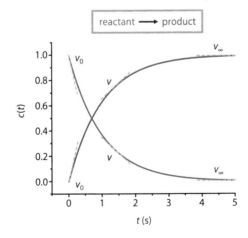

**FIGURE 6.1: Concentration changes and reaction velocity**. The concentration of the reactant (blue) decreases with reaction time, whereas the concentration of product (red) increases. The reaction velocity is the change in concentration with time. Broken lines indicate the initial velocity $v_0$, the velocity $v$ at time $t$, and the velocity $v_\infty$ at the end of the reaction. The reaction velocity decreases from the initial velocity $v_0$ to the final velocity $v_\infty = 0$ during the reaction.

119

For a reaction of A and B to C (Scheme 6.1),

$$A + B \longrightarrow C$$

scheme 6.1

we can express the reaction velocity as the change in concentration of A, B or C as

$$v = \frac{dc}{dt} = -\frac{d[A]}{dt} = -\frac{d[B]}{dt} = \frac{d[C]}{dt}$$

eq. 6.2

In the formulation of reaction velocities, the stoichiometry coefficients of the reactants and products need to be taken into account. For a similar reaction of A and two B to C (Scheme 6.2),

$$A + 2\,B \longrightarrow C$$

scheme 6.2

the reaction velocity is

$$v = \frac{dc}{dt} = -\frac{d[A]}{dt} = -\frac{1}{2} \cdot \frac{d[B]}{dt} = \frac{d[C]}{dt}$$

eq. 6.3

The factor ½ in eq. 6.3 is necessary because for each A that reacts, two molecules of B are consumed. The concentration of B therefore decreases twice as fast as the concentration of A. The reaction velocity, on the other hand, must be identical independent of which change in concentration we use as a reference.

To monitor the progress of chemical reactions and to determine reaction velocities, we need to follow changes in the concentrations of reactants or products. This can be done discontinuously by taking samples at different time points, with subsequent analysis, or continuously by online monitoring of a suitable spectroscopic signal that reports on disappearance of reactants or appearance of products. While some methods provide absolute information on reactant or product concentrations, others provide a signal that is proportional to the concentration and only measure relative changes. Methods to monitor chemical reactions will be treated in Part IV, particularly in Chapter 25.

*Rate laws* describe the dependence of reaction velocities on the concentrations of reactants A, B, C,... . Reaction velocities are proportional to the concentrations of the reactants, and rate laws have the general form

$$v = \frac{dc}{dt} = k[A]^a[B]^b[C]^c\ldots$$

eq. 6.4

The proportionality constant $k$ in eq. 6.4 is the *rate constant* of the reaction. The exponents $a$, $b$, $c$, ... are the *order of the reaction* with respect to the reactants A, B, C, .... They are empirical numbers that are determined experimentally. To determine $a$, the concentration of reactant A is varied, while $[B]$, $[C]$, ... are kept constant, and the reaction velocity is measured as a function of $[A]$. The value of $b$ can be determined by varying the concentration of $[B]$ while keeping $[A]$, $[C]$, ... constant, and so on. With the empirical exponents, a rate law in the form of eq. 6.4 can then be formulated. The overall order of the reaction is the sum of the orders for the individual reactants, $a+b+c+...$ . Reactions whose velocity is independent on the reactant concentration are called *zero-order reactions*. Such a behavior is observed for enzyme-catalyzed reactions at a high excess of substrate (see Section 11.1). *A first-order reaction* depends linearly on the concentration, a *second-order reaction* depends on the square of the concentration. We will treat first- and second-order reactions in more detail in Chapter 7.

The reaction order, overall or with respect to a particular reactant, can be a non-integer number. Many reactions proceed through a sequence of reaction steps that define the *reaction mechanism*. The rate laws of such complex, multi-step reactions cannot be derived from the overall reaction scheme, but depend on the individual steps of the reaction. Rate laws therefore reflect the mechanism of the reaction.

Reactions that occur the way they are written in the reaction scheme and cannot be further sub-divided into a sequence of individual reaction steps are called *elementary reactions*. If the reaction in Scheme 6.3 is an elementary reaction, such that $a$ molecules of A, $b$ molecules of B, $c$ molecules of C, ... collide to form the reaction products

$$aA + bB + cC + ... \xrightarrow{k} \text{products} \qquad \text{scheme 6.3}$$

then we can write the rate law as

$$v = \frac{dc}{dt} = k[A]^a[B]^b[C]^c... \qquad \text{eq. 6.5}$$

This rate law is formally identical to the general rate law in eq. 6.4, but now the exponents are the stoichiometry coefficients $a$, $b$, and $c$. A reaction involving one molecule of A depends linearly on $[A]$, a reaction involving two molecules of A depends on $[A]^2$, and so on. For an elementary reaction, the reaction order is thus an integer number.

The number of molecules that react in an elementary reaction is defined as the *molecularity* of the reaction. The molecularity is always an integer number, there are no half molecules. We call reactions *unimolecular* if one reactant molecule is involved, *bimolecular* for two reactants, and *tri-molecular* for three reactants. A higher molecularity has not been observed, possibly because it is very unlikely that more than three molecules collide with each other in the correct orientation and with sufficient energy for a reaction to occur. For elementary reactions, the molecularity and the overall order of the reaction are equal. However, this is an exception, and in general it is important to distinguish between molecularity and reaction order.

A simple example of a one-step elementary reaction is the bimolecular reaction of hydrogen with iodine to HI (Scheme 6.4):

$$H_2 + I_2 \xrightarrow{k} 2\,HI \qquad \text{scheme 6.4}$$

With the information that the reaction is an elementary reaction, we can write the rate law as

$$v = \frac{dc}{dt} = \frac{1}{2} \cdot \frac{d[HI]}{dt} = k[H_2][I_2] \qquad \text{eq. 6.6}$$

The factor ½ takes into account that the reaction of one $H_2$ and one $I_2$ produces two molecules of HI, and the concentration of HI therefore increases twice as fast as the concentrations of $H_2$ and $I_2$ decrease (see eq. 6.3). Despite the same overall reaction scheme (Scheme 6.5), the situation is different for the corresponding reaction of bromine with hydrogen:

$$H_2 + Br_2 \xrightarrow{k} 2\,HBr \qquad \text{scheme 6.5}$$

This reaction follows a radical mechanism and occurs in several steps (Box 6.1). Experimental determination of the reaction velocity yielded the empirical rate law

$$v = \frac{dc}{dt} = \frac{1}{2} \cdot \frac{d[HBr]}{dt} = \frac{k[H_2]\sqrt{[Br_2]}}{1 + k'\dfrac{[HBr]}{[Br_2]}} \qquad \text{eq. 6.7}$$

The drastically different rate laws for these two formally similar reactions exemplify that the rate law of a reaction cannot be deduced from the overall reaction equation.

**BOX 6.1: MECHANISM OF THE REACTION OF
BROMINE WITH HYDROGEN.**

Bromine and hydrogen react in a multi-step radical reaction. First, $Br_2$ decomposes into two bromine radicals:

$$Br_2 \xrightarrow{k_1} 2\ Br\cdot$$

scheme 6.6

The radicals generated in this starting reaction can then react with molecular hydrogen

$$H_2 + Br\cdot \underset{k_{-2}}{\overset{k_2}{\rightleftharpoons}} HBr + H\cdot$$

scheme 6.7

producing hydrogen radicals. These hydrogen radicals also react with $Br_2$ to give HBr

$$Br_2 + H\cdot \xrightarrow{k_3} HBr + Br\cdot$$

scheme 6.8

leading to regeneration of a bromine radical, and keeping the chain-reaction going. The reaction stops with the recombination of two bromine radicals, which is the reverse reaction of the starting reaction (Scheme 6.6).

$$2\ Br\cdot \xrightarrow{k_{-1}} Br_2$$

scheme 6.9

From the set of elementary reactions that define the reaction mechanism, the rate law for HBr formation can be derived as:

$$v = \frac{1}{2}\cdot\frac{\mathrm{d}\left[HBr\right]}{\mathrm{d}t} = \frac{k_2\sqrt{\dfrac{k_1}{k_{-1}}}\left[H_2\right]\sqrt{\left[Br_2\right]}}{1+\dfrac{k_{-1}}{k_3}\dfrac{\left[HBr\right]}{\left[Br_2\right]}}$$

eq. 6.8

which is in agreement with eq. 6.7 for

$$k = k_2\sqrt{\frac{k_1}{k_{-1}}}$$

eq. 6.9

and

$$k' = \frac{k_{-1}}{k_3}$$

eq. 6.10

The agreement of the empirical rate law with the rate law derived from the mechanism tells us that this reaction mechanism describes the pathway of the reaction.

## QUESTIONS

6.1    Express the reaction velocities of the following reactions in terms of each reactant/product:
(1) $S \rightarrow P$
(2) $E + S \rightarrow E + P$
(3) $ATP + H_2O \rightarrow ADP + P_i$
(4) $NAD^+ + H^+ (+ 2\,e^-) \rightarrow NADH$
(5) $H_2O \rightarrow H^+ + OH^-$
(6) $2M \rightarrow D$

6.2    What is the reaction velocity for a reaction of type $aA + bB \rightarrow cC + dD$?

6.3    Write the rate laws for the forward and reverse reactions:
(1) $S \rightleftarrows P$
(2) $E + S \rightleftarrows E + P$
(3) $ATP + H_2O \rightleftarrows ADP + P_i$
(4) $NAD^+ + H^+ (+ 2\,e^-) \rightleftarrows NADH$
(5) $H_2O \rightleftarrows H^+ + OH^-$
(6) $2M \rightleftarrows D$
The forward rate constant is $k_f$, the rate constant for the reverse reaction is $k_r$.

6.4    What is the order of the reaction of iodine and hydrogen with respect to the reactants, and what is the overall order of the reaction? What is the molecularity? What are the corresponding orders and molecularities for the individual steps of the reaction of bromine with hydrogen?

6.5    The concentration of a reactant A decays exponentially according to $A(t) = A_0 e^{-kt}$ with the rate constant $k$. What is the reaction velocity as a function of time? What is the initial velocity of the reaction?

# Integrated Rate Laws for Uni- and Bimolecular Reactions

We have seen in Chapter 6 that rate laws describing the dependence of reaction velocities on the concentrations of reactants can be formulated directly from the reaction scheme of elementary reactions. By integrating the rate law, the time dependence of the concentrations of reactants or products is obtained, the *integrated rate law*. In the following, we will derive integrated rate laws for common types of elementary reactions. We start by considering a simple unimolecular reaction where a compound A decays to products (Scheme 7.1):

$$A \xrightarrow{k} products$$

scheme 7.1

An example of such a reaction is radioactive decay (Box 7.2). The reaction velocity $v$ of such a reaction is defined as the change in the concentration of A over time:

$$v = \frac{dc}{dt} = -\frac{d[A]}{dt}$$

eq. 7.1

For the unimolecular reaction according to Scheme 7.1, the reaction velocity $v$ at any given point in time is proportional to the actual concentration of A that can react to products:

$$v \propto [A]$$

eq. 7.2

The reaction velocity depends on the first power of the concentration of A: the reaction is a first-order reaction. We now introduce the proportionality constant $k$, the rate constant of the reaction:

$$v = k[A] = -\frac{d[A]}{dt}$$

eq. 7.3

125

The time-dependent change of $v$ as a function of the concentration of A in eq. 7.3 is the rate law for this reaction. We have seen before that the reaction velocity has units M s$^{-1}$. If we now compare the units in eq. 7.3, we can deduce that the rate constant $k$ for a first-order reaction is measured in s$^{-1}$.

Eq. 7.3 is a differential equation, and we can separate the variables A and $t$ by rearranging to

$$\frac{\mathrm{d}[A]}{[A]} = -k\mathrm{d}t \qquad \text{eq. 7.4}$$

To obtain an expression of the concentration of A as a function of time, we now integrate from $A_0$, the starting concentration at time $t = 0$, to the concentration $[A](t)$ at time $t$ on the left-hand side, and from $t = 0$ to $t$ on the right-hand side:

$$\int_{A_0}^{[A](t)} \frac{\mathrm{d}[A]}{[A]} = -k \int_0^t \mathrm{d}t \qquad \text{eq. 7.5}$$

Integration gives

$$\ln\left(\frac{[A](t)}{A_0}\right) = -kt \qquad \text{eq. 7.6}$$

or

$$[A](t) = A_0\, \mathrm{e}^{-kt} \qquad \text{eq. 7.7}$$

The time dependence of the reactant concentration is the *integrated rate law*. According to eq. 7.7, the concentration of A decays exponentially from its starting value $A_0$ at $t = 0$ to zero at the end of the reaction ($t \to \infty$) with the rate constant $k$ (Figure 7.1). The larger $k$, the faster is the reaction, and the faster A is consumed. $A_0$ is the *amplitude* of the exponential function, and often called the amplitude of the reaction. If A is converted into one product B, and we start the reaction from the initial concentration $A_0$ without any B ($B_0 = 0$), we can formulate the mass conservation as

$$A_0 = [A](t) + [B](t) \qquad \text{eq. 7.8}$$

which gives

$$[B](t) = A_0 - [A](t) = A_0 - A_0\, \mathrm{e}^{-kt} = A_0\left(1 - \mathrm{e}^{-kt}\right) \qquad \text{eq. 7.9}$$

Eq. 7.9 describes exponential growth: the concentration of B increases exponentially with the rate constant $k$ from $B_0 = 0$ to $B_\infty = A_0$ (Figure 7.1). The amplitude of the reaction is $A_0$.

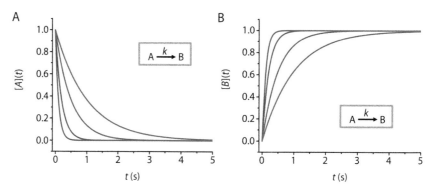

**FIGURE 7.1:   A first-order reaction A→B.** A: The concentration of A decays exponentially from $A_0$ to zero. The rate constant determines how rapidly A is consumed. B: The concentration of the product B increases exponentially from zero to $B_\infty = A_0$. The set of plots corresponds to $k = 1$ s$^{-1}$ (blue), $k = 2$ s$^{-1}$ (violet), $k = 5$ s$^{-1}$ (pink), and $k = 10$ s$^{-1}$ (red).

To determine the rate constant for the reaction from A to products experimentally, we need to measure the concentration [A] as a function of time, either continuously or discontinuously (see Chapter 25 for kinetic methods). The rate constant $k$ for the reaction can then be determined by non-linear regression using eq. 7.7 (Figure 7.2; Box 7.1).

## BOX 7.1: DETERMINATION OF RATE CONSTANTS FROM EXPERIMENTAL DATA.

To determine a rate constant for a chemical reaction, we monitor the concentration(s) of reactant(s) or product(s) as a function of time, either continuously or discontinuously (Figure 7.2). We then describe the experimental data points by a single exponential function, varying the amplitude and the rate constant until we obtain the optimal description of the data points. The rate constant that gives the best fit between curve and data is the rate constant $k$ of the reaction.

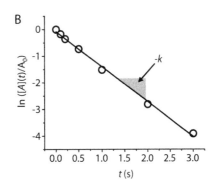

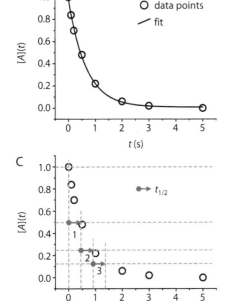

FIGURE 7.2:   **Experimental data and description by a first-order integrated rate law.** A: Fitting curves to experimental data. Experimental data points for the decrease in concentration of the reactant are best described by an exponential decay (eq. 7.7) with the rate constant $k = 1.52$ s$^{-1}$ : the reaction is a first-order reaction with a rate constant $k = 1.52$ s$^{-1}$. B: Linearization according to eq. 7.6 and determination of $k$ from the slope gives $k = 1.31$ s$^{-1}$. C: Determination of the reaction order and the rate constant from half-lives (red arrows) according to eq. 7.13. Three half-lives are approximately 1.4 s, corresponding to $k = 1.49$ s$^{-1}$. The obtained rate constants differ because of different weighting (B) and approximation errors (C).

Alternatively, we can use the linear relationship between $\ln([A]/A_0)$ and the time $t$ according to eq. 7.6 to determine the rate constant of a first-order reaction. We obtain $k$ from the slope of a plot of $\ln([A]/A_0)$ as a function of $t$ (Figure 7.2). However, transformation of the concentrations to the logarithm transforms the experimental errors associated with measuring the concentrations in a non-linear fashion. Therefore, it is better to directly plot the experimentally determined concentrations as a function of time, and to describe the experimental data points by a curve according to eq. 7.7 to extract $k$. In general, linearized plots of converted data values should be avoided due to the error transformation (see Figure 11.4). Finally, first-order rate constants can be calculated from half-lives (eq. 7.13; Figure 7.2).

Often, eq. 7.7 is written in terms of a time constant $\tau$ instead of the rate constant $k$. The time constant $\tau$ of an exponential function is the time $t$ after which the exponential function has decayed by $1/e$, to $A(\tau) = A_0/e$. For eq. 7.7, this translates into

$$[A](\tau) = \frac{A_0}{e} = A_0\, e^{-k\tau}$$
eq. 7.10

Solving this equation for $\tau$ yields

$$\tau = \frac{1}{k}$$
eq. 7.11

The time constant $\tau$ of the reaction has units of s, and is sometimes also called the *lifetime* of A. Fast reactions are characterized by high rate constants $k$ and thus small lifetimes $\tau$, whereas slow reactions have a small $k$ and the lifetime $\tau$ of the reactant is high. The time constant or lifetime $\tau$ describes how long on average the molecules remain as A before they react to products.

In analogy to the time constant or lifetime $\tau$, we can also define a *half-life* $t_{1/2}$, the time at which the starting concentration $A_0$ has halved. From

$$[A](t_{1/2}) = \frac{A_0}{2} = A_0\, e^{-kt_{1/2}}$$
eq. 7.12

we obtain the half-life $t_{1/2}$ as

$$t_{1/2} = \frac{\ln 2}{k} \approx \frac{0.7}{k}$$
eq. 7.13

It is noteworthy that the half-life for a first-order reaction is independent of the starting concentration $A_0$ (Figure 7.3). As a consequence, the first half-life, i.e. the time for the concentration to be reduced from $A_0$ to $A_0/2$, is identical the second half-life (from $A_0/2$ to $A_0/4$), the third half-life (from $A_0/4$ to $A_0/8$), and so on. Because of their constant half-lives, first-order reactions can be used as a reference clock. An example is dating of material by the radiocarbon method (Box 7.2).

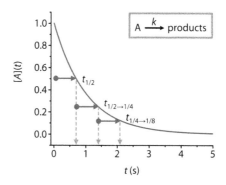

**FIGURE 7.3:**    **Half-lives in a first-order reaction**. For a first-order reaction, the half-life $t_{1/2}$ is independent of the starting concentration $A_0$. Therefore, the first half-life $t_{1/2}$, the time for the concentration to be reduced from $A_0$ to $A_0/2$, is identical to the second half-life ($t_{1/2 \to 1/4}$, from $A_0/2$ to $A_0/4$), the third half-life ($t_{1/4 \to 1/8}$, from $A_0/4$ to $A_0/8$), and so on.

## BOX 7.2: A FIRST-ORDER REACTION: RADIOACTIVE DECAY AND RADIOCARBON DATING.

Radioactive decay is a prominent example of a first-order reaction. At any given time, the velocity $v$ of the decay reaction is proportional to the number $N$ of radioactive molecules present, with the rate constant $k$ as the proportionality constant:

$$v = -\frac{dN}{dt} = kN \qquad \text{eq. 7.14}$$

Eq. 7.14 is analogous to eq. 7.3, and by rearranging and integrating, we obtain

$$N(t) = N_0\,e^{-kt} \qquad \text{eq. 7.15}$$

This integrated rate law is formally identical to the first-order rate law in eq. 7.7. Radioactive isotopes are characterized by their half-lives $t_{1/2}$ (eq. 7.13; Table 7.1).

**TABLE 7.1**
**Half-lives of radioisotopes.**

| isotope | half-life (approx.) | use |
|---------|---------------------|-----|
| $^{14}C$ | 5370 a | radiocarbon dating, radiotracer in biochemistry |
| $^{3}H$ | 12 a | radiotracer in biochemistry |
| $^{35}S$ | 87 d | radiotracer in biochemistry |
| $^{125}I$ | 59 d | nuclear medicine: determination of bone density |
| $^{32}P$ | 14 d | radiotracer in biochemistry |
| $^{131}I$ | 8 d | nuclear medicine: diagnostics of thyroid function |
| $^{18}F$ | 1.8 h | nuclear medicine: radiotracer in positron emission tomography |

Carbon occurs in nature as three different *isotopes*, $^{12}C$ (> 98%), $^{13}C$ (~1%) and $^{14}C$ (< $10^{-10}$%). $^{12}C$ and $^{13}C$ are stable isotopes, whereas $^{14}C$ radioactively decays to $^{14}N$ with a half-life of 5370 years. Living organisms maintain a constant level of $^{14}C$ in their bodies: although $^{14}C$ decays to $^{14}N$, the $^{14}C$ levels are replenished through exchange with the surroundings. Once organisms die, metabolism stops, and $^{14}C$ is not replenished anymore. The $^{14}C$ level then decreases because of its radioactive decay, following a first-order rate law:

$$\left[^{14}C\right](t) = \left[^{14}C\right]_0 e^{-kt} = \left[^{14}C\right]_0 e^{-\frac{\ln 2}{t_{1/2}} \cdot t} \qquad \text{eq. 7.16}$$

In eq. 7.16, $t = 0$ is the time when metabolism stopped. $[^{14}C]_0$ is the constant level of $^{14}C$ in living organisms ($10^{-10}$%) that was present at $t = 0$, $t_{1/2}$ is the half-life of $^{14}C$ (5370 years), and $[^{14}C](t)$ is the concentration of $^{14}C$ at the time $t$ after death of the organism. By determining the $^{14}C$ level of organic material that contains carbon, the time $t$ that has elapsed since the organism died can be calculated. Radiocarbon dating is a standard method in archaeology, and has been used to date wood, seeds, pollen, corals, shells, bones, hair, peat, soil, fabrics, paper, paint, and pottery. After ten half-lives, the amount of $^{14}C$ is below the levels that can be quantified, limiting the age determination to < 53700 years.

We now derive the rate law for a bimolecular reaction. First, we consider a reaction of two (identical) molecules of A to products (Scheme 7.2):

$$A + A \xrightarrow{k} products$$

scheme 7.2

The association of two protein monomers to a dimer is an example of such a type of reaction (see Box 7.4). The definition of the reaction velocity $v$ as the change in concentration of A with time is

$$v = -\frac{1}{2} \cdot \frac{d[A]}{dt}$$

eq. 7.17

The factor ½ in eq. 7.17 takes into account that two molecules of A react to products. Per product formed, the concentration of A thus decays twice as much compared to a unimolecular reaction (Scheme 7.1). When we express the reaction velocity for the bimolecular reaction as a function of the rate constant and the actual concentration of A, we also have to take into account that two molecules of A react with each other. The concentration of A therefore appears twice as a factor in the rate law, and the reaction velocity depends on the square of the concentration of A: the reaction is a second-order reaction.

$$v = -\frac{1}{2} \cdot \frac{d[A]}{dt} = k[A][A] = k[A]^2$$

eq. 7.18

To obtain the reaction velocity $v$ in units of M s$^{-1}$, the product on the right-hand side of eq. 7.18 must also yield M s$^{-1}$. With the two concentration terms (units M), we can derive the unit M$^{-1}$ s$^{-1}$ for the rate constant $k$ of a second-order reaction.

In order to integrate the rate law, we again separate the variables in eq. 7.18:

$$\frac{d[A]}{[A]^2} = -2k \, dt$$

eq. 7.19

By integration from $A_0$ to $A$, and from 0 to $t$, we obtain

$$\int_{A_0}^{A} \frac{d[A]}{[A]^2} = -2k \int_0^t dt$$

eq. 7.20

and

$$-\frac{1}{[A]} + \frac{1}{A_0} = -2kt$$

eq. 7.21

which can be rearranged to:

$$\frac{1}{[A]} = 2kt + \frac{1}{A_0}$$

eq. 7.22

The integrated rate law in eq. 7.22 reveals a linear dependence between the inverse of [A] and $t$. From a linear plot of 1/[A] as a function of $t$, we can determine the second-order rate constant $k$ from the slope (2$k$). Alternatively, we can rearrange to:

$$[A](t) = \frac{A_0}{1 + 2ktA_0}$$

eq. 7.23

and can directly plot [A] as a function of time (Figure 7.4). The concentration of A initially decays more rapidly compared to a first-order reaction, but then approaches zero more slowly than it does for a first-order reaction (Figure 7.4). In other words, we can directly distinguish between a first-order reaction (Scheme 7.1) and a second-order reaction (Scheme 7.2) from the shape of the [A](t) curves (Box 7.3).

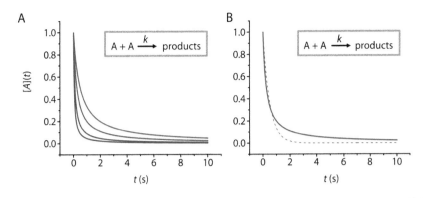

FIGURE 7.4:   **A second-order reaction: A+A→products.** A: Concentration of A (in µM) as a function of time for a second-order rate constant of $k = 1$ µM$^{-1}$ s$^{-1}$ (blue), $k = 2$ µM$^{-1}$ s$^{-1}$ (violet), $k = 5$ µM$^{-1}$ s$^{-1}$ (pink), and $k = 10$ µM$^{-1}$ s$^{-1}$ (red). B: Comparison of $[A](t)$ for a second-order reaction (solid line) with $k = 2$ µM$^{-1}$ s$^{-1}$ and a first-order reaction (dashed line) with the rate constants $k = 2$ s$^{-1}$. The concentration of A decays more rapidly in the beginning than for the corresponding first-order reaction with the same rate constant, but levels off more slowly.

### BOX 7.3: DISTINCTION BETWEEN FIRST- AND SECOND-ORDER REACTIONS.

The dependence of the concentration of the reactant as a function of time differs for first- and second-order reactions. To distinguish between first- and second-order reactions, we plot the experimentally determined concentration of the reactant as a function of time, and try to describe the data with the integrated rate laws for first- and second-order reactions (Figure 7.5). The correct integrated rate law describes the experimental data, while the other rate law yields a curve that systematically deviates from the data.

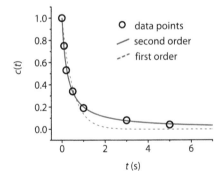

FIGURE 7.5:   **Distinction between first- and second-order reactions.** Experimentally determined concentration (in µM) of the reactant as a function of time (black circles) and best fit of curves for a first-order reaction (dotted line, eq. 7.7, $k = 2$ s$^{-1}$) and a second-order reaction (solid line, eq. 7.23, $k = 2$ µM$^{-1}$ s$^{-1}$). The single exponential curve deviates systematically from the data, indicating that the reaction is not a first-order reaction.

We can calculate the half-life of A for the second-order reaction as

$$[A](t_{1/2}) = \frac{A_0}{2} = \frac{A_0}{1 + 2kt_{1/2}A_0} \qquad \text{eq. 7.24}$$

and obtain

$$t_{1/2} = \frac{1}{2kA_0} \qquad \text{eq. 7.25}$$

Note that in contrast to the first-order reaction, where the half-life is independent of the initial concentration of A, the half-life for the second-order reaction now depends on the initial concentration of A. As a consequence, the first half-life, from $A_0$ to $A_0/2$, is different than the second half-life, from $A_0/2$ to $A_0/4$. By comparing the half-life of a reaction at different starting concentrations we can therefore also distinguish between first- and second-order reactions. From

$$[A](t_{1/4}) = \frac{A_0}{4} = \frac{A_0}{1 + 2kt_{1/4}A_0} \qquad \text{eq. 7.26}$$

we obtain $t_{1/4}$ as

$$t_{1/4} = \frac{3}{2kA_0} \qquad \text{eq. 7.27}$$

The second half-life is the time that has passed from half to a quarter of the starting concentration of A and can be calculated as the difference between $t_{1/4}$ and $t_{1/2}$:

$$t_{1/4} - t_{1/2} = \frac{2}{2kA_0} \qquad \text{eq. 7.28}$$

From a comparison of eq. 7.25 and eq. 7.28 we see that the second half-life thus is twice the first half-life. The third half-life (from $A_0/4$ to $A_0/8$) is twice of the second half life, and so on (Figure 7.6).

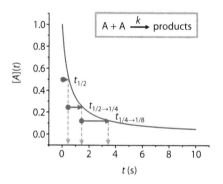

FIGURE 7.6:   **Half-lives in a second-order reaction.** For a second-order reaction, the half-life $t_{1/2}$ depends on the starting concentration $A_0$. As a consequence, the second half-life, the time passing between $A_0/2$ to $A_0/4$ is twice as long as the first half-life (from $A_0$ to $A_0/2$), the third half-life (from $A_0/4$ to $A_0/8$) is twice as long as the second, and so on.

A different type of a bimolecular reaction is a reaction of A with a different molecule B to products with the rate constant $k$ (Scheme 7.3):

$$A + B \xrightarrow{k} products \qquad \text{scheme 7.3}$$

This scheme describes any association of two different molecules, and thus any binary complex formation (Box 7.4). The corresponding rate law is

$$v = -\frac{d[A]}{dt} = -\frac{d[B]}{dt} = k[A][B] \qquad \text{eq. 7.29}$$

The reaction is of first order with respect to A and to B, but second order overall. The unit of $k$ is $M^{-1}\,s^{-1}$. We can now separate the variables

$$-\frac{d[A]}{[A][B]} = k\,dt \qquad \text{eq. 7.30}$$

In eq. 7.30, both $[A]$ and $[B]$ depend on the time $t$. To obtain a form of this equation that we can integrate, we express the current concentration of A as $A_0$ minus the concentration $x$ that has already reacted to products. As one molecule A reacts with one molecule of B, we can express the concentration of B as $B_0$ minus the same concentration $x$. $x$ thus represents the product formed, and the rate law for product formation is

$$\frac{dx}{dt} = k[A][B]$$

eq. 7.31

We thereby reduce the equation to a single integration variable $x$ and its change $dx = -dA$. Eq. 7.30 then becomes

$$\frac{dx}{(A_0 - x)(B_0 - x)} = k\, dt$$

eq. 7.32

which we can now integrate from 0 to $x(t)$, and from 0 to $t$:

$$\int_0^{x(t)} \frac{dx}{(A_0 - x)(B_0 - x)} = \int_0^t k\, dt$$

eq. 7.33

The right-hand side of eq. 7.33 can be integrated to $kt$. To integrate the left-hand side, we need to express the fraction by partial fractions (see Section 29.7) and then integrate the partial fractions individually by substitution (see Section 29.6). We can express the fraction on the left-hand side as

$$\frac{1}{(A_0 - x)(B_0 - x)} = \frac{1}{(A_0 - B_0)} \cdot \left( \frac{1}{(B_0 - x)} - \frac{1}{(A_0 - x)} \right)$$

eq. 7.34

We now have to evaluate the integral

$$\int_0^{x(t)} \frac{1}{(A_0 - B_0)} \cdot \left( \frac{1}{(B_0 - x)} - \frac{1}{(A_0 - x)} \right) dx = kt$$

eq. 7.35

$1/(A_0-B_0)$ is a constant factor, and the two fractions can be integrated separately to

$$\frac{1}{(A_0 - B_0)} \cdot \left( \left[ -\ln(B_0 - x) \right]_0^{x(t)} - \left[ -\ln(A_0 - x) \right]_0^{x(t)} \right) + \text{const.} = kt$$

eq. 7.36

and we obtain

$$\frac{1}{(A_0 - B_0)} \cdot \left( \ln \frac{A_0 - x(t)}{B_0 - x(t)} - \ln \frac{A_0}{B_0} \right) + \text{const.} = kt$$

eq. 7.37

The boundary condition $x(0) = 0$ is fulfilled with an integration constant (const.) of zero. Solving eq. 7.37 for $x(t)$ gives the integrated rate law

$$x(t) = \frac{A_0 - A_0\, e^{k(A_0 - B_0)t}}{1 - \frac{A_0}{B_0}\, e^{k(A_0 - B_0)t}}$$

eq. 7.38

$[A](t)$ and $[B](t)$ can be calculated as $A_0 - x(t)$ and $B_0 - x(t)$ from mass conservation

$$[A](t) = A_0 - \frac{A_0 - A_0\, e^{k(A_0 - B_0)t}}{1 - \frac{A_0}{B_0}\, e^{k(A_0 - B_0)t}}$$

eq. 7.39

and

$$[B](t) = B_0 - \frac{A_0 - A_0\, e^{k(A_0-B_0)t}}{1 - \dfrac{A_0}{B_0}\, e^{k(A_0-B_0)t}}$$

eq. 7.40

The concentrations $[A](t)$, $[B](t)$ and $x(t)$ are depicted in Figure 7.7.

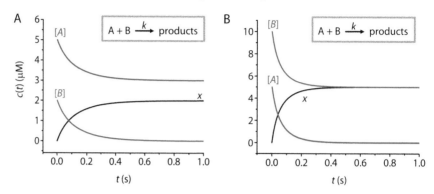

**FIGURE 7.7:** **Second-order reaction A+B → products.** $[A](t)$, $[B](t)$ and $x(t)$ for a bimolecular reaction of A and B to products. A: $A_0 = 5\ \mu M$, $B_0 = 2\ \mu M$ and $k = 2\ s^{-1}$. B: $A_0 = 5\ \mu M$, $B_0 = 10\ \mu M$ and $k = 2\ s^{-1}$.

Note that eq. 7.40 and all derived forms are not defined for $A_0 = B_0$. For this special case, we have derived the rate law before: for $A_0 = B_0$, and a reaction where one molecule A reacts with one molecule of B, the concentrations of A and B are equal at any given point in time, $[A](t)=[B](t)$. Therefore, the rate law can be written as

$$-\frac{d[A]}{dt} = -\frac{d[B]}{dt} = k[A][B] = k[A]^2$$

eq. 7.41

which is identical to eq. 7.18 for the second-order reaction A+A→products, leading to the identical equation for $[A](t)$ (eq. 7.23).

---

### BOX 7.4: SECOND-ORDER REACTIONS: DIMERIZATION AND LIGAND BINDING.

Many enzymes need to associate to dimers to perform their biological functions. Examples are glutathione-S-transferase, alcohol dehydrogenase, transcription factors, and type II DNA restriction enzymes. After biosynthesis and folding of the monomeric forms, two monomers form a homodimer in a second-order reaction according to Scheme 7.2. The formation of heterodimers by two different subunits is a second-order reaction following Scheme 7.3. In fact, any binary complex formation is described by Scheme 7.3, including the association of two complementary strands of DNA or RNA, protein-protein interactions such as the complex formation of Ras-like small G-proteins with their effectors, protein-ligand interactions such as binding of a substrate, inhibitor or activator molecule to an enzyme, as well as protein-nucleic acid interactions.

---

We can apply the same principles we have used for unimolecular and bimolecular reactions to formulate rate laws and derive integrated rate laws, expressions for the concentrations of reaction partners as a function of time, for trimolecular reactions and reactions of higher molecularity.

However, trimolecular reactions and reactions of higher molecularity are very rare (see Section 13.3), and will therefore not be treated.

## QUESTIONS

7.1 You measure the concentration of a reactant as a function of time and obtain the following dataset:

| $t$ (s) | $c(t)$ ($\mu$M) |
|---|---|
| 0 | 10 |
| 1 | 7.1 |
| 2 | 5.0 |
| 3 | 3.5 |
| 4 | 2.5 |
| 5 | 1.7 |
| 6 | 1.3 |
| 8 | 0.6 |
| 10 | 0.3 |

What are the reaction order and the rate constant?

7.2 A reactant shows the following decrease in concentration with reaction time:

| $t$ (s) | $c(t)$ ($\mu$M) |
|---|---|
| 0 | 1 |
| 1 | 0.74 |
| 2 | 0.59 |
| 3 | 0.49 |
| 4 | 0.42 |
| 5 | 0.37 |
| 6 | 0.32 |
| 8 | 0.27 |
| 10 | 0.22 |

Determine the reaction order and estimate the rate constant.

7.3 An old tree trunk is discovered and subjected to radiocarbon dating. The $^{14}$C percentage in the analyzed samples is $10^{-12}$%. What is the age of the tree? $[^{14}C]_0 = 10^{-10}$%, $t_{1/2} = 5730$ a.

7.4 You determine the concentration of a reactant that reacts on its own at time $t = 3.6$ s as $c(t) = 0.27$ mM. The starting concentration was 10 mM. Does the reaction follow the rate law for first- and second-order reactions? What is the rate constant?

7.5 The reactants A and B react to products in an irreversible reaction. What is the concentration of A and B at $t = 5$ s for (1) $A_0 = 1$ $\mu$M, $B_0 = 10$ $\mu$M, (2) $A_0 = 10$ $\mu$M, $B_0 = 1$ $\mu$M, (3) $A_0 = 6$ $\mu$M, $B_0 = 5$ $\mu$M and for (4) $A_0 = 20$ $\mu$M, $B_0 = 29$ $\mu$M? $k = 0.2$ $\mu$M$^{-1}$ s$^{-1}$.

7.6 Derive the integrated rate law for the trimolecular reaction 3 A $\rightarrow$ products. The rate constant is $k$. What is a suitable plot to determine the rate constant by linear regression?

# Reaction Types

I n Chapter 7, we have calculated integrated rate laws for isolated reactions. In the following, we will derive rate laws for reaction schemes comprising two reactions: reversible reactions, parallel reactions and sequential reactions. These reaction types are important elements of more complex reaction schemes. Their treatment will enable us to analyze and quantitatively understand complex biochemical reaction pathways and networks.

## 8.1 REVERSIBLE REACTIONS

We first look at a *reversible reaction* in which A is converted to B in a forward reaction with the rate constant $k_1$, and B is converted to A in a reverse reaction with the rate constant $k_{-1}$ (Scheme 8.1):

$$A \underset{k_{-1}}{\overset{k_1}{\rightleftharpoons}} B$$

scheme 8.1

Examples for reversible reactions are isomerizations, such as the interconversion of dihydroxyacetone phosphate and glyceraldehyde-3-phosphate in glycolysis, protein folding according to the two-state model (see Section 16.3.4), or conformational changes of proteins, e.g. between open and closed states (see Section 4.3). We can express the rate law for a reversible reaction in terms of the forward and reverse reactions as

$$v = \frac{\mathrm{d}c}{\mathrm{d}t} = -\frac{\mathrm{d}[A]}{\mathrm{d}t} = \frac{\mathrm{d}[B]}{\mathrm{d}t}$$

eq. 8.1

The change in concentration of A is brought about by the forward reaction that leads to a consumption of A, and thus a decrease in its concentration. The concentration change over time is the product of the rate constant $k_1$ and the actual concentration of A. This term appears with a negative sign to take into account the decrease of $[A]$ caused by this process. The reverse reaction, from B to A, leads to an increase in the concentration of A. Its velocity is the product of the rate constant $k_{-1}$ and the current concentration of B, and this term enters the balance with a positive sign due to the increase in $[A]$ (see also Box 8.2). The change in $[A]$ over time is thus

$$\frac{d[A]}{dt} = -k_1[A] + k_{-1}[B]$$   eq. 8.2

We know from the law of mass conservation that the total concentration of A ($A_0$) that was present at $t = 0$ will either remain as A, or will have been converted to B. When the reaction starts from A, i.e. the starting concentration of B is zero, eq. 8.3

$$[A] + [B] = A_0$$   eq. 8.3

has to be fulfilled at any given time. We can now solve eq. 8.3 for [B], and substitute [B] in eq. 8.2, and obtain

$$\frac{d[A]}{dt} = -k_1 \cdot [A] + k_{-1}(A_0 - [A]) = -(k_1 + k_{-1})[A] + k_{-1}A_0$$   eq. 8.4

Separating the variables gives

$$\frac{d[A]}{-(k_1 + k_{-1})[A] + k_{-1}A_0} = dt$$   eq. 8.5

We can now integrate eq. 8.5 by substitution: we substitute the denominator for $u$, and obtain

$$\frac{du}{d[A]} = -(k_1 + k_{-1}) \quad \text{or} \quad d[A] = \frac{du}{-(k_1 + k_{-1})}$$   eq. 8.6

The integral then becomes

$$\frac{1}{-(k_1 + k_{-1})} \cdot \int \frac{1}{u} du = \int_0^t dt$$   eq. 8.7

or

$$\int \frac{1}{u} du = \int_0^t -(k_1 + k_{-1}) dt$$   eq. 8.8

and can be evaluated as

$$\ln u = -(k_1 + k_{-1})t$$   eq. 8.9

Substituting back $u$ for the denominator in eq. 8.5 yields

$$\ln u = \left[ \ln\left(-(k_1 + k_{-1})[A] + k_{-1}A_0\right) \right]_{A_0}^{A(t)} = -(k_1 + k_{-1})t$$   eq. 8.10

We can now evaluate the integral on the left from $A_0$ to $A(t)$ to

$$\ln\left(-(k_1 + k_{-1})[A](t) + k_{-1}A_0\right) - \ln\left(-(k_1 + k_{-1})A_0 + k_{-1}A_0\right) = -(k_1 + k_{-1})t$$   eq. 8.11

Solving eq. 8.11 for [A](t) then yields

$$[A](t) = \frac{A_0}{k_1 + k_{-1}}\left(k_{-1} + k_1 \, e^{-(k_1 + k_{-1})t}\right)$$   eq. 8.12

Eq. 8.12 contains a constant term and an exponential function that describes the exponential decay of [A] as a function of time. Note that instead of the rate constant $k$ in the exponent, as in the irreversible reaction (eq. 7.40), now the sum of the rate constants for the forward and reverse reaction, $k_1 + k_{-1}$, appears as an *apparent rate constant* $k_{app}$.

We can calculate $[B]$ as a function of time from the mass conservation as $A_0 - [A](t)$:

$$[B](t) = \frac{A_0 \cdot k_1}{k_1 + k_{-1}} \left(1 - e^{-(k_1 + k_{-1})t}\right)$$

eq. 8.13

Again, the apparent rate constant for the appearance of B is the sum of the individual rate constants $k_1$ and $k_{-1}$. Thus, if we follow the disappearance of A or the appearance of B over time by a suitable spectroscopic signal (see Chapters 19 and 25) and describe the signal as a function of time by a single exponential, we can only determine the sum of the rate constants. To extract values for both $k_1$ and $k_{-1}$, we need to know the absolute amplitudes, i.e. we need to be able to convert our signal into concentrations of A and B.

We know that if a reaction occurs in both directions, an equilibrium between A and B will be established. The rate constant with which the equilibrium is approached is the same apparent rate constant for the consumption of A and the production of B (Figure 8.1). The apparent rate constant is independent of the starting conditions: no matter if we start from A or B, the equilibrium concentrations will be reached with $k_{app} = k_1 + k_{-1}$. In fact, although we have not derived it here, this is also true when we start from any non-equilibrium mixture of A and B.

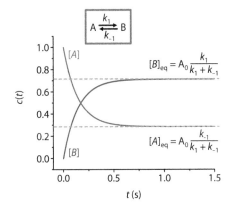

**FIGURE 8.1:    A reversible reaction.** Concentrations $c(t)$ of A (blue) and B (red) as a function of time for rate constants $k_1 = 5$ s$^{-1}$ and $k_{-1} = 2$ s$^{-1}$. A decays exponentially to its equilibrium concentration $[A]_{eq}$ with an apparent rate constant of $k_{app} = k_1 + k_{-1} = 7$ s$^{-1}$, and B grows exponentially to its equilibrium concentration $[B]_{eq}$ with the same apparent rate constant. The equilibrium constant $K = [B]_{eq}/[A]_{eq}$ is the ratio of the two rate constants $k_1$ and $k_{-1}$.

What can we say about the equilibrium concentrations $[A]_{eq}$ and $[B]_{eq}$, once the equilibrium has been established? To answer this question, we have to determine the concentrations of A and B for $t \to \infty$. In eq. 8.12 and eq. 8.13, the exponential term will become zero for $t \to \infty$, leaving us with the constant term

$$[A]_{eq} = \frac{k_{-1} A_0}{k_1 + k_{-1}}$$

eq. 8.14

The concentration of B in equilibrium can be calculated from the mass conservation, or directly from the limit of eq. 8.13 for $t \to \infty$. In both cases, we obtain

$$[B]_{eq} = A_0 - [A]_{eq} = \frac{k_1 A_0}{k_1 + k_{-1}}$$

eq. 8.15

We can now calculate the equilibrium constant K from the ratio of the equilibrium concentrations $[A]_{eq}$ and $[B]_{eq}$ as

$$K = \frac{[B]_{eq}}{[A]_{eq}} = \frac{\dfrac{k_1 A_0}{k_1 + k_{-1}}}{\dfrac{k_{-1} A_0}{k_1 + k_{-1}}} = \frac{k_1}{k_{-1}} \qquad \text{eq. 8.16}$$

Eq. 8.16 illustrates the relationship between thermodynamic data (equilibrium constant) and kinetic data (rate constants). The position of the equilibrium is the result of a *kinetic competition* of forward and reverse reactions: if both reactions occur with the same rate constant, K equals unity. If $k_1$ is e.g. 10-fold higher than $k_{-1}$, then there will be 10-times the amount of product compared to reactant, and *vice versa*.

Note that the rate constants $k_1$ and $k_{-1}$ have different units if we consider the association reaction of two molecules A and B to a complex AB (Scheme 8.2).

$$A + B \underset{k_{-1}}{\overset{k_1}{\rightleftharpoons}} AB \qquad \text{scheme 8.2}$$

In this case, $k_1$ is a second-order rate constant for a bimolecular reaction, with units $M^{-1}\,s^{-1}$, and $k_{-1}$ is the first-order rate constant for a unimolecular reaction, with units $s^{-1}$. Nevertheless, the equilibrium constants $K_a$ for complex formation or $K_d$ for complex dissociation can still be expressed by the ratio of the rate constants. The equilibrium constant for complex formation, $K_a$, in $M^{-1}$ (see Section 3.2; Box 3.2) is

$$K_a = \frac{[AB]_{eq}}{[A]_{eq}[B]_{eq}} = \frac{k_1}{k_{-1}} \qquad \text{eq. 8.17}$$

The dissociation constant $K_d$ of the complex AB is the inverse of the association constant, and thus

$$K_d = \frac{[A]_{eq}[B]_{eq}}{[AB]_{eq}} = \frac{k_{-1}}{k_1} \qquad \text{eq. 8.18}$$

with units M. (Bio)chemical equilibria are often discussed in terms of $K_d$ rather than $K_a$ because of the more convenient units (see Section 3.2).

## 8.2 PARALLEL REACTIONS

We now consider a reaction scheme where A can react either to B or to C in two *parallel reactions* (Scheme 8.3). There are numerous examples for this reaction scheme in metabolism, where many reaction intermediates can enter two or more alternative pathways (see Figure 8.3).

$$A \overset{k_1}{\underset{k_2}{\diagdown}} \begin{array}{c} B \\[4pt] C \end{array} \qquad \text{scheme 8.3}$$

The conversion of A to B is characterized by a rate constant $k_1$, and the conversion of A to C occurs with rate constant $k_2$. We can write the balance for the change in [A] over time by summing up the contributions from reactions consuming A:

$$\frac{d[A]}{dt} = -(k_1 + k_2) \cdot [A] \qquad \text{eq. 8.19}$$

B is produced from A, and the rate law is

$$\frac{d[B]}{dt} = k_1 \cdot [A] \qquad \text{eq. 8.20}$$

Similarly, C is produced from A according to the rate law

$$\frac{d[C]}{dt} = k_2 \cdot [A]$$

eq. 8.21

By integration of eq. 8.19 we obtain the time dependence of A as

$$[A](t) = A_0\, e^{-(k_1+k_2)t}$$

eq. 8.22

Thus, A disappears with an apparent rate constant that is the sum of the individual rate constants, $k_{app} = k_1 + k_2$. We can substitute eq. 8.22 into eq. 8.20 and integrate to the corresponding equation for [B](t) as

$$[B](t) = A_0 \frac{k_1}{k_1+k_2}\left(1 - e^{-(k_1+k_2)t}\right)$$

eq. 8.23

Finally, we can calculate [C](t) from mass conservation as $A_0 - [A](t) - [B](t)$ as

$$[C](t) = A_0 \frac{k_1}{k_1+k_2}\left(1 - e^{-(k_1+k_2)t}\right)$$

eq. 8.24

B and C are thus also produced with an apparent rate constant $k_{app} = k_1 + k_2$ (Figure 8.2).

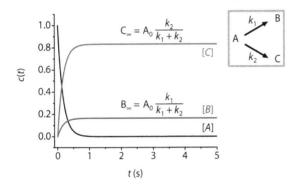

FIGURE 8.2:   **Parallel reactions.** Concentrations $c(t)$ of A, B, and C as a function of time for a parallel reaction of A to B with rate constant $k_1$ (1 s$^{-1}$) and of A to C with rate constant $k_2$ (5 s$^{-1}$). A decays with an apparent rate constant of $k_1 + k_2$, and the concentrations of B and C increase exponentially with the same apparent rate constant. The final ratios of B and C depend on the ratio of $k_1$ and $k_2$. The final concentration of B, $B_\infty$, is $A_0$ multiplied with $k_1/(k_1 + k_2)$, the final concentration of C, $C_\infty$, equals $A_0$ times $k_2/(k_1 + k_2)$. This situation is also called kinetic competition.

According to eq. 8.22–eq. 8.24, A(t), B(t) or C(t) from our kinetic experiment can all be described by a single exponential function, but the apparent rate constant determined from the data is the sum of the two individual rate constants $k_1$ and $k_2$. As discussed for the reversible reaction (Section 8.1), we can only determine the individual rate constants from [B](t) or [C](t) if we also know the absolute amplitude of the exponential function, i.e. $B_\infty$ or $C_\infty$.

What determines the final yield in B and C in such a reaction scheme? We can calculate the final concentrations of B and C for $t \to \infty$ from eq. 8.23 and eq. 8.24, and obtain the ratio of $B_\infty$ and $C_\infty$ as

$$\frac{B_\infty}{C_\infty} = \frac{A_0 \dfrac{k_1}{k_1+k_2}}{A_0 \dfrac{k_2}{k_1+k_2}} = \frac{k_1}{k_2}$$

eq. 8.25

Thus, the ratio of the rate constants $k_1$ and $k_2$ determines the ratio of the two products. Again, we have a kinetic competition between the two reactions.

We find numerous examples for parallel reactions in biochemistry (Figure 8.3). Every time a metabolic intermediate is a branch point and can be channeled into different pathways, the relative entry into the individual pathways (*partitioning*) is determined by kinetic competition. In glycolysis, for example, pyruvate can be further metabolized to acetyl-CoA, and then enter the citric acid cycle, or it can be used for the biosynthesis of alanine. Acetyl-CoA itself can either be channeled into the citric acid cycle, or it can be used for the synthesis of fatty acids. Glyceraldehyde-3-phosphate is metabolized in the glycolytic pathway to 1,3-bisphosphoglycerate, or used for the conversion of sugars in the pentose phosphate pathway. The overall reaction velocity of each reaction sequence after the branch point is determined by the slowest step of the sequence, the rate-limiting step (Chapter 9). The ratio of the rate constants for the rate-limiting step determines the fractions of the metabolic intermediate that will accumulate as the end products of the two alternative pathways.

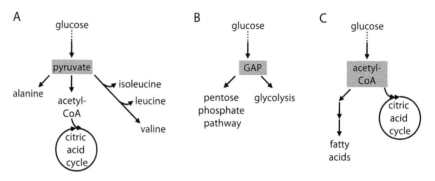

**FIGURE 8.3:  Metabolic branch points and parallel reactions.** A: Pyruvate generated from glucose in glycolysis can be converted to alanine, channeled into the synthesis pathway for the amino acids isoleucine, leucine, and valine, or converted to acetyl-CoA that enters the citric acid cycle. B: Glyceraldehyde-3-phosphate (GAP) is either metabolized *via* the glycolysis pathway or enters the pentose phosphate pathway. C: Acetyl-CoA can either be channeled into the citric acid cycle or enter the synthesis pathway of fatty acids. The distribution over the different pathways depends on the rate constants of the respective processes (kinetic partitioning).

## 8.3  CONSECUTIVE REACTIONS

Finally, we consider a sequence of two reactions from A to B with $k_1$, and from B to C with $k_2$, according to Scheme 8.4:

$$A \xrightarrow{k_1} B \xrightarrow{k_2} C$$

scheme 8.4

Virtually all metabolic pathways, such as the reaction sequence of glycolysis, or the synthesis of fatty acids or nucleobases, are *consecutive reactions*. The rate laws for A, B and C for a reaction sequence A→B→C according to Scheme 8.4 are

$$\frac{d[A]}{dt} = -k_1 \cdot [A]$$

eq. 8.26

and

$$\frac{d[B]}{dt} = k_1 \cdot [A] - k_2 \cdot [B]$$

eq. 8.27

and

$$\frac{d[C]}{dt} = k_2 \cdot [B]$$

eq. 8.28

We can integrate eq. 8.26 to obtain [A](t) as

$$[A](t) = A_0\ e^{-k_1 t}$$

eq. 8.29

The concentration of A thus decays exponentially with the rate constant $k_1$, and shows identical behavior as in an isolated first-order reaction without subsequent reactions (eq. 7.7). The expression for $[A](t)$ may be substituted into eq. 8.27, which can be integrated to

$$[B](t) = A_0 \frac{k_1}{k_2 - k_1}\left(e^{-k_1 t} - e^{-k_2 t}\right)$$

eq. 8.30

Finally, $[C](t)$ can be calculated from the mass conservation as $[C](t) = A_0 - [A](t) - [B](t)$ as

$$[C](t) = A_0\left(1 + \frac{k_1\,e^{-k_2 t} - k_2\,e^{-k_1 t}}{k_2 - k_1}\right)$$

eq. 8.31

The time dependence of $[A]$, $[B]$, and $[C]$ is shown in Figure 8.4. B is only transiently produced as an intermediate. The degree of accumulation of B depends on the ratio of $k_1$ and $k_2$: if $k_1 > k_2$, the reaction from B to C is slower than the reaction from A to B, and B will accumulate. In this case, B is formed with the rate constant $k_1$, and then slowly disappears with rate constant $k_2$.

If $k_1 < k_2$, then the reaction from B to C is faster than the reaction from A to B. Thus, any B that has been formed will rapidly react to C, and B will not accumulate as an intermediate. In this case, B is formed with the rate constant $k_2$, and decomposes with $k_1$. This intrinsically non-intuitive behavior becomes clear if we look at the equation for $[B](t)$ (eq. 8.30), and its behavior for the two cases: the amplitude factor in eq. 8.30 contains the difference $k_2 - k_1$ in the denominator. This difference will be positive for $k_1 < k_2$. Consequently, the first exponential term with $k_1$ will stay positive, and will describe an exponential decay. The second exponential term with $k_2$ will remain negative, describing

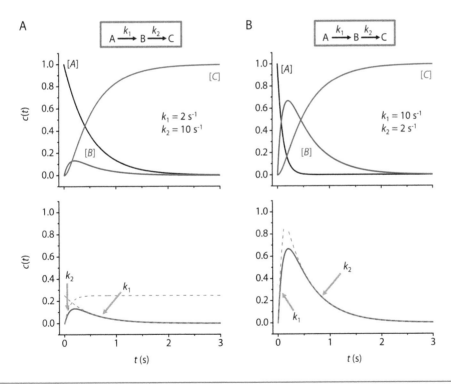

**FIGURE 8.4: Consecutive reactions.** A: Consecutive reactions from A to B to C with rate constants $k_1 < k_2$ ($k_1 = 2\ \text{s}^{-1}$, $k_2 = 10\ \text{s}^{-1}$). The concentration of A decreases exponentially with $k_1$. The concentration of B shows a maximum. The maximum concentration is reached with $k_2$, and decays with $k_1$ (bottom). The maximum concentration of B depends on $A_0$ and the ratio of $k_1/(k_1 + k_2)$. C is formed after a short lag period after A has been converted to B. B: Consecutive reactions with rate constants $k_1 > k_2$ ($k_1 = 10\ \text{s}^{-1}$, $k_2 = 2\ \text{s}^{-1}$). Because the reaction consuming B is now slower than the reaction producing B, B accumulates transiently. The maximum concentration of B is reached with $k_1$, and decays with $k_2$ (bottom).

exponential growth. Thus, B will be formed rapidly with $k_2$, and decay slowly with $k_1$. However, if $k_1 > k_2$, the sign of the denominator becomes negative, converting the first exponential term with $k_1$ into an exponential growth, and the second exponential term with $k_2$ into an exponential decay. Thus, in this case B will be formed rapidly with $k_1$, and decay slowly with $k_2$.

$B_{max}$ can be calculated from the first derivative of eq. 8.30: at the maximum of the curve describing $[B](t)$ the first derivative will be zero (eq. 8.27)

$$\frac{d[B](t)}{dt} = k_1[A] - k_2[B] = 0 \qquad \text{eq. 8.27}$$

or

$$k_1[A] = k_2[B] \qquad \text{eq. 8.32}$$

We can rearrange this equation to obtain

$$k_1\left(A_0 - [B]_{max}\right) = k_2[B]_{max} \qquad \text{eq. 8.33}$$

and solve for $[B]_{max}$, the maximum concentration of B:

$$[B]_{max} = \frac{k_1}{k_1 + k_2}A_0 \qquad \text{eq. 8.34}$$

According to this equation, we can neglect $k_2$ in the denominator for $k_1 \gg k_2$, and $[B]_{max}$ will approach $A_0$ (Figure 8.4). For $k_1 \ll k_2$, we can neglect $k_1$ in the denominator. $[B]_{max}$ will then reach

$$[B]_{max} \approx \frac{k_1}{k_2}A_0 \qquad \text{eq. 8.35}$$

In this case, the ratio of the two individual rate constants determines the amount of B that accumulates (Figure 8.5). The faster the formation of B compared to its decomposition, the higher $[B]_{max}$ (Box 8.1). Using eq. 8.30, we can also determine the time $t_{max}$ when the intermediate B reaches the maximal concentration. The time $t_{max}$ is the ideal time to capture an unstable reaction intermediate, either to characterize it or to study its reaction with other compounds.

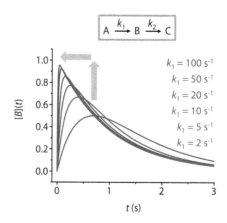

**FIGURE 8.5:** **Accumulation of B as a function of $k_1$ and $k_2$.** The concentration of B as a function of time is shown for $k_1 = 2\ s^{-1}$, $5\ s^{-1}$, $10\ s^{-1}$, $20\ s^{-1}$, $50\ s^{-1}$, and $100\ s^{-1}$. For all curves, $k_2$ is $1\ s^{-1}$. The faster the first reaction, the more B is transiently accumulated, and the maximum concentration of B is reached earlier (gray arrows).

**BOX 8.1: ACCUMULATION OF INTERMEDIATES IN SEQUENTIAL REACTIONS: ENZYMATIC OXIDATION OF ETHANOL.**

Ethanol is metabolized by oxidation (Scheme 8.5). In a first step, alcohol dehydrogenase (ADH) oxidizes ethanol to acetaldehyde. Acetaldehyde is then oxidized to acetic acid by acetaldehyde dehydrogenase (ALDH).

$$\text{ethanol} \xrightarrow[\text{ADH}]{k_1} \text{acetaldehyde} \xrightarrow[\text{ALDH}]{k_2} \text{acetic acid}$$

scheme 8.5

Acetaldehyde is toxic and causes nausea, headaches, and tachycardia. Normally, the intermediate acetaldehyde does not accumulate because its oxidation is faster than its generation by ethanol oxidation. However, in people with mutations in ALDH, acetaldehyde oxidation is slower, and acetaldehyde accumulates to higher levels. ALDH dehydrogenase inhibitors are used in the treatment of alcoholism to increase the negative effects and discourage patients from drinking.

In principle, we have now treated the key elements of complex reaction pathways and networks. While the definition of reaction velocities for complex reaction networks follows the same basic principle as for the building blocks we have treated in Chapters 7 and 8, the derivation of integrated rate laws becomes more and more challenging. From the set of differential equations that describe the reaction network, the temporal behavior of the entire network can be modeled using numerical methods (Box 8.2).

We will see in Chapter 12 how we can apply the principles we have learnt from model reactions to quantitatively understand more complex multi-step reactions.

**BOX 8.2: DIFFERENTIAL EQUATIONS TO DESCRIBE REACTION NETWORKS.**

In reaction networks, the same compound may be a reactant for one reaction, and a product of another reaction. Although this and the lack of directionality pose a problem in defining a reaction velocity, the temporal behavior of such a network can be described quantitatively. Even for the most complicated network, we can express the changes in concentration for each of the reactants in form of differential equations. We only have to consider all processes in which this compound is involved, and separate them into processes towards the compound that cause an increase of its concentration, and all processes that lead away from this compound and cause a concentration decrease (Figure 8.6).

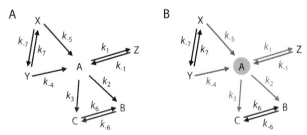

FIGURE 8.6: **A reaction network.** A: The compounds A, B, C, X, Y, and Z are connected through several reversible and irreversible reactions. B: To describe the change in concentration of A over time, we need to take into account all processes that lead away from A and cause a decrease its concentration (blue) with a negative sign, and all processes that lead to formation of A and cause an increase in the concentration of A (red) with a positive sign.

*(Continued)*

We first focus on the balance for compound A (Figure 8.6). All processes that lead away from A contribute to a concentration decrease, and need to be taken into account with a negative sign in the expression for d[A]/dt. In contrast, all processes towards A increase the concentration of A and are taken into account with a positive sign:

$$\frac{d[A]}{dt} = -k_1[A] - k_2[A] - k_3[A] + k_{-1}[Z] + k_{-4}[Y] + k_{-5}[X] \qquad \text{eq. 8.36}$$

Similarly, we can formulate the change in the concentration of B as

$$\frac{d[B]}{dt} = -k_{-6}[B] + k_2[A] + k_6[C] \qquad \text{eq. 8.37}$$

and can express the changes in concentrations for all other compounds by differential equations. This set of differential equations cannot be solved analytically. Nevertheless, the temporal behavior of the reaction network can be calculated from these equations using numerical methods, provided that the individual rate constants are known. This approach is the core of many systems biology approaches to model metabolic or signaling networks. The predictive power for changes in the entire network in response to specific changes depends on how well-determined the rate constants, the connections between individual compounds and their starting concentrations are.

## QUESTIONS

8.1    A reversible reaction A + B $\rightleftarrows$ AB can be monitored by fluorescence because the fluorescence of B increases upon AB formation. Binding is too fast to be measured and already complete within the dead time of your experiment. How can you determine $K_{eq}$, $k_{-1}$ and $k_1$?

8.2    You start from a reactant A and monitor formation of the product B of a reaction as a function of time. $B(t)$ can be described by a single exponential function with a rate constant $k = 0.4$ s$^{-1}$. The end concentration of B is 7.5 µM. What are the possible explanations for this behavior and what is the meaning of the measured rate constant in each case? Which piece of information is needed to distinguish between the different possibilities?

8.3    An enzyme binds and hydrolyzes ATP. You want to populate the enzyme-ATP complex to study its binding to an interaction partner. ATP binding induces a conformational change in the enzyme, and the overall binding reaction is slow with $k_{bind} = 0.1$ s$^{-1}$. The rate constant of ATP hydrolysis is $k_{hyd} = 10^{-3}$ s$^{-1}$. How long do you have to incubate the enzyme with ATP to maximize the concentration of the ATP-bound form?

## REFERENCE

Hald, J. and Jacobsen, E. (1948) A drug sensitizing the organism to ethyl alcohol. *Lancet 2*(6539): 1001–1004.
· treatment of alcoholism with disulfiram/Antabuse

# Rate-Limiting Steps

For a consecutive reaction from A *via* B to C according to Scheme 9.1

$$A \xrightarrow{k_1} B \xrightarrow{k_2} C$$

scheme 9.1

we have derived in Section 8.3 that the formation of C follows eq. 8.31

$$[C](t) = A_0 \left( 1 + \frac{k_1 e^{-k_2 t} - k_2 e^{-k_1 t}}{k_2 - k_1} \right)$$

eq. 8.31

We now consider the limiting case that $k_1 \ll k_2$, i.e. the reaction from A to B is much slower than the reaction from B to C. With this limiting condition, the two exponential terms in eq. 8.31 will contribute very differently, with

$$k_1 e^{-k_2 t} \ll k_2 e^{-k_1 t}$$

eq. 9.1

As an approximation, we can therefore ignore the first exponential term in the numerator. Similarly, we can simplify the denominator as

$$k_2 - k_1 \approx k_2$$

eq. 9.2

Altogether, eq. 8.31 then simplifies to

$$[C](t) \approx A_0 \left( 1 - \frac{k_2 e^{-k_1 t}}{k_2} \right)$$

eq. 9.3

or

$$[C](t) \approx A_0 \left( 1 - e^{-k_1 t} \right)$$

eq. 9.4

Thus, for the limiting condition that the first reaction is much slower than the second reaction in this sequence, C is formed approximately by an exponential function with the rate constant $k_1$

(Figure 9.1). The slow first step in the reaction sequence thus determines the overall reaction velocity from A to C. We therefore call the reaction from A to B with the rate constant $k_1$ the *rate-limiting step* of this sequence of reactions.

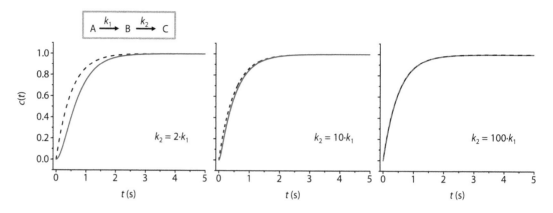

**FIGURE 9.1:**  **Rate-limiting steps.** The slowest step in a consecutive reaction determines the rate of product formation. For two consecutive reactions A→B→C with rate constants $k_1$ and $k_2$, the formation of C can be described by an exponential increase with $k_1$ if $k_1 < k_2$. The dashed lines (black) represent the approximated single exponential growth (eq. 9.4), the solid line (blue) shows the concentration of C calculated according to eq. 8.31 without approximation. The larger $k_2$ is compared to $k_1$, the more accurate is the description of $[C](t)$ by the approximation. $k_1 = 2$ s$^{-1}$.

The rate-limiting step of a reaction sequence is often called the "bottleneck" of the pathway. The existence of a rate-limiting step in a reaction sequence has important implications for the response of the entire pathway to changes in reactant concentrations or to changes in the rate of individual steps. If we increase the rate constant $k_2$ in Scheme 9.1, e.g. by adding a catalyst for this reaction step (see Chapter 14), the overall reaction velocity will not be affected. If we, however, managed to accelerate the first reaction that constitutes the rate-limiting step, we would achieve an increase in the overall reaction velocity, resulting in a faster production of C. Overall reaction velocities are most efficiently altered by changing the velocity of the rate-limiting step(s) of reaction sequences! Hence, rate-limiting steps in biochemical reaction pathways are often subject to tight regulation (Box 9.1).

---

**BOX 9.1: RATE-LIMITING STEPS IN METABOLIC
PATHWAYS AND THEIR REGULATION.**

Metabolic pathways can be regulated efficiently by accelerating rate-limiting steps. In a four-step reaction sequence A→B→C→D→E, the slowest step determines the rate of product generation ("bottleneck"). By up-regulating the enzyme that catalyzes the rate-limiting step, the rate of product formation and the overall flux through the pathway are increased (Figure 9.2). Up-regulation of any of the enzymes that catalyze faster steps has no effect on product formation rate and flux. If more than one of the reactions in the pathway are slow, acceleration of the slowest step only leads to an up-regulation until the second-slowest step becomes rate-limiting for the entire pathway. Efficient regulation of pathways is therefore only achieved by regulating the activities of several enzymes. In branched pathways, up-regulation of individual steps in one branch affects the kinetic partitioning between the two branches (Figure 9.2). Thus, regulation of individual steps in metabolic networks will have complex effects on different pathways and the balance of the entire network.

*(Continued)*

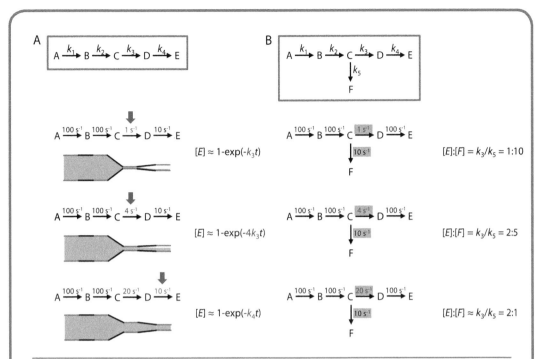

**FIGURE 9.2:** **Rate-limiting steps and flux in metabolic pathways.** A: The slowest reaction in a pathway (here: $k_3$) determines the rate of product formation and the flux (light blue) through the pathway. Acceleration of the rate-limiting step from $k_3 = 1$ s$^{-1}$ to $k_3 = 4$ s$^{-1}$ increases the rate of product formation and the overall flux. Further acceleration to $k_3 = 20$ s$^{-1}$ leads to a further increase in the rate of product formation and to a change in the rate-limiting step (blue arrow). The reaction D→E is now rate-limiting: $k_4$ limits the rate of product formation. Further up-regulation of the enzyme that catalyzes the conversion of C to D does not lead to an increase in the rate of product formation or the overall flux. B: Same reaction sequence with a second branch from C to F. Regulation of the step C→D in the pathway from A to E affects the kinetic partitioning of C into the two alternative pathways, and the ratio of their end products E and F. The relevant rate constants for partitioning are highlighted in orange.

The identification of rate-limiting steps in a reaction sequence is important for the dissection of the reaction mechanism, and to understand the behavior of the overall system to changes in concentrations. Rate-limiting steps in biochemical reactions can be identified by the *kinetic isotope effect*. For example, a C-H bond in a substrate can be changed into a C-D bond (D = deuterium, $^2$H). Due to the larger atomic mass of deuterium, the vibration frequency of the C-D bond is smaller than that of the C-H bond (see Section 19.4.1), and the zero-point energy of the C-D bond is lower than for the C-H bond. This decrease in energy will lead to an increase in activation energy for C-D bond breakage compared to breakage of a C-H bond. The higher activation energy leads to a smaller rate constant for bond breaking (Section 13.2): a C-D bond breaks more slowly than a C-H bond. The exchange of H to D will therefore slow down the overall reaction if breakage of this bond constitutes the rate-limiting step of the reaction (Figure 9.3), which is a *primary kinetic isotope effect*. If breakage of this bond is not the rate-limiting step, the overall reaction velocity will not be affected by the isotopic substitution.

Sometimes a kinetic isotope effect is measured even if the bond where the H-D substitution has been introduced is not the bond that breaks in the rate-limiting step. This effect is called *secondary kinetic isotope effect* (Box 9.2). A secondary isotope effect occurs when the substituted bond and the bond that breaks in the rate-limiting step are vibrationally coupled due to hyperconjugation.

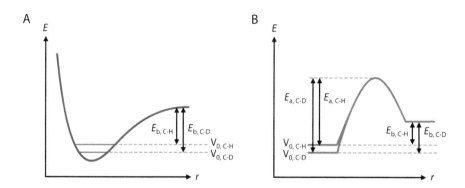

**FIGURE 9.3:   The kinetic isotope effect.** A: Lennard-Jones potential (see Section 15.3) for a C-H and a C-D bond. The vibrational energy of the C-D bond is lower than for the C-H bond because of the larger mass of deuterium compared to hydrogen. B: Because of the lower energy (higher stability) of the C-D bond, the energy difference to the transition state (activation energy $E_a$, see Section 13.2) is higher for the C-D bond than for the C-H bond, and the rate constant for bond breaking is smaller.

The H/D kinetic isotope effect is large because an exchange of H for D doubles the mass, and has a large effect on bond vibration. Kinetic isotope effects based on substitution of $^{12}C$ by $^{13}C$, for example, are much smaller due to the smaller relative change in mass.

---

### BOX 9.2: PRIMARY AND SECONDARY KINETIC ISOTOPE EFFECTS AND ENZYME MECHANISMS.

The degradation of the non-polar amino acids methionine, isoleucine, and valine and of fatty acids with odd numbers of carbon atoms leads to propionyl-CoA. Propionyl-CoA is carboxylated to methylmalonyl-CoA, which then isomerizes to succinyl-CoA. The isomerization reaction is catalyzed by methylmalonyl-CoA mutase (Figure 9.4). Methylmalonyl-CoA mutase is a cobalamin-dependent enzyme. The cofactor 5′-deoxyadenosyl cobalamin (coenzyme $B_{12}$) is transiently converted into a radical during catalysis, and the isomerization follows a radical mechanism, which is very unusual in biochemistry. When the hydrogen in the methyl group of methylmalonyl-CoA is substituted by deuterium, a strong kinetic isotope effect ($k_H/k_D = 3.5$) is observed, pointing to abstraction of a hydrogen radical from this methyl group as the rate-limiting step of the catalyzed reaction. Miller & Richards (1969) *J. Am. Chem. Soc.* 91(6): 1498–1507.

Fumarase catalyzes the dehydration of malate to fumarate, a reaction at the end of the citric acid cycle, when oxaloacetate is regenerated from succinate. Fumarase catalyzes the stereospecific *trans*-addition of water to the C-C double bond (Figure 9.4). The mechanism has been analyzed using malate with a tritium at C2 or C3 as a substrate, and monitoring the enzyme-catalyzed dehydration to fumarate. For both substrates, a secondary kinetic isotope effect ($k_H/k_T \approx 1.14$) was observed, which is consistent with catalysis through a carbenium ion intermediate. Secondary kinetic isotope effects are generally used to provide evidence for carbenium ion intermediates. Schmidt *et al.* (1969) *J. Am. Chem. Soc.* 91(21): 5849–5854.

*(Continued)*

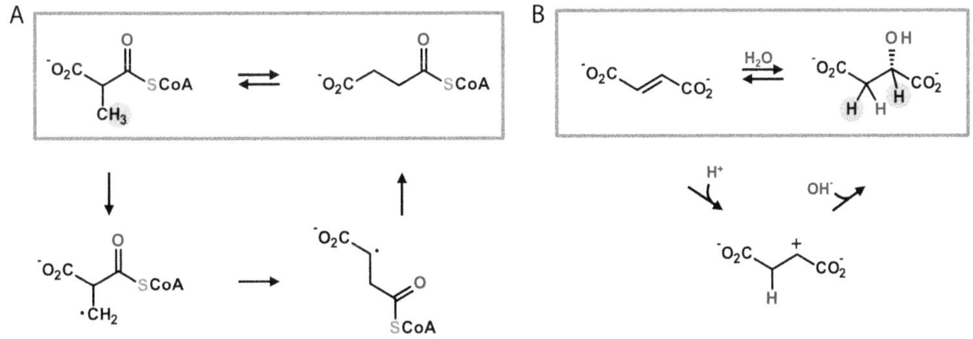

**FIGURE 9.4: Isotope effects and mechanisms of enzymatic reactions: methylmalonyl-CoA mutase and fumarase.** A: Methylmalonyl-CoA-mutase catalyzes the isomerization of methylmalonyl-CoA to succinyl-CoA through a radical mechanism. Substitution of a hydrogen in the methyl group (blue) by deuterium leads to a primary kinetic isotope effect, providing evidence for the rate-limiting abstraction of a hydrogen radical from the methyl group at the beginning of the reaction. B: Fumarase catalyzes the hydration of fumarate to malate through a carbenium ion intermediate. A secondary kinetic isotope effect is observed when the hydrogens at C2 or C3 (blue) are substituted for tritium.

## QUESTIONS

9.1 DNA strands (A, B) associate rapidly to mismatched duplexes (I; $k_{A,B \to I}$) that readily dissociate ($k_{I \to A,B}$) or slowly rearrange to the base-paired duplex (P) with $k_{I \to P}$. What is the reaction velocity for duplex formation as a function of the concentration of the DNA strands A and B and the apparent rate constant of duplex formation?

## REFERENCES

Miller, W. W. and Richards, J. H. (1969) Mechamism of action of coenzyme B12. Hydrogen transfer in the isomerization of methylmalonyl coenzyme A to succinyl coenzyme A. *J Am Chem Soc* **91** (6): 1498–1507.
· kinetic isotope effect to identify the rate-limiting step in the mechanism of methylmalonyl-CoA mutase

Schmidt, D. E., Jr., Nigh, W. G., Tanzer, C. and Richards, J. H. (1969) Secondary isotope effects in the dehydration of malic acid by fumarate hydratase. *J Am Chem Soc* **91** (21): 5849–5854.
· kinetic isotope effect to identify the rate-limiting step in the mechanism of fumarate hydratase

# Binding Reactions
## One-Step and Two-Step Binding

We will now come back to binding reactions. We have seen one example of a binding event in Scheme 7.3 before, describing the irreversible binding of B to A to form products, such as an AB complex:

$$A + B \overset{k_1}{\rightleftharpoons} AB$$

scheme 10.1

The rate law for this second-order binding reaction is

$$\frac{d[AB]}{dt} = -\frac{d[A]}{dt} = -\frac{d[B]}{dt} = k_1[A][B]$$

eq. 10.1

We now consider this binding reaction under conditions where the initial concentration of B is much higher than the concentration of A. Under such circumstances, only a small fraction of B will end up forming a complex with A, whereas all A will eventually be bound to B. The concentration of A thus goes from $A_0$ to zero during the reaction, but the concentration of free B will remain essentially constant. We can therefore approximate the rate law for formation of the AB complex by using the initial concentration of B, $B_0$, in eq. 10.1:

$$-\frac{d[A]}{dt} \approx k_1[A]B_0 = k_{obs}[A]$$

eq. 10.2

with

$$k_{obs} = k_1 B_0$$

eq. 10.3

The rate constant $k_{obs}$ is the observed rate constant for the exponential law that describes the formation of the AB complex over time. As in previous examples, we now separate the variables for integration.

$$\frac{\mathrm{d}[A]}{[A]} = -k_{obs}\mathrm{d}t \qquad\qquad \text{eq. 10.4}$$

and integrate eq. 10.4 over the concentration range $A_0$ at $t=0$ to $[A](t)$ at time $t$ to obtain

$$\ln\left(\frac{[A]}{A_0}\right) = -k_{obs}t \qquad\qquad \text{eq. 10.5}$$

or

$$[A](t) = A_0\,e^{-k_{obs}t} \qquad\qquad \text{eq. 10.6}$$

The rate constant $k_{obs}$ is the observed rate constant for the exponential law that describes the decrease in concentration of A over time. The concentration of the AB complex increases exponentially with $k_{obs}$:

$$[AB](t) = A_0 - [A](t) = A_0\left(1 - e^{-k_{obs}t}\right) \qquad\qquad \text{eq. 10.7}$$

Formation of AB is thus also described by a single exponential function, although complex formation is a second-order reaction. We call this reaction a *pseudo-first-order reaction*, and the conditions under which a reaction behaves as if it were a first-order reaction (here $B_0 \approx [B] \gg A_0$) *pseudo-first-order conditions*.

Following the course of a reaction with one partner provided in large excess is a convenient and widely applicable approach to simplify complex reactions and to deduce reaction mechanisms.

Let us now take a look at a binding equilibrium for AB complex formation according to Scheme 10.2.

$$A + B \;\underset{k_{-1}}{\overset{k_1}{\rightleftharpoons}}\; AB \qquad\qquad \text{scheme 10.2}$$

The rate laws for consumption of A and B are

$$\frac{\mathrm{d}[A]}{\mathrm{d}t} = -k_1[A][B] + k_{-1}[AB] \qquad\qquad \text{eq. 10.8}$$

and

$$\frac{\mathrm{d}[B]}{\mathrm{d}t} = -k_1[A][B] + k_{-1}[AB] \qquad\qquad \text{eq. 10.9}$$

Similarly, we can formulate the rate law in terms of formation of AB as

$$\frac{\mathrm{d}[AB]}{\mathrm{d}t} = k_1[A][B] - k_{-1}[AB] \qquad\qquad \text{eq. 10.10}$$

If we let the reaction proceed under pseudo-first-order conditions, the concentration of B will be constant over the reaction ($[B] \approx B_0$):

$$\frac{\mathrm{d}[B]}{\mathrm{d}t} \approx 0 \qquad\qquad \text{eq. 10.11}$$

and we can simplify the rate laws for A and AB to

$$\frac{d[A]}{dt} = -k_1[A]B_0 + k_{-1}[AB] \qquad \text{eq. 10.12}$$

and

$$\frac{d[AB]}{dt} = k_1[A]B_0 - k_{-1}[AB] \qquad \text{eq. 10.13}$$

According to mass conservation, we can express $[A]$ as $A_0-[AB]$:

$$\frac{d[AB]}{dt} = k_1\left(A_0 - [AB]\right)B_0 - k_{-1}[AB] = k_1 A_0 B_0 + \left(-k_1 B_0 - k_{-1}\right)[AB] \qquad \text{eq. 10.14}$$

and can separate the variables

$$\frac{d[AB]}{k_1 A_0 B_0 - \left(k_1 B_0 + k_{-1}\right)[AB]} = dt \qquad \text{eq. 10.15}$$

The integral on the left can be evaluated by substitution of $u$ for the denominator, with

$$\frac{du}{d[AB]} = -\left(k_1 B_0 + k_{-1}\right) \quad \text{or} \quad d[AB] = \frac{du}{-\left(k_1 B_0 + k_{-1}\right)} \qquad \text{eq. 10.16}$$

giving

$$\frac{1}{-(k_1 B_0 + k_{-1})} \cdot \int \frac{1}{u}\, du = \int_0^t dt \qquad \text{eq. 10.17}$$

and

$$\frac{1}{-(k_1 B_0 + k_{-1})} \cdot \ln u = t \qquad \text{eq. 10.18}$$

Substituting back $u$ and evaluating the integral from 0 to $[AB](t)$ then yields

$$\frac{\ln\left(k_1 A_0 B_0 - \left(k_1 B_0 + k_{-1}\right)[AB]\right) - \ln\left(k_1 A_0 B_0\right)}{-\left(k_1 B_0 + k_{-1}\right)} = t \qquad \text{eq. 10.19}$$

which can be solved for $[AB](t)$:

$$[AB](t) = A_0 \frac{k_1 B_0}{k_1 B_0 + k_{-1}}\left(1 - e^{-(k_1 B_0 + k_{-1})t}\right) \qquad \text{eq. 10.20}$$

The exponential term in eq. 10.20 has the form of a simple growth reaction with the observed rate constant

$$k_{\mathrm{obs}} = k_1 B_0 + k_{-1} \qquad \text{eq. 10.21}$$

Thus, the observed rate constant $k_{\mathrm{obs}}$ for formation of the AB complex will depend on the concentration of $B_0$ (in excess), and on the rate constants $k_1$ and $k_{-1}$ of the elementary reactions. The dependence of the observed rate constant on the concentration $B_0$ is linear, with a slope of $k_1$ and $y$-axis intercept of $k_{-1}$ (Figure 10.1). If we measure the observed rate constant for binding at different concentrations $B_0$, and then plot $k_{\mathrm{obs}}$ as a function of $B_0$, we can therefore calculate $k_1$ and $k_{-1}$, the rate constants for the individual binding and dissociation reactions, by linear regression.

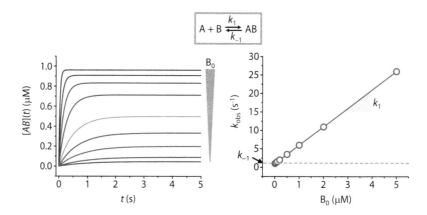

**FIGURE 10.1:** **One-step binding: rate constants**. Left: Kinetic transients for complex formation at different concentrations $B_0$ (0.01, 0.02, 0.05, 0.1, 0.2, 0.5, 1, 2, and 5 μM, from blue to red) under pseudo-first-order conditions ($B_0 \gg A_0$, $[B](t) \approx B_0$). Each transient can be described by a single exponential to extract the observed rate constant $k_{obs}$. Right: For a binding reaction in a single step, $k_{obs}$ depends linearly on the concentration of B. The rate constants $k_1$ and $k_{-1}$ can be determined from the slope and the intercept of the line with the y-axis.

The equilibrium constant $K_d$, the dissociation constant of the AB complex (Section 3.2), can be calculated as $k_{-1}/k_1$ (eq. 3.18). It is interesting to note that the final AB concentration of each kinetic transient, which is the concentration of AB in equilibrium at the respective concentrations of B, can also be used to determine the $K_d$ value of the AB complex: a plot of $[AB]_{eq}$ as a function of $B_0$ represents a binding curve, and the $K_d$ value can be determined using the solution of the quadratic equation that describes the complex concentration in equilibrium as a function of $A_0$ and $B_0$ (eq. 10.22).

$$[AB] = \frac{A_0 + B_0 + K_d}{2} - \sqrt{\left(\frac{A_0 + B_0 + K_d}{2}\right)^2 - A_0 B_0}$$

eq. 10.22

An agreement of this $K_d$ value with the value calculated from the rate constants $k_1$ and $k_{-1}$ serves as an internal control for the validity of the applied model (Figure 10.2).

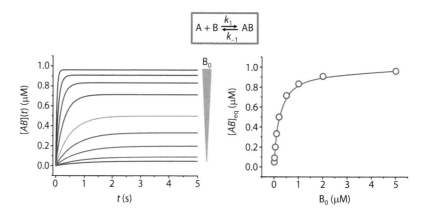

**FIGURE 10.2:** **One-step binding: amplitudes**. Left: Kinetic transients for complex formation at different concentrations $B_0$ (0.01, 0.02, 0.05, 0.1, 0.2, 0.5, 1, 2, and 5 μM, from blue to red) under pseudo-first-order conditions ($B_0 \gg A_0$, $[B](t) \approx B_0$). Right: The final AB concentration $[AB]_{eq}$ of each kinetic transient in panel A can be plotted as a function of the concentration $B_0$ to obtain a binding curve and to determine the $K_d$ value of the AB complex.

Eq. 10.21 only holds if binding is indeed a one-step reaction, and if our approximation ($B_0 \gg A_0$, $[B](t) \approx B_0$) is correct. Should the observed rate constant $k_{obs}$ not show a linear dependence on the concentration $B_0$, our initial model for binding, i.e. binding in one step, will most likely be incorrect, and we have to consider a different model to describe our experimental data.

We can obtain a corresponding expression for the dependence of $k_{obs}$ on the concentration $B_0$ for a two-step binding reaction, consisting of a rapid equilibrium with the equilibrium dissociation constant $K_{d1}$, and a second step in which AB is converted to a different complex AB* with the forward and reverse rate constants $k_2$ and $k_{-2}$ (Scheme 10.3).

$$A + B \underset{K_{d1}}{\rightleftharpoons} AB \underset{k_{-2}}{\overset{k_2}{\rightleftharpoons}} AB^* \qquad \text{scheme 10.3}$$

$K_{d1}$ for the rapid equilibrium is

$$K_{d1} = \frac{[A][B]}{[AB]} \qquad \text{eq. 10.23}$$

Combined with mass conservation for A and pseudo-first-order conditions ($[B](t) \approx B_0$, we can write

$$K_{d1} = \frac{(A_0 - [AB] - [AB^*])B_0}{[AB]} \qquad \text{eq. 10.24}$$

and solve for [AB]

$$[AB] = \frac{A_0 B_0 - [AB^*]B_0}{B_0 + K_{d1}} \qquad \text{eq. 10.25}$$

Now we formulate the rate law for product formation as

$$\frac{d[AB^*]}{dt} = k_2[AB] - k_{-2}[AB^*] \qquad \text{eq. 10.26}$$

and substitute [AB] by the expression in eq. 10.25:

$$\frac{d[AB^*]}{dt} = k_2 \frac{A_0 B_0 - [AB^*]B_0}{B_0 + K_{d1}} - k_{-2}[AB^*] \qquad \text{eq. 10.27}$$

Eq. 10.27 can be integrated by separation of variables and substitution to

$$[AB^*](t) = \frac{k_2 A_0 B_0}{B_0 k_2 + k_{-2}(B_0 + K_{d1})} - \frac{B_0 + K_{d1}}{B_0 k_2 + k_{-2}(B_0 + K_{d1})} e^{-\left(\frac{k_2 B_0}{B_0 + K_{d1}} + k_{-2}\right)t} \qquad \text{eq. 10.28}$$

with the observed rate constant $k_{obs}$ of AB* complex formation

$$k_{obs} = k_{-2} + k_2 \frac{B_0}{B_0 + K_{d1}} \qquad \text{eq. 10.29}$$

Thus, in this case the observed rate constant $k_{obs}$ exhibits a hyperbolic dependence on the concentration $B_0$ (Figure 10.3). We can determine the individual rate constants $k_2$ and $k_{-2}$, as well as the equilibrium constant for the first equilibrium, $K_{d1}$, by measuring AB* complex formation at different concentrations of B. If we keep $B_0 \gg A_0$, then each reaction will be a pseudo-first-order reaction with $[B](t) \approx B_0$, and can be described by a single exponential to calculate $k_{obs}$. The secondary plot of $k_{obs}$ as a function of $B_0$ can then be analyzed with eq. 10.29 to obtain $K_{d1}$, $k_2$ and $k_{-2}$ (Figure 10.3). We will see in Section 12.3 how we can directly arrive at eq. 10.29 by considering transit times.

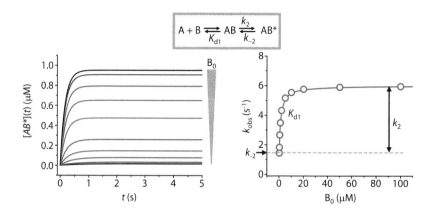

**FIGURE 10.3:** **Two-step binding: rate constants**. Left: Kinetic transients for complex formation at different concentrations $B_0$ (0.1, 0.2, 0.5, 1, 2, 5, 10, 20, 50, and 100 μM from black to blue to red to black) under pseudo-first-order conditions ($B_0 \gg A_0$, $[B](t) \approx B_0$). Each transient can be described by a single exponential to extract $k_{obs}$. Right: For a two-step binding reaction, $k_{obs}$ depends hyperbolically on the concentration of B. The rate constants $k_2$ and $k_{-2}$ can be determined from the amplitude of the hyperbola and its intercept with the $y$-axis. The dissociation constant $K_{d1}$ is determined by fitting a hyperbolic curve to the data using eq. 10.29.

Again, the final value for AB* corresponds to the equilibrium value, and a binding curve is obtained by plotting $[AB^*]_{eq}$ as a function of $B_0$. From this binding curve, an overall $K_d$, $K_{d,tot}$, can be determined using the solution of the quadratic equation (eq. 3.25, eq. 10.22). Agreement of this $K_{d,tot}$ with the overall $K_d$ calculated from $K_{d1}$, $k_2$ and $k_{-2}$ as

$$K_d = K_{d1} \frac{k_{-2}}{k_2}$$

eq. 10.30

serves as an internal control for the validity of the applied model (Figure 10.4). Note that measuring $[AB^*]_{eq}$ as a function of $B_0$ only provides the overall $K_d$ value when the second equilibrium lies on the right-hand side, i.e. $K_{d2} < 1$ ($k_2 > k_{-2}$).

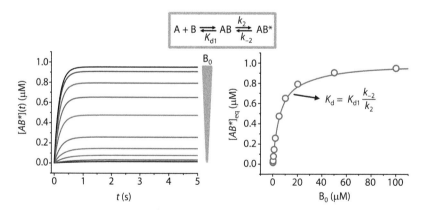

**FIGURE 10.4:** **Two-step binding: amplitudes**. Left: Kinetic transients for complex formation at different concentrations $B_0$ (0.1, 0.2, 0.5, 1, 2, 5, 10, 20, 50, and 100 μM from black to blue to red to black) under pseudo-first-order conditions ($B_0 \gg A_0$, $[B](t) \approx B_0$). Right: The final AB* concentration $[AB^*]_{eq}$ of each kinetic transient in the left panel can be plotted as a function of $B_0$ to obtain a binding curve. The $K_d$ value of the AB* complex is determined by fitting a curve to the data using eq. 3.25.

We have treated two-step binding under the assumption that we measure the formation of the AB* complex, the end product of the binding reaction. Often, binding reactions are monitored using a spectroscopic probe (see Chapter 25). If formation of the AB complex (Scheme 10.2) does not lead to a change in signal, i.e. AB formation is *spectroscopically silent*, and the change in signal is associated with the conversion of AB to AB*, binding under pseudo-first-order conditions follows a single exponential function, and we can describe $k_{obs}$ as a function of $B_0$ by eq. 10.29. However, the change in signal could instead occur with AB formation, and no further change is associated with the conversion to AB*. In this case, the kinetic transients will follow a double exponential function even under pseudo-first-order conditions. The first, more rapid phase reflects formation of AB from A and B, and the second, slower phase reflects the change in the first equilibrium because of the slow conversion of AB to AB*. If the observed rate constants $k_{obs,1}$ and $k_{obs,2}$ of the two phases are sufficiently different in magnitude, the concentration dependence of $k_{obs,1}$ and $k_{obs,2}$ can be analyzed separately to determine the *microscopic rate constants* $k_1$, $k_{-1}$, $k_2$ and $k_{-2}$. If the two phases have similar observed rate constants, their dependence on $k_1$, $k_{-1}$, $k_2$ and $k_{-2}$ is more complex, but it is possible to extract the individual rate constants $k_1$, $k_{-1}$, $k_2$ and $k_{-2}$ from the dependence of the observed rate constants on $B_0$. For a detailed treatment of these types of reactions and the different cases, the reader is referred to more specialized literature.

## QUESTIONS

10.1    The binding of a ligand L to its interaction partner shows a hyperbolic dependence on the ligand concentration under pseudo-first order conditions. $k_{obs}$ as a function of $[L]$ gives $K_{d1} = 100\ \mu M$ and $k_2 = 1\ s^{-1}$. The $y$-axis intercept is not well-defined and $k_{-2}$ cannot be determined accurately from the fit – why? In an equilibrium titration, the overall dissociation constant is determined to $K_{d,overall} = 10\ nM$. Calculate $k_{-2}$. How can $k_{-2}$ be measured directly and why is the experimental determination of $k_{-2}$ difficult?

10.2    You determine the observed rate constants for binding of an ATPase to a non-hydrolyzable ATP analog under pseudo-first-order conditions. The ATPase concentration is 0.1 μM. $k_{obs}$ shows a linear dependence on the concentration of the ATP analog in a concentration range of 1–10 μM. From the slope, you calculate $k_+ = 0.1\ \mu M^{-1}\ s^{-1}$, the intercept is $k_- = 0.001\ s^{-1}$. What are the possible meanings of these rate constants?

10.3    Binding of a substrate analog to an enzyme follows a two-step mechanism with $k_{-2} = 1 \cdot 10^{-5}\ s^{-1}$. You want to determine the overall $K_d$ value in a fluorescence equilibrium titration. How do you perform the experiment?

## REFERENCES

Bernasconi, C. F. (1976) Relaxation kinetics. New York, Academic Press.
· a detailed book about complex kinetics and the extraction of microscopic rate constants from observed rate constants

# Steady-State (Enzyme) Kinetics

I n the previous sections we have treated reaction velocities where we explicitly took into account changes in all reactants over time. We have seen that we can create pseudo-first-order conditions by providing one reactant in large excess, such that its concentration remains constant during the reaction (Chapter 10). We will now derive the quantitative concepts underlying *steady-state reactions*, i.e. reactions where one component is in large excess, such that its concentration does not change with time during the reaction, and the reaction continues without reaching an endpoint. This treatment is particularly useful for enzyme-catalyzed reactions, where the substrate is usually in large excess.

An enzymatic reaction starts with binding of the substrate S to the enzyme E, leading to the formation of an enzyme-substrate complex ES. The equilibrium dissociation constant of the ES complex is $K_S$. The ES complex is converted into an enzyme-product complex EP in a first-order reaction with the rate constant $k_p$. Finally, the EP complex dissociates into enzyme and products, characterized by the equilibrium dissociation constant $K_p$ (Scheme 11.1). The enzyme is regenerated at the end of the reaction, in agreement with its action as a catalyst of the reaction from S to P (see Chapter 14).

$$E + S \underset{K_S}{\rightleftharpoons} ES \xrightarrow{k_p} EP \underset{}{\overset{K_p}{\rightleftharpoons}} E + P$$

scheme 11.1

We now simplify this scheme by assuming that there is only one central complex, namely ES, that directly decomposes into enzyme and product(s) in an irreversible reaction with the rate constant $k_p$. Scheme 11.1 then becomes

$$E + S \underset{K_S}{\rightleftharpoons} ES \xrightarrow{k_p} E + P$$

scheme 11.2

We can also specify the rate constants for the forward and reverse reaction of the first equilibrium:

$$E + S \underset{k_{-1}}{\overset{k_1}{\rightleftharpoons}} ES \xrightarrow{k_p} E + P$$

scheme 11.3

Now we can derive rate laws for product formation under two different conditions. We will first treat the case where formation of the ES complex is rapid, and the rate of product formation is rate-limiting, i.e. $k_1, k_{-1} > k_p$ (the *rapid equilibrium approach*). After that, we derive the rate law for the case where product formation is not rate-limiting, but occurs on a comparable time scale to the equilibration of E and ES.

## 11.1  RAPID EQUILIBRIUM (MICHAELIS-MENTEN FORMALISM)

If the enzyme-substrate complex ES is formed in a rapid equilibrium, we can define the dissociation constant of the ES complex, $K_S$, as

$$K_S = \frac{[E][S]}{[ES]} \qquad \text{eq. 11.1}$$

and obtain

$$[ES] = \frac{[E][S]}{K_S} \qquad \text{eq. 11.2}$$

for the concentration of ES in equilibrium.

The velocity of product formation $v$ is determined by the rate-limiting step, and is

$$v = k_p[ES] \qquad \text{eq. 11.3}$$

The *maximum velocity* $v_{max}$ is reached if all enzyme molecules are bound to a substrate, or, put in other words, under saturation of the enzyme with substrate. In this case, we can substitute $[ES]$ by the total enzyme concentration $[E]_t$:

$$v_{max} = k_p[E]_t \qquad \text{eq. 11.4}$$

We can now divide eq. 11.3 by $[E]_t$. On the right-hand side, we write $[E]_t$ as the sum of the concentrations of the two possible enzyme forms, E (free form) and ES (substrate-bound form).

$$\frac{v}{[E]_t} = \frac{k_p[ES]}{[E]+[ES]} \qquad \text{eq. 11.5}$$

Now we can substitute $[ES]$ by the expression in eq. 11.2,

$$\frac{v}{[E]_t} = \frac{k_p\dfrac{[E][S]}{K_S}}{[E]+\dfrac{[E][S]}{K_S}} \qquad \text{eq. 11.6}$$

Division by $k_p$ and cancelling $[E]$ yields

$$\frac{v}{k_p[E]_t} = \frac{\dfrac{[S]}{K_S}}{1+\dfrac{[S]}{K_S}} \qquad \text{eq. 11.7}$$

The denominator on the left-hand side of the equation above denotes the maximum reaction velocity $v_{max}$ (eq. 11.4). By expanding with $K_S$ on the right, we obtain the *rate* or *velocity equation*

$$\frac{v}{v_{max}} = \frac{[S]}{K_S+[S]} \qquad \text{eq. 11.8}$$

or

$$v = v_{max} \cdot \frac{[S]}{K_S + [S]}$$

eq. 11.9

Eq. 11.9 describes a hyperbola (Figure 11.1). For $[S] = K_S$, the half-maximum velocity is reached. For $[S] \rightarrow \infty$, i.e. saturation of the enzyme with substrate, the velocity reaches $v_{max}$. The constant $K_S$ is often referred to as the *Michaelis-Menten constant* $K_M$. We will use $K_S$ to refer to the true equilibrium constant of ES dissociation, and $K_M$ for apparent equilibrium dissociation constants.

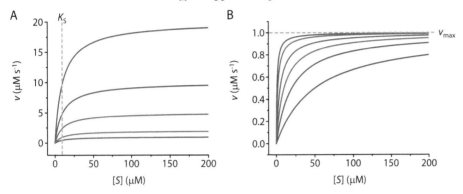

**FIGURE 11.1:** **Reaction velocity as a function of substrate concentration**. A: Hyperbolic dependence of the reaction velocity $v$ on the free substrate concentration $[S]$ for $v_{max} = 1\ \mu M\ s^{-1}$, $2\ \mu M\ s^{-1}$, $5\ \mu M\ s^{-1}$, $10\ \mu M\ s^{-1}$, and $20\ \mu M\ s^{-1}$ (blue to red) and $K_S = 10\ \mu M$. B: Hyperbolic dependence of the reaction velocity $v$ on the free substrate concentration $[S]$ for $v_{max} = 1\ \mu M\ s^{-1}$, and $K_S = 1\ \mu M$, 2, $\mu M$, 5 $\mu M$, 10 $\mu M$, and 20 $\mu M$ (blue to red). The Michaelis-Menten constant $K_S$ is the substrate concentration for which $v$ is $0.5 \cdot v_{max}$.

The hyperbola that describes the dependence of the reaction velocity on substrate concentration has three distinct regions (Figure 11.2): For $[S] \ll K_S$, the reaction velocity shows a linear increase with substrate concentration. In this regime, the reaction behaves as a first-order reaction. In this case, $[S]$ in the denominator of eq. 11.9 can be neglected, and we can approximate the reaction velocity as

$$v \cong v_{max} \frac{[S]}{K_S} = k[S]$$

eq. 11.10

$k$ is the first-order rate constant for the overall reaction. First-order kinetics is observed for $[S] < 0.01 \cdot K_S$. At very high concentrations of $[S]$, the reaction velocity becomes independent of the substrate concentration. In this regime, the reaction follows zero-order kinetics, and $v = v_{max}$. Zero-order kinetics is observed for $[S] > 100 \cdot K_S$. In-between these extreme cases, the reaction velocity exhibits the hyperbolic dependence on substrate concentration expressed by eq. 11.9.

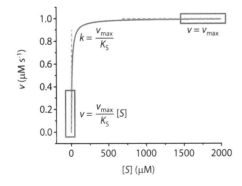

**FIGURE 11.2:** **Zero- and first-order kinetics for high and low substrate concentrations.** Reaction velocity $v$ as a function of substrate concentration $[S]$. For $[S] \ll K_S$, the reaction is first-order, with $k = v_{max}/K_S$. At high concentrations $[S] \gg K_S$, the reaction velocity becomes independent of substrate concentration (zero-order), with $v = v_{max}$. $K_S = 10\ \mu M$, $v_{max} = 1\ \mu M\ s^{-1}$.

## 11.2 STEADY-STATE APPROXIMATION (BRIGGS-HALDANE FORMALISM)

The Michaelis-Menten formalism assumes a true thermodynamic equilibrium between enzyme and substrate and the enzyme-substrate complex ES. This is only true if $k_p$ is much smaller than the rate constant for its dissociation back to enzyme and substrate ($k_{-1}$ in Scheme 11.3). If $k_p > k_{-1}$, then product formation is rapid once the enzyme-substrate complex has been formed. As we have seen before for consecutive reactions (Section 8.3), ES will not accumulate in this case, and never reaches the concentration that would correspond to the thermodynamic equilibrium.

$$\text{E} + \text{S} \underset{k_{-1}}{\overset{k_1}{\rightleftharpoons}} \text{ES} \overset{k_p}{\longrightarrow} \text{E} + \text{P} \qquad \text{scheme 11.3}$$

According to the principle of Le Chatelier (Section 3.1.2), the withdrawal of ES from the equilibrium by the product-forming reaction shifts the equilibrium to the right, further towards ES. If the substrate is in large excess over enzyme, then, after a short period of time, a steady state will be established at which the ES concentration is constant over time:

$$\frac{d[ES]}{dt} = 0 \qquad \text{eq. 11.11}$$

In this *steady-state*, ES is produced with the same velocity as it is converted to E and P. According to Scheme 11.3, the change in ES concentration over time is

$$\frac{d[ES]}{dt} = k_1[E][S] - k_{-1}[ES] - k_p[ES] \qquad \text{eq. 11.12}$$

By combining eq. 11.12 and eq. 11.11, we obtain

$$k_1[E][S] = (k_{-1} + k_p)[ES] \qquad \text{eq. 11.13}$$

which we can solve for [ES]:

$$[ES] = \frac{k_1}{k_{-1} + k_p}[E][S] \qquad \text{eq. 11.14}$$

By defining an apparent (!) equilibrium constant $K_M$ as

$$K_M = \frac{k_{-1} + k_p}{k_1} \qquad \text{eq. 11.15}$$

we can write eq. 11.14 in a similar form as eq. 11.2 for the rapid equilibrium approach:

$$[ES] = \frac{[E][S]}{K_M} \qquad \text{eq. 11.16}$$

The velocity of product formation is

$$v = k_p[ES] \qquad \text{eq. 11.3}$$

We now again divide by the total enzyme concentration $[E]_t$, and express $[E]_t$ as the sum of the concentrations of free enzyme E and enzyme-substrate complex ES (mass conservation). If we now express [ES] as a function of the constant $K_M$ (eq. 11.16), we obtain the equivalent expressions to eq. 11.6:

$$\frac{v}{[E]_t} = \frac{k_p \dfrac{[E][S]}{K_M}}{[E] + \dfrac{[E][S]}{K_M}} \qquad \text{eq. 11.17}$$

and

$$\frac{v}{k_p [E]_t} = \frac{\dfrac{[S]}{K_M}}{1 + \dfrac{[S]}{K_M}} \qquad \text{eq. 11.18}$$

which may be simplified to

$$\frac{v}{v_{max}} = \frac{[S]}{K_M + [S]} \qquad \text{eq. 11.19}$$

or

$$v = v_{max} \cdot \frac{[S]}{K_M + [S]} \qquad \text{eq. 11.20}$$

Thus, for the steady-state approach we obtain a velocity equation that describes the hyperbolic dependence of the reaction velocity on the substrate concentration (Figure 11.1) and is formally identical to the rate equation for the rapid equilibrium approach (eq. 11.9). $K_M$ is the substrate concentration at which the reaction occurs with half-maximal velocity. It is important to note that, despite the same shape of the velocity curves, the equilibrium constants $K_S$ and $K_M$ have different meanings. While $K_S$ is the true equilibrium dissociation constant of the ES complex, $K_M$ reflects a ratio of steady-state concentrations. If product formation is rapid ($k_p \gg k_{-1}$), $K_M$ is larger than $K_S$, and provides an upper limit for $K_S$. Eq. 11.9 and eq. 11.20 become equivalent for $k_p \ll k_{-1}$, i.e. when product formation becomes rate-limiting, and ES can reach its thermodynamic equilibrium instead of the steady-state equilibrium. In this case, $K_M$ equals $K_S$, the dissociation constant of the ES complex.

We can determine $v_{max}$, $k_p$ and $K_M$ for enzyme-catalyzed reactions by measuring the product concentration as a function of time for different concentrations of substrates, such that $S_0 \gg [E]_t$. The rate of product formation $v$ is then plotted as a function of the substrate concentration $[S] \approx S_0$ (Figure 11.1). The maximum reaction velocity $v_{max}$ is obtained as the saturating value of the resulting hyperbolic curve at high substrate concentrations. From the substrate concentration at which the reaction velocity $v$ is half-maximal, $K_M$ or $K_S$ are determined. The rate constant $k_p$, also called turnover number $k_{cat}$, can be calculated according to eq. 11.4 by dividing $v_{max}$ by the total enzyme concentration $[E]_t$.

Historically, eq. 11.20 has been converted into various reciprocal forms such that a linear relationship of some sort between $v$ and $S$ was obtained, and $v_{max}$ and $K_M$ were accessible by linear regression. Examples for these linearized forms are the *Lineweaver-Burk plot*, where the inverse of the reaction velocity, $1/v$, is plotted as a function of $1/[S]$ according to

$$\frac{1}{v} = \frac{1}{v_{max}} + \frac{K_M}{v_{max}[S]} \qquad \text{eq. 11.21}$$

and the *Eadie-Hofstee plot*, where $v$ is plotted as a function of $v/[S]$ according to

$$v = v_{max} - \frac{v \cdot K_M}{[S]}$$

eq. 11.22

(Box 11.1). The reciprocal forms and the linear plots provide a very simple and intuitive representation of different mechanisms of inhibition (Section 11.6). However, conversion of the original data ($v$ as $f([S])$ into reciprocal forms leads to a compression of data points, and alters the associated errors (Figure 11.4). Modern data analysis programs directly describe data points by complex equations without the need for further transformations. The parameters of the fitting equation, here $v_{max}$ and $K_M$, are varied during the fitting procedure, such that deviations between the equation and the experimental data points are minimized and the data are optimally described. Direct analysis of data avoids inappropriate weighting and error transformations associated with linearization.

---

### BOX 11.1: LINEARIZED FORMS OF THE MICHAELIS-MENTEN EQUATION: LINEWEAVER-BURK AND EADIE-HOFSTEE PLOTS.

In a Lineweaver-Burk plot, $1/v$ is plotted as a function of $1/[S]$. The slope of this plot is $K_M/v_{max}$, the intercept with the $y$-axis is $1/v_{max}$ (eq. 11.21). The $x$-axis is intercepted at $-1/K_M$. For a set of enzyme-catalyzed reactions with the same $v_{max}$, but different $K_M$ values, the $y$-axis intercept does not change, but the curves have different slopes and different $x$-axis intercepts. For a set of reactions with the same $K_M$, but different $v_{max}$ values, on the other hand, the $y$-axis intercept and the slope will differ, but the $x$-axis intercepts are the same (Figure 11.3; Section 11.6).

In an Eadie-Hofstee plot, $v$ is plotted as a function of $v/[S]$. The slope is $K_M$, the $y$-axis intercept is $v_{max}$, and the $x$-axis intercept is $v_{max}/K_M$ (eq. 11.22). For a set of enzyme-catalyzed reactions with the same $v_{max}$, but different $K_M$ values, the $y$-axis intercept does not change, but the curves have different slopes and $x$-axis intercepts. For a set of reactions with the same $K_M$, but different $v_{max}$ values, on the other hand, the slope will be the same, but the $x$- and $y$-axis intercepts differ (Figure 11.3; Section 11.6).

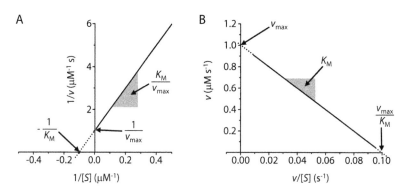

**FIGURE 11.3:** **Reciprocal plots for velocities of enzyme-catalyzed reactions.** Lineweaver-Burk (A) and Eadie-Hofstee plots (B) for enzymatic reactions with the same $K_M$ and $v_{max}$ values ($K_M = 10\ \mu M$, $v_{max} = 1\ \mu M\ s^{-1}$). A: The slope of a Lineweaver-Burk plot is $K_M/v_{max}$, the intercepts with the $y$- and $x$-axis are $1/v_{max}$ and $-1/K_M$. B: The slope of the Eadie-Hofstee plot is $K_M$, the intercepts with the $y$- and $x$-axis are $v_{max}$ and $v_{max}/K_M$.

The conversion of measured values into the reciprocal forms leads to a transformation of the associated errors (Figure 11.4). Whenever possible, conversion of data points should therefore be avoided!

*(Continued)*

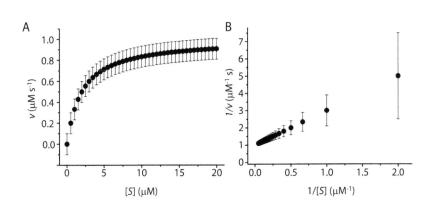

FIGURE 11.4: **Error compression in Lineweaver-Burk plots.** Direct data with associated errors for the reaction velocities (A) and corresponding Lineweaver-Burk plot (B). The error propagation from $v$ to $1/v$ is calculated as

$$\sigma_{\frac{1}{v}} = \sqrt{\sigma_v^2 \cdot \left( \frac{\partial \left( \frac{1}{v} \right)}{\partial v} \right)^2} = \sqrt{\sigma_v^2 \cdot \frac{1}{v^4}} \qquad \text{eq. 11.23}$$

(see Section 29.12). The evenly spaced original data points are compressed into the left bottom corner of the plot, and do not contribute much to the definition of the slope. In addition, the propagated errors show that the reciprocal values cannot be weighted identically.

## 11.3 pH DEPENDENCE

We will discuss the general principles of acid-base catalysis in Section 14.2. Here, we consider a reaction where a basic group on the enzyme needs to be protonated to catalyze a reaction. In this case, the reaction velocity will depend on the fraction of the enzyme that is protonated. The reaction will occur at maximum velocity at pH values where the group is fully protonated, i.e. at pH values at least one unit below the $pK_A$ value of the basic group (Section 3.2). At higher pH values, above the $pK_A$ value, the basic group on the enzyme will become deprotonated, which depletes the active species and reduces the reaction velocity (Figure 11.5).

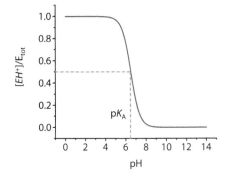

FIGURE 11.5: **Protonated and deprotonated states of an enzyme as a function of pH**. The fraction of protonated enzyme as a function of pH is depicted for a $pK_A$ value of 6.5. At pH values well below the $pK_A$, the enzyme is in the protonated active state, and will show maximal activity. At pH values well above the $pK_A$, the enzyme is in the deprotonated, inactive state and will show no activity.

We now derive the rate equation for an enzymatic reaction where $EH^+$ is the active form that binds to a negatively charged substrate $S^-$. Product formation occurs from the $EH^+S^-$ complex with the rate constant $k_p$ (Scheme 11.4):

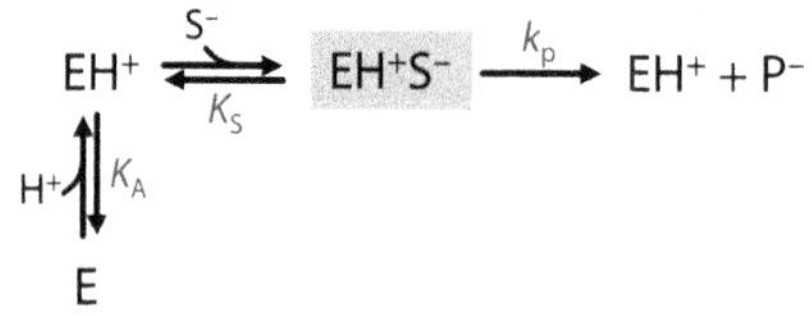

$$EH^+ \underset{K_S}{\overset{S^-}{\rightleftharpoons}} EH^+S^- \xrightarrow{k_p} EH^+ + P^-$$

$$H^+ \updownarrow K_A$$

$$E$$

scheme 11.4

We will follow a widely applicable procedure to derive the rate equation, the reaction velocity as a function of substrate concentration and apparent $v_{max}$ and $K_M$ values, for reactions as a function of pH. The standard sequence of steps to derive these equations is outlined in general terms in Box 11.2. We will apply this procedure to enzymes with multiple substrate binding sites (Sections 11.4, 11.5), and to enzyme-catalyzed reactions in the presence of inhibitors (Section 11.6).

The rate of product formation by the enzyme depends on the rate constant $k_p$ for product formation and the concentration of the active enzyme complex $[EH^+S^-]$ that catalyzes product formation. The reaction velocity $v$ is thus

$$v = k_p \left[ EH^+S^- \right]$$

eq. 11.24

We can now divide by the total enzyme concentration $[E]_t$ on both sides of the equation. On the left-hand side, we do not further specify $[E]_t$ but on the right-hand side we write $[E]_t$ as the sum of the concentrations of all possible complexes the enzyme can form (eq. 11.25)

$$\frac{v}{[E]_t} = \frac{k_p \left[ EH^+S^- \right]}{[E] + \left[ EH^+ \right] + \left[ EH^+S^- \right]}$$

eq. 11.25

We now divide both sides by $k_p$, and obtain

$$\frac{v}{k_p [E]_t} = \frac{v}{v_{max}} = \frac{\left[ EH^+S^- \right]}{[E] + \left[ EH^+ \right] + \left[ EH^+S^- \right]}$$

eq. 11.26

The product $k_p \cdot [E]_t$ corresponds to the maximum velocity of the reaction, $v_{max}$, under conditions when all enzyme is protonated and saturated with substrate. In the next step, we express all concentrations of enzyme complexes on the right-hand side in terms of the equilibrium dissociation constants $K_A$ and $K_S$. The protonated enzyme concentration $[EH^+]$ is

$$\left[ EH^+ \right] = \frac{[E][H^+]}{K_A}$$

eq. 11.27

and the concentration of the protonated and substrate-bound enzyme $[EH^+S^-]$ is

$$\left[ EH^+S^- \right] = \frac{\left[ EH^+ \right]\left[ S^- \right]}{K_S} = \frac{\left[ H^+ \right][E]\left[ S^- \right]}{K_S K_A}$$

eq. 11.28

Substitution of these expressions into eq. 11.26 gives

$$\frac{v}{v_{max}} = \frac{\dfrac{[E]\left[ H^+ \right]\left[ S^- \right]}{K_S K_A}}{[E] + \dfrac{[E]\left[ H^+ \right]}{K_A} + \dfrac{[E]\left[ H^+ \right]\left[ S^- \right]}{K_S K_A}}$$

eq. 11.29

By cancelling [E] on the right-hand side, this expression can be simplified to

$$\frac{v}{v_{max}} = \frac{\dfrac{\left[H^+\right]\left[S^-\right]}{K_S K_A}}{1 + \dfrac{\left[H^+\right]}{K_A} + \dfrac{\left[H^+\right]\left[S^-\right]}{K_S K_A}} \qquad \text{eq. 11.30}$$

The proton concentration [$H^+$] on the right-hand side of eq. 11.30 can be cancelled, and eq. 11.30 can be rearranged to

$$\frac{v}{v_{max}} = \frac{\left[S^-\right]}{\dfrac{K_S K_A}{\left[H^+\right]} + K_S + \left[S^-\right]} = \frac{\left[S^-\right]}{K_S\left(1 + \dfrac{K_A}{\left[H^+\right]}\right) + \left[S^-\right]} \qquad \text{eq. 11.31}$$

Eq. 11.31 can be expressed as

$$\frac{v}{v_{max}} = \frac{\left[S^-\right]}{K_{M,app} + \left[S^-\right]} \qquad \text{eq. 11.32}$$

with

$$K_{M,app} = K_S\left(1 + \frac{K_A}{\left[H^+\right]}\right) \qquad \text{eq. 11.33}$$

From eq. 11.32 and eq. 11.33, we see that $v_{max}$ is not pH-dependent, but $K_{M,app}$ now is an apparent Michaelis-Menten constant that depends on the $K_A$ value of the basic group and on the proton concentration, i.e. the pH. At high concentrations of protons (low pH), $K_{M,app}$ approaches $K_S$, because all enzyme is in the protonated state and can bind the substrate $S^-$. At low concentrations of protons, (high pH), $K_{M,app}$ becomes larger than $K_S$. The increase in $K_{M,app}$ reflects the decrease in substrate binding when only a small fraction of the enzyme is protonated.

---

**BOX 11.2: A STANDARD PROCEDURE TO DERIVE VELOCITY EQUATIONS AND APPARENT $K_M$ AND $v_{MAX}$ VALUES.**

For more complex reaction schemes that comprise more reactions than just substrate binding and product formation, expressions for $v/v_{max}$ as a function of substrate concentration can be obtained by a standard procedure following these steps:

1. Description of the reaction velocity $v$ as a function of the concentration of active complexes of the enzyme that catalyze product formation (eq. 11.24). If more than one complex catalyzes product formation, the concentrations of these complexes have to be added.
2. Division of both sides of the equation by the total enzyme concentration $[E]_t$. On the right-hand side, $[E]_t$ is expressed as the sum of the concentrations of all complexes the enzyme can form (eq. 11.25).
3. Division by $k_p$ gives $v/v_{max}$ on the left-hand side (eq. 11.26).
4. Expression of all concentrations of enzyme complexes in terms of [E], [S], and dissociation constants (eq. 11.27–eq. 11.29) and cancelling [E] from all terms (eq. 11.30).
5. Simplification and rearrangement to obtain an equation in the form of the Michaelis-Menten equation (eq. 11.31, eq. 11.32).

*(Continued)*

6. Relating $v_{\mathrm{max,app}}$ to $v_{\mathrm{max}}$: division of the equation by the factor before [S] in the numerator to obtain an expression for $v_{\mathrm{max,app}}$ (not applicable in this example because $v_{\mathrm{max,app}} = v_{\mathrm{max}}$). Relating $K_{\mathrm{M,app}}$ to $K_{\mathrm{S}}$: summary of all terms added to [S] in the denominator to obtain an expression for $K_{\mathrm{M,app}}$ (eq. 11.33).

This procedure can be used to derive expressions for the reaction velocity as a function of substrate concentration and for apparent $v_{\mathrm{max}}$ and $K_{\mathrm{M}}$ values for enzyme-catalyzed reactions as a function of pH, in the presence of inhibitors (Section 11.6) or activators, and for enzymes with multiple substrate binding sites (Sections 11.4, 11.5).

We can derive the corresponding form of eq. 11.33 for a scenario where the deprotonated enzyme, but not the protonated enzyme, binds substrate (Scheme 11.5).

$$\mathrm{E} \underset{K_{\mathrm{S}}}{\overset{\mathrm{S}^-}{\rightleftharpoons}} \mathrm{ES}^- \xrightarrow{k_{\mathrm{p}}} \mathrm{E} + \mathrm{P}^-$$

$$K_{\mathrm{A}} \updownarrow \mathrm{H}^+$$

$$\mathrm{EH}^+$$

scheme 11.5

In this case, we can derive $v_{\mathrm{max,app}} = v_{\mathrm{max}}$, and

$$K_{\mathrm{M,app}} = K_{\mathrm{S}} \left( 1 + \frac{\left[H^+\right]}{K_{\mathrm{A}}} \right)$$

eq. 11.34

Here, $K_{\mathrm{M,app}}$ becomes equal to $K_{\mathrm{S}}$ for high pH (low [$H^+$]), where the enzyme is mostly deprotonated and binds substrate. For low pH (high [$H^+$]), where most of the enzyme is protonated and does not bind substrate, $K_{\mathrm{M,app}}$ is larger than $K_{\mathrm{S}}$ (Figure 11.6). Formally, this behavior is identical to competitive inhibition, with protons acting as a competitive inhibitor (see Section 11.6.2).

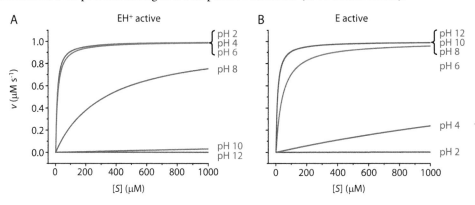

FIGURE 11.6:   **Influence of protolysis equilibria on $K_{\mathrm{M,app}}$.** Reaction velocity $v$ as a function of substrate concentration at different pH. A: The protonated state EH$^+$ is the active species (Scheme 11.4). At low pH, the enzyme is protonated and active. With increasing pH, the enzyme becomes deprotonated, and the $K_{\mathrm{M,app}}$ value increases (eq. 11.33). B: The deprotonated enzyme E is the active species (Scheme 11.5). At high pH, the enzyme is deprotonated and active. With decreasing pH, the enzyme becomes protonated, and $K_{\mathrm{M,app}}$ increases (eq. 11.34). p$K_{\mathrm{A}} = 6.5$ ($K_{\mathrm{A}} = 3.6\ 10^{-7}\ \mathrm{M}^{-1}$).

It is also possible that both enzyme and substrate need to be in the correct protonated/deprotonated state for the reaction to occur, e.g. if the substrate is a weak acid that is only bound in its deprotonated S$^-$ form, and the enzyme carries a basic group in its active site that needs to be protonated for the substrate to bind. The maximum activity would then be observed at pH values where all the enzyme exists in the protonated state as EH$^+$ and all the substrate exists in the deprotonated state as S$^-$. This condition may not be possible if the two p$K_{\mathrm{A}}$ values of the groups involved are too similar (Figure 11.7).

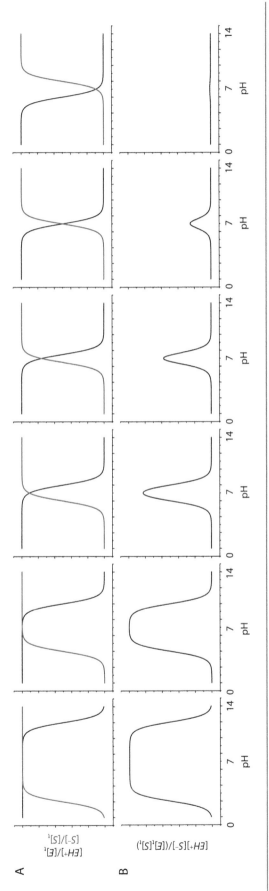

**FIGURE 11.7:** **Protonated and deprotonated states of enzyme and substrate and reaction velocity as a function of pH.** A: Fraction protonated enzyme ($EH^+$, black) and deprotonated substrate ($S^-$, red) as a function of the pH. $pK_A$ values: 12 (enzyme)/2 (substrate), 10/4, 8/6, 7.6/6.5, 7/7, and 6/8. B: The enzyme is active in its protonated state, the substrate has to be deprotonated to be bound. B: The product of the fractions of protonated enzyme and deprotonated enzyme as a function of pH (black) equals the probability of an encounter between $EH^+$ and $S^-$, and is thus a measure of the reaction velocity.

## 11.4 TWO OR MORE NON-INTERACTING ACTIVE SITES

So far, we have limited our considerations to enzymes with one active site. Often, enzymes assemble into dimeric, trimeric, tetrameric, or higher oligomeric states, combining more than one active site in one entity. We first look at a dimeric enzyme that contains two active sites (Scheme 11.6). For simplicity, we now refer to the dimeric enzyme as E, with E binding one or two substrate molecules S. Thus, the enzyme can form an ES, an SE, or an SES complex. We assume first that the active sites are independent, meaning the dissociation of substrates from each site is described by $K_S$, and they all catalyze conversion to products with $k_p$.

scheme 11.6

When we express the reaction velocity in terms of rate constants and concentrations of relevant species, we now have to take into account all enzyme complexes from which product is formed, and the number of active sites leading to product formation in the respective complex. For Scheme 11.6, the complexes ES, SE, and SES catalyze product formation. While ES and SE convert one substrate into the respective product, the SES complex turns over two substrate molecules to two molecules of the product, and generates product twice as rapid as the ES and SE complexes. The overall reaction velocity is thus described by eq. 11.35:

$$v = k_p[ES] + k_p[SE] + 2k_p[SES]$$

eq. 11.35

Again, we divide by the total enzyme concentration $[E]_t$, specify the individual complexes contributing to $[E]_t$ on the right-hand side of the equation only, and obtain

$$\frac{v}{[E]_t} = \frac{k_p[ES] + k_p[SE] + 2k_p[SES]}{[E] + [ES] + [SE] + [SES]}$$

eq. 11.36

The enzyme complexes are expressed as

$$[ES] = [SE] = \frac{[S][E]}{K_S}$$

eq. 11.37

and

$$[SES] = \frac{[S][ES]}{K_S} = \frac{[S]^2}{K_S^2}[E]$$

eq. 11.38

according to the definition of the equilibrium dissociation constant $K_S$. Substitution into eq. 11.36, division of all terms on the right-hand side by the enzyme concentration $[E]$, and rearrangement yields

$$\frac{v}{[E]_t} = \frac{k_p\dfrac{[S]}{K_S} + k_p\dfrac{[S]}{K_S} + 2k_p\dfrac{[S]^2}{K_S^2}}{1 + \dfrac{[S]}{K_S} + \dfrac{[S]}{K_S} + \dfrac{[S]^2}{K_S^2}} = \frac{2 \cdot k_p\dfrac{[S]}{K_S} + 2k_p\dfrac{[S]^2}{K_S^2}}{1 + 2\dfrac{[S]}{K_S} + \dfrac{[S]^2}{K_S^2}}$$

eq. 11.39

and

$$\frac{v}{2k_p[E]_t} = \frac{v}{v_{max}} = \frac{\dfrac{[S]}{K_S} + \dfrac{[S]^2}{K_S^2}}{1 + 2\dfrac{[S]}{K_S} + \dfrac{[S]^2}{K_S^2}}$$
eq. 11.40

with

$$2k_p[E]_t = v_{max}$$
eq. 11.41

The maximum velocity $v_{max}$ is now twice as large as for the same enzyme with one active site, because both active sites in the dimer contribute to product formation.

If we look back at where the individual terms in eq. 11.39 come from in the derivation, we see that the two $[S]/K_S$ terms in the numerator reflect the two species that catalyze product formation, ES and SES. The exponent of each $[S]/K_S$ term corresponds to the number of binding sites occupied (1 for $ES_1$, 2 for $ES_2$). The coefficients of each term contain two components: the rate constant of product formation by the respective complex, and the number of possibilities to generate this species. For two active sites, there are two ways to generate an ES complex (ES and SE, in the following $ES_1$), but only one possibility to generate an $ES_2$ complex. The rate constant of product formation for $ES_1$ is $k_p$. The rate constant for $ES_2$ is $2k_p$ because both active sites generate product. Finally, the three terms in the denominator represent all forms of the enzyme, free enzyme (E), enzyme with one substrate molecule bound ($ES_1$) and enzyme with two substrate molecules bound ($ES_2$). Each enzyme-substrate complex is reflected by an $[S]/K_S$ term with the exponent corresponding to the number of binding sites occupied. Note that $x^0$ equals unity, thus the free enzyme is reflected by $([S]/K_S)^0 = 1$, it is an enzyme-substrate complex with no substrate bound, so to say. The coefficients in front of the $[S]/K_S$ terms correspond to the number of possibilities in which the species can be generated, which is two for $ES_1$, and one for $ES_2$ (Figure 11.8). These coefficients are the *binomial coefficients* that are found in *Pascal's triangle* (see Section 29.3).

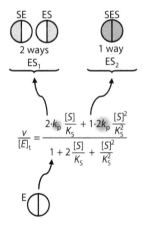

FIGURE 11.8:  **Enzyme-substrate complexes for an enzyme with two substrate binding sites**. In the numerator, each $[S]/K_S$ term represents a substrate-bound enzyme species. The number of occupied substrate binding sites (highlighted in blue) is reflected in the exponent of the $[S]/K_S$ term. Each term is multiplied by the number of different ways to generate the respective complex (2 ways for $ES_1$, one way for $ES_2$), and by the rate constant of product formation for each complex ($k_p$ for $ES_1$, $2k_p$ for $ES_2$, orange). The denominator contains one $[S]/K_S$ term for each species of enzyme present (E, $ES_1$, $ES_2$). Again, the exponent specifies the number of binding sites occupied, giving 1 for the free enzyme ($x^0 = 1$). Each term is multiplied by the number of ways to generate the respective complex (1, 2, and 1 for E, $ES_1$, and $ES_2$). Division by $2k_p$, the rate constant for the $ES_2$ complex that is fully occupied with substrate, then gives $v_{max}$ in the denominator on the left-hand side of the equation.

Recognizing this principle, we can now directly construct the equation for an enzyme with four non-interacting catalytic sites. On the left-hand side we write $v/[E]_t$. On the right-hand side, we have to consider all species that catalyze product formation in the numerator. For an enzyme with four active sites, we have enzyme with one, two, three, or four substrates bound (ES, $ES_2$, $ES_3$, $ES_4$, respectively), represented by $([S]/K_S)$, $([S]/K_S)^2$, $([S]/K_S)^3$, and $([S]/K_S)^4$. The factors for each term depend on the number of possibilities to generate the respective complex. There is only one way to populate an ES or $ES_4$ complex, but there are six different possibilities for $ES_2$, and four for $ES_3$. With these considerations, we have completed the numerator. For the denominator, we have to take into account free enzyme E, ES, $ES_2$, $ES_3$, and $ES_4$ represented by $([S]/K_S)^n$ with $n$ = 0, 1, 2, 3, and 4. The coefficients are the numbers of ways to generate the respective species, 1 for E, 4 for $ES_1$, 6 for $ES_2$, 4 for $ES_3$, and 1 for $ES_4$ (Figure 11.9): the binomial coefficients that are found in Pascal's triangle (see Section 29.3).

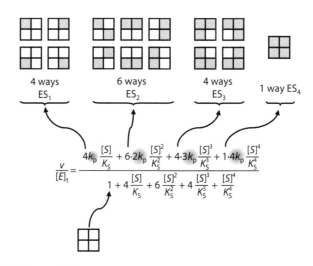

**FIGURE 11.9** **Enzyme-substrate complexes for an enzyme with four substrate binding sites.** In the numerator, each $[S]/K_S$ term represents a substrate-bound enzyme species ($ES_1$, $ES_2$, $ES_3$, $ES_4$, occupied substrate binding sites highlighted in blue). The exponent of the $[S]/K_S$ term is the respective number of occupied substrate binding sites. Each term is multiplied by the number of different ways to generate the respective complex (4 ways for $ES_1$, 6 ways for $ES_2$, 4 ways for $ES_3$, 1 way for $ES_4$), and by the rate constant of product formation for each complex ($k_p$, $2k_p$, $3k_p$, and $4k_p$ (orange) for $ES_1$, $ES_2$, $ES_3$, $ES_4$). The denominator contains one $[S]/K_S$ term for each species of enzyme present (E, $ES_1$, $ES_2$, $ES_3$, $ES_4$), with the exponent equal to the number of binding sites occupied. Each term is multiplied by the number of ways to generate the respective complex (1, 4, 6, 4, 1 for E, $ES_1$, $ES_2$, $ES_3$, $ES_4$). Division by $4k_p$, the rate constant for product formation from the $ES_4$ complex that is fully occupied with substrate, then gives $v_{max}$ in the denominator on the left-hand side of the equation.

Altogether, we obtain eq. 11.42,

$$\frac{v}{[E]_t} = \frac{4k_p\dfrac{[S]}{K_S} + 6\cdot 2k_p\dfrac{[S]^2}{K_S^2} + 4\cdot 3k_p\dfrac{[S]^3}{K_S^3} + 1\cdot 4k_p\dfrac{[S]^4}{K_S^4}}{1 + 4\dfrac{[S]}{K_S} + 6\dfrac{[S]^2}{K_S^2} + 4\dfrac{[S]^3}{K_S^3} + \dfrac{[S]^4}{K_S^4}}$$

eq. 11.42

Dividing eq. 11.42 by $4k_p$ leaves us with

$$v_{max} = 4k_p[E]_t$$

eq. 11.43

in the denominator on the left, and smaller coefficients in the numerator on the right:

$$\frac{v}{v_{max}} = \frac{\dfrac{[S]}{K_S} + 3\dfrac{[S]^2}{K_S^2} + 3\dfrac{[S]^3}{K_S^3} + \dfrac{[S]^4}{K_S^4}}{1 + 4\dfrac{[S]}{K_S} + 6\dfrac{[S]^2}{K_S^2} + 4\dfrac{[S]^3}{K_S^3} + \dfrac{[S]^4}{K_S^4}}$$

eq. 11.44

We can write the sums in the numerator and denominator of eq. 11.44 as binomials (eq. 11.45):

$$\frac{v}{v_{max}} = \frac{\dfrac{[S]}{K_S}\left(1+\dfrac{[S]}{K_S}\right)^3}{\left(1+\dfrac{[S]}{K_S}\right)^4} \qquad \text{eq. 11.45}$$

for four active sites. Eq. 11.45 can be generalized to *n* active sites:

$$\frac{v}{v_{max}} = \frac{\dfrac{[S]}{K_S}\left(1+\dfrac{[S]}{K_S}\right)^{n-1}}{\left(1+\dfrac{[S]}{K_S}\right)^n} \qquad \text{eq. 11.46}$$

By eliminating $(1+[S]/K_S)^{n-1}$, eq. 11.46 reduces to eq. 11.47:

$$\frac{v}{v_{max}} = \frac{\dfrac{[S]}{K_S}}{1+\dfrac{[S]}{K_S}} = \frac{[S]}{K_S+[S]} \qquad \text{eq. 11.47}$$

which is identical to eq. 11.19 for an enzyme with a single active site. Thus, an enzyme with *n* independent active sites shows the same hyperbolic dependence of the reaction velocity on the substrate concentration as an enzyme with only one active site. We therefore cannot derive the number of active sites on the enzyme from a measurement of *v* as a function of [S]. To determine the substrate binding stoichiometry, we have to resort to equilibrium titrations (active site titration; see Section 19.5.5.1).

## 11.5  TWO OR MORE INTERACTING ACTIVE SITES: COOPERATIVITY AND THE HILL EQUATION

The rate equations for enzymes with one or more active sites become different when the individual active sites of the enzyme interact with each other. The interaction between two active sites can be positive or negative: the binding of substrate to one active site can promote or inhibit binding of a second substrate molecule to the second active site. If the equilibrium dissociation constant for the $ES_1$ complex is $K_S$, we can describe the second equilibrium constant for dissociation of one substrate from $ES_2$ ($ES_2 \rightarrow ES_1 + S$) as $\alpha \cdot K_S$ (Scheme 11.7). The parameter $\alpha$ takes into account the interaction between the two active sites, and is a measure of the *cooperativity* of the enzyme. A positive interaction or *positive cooperativity* is expressed by $\alpha < 1$ (corresponding to a smaller dissociation constant $K_S$, or larger binding constant, for substrate 2; Sections 3.2, 3.4): binding of one substrate molecule facilitates binding of the second substrate by $\alpha$-fold. A negative interaction is expressed by $\alpha > 1$ (corresponding to a larger dissociation constant $K_S$, or smaller binding constant, for substrate 2; *negative cooperativity*). In this case, binding of the first substrate leads to an $\alpha$-fold reduction in affinity for the second substrate molecule.

scheme 11.7

In the equation for the reaction velocity, we have to take into account the factor $\alpha$ for the second binding event, i.e. in all terms that represent the SES complex, which leads to eq. 11.48:

$$\frac{v}{[E]_t} = \frac{2k_p\dfrac{[S]}{K_S} + 2k_p\dfrac{[S]^2}{\alpha K_S^2}}{1 + 2\dfrac{[S]}{K_S} + \dfrac{[S]^2}{\alpha K_S^2}}$$

eq. 11.48

or

$$\frac{v}{v_{max}} = \frac{\dfrac{[S]}{K_S} + \dfrac{[S]^2}{\alpha K_S^2}}{1 + 2\dfrac{[S]}{K_S} + \dfrac{[S]^2}{\alpha K_S^2}}$$

eq. 11.49

In eq. 11.49, the sums in the numerator and denominator cannot be expressed as binomial expressions, and the equation therefore does not reduce to a simple hyperbolic form! Thus, the dependence of the reaction velocity $v$ on the enzyme concentration $[S]$ will be different if there is cooperativity between the active sites (Figure 11.10). According to eq. 11.49, the reaction velocity $v$ shows a sigmoidal dependence on the substrate concentration. While the end value at substrate saturation is still $v_{max}$, $K_S$ now does not represent the concentration at which the half-maximal velocity is reached.

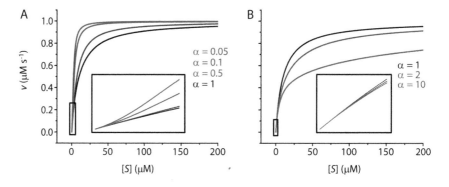

**FIGURE 11.10:** **Cooperativity and sigmoidal dependence of the reaction velocity on substrate concentration**. Reaction velocity $v$ as a function of substrate concentration $[S]$ for an enzyme with two substrate binding sites and cooperativity. A: Positive cooperativity. The cooperativity parameter $\alpha$ is 1 (black), 0.5 (brown), 0.1 (red), and 0.05 (pink). The inset shows the region for small $[S]$ where the differences are most pronounced: with increasing cooperativity (decreasing cooperativity parameter $\alpha$), the curves become more and more sigmoidal, and reach $v_{max}$ at lower substrate concentration. B: Negative cooperativity. The cooperativity parameter $\alpha$ is 1 (black), 2 (dark blue), and 10 (blue). The inset shows the region for small $[S]$: all curves show a hyperbolic dependence on substrate concentration.

Hemoglobin contains four binding sites for its substrate oxygen and is an example of cooperative binding. To describe the binding of four ligand molecules to four cooperative sites, we have to introduce a factor to take into account the interaction of the sites for each binding step, i.e. binding of the first substrate molecule favors/disfavors binding of the second by $\alpha$-fold, binding of the second in turn favors/disfavors binding of the third ligand by $\beta$-fold, and binding of the third ligand favors/disfavors binding of the fourth substrate molecule by $\gamma$-fold (Scheme 11.8). This type of model is called the *sequential model* of cooperativity

scheme 11.8

In this case, eq. 11.49 becomes

$$\frac{v}{v_{\max}} = \frac{\dfrac{[S]}{K_S} + 3\dfrac{[S]^2}{\alpha K_S^2} + 3\dfrac{[S]^3}{\alpha^2 \beta K_S^3} + \dfrac{[S]^4}{\alpha^3 \beta^2 \gamma K_S^4}}{1 + 4\dfrac{[S]}{K_S} + 6\dfrac{[S]^2}{\alpha K_S^2} + 4\dfrac{[S]^3}{\alpha^2 \beta K_S^3} + \dfrac{[S]^4}{\alpha\beta^2\gamma K_S^4}}$$

eq. 11.50

and the dependence of $v$ on $[S]$ will again be sigmoidal (Figure 11.11).

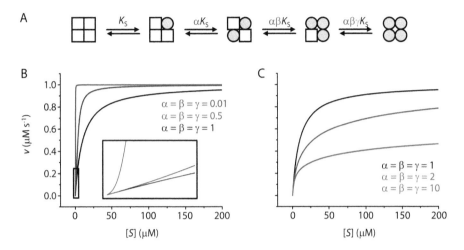

**FIGURE 11.11:    Sigmoidal dependence of reaction velocities on substrate concentration for four cooperative sites**. A: Binding scheme. B: Reaction velocity $v$ as a function of substrate concentration $[S]$ for an enzyme with four substrate binding sites and cooperativity parameters $\alpha$, $\beta$, and $\gamma$ of 1 (no cooperativity, black), 0.5 (brown), and 0.01 (red). The inset shows the region for small $[S]$ where the differences are most pronounced: with increasing cooperativity (decreasing cooperativity parameters), the curves become more and more sigmoidal, and reach $v_{\max}$ at lower substrate concentration. C: Reaction velocity $v$ as a function of substrate concentration $[S]$ for an enzyme with four substrate binding sites and negative cooperativity: $\alpha$, $\beta$, and $\gamma = 1$ (no cooperativity, black), 2 (dark blue), and 10 (blue). All curves are hyperbolic.

In contrast to the sequential model, cooperativity can also be rationalized in terms of a *concerted model*. In the concerted model, binding of the first ligand converts all subunits of the enzyme into a high-affinity state, leading to the same sigmoidal dependence of the reaction velocity on substrate concentration as we derived for the sequential model. The concerted model explains positive cooperativity, but cannot describe negative cooperativity.

Let us now consider a case where the cooperativity is so high that binding of one substrate molecule will immediately lead to saturation of the remaining three binding sites with substrate. In this case, the enzyme only exists in the free form or in the form saturated with four substrate molecules, which will simplify things tremendously. The intermediate states with one, two, or three substrate molecules bound will not be significantly populated. Thus, we do not have to consider them in eq. 11.50, and can simplify this expression to

$$\frac{v}{v_{\max}} = \frac{\dfrac{[S]^4}{\alpha^3\beta^2\gamma K_S^4}}{1 + \dfrac{[S]^4}{\alpha^3\beta^2\gamma K_S^4}} = \frac{\dfrac{[S]^4}{K'}}{1 + \dfrac{[S]^4}{K'}} = \frac{[S]^4}{K' + [S]^4}$$

eq. 11.51

with

$$v_{\max} = 4k_p[E]_t$$

eq. 11.43

and

$$K' = \alpha^3 \beta^2 \gamma K_S^4 \qquad\qquad \text{eq. 11.52}$$

Eq. 11.51 is called the *Hill equation*. The Hill equation can be generalized to *n* interacting sites as

$$\frac{v}{v_{\max}} = \frac{[S]^n}{K' + [S]^n} \qquad\qquad \text{eq. 11.53}$$

with

$$K' = K_S^n \left( \alpha^{n-1} \beta^{n-2} \gamma^{n-3} \ldots \right) \qquad\qquad \text{eq. 11.54}$$

The parameter *n* is called the *Hill coefficient*. For the limiting case of high cooperativity, *n* equals the number of binding sites. If we have intermediate or low cooperativity, we can still use the Hill equation to describe the dependence of the reaction velocity on the substrate concentration (Figure 11.12), and can extract a value for the Hill coefficient *n*. However, in these cases *n* does not correspond to the number of binding sites but merely provides a lower limit for the number of binding sites present (Box 11.3). The Hill equation does not describe negative cooperativity.

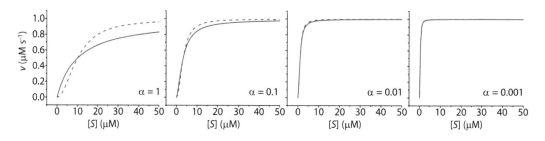

**FIGURE 11.12:** **Approximation by the Hill equation for different levels of cooperativity.** The reaction velocity of an enzyme with two substrate binding sites as a function of the substrate concentration [S] is calculated explicitly (eq. 11.50, blue), or according to the Hill approximation (eq. 11.53 with *n* = 2, blue dashed line). The cooperativity parameter α is 1 (A), 0.1 (B), 0.01 (C) and 0.001 (D). The Hill equation is a good approximation for high cooperativity (α = 0.01, α = 0.001).

---

**BOX 11.3: THE HILL COEFFICIENT AND THE NUMBER OF BINDING SITES.**

For enzymes with multiple binding sites and low to intermediate cooperativity, the dependence of the reaction velocity on the substrate concentration can still be described satisfactorily by the Hill equation. The Hill coefficient retrieved by curve-fitting to experimental data will not be an integer value in these cases, however, and is often called the *apparent Hill coefficient* $n_{app}$. For an enzyme with two low to intermediately cooperative sites (α = 0.5, 0.2, 0.1, 0.01), the extracted Hill coefficients $n_{app}$ are 1.17, 1.37, 1.49, 1.78 (Figure 11.13). For high cooperativity (α = 0.001), a Hill coefficient of 1.97, i.e. close to the number of binding sites, is extracted, whereas in the absence of cooperativity (α = 1), the Hill coefficient reduces to $n_{app} = 1$.

*(Continued)*

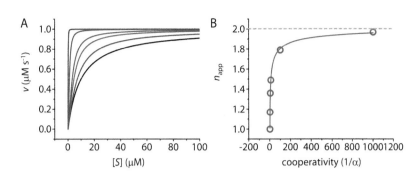

**FIGURE 11.13: The apparent Hill coefficient.** A: Reaction velocity as a function of substrate concentration for an enzyme with two cooperative sites with cooperativity parameter $\alpha = 1, 0.5, 0.2, 0.1, 0.01, 0.001$ (from black over blue and red to brown). Non-linear least-square fitting with the Hill equation (eq. 11.53) gives $n_{app} = 1, 1.17, 1.37, 1.49, 1.78,$ and $1.97$. B: Apparent Hill coefficient $n_{app}$ as a function of the cooperativity ($1/\alpha$). For high cooperativity, $n_{app}$ approaches the number of binding sites.

Even for low to intermediate cooperativity, we can still obtain some information from the apparent Hill coefficient $n_{app}$: the next larger integer $n > n_{app}$ is the minimum number of binding sites. A Hill coefficient of 1.5 can indicate that we are looking at an enzyme with two binding sites that show appreciable cooperativity, at an enzyme with four binding sites and moderate to low cooperativity, or at an enzyme with numerous binding sites and low cooperativity.

Hemoglobin is a tetrameric protein that contains four binding sites for oxygen. It functions in oxygen transport in the blood stream. The four oxygen binding sites of the hemoglobin tetramer show intermediate cooperativity. From the oxygen binding curve, an apparent Hill coefficient of $n_{app} = 2.8$ can be determined.

# 11.6 INHIBITION OF ENZYME ACTIVITY

## 11.6.1 PRODUCT INHIBITION IN REVERSIBLE REACTIONS

Many enzymes are inhibited by the products of the reaction they catalyze, leading to a negative feedback on enzyme activity. Product inhibition is a common mechanism for the regulation of enzyme activity. We assume that enzyme and substrate on one side, and enzyme and product on the other side are in rapid equilibrium, characterized by the dissociation constants of ES and EP, $K_S$ and $K_P$. Forward (product formation) and reverse reactions occur with rate constants $k_f$ and $k_r$ (Scheme 11.9).

$$E + S \underset{K_S}{\rightleftharpoons} ES \underset{k_r}{\overset{k_f}{\rightleftharpoons}} EP \overset{K_p}{\rightleftharpoons} E + P \qquad \text{scheme 11.9}$$

When we consider initial velocities only, i.e. we focus on the initial part of a reaction starting from substrate S, where no product has been formed yet ($[P] = 0$), the forward reaction occurs at the velocity $v_f$

$$v_f = v_{max,f} \frac{[S]}{K_S + [S]} \qquad \text{eq. 11.55}$$

If we start the reaction from the right-hand side by supplying the enzyme with product P instead of substrate S, the initial reaction velocity $v_r$ for the reverse reaction is

$$v_r = v_{max,r} \frac{[P]}{K_P + [P]}$$ 

eq. 11.56

Enzyme and substrate on one side, and enzyme and product on the other side of the reaction are in rapid equilibrium, and $k_f$ and $k_r$ are rate-limiting for the overall forward or reverse reaction. We can express the net reaction velocity $v_{net}$ from substrate S to product P as the difference between the forward (from ES to EP) and the reverse reaction velocity (from EP to ES) (eq. 11.57):

$$v_{net} = k_f[ES] - k_r[EP]$$ 

eq. 11.57

From here on, we follow the standard procedure to obtain a velocity equation (Box 11.2). Division by the total enzyme concentration yields

$$\frac{v_{net}}{[E]_t} = \frac{k_f[ES] - k_r[EP]}{[E] + [ES] + [EP]}$$ 

eq. 11.58

Expressing $[ES]$ and $[EP]$ in terms of the equilibrium dissociation constants $K_S$ and $K_P$ gives

$$\frac{v_{net}}{[E]_t} = \frac{k_f \dfrac{[S]}{K_S}[E] - k_r \dfrac{[P]}{K_P}[E]}{[E] + \dfrac{[S]}{K_S}[E] + \dfrac{[P]}{K_P}[E]}$$ 

eq. 11.59

Elimination of $[E]$ on the right-hand side and multiplication with $[E]_t$ then gives $v_{net}$ as

$$v_{net} = \frac{k_f[E]_t \dfrac{[S]}{K_S} - k_r[E]_t \dfrac{[P]}{K_P}}{1 + \dfrac{[S]}{K_S} + \dfrac{[P]}{K_P}}$$ 

eq. 11.60

The terms $k_f[E]_t$ and $k_r[E]_t$ denote the maximum velocities for forward and reverse reactions, respectively, and eq. 11.60 can be written as

$$v_{net} = \frac{v_{max,f} \dfrac{[S]}{K_S} - v_{max,r} \dfrac{[P]}{K_P}}{1 + \dfrac{[S]}{K_S} + \dfrac{[P]}{K_P}}$$ 

eq. 11.61

From eq. 11.61, we see that with increasing concentrations of product, the second term in the numerator increases, leading to a decrease of the numerator. In addition, the last term in the denominator, and thus the total denominator, increases. Both effects contribute to a decrease of the net reaction velocity $v_{net}$ with increasing product concentration. This mechanism of enzyme inhibition is called *product inhibition*. Product inhibition depends on $K_P$, and is stronger for small values of $K_P$, i.e. when the enzyme binds tightly to the product. Strong product binding is often observed in enzymes that catalyze reactions with chemically similar substrates and products. ATP-hydrolyzing enzymes frequently show product inhibition by ADP. Product inhibition is also very common in self-cleaving ribozymes where substrate and product only differ by one phosphodiester bond (see Chapter 14).

When the reaction reaches equilibrium, $v_{net}$ is zero, and eq. 11.61 becomes

$$0 = v_{max,f} \frac{[S]_{eq}}{K_S} - v_{max,r} \frac{[P]_{eq}}{K_P}$$ 

eq. 11.62

which can be rearranged to

$$K_{eq} = \frac{[P]_{eq}}{[S]_{eq}} = \frac{v_{max,f}K_P}{v_{max,r}K_S}$$

eq. 11.63

$K_{eq}$ is the equilibrium constant for the overall reaction from substrate to product. Note that $K_{eq}$ only depends on the difference in free energy between product and substrate (see Section 3.1), and is not altered by the enzyme (Section 14.1). Using the expression in eq. 11.63 for $K_{eq}$, we can eliminate $v_{max,r}$ from eq. 11.61, and obtain an expression for $v_{net}$ that is a function of $v_{max,f}$ only:

$$v_{net} = v_{max,f} \frac{[S] - \dfrac{[P]}{K_{eq}}}{K_S\left(1 + \dfrac{[P]}{K_P}\right) + [S]}$$

eq. 11.64

From eq. 11.64, we see that accumulated product leads to a higher apparent $K_{M,app} = K_S \cdot (1+[P]/K_P)$, which decreases $v_{net}$. In addition, $v_{net}$ is reduced because the effective substrate concentration is reduced to $[S]-[P]/K_{eq}$ (Figure 11.14). The ratio $[P]/K_{eq}$ is equal to the concentration of the substrate in equilibrium, $[S]_{eq}$, and the difference between the substrate concentration $[S]$ and the ratio $[P]/K_{eq}$ in the numerator thus denotes the deviation of the substrate concentration from the equilibrium value. As long as $[S] > [S]_{eq}$, the net reaction velocity will remain positive, and the reaction will proceed from substrate to product. With ongoing forward reaction, the substrate concentration will approach the equilibrium value. For $[S]_{eq} > [S]$, on the other hand, $v_{net}$ will be negative. In this case, the reaction proceeds from product to substrate, so as to establish the equilibrium concentration. This behavior is thus in accordance with the principle of Le Chatelier. At equilibrium, the numerator in eq. 11.64 becomes zero, and the net velocity is also zero. However, this does not mean that the reaction has stopped! The forward and reverse velocities can still be calculated from eq. 11.55 and eq. 11.56, but they are now equal: the reaction proceeds in both directions at the same speed. For this reason we refer to chemical equilibria as "dynamic".

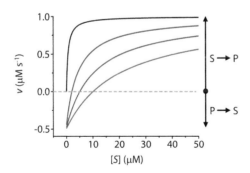

**FIGURE 11.14: Product inhibition.** Reaction velocity as a function of substrate concentration in the absence (black) and presence of 2 µM (red), 5 µM (purple), and 10 µM (blue) product; $K_S = 0.6$ µM, $K_P = 0.3$ µM, $v_{max,f} = 1$ µM s⁻¹, $K_{eq} = 1$. For positive net velocities, the reaction proceeds towards products, for negative velocities, it proceeds towards substrates.

If substrate and product of an enzyme are chemically rather similar, the product efficiently competes with the substrate for the substrate binding site. This type of inhibition is therefore also termed competitive inhibition. We will treat this and other inhibition mechanisms and the effects on $k_{cat}$ and $K_M$ in the following (Sections 11.6.2–11.6.4).

### 11.6.2 COMPETITIVE INHIBITION

A *competitive inhibitor* competes with the substrate for the same binding site on the enzyme. Thus, we have a substrate binding equilibrium ($K_S$), and a competing equilibrium of inhibitor binding ($K_I$). Only the ES complex, but not the enzyme-inhibitor complex EI, can convert the substrate into products (Scheme 11.10).

$$E \; \underset{S}{\overset{K_S}{\rightleftharpoons}} \; \boxed{ES} \; \xrightarrow{k_p} \; E + P$$

$$I \updownarrow K_I$$

$$EI$$

scheme 11.10

As only the enzyme-substrate complex ES contributes to product formation, we can express the reaction velocity as

$$v = k_p[ES] \tag{eq. 11.3}$$

We now follow the standard procedure (Box 11.2) and divide by the total enzyme concentration, and take into account that the enzyme can exist as free enzyme E, as ES complex or as EI complex.

$$\frac{v}{[E]_t} = \frac{k_p[ES]}{[E]+[ES]+[EI]} \tag{eq. 11.65}$$

With [ES] expressed in terms of the equilibrium dissociation constant of ES, $K_S$, and [EI] expressed in terms of the equilibrium constant for dissociation of the EI complex, $K_I$, we obtain

$$\frac{v}{[E]_t} = \frac{k_p \dfrac{[S]}{K_S}[E]}{[E]+\dfrac{[S]}{K_S}[E]+\dfrac{[I]}{K_I}[E]} \tag{eq. 11.66}$$

With $v_{max} = k_p[E]_t$, this equation can be simplified to

$$\frac{v}{v_{max}} = \frac{[S]}{K_S\left(1+\dfrac{[I]}{K_I}\right)+[S]} = \frac{[S]}{K_{M,\,app}+[S]} \tag{eq. 11.67}$$

with

$$K_{M,\,app} = K_S\left(1+\frac{[I]}{K_I}\right) \tag{eq. 11.68}$$

A competitive inhibitor thus does not affect the $v_{max}$ of the reaction, but the apparent Michaelis constant is increased by $(1+[I]/K_I)$ compared to $K_S$ (Figure 11.15). At increasing substrate concentrations, the substrate can compete with the inhibitor for binding, and $v_{max}$ is reached.

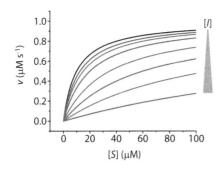

FIGURE 11.15: **Competitive inhibition**. Reaction velocity as a function of the substrate concentration for $K_S = 10\ \mu M$, $v_{max} = 1\ \mu M\ s^{-1}$, and $K_I = 50\ \mu M$ in the absence of inhibitor (black), and in the presence of 1, 10, 20, 50, 100, 200, and 500 μM inhibitor (red: 1, 20, 100, and 500 μM, blue: 10, 50, and 200 μM). A competitive inhibitor interferes with substrate binding, and $K_{M,app}$ is higher than $K_S$. At saturation with substrate, $v_{max}$ is reached.

Many enzymes are inhibited competitively by substrate analogs. One example is succinate dehydrogenase, the enzyme that catalyzes the oxidation of succinate to fumarate in the citric acid cycle, which is inhibited by malonate. Malonate ions bind to the substrate binding site, but are not turned over.

Competitive inhibition can also be exploited therapeutically. The enzyme dihydrofolate reductase (DHFR) catalyzes the reduction of dihydrofolate to tetrahydrofolate, which is the central cofactor for methyl group transfer in the synthesis of purine nucleotides and of thymidylate. The DHFR reaction is particularly important for proliferating cells, such as cancer cells. In chemotherapy, the competitive DHFR inhibitor methotrexate is used to inhibit DHFR in cancer cells. The bacterial DHFR is also a drug target. Bacteria have a different type of DHFR that is competitively inhibited by trimethoprim. This antibiotic has little effect on the human enzyme.

Non-hydrolyzable ATP analogs can act as competitive inhibitors for enzymes that hydrolyze ATP. Novobiocin, an inhibitor of DNA gyrase that is used in the treatment of bacterial infections, inhibits the DNA supercoiling activity (see Section 17.5.3.2) of gyrase by competing with ATP for binding to the ATP binding site.

## 11.6.3 NON-COMPETITIVE INHIBITION

In another mode of enzyme inhibition, the inhibitor I binds to both the free enzyme and to the enzyme-substrate complex, leading to formation of an EI or ESI complex (Scheme 11.11). Here, the inhibitor binds to a different site on the enzyme from that of the substrate. We can assume that $K_S$ is the same for the ES and ESI complexes and that ESI does not catalyze product formation,

$$
\begin{array}{ccccc}
E & \underset{S}{\overset{K_S}{\rightleftharpoons}} & ES & \overset{k_p}{\longrightarrow} & E + P \\[2pt]
I\ \big\updownarrow\ K_i & & K_i\ \big\updownarrow\ I & & \\[2pt]
EI & \underset{S}{\overset{K_S}{\rightleftharpoons}} & ESI & &
\end{array}
$$

scheme 11.11

The reaction velocity again is

$$v = k_p \left[ ES \right]$$

eq. 11.3

leading to a similar expression as for the case of competitive inhibition (eq. 11.65):

$$\frac{v}{\left[ E \right]_t} = \frac{k_p \left[ ES \right]}{\left[ E \right] + \left[ ES \right] + \left[ EI \right] + \left[ ESI \right]}$$

eq. 11.69

The only difference from eq. 11.65 is the appearance of $[ESI]$ in the denominator. With $[ES]$, $[EI]$, and $[ESI]$ expressed in terms of the equilibrium dissociation constants $K_S$ and $K_I$, we obtain

$$\frac{v}{[E]_t} = \frac{k_p \dfrac{[E][S]}{K_S}}{[E] + \dfrac{[E][S]}{K_S} + \dfrac{[E][I]}{K_I} + \dfrac{[E][S][I]}{K_S K_I}} \qquad \text{eq. 11.70}$$

Again, we can cancel $[E]$ from all terms, divide eq. 11.70 by $k_p$, and substitute $k_p[E]_t$ by $v_{max}$. Expansion of the right-hand side with $K_S$ and simplifying the denominator leads to

$$\frac{v}{v_{max}} = \frac{[S]}{K_S \left(1 + \dfrac{[I]}{K_I}\right) + [S]\left(1 + \dfrac{[I]}{K_I}\right)} \qquad \text{eq. 11.71}$$

After transferring the factor $(1+[I]/K_I)$ to the left-hand side, we can express eq. 11.71 in the form

$$\frac{v}{v_{max,\,app}} = \frac{[S]}{K_S + [S]} \qquad \text{eq. 11.72}$$

with

$$v_{max,\,app} = \frac{v_{max}}{\left(1 + \dfrac{[I]}{K_I}\right)} \qquad \text{eq. 11.73}$$

Thus, by sequestering the enzyme in an inactive EI or ESI complex, a *non-competitive inhibitor* reduces $v_{max}$ by $(1+[I]/K_I)$ but does not affect $K_M$ (Figure 11.16). In other words, a non-competitive inhibitor does not affect substrate binding, but only reduces the reaction rate.

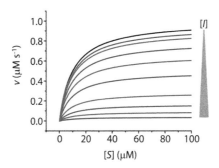

FIGURE 11.16: **Non-competitive inhibition**. Reaction velocity as a function of the substrate concentration for $K_S = 10$ μM, $v_{max} = 1$ μM s$^{-1}$, and $K_I = 50$ μM in the absence of inhibitor (black), and in the presence of 1, 2, 5, 10, 20, 50, 100, 200, and 500 μM of a non-competitive inhibitor (red: 1, 5, 20, 100, and 500 μM, blue: 2, 10, 50, and 200 μM). A non-competitive inhibitor sequesters the enzyme in an inactive complex, leading to a reduced $v_{max}$. The inhibitor does not affect substrate binding, however, and $K_{M,app}$ equals $K_S$.

Non-competitive inhibitors bind to sites on the enzyme distinct from the substrate binding site, and interfere indirectly with substrate binding. This type of inhibition is often found in feedback inhibition, and is referred to as *allosteric* (Greek: another place). An example is the inhibition of pyruvate kinase by alanine. Pyruvate kinase catalyzes the oxidative dephosphorylation of phosphoenol pyruvate to pyruvate, which can be converted to alanine (see Figure 8.3). Alanine acts as a non-competitive inhibitor of pyruvate kinase, ensuring that pyruvate production is reduced when sufficient amounts of alanine are present.

## 11.6.4 Mixed Inhibition

We have derived eq. 11.72 and eq. 11.73 for non-competitive inhibition assuming that $K_S$ for dissociation of substrate from ES and ESI is identical. However, these constants often differ, leading to altered $K_M$ and $v_{max}$. This type of inhibition is called *mixed inhibition*.

One type of mixed inhibition is *uncompetitive inhibition*. In this case, the inhibitor binds to ES, but not to the free enzyme, i.e. the binding site for the inhibitor is created only upon binding of substrate to the enzyme. Scheme 11.11 then simplifies to:

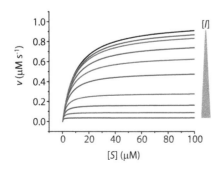

scheme 11.12

We can now express the reaction velocity as the product of $k_p$ and the concentration of active enzyme, i.e. the ES complex, divide the equation by the total enzyme concentration, express all complex concentrations by $[E]$ $[S]$ and $K_S$ or $[E]$, $[I]$ and $K_I$, substitute $k_p[E]_t$ by $v_{max}$, and simplify to

$$\frac{v}{v_{max,app}} = \frac{[S]}{K_{M,app} + [S]}$$

eq. 11.74

with

$$v_{max,\,app} = \frac{v_{max}}{\left(1 + \dfrac{[I]}{K_I}\right)}$$

eq. 11.72

and

$$K_{M,\,app} = \frac{K_S}{\left(1 + \dfrac{[I]}{K_I}\right)}$$

eq. 11.75

An uncompetitive inhibitor thus reduces $v_{max}$ and $K_M$ by a factor of $(1+[I]/K_I)$ (Figure 11.17). The decreased $K_{M,app}$ in non-competitive inhibition corresponds to favored substrate binding when the inhibitor is present. Note that this factor $(1+[I]/K_I)$ is equal to the factor by which $K_{M,app}$ is increased over $K_S$ by a competitive inhibitor that competes with the binding, whereas the uncompetitive inhibitor favors substrate binding by binding to the product of the substrate binding equilibrium ES (principle of Le Chatelier; Section 3.1.2).

**FIGURE 11.17:   Uncompetitive inhibition.** Reaction velocity as a function of the substrate concentration for $K_S = 10\ \mu M$, $v_{max} = 1\ \mu M\ s^{-1}$, and $K_I = 50\ \mu M$ in the absence of inhibitor (black), and in the presence of 1, 2, 5, 10, 20, 50, 100, 200, and 500 μM of a uncompetitive inhibitor (red: 1, 5, 20, 100, and 500 μM, blue: 2, 10, 50, and 200 μM). An uncompetitive inhibitor sequesters the enzyme in an inactive complex, leading to a reduced $v_{max}$. The inhibitor also affects substrate binding by formation of ESI, leading to a decreased $K_{M,app}$ compared to $K_S$.

We have seen in this chapter that different types of inhibition differentially affect $v_{max}$ and $K_M$. The inhibition types cause characteristic differences in reciprocal plots, and from these plots rapid classification of inhibitors is possible (Box 11.4).

---

### BOX 11.4: ENZYME INHIBITION AND LINEAR PLOTS.

Reciprocal plots after Lineweaver-Burk or Eadie-Hofstee (Section 11.2; eq. 11.21, eq. 11.22) are particularly useful to study inhibition (Figure 11.18). Competitive inhibition reduces $K_M$, but does not affect $v_{max}$. Lineweaver-Burk plots for enzymes in the presence of increasing concentrations of inhibitor have the same $y$-axis intercept, but the slope increases with the inhibitor concentration. The same behavior is seen in an Eadie-Hofstee plot. Non-competitive inhibition reduces $v_{max}$, but not $K_M$. This type of inhibition is reflected in an increased slope and $y$-axis intercept with increasing inhibitor concentration in Lineweaver-Burk plots, and a decreasing $y$-axis intercept, but identical slopes, in Eadie-Hofstee plots. For an uncompetitive inhibitor, both $v_{max}$ and $K_M$ are affected, and slopes and intercepts change with increasing inhibitor concentrations in both plots.

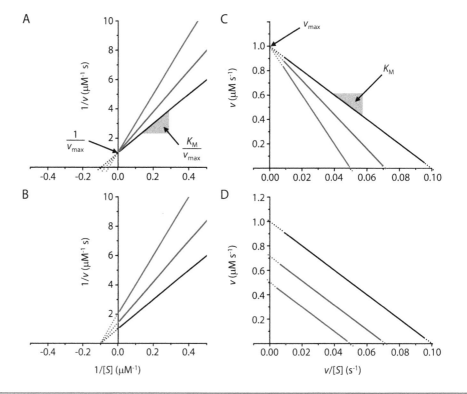

**FIGURE 11.18: Reciprocal plots and inhibition.** Lineweaver-Burk (A,B) and Eadie-Hofstee plots (C,D) for competitive inhibition (A,C) and non-competitive inhibition (B,D) in the absence of inhibitor (black), and in the presence of increasing concentrations of inhibitor (red, blue).

---

Competitive, non-competitive, and uncompetitive inhibition are based on reversible interactions of the inhibitor with the enzyme. Enzymes can also be inhibited irreversibly by molecules that chemically modify the active site, by transition state analogs that bind tightly to the active site, and by suicide inhibitors that are bound and turned over, but not to completion (Box 11.5).

## BOX 11.5: IRREVERSIBLE ENZYME INHIBITION.

Enzymes are inhibited essentially irreversibly by compounds that bind covalently or non-covalently so as to modify their active site. Residues in the active site are typically very reactive, which also makes them prone to reactions with compounds other than the cognate substrate. For example, active site cysteines form complexes with heavy metal ions, leading to inactivation of the enzyme. Active site cysteines also readily react with iodoacetamide. The nerve gas diisopropylfluorophosphate forms an ester bond with a serine in the active site of acetylcholine esterase and irreversibly inhibits the enzyme. Compounds that react with certain residues are called *group-specific inhibitors*. They can be used to map active site residues.

Molecules that bear similarity to the substrate and cause covalent modification of an enzyme are called *affinity labels*. An example is tosyl-L-phenylalanine chloromethyl ketone that binds to the active site of the serine protease chymotrypsin and covalently modifies a histidine and a serine in the active site.

*Transition state analogs* are compounds that have a chemical structure similar to the transition state (Section 13.2) of the catalyzed reaction. They may bind very tightly, essentially irreversibly, to the active site of the enzyme, leading to inhibition. *Transition state analog inhibitors* are very useful to investigate enzyme mechanisms. The neurotoxin sarin (methylfluoro phosphoric acid isopropylester) contains a tetrahedral phosphonate group that mimics the transition state of acetylcholine hydrolysis, and acts as a potent transition state analog inhibitor of acetylcholine esterase.

*Suicide inhibitors* are molecules that bind to the active site of the enzyme and undergo turnover up to a certain step, but are not converted to product. They remain covalently bound to the enzyme, inhibiting its activity. The antibiotic penicillin is an example of a suicide inhibitor. Penicillin is a $\beta$-lactam that binds to the active site of D-alanine transpeptidase, an enzyme involved in cell wall synthesis of bacteria (and not present in humans). The transpeptidase forms crosslinks between D-alanine and glycine residues in the peptidoglycans of the bacterial cell wall. It forms a transient acyl intermediate between a serine in the active site and D-alanine. The acyl-enzyme intermediate then reacts with a terminal glycine to form the crosslinking peptide bond. Penicillin binds the active site and forms a stable acyl intermediate with the serine in the active site that cannot undergo turnover, leading to suicide inhibition. Suicide inhibitors hijack the enzyme mechanism, which makes them very specific. Such *mechanism-based inhibitors* are therefore very attractive for therapeutic applications. For example, suicide inhibitors of the enzyme monoamine oxidase are employed in the therapy of depression and Parkinson's disease. Monoamine oxidase catalyzes the deamination of dopamine and serotonin in the brain. Low levels of dopamine are associated with Parkinson's disease, low serotonin levels with depression. Monoamine oxidase catalyzes the deamination of both of these neurotransmitters and reduces their levels.

## QUESTIONS

11.1    The derivation of the Michaelis-Menten velocity equation is based on the assumption that $[S] \approx S_0$, which requires $[S] \ll K_S$. Derive the velocity equation for $[S] \approx K_S$.

11.2    What is the velocity $v$ relative to $v_{max}$ at $[S] = 2 \cdot K_S, 3 \cdot K_S, 5 \cdot K_S$, and $9 \cdot K_S$? What is the ratio of substrate concentrations for which $v = 0.9 \cdot v_{max}$ and $v = 0.5 \cdot v_{max}$ ($[S]_{0.9}/[S]_{0.5}$)?

11.3    Ethanol (EtOH) is oxidized to acetaldehyde by alcohol dehydrogenase (ADH), using NAD$^+$ as an oxidant. Starting from a concentration of $[EtOH] = 1.0$ g L$^{-1}$ in the body, ADH reduces the ethanol concentration by 10% in 1 h. ADH is saturated with EtOH throughout the reaction, and $NAD^+$ is maintained constant. What is the reaction velocity in $\mu M\ s^{-1}$? $M_m(EtOH) = 40.07\ g\ mol^{-1}$. What are the molar concentrations and the blood alcohol levels in ‰ at the beginning and at the end, assuming equal distribution throughout the body? Is this assumption valid? How long does it take until the blood alcohol level is below 0.3‰?

11.4 The radical reaction of bromine and hydrogen to give HBr occurs according to the following reaction scheme:

$$Br_2 \xrightarrow{k_1} 2\ Br\cdot$$

$$H_2 + Br\cdot \underset{k_{-2}}{\overset{k_2}{\rightleftharpoons}} HBr + H\cdot$$

$$Br_2 + H\cdot \xrightarrow{k_3} HBr + Br\cdot$$

$$2Br\cdot \xrightarrow{k_{-1}} Br_2$$

Derive the rate law for HBr production under steady-state assumptions for $H\bullet$ and $Br\bullet$ radicals. Why is this assumption appropriate?

11.5 The enzymatic activity of an enzyme with $K_S = 2$ mM that converts substrate S into product P is measured at an initial substrate concentration $S_0$ of 10 $\mu M$. After 5 min, the substrate concentration is halved. What is the rate constant $k$, the maximal velocity $v_{max}$ and the concentration of product after 12 min?

11.6 An enzyme has a histidine residue in the active site ($pK_{A,E} = 6.5$) that must be protonated for catalysis. Its substrate is the deprotonated form of a weak acid with $pK_{A,S} = 4.5$. What is the pH optimum of the enzyme-catalyzed reaction?

11.7 An enzyme has an aspartate in the active site that needs to be deprotonated for substrate binding and catalysis. The dissociation constant for the aspartate is $K_A$, the dissociation constant of the enzyme-substrate complex is $K_S$. The rate constant of product formation is $k_p$. Derive the equation for the reaction velocity. What is the effect of $H^+$?

11.8 Calculate the ratio of substrate concentrations at which $v = 0.9 \cdot v_{max}$ and $v = 0.1 \cdot v_{max}$ for an enzyme with the apparent Hill coefficient $n_{app} = 2.6$ and without cooperativity.

11.9 An enzyme shows a sigmoidal dependence of the reaction velocity on substrate concentration. The $[S]_{0.9}/[S]_{0.1}$ ratio is 7.6. What is $n_{app}$? What is the number of substrate binding sites?

11.10 What is the relative distribution of the different enzyme forms and the reaction velocity for an enzyme with two active sites and a cooperativity factor $\alpha = 0.2$ at $[S] = 0.4 \cdot K_S$?

11.11 A dimeric enzyme shows positive cooperativity for substrate binding. Cooperativity can be described by the concerted model. The enzyme exists in two conformations that are in equilibrium: either both protomers are in the T state that does not bind substrate, or both protomers are in the R state that binds substrates with high affinity (dissociation constant $K_S$). The equilibrium constant for the conformational equilibrium is $L = [T]_0/[R]_0$. The rate of product formation is $k_p$ (for each active site). Derive the velocity equation for this enzyme, and generalize the solution to $n$ sites.

11.12 Hexokinase (HK) binds glucose and catalyzes its phosphorylation to glucose-6-phosphate. Fructose binds to the enzyme and inhibits glucose phosphorylation. The dissociation constant of the HK/glucose complex is $K_S$, the dissociation constant of the HK/fructose complex is $K_I$. The rate-limiting step of product formation occurs with $k_p$. Derive the rate equation and compare the apparent $K_M$ and the apparent $v_{max}$ to the values in the absence of inhibition. What are the limiting values for $v_{max,app}$ and $K_{M,app}$ for very low and very high fructose concentration? What happens at [fructose] = $K_I$?

11.13 The activity of an enzyme with $K_S = 50$ μM and $v_{max} = 1$ μM s$^{-1}$ is measured in the presence of 400 μM substrate and 500 μM of an inhibitor with a $K_I$ of 250 μM. What is the degree of inhibition for competitive, non-competitive and uncompetitive inhibition?

11.14 What is the relative inhibition of an enzyme by a competitive inhibitor at $[S] = K_S$ and $[I] = K_I$?

11.15 Why is it effective to treat methanol or ethylene glycol poisoning with large (sub-lethal) doses of ethanol?

11.16 RNase A has two histidines in the active site. All species bind substrate, but for catalysis His12 must be deprotonated, and His119 must be protonated. Deprotonation of His12 favors protonation of His119. Write a reaction scheme and derive the velocity equation as a function of the $[H^+]$ concentration.

## REFERENCES

Segel, I. H. (1975) Enzyme kinetics: behavior and analysis of rapid equilibrium and steady state enzyme systems. New York, Wiley.
· a compendium of steady-state enzyme kinetics

Koshland, D. E., Jr., Nemethy, G. and Filmer, D. (1966) Comparison of experimental binding data and theoretical models in proteins containing subunits. *Biochemistry* **5**(1): 365–385.
· description of the Koshland-Nemethy-Filmer model of cooperativity

Monod, J., Wyman, J. and Changeux, J. P. (1965) On the Nature of Allosteric Transitions: A Plausible Model. *J Mol Biol* **12**: 88–118.
· Monod-Wyman-Changeux model of cooperativity

De La Cruz, E. M., Sweeney, H. L. and Ostap, E. M. (2000) ADP inhibition of myosin V ATPase activity. *Biophys J.* **79**(3): 1524–1529.
· product inhibition of the ATPase myosin by ADP

# Complex Reaction Schemes and Their Analysis

In Chapter 11 we learnt the principles for translating enzymatic reaction schemes into rate equations. This chapter illustrates that the same basic principles allow us to derive rate equations for binding of two substrates, and presents tools to simplify the analysis of quite complex reaction schemes.

## 12.1 BINDING OF TWO SUBSTRATES

In sequential mechanisms, two (or more) substrates are bound to the enzyme first, and then react to give the product. Binding can be random, in any possible sequence, or ordered, with a defined sequence of binding events.

### 12.1.1 RANDOM BINDING

Two substrate molecules can bind to the enzyme in random order, i.e. binding of A followed by binding of B or binding of B followed by binding of A leads to the same ternary complex EAB (Scheme 12.1). From EAB, products are formed with the rate constant $k_p$. The binding equilibria of A and B are described by the equilibrium dissociation constants $K_A$ and $K_B$. Thermodynamic linkage (Section 3.4) can be accounted for by the factor $\alpha$ (Section 3.4):

$$E \underset{K_A}{\overset{A}{\rightleftharpoons}} EA$$

$$K_B \updownarrow B \qquad \alpha K_B \updownarrow B$$

$$EB \underset{\alpha K_A}{\overset{A}{\rightleftharpoons}} EAB \xrightarrow{k_p}$$

scheme 12.1

Product formation from EAB occurs with the reaction velocity

$$v = k_p[EAB] \qquad \text{eq. 12.1}$$

Following the standard procedure (Box 11.2) described several times before, we divide by the total enzyme concentration

$$\frac{v}{[E]_t} = \frac{k_p[EAB]}{[E]+[EA]+[EB]+[EAB]}$$  eq. 12.2

substitute the concentrations of the enzyme complexes in terms of free ligand concentrations [A] and [B] and their corresponding equilibrium constants, rearrange and simplify to

$$\frac{v}{v_{max}} = \frac{\dfrac{[A][B]}{\alpha K_A K_B}}{1 + \dfrac{[A]}{K_A} + \dfrac{[B]}{K_B} + \dfrac{[A][B]}{\alpha K_A K_B}}$$  eq. 12.3

We can now expand eq. 12.3 with $\alpha K_A K_B$ to

$$\frac{v}{v_{max}} = \frac{[A][B]}{\alpha K_A K_B + \alpha K_B[A] + \alpha K_A[B] + [A][B]}$$  eq. 12.4

Cancelling [B] yields

$$\frac{v}{v_{max}} = \frac{[A]}{\alpha K_A\left(1 + \dfrac{K_B}{[B]}\right) + [A]\left(1 + \dfrac{\alpha K_B}{[B]}\right)}$$  eq. 12.5

By division of the denominator on the right-hand side and multiplying the numerator in the left-hand side with $1 + \alpha K_B/[B]$, we can rearrange eq. 12.5 to an equation of the form

$$\frac{v}{v_{max,app}} = \frac{[A]}{K_{M,app} + [A]}$$  eq. 12.6

with

$$v_{max,app} = \frac{v_{max}}{1 + \dfrac{\alpha K_B}{[B]}}$$  eq. 12.7

and

$$K_{M,app} = \frac{\alpha K_A\left(1 + \dfrac{K_B}{[B]}\right)}{\left(1 + \dfrac{\alpha K_B}{[B]}\right)}$$  eq. 12.8

Thus, for a given concentration of A, both $v_{max,app}$ and $K_{M,app}$ depend on the concentration of B. Under the limiting condition that B is saturating, the terms $K_B/[B]$ and $\alpha K_B/[B]$ become small, and both $v_{max,app}$ and $K_{M,app}$ become independent of the concentration of B (Figure 12.3):

$$v_{max,app} \cong v_{max}$$  eq. 12.9

$$K_{M,app} \cong \alpha K_M$$  eq. 12.10

In the absence of cooperativity, for $\alpha = 1$, eq. 12.8 simplifies to $K_{M,app} = K_A$, and so $K_{M,app}$ does not depend on [B], but $v_{max}$ still does (Figure 12.1).

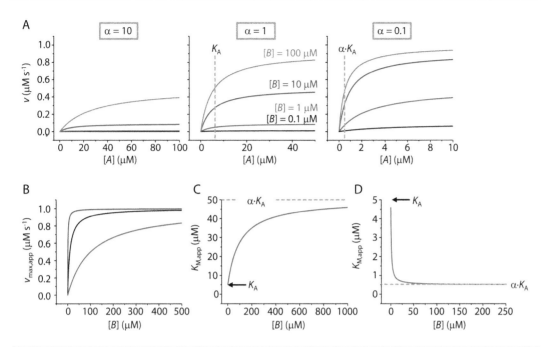

**FIGURE 12.1:   Random substrate binding.** A: Reaction velocity as a function of concentration of substrate A at different concentrations of substrate B in the absence of thermodynamic linkage ($\alpha = 1$), and with positive ($\alpha = 0.1$) and negative ($\alpha = 10$) thermodynamic linkage. $v_{max} = 1$ $\mu$M s$^{-1}$, $K_A = 5$ $\mu$M, $K_B = 10$ $\mu$M, $[B] = 0.1$ $\mu$M, 1 $\mu$M, 10 $\mu$M, 100 $\mu$M. B: $v_{max,app}$ as a function of $[B]$ for no thermodynamic linkage ($\alpha = 1$; black), positive ($\alpha = 0.1$; red) and negative thermodynamic linkage ($\alpha = 10$; blue). $v_{max,app}$ reaches $v_{max}$ when $[B]$ is saturating ($[B] > \alpha K_B$). C: $K_{M,app}$ as a function of $[B]$ for $\alpha = 10$. $K_{M,app}$ increases with increasing $[B]$ and approaches $\alpha K_A$ at saturating concentrations of B. D: $K_{M,app}$ as a function of $[B]$ for $\alpha = 0.1$. $K_{M,app}$ decreases with increasing $[B]$ and approaches $\alpha K_A$ at saturating concentrations of B.

Many enzymes form a ternary complex with two ligands to catalyze their reaction. Examples are found among transferases, such as phosphoryl-, acetyl-, or methyl transferases (although some of them follow a ping-pong mechanism (Section 12.2) with a covalently modified enzyme intermediate). Methionine synthase catalyzes methionine synthesis by transfer of a methyl group from methyltetrahydrofolate to homocysteine. Kinetic and equilibrium studies have shown that both substrates are bound in random order, but with high thermodynamic linkage ($\alpha \approx 30$). Other examples are ATP-dependent enzymes that couple ATP binding and hydrolysis to a certain reaction. RNA helicases unwind RNA duplex regions in an ATP-dependent reaction. Some, but not all of them, bind ATP and the RNA duplex in random order, with different levels of thermodynamic linkage.

## 12.1.2 Ordered Binding

In some enzymes, the two substrates have to be bound in a defined order for the reaction to proceed. *Ordered* or *sequential binding* occurs when binding of the first substrate induces a conformational change in the enzyme that leads to formation of the binding site for the second substrate. Substrates have to be bound in a defined order when the binding sites are arranged on the enzyme such that binding of the second substrate blocks the binding site for the first substrate (Figure 12.2). Similarly, release of the products may be ordered.

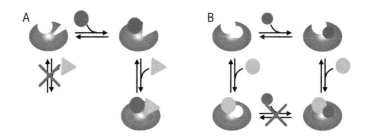

FIGURE 12.2: **Sequential binding.** A: Binding of substrate 1 (blue) to the enzyme causes a conformational change that leads to formation of the binding site for substrate 2 (orange). B: The binding site for substrate 2 (orange) blocks the binding site for substrate 1 (blue).

$$E + A \underset{K_A}{\overset{A}{\rightleftharpoons}} EA$$
$$K_B \updownarrow B$$
$$EAB \xrightarrow{k_p}$$

scheme 12.2

To evaluate the reaction scheme with ordered binding of substrates A and B (Scheme 12.2), we again begin by formulating the rate of product formation from the EAB complex:

$$v = k_p[EAB] \qquad \text{eq. 12.11}$$

When we now divide by the total enzyme concentration, we only have to take into account free enzyme E, the EA complex, and the EAB complex, as the EB complex does not exist (compare eq. 12.2):

$$\frac{v}{[E]_t} = \frac{k_p[EAB]}{[E]+[EA]+[EAB]} \qquad \text{eq. 12.12}$$

We apply the same sequence of operations as before (Box 11.2): by substituting the concentrations of the enzyme complexes in terms of free ligands and equilibrium constants and division by $[E]$ we obtain

$$\frac{v}{v_{max}} = \frac{\dfrac{[A][B]}{K_A K_B}}{1 + \dfrac{[A]}{K_A} + \dfrac{[A][B]}{K_A K_B}} \qquad \text{eq. 12.13}$$

and

$$\frac{v}{v_{max}} = \frac{[A][B]}{K_A K_B + K_B[A] + [A][B]} \qquad \text{eq. 12.14}$$

Canceling $[B]$ gives

$$\frac{v}{v_{max}} = \frac{[A]}{K_A \dfrac{K_B}{[B]} + [A]\left(1 + \dfrac{K_B}{[B]}\right)} \qquad \text{eq. 12.15}$$

Division by $1+K_B/[B]$ and rearrangement of eq. 12.15 leads to

$$\frac{v}{v_{max,app}} = \frac{[A]}{K_{M,app} + [A]} \qquad \text{eq. 12.6}$$

with

$$v_{max,app} = \frac{v_{max}}{\left(1 + \dfrac{K_B}{[B]}\right)} \qquad \text{eq. 12.16}$$

and

$$K_{M,app} = \frac{K_A K_B}{[B] + K_B}$$
eq. 12.17

Thus, for ordered binding, both $v_{max}$ and $K_{M,app}$ depend on the concentration of B. If $[B]$ is saturating, i.e. $[B] \gg K_B$, then $v_{max,app}$ approaches $v_{max}$ and becomes independent of $[B]$, and $K_{M,app}$ becomes $K_A K_B / [B]$ (Figure 12.3).

The expression for $v_{max,app}$ in eq. 12.16 is the same expression we have derived for the reaction with random substrate binding (eq. 12.7), except that the factor $\alpha$ for thermodynamic coupling is missing here. In fact, the cooperativity parameter is implicitly contained in the value of $K_B$. The constant $K_B$ in Scheme 12.2 denotes the equilibrium constant for the dissociation of the EAB complex to give EA and B. No binding of B in absence of A is equivalent to an infinitely high $K_d$ value of the EB complex. $K_B$ in Scheme 12.2 is much lower than this $K_d$, and reflects $\alpha K_B$ in Scheme 12.1 (random binding).

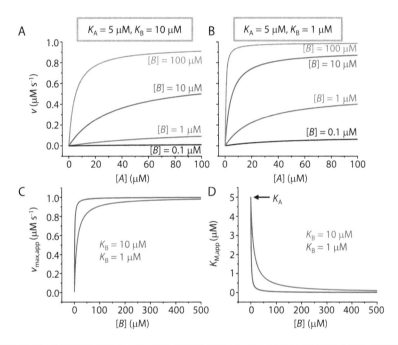

**FIGURE 12.3: Ordered sequential binding.** A, B: Reaction velocity as a function of concentration of substrate A at different concentrations of substrate B. $[B] = 1\,\mu M$, $10\,\mu M$, $100\,\mu M$, $v_{max} = 1\,\mu M\,s^{-1}$. A: $K_A = 5\,\mu M$, $K_B = 10\,\mu M$, B: $K_A = 5\,\mu M$, $K_B = 1\,\mu M$. C: $v_{max,app}$ as a function of $[B]$ for $K_A = 5\,\mu M$, $K_B = 10\,\mu M$ (blue) and $K_A = 5\,\mu M$, $K_B = 1\,\mu M$ (red). For saturating $[B]$, $v_{max,app}$ approaches $v_{max}$. D: $K_{M,app}$ as a function of $[B]$ for $K_A = 5\,\mu M$, $K_B = 10\,\mu M$ (blue) and $K_A = 5\,\mu M$, $K_B = 1\,\mu M$ (red). For saturating $[B]$, $K_{M,app}$ approaches $K_A \cdot K_B / [B]$.

An example of an enzyme with ordered binding of its substrates is serotonin-N-acetyltransferase, an enzyme that catalyzes the transfer of an acetyl group from acetyl-CoA to the terminal amino group of the neurotransmitter serotonin. N-acetyl serotonin is a precursor in the biosynthesis of melatonin, a hormone that regulates the day-night rhythm in humans. Structural studies have shown that binding of acetyl-CoA to the transferase leads to a conformational change of the enzyme and formation of the serotonin binding site. Aspartate transcarbamoylase catalyzes the transfer of a carbamoyl group from carbamoyl phosphate to aspartate, which is converted to dihydroorotate and then orotate, a precursor in pyrimidine biosynthesis. Aspartate transcarbamoylase binds its two substrates in a defined order, first carbamoyl phosphate, and subsequently aspartate. Similar to serotonin-N-acetyltransferase, carbamoyl phosphate binding causes a conformational change of the enzyme that leads to formation of the aspartate binding site. The dissociation of products is also ordered: N-carbamoyl aspartate leaves the enyzme before phosphate.

## 12.2 PING-PONG MECHANISM

In a *ping-pong mechanism*, substrate A-X binds and transfers a group X to the enzyme. The product leaves the enzyme before substrate B binds to the modified enzyme E-X, and receives group X (Scheme 12.3). Examples for these enzymes are aminotransferases, also called transaminases, and biotin-dependent carboxylases.

$$E + A\text{-}X \underset{}{\overset{K_A}{\rightleftharpoons}} E\cdot A\text{-}X \underset{}{\overset{A}{\rightleftharpoons}} \boxed{E\text{-}X} \underset{}{\overset{B}{\rightleftharpoons}} E\text{-}X\cdot B \rightleftharpoons E\cdot B\text{-}X \underset{}{\overset{K_B}{\rightleftharpoons}} E + B\text{-}X$$ scheme 12.3

From initial velocities in the absence of product B-X, we can derive expressions for the apparent $v_{\text{max,app}}$ and $K_{\text{M,app}}$:

$$v_{\text{max, app}} = \frac{v_{\text{max}}}{\left(1 + \dfrac{K_B}{[B]}\right)}$$ eq. 12.18

and

$$K_{M, \text{app}} = \frac{K_A}{\left(1 + \dfrac{K_B}{[B]}\right)}$$ eq. 12.19

The direct analysis of this reaction scheme is rather involved, which is why we do not treat it here. We can more easily gain insight into the overall behavior of such an enzyme using the concept of transit times and net rate constants (Section 12.3). From eq. 12.18 and eq. 12.19, we see that at high concentrations of B, $v_{\text{max,app}}$ approaches $v_{\text{max}}$, and $K_{\text{M,app}}$ approaches $K_A$ (Figure 12.4).

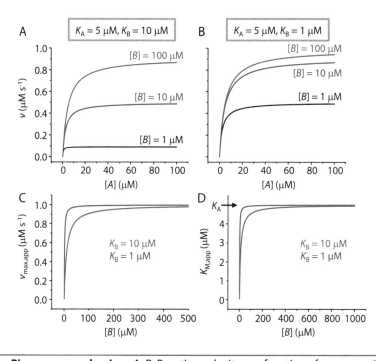

**FIGURE 12.4:** **Ping-pong mechanism.** A, B: Reaction velocity as a function of concentration of substrate A at different concentrations of substrate B. [B] = 1 μM (black), 10 μM (blue), 100 μM (red). $v_{\text{max}}$ is 1 μM s⁻¹. A: $K_A = 5$ μM, $K_B = 10$ μM. B: $K_A = 5$ μM, $K_B = 1$ μM. C: $v_{\text{max,app}}$ as a function of [B] for $K_A = 5$ μM, $K_B = 10$ μM (blue) and $K_A = 5$ μM, $K_B = 1$ μM (red). At saturating concentrations of B, $v_{\text{max,app}}$ approaches $v_{\text{max}}$. D: $K_{\text{M,app}}$ as a function of [B] for $K_A = 5$ μM, $K_B = 10$ μM (blue) and $K_A = 5$ μM, $K_B = 1$ μM (red). At saturating concentrations of B, $K_{\text{M,app}}$ approaches $K_A$.

An example of an enzyme that follows such a ping-pong mechanism is pyruvate carboxylase, which generates oxaloacetate from pyruvate and $CO_2$. $CO_2$ is activated by phosphorylation to carboxyphosphate, $HOCO_2\text{-}PO_3{}^{2-}$, using ATP as a phosphoryl donor. Pyruvate carboxylase then first binds carboxyphosphate (A-X in Scheme 12.3). Carboxylation of the biotin cofactor leads to carboxybiotin (E-X) and release of phosphate (A). The carboxybiotin cofactor is attached to a lever arm that allows it to swing over a large distance to the bound pyruvate (B), and the $CO_2$ is transferred from carboxybiotin to pyruvate (corresponding to B-X in Scheme 12.3) to give oxaloacetate (Q). Other enzymes that form an enzyme intermediate are various transferases such as phospho- or acetyltransferases (X = transferred group), oxidoreductases such as flavin reductase (X = hydride), and serine proteases that form an acyl enzyme intermediate (X = acyl peptide, transferred to $H_2O$).

## 12.3 NET RATE CONSTANTS AND TRANSIT TIMES

Derivation of the rate equation becomes involved for more complex reaction schemes. Sometimes, insight into the behavior of complex reactions can already be gained by applying some principal concepts, and it may not be necessary to derive the full rate equation. We first consider a reaction scheme of two sequential reactions (Scheme 12.4):

$$A \underset{k_{-1}}{\overset{k_1}{\rightleftharpoons}} B \overset{k_2}{\longrightarrow} C$$

scheme 12.4

How can we intuitively describe the reaction velocity for the conversion of A to C? The overall reaction rate depends on the rate of conversion of A to B, multiplied by the probability that B is converted to C instead of being re-converted to A in the reverse reaction. The rate for the conversion of A to B is $k_1[A]$, and the probability of conversion of B to C instead of back to A is the result of a kinetic partitioning between the forward reaction to C with $k_2$ and the reverse reaction to A with $k_{-1}$. The partitioning is described by $k_2/(k_{-1}+k_2)$.

Thus, the *net velocity* $v_{net}$ for the reaction sequence from A to B is the product of these two expressions,

$$v_{net} = \frac{k_1 k_2}{k_{-1} + k_2}[A]$$

eq. 12.20

and the *net rate constant* $k_1'$ for the reaction A→B is

$$k_1' = \frac{k_1 k_2}{k_{-1} + k_2}$$

eq. 12.21

To include the conversion from B to C in our consideration, we first have to introduce the concept of *transit times*. (First-order) rate constants have units $s^{-1}$, and the inverse of the rate constant thus defines a time $t_{transit}$ that is needed for the corresponding step of the reaction:

$$t_{transit} = \frac{1}{k}$$

eq. 12.22

This transit time is equal to the time constant $\tau$ of the reaction step or the lifetime $\tau$ of the reacting species in this step (see Chapter 7). For a four-step reaction sequence from A through B, C, and D to E, with rate constants of $k_1$, $k_2$, $k_3$, and $k_4$ (Scheme 12.5)

$$A \overset{k_1}{\longrightarrow} B \overset{k_2}{\longrightarrow} C \overset{k_3}{\longrightarrow} D \overset{k_4}{\longrightarrow} E$$

scheme 12.5

the overall transit time is the sum of the transit times of each individual step. We can therefore sum up the individual transit times to calculate the total transit time from A to E, and to convert it to the overall rate constant $k$

$$t_{\text{transit}, A \to E} = \frac{1}{k_1} + \frac{1}{k_2} + \frac{1}{k_3} + \frac{1}{k_4} + \frac{1}{k_5} = \frac{1}{k} \qquad \text{eq. 12.23}$$

We can now combine the two concepts to finish the evaluation of Scheme 12.4, and can calculate the total transit time for the reaction from A to C from the sum of the inversed net rate constants $k_1'$ (eq. 12.21) and $k_2$:

$$t_{\text{transit}, A \to C} = \frac{k_{-1} + k_2}{k_1 k_2} + \frac{1}{k_2} = \frac{1}{k} \qquad \text{eq. 12.24}$$

For $k_2 \ll k_1$, the second term in eq. 12.24 is the dominant term which determines the transit time and the overall rate constant of the reaction. In other words: the second step is rate-limiting, and $k_2$ determines the overall rate constant $k$.

The concept of net rate constants and transit times can be applied to directly examine the enzymatic reaction occurring after Scheme 12.6, without deriving the rate law step by step.

$$E + S \underset{k_{-1}}{\overset{k_1}{\rightleftharpoons}} ES \overset{k_2}{\longrightarrow} E + P \qquad \text{scheme 12.6}$$

We analyze the reaction scheme by describing the substrate binding equilibrium by a first-order net rate constant $k_1'$. For the second step, the net rate constant $k_2'$ equals $k_2$:

$$E + S \overset{k_1'}{\longrightarrow} ES \overset{k_2'}{\longrightarrow} E + P \qquad \text{scheme 12.7}$$

According to eq. 12.21, we can calculate the net rate constant for the binding equilibrium as

$$k_1' = k_1[S] \cdot \frac{k_2}{k_{-1} + k_2} \qquad \text{eq. 12.25}$$

The total transit time is then

$$t_{\text{transit}, S \to P} = \frac{1}{k_1'} + \frac{1}{k_2'} = \frac{1}{k_1[S]} \cdot \frac{k_{-1} + k_2}{k_2} + \frac{1}{k_2} \qquad \text{eq. 12.26}$$

The overall rate constant $k$ for the conversion of S to P is the inverse of the total transit time (eq. 12.24), and we obtain

$$k = \frac{1}{\dfrac{1}{k_1[S]} \cdot \dfrac{k_{-1} + k_2}{k_2} + \dfrac{1}{k_2}} = \frac{k_1 k_2[S]}{k_{-1} + k_2 + k_1[S]} \qquad \text{eq. 12.27}$$

By canceling $k_1$ on the right, eq. 12.27 can be expressed in the form

$$k = \frac{k_2[S]}{\dfrac{k_{-1} + k_2}{k_1} + [S]} \qquad \text{eq. 12.28}$$

The rate constant $k$ is the ratio of reaction velocity $v$ and total enzyme concentration $[E]_t$, and we can re-write eq. 12.28 as

$$\frac{v}{[E]_t} = \frac{k_2[S]}{\dfrac{k_{-1} + k_2}{k_1} + [S]}$$

eq. 12.29

Eq. 12.29 is equivalent to the Briggs-Haldane equation (eq. 11.20) with $v_{max} = k_2 [E]_t$ and $K_M = (k_{-1} + k_2/k_1)$, demonstrating the validity of the simplified approach using transit times and net rate constants.

We can apply the concept of transit times to analyze the ping-pong mechanism (Section 12.2; Scheme 12.3). To this end, we re-write Scheme 12.3, and describe the rapid equilibrium by the rate constants $k_1$ and $k_{-1}$, the transfer of group X to the enzyme by $k_2$, and the transfer to substrate B by $k_3[B]$ (Scheme 12.8).

$$E + A\text{-}X \underset{k_{-1}}{\overset{k_1}{\rightleftharpoons}} E\text{·}A\text{-}X \overset{k_2}{\longrightarrow} E\text{-}X \overset{k_3[B]}{\longrightarrow} E + B\text{-}X$$

scheme 12.8

To calculate the overall transit time $t_{transit} = 1/k$, we need to sum up the inverse of the rate constants for each step, $k_1'$, $k_2'$, and $k_3'$. These individual net rate constants are

$$k_3' = k_3[B]$$

eq. 12.30

$$k_2' = k_2$$

eq. 12.31

and

$$k_1' = k_1[A] \cdot \frac{k_2}{k_{-1} + k_2}$$

eq. 12.32

The total transit time for the transfer of the group X from A to B then is

$$t_{transit, A-X \to B-X} = \frac{1}{k_1[A]} \cdot \frac{k_{-1} + k_2}{k_2} + \frac{1}{k_2} + \frac{1}{k_3[B]} = \frac{1}{k} = \frac{[E]_t}{v}$$

eq. 12.33

Compared to a simple enzymatic reaction that consists of formation of the enzyme-substrate complex followed by product formation (eq. 12.26), the only difference in eq. 12.33 is the last term that increases the transit time and thus reduces the overall reaction velocity $k$. At high concentrations of B, this term becomes small and can be neglected. The rate of product formation is then limited by the formation of E-X from EA ($k_2$) that becomes the rate-limiting step (Chapter 9). This is in agreement with the prediction from the explicit equation for $v_{max,app}$ (eq. 12.18; see Figure 12.4): for high concentrations of B, $1+K_B/[B]$ reduces to unity, and $v_{max,app}$ becomes independent of $[B]$ and equals $v_{max}$. Although we have not explicitly derived the rate equation, we can already make predictions for the behavior of such a reaction from these considerations.

We have seen that the net rate constant for an irreversible step equals the microscopic rate constant $k_n$ for this step. For bimolecular reaction steps, the net first-order rate constant is the product of the microscopic second-order rate constant $k_n$ and the concentration of the ligand X, i.e. $k_n [X]$. Net rate constants for reversible steps are calculated from the product of the forward reaction and the partitioning ratio for forward and reverse reactions (eq. 12.21). The evaluation of net rate constants is always performed backwards, moving step by step from the end of the reaction scheme to the first step. For the reaction scheme

$$E \underset{k_{-1}}{\overset{k_1[S]}{\rightleftharpoons}} E_1 \underset{k_{-2}}{\overset{k_2}{\rightleftharpoons}} E_2 \overset{k_3}{\longrightarrow} E$$

scheme 12.9

we can formulate the net rate constants $k_1'$, $k_2'$, and $k_3'$, starting with $k_3'$ as

$$k_3' = k_3 \tag{eq. 12.34}$$

$$k_2' = k_2 \cdot \frac{k_3'}{k_{-2} + k_3'} = k_2 \cdot \frac{k_3}{k_{-2} + k_3} \tag{eq. 12.35}$$

and

$$k_1' = k_1[A] \cdot \frac{k_2'}{k_{-1} + k_2'} = k_1[A] \cdot \frac{k_2 \cdot \dfrac{k_3}{k_{-2} + k_3}}{k_{-1} + k_2 \cdot \dfrac{k_3}{k_{-2} + k_3}} \tag{eq. 12.36}$$

Again, we can calculate the overall net rate constant from the inverse of the sum of the net rate constants for each step (eq. 12.23). The concept of net rate constants and transit times is particularly helpful to analyze complex multi-step reactions.

## QUESTIONS

12.1   From a thermodynamic perspective, the change in free energy for binding of two substrates to an enzyme must be independent of the order of binding. Why can ordered binding be observed?

12.2   An enzyme follows the ping-pong mechanism. In the first step, E reacts with substrate A to give product P and E'. E' then binds substrate B. Conversion of B to product Q regenerates E. Using the steady-state approximation for E', derive the rate equation for the generation of the products.

12.3   An enzyme exists in two conformations, $E_1$ and $E_2$, which are interconverted with the rate constants $k_1$ and $k_{-1}$. Only $E_2$ binds substrate $S$. The bimolecular rate constant for substrate binding is $k_2$, the ES complex dissociates to $E_2$ and S with $k_{-2}$. Product formation occurs with $k_p$. Derive the overall rate constant of product formation using the principle of net rate constants and transit times.

## REFERENCES

Taurog, R. E., Jakubowski, H. and Matthews, R. G. (2006) Synergistic, random sequential binding of substrates in cobalamin-independent methionine synthase. *Biochemistry* **45**(16): 5083–5091.
· analysis of random substrate binding to methionine synthase

Samatanga, B. and Klostermeier, D. (2014) DEAD-box RNA helicase domains exhibit a continuum between complete functional independence and high thermodynamic coupling in nucleotide and RNA duplex recognition. *Nucleic Acids Res* **42**(16): 10644–10654.
· thermodynamic study that shows different order of binding and different thermodynamic linkage in ATP and RNA binding to RNA helicases

Segel, I. H. (1975) Enzyme kinetics : behavior and analysis of rapid equilibrium and steady state enzyme systems. New York, Wiley.
· a compendium of steady-state enzyme kinetics

Hickman, A. B., Namboodiri, M. A., Klein, D. C. and Dyda, F. (1999) The structural basis of ordered sub-strate binding by serotonin N-acetyltransferase: enzyme complex at 1.8 A resolution with a bisubstrate analog. *Cell* **97**(3): 361–369.
· structure of serotonin-N-acetyltransferase in complex with a bisubstrate inhibitor that suggests conformational changes upon acetyl-CoA binding

Porter, R. W., Modebe, M. O. and Stark, G. R. (1969) Aspartate transcarbamylase. Kinetic studies of the catalytic subunit. *J Biol Chem* **244**(7): 1846–1859.

Hsuanyu, Y. and Wedler, F. C. (1988) Kinetic mechanism of catalytic subunits (c3) of *E. coli* aspartate transcarbamylase at pH 7.0. *Biochim Biophys Acta* **957**(3): 455–458.
· kinetic studies that demonstrate ordered binding of substrates to aspartate transcarbamoylase

Wang, J., Stieglitz, K. A., Cardia, J. P. and Kantrowitz, E. R. (2005) Structural basis for ordered substrate binding and cooperativity in aspartate transcarbamoylase. *Proc Natl Acad Sci U S A* **102**(25): 8881–8886.
· structural study that shows conformational changes upon aspartate binding to aspartate transcarbamoylase

Wedler, F. C. and Gasser, F. J. (1974) Ordered substrate binding and evidence for a thermally induced change in mechanism for *E. coli* aspartate transcarbamylase. *Arch Biochem Biophys* **163**(1): 57–68.
· kinetic study that demonstrates ordered release of products from aspartate transcarbamoylase

Tanner, J. J., Lei, B., Tu, S. C. and Krause, K. L. (1996) Flavin reductase P: structure of a dimeric enzyme that reduces flavin. *Biochemistry* **35**(42): 13531–13539.
· structural basis for ping-pong mechanism by flavin reductase

Cleland, W. W. (1975) Partition analysis and the concept of net rate constants as tools in enzyme kinetics. *Biochemistry* **14**(14): 3220–3224.
· a clear description of the concept of net rate constants

# Temperature Dependence of Rate Constants

## 13.1 THE ARRHENIUS EQUATION

Reaction velocities and rate constants depend on temperature. As a rule of thumb, reaction velocities increase by a factor of 2–4 for each increase in temperature by 10°C. Empirically, it has been found that the temperature dependence of rate constants can be described by an exponential law as

$$k(T) = A \cdot e^{-E_a/RT} \qquad \text{eq. 13.1}$$

Eq. 13.1 is called the *Arrhenius equation*. $E_a$ is the activation energy of the reaction, the energy difference between the reactants and an activated complex formed by the reactants that decays to products (see Figure 13.1). By measuring rate constants at different temperatures, we can determine the activation energy for the reaction using the Arrhenius equation. In the following chapters, we will discuss two different concepts, the *transition state theory* (Section 13.2) and the *collision theory* (Section 13.3), that provide different molecular explanations for the temperature dependence of rate constants described by the Arrhenius equation.

## 13.2 TRANSITION STATE THEORY

Reactants and products of a reaction have a free energy $G$, which is the sum of their standard free energies of formation (Section 2.5.3; Figure 2.23). Reactants A and B and products P are separated from each other by an energy barrier, the *activation barrier* (Figure 13.1). The activation barrier reflects the high energy of an activated collision complex $AB^{\neq}$ of the reactants that is ready to dissociate into the products of the reaction. Parameters pertaining to the activated complex, its formation and dissociation, are typically marked by the superscript $\neq$.

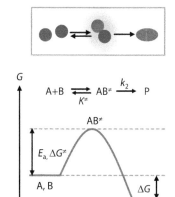

**FIGURE 13.1: Transition state theory.** The reactants A and B and the products P have different free energies $G$. The difference between their free energies is the change in free energy during the reaction, $\Delta G$. Reactants and products are separated by an activation barrier that represents a high-energy collision complex of A and B that dissociates into the products P. The height of the activation barrier relative to the energy of A and B is the activation energy $E_a$ or $\Delta G^{\neq}$.

We now assume that the reactants A and B and the activated complex $AB^{\neq}$ are in rapid equilibrium, and the velocity of the dissociation of $AB^{\neq}$ determines the overall reaction velocity of product formation (Scheme 13.1).

$$A + B \underset{K^{\neq}}{\rightleftharpoons} AB^{\neq} \xrightarrow{k_2} products \qquad \text{scheme 13.1}$$

The reaction velocity then is

$$v = k_2 \left[ AB^{\neq} \right] \qquad \text{eq. 13.2}$$

The concentration of the activated complex can be expressed in terms of the equilibrium constant of activation, $K^{\neq}$, as

$$K^{\neq} = \frac{\left[ AB^{\neq} \right]}{[A][B]} \qquad \text{eq. 13.3}$$

Note that $K^{\neq}$ is the equilibrium constant of $AB^{\neq}$ complex <u>formation</u>. Eq. 13.3 can be rearranged to give the concentration of the activated complex:

$$\left[ AB^{\neq} \right] = K^{\neq} [A][B] \qquad \text{eq. 13.4}$$

We know that equilibrium constants are related to free energy changes under standard conditions. Together with the Gibbs-Helmholtz equation (eq. 2.105), we can therefore write for $K^{\neq}$:

$$K^{\neq} = e^{-\Delta G^{\neq}/RT} = e^{-\Delta H^{\neq}/RT} e^{\Delta S^{\neq}/R} \qquad \text{eq. 13.5}$$

$\Delta G^{\neq}$ is the free energy of activation, $\Delta H^{\neq}$ is the activation enthalpy, and $\Delta S^{\neq}$ is the activation entropy. We can now express the concentration of the activated complex in terms of the activation enthalpy and entropy, and as a function of the concentration of the reactants A and B:

$$\left[ AB^{\neq} \right] = e^{-\Delta H^{\neq}/RT} e^{\Delta S^{\neq}/R} [A][B] \qquad \text{eq. 13.6}$$

Now we have to relate $k_2$ to molecular properties of the transition state. The rate constant $k_2$ for the dissociation of the transition state to the product(s) will depend on the vibrational energy of the scissile bond, the bond that needs to be broken in the reaction from transition state to products. The vibrational energy depends on the thermal energy of the molecule, which is the product of the Boltzmann constant $k_B$ and the temperature $T$ (see Section 19.1 and eq. 19.5). The vibration frequency $\nu$ can be expressed as

$$\nu = \frac{k_B T}{h} \qquad\qquad \text{eq. 13.7}$$

with the Planck constant $h$. We have come across this vibration frequency before when we discussed the kinetic isotope effect (Chapter 9) which is caused by different vibration frequencies of the scissile bond, depending on the masses of the bonded atoms. We can now express the rate constant $k_2$ in Scheme 13.1 by multiplying this frequency with a transmission coefficient $\kappa$

$$k_2 = \kappa \frac{k_B T}{h} \qquad\qquad \text{eq. 13.8}$$

The coefficient $\kappa$ describes the probability of the bond to dissociate in the direction of the product(s). Substituting eq. 13.6 and eq. 13.8 into eq. 13.2 now yields

$$\nu = \kappa \frac{k_B T}{h} e^{-\Delta H^{\neq}/RT} e^{\Delta S^{\neq}/R} [A][B] \qquad\qquad \text{eq. 13.9}$$

Eq. 13.9 is called the *Eyring equation*. It provides an expression for the reaction velocity $\nu$ for a bimolecular reaction of A and B in the form of

$$\nu = k[A][B] \qquad\qquad \text{eq. 13.10}$$

with the rate constant $k$

$$k = A \cdot e^{-\Delta H^{\neq}/RT} \qquad\qquad \text{eq. 13.11}$$

and the pre-exponential factor $A$:

$$A = \kappa \frac{k_B T}{h} e^{\Delta S^{\neq}/R} \qquad\qquad \text{eq. 13.12}$$

These equations relate the *activation energy* $\Delta G^{\neq}$ for a reaction to its rate constant $k$. A high energy barrier corresponds to a small $k$ and a slow reaction, and a low energy barrier corresponds to a high $k$ and a fast reaction. Eq. 13.12 is formally equivalent to the empirical Arrhenius equation (eq. 13.1), with $E_a = \Delta H^{\neq}$. The pre-exponential factor $A$ in eq. 13.11 contains the entropy of activation $\Delta S^{\neq}$.

Note that the activation energies $\Delta G^{\neq}$ for the forward and reverse reactions differ if reactants and products have different free energies. For a spontaneous reaction with $\Delta G < 0$, $\Delta G_f^{\neq}$ for the forward reaction is smaller than $\Delta G_r^{\neq}$ for the reverse reaction, and the forward reaction will be faster than the reverse reaction. The equilibrium constant $K_{kin}$ for product formation, which is the ratio of $k_f$ and $k_r$, is thus larger than unity. Under standard conditions we can also relate the energy difference $\Delta G^0$ to the equilibrium constant $K$, which is larger than unity for exergonic reactions with $\Delta G^0 < 0$. Kinetic and energetic considerations are thus consistent. For an endergonic reaction ($\Delta G^0 > 0$, $K < 1$), $\Delta G_f^{\neq}$ is larger than $\Delta G_r^{\neq}$, and $k_f$ is smaller than $k_r$, consistent with $K_{kin} = k_f/k_r < 1$. For a reaction in equilibrium, we have $\Delta G^0 = 0$ and $K = 1$ on the energy side, and $\Delta G_f^{\neq} = \Delta G_r^{\neq}$, $k_f = k_r$, and $K_{kin} = k_f/k_r = 1$ on the kinetic side (Figure 13.2).

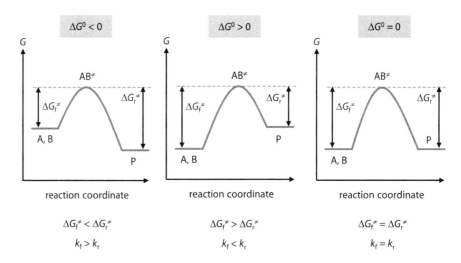

**FIGURE 13.2:** **Activation energy and rate constants for forward and reverse reactions.** Energy diagrams for an exergonic (left, $\Delta G^0 < 0$, $K_{eq} > 1$) and an endergonic reaction (center, $\Delta G^0 > 0$, $K_{eq} < 1$) and for a reaction in equilibrium (right, $\Delta G^0 = 0$, $K_{eq} = 1$). The activation energies for the forward and reverse reaction, $\Delta G_f^{\neq} = \Delta G_r^{\neq}$, are indicated. For the exergonic reaction, $k_f$ exceeds $k_r$, and the equilibrium constant $K_{kin}$ calculated from the rate constants is larger than unity, in agreement with the $K_{eq}$ from $\Delta G^0$. For the endergonic reaction, $k_f$ is smaller than $k_r$, giving $K_{kin} > 1$. In equilibrium, $\Delta G_f^{\neq}$ equals $\Delta G_r^{\neq}$, $k_f$ equals $k_r$, and $K_{kin}$ equals 1. In all three scenarios, the energetic and kinetic considerations are consistent.

## 13.3 COLLISION THEORY

An alternative approach to reaction velocities assumes that the reaction velocity is proportional to the number of collisions $Z$ of the reactants, and to the probability that they collide with sufficient energy for the reaction to occur. The kinetic energy required for a productive collision is called the activation energy, $E_a$. The probability that the kinetic energy of the colliding molecules is equal to the activation energy is $e^{-E_a/RT}$. Multiplication of these two expressions yields the reaction velocity $v$:

$$v = Z \cdot e^{-E_a/RT}$$

eq. 13.13

The number of collisions $Z$ will depend on the concentrations of the reactants, and we can express $Z$ as

$$Z = X \cdot [A][B]$$

eq. 13.14

yielding

$$v = X \cdot [A][B] \cdot e^{-E_a/RT}$$

eq. 13.15

Again, this is an expression for the reaction velocity $v$ in the form of

$$v = k[A][B]$$

eq. 13.16

this time with

$$k = X \cdot e^{-E_a/RT}$$

eq. 13.17

Eq. 13.17 has the same form as the empirical Arrhenius equation (eq. 13.1), with $A = X$. The pre-exponential factor $X$ is a probability factor that takes into account speed, size and orientation of reactants. The units depend on the type of the reaction: for a first-order reaction the units are $s^{-1}$, for a second-order reaction $M^{-1} s^{-1}$. $E_a$ is the activation energy of the reaction ($\Delta G^{\neq}$).

Although eq. 13.9 (transition state theory) and eq. 13.15 (collision theory) have the same form, the molecular meanings of the pre-exponential factors and activation energies are different. Both theories explain the concentration and temperature dependence of rate constants, but describe the effects with different parameters. The increase in rate constants with temperature is interpreted as accelerated breaking of the relevant bond in the transition state because of a higher vibration frequency (transition state theory) or as increased collision frequency due to higher kinetic energy of the reactants (collision theory). The increase in reaction velocities with increased reactant concentration is caused by a shift in the activation equilibrium to the transition state (transition state theory) or a higher frequency of collisions with less distance separating the reactant molecules (collision state theory).

# 13.4 KINETIC AND THERMODYNAMIC CONTROL OF REACTIONS

Thermodynamic considerations allow us to predict if a reaction occurs spontaneously (Section 2.5), but do not provide any information on the rate with which the reaction occurs. The reaction velocity depends on the height of the energy barrier that separates the reactant(s) from the product(s). Therefore, a thermodynamically favorable reaction might occur very slowly if the activation energy is high. Compounds that are stable because they react slowly, although the reaction would be energetically downhill, are called *metastable*. One prominent example of a metastable compound is ATP. We have discussed ATP hydrolysis as a highly exergonic reaction several times. Despite the large negative $\Delta G^0$, ATP hydrolysis is rather slow because of a high activation barrier for hydrolysis, and ATP is quite stable. The situation is different *in vivo*, where ATP hydrolysis is catalyzed by enzymes that reduce the activation barrier and increase the rate of ATP hydrolysis by several orders of magnitude. Peptides are another biochemical example of metastable compounds. Peptide bond formation is an endergonic reaction that is coupled to ATP hydrolysis in the cell. The inverse reaction, peptide bond hydrolysis, is exergonic, but the activation energy is very high, and the spontaneous hydrolysis of peptides is hardly observed.

It is important to appreciate the implications of the interplay between kinetics and thermodynamics for parallel reactions (Section 8.2). If a reactant A can be converted to B or C, and the free energy change $\Delta G_r$ is more negative for the reaction to C than for the reaction to B (Figure 13.3), then C will be the more populated (or even the exclusive) reaction product at the end of the reaction: the reaction is under *thermodynamic control*. However, the time that is necessary until all three compounds have equilibrated depends on the rate constants for individual steps, and these depend on the activation energies. If the activation barrier between A and the stable product C is much higher than the energy barrier that separates A from the less stable product B (Figure 13.3), then A will be converted much more rapidly to B than to C, and B is generated at the beginning of the reaction. The (transient) accumulation of B is a *kinetically controlled* reaction. During the slow equilibration, B can then be re-converted to A and either revert to B or slowly react to C. Once C is generated, the reaction back to A is slow because of the high energy barrier. Nevertheless, after some time, the thermodynamically controlled equilibrium fractions of A, B, and C are reached. In extreme cases, depending on the differences in $\Delta G^0$ and $\Delta G^{\neq}$ for the two branches of the reaction, the less stable products might be exclusively generated in a kinetically controlled reaction, followed by an infinitely slow equilibration to the thermodynamically controlled product. By performing a reaction under kinetic control, thermodynamically disfavored reaction products can be captured.

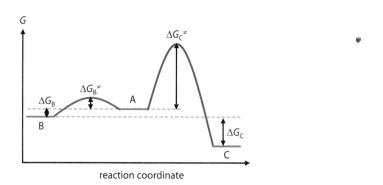

FIGURE 13.3:   **Thermodynamic and kinetic control.** Reactant A can react to products B or C (parallel reaction). The reaction to B has a low activation energy and is rapid. The reaction to C has a high activation energy and is slow. C is the more stable product. Under kinetic control, B will be formed. Under thermodynamic control, C is the dominant product.

## QUESTIONS

13.1   The velocity of a reaction increases two-fold when the temperature is increased from 25°C to 37°C. Calculate the activation energy. How does the activation energy change when the increase in the rate constant $k$ is four-fold?

13.2   A protein folds to the native state that is 30 kJ mol$^{-1}$ more stable than the unfolded state. The activation barrier for folding is $E_a = 40$ kJ mol$^{-1}$. The protein can also form a non-functional intermediate ($\Delta G_r = -20$ kJ mol$^{-1}$). The activation energy is $E_a = 5$ kJ mol$^{-1}$. What are the (relative) rate constants for the reaction from U to I, I to U, U to N and N to U? What are the approximate fractions of I and N under kinetic control, what is the fraction of N under thermodynamic control? Which rate constant determines the time scale of folding into the native state? Assume in a first approximation that the pre-exponential factor of the Arrhenius equation is identical for all four reactions. What are the equilibrium constants and the fractions of N, U and I in equilibrium?

13.3   DNA strands (A,B) associate rapidly to mismatched intermediates that slowly rearrange to the correctly base-paired duplex P. Derive an expression for the activation energy of the overall reaction from A and B to P. What is the reason for a "negative activation energy"?

13.4   Bimolecular reactions can be described by a first step that is the encounter of the reactants, and a second step that is the chemical reaction to products. Derive the overall rate constant and simplify for a diffusion-controlled reaction in which the formation of the encounter complex (diffusion) is rate-limiting, and the reaction to products is rapid, and for an activation-controlled reaction where the reaction to products is rate-limiting.

# Principles of Catalysis

Catalysts accelerate chemical reactions without affecting the equilibrium constant. From the relationship between forward and reverse rate constants and the equilibrium constant (eq. 8.16, eq. 8.18), we can infer that catalysis must accelerate forward and reverse reactions by the same factor. According to transition state theory (Section 13.2), catalysts accelerate chemical reactions by stabilization of the active complex, the transition state (Figure 14.1). As a consequence, the activation barrier is reduced, and the rate constant is increased. The catalyst is regenerated at the end of the reaction, meaning it is not consumed but can catalyze multiple turnovers.

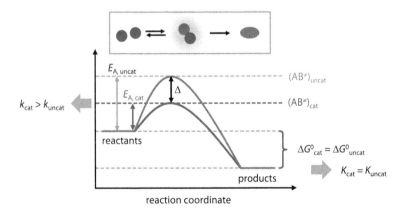

FIGURE 14.1: **Catalysis and the transition state.** A catalyst stabilizes the transition state $AB^{\neq}$ of a reaction. The concomitant reduction of the energy barriers for forward and reverse reactions ($\Delta$) leads to an acceleration ($k_{cat} > k_{uncat}$) of the reaction in both directions. The free energy of the reactants and products is unaltered, and the equilibrium constant $K$ of the reaction is therefore not changed.

## 14.1 ENZYME CATALYSIS

Biomolecules that act as catalysts include enzymes, catalytically active antibodies, and ribozymes. An enzyme-catalyzed reaction starts with formation of the enzyme-substrate complex (Chapters 11, 12). From the enzyme-substrate complex, the transition state is reached, leading to the enzyme-product complex that dissociates to enzyme and product (Figure 14.2; see also Scheme 11.9). The catalytic action of these enzymes is caused by stabilization of the transition state of the reaction: enzymes bind tightly to the transition state of the catalyzed reaction. We have encountered this phenomenon before as the basis for the efficient inhibition of enzyme activity by transition state analogs (Section 11.6; Box 11.5). The stabilization of the transition state leads to a decrease in activation energy, both for the forward reaction from reactants to products and for the inverse reaction from products to reactants. The energies of the reactants (enzyme substrates) and product are unaltered, however, meaning the change in the standard free energy $\Delta G^0$ is identical for the catalyzed and uncatalyzed reaction. Thus, the equilibrium constants of uncatalyzed and enzyme-catalyzed reactions are also identical.

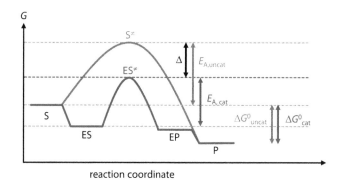

**FIGURE 14.2:    Enzyme catalysis and transition state stabilization.** Enzymes bind their substrate S and form an enzyme-substrate complex (ES). The enzyme-substrate complex is converted to the enzyme-product complex (EP), from which the product P dissociates (see also Scheme 11.9). Enzymes stabilize the transition state ($\Delta$), and achieve an acceleration of the reaction by lowering the energy barrier for both forward (S→P) and reverse reactions (P→S) compared to the uncatalyzed reaction (gray). The energy of the substrate and the product is unaltered, and the standard free energy changes $\Delta G^0_{cat}$ and $\Delta G^0_{uncat}$ of the catalyzed and uncatalyzed reactions are identical. Hence, enzymes do not alter the equilibrium between substrate and product. Note that the energy diagram only reflects the energy levels for the fraction of substrate that is bound to the enzyme. Under catalytic conditions, the enzyme concentration is much lower than the substrate concentration, and almost all substrate molecules which contribute to the overall energy are in their free form.

Note that for enzyme-catalyzed reactions, we have to distinguish between the overall equilibrium constant $K$ between substrate and product ($[P]/[S]$), and the *on-enzyme equilibrium* $K_{enz}$ between ES and EP ($[EP]/[ES]$). Overall equilibrium constant and on-enzyme equilibrium constant can have different values. Nevertheless, under catalytic conditions (i.e. at very low concentrations of enzyme compared to substrate) enzymes do not alter the overall equilibrium constant of the catalyzed reaction: the equilibrium constant $K$ is defined by the difference in free energy between substrate and product.

Interactions of the enzyme with the substrate (see Section 16.2.7) contribute to enzyme specificity. However, we can directly infer from the energy diagram in Figure 14.2 that extremely tight binding of the enzyme to its substrate, i.e. a low energy of the ES complex compared to free substrate, is not a successful strategy for catalysis. The amount to which the energy of the ES complex is reduced directly adds to the activation energy for the enzyme-catalyzed forward reaction. To lower the activation energy, enzymes need to bind the transition state of the reaction much more tightly than the ground state (Figure 14.3). Enzymes are not perfectly complementary to their substrate, but undergo conformational changes that strengthen the interaction with the substrate along the reaction coordinate (induced fit; see Section 16.2.7). The concept of strong binding to transition states as the basis for catalysis has been used to generate catalytic antibodies: antibodies generated against transition state analogs of chemical reactions often catalyze the respective reaction.

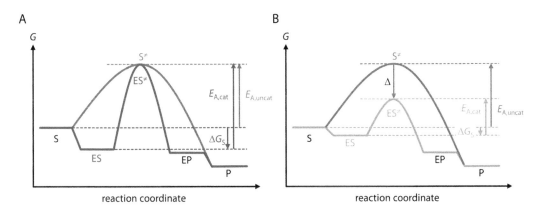

**FIGURE 14.3:** **Effect of substrate and transition state binding on enzyme catalysis.** A: An "enzyme" that binds tightly to the substrate ($\Delta G_S$), but not to the transition state does not lower the activation barrier $E_a$, and does not catalyze the reaction. B: An enzyme that binds weakly to the substrate ($\Delta G_S$), but tightly to the transition state ($\Delta$) lowers the activation barrier $E_a$ for the reaction, and efficiently catalyzes the conversion from substrate to product.

Very tight binding of the enzyme to the products is counterproductive for efficient catalysis because the products need to dissociate for regeneration of the enzyme. Tight binding of the substrate and a low energy of the EP complex would render substrate dissociation endergonic, and would prevent the enzyme from multiple turnovers. This problem is often encountered in catalytic RNAs, such as ribozymes that cleave an RNA substrate. Here, the cleaved RNA strand remains bound to the ribozyme by base pairing and does not readily dissociate. The dissociation then becomes rate-limiting for the overall reaction. The catalytic efficiency of protein enzymes is also reduced when substrate and product are very similar. In this case, the product is also bound tightly, leading to product inhibition (Section 11.6.1). An example is ATP hydrolysis, where the substrate ATP and the product ADP/$P_i$ are chemically not too different. Many ATP-hydrolyzing enzymes, such as the muscle protein myosin, show strong product inhibition by ADP.

In the following, we will briefly discuss the main strategies of enzymes to stabilize transition states and to accelerate biochemical reactions. For more detailed information, the reader is referred to textbooks on catalysis.

## 14.2 ACID-BASE CATALYSIS

*Acid-base catalysis* involves the transfer of a proton to the reactant to stabilize negative charges in the transition state (acid catalysis), or the abstraction of a proton to stabilize a developing positive charge (base catalysis). The overall reaction scheme for catalysis of the reaction of A and B to product P by an acid HX is (Scheme 14.1):

$$A + H\text{-}X \underset{k_{-1}}{\overset{k_1}{\rightleftharpoons}} AH^+ + X^-$$

$$AH^+ + B \xrightarrow{k_2} P \qquad \qquad \text{scheme 14.1}$$

The overall reaction velocity of product formation is

$$v = k_2 \left[ AH^+ \right] \left[ B \right] \qquad \qquad \text{eq. 14.1}$$

Similar to the steady-state approximation for the ES complex in the derivation of the rate equation for enzyme reactions (Section 11.2), we can formulate a steady-state condition for $[AH^+]$:

$$\frac{d\left[ AH^+ \right]}{dt} = k_1 [A][HX] - k_{-1}\left[ AH^+ \right]\left[ X^- \right] - k_2\left[ AH^+ \right]\left[ B \right] \approx 0 \qquad \qquad \text{eq. 14.2}$$

From eq. 14.2, we can derive an expression for $[AH^+]$

$$[AH^+] = \frac{k_1[A][HX]}{k_{-1}[X^-] + k_2[B]}$$

eq. 14.3

and substitute $AH^+$ in the reaction velocity (eq. 14.1) to obtain eq. 14.4:

$$v = \frac{k_1 k_2 [A][HX][B]}{k_{-1}[X^-] + k_2[B]}$$

eq. 14.4

In a first case, we assume that protonation and deprotonation are rapid, and product formation is rate-limiting, with

$$k_{-1}[X^-] \gg k_2[B]$$

eq. 14.5

so that we can neglect the term $k_2[B]$ in the denominator in eq. 14.4, and can approximate the reaction velocity $v$ as

$$v \approx \frac{k_1 k_2 [A][HX][B]}{k_{-1}[X^-]}$$

eq. 14.6

From the definition of the equilibrium constant for the protolysis of HX, $K_A$:

$$K_A = \frac{[H^+][X^-]}{[HX]}$$

eq. 14.7

we obtain

$$\frac{[HX]}{[X^-]} = \frac{[H^+]}{K_A}$$

eq. 14.8

which we can now substitute into eq. 14.6, which gives

$$v \approx \frac{k_1 k_2 [A][B][H^+]}{k_{-1} K_A}$$

eq. 14.9

The rate of the reaction thus depends linearly on the concentration of protons. *Specific acid catalysis* refers to such reactions where protons act as the catalysts. In analogy, *specific base catalysis* refers to reactions catalyzed by hydroxide ions. Examples are the base-catalyzed cleavage of ribonucleic acid, or the acid-catalyzed depurination of DNA (Box 14.1).

**BOX 14.1: SPECIFIC ACID-BASE CATALYSIS.**

The cleavage of RNA at high pH occurs *via* specific base catalysis by hydroxide ions. DNA undergoes depurination under acidic conditions *via* specific acid catalysis by protons (Figure 14.4).

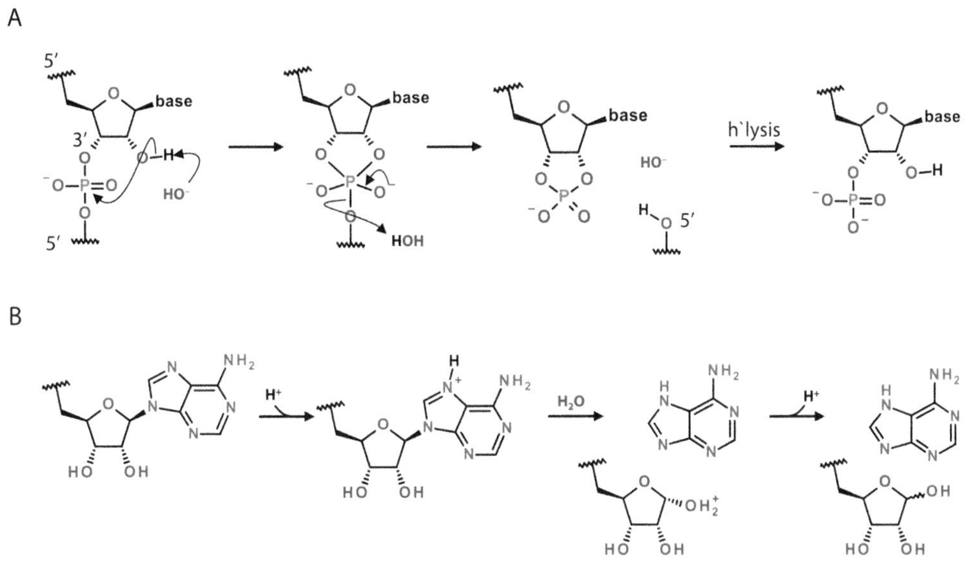

A

B

**FIGURE 14.4:    Specific acid and base catalysis.** A: RNA cleavage by specific base catalysis (OH⁻). B: DNA depurination by specific acid catalysis (H⁺).

Now we treat the case

$$k_{-1}\left[X^-\right] \ll k_2\left[B\right]$$

eq. 14.10

meaning a slow protolysis equilibrium is followed by rapid product formation. In this case we can neglect the first term in the denominator of eq. 14.4, and obtain a different approximation for *v*:

$$v \approx \frac{k_1 k_2\left[A\right]\left[HX\right]\left[B\right]}{k_2\left[B\right]}$$

eq. 14.11

or

$$v \approx k_1\left[A\right]\left[HX\right]$$

eq. 14.12

Here, the rate of the reaction is proportional to the concentration of the acid, not to the proton concentration. This case is called *general acid catalysis*. Specific and general acid catalysis are associated with different energy profiles (Figure 14.5).

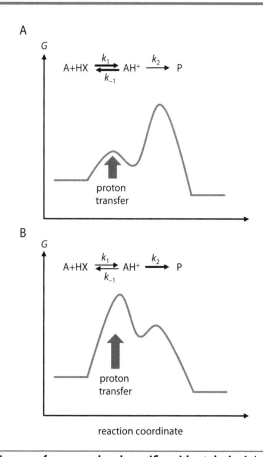

**FIGURE 14.5:** **Energy diagrams for general and specific acid catalysis**. A: In specific acid (base) catalysis, rapid transfer of the proton is followed by a slow conversion of the protonated substrate to product(s). The rate constant of the overall reaction depends linearly on the proton concentration, and is thus pH-dependent. B: In general acid (base) catalysis, proton transfer is slow, and dissociation of the protonated substrate to products is rapid. The rate constant of the overall reaction is proportional to the concentration of the free acid HA.

In general acid-base catalysis, positive or negative charges developing during the reaction, i.e. in a transition state, are stabilized by proton transfer to a base or from an acid, respectively. This is a common mechanism in enzyme-catalyzed reactions (Box 14.2).

---

### BOX 14.2: GENERAL ACID-BASE CATALYSIS.

The enzyme RNase A catalyzes the hydrolysis of phosphodiester bonds in ribonucleic acids (Figure 14.6). In the first step, a histidine side chain (His12) acts as a general base, and abstracts a proton from the 2′-hydroxyl of the scissile bond. The deprotonated 2′-O⁻ then forms a cyclic phosphate by nucleophilic substitution at the phosphorous. The leaving group, the 5′-end of the cleaved RNA, is protonated at the 5′-O⁻ by a second histidine, His 119, which acts as a general acid. In the second step, the inverse reaction takes place with water: His119 now acts as a general base, and accepts a proton from $H_2O$. The hydroxyl then performs a nucleophilic substitution at the phosphorus of the cyclic phosphate. The 2′-O⁻ leaving group is protonated by His12 that now acts as a general acid. Catalytic residues for acid-base catalysis include histidine, serine, and cysteine, but also cofactors, e.g. pyridoxal phosphate.

*(Continued)*

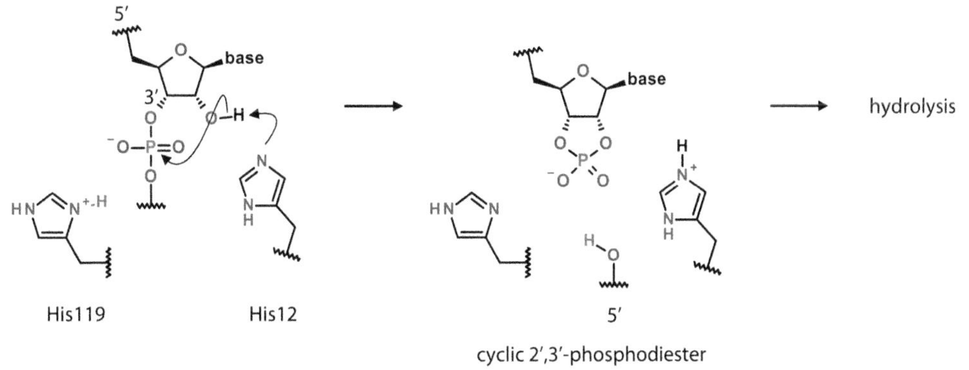

**FIGURE 14.6:** **General acid-base catalysis in the reaction of RNase A**. His12 acts as a general base and abstracts a proton from the 2'-OH of the ribose. The 2'-O⁻ then performs a nucleophilic substitution at the phosphorous. His119 acts as a general acid and protonates the leaving group. The 5'-RNA dissociates from the enzyme, and is replaced by a water molecule. The hydrolysis of the cyclic phosphodiester is achieved by the reverse sequence: His119 acts as a general base and accepts a proton from the water molecule. The OH⁻ performs a nucleophilic substitution at the phosphorous, and His12 acts as a general base and protonates the leaving group, the 2'-O⁻ of the ribose. At the end of the reaction, His12 and His119 are in the same protonation states as before.

## 14.3 ELECTROSTATIC AND COVALENT CATALYSIS

Charges of the transition state do not have to be stabilized by proton transfer involving an acid or base. Positive charges can also be stabilized by electrostatic interactions (see Section 15.3) with negatively charged groups in the vicinity (*nucleophilic catalysis*), and negative charges can be stabilized by positively charged groups or metal ions (*electrophilic catalysis*). As an alternative to transition state stabilization, catalysts sometimes also allow the reaction to follow alternative pathways. A common mechanism for the enzymatic catalysis of biochemical reactions involves *covalent catalysis* (Box 14.3), in which the catalyst, often a prosthetic group or a reactive protein side chain, activates the substrate by formation of a covalent intermediate. Active site residues for covalent catalysis are nucleophilic, and are good leaving groups to ensure turnover of the intermediate to products. Pyridoxal phosphate-dependent enzymes such as aminotransferases are prominent examples for covalent catalysis (Box 14.3). Serine proteases such as trypsin and chymotrypsin also employ covalent catalysis.

### BOX 14.3: ACTIVATION BY FORMATION OF A COVALENT INTERMEDIATE.

Aminotransferases are pyridoxal phosphate-dependent enzymes that transfer amino groups from amino acids to α-keto acids during amino acid degradation (Figure 14.7). Aspartate aminotransferase catalyzes the transfer of the amino group from aspartate to α-ketoglutarate, leading to formation of oxaloacetate and glutamate. In a reaction catalyzed by glutamate dehydrogenase, glutamate then undergoes oxidative deamination, which leads to regeneration of α-ketoglutarate and the formation of ammonium ions. In vertebrates, ammonium ions are converted to urea that is excreted. Aminotransferase contains the prosthetic group pyridoxal phosphate that forms a covalent bond with a lysine residue in the active site (Figure 14.7). When the substrate amino acid binds, a covalent intermediate between the amino acid and pyridoxal phosphate is formed, in which the

*(Continued)*

amino acid replaces the lysine residue. The resulting aldimine intermediate is isomerized to a ketimine intermediate that is hydrolyzed to a α-keto acid and pyridoxamine phosphate. The α-keto acid dissociates from the enzyme. In the second half of the reaction, the binding site is occupied by α-ketoglutarate that forms a ketamine with pyridoxamine phosphate. Now the ketimine is isomerized to an aldimine that hydrolyzes to glutamate and pyridoxal phosphate.

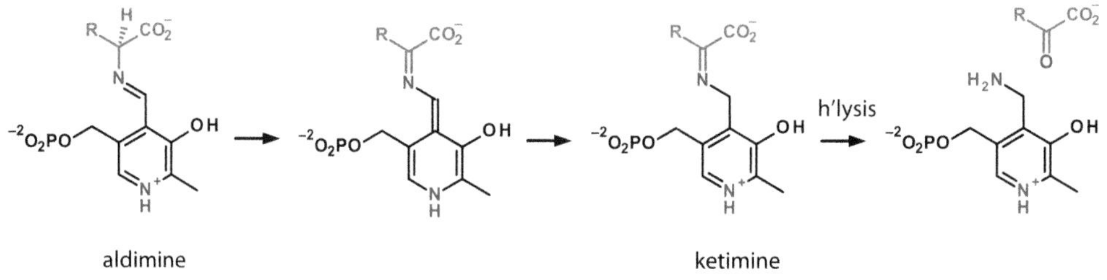

**FIGURE 14.7: Covalent catalysis by aminotransferases.** Aminotransferases use pyridoxal phosphate as a prosthetic group. Pyridoxal phosphate forms a covalent bond with a lysine residue in the active site of the aminotransferase. An amino acid that is bound as a substrate replaces the catalytic lysine, and forms an aldimine with pyridoxal phosphate that is isomerized to a ketamine, and hydrolyzed to the α-keto acid and pyridoxamine phosphate.

## 14.4 INTRAMOLECULAR CATALYSIS AND EFFECTIVE CONCENTRATIONS

*Intramolecular catalysis* has a major advantage in that it reduces the entropic cost of bringing together reactants. This effect relates to the entropic term in the transition state theory (eq. 13.12). When two substrate molecules react to one product molecule, an overall reduction in entropy is observed. This entropy loss is only compensated slightly by increases in entropy caused by different modes of internal rotation and vibration in the product molecule. By using intramolecular catalysis, enzymes avoid this unfavorable entropic term. A reaction with general acid- or base-catalysis has a second-order rate constant, and its rate increases linearly with the concentration of the acid or base (eq. 14.12). Catalysis of the same reaction within an enzyme-substrate complex is a unimolecular reaction, with a first-order rate constant. As the catalytic group and the substrate are part of the same entity, the concentration of the acid or base is difficult to define. An "effective" concentration can be calculated by comparing reaction rates of model compounds that contain the substrate and the catalytic group within the same molecule with rates of the intermolecular reaction at different concentrations of the acid or base. The concentration of external acid or base required to give the same reaction rate as observed for intramolecular catalysis corresponds to the effective concentration of this group in the enzyme. Effective concentrations are typically in the high molar range (general acid catalysis) and can reach values as high as $10^5$ M in nucleophilic catalysis. These high local concentrations contribute to the high catalytic efficiency of enzymes.

## QUESTIONS

14.1 Using a thermodynamic cycle, illustrate why an enzyme needs to bind tightly to the transition state, not the ground state of the reaction, to be an efficient catalyst.

14.2 An enzyme reduces the activation energy $E_a$ of the catalyzed reaction A $\rightleftarrows$ B by 5.7 kJ mol$^{-1}$. What is the acceleration of the forward and reverse reaction at 25°C?

# REFERENCES

Jencks, W. P. (1987) Catalysis in chemistry and enzymology. New York, Dover.
· a compendium about the principles of catalysis

Knowles, J. R. (1991) Enzyme catalysis: not different, just better. *Nature* **350**(6314): 121–124.
· a review article that explains why enzymes are excellent catalysis

Pollack, S. J., Jacobs, J. W. and Schultz, P. G. (1986) Selective chemical catalysis by an antibody. *Science* **234**(4783): 1570–1573.
· description of the first catalytic antibody that tightly binds to a transition state analog

De La Cruz, E. M., Sweeney, H. L. and Ostap, E. M. (2000) ADP inhibition of myosin V ATPase activity. *Biophys J* **79**(3): 1524–1529.
· product inhibition of myosin

Moore, S. A. and Jencks, W. P. (1982) Reactions of acyl phosphates with carboxylate and thiol anions. Model reactions for CoA transferase involving anhydride formation. *J Biol Chem* **257**(18): 10874–10881.
· effective concentrations

Page, M. I. and Jencks, W. P. (1971) Entropic contributions to rate accelerations in enzymic and intramolecular reactions and the chelate effect. *Proc Natl Acad Sci U S A* **68**(8): 1678–1683.
· entropic effects in catalysis

# Part III
# Molecular Structure and Stability

Parts I and II have introduced the energetic and kinetic principles that govern all molecular interactions between molecules and within molecules. In the following Part, we discuss the structural aspects and biophysical consequences of molecular interactions, starting from simple organic entities of a few atoms to increasingly large biological molecules with masses in the MDa range.

# Molecular Structure and Interactions

$\mathbf{W}$hile some basic concepts of organic chemistry are reviewed in the following, the reader should be familiar with the different bond types ($\sigma$,$\pi$), bond energies, hybridization of atomic orbitals to molecular orbitals (sp, sp$^2$, sp$^3$), and the concept of electronegativity.

## 15.1 CONFIGURATION AND CONFORMATION

Since isomers play an important role in biology, we briefly recapitulate the different kinds of isomers that molecules can form (Figure 15.1). *Isomers* are molecules that have the same sum formula, but the order of atoms and the sequence of chemical bonds connecting these atoms differs. The first main class of isomers, the *constitutional isomers*, has fundamentally different arrangements of atoms. Examples for different constitutions are the simple alcohols 1-propanol and 2-propanol. Two important metabolites in glycolysis, the phosphates of dihydroxyacetone and glycerol aldehyde (Figure 15.1) are also constitutional isomers. The second class of isomers is the *stereoisomers*. While stereoisomers have the same constitution, they possess a different spatial arrangement of atoms. The interconversion of constitutional isomers or of stereoisomers is only possible when bonds within the molecule are broken and re-formed. The associated energies for bond breakage are substantial (Table 15.1) and lie in the hundreds of kJ mol$^{-1}$. Enzymes are central to biology because they catalyze breakage and re-formation of chemical bonds at physiological conditions, not only during the interconversion of isomers but for many other metabolic transformations.

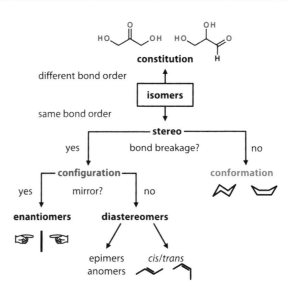

**FIGURE 15.1:** **Classification of isomers.** Molecules with a different bond sequence are constitutional isomers, all others are stereoisomers. The stereoisomers can be grouped into isomers with different conformation (no bond breakage necessary for interconversion) or configuration (bond breakage necessary). Enantiomers are mirror images, whereas diastereomers include both hindered rotation (*cis/trans*) and several stereocenters.

To put the different forms of isomers into perspective, a few illustrative biological examples are given. Ubiquitous in biology are *tautomers*, which are configurational isomers where a hydrogen atom is shifted from a carbon atom next to a keto or aldehyde group to the oxygen atom (Figure 15.2). This *keto/enol tautomerism* is frequently observed in biochemistry, for example in reducing sugars and as part of interconversion of metabolites in many metabolic pathways. Nitrogen atoms in the histidine imidazole (Figure 15.2) and nucleic acid bases (Section 17.1) can also participate in keto/enol tautomerism, in which case the isomers are *imino/enamine tautomers*. The energy difference between different tautomers is comparatively low and isomerization can therefore occur spontaneously.

In *cis/trans isomerization*, a $\pi$-bond is broken and re-established after a 180° rotation about the remaining $\sigma$-bond (Figure 15.2). *Cis/trans* isomerization of retinal is the key step in the visual process (Box 19.1 in Section 19.2.6) with the *trans* isomer usually being more stable than the *cis* isomer. Unsaturated fatty acids also occur as cis or trans isomers. In lipid bilayers, the *cis* isomers cannot pack as tightly as the corresponding *trans*- and saturated fatty acids, and a higher content of *cis*-fatty acids can therefore increase the fluidity of membranes (Section 16.2.8).

Probably the most important configurational isomers in biological molecules are the *enantiomers* (Figure 15.2). These molecules possess a center of mirror symmetry or handedness, which is somewhat confusingly also called *stereogenic center, stereocenter,* or *chiral center*. Molecules that behave like mirror images to each other form an enantiomeric pair. Just like *cis/trans* isomers and tautomers, enantiomers cannot be interconverted by simple rotation and translation operations. Whole classes of molecules, such as amino acids in proteins, nucleotides in nucleic acids, and lipids in membranes, occur only in one enantiomeric form in nature. Most chiral centers in biological molecules are carbon atoms with four different substituents $R_1$–$R_4$. To establish whether a molecule has the *(R)* (lat.: *recte*, right) or *(S)* (lat.: *sinister*, left) configuration, the substituents are prioritized by their atomic numbers, multiplied by the bond order, if applicable. The molecule is turned such that the substituent $R_1$ of lowest priority points to the background. Going from $R_2$ to $R_4$ in a clockwise (right) or counter-clockwise (left) rotation then defines *(R)* and *(S)* configurations. Molecules containing more than one stereocenter are known as *diastereoisomers* or just *diastereomers*. All carbohydrates, nucleotides, and the amino acids isoleucine and threonine belong to this class. The presence of $n$ stereocenters in a molecule gives rise to $2^n$ diastereomers, but in biology usually only a single diastereomer is relevant. Molecules that contain several stereocenters and that differ only in the configuration of a single chiral center are termed *epimers*. For example, mannose is the C2-epimer of glucose, and galactose is the C4-epimer of glucose.

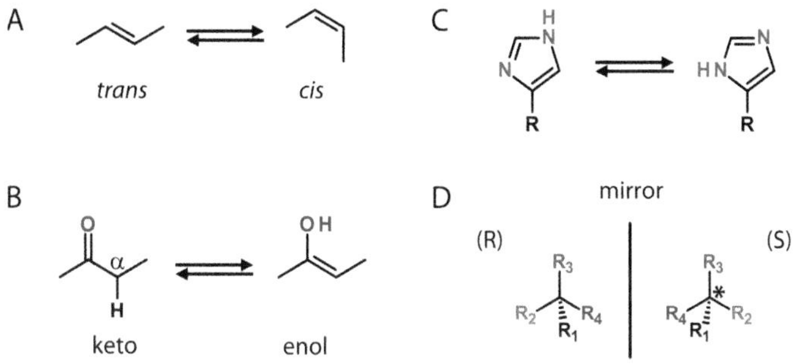

**FIGURE 15.2:** **Common configurational isomers found in molecules.** A: In *cis/trans* isomerism, a π-bond is broken and re-formed after rotation about the remaining σ-bond. B: In keto-enol tautomers, a hydrogen atom is shifted between a carbon atom and an oxygen atom. C: A substituted imidazole, such as histidine, can also exist as two tautomers. Both are observed in proteins, but the one shown on the left is more stable. D: Enantiomers possess mirror symmetry (shown as a vertical line). The most common stereocenters in biology are asymmetric carbon atoms with four different substituents $R_1$–$R_4$.

In contrast to configurations, *conformations* are the different relative orientations of parts of a molecule to one another. Conformations can be interconverted without bond breakage, typically by rotations about single bonds, which can lead to very different three-dimensional shapes of the molecule. Conformational isomers are known as *conformers*.

Changes in conformation do not require bond breakage and usually have low activation barriers. As a consequence, interconversion of conformers is a rapid process, and a molecule can sample thousands of conformations per second (Table 16.9). In the next paragraphs, we will see how such rotations affect the structure and internal energy of molecules.

The total internal energy $E_{total}$ of an isolated molecule of $N$ atoms depends on its intramolecular interactions, which can be divided into covalent and non-covalent contributions:

$$E_{total} = \sum_{i=1}^{N} \left( E_{covalent} + E_{non\text{-}covalent} \right) \qquad \text{eq. 15.1}$$

Covalent or *bonded* interactions occur between atoms that are connected by chemical bonds (Section 15.2). Non-covalent or *non-bonded* interactions include all other types of interactions, including electrostatic attraction or repulsion between ions and dipoles, and steric repulsions (Section 15.3). Every conformation of a molecule corresponds to a certain energy, so a multidimensional plot of all energies as a function of the conformation represents the *energy surface* of the molecule. Minima in the energy surface represent thermodynamically stable conformations. We will revisit energy surfaces when discussing the Ramachandran plot for proteins (Figure 16.16), protein folding (Section 16.3), and the ribose angles in nucleotides (Figure 17.11). To a first approximation, the overall energy of a molecule can be estimated by adding the energies of pairwise interactions that have been measured between small groups of atoms with reference molecules. However, since the conformational energy of molecules has both enthalpic and entropic parts, and the entropy is usually not known, it is difficult to calculate an accurate value for the conformational energy by this summation. Quantum mechanical *ab initio* calculations, often done *in vacuo* to simplify the system, try to estimate the total energy of a molecule based on its electron distributions without inclusion of experimental data. Other methods use empirical *energy potentials* that describe the lowest energy, or ground state of a molecule as a function of distances and angles between atoms (Section 18.2). The energy terms in such functions are derived from the conformational preferences observed in highly precise and accurate crystal structures of small organic and inorganic molecules (collected in the Cambridge Structural Database).

## 15.2 COVALENT INTERACTIONS

Covalent or *bonded* interactions are described by energy terms that depend on the positions of bonded atoms. These energy terms include the atomic bonds, the bond angles, and the torsion angles (eq. 15.2).

$$E_{\text{covalent}} = E_{\text{bond}} + E_{\text{angle}} + E_{\text{torsion}} \qquad \text{eq. 15.2}$$

Bonds include single, double, triple, and the dative bonds of ligands to metal ions. Bond angles or *1,3-angles* are defined by three atoms connected by two bonds. *Torsion angles*, also termed 1,4- or dihedral angles, are defined by four atoms that are linearly connected by three bonds. If viewed along the central bond, the second and third atoms are projected onto each other, leaving three "atoms" that describe the torsion angle in the same way as the 1,3-angle (see Figure 15.5). Figure 15.3 shows the shapes of the energy functions for changes in bond length, 1,3-angles, and torsion angles.

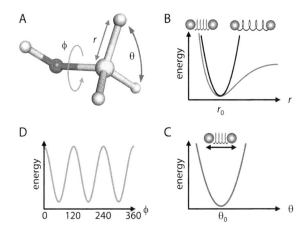

FIGURE 15.3: **Energy terms describing covalent bonds and their dependence on distance and angles.** A: The bond distance $r$, the 1,3-angle $\theta$, and the 1,4-angle (torsion) $\phi$ are indicated in separate colors for the simple molecule methanol ($CH_3OH$). The plots in B-D show the energy variation with distance or angle in the same color code. B: For the equilibrium distance $r_0$, a compression shows a parabolic effect on energy (black curve), but this potential function is inaccurate for large values of $r$. Lengthening of bonds will eventually lead to bond breakage, hence an asymptotic trace of the right arm of the energy function must be included (blue). C: The energy change upon compression or widening of the equilibrium bond angle $\theta_0$ can be described by a parabola (magenta). D: The torsion angle energy of a methyl group follows a sinusoid pattern with minima repeating every 120° (green).

## 15.2.1 Covalent Bonds

Probably the simplest model to describe the energy potential of a covalent bond is the harmonic oscillator: a spring with a spring constant $k_r$ acting between the atoms. This concept is treated in more detail in Section 19.4.1 on infrared spectroscopy. For our discussion here, it is sufficient to note that integration of Hooke's law (eq. 19.54) gives the parabolic potential energy as:

$$E_{bond} = \frac{1}{2} k_r \left( r - r_0 \right)^2$$

eq. 15.3

The minimum bond energy corresponds to the equilibrium bond distance $r_0$. Any deviation from $r_0$ increases the energy. For small deviations of bond length, the energy increase is well described by a parabola. Because the energy is symmetric about $r_0$, eq. 15.3 is a *harmonic* function. However, in this model, the energy would also increase to infinity upon excessive lengthening of the bond, which cannot be true because at some point the bond will break and the energy of the system will then be constant. The potential function therefore is to be modified from a parabola to take into account the finite energy of the system once the bond is broken (Figure 15.3). The energy range of covalent bonds is 200–800 kJ mol$^{-1}$, much higher than any non-covalent interactions, which are in the range of only a few kJ mol$^{-1}$. Table 15.1 lists mean energies and lengths of frequently encountered chemical bonds.

**TABLE 15.1:**
**Mean energies and lengths for bonds found in biological molecules.**

| bond | energy (kJ mol$^{-1}$) | bond length (pm) |
|------|------------------------|------------------|
| C-H | 415 | 109 |
| O-H | 465 | 96 |
| N-H | 389 | 101 |
| S-H | 348 | 143 |
| C-C | 348 | 154 |
| C=C | 611 | 147 |
| C≡C | 837 | 138 |
| C-O | 360 | 143 |
| C=O | ~740[a] | 121 |
| C-N | 306 | 147 |
| C=N | 615 | 128 |
| C≡N | 892 | 114 |
| C-S | 272 | 182 |
| S-S | 251 | 203 |
| S=O | 578 | 144/150[b] |

*Source:* Data from *Handbook of Chemistry and Physics* (hbcpnetbase.com), with permission.
[a] Aldehydes: 737 kJ mol$^{-1}$. Ketones: 749 kJ mol$^{-1}$.
[b] Sulfones and sulfoxides.

## 15.2.2 Bond Angles and Torsion Angles

The widening and compression of a bond angle around its equilibrium position can be described by a harmonic potential similar to the covalent bond. The spring constant is $k_\theta$ and the equilibrium bond angle is $\theta_0$.

$$E_{\text{angle}} = \frac{1}{2} k_{\theta} \left( \theta - \theta_0 \right)^2 = \frac{1}{2} k_{1-3} \left( r - r_0 \right)^2 \qquad \text{eq. 15.4}$$

Alternatively, bond angle variations can be described by the oscillation of a spring with the spring constant $k_{1-3}$ that connects the first and third atom of the three atoms that make up the bond angle (Figure 15.4) and the distance $r$ between these two atoms. The expression of bond angles in terms of distances is useful for speeding up computations.

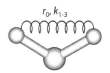

**FIGURE 15.4:** **Alternative description of bond angles as distances.** Measuring the distance between two atoms that are bonded to a third atom is an equivalent way of describing the bond angle.

Torsion angles determine the spatial relation of four consecutively bonded atoms 1-2-3-4. Rotation about the central bond (3-4) changes the relative positions of atoms 1 and 4. The conformational energy of the molecule depends on how close the atoms 1 and 4 come.

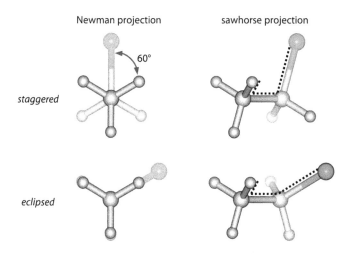

**FIGURE 15.5:** **Staggered and eclipsed conformation of chloroethane.** The top panel shows the staggered conformation of chloroethane. The torsion angle is 60°. On the bottom, the eclipsed conformation is drawn with a torsion angle of 0°. The left-hand side is the Newman projection, the right-hand side is the sawhorse projection. Both projections show the same conformations. The Newman projection reduces the 1,4-angle to the 1,3-angle concept. The sawhorse projection emphasizes the bond about which the substituents are rotated.

Depending on the arrangement of substituents, conformations are called either *staggered* or *eclipsed* (Figure 15.5). In the staggered conformation, the substituents are positioned with the maximum possible distance from each other, which corresponds to an energetic minimum. In contrast, the eclipsed conformation brings the substituents closer together, and steric clashes give rise to a maximum of energy. Staggered and eclipsed conformations are the energetic extremes, and the many conformations between these extremes have intermediate energies. A cosine function describes the periodic change of the energy as a function of the torsion angle (eq. 15.5 and Figure 15.6).

$$E_{\text{torsion}} = \frac{1}{2} k_{\phi} \left( 1 + \cos \left( n\phi - \delta \right) \right) \qquad \text{eq. 15.5}$$

The periodicity $n$ indicates the number of energy minima within a 360° rotation about the bond. The variable δ is the phase shift of the cosine function, and the force constant is $k_{\phi}$. For a tetrahedral carbon atom, *n* equals three and δ is zero (Figure 15.6). The energy minima are therefore at ϕ angles of 60°, 180°, and 300°. A double bond (*n* = 2 and δ = 180°) may be treated as a special case of a torsion angle with distinct energy minima at 180° (*trans*) and 0° (*cis*). A *cis/trans* isomerization requires breaking of the π-bond, which leads to an energy maximum 90° away from planarity (Figure 15.6).

In molecules with different substituents, not all energetic minima and maxima are equal. Upon rotation of the single bond connecting the different substituents, different distances are sampled between them and the torsion potential function will become more complicated (Figure 15.6). *Gauche* are staggered conformations where the larger substituents are 60° apart. The staggered conformation with the lowest energy is termed *anti*. Because torsion angles are so variable, they play an important role in defining the preferred conformations of flexible molecules such as fatty acids (Section 16.2.8), but also of protein and nucleic acid backbones. The energies of the conformers may differ by only a few kJ mol⁻¹, but in the context of large macromolecules, the conformational energies can sum to large values.

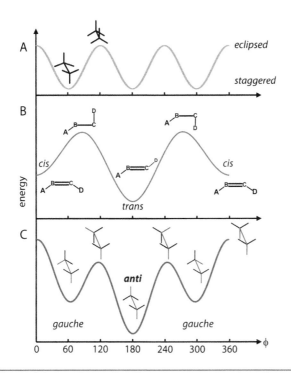

**FIGURE 15.6:** **Periodicity of the conformational energy upon rotation about a dihedral angle.** A: The top panel shows a single bond with all substituents the same, as in ethane. Every 60°, the torsion energy oscillates between the low-energy, staggered conformation and the high-energy, eclipsed conformation. B: Rotation about a double bond has a periodicity of 90° but there are two energy minima, one for *cis* and one for *trans*. Often, the *cis*-configuration is less stable (i.e. has more energy) than the *trans*-configuration because the substituents A and D come close to each other in the *cis*-configuration. The highest energies are reached when the π-bond is broken. C: Example of a molecule with two different substituents (black and magenta bonds), such as in butane. The larger magenta substituents (methyl groups on butane) give rise to local minima and maxima depending on how close they come. The torsion potential function is more complicated and has three cosine terms in this case.

## 15.3 NON-COVALENT INTERACTIONS

Non-covalent or non-bonded interactions are through-space interactions between atoms that are not directly connected to each other by chemical bonds. Table 15.2 gives an overview of the different kinds of non-bonded interactions that will be treated in the following chapters in the order of descending energetic contribution.

**TABLE 15.2:**
**Non-covalent interactions, sorted by strength and including distance ($r$) dependence.**

| type of interaction | distance dependence | energy range (kJ mol⁻¹) | distance range (pm) | example |
|---|---|---|---|---|
| charge-charge | $1/r$ | 12–40 | 200–600 | Arg···Asp |
| charge-dipole | $1/r^2$ | <30 | 320–450 | Lys···O=C |
| dipole-dipole (incl. H-bond, CH-π) | $1/r^3$ | 4–30 | 240–320 | NH···O=C |
| charge-induced dipole (cation-π) | $1/r^4$ | <20 | 450–510 | Arg···Phe |
| dispersion (2x induced dipole) | $1/r^6$ | 2–6 | 330–440 | Met···Phe |

*Note:* For comparison, the energy range of a covalent bond is 200–800 kJ mol⁻¹ and the thermal energy $RT$ at 25°C is 2.5 kJ mol⁻¹. Arg, Asp, Lys, Met, and Phe refer to amino acid side chains (Table 16.1).

All non-covalent interactions are strongly distance-dependent, but to different extents. The attractive terms of the energy potential functions have the general form $1/r^n$, where $n$ is an integer between one and six, and $r$ is the distance between the interacting atoms (Figure 15.7). Interactions with large $n$ are characterized by a steep potential, and the strength of the interaction decreases strongly with increasing distance $r$. Such interactions are termed *short-range*. By contrast, smaller $n$ give rise to less steep potentials, and a smaller decrease in interaction energy with increasing $r$. These interactions are termed *long-range* interactions. In the following, we will see that ionic interactions are strong and long-range ($n = 1$) while van der Waals interactions are weak and short-range interactions ($n \geq 6$).

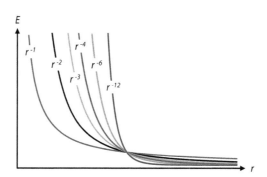

**FIGURE 15.7: Graphs of the form 1/$r^n$.** Ionic interactions follow a potential with the least distance dependence ($n = 1$). At the other extreme with very strong distance dependence are the van der Waals interactions. The potentials with $n = 6$ and $n = 12$ are part of the van der Waals potential (Section 15.3.4).

The different kinds of interactions between atoms and groups of atoms can more or less arbitrarily be divided in polar and non-polar interactions. Polar interactions involve ions ("*monopoles*") or *permanent dipoles* while non-polar interactions occur between neutral atoms and involve *induced dipoles*. Polar interactions are generally stronger than non-polar interactions. Both polar and non-polar interactions are affected by the dielectric constant $\varepsilon_r$ of the medium and by the relative orientation of the dipoles involved.

### 15.3.1 Ionic Interactions

The energy potential $E_{ionic}$ for the pairwise interaction of charges at a distance $r$ is described by Coulomb's law:

$$E_{ionic} = \frac{Z_1 \cdot Z_2 \cdot e^2}{4\pi\varepsilon_0\varepsilon_r \cdot r}$$

eq. 15.6

$Z_i$ is the number of charges of the interacting ions (including the sign of the charge), $e$ is the magnitude of the elementary charge ($1.602 \cdot 10^{-19}$ C), $\varepsilon_0$ is the dielectric constant of vacuum ($8.85 \cdot 10^{-12}$ C V$^{-1}$ m$^{-1}$), and $\varepsilon_r$ is the *relative dielectric constant* of the medium in which the ionic interaction takes place. The value of $\varepsilon_r$ is unity in vacuum, two for paraffin oil, 33 for methanol, and 81 for water. The product $4\pi\varepsilon_0\varepsilon_r$ is the dielectric constant $D$. $\varepsilon_r$ enters eq. 15.6 in the denominator, which means any value larger than unity leads to a weakening of the ionic interaction compared to vacuum. Solvents with high $\varepsilon_r$ thus have a large dampening effect on ionic interactions. For macromolecules in aqueous solutions, ionic interactions are weak on the surface and strong when buried in the non-polar interior of the macromolecule. The hydrophobic interior of proteins is often described by values of $\varepsilon_r$ between 1 and 20.

The dielectric constant $D$ and thus the relative dielectric constant $\varepsilon_r$ occur in the denominator of the energy functions for all other types of polar interactions. Therefore, all polar interactions can be described by similar potential functions to ionic interactions. In contrast to simple ions that are spherical monopoles, so that their interaction energy depends only on distance, all dipolar interactions depend in addition on the relative orientation of the dipoles.

### 15.3.2 Interactions between Ions and Dipoles

A dipole is generated when the electron distribution around an atom or a group of atoms is not spatially uniform. Dipoles can be described as vectors with a direction and a magnitude. The overall magnitude of the dipole depends on the magnitude of the charges and their distance vector $\boldsymbol{l}$ (Figure 15.8). The *dipole moment* $\boldsymbol{\mu}$ is then defined as:

$$\vec{\mu} = \boldsymbol{\mu} = \boldsymbol{l} \cdot \delta^+ \cdot \delta^-$$

eq. 15.7

In eq. 15.7, the dipole moment $\boldsymbol{\mu}$ and the distance vector $\boldsymbol{l}$ are written in boldface, but without the vector symbol. The charges can be integer multiples of the elementary charge or *partial* charges ($\delta^+$ and $\delta^-$). Table 15.3 lists the partial charges of the main-chain protein atoms and some side-chain atoms in proteins.

| residue | atom | partial charge |
|---|---|---|
| **TABLE 15.3:** | | |
| **Partial charges of some atoms in proteins.** | | |
| backbone | N | −0.36 |
| | $H_N$ | +0.18 |
| | $C_\alpha$ | +0.06 |
| | $H_\alpha{}^a$ | +0.02 |
| | C (carbonyl) | +0.45 |
| | O (carbonyl) | −0.38 |
| Ser / Thr | $C_\beta$ | +0.13 / +0.16 |
| | $H_\beta$ | +0.02 / +0.01 |
| | $O_\gamma$ | −0.31 |
| | $H_\gamma$ | +0.17 |
| Tyr | $O_\eta$ | −0.33 |
| | $H_\eta$ | +0.16 |
| Cys | $C_\beta$ | −0.105 |
| | $H_\beta$ | +0.05 |
| | $S_\gamma$ | +0.01 |
| | $H_\gamma$ | +0.01 |
| Lys / Lys⁺ | $N_\zeta$ | −0.39 / −0.32 |
| | $H_\zeta$ | +0.15 / +0.32 |
| Glu/Glu⁻ | $O_{\varepsilon 1}$ | −0.36 / −0.57 |
| | $O_{\varepsilon 2}$ | −0.35 / −0.57 |
| | $H_{\varepsilon 2}$ | +0.21 / n.a. |

*Source:* Data cited from Momany, F.A. et al., *J. Phys. Chem.*, 79(22), 2361–2381, 1975, with permission.

a Pro and Gly have higher charges for $H_\alpha$ of about 0.04 and 0.05, respectively.

For the interaction of an ion with a dipole, the position of the ion relative to the dipole is important because the distance vector **r** changes with ion placement (Figure 15.8).

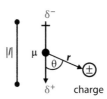

**FIGURE 15.8:** **Charge-dipole interaction.** A dipole with two partial charges (δ⁺ and δ⁻) separated by a distance vector **l** has a dipole moment **μ**. The extent of the interaction of such a dipole with a charge at position **r** is dependent on their relative orientation, expressed by the dot product (**μ·r**), and the sign of the charge. The direction of the dipole is from δ⁻ to δ⁺.

The potential energy of an ion dipole interaction can then be expressed by eq. 15.8.

$$E_{\text{ion-dipole}} = \frac{Z \cdot e \cdot (\boldsymbol{\mu} \cdot \boldsymbol{r})}{D \cdot |\boldsymbol{r}|^2}$$

eq. 15.8

The dot product $(\boldsymbol{\mu} \cdot \boldsymbol{r}) = |\boldsymbol{\mu}| \cdot |\boldsymbol{r}| \cdot \cos \theta$ (Section 29.9.1) takes into account the angle between the distance vector between dipole and the ion and the dipole moment. The ion-dipole interaction follows a $1/r^2$ potential and for a given distance $r$ is weaker than pure ionic interactions. The dielectric constant $\varepsilon_r$ of the medium is again a dampening factor for the interaction.

What happens when we place two permanent dipoles close enough together so they can interact? Figure 15.9 shows several alignments of dipoles to highlight that both the distance between the dipoles and their relative orientation determine whether an attraction or repulsion results.

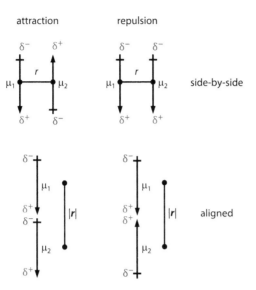

**FIGURE 15.9:    Interactions between two permanent dipoles.** The top panel shows antiparallel and parallel side-by-side placements of dipoles. The bottom panel shows head-to-tail (parallel) and head-to-head (antiparallel) alignment of the dipoles. Placing the same polarities close together leads to repulsion, otherwise the dipoles attract each other.

The potential between two dipoles is given in eq. 15.9.

$$E_{\text{dipole-dipole}} = \frac{\boldsymbol{\mu}_1 \cdot \boldsymbol{\mu}_2}{D \cdot |\boldsymbol{r}|^3} - \frac{3 \cdot (\boldsymbol{\mu}_1 \cdot \boldsymbol{r}) \cdot (\boldsymbol{\mu}_2 \cdot \boldsymbol{r})}{D \cdot |\boldsymbol{r}|^5} \qquad \text{eq. 15.9}$$

Again, the interaction is dependent on the dielectric constant D of the medium. The potential for the interaction between two permanent dipoles follows a $1/r^3$ dependence, and dipole-dipole interactions are thus of shorter range than interactions between dipoles and ions. The dot products ($\boldsymbol{\mu}_i \cdot \boldsymbol{r}$) describe the orientation of the dipole moments relative to the distance vector **r** (Section 29.9.1). In the special case of side-by-side orientation of dipoles, the dipole moments and the distance vector are perpendicular to each other, and $\boldsymbol{\mu}_i \cdot \boldsymbol{r} = 0$, so that eq. 15.9 reduces to

$$E_{\text{dipole-dipole}} = \frac{\boldsymbol{\mu}_1 \cdot \boldsymbol{\mu}_2}{D \cdot |\boldsymbol{r}|^3} \qquad \text{eq. 15.10}$$

For head-to-head or head-to-tail alignment of the dipolar and distance vectors, the dot products $\boldsymbol{\mu}_i \cdot \boldsymbol{r} = |\boldsymbol{\mu}_i \boldsymbol{r}|$, and eq. 15.9 simplifies to

$$E_{\text{dipole-dipole}} = \pm 2 \cdot \frac{\boldsymbol{\mu}_1 \cdot \boldsymbol{\mu}_2}{D \cdot |\boldsymbol{r}|^3} \qquad \text{eq. 15.11}$$

Here, the negative sign corresponds to head-to-tail alignment of the dipoles, which leads to attraction and thus a negative potential. We will see that in α-helices the dipole moments of the peptide bonds are aligned head-to-tail, leading to a permanent *helix dipole* (Figure 16.17).

The presence of an ion can also induce a dipole in an otherwise non-polar molecule. The interaction of an ion with such an induced dipole has a distance dependence of $1/r^4$ (eq. 15.12), which makes this interaction quite short-range.

$$E_{\text{ion-induced dipole}} = \frac{Z \cdot e^2 \cdot \alpha}{2 \cdot D^2 \cdot |r|^4} \qquad \text{eq. 15.12}$$

Z is the number of charges of the ion that induces the dipole in another molecule or molecular group. α is the *molecular polarizability* of the molecular group that generates the induced dipole under the influence of the ion. The *cation-π interaction* (Figure 15.10) between positive charges and the π-electrons of an aromatic system is an example of an ion-induced dipole interaction that is frequently observed in crystal structures of biological molecules. Cation-π interactions are formed between positively charged amino acids and aromatic protein side chains (Chapter 16) or nucleobases (Section 17.1). The positive charge can also be a metal ion, often $Ca^{2+}$ or $K^+$.

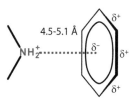

FIGURE 15.10: **Cation-π interaction.** The positively charged nitrogen is centered on the aromatic plane, polarizing the π-electrons. Such interactions are frequently observed between the positively charged side chains of arginine and lysine and the aromatic side chains of phenylalanine, tyrosine, and tryptophan. Bases in nucleic acids can interact with charges in the same manner.

### 15.3.3 HYDROGEN BONDS

In the last section, we discussed the interaction of dipoles and charges with respect to their orientation. Precise alignment of dipoles is also essential for the formation of hydrogen bonds, or *H-bonds*, one of the most important interactions in biology. H-bonds are characterized by two electronegative atoms that share a hydrogen atom in a quite restricted geometry (Figure 15.11). The H-atom is provided by the *donor* group and interacts with a lone pair of the *acceptor* group. The sharing of the H-atom by two electronegative atoms has partial covalent character, which sets the H-bond apart from other dipolar interactions. A full understanding of the H-bond requires quantum-mechanical treatment.

The main factors influencing the strength of an H-bond are the magnitudes of the two dipoles involved and their relative orientation. Electronegativity is a useful concept to judge the propensity of an atom to attract electrons relative to other atoms (see Table 15.4). A hydrogen atom bound to an electronegative atom X (O, N, sometimes S and C in biological molecules) carries a partial positive charge due to shifting of the X-H bond electrons toward the atom X. In water this partial charge is about +0.4 on each H-atom, leaving −0.8 on the O-atom. The larger the electronegativity of X, the larger the partial charges, and the larger the dipole moment of the X-H bond. The strongest H-bonds are formed between dipoles that contain the electronegative oxygen and nitrogen atoms. Much weaker H-bonds with the less electronegative sulfur or carbon atoms are also observed in macromolecules.

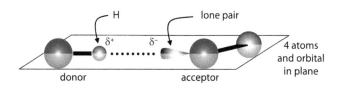

FIGURE 15.11: **H-bond geometry.** An electronegative oxygen atom is shown as a red sphere. Hydrogen and carbon are shown as small and large gray spheres. In the donor dipole, the hydrogen atom carries the positive partial charge δ⁺. The negative partial charge δ⁻ of the acceptor dipole is concentrated in the lone pairs. One of them is shown as a gray lobe. The H-bond is shown as a dashed line. The optimal geometry of an H-bond is in the direction of the lone pair, with the four atoms O–H...O–C in the same plane.

In addition to the magnitude of the partial charges and the strengths of the two dipoles, the geometry is an important determinant of H-bond strengths. In the energetically most favorable orientation, the H-atom of the donor interacts with a lone pair of the acceptor in the direction of the molecular orbital. Thus, the three atoms with the shared H-atom in the middle and the molecular orbital should be collinear. In addition, the four atoms involved – two from the donor and two from the acceptor – should be coplanar (Figure 15.11). This pronounced directionality of H-bonds establishes high selectivity in recognition events both between macromolecules and between macromolecules and ligands, which is why H-bonds are so important in biology. Figure 15.12 shows a few examples of H-bonds that can form in proteins. A similarly pronounced directionality for the strength of interactions is also observed in halogen bonds (Box 15.1), where chlorine, bromine, or iodine atoms accept lone pairs from other atoms.

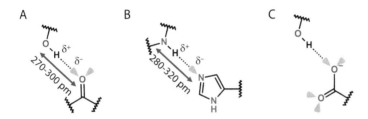

**FIGURE 15.12:**    **Examples of H-bonds.** The H-bond is shown as a dashed line and orbitals are drawn as gray lobes. The distance between the H-atom and the electronegative atom (N or O) is significantly shorter than the sum of atomic radii, indicating significant orbital overlap. Generally, the O-H···O or N-H···N angle is > 150°, with 180° being optimal. A: The hydroxyl group of an alcohol binds to a carbonyl group. B: An NH-group is the donor and the imino nitrogen of imidazole is the acceptor. C: Charged H-bonds are slightly shorter and more stable than non-charged H-bonds.

Distances of H-bonds range between 270–310 pm if the donor is OH or NH and the acceptor is a carbonyl group (Figure 15.12). H-bonds can be even shorter by ca. 10 pm if permanent charges are involved, e.g. between protonated amines and deprotonated carboxylic acids. Somewhat longer H-bonds up to 320 pm are formed if the acceptor is sulfur or an imino-nitrogen. The energy range of H-bonds is large, they range from just 5 kJ mol$^{-1}$ to 40 kJ mol$^{-1}$ *in vacuo*. The dielectric constant $\varepsilon_r$ of the solvent influences the energy of the H-bond in a similar way to the other dipolar interactions. H-bonds exposed to water are weaker than H-bonds that are shielded from water, for example in the hydrophobic interior of proteins. High-resolution crystal structures show that in the core of proteins, most of the H-bond donors and almost all of the H-bond acceptors are saturated, indicating that it is energetically unfavorable to leave H-bonding valences unoccupied. Together with the hydrophobic effect (see Box 16.3), H-bond formation is a major driving force in protein folding (Section 16.2.8.1). If the donor and acceptor groups are within the same molecule, *intramolecular H-bonds* are formed that can dictate the preferred conformation of molecules. *Intermolecular H-bonds* are essential for the specificity of molecular recognition (Section 16.2.7).

Water not only reduces the strength of hydrogen bonds due to its high relative dielectric constant $\varepsilon_r$, but is both a powerful H-bond donor and acceptor by itself. With the exception of membranes, water is the solvent for most biological systems, and therefore effectively competes for H-bond formation with the solute. Thus, an intramolecular H-bond at the surface of a protein has little energetic contribution to stability because it is almost isoenergetic to an H-bond formed with water. However, a <u>network</u> of several H-bonds can have a significantly stabilizing effect. H-bond networks are cage-like structures between several donors and acceptors that are held together by mutual and reciprocal H-bonds. Such networks are ubiquitous in biological macromolecules. The formation of the network is highly cooperative, and as a result it can be very sensitive to disturbances. It is sometimes sufficient to remove a single H-bond by mutation of a side chain, dissociation of a ligand, or hydrolysis of a nucleotide in order to destroy an entire network of H-bonds and drastically change the conformation of a protein. The cooperativity of H-bond networks is used for the regulation of many cellular processes.

---

**BOX 15.1: HALOGEN BONDS.**

The halogen bond is an attractive electrostatic interaction between a halogen atom and a lone pair from another atom (Figure 15.13). The interaction is highly directional, a feature it shares with the H-bond (Figure 15.12) and that enables halogen bonds to establish selectivity for ligand binding. Halogen atoms Cl, Br, and I (but not F) in organic molecules have anisotropic charge distribution: the three lone pairs form a central belt of negative charge around the halogen atom and isolate a region of positive electrostatic potential on the outermost portion of the halogen atom's surface (Figure 15.13). This positively polarized region is termed the σ-*hole* and acts as an acceptor in the halogen bond. The strength of a halogen bond correlates with the size of the σ-hole and the polarizability of the halogen. Electron-withdrawing substituents bound to the halogen increase the σ-hole: halogen bonding is stronger for the sequence Cl < Br < I.

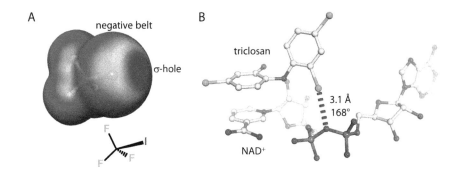

**FIGURE 15.13:**   **Halogen bonding.** A: The molecular surface of trifluoromethyl iodide, $CF_3I$, is colored according to negative (red) and positive (blue) electrostatic potential. The electronegative fluorine atoms pull the electrons of the carbon-iodine σ-bond toward them, resulting in a σ-hole at the extension of the C-I bond (visible as a blue patch). B: Triclosan, a pesticide, bound to the $NAD^+$ cofactor in enoyl-acyl carrier protein reductase. One of the chlorine atoms (green) forms a halogen bond (magenta dashes) with an oxygen atom of $NAD^+$ (PDB-ID 3pjf).

As in H-bonds, the optimal angle for halogen bonding between donor (the C-Hal bond) and the acceptor (the atom carrying a lone pair) is 180°, but smaller values down to 140° have been observed. The distance between the donor and acceptor atom is slightly shorter (ca. 7%) than the sum of their van der Waals radii, and the energy of a halogen bond is on the order of $2-8$ kJ mol$^{-1}$. Only a few hundred halogen bonds have been identified in biological systems, but the use of halogen bonds is becoming popular in structure-based drug design.

---

### 15.3.4 INTERACTIONS BETWEEN INDUCED DIPOLES: VAN DER WAALS INTERACTIONS

Even molecules without a permanent dipole will attract each other electrostatically. The reason is the fluctuation of the electron distribution around an atom, which leads to a transient dipole. Before it vanishes, this transient dipole polarizes electrons from neighboring molecules, thereby inducing another dipole in the second molecule. The induced dipoles are tiny and very short-lived, but they lead to a weak attraction between the molecules. The attraction depends on the inverse 6$^{th}$ power of the distance, making it a very short-range interaction compared to the other dipolar interactions

(Table 15.2). The attractive force has been predicted by quantum mechanics and is also known as the *London dispersion force*. The dispersion potential $E_{\text{London}}$ depends on the distance $r$, the energy term $E = hv$ (where $v$ is the frequency of the charge fluctuation and $h$ is the Planck constant), and the polarizabilities $\alpha_i$ of the interacting atoms:

$$E_{\text{London}} = -\frac{3E\alpha_1\alpha_2}{4r^6}$$

eq. 15.13

The polarizability is a property of an individual atom or of a group of atoms that depends on the volume and the number of electrons contained in that volume (the electron density). For example, xenon atoms (54 electrons within 5.3 $\text{Å}^3$) are more easily polarized than argon atoms (18 electrons distributed over 1.5 $\text{Å}^3$). The stronger dispersion forces between xenon atoms compared to argon atoms lead to a higher boiling point for xenon ($-108°C$ for xenon, $-186°C$ for argon). Although individual London dispersion forces are weak, they sum up to significant contributions to the overall energy of a molecule.

The dispersion potential (eq. 15.13) predicts a stronger attraction the closer the atoms come together. However, below a certain distance the electrostatic repulsion of the electron clouds and, at even closer distance, the repulsion between the positively charged nuclei sets in. The strong repulsive force between atoms at close range is described by an energy term that depends on the inverse 12th power of the distance.

$$E_{\text{vdW}} = \frac{A}{r^{12}} - \frac{B}{r^6}$$

eq. 15.14

The potential function in eq. 15.14 is the *Lennard-Jones potential* or *6–12 potential*, describing the van der Waals energy $E_{\text{vdW}}$. The actual values for the repulsive coefficient A and the attractive coefficient B depend on the type of interacting atoms and are tabulated. The balance of the attractive London dispersion forces and the repulsive forces of the nuclei produces an energy minimum at a distance $r_0$ termed the *van der Waals contact distance* (Figure 15.14). At this distance, the net attractive force between the atoms is maximal.

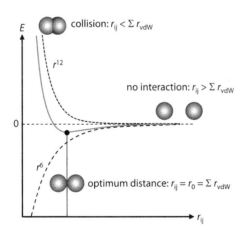

**FIGURE 15.14:    Lennard-Jones potential.** Atoms i and j are at minimum energy at the van der Waals contact distance $r_0$. The shape of the Lennard-Jones potential also applies to H-bonds (Section 15.3.3).

The van der Waals contact distance is the sum of the van der Waals radii of the interacting atoms. Van der Waals radii of atoms and groups of atoms that are important in biology are given in Table 15.4.

| TABLE 15.4: |
| --- |
| **Van der Waals radii (in pm) of selected atoms and groups.** |

| H 120 | C   170–177 | N 155–166 | O   150–152 | F   146–147 |
| --- | --- | --- | --- | --- |
|  | CH$_2$ 180 | P 180–190 | S   180–189 | Cl 175–182 |
|  | CH$_3$ 185 |  | Se 182–190 | Br 183–186 |
|  |  |  |  | I   198–204 |

*Note*: Radii are mean values from small molecules in the Cambridge Structural Database.

Van der Waals interactions between the aliphatic chains of lipids drive the formation of membranes (Section 16.2.8), and those between hydrophobic amino acids, both aliphatic and aromatic, contribute significantly to protein stability (Section 16.3). Aromatic groups in nucleic acids (Figure 17.2), many prosthetic groups (Figure 19.13), and in small molecules are also frequently involved in van der Waals interactions (Figure 15.15). Sulfur is part of cysteine, methionine, and the cofactor S-adenosyl methionine. Its large van der Waals radius and comparatively large polarizability make sulfur a good binding partner for aromatic systems. The mean distances of the atoms in such interactions are about 350–400 pm.

The geometric restraints for van der Waals interactions are less strict than those for dipolar interactions. Favorable geometries between aromatic systems are the face-to-face and the edge-to-face orientations (Figure 15.15). The face-to-face interaction is characteristic for the base stacking of nucleic acids (Section 17.4), and the edge-to-face orientation is found between aromatic groups in protein-DNA interactions. The π-systems of the amide bonds in peptides and proteins can also favorably stack on aromatic systems.

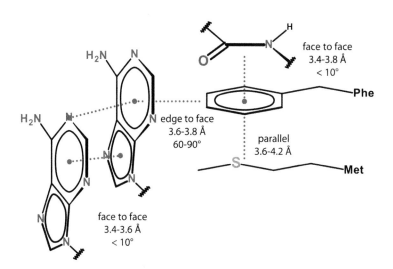

**FIGURE 15.15:    Face-to-face and edge-to-face interactions**. Interactions are illustrated for an amide bond (top), the aliphatic methylene groups of a methionine side chain, and aromatic groups such as the nucleobase adenine and the side chain of phenylalanine. Amide groups stack almost parallel on top of aromatic systems with a tilt of generally less than 10°. Aromatic groups tend to pack either perpendicular or parallel to each other. Often, the planar groups are shifted relative to each other in the parallel orientation.

In the following chapters we will turn our attention to the structure of proteins and nucleic acids where we will encounter all of the different types of interactions.

## QUESTIONS

15.1   The hydrolysis of phosphoenol pyruvate (shown) to pyruvate and inorganic phosphate is thermodynamically strongly favored, liberating ca. $-62$ kJ mol$^{-1}$. What could be the reason?

15.2   The C-N bond in amide groups has partial double bond character. Can this be explained by tautomerism?

15.3   Methionine sulfoxide is a natural oxidation product in antibodies. Is this side chain chiral?

15.4   The structure of glucose is shown. Draw the structures of mannose (C2-epimer) and galactose (C4-epimer).

glucose

15.5   The structure of tartrate is shown. How many stereocenters and how many stereoisomers does this ion have?

15.6   For the potential functions of the form $1/r^n$, what is the change in energy (in %) when the distance $r$ is increased by 10% for $n = 1$ and $n = 6$?

15.7   The short range effect of the $1/r^n$ potentials can be visualized by calculation of the distance change that is necessary to halve the interaction energy. Derive the general formula for the factor by which the distance has to change in order to halve the energy as a function of $n$.

15.8   For the H-bonds shown, assign the possible protein side chains and main-chain parts that can engage in them as donor or acceptor (see Figures 16.1 and 16.3 for help).

15.9   N-methyl acetamide dimerizes in solution by forming H-bonds. Draw the structure of the dimer. The dimerization is favored in non-polar solvents such as $CCl_4$, but disfavored in water. What is the reason?

## REFERENCES

Bissantz, C., Kuhn, B. and Stahl, M. (2010) A medicinal chemist's guide to molecular interactions. *J Med Chem* **53**(14): 5061–5084.
· discusses the preferred conformations of small molecules

Desiraju, G. R. and Steiner, T. (2006) The weak hydrogen bond in structural chemistry and biology. Oxford University Press.
· in-depth treatment of the hydrogen bond

Klostermeier, D. and Millar, D. P. (2002) Energetics of hydrogen bond networks in RNA: hydrogen bonds surrounding G+1 and U42 are the major determinants for the tertiary structure stability of the hairpin ribozyme. *Biochemistry* **41**(48): 14095–14102.
· an example of the sensitivity of H-bond networks, analyzed by double- and triple-mutant cycles

Hassel, O. and Hvoslef, J. (1954) The Structure of Bromine 1,4-Dioxanate. *Acta Chem Scand* **8**(5): 873.
· first description of a halogen bond

Clark, T., Hennemann, M., Murray, J. S. and Politzer, P. (2007) Halogen bonding: the sigma-hole. Proceedings of "Modeling interactions in biomolecules II", Prague, September 5th–9th, 2005. *J Mol Model* **13**(2): 291–296.
· detailed explanation on how the σ-hole is formed

Scholfield, M. R., Zanden, C. M., Carter, M. and Ho, P. S. (2013) Halogen bonding (X-bonding): a biological perspective. *Protein Sci* **22**(2): 139–152.
· review of halogen bonds in biology and their relation to H-bonds

Hardegger, L. A., Kuhn, B., Spinnler, B., Anselm, L., Ecabert, R., Stihle, M., Gsell, B., Thoma, R., Diez, J., Benz, J., Plancher, J. M., Hartmann, G., Banner, D. W., Haap, W. and Diederich, F. (2011) Systematic investigation of halogen bonding in protein-ligand interactions. *Angew Chem Int Ed* **50**(1): 314–318.
· role of halogen bonds in structure-based drug design

Alvarez, S. (2013) A cartography of the van der Waals territories. *Dalton Trans* **42**(24): 8617–8636.
· statistical analysis of van der Waals contacts in molecules

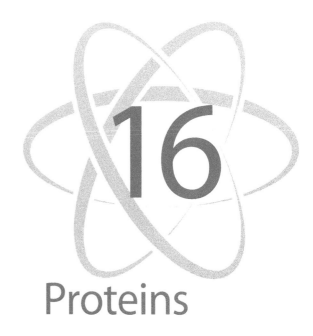

# Proteins

The name protein originates in 1838 from the Greek πρώτειος (*proteios*, primary), indicating an essential substance for life. It was originally believed that one gene codes only for a single protein (*one-gene-one-polypeptide hypothesis*), but in the human genome only 20000–25000 genes code for >100000 proteins that constitute our *proteome*, the sum of all proteins in our body. The observed protein variety is generated by several mechanisms acting on the level of DNA, messenger RNA (mRNA), and the proteins, including gene shuffling, alternative transcription and translation initiation, mRNA splicing, post-translational modification, proteolytic maturation, and somatic mutagenesis in immune cells. While in the early 1900s, proteins were viewed as unstructured colloids akin to wallpaper paste, the first crystal structure of myoglobin in 1958 showed that proteins are well-defined structural entities. Modern structure determination methods are discussed in Part IV (Chapters 20 and 22 and Section 23.2.4) of this book.

We begin our treatment of macromolecular structure by first introducing the building blocks, the amino acids, followed by how these condense into their polymers, the proteins, and then how proteins fold into their functional three-dimensional conformation. A similar approach is taken for membrane assembly from lipids (Section 16.2.8.3).

## 16.1 AMINO ACIDS AND THE PEPTIDE BOND

### 16.1.1 PROPERTIES OF THE TWENTY CANONICAL AMINO ACIDS

Proteins are polymers of amino acids that are linked *via* amide bonds, also termed *peptide bonds* (Section 16.1.2). The twenty canonical amino acids differ in their side-chain properties (Figure 16.1; Table 16.1). The diversity of proteins is further increased by chemical modification of the standard amino acids (phosphorylation or glycosylation), or by incorporation of entirely different, non-canonical amino acids such as selenocysteine or formylglycine (Section 16.1.4). When condensed into a polypeptide chain, the amino acids in proteins are called *residues*.

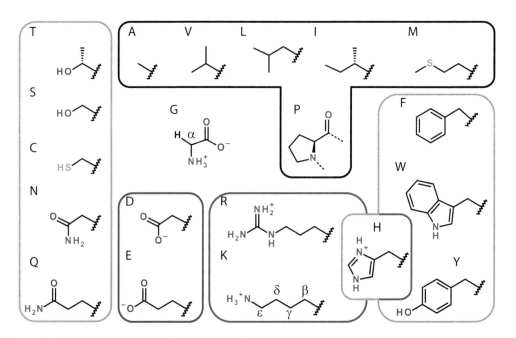

**FIGURE 16.1:**   **The twenty canonical amino acids**. The structure of glycine is shown in its zwitterionic form and the $C_\alpha$ atom labeled. Glycine has only an H-atom as the side chain. The other side chains are grouped according to chemical nature, hydrophobicity, and acidity. Labels are given in single letter code (Table 16.1). Functional groups are charged as they would be at pH 7. Aliphatic side chains are grouped in a black box, but note that the methylene groups of glutamine, glutamate, arginine, and lysine and the methyl group of threonine also have aliphatic properties. Aromatic side chains are grouped in a gray box. Histidine can be charged under physiological conditions, so it belongs to both the hydrophobic and the positively charged groups (blue box). The negatively charged aspartate and glutamate (red box) are placed next to their isosters asparagine and glutamine. Polar side chains are grouped in a green box. Threonine and isoleucine are the only amino acids with chiral side chains. Cysteine is redox active and can form disulfide bonds.

Amino acids are typically abbreviated either by their three-letter or single-letter codes. The single-letter code is useful for protein sequence alignments (Section 18.1.2), but the readability of text is better when using the three-letter code. In this book, both systems are used, depending on the context. Starting from the $C_\alpha$, the atoms in the side chains are usually labeled β, γ, δ, etc.

**TABLE 16.1**
**Properties of the twenty canonical amino acids.**

| name | 3-letter | 1-letter | $pK_A$ side chain | residue mass (Da) | hydrophobicity[a] | H-bond | functionality |
|---|---|---|---|---|---|---|---|
| alanine | Ala | A | | 71.08 | 0.25 / 1.8 | — | aliphatic |
| cysteine | Cys | C | 8.6–9.5 | 103.14 | 0.04 / 2.5 | weak D/A | polar, hydrophobic |
| aspartate | Asp | D | 3.7–4.0 | 115.09 | −0.72 / −3.5 | A | charged |
| glutamate | Glu | E | 4.2–4.5 | 129.11 | −0.62 / −3.5 | A | aliphatic, charged |
| phenylalanine | Phe | F | | 147.18 | 0.61 / 2.8 | — | aromatic, hydrophobic |
| glycine | Gly | G | | 57.05 | 0.16 / −0.4 | — | — |
| histidine | His | H | 6.0–7.0 | 137.14 | −0.40 / −3.2 | D/A | aromatic, polar, charged positive or neutral |
| isoleucine | Ile | I | | 113.16 | 0.73 / 4.5 | — | aliphatic |
| lysine | Lys | K | 10.4–11.1 | 128.17 | −1.10 / −3.9 | D | aliphatic, charged positive |
| leucine | Leu | L | | 113.16 | 0.53 / 3.8 | — | aliphatic |
| methionine | Met | M | | 131.20 | 0.26 / 1.9 | — | aliphatic |
| asparagine | Asn | N | | 114.10 | −0.64 / −3.5 | D/A | carbonyl and –$NH_2$ |
| proline | Pro | P | | 97.12 | −0.07 / −1.6 | — | aliphatic |
| glutamine | Gln | Q | | 128.13 | −0.69 / −3.5 | D/A | aliphatic, carbonyl and –$NH_2$ |
| arginine | Arg | R | 12.3–12.6 | 156.19 | −1.76 / −4.5 | D | aliphatic, charged positive |

| serine | Ser | S | ~16 | 87.08 | −0.26 / −0.8 | D/A | polar, −OH |
| threonine | Thr | T | ~16 | 101.10 | −0.18 / −0.7 | D/A | aliphatic, polar, −OH |
| valine | Val | V | | 99.13 | 0.54 / 4.2 | — | aliphatic |
| tryptophan | Trp | W | | 186.21 | 0.37 / −0.9 | D | aromatic, NH |
| tyrosine | Tyr | Y | 9.6–10.3 | 163.17 | 0.02 / −1.3 | D/A | aromatic, −OH |

*Note:* The p$K_A$ of the N-terminal $NH_3^+$-group is 6.8-8.0, and that of the C-terminal carboxyl group is 3.5-4.3. All p$K_A$ values can be shifted by several units due to the protein environment. The mass of the amino acid residue in a protein (in Da) is the molecular mass minus that of water.

[a] The first number is from Eisenberg et al. (1982) *Faraday Symp. Chem. Soc.* 17(0): 109–120, and the second is from Kyte & Doolittle (1982) *J. Mol. Biol.* 157(1): 105–132.

Amino acid side chains can be grouped according to their biophysical properties, i.e. charged, polar, aliphatic, or aromatic, although there is significant overlap between the categories. For instance, the methylene groups of the polar Lys, Glu, and Gln frequently engage in van der Waals interactions. The NH-group of the aromatic Trp side chain, an indole, frequently acts as an H-bond donor. Cys can engage both in weak H-bonds and in van der Waals interactions. The properties of a side chain may also change as a function of pH, H-bonding environment, temperature, or dielectric constant of the environment: the aromatic imidazole side chain of His can be charged or neutral and, depending on its protonation state, act as H-bond donor, acceptor, or both (Figure 16.2).

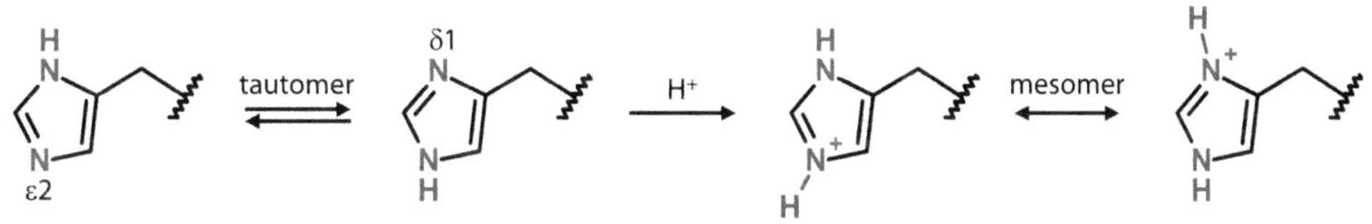

**FIGURE 16.2:    Protonation states of His.** The tautomer with the hydrogen at Nε2 is slightly more stable than the other tautomer with the hydrogen at Nδ1. After protonation of the imidazole side chain, the positive charge is stabilized by resonance. Both mesomeric structures are shown (note the different arrow types for a chemical equilibrium and mesomeric structures).

His is the only amino acid whose p$K_A$ value is close to the physiological pH, and although His is statistically one of the less frequent amino acids in proteins, it is the most frequent residue in enzyme active sites because His is suited for acid-base catalysis (Section 14.2) near neutrality. The side chains of other amino acids such as Lys, Cys, Ser, Tyr, and Glu are also used in enzyme catalysis, but their p$K_A$ values usually need to be shifted to values near 7. In these enzymes, specific local environments around the catalytic amino acid stabilize the protonated or deprotonated form of the side chain so it can be used in catalysis. A well-studied enzyme class with shifted p$K_A$ values is proteases. In Cys-, Ser-, or Thr-proteases, the protein environment makes these side chains more acidic by stabilizing the deprotonated form, which increases their nucleophilicity for cleavage of amide bonds in proteins. Similarly, in protein disulfide isomerases, the protein environment reduces the p$K_A$ of an active site Cys to about 4.5, four orders of magnitude from the normal value. The stabilized thiolate anion acts as a nucleophile to accelerate disulfide exchange, a rate-limiting step in protein folding (Section 16.3.5.2).

## 16.1.2  THE PEPTIDE BOND

The amide group formed upon condensation of amino acids is known as a *peptide group* (Figure 16.3), and the bond connecting the amino acids is the *peptide bond*. The peptide bond has partial double bond character because of conjugation of the π-electrons. Due to the large electronegativity of oxygen, the π-electrons are shifted towards the oxygen atom, establishing a permanent dipole with its positive end at the NH hydrogen atom and the negative end at the oxygen atom of the CO-group. This dipole is especially important for H-bonds that stabilize the secondary structure elements in proteins (Section 16.2).

A polymer of amino acids is termed a *polypeptide* or just *peptide* when it is small, and *protein* when it is larger. The repeating sequence of atoms in a polypeptide, $-[-C_\alpha-C-N-]_n-$, is the *main chain* or *backbone*. While the backbone conformation determines the secondary structure of the polypeptide, the amino acid *side chains* connected to the $C_\alpha$ atoms provide chemical and biophysical properties, and they stabilize the three-dimensional structure of the protein.

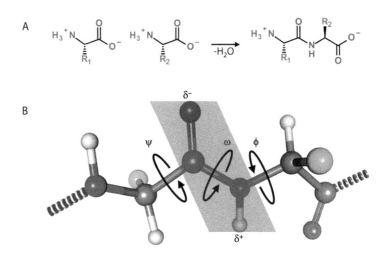

**FIGURE 16.3:** **Formation and properties of the peptide bond.** A: Formation of a dipeptide by condensation of the amino and carboxyl groups of amino acids. B: Overall structure of the polypeptide chain. The planar peptide group is highlighted (blue). The amide is a dipole with its negative part at the carbonyl oxygen ($\delta^-$). The torsion angles $\phi$, $\omega$, and $\psi$ are indicated. $\phi$ and $\psi$ can adopt many values and define the conformation and flexibility of the polypeptide chain. $\omega$ is either 0° or 180°, and determines whether a peptide bond is *cis* or *trans*. Side chains are indicated as green spheres.

The structural context of the peptide bond is described by the torsion angles $\psi$ and $\phi$ (Figure 16.3). $\psi$ describes the rotation around the $C_\alpha$-C bond and $\phi$ the rotation around the N-$C_\alpha$ bond. The torsion angles $\psi$ and $\phi$ can adopt many different values, giving polypeptides considerable conformational flexibility. Not all combinations of $\psi$ and $\phi$ are allowed, however, as some lead to steric clashes between atoms. Certain $\psi$ and $\phi$ combinations are characteristic for the secondary structure of the main chain, as we will see in the next section.

The torsion angle $\omega$ describes the rotation about the peptide bond. In contrast to $\psi$ and $\phi$, $\omega$ is much more restricted, and it clusters predominantly around two values: 0° and 180°, corresponding to the *cis*- and *trans*-configurations of a planar peptide group, respectively. The default configuration in ribosomal protein biosynthesis is *trans*, but many proteins require *cis*-peptide bonds for their function. For *trans-cis* isomerization, the conjugation of the peptide bond needs to be broken. The total rotational energy barrier to remove the conjugation in the peptide bond is about 80 kJ mol$^{-1}$, indicating a very slow isomerization. Rotation of the peptide bond by just 10° out of the plane already requires about 4 kJ mol$^{-1}$. Analysis of high-resolution protein crystal structures showed that in about 10–15% of all *trans*-peptide bonds, $\omega$ deviates from 180° by $\geq$10° with occasional deviation of over 30° from planarity. So while in most cases the peptide bond is planar, some distortion of the peptide plane is normal.

In *cis*-peptide bonds, the orientation of the CO- and NH-groups is inverted with respect to the *trans*-configuration. A *trans*-peptide is more stable than a *cis*-peptide by 2–10 kJ mol$^{-1}$, depending on which amino acids form the peptide bond. The main reason for the lower stability of *cis*-peptide bonds is a steric clash between the side chains in the *cis*-configuration (Figure 16.4). If the C-terminal residue of the peptide bond is Pro, the *cis*-configuration is only slightly less stable than the *trans*-configuration by about 2 kJ mol$^{-1}$.

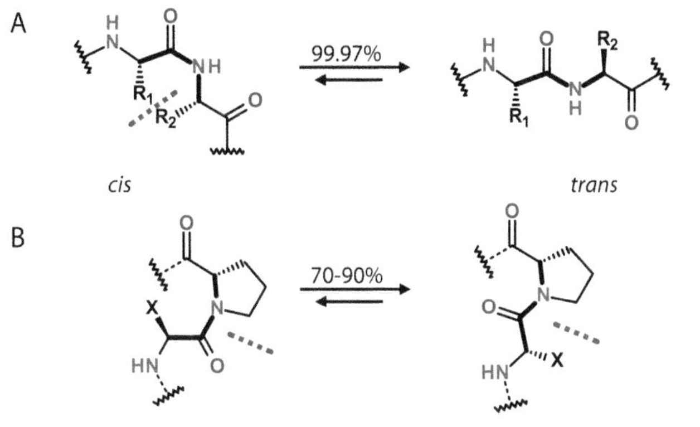

**FIGURE 16.4:    The *cis*-peptide bond.** A: The side chains in a non-prolyl *cis*-peptide clash with each other (dashed red line) but point away from each other in the preferred *trans*-peptide. B: In a prolyl peptide, both *cis*- and *trans*-configurations lead to clashes, so the energy difference between these states is smaller than in the non-prolyl case. The amino acid X preceding the proline can in principle be any amino acid, but the aromatic side chains His, Phe, Tyr, and Trp are preferred. The side chain of X often favorably stacks on top of the aliphatic pyrrole of Pro.

So why is the peptidyl-prolyl peptide bond a special case? In solution, 10–30% of all X-Pro sequences in peptides are found to be *cis*, in agreement with the ratio expected from the $\Delta G$ values. The structural explanation why Pro does not strongly disfavor the *cis*-configuration can be seen from Figure 16.4. In an X-Pro peptide bond either orientation leads to clashes with the pyrrole side chain, which is why both conformations have higher energy compared to non-Pro peptides. The corresponding energy difference between the *cis*- and *trans*-configurations is thus smaller for an X-Pro peptide than for a non-prolyl peptide. Nevertheless, prolyl-isomerization can be a rate-limiting step during protein folding (Section 16.3.5). Most *cis*-peptides involving Pro residues are located at tight turns in loop regions, whereas non-prolyl *cis*-peptides are typically restricted to enzyme active sites or dimerization interfaces.

## 16.1.3 Side-Chain Rotamers

Analysis of high-resolution protein crystal structures with well-defined electron density for the side chains has shown that the amino acid side chains adopt preferred conformations or *rotamers*. Rotamers are described by the torsion angles (Section 15.2.2), $\chi$, of side chains. The $\chi_1$ angle is the torsion angle around the $C_\alpha$–$C_\beta$ bond, $\chi_2$ corresponds to the rotation around the $C_\beta$–$C_\gamma$ bond, etc. The torsion angles of the rotatable single bonds along a side chain are clustered around values corresponding to staggered conformations. In amino acids with short side chains such as Ser, Cys, Thr, and Val, there are only three preferred conformations, each rotated 120° about the $C_\alpha$–$C_\beta$ bond (Figure 16.5). The relative energies of these rotamers are determined by interactions with the main-chain or neighboring side-chain atoms: staggered conformations are preferred over eclipsed conformations. Long side chains also show the basic three staggered conformations around the $C_\alpha$–$C_\beta$ bond seen with the small side chains, but have additional degrees of freedom and are more flexible (Figure 16.5). Arg has 33 preferred rotamers, Lys and Met have 24 and 13 rotamers, respectively.

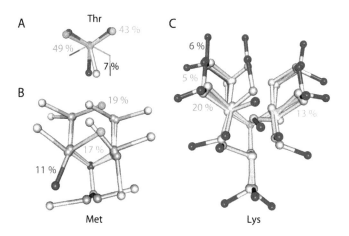

**FIGURE 16.5:   Preferred side-chain conformations in polypeptides.** The view is along the $C_\beta$-$C_\alpha$ bond onto the backbone of the polypeptide, and only the side-chain atoms are shown. The percentages for the most frequently observed rotamers in high-resolution crystal structures are given in the same color as the terminal C-atom of the respective rotamer. A: Short side chains have three preferred rotamers separated by 120°. B: The 13 preferred rotamers for Met. C: Lys has 24 preferred rotamers.

### 16.1.4  POST-TRANSLATIONAL MODIFICATIONS

The chemical and conformational space of proteins is increased by hundreds of post-translational modifications. Post-translational modifications can be covalent or non-covalent, permanent or transient, general for many proteins or specific to a single class of proteins. They occur spontaneously or are introduced by enzymes. Most post-translational modifications are introduced into unfolded proteins in the *endoplasmic reticulum* or the *Golgi apparatus*. Other modification reactions act on folded proteins, for example phosphorylation in the cytoplasm and ADP-ribosylation in the nucleus. Mass spectrometry (Section 26.1) has been invaluable for the identification of post-translational modifications, and the list of known post-translational modifications is continuously increasing. We will limit our survey to a few frequently observed modifications and those that drastically change the biophysical properties of certain amino acids. Table 16.2 gives a short list of common amino acid modifications in proteins.

**TABLE 16.2**
**Summary of frequent post-translational modifications.**

| amino acid | modification | function / effect |
|---|---|---|
| N-terminus | deformylation of Met | removal of N-formyl group after translation |
| | cyclization of Gln | pyro-glutamate in antibodies |
| | myristoylation | membrane anchor |
| Cys | oxidation to disulfide bonds | protein structure (see text), catalysis |
| | oxidation to sulfenic, sulfinic, sulfonic acids | sulfenic acids in catalysis, the others are irreversible oxidation products |
| | oxidation to formylglycine | active site residue of sulfatases |
| | glycosylation, binding of heme | |
| Asp | iso-aspartate | occasional shift of the peptide bond |
| Glu | γ-carboxylation | chelation of $Ca^{2+}$ in blood coagulation factors |
| | ADP-ribosylation | |
| His | methylation, phosphorylation, cofactor attachment | |
| | conversion to diphtamide | |
| | ADP-ribosylation of diphtamide | in active site of translation elongation factor 2 |
| Lys | methylation | mono-, di-, and trimethyl known |
| | hydroxylation and glycosylation | stabilizes and cross-links collagen triple helix |
| | ubiquitination | signal for degradation by proteasome |
| | cofactor binding | binding of pyridoxal phosphate and retinal |
| | reversible histone acetylation | regulation of gene expression |

| Met | oxidation to sulfoxide | unspecific, but two diastereomers exist |
|---|---|---|
| Asn | glycosylation | protein folding and stability |
| | spontaneous deamidation | |
| Pro | hydroxylation plus glycosylation | H-bonds in collagen |
| Gln | spontaneous deamidation | protein aging |
| Arg | methylation, phosphorylation, ADP-ribosylation, and hydrolysis to citrulline and ornithine | |
| Ser / Thr | glycosylation | mucus and extracellular matrix |
| | phosphorylation | signal transduction |
| Tyr | sulfation, phosphorylation | |
| | various phosphoesters with nucleotides | |
| C-terminus | methylation, amidation, acylation? | removal of negative charge and preparation as |
| | glycosyl-phosphatidylinositol | membrane anchor |
| | farnesylation / geranyl-geranylation (Cys) | |

*Note:* The list is by necessity incomplete, only a few examples are listed per amino acid.

### 16.1.4.1 Glycosylation

The addition of carbohydrates occurs at the side chains of Ser, Thr, and Asn residues. Glycosylation at Ser and Thr is termed *O-glycosylation*, whereas glycosylation at Asn residues is referred to as *N-glycosylation*. Asn side chains are glycosylated when they are part of the N-X-[ST(C)] motif, where X is any amino acid except Pro (Figure 16.6). The transfer of single sugars or whole trees of carbohydrates is catalyzed enzymatically, and up to 500 different enzymes are involved in protein glycosylation. The glycans on a protein can easily double the protein mass, and substantially alter the biophysical properties of the protein. Glycosylation increases both the solubility and stability of a protein, but does not alter its structure. It has been estimated that about half of all eukaryotic proteins are glycosylated. Most proteins that are located at the cell surface are glycosylated, creating a dense layer of glycan near the cell surface, the *glycocalyx*. Glycans are involved in intracellular protein transport, protein-protein interactions, and cell-cell contacts *via* their *extracellular matrix*.

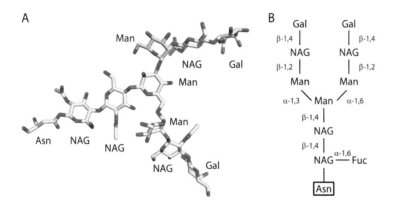

**FIGURE 16.6:  N-glycosylation.** A: A complex glycan is connected to the side chain of Asn (gray). The formal connectivity of the monosaccharides is shown in (B). B: NAG is N-acetylglucosamine, Man is mannose, and Gal is galactose.

Prokaryotes do not have enzymes of the glycosylation pathway, and prokaryotic proteins are not glycosylated. Recombinant production of proteins in *E. coli* that are post-translationally modified by glycosylation in their natural environment often leads to unstable and insoluble proteins.

### 16.1.4.2 Phosphorylation

Phosphorylation, the transfer of a phosphoryl group, occurs at the side chains of Ser, Thr, or Tyr. Phosphoryl group transfer from ATP is catalyzed by enzymes termed *kinases*. The reverse reaction, the hydrolysis of the phosphoester, is catalyzed by *phosphatases* (Figure 16.7). The reversible modification of proteins with phosphate groups functions as a molecular switch: phosphorylation can either activate or inactivate a protein, and conversely dephosphorylation inactivates the protein and switches its activity off, or activates and switches it on. Phosphorylation is one of the most important and best studied post-translational modifications because it regulates a whole set of cellular processes including cell cycle, cell growth and metabolism, apoptosis, and signal transduction.

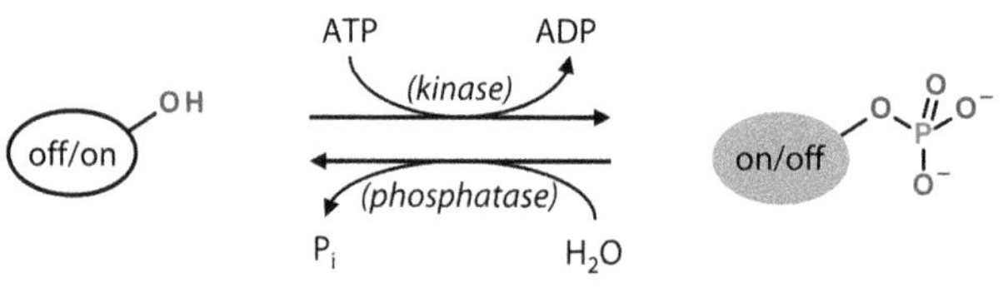

**FIGURE 16.7:    Protein phosphorylation.** Phosphorylation of the hydroxyl groups of Ser, Thr, and Tyr can activate a previously inactive protein or *vice versa*. The phosphoryl donor is almost always ATP. In special cases Asp, His, and Lys are also phosphorylated.

Phosphorylated amino acid residues in proteins are recognized by specific protein domains (Section 16.2.4). For example, phosphotyrosine is recognized by the Src homology 2 (SH2; Figure 16.30) and the phosphotyrosine binding (PTB) domains. Phosphoserine and phosphothreonine are bound by WW (Figure 16.30) and fork head-associated (FHA) domains, respectively. By binding to phosphorylated residues in proteins, these domains mediate the assembly of protein-protein complexes only when their binding partner is phosphorylated, a strategy central to signal transduction across cell membranes.

Phosphorylation is also a means to modulate enzymatic activity. Glycogen phosphorylase becomes activated by phosphorylation and then releases glucose from its storage form glycogen. By contrast, pyruvate dehydrogenase, an enzyme connecting glycolysis with the citric acid cycle, is less active when phosphorylated.

### 16.1.4.3 Hydroxylation

Hydroxylation occurs at Pro and Lys residues (Figure 16.8) and is enzymatically catalyzed. It is a rather specific but important post-translational modification in collagen, where Pro and Lys residues are hydroxylated by the enzymes prolyl-4-hydroxylase, prolyl-3-hydroxylase, and lysyl-5-hydroxylase. Pro in XPG sequences in pro-collagen is hydroxylated at the 4-position to *trans*-4-hydroxyl-Pro, whereas Pro in GP sequences is modified to *trans*-3-hydroxyl-Pro. Lys is hydroxylated at the 5-position when part of an XKG sequence. The hydroxylated residues engage in H-bonds that stabilize the collagen triple helix (Figure 16.21). The collagen hydroxylases are $Fe^{2+}$-dependent and require ascorbate (vitamin C) to keep the iron in the reduced $Fe^{2+}$ state. Scurvy, a disease caused by a lack of vitamin C, leads to less or no hydroxylation of collagen, and a reduced collagen stability in connective tissue because these hydrogen bonds cannot be formed.

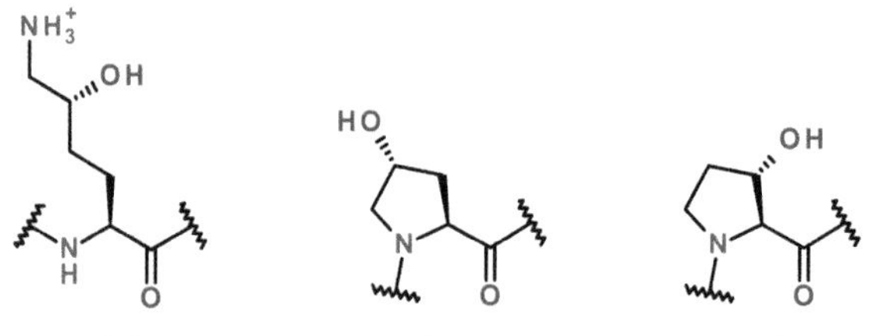

5-HyL     4-HyPro     3-HyPro

**FIGURE 16.8:    Hydroxylated Lys and Pro.** 5-hydroxylysine (5-HyL) can be further oxidized to introduce crosslinks between collagen chains. The methyl group in the hydroxyprolines (HyPro) is *trans* to the carbonyl group.

A hydroxylated serine is present in the non-canonical amino acid formylglycine, which acts as a catalytic residue of sulfatases, important enzymes in the hydrolysis of sulfate esters (Box 16.1).

**BOX 16.1: FORMYLGLYCINE.**

Formylglycine (FGly, Figure 16.9) is the catalytic residue in an enzyme class called *sulfatases*. It is generated by oxidation of a cysteine (in eukaryotes) or serine (in prokaryotes) side chain with molecular $O_2$ while the sulfatase is still in its unfolded state. If sulfatases are not modified this way, they fold into their native structure but are inactive, which results in a variety of severe diseases. The oxidation reaction is catalyzed by the *FGly-generating enzyme* (FGE). The active form of FGly is the hydrated aldehyde, a *geminal diol* that is coordinated to a $Mg^{2+}$ ion in the active site of the sulfatases. FGly can therefore be viewed as a hydroxylated serine, and indeed it acts as the nucleophile for the hydrolysis of sulfate esters, as does a serine in phosphatases. Eukaryotic FGE recognizes the Cys in the sequence context C-[TSAC]-PSR, and oxidizes it to FGly. Attaching this sequence at the N- or C-terminus of a protein, or within a loop region, allows the introduction of a formyl group into the protein that can be coupled with an amine group of a fluorophore, spin label, etc.

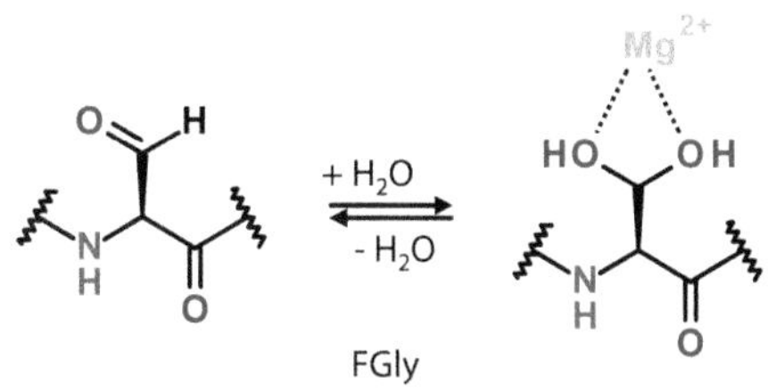

**FIGURE 16.9:    Formylglycine.** In sulfatases, the formyl group is hydrated to a geminal diol that coordinates a $Mg^{2+}$ ion.

#### 16.1.4.4 Carboxylation

An additional carboxyl group at the γ-atom of Glu (Figure 16.10) in proteins of the blood coagulation cascade introduces the ligands for binding of $Ca^{2+}$, which is essential for their function. Carboxylation of several glutamate residues occurs in the blood coagulation factors II (thrombin), VII, IX, and X. The carboxylase that catalyzes carboxylation of Glu to γ-carboxy Glu (Gla) is vitamin K-dependent. Warfarin and dicoumarol competitively inhibit the enzyme vitamin K epoxide reductase, thus depleting the vitamin K pool and ultimately rendering the carboxylase non-functional. These agents thereby inhibit blood clotting and act as anticoagulants. While warfarin was developed as a pesticide and medication against thrombosis, dicoumarol is a natural product.

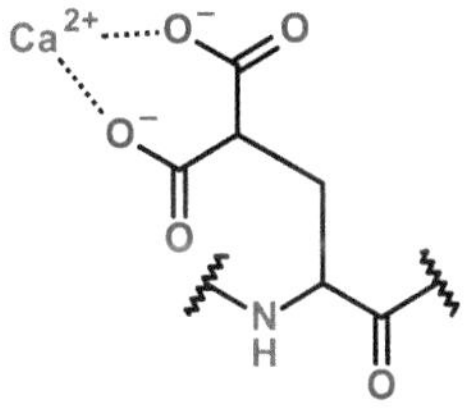

**FIGURE 16.10:    γ-carboxy glutamate (Gla).** $HCO_3^-$ is the carbon donor for the carboxylation. Normally, 1,3-dicarbonic acids are unstable but deprotonation and chelation by $Ca^{2+}$ stabilizes Gla.

### 16.1.4.5  Disulfide Bonds

A disulfide bond or disulfide bridge results from the oxidation of two Cys residues. The cytoplasm is a reducing environment, and therefore cytoplasmic proteins usually do not contain disulfide bonds. In extracellular proteins, disulfide bonds are omnipresent. In eukaryotes, disulfide bonds are synthesized in the oxidizing environment of the endoplasmic reticulum, and in bacteria the corresponding compartment is the periplasmic space. As a covalent interaction, disulfides can have a huge stabilizing effect on protein structure. Such disulfide bonds are usually buried in the interior of the protein, inaccessible for reducing agents or disulfide-reducing enzymes. On the other hand, disulfides at the surface of proteins can be reduced to cysteines. Enzymes such as protein disulfide isomerases (Figure 16.51 in Section 16.3.5.2) have two solvent-exposed Cys residues in their active site that cycle between reduced and oxidized (disulfide-bonded).

The disulfide structures are described by the torsion angles $\chi_i$ when moving along the disulfide bond from one Cys to the other. There are only a few preferred geometries for disulfide bridges, including the left-handed spiral, the right-handed hook, and an extended structure found in immunoglobulins. The preferred $\chi_1$ angles of Cys (the torsion about the $C_\alpha$–$C_\beta$ bond) are around $-60°$ and $180°$ with a strong preference for $\chi_1 = -60°$. The $\chi_3$ torsion angle (around the $S_\gamma$–$S_\gamma{'}$ bond) is in the range of $\pm 90°$, meaning that when viewed along the disulfide-bond, the $C_\beta$–$S_\gamma$ bonds are almost perpendicular to each other. If $\chi_3$ is $-90°$, a left-handed spiral is formed by the atom sequence $C_\alpha$–$C_\beta$–$S_\gamma$–$S_\gamma{'}$–$C_\beta{'}$–$C_\alpha{'}$ (Figure 16.11), which is the predominant structure of disulfides in proteins. If $\chi_3$ is $+90°$, the disulfide is right-handed and has the shape of a hook (Figure 16.11). The third kind of disulfide bond structure stabilizes the immunoglobulin fold (Figure 16.31 in Section 16.2.4) by crosslinking two β-sheets. This disulfide bridge is extended, with both the $\chi_1$ and $\chi_2$ angles close to $180°$, such that the $C_\alpha$ atoms are maximally far apart from each other (up to about 700 pm).

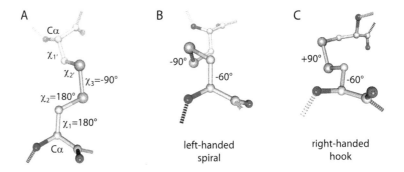

**FIGURE 16.11:**  **Disulfide bonds in proteins.** The predominant disulfide bond structures found in proteins are the left-handed spiral, the right-handed hook, and an extended disulfide bond prevalent in immunoglobulins. The $\chi_i$ torsion definitions are noted for the extended disulfide bond in (A). The spiral (B) and the hook (C) both have $\chi_1$ close to $-60°$.

### 16.1.4.6  Metal Binding

Bound metal ions are often important for protein function. Examples of functional metal ions include the $Mg^{2+}$ in the active site of sulfatases (Box 16.1) and $Ca^{2+}$ in enzymes of the blood coagulation cascade (Section 16.1.4.4). Fe and Cu ions are frequently found at the active sites of redox-active enzymes. Proteins can furthermore be immensely stabilized by metal binding, the second most prominent crosslink after disulfide bonds. The covalent bond between a metal ion and the lone pair of a side chain or a peptide group is a *dative bond* or *coordinate bond*. Dative bonds are longer (210–280 pm) and therefore not quite as strong as a canonical covalent bond (Table 15.1). Metal ions are bound with equilibrium dissociation constants in the range of pM (almost irreversible) to mM (loose interaction). The most common stabilizing metal ions are $Ca^{2+}$ and $Zn^{2+}$ (Figure 16.12), which can bind very tightly to proteins, and $Mg^{2+}$, $Na^+$, and $K^+$ that are less tightly bound. Asp and Glu side chains, but also hydroxide ions and carbonyl groups are frequent ligands for metal ions. Removal of structural metal ions greatly reduces protein stability and may lead to unfolding. For example, chelation of $Ca^{2+}$ in subtilisin or calmodulin by EDTA or EGTA significantly destabilizes these proteins by leaving a negatively charged, destabilizing cavity.

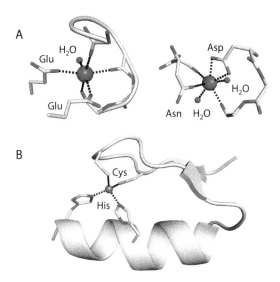

FIGURE 16.12:    **Metal binding in proteins.** A: The two $Ca^{2+}$ binding sites in the FGly-generating enzyme FGE (PDB-ID 1y1e). The coordinate bonds are shown as black dashed lines. Water molecules or hydroxide ions (these are difficult to distinguish) are shown as red spheres. Side chains binding to the $Ca^{2+}$ (magenta spheres) are labeled. All other contacts are made by the carbonyl groups of main-chain peptide bonds. B: Tetrahedral $Zn^{2+}$-coordination by two Cys and two His side chains in the DNA-binding $C_2H_2$ zinc finger Zif268 (PDB-ID 1aay). The $Zn^{2+}$ ion is shown as a gray sphere.

Depending on the type of ion, different coordination geometries are favored. The vast majority of metal ions are coordinated octahedrally with four ligands and the metal ion in a plane and the last two ligands perpendicular to the plane (left $Ca^{2+}$ site in Figure 16.12). Softer, less electronegative ions, such as $K^+$ and $Ca^{2+}$, can accept more than six ligands (right $Ca^{2+}$ site in Figure 16.12) and the ligands may form a heavily distorted octahedron around the metal ion. $Zn^{2+}$, on the other hand, strongly favors tetrahedral over octahedral coordination. If water is bound to $Zn^{2+}$, it has a reduced $pK_A$ value and a higher propensity to form a hydroxide. Such $Zn^{2+}$-bound hydroxides are the nucleophiles in *zinc hydrolases*.

Post-translational modifications occur after the polypeptide chain has left the ribosome. Some non-canonical amino acids are directly encoded in the natural mRNA transcript (Box 16.2).

---

**BOX 16.2: EXPANDED GENETIC CODE.**

Non-canonical amino acids are sometimes directly encoded in the natural mRNA transcript of the gene by stop codons. A triplet of nucleobases in the mRNA, the codon, is recognized by an anticodon of a transfer RNA (tRNA; Section 17.6.2). The stop codons are UAG (*amber*), UAA (*ochre*), and UGA (*opal*). If a stem-loop structure is present close

*(Continued)*

to the UGA codon, a specialized tRNA can decode it as selenocysteine (SeCys). Likewise, the UAG codon can be translated into pyrrolysine (Pyl). SeCys and Pyl (Figure 16.13) have been termed the 21st and 22nd amino acids. The selenol group of SeCys has a low $pK_A$ value of ca. 5.5. It is found in redox-active proteins that require a very negative redox potential for their function such as glutathione peroxidase, thioredoxin reductase, and formate dehydrogenases. Pyl is essentially a lysine with an additional pyrrole ring, a constituent in proteins of methanogenic archaebacteria.

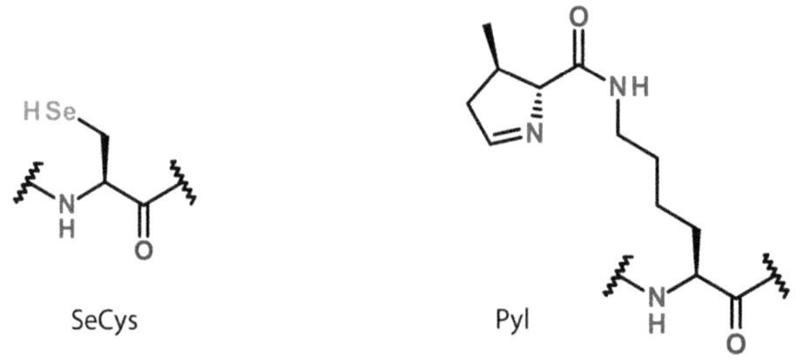

**FIGURE 16.13:**   **Selenocysteine (SeCys) and pyrrolysine (Pyl).**

The translation of stop codons into amino acids is known as *suppression* and has been used with great success to introduce several dozen novel chemical functionalities in proteins. Nowadays, the term *expanded genetic code* solely refers to these chemically altered amino acids. A tRNA that recognizes a stop codon is charged with a non-standard amino acid, either by a chemical reaction or using an aminoacyl-tRNA synthetase. A gene with the stop codon at the desired position is transcribed into mRNA, but translation of the mRNA continues past the stop codon. It is decoded ("suppressed") by the tRNA carrying the cognate anticodon. The new amino acid is incorporated into the nascent protein either by *in vitro* translation or by feeding genetically modified bacteria that make the necessary tRNA and synthetase with the modified amino acid.

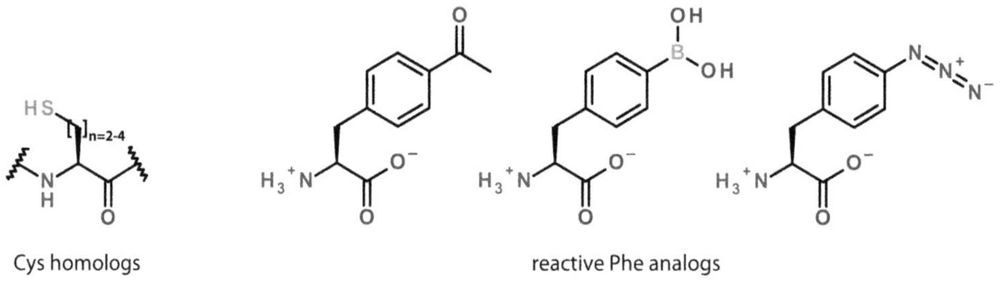

**FIGURE 16.14:**   **Homologs of Cys and Phe analogs.**

Using these novel side chains, protein function can be studied with homologs of natural amino acids. Cys, Gln, and Met have been elongated by adding $CH_2$ groups. By introducing reactive groups such as ketones, boronic acids, or azides, proteins can be modified with NMR-sensitive isotopes, fluorophores, IR-sensitive probes, and spin labels (Figure 16.14). The chemistry of these novel chemical groups is very different (*orthogonal*) to that of the canonical amino acid side chains (hydroxyl, thiol, amino groups) and allows labeling of proteins by two or more chemical groups, for instance in FRET studies (Section 23.1.7.6).

## 16.2 PROTEIN STRUCTURE

The primary structure of proteins is the sequence of amino acids connected by peptide bonds. At least for small proteins, this sequence contains all the information that is necessary to produce the final fold of the protein. During the folding process, the protein sequence adopts *secondary structure* elements such as α-helix, β-strands, and several kinds of loops and turns. The secondary structure elements continue to fold into *super-secondary structures*, also called structural *motifs*. The next level of organization is the *tertiary structure*, which includes globular domains and fibers. Many interactions in domains are between residues located far away in sequence. Proteins that further assemble into oligomers adopt a *quaternary structure*. Sometimes a quaternary structure stabilizes a tertiary structure that otherwise would not be thermodynamically stable on its own. Examples are some ribosomal proteins that only adopt a defined structure when ribosomal RNA (rRNA) is present. Figure 16.15 gives an overview of the different levels of protein structure.

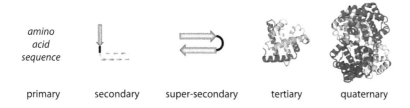

amino acid sequence

primary   secondary   super-secondary   tertiary   quaternary

**FIGURE 16.15:** **Different levels of protein structure.** The primary structure is simply the amino acid sequence. The backbone torsion angles define the secondary structure elements α-helix and β-strand, which are connected by loop regions. These elements can also form super-secondary structures or compact three-dimensional motifs of adjacent secondary structure elements. Super-secondary structures are generally smaller than a whole protein domain, which defines the tertiary structure. Association of several polypeptide chains into a complex is termed quaternary structure.

We have introduced the torsion angles φ and ψ that describe the conformation of the protein main chain (Section 16.1.2). A *Ramachandran plot* represents an energy surface as a function of φ and ψ (Figure 16.16). The energetically favored φ/ψ combinations are characteristic of the major secondary structure elements α-helix, β-strand, and turns, which are discussed in the following sections. The α-helix corresponds to the energy minimum of $\phi = -60 \pm 10°$ and $\psi = -40 \pm 10°$, whereas the β-strand conformation has a large and shallow minimum around $\phi = -120 \pm 20°$ and $\psi = 125 \pm 15°$. Folded proteins are basically a string of secondary structure elements connected by loops and turns that contact each other through space to establish a three-dimensional structure.

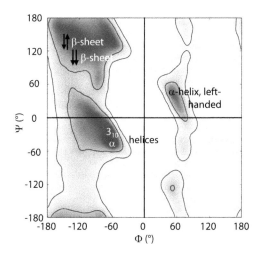

**FIGURE 16.16: Ramachandran plot for non-Gly residues with secondary structure elements labeled.** Darker areas are φ/ψ combinations of lowest energy while white areas indicate high energy and are therefore "forbidden". A left-handed α-helix is not formed to a large extent in nature but sometimes Gly adopts this backbone conformation. The figure was created with the programs Phenix and Molprobity.

Different amino acids prefer different φ/ψ combinations and thus have different secondary structure propensities (Table 16.3). For instance, Ala is much more frequently encountered in α-helices than in any other secondary structure type. Pro and Gly are most frequently encountered in turns, and the small hydrophobic side chains Val and Ile are frequently found in β-sheets. Other residues, such as His, do not have a special propensity for a particular secondary structure. The propensities of different amino acids for certain structures are sometimes used in secondary structure prediction algorithms (Section 18.1.3).

**TABLE 16.3**
**Amino acid propensities for secondary structure.**

| amino acid | α-helix | β-strand | turn |
|---|---|---|---|
| E | **1.59** | 0.52 | 1.01 |
| A | **1.41** | 0.72 | 0.82 |
| L | **1.34** | 1.22 | 0.57 |
| M | **1.30** | 1.14 | 0.52 |
| Q | **1.27** | 0.98 | 0.84 |
| K | **1.23** | 0.69 | 1.07 |
| R | **1.21** | 0.84 | 0.90 |
| H | **1.05** | 0.80 | 0.81 |
| V | 0.98 | **1.87** | 0.41 |
| I | 1.09 | **1.67** | 0.47 |
| Y | 0.74 | **1.45** | 0.76 |
| C | 0.66 | **1.40** | 0.54 |
| W | 1.02 | **1.35** | 0.65 |
| F | 1.16 | **1.33** | 0.59 |
| T | 0.76 | **1.17** | 0.90 |
| G | 0.43 | 0.58 | **1.77** |
| N | 0.76 | 0.48 | **1.34** |
| P | 0.34 | 0.31 | **1.32** |
| D | 0.99 | 0.39 | **1.24** |
| S | 0.57 | 0.96 | **1.22** |

*Source:* Reproduced from Williams et al., *BBA*, 916(2), 1987, 200–204, with permission.

*Note:* Bold numbers represent preferred secondary structure elements for these residues. A value of unity means no significant propensity for this secondary structure.

## 16.2.1 HELICAL SECONDARY STRUCTURE ELEMENTS

### 16.2.1.1 α-helix

The α-helix is a very compact secondary structure element (Figure 16.17) where, arithmetically, 3.6 residues form one turn of a right-handed helix with a *pitch* (height difference per turn) of 540 pm. Thus, each residue is rotated by 100° with respect to the next one. The right-handedness of the helix is a consequence of the $C_\alpha$ chirality of the amino acids. α-helices are stabilized by main-chain H-bonds: the carbonyl group of residue *i* H-bonds to the NH-group of residue *i + 4*. As a consequence of these H-bonds, the peptide planes are aligned parallel to the α-helix axis. The individual dipoles of each peptide bond therefore sum up to a macroscopic *helix dipole* with about half an electronic charge at either end. This macro-dipole often stabilizes anions bound near the helix N-terminus in proteins (see Figure 16.36 in Section 16.2.7). Also because of the *i, i + 4* H-bonds in the helix, the side chains of residues *i* and *i + 4* lie on the same side of the helix. If both are hydrophilic, they form a hydrophilic patch on one side of the helix, if both are hydrophobic, they generate a hydrophobic side. Many helices on the surface of proteins are *amphiphilic* (equivalent to *amphipathic*), meaning that they have hydrophilic residues on one side and hydrophobic residues on the other. Amphipathic helices can bury their hydrophobic part in the interior of proteins or in one leaflet of a membrane, while the hydrophilic side is exposed to water.

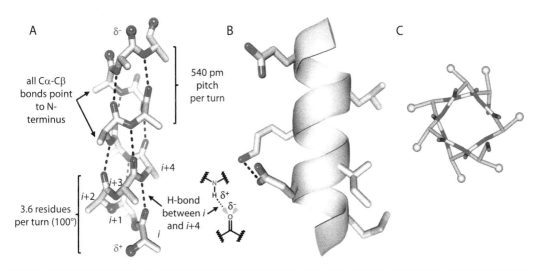

**FIGURE 16.17:    Geometry and properties of a α-helix.** A: Stick representation of a poly-Ala sequence. The dashed lines represent the main chain H-bonds between residues *i, i+4*. Three N-terminal nitrogen atoms and three C-terminal oxygen atoms that do not have an H-bonding partner are shown as spheres. The helix dipole is indicated by the δ+ and δ− labels. B: Ribbon representation of an amphipathic α-helix with polar side chains on the left-hand side, non-polar side chains on the right-hand side. A charged H-bond between a Glu-Lys pair is shown as a dashed line. C: The top view of the α-helix shows staggered side chains.

In α-helices, the first three NH-groups and the last three carbonyl groups have no H-bonding partners (Figure 16.17). For stability reasons, these unsatisfied H-bond donors and acceptors need a counterpart (Section 15.3.3). Free H-bonding valences at the termini of α-helices are often *capped* by amino acids just outside the actual helix. These residues at the position half-in and half-out of an α-helix are termed the N-cap and C-cap, respectively (Figure 16.18).

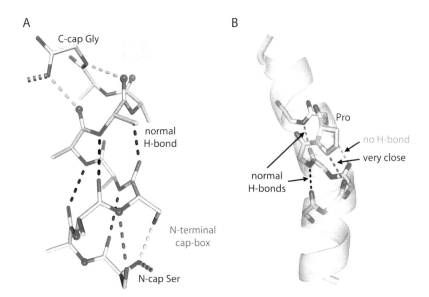

FIGURE 16.18:   **Helix capping and kinking.** A: The Gly at the C-terminus of the α-helix alters the direction of the polypeptide chain. The chain would normally follow the right-handed spiral but escapes to the left. The peptide groups on either side of the Gly are tilted towards the helix axis to form two H-bonds in inverted sequence order (cyan dashes). The N-cap shown here is a Ser whose side chain caps the NH-group of the third residue of the α-helix (magenta dashes). The side chain of this third residue accepts an H-bond from the N-cap residue (yellow dashes), completing the cap-box. PDB-ID 1lmb. B: Pro can kink helices. Pro in the middle of a long α-helix leads to a kink and change of direction of the polypeptide chain. Because of the covalent bond between the Cδ atom of Pro and the main-chain nitrogen, no H-bond is possible (green dashes). The Cδ atom has an unfavorably close contact with the carbonyl group of the $i-4$ residue (red dashes), which induces a kink in the helix. The α-helical H-bonding pattern is reached again at the $i+2$ residue.

All amino acid residues except proline can form α-helices, although some are preferred over others (see Table 16.3). The Pro side chain is a ring (Figure 16.1) that not only interrupts the $i,i+4$ H-bonding pattern but also clashes with the carbonyl oxygen atom of the preceding residue. Thus, helices often terminate at a Pro, and Pro is therefore termed a *helix breaker*. In longer helices, Pro residues do not destabilize the entire helix but introduce a kink (Figure 16.18). Such kinks can be of functional importance and are found in many G-protein coupled receptors (Section 16.2.8.4), membrane proteins of seven transmembrane α-helices that pack together to form a ligand binding site.

An isolated α-helix in solution is unstable. Packing of α-helices against each other stabilizes them by a mechanism that can be described as "knobs-in-holes" where the side chains interdigitate. If two helices are inclined by 20° relative to each other and are overwound slightly to a helical repeat of 3.5 residues, every seventh residue of one helix will fit into a "hole" on the surface of the other. The helices wind around each other in a left-handed superhelix (Figure 16.19). This *coiled-coil* structure can often be predicted from a sequence that contains *heptad repeats*, i.e. a similar sequence every seven residues (two helical turns) that end up on the same side of the helix. The side chains that form the knobs (and the walls of the holes) are mostly hydrophobic. Leu predominates, but other aliphatic side chains, and the larger aromatic side chains are sometimes found in the interface of coiled-coils. Leu-rich regions forming coiled-coils are also known as a *Leu-zipper*, a classical protein dimerization module that is, among others, found in DNA-binding proteins (Section 17.5.4.1).

The helices in a coiled-coil can be parallel or antiparallel, with two, three and four helices forming the bundle. Larger coiled-coils with more helices can be viewed as assemblies of the smaller 2-, 3-, and 4-stranded coiled-coils. A schematic visualization of coiled-coils is the *helix wheel*. A single helix wheel shows the heptad repeats, and several helix wheels next to each other show the interacting residues in the coiled-coil.

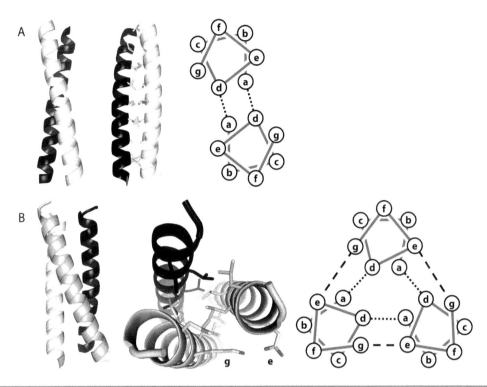

FIGURE 16.19:    **Coiled-coils of α-helices.** A: The Leu-zipper is a classical parallel two-stranded coiled-coil where the slightly distorted α-helices with a helical repeat of 3.5 residues assemble into a left-handed superhelix. The interface is hydrophobic. The schematic drawing on the right is the helix wheel representation of the heptad repeat viewed from the C-terminus. It shows how side chains from residues **a** (most frequently Leu, Ile, or Ala) and **d** (most frequently Leu or Ala) pack against each other in the two-stranded coiled-coil. B: A three-stranded parallel coiled-coil.

In principle, the knob-and-hole pattern also interlocks when α-helices are oriented an angle of about 70°. This almost perpendicular helix-crossing is stabilized by fewer interactions, and is often found as part of a larger structure where it is stabilized by other interactions.

### 16.2.1.2 $3_{10}$-, poly-Pro, and Collagen Helices

Apart from the α-helix discussed above, there are several more helical secondary structure elements that highlight the structural plasticity of the polypeptide chain: the $3_{10}$-helix, the poly-Pro helix, and the collagen helix. The $3_{10}$-helix follows an *i, i + 3* H-bonding pattern along the main chain, one residue less than the α-helix. Therefore, the $3_{10}$-helix is more densely packed than the α-helix and has a smaller radius than the α-helix (Table 16.4). The main chain in a $3_{10}$-helix adopts tight turns and its side chains are eclipsed when viewed along the helix axis (Figure 16.20), leading to conformational strain. In addition, the H-bonds between residues *i* and *i+3* contribute little stabilization energy because the amide groups dipoles are not aligned head-to-tail. As a result, the $3_{10}$-helix is less stable than the α-helix. The longest $3_{10}$-helices in natural proteins have just two turns and are often found at the end of α-helices where they contribute capping residues (Figure 16.18). The $3_{10}$-helix got its name from the fact that the ring closed off by the H-bond (including the hydrogen) involves ten atoms from three residues along the sequence (Figure 16.20). By this definition, the α-helix is technically a $3.6_{13}$-helix, although this nomenclature is rarely used.

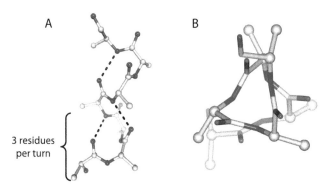

A

B

3 residues
per turn

FIGURE 16.20:   **The 3₁₀-helix.** A: The main chain is tightly wound. The $i, i + 3$ H-bonds (black dashes) are 300 pm long, a distance close to the optimum, but they have sub-optimal geometry. B: The side chains in the 3₁₀-helix emanate every 120° from the helix axis and are almost eclipsed (compare to Figure 16.17). The tightly wound main chain and the close contacts of the side chains lead to conformational strain.

We have seen before (Figure 16.18) that Pro residues do not fit into a regular α-helix but introduce kinks. Longer poly-Pro stretches can form two types of poly-Pro helices. Type I has only *cis*-peptide bonds and forms in organic solvents, but has not been found in proteins. By contrast, type II has all peptide bonds in *trans* and is the predominant form in water. Pro has no H-bond acceptor, so the conformation of the poly-Pro helix is not stabilized by H-bonds but is a result of the conformationally restricted Pro side chains. In contrast to all other types of helices, the poly-Pro-II helix is left-handed (Figure 16.21). It is found in short linker regions connecting individual domains of multi-domain proteins. SH3-domains (Figure 16.27) and WW domains (Figure 16.30) specifically bind these conformations, which is important for the assembly of multi-protein complexes.

Very long poly-Pro-II helices are the basic structural elements in collagen, a ubiquitous framework protein that is part of tendons, ligaments, skin, and cartilage. About a dozen closely related variants of collagen exist, all of which have the basic sequence (Gly-Xaa-Yaa)ₙ, where Xaa and Yaa are often Pro. Many of the Pro and also Lys at position Yaa are post-translationally hydroxylated to 3-hydroxy-Pro (HyPro) and 5-hydroxy-Lys (Section 16.1.4.3). Three poly-Pro-II helices associate into a parallel trimer (Figure 16.21) termed the *collagen helix*. The small Gly residues in the (Gly-Xaa-Yaa)ₙ sequence form cross-strand H-bonds and, importantly, allow the three chains to come close enough to form the triple helix in the first place. This explains why Gly is conserved at every third position in the collagen sequence. The triple helix is stabilized by cross-strand H-bonds from NH-groups of Gly with the CO-groups of Pro. Additional stabilization is provided by crosslinking of the chains with hydroxylysine and by hydroxyproline, although not by direct H-bonds as the OH-group is too far away from other chains.

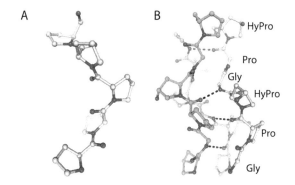

A

B

HyPro

Pro

Gly

HyPro

Pro

Gly

FIGURE 16.21:   **Helical secondary structure elements that involve Pro.** A: Poly-Pro-II helix. The peptide bonds are *trans*, and the helix is left-handed. In a poly-Pro sequence no H-bonds exist between residues along the helix because no H-bond acceptor is present. B: Collagen triple helix made by three chains of poly-Pro-II geometry (PDB-ID 4dmt). H-bonds between the chains are made between Gly of one chain and residue Xaa (not Pro) of another chain in the Gly-Xaa-Yaa repeat.

Table 16.4 summarizes the geometric properties of the helices that we discussed in this section.

| Table 16.4 | | | | |
| --- | --- | --- | --- | --- |
| **Geometric properties of helical polypeptide conformations.** | | | | |
| helix | α | $3_{10}$ | poly-Pro-I | poly-Pro-II |
| handedness | right | right | right | left |
| # residues per turn | 3.6 | 3.0 | 3.3 | 3 |
| pitch (pm) | 540 | 600 | 630 | 930 |
| radius of helix (pm) | 230 | 190 | — | 70[a] |
| preferred φ | −60 | −75 | -83 | −75 |
| preferred ψ | −45 | 0 | 158 | 150 |
| H-bond | i, i+4 | i, i+3 | — | — |
| # of atoms in ring | 13 | 10 | — | — |
| [a] The 70 pm radius of the poly-Pro-II helix refers to the collagen triple helix. | | | | |

## 16.2.2 β-Strands and Their Super-Secondary Structures (β-Sheets)

The second important structural element of proteins is the β-strand, a maximally extended polypeptide (Figure 16.22).

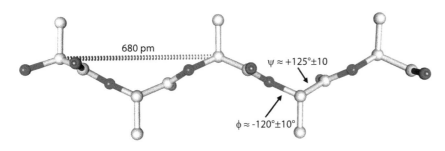

**FIGURE 16.22:   Ideal β-strand conformation.** In a single ideal β-strand, the side chains alternatively point up and down. The main-chain torsion angles are roughly at φ/ψ combinations of −130/130°. This corresponds to a shallow minimum in the upper left corner of the Ramachandran plot (Figure 16.16).

The extended conformation is a preferred structure of a polypeptide and corresponds to a broad energetic minimum in the Ramachandran plot at φ/ψ combinations of roughly −130/130°. There are two regular patterns in a β-strand: subsequent side chains follow an up-down pattern and the peptide groups have a back-and-forth orientation. Thus, every second side chain and every second peptide group point into the same direction. The true repeating unit of a β-strand is therefore a di-peptide with the $C_\alpha$-atoms separated by a distance of ca. 680 pm.

A single β-strand is not stable *per se*. In contrast to the more compact α-helix where local *i, i + 4* interactions stabilize the turn, there are no interactions between nearby atoms in a β-strand. To satisfy the H-bond donors and acceptors in β-strands they assemble into β-*sheets*. In *parallel* β-*sheets*, all β-strands point into the same direction. *Antiparallel* β-sheets have alternating directions of their β-strands, and *mixed* β-sheets have both parallel and antiparallel strands (Figure 16.23).

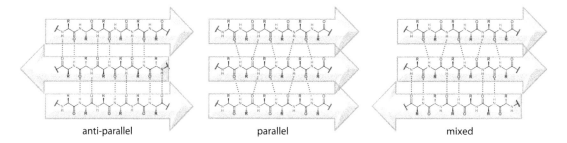

anti-parallel    parallel    mixed

**FIGURE 16.23:** **β-strands associate into antiparallel, parallel, and mixed β-sheets.** The directions of the polypeptides from the N- to the C-terminus are shown as arrows. H-bonds (green dashes) are formed between the peptide groups of adjacent β-strands and have almost perfect geometry. Note that the H-bonds in antiparallel β-sheets are parallel to each other and come in pairs while in parallel β-sheets they are neither parallel nor spaced evenly.

β-sheets are pleated: the $C_\alpha$ atoms of subsequent residues are slightly above and below the plane defined by the sheet. Together with the alternating upwards and downwards pointing side chains, "ripples" perpendicular to the direction of the polypeptide traverse the sheet. The side chains forming these ripples interact *via* inter-strand H-bonds or van der Waals interactions (Figure 16.24).

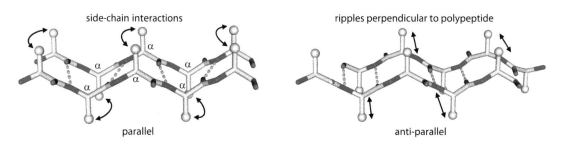

side-chain interactions    ripples perpendicular to polypeptide

parallel    anti-parallel

**FIGURE 16.24:** **Parallel and antiparallel β-sheets.** H-bonds are shown as green dashed lines. H-bonds in parallel β-sheets are angled with respect to each other while antiparallel β-sheets have parallel H-bonds. Common to both β-sheets are side chain interactions between neighboring strands (curved double arrows) and the formation of ripples: the up-and-down path followed by the $C_\alpha$ atoms gives rise to ripples or pleats perpendicular to the polypeptide chains (straight double arrows).

Mathematically, the β-strand can be described as a helix with two residues per turn and a translation of 320–340 pm per residue (Table 16.5).

| TABLE 16.5 Geometric properties of β-strand conformations. | | |
|---|---|---|
| strand | ⇅ β | ⇈ β |
| # residues per turn | 2 | 2 |
| pitch (pm) | 680 | 640 |
| preferred φ | −140 | −119 |
| preferred ψ | 135 | 113 |
| H-bond | long-range | long-range |

Two β-strands connected by a short loop can form a *β-hairpin* (Figure 16.25) that can further assemble into the prevalent super-secondary structures *β-meander*, *Greek key*, and *jellyroll* (Figure 16.25), all of which are antiparallel sheets. The β-meander topology consists of antiparallel β-strands linked by short loops and is found in antiparallel β-barrel and in β-propeller structures (see later in this section). The Greek key motif is a four-stranded antiparallel β-sheet found in immunoglobulin domains (Figure 16.31). The jellyroll is an extension of the Greek key motif with an additional β-hairpin attached to one side. It is prevalent in many spherical virus coat proteins.

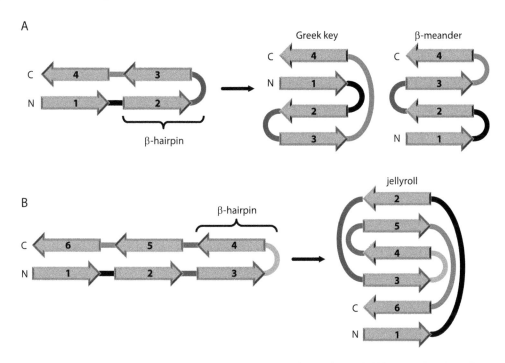

FIGURE 16.25:    **Super-secondary structures from antiparallel β-strands.** A: The β-hairpin forms from two consecutive β-strands that are connected by a short turn. Four β-strands can form the regular up-and-down arrangement of a β-meander, or a Greek key motif (named after Greek ornamental artwork). A total of 24 parallel and antiparallel arrangements with different connections can be drawn from four β-strands. The β-meander and Greek key motif are by far the most frequently observed β-motifs in protein structures. B: Six strands are often found to pack into the jellyroll motif: the three β-hairpins made up by strands 3/4, 2/5, and 1/6 are rolled up into a spiral.

The shortest connections that are possible between the β-strands depend on the nature of the sheet. In antiparallel sheets, these connections can be as short as two residues (a reverse turn, Section 16.2.3). For parallel β-sheets, there are a minimum number of residues connecting the β-strands because the end of the first strand and the start of the second strand are on opposite ends of the sheet. The *crossovers* can be in two different configurations, left- and right-handed. Most β-sheets have a right-handed twist when viewed along the directions of the strands. The right-handedness of the crossover is possibly aided by this inherent right-handed twist (Figure 16.26), and the vast majority of crossovers are right-handed. Instead of a loop (β-loop-β motif), the crossover can also be formed by an α-helix (β-α-β motif). β-loop-β and β-α-β motifs are the building blocks of *Leucine-rich repeat* (see Figure 16.30) and *TIM-barrel* proteins (see Figure 16.31).

Planar β-sheets or β-sheets with twists close to zero can grow infinitely. Small peptides are known to assemble into infinite sheets and form filaments in a number of diseases including Alzheimer's disease (see Table 16.10).

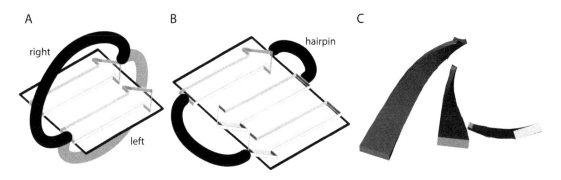

**FIGURE 16.26:**   **β-strand connections and bending of sheets.** A: Crossovers connect β-strands at opposite edges of the sheet. The two strands of a parallel β-sheet and the crossover connecting them form a spiral with a defined handedness. The right-handed crossover is by far the dominant connection. The left-handed crossover (light gray) is very rarely observed. B: Hairpins connect adjoining strands on the same side of the β-sheet. C: Most β-sheets display a right-handed twist when viewed along the direction of the strands. Several more strands may finally close into a β-barrel.

A β-sheet of at least five strands may roll up into a barrel where the strands forming the edges of the sheet H-bond to each other (Figure 16.27). The strands in a barrel are inclined relative to the main barrel axis: the fewer strands a barrel has, the more inclined they tend to be (compare the SH3 and GFP structures in Figure 16.27). Some barrels are stable on their own, while others are part of a larger structure. Most stable β-barrels are entirely antiparallel or are mixed parallel and anti-parallel. Barrels with exclusively parallel β-strands necessarily have long crossovers on their outside and are thus always embedded in a larger structure.

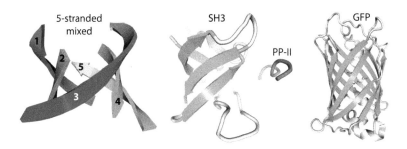

**FIGURE 16.27:**   **β-barrels in proteins.** The smallest barrel is a 5-stranded, antiparallel barrel where only the first and last strands (1 and 5) are oriented parallel to each other. SH3 domains are examples of such small barrels that bind to Pro-rich sequences in the poly-Pro-II conformation (Figure 16.21). The green fluorescent protein (GFP) is an antiparallel 13-mer barrel housing a fluorophore made by post-translational modification of amino acids.

### 16.2.3 REVERSE TURNS

Reverse turns change the direction of a polypeptide by up to 180° over as little as three or four residues. Also known as *tight turns*, *β-turns*, or simply *turns*, reverse turns are the most common type of non-repetitive structure in proteins. Turns can make up anywhere between 24% and 45% of the entire protein structure. Most turns are at the surface of proteins, which explains the predominance of hydrophilic residues in turns (Table 16.3). Of the many different types of turns that have been described, we limit our discussion to the six most frequently observed in protein structures, the type I, I′, II, II′, VIa, and VIb (Figure 16.28).

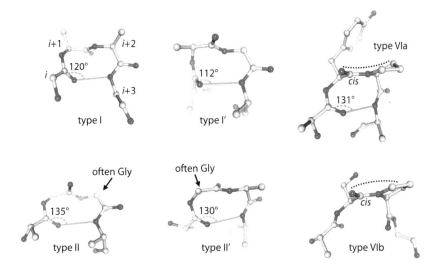

**FIGURE 16.28:   Frequent turns in protein structures.** H-bonds are shown as green lines between residues *i* and *i* + 3. The angle between the H-bond donor (carbonyl) and the acceptor (NH) are also given. The H-bond angles deviate significantly from the optimal geometry of 180°. In fact, H-bonding is not a prerequisite for the existence of a turn. Type VI turns have a *cis*-Pro peptide bond. The subtypes VIa and VIb differ mainly by the main-chain conformation on both sides of the Pro. Type VIa is usually H-bonded, type VIb is not.

Most turn geometries allow formation of an H-bond between the first residue *i* and the *i+3* residue, but this is not a strict requirement. A survey of turns in protein structures found that only about half of them are H-bonded at all.

Type I and type II turns fold the peptide chain around an almost square corner. Both turns bend the polypeptide chain in a way that in principle allows continuation of an antiparallel β-sheet. Type I and type II turns are related to one another by a 180° flip of the central peptide unit. Turns of types I′ and II′ have backbone conformations that are mirror images of types I and II: the φ/ψ angles of the central two amino acids are inverse to those of type I and II (Table 16.6), changing the handedness of the turns as well. Type I′ and II′ turns are well-suited to connect an antiparallel β-sheet because their handedness aligns better with the right-handed twist of the β-strands in the sheet (Figure 16.26).

Turns of type VI have a *cis*-Pro at position *i* + 2. There are two distinct conformations, type VIa and VIb. The Pro in type VIa has near α-helix backbone conformation while the Pro in type VIb has β-conformation. The "plane" made by the *cis*-Pro peptide group and the Pro side chain has a more "concave" orientation relative to the overall curve of the turn in type VIa but is more "convex" in type VIb (Figure 16.28). Whereas type VIa is typically hydrogen-bonded, type VIb is usually not. Pro is also prominent in other turns and fits well at position *i* + 1 of types I and II turns, and at position *i* + 2 of type II′.

α-helices and β-strands are repetitive secondary structures with rather well-defined φ/ψ combinations for each residue that cluster in defined regions in the Ramachandran plot (Figure 16.16). By contrast, reverse turns have different φ/ψ combinations for each residue. In fact, the main-chain torsion angles for the central two residues of a turn, *i* + 1 and *i* + 2, are diagnostic for the type of turn (Table 16.6).

**TABLE 16.6**

**φ/ψ combinations for the central two amino acid residues in turn.**

| turn type | (*i*+1) | (*i*+1) | (*i*+2) | (*i*+2) |
|---|---|---|---|---|
| I | −60 | −30 | −90 | 0 |
| I′ | 60 | 30 | 90 | 0 |
| II | −60 | 120 | 80 | 0 |
| II′ | 60 | −120 | −80 | 0 |
| VIa | −60 | 120 | −90 | 0 |
| VIb | −135 | 135 | −75 | 160 |

Most of the φ/ψ angles for residues in turns also fall into the energetic minima of the Ramachandran plot for regular secondary structures (Figure 16.16), but additional combinations are found for the central two amino acid residues, *i + 1* and *i + 2* (Figure 16.29).

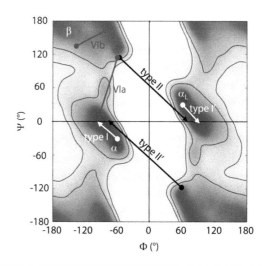

**FIGURE 16.29:   φ/ψ angle combinations in reverse turns.** The Ramachandran plot for Gly shows more and shallower energy minima than those for the other canonical amino acids (Figure 16.16). The arrows start with a circle at the φ/ψ angles for residue *i + 1* and end at those for residue *i + 2*. The Pro-containing type VIa and VIb turns are also shown (magenta arrow heads). Figure created with the programs Phenix and Molprobity.

A short hairpin between adjacent antiparallel β-strands is often a reverse turn where the H-bond formed by the turn is at the same time part of the β-sheet. It is therefore not surprising that not all amino acid residues are equally well-suited for a tight turn. The most frequent amino acids observed in turns are Gly and Pro, followed by the small and hydrophilic Asn, Asp, and Ser. The high preference of Gly in type I′, II, and II′ is due to the dihedral angle restraints for the central positions (*i + 1* and *i + 2*). For instance, in the type II turn, the carbonyl oxygen of the central peptide group between residues *i + 1* and *i + 2* would come very close to a side chain in position *i + 2*, which is avoided when having Gly at this position. Indeed, 61% of type II turns have Gly at this position. A similar clash would exist between the NH-group of the central peptide group and a larger side chain in position *i + 1* in a type II′ turn, which is why Gly is the preferred residue here. The side-chain ring of Pro locks the φ angle at −60°, so its energy minima roughly correspond to those of the α-helix and the β-strand (Figure 16.16). In type VIa turns, the Pro has α-conformation while in type VIb, it has β-conformation.

Type I, I′, II, II′, VIa, and VIb turns represent the vast majority of turns in protein structures. Other classes of turns (termed type III, IV, VII, and γ) are slight structural variations of these themes that are rarely found in protein structures.

### 16.2.4 PROTEIN DOMAINS & TERTIARY STRUCTURE

The next level of protein structure is the connection of secondary structure elements into super-secondary structures, of which we already know β-sheets and coiled-coils. Most super-secondary structures (the coiled-coil is an exception) are not stable *per se* but assemble into larger protein *domains*. A strict definition of a protein domain is an *autonomous folding unit*, i.e. a sequence of amino acids with the intrinsic ability to fold into a compact structure, such as WW, SH2, PH (Figure 16.30), and SH3 domains (Figure 16.27). Domains are the principal building blocks of proteins: they occupy a space separate from other domains, are usually associated with a specific function, and share sequence similarity or structural similarity to other domains. In biology, a more lenient definition of a domain is any more or less globular part of a protein that has a defined function, for instance a catalytic activity or a binding property. Some domains are assembled from repeating super-secondary structures. Although generated from simple super-secondary structure motifs, *repeat proteins* have a broad spectrum of shapes, ranging from planar, crescent-shaped, or circular forms to helical solenoids (Figure 16.30).

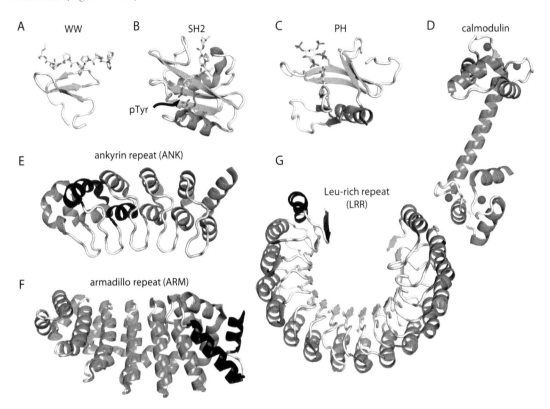

**FIGURE 16.30:   Super-secondary structures in small protein domains and repeat proteins.** A: Similar to the SH3 domain (Figure 16.27), the WW domain recognizes Pro-rich peptides, mediating protein-protein interactions. It is a tiny protein domain of only 40 residues with a three-stranded β-sheet that contains two signature Trp residues (PDB-ID 1upr). Some WW domains bind peptides containing phosphoserine. B: SH2 domains are phosphotyrosine (pTyr) binding domains. Phosphorylation of a substrate protein by a tyrosine kinase triggers SH2 domain binding, often in the context of cellular signaling (PDB-ID 1h9o). C: PH domains are binding modules for phosphoinositide lipids (Table 16.8). Proteins with PH domains often function as adapters that bind and tether other proteins to membranes. The example shown (PDB-ID 1upr) is in complex with inositol 1,3,4,5-tetrakis phosphate, the head group of an important class of phosphoinositide lipids. D: The $Ca^{2+}$-binding protein calmodulin has several EF-hand motifs. The EF-hand is the most common $Ca^{2+}$-binding motif and consists of two short α-helices connected by a loop that binds the ion (PDB-ID 1exr). E–G: The repeating super-secondary structures in three repeat proteins are marked in black. E: The Ankyrin repeat (ca. 33 residues) has two α-helices separated by a long loop (PDB-ID 2rfm). F: The slightly longer (ca. 40 residues) Armadillo repeat consist of two α-helices separated by a turn (PDB-ID 4plr). G: Conserved Leu residues between an α-helix and a β-strand coined the name for the Leu-rich repeat (PDB-ID 1z7x).

Based on their secondary structure content, protein domains can be classified into all-α, α/β, and all-β (Figure 16.30). All-α domains only contain helices connected by loops. α/β-domains contain a mixture of β-barrels and/or twisted β-sheets along with α-helices that pack against the β-structures. All-β domains exclusively contain β-barrels and β-*sandwiches*, all of which are usually formed by antiparallel β-sheets. The β-sandwich is a variant of the β-barrel where two β-sheets lie on top of each other in an almost perpendicular orientation so the outer strands do not interact as intimately as they do in barrels (Figure 16.31). *Ig*- or *immunoglobulin* domains (Figure 16.31) are examples of very stable β-sandwiches, made from two Greek key motifs that pack against each other and that are covalently linked by a disulfide bond (Section 16.1.4.5).

The topology of secondary structure elements in a domain determines the *fold* of the domain. The number of different folds that proteins can adopt is not infinite. There are about 1300 fundamentally different ways of how a polypeptide chain can fold. About ten *superfolds* have been enriched over all others throughout evolution, and they are shared by about 30% of all proteins. Some superfolds, such as the *Rossmann fold* and the *globin fold* are tied to specific biochemical functions, while the functions of others, such as the Ig-fold, TIM-barrels, and β-propellers are quite variable. The Rossmann fold (Figure 16.31) is a mononucleotide-binding domain that is usually embedded in a larger protein. Its structure is a sequence of βαβ-motifs, which generates a central parallel β-sheet with amphipathic α-helices as crossovers. Enzymes with two Rossmann folds bind dinucleotides, such as the cofactors FAD and NAD⁺ (Figure 19.13). The globin fold (Figure 16.31) encompasses eight α-helices that form a pocket for heme to bind oxygen in the transport and storage proteins myoglobin and hemoglobin, and in cytochromes, which are important de-toxifying enzymes in the liver. The Ig-fold is an ubiquitous structural unit in cell surface receptors and used for antigen recognition by antibodies. The β-propeller is a superfold that varies widely. β-propellers are circular arrangements of antiparallel β-sheets and are found in enzymes, structural proteins, and proteins that bind to other proteins. The number of blades in the propeller is between 4 and 9, although 5- and 6-bladed propellers are most common. The triosephosphate isomerase (TIM) barrel is an 8-stranded parallel $(\beta\alpha)_8$ fold that is found in about 10% of all enzymes, including such diverse functionalities as hydrolases, lyases, isomerases, and oxido-reductases. The β-barrel is entirely hydrophobic on its outside and completely shielded from solvent by amphipathic helices.

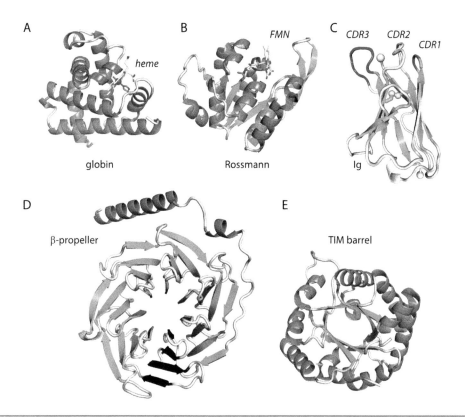

**FIGURE 16.31:   Recurring protein folds.** A: The globin fold is entirely made of α-helices that form a hydrophobic pocket for a heme group (PDB-ID 1a6k). B: The Rossmann fold is a mononucleotide-binding module (PDB-ID 1g5q). C: The immunoglobulin (Ig) domain (taken from PDB-ID 1igt) is a small and very stable β-sandwich. D: β-propellers are circular assemblies of small antiparallel β-sheets. The example shown here is the β-subunit of a heterotrimeric G-protein, which serves as a binding platform for the α- and γ-subunits (PDB-ID 1got). E: In TIM-barrel proteins, the central parallel β-barrel is stabilized by the crossover α-helices on the outside (PDB-ID 1tim).

Larger (>50 kDa) proteins tend to contain several smaller domains, organized like pearls on a string with linkers connecting the domains. Depending on the length and sequence of the linkers, the domains may or may not interact with each other. If the linkers are rather short, the domains may functionally cross-talk. If the linkers are long and/or flexible, often containing sequences rich in Gly, Pro, Ser, or Gln, the individual domains connected by the linkers can adopt very different juxtapositions relative to one another. Such domains may not necessarily functionally interact, but the domain movements enable large conformational changes of the protein, which is important for function. Titin is an extreme example of a multi-domain protein where the domains apparently do not functionally cross-talk. This vertebrate muscle-organizing protein has the largest single polypeptide chain in humans with 27000–38000 residues depending on the titin isoform. Among others, titin harbors a kinase domain, 152 Ig domains, and 132 fibronectin-like domains (β-sandwiches similar to Ig domains). While it is often observed that individual domains can be isolated from multi-domain proteins and are stable in solution, for many other proteins the linker is important for domain stabilization.

## 16.2.5 QUATERNARY STRUCTURE & PROTEIN-PROTEIN INTERACTIONS

Several polypeptide chains that assemble into higher-order oligomers form a quaternary structure. A weak correlation exists between the molecular mass of a protein and its tendency to oligomerize: monomeric proteins generally have rather small molecular masses of around 30 kDa, whereas larger proteins >50 kDa molecular mass tend to form oligomers. Assembly into a quaternary structure is a general strategy to increase protein functionality and stability, or to enable protein regulation. Oligomers frequently undergo cycles of large conformational changes that may not be possible for monomers. Oligomerization can be made irreversible by covalent crosslinks such as disulfide bonds, which results in very stable assemblies. If the assembly is reversible, the oligomer is normally in dynamic equilibrium with its constituent monomers. A survey of crystal structures of oligomeric proteins (Table 16.7) has shown that homo-oligomers are more frequent than hetero-oligomers, small stoichiometries are preferred over large stoichiometries, and even-numbered components are preferred over odd-numbered components.

| TABLE 16.7 Stoichiometries of the most frequent oligomers. | | | | | |
|---|---|---|---|---|---|
| homo $\alpha_n$ | 2 (32) | 3 (3.8) | 4 (8.2) | 5 (0.40) | 6 (2.6) | 7 (0.12) |
| | 8 (1.0) | 10 (0.34) | 12 (0.56) | 24 (0.17) | 48 (0.13) | 60 (0.16) |
| hetero $\alpha_n \beta_m$ | 1,1 (4.6) | 2,1 (0.48) | 2,2 (1.7) | 3,3 (0.29) | 4,4 (0.22) | 4,2 (0.16) |

*Note:* Data are from the protein databank (PDB) as of January 2016 (35460 crystal structures with <90% sequence identity). Only hetero-oligomers >0.1% frequency are listed. Numbers in parentheses are percentages with the values for *n* and *m* (in hetero-oligomers) in front. For comparison, there are 46% monomers. Heterotrimers of the form αβγ constitute 0.89%, and αβγδ heterotetramers contribute 0.28% to all structures.

### 16.2.5.1 Homo-Oligomers

Homo-oligomers are assembled from several copies of the same monomer. They often display internal symmetry (Figure 16.32) and incorporate two-fold, three-fold, etc. rotation axes that relate one monomer to the other(s). Such a symmetric arrangement maximizes the number of interactions between the monomers and is therefore energetically favored compared to the monomers. The energy released by the attractive interactions outweighs the entropic loss during oligomerization. Entropy is not only lost by fixing the subunit but also by large conformational changes in loop regions or N- and C-termini, which are often flexible in the monomer and become fixed in the oligomer (see the tetramers in Figure 16.32). Thus, oligomer formation is a good example of *enthalpy-entropy compensation*, and we will see more examples throughout this section.

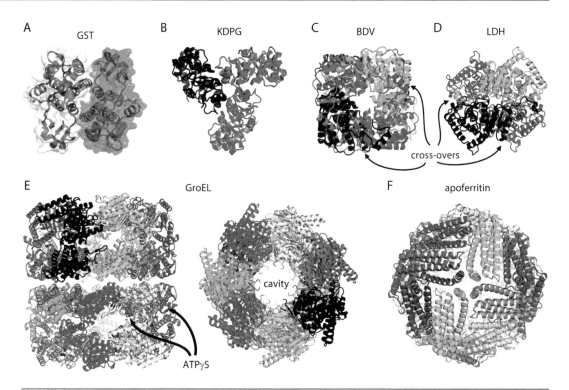

FIGURE 16.32: **Quaternary structure of selected homo-oligomers.** For all examples but the side-view in (E) the rotation axis points towards the reader. A: *Schistosoma mansoni* glutathione S-transferase (GST; PDB-ID 1u31) is a symmetric dimer. The subunits are colored black and red with transparent surfaces to highlight the interface. B: *E. coli* 2-keto-3-deoxy-6-phosphogluconate aldolase (KDPG; PDB-ID 1fq0) is a trimer. C: The Borna disease virus nucleoprotein (BDV; PDB-ID 1n93) is a planar tetramer. D: By contrast, the human muscle L-lactate dehydrogenase (LDH; PDB-ID 4ojn) forms a tetramer in the shape of a tetrahedron. There are large crossovers from one monomer to the other, which helps to stabilize the oligomer. E: The *E. coli* chaperone GroEL is a large 14-mer composed of two seven-membered rings stacked on top of each other. This structure has the non-hydrolyzable ATP analog ATPγS bound (PDB-ID 1sx3). Two orientations rotated 90° from each other are shown. F: Apoferritin is a hollow sphere made of 24 α-helical subunits (PDB-ID 2w0o).

The simplest and most frequent oligomer is the homodimer, found in 32% of all reported crystal structures. Among them is glutathione S-transferase (GST), which is often used as an N-terminal fusion (tag) to purify a protein of interest by affinity chromatography on glutathione sepharose. An example of a trimer is the 2-keto-3-deoxy-6-phosphogluconate aldolase, which forms a planar triangle. While there is only one way to arrange monomers into symmetric dimers and trimers, four monomers can be arranged symmetrically in different ways. A tetramer can either be planar, like the Borna disease virus nucleoprotein, or adopt the form of a tetrahedron with the monomers sitting at the four corners, such as lactate dehydrogenase (Figure 16.32). Hexamers can be planar six-membered rings or dimers of trimers or trimers of dimers. Octamers can be planar rings, dimers of tetramers, or tetramers of dimers, and so on for higher-order oligomers. For uneven numbers of monomers, the most prominent arrangement is a planar ring or a helix (a ring with a pitch). GroEL, for example, is a chaperone composed of two regular seven-membered rings stacked on top of each other that enclose a cavity in which proteins can fold (Figure 16.32 and 16.33). When tubulin monomers assemble into filaments, they form 13-mer rings where each monomer is translated along the filament axis compared to its predecessor, giving rise to an infinite helix. Oligomers composed of many monomers can also from hollow shapes that function as containers. Apoferritin (Figure 16.32) is an example of a container made from 24 identical subunits to store iron in an internal cavity. The concept of storage by formation of hollow spheres is also used by many viruses (see Chapter 17).

### 16.2.5.2 Hetero-Oligomers

*Hetero-oligomers* are formed by association of different subunits. The principles relating to the structure and stability of homo-oligomers also apply to hetero-oligomers. Many hetero-oligomers contain several copies of the same subunit along with other subunits. Other hetero-oligomers have subunits of similar structures that can assemble into *pseudo-symmetric* shapes. This is exemplified by hemoglobin (Figure 16.33), a pseudo-symmetric $\alpha_2\beta_2$ heterotetramer formed by $\alpha$- and $\beta$-subunits with similar, but not identical structures. The $\alpha_6\beta_6$ heterohexamer of aspartate transcarbamoylase is composed of two $\alpha_3$-trimers and three $\beta_2$-dimers. Pseudo-symmetry can repeat many times to form the hollow spheres of capsids that store the viral genome. In foot-and-mouth disease virus (Figure 16.33), three similar subunits form a trimer, which further assembles into the capsid. More examples of hetero-oligomers with at least some pseudo-symmetry include lactose synthase ($\alpha\beta$), tryptophan synthase ($\alpha_2\beta_2$), cholera toxin ($\alpha\beta_5$), ribulose bisphosphate carboxylase ($\alpha_8\beta_8$), troponin ($\alpha\beta\gamma$), phosphorylase kinase ($\alpha_4\beta_4\gamma_5\delta_4$), RNA polymerase ($\alpha_2\beta\beta'$), $F_1$ ATPase ($\alpha_3\beta_3\gamma\delta\varepsilon$), and antibodies.

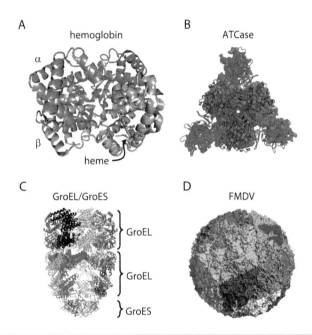

**FIGURE 16.33:    Quaternary structure of selected hetero-oligomers.** A: Fetal deoxy-hemoglobin is an $\alpha_2\gamma_2$ heterotetramer (blue and pink; PDB-ID 2hhb), B: Aspartate transcarbamoylase (ATCase; PDB-ID 1q95) is an $\alpha_6\beta_6$ heterohexamer. The catalytic $\alpha_3$ trimers are colored red and the regulatory $\beta_2$ dimers are shown in blue. C: The small GroES homoheptamer attaches to the GroEL ring and induces large conformational changes in GroEL. D: Foot-and-mouth disease virus (FMDV; PDB-ID 1bbt) is a hollow icosahedral sphere that contains the single-stranded RNA genome of the virus. The icosahedron is assembled from three similar subunits (VP1, VP2, and VP3) that form trimers (colored), which further assemble into the viral capsid.

In other cases, hetero-oligomers are composed of such different monomers that they entirely lack internal symmetry. Many complexes between RNA and proteins belong to this class, for instance ribosomes (Figure 17.37) and spliceosomes, which contain dozens of different and structurally unrelated protein subunits that form intricate networks of non-covalent bonds across molecular interfaces. The following section will discuss the nature of these interactions that form the interfaces of macromolecular complexes.

### 16.2.6 PROTEIN-PROTEIN INTERACTIONS

Macromolecular complexes are held together by non-covalent interactions (Section 15.3): H-bonds, electrostatic, and van der Waals interactions. The main characteristics of the interacting molecular surfaces are shape complementarity, biophysical complementarity, and only few embedded water molecules (Figure 16.34).

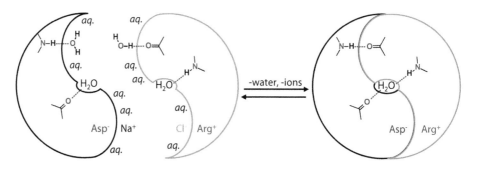

**FIGURE 16.34:    Biophysical and shape complementarity of macromolecular interfaces.** Before subunit association, charges are screened by counter ions (Na$^+$ and Cl$^-$) and the surfaces are covered with water molecules, symbolized by "*aq.*". Upon complex formation, most solvent molecules are displaced from the interface. High-affinity complexes bury large interfaces and have high shape complementarity. Occasionally, some water molecules are trapped in the interface and become an integral part of the complex.

High shape complementarity maximizes the number of interactions in a complex by avoiding cavities within the interface. Biophysical complementarity in addition means a maximum number of attractive interactions between the binding partners across the interface and the absence of unsatisfied H-bonding valences. Upon complex formation, most water molecules and ions that cover the surface of the isolated binding partners in solution are excluded or "squeezed out" from the interface. High-affinity complexes usually have no or just a few water molecules embedded in their interface, but many hydrophobic interactions. If water molecules are locked in the interface, they tend to form a maximum number of H-bonds. Polar or charged groups from the binding partner must form H-bonds or salt bridges across the interface to avoid energetically unfavorable burial of polar atoms in the non-polar environment of the complex. Due to their distinct directionality, H-bonds are important for the specificity of the interaction.

#### 16.2.6.1 Surface Complementarity and Buried Surface Area

High-affinity macromolecular complexes usually have an extensive interface together with high shape complementarity. The better the shapes of the interacting surfaces fit together, the more attractive interactions can be formed. Shape complementarity of a complex can be calculated and expressed as a number between zero and unity, where a value of zero means no fit at all and a value of unity signifies perfect complementary (Figure 16.34). Typical values for protein-protein complexes are in the range of 0.5–0.9.

On average, proteins bury 10–20% of their surface in a complex interface. The *buried surface area* (BSA) is a rough indicator for protein complex stability in solution. If the structure of the complex is known, the BSA can be calculated from the difference in *accessible surface areas* (ASA) of the isolated binding partners and the complex. The ASAs are determined by rolling a sphere with the radius of a water molecule (170 pm) over the surface, and tallying the contact area. The BSA is then the sum of the ASAs of the constituent proteins minus the ASA of the complex. Common sizes of interaction surfaces are in the range of 10–55 nm$^2$, so if the BSA is on the order of 10 nm$^2$ or larger, the complex is probably also stable in solution. However, there is considerable variety and much smaller surface areas have been observed for stable complexes.

A statistical analysis of surfaces in protein-protein complexes has shown that complex interaction surfaces of less than 20 nm$^2$ are usually one continuous patch on the surface of each binding partner. In contrast, larger interfaces tend to be divided into several patches with at least one patch of a larger size, similar to that present in the single-patch interfaces of smaller complexes. The contributions of the different residues to complex stability in a patch also varies. There are interaction *hot spots*: within a single surface patch, a set of buried atoms is responsible for the majority of the interactions. This hot spot is surrounded by atoms that remain solvent-accessible.

### 16.2.6.2 Energetics of Macromolecular Interactions

The energies liberated by protein-protein complex formation can be considerable. For example, while the insulin dimer has a $K_d$ value of only $10^{-5}$ M, the strong trypsin/trypsin-inhibitor interaction has a $K_d$ value of $10^{-13}$ M, corresponding to interaction energies between −28 and −74 kJ mol⁻¹, respectively, at 25°C. About ⅔ of the interaction surfaces is non-polar, and hydrophobic interactions have the largest contribution to association. Occasionally, buried salt bridges can also strongly contribute to binding because of the lower dielectric constant in the interface compared to bulk solvent, which increases the strength of the ionic interaction (Section 15.3.1). Mutagenesis at subunit interfaces and comparison of the complex stability to the unperturbed value is useful to establish the energetic contribution of individual side chains to complex stability.

Removal of a single interaction from a protein-protein interface can cause dissociation or stabilization of an oligomer. For example, a natural mutation in the β-chain of hemoglobin, Tyr35Phe, removes an H-bond to Asp126 in the α-chain. People carrying this hemoglobin "Philly" suffer from mild anemia because loss of the H-bond leads to increased dissociation of the $\alpha_2\beta_2$ tetramer into monomers.

Formation of additional salt bridges across the subunit interface is a general strategy of protein stabilization in thermophiles compared to their mesophilic counterparts. H-bonds, on the other hand, have little energetic contribution to complex stability because the H-bonds with water in the monomeric subunits are replaced by almost isoenergetic H-bonds in the complex.

If we summed the theoretical pairwise enthalpic contributions between atoms in an interface, the resulting energies would be much more negative than those observed experimentally. The enthalpic contributions are counteracted by entropic losses upon complex formation: confinement of the interaction partners in a single particle, conformational changes at the interface, and trapping of water molecules reduces the entropy. As a result, protein-protein interactions release only about 30–70% of the energy expected from summing the individual enthalpic interactions. Especially the water molecules bound to the binding partners prior to complex formation are an important factor for the entropy of complex formation (enthalpy-entropy compensation).

### 16.2.6.3 Role of Water – The Hydrophobic Effect

Water as the solvent will cover all exposed surfaces of a protein, with two important consequences. First, water will saturate all H-bond donors and acceptors present on protein surfaces. Second, water will form ordered structures known as *clathrates* around hydrophobic surfaces such as the aliphatic or aromatic amino acid side chains of a protein, the bases of nucleic acids, or the hydrophobic groups of a ligand. While water in ice forms four H-bonds and in solution forms an average of 3.4 H-bonds, on hydrophobic surfaces water can form only about three H-bonds because part of the water molecule engages in van der Waals interactions (Figure 16.35). The hydrophobic surfaces also restrict the orientation and mobility of water, so its entropy is much reduced compared to bulk solvent. The resulting clathrate structures are the best compromise that water can achieve when bound to hydrophobic surfaces.

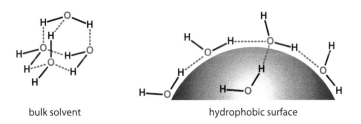

bulk solvent                    hydrophobic surface

**FIGURE 16.35:** **Bulk water and water at hydrophobic surfaces.** Possible H-bonds are shown as dashed blue lines.

Release of water from hydrophobic surfaces increases both the entropy of water and releases energy due to increased H-bonding in bulk water. The positive entropy change contributed by a single water molecule can amount to 40–70 J mol$^{-1}$ K$^{-1}$. This is one manifestation of the *hydrophobic effect* (Box 16.3), where increased interaction of water molecules drives hydrophobic groups out of the aqueous phase. The hydrophobic effect is important whenever hydrophobic groups are solvated by water. Complex formation involving hydrophobic binding interfaces is driven by the hydrophobic effect.

---

### BOX 16.3: THE HYDROPHOBIC EFFECT.

The hydrophobic effect describes the tendency of molecules to exclude water from non-polar surfaces. Bulk water is organized into networks of H-bonds, which rapidly and constantly rearrange. When small, non-polar molecules such as noble gases, acetylene, or methyl sulfide are introduced into water, they disrupt some of these networks and dissolve all to about the same small extent by forming favorable van der Waals interactions with water molecules. The water molecules bound to the non-polar surface engage in fewer H-bonds among each other, but these H-bonds are slightly stronger compared to bulk solvent because solute-bound water molecules are less mobile. At low temperatures, the sum of van der Waals interactions and H-bonds cause the overall $\Delta H$ to be slightly negative. However, at the same time several layers of water molecules become highly organized as cage-like clathrate structures around the non-polar surfaces. This ordering corresponds to a large and negative change in entropy $\Delta S$, and a concomitant increase in heat capacity. Dissolving hydrophobic molecules in water is therefore entropically strongly disfavored, and the overall free energy change $\Delta G$ is positive: $-T{\cdot}\Delta S \gg \Delta H$, so therefore $\Delta G > 0$. At higher concentrations of the non-polar solute, its hydrophobic groups tend to interact with each other rather than with water in order to minimize the decrease in entropy: the solute forms non-polar "islands" or droplets. In biology, the exclusion of water from hydrophobic surfaces is the main driving force for protein folding, protein-protein and protein-ligand interactions, base stacking, formation of vesicles and membranes by lipids, and insertion of membrane proteins into membranes.

Starting from low temperatures, the hydrophobic effect first increases with temperature, which leads to a reduced solubility of hydrophobic solutes in water. With increasing temperatures, the water molecules fixed in the clathrate structures become increasingly mobile and interact less with each other. As a consequence, both the entropic penalty for dissolving a non-polar molecule in water and the favorable change in enthalpy decrease with temperature. In other words, the entropic term becomes less unfavorable, and the enthalpic term becomes less favorable. Because of the different temperature dependence of $\Delta H$ and $\Delta S$ (eq. 2.69, eq. 2.87), $\Delta G$ and thus the hydrophobic effect reach a maximum at the temperature where $\Delta S$ becomes zero (see Figure 2.25). At higher temperatures, $\Delta S$ becomes positive, but is now overcompensated by the large and positive $\Delta H$, leading to a decrease of $\Delta G$ for transferring a non-polar molecule to water, and a decrease in the hydrophobic effect. This decrease of the hydrophobic effect occurs at very high, non-physiological temperatures (>100°C).

The hydrophobic effect can be quantified by measuring the free energy of transfer of a hydrophobic molecule from a reference state (e.g. octanol, approximating the hydrophobic interior of a protein) into a hydrophilic environment (e.g. water). To a first approximation such transfer energies are additive and molecules can be conceptually broken down into individual contributions from their constituent parts (phenyl groups, amide bonds, etc.). For instance, the transfer of methylene groups from the hydrophobic interior of a protein into bulk solvent has been estimated to 3.3 kJ mol$^{-1}$ per methylene. Conversely, complete isolation of water from bulk solvent into a non-polar environment such as a hydrophobic cavity in a protein costs about 26 kJ mol$^{-1}$ due to breaking of H-bonds and the large positive entropy change.

Protein-protein interfaces much resemble the interior of proteins, where the residues are also densely packed with high surface complementarity and few, if any, cavities. Hydrophobic interactions dominate the free energy, while H-bonds and other polar interactions do not contribute much to stability. Protein interiors, however, have less ionic interactions and less buried water molecules per surface area than complex interfaces. The principles of macromolecular interactions also apply to protein-ligand interactions that will be treated next.

### 16.2.7 PROTEIN-LIGAND INTERACTIONS

Small molecule or ligand binding to proteins and other macromolecules (Section 17.5.4.2) follows the same principles as protein-protein interactions, including shape and biophysical complementarity to maximize the number of hydrophobic contacts and H-bonds. The packing between the ligand and the protein is about as tight as the packing in the hydrophobic core of a protein. Similar to protein-protein interactions, the main energetic driving force for protein-ligand interactions is the release of water from those hydrophobic surfaces that become buried in the complex. H-bond donors and acceptors of ligands are paired with those of the protein, which imposes selectivity for the ligand. H-bonds between proteins and ligands can also be mediated by water molecules, in which case water becomes an integral part of the ligand binding site. If charges are present on the ligand, such as in nucleotides, they are usually oriented towards the solvent. Charges that are buried in the interface are neutralized by counter-charges from metal ions or side chains, or by the partial charge of helix dipoles and peptide groups. In some cases, charges are efficiently dispersed, or *solvated*, by a network of H-bonds with oriented peptide dipoles that often connects to a nearby opposite charge. Another way to dissipate charges is to change the $pK_A$ values of the ligand by the protein environment such that the ligand is bound in its neutral form.

All of these ligand-binding principles are summarized in Figure 16.36 using as an example the small GTP-binding protein Ras (from *rat sarcoma*), a molecular switch that is inactive when bound to GDP, but activates a signal transduction pathway when bound to GTP. The selectivity by H-bonds can be appreciated from the Asp side chain that specifically binds to the guanine base of the nucleotide. Any other nucleotide is rejected due to H-bonding incompatibility. Both faces of the guanine base are in van der Waals contact with a hydrophobic Phe side chain and the methylene groups

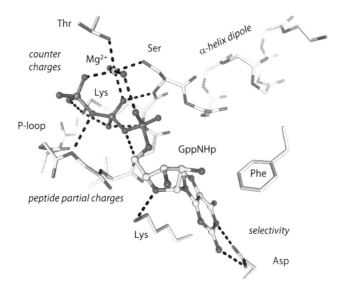

**FIGURE 16.36:** **GTP-binding in Ras.** A non-hydrolyzable GTP-analog, GppNHp, is bound to the human Ras p21 protein (PDB-ID 5p21). The guanine base is sandwiched in van der Waals contacts with Phe and Lys side chains. An Asp side chain establishes selectivity for guanine by two H-bonds. The charges on the α- and β-phosphates are neutralized by a $Mg^{2+}$ ion (green sphere) and a nearby Lys side chain. H-bonds are shown as dashed black lines. The P-loop donates several H-bonds to the phosphates, further dissipating their negative charges. Only the backbone of the α-helix that orients its dipole towards the phosphates is shown.

of a Lys. The charges on the α- and β-phosphates are neutralized by the positive charges of a Lys side chain and a $Mg^{2+}$ ion. Further charge dispersion is achieved by a network of H-bonds from NH-groups of the phosphate-binding *P-loop* and the positive end of a helix dipole immediately following the P-loop.

Binding sites for ligands range from shallow depressions on the surface of the protein to internal pockets buried deep within the protein. Generally, larger ligands, such as oligosaccharides, and those that need to dissociate from the protein (e.g. large substrates in an enzymatic reaction) tend to bind in surface depressions or grooves. These shallow binding pockets facilitate association and dissociation without expenditure of energy for large conformational changes by the protein. In contrast, permanently bound prosthetic groups such as heme, chlorophyll, pyridoxal phosphate (Figure 19.13 and Figure 19.39), but also structural metal ions and FeS clusters, are bound in deep pockets. Removal of such ligands often destabilizes the entire protein, and the ligand has to be considered as an integral part of the protein structure.

The *lock-and-key* model of ligand binding implies a pre-formed binding site on the protein with little flexibility, which consequently has high selectivity for a certain ligand. This model neglects conformational changes of the protein, which are almost invariably required for ligand binding. Two other models, *induced fit* and *conformational selection*, describe the conformational changes of a protein upon ligand binding. In the induced fit model, a ligand weakly binds to a sub-optimal protein conformation and induces a conformational change in the protein to best accommodate the ligand. The protein in the induced fit model is viewed as rather static, switching only between a few conformations. In contrast, the conformational selection model assumes that proteins are rather malleable and change conformations rapidly between isoenergetic states. One of these states is highly complementary to the ligand and hence is selected for binding. However, because the protein samples many different conformations, the one that best fits the ligand is not highly populated. Figure 16.37 compares these extremes and shows that they fulfill the condition of a thermodynamic cycle (Section 2.2.5). The actual binding event for any particular protein is often a mixture of these extremes, although some ligands may prefer one pathway over the other.

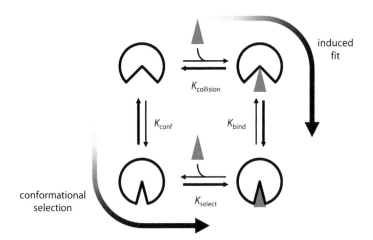

**FIGURE 16.37: Induced fit versus conformational selection.** In the induced fit model, a ligand (blue triangle) forms a weak complex with a protein conformation that does not match the ligand's shape (binding constant $K_{collision} \ll 1\ M^{-1}$). Closure of the protein around the ligand leads to the high-affinity complex, but energy is needed to induce the binding-competent conformation of the protein. In the alternative conformational selection model, intrinsic protein flexibility samples many conformations, and the ligand selects the best-fitting shape.

Most proteins undergo only small conformational changes on the level of side chain and loop movements when binding to a ligand, indicating that the binding site is pre-formed. A certain amount of flexibility, i.e. conformational entropy, is required for fast association of the protein with the ligand. Sizeable surface loop movements exceeding 1 nm are necessary for binding of large co-substrates such as SAM or NADH to enzymes. Larger conformational changes include rigid body domain rearrangements and protein folding around a ligand. An extreme example is folding of unstructured regions into a defined conformation upon DNA binding (Section 17.5.4.1). Rigid body domain movements are triggered by ligands that bind at protein interfaces in oligomers. Ligand binding to *allosteric* sites, i.e. sites remote from the active site of an enzyme, can also induce rigid body domain movements, which in turn change the catalytic efficiency of enzymes (e.g. by non-competitive inhibition; Section 11.6.3).

In addition to protein flexibility, the entropy of the ligand also plays an important role for the overall binding energy. The more flexible the ligand, the higher the entropic cost for binding in a defined conformation. One aim in structure-based drug design is to pre-organize the conformation of a ligand by suitable substituents or intramolecular H-bonds such that the entropy loss upon binding is minimized and the overall affinity is thus increased.

The affinity of permanently bound ligands for their protein is usually very high with $K_d$ values in the nanomolar or picomolar range, while metabolites (substrates and products) usually have very low affinity for their enzymes, indicated by $K_M$ values in the high micromolar or sometimes millimolar range. Enzymes are optimized for tight binding of the transition state (Section 14.1) and high substrate turnover, which is realized by low to moderate affinities for both the substrates and the products. The low affinity for substrates is often due to substrate or enzyme distortion required for catalysis. Part of the binding energy is converted into potential energy in the ES complex, which can then be released during catalysis or product release. An important aspect to mention here is that ligand binding in the cell does not occur under ideal conditions. The millimolar concentrations of proteins in the cell increase the viscosity of the cytoplasm, limiting diffusion and rotation of macromolecules compared to an ideal solution. The ensuing proximity effects, i.e. hindered dissociation of a complex due to molecular crowding, can considerably increase the affinity of a ligand for a protein. The effect is particularly relevant for the two-dimensional diffusion of molecules in membranes.

## 16.2.8 MEMBRANE PROTEINS AND THEIR LIPID ENVIRONMENT

Most of the discussion in the preceding sections involved proteins dissolved in water. Membranes are the solvent for membrane proteins, which have different properties and structures compared to soluble proteins. We will start by taking a look at the membrane constituents, the lipids, the driving forces for membrane formation, and then will focus on the structure of membrane proteins.

### 16.2.8.1 Biological Roles of Lipids and Membranes

The main biological functions of lipids are energy storage, signal transduction, and the formation of membranes. *Triglycerides* or *glycerolipids* (Table 16.8) are highly reduced molecules with a high energy density. Lipids and related molecules also act as hormones and second messengers in signal transduction processes. Sterols such as estrogen, testosterone, and cortisol bind to nuclear hormone receptors and regulate metabolic processes and the blood pressure. Diacylglycerol (DAG) and phosphatidylinositol are second messengers generated by activation of G-protein coupled receptors (GPCRs) and are involved in the activation of protein kinase C and $Ca^{2+}$-mediated signaling, respectively. Lastly, most lipids tend to associate into layers and bilayers, termed *membranes*. Membranes separate cells from the exterior and delimit compartments within the cell. Eukaryotic cells are subdivided into membrane-enclosed organelles including the nucleus, mitochondria, endoplasmic reticulum, the Golgi system, endosomes, lysosomes, etc., which allow separation and thus regulation of biochemical processes.

Membranes establish an almost closed system (Chapter 1) that allows exchange of heat but substantially limit the exchange of matter with its surroundings: while small hydrophobic molecules may passively diffuse through membranes, ions cannot pass. The impermeability of membranes for

**TABLE 16.8**
**Types of lipids according to the LIPID MAPS consortium.**

| type | structure, example, remarks | |
|---|---|---|
| fatty acids |  oleate | |
| glycerolipids |  glycerol, palmitoyl, stearyl, arachidonyl | |
| glycero-phospholipids |  phosphatidyl-choline (lecithin), inositol | $R_1$, $R_2$ can be any fatty acid. Instead of choline the phosphoester can also be made with ethanolamine, inositol, and serine, all of which are found in membranes. In phosphatidylinositol phosphates (PIPs) the inositol can be phosphorylated to various extents |
| sphingolipids |  sphingosine | $R_1$ is a fatty acid. If $R_2$ is H, the lipid is called ceramide. If $R_2$ = phosphatidylcholine it is a sphingomyelin |
| steroids | | R = H: cholesterol. R can also be a fatty acid or a sugar |
| prenyl lipids |  farnesol | Two isoprenyl moieties: geraniol also: retinol (vitamin A), β-carotene (Figure 19.13) |
| saccharolipids |  lipid A, sugars | In Saccharolipids, the glycerol in glycerolipids or phospholipids is replaced by a sugar which forms esters and amides with up to seven fatty acids. Lipid A is part of the lipopolysaccharides in the membrane of Gram-negative bacteria. |
| polyketides |  aflatoxin B1   erythromycin | These are secondary metabolites synthesized by polymerization of acetyl and propionyl subunits. They exhibit extraordinary structural diversity: many are cyclic molecules that can be further glycosylated and methylated. Many are strongly toxic, or have antibiotic, anti-cancer, or anti-parasitic activities. They do not form membranes. |

ions enables the storage of energy in form of proton and other ion gradients (Section 4.2). Exchange of matter between membrane-separated compartments is regulated by special membrane proteins (Section 16.2.8.4) known as channels and transporters (Section 4.3). Signal-transducers transport information across membranes.

### 16.2.8.2  Types of Lipids

Very generally, lipids are amphiphilic molecules with a hydrophobic moiety at one end that is connected to a hydrophilic *head group* at the other end. The hydrophobic part and the head group can vary widely, and more than 40000 different lipids have been reported in the Lipid Maps Structure Database (lipidmaps.org). Variations of the hydrophobic part include chemical properties, chain length, as well as number, position, and stereochemistry of double bonds. The head groups, which are often, but not always, charged, can be as small as a hydroxyl group or as large as a phosphorylated oligosaccharide. Based on the chemical classes of hydrophobic and head groups, eight principal types of lipids have been defined from which all other lipids observed in nature are derived (Table 16.8).

Glycerophospholipids, or just phospholipids, exhibit the largest diversity of any lipid family. Phospholipids are the major component of lipids in mammalian membranes (70–90%), followed by sterols, mainly cholesterol (10–20%), and sphingomyelin (5–10%).

A variation on the concept of lipids is *detergents*, also known as *surfactants* or *surface-active agents*. Detergents are synthetic amphiphilic molecules with similar biophysical properties to lipids. In contrast to many lipids, detergents have at least some solubility as monomers in water and are used to gently extract and stabilize membrane proteins (Section 16.2.8.4). Both lipids and detergents behave similarly when mixed with polar and non-polar solvents by formation of monolayers, micelles, membranes, and vesicles. We will use *amphiphiles* as a collective term for lipids and detergents when discussing their biophysical properties.

### 16.2.8.3  Super-Structures Formed by Lipids and Detergents

When placed in water, amphiphiles partition as monolayers at the water/air interface, orienting their hydrophilic head groups towards the water and the hydrophobic parts towards the non-polar air (Figure 16.38). The common property of amphiphiles is to reduce the surface tension at a polar/hydrophobic interface, e.g. water/air or water/oil. If the concentration of amphiphiles is increased above the *critical micelle concentration* (CMC), they form micelles, bilayers, and vesicles (Figure 16.38). The main driving forces for this assembly are the hydrophobic effect (Section 16.2.6.3) and, in charged amphiphiles, electrostatic repulsion of the head groups. Expulsion of ordered water molecules from the hydrophobic surfaces of amphiphiles upon assembly into micelles or membranes increases the entropy of water. At the same time, new van der Waals interactions can be formed between the hydrophobic parts, which decreases the overall enthalpy. Exactly which type of assembly is formed not only depends on the chemical nature and concentration of the amphiphile, but also on the osmolarity of the solvent, the ionic strength, and the temperature. The *detergent packing parameter P* of an amphiphile is helpful in predicting the preferred shape of the lipid assembly:

$$P = \frac{V}{A \cdot L}$$

eq. 16.1

$V$ and $L$ are the volume and length, respectively, of the amphiphile chain, and $A$ is the cross-sectional area of the head group. Lipids with $P$-values $<\frac{1}{3}$ tend to form spherical micelles, those with $\frac{1}{3} < P < \frac{1}{2}$ prefer cylindrical micelles, and very hydrophobic lipids assemble into more lamellar aggregates, or membranes. For example, small amphiphiles such as lipids with only a single fatty acid chain form spherical micelles, and larger hydrophobic groups such as those in glycerolipids favor formation of membranes.

Micelles are unilamellar objects of 2–20 nm diameter and masses of typically less than 100 kDa that contain between 50 and 100 amphiphile molecules. The CMC values of amphiphiles range from a few nanomolar for large hydrophobic chains to many millimolar for short chains. For example, in phosphatidylcholines that carry a single fatty acid, the CMC depends exponentially on the length of the fatty acid chain: five, six, and seven carbon atoms (pentanoic, hexanoic, and heptanoic acid) lead to CMC of 90 mM, 15 mM, and 1.4 mM, whereas phosphatidylcholines with 12, 14, and 16 carbon atom groups (lauryl, myristoyl, and stearyl) have CMC of 90 nM, 6 nM, and 0.5 nM, respectively. This shows that micelles are more easily formed by lipids with increasing chain length. As detergents are frequently used for the purification of membrane proteins, the CMC of the detergent needs to be known to avoid unnecessary micelle formation.

Membranes are lipid bilayers where the hydrophobic parts of amphiphiles form a non-polar center that excludes water, whereas the hydrophilic head groups are in contact with water. In contrast to the comparatively small micelles, membrane bilayers can span large distances of up to a millimeter (Figure 16.38). Biological membranes have an overall thickness of about 4 nm with the two hydrophobic leaflets covering about 3 nm and the hydrophilic part spanning 0.5–1 nm. A layer of water molecules and ions is associated with the polar surface of membranes. The negative charges on biological membranes are the reason for their insulating properties, and their non-conductivity for ions or negatively charged metabolites (Box 16.4). By contrast, hydrophobic molecules with positive charges have some capacity to overcome membrane barriers, which is of advantage for the diffusion of pharmaceuticals to the interior of cells. Gases are small and non-polar, so they have little difficulty dissolving in and diffusing across membranes.

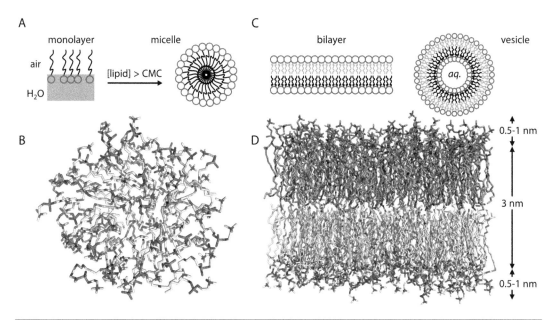

**FIGURE 16.38:   Assembly of lipids and detergents into monolayers, micelles and bilayers.** A: At a water/air or water/oil interface, amphiphiles first partition into a monolayer followed by unilamellar micelles when the concentration is larger than the CMC. The calculated micelle assembly consists of 65 molecules of n-dodecylphosphocholine (DPC), a lipid with only a single alkyl chain. B: Amphiphiles with larger hydrophobic chains organize into membrane bilayers that bury the hydrophobic parts and expose the hydrophilic head groups toward the aqueous phase. Vesicles form when the bilayers curve into water-filled hollow spheres. The simulated gel-like assembly is colored yellow and gray to highlight the individual leaflets.

## BOX 16.4: ELECTROSTATIC REPULSION OF METABOLITES BY MEMBRANES.

The non-permeability of membranes for negatively charged metabolites is used to keep neutral molecules within the cell by conferring negative charges onto them. This general principle in biology is exemplified by the phosphorylation of nutrients after their uptake into the cell. Glucose for example is transported by the glucose transporter from the extracellular milieu across the membrane into the cytoplasm and then rapidly converted into the negatively charged glucose-6-phosphate, a reaction catalyzed by hexokinase at the expense of ATP. Other kinases that serve to phosphorylate metabolites are fructokinase and galactokinase. In addition, many metabolites carry negative charges not only in the form of phosphoryl groups but also carboxyl groups. Examples include fatty acids, pyruvate, oxaloacetate, citrate, and many more. Intermediates of the biosynthesis of amino acids and nucleotides are also negatively charged. Electrostatic repulsion of these negative charges by membranes ensures that these molecules stay within their compartment and do not diffuse away.

Membranes are highly dynamic structures. The lipids in a membrane diffuse laterally at a high velocity of about 2 $\mu$m s$^{-1}$. This means that a lipid in the membrane of a typical bacterium can travel from one end to the other in just a second. Lateral diffusion velocities can be measured by fluorescence methods such as FRAP (Section 19.5.5.3) or single-molecule tracking (Section 23.1.7.3). By contrast, transverse diffusion or *flip-flop* of a lipid from one layer to the other is about nine orders of magnitude slower than lateral diffusion because the polar head group has to traverse the hydrophobic layer of the membrane, which is energetically disfavored. Thus, the contents of the outer and inner leaflets of the bilayer do not mix appreciably, which explains why biological membranes can maintain asymmetry of the leaflet contents. For example, the outer leaflet of erythrocyte membranes has a high content of sphingomyelin and phosphatidylcholine, whereas the inner leaflet is rich in phosphatidylserine. Rupture of an erythrocyte exposes phosphatidylserine to the extracellular medium, thus signaling the body that the cell is dying. Membranes in contact with water tend to form vesicles in order to further minimize exposure of hydrophobic surfaces.

### 16.2.8.4 Properties and Structure of Membrane Proteins

Membrane proteins are permanently or at least most of the time associated with a membrane. About 20–30% of all proteins are membrane proteins, but they are difficult to study outside their lipidic environment. The general problem is the exposure of hydrophobic surfaces once they are removed from their membrane, which leads to strong destabilization of the proteins due to the hydrophobic effect (Figure 16.35). Lipids are an integral part of the structure of membrane proteins, just as water is for soluble proteins, and both solvents have a decisive effect on the structure of the solute. In essence, membrane proteins dissolved in a membrane are a two-dimensional solution of hydrophobic proteins in a hydrophobic solvent. Some membrane proteins diffuse laterally just as quickly as lipids, while others are tethered to stationary structures outside the membrane, such as the cytoskeleton. Polar groups in membrane proteins are either solvent-exposed, pointing away from the membrane, or they bind to the polar head groups of the lipids. Any polar groups or charges that are buried within the membrane must interact with other polar groups or charges, similar to polar groups in the hydrophobic core of soluble proteins or within protein-protein interfaces.

The classification of membrane proteins is based on the extent to which they are associated with, or integrated into the membrane (Figure 16.39). *Monotopic* or *peripheral* membrane proteins only interact with one leaflet of the lipid bilayer. *Integral* membrane proteins traverse the membrane. Depending on the number of crossings of the polypeptide chain through the lipid bilayer, integral membrane proteins are subdivided into *bitopic* (single pass) or *multi-topic* (multi-pass).

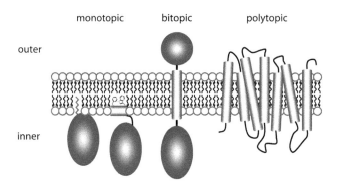

**FIGURE 16.39: Classification of membrane proteins.** Monotopic proteins are attached to a single leaflet of the membrane. Bitopic proteins have a transmembrane helix as the membrane anchor that passes both leaflets. Polytopic proteins have several passes of the protein structure through the membrane.

Some monotopic membrane proteins interact electrostatically with the polar head groups of the lipids e.g. by $Ca^{2+}$ ions, but most protein-membrane interactions are hydrophobic. Amphipathic α-helices, hydrophobic loops, and lipids post-translationally attached to proteins are ways to maximize hydrophobic interaction of proteins with a single leaflet of the membrane. For example, the small GTP-binding protein Arf is anchored in the membrane by an amphipathic α-helix and an N-terminal myristoyl group. Ras and Rab proteins have C-terminal farnesyl and geranylgeranyl groups, respectively, covalently bound to a Cys side chain. Other monotopic proteins have a C-terminal glycophosphatidylinositol (GPI) anchor, a lipid with the same core structure as phosphatidylinositol phosphates (PIPs; Table 16.8) plus several saccharides, phosphates, and ethanolamine groups attached to the inositol. GPI anchors are attached to the C-terminus of enzymes such as choline esterases and phospholipases, but also to adhesion molecules, complement regulatory proteins, and receptors. The connection of monotopic proteins with the membrane is not irreversible. The activity of Rab proteins is regulated *via* removal from the membrane by regulatory proteins.

In contrast to monotopic proteins, bitopic membrane proteins have a single transmembrane α-helix. The residues in the center of the helix are all hydrophobic to interact with the lipids of the membrane. The transmembrane helix permanently tethers the protein to the membrane and divides it into a cytoplasmic part and an extracellular part. Many cell surface receptors and protein kinases transduce external signals across the membrane into the cytoplasm and belong to this group of membrane proteins. The transmembrane helix itself is quite stable because its H-bonds are stronger in membranes due to the lower dielectric constant compared to aqueous solution, and because there is no water to compete for H-bonding. H-bonding is thus a strong driving force for solvation of peptide groups in a membrane environment. A stretch of hydrophobic amino acids will readily form α-helices in any hydrophobic environment because all polar main-chain atoms are tied in H-bonds and no unfavorable contacts between polar and non-polar atoms are left. A continuous stretch of ca. 20 hydrophobic residues in the sequence of an unknown protein is often diagnostic of a transmembrane helix.

The most complex type of membrane proteins are *polytopic* proteins, which pass several times through the membrane (Figure 16.40). Important classes of polytopic membrane proteins are enzymes, transporters, pores, and channels.

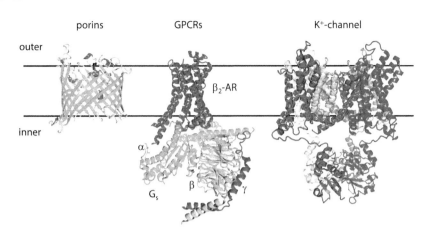

FIGURE 16.40:    **Examples of membrane protein structures.** Bacterial porins usually form trimers in the membrane but for clarity only a monomer is shown (PDB-ID 1a0s). The $\beta_2$-adrenergic receptor ($\beta_2$-AR) is a G-protein coupled receptor (GPCR) in eukaryotes that relays extracellular signals to the cytoplasm by docking to the heterotrimeric G-protein $G_s$ (PDB-ID 3sn6). The $K^+$ channel is a tetramer with a central polar path for the ions. Only the membrane-bound $\alpha_4$-subunit is shown (PDB-ID 4jta). Black lines show the membrane boundaries.

Porins are water-filled channels in the outer membrane of Gram-negative bacteria that allow small (<600 kDa) and polar or charged molecules such as carbohydrates, ions, and amino acids to pass across the outer membrane by passive diffusion. Each member of the large class of G-protein coupled receptors (GPCRs) binds to an extracellular ligand, e.g. a hormone, and then relays the signal across the cell membrane. For example, the $\beta_2$-adrenergic receptor ($\beta_2$-AR) is activated by a small molecule and then binds to the stimulatory G protein ($G_s$) (Figure 16.40). In the cell, this complex activates adenylyl cyclase, which in turn produces the second messenger cAMP for signal transduction. The $K^+$ channel is an example of functional oligomerization within a membrane. A central ion-conducting channel is formed from four monomers arranged as a planar tetramer.

The membrane-buried part of the proteins is either all-$\alpha$ or all-$\beta$. No integral membrane proteins with $\alpha/\beta$-structure have been described. The range of membrane protein architectures is consequently more limited compared to soluble proteins. There are currently about 2600 structures (~3% of all known structures) of transmembrane proteins known, of which ca. 2300 are all-$\alpha$ and ca. 300 are all-$\beta$ structures. The part of the protein that is embedded in the membrane is formed by helix bundles (GPCRs, $K^+$ channel) and $\beta$-barrels (porins). Similar to the formation of single-pass transmembrane helices, $\beta$-barrel and helix bundle formation in polytopic membrane proteins is driven by maximizing the number of H-bonds within the membrane. This would be difficult in $\alpha/\beta$ structures that have open $\beta$-sheets where the first and last strands expose polar peptide groups to the lipids. The maximization of H-bonds is the reason for the limited structural variety of membrane proteins: $\beta$-barrels in membrane proteins can be entirely antiparallel (even number of strands) or have two parallel strands (the first and last in odd numbers of strands). Polar loop regions connecting the $\beta$-strands protrude from the membrane on both sides. Regardless of the topology, the face of the barrel that contacts the membrane is invariably made of hydrophobic side chains. Depending on the function of the protein, the inner face of the barrel can be either hydrophobic or hydrophilic (as in porins).

## 16.3 FOLDING AND STABILITY

The thermodynamic stability of proteins under physiological conditions is small: the free energy $G_N$ for the native state is often just a few kJ mol$^{-1}$ more negative than $G_U$ for the unfolded state. The native state has many attractive interactions but low entropy, whereas the unfolded state has few attractive interactions but high entropy (see Box 2.9). Therefore, both $G_N$ and $G_U$ are large negative numbers. The net stability of the protein under specific conditions is the small difference $G_U-G_N$, which is typically around 20–40 kJ mol$^{-1}$. This corresponds to a small negative $\Delta G$ for folding of proteins which is a result of entropy-enthalpy compensation (see Box 2.7). The enthalpic change during protein folding is dominated by differential interactions between the side chains in the folded state and of the side chains with water in the unfolded state. The entropic change is a balance between the reduced entropy of the folded state and the increased entropy of the water that is released upon folding. The folding equilibrium between native and unfolded states is easily shifted as a function of temperature and solvent composition. There are two possible strategies to destabilize a protein, either by destabilizing the native state or, maybe less intuitively, by stabilizing the unfolded state (Figure 16.41).

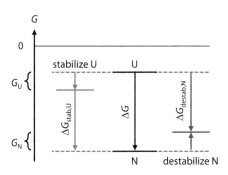

FIGURE 16.41:    **Dependence of protein stability on the energies of the folded and unfolded states.** Black lines indicate the energy levels of a protein with the energy difference $\Delta G$ between the folded state N and the unfolded state U. The red lines show protein destabilization by destabilizing the folded state ($\Delta G_{destab,N}$). The blue lines show protein destabilization by stabilization of the unfolded state ($\Delta G_{stab,U}$). The protein is destabilized in both cases.

## 16.3.1 Driving Forces for Protein Folding

The tendency of non-polar molecules to aggregate in polar solvents is known as the hydrophobic effect (Box 16.3). We have encountered this effect when discussing water structures on hydrophobic surfaces (Section 16.2.6.3) and the formation of vesicles and membranes in aqueous environments (Section 16.2.8). The same principle of water exclusion from hydrophobic surfaces applies to the folding of polypeptides: the hydrophobic effect drives the clustering of hydrophobic side chains to form the tightly packed hydrophobic core of the native state.

Hydrophobic and hydrophilic amino acids (Table 16.1) have different contributions to the free energy of folding. The transfer of non-polar side chains from the polar solvent water to the non-polar environment of the protein core has both favorable enthalpic and entropic contributions. The enthalpic contribution comes from van der Waals interactions between the amino acid side chains. Ordered water molecules on the surface of the hydrophobic side chains are released into bulk solvent, which is entropically beneficial. However, an enthalpic penalty has also to be paid because the strong H-bonds between the ordered water molecules on hydrophobic surfaces (Figure 16.35) first have to be broken before they can re-form weaker H-bonds in bulk solvent. Upon unfolding, the hydrophobic side chains become again solvent-exposed, leading to an increase in the heat capacity $C_p$, similar to what is observed when small non-polar molecules are dissolved in water (Box 16.3). It is this balance between entropic and enthalpic contributions in the folded *versus* the unfolded protein that determines the net free energy of the folded protein. A stable monomeric protein has few hydrophobic side chains on its surface. If present, they are mostly single and surrounded by hydrophilic side chains. If clusters of hydrophobic residues occur at the surface of a protein they are usually part of a protein-protein interaction site (Section 16.2.6).

Compared to the burial of hydrophobic residues in the protein core, hydrophilic side chains have less effect on protein stability. Hydrophilic amino acids decorate the solvent-exposed surface of soluble proteins where they form almost isoenergetic H-bonds among themselves or with water. This is different for H-bonds formed by main-chain and side-chain groups that are buried in the hydrophobic core of the protein: buried H-bonds have a large energetic contribution to stability because there is no competition by H-bond formation with solvent molecules and the interaction between donor and acceptor is stronger in environments of low dielectric constant (Section 15.3.3). Buried H-bonds are also responsible for the *cooperativity* of protein folding (Section 16.3.4).

Although most of the water is removed from hydrophobic surfaces upon folding, water may still end up in a non-polar pocket of the protein interior. About 1% of the protein core is not completely packed with side chains. Sub-optimal packing of hydrophobic cores reduces van der Waals contacts between side chains and at the same time increases their entropy. For these energetic reasons, nature "abhors a vacuum" (*horror vacui*) in the protein interior, and one of the strategies by which proteins are stabilized in thermophilic organisms is exquisite packing of the hydrophobic core with few or no cavities. However, if a cavity is larger than the volume of water (>30 $\mathring{A}^3$ or 0.03 nm$^3$), water being the solvent will have to fill it. A single water molecule in a hydrophobic environment is frustrated because it cannot form any H-bond. A minimum of three water molecules must be trapped to form two H-bonds each, and indeed over 90% of water molecules in hydrophobic cavities form 3–4 H-bonds.

Other components in the solvent also affect the stability of the protein. Salts can stabilize the native state of a protein or the unfolded state, leading to a stabilization or destabilization, respectively. The observed effect of salts on protein solubility and stability led to the Hofmeister series (Box 16.5).

---

### BOX 16.5: THE HOFMEISTER SERIES.

The effect of salts on protein solubility was tested ~1880, leading to the empirical *Hofmeister series* of cations and anions (Figure 16.42).

| | salting out | salting in | |
|---|---|---|---|
| hydrophobic effect ↑ | | | hydrophobic effect ↓ |
| surface tension ↑ | $NH_4^+ > K^+ > Na^+ > Li^+ > Mg^{2+} > Ca^{2+} >$ guanidinium$^+$ | | surface tension ↓ |
| stability ↑ | $SO_4^{2-} > HPO_4^{2-} > CH_3CO_2^- > Cl^- > NO_3^- > Br^- > I^- > SCN^-$ | | stability ↓ |
| solubility ↓ | | | solubility ↑ |

FIGURE 16.42:   **Effect of salts on protein solutions.**

Salts from the combinations of ions on the left of the series (such as ammonium sulfate) were found to strongly reduce protein solubility (*salting out*) while those from the right-hand side of the series (such as guanidinium thiocyanate) strongly solubilize proteins (*salting in*). The effects on solubility are macroscopically related to an increase or reduction in the surface tension of water, and microscopically related to an increase or reduction of the hydrophobic effect. Salts like ammonium sulfate increase both the surface tension of water and the hydrophobic effect and concomitantly decrease the solubility of proteins. The net result is the stabilization of the native state for the price of reduced protein solubility. Ammonium sulfate both stabilizes and precipitates proteins from aqueous solutions and therefore is often used for protein purification, concentration, crystallization, and long-term storage of precipitated proteins. By contrast, solubilizing salts from the right-hand side of the Hofmeister series decrease the surface tension of water, which decreases the hydrophobic effect and promotes solubilization of non-polar surfaces in water. However, non-polar surfaces are a hallmark of unfolded proteins, so the net result is a destabilization of the protein by stabilization (solubilization) of the unfolded state. Ions of such *chaotropic* salts directly bind to the hydrophobic surfaces exposed in

*(Continued)*

the unfolded state. Guanidinium ions for instance (Figure 16.43) stack against peptide groups and aromatic side chains *via* cation-π interactions (Section 15.3.2). Basically any organic solvent or any substance that efficiently competes for H-bonding can act as a chaotrope. Some of these chemicals are used in the laboratory to remove proteins from solutions of nucleic acids (e.g. in phenol extraction and ethanol precipitation) or to solubilize denatured proteins (e.g. SDS for polyacrylamide gel electrophoresis). Urea (Figure 16.43) is a chaotrope that forms H-bonds with the peptide group, possibly competing with the native H-bonding in proteins, and thus contributes to solvation of the unfolded state and induces protein unfolding. The alkyl chain of SDS (Figure 16.43) interacts with hydrophobic surfaces of aliphatic and aromatic amino acids, again solubilizing (and stabilizing) the unfolded state. While binding of SDS to hydrophobic side chains leads to destabilization of soluble proteins, membrane proteins are stabilized by small concentrations of detergents such as SDS because they interact with hydrophobic surfaces present in the folded state (Section 16.2.8.4).

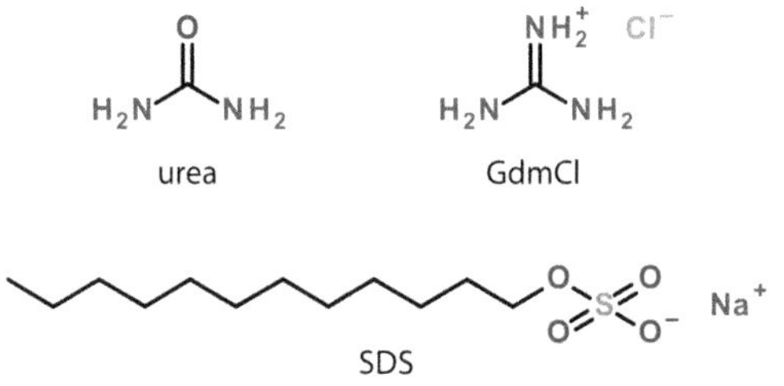

**FIGURE 16.43: Structures of the common chemical denaturants.** Shown are urea, guanidinium chloride (GdmCl), and sodium dodecylsulfate (SDS).

### 16.3.2 First Folding Experiments and the Levinthal Paradox

In 1956, Christian Anfinsen studied the dependence of RNase A activity as a function of reducing agent. RNase A is a small protein of 124 amino acids with four disulfide bonds that is deactivated by high concentrations of the denaturants urea and guanidinium chloride. Anfinsen unfolded RNase in urea and reduced the disulfide bonds with β-mercaptoethanol. When this inactive form was exposed to air, disulfide bonds re-formed in a random fashion. After removal of urea by dialysis, this *scrambled* RNase was almost inactive, retaining just 1% of its initial activity. However, upon addition of just a catalytic amount of reducing agent, the RNase regained its full activity through the process of *disulfide shuffling* (Figure 16.44).

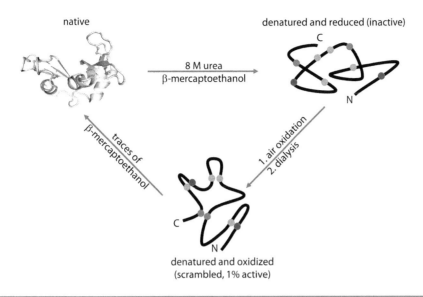

native                                                          denatured and reduced (inactive)

8 M urea
β-mercaptoethanol

traces of
β-mercaptoethanol

1. air oxidation
2. dialysis

denatured and oxidized
(scrambled, 1% active)

FIGURE 16.44:    **Anfinsen's experiment.**

It was apparent that the unfolded form of RNase A was trapped in an almost inactive state after air oxidation, indicating that incorrect disulfide bonds had been formed. The fact that only a catalytic amount of reducing agent was necessary to reactivate RNase A shows that the driving forces for this autonomous, self-assembling folding process must be encoded in the primary structure (the sequence) of the polypeptide. This impressive result is corroborated by a simple calculation: when randomly combining the 20 common amino acids for a typical protein of 200 residues, the resulting number of possible sequences is $20^{200}$ or $1.6 \cdot 10^{160}$, which far exceeds the estimated number of atoms in the universe (ca. $10^{80}$). However, the estimated number of different protein sequences in nature is much lower, only $10^{10}$–$10^{13}$, and many of these sequences are closely related. Most of the statistically possible sequences lead to polypeptides that are not folding-competent. Anfinsen was awarded the 1972 Nobel Prize in Chemistry for the "clarification of the relationship between the structural properties of proteins and their biological functions".

A simple Gedankenexperiment shows that it is impossible for the polypeptide chain to sample all possible conformations and find the native state by a random search. Let us assume that each amino acid residue in a polypeptide can sample ten different conformations. For N residues, we then obtain $10^{N}$ different conformations. The average rotation frequency around a chemical bond is ca. $10^{14}$ s$^{-1}$, so sampling of all conformations in just a 40-residue polypeptide would require $3 \cdot 10^{18}$ years (the universe is about $10^{10}$ years old), which is clearly longer than the few milliseconds to seconds that a typical protein takes to fold into its native state. This discrepancy between statistical sampling of the folding space and the true time frame of protein folding is known as the *Levinthal paradox*, formulated in 1969. The kinetics of protein folding must be guided by additional information contained in the primary structure, but exactly how this works remains a conundrum. We will see in Chapter 18 on computational biology that protein models can be routinely calculated if a high-resolution structure of a homolog is known, but *de novo* calculation of a protein structure remains impossible for any but the smallest proteins. The *protein folding code* remains un-deciphered, not least because it is highly redundant: proteins of very different sequence can adopt very similar structures, and some proteins even adopt different structures under different conditions.

## 16.3.3 Energy Landscapes for Protein Folding

The Levinthal paradox assumes that all conformations accessible to the polypeptide chain are sampled with equal probability until the native state, the conformation of lowest free energy, is found. The associated energy landscape is a flat surface with a single minimum, akin to the putting green in a golf course (Figure 16.45). This scenario is unrealistic not only for reasons of time, but also for energetic reasons. Random sampling assumes that all conformations have equal energy, but the Ramachandran energy surface (see Figure 16.16) shows that certain conformations have lower energies than others.

A more realistic energy landscape for folding is a *folding funnel*. In its idealized, smooth shape, all starting points at a perimeter have the same energy. This model correctly predicts that with each native-like contact that is established, $\Delta G$ becomes more and more negative, and the structure is closer to the native state. However, the smooth funnel model lacks local energy minima where folding intermediates can reside. Such intermediates have been observed experimentally for proteins that contain disulfide bonds, *cis*-peptides, or several domains. Probably the best model to visualize protein folding is a *rugged funnel*, an extension of the smooth funnel. The energy landscape has local minima with activation barriers between them. Proteins that reach a local minimum with a high energy barrier will need time to overcome this barrier and to reach the native state. Local minima therefore lock the molecule in a *kinetic trap*, which has been evidenced experimentally for some proteins. In the rugged funnel model the folding of a protein may proceed without detectable intermediates or through one or more intermediates. There are several alternative pathways toward the native state, making the rugged folding funnel the most realistic scenario to describe the energy landscape of protein folding.

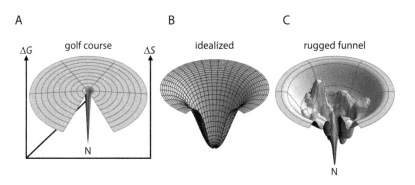

**FIGURE 16.45:   Energy landscapes of protein folding.** A: In the "golf course putting green" landscape from Levinthal's paradox, all folding intermediates have equal energy, an impossible scenario. B: The smooth funnel landscape is more realistic because folding can start from any point on the energy surface and should move in the correct direction. C: The rugged energy landscape includes local minima for folding intermediates that can act as kinetic traps on the way to the folded state. (Panels A and C reprinted from Dill & Chan [1997], *Nat. Struct. Biol.* 4(1): 10–19, with permission.)

Although the funnel analogy is useful for visualizing protein folding and its energetics, the funnels must not be taken literally: folding funnels give the wrong impression that the shape of the energy landscape guides the protein towards the folded state. As a consequence, the folding process should accelerate towards the native state as the walls of the funnel get steeper, but this is not observed experimentally. From a thermodynamic point of view, the conformational entropy decreases toward the native state ($T\Delta S < 0$), while at the same time attractive interactions are formed that decrease the enthalpy ($\Delta H < 0$). Both $\Delta H$ and $T\Delta S$ are large energies, making the net decrease in $\Delta G$ very small.

The marginal stability of the native state of a protein often goes along with high flexibility, which is important for ligand binding, catalysis, and protein-protein recognition. *Intrinsically disordered proteins* are proteins without a regular structure, i.e. they have no "native" state, but are nevertheless functional (Box 16.6).

---

**BOX 16.6: INTRINSICALLY DISORDERED PROTEINS.**

Until the 1990s it was assumed that a functional protein needs a defined three-dimensional structure. The few exceptions of functional but unfolded proteins were limited to small peptide hormones such as glucagon, which were thought to fold upon binding to their

*(Continued)*

receptors. The unfolded state of proteins was mainly considered non-functional, populated only during synthesis and chaperone-assisted folding, and before degradation. A sequence database survey conducted in 1994 showed that about half of all proteins known at that time contain stretches of >40 residues that are of low complexity, i.e. contain sequences that will likely not fold into any defined structure. Eukaryotic genomes were found to encode more proteins with intrinsic disorder than those from prokaryotes. Nowadays a few dozen proteins are known to be entirely unfolded or *intrinsically disordered* under physiological conditions. Compared to folded proteins, intrinsically disordered proteins have a higher hydrophobicity and lower net charge, and are therefore more aggregation-prone. They display little, if any, secondary structure, and usually do not adopt a globular shape. Some of them, for instance α-synuclein in nerve cells, adopt helical secondary structure when in contact with phospholipids, while others, including the zinc finger-containing transcriptional repressor RYBP, have an unfolded DNA-binding domain that adopts a regular structure upon engagement with its target sequence.

### 16.3.4 MATHEMATICAL DESCRIPTION OF THE TWO-STATE MODEL

The simplest conceptual model for protein folding is the two-state model that assumes a thermodynamic equilibrium between only two states, the native state N and the unfolded state U (Figure 16.46). By native state, we usually mean the functional form of the protein, often a single conformation for small proteins or a set of defined conformations for larger proteins exhibiting domain motions. The native state is characterized by contacts between residues far away in sequence that bring together secondary structure elements. By contrast, the unfolded state is an ensemble of conformations that exhibit only local interactions. These conformations rapidly interconvert and have similar energies.

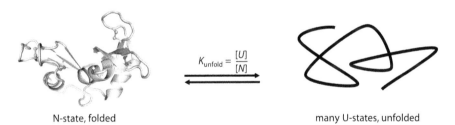

$$K_{\text{unfold}} = \frac{[U]}{[N]}$$

N-state, folded                    many U-states, unfolded

**FIGURE 16.46:   The two-state model of protein folding.** *RNase* A (PDB-ID 1rat) is shown with its disulfide bonds colored as in Figure 16.44. The ratio of the concentrations of the unfolded state U and the folded state N defines a constant $K_{\text{unfold}}$ that is dependent on experimental conditions such as temperature or denaturant concentration.

The thermodynamic equilibrium between native state N and unfolded state U can be described by the equilibrium constant $K_{\text{unfold}}$:

$$K_{\text{unfold}} = \frac{[U]}{[N]} = \frac{1-\alpha}{\alpha} \qquad \text{eq. 16.2}$$

Since the total protein concentration is the sum of folded and unfolded protein, $K_{\text{unfold}}$ can also be described as a function of the folded fraction α (eq. 16.2). The relation $\Delta G^0 = -RT\ln K_{\text{unfold}}$ is used to calculate the free energy of unfolding under standard conditions. A typical two-state unfolding transition, induced by increasing concentrations of denaturants or increasing temperatures, is shown in Figure 16.47.

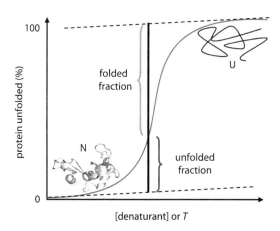

**FIGURE 16.47:   A two-state unfolding transition.** The thermal or chemical denaturation of a protein is followed by a spectroscopic signal. The baselines before and after the transition represent the dependence of the measured signals for the folded and unfolded states on the concentration of denaturant or temperature. By extrapolation of these baselines, the U/N ratios can be determined that correspond to the equilibrium constant $K_{unfold}$ at any given concentration of denaturant or temperature. The inflection point of the transition corresponds to $K = 1$ (50% folded). The gradient in the midpoint of the transition is a measure of the cooperativity of unfolding.

The unfolding transition can be monitored by any signal that changes during the experiment, i.e. that differs between native and unfolded states of the protein studied, provided that it can be performed under all required experimental conditions (high temperature or denaturant concentrations, extremes of pH). Changes in UV-VIS absorbance (Section 19.2) or fluorescence (Section 19.5) use aromatic amino acids as chromophores and report on changes in tertiary structure. Far-UV circular dichroism (Section 19.3), on the other hand, measures absorption properties of the protein backbone and reports on changes in secondary structure. Nuclear magnetic resonance (Section 20.1) and heat capacity (differential scanning calorimetry, Section 27.2) are also frequently used probes. The baselines of the unfolding curve represent the change in signal (if any) as a function of denaturant concentration or temperature, but have nothing to do with the actual folding/unfolding event. Each point in the transition region corresponds to an equilibrium between folded and unfolded states. The inflection point is at $K = 1$, or 50% folded/unfolded. The corresponding temperature $T_m$ is the melting temperature of the protein. The steeper the almost linear middle part of the transition, the more cooperative is the (un)folding. A quantitative thermodynamic treatment of protein folding according to eq. 16.2 is only possible if the unfolding is reversible, a prerequisite that is often neglected when protein stability *per se* is the focus (see Box 16.7).

A mathematical description of a reversible two-state unfolding transition was introduced in 1988 by Santoro and Bolen. Denaturant-induced unfolding by urea, guanidinium chloride, or acid typically results in a baseline of the measured signal ($y$ in the following equations). The baselines for the folded state N and the unfolded state U are expressed as a linear function of the denaturant concentration $[D]$ with the slope $m$ (Figure 16.47)

$$y_U = y_U^0 + m_U[D] \quad \text{and} \quad y_N = y_N^0 + m_N[D]$$

eq. 16.3

Using the law of mass action, we can expand eq. 16.2 to replace the actual concentrations of $[N]$ and $[U]$ with the measured signal $y_{obs}$.

$$K_{unfold} = \frac{[U]}{[N]} = \frac{y_{obs} - y_N}{y_U - y_{obs}}$$

eq. 16.4

The free energy of folding $\Delta G_{stab}$ is the difference between the free energies of the folded ($G^0_N$) and unfolded states ($G^0_U$), linearly extrapolated to zero denaturant concentration ($[D] = 0$). The experimentally observed $\Delta G_{unfold}$ in an unfolding transition is a function of $\Delta G_{stab}$, the denaturant concentration, and the cooperativity parameter $m_G$:

$$\Delta G_{\text{unfold}} = \Delta G_{\text{stab}} + m_G \cdot [D] = -RT \ln K_{\text{unfold}} \qquad \text{eq. 16.5}$$

The linear dependence of unfolded protein on denaturant concentration in the equation above is known as the Tanford model, which applies to many proteins. The cooperativity parameter $m_G$ is defined as

$$m_G = \frac{\partial \Delta G_{\text{unfold}}}{\partial [D]} \qquad \text{eq. 16.6}$$

The more steeply the $\Delta G_{\text{stab}}$ changes with denaturant concentration, the more cooperative the folding event, and the larger the cooperativity parameter. Solution of eq. 16.4 for $y_{\text{obs}}$ and replacement of $y_N$ and $y_U$ (from eq. 16.3) and $K_{\text{unfold}}$ (from eq. 16.5) yields the equation needed to describe the observed spectroscopic parameter $y_{\text{obs}}$ as a function of denaturant concentration $[D]$ with the fit parameters $\Delta G_{\text{stab}}^0$ and $m_G$ (if we fix the parameters for the baselines):

$$y_{\text{obs}} = \frac{y_N^0 + m_N [D] - y_U^0 - m_U [D]}{1 + e^{-\frac{\Delta G_{\text{stab}}^0 + m_G [D]}{RT}}} + y_U^0 + m_U [D] \qquad \text{eq. 16.7}$$

The procedure to derive the corresponding expression for temperature-induced reversible unfolding events is similar. Here, we start from the Gibbs-Helmholtz equation for protein folding:

$$\Delta G_{\text{stab}} = \Delta H_{\text{stab}} - T \Delta S_{\text{stab}} \qquad \text{eq. 16.8}$$

The enthalpy and entropy terms in eq. 16.8 are temperature-dependent and include the change in heat capacity $\Delta C_p$, which is the difference in heat capacity between the unfolded and the folded states. As derived in the section on thermodynamics (Section 2.3), we know that

$$\Delta H_{\text{stab}}(T) = \Delta H_{\text{stab}}(T_m) + \Delta C_p (T - T_m) \qquad \text{eq. 16.9}$$

$T_m$, the melting temperature of the protein, is taken as the reference temperature. The temperature-dependence of the entropy (Section 2.4.3) is

$$\Delta S_{\text{stab}}(T) = \Delta S_{\text{stab}}(T_m) + \Delta C_p \cdot \ln \frac{T}{T_m} \qquad \text{eq. 16.10}$$

At the temperature $T_m$, we have the same concentrations of native and unfolded protein ($K_{\text{obs}} = 1$ and $\Delta G = 0$). We can use this information to eliminate $\Delta S_{\text{stab}}(T_m)$ from eq. 16.10.

$$\Delta G_{\text{stab}}(T_m) = 0 = \Delta H_{\text{stab}}(T_m) - T \Delta S_{\text{stab}}(T_m) \qquad \text{eq. 16.11}$$

It follows that

$$\Delta S_{\text{stab}}(T_m) = \frac{\Delta H_{\text{stab}}(T_m)}{T} = \frac{\Delta H_m}{T} \qquad \text{eq. 16.12}$$

We can now combine eq. 16.8, eq. 16.9, eq. 16.10, and eq. 16.12 to express the free energy $\Delta G_{\text{obs}}$ as a function the temperature $T$, the change in heat capacity $\Delta C_p$, and the "melting" enthalpy $\Delta H_m$ at the melting temperature $T_m$.

$$\Delta G_{\text{stab}}(T) = -RT \ln K_{\text{unfold}} = \Delta H_m \left(1 - \frac{T}{T_m}\right) + \Delta C_p \left(T - T_m - T \cdot \ln \frac{T}{T_m}\right) \qquad \text{eq. 16.13}$$

The baselines of the experimentally observed unfolding transition are now a linear function of temperature.

$$y_U = y_U^0 + m_U T \quad \text{and} \quad y_N = y_N^0 + m_N T \qquad \text{eq. 16.14}$$

Combination of eq. 16.4, eq. 16.13, and eq. 16.14 then yields an expression for the observed signal $y_{obs}$ of the same format as eq. 16.7:

$$y_{obs} = \frac{y_N^0 + m_N T - y_U^0 - m_U T}{1 + e^{-\frac{\Delta H_m}{R}\left(\frac{1}{T} - \frac{1}{T_m}\right) - \frac{\Delta C_p}{R}\left(1 - \frac{T_m}{T} - \ln\frac{T}{T_m}\right)}} + y_U^0 + m_U T \qquad \text{eq. 16.15}$$

Eq. 16.7 and eq. 16.15 are the general expressions to describe the signal $y_{obs}$ used to monitor unfolding transitions in order to extract the thermodynamic parameters of (un)folding. In practice, temperature-induced unfolding experiments have the advantage that the measured data are directly related to thermodynamic quantities ($\Delta H$, $\Delta G$, $\Delta C_p$), whereas chemically induced unfolding introduces additional parameters such as the dependence of $\Delta G$ on the denaturant concentration, which has to be taken into account to retrieve thermodynamic values for physiological conditions.

An interesting prediction of eq. 16.13 is *cold denaturation*. Proteins can also unfold when the temperature is lowered, which may be counterintuitive at first. A plot of $\Delta G_{stab}$ as a function of temperature, termed the protein stability curve, has a bell shape and shows that $\Delta G_{stab}$ reaches zero at two different temperatures (Figure 16.48; see Section 2.5.2 ), so the protein has actually two melting points. The temperature at which the protein has its maximum stability lies between the two melting temperatures. The overall curvature of the stability curve, and therefore the temperature range where the protein is stable, is determined by the values of $\Delta C_p$ and $\Delta H_m$. The origin of $\Delta C_p$ is mostly related to the hydrophobic effect and comes from the increase in solvent-exposed non-polar surface area when the protein unfolds. $\Delta C_p$ is usually positive in the range 0–30 kJ mol$^{-1}$ K$^{-1}$, and lies at around 8 kJ mol$^{-1}$ K$^{-1}$ for most proteins. The melting enthalpy $\Delta H_m$ is also positive and in the range between almost zero to a few hundred kJ mol$^{-1}$.

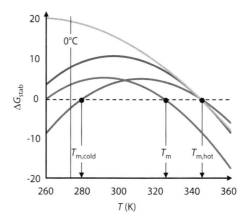

**FIGURE 16.48:** **Protein stability curves.** The freezing point of water is marked by the line at 0°C (273 K). A mesophilic (blue curve) protein is shown with a $T_m$ of 323 K. The $T_m$ for its cold denaturation is at 260 K, below the freezing point of water. The three stability curves of thermostable proteins (magenta, red, and orange) have the same $T_{m, hot}$ for heat denaturation at 343 K but different $T_{m, cold}$ for cold denaturation. A protein can be stabilized in several ways: the magenta curve has the same $\Delta H_m$ and $\Delta C_p$ as the mesophilic protein but a 20°C higher $T_m$ and is just right-shifted from the mesophilic curve. The shift makes the cold denaturation temperature $T_{m, cold}$ at 280 K experimentally accessible. The red curve has a more positive $\Delta H_m$ but the same $\Delta C_p$ as the magenta and blue curves. The orange curve has a smaller $\Delta C_p$ than the other curves.

For a mesophilic protein, cold denaturation is difficult to observe when the $T_m$ for cold denaturation is below the freezing point of the buffer system. For thermophilic proteins, the stability curve may be shifted to higher temperatures so that cold denaturation can be observed experimentally (Figure 16.48). Cold denaturation is enthalpy-driven while heat denaturation is entropy-driven (see Box 2.7). If $\Delta C_p$ is large, the stability curve is strongly bent. In such a case, the melting temperature for cold denaturation can become experimentally accessible.

If the unfolding process is not reversible, the midpoint of the transition still provides information about the stability of the protein against the denaturant, but no thermodynamic parameters can be extracted from the data. In comparative studies of protein stabilities, it is often enough to measure the unfolding transition and compare their midpoints. Such a ranking is used in the *Thermofluor®* assay (Box 16.7).

---

### BOX 16.7: THERMOFLUOR® ASSAY FOR PROTEIN STABILIZATION AND LIGAND BINDING.

The exposure of hydrophobic surfaces upon protein unfolding can be used to estimate the relative stabilities of protein variants or to rank the affinities of ligands that bind to, and hence stabilize, the protein. Protein variants or protein-ligand complexes that are more stable will have a higher melting temperature $T_m$. The magnitude of the shift in $T_m$ is a measure of the extent of protein stabilization (Figure 16.49). The exposure of hydrophobic surfaces upon temperature-induced unfolding is measured as an increase in fluorescence of the fluorescent dye SYPRO Orange that binds to these surfaces, which causes an increase in its quantum yield. The measurement can be automated and conducted in micro-titer plates such that dozens of protein variants or protein-ligand complexes can be ranked in a matter of a few hours.

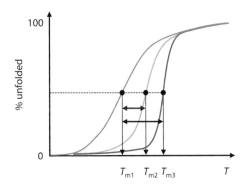

FIGURE 16.49:    **Thermofluor assay.** A shift in melting temperature $T_m$ is the readout for stabilization of the protein by binding of the respective ligand.

---

It may not be immediately obvious that the stabilization of the folded state of the protein is a consequence of ligand binding. In fact, any ligand that binds more strongly to the native than to the unfolded state of a protein causes an increase in protein stability (Box 16.8).

---

### BOX 16.8: LIGAND BINDING STABILIZES THE NATIVE STATE OF A PROTEIN IN GENERAL.

We consider a ligand L that can in principle bind both to the native state N and the unfolded state U of a protein. The thermodynamic cycle of this system is given by

$$
\begin{array}{ccc}
\text{N} & \underset{K_{fold}}{\rightleftharpoons} & \text{U} \\
K_{d,N} \updownarrow & & K_{d,U} \updownarrow \\
\text{NL} & \rightleftharpoons & \text{UL}
\end{array}
$$

scheme 16.1

*(Continued)*

The equilibrium constant for protein folding is

$$K_{\text{fold}} = \frac{[N]}{[U]}$$

eq. 16.16

and the equilibrium dissociation constants of the ligand complexes are

$$K_{d,N} = \frac{[N][L]}{[NL]} \quad \text{and} \quad K_{d,U} = \frac{[U][L]}{[UL]}$$

eq. 16.17

We define an apparent equilibrium constant $K_{\text{app}}$ (see Section 3.4) as the ratio of all folded and all unfolded species:

$$K_{\text{app}} = \frac{[N]+[NL]}{[U]+[UL]}$$

eq. 16.18

Substituting $K_{d,U}$ and $K_{d,N}$ into the definition for $K_{\text{app}}$ gives

$$K_{\text{app}} = \frac{[N]+\dfrac{[N][L]}{K_{d,N}}}{[U]+\dfrac{[U][L]}{K_{d,U}}} = \frac{[N]}{[U]} \cdot \left( \frac{1+\dfrac{[L]}{K_{d,N}}}{1+\dfrac{[L]}{K_{d,U}}} \right) = K_{\text{fold}} \cdot \left( \frac{1+\dfrac{[L]}{K_{d,N}}}{1+\dfrac{[L]}{K_{d,U}}} \right)$$

eq. 16.19

Now, if the ligand stabilizes the native state, then the condition

$$K_{\text{app}} > K_{\text{fold}}$$

eq. 16.20

must be true. This means that

$$1+\frac{[L]}{K_{d,N}} > 1+\frac{[L]}{K_{d,U}}$$

eq. 16.21

which it true for

$$K_{d,U} > K_{d,N}$$

eq. 16.22

Smaller equilibrium dissociation constants mean tighter binding. The result of this exercise is that as long as the ligand binds more strongly to the native state N than to the unfolded state U, the native state is stabilized compared to the absence of ligand. This is true for all ligand concentrations. If a "ligand" binds more strongly to the unfolded state, the protein is destabilized. In this case the ligand is called a denaturant.

### 16.3.5 FOLDING PATHWAYS AND MECHANISMS OF PROTEIN FOLDING

The complete folding pathway, including thermodynamic parameters, microscopic rate constants, and intermediates (if present), has been elucidated for a number of proteins such as ubiquitin, lysozyme, RNase, myoglobin, and the cold shock protein CspB. The best understood examples in terms of kinetics and thermodynamics are small, autonomously folding single-domain proteins with no folding intermediates that display a two-state folding transition (Figure 16.47). However, this direct pathway seems to be the exception rather than the rule. Many proteins populate one or more folding intermediates, use different folding pathways, or fall into kinetic traps from which they need to be rescued. Furthermore, oligomers need guided assembly while at the same time competing aggregation has to be excluded.

The folding of a nascent polypeptide begins as soon as it exits the ribosome, which will be a fundamentally different pathway from that of an already synthesized protein that undergoes cycles of unfolding and refolding. Figure 16.50 gives a schematic overview of the competing processes from a polypeptide to higher order structures.

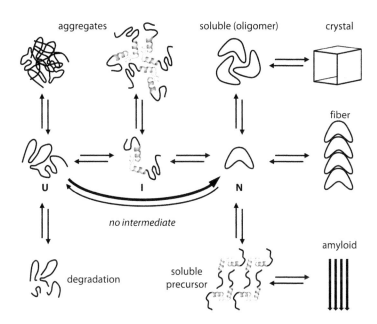

**FIGURE 16.50: Overview of protein folding, assembly, and aggregation.** The result of ribosomal peptide synthesis is an unfolded protein U, which can either fold directly or *via* intermediates (labeled I) to the native state N. The native state can proceed to assemble into oligomers, fibers, and sometimes even crystals. Folding-incompetent proteins, such as variants or protein fragments from premature translation termination, are degraded. Chaperones are vital for the folding of large multi-domain proteins and to avoid aggregation. Because $\Delta G_{fold}$ is small for most proteins and the energy landscape of folding is rather flat, it is possible for a native protein to convert into a soluble precursor of toxic amyloids (Section 16.3.6).

Several models have been proposed to explain how proteins can fold in a matter of milli- or sometimes microseconds, and most of these models separate secondary structure formation from tertiary structure formation. The *nucleation model* assumes that a group of residues close in sequence forms a secondary structure element (helix, loop, or β-hairpin), from which new secondary structure is propagated along the chain. Tertiary structure forms from interactions of the string of secondary structure elements along the polypeptide chain. By contrast, in the *diffusion-collision model*, secondary structure elements form independently at several locations throughout the polypeptide. These elements then diffuse, collide, and bind to each other until the polypeptide chain coalesces into the native tertiary structure. Finally, the *hydrophobic collapse model* proposes that the hydrophobic effect drives compaction of the polypeptide chain, which then enables formation of secondary structure. A hydrophobic collapse imposes steric restrictions to movements, and also much reduces the number of water molecules around the polypeptide chain that compete with NH-and CO-groups for H-bonds. Limited movement and desolvation thus promote secondary structure formation.

The first two models propose that secondary structure forms first and enables formation of tertiary structure. The hydrophobic collapse model is the reverse: secondary structure forms only after a three-dimensional collapse. The marriage of these models is the *nucleation-condensation model* where secondary and tertiary structure is formed in parallel: a nucleus of neighboring residues forms a native-like secondary structure that is subsequently stabilized by tertiary interactions. Since no two proteins are alike, the folding mechanisms of different proteins will bear more or less similarity to one or the other model. For example, helical membrane proteins are thought to fold sequentially, following the nucleation and the diffusion collision models. α-helices are formed within the membrane, which is a thermodynamically favored event due to the absence of water (Section 16.2.8.4). The helices then diffuse laterally to form the helical bundles of the native structure. By contrast, small globular proteins such as SH3 domains and the chymotrypsin inhibitor fold according to the nucleation-condensation model.

The fastest steps in folding are formation of secondary structure elements and the hydrophobic collapse of the polypeptide chain to generate a compact, *hydrated intermediate* known as a *molten globule*. While for two-state transitions the molten globule represents a transition state that cannot be observed, it is an experimentally tractable folding intermediate for other proteins (Section 16.3.5.4). After the fast collapse, slower rearrangement and dehydration of the interior of the molten globule produces the native state of the protein. In the case of oligomers, this process is followed by subunit assembly. The sequence secondary-tertiary-quaternary structure formation during folding sounds logical but is by no ways strict: the hydrophobic collapse of secondary structure elements drives further secondary structure formation by exclusion of water, and re-shuffling of secondary structure elements is possible even after the quaternary structure has been established. In oligomers of multi-domain proteins, folding and assembly can occur simultaneously. Table 16.9 summarizes the time scales for different molecular processes relevant to protein folding.

**TABLE 16.9**
**Time scales for molecular processes in protein folding.**

| time range | extent | scale | action |
|---|---|---|---|
| fs – ps | <10 pm | local | bond vibrations (stretch and bend), source of IR signal |
| ~ 1 ps | | local | rotation correlation time of water molecules; flipping of aromatic rings |
| <1 ns | | local | free side-chain rotation, lifetime of H-bonds |
| 1 ns – 1 μs | <0.5 nm | medium | movement of protein loops. |
| | | | quasi-harmonic twisting of β-sheets and flexing of α-helices |
| many ns – 1 μs | a few nm | medium | small peptide folding, helix/coil transition |
| many μs – 1 ms | <1 nm | medium | subunit (allostery) and domain (hinge) motions |
| 1 μs – 1 ms | >1 nm | large | membrane transfer of proteins |
| ms – many min | >10 nm | large | protein subunit association |
| 1 ms – >1 s | >10 nm | large | protein folding and unfolding |

### 16.3.5.1 Fast Steps in Protein Folding: Secondary Structure Formation

The formation of α-helices in solution has been studied in considerable detail and reveals many of the key requirements for any secondary structure to form. An isolated α-helix in solution is only marginally stable: it is in dynamic equilibrium with the unfolded (random coil) state with rate constants of about $10^5$–$10^7$ s$^{-1}$. While the rate of helix formation is not dependent on the final length of the helix, the unfolding of helices is strongly length-dependent: the longer the helix, the slower the unfolding. Two models have been developed to explain the kinetics of helix-random coil interconversion. The *Zimm-Bragg model* looks at H-bond formation in the helix and helix extension, whereas the *Lifson-Roig model* defines helicity by φ/ψ angles, not by H-bonding. This model can include Pro residues (that cannot form hydrogen bonds in the helix) and other discontinuities of the helix. In both models, the rate-limiting step is helix initiation, which can in principle occur anywhere in the sequence but is slower than helix extension by a factor of 300–500. The slow initiation is explained by the fact that at least five residues must adopt α-helical conformation to form just a single *i, i + 4* H-bond that spatially fixes the intervening three residues. By contrast, helix extension only requires the fixing of the conformation of one additional residue at a time, and this residue is the next in the sequence and thus close by. Taken together, the entropic penalty for helix initiation is much higher than for helix extension. There are also enthalpic arguments to explain the slow helix initiation. The dipoles of the peptide bonds are aligned parallel to each other in the α-helix (see Figure 16.17). For the first helix-initiating turn, the dipoles are placed side-by-side, without a displacement along the yet to be established helix axis, which is not a favorable arrangement (see Figure 15.9). Only after the helix extends to more than one turn do the dipoles interact favorably in a head-to-tail orientation from one turn to the next, creating the helix macro-dipole.

In the Zimm-Bragg and Lifson-Roig models, all amino acids have the same probability of incorporation into the helix. This is not the case for the charged amino acids Glu and Asp that interact with the helix dipole: Glu and Asp are about four times more likely to extend the helix at the N-terminus than at the C-terminus. Also, the growing α-helix zips up in both directions until a breakpoint such as Pro or a capping residue (Gly as a C-cap for example) is reached. At this point, the probability of incorporation increases for any residue at the opposing end of the growing α-helix. Furthermore, polar side chains (Ser, Thr, Asn) can H-bond to the backbone and thus interfere with helix elongation. The Zimm-Bragg and Lifson-Roig models also do not include interactions between side chains, which can significantly influence helix stability. The propensities of individual residues to form α-helices have been studied by changing individual positions in short peptides to all 20 amino acids and determining the helix content using CD spectroscopy (see Table 16.3).

The zipper-like extension of secondary structure also applies to super-secondary structures such as the helical coiled-coils and the collagen triple helices. β-sheets also form in a similar manner to helices. They are initiated by a minimum of three residues in the extended conformation (see Figure 16.22) and then extend rapidly in both directions until a break is reached, e.g. Pro, that folds into a turn. It has been suggested that turns (see Figure 16.28) are important decision points in protein folding because they can define the direction that the polypeptide chain takes. The strongly preferred residues Pro and Gly in turns can therefore have strong guidance for folding of the entire protein.

### 16.3.5.2 Rate-Limiting Steps and Protein Folding *In Vivo*

Among the slowest steps in protein folding are prolyl isomerization, disulfide bond formation, and folding of large domains that are prone to aggregate. *In vivo*, these reactions are catalyzed by specific sets of enzymes, the *prolyl isomerases* (PPIs), *protein disulfide isomerases* (PDIs), and *chaperones*. We will only outline the principles of action of these enzymes.

Although ribosomal protein synthesis defaults to the *trans*-configuration, *cis*-peptide bonds (see Figure 16.4) are quite common in proteins. *Cis*-peptides are formed slowly at physiological pH with rate constants of only 0.1–0.01 s$^{-1}$ (the reaction is much accelerated at extremes of low and high pH), so spontaneous prolyl isomerization can be the rate-limiting step during protein folding. PPIs are very efficient enzymes with $k_{cat}/K_M$ values of $10^5$–$10^6$ M$^{-1}$s$^{-1}$, accelerating the *trans/cis* isomerization by a factor of $>10^6$. Proteins with *cis*-peptides fold significantly faster in the presence of PPI, because PPIs accelerate this rate-limiting step. Folded forms of proteins with non-native *trans*-prolyl peptide bonds do exist and have a native-like structure, and in the case of enzymes these may have partial activity. An example is RNase A, which has four Pro residues and two *cis*-prolyl peptide bonds in its native state. A folding intermediate with a *trans*-prolyl peptide bond at Pro93 has a compact native-like structure as detected by NMR hydrogen exchange (Section 16.3.5.3) and it has some RNase activity. The complete folding scheme of RNase A encompasses all 16 possible *cis/trans* isomers.

We have encountered the important role of disulfide bond exchange for protein folding when discussing the Anfinsen experiment (Section 16.3.2). The number of possible disulfide pairings increases with the number of Cys residues. Not all Cys may be tied up in disulfide bonds and only a single set of disulfide bonds corresponds to the native state. PDIs both facilitate the (*de novo*) introduction of disulfide bonds and the exchange of incorrectly formed disulfide bonds. In eukaryotes, disulfide bond formation takes place in the endoplasmic reticulum, which has a less negative redox potential (– 190 mV) than the cytoplasm (–230 mV), and thus facilitates Cys oxidation (see Section 5.3). PDIs display broad substrate specificity, including multi-domain proteins and oligomers, and have no sequence specificity, which is important so that any folding protein can settle into its thermodynamically favored disulfide-bonded pattern. The active sites of PDIs contain a reactive pair of Cys in a CXXC sequence that forms the first turn of a long α-helix. The influence of the helix dipole causes unusually low $pK_A$ values of these Cys residues. Introduction of disulfide bonds starts from the oxidized PDI, leaving the PDI in a reduced state. In eukaryotes, the reduced PDI is re-oxidized by endoplasmic reticulum oxidoreductin 1 (Ero1). Proteins that spontaneously fold into a near-native state will bring the correct Cys residues together, so only oxidation needs to be catalyzed by PDI. For incorrectly formed disulfide bonds, catalysis of disulfide exchange can proceed *via* all possible intermediates, regardless of whether they involve conformational changes in the protein substrate or not (Figure 16.51).

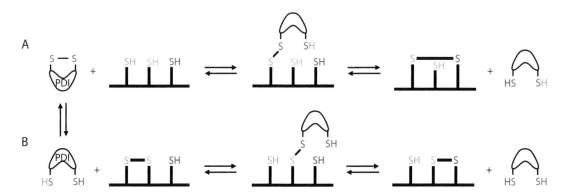

**FIGURE 16.51:    Disulfide bond formation and exchange during protein folding.** A: Disulfide bond formation at the expense of oxidized PDI. Oxidized PDI forms a mixed disulfide bond in an enzyme-substrate intermediate. Intramolecular disulfide bond formation during protein folding releases reduced PDI. B: PDI-catalyzed disulfide exchange *via* a mixed disulfide intermediate. PDI-catalyzed disulfide bond formation and exchange (shuffling) can accelerate the folding by a factor of 100. In bacteria, separate enzymes are used for introduction and shuffling of disulfide bonds.

"Chaperones" is the collective term for a family of unrelated proteins that aid the folding and assembly of other proteins but that are not part of their final native state. In this respect chaperones are true catalysts, but they also shift the equilibrium towards the folded state by expending energy in the form of ATP hydrolysis. Proteins that are associated with chaperones and regulate their activity but that do not hydrolyze ATP are termed co-chaperones. Chaperones bind to the solvent-exposed hydrophobic surfaces typical for unfolded proteins and folding intermediates in order to suppress potentially disease-causing aggregate formation (Section 16.3.6).

The main chaperone systems, Hsp60 and Hsp70, have been defined based on the molecular mass of their heaviest component. Heat shock proteins (Hsps) are chaperones produced as a stress response at elevated temperatures. GroEL, a tetradecamer of two seven-membered rings stacked on top of each other (see Figure 16.32), is the main chaperone of the Hsp60 group in *E. coli*. Together with its co-chaperone GroES, GroEL cordons off a large hydrophobic space to isolate aggregation-prone unfolded proteins from the remainder of the cellular proteins. In an ATP-dependent reaction, the GroEL-GroES complex releases the folding intermediate inside the complex for another attempt at folding, i.e. to lift the folding intermediate out of its local energetic minimum. If the protein has folded, it will have polar surfaces with low affinity for the GroEL-GroES complex, and be released. If the protein still exposes hydrophobic surface the chaperone will re-bind for another round of ATP hydrolysis and protein release. Several binding and release cycles may be required until the native state is reached, or the protein is eventually degraded by proteases, many of which (Lon, Clp, HtrA) are also heat shock proteins.

The *E. coli* chaperone DnaK is a representative of the Hsp70 family of chaperones that acts in conjunction with the co-chaperones DnaJ and GrpE. Similar to GroEL, DnaK binds and hydrolyzes ATP. DnaJ stimulates the ATPase activity, and GrpE is a nucleotide exchange factor that accelerates ADP/ATP exchange after hydrolysis. The ATP state of DnaK binds hydrophobic regions rapidly, but with low affinity. Hydrolysis of ATP to ADP and phosphate generates a high-affinity state, and the peptides remains stably bound until the ADP is exchanged for ATP. Hsp70 proteins promote assembly and disassembly of oligomers, and keep newly synthesized proteins in an unfolded state that can be translocated through membranes.

### 16.3.5.3 Kinetics of Protein Folding

In Section 16.3.4, the two-state protein folding process with the native state N and the unfolded state U was treated from the equilibrium point of view. Let us now consider the same system from a kinetic perspective. The equilibrium between the N and U states is characterized by the microscopic rate constants $k_{fold}$ and $k_{unfold}$ (Scheme 16.2).

$$N \underset{k_{fold}}{\overset{k_{unfold}}{\rightleftarrows}} U$$

scheme 16.2

The description of a reversible reaction according to Scheme 16.2 is discussed in detail in Section 8.1. Here, we apply this knowledge to protein folding. The equilibrium constant $K_{unfold}$ is described by the ratio of the folded and unfolded protein concentrations or by the ratio of the rate constants:

$$K_{unfold} = \frac{[U]}{[N]} = \frac{k_{unfold}}{k_{fold}}$$

eq. 16.23

When we start from a solution of a native protein and rapidly dilute it into a high concentration of denaturant, we can follow the unfolding reaction as a function of time, using any of the spectroscopic probes that are also suitable for measuring equilibrium protein unfolding. Similarly, we can follow the folding reaction if we start from a solution of unfolded protein that we rapidly dilute into native conditions. In either case, the observed rate $k_{obs}$ at each denaturant concentration will be an apparent rate constant that is a composite of $k_{fold}$ and $k_{unfold}$ at that particular denaturant concentration:

$$k_{obs} = k_{fold} + k_{unfold}$$

eq. 16.24

$k_{obs}$ is obtained by fitting an exponential function (Chapter 7) to each unfolding trace. Plotting the logarithms of the $k_{obs}$ as a function of denaturant concentration results in a V-shaped plot that has been named the *chevron plot* (Figure 16.52), from which the true microscopic rate constants $k_{fold}^0$ and $k_{unfold}^0$ in the absence of any denaturant can be extracted.

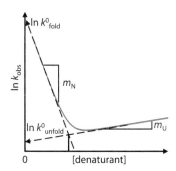

**FIGURE 16.52:    Chevron plot.** A plot of the logarithm of the apparent rate constant of folding $\ln(k_{obs})$ as a function of denaturant concentration has a V-shape. Extrapolations of the linear regions to zero denaturant yield the logarithms of the microscopic rate constants $k^0_{fold}$ and $k^0_{unfold}$. The intersection of the linear extrapolations is the denaturant concentration at the midpoint of the transition.

The chevron plot is V-shaped because both microscopic folding and unfolding rate constants are affected by the denaturant, an effect that has nothing to do with the folding process itself. Specifically, the logarithms of the rate constants $k_{fold}$ and $k_{unfold}$ change linearly with the denaturant concentration:

$$\ln k_{fold} = \ln k_{fold}^0 - m_N \cdot [D]$$

eq. 16.25

$$\ln k_{unfold} = \ln k_{unfold}^0 + m_U \cdot [D]$$

eq. 16.26

The values of the slopes $m_N$ and $m_U$ describe how strongly the rate constants depend on the denaturant D. The intersections of these lines with the $y$-axis are the logarithms of the microscopic rate constants in the absence of denaturant. If we combine equations eq. 16.24–eq. 16.26, we obtain the general expression for the chevron plot.

$$\ln k_{obs} = \ln\left(k_{fold} + k_{unfold}\right) = \ln\left(k_{fold}^0 \cdot e^{-m_N \cdot [D]} + k_{unfold}^0 \cdot e^{m_U \cdot [D]}\right)$$    eq. 16.27

For low denaturant concentrations, $\ln(k_{obs})$ approaches $\ln(k_{fold}^0 + k_{unfold}^0)$, meaning the sum of the microscopic rate constants is obtained from the $y$-axis intercept. For high denaturant concentrations, $\ln k_{obs}$ is dominated by $\ln(k_{unfold}^0 \cdot e^{m_U \cdot [D]})$, which is equal to $\ln(k_{unfold}^0 \cdot m_U \cdot [D])$, and the microscopic rate constant for folding can be obtained from extrapolation of the linear dependence of $k_{obs}$ on $[D]$ to the $y$-axis (Figure 16.52).

Jumping from native to unfolding conditions or *vice versa* is a general technique to study protein folding. Solutions of folded proteins can be diluted into high denaturant concentrations to induce unfolding. Laser flashes can be used to induce defined temperature jumps (*T*-jumps) in protein solutions (Section 25.4). Similarly, pressure jumps (*p*-jumps) of several thousand bar can be applied to modulate protein refolding (see Box 25.3): since the unfolded state has a larger volume than the native state (principle of Le Chatelier), pressure increase can induce folding while a reduction in pressure may promote unfolding. Slow folding reactions with half-lives of >10 s can be initiated by manual mixing and then be followed in standard spectrometers. Fast reactions, however, require stopped-flow (Section 25.1) or quench flow systems (Section 25.2) to rapidly initiate and follow the reactions on the millisecond time scale. With relaxation methods, reactions with half-lives in the microsecond range can be monitored (see Box 25.3).

The rapid generation of secondary structures and the slower subsequent steps of prolyl isomerization, disulfide bond formation, etc. can be measured by the transient kinetic methods detailed in Chapter 25. If the rate constants of each step are sufficiently different, the kinetic folding transition will have several phases that can be evaluated individually by a multi-exponential curve fit. An example is folding of proteins that require the slow *cis/trans* isomerization of peptide bonds. It is not uncommon for the unfolded state to have both *cis*-prolyl and *trans*-prolyl peptide bonds. If the native state contains a *cis*-prolyl peptide, the folding kinetics will be biphasic. The fast phase (*burst phase*) corresponds to those polypeptides that already have the correct *cis*-peptide, and the slow phase to those that need to first form a *cis*-peptide before fast folding can take place.

While all these methods monitor global folding of the protein, *local protein folding* can be monitored by NMR spectroscopy in H/D exchange experiments (see Section 20.1.7). An unfolded protein is diluted in a native buffer containing $D_2O$, and the exponential decay of resonances from amide protons is monitored as a function of time. Amide protons from the peptide groups and the polar side chains can be replaced by a deuteron from the solvent, which is NMR-silent. Amide proton exchange rates are a measure of the accessibility of the protein backbone. Peptide groups whose amide protons exchange rapidly are not protected by the structure of the protein or a folding intermediate. Amide protons of peptide groups that do not or only very slowly exchange with deuterons are probably part of the hydrophobic core or of an early folding intermediate (see Section 20.1.7). Different peptide groups can thus be classified into early and late folding. The only requirement for these jump-techniques to work is that the folding/unfolding is reversible under all conditions.

### 16.3.5.4 Folding Intermediates in Monomers and Oligomers

Nearly all proteins will fold through intermediates. For small proteins such as SH3 domains, the intermediates are often too short-lived and are not detected: these proteins show a simple two-state folding behavior and single exponential folding and unfolding reactions. Biphasic folding reactions, on the other hand, are not in agreement with the canonical two-state model, and point to the presence of intermediates along the folding pathway. A common property of folding intermediates is exposure of hydrophobic surfaces to the solvent, which can lead to aggregation as a competing reaction of folding. The nucleation-condensation model proposes that a small number of residues nucleate some key secondary structure elements, which is followed by a collapse of the polypeptide chain to a molten globule. A molten globule is flexible ("molten") and adopts a condensed globular shape. Molten globules have substantial native-like secondary structure content with a topology close to the native state, but they still lack many specific tertiary interactions and are thus not functional (enzymes are inactive). The key stabilizing interactions of molten globules are non-specific hydrophobic interactions between non-polar side chains that are concentrated at the core of the molten globule. Water is present throughout the entire structure, which results in a Stokes radius (see Section 26.2.2) that is 10–15% larger than that of the corresponding folded protein. Only during later stages of folding is the water completely excluded from the hydrophobic core, and the tertiary interactions characteristic for the folded state are established. In multi-domain proteins, each domain may fold individually directly or *via* a molten globule and then associate with other domains to form a compact structure.

The folding of oligomers usually proceeds *via* n-mer intermediates, not sequentially by addition of one monomer to another (Scheme 16.3). A tetramer T, for instance, will have dimers (D) along its folding pathway, possibly with different conformations (D') than those present in the folded tetramer.

$$4\,M_U \rightleftharpoons 4\,M_f \rightleftharpoons 2\,D' \rightleftharpoons 2\,D \rightleftharpoons T' \rightleftharpoons T \qquad \text{scheme 16.3}$$

Similarly, hexamers, dodecamers, and 24-mers have dimers, trimers, tetramers, and hexamers as intermediates. Folding and assembly of oligomers bears the danger of aggregation, which is usually a reaction of higher order than oligomerization. At high monomer concentrations, aggregation is therefore favored over oligomerization.

A beautiful connection between folding of a monomeric protein and the assembly of oligomers is that both processes bury about the same percentage of surfaces. The BSA upon folding is the difference between the ASAs of the unfolded and the folded states (Section 16.2.6.1), and can be approximately calculated from the molecular mass (eq. 16.28).

$$A_{\text{buried}} = 0.0145 \cdot M_M - 0.111 \cdot M_M{}^{2/3} \qquad \text{eq. 16.28}$$

$A_{\text{buried}}$ amounts to about 0.9 nm$^2$ per residue for a protein of 50 kDa molecular mass, and about 1.2 nm$^2$ per residue for a 35 kDa protein. Oligomers also bury about 1.2 nm$^2$ per residue, which means that oligomerization is a way to achieve the same degree of surface burial as that observed in small monomeric proteins.

## 16.3.6 PROTEIN FOLDING DISEASES

Proteins with a small negative value for $\Delta G_{\text{fold}}$ are prone to unfold at slightly elevated temperatures. Mutations in proteins can make $\Delta G_{\text{fold}}$ less negative, so that a substantial fraction of the protein may be in the unfolded state at physiological temperatures. The presence of unfolded proteins in the cell is often linked to disease. There are two possible reasons for these detrimental effects: loss of function of the folded state or gain of function of the unfolded state. The diseases are more severe in homozygous patients where both copies of the affected genes are mutated, and only the destabilized protein variant is synthesized.

Loss of function often results from a mutation in the hydrophobic core that destabilizes the whole protein. An example is the Leu444Pro mutation in glucocerebrosidase, which leads to Gaucher disease. When glucocerebrosidase is mutated (or absent), its substrates (glycosphingolipids) accumulate in the lysosome. Together with Fabry disease and Tay-Sachs disease, where degradation of glycosphingolipids is also impaired, Gaucher disease belongs to the class of *lysosomal storage diseases*. There are about 200 mutations known to cause Gaucher disease, most of which affect folding of glucocerebrosidase to different extents, which in turn leads to differently severe outcomes.

The tumor suppressor p53 is a transcription factor that can activate DNA repair, initiate apoptosis, and arrest the cell cycle. p53 is involved in so many cellular processes that it has been dubbed the *guardian of the genome*. Its rather low melting temperature of around 45°C is beneficial for rapid regulation of p53 activity. On the other hand, the marginal stability of p53 has the downside that mutations readily lead to temperature-sensitive variants of p53 that are unstable and unfolded to different extents at the physiological temperature of 37°C. About 30% of all cancer-related p53 mutations lead to temperature-sensitive variants. In principle, it could be possible to "back-stabilize" the slightly destabilized mutated proteins using a ligand that binds to and stabilizes the folded state of p53, which would be an approach towards treatment of these diseases. This has been attempted for the cancer-causing Y220C mutation in p53 where a cavity on the surface of p53 created by the mutation was filled by a small molecule.

A third example of folding diseases is caused by mutations of collagen, which we know from our treatment of helical secondary structures (Figure 16.21). *Osteogenesis imperfecta* is a disease caused by mutations in collagen that hinder the close approach of the three strands forming the collagen helix. In principle, scurvy is a type of a collagen folding disease, where a lack of vitamin C leads to insufficient Pro-hydroxylation and destabilization of the triple helix due to the lack of inter-strand H-bonds.

Gain of function, on the other hand, can be observed when unfolded proteins or their degradation products adopt new functions. Molecular crowding favors protein (re)folding but it also favors aggregation. If the concentration of an unfolded protein or one of its degradation products reaches a critical level, it may aggregate into regular structures and form *amyloid plaques*. In Alzheimer's disease for instance, the combined action of the two proteases BACE-1 and γ-secretase on the amyloid precursor protein APP liberates a 4 kDa peptide of just forty residues termed Aβ. The peptide can assemble into amyloids, which are infinite β-sheets of stacked β-strands that have a superhelical twist. The plaques have been implicated in Alzheimer disease progression. A few more examples are collected in Table 16.10.

**TABLE 16.10**
**Examples of protein folding diseases.**

| disease, protein | remarks |
|---|---|
| hypercholesterolemia, low-density lipoprotein receptor LDLR | LDL is a major cholesterol-carrying lipoprotein. Non-functional LDLR (>200 mutations known) leads to high blood levels of cholesterol. |
| cystic fibrosis. cystic fibrosis transmembrane regulator | Integral membrane Cl⁻ channel. ~1500 gene mutations known, ⅔ of all patients have ΔF508. |
| phenylketonuria. phenylalanine hydroxylase | >400 gene mutations known. |
| Marfan syndrome, fibrillin | Fibrillins are part of the connective tissue between cells and regulate transforming growth factor beta (TGF-β) bioavailability. Mutants do not bind to TGF-β. |
| osteogenesis imperfecta and scurvy, procollagen | See text. |
| sickle cell anemia, hemoglobin | $\alpha_2\beta_2$ heterotetramer. The β-chain Glu6Val mutant polymerizes at low $pO_2$ |
| Tay-Sachs disease, β-hexosaminidase | Lysosomal storage disease. The enzyme can no longer metabolize GM2 gangliosides. |
| *retinitis pigmentosa*, rhodopsin | >100 gene mutations known to cause this autosomal dominant disease. |
| cancer, p53 | See text. |
| Alzheimer's disease, amyloid precursor protein APP | APP is a cell surface receptor on neurons that is degraded to amyloid-β peptides Aβ (40–42). Extracellular aggregation of the peptides leads to amyloid plaques that can cause synaptic disconnection. Also, intracellular accumulation of phosphorylated tau protein leads to paired helical filaments and neurofibrillary tangles. |
| cataracts, crystallins | Protein aggregates in the lens, but unclear if these are amyloids. Proteins have aged (deamidation, disulfide-bond formation). |
| TTR amyloidosis Type I, transthyretin | Carrier for thyroid hormones and retinol. Homotetramer. >80 mutations described, most of which are related to amyloid deposition. |
| Huntington's disease, huntingtin | Possible role in microtubule-mediated transport or vesicle function. Genetic CAG-expansion leads to poly-Gln huntingtin. The Gln-rich sequences can aggregate to infinite β-sheets. |

Not all mutations will cause protein folding problems. Many affect gene transcription, mRNA splicing, protein maturation and transport. In several cases more than one gene can cause the disease when mutated. For instance, hypercholesterolemia can be caused by mutations in the APOB, LDLR, LDLRAP1, and PCSK9 genes.

## QUESTIONS

16.1   Calculate the concentration of neutral Gly in a 1 M solution at pH = pI. The $pK_A$ values for the carboxyl and ammonium groups are $pK_{A1} = 2.3$ and $pK_{A2} = 9.8$, respectively. Is it a good idea to draw the neutral form of free amino acids?

16.2   Calculate the expected percentage of prolyl *cis*-peptide bonds ($\Delta G = -2$ kJ mol$^{-1}$) and non-prolyl *cis*-peptide bonds ($\Delta G = -10$ kJ mol$^{-1}$) at 25°C.

16.3   The three most prominent rotamers of Thr are shown. The rotamer in cyan accounts for 43% in Thr, but the same conformation accounts for 73% of all rotamers in Val. What could be the explanation for this high propensity?

16.4   Protonated 1,3-dicarbonic acids and also 3-ketoacids are unstable. They lose a molecule of $CO_2$ by a pericyclic reaction. Draw the movement of electrons and the intermediates of the decarboxylation reaction. How is this decarboxylation avoided *in vivo*?

16.5   Draw the two alternative Greek key motifs.

16.6   The monomer-dimer equilibrium $2M \rightleftarrows D$ is characterized by the equilibrium constant $K_a$. The total monomer concentration is $M_0$. Derive a formula that relates the mole fraction $M/M_0$ of monomer M to the equilibrium constant $K_a$. Draw a graph $M/M_0$ as a function of $M_0$ with logarithmic scale for $M_0$ and $K_a$ values of 1000 M$^{-1}$ and 10$^6$ M$^{-1}$. Show that the mole fraction is 0.5 for $M_0 = 1/K_a$ ($= K_d$). Using l'Hôpital's rule, show that the mole fraction approaches unity (complete complex dissociation) for $M_0 \rightarrow 0$ and approaches zero (100% complex) for $M_0 \rightarrow \infty$.

16.7   In hydrophobic environments, a set of five water molecules often forms a planar ring. What could be the explanation? The H-O-H angle in water is 104.5°.

16.8   GppNHp is a non-hydrolyzable GTP analog where the oxygen atom bridging the β- and γ-phosphate is replaced by a NH group. In the figure below, an H-bond is drawn between a peptide NH group of the P-loop and the N-atom of GppNHp. Why is this allowed?

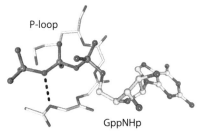

16.9   Given is the scheme for alternative ligand binding by induced fit and conformational selection:

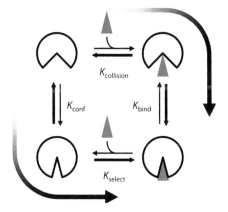

Express the binding constant $K_{\text{bind}}$ by the other three equilibrium binding constants. Why is $K_{\text{select}}$ much larger than $K_{\text{collision}}$? $K_{\text{conf}}$ is dominated by thermal energy, which interconverts the conformations. This energy is then kept upon ligand binding. Would you expect $K_{\text{conf}}$ to be larger or smaller than unity?

16.10   When an amphiphilic molecule is mixed with a little water and a large excess of oil, another type of micelles, *inverted micelles*, is formed. What would be the general structure of such a micelle and what are the driving forces for its formation?

16.11   Calculate the minimum number of residues for an $\alpha$-helix to perpendicularly traverse a membrane of 3 nm mean thickness.

16.12   Why can a single $\beta$-strand not exist in a membrane? What would be the shortest $\beta$-barrel to traverse a membrane?

16.13   Why is guanidinium chloride less destabilizing than guanidinium thiocyanate?

16.14   Guanidinium sulfate stabilizes proteins, what does this tell us about the relative effect of anions *versus* cations in the Hofmeister series?

16.15   If RNase A is (1) denatured and reduced, (2) re-oxidized by air in the presence of denaturant, and (3) dialyzed into native conditions, the resulting scrambled RNase A has 1% activity compared to the starting material. What could be the explanation? Remember that native RNase A has eight Cys tied in four disulfide bonds.

16.16   Why is it not always advisable to purify proteins from thermophilic microorganisms at $4°C$?

16.17   The stability of a protein is given by

$$\Delta G_{\text{stab}}(\text{T}) = -RT \ln K_{\text{unfold}} = \Delta H_m\left(1 - \frac{T}{T_m}\right) + \Delta C_p\left(T - T_m - T \cdot \ln \frac{T}{T_m}\right)$$

Using analytical calculus, show that the temperature of maximum stability is at

$$T = T_m \cdot \exp\left(-\frac{\Delta H_m}{T_m \cdot \Delta C_p}\right)$$

## REFERENCES

Creighton, T. E. (1992) Proteins: Structures and Molecular Properties, Freeman.
· a classic book on protein structure and interactions

Creighton, T. E., Ed. (1992) Protein folding. New York, Freeman.

Pain, R. H., Ed. (1994) Mechanisms of Protein Folding. Oxford, IRL Press.
· comprehensive references on all aspects of protein folding

Liu, C. C. and Schultz, P. G. (2010) Adding new chemistries to the genetic code. *Annu Rev Biochem* **79**: 413–444.
· review on the introduction of modified amino acids in proteins

Chothia, C. (1984) Principles that determine the structure of proteins. *Annu Rev Biochem* **53**: 537–572.
· early review on protein structure

Berkholz, D. S., Driggers, C. M., Shapovalov, M. V., Dunbrack, R. L., Jr. and Karplus, P. A. (2012) Nonplanar peptide bonds in proteins are common and conserved but not biased toward active sites. *Proc Natl Acad Sci U S A* **109**(2): 449–453.
· statistical analysis of distorted peptide bonds in protein structures

Eisenberg, D., Weiss, R. M., Terwilliger, T. C. and Wilcox, W. (1982) Hydrophobic moments and protein structure. *Faraday Symposia of the Chemical Society* **17**(0): 109–120.

Kyte, J. and Doolittle, R. F. (1982) A simple method for displaying the hydropathic character of a protein. *J Mol Biol* **157**(1): 105–132.
· two hydropathy scales

Crick, F. H. C. (1953) The Packing of α-Helices: Simple Coiled-Coils. *Acta Cryst* **6**: 689–697.
· first description of the packing of α-helices - this work helped Crick later in the determination of the helical DNA structure

Lupas, A. N. and Gruber, M. (2005) The structure of α-helical coiled coils. *Adv Protein Chem* **70**: 37–78.
· review on α-helical coiled-coils

Richardson, J. S. and Richardson, D. C. (1988) Amino acid preferences for specific locations at the ends of α-helices. *Science* **240**(4859): 1648–1652.
· review on helix capping

Chou, P. Y. and Fasman, G. D. (1977) β-turns in proteins. *J Mol Biol* **115**(2): 135–175.
· survey of turns in protein structures

Sibanda, B. L. and Thornton, J. M. (1985) β-hairpin families in globular proteins. *Nature* **316**(6024): 170–174.
· review on β-hairpins in proteins

Bhaskara, R. M. and Srinivasan, N. (2011) Stability of domain structures in multi-domain proteins. *Scientific reports* **1**: 40.
· linkers in proteins

Copley, R. R. and Bork, P. (2000) Homology among (βα)$_8$ barrels: implications for the evolution of metabolic pathways. *J Mol Biol* **303**(4): 627–641.
· review on TIM-barrels

Lawrence, M. C. and Colman, P. M. (1993) Shape complementarity at protein/protein interfaces. *J Mol Biol* **234**(4): 946–950.
· introduces the surface complementarity coefficient

Chakrabarti, P. and Janin, J. (2002) Dissecting protein-protein recognition sites. *Proteins* **47**(3): 334–343.
· review on protein-protein recognition sites

Chothia, C. and Janin, J. (1975) Principles of protein-protein recognition. *Nature* **256**(5520): 705–708.
· early review on protein-protein interactions

Jaenicke, R. and Böhm, G. (1998) The stability of proteins in extreme environments. *Curr Opin Struct Biol* **8**(6): 738–748.
· review on the stability of proteins from extremophiles

Dunitz, J. D. (1994) The entropic cost of bound water in crystals and biomolecules. *Science* **264**(5159): 670.
· entropy of bound water molecules

Davis, B. D. (1958) On the importance of being ionized. *Arch Biochem Biophys* **78**(2): 497–509.
· a classic read on compartmentalization of metabolites in cells

Wright, P. E. and Dyson, H. J. (1999) Intrinsically unstructured proteins: re-assessing the protein structure-function paradigm. *J Mol Biol* **293**(2): 321–331.
· review on intrinsically unfolded proteins

Go, Y. M. and Jones, D. P. (2008) Redox compartmentalization in eukaryotic cells. *Biochim Biophys Acta* **1780**(11): 1273–1290.
· review on disulfide formation *in vivo*

Pace, C. N. (1995) Evaluating contribution of hydrogen bonding and hydrophobic bonding to protein folding. *Meth Enzymol* **259**: 538–554.
· contribution of H-bonds to protein folding

Matthews, B. W. and Liu, L. (2009) A review about nothing: are apolar cavities in proteins really empty? *Protein science : a publication of the Protein Society* **18**(3): 494–502.
· review on non-polar cavities in proteins

Bolen, D. W. and Santoro, M. M. (1988) Unfolding free energy changes determined by the linear extrapolation method. 2. Incorporation of $\Delta G°_{N-U}$ values in a thermodynamic cycle. *Biochemistry* **27**(21): 8069–8074.

Santoro, M. M. and Bolen, D. W. (1988) Unfolding free energy changes determined by the linear extrapolation method. 1. Unfolding of phenylmethanesulfonyl α-chymotrypsin using different denaturants. *Biochemistry* **27**(21): 8063–8068.
· first mathematical description of protein folding and unfolding

Schmid, F. X. (2001) Prolyl isomerases. *Adv Protein Chem* **59**: 243–282.
· review on prolyl isomerases

Jaenicke, R. (1987) Folding and association of proteins. *Prog Biophys Mol Biol* **49**(2–3): 117–237.
· folding of oligomers

Basse, N., Kaar, J. L., Settanni, G., Joerger, A. C., Rutherford, T. J. and Fersht, A. R. (2010) Toward the rational design of p53-stabilizing drugs: probing the surface of the oncogenic Y220C mutant. *Chem Biol* **17**(1): 46–56.
· a destabilized, cancer-causing variant of p53 was stabilized by a small molecule - stabilization of destabilized proteins by small molecules could be a way to treat disease in general

## ONLINE RESOURCES

pir.georgetown.edu/resid
· database of post-translational modifications

people.ucalgary.ca/~tieleman/download.html
· structural models of micelles

heller.userweb.mwn.de/membrane/membrane.html
· structural models of membranes

www.abren.net/protherm
· protherm database

# 17

# Nucleic Acids

T here are two types of nucleic acid, *ribonucleic acid* (RNA) and *deoxy-ribonucleic acid* (DNA), named after their predominant localization (in the nucleus) and their acidic character. The main function of DNA is to store genetic information, although some viruses use RNA for this purpose. RNA has a wider range of functions as messenger RNA (mRNA) and transfer RNA (tRNA) during gene expression and translation. *Ribozymes* are enzymatically active RNA molecules, probably the first enzymes on the planet that existed billions of years ago before any proteins were around. RNAs also function as structural and catalytic components in *ribonucleoprotein* complexes such as ribosomes, spliceosomes, and telomerases. Over the last decade, many non-coding or *regulatory RNAs* have been discovered that regulate genome organization and gene expression.

Nucleic acids are polymers of nucleotides or in short *polynucleotides*. Similar to amino acids in proteins, it is the nucleotide sequence that determines the final structure, function, and information content of nucleic acids. We will begin this chapter with a chemical and geometric view on the constituents of nucleic acids, their assembly into secondary and tertiary structures, and finish with a brief description on the interconversion of DNA topoisomers and the folding of RNA (Figure 17.1).

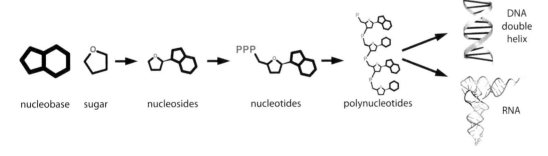

FIGURE 17.1: **From simple organic groups to nucleic acids.** Nucleobases combine with sugars to nucleosides, which are phosphorylated to nucleotides and then polymerized into polynucleotides. Two strand of a polynucleotide can form a double helix or, in the case of RNA, a single strand can fold into a complex three-dimensional shape.

## 17.1 NUCLEOBASES, NUCLEOSIDES AND NUCLEOTIDES

The number of canonic building blocks that make up nucleic acids is smaller than for proteins: just four nucleobases compared to the twenty standard amino acids. A nucleobase is a flat, aromatic heterocycle, some of which have weakly basic character, hence the name nucleo*base*. The four nucleobases found in DNA are adenine (A), guanine (G), cytosine (C), and thymine (T). RNA contains uracil (U) instead of T (Figure 17.2). U and T are identical but for the extra methyl group in T. The $pK_A$ values of protonated G and C are in the range of carbonic acids (acetic acid has $pK_A$ = 4.7). Protonated G and C are not observed in solution but do occur in base triplets that are shielded from solvent (Section 17.5.2).

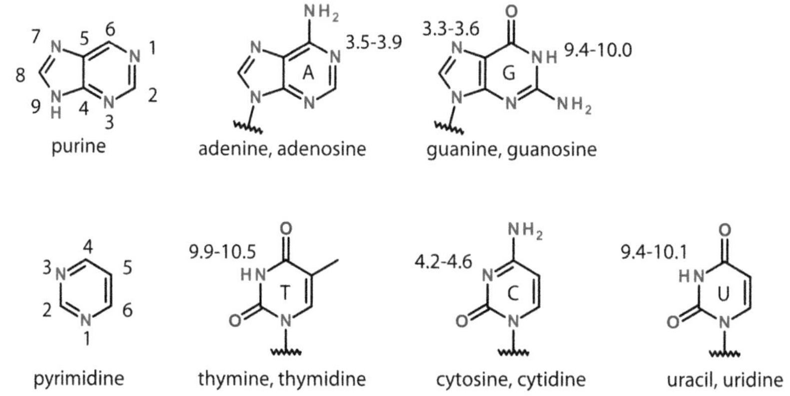

FIGURE 17.2: **Structure and numbering of nucleobases.** Adenine and guanine are derivatives of purine. Using the purine numbering system, adenine is 6-amino purine. Thymine, cytosine, and uracil are derivatives of pyrimidine. Note that uracil is thymine that lacks the methyl group at the 5-position. The most stable tautomers of the bases are shown. The wavy line indicates the glycosidic bond to the (deoxy)ribose in the nucleosides adenosine, guanosine, thymidine, cytidine, and uridine. Number ranges next to the nitrogen atoms in rings are $pK_A$ values found in nucleosides and nucleotides.

The unit comprised of a nucleobase and a (deoxy)ribose is termed *nucleoside*. In purines it is the N9-atom and in pyrimidines the N1-atom that forms a *glycosidic bond* with the C1′-atom of a ribose sugar (Figure 17.3). *Nucleotides* are phosphorylated nucleosides. For instance, ATP is adenosine 5′-triphosphate, an adenosine nucleoside that has a string of three phosphoryl groups attached to the 5′-hydroxyl group of ribose. Figure 17.3 illustrates the blueprint of nucleosides and nucleotides

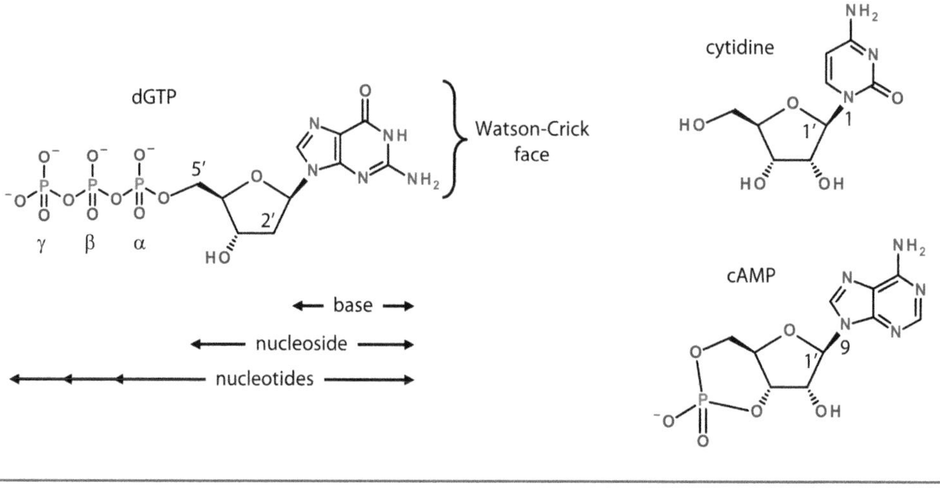

FIGURE 17.3: **Nucleobases, nucleosides, and nucleotides.** The ribose condenses with a nucleobase to form a nucleoside. A nucleotide is a phosphorylated nucleoside. dGTP is 2′-deoxy guanosine 5′-triphosphate (dGTP) and cAMP is cyclic 2′,5′ adenosine monophosphate. For the numbering of the ribose see Section 17.2. Phosphate groups are labeled with Greek letters. The Watson-Crick face contains the H-bond donors and acceptors in nucleobases that form standard base pairs.

with the nucleoside cytidine (no phosphoryl group) and the nucleotides dGTP and cAMP as examples. The nucleotides dGTP, dCTP, dTTP, and dATP with 2′-deoxyribose as the sugar are building blocks of DNA. In contrast, ribose is the sugar in RNA. The phosphoryl group in cAMP, a second messenger in signal transduction, forms a cyclic phosphodiester with two hydroxyl groups of ribose.

The $pK_A$ of the phosphate group is around 2, so all nucleotides are strong acids and are negatively charged under physiological conditions. The term nucleic *acid* stems from this acidic property of the nucleotides. Nucleotide triphosphates have 3–4 negative charges that strongly attract cations. Most nucleotides and also nucleic acids are bound to metal ions ($Mg^{2+}$, $Na^+$) in solution.

The *Watson-Crick* face in nucleobases is the side that forms H-bonds in the standard Watson-Crick base pairs, which are the dominant pairings in all double-helical nucleic acid structures (Figure 17.4). There are two H-bonds in A-T and three in G-C pairs, which makes the G-C pair the more stable one. The H-bonding pattern establishes selectivity for the base pairings.

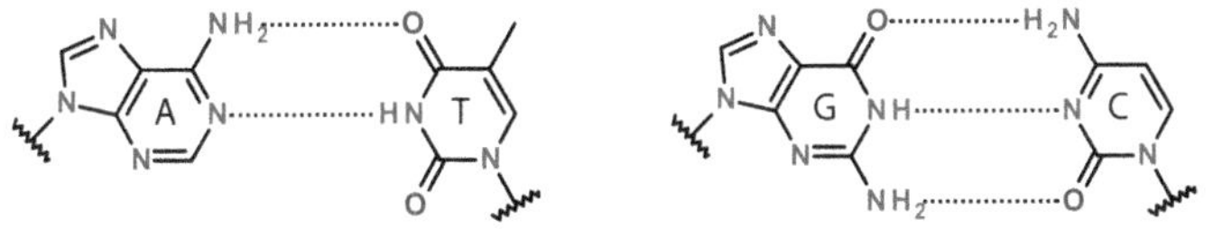

**FIGURE 17.4:    Standard Watson-Crick base pairs.** An AT base pair forms two H-bonds. The more stable GC base pair has three H-bonds. The two pairs are isosteric, i.e. they have roughly the same overall size, which is important when they stack to form a double helix in DNA and RNA. The biologically relevant tautomers of the nucleobases are shown (Box 17.1).

---

### BOX 17.1: NUCLEOBASE TAUTOMERS.

The nucleobase structures shown in Figure 17.2 show the most stable tautomer, which is the configuration that engages in base pairing. Determination of which tautomers of the nucleobases are the physiologically relevant ones was not straightforward. Nucleobases were first drawn in their enolic forms, but these would not align into base pairs. The solution came with the analysis of DNA composition in the 1940s by Erwin Chargaff, who showed that the molar ratio of adenosine and thymidine in double-stranded DNA is always unity ($[A] = [T]$), and the same is true for the ratio of guanosine and cytidine ($[G] = [C]$). These *Chargaff rules* suggested A-T and G-C base pairs. When James Watson realized that A would pair with T and G would pair with C when all nucleobases are formulated in their keto forms, he provided the explanation of the Chargaff rules... (Figure 17.4).

---

## 17.1.1 NON-STANDARD NUCLEOBASES IN DNA

Similar to the post-translational modifications of side chains in proteins, the nucleobases in DNA and RNA are subject to modification, which expands the information content of nucleic acids. One of the most frequent modifications in DNA is methylation (Figure 17.5), as in 5-methyl-cytosine, 5-hydroxymethyl-cytosine, and $N^6$-methyl-adenine. In DNA of higher eukaryotes, the amount of cytosine methylation depends on the organism, its developmental stage, and the tissue. It is estimated that 1.3–1.5% of all cytosine nucleotides in mouse and human brain DNA are methylated. In mammals, the degree of methylation is high during neurogenesis and decreases with age. This additional, sequence-independent level of information encoded in eukaryotic DNA is part of *epigenetics*. The degree of methylation is controlled by DNA methyl transferases. Bacteria use DNA methylation to distinguish their own DNA from foreign DNA, e.g. from bacteriophages. Methylation is also used in bacteria to distinguish parent DNA from recently replicated DNA. If a mismatch occurs during DNA replication, the error is repaired taking the sequence of the methylated parent DNA as the reference.

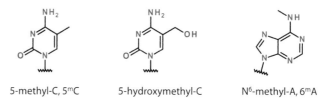

5-methyl-C, 5^mC          5-hydroxymethyl-C          N^6-methyl-A, 6^mA

**FIGURE 17.5:** **Modified bases in DNA.** 5-methyl-C is by far the most frequently modified nucleobase in DNA followed by 5-hydroxymethyl-C and N^6-methyl-A. None of these modifications change the base-pairing abilities of the nucleobase.

## 17.1.2 NON-STANDARD NUCLEOBASES IN RNA

RNA shows much more variation of nucleobases than DNA due to a large variety of post-transcriptional modifications. They are introduced by dedicated RNA-modifying enzymes, either by editing the nucleobases *in situ* or by replacement of the entire nucleobase. The Modomics database lists over 160 different post-transcriptional RNA modifications, most of them in tRNAs (see Section 17.6.2). Modifications that for structural reasons are always present in tRNA include dihydrouracil, pseudouracil, and thymine (Figure 17.6). The saturated C5 and C6 atoms in dihydrouracil lead to nonplanarity (puckering) of the pyrimidine ring. In pseudo-uracil, the N1 and C5 atoms are formally exchanged, leading to an unusual C-glycosidic bond. Thymine is a regular component of DNA, but also occurs in tRNA where it is bound to ribose as ribothymidine. 4-thiouridine has a sulfur atom in place of the usual oxygen. This fluorescent nucleobase happens to be quite reactive and is often used for fluorescent and spin labeling, as well as crosslinking studies.

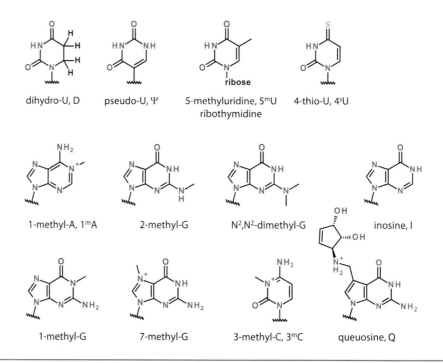

dihydro-U, D    pseudo-U, Ψ    5-methyluridine, 5^mU    4-thio-U, 4^sU
                              ribothymidine

1-methyl-A, 1^mA    2-methyl-G    N^2,N^2-dimethyl-G    inosine, I

1-methyl-G    7-methyl-G    3-methyl-C, 3^mC    queuosine, Q

**FIGURE 17.6:** **Modified bases in tRNA.** These modified bases are present in tRNA where they stabilize the structure by conserved interactions or help decoding mRNA. Methyl groups are indicated as "Me". Some methylations increase the $pK_A$ value of the nucleobase above 7 so that at physiological pH the nucleobase is protonated. If present at the third (*wobble*) position of the anticodon 5'-IAU-3', inosine can pair with A, U, and C and thus decode the mRNA triplets AUA, AUU, and AUC, which all code for Ile. Queuosine replaces the G at the wobble position of tRNAs, with GUN anticodons coding for Asp, Asn, His, and Tyr.

Methylation is another very common modification in tRNA, which can add positive charges, change p$K_A$ values and H-bonding capabilities, and change the stacking properties of the nucleobases. Adenine can be methylated at N1 and N6 (as in DNA), guanine at N1, N2, and N7, cytosine at N3, and uracil at C5 (Figure 17.6). Methylation of adenosine to 1-methyl-adenosine (1$^m$A) increases the p$K_A$ value from 3.5 to 8.3, a decrease in acidity of almost five orders of magnitude. 1$^m$A is therefore charged under physiological conditions and engages in electrostatic and strong stacking interactions. At the same time, the Watson-Crick face cannot form the standard base pair with thymine due to steric repulsion. In addition to tRNA, ribosomal RNA (rRNA) also contains dozens of modifications (mostly methylation), which cluster at the catalytic sites for peptide bond formation and translocation in the ribosome. Eukaryotic mRNA carries a cap-structure at its 5′-end, a methylated G with a 5′-5′-triphosphate that is recognized during the initiation of ribosomal translation.

Several more radically modified nucleobases are specific for particular tRNAs. For example queuosine (Q) bases have a pentenyl ring attached to the N7-atom of guanosine *via* a methyl-amino linker. Q bases are present at the third position of the tRNA anticodon where they pair with U and sometimes also with C. The alternative pairings enable the same tRNA to recognize different codons on the mRNA. The fact that there are less tRNAs in a cell (maximum of 41) than mRNA codons for amino acids (61) has been termed the *wobble hypothesis*. We will return to this hypothesis when we look at the various base-pairing possibilities in Section 17.4.

## 17.2 RIBOSE AND NUCLEOBASE CONFORMATIONS

The nucleobase in nucleotides is bound to a sugar by an N-glycosidic bond. The sugar has a 5-membered ring that resembles the heterocycle furan, hence such sugars are termed *furanoses*. Ribose in RNA and 2′-deoxyribose in DNA are furanoses that only differ by the hydroxyl group at the second carbon atom (Figure 17.7). To distinguish the atoms of the nucleobases from those of the ribose, a prime (′) is used for the sugar numbering.

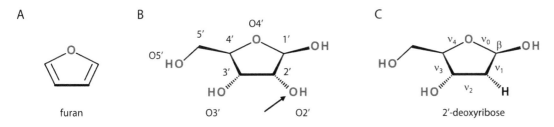

FIGURE 17.7:    **Ribose nomenclature.** A: The saturated furan derivative tetrahydrofuran makes up the framework for furanoses. B: Ribose and 2′-deoxyribose differ only by the hydroxyl group at C2′ (arrow). C: Nucleic acids exclusively contain the diastereomer shown, the β-anomer. The torsion angles within the ribose ring are labeled $v_0$ to $v_4$.

The C1′ atom of ribose is the *anomeric* carbon, which gives rise to two diastereomers termed *anomers*. α-anomers have the 1′-hydroxyl group below the plane of the furanose ring (as drawn in Figure 17.7). Only the β-anomer with the substituent above the furanose ring occurs in nucleic acids. The β-anomeric configuration is fixed in nucleosides by the nucleobase. The torsion angles $v_0$ to $v_4$ in ribose are not as tightly restricted as the peptide backbone angles, but they are still highly correlated: changing one torsion angle automatically induces concomitant rotations about the other four bonds.

### 17.2.1 SUGAR PUCKER

The most energetically favorable conformations of the ribose are the *envelope* and *twist*. Both conformations are non-planar, and the degree of non-planarity is termed *puckering*. A planar 5-membered ring would have all carbon atoms eclipsed, which corresponds to maximal torsional energy (Section 15.2.2). Puckering is a general way to reduce the torsional energy in a ring. In the envelope conformation, four ring atoms are almost co-planar and the fifth atom is above (*endo*) or below (*exo*) this plane. Nucleobases in nucleosides and nucleotides always lie in the *endo* face of the ribose. The *twist* conformation has one atom above and one below the plane defined by atoms C1′, C4′, and O4′ (Figure 17.8). Twist conformations are somewhat less stable than the envelope conformation and sometimes occur in tRNA.

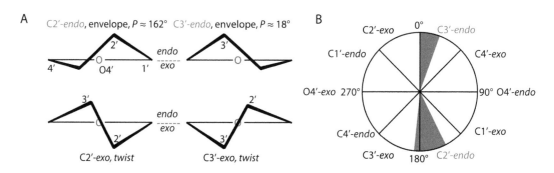

**FIGURE 17.8:** **Ribose conformations.** A: The conformations are defined relative to the plane that is formed by the atoms C1′, C4′, and O4′. The view is side-on to the plane so that the three atoms appear as a line. The conformations are drawn in an exaggerated fashion to highlight their differences. Envelope conformations have four atoms in a plane while in exo-conformations only three out of the five ribose atoms are in plane. B: The circular plot shows how the phase angle of the ribose ring corresponds to the various *endo*- and *exo*- conformations. The energetically preferred C2′-*endo* and C3′-*endo* regions are marked in blue.

The *sugar pucker* is a convenient way to quantify the degree of puckering in nucleic acids by describing the overall conformation of the sugar. It is defined by the five torsion angles $v_0$ to $v_4$ of the ribose ring (Figure 17.7):

$$\text{pucker} = \frac{v_2}{\cos P} \qquad \text{eq. 17.1}$$

The *pseudo-rotation or phase angle P* is defined by the five individual torsion angles as

$$\tan P = \frac{(v_4 + v_1) - (v_3 + v_0)}{2v_2(\sin 36° + \sin 72°)} \qquad \text{eq. 17.2}$$

The most important sugar pucker in nucleic acids is the deviation of the C2′ and C3′ atoms from a plane defined by the other three atoms. C2′-*endo* conformations have phase angles between 144° and 190° and are characteristic for B-DNA (Section 17.5.1). Angles between zero and 18° define C3′-*endo* conformations, which are found in A-DNA and helical RNA structures (Section 17.6). These phase angles correspond to sugar puckers in the 25°−45° range.

### 17.2.2 Syn- and Anti-Conformations

The orientation of the nucleobase with respect to the ribose is a major energetic factor for nucleotide conformation. The torsion angle $\chi$ defines this rotation about the glycosidic bond. In the overwhelming majority of all nucleobases, $\chi$ is in the ranges 90° to 180° and −90° to −180°, the *anti*-conformation. The energetically most favored range is $-180° < \chi < -115°$ for pyrimidines and $-180 < \chi < -60°$ for purines. In the *anti*-conformation, there are only minimal clashes between atoms of the nucleobase and the ribose. $\chi$ angles between −90° and 90° are characteristic of the *syn*-conformation, with the nucleobase above the ring (Figure 17.9).

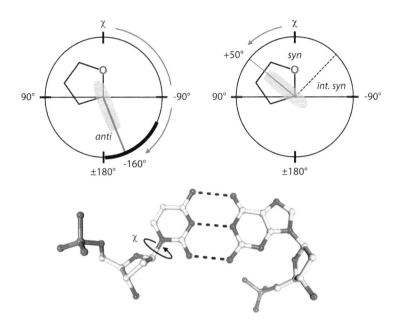

**FIGURE 17.9:    Nucleobase conformations.** The ribose and nucleobase are abbreviated as a tetrahydrofuran and a gray rod, respectively. Rotation about the glycosidic bond can place the nucleobase away from the ribose (*anti*) or on top of it (*syn*). Intermediate *syn*-conformations are the sub-category with $-45° > \chi \geq -90°$. The lower part of the figure shows a CG base pair in Z-DNA conformation. Cytosine is *anti* and guanine is *syn*.

The *syn*-conformation is more compact than the *anti*-conformation but will inevitably place atoms of the nucleobase close to the ribose. Therefore, *syn*-conformations are of higher energy than *anti*-conformations and are much less prevalent than the *anti*-conformations. The clashes between nucleobase and ribose are not too severe for purine nucleotides because of their relatively small 5-membered ring above the ribose. Pyrimidine nucleotides with their larger six-membered ring clash more severely with the ribose in the *syn*-conformation, especially the exocyclic carbonyl oxygen at C4. For these reasons, *syn*-conformations are four times more frequent in purine nucleotides than in pyrimidine nucleotides. Many *syn*-nucleotides participate in tertiary interactions, interact with ligands, or contribute to ribozyme active sites. The *syn/anti*-conformation may even define the overall conformation of DNA: a stretch of guanine nucleotides in the *syn*-conformation can be placed into a left-handed polynucleotide helix, the Z-DNA (Figure 17.17).

## 17.3 PRIMARY STRUCTURE OF NUCLEIC ACIDS

Nucleotides are joined together to form *single-stranded* nucleic acids by phosphodiester bonds that connect the 3′- and 5′-hydroxyl groups of neighboring nucleotides. The sequence of nucleobases defines the primary structure of a nucleic acid, just like the sequence of amino acids defines the primary structure of proteins. The only differences between DNA and RNA are the nucleobase thymine (T) and 2′-deoxyribose in DNA compared to the nucleobase uridine (U) and ribose in RNA. The *polarity* of the strand is given in the 5′→3′–direction.

In single-stranded nucleic acids, adjacent nucleobases engage in hydrophobic *stacking interactions*. The extent of the aromatic system and hence the nature of the nucleobases determines the stability of the stacking interaction: interactions between purines are stronger than between purines and pyrimidines, which in turn are stronger than stacking between two pyrimidines. We will revisit base pair stacking in the context of the DNA double helix (Section 17.5.1).

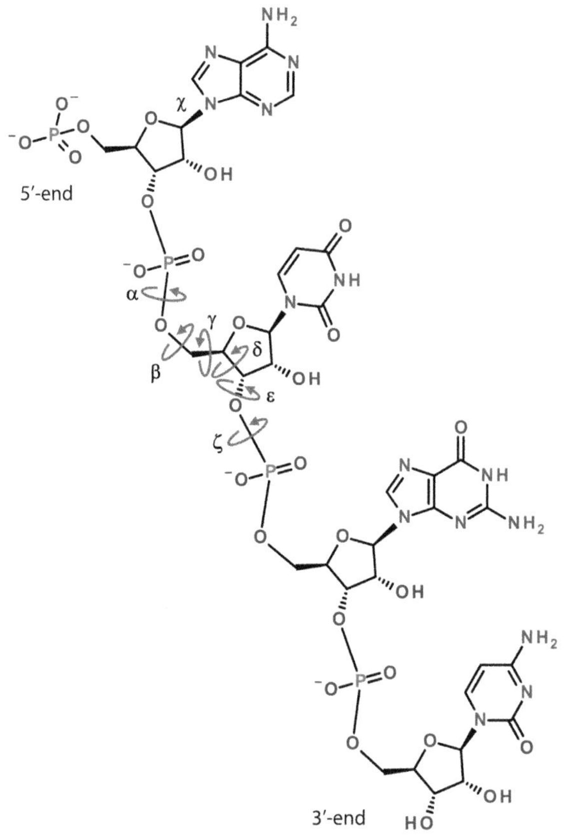

**FIGURE 17.10:** **Primary structure of nucleic acids.** Shown is an RNA tetranucleotide of sequence 5′-AUGC-3′. Sequences are generally written in the 5′→3′–direction. The 5′- and 3′ -ends are chemically distinct and establish strand polarity. The six torsion angles along the phosphodiester backbone are labeled. Note that the δ-angle is identical to the $\nu_3$-angle in Figure 17.7.

We have seen in Section 16.1.2 that we can describe the conformation around a peptide bond and the resulting energy potential by the torsion angles $\Phi$ and $\Psi$. These torsion angles adopt values that correspond to low energies in the Ramachandran energy surface (Figure 16.16). The equivalent potential for nucleic acids is more complicated because there are eleven rotatable bonds per nucleotide, distributed along the phosphodiester backbone and within the ribose (Figure 17.10). Single-stranded nucleic acids are therefore inherently much more flexible than proteins. The torsion angles in nucleic acids adopt values that correspond to energetically favorable conformations. β-angles, for instance, center around $180 \pm 20°$, which corresponds to the energetically favored staggered conformation (Figure 15.5) of the hydrogen atoms at C5′ and the lone pairs at O5′. Some of the torsion angles are also strongly correlated. α-angles center around $300 \pm 20°$ and correlate with a narrow γ-angle distribution of around $50 \pm 10°$. Likewise, the ε-angles are about $190 \pm 20°$ and correlate well with ζ-angles in the $270 \pm 20°$ range. Since the χ-angle of the nucleobase is just opposite to the γ-angle, these two are also highly correlated: the *syn*-conformation correlates with a γ-angle of $180 \pm 30°$.

We have seen in Section 17.2.1 that the set of torsion angles in the ribose ring can be conveniently summarized by a pseudo-rotation angle P (eq. 17.2) with its prominent energetic minima at the *endo-* and *exo*-conformations. This sugar pucker again is correlated with the glycosidic angle χ (*syn-* and *anti*-conformations) of the nucleobases in the double-helical nucleic acid structures (Sections 17.5 and 17.6). Figure 17.11 shows a schematic energy surface for the various P/χ angle combinations, which uses the same concept of two angles as the Ramachandran plot for proteins (Figure 16.16).

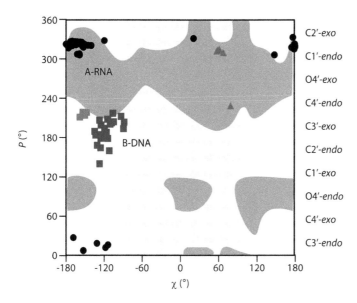

**FIGURE 17.11:    Schematic energy potential for the pseudo-rotation angle *P versus* the glycosidic angle χ.** This kind of representation is analogous to the Ramachandran plot for proteins. The potentials are different for purines and pyrimidines, and also for ribose and 2′-deoxyribose. The yellow areas are the forbidden angle combinations for a 2′-deoxy purine nucleoside where the calculated energy exceeds 24 kJ mol⁻¹. Adapted from Saran *et al*. (1973). *Theor. Chim. Acta* 30: 31–44. Different ribose conformations are marked on the right-hand side. Note that the C2′-*endo* and C3′-*exo* conformations have similar pseudo-rotation angles of ($P \approx 180 \pm 20°$). The same is true for the C2′-*exo* and C3′-*endo* conformations ($P \approx 0 \pm 20°$). Black circles represent the RNA geometry of a riboswitch with mostly A-RNA geometry (PDB-ID 4rzd). RNA can access conformations that are forbidden for DNA. Red squares represent angles found in regular B-DNA (PDB-ID 1bna). Z-DNA (PDB-ID 4ocb) is represented by pyrimidine nucleotides in *anti*-conformation (blue) and purine nucleotides in *syn*-conformation (magenta). The area shaded in gray is the *syn*-conformational space of $-90° < \chi < 90°$.

In summary, the correlation between the torsion angles in nucleotides leads to energetically preferred conformations of the entire polymer (see Table 17.1). While the low-energy torsion angle combinations ensure a minimum of steric clashes between atoms, most of the stabilization energy in nucleic acids structures comes from H-bonds between bases and the stacking of base pairs, which we will take a look at next.

# 17.4  BASE PAIRING AND STACKING

## 17.4.1  H-BONDS BETWEEN NUCLEOBASES

In addition to the canonical GC and AT Watson-Crick base pairs, several other nucleotide orientations allow formation of H-bonds, among them *mismatched* base pairs, reverse Watson-Crick base pairs, Hoogsteen base pairs, and reverse Hoogsteen base pairs (Figure 17.12). In reverse base pairs, the chains of the two strands are parallel (same polarity), putting the glycosidic bonds on opposite (*reverse*) sides of the base pair.

Mismatched base pairs use the Watson-Crick face of the nucleobases but pair with the "wrong" nucleobase. These include G-G, G-A, G-T, A-A, A-C, T-T, T-C, C-C, and G-U. Mismatched base pairs have less or weaker H-bonds than usual base pairs and can lead to replication errors and mutations if they occur in DNA. Some mismatched base pairs are important for the decoding of mRNA by tRNAs. The mismatched G-U pair is particularly important because it allows the same tRNA to read out two different codons (see Figure 17.36). In G-U pairs the U is shifted in the plane of the base pair to match the H-bond donors and acceptors of G and to form two H-bonds (Figure 17.12).

Hoogsteen pairing always involves the N7 atom of a purine in the *Hoogsteen face* (Figure 17.12) and the Watson-Crick face of a pyrimidine. Smaller distances between the C1′-atoms in Hoogsteen pairs compared to Watson-Crick pairs make Hoogsteen pairs unsuited for regular DNA structures. Instead, Hoogsteen pairs are prominent in RNA tertiary structure, Z-DNA, and DNA structures containing more than two strands (Section 17.5.2). Reverse base pairs, particularly reverse Hoogsteen base pairs, are present in DNA triplexes (Section 17.5.2.1) and frequently observed in folded RNA. In reverse base pairs the glycosidic bonds of the nucleotides have opposite directions (Figure 17.12).

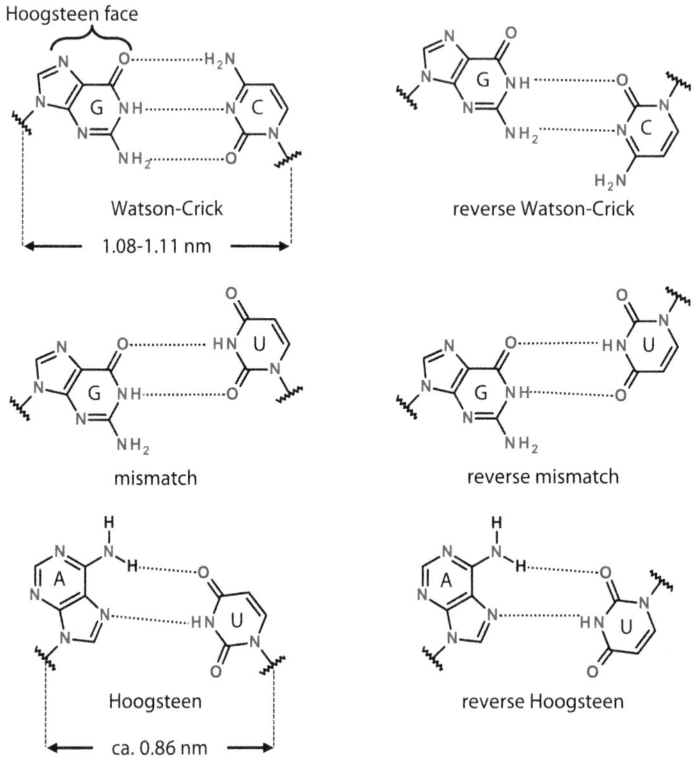

Hoogsteen face

Watson-Crick

1.08-1.11 nm

reverse Watson-Crick

mismatch

reverse mismatch

Hoogsteen

ca. 0.86 nm

reverse Hoogsteen

**FIGURE 17.12: Base-pairing geometries.** Base pairing is dictated by H-bonds between bases with matching donors and acceptors. The distances between the C1′ atoms (wiggly lines) in Watson-Crick base pairs range between 1.08–1.1 nm. The nucleobases are on the same side in a *cis*-like orientation of their glycosidic bonds, as opposed to reverse base pairs (right-hand column) where the glycosidic bonds are oriented antiparallel. Mismatched base pairs such as G-U are frequently observed in tRNA. The purine in Hoogsteen pairs has *syn*-conformation and the C1′-C1′ distance is only 0.86 nm.

Calculations for the *in vacuo* H-bond strengths of base pairs showed that the G-C base pair is the most stable, as it has three H-bonds. In line with the vacuum calculations, base pairing is favored in the non-polar solvent $CHCl_3$ compared to the polar solvent dimethyl sulfoxide (DMSO). However, under physiological conditions water competes with the bases for H-bonds and thus decreases their energetic contribution to base pair stability: in order to form a single H-bond between nucleobases, two H-bonds with two water molecules need to be broken. The net energy contribution of H-bonds may therefore be quite small, similar to what is observed with proteins. This leaves base stacking as the main energetic contributor to nucleic acid stability.

## 17.4.2 IMPORTANCE OF BASE PAIR STACKING FOR DOUBLE HELIX FORMATION

The sequence of nucleotides in a single-stranded nucleic acid establishes a unique H-bonding pattern at the Watson-Crick faces of the nucleobases. Thus, two sequences of complementary nucleobases can associate into a duplex, a spontaneous process that results in a stable helical structure held together by H-bonds between the bases and, more importantly, by van der Waals interactions between base pairs stacked on top of each other. The base stacking in double helices provides a higher energetic contribution than in single-stranded nucleic acids because the double helix has a hydrophobic core where the nucleobases are shielded from solvent. As with proteins, the hydrophobic effect is an important driving force for the assembly of polynucleotide single-strands into a double helix because it minimizes solvent-exposed hydrophobic surface areas of the nucleobases. At the same time, H-bonding is stronger in the hydrophobic environment due to the smaller $\varepsilon_r$, and the electrostatic repulsion of the charges in the phosphate backbone on the outside of the helix is also minimized. These principles go beyond double-stranded DNA and are also important for triple and quadruple DNA helices, and for folding of RNA molecules. Just like helices in proteins, the helix of nucleic acids is described by general characteristics, including its diameter, axial rise, and pitch (Figure 17.13).

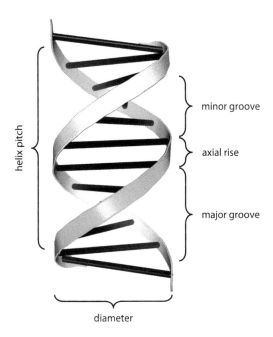

**FIGURE 17.13:    Schematic depiction of the DNA double helix.** The helix is composed of two antiparallel and complementary strands (green and gray with base pairs as black rods) that bury the base pairs in the center and expose the phosphate backbone to solvent. A major and a minor groove run parallel to the strands. Helical parameters including the pitch, the diameter, the number of nucleotides for a full rotation, and the width of the major and minor grooves depend on the actual conformation of the helix (see Table 17.1).

A good measure for the contribution of base pair stacking to the stability of nucleic acids is the melting temperature $T_m$ (see Figure 19.20) of a double helix made of defined oligonucleotides. Energy functions have been derived for the stacking of the various base pair doublets. Experimentally determined $T_m$ for large double-stranded B-DNA correlate well with the calculated base stacking energy (Figure 17.14).

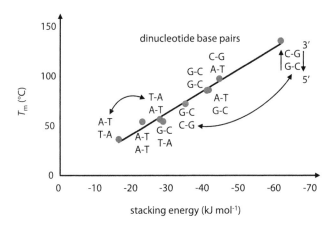

**FIGURE 17.14:** **Linear correlation of calculated stacking energies with melting temperatures.** $T_m$ is calculated from the experimental melting temperature of oligonucleotides containing specific dinucleotides. The stacking energies were calculated using a potential function optimized for nucleic acids. A dash between bases marks a base pair that stacks onto another base pair. The arrows in the top right dinucleotide point from the 3'- to the 5'-end. $T_m$ strongly depends on whether the purine is at the 5'- or 3'-end (curved arrows). Adapted from Ornstein & Rein (1978). Biopolymers 17(10): 2341–2360 and Gotoh & Tagashira (1981). Biopolymers 20(5): 1033–1042.

In general, because the AT-rich sequences have the lowest stacking energy, an AT-rich duplex (or AU-rich for RNA) melts at lower $T_m$ compared to GC-rich duplexes. However, the stability of a double-stranded DNA depends not only on the composition (type of nucleobase) but also on the sequence of the base pairs (Figure 17.14). The most stable stacking interaction is calculated between a C-G pair and an adjacent G-C pair. If the order is changed to G-C packing on C-G, the stacking energy is almost halved, which is explained by the larger overlap of the nucleobases in a $3' \rightarrow 5'$ purine-pyrimidine stack compared to a pyrimidine-purine stack. Similarly, if two G-C pairs stack onto each other, the stacking interaction is weaker than for the G-C and C-G pairs. On the other end of the energy scale, the trend is the same for A-T base pairs. A repetitive 5'-AT-3' sequence may form a "hinge" site in DNA: the weak stacking energy of this sequence provides more flexibility to the duplex, enabling it to kink at such sequences (Section 17.5.3.3).

## 17.4.3 BASE PAIR GEOMETRIES

A strictly planar base pair stacking on top of another base pair should be the energetically most favorable situation because this way the largest surface is available for hydrophobic interactions. The optimal distance between base pairs is close to the sum of the van der Waals radii of the atoms making up the base pairs, about 340 pm. However, depending on the type of helical structure, the bases in a pair can be slightly twisted, sheared, opened, buckled, and staggered relative to each other. Some of these conformational adaptations are absolutely necessary to form the experimentally observed A-, B-, and Z-forms of DNA (see Section 17.5.1). Figure 17.15 summarizes the main structural parameters for base pairs.

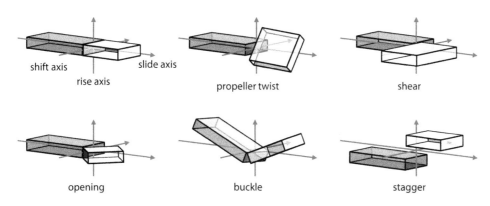

**FIGURE 17.15:** **Base pair geometries.** The helix axis is the rise axis, the long axis in the base pair is the slide axis, and the shift axis is perpendicular to the first two. The movements are exaggerated for clarity. Bases can be laterally displaced, rotated, or tilted. A lateral displacement along the helix axis is the rise, displacements in the plane of the base pairs are shifts when they occur sideways with respect to the base pair or slide when they increase the distance between the bases of a pair. Shearing is a shifting of the bases, stagger is a change in rise. Rotations can occur around the helical axis (opening) or around the long axis of the base pair (propeller twist). A tilt out of the base plane (perpendicular to the helical axis) is a buckle.

The stacking of base pairs in different forms of DNA and RNA helices sometimes requires variations to the parallel orientation of base pairs, described by rise, slide, shift, twist, tilt, and roll (Figure 17.16). These local structural adaptations of base pair conformations translate into fundamentally different structures of the various double helices.

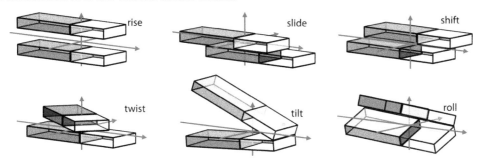

**FIGURE 17.16:** **Relative orientations of base pairs.** Rise, slide, and shift are displacements of one base pair relative to another along the same-named axes. Twist, tilt, and roll are rotations about the rise, shift, and slide axes, respectively.

## 17.5 DNA STRUCTURES AND CONFORMATIONS

The archetype of the DNA structure is the double helix with two complementary strands running antiparallel to each other. The strong base pair stacking causes the duplex to behave essentially as a single molecule. However, DNA and RNA are far from being static molecules: unpaired regions in the middle (*breathing*) or at the ends (*fraying*) of a duplex may form spontaneously or can be induced by higher temperature or low ionic strength. DNA structures with three and four strands also exist, although their biological roles are not entirely clear. In addition to Watson-Crick base pairs, triplex and quadruplex DNA helices use the Hoogsteen face of nucleobases to form their triple and quadruple base pairs (Section 17.5.2).

### 17.5.1 DNA DOUBLE HELICAL STRUCTURES

Double-stranded DNA is polymorphic and can adopt A-, B-, and Z-forms depending on the sequence and solvent conditions. X-ray diffraction studies on hydrated DNA fibers established that the different forms of DNA are dependent on humidity. The canonical B-form of the DNA double helix is stabilized by high humidity while salt, alcohols, and low humidity stabilize the other two forms. A- and B-form helices are right-handed, but the Z-form is left-handed. The geometric parameters of A-,B-, and Z-DNA are summarized in Table 17.1. Figure 17.17 shows the three DNA double helix structures side-by-side.

**TABLE 17.1**
**Helical parameters of double-strand nucleic acids.**

| structure | A | B | Z |
|---|---|---|---|
| helicity | right | right | left |
| helix diameter (nm) | 2.3 | 2.0 | 1.8 |
| mean # bp per turn, $\langle c \rangle$ | 11 | 10–10.5 | 12 (2 bp repeat) |
| twist per bp (°) | 33 | 34–36 | –9 and –51 |
| displacement per bp (nm) | 0.45 | –0.02 to –0.18 | –0.2 to –0.3 |
| helix pitch (nm) | 2.8 | 3.4 | 4.5 |
| rise per bp (nm) | 0.26 | 0.33–0.34 | 0.37 |
| tilt of bp (°) | 20 | –6 | –7 |
| propeller twist (°) | 18 | 16 | 0 |
| helix diameter (nm) | 2.3 | 2.0 | 1.8 |
| glycosidic bond | *anti* | *anti* | *anti* (C) and *syn* (G) |
| sugar pucker | C3'–*endo* | C2'–*endo* | C2'–*endo* (*anti*) and C3'–*endo* (*syn*) |
| $P_n – P_{n+1}$ distance (nm) | 0.56 | 0.66 | 0.56 and 0.64 |
| C1'–C1' distance (nm) | 0.48 | 0.49 | 0.51 and 0.59 |
| major / minor groove width (nm) | 0.27 / 1.1 | 1.17 / 0.57 | 0.88 / 0.2 |
| major / minor groove depth (nm) | 1.35 / 0.28 | 0.88 / 0.75 | 0.37 / 1.38 |
| average $\alpha$ and $\zeta$ (°) | –50 / –78 | –41 / –157 | –137 / 80 (Py) and 47 / –69 (Pu) |
| average $\chi$ (°) | –154 | –102 | –159 (Py) and 68 (Pu) |

*Note:*   Py and Pu denote pyrimidines and purines, bp is short for base pair.

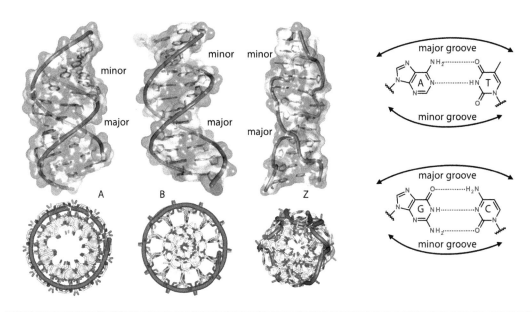

**FIGURE 17.17:    Double-stranded helices in A-, B-, and Z-form.** A dodecamer sequence is shown for A-DNA, B-DNA, and Z-DNA (from left to right). A- and B-DNA are right-handed helices of 2.3 nm and 2.0 nm diameters, respectively. The narrow poly-CG Z-DNA (PDB-ID 4ocb) is left-handed. It earns its name from the zigzag path of the phosphate backbone. The major and minor grooves are indicated both on the overviews on the left and on the base pairs on the right. The upper (Hoogsteen face) and the lower parts of the base pairs lie in the major and minor groove, respectively. Note that in A-DNA, the major groove is actually less wide than the minor groove. A-, B-, and Z-DNA are all antiparallel and are interconvertible under certain buffer conditions.

All three main DNA conformations have been studied by fiber diffraction, NMR spectroscopy, and X-ray crystallography. The first atomic model of a DNA molecule was the B-DNA, built by James Watson and Francis Crick in 1953 using information from fiber diffraction photographs taken by Rosalind Franklin. B-DNA is the predominant conformation at high humidity and in solution, accounting for the vast majority of genomic sequences. The bases in B-DNA are not very much displaced from the helix axis (negligible slide) and almost perpendicular to it (tilt almost zero). Ten base pairs make up a full turn of the helix with a pitch of 3.4 nm. Thus, the stacked bases in B-DNA have van der Waals distance (0.34 nm), enabling optimal interactions. The numbers of bases for a full turn can vary between 10 and 10.5. All bases are in the *anti*-conformation and the deoxyribose is C2′-*endo*. The surface of B-DNA shows two depressions known as *major and minor grooves*. The Hoogsteen faces of the nucleobases lie at the bottom of the wide and shallow major groove. H-bond donors and acceptors close to the glycosidic bond that are not involved in base pairing lie in the narrow minor groove (Figure 17.17). X-ray crystallography and NMR have revealed a chain of water molecules bound in the minor groove of B-DNA, which is known as the *spine of hydration*. Both the major and minor grooves in B-DNA are sites of interactions with proteins and small molecules (Section 17.5.4).

A-DNA is the dominating conformation at low (75%) humidity and also the main conformation of double-stranded RNA. Similar to B-DNA, the base pairs in A-DNA and A-RNA are of Watson-Crick type with all glycosidic $\chi$-angles in the *anti*-conformation. In contrast to B-DNA, the ribose conformation is C3′-*endo*, not C2′-*endo*. In addition, the bases in the A-form are strongly displaced from the helical axis by about 0.45 nm, leaving a void along the center of the helix. This slide makes the minor groove very shallow and the major groove deep and narrow. Indeed, the minor groove in A-DNA and A-RNA is almost as broad as the major groove of B-DNA (Figure 17.17). The A-form has eleven base pairs in a single turn of 2.8 nm pitch. The vertical rise is thus only 0.26 nm per base pair, which is much smaller than van der Waals contact distance (0.34 nm). To avoid clashes, the bases in the base pairs are strongly tilted by ca. 20° relative to the helix axis.

Z-DNA has a repeating unit of two nucleotides, usually pyrimidine-purine, that regularly alternate in their glycosidic $\chi$-angle between *anti* for the pyrimidine and *syn* for the purine. Along with the $\chi$-angle, the sugar pucker alternates between C2′-*endo* (*anti*) and C3′-*endo* (*syn*), imposing a zigzag path on the phosphate backbone that gave the Z-form its name. Although twelve nucleotides make up a full turn, the Z-form is therefore better described as a repeating unit of six dinucleotides. CD- and [31]P-NMR spectroscopy showed that Z-DNA is stabilized by high salt concentrations (4 M NaCl) and the ammonium ions of spermine and spermidine. This indicates that the transition from B-DNA to Z-DNA is possible when reducing the electrostatic repulsion between opposite DNA strands by counter ion binding. Certain chemically modified sequences containing 5-methylcytosine or 5-bromocytosine can form stable Z-DNA against which specific antibodies have been raised. Such antibodies have been shown to bind to the polytene chromosomes of enhanced transcription in *Drosophila* and near promoter sites of the c-MYC gene in mammalian nuclei, suggesting a biological role for Z-DNA. Indeed, stretches of Z-DNA may form to relax negatively supercoiled DNA (Section 17.5.3.2) during transcription of certain genes.

Base-paired duplex structures different from A-, B-, and Z-DNA are useful tools in biological and biophysical applications, ranging from gene silencing to surface attachment of molecules. Box 17.2 describes the building blocks of two of those altered nucleic acids, the peptide nucleic acids (PNAs) and locked nucleic acids (LNAs). These DNA-like molecules form very stable duplexes and are resistant to nucleases.

## BOX 17.2: PEPTIDE NUCLEIC ACIDS (PNAs) AND LOCKED NUCLEIC ACIDS (LNAs).

PNAs are hybrid molecules of nucleobases connected to a polypeptide backbone (Figure 17.18). The whole phosphodiester backbone of DNA is replaced by a polymer of N-(2-aminoethyl)-glycine monomers or a similar molecule. There are two amide bonds per monomer. One is located close to the former position of the phosphodiester group, the other links a nucleobase to the polyamide backbone. PNAs can form exceptionally stable Watson-Crick base pairs and heteroduplexes with complementary single DNA and RNA strands. Even triplexes can form between double-stranded DNA and a PNAs strand. Because PNAs are uncharged they penetrate cells more easily than DNA or RNA and can be used to deliver drugs conjugated to them. PNAs are also not sensitive to nucleases, making them attractive candidates for sequence-specific DNA silencing and RNA interference.

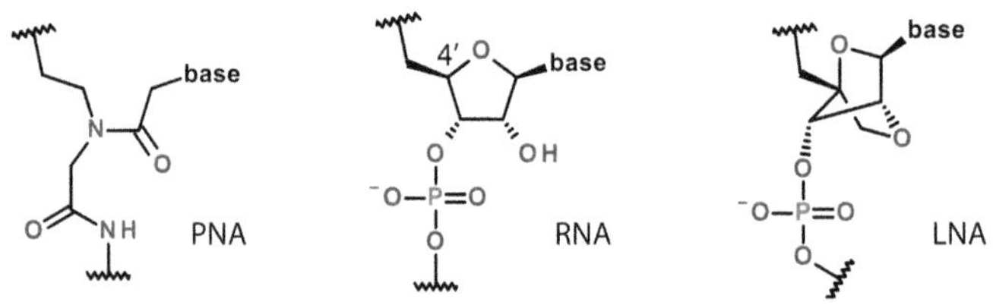

**FIGURE 17.18:**  **Comparison of peptide nucleic acids (PNAs) and locked nucleic acids (LNAs) with regular RNA.**

Many of the properties of PNAs apply to another type of synthetic nucleic acid, the locked nucleic acids (Figure 17.18). LNAs are stable, nuclease-resistant RNA derivatives where the 2′-hydroxyl group is connected by a methylene (CH$_2$) group to the C4′-atom. The connection creates a bicyclic ring system that locks the furanose into the C3′-*endo* conformation. In contrast to PNAs, LNAs retain a negatively charged backbone. LNAs bind to complementary RNA and DNA strands with very high affinity, making them ideal research tools for gene silencing.

## 17.5.2 TRIPLE AND QUADRUPLE DNA HELICES

### 17.5.2.1 Triplexes

The major groove of B-DNA is wide enough to accommodate a third strand of DNA to form a triple helix or *triplex*, especially when one strand of the duplex is rich in purines. Depending on the type of triplex, the nucleobases of the third strand form either Hoogsteen or reverse Hoogsteen interactions with the Watson-Crick base pairs of the duplex, leading to *base triplets* (Figure 17.19). Although they have been detected *in vivo*, it is uncertain whether triple helices have a biological function. There are three types of DNA triple helices based on the sequence composition and the orientation of the third strand's backbone relative to that of the poly-purine strand of the duplex. In the *TC triplex*, a pyrimidine-rich third strand runs parallel to the purine-rich strand of the duplex, forming T·A-T and C⁺·G-C Hoogsteen triplets (Figure 17.19). Triplets involving C⁺·G-C pairings require low pH to protonate the N3-atom of cytosine (Figure 17.2). In a *GT triplex*, the third strand can be oriented either parallel or antiparallel to the poly-purine strand of the duplex, forming Hoogsteen or reverse Hoogsteen triplets, respectively, of the type G·G-C and T·A-T. In the *GA triplex*, the purine-rich third strand runs antiparallel to the purine-rich strand of the duplex, forming reverse Hoogsteen G·G-C and A·A-T triplets.

A

B

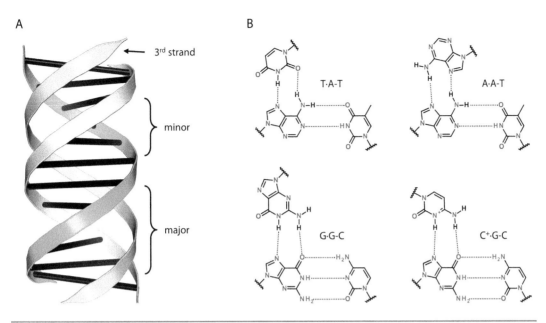

**FIGURE 17.19:     DNA triple helix.** A: In the DNA triplex, an extra strand is bound in the major groove of B-DNA (colored yellow). B: Examples of base triplets observed in triple helices. Watson-Crick pairing is at the bottom, and (reverse) Hoogsteen pairings are at the top of the triplet. When cytosine forms a triple with a G-C pair (lower right), it must be protonated at the N3-atom. Only if there are two H-bond donors can a Hoogsteen pair with the O6- and N7-atoms of guanine be formed.

Intermolecular triple helices have attracted much interest for their therapeutic potential. For instance, oligonucleotides that are able to form triplexes with promotor regions in genomic DNA can be used for gene silencing. The triple helix approach can be used to target chemicals or proteins to specific sites on the DNA. To do so, crosslinking agents, nucleases, or transcription factors are fused to triplex-forming oligonucleotides. Once the triplex is formed, the cargo delivered by the oligonucleotide will cross-link, mutate, or hydrolyze the target DNA. Triple helices are also useful to immobilize double-stranded DNA at surfaces (e.g. for TIRF microscopy; Section 23.1.4) *via* triplex-forming oligonucleotides that are covalently coupled to a surface.

### 17.5.2.2 Quadruplexes and Telomeres

DNA quadruplexes are formed by four G-rich DNA strands using both the Watson-Crick and the Hoogsteen faces. The O6-atoms of the guanines in the planar *G-quartet* point towards the center of the plane and can coordinate metal ions, which further stabilize the quadruplex. The metal ions are sandwiched between two G-quartet planes and coordinated by eight O6-atoms from the guanines. Alkali ions such as $K^+$ and $Na^+$ are bound at the center of the quadruplex, but $Sr^{2+}$ and $Tl^+$ ions can also bind *in vitro*. The direction of the strands in a quadruplex can be parallel or antiparallel with different connectivities when the quadruplex is made up of one, two, or four DNA molecules (Figure 17.20).

Telomeres are G-rich DNA repeat sequences (>2000 times TTAGGG in vertebrates) at the ends of chromosomes that form quadruplexes which have been detected using antibodies. Telomeres have regulatory functions, but their primary function is to provide primers for DNA replication. Telomeres continue to shorten throughout life and have been implicated in the development of cancer.

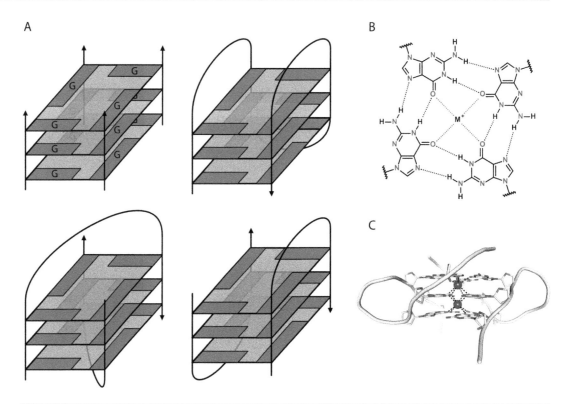

**FIGURE 17.20:   DNA quadruplexes.** A: In the simplest case, four independent poly-G sequences form an intermolecular parallel quadruplex (top left). The quadruplex can also be formed by a single DNA strand (top right) or by two hairpin strands that either cross over (bottom left) or are parallel to each other (bottom right). B: Typical Hoogsteen G-quartet with a central metal ion M+ (Na+ or K+). C: The side-view of a quadruplex made from two hairpins shows how K+ is bound between neighboring quadruplets (PDB-ID 4p1d).

### 17.5.3 HIGHER ORDER DNA STRUCTURES

DNA is often kinked, bent, extruded into *cruciform structures*, or curled up into *topoisomers*. The extent of deviation from the standard, linear B-DNA conformation is dependent on the actual sequence, the binding of protein partners, and the particular environment of the DNA.

#### 17.5.3.1 Helix Junctions

*Inverted repeat* or *palindromic* sequences are self-complementary (Figure 17.21). Under certain solution conditions, inverted repeats can form cruciform structures where the DNA is extruded from its linear structure into a cross-shaped form. The biological significance of cruciform structures is still debated because some of the nucleobases are unpaired, making the cruciform less stable than fully base-paired B-DNA. A biologically important structure that very much resembles a cruciform is the *Holliday junction* (Figure 17.21), named after Robin Holliday. This junction is an important intermediate in *homologous recombination*, a process used in cells to exchange genetic material. The Holliday junction is recognized by the RuvAB helicase, an ATP-dependent motor protein in bacteria that can move the junction along the DNA and determines the outcome of the recombination event.

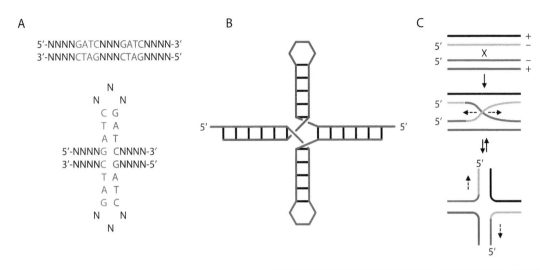

FIGURE 17.21: **DNA cruciform structures**. A: The short palindromic sequence GATC is separated by three nucleotides and flanked by four nucleotides. All nucleotides form base pairs in standard DNA, but here have been abbreviated as N for simplicity. Extrusion of the palindromic sequence into a cruciform leaves the bases between the inverted repeats unpaired. B: A four-way junction is formed by two DNA strands (red and blue). Base pairings are indicated by black lines. C: Formation of a Holliday junction during homologous recombination (marked by an X). The Holliday junction is made up of four DNA strands (different colors). After strand nicking and swapping of the nicked strand, the assembly can be re-drawn to resemble a cruciform structure but without unpaired bases as in (A) and (B). The amount of DNA exchanged during the process depends on how far the junction moves along the sequence (indicated by dashed arrows).

Whereas the cruciform structure is formed by two DNA strands, the Holliday junction contains four DNA strands, which avoids high-energy loop regions with unpaired nucleobases. The Holliday junction is not the only junction prevalent in biology. Depending on the number of helices that meet at the junction, there are four-way, three-way, and two-way junctions. The common theme in junctions is to maximize the interactions by co-axial stacking of two or more helices (Figure 17.22).

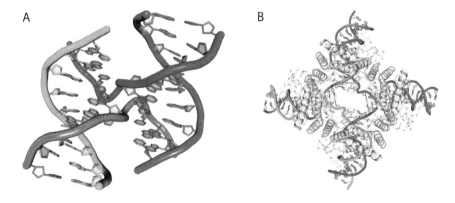

FIGURE 17.22: **Crystal structures of Holliday junctions.** The mini-junction (PDB-ID 467D) on the left is made of the decamer sequence 5′-CCGGGACCGG-3′, which is self-complementary but for the central GA sequence. These bases form non-standard base pairs at the center of the junction. The four helical arms of the junction form two duplexes by pairwise co-axial stacking. By contrast, the Holliday junction of the LoxP sequence 5′-ATAACTTCGTATA-NNNTANNN-TATACGAAGTTAT-3′ on the right is a planar cruciform-like structure stabilized by the Cre protein.

The sequence at the junction determines the relative arrangement of the helical arms in helical junctions. The global structure and conformational changes of junctions can be studied by FRET using fluorescent labels introduced at the ends of the oligonucleotides (Section 19.5.4).

The replication fork is an example of a three-way junction: the DNA that has not yet been replicated forms one helical arm, the two copies of newly synthesized DNA form the other two arms.

If DNA replication stalls due to a DNA lesion, there are two mechanisms for repair: homologous recombination or transient formation of the *chicken foot structure*. In this four-way junction, the newly synthesized strands serve as their own templates until enough DNA is synthesized to by-pass the original lesion. Reversal of the chicken foot structure to a three-way junction restores the replication fork (Figure 17.23).

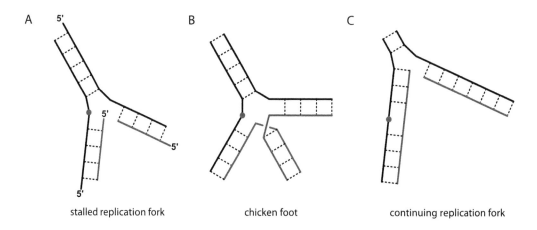

stalled replication fork            chicken foot            continuing replication fork

**FIGURE 17.23:** **Junctions in DNA replication.** The parent strands are shown in black with base pairings as dashed lines. A DNA lesion is marked by a magenta circle. A: Intact replication fork. The 5'-ends of the parent DNA, the continuously synthesized strand (red), and the Okazaki fragment (blue) are indicated. Replication stops at the lesion. B: The newly synthesized strands use each other as new template by transiently forming a structure that resembles a chicken foot. C: The chicken foot structure is reverted into a replication fork, and the DNA lesion has been by-passed.

Helical junctions are also frequent structural elements in RNA, and all aspects discussed here for DNA junctions also pertain to RNA junctions. We will see an example of an RNA four-way junction in the chapter about tRNA (Section 17.6.2).

### 17.5.3.2 DNA Supercoiling

In the schematic depiction of the DNA strands in a Holliday junction in Figure 17.21, and also for the replication fork in Figure 17.23, we have neglected the helical structure of DNA. The double-stranded helical structure of DNA is an impediment to essentially all biological processes that require strand separation. When DNA is replicated, the two strands have to be separated to form the replication bubble. The movement of the replicating DNA polymerase along the DNA and the concomitant separation of the two strands of the double helix require the two DNA strands to rotate. However, this rotation is not feasible in cellular DNA that is packed very tightly. In circular covalently closed DNA such as plasmids and bacterial chromosomes, the ring shape fixes the ends of the DNA and prevents their rotation. In eukaryotes, DNA is bound to nucleosomes (see Figure 17.29) that form higher order structures. Chromosomal DNA can therefore be regarded as an assembly of circular domains, loops whose ends are fixed and cannot rotate freely. The same is true for very long linear DNA, which is too inert to freely rotate about its long axis.

As the DNA cannot freely rotate, movement of the replication fork along the DNA double helix generates torsional stress. This torsional stress leads to overwinding of the DNA ahead of the replication fork, and underwinding behind it. The same is observed when the RNA polymerase moves along the DNA in the transcription bubble. Overwinding and underwinding of the double helix leads to coiling of the helix around itself, and to the formation of supercoils. DNA that is overwound forms positive supercoils that stabilize DNA in its double-stranded form and counteract strand separation. Overwinding ahead of the replication fork or the transcription bubble thus impedes fork progression and replication, and needs to be removed. Underwound DNA forms negative supercoils. Underwinding favors strand separation, which is the reason for the higher propensity of negatively supercoiled DNA to form cruciform structures *in vitro*.

The helical nature of the DNA also has consequences at the end of replication: the two DNA circles generated by replication of a circular DNA are linked like two members of a chain (Figure 17.24). These linked circles are called *catenanes*. For segregation of one copy of the DNA to each daughter cell during cell division, catenanes have to be resolved. Other knotted forms of DNA are also observed, but will not be treated here.

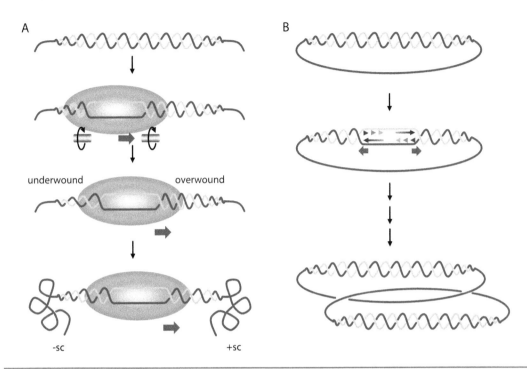

**FIGURE 17.24: Movement of RNA polymerase and transcription bubbles generates supercoiling.**
A: DNA transcription by RNA polymerase (orange) requires unwinding of the DNA double helix and genera-tion of a transcription bubble. During transcription, movement of the transcription bubble requires unwinding of the DNA in the direction of RNA polymerase movement, and rewinding behind it. Because the ends of a circular DNA cannot rotate freely, the torsional stress leads to positive supercoiling of the DNA ahead of the RNA polymerase, and negative supercoiling in its wake. The positive supercoils ahead of the moving RNA poly-merase are an impediment for its progression, and need to be removed by topoisomerases. B: DNA replication of covalently linked DNA circles leads to the production of catenanes, two linked double-stranded DNA rings. The two strands of the template DNA are depicted in blue, the newly synthesized DNA in gray. The arrow rep-resents leading strand synthesis, and the triangles indicate Okazaki fragments in lagging strand synthesis. For segregation of the two DNA molecules to the daughter cells, the catenanes are unlinked by topoisomerases.

To describe supercoiling quantitatively, we first define the *linking number* ($Lk$) of a DNA molecule. We have seen in Section 17.5.1 that the average number of base pairs per turn in B-DNA is $\langle c \rangle$ = 10–10.5. The exact number depends on temperature, ionic strength, and pH. B-DNA with 63 base pairs thus forms six helical turns. We can directly link the ends of this DNA into a covalent circle. The linking number of the circular DNA describes how often the strands are linked within this circle. In this case, the linking number is simply the number of base pairs divided by the number of base pairs per turn, i.e. the number of helical turns ($Lk$ = 6). Such a DNA is called *relaxed* DNA, and its linking number is also referred to as $Lk^0$. The linking number is always an integer number.

We now take the same piece of DNA with six helical turns, keep one end fixed and rotate the other end in the rotational sense of the helix, such that one helical turn is removed (Figure 17.25). The number of helical turns then reduces to 5, and the number of base pairs per turn increases: the DNA is said to be *underwound*. Conversely, we can *overwind* this DNA by rotating one end against the rotation sense of the helix, such that the number of helical turns increases. Overwinding and underwinding DNA cause torsional strain in the DNA. When the overwound or underwound DNA is covalently closed, this strain can be accommodated in two ways: either by changes in twist $Tw$, or by changes in writhe $Wr$. The twist $Tw$ corresponds to the number of turns of the double helix, the writhe $Wr$ quantifies how often the double helix is coiled (Figure 17.25). In other words, the writhe quantifies the number of supercoils present. $Wr$ can be positive or negative, reflecting the number of positive or negative supercoils present in the DNA. The double-stranded crossovers in positively supercoiled DNA are positive, and the resulting superhelical turn is left-handed. Negatively super-coiled DNA has negative crossovers and right-handed superhelical turns (Figure 17.29).

The twisting number $Tw$ and the writhe $Wr$ add up to the linking number $Lk$:

$$Lk = Tw + Wr \qquad \text{eq. 17.3}$$

While $Lk$ is always an integer number, $Tw$ and $Wr$ can adopt any value. Relaxed B-DNA is described by $Wr = 0$ and $Tw = Lk = N/10.5$, where $N$ is the number of base pairs. Chemically identical DNA molecules that differ only by their $Lk$ are called *topoisomers* and are interconverted by enzymes termed *topoisomerases*.

So far we have restricted our considerations to DNA molecules with integer numbers of helical turns. What happens when the number of base pairs $N$ is not a multiple of 10.5 base pairs per turn, and the ratio $N/\langle c \rangle$ gives a non-integer value? In this case, covalent joining of the ends into a circle is not possible by mere bending of the DNA, but requires some torsion of the DNA to align the ends: we have to slightly over- or underwind the DNA double helix for ring closure. The linking number of the relaxed DNA circle will be the closest integer to the calculated value ratio $N/\langle c \rangle$. For a 108 base pair DNA, $N/\langle c \rangle = 10.3$. This DNA will form a covalently closed circle with $Lk = 10$. The deviation of the linking number $Lk$ from $Lk^0 = N/\langle c \rangle$ gives rise to some torsional stress even in the relaxed form. As a consequence, the most relaxed topoisomer of this DNA already has some twist and/or writhe. While this torsional stress is small for large DNA molecules, it becomes more and more significant for small DNA circles.

The linking number of covalently closed DNA is constant. $Lk$ can only be changed by transient breakage of the phosphodiester backbone. A change in linking number generates a new topoisomer that will adjust to its different linking number either by over- or underwinding (twist), or by super-coiling $Wr$, or both. Starting from the relaxed DNA circle with $Wr = 0$ and $Tw = Lk = 6$ (see above), a negative supercoil can be introduced by cleavage of both strands, introducing underwinding

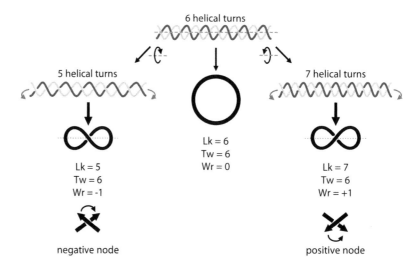

**FIGURE 17.25: DNA supercoiling.** A regular B-form DNA with 6 helical turns can be circularized to a covalently closed, relaxed DNA (middle, circle). The linking number of the circular DNA is $Lk = 6$. Underwinding (left) or overwinding (right) before circularization introduces torsional stress that is distributed into twist $Tw$ and writhe $Wr$, and generates negatively and positively supercoiled DNA. The dotted gray line marks the double-helical and superhelical axis, respectively. Negative supercoils form negative crossovers (alignment of the upper double helix with the bottom double helix requires a rotation in clockwise direction). Positive supercoils form positive crossovers (alignment of the upper double helix with the bottom double helix requires a rotation in counter-clockwise direction).

(*Tw* = 5 turns) and re-ligation of both strands (*Lk* = 5). The twist *Tw* either remains at 5, or increases to 6 upon formation of one negative supercoil (*Wr* = −1). Conversely, the relaxed DNA can be positively supercoiled by cleavage, overwinding, re-ligation, and formation of one positive supercoil (*Lk* = 7, *Tw* = 6, *Wr* = +1).

Topoisomers have different hydrodynamic radii and can therefore be resolved by gel electrophoresis (Figure 17.26). Negative and positive topoisomers of the same linking number have different propensities to interact with intercalators, which can be used for their separation by gel electrophoresis in two dimensions. Intercalators are planar, aromatic molecules that bind to DNA by insertion between two adjacent base pairs (Section 17.5.4.2). A well-known intercalator is ethidium bromide (see Figure 17.34) that becomes fluorescent upon DNA binding and is used to visualize DNA in gels. Intercalation leads to local unwinding of the DNA and a decrease in *Tw*. For circular DNA, *Lk* is constant, and a decrease in *Tw* inevitably leads to an increase in *Wr* (eq. 17.3). A negatively supercoiled plasmid thus becomes more and more relaxed when intercalators bind. Once it is fully relaxed, further binding of the intercalator leads to accumulation of positive supercoils until the maximum number of supercoils is reached, and the DNA is saturated with intercalating molecules. The more negatively supercoiled the plasmid is, the more intercalating molecules it can bind. A positively supercoiled plasmid also becomes more and more positively supercoiled with each molecule that intercalates, and reaches the limiting number of supercoils with fewer intercalating molecules bound. As a consequence, negatively supercoiled DNA has a lower electrophoretic mobility than positively supercoiled DNA in the presence of intercalators. If the second dimension of gel electrophoresis is performed in the presence of intercalators such as chloroquine, negatively supercoiled DNA lags behind, and can be separated from positively supercoiled DNA with the same *Wr*. A ladder of different topoisomers is separated into an arch with a left branch corresponding to the negatively supercoiled species, and a right branch formed by the positively supercoiled DNA.

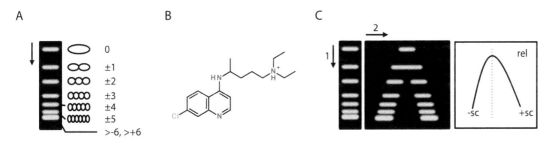

**FIGURE 17.26:** **Separation of DNA topoisomers.** A: The higher the number of supercoils the more compact the molecule, and the faster it migrates on an agarose gel. B: Chloroquine is a DNA intercalator that causes untwisting of the DNA upon binding. C: Negative and positively supercoiled DNA can be separated by two-dimensional gel electrophoresis. After separation in the first dimension according to the hydrodynamic radius and thus to the number of supercoils, the separation in the second dimension is performed in the presence of an intercalator such as chloroquine. The more negatively supercoiled the DNA, the more intercalating molecules are bound, leading to a decrease in electrophoretic mobility in the second dimension. The more positively supercoiled the DNA, the fewer intercalating molecules are bound, and the electrophoretic mobility remains high. Separation of DNA with the same number of negative and positive supercoils in the second dimension leads to an overall arch-like band pattern. The left branch corresponds to negatively supercoiled DNA, the right branch to positively supercoiled DNA.

Supercoiled DNA has a higher free energy than relaxed DNA because of its torsional energy (Figure 17.27). DNA supercoiling is thus an endergonic reaction. Relaxation, on the other hand, is an exergonic process and thus in principle a spontaneous reaction. Nevertheless, the thermodynamically favorable relaxation of DNA by rotating about its ends is not possible if the DNA is long and too inert to rotate, or if the DNA is circular covalently closed. In the cell, the interconversion of topoisomers is catalyzed by topoisomerases that transiently cleave the DNA to change its linking number (Figure 17.27).

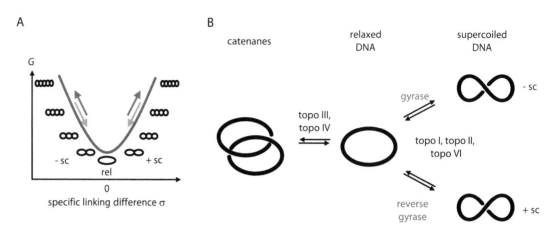

**FIGURE 17.27:** **Interconversion** of topoisomers by DNA topoisomerases. A: Energy diagram. The torsional energy of supercoiled DNA and the free energy *G* increase with the number of supercoils. Relaxation of DNA is an energetically favorable process (green), whereas the introduction of supercoils is endergonic (red). B: DNA topoisomerases interconvert topoisomers. Topo I, II, and reverse gyrase are type I topoisomerases that cleave one strand of their double-stranded DNA substrate. Topo II, IV, VI, and gyrase are type II topoisomerases that cleave both strands of the DNA (see Figure 17.28). Topo II and IV catalyze DNA catenation and decatenation. Gyrase and reverse gyrase introduce negative and positive supercoils into DNA, respectively, at the expense of ATP hydrolysis (red). Topo I, II, and IV catalyze DNA relaxation. Supercoiled DNA can also be catenated/decatenated (not shown).

There are two classes of DNA topoisomerases. Type I topoisomerases cleave only one DNA strand of the duplex and change *Lk* in steps of one while type II topoisomerases cleave both strands and change *Lk* in steps of two. DNA cleavage is carried out by one (type I) or two (type II) catalytic tyrosine residues that perform a nucleophilic substitution at the scissile phosphate. The resulting 3′- (type IB) or 5′-phosphotyrosyl ester (type IA) of the cleaved DNA strand is a covalent intermediate. As an energetically favorable reaction, relaxation is spontaneous and does not require ATP hydrolysis. Type I topoisomerases achieve DNA relaxation either by passage of the uncleaved strand through the gap in the cleaved strand of the double helix (type IA), or by controlled rotation of the free end on the enzyme (type IB, Figure 17.28). At the end of the reaction, re-ligation of the DNA is mediated through nucleophilic substitution by the free 3′- (type IA) or 5′-hydroxyl group (type IB) of the DNA at the phosphotyrosyl intermediate. Reverse gyrase is a type IA topoisomerase that introduces positive DNA supercoils into DNA. The introduction of supercoils is an endergonic reaction and is therefore coupled to ATP hydrolysis.

Type II topoisomerases catalyze DNA relaxation and supercoiling in an ATP-dependent reaction by passing a double-stranded region through the gap in the cleaved duplex (Figure 17.28). If the transported duplex is part of the same DNA as the cleaved DNA (as in plasmids), this reaction leads to DNA relaxation or the introduction of supercoils. If the two DNA regions belong to different molecules, DNA is catenated or decatenated. Eukaryotic topoisomerase II relaxes DNA in an ATP-dependent reaction. Although ATP hydrolysis is not required for the removal of supercoils itself, it is coupled to conformational changes of the enzyme that drive strand passage. The type II topoisomerase topo IV catalyzes ATP-dependent DNA decatenation. DNA gyrase is a bacterial type II topoisomerase that introduces negative supercoils into DNA in an ATP-dependent reaction.

A

type IA

B

type IB

C

type II

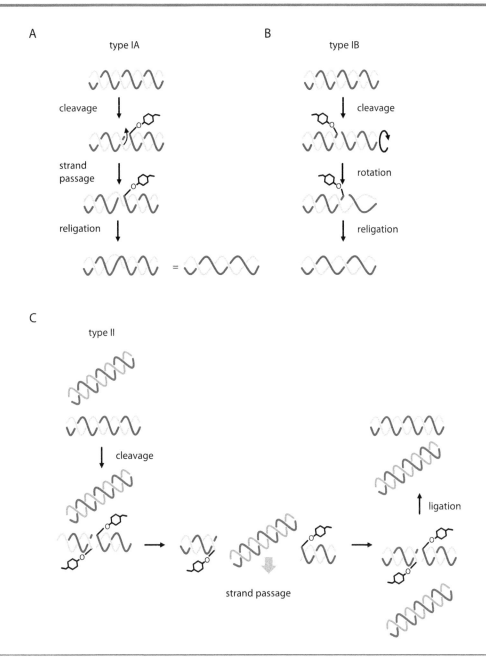

**FIGURE 17.28:    Supercoiling, relaxation, and catenation/decatenation by DNA topoisomerases.** A: Type IA topoisomerases cleave a single strand but form a phosphotyrosyl ester with its 5′-end. Passage of the uncleaved strand (light blue) through the gap in the cleaved strand (dark blue), followed by re-ligation, leads to DNA relaxation or decatenation. Reverse gyrase uses the inverse reaction sequence to catalyze the ATP-dependent introduction of positive supercoils into DNA. B: Type IIB topoisomerases cleave one strand of their DNA substrate and remain bound to the 3′-end of the cleaved strand. Controlled rotation of the DNA (arrow) leads to relaxation. C: Type II topoisomerases cleave both strands of their DNA substrate, and remain covalently attached to the 5′-ends. Passage of a second double-stranded DNA through the gap in the cleaved duplex, followed by re-ligation, changes the linking number by 2. Intramolecular strand passage leads to DNA supercoiling or relaxation, intermolecular strand passage to catenation/decatenation.

*In vivo*, topoisomerases constantly introduce and remove supercoils. The steady-state supercoiling level of cellular DNA is determined by the opposing activities of the set of topoisomerases present in the organism. The overall degree of supercoiling is quantified by the *specific linking difference* σ of the DNA.

$$\sigma = \frac{Lk - Lk^0}{Lk^0} = \frac{\Delta Lk}{Lk^0}$$

eq. 17.4

$Lk^0$ is the linking number of the relaxed state of the DNA ($N$/10.5), $Lk$ is the actual linking number, and $\Delta Lk$ the linking difference. The parameter σ is also called the *superhelical density*, although it does not specify the distribution of twist and writhe. The *specific linking difference* is normalized to the number of helical turns ($Lk^0$) and thus the size of the DNA, and allows comparing the superhelicity of different DNAs. Positive values of σ correspond to positively supercoiled/overwound DNA, negative values to negatively supercoiled, underwound DNA. Using eq. 17.4, we can calculate the effect of deviations between $Lk$ and $Lk^0$ for DNAs whose number of base pairs is not an integer multiple of 10.5: for a 108 bp DNA, $Lk^0$ is 108/10.5 = 10.3. The covalent circle has a linking number $Lk$ = 10, which gives $\Delta Lk$ = –0.3, and σ = –0.3/10.3 = –0.03. The same linking difference for a 3 kb plasmid corresponds to σ = –0.001, and the torsional stress of the most relaxed species is negligible. Most organisms maintain their DNA moderately negatively supercoiled. The local supercoiling density is a major determinant for gene expression.

### 17.5.3.3 DNA Bending and Kinking

Both bending and kinking change the direction of the axis of the DNA double helix. DNA kinking is brought about by a change in local backbone conformation at a specific base pair. In contrast, bending is defined as a curvature over a stretch of several base pairs. Short poly-A sequences (*A-tracts*) of 4–6 adenosines can induce a curvature of ca. 18° ± 2° in B-DNA. If the A-tracts repeat in phase with the rise of the B-DNA (10-11 base pairs), they are all on the same side of the helix and the curvatures induced by each A-tract sum up to a global curving of the DNA. Curvature can be detected by methods that are sensitive to the hydrodynamic radius, including light scattering, analytical ultracentrifugation (Section 26.2), gel permeation chromatography, and electrophoresis. Poly-A tracts are often found in regulatory elements of DNA, suggesting a role in gene transcriptional regulation, replication, or recombination. *In vitro*, intrinsic bending of DNA by phased A-tracts can be used to create small DNA rings of about 100 base pairs. Analysis of poly-A DNA crystal structures shows a tendency towards a narrow minor groove and a large propeller twist of the base pairs such that bifurcated H-bonds are possible, where one nucleobase H-bonds at the same time to two nucleobases of the opposing strand.

Functional bending and kinking of DNA can also be induced or strengthened by protein binding. An example is the cellular storage form of DNA, *chromatin*, where DNA is wrapped around a core particle of eight histones to form a *nucleosome*. The histone octamer wraps a stretch of 146 base pairs of regular B-DNA on its outer surface in almost two complete turns of a left-handed superhelix (Figure 17.29). The driving force for bending is the electrostatic attraction between basic residues on the surface of histones and the negatively charged phosphate backbone of DNA. Bending of the DNA around the nucleosome compresses the minor groove, leading to an increase of negative charge density (*electrostatic focusing*) that is neutralized by Arg side chains on the surface of the histones.

An example of protein-induced DNA kinking is the TATA-box binding protein (TBP) bound to the TATA box in a DNA double helix. An estimated 10–20% of eukaryotic genes have a TATA sequence upstream of their promoter. TBP is a transcription factor that assists in positioning the RNA polymerase II for initiation of transcription. Binding of TBP to the TATA box kinks the DNA twice, once at each TA sequence, by a total of about 115° (Figure 17.29). DNA unwinding and kinking is facilitated by the weak stacking interaction of the 5′-TA-3′ sequence (Figure 17.14). The large concave 10-stranded β-sheet of TBP covers the minor groove using mainly hydrophobic interactions between protein side chains and the ribose furan rings. The extreme widening of the minor groove upon TBP binding changes the overall conformation of the B-DNA to A-DNA. We will encounter more examples of DNA bending and kinking upon protein binding when we discuss the structural principles of DNA-protein recognition (Section 17.5.4.1).

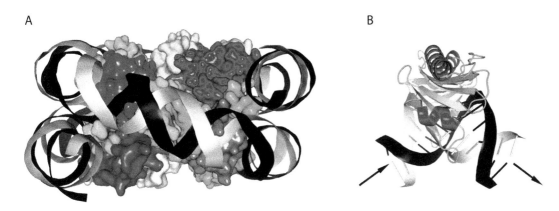

A                                                B

**FIGURE 17.29:**    **Protein-induced DNA bending.** A: In the nucleosome, the histone octamer wraps a 146 base pair DNA (PDB-ID 1kx5). B: Strong kinking of DNA by the TATA-box binding protein (PDB-ID 1cdw). Arrows give the direction of the DNA helix axis before and after kinking.

## 17.5.4  DNA INTERACTIONS WITH PROTEINS AND LIGANDS

### 17.5.4.1  DNA Recognition by Proteins

Binding of proteins such as transcription factors and other regulators of gene expression to DNA can be sequence-specific or unspecific. Sequence-specific binding usually involves direct H-bonding between the nucleobases and a protein, a recurring principle of molecular recognition. Sequence-independent or indirect DNA recognition, on the other hand, does not rely on direct contact with base pairs but on variations in the electrostatic potential and the shape of the DNA. An excellent example of sequence-independent DNA binding by proteins is the nucleosome (Figure 17.29). Other proteins that bind DNA (and RNA) independent of its sequence often have intrinsically disordered regions with positively charged side chains. These extended regions may act as a "lasso" that helps to capture DNA. Once in complex with DNA, the extended region can fold into a defined structure. Z-DNA can be recognized in a sequence-independent manner *via* its peculiar zigzag electrostatic potential: a domain in the ADAR1 protein (double-stranded RNA adenosine deaminase) recognizes the charge and shape of five consecutive phosphodiester groups along the Z-DNA backbone.

Single-stranded nucleic acids can be recognized sequence-specifically by H-bonds of protein main-chain and side-chain groups to the Watson-Crick faces of nucleobases. Such single-stranded regions may form upon bending, kinking, or partial unwinding of double helices. Sequence-specific recognition of intact double helices is possible through H-bonds to groups in the major groove. The side chains of proteins that bind to the major groove bind the Hoogsteen faces of the nucleobases for H-bonding partners. Regardless of their architecture, sequence-specific DNA binding proteins use common themes to read out the nucleotide sequence (Figure 17.30).

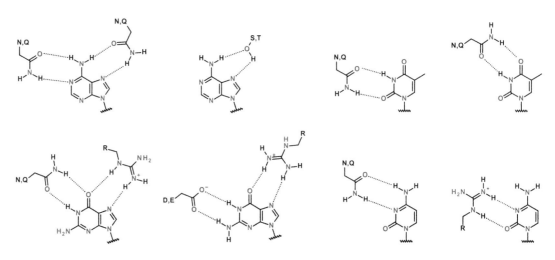

**FIGURE 17.30:    H-bonds between nucleobases and protein side chains.** Amino acids are given in single letter notation. Both the Watson-Crick and the Hoogsteen faces are used for H-bonding and recognition. For arginine, several orientations of the guanidinium side chain relative to the nucleobase are possible.

An α-helix binding to the major groove is arguably the most prevalent motif of sequence-specific DNA recognition. The side chains on one side of the α-helix recognize the DNA sequence by binding to the Hoogsteen face of a stretch of nucleobases in the major groove. This binding mode is found in *helix-turn-helix*, *basic leucine zipper*, and the *zinc finger* proteins (Figure 17.31). A minimal helix-turn-helix or HTH motif has about 20 residues and folds into two α-helices connected by a β-turn. Variations of the HTH motif have a longer loop, and are termed *helix-loop-helix motif*. In both motifs the two helices are almost perpendicular to each other. The second helix, termed the *recognition helix*, binds into the major groove of DNA in a sequence-specific manner. The HTH motif on its own is not stable, but is always part of a larger protein that may contribute additional residues for DNA binding.

The basic leucine zipper (bZIP) motif consists of two amphipathic α-helices forming a left-handed coiled-coil super-secondary structure (see Figure 16.19) at one end, and extend uninterrupted into the major groove of DNA at the other end. bZIP proteins are therefore examples where a single α-helix has two different functions, dimerization and DNA binding (Figure 17.31). Transcription factors with this motif can be homo- or heterodimeric. Interestingly, in the Jun/Fos heterodimeric transcription factor, the monomers are unfolded but form a Leu-zipper in the heterodimer. The basic DNA-binding sequences, too, are unfolded unless they bind to DNA.

Zinc fingers (ZIFs) are small domains of up to 90 but generally less than 50 residues that are stabilized by a $Zn^{2+}$ ion. Eight types of zinc fingers have been identified based on structural differences. C2H2-type zinc fingers tetrahedrally coordinate a $Zn^{2+}$ ion between a two-stranded β-sheet and an α-helix using two Cys and two His side chains, with the α-helix binding to the major groove. Up to sixty copies of this type of zinc finger are found in transcription factors. Because each helix recognizes 3–4 adjacent base pairs, each additional zinc finger increases DNA specificity. Artificial concatenation of zinc fingers with different sequence specificities allows generation of DNA-binding modules that recognize only a single site in the human genome, which is used for gene editing (Box 17.3). Zinc fingers with three Cys and one His, or with four Cys are also found.

DNA binding by HTH, bZIP, and ZIF proteins generally does not disrupt base pairing, but the DNA is usually bent to different extents (Figure 17.31).

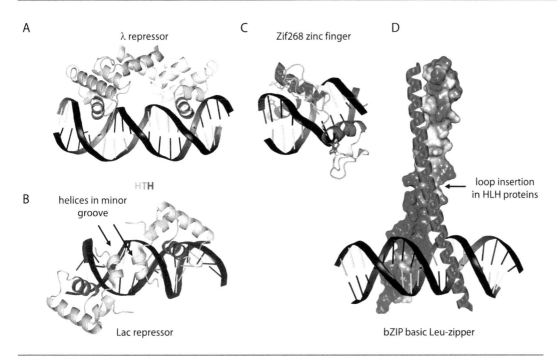

**FIGURE 17.31:** **α-helical DNA binding motifs.** A: The λ-repressor (PDB-ID 1lmb) is a dimeric protein that binds to the operator of the *cro* gene of the viral DNA to maintain a lysogenic (silent, non-virulent) life cycle. The recognition helices (red) are 3.4 nm apart, corresponding to a full turn of B-DNA. B: The Lac repressor, which shuts down transcription of the *lac* operon when bound to its operator DNA (PDB-ID 1efa), not only binds to the major groove with an HTH motif, but strongly distorts the minor groove of the DNA by additional α-helices (yellow). C: In the eukaryotic transcription factor Zif268, three consecutive zinc fingers, spaced at intervals of 3 base pairs, bind the major groove with α-helices (PDB-ID 1aay). D: bZIP proteins also recognize DNA by way of an α-helix in the major groove. One of the amphipathic helices of the dimeric viral Jun protein is shown with its electrostatic surface potential to highlight the basic DNA-binding region and the hydrophobic leucine zipper (PDB-ID 2h7h).

In contrast to the major groove, the diversity of H-bond donors and acceptors in the minor groove is not high enough to enable sequence-specific binding of proteins. In rare cases, the minor groove can also host an α-helix from an HTH motif or another DNA-binding motif. The Lac repressor is such an example where the minor groove is distorted so much that two α-helices (not belonging to an HTH motif) fit into it side-by-side (Figure 17.31). The minor groove of DNA either interacts with a single β-strand or loop or, after widening and DNA bending or kinking, with a whole β-sheet as in the case of TBP (Figure 17.29).

---

### BOX 17.3: GENE EDITING BY PROGRAMMABLE ZINC FINGER NUCLEASES.

Many human diseases are the result of point mutations in genes. Traditional gene therapy seeks to repair the lost function of the mutated gene by adding another copy, often using a virus as a vector. The best way to treat a genetic disease would be to mutate the affected gene in the germline back to its wild-type sequence. This would work if the DNA double-strand of the human genome ($3.2 \cdot 10^9$ base pairs) could be cleaved at precisely the location of the mutation, and the repair mechanisms of the cell would then eliminate the mutation. A novel technique using *programmable zinc finger nucleases* (ZFN) promises to achieve just that. ZFN are chimeric enzymes with a specific DNA-recognition part made of a sequence of several zinc fingers fused to the DNA-cleaving part from the unspecific bacterial restriction endonuclease FokI. Zinc fingers of different specificities can be concatenated to recognize virtually any DNA sequence, which is important for cleaving the genome only once. The FokI nuclease is a monomer in solution but active as a dimer when bound to DNA, so two different sets of zinc finger

*(Continued)*

concatemers can be fused to FokI and then mixed to obtain a functional dimer with eight zinc fingers. The zinc fingers anchor FokI at a specific location on the DNA, which is then cleaved by the unspecific nuclease 9 and 13 bases away from the recognition site dictated by the zinc fingers. A series of four zinc fingers per FokI is sufficient to establish a unique address in the human genome: four zinc fingers recognizing three base pairs each, times two for the FokI dimer, is $4^{24} = 2.8 \cdot 10^{14}$, much larger than the human genome. The result of the asymmetric 9/13 cleavage by FokI is a double-strand break with overhangs. If an oligonucleotide with the healthy gene sequence is provided together with the ZFN, homologous recombination by the cell can replace the mutated region with the correct one. The technique has been used for gene editing (mutation, deletion, insertion) in animals and plants. A variation of this approach uses the CRISPR-Cas9 system (clustered regulatory interspace short palindromic repeats): the bacterial nuclease Cas9 is used for cleavage, and an RNA complementary to the target site is used for specificity.

Homoeobox proteins have a canonical HTH motif but, in addition, bind into the minor groove with a loop region extending away from the protein. The MetJ repressor (Figure 17.32) is the prototypical representative of the β-hairpin/ribbon class of DNA-binding proteins. These repressor proteins fall into six families of divergent structures, but all use β-sheets for DNA recognition. MetJ is a homodimer with a protein-protein interface constructed by a helix bundle and a single β-strand. The resulting two-stranded antiparallel β-sheet is placed into the major groove almost perpendicular to the edges of the nucleobases so that the side chains of only one side of the β-sheet scan the DNA sequence for H-bonding partners. The integration host factor IHF dramatically bends its target DNA by almost 160° (Figure 17.32). IHF is an *E. coli* protein that is used by the bacteriophage λ to integrate the viral genome into the bacterial chromosome. IHF is a heterodimer with a β-hairpin extending from each monomer into the minor groove. A single β-strand of the hairpin intercalates its side chains, most importantly a conserved Pro, into the groove, thereby kinking the DNA. The two kinks induced by the two monomers add to an almost 160° bend of the DNA towards the IHF dimer. All six α-helix dipoles in IHF are oriented with their N-terminus towards the phosphodiester backbone for favorable electrostatic interactions (see Figure 16.17).

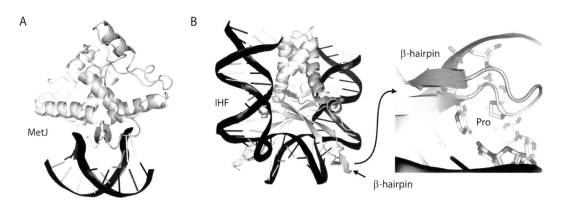

**FIGURE 17.32:**   β-strand DNA binding motifs. A: Two β-strands (yellow) in the major groove are used for sequence-specific DNA recognition by the MetJ repressor (PDB-ID 1cma). B: The minor groove is the binding site for a single β-strand of the integration host factor (PDB-ID 1ihf). A two-stranded β-sheet inserts the edge of a single strand into the minor groove. The loop region connecting the strands has a conserved Pro residue that intercalates between two base pairs, contributing to the DNA kinking.

### 17.5.4.2 Small Molecule Binding to DNA

Binding of small molecules to DNA uses the same kinds of interactions as in protein-protein and protein-nucleic acid interactions, i.e. H-bonds, ionic, and van der Waals interactions, but can take place in ways that are not accessible to macromolecules. Three main mechanisms for small molecule binding can be distinguished, *ion condensation*, minor groove binding, and *intercalation*.

Ion condensation is the unspecific binding of metal ions and other positively charged ions, such as spermine and spermidine to the phosphate backbone of nucleic acids. On average, 0.8–0.9 positive charges per phosphate group are associated with DNA. Metal ions, particularly $Na^+$ and $Mg^{2+}$ *in vivo*, are required for double helix formation because they counteract electrostatic repulsion of the negative charges of the phosphate backbone: the charges would otherwise be too close to allow formation of a stable duplex. The large entropic loss of ion condensation is over-compensated by the favorable enthalpic contribution of strong ion pair formation. The ions are not fixed to specific binding sites: they can slide along the backbone. Ions often bind to the phosphate backbone with their hydration shells intact.

Unlike most proteins, which bind to the major groove of DNA, small molecules other than metal ions either bind to the minor groove of DNA or they intercalate between base pairs. *Intercalation*, the insertion of a hydrophobic moiety between two base pairs, is driven by the hydrophobic effect, while minor groove binding is dominated by both H-bonds and van der Waals interactions. The small molecules shown in Figure 17.33 are planar, positively charged molecules with an overall crescent shape that H-bond into the minor groove.

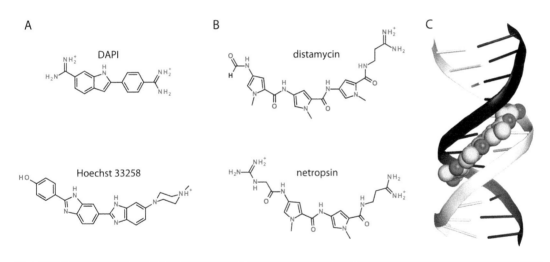

**FIGURE 17.33:** **Molecules binding to the minor groove of DNA.** A: DAPI (4',6-diamidino-2-phenyl-indole) and Hoechst 33258 are used as tools in fluorescence microscopy. Both molecules bind to AT-rich sequences in double-stranded DNA, accompanied by an increase in their blue fluorescence at 460 nm. B: Netropsin and distamycin are small polymers of methyl-pyrrole connected by amide bonds. Their curvature follows that of the minor groove, as can be seen from the distamycin/DNA complex in panel C (PDB-ID 267d). All molecules have positive charges as amidinium or ammonium groups, which electrostatically interact with the phosphodiester backbone.

DAPI and the Hoechst dyes are minor groove binders that are used as tools for fluorescent DNA staining. The polyamides netropsin and distamycin are natural products with antibiotic and antiviral activity. Their curved shapes follow that of the DNA minor groove. Distamycin contacts up to four consecutive base pairs. Minor groove binding small molecules form H-bonds with their NH-groups to the N3-atom of adenine and O2-atom of thymine. Hydrophobic interactions are formed between the aromatic groups and the nucleobases. The water molecules normally located in the minor groove as the spine of hydration are displaced by the small molecule, so that binding is entropically favored. All of the minor groove binders in Figure 17.33 prefer AT-rich sequences. Binding to GC-rich sequences is disfavored because the 2-amino group of guanine bases that points into the minor groove (see Figure 17.17) sterically hinders the binding of the small molecule.

Molecules with large aromatic planes that have similar sizes to base pairs can intercalate between them without interrupting the base pairing. The strong stacking interaction between the base pairs and the intercalator releases more energy than is required to first un-stack the base pairs and locally unwind the DNA to allow binding of the intercalator. Each intercalated molecule lengthens the DNA by about 0.3 nm. A linear extension of the DNA would require lengthening by the van der Waals distance between aromatic systems (ca. 0.36 nm). As a result, the DNA is strongly unwound and rigidified by intercalation (Section 17.5.3.2), which can interfere with the action of topoisomerases to such an extent that DNA replication and transcription become impossible. Some intercalators such as the *Streptomyces*-derived anthracyclines are therefore efficient anti-cancer drugs (Figure 17.34).

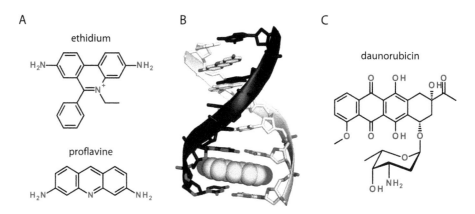

**FIGURE 17.34:** **DNA intercalating molecules.** A: Amino groups of ethidium and proflavine H-bond to the phosphodiester backbone while the aromatic parts intercalate into DNA. B: Two proflavines are bound to the self-complementary DNA hexamer 5′-CGATCG-3′ (PDB-ID 3ft6) and are rendered as sticks and as space-filling model. C: Daunorubicin is a prototypic anthracycline. In addition to the intercalating aromatic part, anthracyclines contain additional sugar or peptidic groups that can bind to the minor groove.

## 17.6  RNA STRUCTURE

Many of the principles we discussed for DNA with respect to base pairing and stacking, H-bonding to protein side chains, and ligand binding also apply to RNA. The main differences between DNA and RNA are in their preferred double helix conformations, their propensity to fold into complex shapes, and their chemical stability. All of these differences can be traced back to the presence of the 2′-hydroxyl group in RNA.

The presence of the 2′-hydroxyl group in RNA strongly disfavors the B-form: in B-form RNA, the 2′-hydroxyl group would be in the C2′-*endo* conformation and clash with the nucleobase of the next nucleotide. Thus, the presence of the 2′-hydroxyl group changes the ribose pucker from C2′-*endo* to C3′-*endo* (see Figure 17.8). As small as this structural change may seem, it has huge consequences on the conformation of the phosphodiester backbone: the torsion angles of the backbone and the ribose ring are highly correlated, so their preferred values depend on the presence or absence of the 2′-hydroxyl group. The preferred helical structure of RNA is therefore the A-form, where the 2′-hydroxyl group does not clash. Instead, the 2′-hydroxyl group is within H-bonding distance to the O4′ oxygen atom of the next ribose in the 3′-direction, which may contribute to A-form stabilization. The B-form is disfavored to such an extent in RNA that even DNA/RNA heteroduplexes have structurally more in common with the A-form than the B-form. The structural differences between DNA A- and B-forms (Section 17.5.1) also apply to the RNA duplex: compared to B-DNA, the RNA double helix has much more tilted base pairs, and they are displaced away from the center of the helix toward the outside (see Figure 17.17). As a result, the minor groove of double-stranded RNA is wide and shallow while the major groove is narrow and deep. RNA recognition by proteins thus requires other structural elements than DNA recognition (Section 17.5.4.1).

The additional H-bonding capacity from the 2′-hydroxyl group strongly increases the structural variability of RNA compared to DNA. In fact, single-stranded RNA can fold into compact structures, much like a protein, where local secondary structure elements are connected by interactions between nucleotides far away in sequence. Aside from the usual Watson-Crick base pairs, there are mismatched, Hoogsteen, purine-purine, pyrimidine-pyrimidine, and reversed base pairs (see Section 17.4) in RNA that are not part of the usual structural repertoire of DNA. A number of unusual nucleobases that are post-transcriptionally introduced into RNA further increase the chemical, and thus structural, diversity of RNA compared to DNA (see Section 17.1.2).

### 17.6.1 RNA Secondary Structure

Complementary sequences in RNA molecules will pair into duplexes, even if the resulting structure leaves a few unpaired bases in *loops* and *bulges* (Figure 17.35). Three or four consecutive base pairs are sufficient to form a *stem-loop* structure. A *hairpin* is a stem-loop formed by a single RNA molecule where the complementary sequences that form the duplex (stem) necessarily have opposite directions. The loop in a hairpin is closed by a single base pair. This loop can, in principle, have any number of nucleobases but the minimum is the *tetraloop* (Figure 17.35) with prevalent consensus sequences CUUG, GNRA, and UNCG (N being any nucleotide and R is any purine). Tetraloops adopt defined conformations that are stabilized by H-bond networks and that provide docking sites for other RNA structural elements during RNA folding. In general, unpaired bases in loops can engage in H bonds with unpaired bases from other elements.

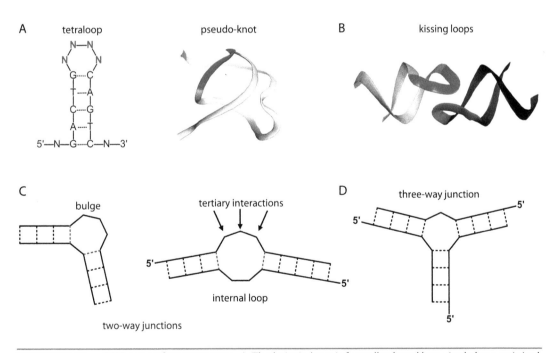

**FIGURE 17.35:    RNA secondary structure.** A: The hairpin loop is formally closed by a single base pair (red dashes). "N" represents any nucleotide and dashed lines symbolize base pairings. A pseudo-knot is formed when a single strand of RNA contacts base pairs with the tetraloop in the same molecule. B: Two-way junctions have unpaired regions between helical parts. Bulge loops have unpaired nucleotides in just one strand, whereas internal loops have unpaired nucleotides in both strands. Internal loops can be symmetric or asymmetric with the same or different number of unpaired nucleotides in the two strands. In contrast to hairpins, bulges and interior loops are closed by two base pairs (red dashes). D: Three-way junction. Four-way junctions are also possible (see the tRNA structure in the next section).

A single strand contacting the tetraloop in a hairpin to form base pairs is the simplest conceivable *pseudo-knot*, termed pseudo because pulling at the RNA strands would completely unravel the structure. The *kissing loop* motif is a structure of two hairpin loops that contact each other to form a co-axial, pseudo-continuous RNA duplex (Figure 17.35). Similar to DNA helical junctions, RNA also forms two-way, three-way, and four-way junctions. Bulges and *interior loops* are two-way junctions formed by one or two RNA strands. In contrast to the terminal hairpin loop, the internal loops in two-way junctions are located between two helical structures and are closed off by two base pairs.

Helical junctions are ubiquitous elements of RNA structure. A varying number of unpaired nucleotides directly at the junction or as internal loops in the helical parts provides docking sites for tertiary interactions by Watson-Crick and Hoogsteen base pairings, leading to pseudo-knots and kissing loops.

## 17.6.2 RNA Tertiary Structure

The tertiary structure of tRNA explains many principles of how RNA molecules fold. In its standard *cloverleaf* representation, tRNA consists of four base-paired regions, three of which are hairpins and one is the acceptor stem where the 5′- and 3′-ends are found. Base triplets are formed between the dihydrouridine (DHU) and pseudouridine (TΨC) loops, transforming the cloverleaf into a compact *L-shape*, which is typical for all tRNAs (Figure 17.36). The base-paired regions in the acceptor stem, anticodon stem, D-arm, and T-arm adopt standard A-RNA structures of variable length. In addition, the anticodon stem and D-arm helices co-axially stack on top of each other, as do the helices of the acceptor stem and the T-arm. The overall tRNA structure thus comprises two A-RNA helices of about 10–12 base pairs that are perpendicular to each other. The anticodon loop, which decodes the information contained in the mRNA, is located at one end of the L-shape. At the other end of the L-shape is the unpaired CCA-sequence that is esterified with an amino acid by an *aminoacyl-tRNA synthetase*.

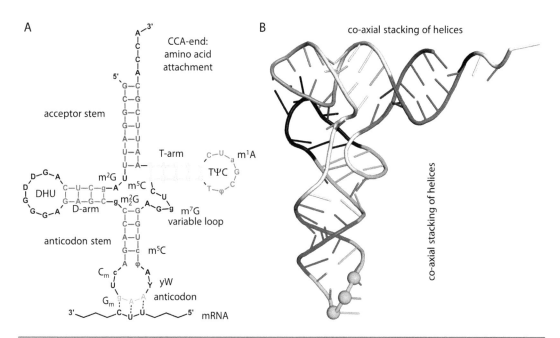

**FIGURE 17.36:** **Structure of tRNA.** A: cloverleaf representation of tRNA. B: the L-shape of tRNA is due to interaction of nucleobases from the TΨC- and DHU-loops. Two perpendicular RNA double helices are formed by several RNA pieces. The same color code is used for both representations. The sizes of the TΨC- and DHU-loops and the number of base pairs in the stems and arms are quite variable, but the total number of stacked base pairs in all tRNAs mostly ranges between 10 and 12 so that the overall shape of the molecule is conserved. The anticodon triplet is marked by three spheres. These unpaired bases stack on top of each other, ready to decode mRNA.

Similar to protein folding, folding of tRNAs minimizes solvent-exposed hydrophobic areas and at the same time maximizes polar and non-polar interactions. In tRNA, this is achieved by formation of base triplets, co-axial stacking of short RNA helices, and formation of kissing loops and pseudo-knots. Co-axial stacking results in a longer helix that is quasi-continuous with uninterrupted base pair stacking along its entire length, which is energetically favorable. Co-axial stacking is the most frequently observed motif for RNA structure stabilization. The structural elements found in tRNA are present in many other RNA molecules. In addition, triple helices are observed, where a single strand of RNA lies in the minor groove of an A-RNA duplex. This is possible because the minor groove in A-RNA is larger than the major groove (see Table 17.1). By contrast, the narrow and deep major groove can host small molecules.

The distinct tertiary structure of RNA arranges nucleobases into the catalytic sites of *ribozymes* and the specific binding sites for ligands in *riboswitches*. Riboswitches are small structural domains in the 5′-untranslated regions of mRNAs that sense the concentration of a metabolite or second messenger. Upon binding of their ligand, they undergo a conformational change (switch) that may regulate the transcription and/or the translation of the mRNA carrying the riboswitch. Riboswitches therefore provide a feedback mechanism for the regulation of biosynthetic pathways including enzyme cofactors such as cobalamin, flavin mononucleotide (FMN), tetrahydrofolate (THF), thiamine pyrophosphate (TPP), S-adenosyl methionine (SAM), and the molybdenum cofactor Moco, as well as amino acids (Gly, Gln, Lys) and purine nucleobases. Four different classes of SAM riboswitches have been described, which bind S-adenosyl methionine, an intermediate of Met biosynthesis, and regulate the methionine biosynthetic pathway. The structure of a class II SAM-riboswitch in complex with SAM is shown in Figure 17.37. It consists of two short regular A-RNA helices and two loop regions. The central part of the riboswitch is a triple helix where the triplets use both the Watson-Crick and Hoogsteen faces of their nucleobases, similar to DNA triple helices. All three helical segments stack on top of each other to form an overall linear pseudo-helical structure. The SAM binding site is in the major groove of the triple helix. The adenine moiety of SAM participates in a base triplet. In absence of SAM, the riboswitch adopts a different structure.

Ribozymes are RNA molecules with catalytic activity, which is an interesting property for nucleic acids because, prior to their discovery in the 1980s, it was assumed that only proteins could act as catalysts. Some ribozymes, such as self-splicing introns, work as naked RNA while others are parts of *ribonucleoprotein complexes*. The ribosome (Figure 17.37) is arguably the most complicated ribozyme and ribonucleoprotein complex. Its 23S rRNA (prokaryotes) or 28S rRNA (eukaryotes) on the large subunit harbors the peptidyl transferase activity for protein biosynthesis. The catalytic nucleobases of the RNA are arranged as a result of RNA folding and formation of the ribonucleoprotein complex.

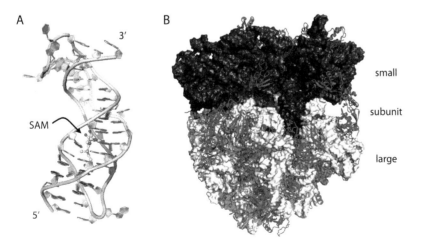

**FIGURE 17.37:** **Riboswitches and RNA-protein complexes**. A: SAM riboswitch (PDB-ID 2qwy). Watson-Crick base pairs are shown in gray. The magenta-colored stretch forms the triple helix where SAM (yellow) binds. Green nucleotides form non-standard base pairs and cyan nucleotides have solvent-exposed nucleobases. B: The 70S ribosome of the thermophilic bacterium *Thermus thermophilus* (PDB-ID 4w2f) shows that ribosomes are essentially large masses of RNA (several MDa, shown as surfaces) held together by a few (<100) proteins.

### 17.6.3 RNA FOLDING

The folding of RNA molecules into their three-dimensional structure follows similar rules to protein folding. The hydrophobic effect, which excludes water from the non-polar surfaces of the nucleobases, drives optimal stacking of base pairs, both within a single helix or between co-axially stacked helices. Buried H-bond donors and acceptors are energetically unfavorable if unpaired, and hence most of the H-bond donors and acceptors in RNA structures are saturated. A special driving force for RNA folding is the electrostatic repulsion of the negative charges on the phosphate backbone, which are solvent-exposed in the folded structure. Tertiary structure formation in RNA leads to close approach of the negatively charged phosphate backbone. Hence, RNA folding is highly dependent on the ionic strength: metal ions bind to and neutralize the negative charges of the phosphate backbone.

RNA folding already starts during transcription as soon as the RNA transcript leaves the RNA polymerase, leading to sequential folding of RNA. Intermediates with stable secondary structures form before transcription of the correct interaction sequence is finished. The intermediates from sequential folding are much more stable than protein folding intermediates and difficult to resolve. Thus, RNA folding may get trapped in local energetic minima corresponding to conformations that may be very different from the native structure. The local minima in the RNA energy landscape are separated by high activation barriers, and thus these minima are very effective *kinetic traps*. A progressive movement towards the global energy minimum as with proteins is therefore less likely with RNA. Instead, intermediate RNA structures have to be dissolved and re-annealed in order to allow them to fold into a more stable structure. RNA helicases are enzymes that use the energy from ATP to dissolve local RNA structure. They can act as "RNA chaperones" and catalyze structural rearrangements of misfolded RNA intermediates.

## QUESTIONS

17.1 Draw the structure of 2′-deoxy-GTP at physiological pH including all stereo centers. Assign the two different $pK_A$-values (about 2 and 7) to the phosphate oxygen atoms. How many charges does dGTP carry under physiological conditions? Explain the difference between a phosphoric acid anhydride and a phosphoester from this structure.

17.2 Draw the structure of deprotonated 1-methyl-adenosine. Can you suggest why the protonated form is much more stable? What about 3-methyl-cytosine?

17.3 Free riboses are hemi-acetals and can open to an aldehyde. The closed furanose is much more stable than the open aldehyde. Upon closure of the aldehyde, two anomeric configurations can form that have different chirality at C1′. Draw the aldehyde form of ribose and explain how the two anomers can be formed upon ring closure.

17.4 Bromination of guanosine at C8 favors the *syn*-conformation. What could be the explanation?

17.5 At room temperature and neutral pH, RNA spontaneously hydrolyzes about 100 times more quickly than DNA. The rate of RNA hydrolysis is considerably increased in alkaline solutions. What might be the reason and which chemical reaction is responsible for RNA instability?

17.6 Draw the G-C Hoogsteen base pair. Why is a protonated cytosine required?

17.7 The A-T base pair is more stable than the A-U base pair. Can you explain?

17.8 Draw the enol forms of G and T. Why can they not base-pair with C and A, respectively?

17.9 The G-U Watson-Crick base pair is shown. Why can 2-thiouracil not base-pair with G but only with A?

**17.10** Inosine (shown) is often present at the first position of the anticodon of tRNA (the wobble position) where it recognizes U, C, and A. Draw the three Watson-Crick base pairs for inosine.

**17.11** The codon AUA codes for isoleucine, while AUG codes for methionine. How can inosine distinguish between the two codons?

**17.12** A plasmid of 1000 bp has a superhelical density of $\sigma = -0.05$. Calculate $Lk$, $Lk^0$, and $\Delta Lk$. How does $\sigma$ change after 5 catalytic cycles of DNA supercoiling by gyrase?

**17.13** Why is DNA relaxation an exergonic reaction although the DNA backbone needs to be cleaved?

**17.14** What is the overall bend of a DNA where 4 poly-A tracts are each spaced by 5 random base pairs?

**17.15** Acidic side chains can form a bi-dentate H-bond to the Watson-Crick face of guanine bases. An Asp side chain binds to the guanine in Ras, as shown in the figure.

A variant of Ras, Asp119Asn, shows much weaker affinity for guanine nucleotides but strongly binds to xanthine nucleotides. Xanthine differs from guanine by substitution of an oxygen atom for the exocyclic amino group. How could the switch in nucleotide specificity be explained?

## REFERENCES

Egli, M. and Saenger, W. (1988) Principles of Nucleic Acid Structure, Springer.

Blackburn, G. M. (2006) Nucleic acids in chemistry and biology. Cambridge, UK, RSC Pub.

Sinden, R. R. (1994) DNA structure and function. San Diego, Academic Press.
· comprehensive treatment of all aspects of nucleic acid chemistry and structure

Ornstein, R. L. and Rein, R. (1978) An optimized potential function for the calculation of nucleic acid interaction energies I. base stacking. *Biopolymers* **17**(10): 2341–2360.

Gotoh, O. and Tagashira, Y. (1981) Stabilities of nearest-neighbor doublets in double-helical DNA determined by fitting calculated melting profiles to observed profiles. *Biopolymers* **20**(5): 1033–1042.
· base stacking

Sokoloski, J. E., Godfrey, S. A., Dombrowski, S. E. and Bevilacqua, P. C. (2011) Prevalence of *syn* nucleobases in the active sites of functional RNAs. *RNA* **17**(10): 1775–1787.
· prevalence of *syn* nucleobases in the active sites of functional RNAs

Rich, A. and Zhang, S. (2003) Timeline: Z-DNA: the long road to biological function. *Nat Rev Genet* **4**(7): 566–572.
· review on Z-DNA

Haran, T. E. and Mohanty, U. (2009) The unique structure of A-tracts and intrinsic DNA bending. *Q Rev Biophys* **42**(1): 41–81.
· review on DNA bending

Potratz, J. P., Del Campo, M., Wolf, R. Z., Lambowitz, A. M. and Russell, R. (2011) ATP-dependent roles of the DEAD-box protein Mss116p in group II intron splicing *in vitro* and *in vivo*. *J Mol Biol* **411**(3): 661–679.
· ATP-dependent unwinding to help RNA folding

Pan, C., Potratz, J. P., Cannon, B., Simpson, Z. B., Ziehr, J. L., Tijerina, P. and Russell, R. (2014) DEAD-box helicase proteins disrupt RNA tertiary structure through helix capture. *PLoS Biol* **12**(10): e1001981.
· RNA tertiary structure disruption by an RNA helicase

Kouzine, F. and Levens, D. (2007) Supercoil-driven DNA structures regulate genetic transactions. *Front Biosci* **12**: 4409–4423.

Kouzine, F., Sanford, S., Elisha-Feil, Z. and Levens, D. (2008) The functional response of upstream DNA to dynamic supercoiling *in vivo*. *Nat Struct Mol Biol* **15**(2): 146–154.

Kouzine, F., Gupta, A., Baranello, L., Wojtowicz, D., Ben-Aissa, K., Liu, J., Przytycka, T. M. and Levens, D. (2013) Transcription-dependent dynamic supercoiling is a short-range genomic force. *Nat Struct Mol Biol* **20**(3): 396–403.
· supercoiling *in vivo*

Krishna, S. S., Majumdar, I. and Grishin, N. V. (2003) Structural classification of zinc fingers: survey and summary. *Nucleic Acids Res* **31**(2): 532–550.
· review on zinc fingers

Devi, G., Zhou, Y., Zhong, Z., Toh, D. F. and Chen, G. (2015) RNA triplexes: from structural principles to biological and biotech applications. *Wiley interdisciplinary reviews RNA* **6**(1): 111–128.
· review on RNA triplexes

Gilbert, S. D., Rambo, R. P., Van Tyne, D. and Batey, R. T. (2008) Structure of the SAM-II riboswitch bound to S-adenosylmethionine. *Nat Struct Mol Biol* **15**(2): 177–182.
· structure of the SAM riboswitch bound to SAM

## ONLINE RESOURCES

modomics.genesilico.pl/modifications
· database of RNA modifications

# Computational Biology

*B*  *ioinformatics* became a research area in the late 1970s when computer power started to increase, and more and more data could be handled and stored. Statistical methods were developed to analyze data that were stored and curated in databases. The first databases stored sequences for proteins and DNA, and with the advent of whole genome sequencing and high throughput mass spectrometric protein sequencing, these databases continue to be the largest that we have. Genome annotation remains one of the most fruitful activities in bioinformatics, identifying and ascribing likely biological functions to genes. However, it quickly became clear that the number of genes in a genome is too small to explain the large variety of proteins and functions in the cell. Recent databases contain content from other "*omics*" fields such as "transcriptomics" (the sum of all mRNAs), "proteomics" (all of the proteins in the cell), "metabolomics" (the pool of metabolites and their interconversion in cells), "glycomics" (the glycosylation patterns in organisms), and "interactomics" (the sum of all interactions between molecules in the cell). In line with the term *genome* for the genetic material, the new datasets are aptly named *transcriptome, proteome, metabolome, glycome*, and *interactome*, respectively. A survey of the methods in bioinformatics is given for example at www.expasy.org and www.click2drug.org. Here, we will look at only a small subset of methods used by scientists from all disciplines on a regular basis.

## 18.1 SEQUENCE ANALYSIS

### 18.1.1 Sequence Composition, Global Properties, and Motifs

Sequences may be analyzed in isolation without any other source of information, or in conjunction with other sequences. Analysis of a single protein or nucleic acid sequence allows prediction of biophysical properties of the molecule and to identify functional elements. Comparison of several sequences reveals conserved regions that can be related to function. Whether a protein is folded or unfolded, or soluble at a certain pH, can often be predicted from its sequence. Nucleic acid sequences contain many functional signatures such as splice sites (RNA), restriction sites (DNA), and binding sequences for transcription factors and other proteins (both). Non-coding RNA sequences can be analyzed for inverted repeats, sequences involved in transcriptional regulation, and riboswitches.

### 18.1.1.1 DNA Sequences

One interesting aspect of a single gene or mRNA sequence is the *codon usage*. The genetic code is degenerate, and with the exception of Met and Trp, most amino acids are represented by 2–6 codons. Usage of these codons is a key factor for the efficient translation of an mRNA from one organism in another organism. Fast-growing organisms such as *S. cerevisiae* or *E. coli* have codon usages that reflect their tRNA concentrations. In order to maximize the translation of an mRNA from one organism (e.g. human) in another (e.g. *E. coli*), the sequence of the transcribed gene can be optimized to match the codon preferences of the target organism. Several online servers are available to optimize the cDNA sequence or generate an optimized synthetic gene from a protein sequence. Table 18.1 compares the codon usage of human and *E. coli* genes.

**TABLE 18.1**
**Codon usage (in %) in *E. coli* and *H. sapiens*.**

| | | | | | | | | | | | |
|---|---|---|---|---|---|---|---|---|---|---|---|
| UUU | F | 58/46 | UCU | S | 14/19 | UAU | Y | 57/44 | UGU | C | 45/46 |
| UUC | F | 42/54 | UCC | S | 15/22 | UAC | Y | 43/56 | UGC | C | 55/54 |
| UUA | L | **13**/8 | UCA | S | 14/15 | UAA | * | **59**/30 | UGA | * | 33/47 |
| UUG | L | 13/13 | UCG | S | **15**/5 | UAG | * | 8/**23** | UGG | W | 100/100 |
| | | | | | | | | | | | |
| CUU | L | 11/13 | CCU | P | 17/29 | CAU | H | 58/42 | CGU | R | **36**/8 |
| CUC | L | 10/**20** | CCC | P | 13/**32** | CAC | H | 42/58 | CGC | R | **37**/18 |
| CUA | L | 4/7 | CCA | P | 19/28 | CAA | Q | 33/27 | CGA | R | 7/11 |
| CUG | L | 49/40 | CCG | P | **51**/11 | CAG | Q | 67/73 | CGG | R | 11/20 |
| | | | | | | | | | | | |
| AUU | I | 50/36 | ACU | T | 17/25 | AAU | N | 47/47 | AGU | S | 16/15 |
| AUC | I | 40/47 | ACC | T | 41/36 | AAC | N | 53/53 | AGC | S | 27/24 |
| AUA | I | 9/17 | ACA | T | 15/28 | AAA | K | **76**/43 | AGA | R | 5/**21** |
| AUG | M | 100/100 | ACG | T | **27**/11 | AAG | K | 24/**57** | AGG | R | 3/**21** |
| | | | | | | | | | | | |
| GUU | V | 26/18 | GCU | A | 16/27 | GAU | D | 63/46 | GGU | G | **33**/16 |
| GUC | V | 21/24 | GCC | A | 27/40 | GAC | D | 37/54 | GGC | G | 39/34 |
| GUA | V | 16/12 | GCA | A | 22/23 | GAA | E | **68**/42 | GGA | G | 12/**25** |
| GUG | V | 37/46 | GCG | A | **35**/11 | GAG | E | 32/58 | GGG | G | 16/25 |

*Note:*   Data (in %) were generated using the Codon Usage Database. The first number represents *E. coli*, the second the human codon. The coded amino acid is given in single letter code.
* denotes a stop codon. Codons with >50% difference in usage between *E. coli* and human are bold.

### 18.1.1.2 RNA Secondary Structure Prediction

Using the conserved arrangements of base pairings in RNA, secondary structures can be predicted by RNA folding algorithms, and relative energies may be assigned to the structures. The result is a schematic plot of the predicted secondary structure with lowest energy, similar to the drawing shown in Figure 17.36. Base-paired regions and stem-loop structures that are close in sequence are predicted with up to 75% accuracy. Different programs will calculate similar lowest energy structures, but may assign different energies to them due to different treatment of ionic strength or differences in the energy terms used for RNA folding. The most widely used energy terms for RNA folding programs are based on the *Turner1999* potential. Inaccuracies in the predicted structures can result from base pair stacking in larger loops or junctions that require different energy parameters compared to standard A-RNA. The larger the RNA sequence, the more energetically similar secondary structures are predicted, and the more difficult it is to find the structure that in solution is most prevalent. The accuracy of the folded structure is increased when several sequences of functionally related RNAs are available that are expected to have the same secondary structure. Alignment of these sequences will identify conserved regions, which usually form the same secondary structures in different RNAs.

Nevertheless, there will be many predicted RNA secondary structures with comparable energy, so the lowest energy arrangement of base pairs represents only a small amount of all the structures expected in solution. RNA folding algorithms usually calculate the *partition function*, i.e. the expected distribution of the predicted lowest energy structures. Most algorithms start from the complete RNA sequence void of any local structure, but *in vivo*, RNA will start to fold as soon as it leaves the RNA polymerase. Co-translational RNA folding is sequential (Section 17.6.3) and will install secondary structure between bases that are close in sequence. As a result, the calculated lowest energy arrangements using RNA folding algorithms may represent only a small fraction (<10%) of the predominant RNA structures *in vivo*. The difficulties with RNA secondary structure predictions extend to tertiary structure prediction algorithms. Currently, no program can accurately predict three-dimensional structures of RNA, probably because the torsional flexibility of RNA is even larger than that of proteins. In addition, cation binding has a large effect on RNA structure, but is difficult to model. Coarse-grained modeling (Section 18.2.4.4), which substitutes pseudo-atoms for groups of atoms in a macromolecule and thus simplifies the computational problem, has been used to predict the three-dimensional structure of small (<40 nucleotides) RNA molecules to an accuracy of 0.2–0.5 nm rmsd to their experimentally determined structures.

### 18.1.1.3 Protein Sequence Composition and Properties

The N-terminus of a protein sequence has predictive power on the estimated half-life of the protein in human, yeast, and *E. coli* cells, called the *N-end rule*. If the N-terminal residue happens to be R, L, K, F, W, or Y, then the half-life of the protein in *E. coli* is in the range of minutes. For recombinant protein production purposes in *E. coli*, a different N-terminal residue type should therefore be chosen, potentially extending the half-life to >10 h. Among the parameters that can be calculated from the entire sequence are the molecular mass, the extinction coefficient, the instability index, the distribution of amino acid residues, and the expected isoelectric point (pI). If the calculated molecular mass does not match the results from mass spectrometry (Section 26.1), post-translational modifications (Section 16.1.4) may be deduced from the mass differences. By counting the number of absorbing amino acids, the expected molar extinction coefficient of the unfolded polypeptide chain can be calculated to an error of <10% from the true value. More accurate values require the Gill & von Hippel method (Section 19.2.7.1). The instability index is based on the occurrence of certain dipeptides in the sequence of unstable proteins. Comparison of the frequency of all 400 possible dipeptides in stable and unstable proteins produced a table of dipeptide instability weight values (DIWV). The instability index of a novel protein sequence of length $L$ is then obtained by summing all DIWVs over all dipeptides $x_i x_{i+1}$ at position $i$ along the sequence (eq. 18.1). Values <40 predict a protein to be stable.

$$\text{instability index} = \frac{10}{L} \cdot \sum_{i=1}^{L-1} \text{DIWV}\left(x_i x_{i+1}\right) \qquad \text{eq. 18.1}$$

The frequency of the twenty amino acid residues in a protein can be compared to the mean distribution in a large set of proteins to gather information on its solubility and stability. If the frequency of hydrophobic amino acids is drastically different from the average, this might point to a membrane protein. For instance, bacteriorhodopsin from *Halobacterium halobium* has considerably more Leu/Gly and much less Glu/Lys than the average protein (Figure 18.1).

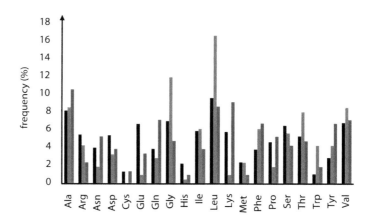

FIGURE 18.1:    **Residue frequency in a soluble and a membrane protein.** The black bars are the average frequencies of the twenty amino acids in the UniProt database from 2013. The membrane protein *H. halobium* rhodopsin (blue) has more Gly and Leu residues than average, while the soluble *D. melanogaster* glutathione S-transferase (GST, red) more closely matches the average distribution.

The number of charges of the protein sequence can be calculated as a function of pH. Strictly, this *in silico* titration is only valid for an unfolded protein where all ionizable side chains (Asp, Glu, His, Arg, Lys, termini) are fully solvent-exposed and their $pK_A$ values (see Table 16.1) are not perturbed by the protein environment or the solvent. Since most of the charges of a protein are located at the surface, the pI of the folded protein is often within ±0.5 pH units of the calculated pI. The isoelectric point of the protein sequence is the pH at which the overall charge is zero (Figure 18.2) and the solubility of the protein is minimal. This information can be used to keep proteins from aggregating during their purification and concentration by buffering at a pH very different from the pI.

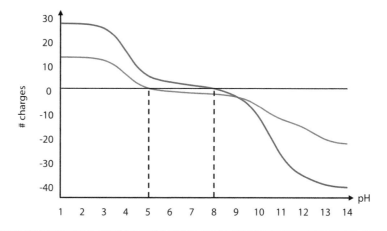

FIGURE 18.2:    **Titration of polypeptides.** GST from *D. melanogaster* (red) and bacteriorhodopsin from *H. halobium* (blue) have quite different pI values (dashed lines at pH 8 for GST and pH 5 for bacteriorhodopsin), and different steepness of the titration curves around the pI. The membrane protein has a lower number of total charges than the soluble protein, especially in the physiological pH range.

The steepness of the titration around the pI can be a guide for how far away from the pI and in which direction the pH of the protein buffer should be. For example, *D. melanogaster* GST (Figure 18.2) has a pI of 8 and will accumulate fewer charges when shifting the pH to 6 (ca. −6 charges) than when shifting the pH to 10 (ca. +10 charges). Note, however, that counter ions may screen surface charges on the proteins. These are not taken into consideration when calculating the charges as a function of pH, so the ionic strength of the buffer is an additional parameter to consider.

In addition to charges, exposed hydrophobic surfaces also determine the solubility of proteins. The Kyte-Doolittle and Sweet-Eisenberg hydropathicity schemes assign each amino acid a score expressing the relative hydrophobicity of its side chain (see Table 16.1). In the Kyte-Doolittle scale, Arg is the most polar side chain with a value of −4.5 and Ile is the most non-polar with a value of +4.5. Values for other schemes are different, but the relation between the amino acids is very similar. The hydropathicity scores are averaged over a small sequence window and plotted as a function of sequence position (Figure 18.3). A continuous stretch of hydrophobic residues is a good indication that this part of the sequence is either buried in the hydrophobic core of the protein or embedded in a membrane.

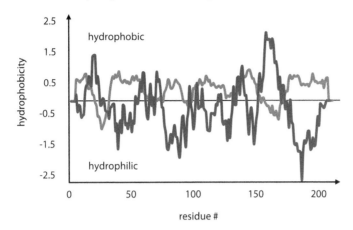

**FIGURE 18.3:    Kyte-Doolittle hydropathicity plot.** The same examples as in Figure 18.1 are shown. Hydrophobicity values from Table 16.1 are assigned to each amino acid and averaged over a window of nine residues. Clearly, the membrane protein bacteriorhodopsin (blue) is overall more hydrophobic than the soluble GST (red). The hydrophilic parts of bacteriorhodopsin are the loops that extend away from the membrane on the cytosolic or extracellular side. Long hydrophobic stretches in GST belong to the hydrophobic core.

Some regions in a protein sequence may be composed of only a single or a few types of amino acids over stretches as long as 100 residues. Such biased amino acid compositions are called *low complexity regions* (LCRs). For example, several intrinsically disordered proteins (see Box 16.6) contain LCRs. Among the repeating residues found in LCRs are Gln in the Parkinson's disease protein huntingtin (Table 16.10), but repeats of Ser, Gly, and Pro can also signify unstructured regions. Proteins with many LCRs seem to have more binding partners than those lacking LCRs. An estimated 50% of all eukaryotic proteins is predicted to have disordered sequences of >40 residues, while only 17% of prokaryotic proteins have such long disordered regions.

Probable sites for ion binding and post-translational modifications, including phosphorylation, glycosylation, and lipidation, can be identified by searching for characteristic short sequence patterns (motifs). A novel protein sequence can be quickly scanned for such motifs, which may tell about the protein's function. A few examples for motifs are given in the following: (1) cAMP-dependent kinase phosphorylates the Ser in the $RRX_nS$ motif ($n$ is usually 1, but 0 and 2 are also recognized). (2) The Tyr in [D/E]XXY motifs is phosphorylated by the anaplastic lymphoma kinase. (3) Ras-like GTP-binding proteins are farnesylated or geranylgeranylated for membrane attachment at the Cys of their C-terminal CaaX box, where the Cys (C) is followed by two aliphatic (aa) and an arbitrary amino acid (X). (4) The phosphate-binding loop or Walker A motif of canonic sequence GXXXXGK(T/S) is frequently found in nucleotide binding proteins. It forms a tight loop that orients its NH-groups towards the phosphates of nucleotides (see Figure 16.36) for H-bonding. The Lys side chain in the motif helps in neutralizing the phosphate charges, and the hydroxyl group of Thr or Ser binds to a $Mg^{2+}$ ion, which in turn complexes the β- and γ-phosphates. Combinations of several motifs in the correct order within a sequence can also help identify protein function. DEAD-box RNA helicases contain a Q-motif that binds the adenine base in ATP, followed by a phosphate-binding loop and a DEAD-sequence. Once this sequence of motifs is identified in a novel protein sequence, it is very likely an RNA helicase.

## 18.1.2 SEQUENCE ALIGNMENT

Sequences of new genes (or proteins) are compared to databases to find whether a homologous gene of known function (a *homolog*) has been identified previously. Genes with the same function but from different species are called *orthologs*, whereas *paralogs* are genes related by gene duplication within a genome that have adopted new functions. Many protein databases are categorized by organism, certain organelles, or diseases. DNA databases may be subdivided into coding and non-coding regions, and specific RNA databases exist for rRNA, tRNA, or plant RNA. Sequence searches over entire protein databases are done using online servers that run a variety of alignment algorithms. The first alignment algorithms attempted to match all sequences over their entire length (Needleman-Wunsch) or to find local alignments by allowing gaps (Smith-Waterman). As the sequence databases became larger, less time-consuming algorithms using heuristic methods such as BLAST (Basic Local Alignment Search Tool) and FASTA were developed. Heuristic methods can come to probable conclusions using incomplete information (such as a sensible alignment from distantly related sequences), but because these methods yield several possibilities, they need a powerful scoring method. For matching and mismatching of nucleic acid sequences, a simple scoring of $\pm 1$ (nucleotides are the same or not between two sequences) may be sufficient. Scoring of protein alignments is more complex and nowadays includes biophysical properties of the amino acid side chains such as polarity, hydrophobicity, charge, etc. (see Figure 16.1). The PAM250 (point-accepted mutation per 100 residues) scoring matrix is a $20 \times 20$ scoring table that assigns a probability to each naturally occurring amino acid exchange after 250 cycles of "evolution": one PAM means about a 1% change in all amino acid positions, and thus can be thought of as a measure of *evolutionary distance*. To obtain the values in the PAM matrix, a few closely related sequences were compared and the amino acid substitutions were counted. BLOSUM (block substitution) is an alternative matrix that uses less homologous sequences and thus may deal better with alignments of distantly related sequences. The matrices are references against which the alignments are scored. The advantage of this kind of scoring alignments is that it includes evolutionary information in the form of different probabilities of amino acid exchanges: according to either matrix, Lys/Arg exchanges have high probability whereas Lys/Leu substitutions are strongly disfavored, which reflects the biophysical properties of these side chains. The more distant the proteins are from an evolutionary point of view, the higher the probability for unlikely amino acid exchanges in their sequences will be.

Many related proteins differ by insertions and deletions, differences in surface loop lengths, and insertions of whole domains. Such sequences cannot be aligned without introduction of gaps. If a gap has to be introduced into a sequence to optimize the alignment, a penalty is given. The value of the penalty may depend on where in the sequence the gap has to be introduced or how large it needs to be. Introducing a new gap is penalized heavier than extending an existing gap. Conserved sequences will form the same structures, while non-conserved sequences are often found in dissimilar structures. Thus, introducing a gap in a conserved region is penalized more than in a non-conserved region of the sequence. While for low homology sequences different programs may result in different alignments, a consensus can often be reached when the results of several algorithms are compared. A very stringent way to align multiple sequences is to include the position of secondary structure elements, disulfide bonds, etc. in the alignment, provided that NMR or crystal structures for some of the proteins are available. Such structure-based alignments can be very different from alignments that are based on sequences only (Figure 18.4).

A variant of BLAST, the position-specific iterated BLAST (PSI-BLAST) has higher predictive success than a simple sequence alignment by using a scoring matrix that is iteratively customized to the query. A first BLAST run returns a multiple sequence alignment, from which a *specialized position-specific scoring matrix* (PSSM) is extracted, highlighting some residue positions that show homology above a threshold. The PSSM is then used as the new query in another PSI-BLAST run and the process is repeated several times. The method can identify remote homologs of the query that would not turn up with pairwise alignments. In general, multiple alignments of more than three sequences are more informative than pairwise alignments because recurring motifs can be identified. These motifs help in establishing phylogenetic relationships. Pairwise differences between sequences are recorded in a matrix of evolutionary distances. The differences

A

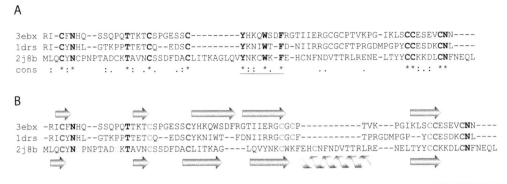

```
3ebx  RI-CFNHQ--SSQPQTTKTCSPGESSC---------YHKQWSDFRGTIIERGCGCPTVKPG-IKLSCCESEVCNN----
1drs  RI-CYNHL--GTKPPTTETCQ--EDSC---------YKNIWT-FD-NIIRRGCGCFTPRGDMPGPYCCESDKCNL----
2j8b  MLQCYNCPNPTADCKTAVNCSSDFDACLITKAGLQVYNKCWK-FE-HCNFNDVTTRLRENE-LTYYCCKKDLCNFNEQL
cons  : *:*       :.  *: .*.   .:*       *:: *. *         ..      .    **:.: **
```

B

```
3ebx  -RICFNHQ--SSQPQTTKTCSPGESSCYHKQWSDFRGTIIERGCGCP-----------TVK---PGIKLSCCESEVCNN----
1drs  -RICYNHL--GTKPPTTETCQ--EDSCYKNIWT--FDNIIRRGCGCF-----------TPRGDMPGP--YCCESDKCNL----
2j8b  MLQCYNCPNPTADCKTAVNCSSDFDACLITKAG----LQVYNKCWKFEHCNFNDVTTRLRE---NELTYYCCKKDLCNFNEQL
```

**FIGURE 18.4:    Multiple sequence alignment including structural data.** Three proteins that belong to the snake toxin-like superfamily were aligned either just using their sequences (A) or including structural information (B). PDB-ID 3ebx is sea snake erabutoxin, PDB-ID 1drs is mamba dendroaspin, and PDB-ID 2j8b is an extracellular domain of the human cell surface receptor CD59. A: The snake toxins have quite similar sequences. The CD59 sequence has an insertion compared to the toxins that is placed by the alignment algorithm at a position to maximize the number of identical (*), very similar (:), and similar (.) residues. A particularly homologous region is underlined. B: All structures belong to the disulfide-rich fold with nearly all-β geometry. If this structural information is used, the alignment shifts significantly. The underlined homologous region from the top panel is gone, the insertion has shifted, and the pairs of Cys (colored) that form disulfide bonds can be assigned, showing that seven Cys positions, not six as in the top panel, are conserved. The secondary structure elements of the snake toxins and CD59, shown on the top and bottom, respectively, coincide very well. If shifts are present, they are only by one or two residues.

are scored by sophisticated methods that compare physico-chemical properties of the side chains and take into account insertions and deletions, preferably mapped to crystal structures, if available. The results of the distance matrix are displayed in the form of a dendrogram that shows the probable diversification of the sequences during evolution. The dendrogram may or may not originate from a common ancestor. If no ancestor is known the dendrogram is represented as a *rootless tree* (Figure 18.5).

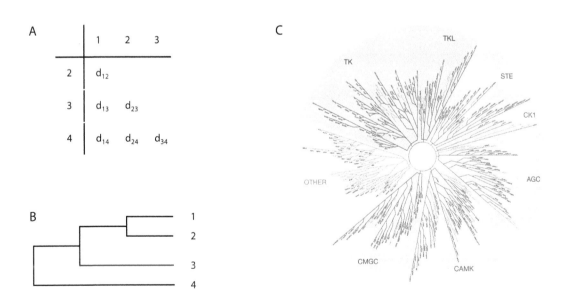

**FIGURE 18.5:    Phylogenetic relations between protein sequences.** A: A distance matrix contains the similarities of all pairwise alignments. These distances can be visualized as a tree (B). C: More than 500 known kinases are grouped into eight families in a tree that has no common ancestor, a rootless tree. Image generated using TREEspot™ Software Tool and reprinted with permission from KINOMEscan®, a division of DiscoveRx Corporation, © DISCOVERX CORPORATION 2015.

### 18.1.3 SECONDARY STRUCTURE PREDICTION

We saw before that different amino acids have different propensities for α-helices, β-sheets, and turns (see Table 16.3). This statistical information from proteins of known structure can be used to predict secondary structure elements from sequence analysis. The first algorithms used single sequences, but better results with >70% accuracy were later obtained by using multiple sequence alignments to determine the conserved regions in protein families that are expected to have conserved secondary and tertiary structure. One of the first algorithms was the *Chou-Fasman method*, which assigns a probability for each amino acid to occur in helices, sheets, or turns. A window of 5–10 amino acids is shifted along the sequence, and the propensities for the secondary structures are tallied. Consistently high values for a particular secondary structure element indicate that the sequence may form this secondary structure. However, a single prediction is still not very accurate, and it is best to combine the results of several different algorithms. Different methods tend to agree on the position of helices, loops, and strands, but may predict their extent differently. Exclusion or disfavoring of certain amino acids in certain secondary structure elements can be an equally powerful approach. For instance, while Pro and Gly are both disfavored in shorter strands and helices, a Pro-Gly or Gly-Pro sequence can be diagnostic for a turn.

Since α-helices are characterized by local interactions, they can be predicted more faithfully than β-strands. Many α-helices on the surface of proteins are amphipathic, with one side contributing to the hydrophobic core and the other exposed to solvent. Two shifted $i$, $i + 4$ patterns of hydrophobic and hydrophilic residues are characteristic for such helices. Transmembrane helices may also be predicted: a membrane is ca. 2 nm thick, and the α-helix has a 540 pm pitch, so it takes 4–5 turns or 16–20 residues for the shortest α-helix to traverse the membrane. Figure 18.6 shows the secondary structure prediction of bacteriorhodopsin together with the hydropathicity data from Figure 18.3. Combination of these complementary methods can give more validity to the prediction of transmembrane regions than either alone. A protein crystal or NMR structure determined later informs on the accuracy of the initial prediction, which allows adjustment of the scoring functions.

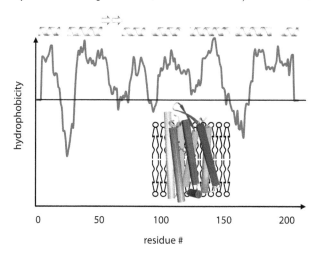

**FIGURE 18.6: Secondary structure prediction.** The structure of *Haloarcula marismortui* (PDB-ID 4pxk; same sequence) is shown as an inset with a sketched membrane. The true helix segments are indicated above the hydrophobicity data from Figure 18.3. All secondary structure elements were correctly predicted by JPRED (top), including the two short β-strands (magenta in the structure). The predicted helices align roughly, but not perfectly, with the hydropathicity data.

## 18.2 MOLECULAR MODELING

Molecular modeling is the umbrella term for numerically representing molecules and simulating their structures and time-dependent conformational changes. The simulations are approximate, and their accuracy must be validated by comparison of the results with experimental data. Molecular modeling is applied in many areas of research, from finding the global energy minimum of small molecule conformations to folding of entire proteins.

Molecular modeling can be performed using the laws of molecular mechanics (MM) and/or quantum mechanics (QM). MM and *molecular dynamics* (MD) calculations adequately describe many aspects of molecular conformation, conformational changes, and rigid body movements, but some molecular properties require the more accurate QM treatment. These include drastic changes in electronic structure, such as fission and formation of chemical bonds, and the conformational description of small molecules using molecular orbitals. While QM treatment of small molecule geometry is becoming routine practice in applications such as crystallographic structure refinement, the problem with QM methods is that they are computationally very demanding. Pure QM calculations on systems with more than 1000 atoms are so time-consuming that is often necessary to treat a small part of the system quantum mechanically, e.g. a ligand in a protein-ligand complex, while the laws of classical mechanics are applied to the rest of the system by using empirical force fields and MM/MD methods.

A *force field* expresses the physico-chemical information that we have about molecules as a sum of energy terms in a *potential energy function*, which allows calculation of the energy of a system (a lipid bilayer, a hydrated protein, a protein-nucleic acid complex, etc.). Thus, a crucial aspect of MM/MD methods is the parametrization of the force field because this will have a strong effect on the calculated energy. Although force fields cannot take quantum effects into account, their constituting energy terms are weighted against each other in such a way that the result agrees with quantum mechanical calculations on small molecule reference compounds. Among the force fields that have been developed to study the structure and dynamics of proteins and nucleic acids are CHARMM, AMBER, GROMOS, MOLOC, and ROSETTA. In the following, we will take a look at force fields and how they are used in MM/MD studies.

## 18.2.1 FORCE FIELDS

There are two main contributors to the potential energy of a molecule, interactions *via* covalent bonds and non-covalent interactions through space (Chapter 15):

$$E_{\text{molecule}} = \sum \left( E_{\text{covalent}} \right) + \sum \left( E_{\text{non-covalent}} \right)$$

eq. 18.2

Covalent interactions include bond distances, 1,3-angles, and torsion (1,4-) angles plus additional *improper dihedral* restraints that maintain flatness of planes, such as peptide bonds and nucleobases, and keep the chirality of stereocenters. Non-covalent energy terms are chiefly the van der Waals interactions and all interactions between permanent dipoles, including H-bonds. The overall non-covalent energy can be implemented as the sum of many individual energy terms (Section 15.3) that are weighted against each other to reflect their relative importance. The ROSETTA force field used in the prediction of protein structures, for example, contains specific energy terms for atom solvation in the folded state and amino acid residue desolvation upon folding. It divides H-bonds in short-range and long-range, and whether they are formed between backbone or side-chain atoms. There can even be different energy terms for H-bonds, depending on the hybridization state of the contributing atoms. The possibilities of shaping the force field by adding specialized energy terms appears endless. However, for a molecule of $N$ atoms, each additional energy term increases the number of energies that have to be calculated by $N \cdot (N-1)$, which slows down the calculations. The force field can be simplified again by incorporating known facts about the molecule under study, such as the presence of cofactors, covalent modifications of residues, or conserved protein regions that should not move during the simulation. These simplifications reduce the number of degrees of freedom of the system, speeding up the calculations. The ROSETTA force field, to keep the example, is simplified by inclusion of empirical information on torsion angles as a Ramachandran potential for the protein backbone (Figure 16.16 in Section 16.2), and preferred rotamers for the side chains (Figure 16.5 in Section 16.1.3). This "fine-tuning" of force fields highlights an important aspect–they can incorporate prior knowledge and experimental data from many different sources to increase the confidence in the energy minimization. This approach is termed *hybrid methods*, and is now routinely applied to large and flexible macromolecular complexes for which one particular method yields only incomplete data (Box 18.1).

**BOX 18.1: HYBRID OR INTEGRATIVE METHODS.**

Hybrid methods integrate the information from complementary structural methods to increase the data-to-parameter ratio and to arrive at a comprehensive structural description of the molecule of interest. Hybrid methods may combine geometric and distance restraints from nuclear magnetic resonance (NMR; Section 20.1), X-ray diffraction (Chapter 22), Förster resonance energy transfer (FRET; Section 19.5.10), and electron paramagnetic resonance (EPR; Section 20.2). These methods are discussed in Part IV. The experimental data from each method are represented by additional energy terms, integrated into a single force field that is then used for energy minimization and structure calculation. This approach is particularly useful for combining short range distance restraints from NMR with long-range distance restraints from EPR (Box 20.10), restraints from NMR and FRET, or EPR and FRET. The hybrid concept also finds applications in the structural modeling of large macromolecular complexes: envelopes from small angle X-ray scattering (SAXS, Section 21.2.4) or cryo-electron microscopy (EM) maps (Section 23.2) can be filled by rigid body modeling of NMR and crystal structures to gain insight into the complex interfaces, and conformational changes monitored by scattering, fluorescence, or NMR can be compared to the models from MD simulations. The accuracy of envelope fitting and modeling procedures can be verified if biochemical information about protein-protein interactions is available by mutagenesis or crosslinking data from mass spectrometry. Using a combination of EM and crosslinking mass spectrometry data, together with docking of X-ray crystal structures and homology modeling, the structure of a large part of the nuclear pore complex could recently be visualized. Bui *et al.* (2013) *Cell* 155(6): 1233–1243.

## 18.2.2 ENERGY MINIMIZATION

Energy minimization is the search for the minimum of the potential energy function by gradually changing the Cartesian coordinates $r_i$ of the atoms. Calculus of analytical functions tells us that in order to find their extrema, the first derivative should be zero. The force $F$ is the first derivative of the potential energy $E$ with respect to the atom coordinates, hence if we find the minimum of the energy function, there are no net forces acting on the atoms.

$$-\frac{\mathrm{d}E}{\mathrm{d}r} = -\nabla E = F = 0 \qquad \text{eq. 18.3}$$

Here, $r$ in bold face is the vector describing the atomic positions $r_i$ and $\nabla$ is the Nabla operator signifying the vector differential, or total differential of the energy function for all variables $r_i$. Energy minimization is usually carried out in Cartesian space, so a set of $N$ atoms is described by $3N$ variables. In contrast to analytical functions where the exact minimum can be found using calculus, molecular motion follows force fields that are too complex to solve analytically, but the derivative of $E$ has to be found numerically. In numerical methods, gradual changes of the atomic positions are introduced, $E$ is re-calculated, and the new set of coordinates is only kept if the energy is lower than the starting energy. After a few thousand iterations, the procedure converges at a conformation of low energy.

Several protocols are in use for energy minimization, among them the *steepest descent*, the *conjugate gradient*, and the *Newton-Raphson* algorithm, all of which use derivatives of the potential energy function to find the minimum, but differ in their paths to approach the minimum. The steepest descent method calculates the force as the first derivative of *E*. Atoms are then moved in the direction of the force. Each movement is perpendicular in direction to the previous step, by definition. The decision how far the atoms are moved can be chosen to depend on the magnitude of the force, reflecting the steepness of the energy surface. Alternatively, the atoms can be moved by an arbitrary step size. Flat energy surfaces warrant larger step sizes while sampling of narrow valleys needs smaller step sizes, but the shape of the surface is not known *a priori*. At the new atomic positions, the energy and its derivative are re-calculated. The derivative provides the new force and its new direction perpendicular to the previous step, and so on. The energy is compared to the energy of the previous conformation. If the energy decreases, the step size is increased by a factor and the procedure repeated until the energy stops decreasing. If after some iterations the energy increases again, the molecule has leapt over an energy minimum, moving uphill on the energy surface. In this case, the system is moved back into the direction of the minimum by a *line search* where a line is drawn between the last three sampled points (representing molecular conformations) on the energy surface. As the energy has increased in the second step compared to the first step, the minimum must be between the first and the third point. It must then be decided where between the first and third point to move the atoms to get the system back on track. A computationally efficient way is to approximate the (curved) line on the energy surface by a quadratic function that is fit to the three points. Its minimum, which is close to but not necessarily identical with the second point, is used as the new starting coordinates. The steepest descent tends to overshoot in narrow energy valleys and tends to become trapped in local energy minima, where it oscillates about the energy minimum.

The conjugate gradient method avoids the oscillating behavior of steepest descent by a different choice of direction. The first iteration is the same as in the steepest descent method, but starting from the second iteration the direction of the atom movement is not necessarily perpendicular to the previous step. Instead, the new direction vector $\boldsymbol{d}_{new}$ is a conjugate between the gradient $\boldsymbol{g}_{new}$ at that starting point and the direction vector $\boldsymbol{d}_{previous}$ of the previous step.

$$\boldsymbol{d}_{new} = -\boldsymbol{g}_{new} + \gamma_{new} \cdot \boldsymbol{d}_{previous} \qquad \text{eq. 18.4}$$

The scaling constant $\gamma_{new}$ is a scalar calculated from the ratio of the following scalar products:

$$\gamma_{new} = \frac{\boldsymbol{g}_{new} \cdot \boldsymbol{g}_{new}}{\boldsymbol{g}_{previous} \cdot \boldsymbol{g}_{previous}} \qquad \text{eq. 18.5}$$

$\gamma_{new}$ in the equation above corresponds to the *Fletcher-Reeves algorithm*, but other algorithms define $\gamma_{new}$ differently. Just as with the steepest gradient method, step sizes for the conjugate gradient method are either determined by a line search or are set arbitrarily.

An algorithm that not only uses the first but also the second derivative of the potential energy function to locate the energy minimum is the Newton-Raphson algorithm. The second derivative provides information on the curvature of the energy surface. The Newton-Raphson algorithm approximates the potential energy function by a Taylor series around the point at coordinates $r_i$:

$$E(\boldsymbol{r}) = E(r_i) + E'(r_i) \cdot (r - r_i) + \frac{E''(r_i)}{2} \cdot (r - r_i)^2 + \cdots \qquad \text{eq. 18.6}$$

Just as with the line search procedure above, the energy function is approximated by a quadratic function and the expansion is therefore stopped at the quadratic term. The first and second derivatives of the potential energy function are then

$$E'(\boldsymbol{r}) = E'(r_i) + r \cdot E''(r_i) - r_i \cdot E''(r_i) \qquad \text{eq. 18.7}$$

and

$$E''(\boldsymbol{r}) = E''(r_i) = \text{const.} \qquad \text{eq. 18.8}$$

Thus, the curvature of the energy surface is constant with this quadratic approximation. The energy is minimal for the condition $E'(r) = 0$, and hence

$$E'(\boldsymbol{r}) = 0 \Rightarrow r = r_i - \frac{E'(r_i)}{E''(r_i)} \qquad \text{eq. 18.9}$$

For a multidimensional function, the second derivative $E''(\mathbf{r})$ transforms into a huge *Hessian* matrix that must be inverted to solve eq. 18.9 for $r$. The algorithm is powerful and for a quadratic potential may find the energy minimum in a single step, regardless of the starting value. In practice, the more complex potential energy functions formulated for molecules require several steps. At each of them matrix inversion has to be performed, which limits the size of the molecules that can be handled to a few hundred atoms. Variants of the Newton-Raphson algorithm seek to avoid the time-demanding matrix inversion to speed up the calculations. For example, the same Hessian matrix can be used unaltered for a number of steps and only $E'(\mathbf{r})$ is re-calculated at each step.

### 18.2.3  Molecular Mechanics and Dynamics

Energy minimization is just concerned with the end result, the lowest possible energy state, but does not care about the way to achieve this goal. By contrast, in *molecular dynamics* (MD), the *trajectory* from one conformational state to the other is of interest, and each conformational state along the trajectory may contain useful biophysical information. MD thus extends the concept of energy minimization to the simulation of molecular motion in a time-dependent manner. Such motions can be anything from simple vibrations (bond stretching and angle bending that give rise to IR spectra; Section 19.4) to the large-scale conformational transitions in protein and RNA folding. In a classical MM approach, Newton's law is used to describe the time-dependent energy function:

$$-\frac{\mathrm{d}E}{\mathrm{d}r_i} = -\nabla E = F_i(t) = \mathrm{m}_i \cdot a_i(t) = \mathrm{m}_i \cdot \frac{\mathrm{d}^2 r_i(t)}{\mathrm{d}t^2} = \mathrm{m}_i \cdot \frac{\mathrm{d}v_i(t)}{\mathrm{d}t} \qquad \text{eq. 18.10}$$

We have seen in the previous section that the force acting on an atom $i$ is numerically calculated from the change in energy between its position $r_i$ and its position a small distance away, $r + \mathrm{d}r_i$. If we know the masses $m_i$ of the atoms, we can calculate their accelerations using Newton's law $F = m_i \cdot a_i$. The acceleration is (1) the second time derivative of the position of the atom and (2) the first time derivative of the velocity (eq. 18.10). We can therefore calculate both the velocity and the position of the atoms from their acceleration by single and double integration over time.

$$a_i(t) = \frac{\mathrm{d}v_i(t)}{\mathrm{d}t} \Rightarrow \int_{t=0}^{t} a_i(t) \cdot \mathrm{d}t = v_i(t) - v_{i0} \qquad \text{eq. 18.11}$$

where $v_{i0}$ is the velocity at time $t = 0$. Thus, we have

$$v_i(t) = a_i(t) \cdot t + v_{i0} \qquad \text{eq. 18.12}$$

Furthermore,

$$v_i(t) = \frac{\mathrm{d}r_i(t)}{\mathrm{d}t} \Rightarrow \int_{t=0}^{t} v_i(t) \cdot \mathrm{d}t = r_i(t) - r_{i0} \qquad \text{eq. 18.13}$$

Substitution of eq. 18.12 into eq. 18.13 and integration yields

$$r_i(t) = r_{i0} + \int_{t=0}^{t} v_i(t) \cdot \mathrm{d}t = r_{i0} + \int_{t=0}^{t} \left[ a_i(t) \cdot t + v_{i0} \right] \cdot \mathrm{d}t = \frac{1}{2} a_i(t) \cdot t^2 + v_{i0}\mathrm{t} + r_{i0} \qquad \text{eq. 18.14}$$

This equation shows that the positions $r_i(t)$ of the atoms can be calculated from the known set of starting positions $r_{i0}$, the known set of starting velocities $v_{i0}$, and the accelerations $a_i(t)$ derived from the differential of the energy function. Before briefly discussing some of the actual integration algorithms used in MD, we will take a look at necessary boundary conditions of an MD simulation.

### 18.2.3.1 Boundary Conditions and Solvation

To prepare the simulation, all atoms of the starting structure are assigned initial velocities, which effectively adds a kinetic energy term to the potential energy function, so the sum of potential and kinetic energy represents the overall energy of the system. The velocities of the atoms, and hence the kinetic energy of the system, are determined by the temperature (Section 13.3).

$$E_{kin} = \frac{3}{2}RT = \frac{1}{2} \cdot \sum_i m_i v_i^2$$

eq. 18.15

The kinetic energy is applied by heating the system to the simulation temperature, whereupon the atoms start to oscillate about their equilibrium positions. The system is then thermodynamically isolated to conserve the total energy, and allowed to evolve. During this process, different atomic positions are calculated by solving the equations of motion. The structures obtained correspond to different energy states, and the movement of the atoms corresponds to different forces, which are calculated in time increments $\delta t$ on the order of 1–2 femtoseconds ($10^{-15}$ s). The volume of the system and the number of atoms in that volume are kept constant using boundary conditions (Figure 18.7): no atoms may leave or enter the volume or else the system is not isolated any more. This is a problem for hydrated macromolecules, because water may attach or detach from the macromolecule and diffuse out of the MD volume. Inclusion of water is crucial for realistic MD simulations because the large dielectric constant of water ($\varepsilon_r = 80$) has a huge dampening effect on dipolar interactions (Section 15.3), and thus on the total energy. To prevent water molecules from diffusing out of the MD volume, *periodic boundary conditions* are applied. The macromolecule is placed in a water-filled cubic box, which is then copied in all three dimensions. The coordinates of the atoms in the central box evolve according to the Newtonian equations and are related to all other boxes by simple translations. Each water molecule leaving the central box at one side is replaced by an exact copy entering the box from the opposite side. The net effect is that the water experiences the same interactions as in bulk solvent (it may diffuse freely), but the number of water molecules per box stays the same (Figure 18.7).

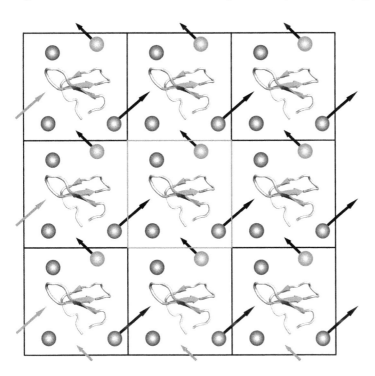

FIGURE 18.7:    **Periodic boundary conditions for explicit solvent molecules.** The protein of interest is at the center of the box, surrounded by solvent molecules (spheres). The number of atoms in the green box stays constant because as soon as one molecule leaves the central box, it is replaced by an identical atom entering from a neighboring box on the opposite side.

A hydrated macromolecule can also be implemented by including extra energy terms describing *implicit solvent models* in the potential energy function or by using a very large number of water molecules and keeping them in place by a *cavity potential function*. However, periodic boundaries with explicit solvent are a computationally more efficient approach because no extra adjustments of the force field are needed and the number of parameters is kept manageable.

### 18.2.3.2 Integration of the Newtonian Equations

Integration of the Newtonian equations of motion cannot be performed analytically but, as in energy minimization, has to be solved numerically. The integration is broken down into a series of small time intervals $\delta t$. At each time point $t, t + \delta t, t + 2\delta t$, etc., the force is calculated, which yields the acceleration, from which then the velocity and the new coordinates at $t+\delta t$ are derived. The force is assumed to be constant during each time step but it may change throughout the simulation due to re-partitioning of potential energy and kinetic energy between the atoms. The Newtonian equations are evaluated using different algorithms, among them the *Verlet integrator*, the *leap-frog* method, and the *velocity Verlet*. Some of the integration algorithms approximate the Newtonian equations of motion as Taylor series:

$$r(t+\delta t)=r(t)+r'(t)\cdot\delta t+\frac{r''(t)}{2!}\cdot\delta t^{2}+\cdots=r(t)+v(t)\cdot\delta t+\frac{a(t)}{2}\cdot\delta t^{2}+\cdots \qquad \text{eq. 18.16}$$

$$v(t+\delta t)=v(t)+a(t)\cdot\delta t+\frac{a'(t)}{2}\cdot\delta t^{2}+\cdots \qquad \text{eq. 18.17}$$

$$a(t+\delta t)=a(t)+a'(t)\cdot\delta t \qquad \text{eq. 18.18}$$

where $r(t)$, $v(t)$, $a(t)$, and $a'(t)$ are the coordinates, velocity, acceleration, and the first derivative of the acceleration, respectively, for every atom at the time $t$. As in energy minimization energy minimization, the Taylor series expansion is often truncated at the quadratic term. The popular Verlet integrator for example uses the following Taylor series approximations for the atomic positions at $t \pm \delta t$:

$$r(t+\delta t)=r(t)+v(t)\cdot\delta t+\frac{a(t)}{2}\cdot\delta t^{2} \qquad \text{eq. 18.19}$$

$$r(t-\delta t)=r(t)-v(t)\cdot\delta t+\frac{a(t)}{2}\cdot\delta t^{2} \qquad \text{eq. 18.20}$$

Addition of eq. 18.19 and eq. 18.20 eliminates the velocities and yields an equation describing the new atomic positions at time $t + \delta t$:

$$r(t+\delta t)=2r(t)-r(t-\delta t)+\frac{a(t)}{2}\cdot\delta t^{2} \qquad \text{eq. 18.21}$$

Thus, the Verlet integrator uses the positions and accelerations at time point $t$ together with the positions at time point $t - \delta t$ from the previous step to calculate the new positions at time point $t + \delta t$. Velocities are not explicitly used in the algorithm but can be calculated by subtraction of eq. 18.19 and eq. 18.20:

$$v(t)=\frac{r(t+\delta t)-r(t-\delta t)}{2\delta t} \qquad \text{eq. 18.22}$$

By contrast, the *leap-frog* algorithm does explicitly include the velocities in the calculations and assumes the following relations:

$$r(t+\delta t)=r(t)+v\left(t+\frac{\delta t}{2}\right)\cdot\delta t \qquad \text{eq. 18.23}$$

$$v\left(t+\frac{\delta t}{2}\right)=v\left(t-\frac{\delta t}{2}\right)+a(t)\cdot\delta t \qquad \text{eq. 18.24}$$

The acceleration is again obtained from the force, then the velocity is calculated at the intermediate time step $t + \delta t/2$ using the velocity from the previous time step $t–\delta t/2$. Finally, the atomic positions are updated by modifying the previous positions $r(t)$ using the intermediate velocity and the next iteration can be started. The name leap-frog stems from the ½ time step intervals at which velocity and position alternate. A disadvantage of the method is that the total energy of the system at a given time cannot be calculated because position and velocity are not synchronized.

Finally, the *velocity Verlet* algorithm returns $r(t)$, $v(t)$, and $a(t)$ at the same time with high precision but at the price of longer computation time using the following relationships:

$$r(t+\delta t)=r(t)+v(t)\cdot\delta t+\frac{1}{2}a(t) \qquad \text{eq. 18.25}$$

$$v(t+\delta t)=v(t)+\frac{1}{2}\big[a(t)+a(t+\delta t)\big]\cdot\delta t \qquad \text{eq. 18.26}$$

Whichever integration method is used, the stability of the coordinates as a function of time (the trajectory) is gauged against an analytical (exact) reference trajectory calculated from the Gibbs free energy. The reference trajectory strictly conserves energy as is required for any MD calculation performed at constant temperature. If the two trajectories diverge, this is an indication of failure to conserve kinetic and/or potential energy during the simulation, and the results will not be useful. The stability of the integration process is improved by using small time increments $\delta t$ for the integration, typically 1–2 fs for protein simulations.

### 18.2.3.3 Trajectory Analysis

The trajectory describing the motion of the molecular system is the series of conformations calculated over the total time of the simulation, including the coordinates, the velocities, and the accelerations of the atoms. If the coordinates change substantially during the simulation, the starting structure was of high energy. This is the case for *de novo* protein folding and NMR structure determinations, which start from an unfolded polypeptide chain and end at a folded, compact structure. Sizeable changes in atom position are also observed for crystal structures that contain errors or for proteins that have several low-energy conformations. In these cases, energy minimization of the starting structure is essential prior to starting the MD simulation. The trajectory is also analyzed in terms of time and ensemble average structures, which are compared to the starting structure. Time and ensemble averages are an important concept in MD because it connects the microscopic energies of atoms to the macroscopic energy of the system. If the MD simulation is done long enough (many picoseconds), and at each time point $t+\delta t$ the system is allowed to reach equilibrium, the trajectory contains all possible positions and velocities accessible to the system, i.e. the molecule samples the entire conformational space. Under these conditions, the *ergodic theorem* applies, stating that an ensemble average is equivalent to a time average over the "infinitely long" trajectory. The theorem also applies *vice versa*: a sufficiently large number of measurements will have a time average that is a good approximation to the equilibrium ensemble average. Importantly, the energies calculated from time and ensemble average structures can be compared to experimentally determined energies, e.g. from isothermal titration calorimetry (Section 27.1), providing a link between theory and experiment.

The end point of an MD simulation is not fixed but only limited by the available computational resources. In practice a few nanoseconds ($10^{-9}$ s) are achieved using several weeks of computation time, which unfortunately is not long enough for many biologically interesting conformational changes (see Table 16.9). In order to speed up the simulations, either the force field or the system may be simplified. Since the most time consuming part of a MD simulation is calculation of the non-bonded energy terms, a cut-off distance can be introduced into the function to disregard all interactions between two atoms sufficiently separated. Experimental data can be included in the potential energy function to serve as powerful restraints. The number of variables in the system itself may be drastically reduced by coarse-grained modeling (Section 18.2.4.4). While these adaptations make the simulations computationally feasible on the microsecond scale, slow conformational transitions in proteins such as allosteric changes, oligomer assembly, and protein folding currently still remain inaccessible to all-atom MD simulation.

### 18.2.4 Applications of Molecular Modeling to Macromolecules

The Anfinsen experiments on RNase showed that the three-dimensional structure is an inherent property of the protein sequence at given environmental conditions (Section 16.3.2). An early dream in protein science has been and still is to calculate the structure of a protein from its sequence alone, to finally decipher the protein code. Since structures of folded proteins are states of low energy, any simulation method must describe the energy of the protein, minimize the protein structure, and then assess which of the candidate structures (often of similar energies) is the most likely one. *De novo* structure prediction of large proteins >50 kDa is still computationally too demanding because there are too many conformations to be assessed. However, it might be possible that the pattern-recognition and puzzle-solving abilities of humans are more efficient at folding tasks than the current computer programs (Box 18.2).

---

**BOX 18.2: A PROTEIN FOLDING GAME - FOLD.IT.**

The protein folding problem has been put into a game that everyone, including children, can play at home. The task is to generate a biophysically sensible protein structure starting from the amino acid sequence by applying some simple rules: the structure should be compact, have no clashing atoms, the hydrophobic residues should be buried, and the hydrophilic residues should point towards the water. This sometimes means moving large parts of the protein sequence, for example to insert a loop as a new β-strand into an existing β-sheet (Figure 18.8). Such drastic changes are often outside the convergence radius of current computer programs but readily seen by the human eye. The folding strategies that the players take can then be incorporated into existing computer programs, for example Rosetta, which is developed by the same research group as Foldit.

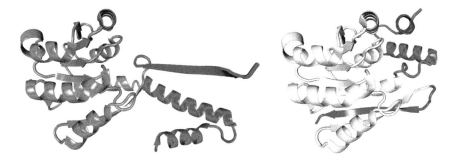

FIGURE 18.8: **Large conformational changes easily recognized by the human eye.** The two structures are from the same protein, but only the right-hand structure is compact. For humans, it is clear how the β-strand (blue) and the two α-helices (red) have to be moved starting from the left-hand structure.

### 18.2.4.1 Fold Recognition

One of the first questions when a new protein structure has been determined by NMR spectroscopy or X-ray crystallography is whether it is a novel fold or has structural homologs. The coordinates are often submitted to the Distance mAtrix aLIgnment (DALI) server to assess if there are structures of similar fold. The DALI method is based on the fact that the atoms in similar structures have similar distances, which surprisingly often finds a common fold, although the sequences may be completely unrelated. Indeed, the number of novel folds added to the protein database (PDB) in recent years is very small. According to the CATH 4.0 database release in 2013, there are only 1375 different folds. CATH classifies protein structures according to secondary structure content (Class), structural similarity in the absence of sequence similarity (Architecture), Topology, and evolutionary relationship (Homologous superfamily). Another database that uses purely structural considerations is the SCOP (structural classification of proteins) database. In February 2016 this database defined 1221 folds, with only 13 new folds added during 2015. There is a fair chance that any new protein sequence is already represented by a known structure. This limited set of folds is the basis of another three-dimensional structure prediction method called *threading*, or *fold recognition*.

Fold recognition is used in cases where no structural information is available for the sequence under study. It is a template-based method where a query sequence is fitted (*threaded*) onto a likely template structure. The major problem is to find the best templates and how to rank the results. In principle, all 1375 different folds known to date can be used for the threading exercise, but often a most likely template can be found by limited sequence homology to a known structure, an annotated biological function for the query sequence, or with the help of biological data such as the position of active site residues in an enzyme or residues essential for binding to another interaction partner. Once a template is selected, its backbone coordinates serve as a three-dimensional framework onto which the query sequence is threaded, allowing for insertions and deletions. The resulting structures naturally have the fold of the template, but in the case that the wrong template has been selected, there will be few favorable and many unfavorable contacts between the side chains of the query sequence. The contacts are summed up to a pseudo-energy score that also includes a solvation term to measure the amount of hydrophobic surface exposed to solvent, which should be minimal for a soluble protein. The structure with the lowest pseudo-energy can be considered the best approximation to the true structure. In the CASP competition (Box 18.3), different algorithms are used to predict structures from the sequence. The results are then compared to the respective NMR and crystal structures that have not yet been published.

---

**BOX 18.3: CASP (CRITICAL ASSESSMENT OF STRUCTURE PREDICTION; PREDICTIONCENTER.ORG).**

The CASP challenge is an biennial competition where, in 2014 (CASP11), 123 research groups and 84 servers competed for the best three-dimensional predictions of the structures of 137 targets using both template-based and template-free methods. The tertiary structures are predicted for sequences of proteins whose structures are known but that have not yet been published (mostly from structural genomics consortia). A very successful case for template-based approaches is a phosphoethanolamine transferase from *Neisseria meningitidis*, which was modeled at an rmsd (root mean square distance) of 43 pm over 332 residues with respect to the true structure (PDB-ID 4kav). Algorithms for template-free modeling deliver less accurate models, but are improving continuously. A typical result is target T0760, a protein from a *Bacteroides* species (Figure 18.9). It superimposes with an rmsd of 170 pm with the crystal structure (PDB-ID 4pqx; colored). Such models are often accurate enough to be used as search models for molecular replacement phasing of diffraction data (Section 22.7.3).

*(Continued)*

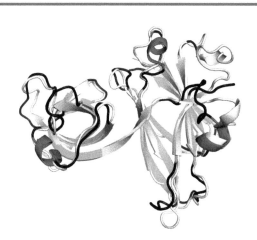

FIGURE 18.9:  **Successful example of threading.** The modeled structure (gray) is virtually indistinguishable from the crystal structure (colored red, green, and black).

### 18.2.4.2 Homology Modeling

Currently, the best approach to predict structures from sequences is homology modeling. A target sequence is aligned with sequences of close homologs, at least one of which must have its crystal structure determined. This *parent structure* or superimposed group of parent structures is then used as a framework for the new structure. If there are several related structures, a conserved core can be extracted that is more reliable to build upon than a single structure. Those residues that align well between the sequences can be assumed to be modeled with high confidence, sometimes on the level of side-chain rotamers. The side chains of all other residues are placed in energetically favorable conformations. Any clashes between the side chains are then removed by energy minimization (Section 18.2.2). For query sequences larger than the template, the last step can be *ab initio* modeling of surface loops for which the parent structure has no counterpart. For very short loops, the conformations of canonical turns (Section 16.2.3) can be tried while for longer loops of eight and more residues, either template-based techniques or *de novo* modeling are useful. Suitable template loops in this *spare parts approach* have similar length, similar connections to the protein, or similar sequence. Loop databases are used to look up the conformations of whole loops or smaller parts thereof. In *de novo* modeling, the loops may be grown residue by residue from the ends, which serve as defined anchor points, until a closure of the loop can be found. An alternative route is to create a loop by adding dipeptides of sensible stereochemistry followed by refinement, or to deconstruct a larger loop until it fits the length, sequence, and anchor points of the target. In all approaches, Ramachandran restraints (Figure 16.16) can be used to ensure a sensible backbone conformation of the modeled loop. Homology models for which a crystal structure became available later showed that the accuracy can vary substantially from ca. 0.1 nm rmsd for ultra-short loops of three residues to about 0.4 nm rmsd for loops of eight residues or more.

### 18.2.4.3 Simulated Annealing

The conformational energy landscape of a macromolecule is rugged with many local minima, separated by energy barriers. Energy minimization methods run the risk of getting trapped in a local energy minimum. To find the global energy minimum, a method is required that allows a molecule to overcome the energy barrier separating one energy minimum from the next. *Simulated annealing* is an application of MD to refine crystal and NMR structures that are far away from their global energetic minimum. The macromolecule is heated *in silico* to the desired starting temperature of several thousand K. This is equivalent to assigning large random velocities velocities to the atoms so the molecule can adopt high-energy conformations to overcome energy barriers between low-energy conformational states (local energy minima). It also means that the radius of convergence is larger for simulated annealing compared to other energy minimization methods: simulated annealing may find the global energy minimum for starting structures far away from it. Other than increasing the likelihood of overcoming local energy barriers, the temperature in simulated annealing has no physical meaning. Upon slow cooling, the atoms move more slowly and can engage in more favorable interactions than in previous conformations at higher temperature. Lower energy conformations become more probable, in accordance with the Boltzmann distribution (eq. 19.7). The annealing process is guided by force fields and experimental data to avoid divergence of the structure and to ensure a chemically sensible result. To find the global energy minimum, an infinitely large number of temperature steps would be required. Simulated annealing is a compromise between computation time and accuracy, and it is uncertain whether the global energy minimum has been found. In practice, this issue is addressed by performing several cycles of simulated annealing at different starting temperatures and with different cooling rates. If the results are identical across different runs, it is likely that the global minimum has been reached. Simulated annealing may also return a series of different structures of similar energy, which is typical for NMR.

### 18.2.4.4  Coarse-Grained Modeling

To calculate the energies for large systems or to increase the time window for MD simulations using QM methods, the number of parameters or degrees of freedom has to be reduced. This is the concept behind coarse-grained methods, which group atoms as *pseudo-atoms* or *beads*. Depending on the required accuracy and the size of the system, a bead can be a whole protein domain or just a few atoms. If relative domain movements are studied, each domain may be represented by a single bead. Large conformational changes may require one bead per secondary structure element. For finer details, suitable groups of atoms are chosen. If each amino acid residue or nucleotide is approximated by a bead, a *chain-compatible* bead model results where the distance of the beads is the same as the characteristic distances ($C_\alpha$-$C_\alpha$ or phosphate-phosphate) in the authentic molecule. Higher precision modeling for proteins is approximated by separate beads for main chain and side chain. Similarly, nucleotides can be broken down into sugar, phosphate, and nucleobase, and lipid alkyl chains can be simplified as sets of 2-4 methylene groups. The solvent can be treated similarly: several water molecules may be merged into a "super-water", and an ion plus its hydration shell becomes an "ion bead". Beads are assumed to behave similar to atoms as point-like masses obeying Newtonian mechanics. Bonds and angles between connected beads are represented by harmonic springs and harmonic angular potentials. The Lennard-Jones and Coulomb potentials are used to represent long-range interactions. The fewer number of parameters of the bead models not only speed up the calculations, but the energy landscape is also smoother so the energetic minimum is found in fewer steps. Coarse-grained modeling enables MD calculations on the formation of protein-lipid complexes, diffusion of proteins in membranes, membrane bending, deformation and dynamics of virus capsids, and large assemblies such as the nuclear core complex. Using coarse-grained models, MD simulations on the millisecond time scale can be realized, which is the time range accessible to fast transient kinetic experiments (Chapter 25), and these experimental results may be used to validate the simulations. Techniques that deliver low-resolution data, such as electron microscopy or small angle scattering, often use coarse-grained modeling together with the experimental data (Box 18.1).

## QUESTIONS

18.1   What is the concentration ratio between the lowest energy and next best predicted RNA secondary structure if they differ by $\Delta\Delta G = 5.7$ kJ mol$^{-1}$ at 37°C?

18.2   Jean-Baptiste Lamarck was a French naturalist. Assuming equal probabilities for all amino acids, what is the probability of finding the LAMARCK sequence by chance? Select the protein BLAST option at http://blast.ncbi.nlm.nih.gov/Blast.cgi and search for this sequence in proteins. How many 100% hits do you find? Select the ABC transporter. Can you find the P-loop motif for nucleotide binding? Using the ProtParam tool at http://web.expasy.org/protparam, calculate the molecular mass, the theoretical pI, and the extinction coefficients at 280 nm for reduced and oxidized Cys residues. Is the difference in extinction coefficients significant? If you do a secondary structure prediction with psipred at http://bioinf.cs.ucl.ac.uk/psipred, what secondary structure is predicted for the LAMARCK sequence? To verify the prediction, search for crystal structures of proteins using the entire sequence of the ABC transporter at www.pdb.org using the "advanced search" option and take a look at the structure of the corresponding sequence with e.g. Pymol or Coot. Is there a difference to the prediction?

## REFERENCES

Leach, A. R. (2001) Molecular modeling: principles and applications. Harlow, England; New York, Prentice Hall.
· comprehensive description of molecular modeling from a medicinal chemist's perspective

Mathews, D. H., Sabina, J., Zuker, M. and Turner, D. H. (1999) Expanded sequence dependence of thermodynamic parameters improves prediction of RNA secondary structure. *J Mol Biol* **288**(5): 911–940.
· energy parameters for RNA secondary structure predictions

Guruprasad, K., Reddy, B. V. and Pandit, M. W. (1990) Correlation between stability of a protein and its dipeptide composition: a novel approach for predicting *in vivo* stability of a protein from its primary sequence. *Protein Engineering* **4**(2): 155–161.
· generation of a table for all 400 dipeptide instability weight values (DIWV) for the calculation of the overall instability index of a protein sequence.

Coletta, A., Pinney, J. W., Solis, D. Y., Marsh, J., Pettifer, S. R. and Attwood, T. K. (2010) Low-complexity regions within protein sequences have position-dependent roles. *BMC Syst Biol* **4**: 43.
· analysis of LCR positions within sequences

Zhang, T., Faraggi, E., Xue, B., Dunker, A. K., Uversky, V. N. and Zhou, Y. (2012) SPINE-D: accurate prediction of short and long disordered regions by a single neural-network based method. *Journal of biomolecular structure & dynamics* **29**(4): 799–813.
· LCR detection by a neural-network method

Vucetic, S., Brown, C. J., Dunker, A. K. and Obradovic, Z. (2003) Flavors of protein disorder. *Proteins* **52**(4): 573–584.
· discusses different types of protein disorder

Khatib, F., Cooper, S., Tyka, M. D., Xu, K., Makedon, I., Popovic, Z., Baker, D. and Players, F. (2011) Algorithm discovery by protein folding game players. *Proc Natl Acad Sci U S A* **108**(47): 18949–18953.
· description of the FoldIt program

## ONLINE RESOURCES

genomes.urv.es/OPTIMIZER
· server to optimize codon usage for mRNA

www.kazusa.or.jp/codon
· codon usage database

web.expasy.org/protparam
· calculate biophysical properties from protein sequences.

http://www.compbio.dundee.ac.uk/jpred
· predicts the secondary structure based on a multiple sequence alignment.

rna.tbi.univie.ac.at/cgi-bin/RNAfold.cgi
· program rnafold for RNA secondary structure prediction.

en.wikipedia.org/wiki/List_of_RNA_structure_prediction_software
· list of RNA structure prediction software.

sparks-lab.org/SPINE-D
· SPINE D server to analyze sequences for potential disorder

www.hprd.org/PhosphoMotif_finder
· database of phosphorylation motifs

www.oxfordjournals.org/our_journals/nar/database/cat/3
· a list of DNA, RNA, and protein databases

www.uniprot.org
· UniProt database for proteins.

blast.ncbi.nlm.nih.gov/Blast.cgi?PAGE_TYPE=BlastSearch
· non-redundant nucleotide database

phylomedb.org
· database of gene phylogenies

ekhidna.biocenter.helsinki.fi/dali_lite
· DALI server

cathdb.info
· CATH database of protein folds

scop.berkeley.edu
· SCOP database of protein folds

# Part IV
# Methods

Parts I and II introduced the thermodynamic and kinetic concepts relevant for biochemistry. In Part III, we have connected these concepts to structure formation by and stabilization of macromolecules. Part IV introduces the biophysical techniques most frequently used to study macromolecules in solution, in crystals, and on surfaces.

# Optical Spectroscopy

Spectroscopy is based on the interaction of electromagnetic radiation with matter. Molecules may absorb energy from the incident light, which converts them into a state of higher energy, the excited state. From the excited state, they return to the ground state by releasing the energy in form of heat or radiation. The different techniques of optical spectroscopy either measure absorption (absorption spectroscopy, Sections 19.2–19.4) or emission of radiation (emission spectroscopy, Section 19.5). Before we discuss the individual spectroscopic techniques in detail, we will begin by reviewing the properties of light and its interaction with molecules that are relevant for spectroscopy.

## 19.1 INTERACTION OF LIGHT AND MATTER

### 19.1.1 LIGHT AS AN ELECTROMAGNETIC WAVE

Light is an electromagnetic wave that consists of oscillating electric and magnetic fields and propagates in a certain direction (Figure 19.1). The electric and magnetic field vectors undergo sinusoidal oscillations vertical to the propagation direction of the light. At any given time, the electric and magnetic components are vertical to each other and in phase, meaning they oscillate in synchrony, such that their maxima, minima, and zero transitions coincide.

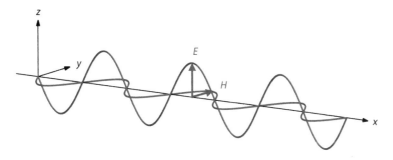

FIGURE 19.1: **Electric and magnetic components of an electromagnetic wave**. The electric (red) and magnetic field components (blue) of a linearly polarized electromagnetic wave propagating in the *x*-direction. The electric field vector (*E*) oscillates in the *xz*-plane, the magnetic field vector (*H*) in the *xy*-plane. *E*-field and *H*-field vectors are always perpendicular to each other.

The oscillations of the electric ($E$) and magnetic ($H$) field vectors in time ($t$) and space ($x$) are described by eq. 19.1 and eq. 19.2:

$$E(t,x) = E_0 \cos\left(2\pi\left(\nu t - \frac{x}{\lambda}\right)\right)$$

eq. 19.1

$E_0$ is the amplitude of the electric field (in V m$^{-1}$), $\nu$ is the *frequency*, i.e. the number of oscillations of the electric and magnetic field vectors per second, and $\lambda$ is the *wavelength*.

$$H(t,x) = H_0 \cos\left(2\pi\left(\nu t - \frac{x}{\lambda}\right)\right)$$

eq. 19.2

$H$ is the strength of the magnetic field (in A m$^{-1}$), $H_0$ is the amplitude.

Spectroscopy is based on the interaction of molecules with the electric and/or the magnetic component of the electromagnetic wave. Most spectroscopic techniques, such as absorption, fluorescence, and infrared spectroscopy, use the interaction of the molecule with the electric component. Circular dichroism (see Section 19.3.2) provides a rare example of a spectroscopic method that is based on the interaction of molecules with both electric and magnetic components of electromagnetic radiation.

The energy $E$ of an electromagnetic wave is proportional to its frequency $\nu$. The proportionality constant is the Planck constant $h$ (6.62607·10$^{-34}$ J s):

$$E = h\nu$$

eq. 19.3

The frequency of electromagnetic radiation can be expressed as the ratio of the speed of light $c$ (3·10$^8$ m s$^{-1}$) and the wavelength $\lambda$, giving:

$$E = h\frac{c}{\lambda}$$

eq. 19.4

The energy of an electromagnetic wave is thus proportional to its frequency, but inversely proportional to its wavelength.

Only a small fraction of the electromagnetic spectrum (Figure 19.2) is visible to the human eye. The visible (VIS) region covers the spectral range from 400 nm (blue light) to 800 nm (red light). This range is flanked by ultraviolet (UV) and X-ray regions towards lower wavelengths (higher energy), and by infrared (IR), microwave, and radio wave regions towards higher wavelengths (lower energy).

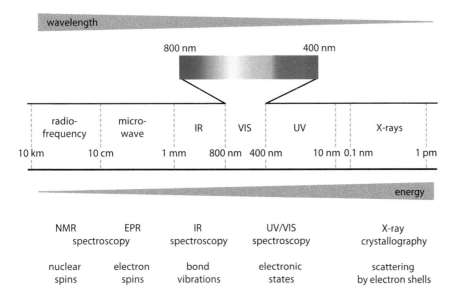

FIGURE 19.2:    **Electromagnetic spectrum and associated spectroscopies**. The spectral region of
λ = 400–800 nm is visible (VIS) for the human eye. Infrared radiation (IR), microwaves, and radio waves are of
lower energy (higher wavelength), ultraviolet (UV) radiation and X-rays are of higher energy (lower wavelength). Nuclear magnetic resonance (NMR) spectroscopy induces transitions of nuclear spins (Section 20.1),
electron paramagnetic resonance (EPR) spectroscopy induces transitions of electron spins (Section 20.2). IR
spectroscopy excites bond vibrations (Section 19.4), UV/VIS spectroscopy (Sections 19.2 and 19.3) induces
transitions between electronic states of molecules. X-rays are scattered by the electron shell of molecules, a
process that does not involve absorption, but is important for structure determination (Chapter 22).

## 19.1.2 PRINCIPLES OF SPECTROSCOPY: TRANSITIONS IN TWO-STATE SYSTEMS

Spectroscopy is based on transitions between different energy levels of a molecule that are induced
by the energy of the incident electromagnetic radiation. Such states can be different electronic states
(UV/VIS absorption and fluorescence spectroscopy; Sections 19.2, 19.3 and 19.5), vibrational states
(IR spectroscopy; Section 19.4), rotational states, or spin states of nuclei (NMR; Section 20.1) or
electrons (EPR; Section 20.2). The energy of these states is quantized: only defined states can be
populated, leading to a set of discrete energy levels.

Incident light can provide the energy to induce transitions from the lower energy level, the
ground state, to a higher energy level, the excited state. To do so, the energy $E$ of the electromagnetic wave (eq. 19.3, eq. 19.4) has to match the energy difference $\Delta E$ between the two states involved
(Figure 19.3). It is important to note that the energy difference $\Delta E$ refers to the difference between
the two states in a single molecule, resulting in very small numbers. Electronic transitions can typically be induced by UV/VIS light in the range of 200–800 nm, corresponding to energies on the
order of $10^{-19}$ J. Sometimes these energies are given in electron volts (eV). An electron volt corresponds to the work associated with moving an electron (with the elementary charge of $1.602 \cdot 10^{-19}$ C)
against a potential difference of 1 V. One electron volt thus equals $1.602 \cdot 10^{-19}$ J. Alternatively, molar
energies are given: $10^{-19}$ J per molecule add up to 16 kJ mol$^{-1}$.

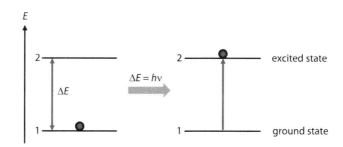

**FIGURE 19.3:** **Light-induced transitions in a two-state system.** A molecule in its lower energy ground state (1, left) can undergo a transition to the excited state (2, right, blue arrow) upon irradiation with light of the frequency ν if the energy of the photons ($h\nu$) equals the energy difference $\Delta E$ that separates the two states.

The population of the two (electronic or other) states in the absence of irradiation depends on the energy difference $\Delta E$ separating the two states and the *thermal energy* of the molecules. For single molecules, the thermal energy is the product of the Boltzmann constant $k_B$ and the temperature $T$:

$$E = k_B T \qquad\qquad \text{eq. 19.5}$$

For one mol of molecules, the thermal energy is

$$E_{\text{molar}} = E \cdot N_A = N_A k_B T = RT \qquad\qquad \text{eq. 19.6}$$

$N_A$ is the Avogadro constant, $R$ the general gas constant. The Boltzmann constant $k_B$ is a microscopic constant, the gas constant $R$ is a macroscopic constant (see Section 2.2.3). At room temperature (298 K), the thermal energy of a molecule, $k_B T$, is ~$4.1 \cdot 10^{-21}$ J, the thermal energy of one mol of molecules, $RT$, is ~2.5 kJ mol$^{-1}$.

The population of two states separated by $\Delta E$ is described by the *Boltzmann distribution* (eq. 19.7). The probability $p_2$ that a molecule is in the high-energy state 2 relative to the probability $p_1$ that it is in the lower energy state 1 is

$$\frac{p_2}{p_1} = \exp\left(-\frac{\Delta E}{k_B T}\right) \qquad\qquad \text{eq. 19.7}$$

For an ensemble of molecules, the number of molecules $n_2$ in the higher energy state relative to the molecules $n_1$ in the lower energy state is

$$\frac{n_2}{n_1} = \exp\left(-\frac{\Delta E_{\text{molar}}}{RT}\right) \qquad\qquad \text{eq. 19.8}$$

The product of the general gas constant $R$ and the temperature $T$ is the thermal energy of one mol of molecules. The higher the energy difference between states 1 and 2, the lower the probability that the molecules are in the high-energy state. The higher the temperature and the thermal energy of the molecules, the higher the probability that the molecules populate the high-energy state.

The number of light-induced transitions from state 1 to 2 (1→2) is determined by the product of the population of state 1 and the transition probability (1→2). The incident light also induces transitions from state 2 to 1. The number of these 2→1 transitions is related to the product of the population of state 2 and the transition probability (2→1). For a simple two-state system, the probabilities for a transition 1→2 and for the reverse transition 2→1 are equal. At the beginning of the illumination, state 1 is more populated than state 2, and by irradiation with light we will be able to induce more transitions from 1→2 than from 2→1. With increasing numbers of molecules in state 2, reverse transitions will start to pick up. Eventually, we will reach equal populations of states 1 and 2. Now the numbers of transitions 1→2 and 2→1 are equal: the transition is saturated, and absorption is no longer observed (Figure 19.4).

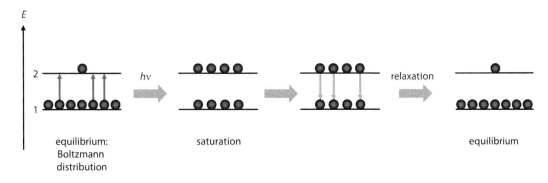

**FIGURE 19.4:   Transitions in a two-state system: Boltzmann distribution, excitation, saturation, and relaxation.** In equilibrium, the lower energy ground state (1) will be more populated than the excited state (2) according to the Boltzmann distribution (eq. 19.7, eq. 19.8). Irradiation with light leads to transitions from the ground state to the excited state (blue arrows). The transition is saturated when both states are equally populated. In the absence of light, molecules will relax from the excited state to the ground state (green arrows), restoring the equilibrium distribution.

## 19.2  ABSORPTION

### 19.2.1  ELECTRONIC, VIBRONIC, AND ROTATIONAL ENERGY LEVELS

As outlined in the previous section, a molecule can absorb light when the energy of a photon corresponds to the energy separating its *electronic ground state* ($S_0$) from its *electronically excited state* $S_1$ (or $S_2$, or higher $S_n$ states). The "S" refers to singlet states, in which electron spins (see Section 20.2) are paired. Absorption of light converts the absorbing molecule into the excited state, in which electrons are more separated from the nuclei compared to the ground state. The energy of the absorbed light is required to overcome the electrostatic attraction that counteracts the separation of the electrons from the nuclei. In addition to electronic excitation, electromagnetic radiation can also induce transitions between *vibronic or rotational states* of the molecule. The energy differences between vibronic states are about one order of magnitude smaller than for electronic states, and the differences between adjacent rotational states are about two orders of magnitude smaller. The electronic, vibronic, and rotational energy levels of a molecule can be depicted schematically in an energy diagram, the *Jablonski diagram* (Figure 19.5). Such a diagram is useful to illustrate processes occurring in spectroscopy, and we will come back to this simplified, but clear depiction several times in the following.

At room temperature (298 K), the thermal energy $RT$ is about 2.5 kJ mol$^{-1}$ (eq. 19.6). This value is on the order of energetic differences between rotational energy levels. Transitions between rotational states can therefore be excited by thermal energy. In contrast, the thermal energy is much smaller than the energy differences between vibronic or electronic states: transitions between vibronic or electronic states cannot be induced by thermal energy at ambient temperatures. Spectroscopy at room temperature therefore starts from the lowest electronic and vibronic energy levels of the molecule, $S_0, V_0$ (Figure 19.5) but not from the lowest rotational level.

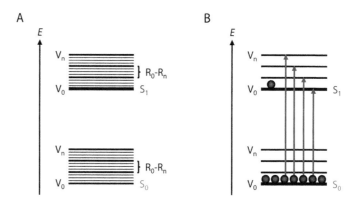

**FIGURE 19.5: Jablonski diagram.** A: A molecule can exist in two electronic states, $S_0$ and $S_1$ (thick lines) $S_0$ is the ground state, corresponding to the lowest energy electron configuration. $S_1$ is the excited state in which electrons are more separated from the nuclei compared to the ground state. The $S_0$ and $S_1$ states are subdivided into different vibrational states, $V_0$–$V_n$ (medium lines) and rotational states ($R_0$–$R_n$, thin lines). The energy differences between adjacent vibronic energy levels are about one order of magnitude smaller than the energy differences between electronic states; the energy differences between neighboring rotational energy levels are two orders of magnitudes smaller. B: According to the Boltzmann distribution, spectroscopy at room temperature starts from $S_0$, $V_0$. Irradiation with light can induce transitions (blue arrows) from $S_0$, $V_0$ to different vibronic levels of $S_1$.

### 19.2.2 TRANSITIONS AND TRANSITION DIPOLES

Generally, transitions between two states of a molecule can be induced by perturbations, i.e. by external electric or magnetic fields. The effectiveness of the perturbation to induce a transition depends on the extent to which it can deform the initial state such that it resembles the final state. Incident light induces transitions between electronic states by altering the charge distribution of the molecule. This change in charge distribution is connected to an induced dipole in the molecule. According to classical mechanics, the dipole **μ** induced by an electric field $E$ depends on the polarizability $\alpha$ of the molecule:

$$\boldsymbol{\mu} = \alpha \cdot \boldsymbol{E} \qquad \text{eq. 19.9}$$

The *transition dipole* μ quantifies the direction and magnitude of changes in the charge distribution within a molecule. The direction of the dipole moment is a property of the molecule, and is fixed with respect to its geometry. The transition dipole moment is a measure of the capacity of the incident light to distort the initial state towards the final state. Therefore, the magnitude of the transition dipole $|\boldsymbol{\mu}| = \mu$ determines the probability for a transition between these states upon irradiation. Transitions with $\mu \neq 0$ are called *allowed transitions*, whereas transitions associated with a transition dipole of zero are *forbidden*. Selection rules tell us if a particular transition between two states is allowed in a quantum-mechanical system. Transitions that are inconsistent with these selection rules are called forbidden. Forbidden transitions can be observed experimentally, but the transition probabilities and absorption intensities are very low.

In Section 19.1.2, we have considered the behavior of two-state systems independent of the molecular properties. Molecules consist of atoms that are linked by covalent bonds. The dependence of the potential energy of two atoms on their distance is described by the *Lennard-Jones potential* (Figure 15.14; see Section 15.3) that reflects the balance of the attractive and repulsive forces between the two atoms. At the distance $r_{min}$, the van der Waals distance of the two atoms, the potential energy reaches a minimum. In electronically excited states, electrons usually occupy non-bonding and anti-bonding orbitals, resulting in less electron density between the nuclei and higher repulsion forces between their positive charges, and an increased bond length $r_{min}$ in the electronically excited state compared to the ground state. The electronic ground state and the first electronically excited state of a molecule can therefore be depicted as two Lennard-Jones potential curves, shifted along the $r$-axis (Figure 19.6). Within these potential energy curves, we can now include the vibronic levels as horizontal lines for ground state and excited state. Absorption is an inherently rapid process, and occurs on the time scale of $10^{-15}$ s. This time scale is too short for bonds to vibrate or for conformational changes of the absorbing molecule to occur. Hence, the molecule is static on the time scale of absorption, and bond lengths do not change. As a consequence, absorption can be depicted as a vertical transition in Lennard-Jones potential diagrams (Figure 19.6). This principle is known as the *Franck-Condon principle*, or the *principle of vertical transitions*.

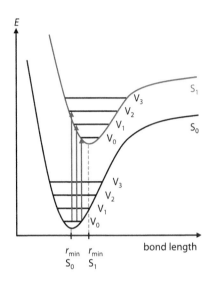

**FIGURE 19.6:    Vertical transitions in the Lennard-Jones potential.** On the time scale of absorption, the geometry of molecules does not change, and bond lengths remain constant. As a consequence, transitions between the electronic ground state $S_0$ and the excited state $S_1$ can be depicted as vertical arrows (blue) in the Lennard-Jones potential. The vibronic levels $V_0$–$V_3$ for both states are indicated.

At room temperature, absorption of light causes transitions from the electronic and vibronic ground state, $S_0,V_0$ (see Section 19.1.2), to the electronically excited state $S_1$. Depending on the energy of the excitation light, the molecules can be excited to higher vibronic levels $V_x$ of the $S_1$ state. An *absorption spectrum* is the dependence of absorption by a molecule, a *chromophore*, as a function of the wavelength of the irradiating light (Figure 19.7). Such a spectrum typically consists of a set of lines. Each set corresponds to a particular $S_n \rightarrow S_{n+1}$ transition, and the individual lines of the set correspond to the individual vibronic transitions involved. In solution, the energies involved in these transitions of the dissolved molecule, the solute, show some variation due to interactions with the solvent. Hence, the measured absorption spectrum of chromophores in solution is the envelope of all vibronic transitions of one set of electronic transitions. The lines underneath the broad absorption bands are called the *fine structure* of the spectrum.

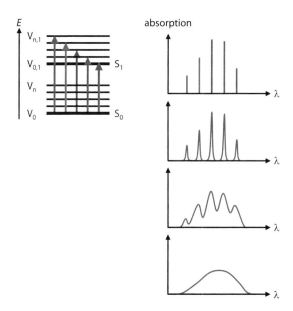

**FIGURE 19.7: From line spectra to continuous spectra.** Left: Jablonski diagram for a molecule that can undergo transitions from $S_0, V_0$ to different vibrational levels of $S_1$ (arrows). The transitions in the Jablonski diagram predict a set of absorption lines at the wavelength corresponding to the energy of each transition (right, top). Such a line spectrum is observed in the absence of collisions with other molecules or with solvent molecules. In the presence of collisions, the discrete energy levels become more degenerate, and the lines in the spectrum broaden and overlap until a continuous spectrum is observed (right, bottom).

An absorption band is characterized by the wavelength of maximum absorption ($\lambda_{max}$) and the *extinction coefficient* $\varepsilon_\lambda$ (in M$^{-1}$ cm$^{-1}$) at this wavelength (see Section 19.2.3), which is a measure of the absorption intensity. The absolute absorption intensity, i.e. the area under the absorption band, is determined by the *dipole strength D*. The dipole strength $D_{i \to f}$ (in the unit debye$^2$, with 1 debye = 3.336·10$^{-30}$ C·m) of the transition from the initial state $i$ to the final state $f$ is related to the measured spectrum by eq. 19.10:

$$D_{i \to f} = 9.18 \cdot 10^{-3} \int \frac{\varepsilon}{v} \, dv$$

eq. 19.10

Thus, the dipole strength can be obtained by integration of the absorption band in a spectrum that has been converted into the molar extinction coefficient $\varepsilon$ (see Section 19.2.3) as a function of the frequency $v$. The dipole strength is proportional to the square of the transition dipole.

### 19.2.3 THE LAMBERT-BEER LAW

One of the most important applications of absorption spectroscopy is the determination of concentrations. To derive a relationship between absorption and concentration, we consider a solution containing absorbing molecules (Figure 19.8) that is illuminated by light of the intensity $I_0$. The light path through the sample can be divided into slices with the incremental thickness dx. The light entering one of these slices will have the intensity $I$. The intensity of the light exiting from the slice, the transmitted light, will be reduced to $I-dI$ due to absorption by the molecules within this slice. The change in intensity d$I$ is proportional to the intensity $I$ of the incident light, and to the concentration $c$ of the absorbing molecules. We name the proportionality constant $a$. This constant is a measure of the absorption strength of the molecule. The reduction in light intensity due to absorption is accounted for by the negative sign in eq. 19.11:

$$dI = -a \cdot c \cdot I \, dx$$

eq. 19.11

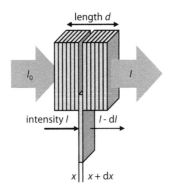

FIGURE 19.8:  **Lambert-Beer law.** A sample in a cuvette of path length *d* is illuminated by light. Within a slice of the depth *dx* (bottom) absorption leads to a reduction of the light intensity from *I* (intensity of the incident light) to *I*−d*I* (intensity of the transmitted light). By integrating over the path length *d* and the light intensity, the Lambert-Beer law can be derived.

Separation of variables in eq. 19.11 gives

$$\frac{\mathrm{d}I}{I} = -a \cdot c\,\mathrm{d}x \qquad \text{eq. 19.12}$$

Eq. 19.12 can then be integrated over all slices of the sample, from $x = 0$ to the path length $d$ and from the initial light intensity $I_0$ to the intensity $I$:

$$\int_{I_0}^{I} \frac{\mathrm{d}I}{I} = -a \int_{0}^{d} c\,\mathrm{d}x \qquad \text{eq. 19.13}$$

The concentration $c$ of the absorbing molecule is homogeneous over the sample. Therefore, $c$ does not depend on the location of $x$, and can be considered as constant. We can evaluate the integrals to

$$\ln \frac{I}{I_0} = -a \cdot c \cdot d \qquad \text{eq. 19.14}$$

Conversion of the natural logarithm in eq. 19.14 into the decadic logarithm (see Section 29.5) according to

$$\ln \frac{I}{I_0} = \ln 10 \cdot \log \frac{I}{I_0} = -\ln 10 \cdot \log \frac{I_0}{I} \qquad \text{eq. 19.15}$$

and definition of the constant $a/\ln 10$ as the molar *extinction coefficient* ε gives

$$\log \frac{I_0}{I} = \varepsilon c\, d \qquad \text{eq. 19.16}$$

The logarithm of the intensity of the incident light $I_0$, divided by the intensity of the transmitted light $I$, is called the absorption $A$. We thereby obtain the *Lambert-Beer law* as

$$A = \varepsilon c\, d \qquad \text{eq. 19.17}$$

The Lambert-Beer law is central for the photometric determination of the concentration of absorbing molecules. If the extinction coefficient for a molecule of interest is known, we can calculate its concentration in solution from the measured absorbance (see Section 19.2.7). It is important to keep in mind that absorption is defined on a logarithmic scale (see Section 19.3.6).

### 19.2.4 SOLVENT EFFECTS AND INFLUENCE OF THE LOCAL ENVIRONMENT

Absorption is linked to a change in charge distribution: the electronically excited state has a different charge distribution than the ground state. In solution, the configuration of solvent molecules around the solute will be such that the solvent stabilizes the electronic state of the solute molecule. This effect is particularly important for aqueous solutions because water molecules are strong permanent dipoles. Ordering of water molecules around aromatic moieties is the molecular basis for the hydrophobic effect (Section 16.3.1 and Figure 16.35). Charged solutes are stabilized in aqueous solutions by hydration, an energetically favorable orientation of the surrounding water molecules (Figure 19.9). When the solute molecule absorbs light and undergoes a vertical transition (Figure 19.6) to the electronically excited state, its charge distribution changes on the $10^{-15}$ s time scale of the absorption process. On this time scale, not only the chromophore and its geometry, but also the surrounding water molecules are unchanged. However, the arrangement of water molecules that was optimal for the stabilization of the ground state is not the energetically favorable geometry for hydration of the chromophore in its excited state. Therefore, this geometry of water molecules now destabilizes the excited state relative to a non-polar solvent, and increases its energy (Figure 19.9). Consequently, the energy difference between $S_0$ and $S_1$ levels is increased compared to a non-polar solvent, and the absorption wavelength will be changed accordingly. A shift in the absorption wavelength towards the blue end of the spectrum (higher energy) is called a *hypsochromic effect*, a shift towards the red (lower energy) is called a *bathochromic effect*.

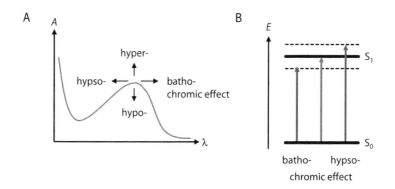

**FIGURE 19.9: Solvent effects on electronic transitions.** A: The environment of a chromophore affects its absorption spectrum, and can shift the maximum of the absorption to lower wavelengths (blueshift, hypsochromic effect), or to longer wavelengths (redshift, bathochromic effect). An increase in the extinction coefficient corresponds to a hyperchromic effect, a decrease to a hypochromic effect. B: Energy diagram for batho- and hypsochromic effects of solvents.

In contrast to changes in the absorption wavelength, changes in the extinction coefficient are called *hypochromic* (decrease) or *hyperchromic effects* (increase; Figure 19.9). These effects are caused by effects of the solvent on the transition dipole and thus on the dipole strength (eq. 19.10) of the molecule.

Similar to the effects of the solvent on electronic transitions of a solute molecule, the local environment of a chromophoric group within a larger molecule such as a protein or a nucleic acid influences its absorption. For example, the absorption of chromophoric side chains of amino acids depends on their chemical environment within the three-dimensional structure of the protein. Therefore, the absorption of a protein in the folded state under native conditions differs from its absorption under denaturing conditions, where it is unfolded and all side chains are solvent-exposed. Absorbance is therefore a powerful spectroscopic probe to monitor protein folding and unfolding (see Section 19.2.7). The direction of the change cannot be predicted, and hyper- or hypochromic as well as batho- and hypsochromic effects upon folding are observed.

The absorption of the nucleobases in the DNA also depends on the environment. The absorbance of the bases in the DNA is influenced by their neighbors within the same strand, and by their pairing partners on the opposite (complementary) strand. DNA shows a strong hyperchromic effect upon dissociation of the double helix into single strands, and absorption is therefore routinely used as a sensitive spectroscopic probe to monitor DNA melting (see Section 19.2.7).

### 19.2.5 INSTRUMENTATION

Absorption is measured in spectrometers that consist of a light source, typically a deuterium arc lamp for UV- and tungsten lamp for visible light or a Xenon arc lamp (UV/VIS), a monochromator to select a certain wavelength of the incident light, a cuvette holder, and a photomultiplier that detects the light transmitted by the sample (Figure 19.10). The photons that enter the photomultiplier generate electrons by the photoelectric effect. These secondary electrons are accelerated in an electric field and amplified to a measureable photocurrent that is proportional to the light intensity. This simplest type of absorption spectrometer is termed single-beam photometer. The absorption of the sample, typically a buffered solution of a protein or nucleic acid, is calculated from the intensities of the incident and the transmitted light. Solvent contributions to the absorption of the sample need to be measured in a second measurement of a sample with the solvent only, and are subtracted from the absorption of the sample. If the wavelength of the incident light is varied, an *absorption spectrum*, $A(\lambda)$, is obtained.

In a double-beam photometer, the incident light is split before the sample. One part is passed through the sample, and the second part is passed through a reference cuvette containing the solvent only. The difference in absorption between sample and reference cell then directly reflects the absorption of the solute in the sample cell. Using a double-beam photometer *difference spectra* may be measured directly. In this case, both cuvettes contain samples, e.g. a protein in the native state in one cuvette and the same protein in the denatured state in the second cuvette.

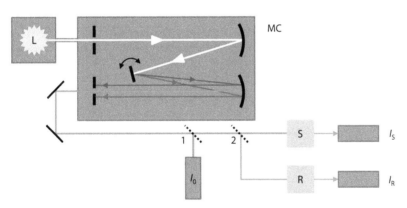

**FIGURE 19.10:    Single- and double-beam photometer.** A light source (L, yellow) emits white light. The monochromator (MC) selects the desired wavelength. The selection depends on the position of an adjustable diffraction grating (double-headed arrow). A small fraction of the light exiting the monochromator is reflected by a partially transmitting mirror (1) to measure the intensity of the incident light $I_0$. The remaining light passes the sample (S) in a cuvette. The transmitted light $I_S$ is detected by a photomultiplier. In a double-beam spectrometer, the light is split a second time (2). One part passes the sample, the second a reference cuvette (R) that contains solvent only. The absorption of the solute molecule in the sample is then calculated from the difference in the intensities of transmitted light, $I_S - I_R$.

### 19.2.6 BIOLOGICAL CHROMOPHORES

Proteins typically show absorption in the UV range, caused predominantly by the aromatic side chains of Tyr and Trp due to $\pi \rightarrow \pi^*$ transitions (Figure 19.11). Tyr and Trp are not very common amino acids, with a relative abundance of 3.2% and 1.3%, respectively, and proteins therefore only contain a small number of chromophores. The extinction coefficients at 280 nm for Tyr and Trp are around 1200 and 5600 $M^{-1}$ $cm^{-1}$. Both Tyr and Trp also exhibit weak absorption below 220 nm. Despite the lower extinction coefficient, tyrosines contribute significantly to protein absorption due to their higher abundance. An average protein of 100 residues would thus typically have an extinction coefficient on the order of 10000 $M^{-1}$ $cm^{-1}$. Although Phe contributes to protein absorption near 260 nm because of a symmetry-forbidden $\pi \rightarrow \pi^*$ transition ($\varepsilon \approx 400\ M^{-1}\ cm^{-1}$), this weak band is usually not observed experimentally in proteins that also contain Trp and Tyr.

The peptide bond shows two absorption bands in the UV range, one around 210–220 nm as the result of a forbidden $n \rightarrow \pi^*$ transition ($\varepsilon \approx 100\ M^{-1}\ cm^{-1}$), and a much stronger band at around 190 nm ($\varepsilon \approx 7000\ M^{-1}\ cm^{-1}$) because of a $\pi \rightarrow \pi^*$ transition. The side chains of some amino acids, such as Asp, Glu, Asn, Gln, Arg, His, and Cys, and Tyr and Trp, absorb in the same spectral range. Their contribution can usually not be measured in the context of proteins, however, because it is very small, and because these residues are always outnumbered by the peptide bonds. Disulfides (cystines) show red-shifted absorption at 250–270 nm ($\varepsilon \approx 300\ M^{-1}\ cm^{-1}$), which is usually too small to be exploited in absorption measurements.

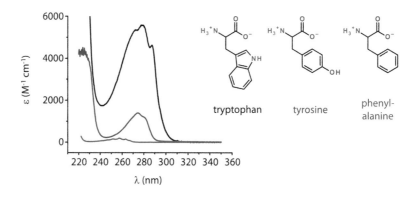

FIGURE 19.11:    **Absorption of aromatic amino acids.** Absorption spectra of tryptophan (black) tyrosine (red) and phenylalanine (blue), converted to the molar extinction coefficient. Dixon *et al.* (2005) *Photochem. Photobiol.* 81(1): 212–213. The structures of tryptophan, tyrosine and phenylalanine are depicted on the right.

The nucleobases of DNA and RNA absorb strongly in the UV as a result of several $n \rightarrow \pi^*$- and $\pi \rightarrow \pi^*$-transitions with maxima between 255 and 275 nm (Figure 19.12). The extinction coefficients at 260 nm are about 8000 $M^{-1}$ $cm^{-1}$ for pyrimidines, and 15000 $M^{-1}$ $cm^{-1}$ for purines. The absorbance spectrum of DNA shows a broad maximum at around 260 nm, with an average extinction coefficient of about 10000 $M^{-1}$ $cm^{-1}$ per nucleotide. In contrast to proteins, each building block provides a chromophore, and the average extinction coefficients for nucleic acid molecules are very high. Therefore, absorption at 260 nm provides a highly sensitive method to determine nucleic acid concentrations.

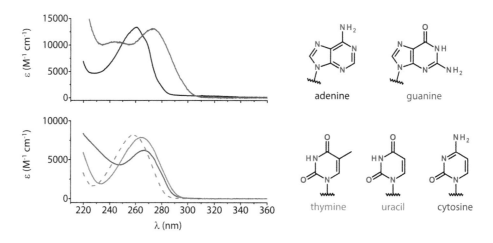

FIGURE 19.12:    **Absorption of bases in nucleic acids.** Absorption spectra of adenine (black) and guanine (blue, top), and cytosine (red), thymine (green) and uracil (green, dotted line, bottom). All spectra are normalized to the molar extinction coefficient. Dixon *et al.* (2005) *Photochem. Photobiol.* 81(1): 212–213. The structures of the nucleobases are depicted on the right.

Several cofactors and prosthetic groups also absorb light in the UV and visible regions (Figure 19.13). Examples include NADH, an electron donor in redox reactions (Chapter 5 and Section 5.6) that absorbs at 340 nm due to an n→π* transition of the conjugated CO group in the nicotinamide ring, and flavins (FMN, FAD) that absorb at 445–465 nm. Flavonoids, such as anthocyanins and flavones, are responsible for the yellow, red, and blue colors of flowers and fruits. Tetrapyrroles are wide-spread chromophores of four pyrrole rings that can either form a macrocyclic, conjugated ring system such as in the porphyrins, or linear structures such as in the bile pigment biliverdin. Heme, the prosthetic groups in hemoglobin, myoglobin, and cytochromes, is a redox-active porphyrin-$Fe^{2+}/Fe^{3+}$ complex. Their intense absorption at 400–500 nm, the "Soret" band, is caused by a π→π* transition in the extensive aromatic system with electron delocalization over all four pyrrole rings. The photosynthesis pigment chlorophyll (Section 5.6.2) is a $Mg^{2+}$ complex of a modified porphyrin. Carotenoids (carotenes and xanthophylls; Figure 19.13) are also involved in photosynthesis and show strong absorption between 450 and 550 nm. The absorption of UV light (at 335 nm) by melanin, generated by the oxidation and polymerization of tyrosines, protects our skin from UV damage and cancer.

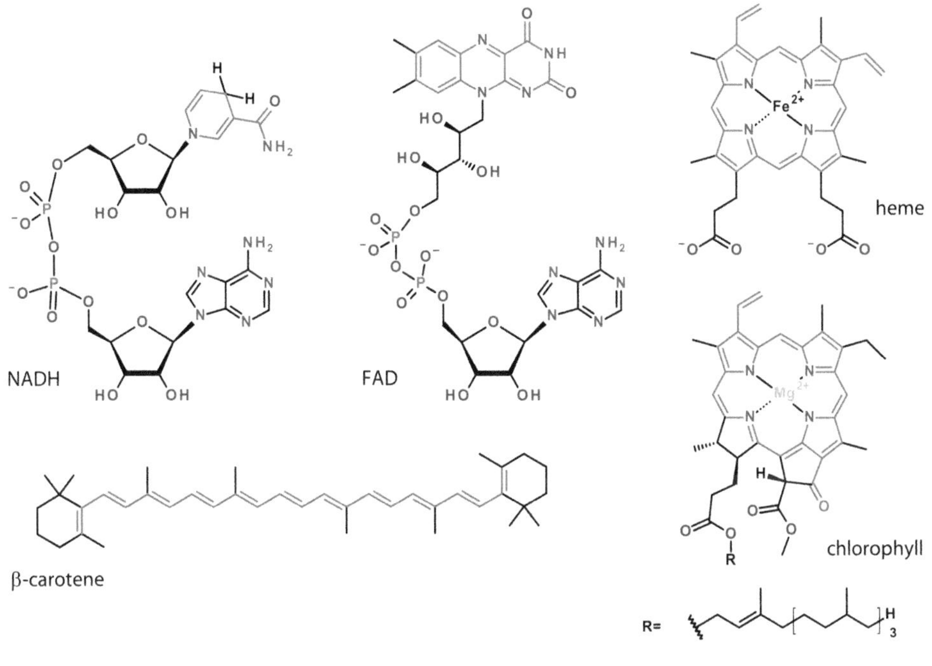

**FIGURE 19.13:    Cofactors and prosthetic groups that show UV absorption.** Nicotinamide adenine dinucleotide (NADH; reduced form), flavin adenine dinucleotide (FAD; oxidized form), the porphyrins heme and chlorophyll, and the carotenoid β-carotene. The delocalized π-electron system responsible for the absorption is highlighted in orange.

Vitamin A, the precursor of the vision pigment retinal, and retinal itself absorb at 370 nm (Figure 19.14). The absorption of retinal is shifted to 500 nm in complex with the protein opsin. It further changes during the *cis→trans* isomerization (Section 15.1) of retinal, the central reaction in the vision process (Box 19.1). Retinal is also the chromophoric unit in light-driven proton pumps and ion channels (bacteriorhodopsin, channelrhodopsin and halorhodopsin; Box 19.2, see Section 4.3).

## BOX 19.1: THE VISION PROCESS.

Retinal (Figure 19.14; $\lambda_{max}$ = 370 nm) is the central chromophore in our vision process. Upon binding of 11-*cis* retinal to the seven-helix transmembrane protein opsin, the absorption maximum undergoes a huge bathochromic shift to $\lambda_{max}$ = 500 nm. 11-*cis* retinal has an extinction coefficient of $\varepsilon$ = 40000 M$^{-1}$ cm$^{-1}$. When 11-*cis* retinal absorbs light, it isomerizes to all-*trans* retinal. The affinity of opsin for all-*trans* retinal is low, and retinal dissociates from rhodopsin. Opsin is a G-protein coupled receptor that interacts with the G-protein transducin. Transducin is a heterotrimeric G-protein. Upon activation by opsin, its $\alpha$-subunit exchanges bound GDP for GTP and activates a phosphodiesterase that hydrolyzes cyclic GMP (cGMP) to GMP. cGMP maintains ligand-gated sodium channels in the plasma membrane in the closed form. Its hydrolysis accordingly leads to opening of these channels. The resulting hyperpolarization (see Section 5.7) of the rod cells affects the transmission of signals from the rod cells to bipolar cells to the ganglia of the optical nerve. Absorption of a single photon leads to isomerization of one retinal molecule, but each opsin activates many transducins and leads to opening of numerous sodium channels. This signal amplification renders our eyes very sensitive optical instruments!

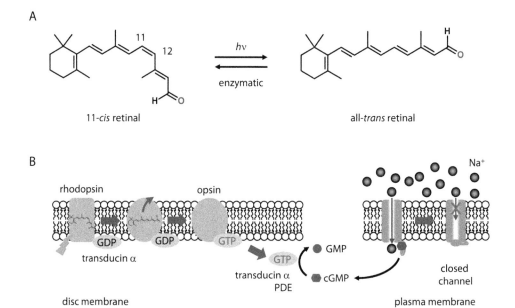

FIGURE 19.14:   **Retinal, rhodopsin, and the vision process. A:** 11-*cis* retinal is converted to all-*trans* retinal upon illumination (*hν*). All-*trans* retinal is enzymatically regenerated to 11-*cis* retinal. **B:** 11-*cis* retinal is bound to the protein opsin, an integral membrane protein in the disc membrane of rod cells. Illumination causes isomerization of the retinal to the all-*trans* form that dissociates from opsin (red arrow). The resulting activated opsin converts the $\alpha$-subunit of the G-protein transducin into the GTP form that has phosphodiesterase (PDE) activity. The PDE hydrolyzes cyclic GMP (cGMP) to GMP, leading to the closure of cGMP-gated sodium channels in the plasma membrane and hyperpolarization of the rod cell.

## BOX 19.2: LIGHT-DRIVEN PROTON PUMPS AND LIGHT-GATED ION CHANNELS.

Bacteriorhodopsin is a transmembrane protein in the purple membrane of Archaea that acts as a light-driven proton pump (Figure 19.15). Its chromophore all-*trans* retinal is covalently bound to a lysine *via* a protonated Schiff base, and separates the pore into two half-channels. Upon absorption of light ($\lambda_{max}$ = 568 nm), retinal isomerizes to 13-*cis* retinal. The resulting conformational change reduces the polarity around the Schiff base, leading to deprotonation and release of the proton to the outside of the cell. A subsequent conformational change in the photocycle of bacteriorhodopsin restores the polarity around the Schiff base, which becomes reprotonated from the inside. One absorption event is thus coupled to the net transport of one proton from the inside of the bacterium to the outside. The light energy is thereby converted into a proton-motive force across the plasma membrane (Section 4.2) that can be exploited to drive the synthesis of ATP. Halorhodopsin is a light-gated chloride pump in halobacteria. Both bacteriorhodopsin and halorhodopsin use light as the energy source for active transport. Channelrhodopsins are light-gated ion channels (Section 4.3) in algae. The discovery of channelrhodopsins and their application for the manipulation of membrane potentials in living cells and animals has opened up the field of optogenetics, and has provided a huge impulse to the field of neurophysiology.

A

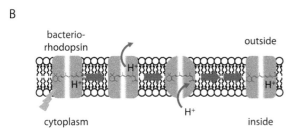

all-*trans* retinal          13-*cis* retinal

B

bacterio-
rhodopsin                          outside

H⁺
H⁺                    H⁺

H⁺

cytoplasm                    inside

**FIGURE 19.15:    Light-gated ion pumps and channels.** A: Light-driven ion pumps use the chromophore all-*trans* retinal that isomerizes to 13-*cis* retinal upon absorption of light. B: Bacteriorhodopsin is a light-driven proton pump that uses the energy of light to generate a proton-motive force (see Section 4.2). The chromophore all-*trans* retinal is bound to the protein by a protonated Schiff base and blocks the proton channel. Upon absorption and isomerization of retinal, and a conformational change of the protein, the Schiff base becomes deprotonated. A proton is released to the outside. Regeneration of the original state of bacteriorhodopsin, i.e. the all-*trans* form of retinal and the original conformation of the opsin, are coupled to the uptake of a proton from the cytoplasm. Overall, one absorption event is coupled to transport of one proton from the cytoplasm to the outside. Halorhodopsin and channelrhodopsins use a similar mechanism to couple the transport of chloride and other ions to the absorption of light.

### 19.2.7 APPLICATIONS

#### 19.2.7.1 Concentration Determination

According to the Lambert-Beer law (eq. 19.17), absorption is proportional to chromophore concentration. Absorption is therefore a useful tool to determine concentrations in solution, provided its extinction coefficient in the solvent used is known. In the case of proteins and DNA, the effect of the local environment has to be taken into account. For an unfolded protein, with solvent-exposed side chains, the extinction coefficient at 280 nm can be calculated by simply adding the extinction coefficients for all chromophoric side chains present. Typically, the aromatic side chains of Tyr and Trp have to be taken into account, as well as cystine groups (cysteines forming disulfide bonds). The absorption differences between denatured and native states of a protein typically do not exceed 10% (Figure 19.19). As a consequence, the calculated $\varepsilon_{280}$ for the unfolded protein is a good first approximation for the extinction coefficient of the native protein for many qualitative purposes. More quantitative approaches require determination of the exact extinction coefficient of the native protein. We can determine the extinction coefficient for the protein in its folded state by comparing the absorbance of a protein solution at the same concentration under denaturing and native conditions (Box 19.3). From the Lambert-Beer law (eq. 19.17) it is clear that the ratio of the absorption $A_{native}$ and $A_{denatured}$ equals the ratio of the extinction coefficients $\varepsilon_{native}$ and $\varepsilon_{denatured}$ when path length and concentrations are identical.

---

**BOX 19.3: DETERMINATION OF EXTINCTION COEFFICIENTS FOR NATIVE PROTEINS ACCORDING TO GILL & VON HIPPEL.**

Gill & von Hippel developed a straightforward procedure for the determination of protein extinction coefficients under native conditions. The only information needed is the protein sequence (or at least the number of the chromophoric groups tyrosine, tryptophan and cystines), and a measurement of the absorption of the protein under native and denaturing conditions. The extinction coefficient for the denatured protein can be calculated from the number of tryptophans ($n_{Trp}$), tyrosines ($n_{Tyr}$) and cystines ($n_{Cystine}$), and the extinction coefficients $\varepsilon_{Trp}$, $\varepsilon_{Tyr}$, and $\varepsilon_{Cystine}$:

$$\varepsilon_{denatured} = n_{Trp} \cdot \varepsilon_{Trp} + n_{Tyr} \cdot \varepsilon_{Tyr} + n_{Cystine} \cdot \varepsilon_{Cystine} \qquad \text{eq. 19.18}$$

Gill & von Hippel determined the extinction coefficients $\varepsilon_{Trp}$, $\varepsilon_{Tyr}$ and $\varepsilon_{Cystine}$ in a 20 mM phosphate buffer, pH 6.5, 6 M guanidinium chloride, using N-acetyl-tryptophanamide as a reference substance for tryptophan, a Gly-Tyr-Gly peptide for tyrosine, and cystine for cysteine (Table 19.1).

**TABLE 19.1**
**Extinction coefficients for model compounds.**

| compound | $\varepsilon_{280\,nm}$ (M$^{-1}$ cm$^{-1}$) |
|---|---|
| N-acetyl-tryptophanamide | 5690 |
| Gly-Tyr-Gly | 1280 |
| cystine | 120 |

Values from Gill & von Hippel (1989) Anal. Biochem. 182(2): 319–326.

The absorption of the denatured protein, $A_{denatured}$, and of the native protein, $A_{native}$, can be described by the Lambert-Beer law as

$$A_{denatured} = \varepsilon_{denatured} \cdot c \cdot d \qquad \text{eq. 19.19}$$

and

*(Continued)*

---

$$A_{\text{native}} = \varepsilon_{\text{native}} \cdot c \cdot d \qquad \text{eq. 19.20}$$

For identical concentrations $c$ and path lengths $d$, the extinction coefficient of the native protein can then be calculated from the absorbances under native and denaturing conditions, $A_{\text{native}}$ and $A_{\text{denatured}}$, according to

$$\varepsilon_{\text{native}} = \frac{A_{\text{native}}}{A_{\text{denatured}}} \cdot \varepsilon_{\text{denatured}} \qquad \text{eq. 19.21}$$

This extinction coefficient $\varepsilon_{\text{native}}$ of the native protein can then be used in concentration determinations of the protein under native conditions, without the need for prior denaturation. It is important to note that the calculation of the extinction coefficients relies on identical concentrations of native and denatured proteins. It is therefore advisable to repeat the absorption measurements with independently diluted samples.

Sometimes a protein of interest does not contain chromophoric amino acids. In this case, the photometric determination of protein concentrations following Ehresmann provides an alternative. This method measures the absorption of the protein backbone and is thus independent of the protein sequence. The relation

$$c = \frac{A_{228.5} - A_{234.5}}{3.14} \qquad \text{eq. 19.22}$$

was established empirically to determine protein concentrations in plant extracts in the presence of nucleic acids. $A_{228.5}$ and $A_{234.5}$ denote the absorption at 228.5 nm and 234.5 nm, respectively. The extinction coefficient of nucleic acids is identical at these wavelengths, and contributions from nucleic acids thus cancel. The factor 3.14 was determined empirically and converts the absorption difference into a concentration in mg/mL (which can be converted into molarity using the molecular mass of the protein). An advantage of the Ehresmann method is its high sensitivity because of the high extinction coefficient and the large number of peptide bonds, and the small amount of sample needed.

Due to the high extinction coefficients of nucleic acids, absorption is also a sensitive means to determine the concentrations of DNA or RNA. Concentration determination of nucleic acids is hampered by the strong hypochromic effect upon duplex formation, and by the dependence of the extinction coefficient of the bases on their neighbors. For larger DNA with an approximately equal distribution of A, G, C, and T, one absorption unit at 260 nm ($d = 1$ cm) corresponds to approximately 55 µg mL$^{-1}$ double-stranded DNA, or 33 µg mL$^{-1}$ single-stranded DNA. For RNA, one absorption unit is equivalent to a concentration of approximately 40 µg mL$^{-1}$. If the sequence (or base composition) of the RNA is known, the concentration can be determined from the absorption of the nucleotide solution after alkaline hydrolysis of the RNA.

### 19.2.7.2 Spectroscopic Assays for Enzymatic Activity

The facile conversion of absorption into concentrations also renders it a sensitive probe for measuring enzymatic activity when the natural substrate and the product of an enzymatic reaction differ in absorption. Examples are enzymes that catalyze redox reactions and transfer electrons to or from flavins or NAD(P)/NAD(P)H, such as oxidases and dehydrogenases. In other cases, *chromogenic substrates* can be used. Chromogenic substrates do not absorb, but enzymatic turnover generates a product that then shows absorption (Figure 19.16). Simple examples are the blue/white screening for expression of β-galactosidase in bacteria using the galactose-derivative 5-bromo-4-chloro-3-indolyl-β-D-galactopyranoside (X-Gal), and the ortho-nitrophenyl-β-galactoside (ONPG) test for β-galactosidase activity. Chromogenic peptides are frequently used as substrates to monitor peptidase or proteinase activity.

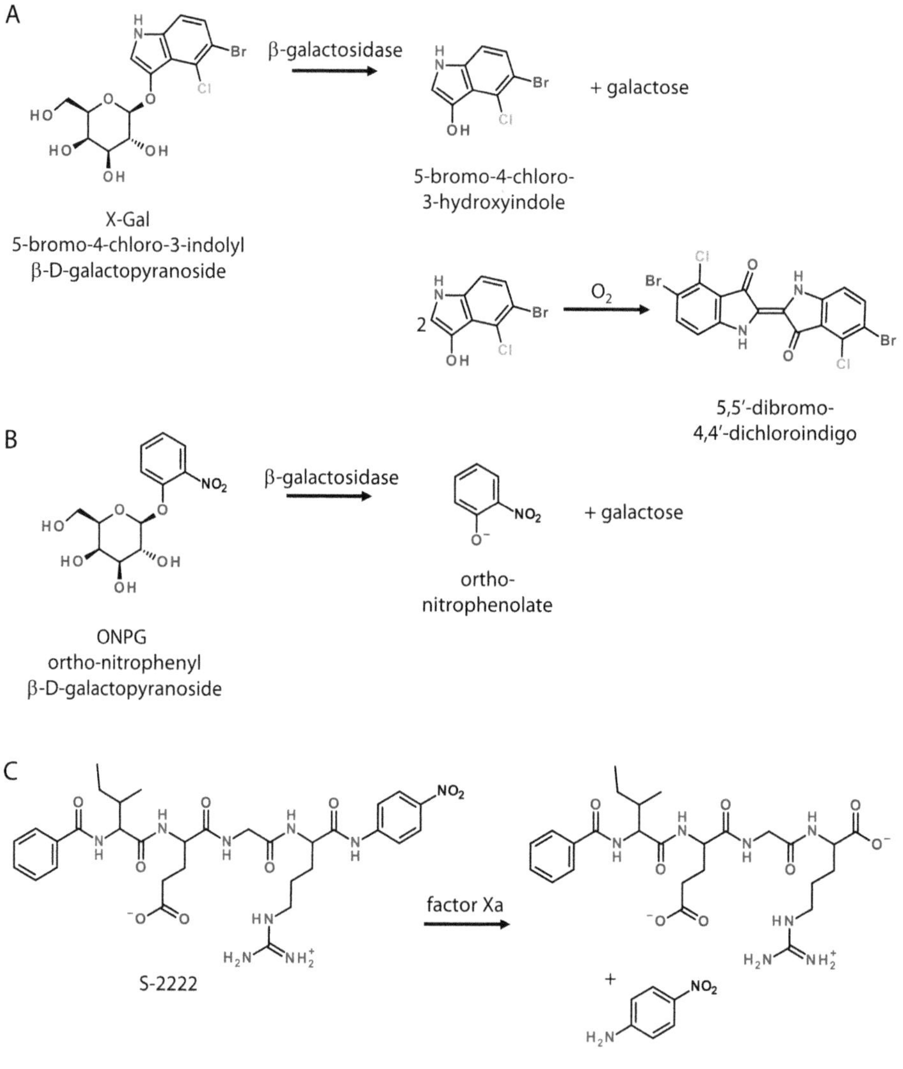

**FIGURE 19.16:** **Chromogenic substrates and their reactions.** A: 5-bromo-4-chloro-3-indolyl-β-D-galactopyranoside (X-Gal) is a chromogenic substrate for the enzyme β-galactosidase. β -galactosidase cleaves the glycosidic bond in X-Gal, and generates 5-bromo-4-chloro-3-hydroxyindole. Two molecules of 5-bromo-4-chloro-3-hydroxyindole are oxidized to 5,5′-dibromo-4,4′-dichloroindigo, a blue insoluble pigment. B: ortho-nitrophenyl-β-D-galactopyranoside is cleaved to ortho-nitrophenolate and galactose by β-galactosidase. The reaction can be monitored by the strong absorption of ortho-nitropheno-late at 420 nm. C: The activity of peptidases and proteinases can be monitored using chromogenic peptides. S-2222 is a tetrapeptide of the sequence Ile-Glu-Gly-Arg, covalently linked to ortho-nitro-aminobenzene by a peptide bond. Factor Xa, a factor of the blood coagulation cascade, cleaves the terminal peptide bond. Factor Xa activity can be monitored *via* the absorption of para-nitro-aminobenzene at 380 nm.

If the enzyme of interest does not turn over substrates with an accompanying change in absorption, the catalyzed reaction can be coupled to other reactions that lead to a spectroscopically measureable output. A *coupled enzymatic assay* is frequently employed to monitor ATP hydrolysis by ATPases (Figure 19.17). The absorption spectra of ATP and ADP are identical, and the hydrolysis of ATP to ADP and inorganic phosphate can therefore not be monitored directly *via* a change in absorption. However, the reaction can be coupled to the oxidation of NADH to $NAD^+$. The nicotinamide ring of this cofactor absorbs at 340 nm in the reduced, but not in the oxidized form. The decrease in absorption at 340 nm therefore provides a sensitive spectroscopic probe for reactions that can be coupled enzymatically to the oxidation of NADH. To couple the hydrolysis of ATP to the oxidation of NADH, the ADP that is generated by the ATPase of interest is converted back to ATP by the enzyme pyruvate kinase that uses phosphoenol pyruvate as a phosphoryl group donor. Phosphoenol pyruvate is converted to pyruvate, which is then reduced to lactate by lactate dehydrogenase, an enzyme that uses NADH as the electron donor. Over all three reactions, one ATP hydrolysis event is coupled to the oxidation of one NADH molecule. With the known extinction coefficient of NADH, the observed decrease in absorption at 340 nm over time can be converted into a change in concentration (of NADH and thus of ATP) over time. This change in ATP concentration directly represents the reaction velocity $v$ (in M s$^{-1}$). In coupled enzymatic assays, it is essential to adjust enzyme concentrations and experimental conditions such that the overall reaction velocity is determined by the enzyme of interest, and not limited by the helper enzymes in the coupled reactions (see Chapter 9).

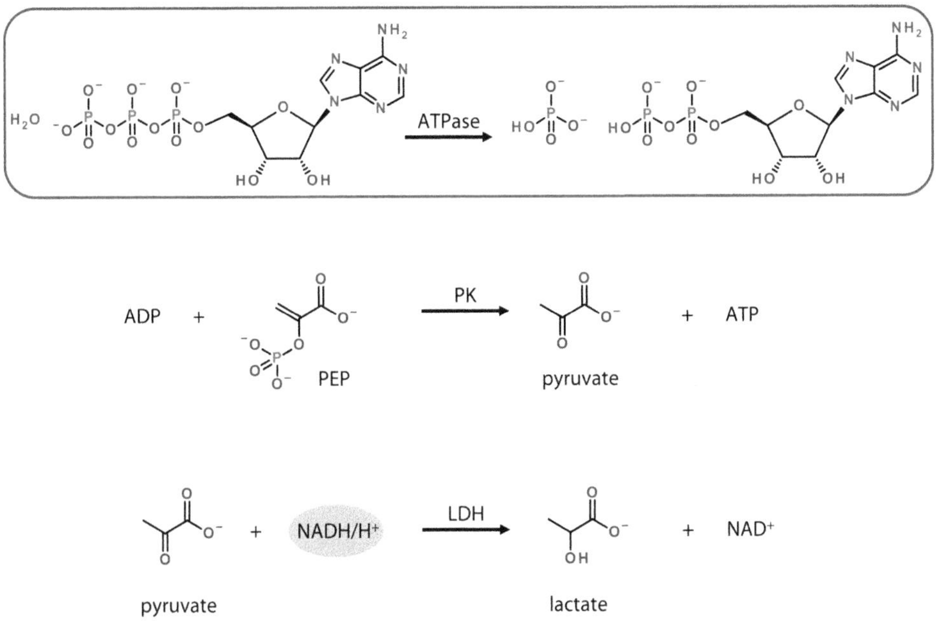

FIGURE 19.17:    **Coupled enzymatic ATPase assay.** ATPases catalyze the hydrolysis of ATP to ADP and phosphate (top). The reaction is spectroscopically silent because ATP and ADP show the same absorption at 260 nm. In a coupled enzymatic assay, the hydrolysis of ATP is coupled to the oxidation of NADH: the ADP generated by ATP hydrolysis is phosphorylated to ATP by pyruvate kinase (PK) using phosphoenol pyruvate (PEP) as a phosphoryl group donor. The pyruvate generated by PK is then reduced to lactate by lactate dehydrogenase (LDH), with the concomitant oxidation of NADH/H$^+$ to NAD$^+$. NADH absorbs at 340 nm, but NAD$^+$ does not. Therefore, NADH consumption can be followed as a decrease in the absorption at 340 nm, and the concentration decrease can be calculated from the change in absorbance using the Lambert-Beer law. Per ATP hydrolyzed, one NADH is oxidized. The concentration decrease of NADH thus directly corresponds to the concentration decrease of ATP.

### 19.2.7.3 Spectroscopic Tests for Functional Groups

Chromogenic reagents can also be used to test for the presence and reactivity of functional groups. One example is the counting of cysteines in proteins using the disulfide 5,5'-dithiobis-2-nitrobenzoic acid (DTNB, *Ellman's reagent*; Figure 19.18). DTNB forms mixed disulfides with solvent-accessible cysteines, which is associated with the release of the TNB$^-$ ion which absorbs light at 412 nm. Per disulfide formed, one molecule of TNB$^-$ is released. The molar amount of TNB$^-$ that is generated can be calculated according to the Lambert-Beer law (eq. 19.17) from the increase in absorption upon incubation of the protein of interest with DTNB. The final concentration of TNB$^-$ directly corresponds to the concentration of reactive cysteines. With the protein concentration known, the ratio of the concentration of free cysteines to protein gives the number of solvent-accessible cysteines per protein. If the experiment is performed under denaturing conditions, the total number of cysteines is determined. The difference between the two numbers (total-accessible) gives the number of cysteines buried in the protein interior. Cysteines are commonly used for site-specific chemical modification of proteins. The reaction of cysteines with maleimides of fluorescent dyes or spin labels is widely used to introduce labels for fluorescence (Section 19.5) or electron paramagnetic resonance (EPR) spectroscopy (Section 20.2). Solvent-accessible cysteines that are suitable for modification can be identified by their reactivity with DTNB. Comparison of the number of reactive cysteines in a DTNB test under reducing and oxidizing conditions, on the other hand, yields information on disulfide bonds in the protein of interest.

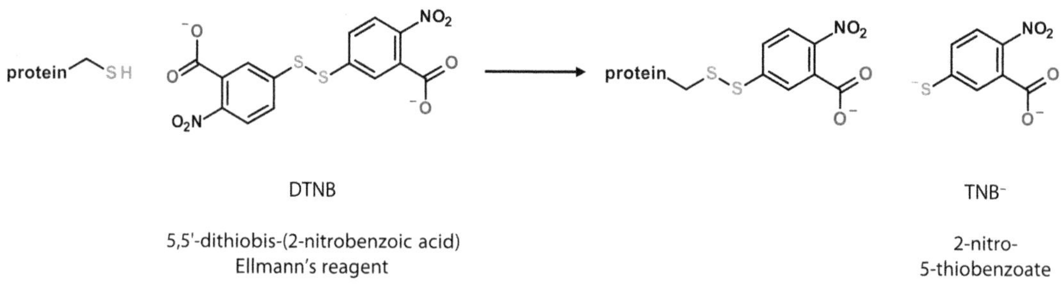

DTNB

5,5'-dithiobis-(2-nitrobenzoic acid)
Ellmann's reagent

TNB⁻

2-nitro-
5-thiobenzoate

**FIGURE 19.18:   DTNB test for accessible cysteines.** DTNB (5,5'-dithiobis-(2-nitrobenzoic acid); shown is the anion), also called Ellman's reagent, forms mixed disulfides with cysteines. The 2-nitro-5-thiobenzoate (TNB⁻) shows absorption at 412 nm. From the increase in absorption at 412 nm, the concentration of TNB⁻ can be calculated. Comparison with the protein concentration gives the number of free cysteines.

### 19.2.7.4 Absorption as a Probe for Structural Changes

Absorption can also be used to monitor structural changes in proteins and nucleic acids because the absorption depends on the local environment of a chromophore (Section 19.2.4). The different absorption of folded and unfolded proteins can be exploited to follow protein folding and unfolding spectroscopically, and to determine the thermodynamic stability of proteins (Box 19.4).

### BOX 19.4: ABSORPTION AS A SPECTROSCOPIC PROBE FOR PROTEIN UNFOLDING.

The absorption of aromatic amino acids depends on their chemical environment. In unfolded proteins, all aromatic amino acids are exposed to the aqueous solvent. Upon folding, at least some of these side chains become buried in the hydrophobic core of the protein. Therefore, the absorption of folded (native) and unfolded (denatured) protein differs. Absorption can thus be used as a spectroscopic probe to follow folding or unfolding of proteins and to determine the thermodynamic stability of proteins (Figure 19.19). An analysis of the equilibrium unfolding curves according to the two-state model of protein folding (Section 16.3.4; Figure 16.47) yields the thermodynamic parameters for folding. The general derivations of the relationships between the measured absorption $A_{obs}$ and the thermodynamic parameters for protein folding for thermal and denaturant-induced unfolding are described in Section 16.3.4.

*(Continued)*

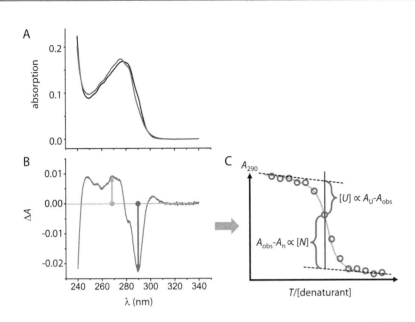

**FIGURE 19.19: Protein unfolding monitored by absorption.** A: The absorption of a native protein (black) differs from the absorption of the unfolded protein (red). Usually, a blueshift is observed upon unfolding. B: From the difference spectrum (denatured – native), the optimal wavelength for monitoring folding/unfolding can be determined (gray: maximum signal increase, red: maximum decrease). C: At 290 nm, the absorption decreases upon unfolding with increasing temperature (thermal denaturation) or denaturant concentration (chemical denaturation, Section 16.3 and Figure 16.43). For a two-state transition, the fraction of native and unfolded protein at each temperature or denaturant concentration can be determined from the difference of the measured absorption ($A_{obs}$) and the absorption of the native or unfolded protein ($A_N$, $A_U$; see Section 16.3.4 for the general derivation of a spectroscopic signal $y_{obs}$ as a function of thermodynamic parameters for protein folding).

The strong hypochromic effect of DNA upon double helix formation provides a sensitive spectroscopic probe for DNA duplex stability. DNA melting curves, measured as the increase in the absorption at 260 nm as a function of temperature, define the melting temperature $T_m$ at which 50% of the double-helical regions are disrupted (Box 19.5). A quantitative thermodynamic analysis requires taking into account the molecular mechanism of dissociation of the duplex into two single-strands.

---

**BOX 19.5: ABSORPTION AS A PROBE FOR DNA DENATURATION.**

Dissociation of a DNA duplex into two single-strands is accompanied by an increase in absorption (hyperchromic effect). The increase in absorption serves as a spectroscopic probe for duplex dissociation, and can be used to compare melting temperatures and duplex stabilities (Figure 19.20).

*(Continued)*

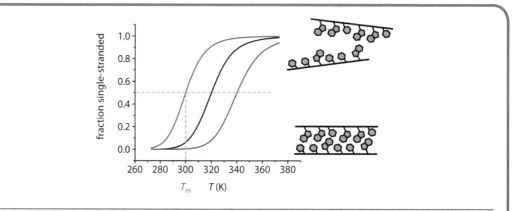

**FIGURE 19.20: DNA melting curve.** Melting curves for three DNA duplexes with increasing thermal stability (blue→black→red). With increasing temperature, the absorption of DNA increases because of dissociation of the duplex into the single strands and the associated hypochromic effect. The inflection point where 50% of the DNA has dissociated is defined as the melting temperature $T_m$.

### 19.2.8 POTENTIAL PITFALLS

Absorption measurements are in principle very straightforward to perform using standard absorption photometers and are easy to analyze. However, the dynamic range of photometers is typically limited to absorption values between 0.001 and 3. One reason for this limitation lies with the sensitivity of the detector: at very low absorption, a tiny difference between the intensities of the incident and the transmitted light has to be determined accurately. If the absorption is high, on the other hand, the little transmitted light will ultimately be below the detection limit. At $A = 3$ and $d = 1$ cm, only 0.1% of the incident light reaches the detector!

At high chromophore concentrations, a non-linearity between absorption and concentration will be observed. Chromophores at the beginning of the light path will experience the full intensity $I_0$ of the irradiating light. Because of the absorption of a significant fraction of the light by these chromophores, molecules further down the pathway will experience a lower light intensity. This phenomenon is called *inner filter effect* (Figure 19.21). Because of the inner filter effect, the measured absorption $A_{obs}$ is lower than the real absorption A according to

$$A_{obs} = A \cdot 10^{-A/2}$$

eq. 19.23

If the inner filter effect cannot be avoided, eq. 19.23 can be used to correct the measured absorption $A_{obs}$.

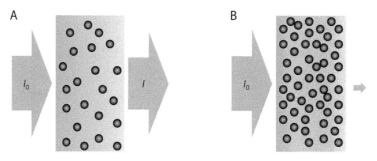

**FIGURE 19.21: Inner filter effect.** A: In a dilute solution, absorption is low, and the chromophores (blue) at the end of the light path through the sample experience nearly the same light intensity as the chromophores at the beginning of the light path. B: In highly concentrated solutions, absorption of light by chromophores at the beginning of the light path reduces the light intensity experienced by chromophores at the end of the light path. As a consequence, the molecules at the end of the light pass will absorb less. The linearity of absorption and concentration according to the Lambert-Beer law breaks down at high concentrations.

Problems can also result from large particles in the sample, such as aggregates, that scatter the incident light (see Section 21.1). Scattering increases with the inverse of $\lambda^4$, and particularly distorts the characteristic protein spectrum in the lower wavelength region (Figure 19.22). Apparent absorption above 300 nm is a strong indicator for the presence of scattering because aromatic residues in proteins do not show absorption in this spectral range.

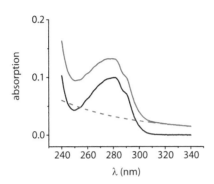

**FIGURE 19.22:**    **Scattering distorts absorption spectra.** The intensity of scattered light (blue, broken line) increases with $1/\lambda^4$. When scattering particles are present, the measured spectrum of a protein (blue) will be distorted from the real protein absorption spectrum (black). Strong scattering not only changes the intensity of absorption, but also the apparent maximum.

## 19.3  LINEAR AND CIRCULAR DICHROISM

### 19.3.1  LINEARLY POLARIZED LIGHT AND LINEAR DICHROISM

Absorption is based on an interaction of the dipole of the absorbing molecule with the electric field vector of the light. The strength of this interaction depends on the relative orientation of the electric field vector and the transition dipole of the molecule. In linearly polarized light, the oscillation of the electric field vector is confined to a plane, the *polarization plane* (Figure 19.23). Molecules oriented such that their transition dipoles are parallel to the polarization direction of the incident light will then preferentially absorb. The probability of a molecule to absorb polarized light depends on the magnitude of the vector component of the transition dipole that is parallel to the polarization direction, and thus on the angle $\theta$ between polarization direction and dipole moment (Figure 19.23). The absorption probability is proportional to $\cos^2\theta$ (Box 19.6). In solution, molecules can freely rotate, and have a random orientation of their transition dipoles. The number of molecules with an orientation between $\theta$ and $d\theta$ is proportional to $\sin\theta \, d\theta$. The overall *absorption probability* $f(\theta)$ then depends on the angle $\theta$ of the transition dipole and the polarization direction according to eq. 19.24:

$$f(\theta)d\theta = \cos^2\theta\sin\theta \, d\theta \qquad \text{eq. 19.24}$$

This angular dependence of the absorption probability leads to the preferential excitation of molecules whose transition dipoles have a large component parallel to the polarization of the incident light. In solution, only those molecules are selectively excited, a phenomenon called *photoselection* (Figure 19.23).

FIGURE 19.23: **Photoselection.** A: In linearly polarized light, the electric field vector is confined to the polarization plane (here: *xz*-plane). B: Left: The absorption probability depends on the angle θ between the transition dipole (blue arrow) and the polarization direction of the incident light (along the *z*-axis). Right: The angular dependence of the absorption probability is qualitatively represented by the intensity of the blue arrows that represent the transition dipoles. The orientations of dipole moments that lead to preferred absorption correspond to a dumbbell shape (gray). C: Chromophores in solution are randomly oriented (left). Their transition dipoles (arrows) are fixed with respect to the molecular geometry, and show random orientation. Upon excitation with light that is linearly polarized in the *z*-direction, those molecules whose transition dipole is oriented parallel to the *z*-axis absorb preferentially (red). As a consequence, the transition dipoles of the excited molecules are aligned. This process is called photoselection. Note that for simplicity excitation is shown only for molecules that are perfectly aligned with the polarization direction of the excitation light.

The dependence of the absorption on the polarization direction of the incident linearly polarized light is called *linear dichroism* (LD):

$$LD(\lambda) = A_v(\lambda) - A_h(\lambda)$$

eq. 19.25

$A_v(\lambda)$ denotes the absorption of light polarized in the *z*-direction (vertical), and $A_h(\lambda)$ is the absorption of light polarized horizontally (along the *y*-axis). Often, the linear dichroism is converted to the *reduced linear dichroism* RD by normalization with the total absorption at this wavelength

$$RD(\lambda) = \frac{LD(\lambda)}{A(\lambda)}$$

eq. 19.26

This expression is also called the *dichroic ratio*. Note that transition dipoles will be randomly oriented in solution (Figure 19.23). The total absorption over all molecules in solution will therefore not depend on the polarization direction of the incident light. Linear dichroism can only be observed for oriented molecules. The orientation of elongated molecules in solution can be achieved by applying flow. Polyelectrolytes or molecules with permanent electric dipoles, such as DNA double helices or helical peptides, can also be oriented in an electric field.

LD measurements are useful to examine whether an absorption band is caused by a single electronic transition between two states a and b, or if it is the result of two overlapping transitions, a→b and a→c. It is unlikely that the transition dipoles for these two transitions have identical directions. Therefore, the dichroic ratio is independent of the wavelength if only one transition is involved, but will depend on the wavelength for two overlapping transitions.

If the direction of the transition dipole is known, linear dichroism measurements can provide structural information. For example, the dipole of aromatic rings usually lies in the plane of the conjugated system. The orientation of aromatic rings within the structure can then be analyzed. An example is the orientation of nucleobases within a DNA duplex. In double helical DNA, the nucleobases are oriented nearly perpendicular to the helical axis, and show strong absorption of light that is polarized in this direction, but no absorption of light that is polarized parallel to the helical axis. In addition, each base pair is rotated relative to the next, and shows different absorption of light polarized in different directions within the plane perpendicular to the helical axis (Figure 19.24). Helical peptides show linear dichroism in the infrared range (see Section 19.4.4).

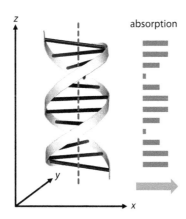

**FIGURE 19.24:**    **Linear dichroism of DNA.** A DNA double helix with its helical axis (red) along the z-direction will only absorb light polarized in the xy-plane, the plane of the bases and their transition dipoles. Because of the rotation of each base pair by 360°/10.5 with respect to the neighboring base pair, the component of the transition dipole along the x-axis changes periodically. The resulting periodic pattern of absorption intensities for linearly polarized light with polarization along the x-axis is depicted on the right (blue bars).

Often, bases are tilted from their orientation exactly perpendicular to the helical axis, giving rise to a small component of the transition dipole parallel to the DNA helical axis (Figure 17.16). The inclination angle can be determined from LD measurements of DNA oriented in an electric field. Conversely, if structural information is available, LD measurements provide information on the direction of the transition dipole. This is possible in crystals with identical orientation of the chromophores, or with chromophores immobilized on surfaces or embedded in a matrix (Box 19.6).

---

**BOX 19.6: DETERMINATION OF TRANSITION DIPOLE DIRECTIONS.**

The probability of absorption is related to the square of the transition dipole induced in a molecule. Let us assume a molecule is oriented with its transition dipole at an angle $\theta$ with respect to the z-axis, and we excite this molecule with linearly polarized light along the z-axis. The dipole induced in the molecule is proportional to $\cos\theta$ (Figure 19.25), and the resulting probability for a transition is proportional to $\cos^2\theta$. If we now rotate the polarization direction of the incident light in the xz-plane, the absorption of the molecule will be modulated by $\cos^2\theta$. From this modulation of absorption, the angle $\theta$ and thus the direction of the transition dipole can be determined. Using this approach, the relative orientations of chlorophyll molecules in a bacterial photosynthetic light harvesting complex have been determined. Hofmann *et al.* (2003) *Phys. Rev. Lett.* 90: 013004.

*(Continued)*

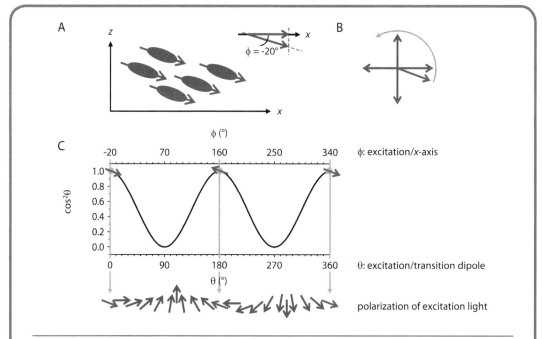

**FIGURE 19.25:** **Determination of transition dipole directions**. A: The transition dipole direction can be determined for aligned molecules (or for single molecules, see Section 23.1.7). B: The polarization direction of the excitation light is rotated, and the fluorescence emission from the molecules is measured for each polarization direction. C: The absorption and thus the fluorescence emission depends on the square of the cosine of the angle between the transition dipole and the polarization direction of the excitation light. From the modulation of the intensity with the direction of the polarization, the direction of the transition dipole can be determined.

### 19.3.2 CIRCULARLY POLARIZED LIGHT AND CIRCULAR DICHROISM

*Optical activity* is a property that is linked to molecular asymmetry. The most common examples are asymmetric carbons in biomolecules, such as the $C_\alpha$ in amino acids and proteins (Section 16.1), or in sugars and nucleic acids. These asymmetric carbons have effects on chromophores in their vicinity. Optically active molecules alter the properties of light, leading to the phenomena of circular birefringence and optical rotation, or circular dichroism and ellipticity. Before we take a closer look at optical activity, we need to understand the concept of *circularly polarized light*, and its relationship to linearly polarized light. Conceptually, we can regard linearly polarized light as the superposition of a *left-handed and a right-handed circularly polarized wave* of the same wavelength and phase (LHCPL, RHCPL; Figure 19.26). At any given point in time, the two vectors of the two circularly polarized waves will superimpose to a vector in the polarization plane of the resulting linearly polarized light.

In circularly polarized light, the tip of the electric field vector describes a helix in time and space (Figure 19.26), which is left-handed for left-handed circularly polarized light, and right-handed for right-handed circularly polarized light. This helical movement is the result of the superposition of the circular change in polarization direction and the propagation direction of the wave. Note that for linearly and circularly polarized light, the spatial repetition unit is the wavelength, and the repetition unit in time is the frequency. In linearly polarized light, the magnitude of the electric field vector is modulated, but its direction is constant. For circularly polarized light, the magnitude is constant, but the direction is modulated.

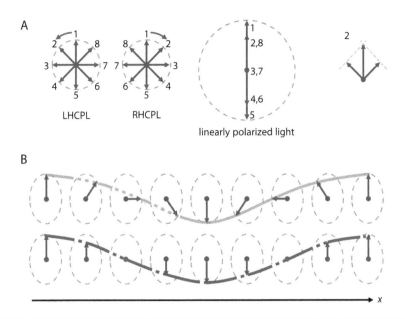

**FIGURE 19.26:** **Circularly polarized light.** A: In linearly polarized light, the oscillation of the electric field vector is confined to one direction, here in the vertical direction. At any given time (from 1 to 8), the electric field vector can be generated by vector addition of two circularly polarized components of equal size, phase and velocity, but opposite orientations. An example for time point 2 is shown on the right. One set of these vectors rotates in clockwise direction, the second set in counter-clockwise direction. The set of vectors whose tip rotates clockwise (right-handed) forms a right-handed circularly polarized light (RHCPL), the other set forms left-handed circularly polarized light. Thus, linearly polarized light can be regarded as a superposition of RHCPL and LHCPL. B: In circularly polarized light, the electric field vector is modulated in direction, not in amplitude. Superimposed with the propagation in the x-direction, the tip of the vector describes a helix (yellow). In linearly polarized light, the direction of the electric field vector is fixed, but the amplitude is modulated. Superimposed with the propagation along the x-axis, its tip describes a cosine function (blue).

Experimentally, circularly polarized light can be generated by superposition of two linearly polarized waves of the same wavelength $\lambda$, but with a 90° angle between their polarization planes and a $\lambda/4$ phase shift. Depending on the direction of the phase shift ($-\lambda/4$ or $+\lambda/4$), superposition generates left- or right-handed circularly polarized light.

If molecules interact differently with the two circularly polarized components of linearly polarized light, the light transmitted by the sample will not remain linearly polarized. There are two possible effects. If the refractive indices for RHCPL and LHCPL are different ($n_L \neq n_R$), the light leaving the sample is still linearly polarized, but the polarization plane is rotated with respect to the polarization of the incident light by an angle $\phi$. This phenomenon is called *circular birefringence* or *optical rotation dispersion* (ORD). The term dispersion takes account of the wavelength dependence of this effect. ORD is measured in degrees, i.e. the angle by which the polarization plane has been rotated by the sample. If RHCPL and LHCPL are absorbed differently ($A_L \neq A_R$ and $\varepsilon_L \neq \varepsilon_R$), the light leaving the sample is elliptically polarized: the tip of its vector traces an ellipse. This phenomenon is called *circular dichroism* (CD). CD is also measured in degrees (eq. 19.27). CD can only be observed at wavelengths where the sample absorbs ($\varepsilon_R \neq \varepsilon_L$). In contrast, ORD can be observed in regions outside absorption bands, and can be measured at high concentrations without inner filter effects (Section 19.2.8). CD and ORD can be interconverted using the Kronig-Kramers transformation. We will only treat CD in the following because of its higher relevance for structural analysis of biomolecules.

The difference in absorption of RHCPL and LHCPL, $\Delta\varepsilon$, that leads to circular dichroism is typically < 10 $M^{-1}$ $cm^{-1}$, whereas $\varepsilon$ of the respective molecule easily exceeds 20000 $M^{-1}$ $cm^{-1}$. Because of the different absorption, the E-field vectors of RHCPL and LHCPL have different amplitudes when they leave the sample. We can formally construct the elliptically polarized light leaving a sample that exhibits circular dichroism by adding these vectors at each time point. The ellipse that is described by the tip of the E-field vector is characterized by the larger and smaller half-axes, $a$ and $b$. Their ratio b/a is called the *elliptic ratio*. CD is measured as ellipticity $\theta$ (in degrees), which is defined as the angle whose tangent is the elliptic ratio:

$$\theta = \arctan\left(\frac{b}{a}\right)$$

eq. 19.27

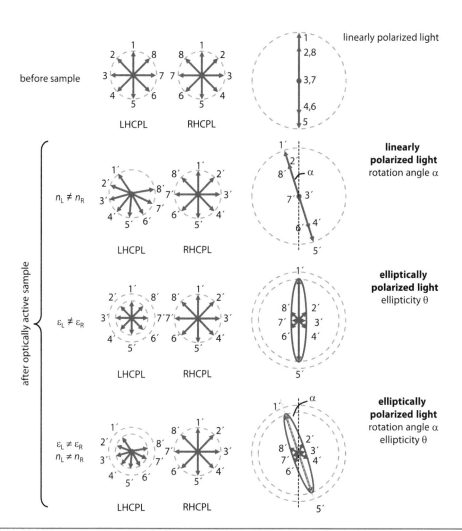

**FIGURE 19.27: Circular dichroism and circular birefringence.** An optically active sample is illuminated with linearly polarized light (top). If the refractive indices for the RHCPL and LHCPL components are different ($n_L \neq n_R$), superposition of the RHCPL and LHCPL components at time points 1 to 8 after the optically active sample gives linearly polarized light, but the polarization plane is rotated by the angle $\alpha$ (optical rotation dispersion, ORD). If the RHCPL and LHCPL components are absorbed to different extents ($\varepsilon_L \neq \varepsilon_R$; here, the RHCPL is not absorbed, but the LHCPL is), superposition of RHCPL and LHCPL after the sample gives elliptically polarized light. The ellipticity $\theta$ is defined by the ratio of the axes of the ellipse. If both $n_L \neq n_R$ and $\varepsilon_L \neq \varepsilon_R$, the light is elliptically polarized, and the plane of the ellipse is rotated by the angle $\alpha$.

To compare ellipticities between different samples, the measured values are converted into the *molar ellipticity* $\theta_{molar}$ that is directly proportional to the difference of the extinction coefficients $\Delta\varepsilon$ for the RHCPL and LHCPL components:

$$[\theta]_{molar} = \frac{100 \cdot \theta}{c \cdot d} \qquad \text{eq. 19.28}$$

$c$ is the molar concentration, and $d$ is the path length of the cuvette. The factor 100 is present for historical reasons and gives a unit for $[\theta]_{molar}$ of deg cm$^2$ dmol$^{-1}$. The ellipticities $\theta$ and $\theta_{molar}$ are proportional to the difference in extinction coefficient $\Delta\varepsilon$. For the molar ellipticity, the relationship is

$$[\theta]_{molar} \approx 3300 \cdot \Delta\varepsilon \qquad \text{eq. 19.29}$$

(derivation in Box 19.7).

---

**BOX 19.7: DERIVATION OF THE RELATIONSHIP BETWEEN THE ELLIPTICITY $\theta$ AND THE ABSORPTION DIFFERENCE $\Delta\varepsilon$.**

We can express the half axes of the ellipse by the electric field vectors of LHCPL and RHCPL, $E_L$ and $E_R$ (Figure 19.27). The long axis $a$ is given by $E_L + E_R$, the short axis $b$ by $E_L - E_R$:

$$\theta = \arctan\frac{E_L - E_R}{E_L + E_R} \qquad \text{eq. 19.30}$$

Measured values for $\theta$ are very small, usually in the range of millidegrees, and so $\arctan\theta$ can be approximated by $\theta$. Furthermore, the intensities $I_L$ and $I_R$ are proportional to the square of the electric field vectors, and we can substitute

$$\theta = \frac{\sqrt{I_L} - \sqrt{I_R}}{\sqrt{I_L} + \sqrt{I_R}} \qquad \text{eq. 19.31}$$

According to the definition of the absorption (eq. 19.16 and eq. 19.17), we can express the intensities $I_L$ and $I_R$ in terms of $A_L$ and $A_R$ and $I_0$:

$$\log\frac{I_0}{I_L} = A_L = -\log\frac{I_L}{I_0} \qquad \text{eq. 19.32}$$

Conversion of the decadic logarithm to the natural logarithm gives

$$\log\frac{I_0}{I_L} = A_L = -\log\frac{I_L}{I_0} = -\frac{\ln\dfrac{I_L}{I_0}}{\ln 10} \qquad \text{eq. 19.33}$$

and

$$I_L = I_0 \exp(-A_L \ln 10) \qquad \text{eq. 19.34}$$

*(Continued)*

Substituting these expressions for $I_L$ and $I_R$ into eq. 19.31 yields

$$\theta = \frac{I_0 \exp\left(-\dfrac{A_L \ln 10}{2}\right) - I_0 \exp\left(-\dfrac{A_R \ln 10}{2}\right)}{I_0 \exp\left(-\dfrac{A_L \ln 10}{2}\right) + I_0 \exp\left(-\dfrac{A_R \ln 10}{2}\right)} \qquad \text{eq. 19.35}$$

from which $I_0$ cancels. We can now expand the expression

$$\theta = \left(\frac{\exp\left(-\dfrac{A_L \ln 10}{2}\right) - \exp\left(-\dfrac{A_R \ln 10}{2}\right)}{\exp\left(-\dfrac{A_L \ln 10}{2}\right) + \exp\left(-\dfrac{A_R \ln 10}{2}\right)}\right) \cdot \frac{\exp\left(-\dfrac{A_L \ln 10}{2}\right)}{\exp\left(-\dfrac{A_L \ln 10}{2}\right)} \qquad \text{eq. 19.36}$$

and obtain

$$\theta = \frac{1 - \exp\left(-\dfrac{A_R \ln 10}{2} + \dfrac{A_L \ln 10}{2}\right)}{1 + \exp\left(-\dfrac{A_R \ln 10}{2} + \dfrac{A_L \ln 10}{2}\right)} = \frac{1 - \exp\left(\dfrac{\Delta A \ln 10}{2}\right)}{1 + \exp\left(\dfrac{\Delta A \ln 10}{2}\right)} \qquad \text{eq. 19.37}$$

with $\Delta A = A_L - A_R$. $\Delta A$ is typically small, and we can approximate the exponential function $e^x$ as $1+x$:

$$\theta \approx \frac{1 - \left(1 + \dfrac{\Delta A \ln 10}{2}\right)}{1 + \left(1 + \dfrac{\Delta A \ln 10}{2}\right)} \qquad \text{eq. 19.38}$$

Because of the small values for $\Delta A$, the term in the denominator can be approximated as 1, leading to

$$\theta \approx \frac{\dfrac{\Delta A \ln 10}{2}}{2} = \frac{\Delta A \ln 10}{4} \qquad \text{eq. 19.39}$$

The last step is the conversion of $\theta$ in rad to degrees, which gives the ellipticity $\theta$ as

$$\theta = \frac{\Delta A \ln 10}{4} \cdot \frac{360}{2\pi} \approx 33 \Delta A \qquad \text{eq. 19.40}$$

and the molar ellipticity $[\theta]_{\text{molar}}$ as

$$[\theta]_{\text{molar}} = \frac{100 \cdot \Delta A}{c \cdot d} \cdot \frac{\ln 10}{4} \cdot \frac{360}{2\pi} \approx 3300 \Delta \varepsilon \qquad \text{eq. 19.41}$$

Thus, ellipticity and molar ellipticity $[\theta]_{\text{molar}}$ are proportional to the difference in extinction coefficient $\Delta \varepsilon$, and therefore proportional to the concentration of the chromophore.

Circular dichroism is the result of an interaction of a chromophore with the electric and magnetic components of light. The absorption intensity of a chromophore depends on the dipole strength, related to the square of the electric transition dipole (Section 19.2.2). The intensity of a CD band is linked to the *rotational strength*, related to the product of the electric and the magnetic transition dipoles. Similar to the dipole strength, the rotational strength for a transition from initial state i to final state f can be determined from a measured CD spectrum by integration over the band.

### 19.3.3 INSTRUMENTATION

A standard CD spectrometer consists of a light source, a monochromator, and a polarizer to generate monochromatic, linearly polarized light (Figure 19.28). Because $\Delta\varepsilon$ is so small it is not possible to measure the absorption of LHCPL and RHCPL separately and then calculate the small difference by subtraction. Instead, a piezo-elastic element or an electro-optic crystal is used to periodically modulate the excitation light, such that LHCPL and RHCPL components are produced in an alternating manner. The transmitted light gives rise to a photocurrent by the detector. The average transmitted light causes a direct current (DC) component, and the differential absorption of LHCPL and RHCPL by an optically active compound is detected as a modulation of the transmitted light, leading to a superimposed small alternating current (AC) signal. This small signal is accumulated over many modulation periods, so very small absorption differences can be determined. A phase-sensitive detector separates the AC and DC components. The ratio (AC/DC) then yields the absorption difference $\Delta A = A_L - A_R$.

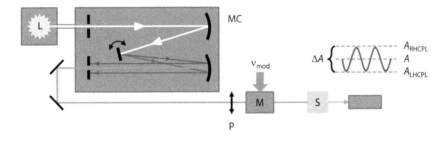

**FIGURE 19.28:    Principle of a CD spectrometer.** The lamp (L) emits white light. The monochromator (MC) selects the desired excitation wavelength (green). After passing the polarizer (P), the light is linearly polarized. A modulator (M) generates an alternating excess of left- and right-handed circularly polarized light, alternating with the frequency $v_{mod}$. Differential absorption of LHCPL and RHCPL then leads to a periodically modulated photocurrent on the detector. The DC component corresponds to the mean absorption $A$, the AC component reflects the difference in absorption $\Delta A$ of RHCPL and LHCPL.

### 19.3.4 BIOLOGICAL CHROMOPHORES THAT SHOW CIRCULAR DICHROISM

Circular dichroism is observed with optically active, i.e. chiral molecules (Section 15.1). Asymmetric carbon atoms are abundant in biologically active molecules such as amino acids and nucleotides. Secondary structures formed by their respective polymers, proteins and nucleic acids, are also chiral. The most important chromophores in proteins are the peptide bonds and aromatic amino acid side chains. The peptide bond exhibits circular dichroism in the spectral region of its absorbance, which is the *far-UV* (190–250 nm; Figure 19.29). Although the peptide bond is not chiral on its own, it shows CD because it is embedded into an asymmetric environment caused by the secondary structure in a folded protein. The magnitude of this dichroism depends on the type of secondary structure element: each secondary structure has a characteristic CD spectrum. CD spectra from proteins of unknown structure can be described as a linear combination of such reference spectra to determine the fractions of secondary structure elements.

Aromatic amino acids in an asymmetric environment, i.e. in the interior of proteins where they are not free to rotate, show a CD signal in the *near-UV* spectral region (230–320 nm; Figure 19.29). In contrast to CD in the far-UV, near-UV CD is thus a useful probe for tertiary structure that immobilizes amino acid side chains. Nucleotides show a rather weak CD signal around 260 nm, which is typically positive for pyrimidine nucleotide and negative for purine nucleotides. The circular dichroism of DNA is caused by the periodic arrangement of the nucleobases in the double helix. Different helical forms, e.g. the A-forms and B-forms of DNA (see Section 17.5.1), give rise to different CD spectra. RNA exhibits CD characteristics similar to A-form DNA (Figure 19.29).

Similar to the near-UV CD of aromatic amino acid side chains, small non-optically active molecules can show induced CD upon binding to a macromolecule in an asymmetric environment. An example is provided by the induced CD of pyridoxal phosphate when it is bound to aspartate aminotransferase.

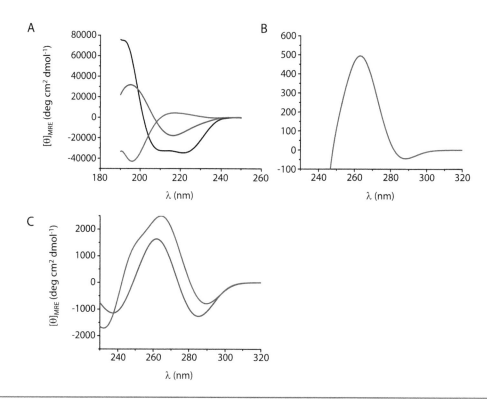

**FIGURE 19.29:** **Circular dichroism of proteins and nucleic acids.** A: Far-UV CD spectra of proteins. The CD spectrum of the peptide bond depends on the secondary structure. Black: α-helix, red: β-sheet, blue: random coil. Chou & Fasman (1974) *Biochemistry* 13(2): 211–222. B: Near-UV CD spectrum of a tryptophan-rich protein. Fixation of the tryptophan side chain in the asymmetric environment within the tertiary structure of the protein gives rise to circular dichroism. C: CD of DNA. The CD spectrum of DNA depends on the helix geometry (red: B-DNA, blue: A-DNA, see Section 17.5.1). Spectra in B and C are calculated using dichrocalc. Bulheller & Hirst (2009) *Bioinformatics* 25(4): 539–540.

## 19.3.5 APPLICATIONS

CD spectra of the peptide bonds of folded proteins depend strongly on the secondary structure (Figure 19.29). Therefore, CD measurements are commonly used to determine the fraction of secondary structure elements in a protein of unknown structure. To this end, a CD spectrum of the protein of interest is measured, and described as the weighted sum of reference spectra. In the simplest case, the reference spectra are from model peptides that adopt a defined secondary structure (α-helix or β-sheet) or form a random coil (rc). The measured spectrum $x(\lambda)$ can then be described as

$$x(\lambda) = f_\alpha \cdot \alpha(\lambda) + f_\beta \cdot \beta(\lambda) + f_{rc}(\lambda)$$

eq. 19.42

where $f_\alpha$, $f_\beta$, and $f_{rc}$ are the fractions of the protein that adopt α-helical or β-sheet structure or form a random coil, and a(λ), b(λ), and rc(λ) denote the corresponding reference spectra.

This rather simple procedure assumes that the entire structure of a protein can be represented by three types of secondary structure (α, β, or lack of regular structure, i.e. random coil), and that spectra of peptides appropriately reflect the CD properties of authentic proteins. However, the CD signal (per amino acid) of an α-helix depends on its length, and the CD spectrum of β-sheets is highly dependent on the environment. It differs for parallel and antiparallel β-sheets, and is affected by the commonly observed twist in β-sheets. Also, other secondary structure elements such as β-turns contribute to the CD spectrum of proteins. Finally, non-regularly structured elements in proteins are often still ordered and do not represent random coils. Altogether, the approach underlying eq. 19.42 is thus too simplistic. Modern programs therefore use a database of reference spectra that has been obtained from a principal component analysis of measured CD spectra of proteins with known structures.

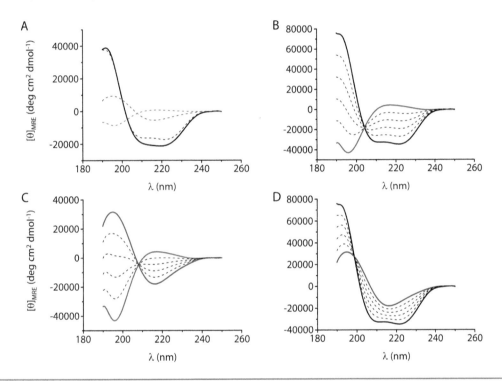

**FIGURE 19.30:** **CD to analyze the secondary structure content of proteins.** A: The CD spectrum of a protein of unknown secondary structure (black) can be described as a linear combination of the reference spectra (dotted lines) for α-helix (black), β-sheet (red), and random coil (blue). The reference spectra are multiplied by the fractions $f_\alpha$ = 0.5, $f_\beta$ = 0.3, and $f_{rc}$ = 0.2. B: CD spectra for a helix→coil transition. The extreme spectra correspond to 100% α-helix (black) and 100% random coil (blue). The dotted spectra in-between reflect 80% α-helix/20% random coil, 60% α-helix/40% random coil, 40% α-helix/60% random coil, and 20% α-helix/80% random coil. C: CD spectra for a β-sheet→coil transition. The extreme spectra correspond to 100% β-sheet (red) and 100% random coil (blue). The dotted spectra in-between reflect 80% β-sheet/20% random coil, 60% β-sheet/40% random coil, 40% β-sheet/60% random coil, and 20% β-sheet/80% random coil. D: CD spectra for an α-helix→β-sheet transition. The extreme spectra correspond to 100% α-helix (black) and 100% β-sheet (red). The dotted spectra in-between reflect 80% α-helix/20% β-sheet, 60% α-helix/40% β-sheet, 40% α-helix/60% β-sheet, and 20% α-helix/80% β-sheet.

Circular dichroism is also a useful probe to follow protein folding or unfolding as a gain or loss in CD signal (Figure 19.30). Typically, the far-UV CD signal at 222 nm is used to follow the disappearance of α-helical structures upon protein unfolding. α-helix→β-sheet transitions can also be readily followed by CD (Figure 19.30). The near-UV CD signal (230–310 nm) originates from aromatic amino acid side chains in an asymmetric environment and is sensitive to changes in tertiary structure. CD has also been used to monitor structural changes of DNA and RNA (Figure 19.30), including thermal stability, metal ion binding, intercalation, and interactions with proteins.

### 19.3.6 POTENTIAL PITFALLS

CD spectroscopy is a rather insensitive method that measures a small difference in absorption between LHCPL and RHCPL on top of a large mean absorption. To increase the signal-to-noise ratio, spectra are often accumulated $n$-fold, which leads to an increase in signal by $n$, while the noise only increases by $\sqrt{n}$. Measurements of CD are based on accurate measurements of transmission, and the overall absorption should therefore be limited. We have seen before that the inner filter effect reduces the measured absorption $A_{obs}$ to $A \cdot 10^{-A/2}$ (eq. 19.23). The measured absorbance $A_{obs}$, and hence the signal-to-noise ratio, is maximal for $A = 0.8686$. Many common buffers including Tris show strong absorption near 200 nm, and buffers have to be selected carefully for far-UV measurements.

## 19.4 INFRARED SPECTROSCOPY

We have seen before (Section 19.2.1) that each electronic state comprises a set of vibronic states. Similar to electronic excitation, vibrations can be excited by radiation of the energy separating the two vibrational states involved. The energetic difference between two vibrational states is typically one order of magnitude smaller than the energy separating two electronic states. Therefore, vibronic excitation can be achieved with *infrared (IR) light*, at wavelengths of 1–10 μm. It is important to note that the energetic difference between the vibronic ground state and the first excited state is larger than the thermal energy, meaning that IR spectroscopy at room temperature starts from $V_0$. While Raman spectroscopy probes vibrations indirectly as scattering (Section 21.1.3), IR spectroscopy directly measures transitions between vibrational energy levels of molecules due to absorption.

### 19.4.1 BOND VIBRATIONS: THE HARMONIC OSCILLATOR

A chemical bond can be described as a *harmonic oscillator* (Figure 19.31; Section 15.2). We imagine a sphere attached to the ceiling by a spring. If we drop the sphere, it will fall until it has extended the spring, and will then be pulled back up again as the spring re-contracts. The force $F$ on the sphere depends on the *stiffness k* of the spring and the current position $x$ of the sphere according to *Hooke's law*:

$$F = -kx \qquad \text{eq. 19.43}$$

This force causes an acceleration $a$ of the particle of mass $m$ that is

$$F = ma \qquad \text{eq. 19.44}$$

The acceleration is the derivative of the velocity $v$, which in turn is the derivative of the position $x$:

$$a = \frac{dv}{dt} = \frac{d^2x}{dt^2} \qquad \text{eq. 19.45}$$

By substituting eq. 19.45 into eq. 19.43 and then equating eq. 19.43 with eq. 19.44, we obtain

$$-kx = m\frac{d^2x}{dt^2} \qquad \text{eq. 19.46}$$

Eq. 19.46 contains $x$ and its second derivative, $d^2x/dt^2$. A possible solution for $x$ is a sine function

$$x = b \sin \omega t \qquad \text{eq. 19.47}$$

with

$$\frac{d^2x}{dt^2} = -b\,\omega^2 \sin\omega t = -\omega^2 x$$

eq. 19.48

We can substitute this expression into eq. 19.46, and obtain

$$-kx = m\left(-\omega^2 x\right)$$

eq. 19.49

and the frequency of the oscillation as

$$\omega = \sqrt{\frac{k}{m}}$$

eq. 19.50

or

$$\nu = \frac{1}{2\pi}\sqrt{\frac{k}{m}}$$

eq. 19.51

The potential energy of the sphere is then

$$E = \int F dx = k\int x dx = \frac{k}{2}x^2$$

eq. 19.52

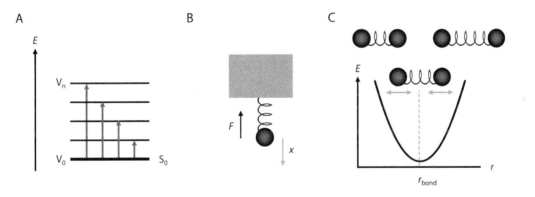

FIGURE 19.31: **Vibration spectroscopy.** A: Absorption of infrared light leads to transitions between vibrational states of a molecule in the electronic ground state $S_0$ (blue arrows). B: Oscillations of chemical bonds around their mean bond length $r_{bond}$ can be described as harmonic oscillators. The potential energy of the harmonic oscillator increases parabolically with increasing deviations from the bond length (eq. 19.52, eq. 19.53).

We can now transfer the concept of a harmonic oscillator to the oscillation of a chemical bond (Section 15.2). Bond vibration can be described quantitatively by restricting the considerations to the area of the Lennard-Jones potential close to the distance where the potential energy is minimal. In this region, the potential can be approximated by a parabola (Figure 19.31), and we can express the potential energy $E_{pot}$ as

$$E_{pot} = \frac{1}{2}k\left(r - r_{bond}\right)^2$$

eq. 19.53

with the force constant $k$, the actual distance $r$, and the bond length $r_{bond}$. Note that the term $(r - r_{bond})$ corresponds to the displacement $x$ in eq. 19.43. The force constant $k$ is a measure of the steepness of the walls of the potential, and the stiffness of the bond. The force during bond vibration follows Hooke´s law:

$$F = k\left(r - r_{\text{bond}}\right)$$

eq. 19.54

The *resonance frequency* $\omega$ of the harmonic oscillation of the bond depends on the reduced mass $\mu$ of the two atoms as

$$\omega = \sqrt{\frac{k}{\mu}}$$

eq. 19.55

or

$$\nu_0 = \frac{1}{2\pi}\sqrt{\frac{k}{\mu}}$$

eq. 19.56

The reduced mass $\mu$ of the oscillator is defined as

$$\mu = \frac{m_1\, m_2}{m_1 + m_2}$$

eq. 19.57

with the masses $m_1$ and $m_2$ of the two atoms. We can infer that bonds between atoms with larger masses absorb at lower frequencies. As an example, the IR absorption of a C-D bond appears at a lower frequency than the vibration of a C-H bond (isotope effect; see Chapter 9, Box 9.2). Stronger bonds with higher stiffness $k$ show IR absorption at higher frequencies. Triple carbon-carbon bonds therefore absorb at higher frequencies than double or single bonds.

The energy of a harmonic oscillator is quantized, and the possible vibrational energy levels $E_n$ depend on the vibrational quantum number $n$ (1, 2, ...):

$$E_n = h\nu_0\left(n + \frac{1}{2}\right)$$

eq. 19.58

According to eq. 19.58, neighboring vibration levels are separated by

$$\Delta E = h\nu_0$$

eq. 19.59

and infrared absorption occurs when the frequency of the incident radiation equals the resonance frequency $\nu_0$ of the bond. IR absorption is usually not given in frequency or wavelength, but in *wavenumbers*. The wavenumber $\tilde{\nu}$ is the inverse of the wavelength $\lambda$

$$\tilde{\nu} = \frac{1}{\lambda} = \frac{\nu}{c}$$

eq. 19.60

and its dimension is typically $cm^{-1}$. Thus, it shows how many periodic waves occur over one centimeter. For example, the three water vibrations are at 3685 $cm^{-1}$, 3605 $cm^{-1}$, and 1885 $cm^{-1}$.

The advantage of wavenumbers lies with their direct proportionality to the energy of the radiation instead of the inverse proportionality between energy and wavelength (eq. 19.61)

$$E = h\frac{c}{\lambda} = hc\tilde{\nu}$$

eq. 19.61

with the Planck constant $h$, and the speed of light $c$.

## 19.4.2  MOLECULE GEOMETRY, DEGREES OF FREEDOM, AND VIBRATIONAL MODES

Vibration modes are only *IR-active*, i.e. give rise to an IR absorption band, when they are associated with a transition dipole moment ($\Delta\mu \neq 0$). Vibrations that are not associated with a transition dipole ($\Delta\mu \neq 0$) are *IR-inactive*. A vibration leads to a transition dipole if the permanent dipole of the molecule changes during the vibration. The number of possible vibration modes depends on the size and shape of the molecule. An atom has three *degrees of freedom*, corresponding to its translational movement along the three axes of a Cartesian coordinate system. A molecule that consists of $N$ atoms has $3N$ degrees of freedom. Three of these degrees of freedom correspond to the translational movement of the entire molecule (i.e. of all atoms simultaneously) in the three dimensions. Molecules can also rotate: for a linear molecule, two degrees of freedom are assigned to rotation (around its long and its short axis), whereas a non-linear molecule can rotate around three different axes, corresponding to three degrees of freedom. Altogether, a linear molecule of $N$ atoms has $3N$-5 degrees of freedom that are correlated with different types of vibrations (modes), and a non-linear molecule of $N$ atoms has $3N$-6 different modes of vibration. Although the degrees of freedom increase tremendously with the size of the molecule, infrared spectroscopy is simplified by the fact that functional groups show distinct vibration patterns and can be assigned even in spectra of rather complex molecules. We will therefore illustrate the different vibration modes with simple examples. First, we consider a diatomic molecule, such as $H_2$. The $H_2$ molecule can vibrate such that the two hydrogen atoms periodically come closer and move further apart. Such a vibration that leads to oscillations of the bond length is called stretching, and this mode of vibration is a *stretching mode*. It is the only vibration mode possible for this molecule ($3N-5 = 1$). $H_2$ does not show IR absorption, however, because its only vibrational mode is not associated with changes in the permanent dipole.

Now we consider two different molecule of three atoms, the linear $CO_2$ molecule ($3N-5 = 4$ vibrational modes), and the non-linear $H_2O$ molecule ($3N-6 = 3$ vibrational modes; Figure 19.32).

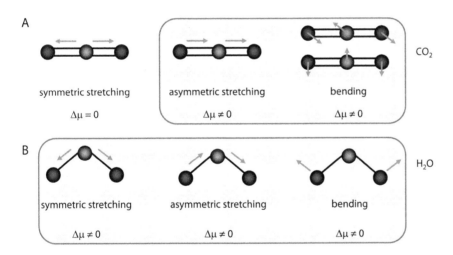

**FIGURE 19.32:    Vibrational modes of $CO_2$ and $H_2O$.** A: $CO_2$ as a linear molecule has four vibrational modes. The four possible modes are the symmetric and asymmetric stretching vibrations, and two bending vibrations, one with the atoms moving into and out of the plane of the diagram, and a second with the atoms moving up and down. The symmetric stretching does not alter the dipole moment ($\Delta\mu = 0$) of the molecule and is thus IR-inactive. B: $H_2O$ is a non-linear molecule with three vibrational modes. It undergoes symmetric and asymmetric as well as a bending vibration. All three modes lead to a change in the dipole moment ($\Delta\mu \neq 0$), and are thus IR-active.

In $CO_2$, the two bonds between the carbon and oxygen atoms can vibrate such that the two oxygen atoms move away and towards the central carbon in synchrony (*symmetric stretching* mode; Figure 19.32) or anti-correlated, with one oxygen moving towards the carbon while the second moves away from the carbon (*asymmetric stretching* mode). Symmetric stretching does not change the dipole moment of the molecule, and thus does not lead to IR absorption: it is IR-inactive. In contrast, the asymmetric stretching is accompanied by a change in the dipole moment of the $CO_2$ molecule, and this vibration gives rise to an IR band: it is IR-active. In addition to the stretching modes, $CO_2$ can undergo *bending* vibrations, where the two oxygen atoms are displaced into the same direction, such that the molecule is deformed from its linear shape, and the bending angle oscillates. A deviation from linearity alters the dipole of the molecule, and the bending vibration is therefore IR-active. $CO_2$ shows two possible modes of bending vibrations, with a relative displacement of carbon and oxygen in two different planes. These modes are isoenergetic and lead to one band in the IR spectrum with twice the intensity compared to the other asymmetric stretching mode.

In the non-linear water molecule, the three possible vibration modes are the symmetric stretching vibration, the asymmetric stretching vibration, and the bending vibration (Figure 19.32). All three vibration modes are associated with a change in the dipole moment of the $H_2O$ molecule. Therefore, all three vibrations are IR-active, and give rise to absorption at 3685 cm$^{-1}$ (symmetric stretching), 3605 cm$^{-1}$ (asymmetric stretching), and 1885 cm$^{-1}$ (bending).

Because of the requirement for changes in the molecular dipole for infrared absorption, highly polarized bonds such as CO-, CN-, PO-, and NH-bonds give rise to intense IR bands. The resonances from vibrations of these bonds are found in the spectral region above 1500 cm$^{-1}$, and can be assigned to the individual bonds. Below wavenumbers of 1500 cm$^{-1}$, a large number of resonances corresponding to vibrations of the entire molecule are observed. This region is therefore called the *fingerprint region*. The fingerprint region is complex and very sensitive to the chemical identity of the molecule. Comparison of the fingerprint region with reference spectra can therefore be a very powerful method to identify molecules.

### 19.4.3 INSTRUMENTATION

An IR spectrometer consists of a lamp that emits radiation in the infrared spectral range, monochromators for selection of excitation and emission wavelength, and a detector. Quartz-tungsten-halogen lamps cover the spectrum up to wavelengths of 5 µm (2000 cm$^{-1}$), whereas heated resistive ceramic wires emit in the range from 1 µm up to several hundred micrometers (from 9500 cm$^{-1}$ to tens of cm$^{-1}$). The wavelength is scanned by moving the diffraction grating of a monochromator.

IR spectra are typically depicted as % transmission as a function of the wavenumber of the incident light. IR absorption can be measured on solid, liquid, or gaseous material. Solids are usually mixed with KBr and compressed into a tablet. Liquids are spread between two NaCl plates, or are dissolved and measured in solution in NaCl cells. NaCl is transparent for infrared light. Special cells sealed with NaCl plates are used for measurements in the gas phase. Spectra are typically measured by ratio-recording (Figure 19.33), where the transmission of the sample is measured relative to the transmission of a reference cell. The *n*-fold accumulation of spectra increases the signal-to-noise ratio.

Modern IR spectrometers excite all frequencies simultaneously and measure an *interferogram* from which the spectrum is obtained by *Fourier transformation* (FT; Section 29.14). The heart of an FT-IR instrument that enables simultaneous measurement of IR absorption at all frequencies is the *Michelson interferometer* that takes the place of the diffraction grating (Figure 19.33). In an FT-IR spectrometer, the excitation light first passes an interference plate. This plate functions as a beam splitter, and reflects half of the light to a fixed mirror, while the other half is transmitted towards a moveable mirror. The reflected light from both mirrors is superimposed on the interference plate, and then passes sample and reference. The light intensity as a function of the position of the moveable mirror, the *interferogram*, is detected by the detectors. To extract the spectrum, i.e. the intensity of the transmitted light as a function of the wavenumber, the interferogram has to be Fourier transformed from intensity as a function of distance into intensity as a function of inverse distance (wavenumbers). In NMR (Section 20.1.5), FT is used to convert a signal as a function of time into a signal as a function of frequency (inverse of time), in X-ray crystallography (Chapter 22), it is employed to convert intensities in reciprocal space (inverse distance) to electron densities in real space (distance). With FT-IR, IR spectra are measured faster (within seconds), with higher signal-to-noise ratio, and with high precision in wavenumbers. Even faster processes can be followed by *step-scan FT-IR*. In this type of experiment, the reaction is triggered in the IR sample cell, for example by laser flash photolysis of a caged compound (see Section 25.3), and the time dependence of the change in IR absorption is measured at a fixed position of the moveable mirror in the Michaelson interferometer. The mirror is moved by a small step, and a new reaction is started to monitor the time dependence of the IR signal at this position of the mirror. This cycle is repeated until the entire interferogram has been sampled step by step. Individual time-dependent measurements are then combined to obtain a time-dependent IR spectrum with a time-resolution of as low as µs–ns, limited by the time needed to start the reaction and by the response time of the detector.

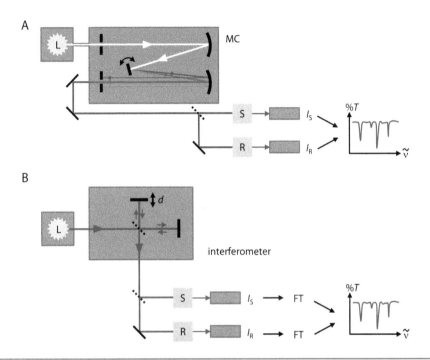

**FIGURE 19.33:**    **IR spectrometers.** A: Ratio-recording IR spectrometer. The excitation wavelength is selected by passing the light from the light source (L) through a monochromator (MC) before irradiating sample (S) and reference (R) cell. The intensity of the transmitted light is measured as a function of the wavelength of the incident light. B: FT-IR spectrometer. The light from the light source (L) enters an interferometer. An interference plate reflects half of the light towards a moveable mirror (double-headed arrow, $d$), the other half passes towards a fixed mirror. The light reflected from both mirrors is recombined at the interference plate. Depending on the distance $d$ the mirror has moved, the two wave trains have traveled different distances, and are phase-shifted relative to each other. Their superposition creates an interferogram. The light exiting the interferometer is then passed through sample (S) and reference cell (R). The absorbance of the sample alters the shape of the interferogram of the transmitted light. From the transmission as a function of the movement $d$ of the mirror, the spectrum (transmission as a function of wavenumbers) can be calculated by Fourier transformation (FT).

## 19.4.4 APPLICATIONS

To monitor changes in vibrations for individual bonds by IR spectroscopy, a particular resonance has to be assigned to a specific group. IR spectra of proteins are very crowded, but assignment of bands can be achieved by site-directed mutagenesis: when the amino acid whose side chain gives rise to a particular resonance is replaced by a different type of amino acid, the corresponding resonance will disappear, enabling assignment. The effect of isotope substitution on the vibration frequency can also be exploited for assignment. In studies of the mechanism of GTP hydrolysis by the oncoprotein Ras, GTP with $^{18}O$ substitutions at the α-, β-, or γ-phosphate has been used to assign phosphate bands to individual phosphates. Isotope-labeling of proteins is achieved by supplying minimal media for recombinant protein production with isotopically labeled amino acids. Using this approach, isotope-labeled tyrosines have been incorporated into photoactive yellow protein, a bacterial blue light sensor, to record isotope-edited IR spectra.

Ligation by metal ions strongly polarizes bonds, and leads to an increase in their IR absorption. The interaction of CO, NO, or $O_2$ with the central iron in heme, for example, strongly affects their vibration frequencies, enabling studies on ligand binding to heme by IR spectroscopy.

The IR spectrum of proteins can also be used to quantify secondary structure elements: the CO stretching vibration of the peptide bond depends on the hydrogen bonding pattern and thus on the type of secondary structure element it is part of. The CO resonance of a protein can be described by a linear combination of the absorption bands for the CO stretching vibration in $\alpha$-helices or $\beta$-sheets to quantify their fractions, similar to the secondary structure determination from far-UV CD spectra (Section 19.3.5). While CD is very sensitive for $\alpha$-helices, IR spectroscopy provides stronger signals for $\beta$-sheets, and the methods are complementary.

The stretch vibration of backbone carbonyl groups (1652 cm$^{-1}$) in the peptide bond exhibits a transition dipole in the direction of the CO double bond. In helical peptides, this dipole is parallel to the helical axis, and helical peptides therefore show linear dichroism (Section 19.3.1) in the infrared with a higher absorbance for infrared light polarized parallel to the helical axis. In contrast, the NH bending vibration is mixed with the CN stretch vibration (1549 cm$^{-1}$), leading to a transition dipole almost perpendicular to the helical axis. Here, a stronger absorption of light polarized perpendicular to the helical axis is observed.

IR spectroscopy can also be used to characterize cold-trapped reaction intermediates (cryo-FT-IR), for example in protein folding. Time-dependent changes in IR spectra are often difficult to interpret because of the large number of signals. Difference spectroscopy simplifies spectra significantly and allows one to focus on those signals that change in intensity upon conversion of one species into another. Maxima in a difference spectrum (product minus reactant) corresponds to a signal that is built up during the reaction, whereas minima (negative peaks) reflect signals that are present in the reactant but not in the product of the reaction. Step-scan FT-IR measures time-dependent changes in IR signals over the entire spectrum and has been used to identify structural changes in bacteriorhodopsin during the photocycle, and to dissect the phosphoryl transfer mechanism during GTP hydrolysis. by the small GTP-binding protein Ras in presence of its GTPase activating protein.

## 19.5 FLUORESCENCE

### 19.5.1 GENERAL CONSIDERATIONS

Fluorescence is emitted when a molecule returns from the excited state to the ground state. While absorption starts from the ground state of a molecule, and therefore reports on molecular properties in the ground state, fluorescence is sensitive for all processes that occur while the molecule is in the excited state. Absorption and emission spectroscopy therefore yield complementary information.

Fluorescence is generally independent of the excitation wavelength. The reason for this lies in the time scale of the processes involved. We have seen before that absorption is a very fast process that occurs on the order of 10$^{-15}$ s (Section 19.2.2). Depending on the energy of the excitation light, the molecule is excited from the singlet ground state $S_0$ to the $S_1$ state (or the higher electronic states $S_2$, etc.; Figure 19.34), and to different vibronic sub-states of $S_1$. Higher electronically excited states will relax to $S_1$ within 10$^{-12}$ s by *internal conversion* (IC). On the same time scale, higher vibronic states of $S_1$ will relax to the $V_0$ level by *vibronic relaxation* (VR). The $S_1, V_0$ state is very long-lived (1–10 ns), and the fluorescence that occurs from this state is therefore independent of the wavelength of the excitation light. Alternatively, molecules can relax to the ground state in non-radiative transitions. Molecules can also enter the triplet state $T_1$ from $S_1$ by *inter-system crossing* (ISC). In this transition, the electron spin is inverted, and the transition is spin-forbidden. From $T_1$, the molecule returns to $S_0$ in a second spin-forbidden transition by emitting *phosphorescence*. Transitions between energy levels of fluorophores are often depicted schematically in a *Jablonski diagram* (Figure 19.34).

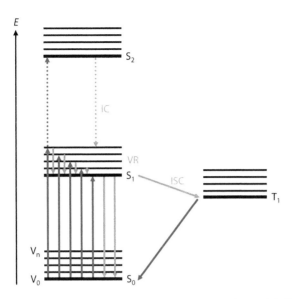

**FIGURE 19.34:    Jablonski diagram for fluorescence.** Excitation from $S_0,V_0$ (blue arrows) leads to the population of different vibronic states $V_0$-$V_n$ of the electronically excited $S_1$ state. Higher vibronic states of $S_1$ relax to the vibrational ground state $S_1,V_0$ by vibronic relaxation (VR, gray). Irradiation with light of higher energy (blue dotted line) can lead to the excitation of higher energy electronic states ($S_2$), followed by non-radiative relaxation to $S_1$ (internal conversion, IC; gray dotted arrow). Molecules return to the ground state by emitting fluorescence (green) or by non-radiative transitions (gray), or enter the triplet state $T_1$ from $S_1$ by inter-system crossing (ISC) and return to $S_0$ by emitting phosphorescence (red).

Because excitation leads to higher-energy vibronic and electronic states, but the return to the ground state occurs from $S_1,V_0$, fluorescence emission is always of longer wavelengths (red-shifted) than absorption/excitation. This difference between absorption and emission maxima is the *Stokes shift*. The solvent can also contribute to the Stokes shift (Figure 19.35). Polar solvent molecules such as water form a solvation shell around the solute to stabilize its charges. Absorption is a fast process during which bond lengths and molecule geometries do not change (Section 19.2.2; Figure 19.6). Solvating water molecules thus maintain their arrangement around the solute upon excitation, but the excited state is more polar than the ground state, and the water is not optimally arranged for stabilization of the excited state charges (Section 19.2.4). Immediately after excitation, the energy of the excited state is therefore higher in aqueous solutions compared to non-polar solvents. During the nanosecond lifetime of the excited state, the water molecules have ample time to re-orient and to optimally stabilize the excited state of the fluorophore. This process is called *solvent relaxation*. Fluorescence then occurs from the excited state that is optimally stabilized by the solvent. Again, the transition is very rapid, and the water molecules maintain their orientation during the electronic transition. As a consequence, the fluorophore returns to an electronic ground state of higher energy than in non-polar solvents. Solvent relaxation then leads to the original, solvated ground state. Polar solvents thus increase the energy difference between excitation and emission wavelengths and contribute to the Stokes shift.

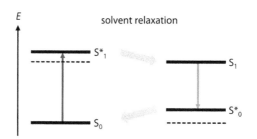

**FIGURE 19.35:** **Solvent relaxation and Stokes shift.** Polar solvents stabilize the charge distribution of solutes. Upon excitation, the charge distribution changes, but the arrangement of solvent molecules around the fluorophore is unaltered. The resulting excited state $S^*_1$ has a higher energy than in a non-polar solvent. Solvent relaxation (gray arrow) is the rearrangement of the solvent molecules to the energetically most favorable geometry for the electron distribution of the excited state ($S_1$). Relaxation of the fluorophore from $S_1$ to the ground state reverts the electron distribution. Again, the solvent molecules remain fixed, and the molecule reaches the $S^*_0$ state that is of higher energy than the original $S_0$ state. Solvent relaxation (gray) restores the original solvated ground state.

### 19.5.2 INSTRUMENTATION

In a standard fluorescence experiment using a conventional fluorescence spectrometer, the fluorophores in the sample are excited by continuous illumination, and the steady-state fluorescence emission intensity is measured. Two types of spectra can be measured, excitation and emission spectra. For *emission spectra*, the excitation wavelength is fixed, and the emission is recorded as a function of wavelength. *Excitation spectra* are measured the other way around: the excitation wavelength is varied and emission is recorded at a fixed wavelength, usually the emission maximum. A fluorescence excitation spectrum is equivalent to the absorption spectrum.

A standard fluorescence spectrometer (Figure 19.36) has a xenon lamp as a light source. An excitation monochromator selects the excitation wavelength. A second monochromator or filter selects the emission wavelength or wavelength range. Emitted light is detected by a photomultiplier. The emission/detection path is oriented at 90° with respect to the excitation path to avoid background from excitation light on the detector. This 90° configuration prevents the construction of double-beam fluorimeters. Measured fluorescence spectra are therefore always corrected for the spectrum of the solvent, recorded in a separate reference experiment. Fluctuations in the intensity of the lamp are detected by a reference diode and corrected by the instrument. The sensitivity of detectors is wavelength-dependent. Most commercial spectrometers automatically correct the spectra for this wavelength-dependent detection efficiency using an internal calibration file. For quantitative comparisons of spectra, it is important to be aware if these corrections are being made by the instrument used.

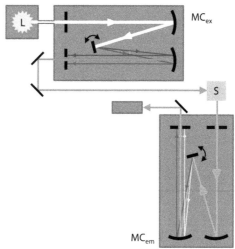

**FIGURE 19.36:** **Fluorimeter.** A light source (L) emits white light. The excitation monochromator ($MC_{ex}$) selects the desired excitation wavelength. Fluorescence light emitted from the sample (S) in a cuvette is detected at right angles. The emission light is split into different wavelengths by the emission monochromator ($MC_{em}$), and is detected by a photomultiplier.

### 19.5.3 Quantum Yield and Lifetime

The *quantum yield* $\phi$ of a fluorophore is the ratio of emitted photons relative to absorbed photons. $\phi$ depends on the kinetic competition (Section 8.2) of fluorescence with other events that lead to the depopulation of the excited state without emission of light. The quantum yield can be calculated from the ratio of the rate constant for fluorescence, $k_f$, and the sum of all rate constants that lead away from the excited state, including those for non-radiative processes summarized here as $k_{nr}$:

$$\phi = \frac{k_f}{k_f + k_{nr}} \qquad \text{eq. 19.62}$$

In the absence of non-radiative processes, the quantum yield is $\phi = 1$, i.e. each excitation event leads to one fluorescence event. In practice, quantum yields will be lower due to the occurrence of non-radiative processes, such as internal conversion and inter-system crossing (Figure 19.34), quenching (Section 19.5.7), or energy transfer (Section 19.5.10). If these non-radiative processes are the dominant pathways for a system to relax to $S_0$, fluorescence may be totally suppressed ($\phi = 0$).

The *fluorescence lifetime* is the mean time the molecule spends in the $S_1$ state before returning to the ground state $S_0$ by emitting fluorescence. The lifetime is the inverse of the sum of all rate constants leading away from the $S_1$ state:

$$\tau = \frac{1}{k_f + k_{nr}} \qquad \text{eq. 19.63}$$

The *intrinsic fluorescence lifetime* $\tau_0$ is defined as the lifetime in the absence of non-radiative processes, and is the inverse of the rate constant for fluorescence, $k_f$:

$$\tau_0 = \frac{1}{k_f} \qquad \text{eq. 19.64}$$

Combining eq. 19.62–19.64, we obtain the relationship between quantum yield and lifetime as

$$\phi = \frac{\tau}{\tau_0} \qquad \text{eq. 19.65}$$

Quantum yields can be determined relative to reference substances (Box 19.8). Fluorescence lifetimes are measured in a lifetime spectrometer (see Section 19.5.9). We will see later that fluorescence lifetime and steady-state fluorescence emission are related, but the measurement of lifetimes provides additional information.

---

**BOX 19.8: DETERMINATION OF QUANTUM YIELDS.**

Quantum yields can be determined from fluorescence emission spectra in comparison with reference substances. To this end, the fluorescence emission spectrum of the reference and the sample are measured using the same instrument settings (excitation wavelength, detection and emission band width, photomultiplier voltage), and corrected for the instrument-specific wavelength dependence of sensitivity. From the areas under the emission band and the extinction coefficients of reference and sample at the excitation wavelength, the fluorescence quantum yield $\phi_s$ of the sample can be calculated (Figure 19.37). The quantum yield of fluorescein has been determined to high accuracy using thermal methods, and fluorescein in 0.1 M NaOH ($\phi = 0.95$) is often used as a reference. Magde *et al.* (2002) *Photochem. Photobiol.* 75: 327–334.

*(Continued)*

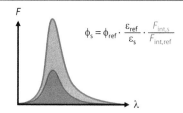

**FIGURE 19.37:   Determination of quantum yields.** From the ratios of the integrated fluorescence $F_{int}$ of a reference substance (ref, blue) and the sample (s, red), the quantum yield of the sample $\phi_s$ can be calculated according to the equation given. $\varepsilon_{ref}$ is the extinction coefficient of the reference substance at the excitation wavelength, $\varepsilon_s$ the extinction coefficient of the sample. $\phi_{ref}$ is the known quantum yield of the reference substance.

### 19.5.4  FLUOROPHORES AND FLUORESCENT LABELING

### 19.5.4.1 Biological Fluorophores

Intrinsic fluorescence of biomolecules is restricted to aromatic amino acids, some cofactors, and unusual nucleobases. In proteins, the aromatic side chains of Trp, Tyr, and Phe absorb and emit in the UV spectral range (Figure 19.38). The quantum yields are ~0.2 (Trp), 0.14 (Tyr), and 0.04 (Phe). The extinction coefficients at the absorption maximum are ~5600 M$^{-1}$ cm$^{-1}$ for Trp ($\lambda_{ex}$ = 280 nm), 1200 M$^{-1}$ cm$^{-1}$ for Tyr ($\lambda_{ex}$ = 274 nm), and 200 M$^{-1}$ cm$^{-1}$ for Phe ($\lambda_{ex}$ = 257 nm), and are a measure of the excitation efficiency (Table 19.2). The product of the extinction coefficient and the quantum yield is a measure of the relative fluorescence "strength", and is highest for Trp (1120 M$^{-1}$ cm$^{-1}$), followed by Tyr (168 M$^{-1}$ cm$^{-1}$) and Phe (8 M$^{-1}$ cm$^{-1}$). For fluorescence studies, Trp is therefore the most important and most useful amino acid. Trp shows the largest Stokes shift (70 nm, $\lambda_{em}$ = 350 nm in aqueous environment). Trp also shows the largest dependence of its fluorescence on the environment, with large changes in quantum yield and strong shifts in emission wavelength to 320 nm and below in hydrophobic surroundings such as the interior of a protein. Therefore, Trp fluorescence is a powerful probe to follow protein folding/unfolding reactions (see Figure 19.55). In contrast, Tyr fluorescence shows a small Stokes shift of about 30–35 nm which is almost independent of the polarity of the surroundings ($\lambda_{em}$ = 305–310 nm). The quantum yield is also dependent on the environment, and Tyr fluorescence is therefore a suitable spectroscopic probe for protein folding and unfolding. Because of its low fluorescence compared to Trp and Tyr and its emission at 260 nm, Phe cannot be used as a fluorescence probe when Trp and/or Tyr are also present.

| TABLE 19.2 | | | |
|---|---|---|---|
| **Absorption and fluorescence of aromatic amino acids.** | | | |
| **amino acid** | **extinction coefficient $\varepsilon$ (M$^{-1}$ cm$^{-1}$)** | **quantum yield $\phi$** | **sensitivity $s = \varepsilon \cdot \phi$ (M$^{-1}$ cm$^{-1}$)** |
| tryptophan | 5600 (280 nm) | 0.2 | 1120 |
| tyrosine | 1200 (274 nm) | 0.14 | 168 |
| phenylalanine | 200 (254 nm) | 0.04 | 8 |

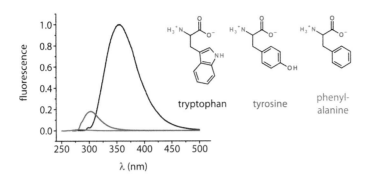

FIGURE 19.38:    **Fluorescence emission spectra of the aromatic amino acids tryptophan, tyrosine and phenylalanine.** Fluorescence emission spectra for tryptophan ($\lambda_{ex}$ = 270 nm, $\phi$ = 0.12), tyrosine ($\lambda_{ex}$ = 260 nm, $\phi$ = 0.13), and phenylalanine ($\lambda_{ex}$ = 240 nm, $\phi$ = 0.022) in aqueous solution (arbitrary units). The structures of the amino acids are shown on the right. Data from Dixon *et al.* (2005) Photochem. Photobiol. 81(1), 212–213.

Several cofactors (see Figure 19.13) contain large aromatic systems that exhibit fluorescence. Examples are NADH ($\lambda_{ex}$ = 340 nm, $\lambda_{em}$ = 460 nm), with the oxidized form NAD$^+$ being non-fluorescent. In NADH, the fluorescence of the nicotinamide moiety is partially quenched due to stacking with the adenine ring (static quenching, see Section 19.5.7). NADH fluorescence often increases upon binding to proteins: binding of the dinucleotide in an extended conformation alleviates quenching, and energy can be transferred from Trp to the nicotinamide moiety.

Flavins and their derivatives FMN and FAD also show fluorescence ($\lambda_{ex}$ ca. 450 nm, $\lambda_{em}$ ca. 525 nm). As for NADH, the fluorescence is partially quenched in FMN and FAD. Notably, flavins show fluorescence in their oxidized form, and are non-fluorescent in the reduced state.

The fluorescent pyridoxal phosphate (PLP; Figure 19.39) is an important cofactor in a multitude of reactions of amino acids (see Box 14.3). The absorption and emission maxima of PLP depend on its ligation state. During transamination reactions, the aldehyde group in PLP forms a Schiff base with the amino group of the substrate, which alters its photophysical properties.

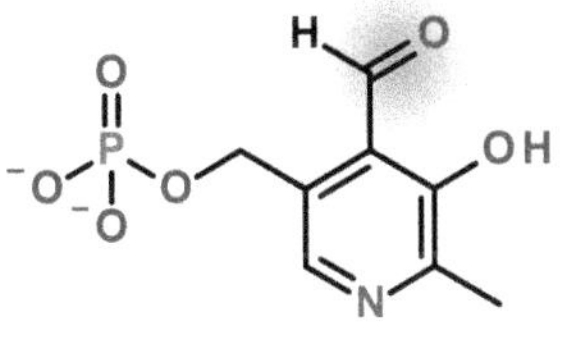

FIGURE 19.39:    **Pyridoxal phosphate (PLP).** PLP is a common cofactor in transamination reactions. During the transamination reaction, an amino group from the first substrate is transferred to PLP and forms a Schiff base with the aldehyde group (orange). In the second step, this amino group is transferred to the second substrate of the reaction. The photophysical properties of PLP change during the catalytic cycle of transaminases.

Nucleotides and nucleic acids do not show fluorescence. Despite their strong absorbance, the canonical nucleobases have an extremely low quantum yield (on the order of $10^{-4}$), resulting in fluorescence "strengths" of 0.04–0.3 M$^{-1}$ cm$^{-1}$. Exceptions are some rare, modified bases such as 4-thiouridine and the Y base that occur in certain tRNAs and that show fluorescence in the UV range. Lipids and membranes lack conjugated electron systems and thus do not exhibit fluorescence.

### 19.5.4.2 Extrinsic Fluorophores and Their Introduction into Proteins, Nucleic Acids, and Lipids

If the biomolecule of interest does not contain a fluorescent group, or if the fluorescent group is not in a suitable location to serve as a spectroscopic probe, fluorescent reporters can be introduced into proteins, nucleic acids, and lipids at the desired position by chemical modification. In addition to their limited occurrence, most intrinsic fluorophores exhibit fluorescence in the UV range, which is experimentally less well accessible, and the extinction coefficients and quantum yields are typically moderate. For highly sensitive fluorescence measurements, fluorophores that absorb in the visible range of the spectrum and have high extinction coefficients and improved quantum yields are frequently employed. These range from rather small organic dyes to quantum dots and entire fluorescent proteins such as the green fluorescent protein (GFP; see Box 19.9). Fluorescent moieties that are introduced into the molecule of interest as reporter groups are also called *extrinsic fluorophores*.

Small organic fluorophores are often covalently attached to proteins using the thiol group of cysteines. A large number of organic dyes have been developed that cover the whole visible spectrum and show high extinction coefficients and quantum yields (Table 19.3) and absorption maxima that coincide with common laser lines. Many of these dyes are commercially available as maleimide or iodoacetamide derivatives that specifically react with thiols (Figure 19.40). Alternatively, succinimidyl ester or isothiocyanate derivatives are available to label the amino terminus or Lys residues.

**TABLE 19.3**
**Absorption and fluorescence properties of organic fluorophores.**

| fluorophore | $\lambda_{max}$ (nm) | $\varepsilon$ ($M^{-1}$ $cm^{-1}$) | $\lambda_{em}$ (nm) | $\phi$ |
|---|---|---|---|---|
| fluorescein | 495 | 75000 | 521 | 0.90 |
| AlexaFluor488 | 490 | 73000 | 524 | 0.92 |
| Atto488 | 500 | 90000 | 520 | 0.80 |
| Cy3 | 555 | 150000 | 570 | 0.31 |
| tetramethylrhodamine | 546 | 90000 | 580 | 0.70 |
| AlexaFluor546 | 556 | 112000 | 573 | 0.79 |
| Cy5 | 646 | 250000 | 662 | 0.20 |
| Alexa647 | 650 | 270000 | 665 | 0.33 |
| Atto647 | 647 | 120000 | 667 | 0.20 |
| Cy7 | 750 | 199000 | 773 | 0.30 |

Note that values are approximate only. Exact values depend on the solvent and the local environment of the fluorophores.

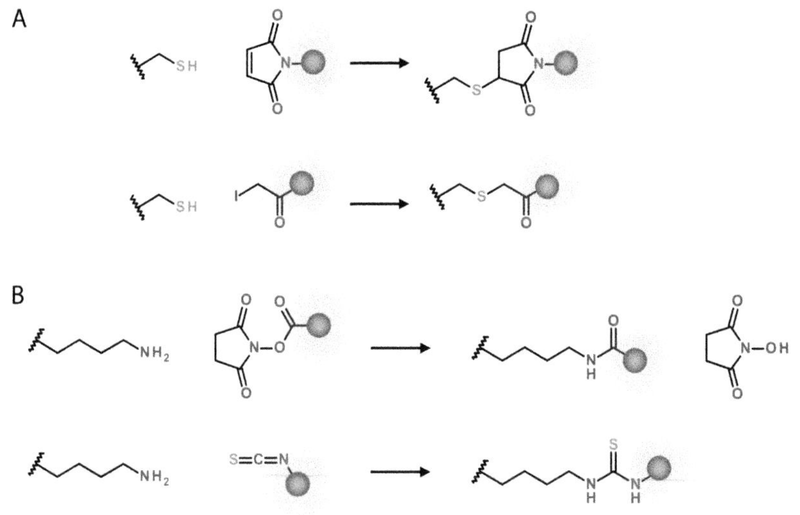

FIGURE 19.40:   **Fluorescent labeling of cysteines and lysines.** A: Cysteines react with maleimides by addition to the double bond (top) and with iodacetamides (bottom) by nucleophilic substitution. Maleimide or iodoacetamide derivatives of fluorescent dyes (green) are used for covalent attachment of fluorophores. B: Lysines can be covalently modified by reaction with succinimidyl esters (top) or isothiocyanates (bottom).

Because of the high natural abundance (~6%), labeling of lysines leads to the introduction of multiple fluorophores into the protein of interest. This approach is frequently used to generate fluorescently labeled antibodies. However, self-quenching of fluorescence (see Section 19.5.7) at high fluorophore densities influences their brightness. The optimal ratio of fluorophores per protein at which the fluorescence signal is maximal depends on the fluorophore.

The incorporation of unnatural amino acids by suppression of stop codons affords an increasing number of bio-orthogonal functional groups, such as keto or azido groups for site-specific modifications of proteins, using simple and efficient reactions (click chemistry, Figure 19.41). Functional groups for subsequent labeling can also be introduced enzymatically. One example is the formylglycine generating enzyme that can be used to genetically encode formyl groups (see Box 16.1). Farnesyl transferases have been used for the introduction of alkyne groups to a four amino acid sequence that can then be modified using click chemistry.

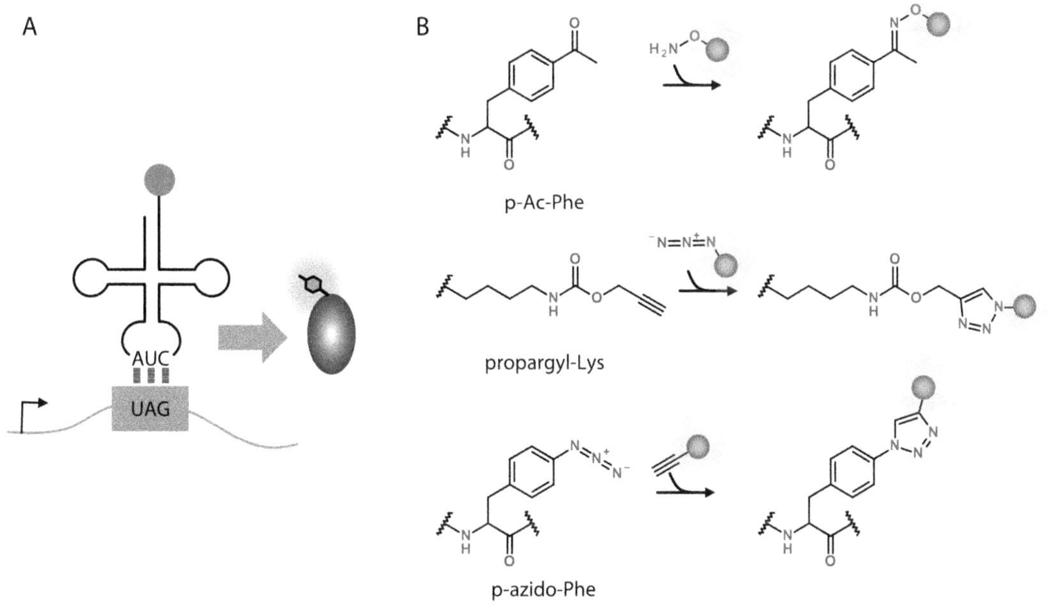

**FIGURE 19.41:**  **Click chemistry and labeling of unnatural amino acids.** A: Internal stop codons are recognized by suppressor tRNAs. The suppressor tRNA can be charged with an unnatural amino acid (orange) that will be incorporated into the protein (blue, right). The unnatural amino acid contains a functional group that can be coupled to a fluorophore in a bio-orthogonal reaction. B: Examples for click chemistry reactions of unnatural amino acids: p-acetyl phenylalanine (p-Ac-Phe) reacts with hydroxylamine derivatives of dyes (green). Propargyl lysine and azido phenylalanine can undergo Cu-mediated cycloadditions of azide- or alkyne-modified dyes, respectively.

A different approach for fluorescent labeling is based on the introduction of N- or C-terminal tags that are subsequently modified with a fluorophore *in vitro* or *in vivo* (Figure 19.42–Figure 19.44). Non-covalent fluorescent labeling is possible by complexation of a hexahistidine-tag with a fluorescent variant of $Ni^{2+}$-nitrilotriacetic acid (Figure 19.42). Similarly, tetracysteine motifs form non-covalent biarsenic complexes with fluorophores. Tags can also be used to introduce covalently attached fluorescent labels. The SNAP tag is a 20 kDa fragment of $O^6$-alkyl-guanine alkyltransferase, involved in DNA repair *in vivo*. The SNAP fragment uses benzylguanine as a substrate. Dyes (derivatized as N-hydroxy-succinimidylesters) can be coupled to its amino group to create a fluorescent benzylguanine derivative, which is then covalently coupled to a Cys of the SNAP tag (Figure 19.43). Similarly, the Halo tag® is derived from a modified haloalkane dehalogenase that has been engineered to accept synthetic chloroalkane ligands. The ligands are modified with fluorophores (or other reporters) and form a covalent bond with the Halo tag fused to the protein of interest. Alternatively, a transglutaminase recognition sequence (Q tag) can be introduced into the protein of interest. Transglutaminase will then transfer a fluorescent probe with an amino group to a conserved Gln within this sequence. Finally, sortase is a bacterial cysteine peptidase that cleaves a C-terminal sorting signal from cell-wall-linked proteins. The five-amino-acid sorting signal (the sortase tag) can be coupled to fluorophores. Sortase recognizes the sortase tag and links it to the N-terminus of the target protein (Figure 19.44). Sortase can also be used for internal or C-terminal labeling. Many of these approaches are also employed for surface immobilization of biomolecules in surface plasmon resonance (Section 26.3.4) or fluorescence microscopy (Sections 23.1.4, 23.1.7).

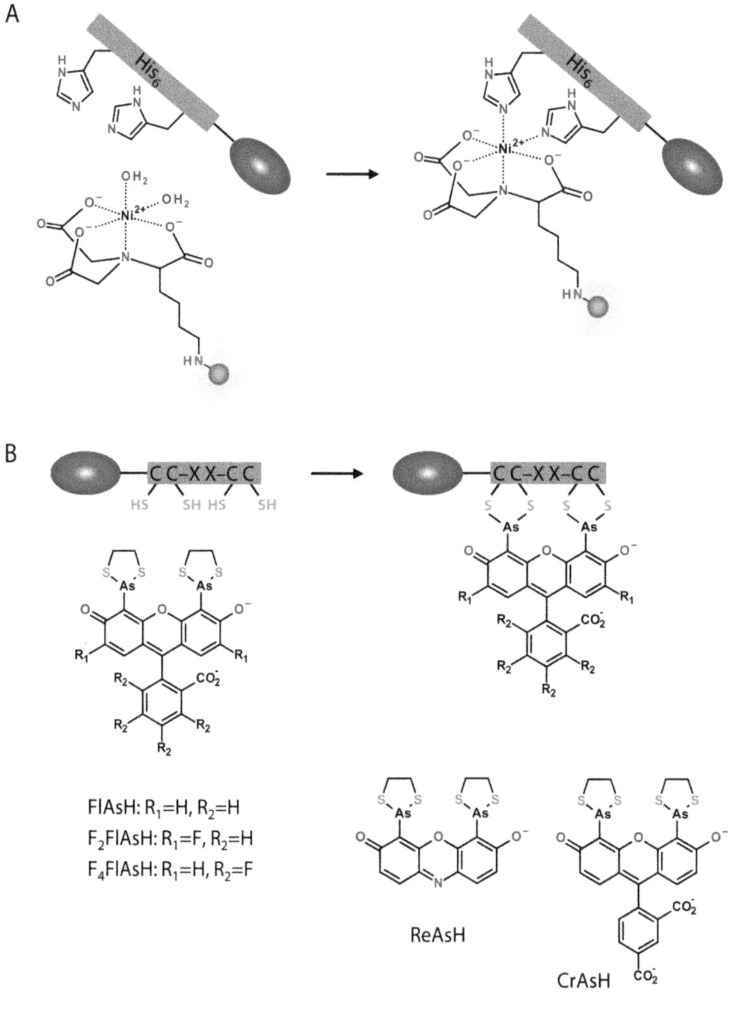

FIASH: $R_1$=H, $R_2$=H
$F_2$FIAsH: $R_1$=F, $R_2$=H
$F_4$FIAsH: $R_1$=H, $R_2$=F

ReAsH

CrAsH

**FIGURE 19.42: Introduction of tags and non-covalent fluorescent labeling. A:** A His-tag (gray) fused to the protein of interest (orange) forms a non-covalent complex with $Ni^{2+}$-nitrilotriacetic acid (NTA). Using fluorescently modified NTA-derivatives, this interaction can be used to couple a fluorophore (green sphere) to a His-tagged protein. **B:** Motifs containing four cysteines (CCXXCC, gray) form non-covalent complexes with biarsenic fluorophore derivatives. The ethane dithiol forms of the biarsenic compounds are non-fluorescent, but complexation by the tetracysteine motif renders them fluorescent. Adding biarsenic complexes to the protein of interest (orange), fused to a tetracysteine motif, allows non-covalent fluorescent labeling of proteins. A number of different biarsenic compounds based on fluorescein or resorufin are available.

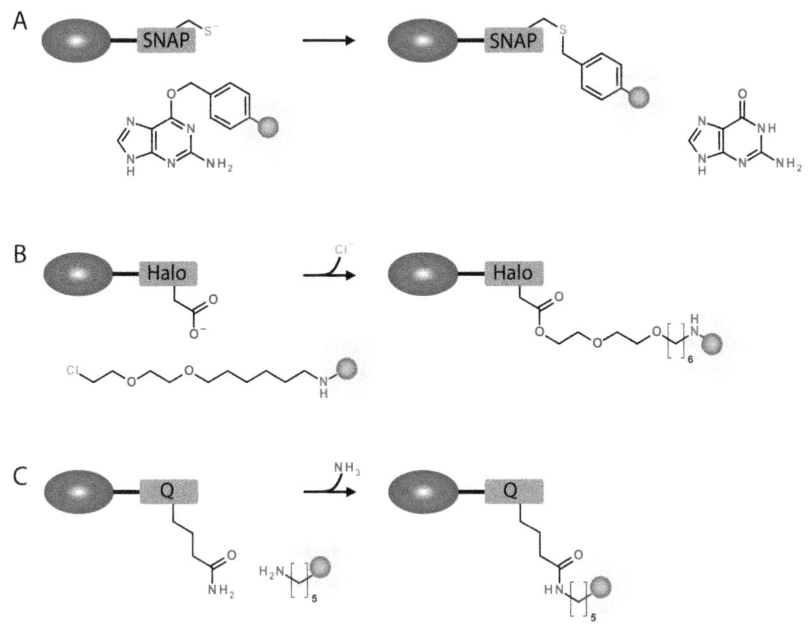

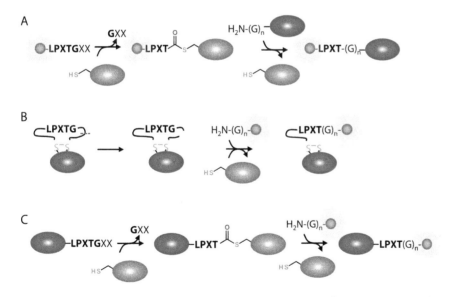

**FIGURE 19.43:** **Introduction of tags for covalent fluorescent labeling.** A: The SNAP tag (gray) is a 20 kDa fragment of $O^6$-alkyl-guanine alkyltransferase. The reaction of a cysteine within the tag with fluorescently labeled (green) benzylguanine can be exploited for covalent fluorescent labeling. B: The Halo tag® (gray) is derived from a haloalkane dehalogenase. An aspartate within the Halo tag forms a covalent adduct with fluorescently modified (green) chloroalkane ligands. C: The Q tag (gray) is a transglutaminase recognition sequence. Transglutaminase transfers fluorescently labeled (green) amines to a conserved glutamine within the tag. The target protein is represented in orange.

**FIGURE 19.44:** **Sortase for N-terminal, internal, or C-terminal fluorescent labeling**. A: Sortase (gray sphere), a bacterial cysteine peptidase that cleaves a C-terminal sorting signal from cell-wall-linked proteins, recognizes the five-amino-acid sortase tag (LPXTG, bold) and links it to the N-terminus of target proteins (orange). This reaction can be used for covalent fluorescent labeling (green sphere) of proteins at the N-terminus. B: For internal labeling, the sortase tag has to be introduced into the protein of interest (orange) within a loop that is closed by a disulfide bond. Cleavage by sortase generates a C-terminal sortase tag. Sortase then transfers this tag, including the protein of interest, to the N-terminus of a fluorescently modified (green) linker peptide. C: For C-terminal labeling, the sortase tag is fused to the C-terminus of the protein of interest (orange). Sortase cleaves the sortase tag between T and G, and couples the tagged protein of interest to the N-terminus of a fluorescently modified (green) linker peptide.

*Expressed protein ligation* is a procedure derived from the reaction sequence in protein splicing, in which a protein that is modified with a thioester at the C-terminus can react with a peptide (or protein) carrying an N-terminal cysteine (Figure 19.45). Thioester-modified proteins and proteins with N-terminal cysteines can be generated by production as fusion proteins with intein sequences (protein introns) and subsequent cleavage of the fusion protein. At the end of the ligation reaction, the protein is linked to the protein *via* an authentic peptide bond. With a fluorophore-labeled peptide, this reaction can be used to introduce a fluorescent label into the protein of interest.

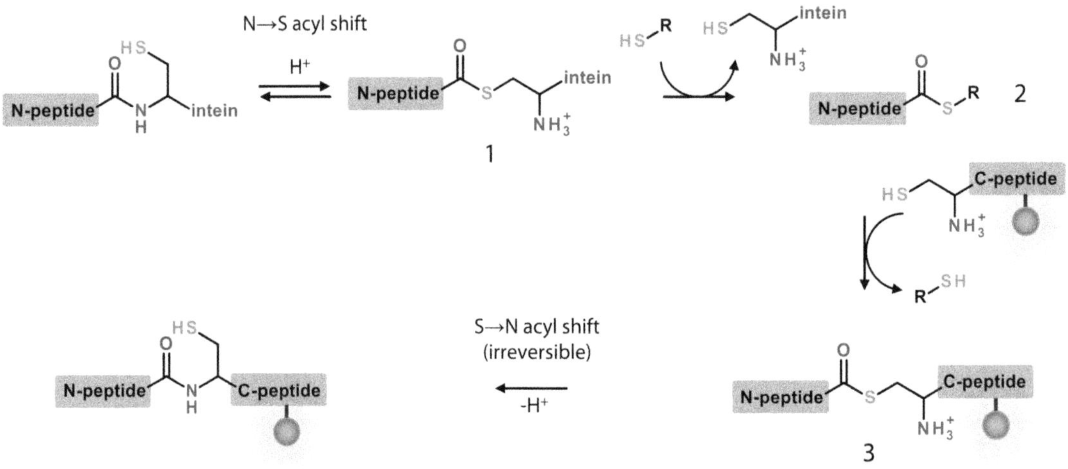

**FIGURE 19.45: Expressed protein ligation.** Two peptides (N-peptide, C-peptide) can be linked by a peptide bond *via* expressed protein ligation. The N-peptide is fused to an intein that has a cysteine at the N-terminus. The nucleophilic attack of the thiol group at the carbonyl carbon of the peptide bond causes an N→S acyl shift to a thioester (2). This thioester is activated by addition of a thiol (R-SH) with a good leaving group R. This thioester (3) reacts with the C-peptide that carries a cysteine at the N-terminus. An irreversible S→N acyl shift then leads to the final product, the two peptides linked by a peptide bond. Ligation of a recombinantly produced fusion protein of N-peptide with a fluorescently labeled synthetic C-peptide allows the introduction of fluorophores (green sphere) into larger proteins.

If their large size is not an issue, proteins can be fluorescently labeled by fusion with fluorescent proteins. Starting from natural fluorescent proteins such as the green fluorescent protein (GFP) from the jellyfish *Aequorea victoria*, a full spectrum of fluorescent proteins has been developed with emission wavelengths that cover the whole visible spectrum (Box 19.9). While these fluorophores are genetically encoded and useful for fluorescence microscopy and *in vivo* localization of biomolecules, their large size of ~25 kDa and unstable fluorescence emission limit their applications.

**BOX 19.9: FLUORESCENT PROTEINS.**

Green fluorescent protein (GFP) is a 27 kDa protein from the jellyfish *Aequorea victoria* that exhibits green fluorescence after excitation with UV or blue light. GFP folds into a β-barrel structure (Figure 19.46). Air oxidation generates the fluorophore, formed by the side chains of a Ser-Tyr-Gly sequence. The fluorophore is located in the center of the protein, protected from solvent. Numerous GFP variants have been generated since its discovery. The first variant was the S65T variant (eGFP) that showed enhanced fluorescence

*(Continued)*

compared to wild-type GFP. Substitution of Y66 generated the cyan fluorescent protein (CFP), with a blueshift in absorbance and emission wavelength. More complex modifications of the fluorophore surroundings led to the yellow fluorescent protein (YFP). CFP and YFP are commonly used as FRET pairs (see Section 19.5.10) for *in vivo* studies. Naturally occurring red fluorescent proteins such as the dimeric DsRed isolated from the sea anemone *Discosoma striata* were the basis for a number of monomeric red fluorescent proteins with improved photophysical properties that cover the red spectral region.

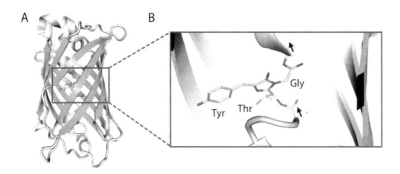

FIGURE 19.46:   **Structure of green fluorescent protein and its fluorophore**. A: Green fluorescent protein (GFP) from the jellyfish *Aequorea victoria*. GFP forms a β-barrel structure, with the fluorophore buried inside the barrel. Upon excitation with blue light, GFP shows green fluorescence. B: The fluorophore in GFP is formed by oxidation and ring formation of the side chains of the three consecutive amino acids Ser65, Tyr66, and Gly67 (in wild-type GFP). The structure depicted here is the enhanced version with an S65T mutation.

Split versions of GFP can be used in complementation assays to investigate protein-protein or other biomolecular interactions. In these assays, a GFP fragment is fused to one potential interaction partner, the remainder of GFP is fused to the second partner. While the individual protein fragments do not exhibit fluorescence, complex formation *in vivo* allows the GFP fragments to recombine to a complete GFP that emits fluorescence. Tripartite GFP versions have also been used in *in vivo* interaction assays. The importance of the discovery and development of GFP into a versatile genetic tool was recognized with the award of the Nobel Prize in Chemistry in 2008 to Osamu Shimomura, Martin Chalfie, and Roger Tsien.

Organic fluorophores, although superior to intrinsic fluorophores, still pose problems for fluorescence measurements. A major drawback is their limited brightness and photostability and their tendency to undergo irreversible photobleaching (see Section 23.1.7.1; Box 23.7). *Quantum dots* are semiconductor nanocrystals whose diameter is similar in magnitude to their de Broglie wavelength, leading to a phenomenon called quantum confinement, and to special molecular properties. As a result of this quantum confinement, the bonding and anti-bonding orbitals of quantum dots are separated by a *band gap* that decreases with the size of the quantum dot (Figure 19.47). Absorption causes the transition of an electron from the bonding to the anti-bonding orbital, leaving a hole in the bonding orbital. Relaxation of the electron to the bonding orbital, and recombination with the hole is then accompanied by fluorescence emission. As a consequence of the size-dependent band gap, the emission wavelength of quantum dots made from the same material depends on their size: larger quantum dots show red-shifted emission, the emission of small quantum dots is blue-shifted.

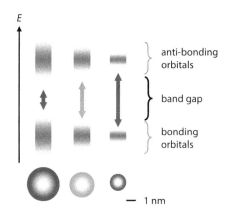

FIGURE 19.47:   **Quantum dots and the size dependence of the band gap.** The bonding and anti-bonding orbitals of semiconductor nanocrystals (red, green, and blue spheres) are separated by the band gap. The larger the nanocrystal, the broader the energy band for each type of orbital, and the lower the energy required for excitation of electrons from a bonding to an anti-bonding orbital (double-headed arrows).

Nanocrystals can be coated with polymers, functionalized with streptavidin, and coupled to biotinylated molecules of interest. Compared to organic fluorophores, quantum dots show an increased brightness because of their higher extinction coefficients and comparable fluorescence quantum yields. They are photostable and show less photobleaching. However, quantum dots show frequent *blinking*, short excursions to a non-fluorescent state such as the triplet state. Blinking can be suppressed by adding Trolox, a vitamin E analog that acts as a triplet state quencher and rescues fluorophores from the non-fluorescent triplet state (see Box 23.7). The toxicity of semiconductor nanocrystals limits *in vivo* applications. Also, the name quantum dot is a bit deceiving when it comes to size. Typically, a functionalized quantum dot has a diameter of a few nanometers, and so is of the size of a 30–50 kDa protein.

Fluorescent nucleotide analogs are powerful tools to study the interaction of proteins with nucleotides. In accordance with their wide-spread function in the cell, a wide range of nucleotides coupled to fluorescent reporter groups are commercially available. Examples are mant nucleotides that carry a N-methylanthraniloyl (mant) group that forms an ester bond with the 2´- or 3´- hydroxyl group of the ribose or deoxyribose (Figure 19.48). Similar to the static quenching (see Section 19.5.7) in NADH und FAD, the mant fluorescence of mant nucleotides in solution is quenched by the

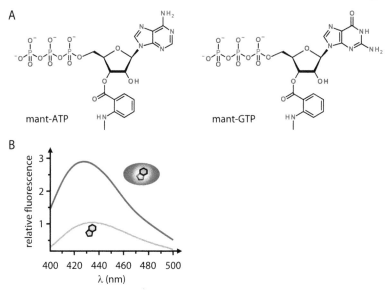

FIGURE 19.48:   **mant nucleotides as a probe for nucleotide binding.** A: mant-ATP and mant-GTP are frequently used to study nucleotide binding to ATP- or GTP-binding proteins. B: Fluorescence of the mant group can be excited at 360 nm. The emission maximum is 440 nm. Upon binding to a protein, the mant fluorescence increases by 2-3-fold due to the alleviation of intermolecular quenching by the adenine/guanine base.

nucleobase that stacks onto the mant group. After specific binding of the nucleobase by a protein, stacking is no longer possible and the mant fluorescence increases, often by a factor of 2–3.

Fluorescent labeling of nucleic acids can be achieved by introducing organic fluorophores during solid-phase synthesis using modified building blocks. This approach is only possible for fluorophores that are stable under the conditions of the coupling and deprotection reactions. One example of such a probe is 2-aminopurine (2-AP; $\lambda_{ex}$ = 310 nm, $\lambda_{em}$ ca. 370 nm), an adenine analog that base-pairs with thymine, but also with cytosine. By incorporating 2-AP in place of adenine, a chemically very similar fluorescent reporter can be introduced into DNA. 2-AP shows strong fluorescence in solution, but its fluorescence is partially quenched due to base stacking when it is part of a DNA duplex. Hence, 2-AP serves as a spectroscopic probe for DNA structure and dynamics, and for the accessibility of nucleotides in nucleic acids. 2-AP fluorescence has been used to monitor base-flipping by DNA-modifying enzymes, or unwinding of DNA or RNA duplexes by helicases (Figure 19.49).

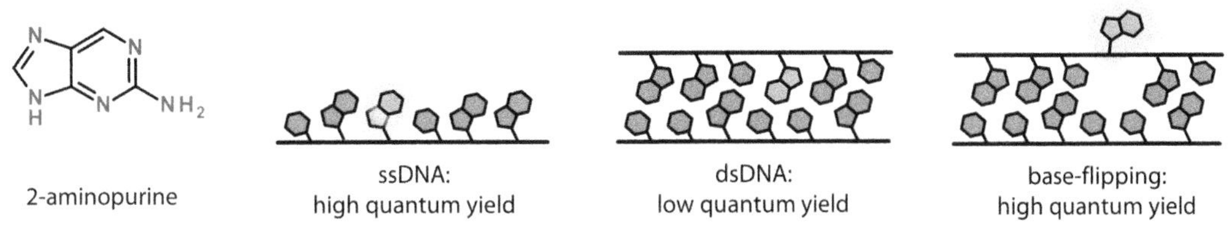

**FIGURE 19.49:** **2-aminopurine fluorescence as a probe for nucleic acid structure and dynamics.** The fluorescence of 2-aminopurine is quenched by base stacking. Its fluorescence is higher in single-stranded than in double-stranded DNA. Therefore, the increase in fluorescence upon duplex separation can be exploited as a spectroscopic probe to monitor unwinding by DNA (or RNA) helicases. Some DNA modification enzymes modify bases while they are flipped out from the duplex. Base flipping alleviates stacking, and is associated with an increase in 2-AP fluorescence.

Fluorescent nucleotides have been generated by organic synthesis. Cytosine analogs with UV absorbance and fluorescence have been developed and introduced into DNA by chemical synthesis (Figure 19.50).

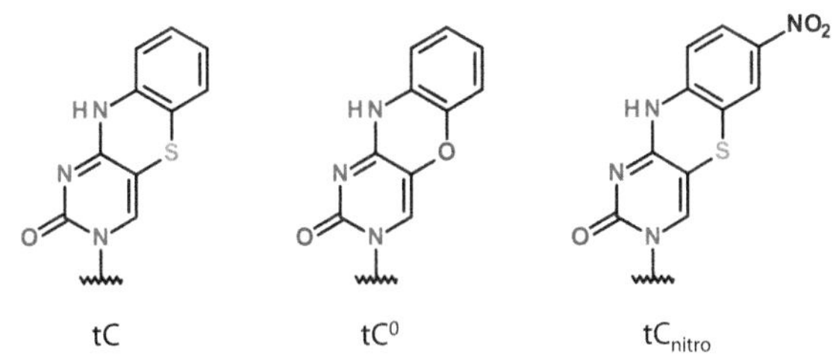

**FIGURE 19.50:** **Fluorescent cytosine analogs.** The cytosine analogs tC, tC$^0$, and tC$_{nitro}$ can be incorporated into DNA by solid-phase synthesis for fluorescence studies. tC and tC$^0$ exhibt fluorescence, whereas tC$_{nitro}$ is non-fluorescent, but can be used as an acceptor in FRET experiments (see Section 19.5.10).

As an alternative to introducing fluorophores into DNA directly during solid-phase synthesis, reactive groups can be introduced instead and modified with fluorophores after synthesis. This procedure allows the incorporation of fluorophores that would not withstand the harsh conditions during DNA synthesis. Longer DNA molecules (> 200 nucleotides) are not amenable to chemical synthesis. In this case, fluorophores can be introduced during PCR reactions with fluorescently modified primers (generated by solid-phase synthesis). Similarly, fluorescently labeled RNA can be generated enzymatically by *in vitro* transcription from a DNA template, using fluorescently modified starter oligonucleotides. Ligation techniques enable the attachment of DNA and RNA fragments, chemically synthesized or enzymatically generated, to large nucleic acid molecules with fluorophores in the desired position(s). In structured RNAs, fluorescent labels have successfully been introduced by hybridizing fluorescently labeled oligonucleotides to unpaired bases in loop regions.

A green fluorescent RNA molecule, called spinach, can be used as a genetically encoded green fluorescent tag for RNA labeling *in vivo*, in the same way as GFP for proteins.

Lipids can be modified with hydrophilic dyes at the polar head group, or with hydrophobic dyes introduced into the hydrophobic side chain without interfering with membrane formation, enabling fluorescent studies on lipid bilayers, micelles, and vesicles. A variety of such lipids is commercially available.

### 19.5.5 Applications

#### 19.5.5.1 Fluorescence as a Probe for Binding: Equilibrium Titrations

Fluorescence is frequently used as probe to monitor binding. In some cases, changes in Trp fluorescence can be employed to monitor ligand binding to proteins, or protein-protein interactions. More often, ligands labeled with extrinsic fluorophores are used as reporters for binding. $K_d$ values for the dissociation of macromolecule-ligand complexes can be determined in *equilibrium titrations* of the fluorescently labeled partner (typically the ligand) with the unlabeled macromolecule (Figure 19.51). For a simple binding equilibrium of a fluorescent ligand B that binds to a biological macromolecule A to form an AB complex

$$A + B \underset{K_d}{\rightleftharpoons} AB \qquad \text{scheme 19.1}$$

we can formulate the equilibrium constant and the mass conservation, and obtain the concentration of the AB complex as a function of the total concentrations $A_t$ and $B_t$ is

$$[AB] = \frac{A_t + B_t + K_d}{2} - \sqrt{\left(\frac{A_t + B_t + K_d}{2}\right)^2 - A_t B_t} \qquad \text{eq. 19.66}$$

(see Section 3.2). We now have to relate the concentration of the AB complex to our spectroscopic signal. Let us assume that we measure the fluorescence of B, which changes upon AB complex formation. To determine the dissociation constant of the AB complex, we use a fixed concentration $B_0$ in the range of the $K_d$ value we expect. In a *titration* experiment, we increase the concentration of A in a stepwise manner, and obtain a binding curve from the fluorescence of B as a function of $A_t$ (Figure 19.51). The change in fluorescence of B upon formation of the AB complex is proportional to the degree of binding ν:

$$\nu = \frac{[AB]}{B_t} \qquad \text{eq. 19.67}$$

We express the total change in fluorescence $\Delta F$ for each step of the titration as the product of the degree of binding and the maximum fluorescence change $\Delta F_{max}$ upon saturation (ν = 1) at the end of the titration:

$$\Delta F = \Delta F_{max} \frac{[AB]}{B_t} \qquad \text{eq. 19.68}$$

Our fluorescence signal $F$ at any given concentration of A then becomes

$$F = F_0 + \Delta F_{max} \frac{[AB]}{B_t} \qquad \text{eq. 19.69}$$

$F_0$ is the fluorescence of free B in the absence of A, and thus the starting value of the titration. In combination with eq. 19.66 we obtain

$$F = F_0 + \Delta F_{max} \cdot \frac{\dfrac{A_t + B_t + K_d}{2} - \sqrt{\left(\dfrac{A_t + B_t + K_d}{2}\right)^2 - A_t B_t}}{B_t} \qquad \text{eq. 19.70}$$

Eq. 19.70 thus describes the fluorescence $F$ as a function of the total concentrations $B_t$ and $A_t$ and the dissociation constant $K_d$. Since we know $A_t$ and $B_t$ for each point of our titration, we can use eq. 19.70 to analyze the titration data and to determine $K_d$ by non-linear least squares fitting. It is important to note that equilibrium titrations can be performed using any spectroscopic signal that changes linearly with complex formation, and $K_d$ values can be determined from these titration curves by the same principle outlined here.

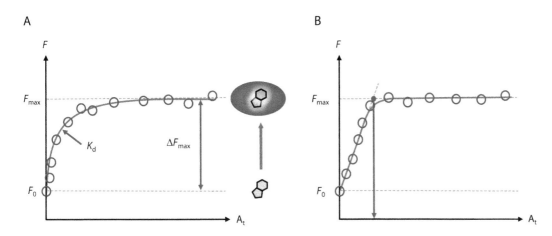

**FIGURE 19.51:   Fluorescence equilibrium titration.** A: Fluorescence equilibrium titration: A fluorescent or fluorescently labeled ligand, provided at a concentration in the range of the expected $K_d$ value, is titrated with its binding partner A. The fluorescence increase upon complex formation serves as a spectroscopic probe for binding. The binding curve obtained (blue circles) can be described by eq. 19.70 (gray curve) to determine the dissociation constant $K_d$ of the complex. B: Active site titration. The fluorescently labeled ligand is provided at a concentration more than 10-fold above the $K_d$ value of the complex, such that nearly all of the macromolecule that is added in each titration step is bound. From the equivalence point (red arrow) of the binding curve, the stoichiometry of binding can be determined.

If the ligand is provided at concentrations above the $K_d$, the binding equilibrium will be strongly on the side of the complex in each step of the titration because of the high ligand concentration. Virtually all of the added macromolecule will bind to a ligand molecule, giving rise to a nearly linear change in fluorescence (Figure 19.51). Saturation is reached when all ligand is bound, and further addition of macromolecule does not change the fluorescence. From the equivalence point, the stoichiometry of binding can be determined. Such an *active site titration* is a convenient and often much more sensitive alternative to the determination of binding stoichiometries by isothermal titration calorimetry (Section 27.1).

In a fluorescence titration, we determine the $K_d$ value for a complex of the fluorescently labeled ligand and its binding partner. What we really want to know is the $K_d$ value of the complex with a non-fluorescent, natural ligand, however. Again, fluorescence equilibrium titrations are a useful approach. Once we have determined the $K_d$ value of the complex with the fluorescent ligand, for example a fluorescently labeled nucleotide, we can perform a *competitive* or *displacement titration*, in which the fluorescent ligand and the non-fluorescent, authentic ligand are present at the same time and compete for the same binding site. We can formally distinguish between competitive and displacement titrations (Figure 19.52). In competitive titrations, the fluorescent ligand is titrated with the molecule of interest in the absence and presence of different concentrations of the non-fluorescent ligand. From the effect of the non-fluorescent ligand on the apparent $K_d$ value, the true $K_d$ value for the complex of the molecule with the non-fluorescent ligand is determined. Displacement titrations start from the complex of the molecule of interest with the fluorescent ligand. The fluorescent ligand is then displaced by adding increasing concentrations of the non-fluorescent ligand. As the fluorescent ligand becomes liberated, its fluorescence will return to the characteristic value of the unbound form. The result of these two approaches should be identical, provided that the reaction is in equilibrium for each titration point.

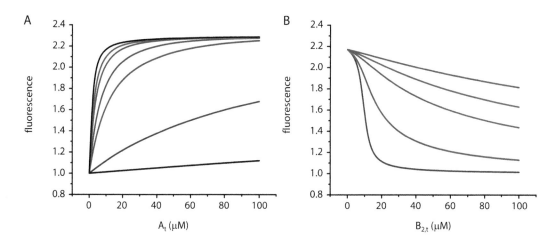

**FIGURE 19.52:   Fluorescence equilibrium competitive and displacement titration.** A: Competitive titrations of a fluorescent ligand $B_1$ with a macromolecule A in the absence of competitor (black), and in the presence of increasing concentrations of a non-fluorescent competitor $B_2$. $B_{1,t}$ = 1 µM, $K_{d1}$ = 1 µM, black (top): no competitor, blue: 10 µM, purple: 20 µM, pink: 50 µM, red: 100 µM, brown: 1 mM, black: 10 mM competitor. B: Displacement titrations of complexes of the macromolecule A and the fluorescent ligand $B_1$ with increasing concentrations of the non-fluorescent ligand $B_2$. $K_{d1}$ = 1 µM, $A_t$ = 10 µM, $B_{1,t}$ = 1 µM, brown: $K_{d2}$ = 0.1 µM, red: $K_{d2}$ = 1 µM, purple: $K_{d2}$ = 5 µM, blue: $K_{d2}$ = 10 µM, dark blue: $K_{d2}$ = 20 µM.

In competitive and displacement titrations two equilibria have to be considered at each point of the titration. The first is binding of the fluorescently labeled ligand $B_1$ with its dissociation constant $K_{d1}$:

$$A + B_1 \underset{K_{d1}}{\rightleftarrows} AB_1$$

scheme 19.2

The second equilibrium is binding of the non-fluorescent ligand $B_2$ with the dissociation constant $K_{d2}$:

$$A + B_2 \underset{K_{d2}}{\rightleftarrows} AB_2$$

scheme 19.3

The equilibrium constants $K_{d1}$ and $K_{d2}$ are defined as

$$K_{d1} = \frac{[A][B_1]}{[AB_1]}$$

eq. 19.71

and

$$K_{d2} = \frac{[A][B_2]}{[AB_2]}$$

eq. 19.72

Both eq. 19.71 and eq. 19.72 have to be fulfilled at each point of the titration. To derive an expression for the fluorescence as a function of the concentration of the non-fluorescent ligand $B_2$, we again need to combine the laws of mass action with the mass conservation for $B_1$ and $B_2$. From these equations, we obtain an expression for the $AB_1$ complex as a function of the total concentrations $A_t$, $B_{1,t}$, and $B_{2,t}$. This time, the result is a cubic equation for the concentration of the $AB_1$ complex that provides our spectroscopic signal (Box 19.10). The solutions of the cubic equation are used to describe the fluorescence signal as a function of $A_t$, $B_{1,t}$ and $B_{2,t}$, and $K_{d1}$ and $K_{d2}$ in analogy to eq. 19.69 and eq. 19.70. Using this expression, $K_{d2}$ is determined from displacement curves by non-linear least squares fitting. $K_{d1}$ is determined independently in a separate titration experiment and is kept constant during the fitting procedure.

## BOX 19.10: DETERMINING $K_d$ VALUES FROM DISPLACEMENT TITRATIONS.

To derive an expression for the fluorescence signal as a function of the concentrations of A, $B_1$ and $B_2$, and the two $K_d$ values, we again have to combine the mass conservation for $B_1$ and $B_2$ and the equilibrium constants $K_{d1}$ and $K_{d2}$. A cubic equation for the concentration of the $AB_1$ complex is obtained:

$$0 = [AB_1]^3 + a \cdot [AB_1]^2 + b \cdot [AB_1] + c \qquad \text{eq. 19.73}$$

with the coefficients

$$a = \frac{A_t(K_{d2} - K_{d1}) + B_{1,t}(2K_{d2} - K_{d1}) + B_{2,t}K_{d1} - K_{d1}^2 + K_{d1}K_{d2}}{K_{d1} - K_{d2}} \qquad \text{eq. 19.74}$$

$$b = \frac{A_t B_{1,t}(K_{d1} - 2K_{d2}) - B_{1,t}^2 K_{d2} - B_{1,t}K_{d1}(B_{2,t} + K_{d2})}{K_{d1} - K_{d2}} \qquad \text{eq. 19.75}$$

and

$$c = \frac{A_t B_{1,t}^2 K_{d2}}{K_{d1} - K_{d2}} \qquad \text{eq. 19.76}$$

This cubic equation has three solutions. For

$$Q \equiv \frac{a^2 + 3b}{9} \qquad \text{eq. 19.77}$$

and

$$R \equiv \frac{2a^3 - 9ab + 27c}{54} \qquad \text{eq. 19.78}$$

and the condition

$$Q^3 - R^2 \geq 0 \qquad \text{eq. 19.79}$$

the three solutions are

$$[AB_1]_1 = -2\sqrt{Q}\cos\left(\frac{\theta}{3}\right) - \frac{a}{3} \qquad \text{eq. 19.80}$$

$$[AB_1]_2 = -2\sqrt{Q}\cos\left(\frac{\theta + 2\pi}{3}\right) - \frac{a}{3} \qquad \text{eq. 19.81}$$

and

$$[AB_1]_3 = -2\sqrt{Q}\cos\left(\frac{\theta + 4\pi}{3}\right) - \frac{a}{3} \qquad \text{eq. 19.82}$$

with

$$\theta = \arccos\left(\frac{R}{\sqrt[3]{Q}}\right) \qquad \text{eq. 19.83}$$

*(Continued)*

For

$$Q^3 - R^2 < 0 \qquad \text{eq. 19.84}$$

Eq. 19.73 has only a single solution

$$[AB_1] = -\text{sgn}(R)\left[\left(\sqrt{R^2 - Q^3} + |R|\right)^{\frac{1}{3}} + \frac{Q}{\left(\sqrt{R^2 - Q^3} + |R|\right)^{\frac{1}{3}}}\right] - \frac{a}{3} \qquad \text{eq. 19.85}$$

The solutions for $[AB_1]$ are used to express the fluorescence signal as a function of $A_t$, $B_{1,t}$, $B_{2,t}$, $K_{d1}$, and $K_{d2}$ based on eq. 19.69:

$$F = F_0 + \Delta F_{max} \frac{[AB_1]}{B_{1,t}} \qquad \text{eq. 19.86}$$

The resulting equations can be implemented in data analysis software to determine $K_{d2}$ from displacement curves by non-linear least squares fitting. $K_{d1}$ is kept constant during the fitting procedure.

Fluorescence can also be measured as a function of time upon mixing of binding partners, and used as a spectroscopic probe to measure association and dissociation rates (see Chapter 25). From these experiments, macroscopic and microscopic rate constants can be extracted (Chapters 7–8, 10).

### 19.5.5.2 Fluorescence as a Probe for the Chemical Micro- and Macro-Environment

Fluorescence reports on the micro-environment of the fluorophore within a biomolecule, and can therefore be used as a probe for structural changes. Due to its high sensitivity for the local chemical environment, Trp fluorescence is a valuable probe to monitor folding and unfolding of proteins (Figure 19.53). 1-anilinonaphtalene-8-sulfonate (ANS) binds to exposed hydrophobic regions of proteins, which is accompanied by an increase in its fluorescence. ANS has been used to monitor their folding and unfolding and to detect partially folded intermediates. The nucleobase 2-aminopurine (2-AP) is a probe for the local environment in the DNA, with high fluorescence when it is part of a single-strand, and low fluorescence within a double-stranded DNA (Figure 19.49). 2-AP fluorescence reports on structural changes of nucleic acids, and has been used to study folding of RNA molecules and to monitor structural changes in RNA and DNA.

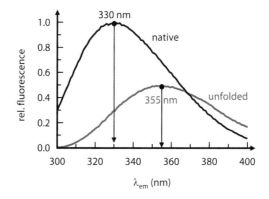

**FIGURE 19.53: Tryptophan fluorescence as a probe to monitor protein unfolding.** Fluorescence emission spectra of a native protein (black) and the same protein in the unfolded state (red). The fluorescence signal decreases due to collisional quenching by water, and the maximum shifts from 330 nm to 355 nm due to the solvent exposure of tryptophans upon unfolding and the resulting more polar environment (see Figure 19.9).

Fluorescence can also be used as a probe for the macroscopic environment. Fluorescence sensors bind to an analyte and respond to its presence by a change in fluorescence. They thereby report on pH values or concentrations of ions and metabolites (Box 19.11).

---

**BOX 19.11: FLUORESCENCE SENSORS.**

Fluorescein shows a fluorescence emission that is dependent on the pH, and can therefore function as a pH sensor. Fluorescein exists in two fluorescent protonation states, the mono-anionic and the di-anionic forms that equilibrate with a $pK_A$ value of ~6.5. The two forms show different absorption and emission wavelengths, and fluorescein can therefore be used as a ratiometric pH probe by measuring emission after excitation at 450 nm (maximum of the mono-anion) and 495 nm (maximum of the di-anionic form). Fluorescence intensities depend on the concentration of the fluorophore, which is difficult to control in *in vivo* applications. Ratiometric approaches, in which the concentration-independent ratio of the fluorescence emission at two wavelengths is measured, are therefore often preferred. Fluorescence sensors for $Ca^{2+}$ and $Mg^{2+}$ ions are chelating molecules that form a high-affinity complex with the respective ion, accompanied by a change in fluorescence. One example is the dye Fura-2, which exhibits a shift in excitation wavelength in the presence of $Ca^{2+}$ and is used as a ratiometric $Ca^{2+}$ sensor. Other fluorescence properties such as fluorescence lifetimes (Section 19.5.9), anisotropy (Section 19.5.8), quenching (Section 19.5.7), or FRET (Section 19.5.10) can also be used for sensing.

---

### 19.5.5.3 Fluorescence and Imaging: Fluorescence Recovery after Photobleaching

Fluorescence is commonly used in imaging applications. Fluorescence microscopy, which will be discussed in detail in Section 23.1, allows fluorescent or fluorescently labeled biomolecules to be localized and tracked over time (see Sections 23.1.2, 23.1.7). Fluorescence microscopy can be employed to monitor diffusion of fluorescently labeled proteins or lipids in a membrane. In *fluorescence recovery after photobleaching* (FRAP) experiments, a membrane area (about 1 or 2 µm in diameter) containing fluorescently labeled lipids or proteins is illuminated with laser light in a microscope, leading to fluorophore excitation and steady-state emission at a constant intensity. At time $t_0$, the light intensity is increased by several orders of magnitude, leading to irreversible photodestruction (bleaching) of the fluorophore and a drop in emitted fluorescence. After the bleaching pulse, the excitation intensity is reduced again to its previous level. The fluorescence signal in the bleached area will now increase over time until it reaches nearly the original value. Fluorescence recovery is caused by fluorescent molecules that re-enter the observed membrane area by lateral diffusion (Figure 19.54). The increase in fluorescence $F(t)$ follows a single exponential time-course:

$$F(t) = F(t_0) + (F(t_\infty) - F(t_0))\exp(-k_{diff}t) \qquad \text{eq. 19.87}$$

where $F(t_0)$ is the initial fluorescence after the bleaching pulse, and $F(t_\infty)$ is the final recovered fluorescence value that equals the fluorescence before the bleaching pulse. From the rate constant $k_{diff}$, the lateral diffusion constant $D_{lateral}$ of the molecule of interest can be determined:

$$D_{\text{lateral}} = \frac{r^2}{4} \cdot k_{\text{diff}}$$

eq. 19.88

$r$ is the radius of the bleached area, and $k_{\text{diff}}$ is the time constant of the exponential fluorescence recovery. With this method, diffusion coefficients between $10^{-7}$ and $10^{-11}$ cm$^2$ s$^{-1}$ (Section 4.1) can be measured.

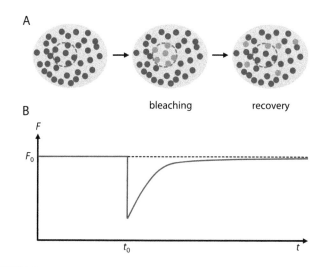

**FIGURE 19.54:**   **Fluorescence recovery after photobleaching (FRAP). A:** Fluorescent proteins (or lipids, red) are evenly distributed in a membrane. At $t_0$, the area outlined in gray is irradiated with laser light to bleach the fluorophores in this region (gray). Over time, the fluorescently labeled molecules will re-distribute evenly due to diffusion. **B:** The diffusion constant of the molecules can be determined from the time dependence of fluorescence recovery.

### 19.5.6 POTENTIAL PITFALLS

Fluorescence experiments require a careful selection of measurement parameters such as excitation and emission wavelength, bandwidth of excitation (and emission), and fluorophore concentration. The emission is recorded at the wavelength with the maximum change in intensity during the process under investigation. This wavelength can be determined from difference spectra that are calculated by subtracting the emission spectrum of the initial state from the spectrum of the final state (Figure 19.55). Maxima in the difference spectrum correspond to wavelengths with maximum increase in fluorescence, minima occur at the wavelength with maximum signal decrease.

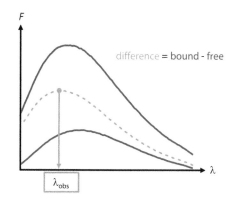

**FIGURE 19.55:**   **Difference spectra and signal maximization.** Before performing an equilibrium fluorescence titration, the signal should be optimized. The maxima and minima of a difference spectrum (orange) between the final state (bound ligand, red) and the initial state (free ligand, blue) show at which wavelength the maximal increase/decrease in fluorescence is observed.

Photobleaching of fluorophores, leading to a decrease in the fluorescence signal with time, is a common problem in fluorescence experiments. An observed fluorescence decrease because of photobleaching can easily be misinterpreted as binding or as a structural change. Wherever possible, fluorescence experiments should preferentially be designed such that an increase in fluorescence can be monitored. If this is not possible, photobleaching can be minimized by using the lowest possible setting for the width of the excitation slit and thus for the excitation bandwidth, and by closing the shutter between individual measurements. Control experiments without addition of binding partners help to distinguish photobleaching from signal changes due to binding.

Fluorescence measurements have to be performed at sufficiently low concentrations to avoid inner filter effects (Section 19.2.8; Figure 19.21). Otherwise, fluorescent molecules further down the excitation pathway within the sample will experience a lower light intensity and show less fluorescence than molecules that are at the beginning of the light path and experience the full excitation intensity $I_0$. If an inner filter effect occurs, the fluorescence intensity does not increase linearly with concentration of the fluorophore, which is a key requirement for many biophysical applications of fluorescence spectroscopy.

Fluorescent impurities in buffers and other reagents will contribute to the overall fluorescence signal and should be avoided by using only the purest chemicals. Reference spectra of the solvent should always be recorded to ensure low fluorescence background. Particles in the solution will scatter the incident light and lead to fluctuations of the readout. Scattering of light also distorts the measured fluorescence spectrum. Rayleigh scattering (Section 21.1.1) occurs at the excitation wavelength. If spectra are measured over a large range of emission wavelengths, the second harmonic of the Rayleigh scattering will be detected at twice the excitation wavelength. For instance, if Trp is excited at 280 nm, the emission spectrum should be measured only above ~290 nm, and not beyond 2·280 nm = 560 nm to avoid the first and second harmonic of the Rayleigh scattering. Raman scattering (Section 21.1.3) of the excitation light leads to excitation of water vibrations and a scattering peak that might distort the spectrum of the molecule of interest (Box 19.12). Correction of fluorescence spectra for the buffer spectrum (that also contains the Raman peak of water) will eliminate this problem. For time-dependent fluorescence measurements the contribution from the Raman peak should be eliminated by using suitable cut-off filters.

---

### BOX 19.12: SCATTERING IN FLUORESCENCE SPECTRA.

Rayleigh scattering (Section 21.1.1) of the excitation light superimposes with the fluorescence emission spectrum at the excitation wavelength and twice the excitation wavelength (second harmonic). The intensity for second and higher harmonics is strongly decreased. Because of this and the limited Stokes shift of common fluorophores, higher than second harmonics are typically not detected in fluorescence spectroscopy. Raman scattering (Section 21.1.3) due to the excitation of water vibrations is a common phenomenon in fluorescence spectroscopy of samples in aqueous solutions (Figure 19.56). The position of the Raman peak depends on the excitation wavelength. The wavenumber for the water vibration is 3600 cm$^{-1}$ (Section 19.4). This energy is removed from the incident light, and the scattered light has a higher wavelength (redshift) than the excitation light. We can calculate the position of the Raman peak in the fluorescence spectrum by converting the

*(Continued)*

excitation wavelength into wavenumbers, subtracting the wavenumber for the water excitation, and re-converting the result to wavelength:

$$\frac{1}{\lambda_{ex}} - 3600 \text{ cm}^{-1} = \frac{1}{\lambda_{Raman}}$$

eq. 19.89

Solving for the wavelength of Raman scattering, $\lambda_{Raman}$, gives

$$\lambda_{Raman} = \frac{\lambda_{ex}}{1 - 3600 \text{ cm}^{-1} \cdot \lambda_{ex}}$$

eq. 19.90

By converting 3600 cm$^{-1}$ into $36 \cdot 10^{-5}$ nm$^{-1}$, we obtain an equation that gives $\lambda_{Raman}$ in units of nm when we enter the excitation wavelength $\lambda_{ex}$ in nm:

$$\lambda_{Raman} = \frac{\lambda_{ex}}{1 - 36 \cdot 10^{-5} \cdot \lambda_{ex}}$$

eq. 19.91

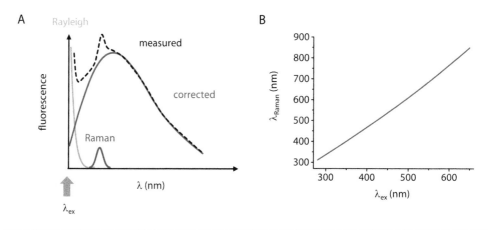

FIGURE 19.56:   **Scattering in fluorescence emission spectra**. A: Rayleigh scattering (orange) is elastic scattering of excitation light. In addition to the first harmonic with $\lambda = \lambda_{ex}$, the second harmonic is observed at $\lambda = 2 \cdot \lambda_{ex}$ when fluorescence spectra are measured over a wide range of wavelengths. The Stokes band of Raman scattering (blue) is found at wavelengths above $\lambda_{ex}$. B: Position of the Raman peak (Stokes band, $\lambda_{Raman}$) as a function of the excitation wavelength $\lambda_{ex}$ (eq. 19.91).

## 19.5.7 FLUORESCENCE QUENCHING

Fluorescence quenching is a general term for all processes that cause a decrease in fluorescence intensity. Quenching can be static or caused by collisions (Figure 19.57). In *collisional quenching*, a fluorophore collides with a quencher molecule during the lifetime of the excited state, and returns to the ground state without emitting a photon. Collisional quenchers provide an alternative process for the depopulation of the excited state, and lead thereby to a decrease in fluorescence lifetime and quantum yield (see Section 19.5.3). In contrast, *static quenching* is caused by formation of a non-fluorescent complex of the quencher and the fluorophore. Intramolecular static quenching occurs in NADH and FAD, where the adenine group stacks on the nicotinamide or flavin moiety (see Section 19.5.4). A static quencher effectively reduces the concentration of the fluorescent species, but does not provide an alternative pathway for relaxation from the excited state to the ground state. Static quenchers therefore do not alter the fluorescence lifetime or the quantum yield.

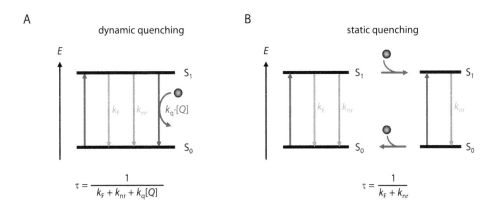

A    dynamic quenching

B    static quenching

$$\tau = \frac{1}{k_F + k_{nr} + k_q[Q]}$$

$$\tau = \frac{1}{k_F + k_{nr}}$$

**FIGURE 19.57:** **Dynamic and static quenching, principle and effect on lifetime.** A: Dynamic quench-ers (yellow) quench fluorescence by collisions with the excited state of the molecule and provide an additional path for relaxation of the excited state. The lifetime $\tau$ is the inverse sum of the rates for all processes that leads to depopulation of the excited state. Collisional quenching leads to a reduction of the fluorescence lifetime. B: A static quencher binds to the excited state of the fluorophore. The fluorophore-quencher complex relaxes to the ground state by non-radiative processes only. A static quencher reduces the concentration of excited molecules that can undergo fluorescence, but does not provide an extra pathway for depopulation of the $S_1$ state. As a consequence, the fluorescence lifetime is unaffected by its presence.

One example of a ubiquitous collisional quencher is molecular oxygen. For accurate measure-ments of fluorescence lifetime and intensity, molecular oxygen dissolved in the samples should be removed. In practice, oxygen levels are often reduced by degassing buffers. The quenching properties of oxygen are commonly ascribed to its paramagnetic character that allows inter-system crossing to the triplet state. The reduction in fluorescence intensity because of collisional quenching by oxy-gen can be exploited for oxygen sensing. Electron-rich and polarizable compounds such as halide ions ($Br^-$, $I^-$) or halogenated compounds and acrylamide also lead to collisional quenching. Trp fluorescence is efficiently quenched by acrylamide.

Collisional quenching can be described quantitatively by the *Stern-Volmer equation*. In the absence of quencher, continuous illumination of the sample leads to a steady-state concentration of fluorophores in the excited state, X*. The differential equation for this steady-state concentra-tion is

$$\frac{d\left[ X^* \right]}{dt} = f\left(t\right) - \left(k_f + k_{nr}\right) \cdot \left[ X^* \right]_0 = 0 \qquad \text{eq. 19.92}$$

where $f(t)$ is the excitation function that quantifies transitions from the ground state to the excited state under constant illumination, and the second term summarizes the depopulation of the excited state with the rate constant $k_f$ for transitions involving fluorescence, and $k_{nr}$ for non-radiative deac-tivation. Note that this differential equation is essentially a rate law and follows the same basic rules (Chapter 7–8, Box 8.2): processes leading to a species (here: a state) enter the balance with a posi-tive sign, processes leading away from this species/state enter with negative signs. Velocities are the products of the respective rate constant and the concentration(s) of the relevant species/state(s). The index "0" in $[X^*]_0$ refers to the absence of quencher.

In the presence of quencher, a third term enters the differential equation to account for the quenching process (eq. 19.93). The velocity of this collisional process is proportional to the quencher concentration $[Q]$ and occurs with a rate constant $k_q$. We therefore obtain

$$\frac{d\left[X^*\right]}{dt} = f(t) - \left(k_f + k_{nr} + k_q[Q]\right) \cdot \left[X^*\right]_0 = 0 \qquad \text{eq. 19.93}$$

By equating eq. 19.92 and eq. 19.93, we obtain

$$\frac{\left[X^*\right]_0}{\left[X^*\right]} = \frac{k_f + k_{nr} + k_q[Q]}{k_f + k_{nr}} = 1 + \frac{k_q}{k_f + k_{nr}}[Q] = 1 + K_{sv}[Q] \qquad \text{eq. 19.94}$$

The steady-state fluorescence emission $F_0$ in the absence or $F$ in the presence of quencher is proportional to the concentration of the excited state of the fluorophore ($[X^*]_0$, $[X^*]$), and we can therefore write

$$\frac{F_0}{F} = 1 + K_{SV}[Q] \qquad \text{eq. 19.95}$$

which is the *Stern-Volmer equation*. $F_0$ and $F$ are the fluorescence intensities in the absence and presence of quencher, and $K_{SV}$ is the *Stern-Volmer constant* with units $M^{-1}$. According to the Stern-Volmer equation, the ratio of the fluorescence in absence and presence of quencher depends linearly on the quencher concentration (Figure 19.58). $K_{SV}$ is the proportionality constant that can be derived from the slope of the Stern-Volmer plot. It is a measure of the solvent-accessibility of a fluorophore. A high value for $K_{SV}$ indicates a strong dependence of the fluorescence on the quencher concentration. In this case, the fluorophore is solvent-exposed and accessible to the quencher, which accordingly can very effectively quench the fluorescence. By contrast, a small $K_{SV}$ indicates low accessibility of the fluorophore.

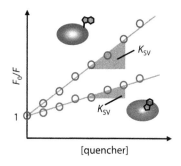

**FIGURE 19.58:** **Stern-Volmer plot.** A plot of the ratio $F_0/F$ (fluorescence in absence of quencher divided by fluorescence at the current quencher concentration) as a function of the quencher concentration is linear (eq. 19.95). From the slope of the plot, the Stern-Volmer constant $K_{SV}$ can be determined. High values for $K_{SV}$, i.e. high sensitivity of the fluorescence on the quencher concentration, are indicative of a solvent-exposed fluorophore (blue: protein with solvent-exposed tryptophan). Low values for $K_{SV}$, i.e. low sensitivity of the fluorescence for the quencher concentration, are indicative of buried fluorophores (red: protein with buried tryptophan).

Fluorescence quenching by small molecule quenchers is therefore a sensitive probe to determine the solvent-accessibility of a fluorophore. For example, surface-exposed tryptophans in proteins will be highly susceptible to quenching, whereas buried tryptophans will not be affected by the addition of quencher (Figure 19.58). In a similar way, quenching can be used to probe the position of a fluorophore in a membrane protein (Figure 19.59). Such an experiment can also provide information on the membrane permeability of the quencher.

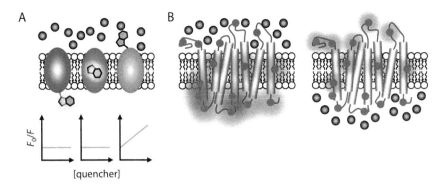

**FIGURE 19.59:**   **Fluorescence quenching and the orientation of membrane proteins.** A: The exposure of tryptophans (black) towards one side of the membrane can be probed by adding quencher (orange). Fluorescence from tryptophans pointing to the same side will be quenched with increasing quencher concentrations, whereas tryptophans in the interior of the protein and/or the membrane or tryptophans pointing towards the distal compartment do not respond to the addition of quencher (bottom). B: By introducing fluorophores (red) at different positions of the membrane protein and testing their reaction to quencher from both sides of the membrane, the topology of a membrane protein can be determined. The red halo indicates emitting fluorophores, red spheres without halo represent fluorophores whose fluorescence is quenched.

## 19.5.8  FLUORESCENCE ANISOTROPY

### 19.5.8.1  Principle of Fluorescence Anisotropy

So far we have only considered fluorescence emission after excitation of the fluorophores by non-polarized light. When fluorophores are illuminated by linearly polarized light, only those with an orientation that aligns the direction of their transition dipole with the polarization direction of the incident light will be efficiently excited. In solution, the fluorophores have a random orientation, and the linearly polarized light will preferentially excite those that have their transition dipole aligned at the time of excitation. We have treated this photoselection process in the context of linear dichroism (see Section 19.3.1; Figure 19.23).

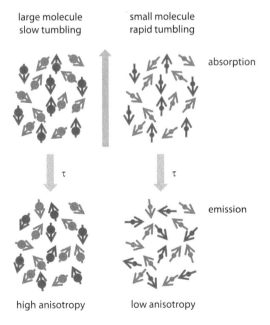

**FIGURE 19.60:**   **Photoselection and randomization of the orientation.** Excitation with linearly polarized light (orange arrow) preferentially excites those fluorophores whose transition dipoles are aligned with the polarization direction of the incident light (top, red). During the lifetime $\tau$ of the excited state, large molecules (left) have rotated very little, and the transition dipoles of the excited fluorophores are still mostly vertically oriented. The polarization direction of the emitted light is therefore the same as that of the incident light. Small molecules (right) rotate much faster, and the orientation of the transition dipoles of the excited fluorophores is randomized during the fluorescence lifetime. Emitted light is therefore not polarized. For simplicity, photoselection is depicted as exclusive excitation of molecules with perfectly aligned transition dipoles.

What consequence does photoselection have for fluorescence emission? If the fluorophores were to emit fluorescence immediately after excitation, the emitted light would also be polarized in the same plane as the excitation light. However, during the nanosecond lifetime of the excited state, mobile fluorophores have enough time to undergo rotational diffusion (Table 16.9), which eventually randomizes the orientation of the excited fluorophores and their transition dipoles, and the fluorophores emit unpolarized light (Figure 19.60). The velocity of rotational diffusion depends on the size of the rotating moiety. If the rotation is too slow for complete randomization of the fluorophore orientation during the excited state lifetime, the emitted light will not be completely depolarized at the time of emission. Instead, it will contain a component polarized parallel to the polarization plane of the excitation light and a component that is polarized perpendicularly with respect to the excitation light. This phenomenon is called *polarization* or *anisotropy*. In the following, we refer to excitation light that is polarized in the vertical direction with respect to the plane of the instrument table. The component of the emitted light with a polarization parallel to the excitation light is $I_v$ (vertical), the component with perpendicular polarization is $I_h$ (horizontal; Figure 19.61). The polarization $p$ is defined as

$$p = \frac{I_v - I_h}{I_v + I_h}$$

eq. 19.96

The fluorescence anisotropy $r$ is defined as

$$r = \frac{I_v - I_h}{I_v + 2I_h}$$

eq. 19.97

Anisotropy and polarization can be interconverted and provide equivalent information. The definition of the anisotropy scales the difference between the two components $I_v$ and $I_h$ of the emitted light by the total intensity ($I_{tot} = I_v + 2I_h$). The horizontally polarized component has to be considered twice because of the two possible orientations perpendicular to the polarization direction of the incident light (Figure 19.61). The anisotropy is therefore independent of the total fluorescence intensity, and additive for mixtures of species with different anisotropies. In contrast, polarization is not additive.

To measure fluorescence anisotropy, a standard steady-state fluorimeter equipped with excitation and emission polarizers is needed (Figure 19.61). The excitation polarizer establishes the vertical polarization direction of the excitation light. The polarizer in the emission path can be aligned in parallel and perpendicular to the excitation polarization to measure $I_v$ and $I_h$, respectively. From these components, the anisotropy can be calculated according to eq. 19.97.

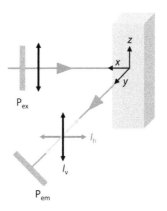

**FIGURE 19.61: Measurement of polarization and anisotropy.** An excitation polarizer ($P_{ex}$) selects excitation light polarized along the *z*-axis (green, black double-headed arrow). The components of the emitted light (orange) polarized parallel ($I_v$, in the *z*-direction) and perpendicular ($I_h$, in the *x*-direction) to the polarization direction of the incident light are measured by placing an emission polarizer ($P_{em}$) in front of the detector. Note that the emitted light has a second perpendicular, horizontally polarized component (light gray double-headed arrow), polarized along the *y*-axis, that cannot be measured in this geometry of the instrument. The two horizontally polarized components are identical.

Monochromators and detectors show different sensitivities for vertically and horizontally polarized light. This instrument non-ideality is expressed by the *G-factor*. In the following, we specify the polarization direction of excitation and emission light by two indices. The first letter of the subscript defines the direction of the excitation polarizer, the second letter specifies the direction of the emission polarizer. The *G*-factor of the instrument is defined as the ratio of the sensitivity for the vertical and parallel components. It is determined by measuring the vertical and horizontal components of the emitted light upon excitation with horizontally polarized light, $I_{hv}$ and $I_{hh}$, according to

$$G = \frac{I_{hv}}{I_{hh}}$$

eq. 19.98

The anisotropy $r$ can then be calculated from the measured intensities as:

$$r = \frac{I_{vv} - G \cdot I_{vh}}{I_{vv} + 2G \cdot I_{vh}}$$

eq. 19.99

### 19.5.8.2 Applications

Anisotropy is a useful spectroscopic probe to monitor processes in which the tumbling time of the entity carrying the fluorophore changes, typically when a small molecule binds to a macromolecule. The anisotropy increases linearly with concentration, therefore an increase in fluorescence anisotropy is frequently used as a probe in equilibrium titration experiments. To maximize the change in anisotropy, the small binding partner (B) is fluorescently labeled, and the binding partner with the high molecular mass (A) is titrated to this small fluorescent ligand (Figure 19.62). The overall tumbling time of the fluorescently labeled partner increases upon complex formation (AB), giving rise to an increase in fluorescence anisotropy. From the corresponding titration curve, the equilibrium dissociation constant of the complex is determined by the same procedure as outlined before (Section 19.5.5.1). When the fluorescence intensity of the fluorescent moiety does not change upon binding, the measured anisotropy $r$ is the population-weighted average of the anisotropy of the free fluorescently labeled molecule, $r_{free}$, and the bound form, $r_{bound}$:

$$r = f_{free} \cdot r_{free} + f_{bound} \cdot r_{bound}$$

eq. 19.100

$f_{free}$ and $f_{bound}$ denote the fractional populations of free and bound forms. The anisotropy can also be expressed in terms of the binding degree, $[AB]/B_t$, as

$$r = r_{free} + \Delta r_{max} \frac{[AB]}{B_t} = r_{free} + \left(r_{bound} - r_{free}\right) \cdot f_{bound}$$

eq. 19.101

Eq. 19.101 is formally identical to eq. 19.69 that describes the fluorescence signal in a fluorescence (intensity) titration as a function of the degree of binding. Hence, anisotropy titrations of a ligand B with a macromolecule A can be described by an equation analogous to eq. 19.70 to extract the $K_d$ value:

$$r = r_{free} + \Delta r_{max} \frac{\frac{A_t + B_t + K_d}{2} - \sqrt{\left(\frac{A_t + B_t + K_d}{2}\right)^2 - A_t B_t}}{B_t}$$

eq. 19.102

Note that eq. 19.100 and consequently also eq. 19.102 do not hold for polarization values because of the non-linear relationship between anisotropy $r$ and polarization $p$.

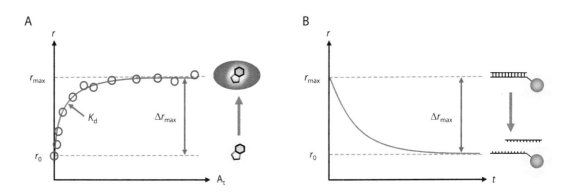

**FIGURE 19.62:   Fluorescence anisotropy to monitor binding and dissociation.** A: A fluorescent ligand, such as a fluorescently labeled nucleotide (yellow) is titrated with a protein (red). The small ligand rotates rapidly, and has a low fluorescence anisotropy ($r_0$). Upon complex formation with the protein, the mobility of the fluorescent ligand decreases, leading to an increase in anisotropy. At saturation, the anisotropy of the complex, $r_{max}$, is reached. The dissociation constant $K_d$ of the complex can be determined by analyzing the data points (blue) with eq. 19.102. B: The dissociation of a fluorescently labeled DNA duplex (black/green) can also be monitored *via* fluorescence anisotropy. The fluorophore, attached to one of the strands of the duplex, shows a high fluorescence anisotropy due to the overall slow tumbling of the duplex. Addition of a helicase at $t = 0$ will lead to unwinding of the duplex, and dissociation of the fluorescently labeled strand. The single strand gives a lower fluorescence anisotropy signal because of its more rapid tumbling. The rate constant for duplex unwinding can be determined by describing the observed anisotropy decrease with time by a single exponential.

The steady-state anisotropy of fluorophores generally reports on their mobility on the time scale of the lifetime of the fluorophore (see Section 19.5.9). Membrane-inserted fluorophores such as 1,6-diphenyl-1,3,5-hexatriene (DPH) have been used as probes to study membrane dynamics.

### 19.5.8.3  Potential Pitfalls of Polarization/Anisotropy Measurements
Scattered light is highly polarized, and will lead to overestimation of the actual fluorescence polarization or anisotropy. Scattered light should therefore be carefully avoided in anisotropy experiments, or rejected by filters.

Anisotropy and polarization are widely used probes to measure binding. Anisotropy of a mixture of two (or more) species can be calculated as the population-weighted average of the individual anisotropy values because anisotropy values are weighted by the total fluorescence intensity and therefore are additive (eq. 19.100). In contrast, polarization values are not additive, and the calculation of the polarization of mixtures cannot be calculated from the polarization of the components. The deviations from additivity are small, but it is important to be aware of this difference when polarization is used as a probe for binding in equilibrium titrations. An increase in anisotropy that is not caused by scattering is a reliable reporter for binding. However, a lack of change in anisotropy during a titration cannot be taken as a proof that the two molecules do not interact! If the fluorophore is flexibly attached to the one of the partners, the orientation of its transition dipole is largely uncoupled from the orientation of the biomolecule itself (see Section 19.5.9.2). In this case, the fluorophore has a low anisotropy because of its rapid local motion relative to the biomolecule both in the free and bound states. A flexibly attached fluorophore therefore does not report on tumbling of the biomolecule, and the anisotropy remains low even if binding occurs. For anisotropy experiments, a rigid attachment of the fluorophore to the molecule of interest is therefore desired. Apart from the local flexibility, the fluorescence lifetime of the fluorophore (see Section 19.5.9.1) also limits its suitability to report on binding. The fluorescence lifetime needs to be on the order of the tumbling times of the free and bound ligand, otherwise the anisotropy is not sensitive to changes in tumbling and molecular mass (see Section 19.5.9.3).

Analysis of anisotropy binding curves for $K_d$ values becomes more complicated when the quantum yield, and thus the fluorescence intensity, changes upon binding. A change in quantum yield means that free and bound forms contribute differently to the measured fluorescence intensities. In this case, the measured overall anisotropy is not a population-weighted but an intensity-weighted average. The change in quantum yield has to be taken into account in data analysis (Box 19.13). If the fluorescence intensity changes upon binding, it is advisable to directly use the intensity rather than fluorescence anisotropy as a probe for binding.

---

**BOX 19.13: ANISOTROPY TITRATIONS WITH CHANGES IN QUANTUM YIELD.**

Eq. 19.100 for the calculation of the total anisotropy for a mixture of free and bound forms with $r_{free}$ and $r_{bound}$ only holds if both species contribute identically to the fluorescence intensity. If the quantum yield and thus the fluorescence changes upon binding, however, the species with the higher quantum yield contributes more to the overall intensity than the species with the lower quantum yield. For an $R$-fold change in quantum yield upon binding

$$R = \frac{\phi_{bound}}{\phi_{free}}$$

eq. 19.103

with the quantum yields $\phi_{free}$ and $\phi_{bound}$ for the free and bound forms of the fluorescent binding partner, the contribution $a_{free}$ of the free form to the fluorescence intensity is

$$a_{free} = \frac{f_{free}}{f_{free} + R \cdot f_{bound}}$$

eq. 19.104

$f_{free}$ and $f_{bound}$ refer to the fractional populations of the free and bound forms. The contribution of the bound form is

$$a_{bound} = \frac{R \cdot f_{bound}}{f_{free} + R \cdot f_{bound}}$$

eq. 19.105

These relative contributions $a_{free}$ and $a_{bound}$ are the weighting factors for the calculation of the overall anisotropy $r$ from the individual anisotropy values $r_{free}$ and $r_{bound}$:

$$r = \frac{f_{free}}{f_{free} + R \cdot f_{bound}} \cdot r_{free} + \frac{R \cdot f_{bound}}{f_{free} + R \cdot f_{bound}} \cdot r_{bound}$$

eq. 19.106

We can now substitute $f_{free}$ by $1 - f_{bound}$. For a binary equilibrium of A and B with the AB complex, $f_{bound}$ can be expressed in terms of $A_t$, $B_t$, and $K_d$ by the solution of the quadratic equation (eq. 19.66 and eq. 19.67). We thereby obtain an equation for the anisotropy $r$ as a function of $A_t$, $B_t$, $K_d$, $r_{free}$, $r_{bound}$, and $R$ that we can use to describe an experimental binding curve. The increase in quantum yield $R$ is typically determined from reference spectra of the free and bound species, and kept constant during the fit.

---

## 19.5.9 TIME-RESOLVED FLUORESCENCE

In the previous sections, we have discussed fluorescence emission during steady-state illumination of the sample. What happens if we illuminate a fluorophore only briefly, with a short light pulse, and then monitor the emitted fluorescence? During illumination, the excited state of the fluorophore, X*, becomes populated. The fluorophores in the excited state will then relax to the ground state. The fluorescence $F(t)$ at any given time is proportional to the concentration of molecules in the excited state $[X^*]$

$$F(t) \propto \left[ X^* \right]$$

eq. 19.107

The velocity of the relaxation depends on the rate constant $k$ for processes leading away from the excited state

$$\frac{d\left[X^*\right]}{dt} = k \cdot \left[X^*\right]$$

eq. 19.108

The rate constant $k$ is the sum of the rate constants for radiative and non-radiative deactivation processes of the excited state. The inverse of $k$ is the fluorescence lifetime $\tau$, the mean time a fluorophore spends in the excited state before returning to the ground state (see Section 19.5.3). The relaxation, accompanied by the emission of fluorescence, will depopulate the excited state and the fluorescence will decay to zero. By separating the variables in eq. 19.108 and integrating over the time $t$, we obtain an expression for the concentration of the fluorophore in the excited state as a function of time:

$$\left[X^*\right](t) = \left[X^*\right]_0 \cdot \exp(-kt)$$

eq. 19.109

Eq. 19.109 can be combined with eq. 19.107 to arrive at the exponential function that describes the *fluorescence decay F(t)*:

$$F(t) = F_0 \cdot \exp(-k \cdot t) = F_0 \cdot \exp\left(-\frac{t}{\tau}\right)$$

eq. 19.110

$F_0$ is the fluorescence intensity at time $t = 0$, and $\tau$ is the *fluorescence lifetime*, the time at which the fluorescence intensity has decayed to $F_0/e$. If different fluorescent species are present, the fluorescence decay is described by a sum of exponentials with individual lifetimes $\tau_i$ and corresponding amplitudes $\alpha_i$:

$$F(t) = \sum_i \alpha_i\, e^{-t/\tau_i}$$

eq. 19.111

The amplitude values $\alpha_i$ correspond to the fraction of the species $i$ in the mixture (Figure 19.63). Multiple fluorescence lifetimes are observed for mixtures of fluorophores with different fluorescence lifetimes. One type of fluorophore can also give rise to more than one lifetime if it can exist in different chemical environments (free or bound, solvent-exposed or buried). In this case, fluorescence lifetime measurements provide information on the relative abundance of the different species, which is proportional to the amplitude $\alpha_i$ of the lifetime component $\tau_i$. In fluorescence intensity measurements under steady-state illumination, this information is averaged out.

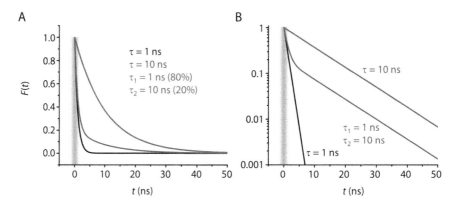

**FIGURE 19.63:    Fluorescence decay.** A: When fluorophores are excited by a short light pulse at $t = 0$ (orange), their fluorescence decays exponentially. After the lifetime $\tau$, the fluorescence has decreased to $1/e$ of its initial value. Fluorophore A has a lifetime of 1 ns (black), fluorophore B of 10 ns (red). A mixture of 80% A and 20% B shows a biexponential decay (blue). Biexponential decays are also obtained for the same fluorophore in different environments (e.g. free and bound). B: Same fluorescence decays as in A plotted with a logarithmic y-axis.

### 19.5.9.1 Measurement of Fluorescence Lifetimes

Fluorescence lifetimes are measured by exciting fluorophores with a very short laser pulse, typically a few to a few hundred ps, and then following the decay as a function of time. For technical reasons, the fluorescence cannot be measured directly on the nanosecond time scale. Therefore, a technique called *time-correlated single photon counting* is used (Figure 19.64). Time-correlated single photon counting measures the arrival time of the first photon emitted from the sample after the excitation pulse. The arrival times of photons measured in a large number of excitation-emission cycles are histogrammed. From the envelope of these arrival time histograms, the fluorescence decay is obtained.

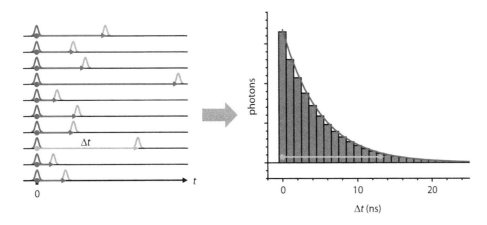

**FIGURE 19.64:**    **Measurement of fluorescence lifetimes by time-correlated single photon counting:** From histograms to decay curves. Fluorescence is excited by a short laser pulse (blue) at $t = 0$, and the first photon emitted from the sample is detected (green). The time between excitation and emission (red arrow) is measured. The measurement is repeated, and the determined time differences between excitation pulse and emission events are histogrammed (right). The envelope of such a histogram is the fluorescence decay (red). The orange arrow indicates the time difference $\Delta t$ for one single excitation/emission cycle (left), and its position in the histogram (right).

Strictly speaking, time-correlated single photon counting yields fluorescence decays and lifetimes directly only when the excitation pulse is much shorter than the fluorescent lifetime. However, this is not always the case. Due to the finite length of the excitation pulse, a series of fluorescence decays is initiated over the duration of the excitation pulse (Figure 19.65). The amplitude of each individual decay is proportional to the intensity of the excitation light at that time point. The measured fluorescence decay is the sum of all of these individual decays, with different amplitudes and starting times. Mathematically, the measured decay is a *convolution* (see Section 29.15) of the *instrument response* (or lamp function) and the true fluorescence decay (Box 19.14). The instrument response is the excitation pulse filtered by the detection electronics. It is measured in a reference measurement with a scattering solution as a sample. The true fluorescence decay can then be retrieved from the measured fluorescence decays obtained by time-correlated single photon counting by deconvolution with the instrument response function.

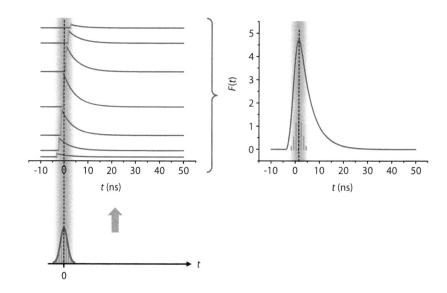

**FIGURE 19.65: Instrument response and measured decay.** The excitation pulse used in time-resolved fluorescence measurements is not an ideal, infinitely short pulse at $t = 0$ but can often be described by a Gaussian intensity distribution, centered at $t = 0$ (bottom left, blue). The exact shape of this lamp function depends on the excitation source used. The excitation pulse consists of a number of infinitely short pulses (gray bars). Each of these individual pulses elicits an impulse response in the sample, i.e. a fluorescence decay with the amplitude of the excitation light intensity at this point in time. Therefore, a finite pulse generates a series of instrument responses with different amplitudes and different starting times (top left). The sum of these individual responses is the measured fluorescence intensity (right). Mathematically, the measured fluorescence intensity is the convolution of the lamp function and the true fluorescence decay (Box 19.14).

---

### BOX 19.14: CONVOLUTION AND DECONVOLUTION.

The measured fluorescence intensity in lifetime experiments is the convolution of the lamp function and the true fluorescence decay. The lamp function can be described as a series of $\delta$-functions. A $\delta$-function is defined as a function that has a certain value for one $x$-value, but has a value of zero for all other $x$-values. To calculate the impulse response $I_k(t)$ to each $\delta$-function, we have to shift the true fluorescence decay $I(t)$ along the time axis by $t_k$. The shifted decay $I(t-t_k)$ is then multiplied by the height $L(t_k)$ of this $\delta$-function and by the time difference $\Delta t$ to the next $\delta$-function:

$$I_k(t) = L(t_k) \cdot I(t - t_k) \Delta t \qquad \text{eq. 19.112}$$

The measured fluorescence intensity $F(t_k)$ is the sum of all these individual impulse responses $I_k(t)$:

$$F(t_k) = \sum_{t=0}^{t=t_k} L(t_k) \cdot I(t - t_k) \Delta t \qquad \text{eq. 19.113}$$

For infinitely small intervals, i.e. small $\Delta t$ values, we can replace the sum by an integral and integrate from $t = 0$ to $t'$ (corresponding to $t_k$):

$$F(t) = \int_0^{t'} L(t') \cdot I(t - t') \mathrm{d}t' \qquad \text{eq. 19.114}$$

By replacing $t'$ with $t-\mu$, we obtain

$$F(t) = \int_0^{t} L(t - \mu) \cdot I(\mu) \mathrm{d}\mu \qquad \text{eq. 19.115}$$

Eq. 19.115 is the mathematical definition of the convolution of two functions $L(t-\mu)$ and $I(t)$, also written as

*(Continued)*

$$F(t) = L(t) \otimes I(t) \qquad \text{eq. 19.116}$$

Convoluting two functions with each other means shifting the mirror image of one (here L(t)) with respect to the $y$-axis relative to the other (here $I(t)$) by µ, and calculating the product. We will see later that this definition is different from the definition of the auto- and cross-correlation (Section 21.1.2, eq. 21.17; Sections 23.1.6.1 and 23.1.6.3, eq. 23.23 and eq. 23.32), where the first function is not inverted (Figure 19.66). We will revisit the convolution in X-ray crystallography, where a macromolecular crystal can be described as the convolution of a lattice function and the function for the entity that is found at each lattice point (Section 22.5).

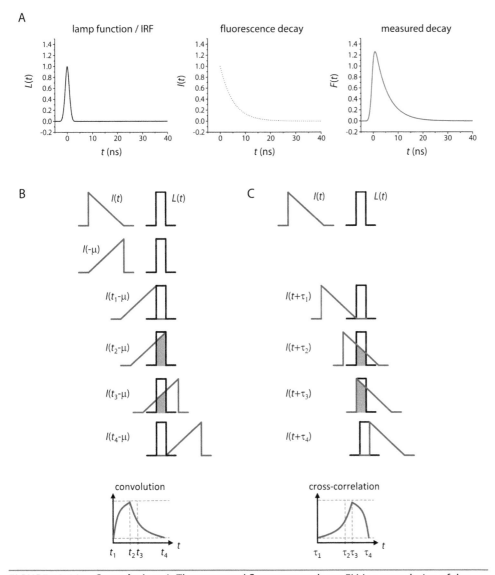

**FIGURE 19.66:    Convolution**. A: The measured fluorescence decay $F(t)$ is a convolution of the lamp function $L(t)$, also called instrument response function, and the true fluorescence decay $I(t)$. B: The convolution of two functions can be constructed by inverting one function, shifting it relative to the second function, and calculating the area under the two curves. This is illustrated for the convolution of $I(t) \otimes L(t)$: the fluorescence decay ($I(t)$, blue) is inverted to $I(-µ)$ and shifted relative to the lamp function ($L(t)$, black). The area under both curves is highlighted in orange for different time points $t_1-t_4$. The convolution $I(t) \otimes L(t)$ (red) is identical to the convolution $L(t) \otimes I(t)$. C: In comparison, auto- and cross-correlation are calculated by shifting one function relative to itself (auto-correlation) or to the second function (cross-correlation), and calculating the area under the curves. Here, the cross-correlation of $L(t)$ and $I(t)$ is illustrated (red). Note that $I(t)$ is not inverted to calculate the cross-correlation. Auto- and cross-correlation are treated in Section 23.1.6.

### 19.5.9.2 Fluorescence Anisotropy Decays and Rotational Correlation Times

Instead of measuring the decay of the (isotropic) fluorescence intensity, fluorophores can also be excited by a pulse of linearly polarized light. From the time-dependent fluorescence emission with parallel and perpendicular polarization with respect to the polarization of the excitation light, the decay $r(t)$ of the fluorescence anisotropy can be calculated from the general definition of anisotropy (Section 19.5.8) as

$$r(t) = \frac{I_v(t) - I_h(t)}{I_v(t) + 2I_h(t)} \qquad \text{eq. 19.117}$$

Note that eq. 19.117 refers to excitation with light polarized in the vertical direction with respect to the plane of the instrument table, and $I_v(t)$ is the parallel component, $I_h(t)$ the perpendicular component. Anisotropy decays can then be described as sums of exponential functions, with amplitude factors $\beta_i$ and time constants that are now called *rotational correlation time* $\tau_{c,i}$.

$$r(t) = \sum_i \beta_i\, e^{-t/\tau_{c,i}} \qquad \text{eq. 19.118}$$

The fluorescence anisotropy decays of fluorescently labeled (bio-)macromolecules can often be described by a sum of two exponentials, with a high rotational correlation time corresponding to the overall tumbling of the fluorophore with the macromolecule, and a small correlation time that reflects the *segmental mobility* of the fluorophore relative to the macromolecule (Figure 19.67). The segmental mobility can be approximated by the "wobbling in a cone" model, in which the linker connecting the fluorophore to the macromolecule can access the inside of a cone. From the amplitude of the two phases in the anisotropy decay, the order parameter S can be calculated according to eq. 19.119:

$$S = \sqrt{\frac{\beta_2}{\beta_1 + \beta_2}} \qquad \text{eq. 19.119}$$

S is related to the half-cone angle $\theta$ within which the fluorophore can move by eq. 19.120:

$$\theta = \arccos\left(\frac{1}{2}\left(\sqrt{1+8S} - 1\right)\right) \qquad \text{eq. 19.120}$$

The larger the half-cone angle, the larger is the conformational space accessible to the fluorophore. Fluorescence anisotropy decays are therefore a useful tool to probe the rigidity of a fluorophore-macromolecule linkage.

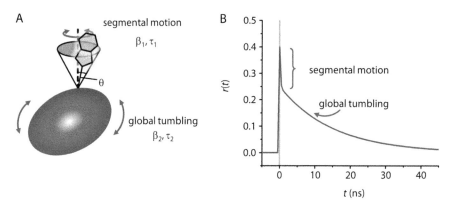

**FIGURE 19.67:** **"Wobbling in a cone" model: segmental flexibility gives rise to two rotational correlation times.** A: A fluorophore (yellow) attached to a macromolecule (orange) by a linker can move within a cone (black) because of the segmental flexibility relative to the macromolecule. B: The anisotropy decay (blue) of this fluorophore is described by two rotational correlation times: a small rotational correlation time $\tau_1$ accounts for the rapid segmental motion of the fluorophore relative to the macromolecule. A larger rotational correlation time $\tau_2$ represents the global tumbling of the fluorophore with the macromolecule. From the amplitudes $\beta_1$ and $\beta_2$ of the two components of the anisotropy decay, the half-cone angle $\theta$ can be calculated (eq. 19.120).

### 19.5.9.3 Rotational Correlation Time and Molecular Size

The rotational correlation time depends on the size of the fluorescent molecule and the viscosity of the medium. Simply speaking, a small molecule will rotate rapidly, and show a small rotational correlation time, i.e. the anisotropy decays rapidly. For large molecules, tumbling is slow, the rotational correlation time is large, and the anisotropy will decay slowly. High viscosities lead to slow tumbling, low viscosities allow rapid tumbling.

Similar to translational motion (see Sections 4.1, 21.1.2), rotational motion can be described by a rotational diffusion coefficient $D_{rot}$. The diffusion coefficient for rotational motion depends on the rotational friction coefficient $f_{rot}$ of the molecule according to eq. 19.121:

$$D_{rot} = \frac{k_B T}{f_{rot}} = \frac{k_B T}{6 V_h \eta}$$

eq. 19.121

with the Boltzmann constant $k_B$, the temperature $T$, the hydrated molecular volume $V_h$, and the viscosity $\eta$. The rotational correlation time $\tau_c$ is defined as

$$\tau_c = \frac{1}{6 D_{rot}} = \frac{V_h \eta}{k_B T}$$

eq. 19.122

$\tau_c$ thus depends linearly on the ratio $\eta/T$. Measurements of $\tau_c$ as a function of viscosity at constant temperature will give a linear plot with the hydrated molecular volume $V_h$ as the slope. The specific hydrated volume for proteins is typically $1 \text{ cm}^3 \text{ g}^{-1}$. To convert the specific hydrated volume into the hydrated volume for a single molecule we need to multiply by the molecular mass and divide by the Avogadro constant $N_A$. From eq. 19.122, we can then estimate a rotational correlation time of 1 ns per 2800 Da molecular mass at 25°C. This rule of thumb only holds for spherical molecules.

The steady-state anisotropy (under constant illumination) can be calculated from the fluorescence decay $I(t)$ and the anisotropy decay $r(t)$ according to

$$r = \frac{\int_0^\infty I(t) r(t) \, dt}{\int_0^\infty I(t) \, dt}$$

eq. 19.123

For a spherical molecule, the anisotropy decay will follow a single exponential:

$$r(t) = r_0 \cdot e^{-t/\tau_c}$$

eq. 19.124

Assuming that the intensity decay also follows a single exponential, we can express $r(t)$ and $I(t)$ as exponential functions and evaluate the integral, and obtain

$$r = \frac{r_0}{1 + \dfrac{\tau}{\tau_c}}$$

eq. 19.125

For static molecules ($\tau_c \to \infty$), the steady-state anisotropy $r$ thus equals $r_0$, the limiting anisotropy. According to eq. 19.125, the anisotropy $r$ depends on the ratio of the fluorescence lifetime $\tau$ and the rotational correlation time $\tau_c$. For $\tau \ll \tau_c$, the anisotropy is $r \approx r_0$, and for $\tau \gg \tau_c$, the anisotropy $r$ will be very low and almost independent of the tumbling time $\tau_c$. A fluorophore thus only reports on changes of the tumbling time when its lifetime is on the order of the rotational correlation time of the species studied (see Section 19.5.8.3).

Eq. 19.125 can be rearranged to

$$\frac{1}{r} = \frac{1}{r_0} + \frac{\tau}{r_0 \tau_c}$$

eq. 19.126

This equation is called the *Perrin equation*. Using the relationship in eq. 19.122, we can also express the Perrin equation as

$$\frac{1}{r} = \frac{1}{r_0} + \frac{\tau k T}{r_0 V_h \eta}$$

eq. 19.127

According to this equation, a plot of $1/r$ as a function of $T/\eta$ (or $\tau$) is linear (Figure 19.68). From the intercept, $r_0$ can be determined. The slope of the plot yields $V_h$, the hydrated volume of the molecule of interest. For fluorescently labeled biomolecules, segmental mobility of the chromophore often causes an increase of the observed $1/r$, i.e. a decrease in the anisotropy $r$. The extrapolated $r_0$ is then also smaller than the true $r_0$. In fluorescence anisotropy decays, this segmental motion leads to a second, smaller rotational correlation time in anisotropy (Figure 19.67).

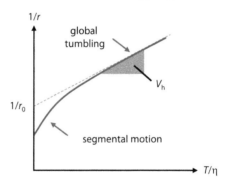

**FIGURE 19.68:   Perrin plot for macromolecules and chromophores.** A plot of $1/r$ as a function of $T/\eta$ is linear (eq. 19.126). From the slope, the hydrated volume $V_h$ of the molecule can be determined (eq. 19.127). If a fluorophore is flexibly attached to a macromolecule, its segmental motion will cause a deviation from linearity at low temperatures/high viscosities, and the limiting anisotropy extrapolated from the linear region of the plot (gray broken line) is lower than the true value.

### 19.5.9.4 Applications

Fluorescence lifetimes report on the local environment of the fluorophore. 2-aminopurine is a fluorescent nucleotide analog whose lifetimes are a sensitive probe for nucleic acid structure and dynamics. In solution, 2-AP fluorescence decays with a single lifetime of about 10 ns. 2-AP fluorescence is quenched within a DNA duplex due to hydrogen bonding and because of stacking with neighboring bases. Quenching is alleviated when stacking or base pairing are perturbed. When 2-AP is incorporated into dsDNA, its fluorescence decay is complex, and multiple lifetimes are observed. Although these lifetimes are difficult to assign to a particular structural change, changes in observed lifetimes and amplitudes are a sensitive qualitative probe for changes in the local environment. Fluorescence lifetimes are also measured in time-resolved FRET experiments to resolve multiple species with different donor-acceptor distances (see Section 19.5.10).

Fluorescence lifetimes are also used for fluorescence sensing. A fluorescence lifetime sensor provides a concentration-independent read-out of changes in steady-state intensities and fluorescence quenching. An example is the fluorescein-coupled chelator Calcium Green, whose fluorescence lifetime increases upon $Ca^{2+}$ binding. Calcium Green is used in live cell fluorescence lifetime imaging (see Section 23.1.5).

Measurement of rotational correlation times can provide unambiguous evidence for binding, even in cases where no increase in steady-state anisotropy is observed. A large segmental flexibility of the fluorophore leads to a low steady-state fluorescence, and may preclude changes in steady-state anisotropy upon binding to a higher molecular mass partner. Anisotropy decays, on the other hand, directly provide rotational correlation times for segmental flexibility and for global tumbling and the associated amplitudes. An increase in the rotational correlation time associated with global tumbling in the presence of a potential binding partner is clear evidence for binding. Rotational correlation times are also measured to quantify the local flexibility of fluorophores attached to biomolecules (eq. 19.119, eq. 19.120), for example as controls in Förster resonance energy transfer experiments (Section 19.5.10).

### 19.5.10 Förster Resonance Energy Transfer

A single fluorophore absorbs excitation light and then emits fluorescence at its characteristic wavelength and with a characteristic fluorescence lifetime. If a second fluorophore is present in the vicinity, the transition dipoles are not independent of each other, but interact in a distance-dependent manner. Due to this dipolar coupling, the first fluorophore can transfer its excitation energy to the second fluorophore. As a result, the first fluorophore will return to the ground state in a radiation-less process, and the second fluorophore undergoes a transition to the excited state. Upon relaxation to the ground state, the second fluorophore will emit fluorescent light. Because the energy is donated by the first fluorophore it is called the *donor*. The second fluorophore taking up the energy is called the *acceptor*. The whole process of radiation-less energy transfer from a donor to an acceptor fluorophore is called *Förster resonance energy transfer* (FRET), after Theodor Förster who described the process quantitatively in 1948.

#### 19.5.10.1 Principle of FRET

FRET is the result of a dipolar interaction and depends on the inverse 6th power of the distance (see van der Waals interactions in Section 15.3.4 and the nuclear Overhauser effect in NMR spectroscopy, Section 20.1.3). Because of its strong distance dependence, FRET can be used as a molecular ruler on the relevant length scales for biomolecules. The rate constant $k_t$ for FRET depends on the inter-fluorophore distance $r$:

$$k_t = \frac{1}{\tau_D} \cdot \left( \frac{R_0}{r} \right)^6$$

eq. 19.128

In eq. 19.128, $R_0$ is the *Förster distance* or *characteristic transfer distance*. $R_0$ depends on the spectral properties of the two fluorophores between which FRET occurs (see eq. 19.132).

The efficiency of the energy transfer $E$ from the donor to the acceptor is determined by the kinetic competition (Section 8.2; Figure 19.69) between energy transfer ($k_t$) and all other processes leading to a depopulation of the excited state, including donor fluorescence ($k_{D,f}$) or non-radiative donor relaxation ($k_{D,nr}$):

$$E = \frac{k_t}{k_{D,f} + k_{D,nr} + k_t}$$

eq. 19.129

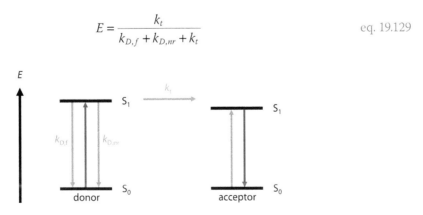

**FIGURE 19.69: Jablonski diagram for FRET.** The donor fluorophore is excited (blue arrow). From the excited state $S_1$, it can either relax to the ground state by radiative ($k_{D,f}$, green) or non-radiative ($k_{D,nr}$, gray) transitions, or transfer its energy to a suitable acceptor. Due to this non-radiative energy transfer ($k_t$, orange), the donor relaxes back to the ground state $S_0$ without fluorescence emission, and the acceptor transitions to the excited state. Once the acceptor relaxes to the ground state, its fluorescence can be observed (red arrow).

We can express the rate constants for relaxation of the donor *via* radiative and non-radiative processes as the inverse of the donor fluorescence lifetime $\tau_D$ as

$$\frac{1}{\tau_D} = k_{D,f} + k_{D,nr}$$

eq. 19.130

By combining eq. 19.128 through eq. 19.130, we obtain the dimensionless FRET efficiency as

$$E = \frac{R_0^6}{R_0^6 + r^6}$$

eq. 19.131

The Förster distance is the distance at which the energy transfer efficiency is 0.5 or 50%. The efficiency of the energy transfer from the donor to the acceptor fluorophore depends on the inverse 6th power of the distance separating the dyes. Due to this prominent distance dependence, the FRET efficiency can be used as a molecular ruler.

The Förster distance $R_0$ is defined by the spectral properties of donor and acceptor fluorophores. It can be calculated according to the relationship

$$R_0{}^6 = 8.785 \cdot 10^{-11} \frac{\kappa^2 \phi_D J}{n^4}$$

eq. 19.132

In eq. 19.132, $\kappa^2$ is the orientation factor, $\phi_D$ is the quantum yield of the donor, $J$ is the overlap integral and $n$ is the refractive index of the medium. Both $J$ and $\phi_D$ can be determined experimentally. The quantum yield $\phi_D$ of the donor can be determined in comparison to reference fluorophores (see Box 19.8). The *overlap integral J* is a measure of the spectral overlap of the fluorescence emission of the donor and the absorption of the acceptor. It can be determined from the fluorescence emission spectrum of the donor, $F_D(\lambda_i)$ normalized to a total intensity of 1, and the absorption spectrum of the acceptor, normalized to the molar extinction coefficient $\varepsilon_A(\lambda_i)$, according to eq. 19.133:

$$J = \int_0^\infty F_D\left(\lambda\right) \cdot \varepsilon_A\left(\lambda\right) \cdot \lambda^4 d\lambda \approx \sum_i F_D\left(\lambda_i\right) \cdot \varepsilon_A\left(\lambda_i\right) \cdot \lambda_i^4$$

eq. 19.133

Typically, the two spectra are measured with a step size of 1 nm, and the overlap integral is then approximated by the sum in eq. 19.133 (Figure 19.70).

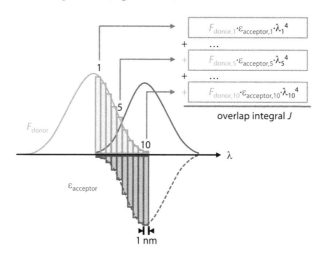

**FIGURE 19.70:** **Calculation of the overlap integral.** The overlap integral *J* can be calculated from the fluorescence emission spectrum of the donor, normalized to an area under the emission peak of 1 (green) and the absorption spectrum of the acceptor in terms of the extinction coefficient ε (red, dotted red line: inverted spectrum for better visibility). The fluorescence and absorption values, measured in 1 nm steps (green and red bars) are multiplied for each wavelength λ in the overlapping region, and each value is weighted by λ⁴, shown exemplarily for the values 1, 5, and 10. The sum of all values is the overlap integral *J* (eq. 19.133).

The *orientation factor* $\kappa^2$ is a measure of the relative orientation of the transition dipoles of donor and acceptor (Figure 19.71). It depends on the angles $\theta_D$ and $\theta_A$ of the donor and acceptor transition dipoles with the vector joining them, and the angle $\theta_{DA}$ between donor and acceptor transition dipoles. The angle $\theta_{DA}$ is the dihedral angle of the two planes defined by the joining vector and the donor and acceptor transition dipole, respectively.

$$\kappa^2 = \left(\cos\theta_{DA} - 3\cos\theta_D \cdot \cos\theta_A\right)^2 \qquad \text{eq. 19.134}$$

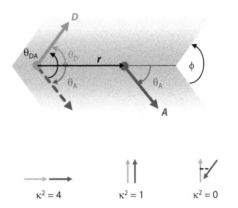

**FIGURE 19.71:** **The orientation factor $\kappa^2$.** The orientation factor $\kappa^2$ describes the relative orientation of the transition dipoles of donor and acceptor fluorophores (eq. 19.134). It depends on the angle $\theta_D$ of the donor transition dipole (green) with the vector $r$ (black) joining the transition dipoles of donor (green) and acceptor (red), the angle $\theta_A$ of the acceptor transition dipole (red) with the vector $r$, and the angle $\theta_{DA}$ between donor and acceptor transition dipoles. This angle $\theta_{DA}$ is the dihedral angle $\phi$ of the two planes defined by the vectors $r$ and the donor and acceptor transition dipole, respectively.

$\kappa^2$ can cover a range of values from zero to four. Clearly, its value has to be known if changes in FRET efficiency are to be interpreted in terms of distance changes. Unfortunately, transition dipole directions are not straightforward to determine, and often unknown. $\kappa^2$ adopts a value of ⅔ when both donor and acceptor fluorophore can sample all possible relative orientations on the time scale of the experiments. To ensure that this condition is met, fluorophores for FRET are ideally attached to the molecule of interest by flexible linkers, such that they can almost freely rotate relative to the molecule of interest. A flexible attachment of fluorophores also minimizes interactions with the labeled biomolecule that may lead to fluorescence changes, and prevents interference with its function. The segmental mobility of fluorophores attached to a macromolecule can be estimated from fluorescence anisotropy decay parameters (see Section 19.5.9).

### 19.5.10.2 Experimental Determination of FRET Efficiencies

The *FRET efficiency E* can be determined experimentally from the donor and acceptor fluorescence intensities $F_D$ and $F_A$ that are measured after excitation of the donor fluorophore (eq. 19.135):

$$E = \frac{F_A}{F_D + F_A} \qquad \text{eq. 19.135}$$

The donor and acceptor contributions can be measured at fixed wavelengths for donor and acceptor emission. More accurately, they are determined from steady-state emission spectra $F_{DA}(\lambda)$ of the molecule that carries both fluorophores (Figure 19.72). Such a spectrum is a linear combination, or weighted sum, of the emission spectra of the donor $F_D(\lambda)$ and of the acceptor $F_A(\lambda)$. The reference spectra are obtained by measuring the fluorescence emission spectrum of the molecule of interest that is labeled with only the donor or only the acceptor.

$$F_{DA}(\lambda) = f_D \cdot F_D(\lambda) + f_A \cdot F_A(\lambda)$$

eq. 19.136

From the weighting coefficients $f_D$ and $f_A$, the FRET efficiency can be calculated (eq. 19.137):

$$E = \frac{f_A}{f_D + f_A}$$

eq. 19.137

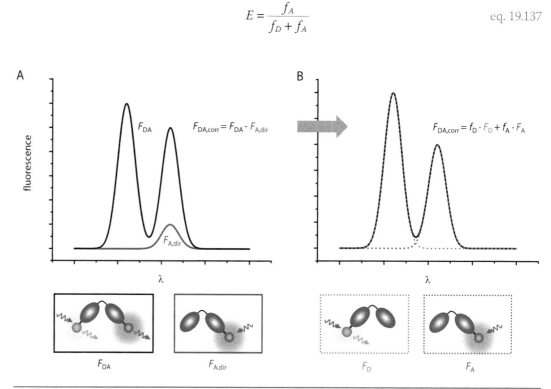

**FIGURE 19.72: Determination of the FRET efficiency from donor and acceptor intensities.** A: The fluorescence emission spectrum of the donor/acceptor-labeled molecule upon donor excitation is recorded ($F_{DA}$, black) and corrected for the contributions from the direct excitation of the acceptor at the donor excitation wavelength ($F_{A,dir}$, red). B: The corrected spectrum $F_{DA,corr}$, calculated by subtracting the measured spectra in A, is described as a weighted sum (linear combination) of the donor fluorescence ($F_D$, green) and the acceptor fluorescence ($F_A$, red). The weighting coefficients $f_D$ and $f_A$ are used to calculate the FRET efficiency $E_{FRET}$ using eq. 19.137.

Methods that calculate the FRET efficiency from measurements of donor and acceptor fluorescence are called *ratiometric methods*. Measuring both donor and acceptor fluorescence provides the inherent control of anti-correlated changes in donor and acceptor fluorescence when FRET occurs. Any change in donor or acceptor fluorescence without the corresponding opposite change in fluorescence of the second dye can be attributed to other events than FRET, such as donor or acceptor quenching. Ratiometric approaches are also not affected by errors in concentration of the donor/acceptor-labeled molecule, or by instabilities of the excitation source. Alternatively, the FRET efficiency can be determined by only measuring the effect of the acceptor on the donor fluorescence (donor quenching), or the effect of the presence of the donor on the acceptor fluorescence (acceptor sensitization). To calculate the FRET efficiency from donor quenching, the donor fluorescence in the absence of acceptor, $F_D$, is determined in a fluorescence measurement of the molecule of interest that only carries a donor fluorophore. The donor fluorescence in the presence of the acceptor,

$F_{DA}$, is measured with the molecule of interest that contains both donor and acceptor fluorophore (Figure 19.73). The FRET efficiency $E$ is then calculated as

$$E = 1 - \frac{F_{DA}}{F_D}$$ 

eq. 19.138

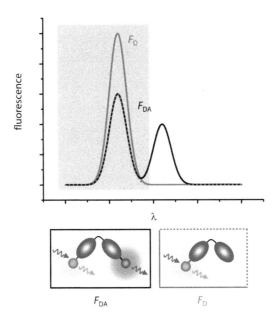

**FIGURE 19.73:    FRET from donor quenching.** The fluorescence emission spectra of the molecule carrying only a donor ($F_D$, green) and of the molecule carrying donor and acceptor ($F_{DA}$, black) after excitation of the donor are measured. To extract the donor contribution from the spectrum of the donor/acceptor-labeled molecule, the spectrum is described as the weighted sum of the donor spectrum (green dotted line) and the acceptor spectrum (not shown). The FRET efficiency can be calculated from the integrated intensities $F_{DA}$ (area under the dotted green curve) and $F_D$ (area under the green curve; eq. 19.138).

To determine the FRET efficiency from sensitized acceptor emission, the fluorescence intensity of the acceptor in the absence of donor, $F_A$, is measured from molecules carrying the acceptor fluorophore only (Figure 19.74). The fluorescence intensity of the acceptor in the presence of donor, $F_{AD}$, (not to be confused with $F_{DA}$, the fluorescence intensity of the donor in the presence of the acceptor, in eq. 19.138!) is measured from molecules carrying both fluorophores:

$$E = \frac{1}{f_D} \cdot \frac{\varepsilon_A}{\varepsilon_D} \left( \frac{F_{AD}}{F_A} - 1 \right)$$ 

eq. 19.139

Here, $f_D$ is the fraction of molecules carrying a donor moiety, $\varepsilon_A$ and $\varepsilon_D$ are the extinction coefficients of the acceptor and the donor fluorophores at the excitation wavelength (of the donor), and $F_A$ and $F_{AD}$ are the fluorescence intensities of the acceptor in the absence and presence of the donor fluorophore. The emission of the acceptor fluorophore needs to be corrected for contributions from the fluorescence of the donor, which can be determined from the reference spectrum of molecules carrying a donor only.

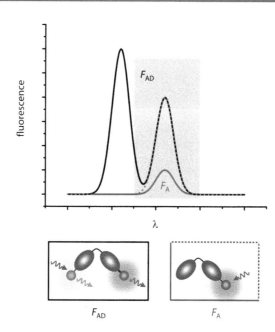

**FIGURE 19.74:** **FRET from acceptor sensitization.** To determine the FRET efficiency from the intensity of sensitized acceptor emission, the fluorescence spectra of the donor/acceptor-labeled molecule ($F_{DA}$, black) and of the molecule carrying only an acceptor ($F_A$, red) are measured. The acceptor contribution is extracted from the spectrum of the donor/acceptor-labeled molecule by describing it as the weighted sum of the donor spectrum (not shown) and the acceptor spectrum (red dotted line). The FRET efficiency can then be calculated from the integrated intensities $F_{AD}$ (area under the dotted red curve) and $F_A$ (area under the red curve) according to eq. 19.139.

Regardless of the method of calculation, fluorescence intensities of ensembles of molecules measured under steady-state illumination are averages over all molecules present. Naturally, FRET efficiencies calculated from average fluorescence intensities represent an average FRET efficiency. Distances obtained from these values by eq. 19.131 may not necessarily represent a true intramolecular distance but an average of all distances present in the sample over the time interval used for the measurement.

### 19.5.10.3 Applications

Intramolecular FRET can be used as a probe for conformational changes of biomolecules. FRET has been employed to study protein folding, ligand-induced conformational changes, and to follow secondary and tertiary structure formation of nucleic acids (Figure 19.75). Intermolecular FRET serves as a probe for binding (Figure 19.75; Box 19.15).

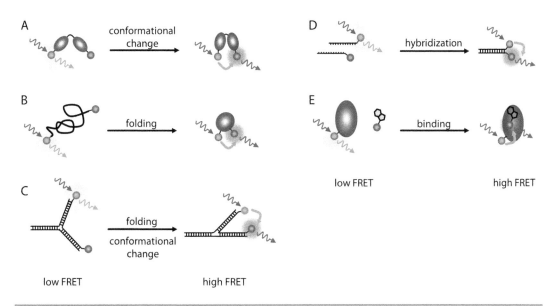

FIGURE 19.75:    **Applications of FRET**. FRET can be used to monitor conformational changes and folding. Intramolecular FRET follows changes in FRET between donor (green) and acceptor (red) dyes that are attached to the same molecule (A–C). Intermolecular FRET follows changes in FRET between dyes that are attached to two binding partners (D, E). A: Protein conformational changes, B: Protein folding, C: Nucleic acid folding and conformational changes, D: DNA hybridization, E: Ligand binding.

---

**BOX 19.15: USING ENERGY TRANSFER FROM TRYPTOPHANS TO MANT AS A PROBE FOR BINDING OF MANT NUCLEOTIDES.**

The emission spectrum of Trp ($\lambda_{max}$ = 320–350 nm) overlaps with the absorption spectrum of mant nucleotides ($\lambda_{max}$ = 360 nm). As a consequence, Trp can serve as a donor for energy transfer to the mant group. FRET from Trp to the mant group provides a highly selective probe to monitor nucleotide binding. In such an experiment, Trp fluorescence is excited at 280–295 nm, where mant does not show absorption. Mant fluorescence emission is monitored at 440 nm. Under these conditions, the observed fluorescence signal is only caused by protein-nucleotide complexes, with no contribution from free mant nucleotide.

FRET can also be used for fluorescence sensing. FRET is a ratiometric method because both donor and acceptor emission are measured, and is thus concentration-independent. Calmodulin, a $Ca^{2+}$-binding protein, undergoes a conformational change upon $Ca^{2+}$ binding. By coupling the fluorescent proteins CFP and YFP to the termini of Calmodulin, this conformational change can be detected by a change in the FRET efficiency between the fluorescent proteins. This fluorescence sensor offers the advantage that it can be genetically encoded and produced directly in the cell. Numerous FRET sensors have been developed to sense ions, metabolites, and signaling molecules.

In addition to using FRET as a qualitative probe, FRET can be employed for the determination of distances according to eq. 19.131. These distances can then be used for mapping of conformational states of proteins, or to determine the architecture of large protein complexes (Box 19.16).

## BOX 19.16: DETERMINING THE STRUCTURAL ORGANIZATION OF THE RNA POLYMERASE-PROMOTOR OPEN COMPLEX BY FRET.

The architecture of the bacterial RNA polymerase holoenzyme (RNA polymerase and initiation factor $\sigma^{70}$) and of the promotor open complex (RNA polymerase/$\sigma^{70}$/DNA template) was defined in FRET experiments by determination of individual donor-acceptor distances and triangulation and distance-constrained docking of the individual structures. RNA polymerase was labeled with the donor fluorescein in four different positions, using intein-mediated C-terminal labeling of the polymerase subunits: N-terminal fragments of the subunit up to the amino acid where the label was to be introduced were produced as C-terminal intein fusions, cleaved to generate a C-terminal thioester, and coupled to cysteinylamido-acetamido-fluorescein. The C-terminal part of the subunit was provided as a separate polypeptide during reconstitution of RNA polymerase. The positions for labeling were distant from each other, chosen to ensure that the position of an acceptor dye on $\sigma^{70}$ would be well-defined by the triangulation procedure. The reconstituted labeled polymerase variants were tested for interaction with $\sigma^{70}$, and their transcription activity was confirmed. The acceptor fluorophore tetramethylrhodamine (TMR) was introduced at 18 different sites in $\sigma^{70}$ by Cys-specific labeling with a TMR-maleimide derivative. All labeled $\sigma^{70}$ variants showed interaction with RNA polymerase and were used in FRET experiments of the holoenzyme. 15 of 18 labeled $\sigma^{70}$ variants still fully supported transcription, and were used for FRET experiments of the promotor open complex. Labeled DNA was generated by synthesis of Cy5-labeled oligonucleotides that were used as primers in PCR reactions to generate a DNA template with Cy5 in the +15, +20, and +25 position from the transcription start site. Holoenzyme and promotor open complexes were assembled and purified by native polyacrylamide gel electrophoresis. 66 and 105 pairwise FRET efficiencies between fluorescein and TMR and between TMR and Cy5 were determined in steady-state fluorescence experiments of holoenzyme or promotor open complex in gel slices. Steady-state and time-resolved fluorescence anisotropy measurements were performed to examine the local mobility of the dyes and to confirm that the dyes reorient on the time scale of the excited state and that the $\kappa^2 = \frac{2}{3}$ approximation is valid. From the measured FRET efficiencies, mean inter-dye distances were calculated, and the distances were used as restraints for docking of crystal structures and homology models of individual subunits and fragments thereof to generate structural models for the holoenzyme (RNA polymerase/$\sigma^{70}$) and the promotor open complex (RNA polymerase/$\sigma^{70}$/DNA template).

The structural models that were obtained from FRET-constrained docking were well-defined, with low root mean square deviations between individual models. When 10% of the restraints were omitted in the docking procedure, the structures that were obtained were very similar, indicating the robustness of the procedure. The structural models rationalize biochemical and genetic data on the interaction between $\sigma^{70}$ and the polymerase core, and between $\sigma^{70}$ and the DNA template in the promotor open complex. Furthermore, they suggest a mechanism for the reduced $\sigma^{70}$ affinity for the polymerase core after synthesis of a 9-11mer RNA. The R3.2 domain of $\sigma^{70}$ occupies the RNA exit channel of the polymerase. Possibly, when a 9-11 nucleotide RNA has been synthesized it competes with this domain for binding to the exit channel, displaces the R3.2 domain and thereby causes the reduced affinity. Mekler *et al.* (2002) *Cell* 108: 599–614.

FRET is also used in imaging applications (Section 23.1.7.6). While fluorescence microscopy is typically limited to the localization of biomolecules, FRET provides additional information. Intramolecular FRET serves as a probe for the conformational state of the molecule of interest. Intermolecular FRET, on the other hand, is an indication for colocalization and/or interaction of the labeled molecules. FRET microscopy on the single-molecule level (Section 23.1.7.6) provides information on the distribution of conformational states in equilibrium, and on the rates of their interconversion.

### 19.5.10.4 Potential Pitfalls

FRET is a robust qualitative probe for changes in distance between the donor and acceptor fluorophores due to conformational rearrangements or binding. It is advisable to always measure changes in both donor and acceptor fluorescence. Anti-correlated changes of donor and acceptor fluorescence are a clear indication of a change in FRET, whereas changes that are not anti-correlated point to effects other than FRET on donor and acceptor fluorescence.

Preparation of donor/acceptor-labeled molecules is the central challenge in intramolecular FRET studies. Ideally, donor and acceptor fluorophores should be introduced into the same molecule site-specifically. Although a variety of different labeling techniques have been established (Section 19.5.4), the combination of orthogonal methods for the introduction of donor and acceptor fluorophores is not straightforward, and the optimization of labeling reactions can be tedious. A common approach for donor/acceptor-labeling of proteins is statistical labeling of a variant with two solvent-exposed cysteines with a mixture of donor- and acceptor maleimides. Such a labeling reaction yields a mixture of donor/donor-, donor/acceptor-, acceptor/donor-, and acceptor/acceptor-labeled molecules. Incomplete labeling further increases the number of different species (D/-, -/D, A/-, -/A). Only the donor/acceptor-labeled species provide FRET data. All other species are unwanted, but contribute to the measured intensities and affect the experimentally determined FRET efficiency. The species carrying donor and acceptor can in principle be isolated by chromatographic purification, although the separation may be challenging as the chromatographic behavior of all species becomes more and more similar the larger the protein under investigation.

The analysis of FRET experiments requires careful control experiments to determine the effect of the local environment on donor quantum yield and on donor and acceptor spectral properties. To convert FRET efficiencies into inter-dye distances, the Förster distance for the particular donor-acceptor pair has to be determined. The position of dye attachment, as well as the presence of binding partners can affect the photophysical properties of the dyes and their mobility, and thus the Förster distance. Restriction of the local mobility of the dye influences the orientation factor $\kappa^2$ and thus the Förster distance. The local mobility of the dye can be determined from fluorescence anisotropy decays (Section 19.5.9.2) to verify that the $\kappa^2 = \frac{2}{3}$ approximation holds. In FRET experiments aiming at distance determination, the Förster distance therefore has to be determined for each labeling configuration, and under the exact conditions (solvent, temperature, presence of interaction partners) of the FRET experiment. Even careful controls cannot overcome the inherent limitation of ensemble FRET efficiencies, the averaging over all molecules present in the sample.

### 19.5.10.5 FRET Efficiencies from Lifetimes

A major drawback of ensemble FRET experiments lies with the averaging over the observed ensemble. For heterogeneous samples, such as mixtures of two conformations, the observed intensities and thus the FRET efficiencies are averaged over all molecules in the sample. Distances calculated from these FRET efficiencies then represent a population-weighted average, and do not reflect a true intramolecular distance. This limitation can be overcome by measuring the fluorescence lifetime of the donor in the absence and presence of the acceptor. FRET provides an additional pathway for depopulation of the excited state of the donor, and therefore reduces the fluorescence lifetime of the donor. Therefore, FRET efficiencies can also be calculated from fluorescence lifetimes of the donor in the absence ($\tau_D$) and presence ($\tau_{DA}$) of the acceptor:

$$E = 1 - \frac{\tau_{DA}}{\tau_D}$$

eq. 19.140

In eq. 19.140, the lifetime of the fluorophore in the absence and presence of the acceptor is a measure of the steady-state fluorescence intensity (eq. 19.138). Eq. 19.140 can be applied if the donor shows a single fluorescence lifetime. If more than one lifetime is present, the (intensity-weighted) average lifetime has to be used to calculate the FRET efficiency.

Time-resolved fluorescence spectroscopy offers an additional advantage over the calculation of FRET efficiencies from intensities: the duration of the experiment, i.e. from the excitation pulse to the first emitted photon, is very short, on the order of nanoseconds. On this time scale, molecules are essentially static, and different conformers do not interconvert. As a consequence, the fluorescence decay contains contributions from all species present, and thus conserves the information on different donor-acceptor distances in different molecules. The distance distribution can be extracted from these decays. First, the donor fluorescence decay in the absence of the acceptor is measured. A multi-exponential decay of the donor fluorescence can be described by eq. 19.141.

$$F_D(t) = \sum_i \alpha_i \exp\left(-\frac{t}{\tau_i}\right)$$

eq. 19.141

In a second experiment, the fluorescence decay of the donor in the presence of the acceptor is measured. By equating eq. 19.131 and eq. 19.140 and solving for $\tau_{DA}$ we see that the lifetime $\tau$ is shortened by $(1 + R_0^6/r^6)$ because of energy transfer:

$$\tau_{DA} = \frac{\tau_D}{\left(1 + \dfrac{R_0^6}{r^6}\right)}$$

eq. 19.142

We can now use this expression for each $\tau_i$ in eq. 19.141 to describe the donor fluorescence decay in the presence of the acceptor, $F_{DA}(t)$ as

$$F_{DA}(t) = \sum_i \alpha_i \exp\left(-\frac{t}{\tau_i} \cdot \left(1 + \frac{R_0^6}{r^6}\right)\right)$$

eq. 19.143

Eq. 19.143 is valid if donor and acceptor are separated by a discrete distance $r$. If we now assume that the molecules are somewhat flexible, allowing the donor-acceptor distance to fluctuate around a mean value, we can describe the distance distribution $P(r)$ as a three-dimensional Gaussian distribution according to

$$P(r) = 4\pi r^2 c \cdot e^{-a(r-b)^2}$$

eq. 19.144

Here, the parameter $a$ corresponds to the width of the distance distribution, $b$ marks its center, and $c$ is an amplitude factor (Figure 19.76). We can then substitute the discrete distance $r$ in eq. 19.143 with the *distance distribution* $P(r)$, and have to integrate over all distances $r$ of the distribution, leading to eq. 19.145:

$$F_{DA}(t) = \int_0^\infty P(r) \sum_i \alpha_i \exp\left(-\frac{t}{\tau_i} \cdot \left(1 + \frac{R_0^6}{r^6}\right)\right) dr$$

eq. 19.145

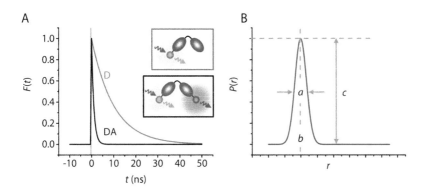

**FIGURE 19.76:** **FRET from fluorescence lifetime measurements.** A: The donor fluorescence lifetime $\tau_D$ (green) is shortened to $\tau_{DA}$ when FRET occurs, and the donor fluorescence decays faster in the presence of the acceptor (black). B: The donor-acceptor distance distribution over all molecules in the solution can be described by a three-dimensional Gaussian distribution (red, eq. 19.144). The parameters $a$, $b$, and $c$ describe the width, the center, and the amplitude (area under the curve) of the distribution.

Eq. 19.145 holds for a single distance distribution only. For heterogeneous samples with different conformational states and different distance distributions $P_n(r)$, we can generalize eq. 19.145 by summing over $n$ distance distributions with a fractional population $f_n$:

$$F_{DA}(t) = \sum_n f_n \int P_n(r) \sum_i \alpha_i \exp\left( -\frac{t}{\tau_i} \cdot \left(1 + \frac{R_0^6}{r^6}\right)\right)dr \qquad \text{eq. 19.146}$$

Thus, by measuring donor fluorescence decays in the absence and presence of acceptor, we can determine the number and fractional populations of the species present (Figure 19.77). The position and width of the Gaussian distribution describe the variations in the distance within one species. From the fractional population of two species in equilibrium, the equilibrium constant can be calculated (Box 19.17). Using time-resolved FRET, it is therefore possible to determine energetic contributions of individual interactions to the stabilization of a conformer by assessing their effect on the conformational equilibrium (Box 19.17).

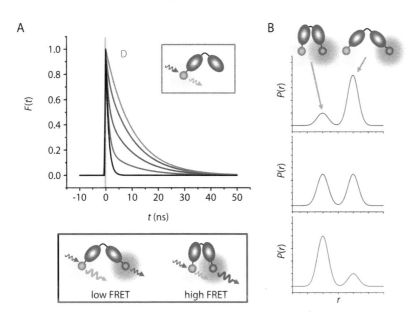

**FIGURE 19.77:** **Conformational states and their population from fluorescence lifetime measurements:** Donor fluorescence decays in the presence of the acceptor (red, purple, blue, black) are faster when FRET occurs than in the absence of FRET (green). By description of the measured decays with eq. 19.146 the distance distribution $P(r)$ may be extracted. The two Gaussian distributions represent the donor-acceptor distance in the low FRET species and the high FRET species. From the areas under the curve, the relative population of the two species can be determined. Red: 20% high FRET, purple: 50% high FRET, blue: 80% high FRET, black: 100% high FRET.

## BOX 19.17: EQUILIBRIUM CONSTANTS FOR CONFORMATIONAL EQUILIBRIA FROM TIME-RESOLVED FRET EXPERIMENTS.

From the donor fluorescence decay in the presence of acceptor, we can calculate distributions of donor-acceptor distances in two (or more) species with different FRET efficiencies. The fractional populations of two conformers in equilibrium, such as a folded and an unfolded state, can thereby be determined (Figure 19.78). From the fractional populations, the equilibrium constant can be calculated, which in turn is linked to $\Delta G^0$ of the reaction (Section 3.1). Energetic contributions of putative interactions that stabilize one conformer over the other can be probed in time-resolved FRET experiments. Donor fluorescence decays for the donor/acceptor-labeled unmodified protein or nucleic acid provide the unperturbed equilibrium constant $K_1$ and the energetic difference $\Delta G_1^0$ between the two conformers. In a second experiment, the donor decay is measured for a modified version in which a group that is hypothesized to be involved in a stabilizing interaction has been removed. For example, the possible energetic contribution of a hydrogen bond formed by a tyrosine hydroxyl group to the overall protein stability can be probed by replacement of this tyrosine by a phenylalanine. The equilibrium constant $K_2$ of folding for this variant is determined by time-resolved FRET experiments, and $\Delta G_2^0$ is calculated. If $K_2$ differs from $K_1$, the energetic contribution of the tyrosine for stability can be quantified as a $\Delta\Delta G^0$ value, the difference between $\Delta G_1^0$ and $\Delta G_2^0$ (see Section 3.4; Box 3.3).

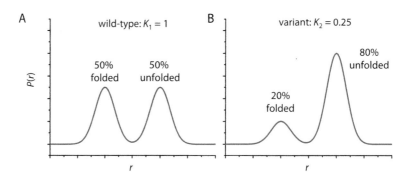

**FIGURE 19.78: Energetic contributions of interactions from time-resolved FRET experiments.** A. Distance distributions allow quantification of folded (low inter-dye distance) and unfolded forms (high inter-dye distance) of a protein or nucleic acid in equilibrium. The areas under the curve reflect the fractional populations, from which equilibrium constants and $\Delta G^0$ values for folding are calculated. $K_1$ for folding is 1, the corresponding $\Delta G_1^0$ is 0 kJ mol⁻¹. B: A destabilized variant is only 20% folded under the same conditions. $K_2$ for folding is 0.25, the corresponding $\Delta G_2^0$ is 3.4 kJ mol⁻¹. The energetic contribution of the group that has been removed in this variant to the overall stability of the folded state is $\Delta\Delta G^0 = \Delta G_2^0 - \Delta G_1^0 = 3.4$ kJ mol⁻¹.

### 19.5.10.6 FRET Efficiencies from Single Molecules

When donor/acceptor-labeled molecules are studied by single-molecule fluorescence microscopy, FRET efficiencies can also be determined for single molecules (see Section 23.1.7). The FRET efficiency for donor/acceptor-labeled single molecules, either in solution or immobilized on a surface, can be calculated from the measured intensities of the donor and acceptor fluorescence, $F_D$ and $F_A$, of a single molecule according to eq. 19.135. FRET efficiencies calculated for a large number of single molecules, one after the other (in solution) or in parallel (on surfaces), can be assembled in a FRET histogram that shows how frequently a certain FRET efficiency is encountered. These FRET histograms directly reveal the different populations present, and their fractions, without any assumptions on the shape of the distribution. For surface-immobilized molecules, the donor and acceptor intensities can be measured over seconds to minutes. Changes in the FRET efficiencies in this time window can be followed directly, which enables direct monitoring of conformational changes over time. Single-molecule FRET experiments avoid averaging over ensembles of molecules, and provide important insight into heterogeneous systems with different species of different donor-acceptor distances. They offer the possibility to monitor events that cannot be synchronized. Single-molecule FRET and its applications are discussed in more detail in Section 23.1.7.

## QUESTIONS

19.1 A solution of ATP has a concentration of 100 μM. What is the absorption and transmission ($d$ = 1 cm, (ATP) = 15400 M$^{-1}$ cm$^{-1}$)?

19.2 Proteins typically have an absorption ratio of $A_{280}/A_{260}$ = 2, whereas the ratio for nucleic acids and nucleotides is $A_{280}/A_{260}$ = 0.5. A purified ATP-binding protein with a molar extinction coefficient $\varepsilon_{280}$ = 12000 M$^{-1}$ cm$^{-1}$, has an absorption of $A_{280}$ = 0.5, $A_{260}$ = 0.35. What is the fraction of ATP (or ADP) that has been co-purified ($\varepsilon_{260}$ (ATP) = 15400 M$^{-1}$ cm$^{-1}$, $d$ = 1 cm)? Calculate the $A_{280}/A_{260}$ ratio for a nucleotide contamination of 1%.

19.3 Fluorescein absorbs light of $\lambda$ = 495 nm. What is the energy difference $\Delta E$ of ground state $S_0$ and excited state $S_1$ per molecule and per mol? The extinction coefficient is $\varepsilon$ = 95000 M$^{-1}$ cm$^{-1}$. How much energy is necessary to excite all molecules in one mL of a fluorescein solution with $A$ = 1.2 ($d$ = 1 cm)? What fraction of the incident light reaches the detector?

19.4 NADH has an extinction coefficient $\varepsilon_{340}$ of 6300 M$^{-1}$ cm$^{-1}$; NAD$^+$ does not absorb at this wavelength. At 260 nm, NAD$^+$ absorbs more strongly, with $\varepsilon_{260}$ (NAD$^+$) = 18000 M$^{-1}$ cm$^{-1}$, and $\varepsilon_{260}$ (NADH) = 15000 M$^{-1}$ cm$^{-1}$. You find a very old stock of NADH in the freezer, and prepare a solution that has an absorption of $A_{260}$ = 0.31, and $A_{340}$ = 0.11 ($d$ = 0.2 cm). What is the fraction of NAD$^+$?

19.5 A solution contains glucose-6-phosphate and glucose-1-phosphate. When an excess of NADP$^+$ and glucose-6-phosphate dehydrogenase is added, a final absorption at 340 nm of $A_{340}$ = 0.60 is measured. Subsequent addition of phosphoglucomutase leads to an increase in absorption to $A_{340}$ = 0.72. What are the glucose-6-phosphate and glucose-1-phosphate concentrations in the solution? $\varepsilon_{340}$(NADPH) = 6300 M$^{-1}$ cm$^{-1}$, $d$ = 1 cm.

19.6 Addition of Ellman's reagent to a 15 μM protein solution leads to an increase in $A_{412}$ by 0.22 units. In denaturing buffer, the increase is 0.43 units. How many surface-exposed and buried cysteines does the protein contain? Does the protein contain disulfide bridges? $\varepsilon$(TNB$^-$) = 14000 M$^{-1}$ cm$^{-1}$.

19.7 An excess of Ellman's reagent is added to a 10 μM solution of an enzyme. Under native conditions, the absorption increases by 0.15, under denaturing conditions it increases by 0.29. Calculate the number of accessible cysteines and explain the result. Repeating the experiment in the presence of the substrate, the absorption increases by 0.04 under native, by 0.30 under denaturing conditions. What is the reason for this behavior? What is the degree of binding? $\varepsilon$(TNB$^-$) = 14000 M$^{-1}$ cm$^{-1}$.

19.8 How can you determine the concentration of a structured RNA molecule of 154 nucleotides, consisting of 54 A, 35 G, 29 C, and 36 U, most accurately by absorption? How can you determine the extinction coefficient of the folded RNA molecule? Why is the procedure not applicable to DNA?

19.9 A CD measurement of the ellipticity at 222 nm gives $\theta_A$ = 1.246 mdeg for a 0.9 μM solution of protein A, a 23.4 kDa protein of 210 amino acids, and $\theta_B$ = 2.345 mdeg for a 1.15 μM solution of protein B, a 43.2 kDa protein of 393 amino acids. Which protein has a higher α-helical content? $d$ = 0.1 cm.

19.10 A protein folding intermediate shows a high far-UV CD signal, no near-UV CD, and tryptophan fluorescence at 330 nm wavelength. What are the characteristics of the intermediate?

19.11 CD spectra for a protein in its folded and unfolded form are given.

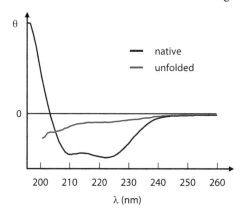

Which wavelength is the most suitable for following unfolding of the protein by CD spectroscopy? Why is the wavelength with maximal signal difference not preferable?

19.12 The heat-induced unfolding of a protein is measured using far-UV CD and tryptophan fluorescence as spectroscopic probes. The melting temperature determined from the transition observed by CD is 5 degrees higher than the melting temperature determined by fluorescence. What is the reason?

19.13 Calculate the stiffness $k$ for CO-, NH-, CH-, and CD- bonds from the harmonic oscillator model. The wavenumbers are 1720 cm$^{-1}$ (CO), 3400 cm$^{-1}$ (NH), 2900 cm$^{-1}$ (CH), and 2200 cm$^{-1}$ (CD). Calculate the average change in bond lengths $\Delta x$ during the stretch vibrations. Is the isotope effect for CH and CD only caused by the difference in (reduced) mass?

19.14 The wavenumber for the C-F stretch vibration is 1200 cm$^{-1}$, the wavenumbers for the C-Cl, CBr, and C-I stretch vibrations are 700 cm$^{-1}$ (C-Cl), 550 cm$^{-1}$ (C-Br), and 500 cm$^{-1}$ (C-I). Can this effect be explained by the different masses of the halogens?

19.15 Why is it necessary to calculate difference spectra of free and bound forms of a fluorescent ligand to select the emission wavelength at which the signal change for a fluorescence titration is maximal? Under what conditions is the observation at the maximum of the emission spectrum optimal?

19.16 Calculate the equations for conversion of anisotropy $r$ into polarization $p$ and *vice versa*.

19.17 A fluorescent ligand in solution has an anisotropy of 0.05. The anisotropy increases to 0.25 upon binding of the ligand to a protein. Calculate the anisotropy and polarization of a solution with 40% saturation. What is the deviation (in %), if a linear dependence of polarization on concentration of bound and free forms is assumed?

19.18 A fluorophore has a quantum yield of 0.2 and a fluorescence lifetime of $\tau$ = 12 ns. Calculate the rate constants $k_f$ and $k_{nr}$ for radiative and non-radiative transitions from $S_1$ to $S_0$ and the intrinsic lifetime $\tau_0$. Upon addition of 0.25 M acrylamide, the lifetime is reduced to 10 ns. What is the Stern-Volmer constant $K_{SV}$ for this quencher? Calculate the $K_d$ value of the fluorophore complex with a quencher that would achieve the same quenching effect by static quenching. The concentration of the fluorophore is 1 μM.

19.19 A protein is very difficult to purify and tends to aggregate and precipitate when it is titrated to a fluorescent ligand. To be able to determine the $K_d$ value of the protein-ligand complex, the titration is performed in reverse, by supplying the protein at a low concentration, and titrating with increasing concentrations of the fluorescent ligand. Derive the equation for the measured fluorescence $F$ as a function of total ligand and protein concentration and the $K_d$. Why is the binding curve obtained more difficult to analyze than a direct titration of fluorescent ligand with the protein?

19.20  A double-cysteine variant of a protein ($\varepsilon_{protein,\,280\,nm}$ = 14 800 $M^{-1}$ $cm^{-1}$) has been labeled with a mixture of donor and acceptor maleimides for FRET experiments ($\varepsilon_{donor,\,495\,nm}$ = 72500 $M^{-1}$ $cm^{-1}$, $\varepsilon_{acceptor,\,540\,nm}$ = 95000 $M^{-1}$ $cm^{-1}$). Both donor and acceptor fluorophores show absorption at 280 nm ($\varepsilon_{donor,\,280\,nm}$ = 9300 $M^{-1}$ $cm^{-1}$, $\varepsilon_{acceptor,\,280\,nm}$ = 26500 $M^{-1}$ $cm^{-1}$), and the acceptor absorbs at the excitation wavelength of the donor ($\varepsilon_{acceptor,\,495\,nm}$ = 18600 $M^{-1}$ $cm^{-1}$). Determine the labeling efficiencies for donor and acceptor from the absorption values at 280 nm, 495 nm, and 540 nm ($d$ = 1 cm), $A_{280}$ = 0.25, $A_{495}$ = 0.31, $A_{540}$ = 0.49. What is the fraction of unlabeled molecules and of protein molecules that carry one donor, two donors, one acceptor, two acceptors, one donor and one acceptor, or no label at all?

19.21  A colleague determines the $K_d$ value of a DNA-protein complex by titration of the protein with a 10-base-pair DNA, and follows binding by measuring the fluorescence of intrinsic tryptophan residues. After an initial increase in tryptophan fluorescence, the fluorescence decreases linearly until it reaches background intensity. What is the reason for this behavior? How can the measured fluorescence values be corrected to obtain a binding curve?

19.22  A globular protein is labeled with fluorophore A that has a lifetime $\tau_A$ = 1 ns, or fluorophore B with a lifetime $\tau_B$ = 10 ns. Which of the labeled proteins has the higher steady-state anisotropy and why?

19.23  Calculate the steady-state anisotropies for fluorophores with lifetimes $\tau$ = 0.1 ns, 0.5 ns, 1 ns, 5 ns, 10 ns, 50 ns, and 100 ns that are rigidly attached to a molecule with $\tau_c$ = 1 ns. Binding of the molecule to a protein causes an increase in rotational correlation time to $\tau_c$ = 10 ns. Which fluorophore is best suited to follow binding, which fluorophores do not report on binding?

19.24  Two proteins of 25 kDa (A) and 45 kDa (B) form a heterodimer. One protein has no tryptophan, the other contains a single tryptophan. What are the prerequisites for measuring binding by a change in tryptophan anisotropy?

19.25  Fluorescein has a lifetime of $\tau_D$ = 5.2 ns when attached to a DNA. When tetramethylrhodamine is attached at the other end of the DNA, the lifetime decreases to $\tau_{DA}$ = 3.6 ns. Upon binding of a protein to the DNA, the lifetime of the donor decreases to 4 ns, the lifetime of the donor in presence of acceptor decreases to 1.2 ns. What is the energy transfer efficiency, the rate constant $k_t$ for energy transfer, and the donor-acceptor distance for the free DNA and for the protein-DNA complex? $R_0$ = 5.4 nm.

19.26  A Cy3-Cy5-labeled protein occurs in three different conformational states with inter-dye distances of 1 nm (50%), 2 nm (10%), and 5 nm (40%). The percentages are the fraction of each conformer in equilibrium. The Förster distance is $R_0$ = 5.2 nm. Calculate the FRET efficiency for each conformer, the mean FRET efficiency, and the average distance over all conformers. What is the mean distance that would be determined from an ensemble FRET experiment? What additional information would be obtained by time-resolved FRET?

19.27  A fluorescein-TMR-labeled enzyme is analyzed with time-resolved FRET. The donor lifetime is 5 ns in the absence and 4 ns in the presence of acceptor. When substrate is added, the donor lifetime is 5 ns in the absence and 1 ns in the presence of the acceptor. The Förster distance is $R_0$ = 5.2 nm. Calculate $E_{FRET}$ and $r$ for both cases.

# REFERENCES

Cantor, C. R. and Schimmel, P. R. (1980) Biophysical Chemistry (Part II) Techniques for the study of biological structure and function. New York, Freeman.
· a comprehensive book on biophysical techniques including optical spectroscopy, with a strong focus on physical background and quantitative derivations

Fasman, G. D. (1996) Circular dichroism and the conformational analysis of biomolecules. New York, Plenum press.
· a comprehensive book about theory and applications of circular dichroism of proteins and other biomolecules

Johnson, W. C., Jr. (1990) Protein secondary structure and circular dichroism: a practical guide. *Proteins* **7**(3): 205–214.
· introductory review on practical aspects in the application of circular dichroism to proteins

Lakowicz, J. R. (1999) Principles of fluorescence spectroscopy. New York Kluwer Academic/Plenum Publishers.
· a compendium of fluorescence spectroscopy

Toseland, C. and Fili, N. (2014) Fluorescent methods applied to molecular motors: from single molecules to whole cells. Basel, Springer.
· a collection of articles on different fluorescent methods and their application to ATP-dependent enzymes

Dixon, J. M., Taniguchi, M. and Lindsey, J. S. (2005) PhotochemCAD 2: a refined program with accompanying spectral databases for photochemical calculations. *Photochem Photobiol* **81**(1): 212–213.
· a program for spectral calculations with an accompanying database of absorption and fluorescence spectra

Nagel, G., Ollig, D., Fuhrmann, M., Kateriya, S., Musti, A. M., Bamberg, E. and Hegemann, P. (2002) Channelrhodopsin-1: a light-gated proton channel in green algae. *Science* **296**(5577): 2395–2398.
· discovery of channelrhodopsins, light-gated ion channels in algae

Cohen, A. E. (2016) Optogenetics: Turning the Microscope on Its Head. *Biophys J* **110**(5): 997–1003.
· review on current state of the field of optogenetics

Gill, S. C. and von Hippel, P. H. (1989) Calculation of protein extinction coefficients from amino acid sequence data. *Anal Biochem* **182**(2): 319–326.
· photometric method to determine the molar extinction coefficient of folded proteins

Ehresmann, B., Imbault, P. and Weil, J. H. (1973) Spectrophotometric determination of protein concentration in cell extracts containing tRNAs and rRNAs. *Anal Biochem* **54**(2): 454–463.
· photometric determination of protein concentrations in the presence of nucleic acids

Ellman, G. L. (1959) Tissue sulfhydryl groups. *Arch Biochem Biophys* **82**(1): 70–77.
· original paper describing the quantification of cysteines by photometry using DTNB

Birdsall, B., King, R. W., Wheeler, M. R., Lewis, C. A., Jr., Goode, S. R., Dunlap, R. B. and Roberts, G. C. (1983) Correction for light absorption in fluorescence studies of protein-ligand interactions. *Anal Biochem* **132**(2): 353–361.
· correction of absorption values for inner filter effects

Dafforn, T. R. and Rodger, A. (2004) Linear dichroism of biomolecules: which way is up? *Curr Opin Struct Biol* **14**(5): 541–546.
· review article on application of linear dichroism to biomolecules

Hofmann, C., Ketelaars, M., Matsushita, M., Michel, H., Aartsma, T. J. and Kohler, J. (2003) Single-molecule study of the electronic couplings in a circular array of molecules: light-harvesting-2 complex from *Rhodospirillum molischianum*. *Phys Rev Lett* **90**(1): 013004.
· determination of the relative orientation of chlorophyll molecules in a bacterial photosynthetic light harvesting complex through their transition dipole moments

Bayley, P. M. and Harris, H. E. (1975) Conformational properties of pig-heart cytoplasmic aspartate aminotransferase. Circular-dichroism and absorption-spectroscopic study of dicarboxylate binding. *Eur J Biochem* **56**(2): 455–465.
· study on the induced circular dichroism of pyridoxal phosphate bound to aspartate aminotransferase

Greenfield, N. J. (2006) Using circular dichroism spectra to estimate protein secondary structure. *Nat Protoc* **1**(6): 2876–2890.
· review article focusing on practical aspects of CD measurements and the interpretation of CD spectra for secondary structure prediction of proteins

Bulheller, B. M. and Hirst, J. D. (2009) DichroCalc-circular and linear dichroism online. *Bioinformatics* **25**(4): 539–540.
· description of the program DichroCalc for calculations of circular and linear dichroism spectra

Chou, P. Y. and Fasman, G. D. (1974) Conformational parameters for amino acids in helical, β-sheet, and random coil regions calculated from proteins. *Biochemistry* **13**(2): 211–222.
· reports how the CD spectra of proteins depend on the secondary structure

Cepus, V., Scheidig, A. J., Goody, R. S. and Gerwert, K. (1998) Time-resolved FTIR studies of the GTPase reaction of H-ras p21 reveal a key role for the β-phosphate. *Biochemistry* **37**(28): 10263–10271.
· study of the mechanism of GTP hydrolysis by the oncoprotein Ras by Fourier transform infrared spectroscopy using GTP with $^{18}$O substitutions at the α-, β-, or γ-phosphate

Rathod, R., Kang, Z., Hartson, S. D., Kumauchi, M., Xie, A. and Hoff, W. D. (2012) Side-chain specific isotopic labeling of proteins for infrared structural biology: the case of ring-D4-tyrosine isotope labeling of photoactive yellow protein. *Protein Expr Purif* **85**(1): 125–132.
· incorporation of isotope-labeled Tyr into photoactive yellow protein for isotope-edited FTIR spectroscopy

Jung, C. (2000) Insight into protein structure and protein-ligand recognition by Fourier transform infrared spectroscopy. *J Mol Recognit* **13**(6): 325–351.
· review of applications of FT-IR spectroscopy to proteins, with a focus on heme proteins

Gerwert, K., Souvignier, G. and Hess, B. (1990) Simultaneous monitoring of light-induced changes in protein side-group protonation, chromophore isomerization, and backbone motion of bacteriorhodopsin by time-resolved Fourier-transform infrared spectroscopy. *Proc Natl Acad Sci U S A* **87**(24): 9774–9778.
· application of step-scan FT-IR spectroscopy to investigate structural changes in bacteriorhodopsin during its photocycle

Allin, C., Ahmadian, M. R., Wittinghofer, A. and Gerwert, K. (2001) Monitoring the GAP catalyzed H-Ras GTPase reaction at atomic resolution in real time. *Proc Natl Acad Sci U S A* **98**(14): 7754–7759.
· application of step-scan FT-IR spectroscopy to the mechanism of GTP hydrolysis by Ras in presence of the Ras GTPase-activating protein

Kotting, C. and Gerwert, K. (2005) Proteins in action monitored by time-resolved FTIR spectroscopy. *Chemphyschem* **6**(5): 881–888.
· review article on time-resolved FT-IR spectroscopy

Magde, D., Wong, R. and Seybold, P. G. (2002) Fluorescence quantum yields and their relation to lifetimes of rhodamine 6G and fluorescein in nine solvents: improved absolute standards for quantum yields. *Photochem Photobiol* **75**(4): 327–334.
· calorimetric study to determine rates of radiative and non-radiative decay for fluorescein and rhodamine 6G in different solvents with high precision

Brannon, J. H. and Magde, D. (1978) Absolute quantum yield determination by thermal blooming. Fluorescein. *The Journal of Physical Chemistry* **82**(6): 705–709.
· determination of fluorescein quantum yield in 0.1 M NaOH by thermal analysis

Noren, C. J., Anthony-Cahill, S. J., Griffith, M. C. and Schultz, P. G. (1989) A general method for site-specific incorporation of unnatural amino acids into proteins. *Science* **244**(4901): 182–188.
· incorporation of unnatural amino acids by suppression of stop codons

Liu, C. C. and Schultz, P. G. (2010) Adding new chemistries to the genetic code. *Annu Rev Biochem* **79**: 413–444.
· review article about incorporation of unnatural amino acids into proteins in *E. coli*, yeast and mammalian cells.

Appel, M. J. and Bertozzi, C. R. (2015) Formylglycine, a post-translationally generated residue with unique catalytic capabilities and biotechnology applications. *ACS Chem Biol* **10**(1): 72–84.
· review about formylglycine generating enzyme that can be used to genetically encode formyl groups

Rabuka, D., Rush, J. S., deHart, G. W., Wu, P. and Bertozzi, C. R. (2012) Site-specific chemical protein conjugation using genetically encoded aldehyde tags. *Nat Protoc* **7**(6): 1052–1067.
· protocol for incorporation of aldehyde groups into proteins using the formylglycine generating enzyme

Carrico, I. S., Carlson, B. L. and Bertozzi, C. R. (2007) Introducing genetically encoded aldehydes into proteins. *Nat Chem Biol* **3**(6): 321–322.
· original paper that reports the genetically encoded incorporation of formylglycine into proteins

Gauchet, C., Labadie, G. R. and Poulter, C. D. (2006) Regio- and chemoselective covalent immobilization of proteins through unnatural amino acids. *J Am Chem Soc* **128**(29): 9274–9275.
· use of farnesyl transferases to introduce four amino acid sequences with alkyne groups into proteins and modification by click chemistry.

Griffin, B. A., Adams, S. R. and Tsien, R. Y. (1998) Specific covalent labeling of recombinant protein molecules inside live cells. *Science* **281**(5374): 269–272.
· original paper describing *in vivo* labeling of proteins using a genetically encoded tetracysteine motif that forms a biarsenic complex with fluorescein

Griffin, B. A., Adams, S. R., Jones, J. and Tsien, R. Y. (2000) Fluorescent labeling of recombinant proteins in living cells with FlAsH. *Methods Enzymol* **327**: 565–578.
· method review of the use of FlAsH tags for fluorescent labeling *in vivo*

Keppler, A., Gendreizig, S., Gronemeyer, T., Pick, H., Vogel, H. and Johnsson, K. (2003) A general method for the covalent labeling of fusion proteins with small molecules *in vivo*. *Nat Biotechnol* **21**(1): 86–89.
· description of the SNAP tag for fluorescent labeling of proteins *in vivo*

Gautier, A., Juillerat, A., Heinis, C., Correa, I. R., Jr., Kindermann, M., Beaufils, F. and Johnsson, K. (2008) An engineered protein tag for multiprotein labeling in living cells. *Chem Biol* **15**(2): 128–136.
· description of the CLIP tag for fluorescent labeling of proteins *in vivo*

Los, G. V., Encell, L. P., McDougall, M. G., Hartzell, D. D., Karassina, N., Zimprich, C., Wood, M. G., Learish, R., Ohana, R. F., Urh, M., Simpson, D., Mendez, J., Zimmerman, K., Otto, P., Vidugiris, G., Zhu, J., Darzins, A., Klaubert, D. H., Bulleit, R. F. and Wood, K. V. (2008) HaloTag: a novel protein labeling technology for cell imaging and protein analysis. *ACS Chem Biol* **3**(6): 373–382.
· description of the Halo tag for fluorescent labeling

Taki, M., Shiota, M. and Taira, K. (2004) Transglutaminase-mediated N- and C-terminal fluorescein labeling of a protein can support the native activity of the modified protein. *Protein Eng Des Sel* **17**(2): 119–126.
· description of fluorescent labeling by transglutaminase-mediated modification of a glutamine

Popp, M. W., Antos, J. M., Grotenbreg, G. M., Spooner, E. and Ploegh, H. L. (2007) Sortagging: a versatile method for protein labeling. *Nat Chem Biol* **3**(11): 707–708.
· description of the use of sortase to attach fluorescent labels to proteins *in vitro*, in cell lysates and on the cell surface

Ritzefeld, M. (2014) Sortagging: a robust and efficient chemoenzymatic ligation strategy. *Chem Eur J* **20**(28): 8516–8529.
· review on applications of sortase-mediated tagging

Theile, C. S., Witte, M. D., Blom, A. E., Kundrat, L., Ploegh, H. L. and Guimaraes, C. P. (2013) Site-specific N-terminal labeling of proteins using sortase-mediated reactions. *Nat Protoc* **8**(9): 1800–1807.
· protocol for N-terminal labeling of proteins using sortase

Guimaraes, C. P., Witte, M. D., Theile, C. S., Bozkurt, G., Kundrat, L., Blom, A. E. and Ploegh, H. L. (2013) Site-specific C-terminal and internal loop labeling of proteins using sortase-mediated reactions. *Nat Protoc* **8**(9): 1787–1799.
· protocol for C-terminal and internal loop labeling of proteins using sortase

Muir, T. W., Sondhi, D. and Cole, P. A. (1998) Expressed protein ligation: a general method for protein engineering. *Proc Natl Acad Sci U S A* **95**(12): 6705–6710.
· original paper reporting the use of expressed protein ligation for protein engineering

Muralidharan, V. and Muir, T. W. (2006) Protein ligation: an enabling technology for the biophysical analysis of proteins. *Nat Methods* **3**(6): 429–438.
· review article about applications of expressed protein ligation in biophysical analyses

Ormo, M., Cubitt, A. B., Kallio, K., Gross, L. A., Tsien, R. Y. and Remington, S. J. (1996) Crystal structure of the *Aequorea victoria* green fluorescent protein. *Science* **273**(5280): 1392–1395.
· crystal structure of green fluorescent protein

Magliery, T. J., Wilson, C. G., Pan, W., Mishler, D., Ghosh, I., Hamilton, A. D. and Regan, L. (2005) Detecting protein-protein interactions with a green fluorescent protein fragment reassembly trap: scope and mechanism. *J Am Chem Soc* **127**(1): 146–157.
· use of split versions of GFP to detect protein-protein interactions

Cabantous, S., Terwilliger, T. C. and Waldo, G. S. (2005) Protein tagging and detection with engineered self-assembling fragments of green fluorescent protein. *Nat Biotechnol* **23**(1): 102–107.
· generation of self-assembling GFP fragments

Cabantous, S., Nguyen, H. B., Pedelacq, J. D., Koraichi, F., Chaudhary, A., Ganguly, K., Lockard, M. A., Favre, G., Terwilliger, T. C. and Waldo, G. S. (2013) A new protein-protein interaction sensor based on tripartite split-GFP association. *Scientific reports* **3**: 2854.
· generation of tripartite GFP to study protein-protein interactions

Kirchhofer, A., Helma, J., Schmidthals, K., Frauer, C., Cui, S., Karcher, A., Pellis, M., Muyldermans, S., Casas-Delucchi, C. S., Cardoso, M. C., Leonhardt, H., Hopfner, K. P. and Rothbauer, U. (2010) Modulation of protein properties in living cells using nanobodies. *Nat Struct Mol Biol* **17**(1): 133–138.
· generation of nanobodies that bind to green fluorescent protein (GFP), structural basis of the reduction or increase in GFP fluorescence by these nanobodies (GFP minimizer and GFP maximizer), and application in expression and localization studies

Paige, J. S., Wu, K. Y. and Jaffrey, S. R. (2011) RNA mimics of green fluorescent protein. *Science* **333**(6042): 642–646.
· development of the green fluorescent RNA aptamer "spinach"

Strack, R. L. and Jaffrey, S. R. (2013) New approaches for sensing metabolites and proteins in live cells using RNA. *Curr Opin Chem Biol* **17**(4): 651–655.
· spinach derivatives for metabolite sensing in living cells

Vogelsang, J., Kasper, R., Steinhauer, C., Person, B., Heilemann, M., Sauer, M. and Tinnefeld, P. (2008) A reducing and oxidizing system minimizes photobleaching and blinking of fluorescent dyes. *Angew Chem Int Ed* **47**(29): 5465–5469.
· suppression of photobleaching and blinking of fluorophores by reducing/oxidizing systems

Cordes, T., Vogelsang, J. and Tinnefeld, P. (2009) On the mechanism of Trolox as antiblinking and antibleaching reagent. *J Am Chem Soc* **131**(14): 5018–5019.
· suppression of fluorophore photobleaching and blinking by the vitamin E analog Trolox

Jones, A. C. and Neely, R. K. (2015) 2-Aminopurine as a fluorescent probe of DNA conformation and the DNA-enzyme interface. *Q Rev Biophys* **48**(2): 244–279.
· review on applications of 2-aminopurine to study nucleic acid structure and dynamics, with a focus on time-resolved fluorescence

Walter, N. G., Harris, D. A., Pereira, M. J. and Rueda, D. (2001) In the fluorescent spotlight: global and local conformational changes of small catalytic RNAs. *Biopolymers* **61**(3): 224–242.
· review on fluorescence probes for large scale and local conformational changes of RNA, including 2-aminopurine and FRET

Andreou, A. Z. and Klostermeier, D. (2014) eIF4B and eIF4G jointly stimulate eIF4A ATPase and unwinding activities by modulation of the eIF4A conformational cycle. *J Mol Biol* **426**(1): 51–61.
· use of 2-aminopurine to monitor RNA unwinding by a helicase

Holz, B., Klimasauskas, S., Serva, S. and Weinhold, E. (1998) 2-Aminopurine as a fluorescent probe for DNA base flipping by methyltransferases. *Nucleic Acids Res* **26**(4): 1076–1083.
· investigation of base-flipping by the methyltransferase M.HhaI using 2-AP fluorescence as a probe

Wilhelmsson, L. M. (2010) Fluorescent nucleic acid base analogues. *Q Rev Biophys* **43**(2): 159–183.
· review article summarizing the structures and properties of fluorescent nucleotide analogs and possible applications

Lee, M. H., Kim, J. S. and Sessler, J. L. (2015) Small molecule-based ratiometric fluorescence probes for cations, anions, and biomolecules. *Chem Soc Rev* **44**(13): 4185–4191.
· review article on ratiometric fluorescence sensors

Klonoff, D. C. (2012) Overview of fluorescence glucose sensing: a technology with a bright future. *J Diabetes Sci Technol* **6**(6): 1242–1250.
· review about glucose sensing by fluorescence sensors

Moore, M. J. and Query, C. C. (2000) Joining of RNAs by splinted ligation. *Methods Enzymol* **317**: 109–123.
· description of RNA ligation techniques with a focus on practical aspects

Smith, G. J., Sosnick, T. R., Scherer, N. F. and Pan, T. (2005) Efficient fluorescence labeling of a large RNA through oligonucleotide hybridization. *RNA* **11**(2): 234–239.
· introduction of donor and acceptor fluorophores into RNase P RNA by hybridization of fluorescently labeled oligonucleotides to unpaired bases in loop regions

Thrall, S. H., Reinstein, J., Wohrl, B. M. and Goody, R. S. (1996) Evaluation of human immunodeficiency virus type 1 reverse transcriptase primer tRNA binding by fluorescence spectroscopy: specificity and comparison to primer/template binding. *Biochemistry* **35**(14): 4609–4618.
· solution of the cubic equation and analysis of displacement titrations

Lentz, B. R. (1993) Use of fluorescent probes to monitor molecular order and motions within liposome bilayers. *Chem Phys Lipids* **64**(1–3): 99–116.
· review article on the use of fluorescent probes to study membrane dynamics and order

Förster, T. (1948) Zwischenmolekulare Energiewanderung und Fluoreszenz. *Annalen der Physik* **437**(1): 55–75.
· original description of the phenomenon of Förster Resonance Energy Transfer

Stryer, L. and Haugland, R. P. (1967) Energy transfer: a spectroscopic ruler. *Proc Natl Acad Sci U S A* **58**(2): 719–726.
· experimental demonstration of fluorescence resonance energy transfer as a spectroscopic ruler using poly-lysine peptides of different lengths

Andreou, A. Z. and Klostermeier, D. (2014) Fluorescence methods in the investigation of the DEAD-box helicase mechanism. *EXS* **105**: 161–192.
· review article on the application of fluorescence techniques to DEAD-box helicases

Sustarsic, M. and Kapanidis, A. N. (2015) Taking the ruler to the jungle: single-molecule FRET for understanding biomolecular structure and dynamics in live cells. *Curr Opin Struct Biol* **34**: 52–59.
· review article on quantitative *in vivo* applications of FRET

Hochreiter, B., Garcia, A. P. and Schmid, J. A. (2015) Fluorescent proteins as genetically encoded FRET biosensors in life sciences. *Sensors* **15**(10): 26281–26314.
· review on the principle of FRET sensors

Mekler, V., Kortkhonjia, E., Mukhopadhyay, J., Knight, J., Revyakin, A., Kapanidis, A. N., Niu, W., Ebright, Y. W., Levy, R. and Ebright, R. H. (2002) Structural organization of bacterial RNA polymerase holoenzyme and the RNA polymerase-promoter open complex. *Cell* **108**(5): 599–614.
· structural modeling of the RNA polymerase holoenzyme and the promotor open complex using 66 and 105 FRET-derived distance restraints

# Magnetic Resonance

Atoms consist of nuclei and electrons that carry positive and negative charges, respectively. Movement of electric charges generates a magnetic field. Electrons and nuclei can therefore have a magnetic moment, originating from a circular electric current, and can be regarded as tiny magnetic dipoles. This magnetic moment is related to the quantum phenomenon called *spin*. The spins of electrons and nuclei are quantized, and are characterized by the *spin quantum number*. In the presence of external magnetic fields, the magnetic moments adopt preferred orientations relative to the field of different energies. In magnetic resonance methods, transitions between these states are induced by electromagnetic radiation. Magnetic resonance methods provide information on the local environment and mobility of the probes and on inter-probe distances.

## 20.1 NUCLEAR MAGNETIC RESONANCE

Nuclear magnetic resonance (NMR) is based on the inherent magnetic properties of nuclei. Therefore, we will focus on nuclear spins in the following chapters, and will return to electron spins when we discuss electron paramagnetic resonance (EPR, Section 20.2).

### 20.1.1 NUCLEAR SPINS AND THE ZEEMAN EFFECT

The positive charge of nuclei is located on the protons, and all nuclei except hydrogen contain more than one proton. The spin quantum numbers $I$ of nuclei are multiples of ½, i.e. 0, ½, 1, ³⁄₂, and so on. In the absence of a magnetic field, the orientation of the magnetic moment of the nuclei is random, and the nuclei are energetically equivalent (degenerated states). In the presence of an external magnetic field, however, the magnetic dipole of a nucleus adopts certain orientations relative to the direction of the field. A nucleus with the spin quantum number $I$ can be found in $(2I+1)$ orientations. NMR experiments on biomolecules are often performed with nuclei with $I = ½$, such as $^1H$, $^{13}C$, $^{15}N$, or $^{31}P$, whose magnetic moment can adopt only two different orientations relative to the magnetic field. In the simplified image of nuclei as tiny magnets, the nuclear spins can be oriented such that the magnetic dipole is aligned parallel or antiparallel to the external magnetic field. These two orientations are of different energies, and the states are separated by an energy difference $\Delta E$. In general, the orientation parallel to the magnetic field is the state with the lower energy, the antiparallel orientation is the higher energy state. The splitting of the two spin states into two levels of different energy by an external magnetic field is known as the *Zeeman effect* (Figure 20.1). The Zeeman effect thus generates the two-state system required for NMR spectroscopy that are inherently present in the molecules in other types of spectroscopy: in NMR (and EPR, see Section 20.2) the energy states that are later used for measurements have first to be created by applying a magnetic field.

In the presence of a magnetic field, we can induce transitions between the lower and the higher energy level by irradiating the sample with electromagnetic radiation of the correct energy to provide the energy difference $\Delta E$. The difference between the energy levels depends on the strength of the external magnetic field. In an external field of strength $B_0$ in the z-direction, the energy difference for a spin ½ nucleus is

$$\Delta E = 2m_z B_0 \qquad \text{eq. 20.1}$$

where $m_z$ is the component of the magnetic moment of the nucleus that is parallel to the external field. The magnetic moment in the z-direction is dependent on the *magnetogyric ratio $\gamma$* of the particular type of nucleus according to

$$m_z = \frac{h\gamma}{4\pi} \qquad \text{eq. 20.2}$$

The magnetogyric ratio depends on the charge and the mass of the nucleus (Table 20.1). It is positive for $^1$H and $^{13}$C, but negative for $^{15}$N. The sign of the magnetogyric ratio determines which orientation of spins is the lower energy state: for nuclei with positive magnetogyric ratio $\gamma$, the orientation parallel to the external field is energetically more favorable, for nuclei with negative $\gamma$, the antiparallel orientation is the lower energy state. By combining eq. 20.1 and eq. 20.2, we obtain

$$\Delta E = \frac{h\gamma}{2\pi} B_0 \qquad \text{eq. 20.3}$$

Even with the strongest magnetic fields in commercially available spectrometers today (23.5 T, 1 GHz NMR spectrometer), the energy difference $\Delta E$ between the two states is tiny compared to the energy differences between electronic or vibrational states. According to the Boltzmann equation (eq. 19.7, eq. 19.8), the different energy levels are therefore populated almost equally, with just a slight excess of nuclei in the lower energy state. With stronger magnetic fields, the energy gap and the population difference increase, leading to higher sensitivity of NMR experiments with increasing magnetic fields.

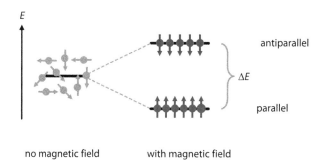

**FIGURE 20.1:** **The Zeeman effect**. Nuclei with spin = ½ can exist in two different energetically different states in the presence of an external magnetic field (red, blue), a phenomenon called Zeeman effect. Due to the small energy differences $\Delta E$, the two states are almost equally populated.

With the relation $\Delta E = h\nu$, we can calculate the frequency $\nu$ of the electromagnetic radiation that is required to induce transitions between two spin states as

$$\nu = \frac{\gamma}{2\pi} B_0 \qquad\qquad \text{eq. 20.4}$$

For a magnetogyric ratio of $268 \cdot 10^6$ $T^{-1}$ $s^{-1}$ (hydrogen, Table 20.1) and typical magnetic field strengths of NMR spectrometers in the 10 T range, we need radiation with a frequency of several 100 MHz or a wavelength of about 1 m, i.e. radiofrequency, to induce transitions of hydrogen spins between the ground state and the excited state in NMR. This frequency is related to the *Larmor frequency* (Section 20.1.2) of the hydrogen nucleus. In fact, NMR spectrometers are named according to the resonance frequency of protons at their respective magnetic field strengths: A 500 MHz spectrometer has a magnetic field of 11.8 T, which translates into a Larmor frequency of 500 MHz for protons. The 23.5 T magnetic field of a 1 GHz spectrometer leads to a Larmor frequency for protons of 1 GHz.

**TABLE 20.1**
**NMR parameters of biologically relevant nuclei.**

| nucleus | spin quantum number | relative abundance (%) | magnetogyric ratio $\gamma$ ($10^6$ $T^{-1}$ $s^{-1}$) |
|---|---|---|---|
| $^1H$ | ½ | > 99.9 | 268 |
| $^2H = D$ | 1 | 0.02 | 41 |
| $^{13}C$ | ½ | 1.1 | 67 |
| $^{15}N$ | ½ | 0.37 | −27 |
| $^{19}F$ | ½ | 100 | 25 |
| $^{31}P$ | ½ | 100 | 108 |

## 20.1.2 A One-Dimensional NMR Spectrum: Larmor Frequency, Chemical Shift, *J*-Coupling, and Multiplicity

### 20.1.2.1 The Larmor Frequency

The application of the external magnetic field not only causes the generation of a two-state system and the alignment of the nuclear spins in the applied magnetic field. The interaction of the magnetic momentum of the nucleus with the magnetic field also leads to precession of the spin around the axis that is defined by the direction of the magnetic field. The frequency of this precession movement is the *Larmor frequency* $\nu_0$. It depends on the magnetogyric ratio $\gamma$ of the nucleus and on the strength $B_0$ of the external magnetic field:

$$\nu_0 = -\frac{\gamma}{2\pi} B_0 \qquad\qquad \text{eq. 20.5}$$

This frequency is of the same magnitude as the frequency of electromagnetic radiation that is needed to induce transitions between the two spin orientations (eq. 20.4). The Larmor frequency $\nu_0$ is therefore also called the resonance frequency of the particular nucleus.

### 20.1.2.2 The Local Magnetic Field and the Chemical Shift

Nuclei of the same chemical identity have the same magnetogyric ratio $\gamma$ and the same Larmor frequency. Therefore, we would not be able to distinguish these nuclei within a large molecule, such as different protons in a protein. What makes NMR spectroscopy so powerful is the effect of the chemical environment of the nucleus on the Larmor frequency. The Larmor frequency of a specific nucleus in a larger molecule depends on the local magnetic field that this nucleus experiences. The electrons of the molecule shield its nuclei from the external magnetic field, rendering the local magnetic field slightly smaller than the external field. The effect of the electrons on the local magnetic field $B_{local}$ can be described by the *shielding constant* $\sigma$:

$$B_{local} = (1 - \sigma) B_0 \qquad\qquad \text{eq. 20.6}$$

Particularly strong effects on the local magnetic field come from electronic ring system currents of aromatic groups. The external magnetic field induces ring currents in aromatic systems, and the current in turn generates an additional local magnetic field (Figure 20.2). This induced field is parallel to the external field outside the ring system, but antiparallel inside the ring system. Hence, nuclei in the plane of the ring on the outside experience a higher local magnetic field, they are deshielded. By contrast, nuclei above or below the plane of the ring system experience a lower local magnetic field: they are shielded from the external field. Paramagnetic ions, such as $Fe^{2+}$, $Fe^{3+}$, or $Mn^{2+}$, that are present in the active sites of many enzymes also strongly affect the local magnetic field. It is this effect of electrons on the local magnetic field that allows distinction of the same type of nuclei by NMR.

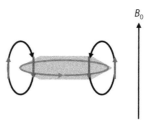

**FIGURE 20.2:** **Local magnetic fields around aromatic ring systems.** The external magnetic field ($B_0$) induces ring currents (gray circular arrow) in the aromatic system (gray hexagon). These ring currents are associated with a magnetic field (black circular arrows) in the direction of the external field (red) outside the aromatic ring, and antiparallel to the external magnetic field inside of the ring (blue). The local magnetic field and the Larmor frequency of a nucleus thus depends on its chemical environment that leads to shielding (blue) or deshielding (red) of nuclei.

As a consequence of the shielding effect of electrons, the same nuclei in different electronic environments within the molecule have slightly different Larmor frequencies. This effect is called the *chemical shift*, because the Larmor frequency is shifted from the expected value (eq. 20.5). The chemical shift δ is expressed in parts per million (ppm) of the unperturbed resonance frequency $\nu_0$ at the external field strength $B_0$. δ is calculated relative to a reference substance with the Larmor frequency $\nu_{ref}$:

$$\delta = \frac{\nu_s - \nu_{ref}}{\nu_0}$$

eq. 20.7

where $\nu_s$ is the measured Larmor frequency at the local magnetic field strength $B_{local}$, and $\nu_0$ is the Larmor frequency at the external field strength $B_0$. While the Larmor frequencies depend on the strength of the external magnetic field (eq. 20.5), the chemical shift δ is independent of the field strength. A common reference substance for proton NMR is tetramethylsilane (TMS). The twelve protons in TMS are in an identical chemical environment and provide a strong reference signal. Shielded nuclei experience a magnetic field $B_{local} < B_0$, and their Larmor frequency is $\nu_s < \nu_0$. The chemical shift δ of shielded nuclei is therefore small. Deshielded atoms experience a magnetic field $B_{local} \approx B_0$, and the Larmor frequency is $\nu_s \approx \nu_0$, which corresponds to larger chemical shifts. In continuous wave NMR experiments where the frequency of the electromagnetic radiation is fixed and the spectrum is measured by changing the external magnetic field $B_0$ (Section 20.1.2.5), a higher external field $B_0$ is required for shielded nuclei than in the absence of shielding. The region of the NMR spectrum where resonances from highly shielded nuclei are found is therefore also called the high-field or upfield region. Resonances from deshielded nuclei populate the low-field or downfield region of the spectrum.

It is not possible to assign a particular nucleus to a specific chemical group just on the basis of its chemical shift, but certain types of nuclei show resonances in characteristic regions of an NMR spectrum. Aliphatic protons, for example, typically exhibit low chemical shifts, whereas protons in aromatic systems show higher chemical shifts (Figure 20.3).

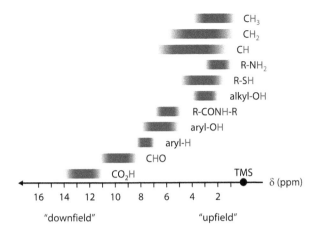

**FIGURE 20.3:    Chemical shifts for protons in different groups.** The chemical shift relative to the reference substance TMS depends on the chemical environment of the proton, resulting in characteristic chemical shifts for protons within functional groups.

Chemical shifts are very sensitive to the local environment, and serve as indicators for secondary structure formation in proteins (Box 20.1) or can be used to map interaction sites (Box 20.2).

---

**BOX 20.1: CHEMICAL SHIFTS AND SECONDARY STRUCTURE.**

The chemical shift of $C_\alpha$-protons in proteins is sensitive to the conformation, and depends on the $\phi$- and $\psi$-angles of the polypeptide chain (Section 16.1.2 and Figure 16.16). $C_\alpha$-protons within $\alpha$-helices typically show an upfield shift (lower chemical shift) compared to random coil, whereas $\beta$-sheet formation is associated with a downfield shift (higher chemical shift), leading to a chemical shift difference $\Delta\delta = \delta_{observed} - \delta_{random\ coil}$. Residues that show $\Delta\delta > 0.1$ are assigned a chemical shift index of 1, whereas $\Delta\delta < -0.1$ corresponds to a chemical shift index of $-1$. Otherwise, the chemical shift index is zero. Protein regions that adopt $\alpha$-helical or $\beta$-sheet structure can be identified by plotting the *chemical shift index* as a function of the position within the amino acid sequence. At least four consecutive residues with a chemical shift index of $-1$ indicate $\alpha$-helix formation. At least three consecutive residues with a chemical shift of 1 point to $\beta$-sheet formation. The remaining regions are classified as random coil. Ends of secondary structure elements are marked by a chemical shift index of opposite sign, or by two residues with a chemical shift index of zero. Wishart *et al.* (1992) *Biochemistry* 31(6): 1647–1651.

---

**BOX 20.2: CHEMICAL SHIFT PERTURBATION
TO MAP INTERACTION SITES.**

The sensitivity of the chemical shift to the environment can be exploited to map binding sites for ligands in chemical shift perturbation experiments. To map the ligand binding site on a protein, an NMR spectrum of the free protein is measured and the resonances of the amide protons are assigned. In a second step, the NMR spectrum is recorded in the presence of the ligand. Some amide proton resonances will undergo a large change in chemical shift upon ligand binding, some show smaller changes, and some do not change

*(Continued)*

at all. In NMR titrations, chemical shifts are plotted as a function of the concentration of the binding partner to determine the $K_d$ value of the complex. This procedure is only possible when binding and dissociation of the ligand are rapid, and the measured chemical shift is a population-weighted average of the chemical shifts of free and bound states (fast-exchange regime, see Section 20.1.7; Box 20.7). Mapping the residues whose amide protons exhibit a large chemical shift perturbation on the structure of the protein reveals the relative location of these residues. A clustering of these amino acids often reveals the binding site for the ligand. However, it has to be noted that chemical shift perturbation may also result from conformational changes. A large chemical shift perturbation therefore cannot directly be taken as a proof for the contribution of a particular amino acid in binding.

### 20.1.2.3  Scalar Coupling and Multiplets

A single nucleus, e.g. a proton, yields one defined resonance in an NMR spectrum, with the spectral position depending on its electronic environment. In a larger molecule, this nucleus (nucleus 1) is bonded to other nuclei by bonding electrons (Figure 20.4), and these nuclei affect the local magnetic field experienced by nucleus 1. The spin of a bonded nucleus 2 can assume two different orientations. Depending on the orientation of spin 2, nucleus 1 has two slightly different Larmor frequencies, and two resonances for nucleus 1 will be detected in an NMR experiment. This phenomenon is called *scalar* or *J-coupling*. The split resonance resulting from *J*-coupling is called a *doublet*. The separation of the two resonances (measured in Hz) is called the *J-coupling constant*.

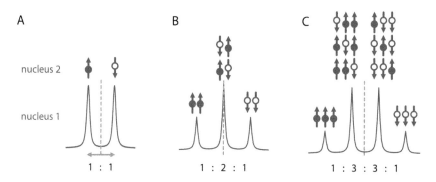

**FIGURE 20.4:    Interaction between two nuclei by scalar coupling.** A: When two nuclei 1 and 2 are connected by chemical bonds, the orientation of the magnetic moment of nucleus 2 (red) is experienced by nucleus 1 (blue), and influences the energy levels of nucleus 1. Coupling of nucleus 1 to nucleus 2 splits the single resonance of nucleus 1 into two resonances of equal intensity, corresponding to transitions of nucleus 1 with the spin of nucleus 2 in either of the two possible orientations (left). The scalar coupling constant *J* is the difference between the frequencies (in Hz) of the two resonances (gray double-headed arrow). B: If nucleus 1 is coupled through chemical bonds to two nuclei 2, the resonance of nucleus 1 is split into a triplet, with the intensities 1:2:1. The orientations of the two nuclei 2 are indicated above the lines of the triplet. C: Coupling to three nuclei 2 splits the resonance of nucleus 1 into a quartet, with an intensity ratio of the four lines of 1:3:3:1, corresponding to the possible orientations of the spins of the three nuclei 2 that are indicated above the line of the quartet (for $\gamma > 0$).

In NMR spectroscopy, *J*-coupling through up to three bonds is typically analyzed. The *Karplus equation* relates the value of *J* for two nuclei coupled through three bonds to the dihedral angle $\phi$ that is defined by the bonds (see Figure 15.5):

$$J = a\cos^2\phi + b\cos\phi + c \qquad \text{eq. 20.8}$$

The constants $a$, $b$, and $c$ depend on the nuclei and the connecting bonds. *J*-coupling constants are between 0.1 and 10 Hz for protons, but can exceed these values for other nuclei. A typical example of *J*-coupling in proton NMR is the splitting of resonances for methyl protons in acetaldehyde into doublets by the neighboring carbonyl proton (Figure 20.5). Note that splitting of the resonance of nucleus 1 into a doublet by nucleus 2 also leads to splitting of the resonance for nucleus 2 by nucleus 1, with the same *J*-coupling constant. Identical coupling constants for two doublet resonances therefore serve as an indication for which nuclei are coupled, i.e. are connected by chemical bonds.

If nucleus 1 has two or three identical neighbors, its resonances are split into a triplet or quadruplet. The area under the resonance lines is proportional to the number of relative spin orientations leading to the particular resonance (Figure 20.6). For acetaldehyde, the resonances of the carbonyl proton are split into a quadruplet due to the four possible orientations of the three spins of the methyl protons (Figure 20.5).

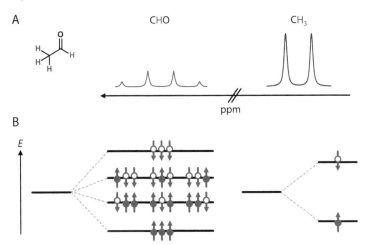

**FIGURE 20.5:    Scalar coupling in acetaldehyde.** A: Acetaldehyde contains one proton in the carbonyl group (blue) and three protons in the methyl group (red). B: The resonance of the three methyl protons is split into a doublet because of coupling with the carbonyl proton whose magnetic moment can adopt two different configurations with different energies. The resonance for the carbonyl proton is split into a quartet because of coupling with the three methyl protons whose magnetic moments can have different combinations of orientation, giving rise to four energy levels.

### 20.1.2.4 Shape of NMR Lines

In emission spectroscopy, resonance lines are *Lorentzian-shaped*. The intensity *I* as a function of the frequency $\omega$ is

$$I(\omega) = I_0 \cdot \frac{\Gamma / 2}{(\omega - \omega_0)^2 + (\Gamma / 2)^2}$$

eq. 20.9

$I_0$ is the amplitude, i.e. the intensity at the resonance frequency $\omega_0$, and $\Gamma$ is the sum of all rate constants that lead away from the excited state. $\Gamma$ corresponds to the full width at half maximum of the Lorentzian line. For NMR transitions, the relaxation from the excited to the ground state occurs with the *transverse relaxation time* $T_2$ of spin inversion (see Section 20.1.4). In the absence of other relaxation processes, $\Gamma$ is

$$\Gamma = \frac{1}{T_2}$$

eq. 20.10

and we can directly determine the relaxation time $T_2$ from the width of an NMR line.

Overall, we thus obtain multiple pieces of information from a one-dimensional NMR spectrum (Figure 20.6): the chemical shift is indicative of the chemical environment of a particular proton. The area under the resonance line is proportional to the number of nuclei with the same chemical shift. The line width depends on the relaxation time. The multiplicity of a particular resonance indicates the number of spins with which the respective nucleus shows scalar coupling. The coupling constant is represented by the distance of the individual lines of a multiplet.

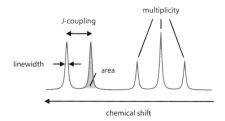

FIGURE 20.6:   **Information from an NMR spectrum.** The chemical shift is characteristic of the local magnetic field experienced by a nucleus. The multiplicity provides information on the number of nuclei that are coupled *via* chemical bonds, and the *J*-coupling is a measure of the coupling strength. The relaxation time (see Section 20.1.4) is encoded in the line width of the resonance, and the area under the resonance or the lines of a multiplet is proportional of the number of nuclei that contribute to this resonance.

### 20.1.2.5 Instrumentation

A simple NMR spectrometer contains a sample compartment within an electromagnetic or permanent magnet that generates the external magnetic field $B_0$ (Figure 20.7). This magnetic field establishes the two-state systems by way of the Zeeman effect. A high-frequency radio transmitter provides the radiation for excitation of spins in the sample. When the resonance condition is fulfilled, spin inversion will be detected by a receiver, an electromagnetic coil in which the change in magnetization of the sample induces a current that is measured. Scanning of excitation frequencies can be achieved in two different ways: either, in an approach analogous to wavelength scanning by means of a monochromator in optical spectroscopy, the magnetic field $B_0$ is kept constant and the frequency $\nu$ of the excitation radiation is varied, or the frequency of the excitation is kept constant and the $B_0$-field is scanned. In this case, the energy differences between states are changed, and the resonance condition for different nuclei is met at different field strengths $B_0$. Both of these *continuous wave* (cw) methods measure the absorption lines for individual transitions separately, which together give the NMR spectrum.

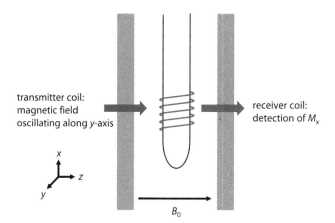

FIGURE 20.7:   **NMR spectrometer.** A static magnetic field $B_0$ in the z-direction is generated by a superconducting magnet (gray) surrounding the sample compartment (black). A transmitter coil (not shown) provides the oscillating magnetic field along the y-axis that induces transitions between the two spin states. The oscillating magnetization in the xy-plane induces an oscillating current in the receiver coil (red) that is detected as the NMR signal.

### 20.1.3 The Nuclear Overhauser Effect: Distance Information

*J*-coupling is mediated by bonding electrons. In contrast, the *nuclear Overhauser effect* (NOE) is caused by an interaction of two spins through space. This dipolar coupling leads to the transfer of magnetization by *cross-relaxation*. Dipolar coupling is strongly distance-dependent: its efficiency depends on the inverse sixth power of the inter-nuclear distance. We have encountered this distance dependence for the interaction of dipoles before in the distance dependence of London forces (eq. 15.13), and in the distance dependence of Förster resonance energy transfer (eq. 19.128). Because of its strong distance dependence, the NOE provides a powerful tool to obtain intramolecular distance information.

NOEs can be detected by changes in the intensity of a particular resonance, when a second resonance is saturated by continuous irradiation, such that the populations of the two states are equal. The measured intensity I can increase (positive NOE) or decrease (negative NOE) relative to the intensity $I_0$ without saturation of the second resonance. The nuclear Overhauser enhancement is defined as

$$NOE = \frac{I - I_0}{I_0}$$

eq. 20.11

To understand the cross-relaxation that leads to the NOE, we consider the energy levels of the two nuclei and the effect of saturation of one transition on the population differences (Figure 20.8). In the unperturbed system of two nuclei A and B, the energetic differences between the energy levels determine the population of the individual states. The lowest energy state, with both spins in an orientation parallel to the external field (for nuclei with $\gamma > 0$), is the state with the highest population. The two possible states with one nucleus in parallel and the second in antiparallel orientation are isoenergetic and have a lower population. The highest energy level, corresponding to both nuclei in an antiparallel orientation to the external magnetic field, is even less populated. Note that the energy differences between the lowest and the intermediate state and between the intermediate and highest energy level are identical. We assume we have $n$ molecules in the states with intermediate energy, where the spins of $A$ and $B$ have opposite orientations. The Boltzmann distribution tells us that in equilibrium we have $n + x$ molecules in the state with the lowest energy level, and $n - x$ molecules in the highest energy level. Irradiation at the Larmor frequency of nucleus A will now alter the populations. Upon saturation of A, the populations of A in the spin-up and spin-down state (at a constant orientation of spin B) become equal (Figure 20.8). Effectively, the saturation of nucleus A leads to the transfer of $x/2$ molecules from the lower to the higher energy level of each A transition. The populations of the two states that are relevant for the transitions of nucleus B are now $n + x/2$ – $(n - x/2)$. The population difference therefore still equals $x$, and the intensity of the resonance of B is not affected. The intensity of this resonance will only be altered if the population differences can be reduced by cross-relaxation in two ways: (1) by transitions between the two states of intermediate energy ($W_0$ transition), and (2) by transitions between the highest and lowest energy level ($W_2$ transition). $W_0$ transitions involve no net spin inversion, whereas $W_2$ transitions invert both spins. Both $W_0$ and $W_2$ transitions are spin-forbidden because they involve changes of two spin states. These transitions are not observed directly by NMR, but can occur during relaxation. $W_0$ and $W_2$ transitions are only possible if the respective nucleic are closer than ca. 0.5 nm. The outcome depends on which of these cross-relaxation pathways dominates. If $W_0$ is the dominant (faster) process, the population difference for the B transitions will be decreased from $x$ to $x - a$. As a consequence, the intensity of the resonance line for B is reduced (negative NOE, Figure 20.8). If, on the other hand, the cross-relaxation *via* $W_2$ is faster, the population difference for B will increase to $x + a$. In this case, the intensity of the resonance of B increases (positive NOE, Figure 20.8).

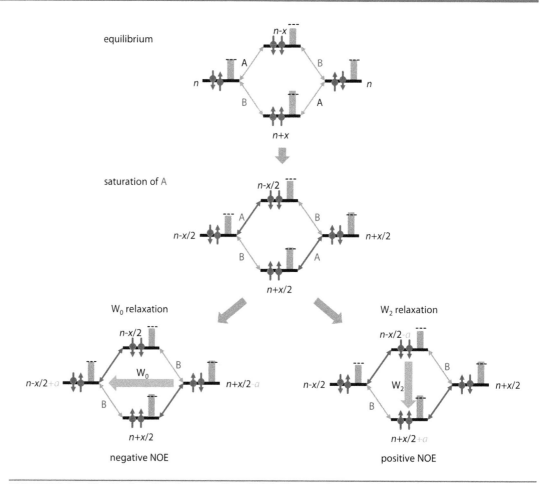

equilibrium

saturation of A

$W_0$ relaxation

$W_2$ relaxation

negative NOE

positive NOE

**FIGURE 20.8: The Nuclear Overhauser effect (NOE).** The populations of the possible energy levels of two spins A (blue) and B (red) according to the Boltzmann distribution. We assume that the population of the two intermediate energy levels is $n$. The lower energy level with the two spins parallel to the external field is more populated ($n + x$), and the higher energy level with both spins antiparallel to the magnetic field is less populated ($n - x$). The gray bars indicate the populations, the dotted black line marks $n$. When we irradiate the sample at the Larmor frequency of nucleus A, the A transitions (red) will be saturated, i.e. half of the A-spins in the ground state will be excited, leading to equal populations of the states involved ($n - x/2$ or $n + x/2$). If the system can now cross-relax by $W_0$ transitions between the intermediate energy levels, involving a spin flip-flop and no change in overall spin, the population difference for nucleus B will be reduced from $x$ to $x - a$, and the resonance for B is smaller. This effect is called the negative nuclear Overhauser effect. If, on the other hand, cross-relaxation by $W_2$ transitions between the highest and the lowest level take place, and both spins flip orientation in the same directions, the population difference for nucleus B will be increased from $x$ to $x + a$. In this case, we observe a larger resonance line and a positive nuclear Overhauser effect.

We will see in Section 20.1.6 that the NOE is an important tool for the determination of molecular structures by two-dimensional NMR spectroscopy. The nuclear Overhauser effect can lead to a transfer of magnetization from one molecule to its binding partner. This effect can be exploited to probe binding, and to screen libraries of molecules for binding to pharmaceutical targets (Box 20.3).

---

### BOX 20.3: SATURATION TRANSFER DIFFERENCE-NMR (STD-NMR) TO SCREEN LIGAND LIBRARIES FOR BINDING.

STD-NMR is based on magnetization transfer from a macromolecule to its ligand by the nuclear Overhauser effect. To investigate protein-ligand binding, a resonance of the protein is selectively saturated by electromagnetic radiation at the corresponding Larmor

*(Continued)*

frequency, and the 1D-proton NMR spectrum is measured (on-resonance spectrum). If the ligand binds to the protein, cross-relaxation at the ligand-protein interface allows for transfer of saturation from the protein to the ligand, which leads to a reduced population difference between the two spin states of the respective proton, and to a reduced intensity in the NMR spectrum. In a control experiment, the saturation is performed with radiation frequency in a spectral region without protein resonances (off-resonance spectrum). The off-resonance spectrum contains resonance lines for all protons in the ligand mixture. The difference spectrum (off-resonance minus on-resonance) then shows the saturation difference, and contains only resonances from ligands to which saturation has been transferred (Figure 20.9).

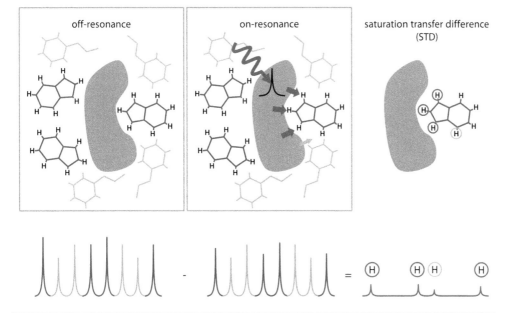

**FIGURE 20.9:    Principle of STD-NMR.** In the off-resonance spectrum, resonances for all protons are recorded. The on-resonance spectrum is taken after selective saturation of a protein resonance. Due to magnetization transfer from the protein to bound ligands, the resonances of protons involved in binding are reduced in intensity. The saturation transfer difference (STD) spectrum shows only resonances from the ligand that binds. The closer the proton to the protein-ligand interface, the higher the intensity of its resonance line.

STD-NMR relies on exchange between bound and free ligand during the experiment. Saturation transfer occurs while the ligand is bound, and the more ligand molecules bind and dissociate from the binding site during the experiment, the higher the transfer difference. Therefore, STD-NMR is ideally suited to detect low-affinity interactions, with $K_d$ values in the low micromolar to the millimolar range. The method can be used for screening and for the identification of lead structures in pharmaceutical research. STD-NMR also provides information on the functional groups of the ligand that mediate binding. Magnetization transfer is achieved through the strongly distance-dependent nuclear Overhauser effect, and saturation is therefore most efficient for ligand protons that are closest to the binding surface. The relative intensity of ligand resonances in the STD spectrum provides information on the orientation of the ligand relative to the binding site and on the protons involved in binding.

### 20.1.4 Magnetization and Its Relaxation to Equilibrium: Fourier Transform-NMR and the Free Induction Decay

In the absence of a magnetic field, the spins of the individual nuclei have random orientations relative to each other, and their magnetic moments cancel over the whole sample. Once an external magnetic field is applied, however, quantum mechanics restricts the spin of each nucleus to two different orientations. Each individual spin precesses around the $z$-axis (of the external magnetic field), but the precession of the individual spins is not correlated. Consequently, the vector components of the magnetization in the $xy$-plane cancel. In contrast, the magnetization in the $z$-direction does not cancel because the equilibrium population of spins parallel to the external field is slightly higher than the population with an antiparallel orientation. The magnetic moments of all spins therefore add up to a macroscopic *magnetization* in the $z$-direction (Figure 20.10).

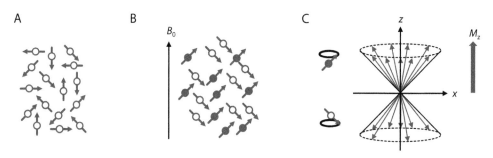

**FIGURE 20.10:** **Generation of macroscopic magnetization.** A: In the absence of an external magnetic field, the orientations of the nuclear spins are random, and their magnetic moments (arrows) cancel. B: In the presence of an external magnetic field, the spins are oriented either parallel (filled circle) or antiparallel (open circle) to the $B_0$-field. C: The tip of the magnetization vector (arrow) precesses around the $z$-axis. The precession of the individual spins is not correlated, and the vector component of the magnetization in the $xy$-plane cancels. Because of the slight excess of spins in the orientation parallel to the external field, their components add up to a magnetization $M_z$ in the $z$-direction.

Now imagine we apply a short pulse of electromagnetic radiation at the Larmor frequency of our nuclei to induce transitions between the two energy levels. For equal populations of both states, the $z$-magnetization will disappear, for population inversion the net magnetization in the $z$-direction will be inverted. What happens to the magnetization in the $xy$-direction? At the beginning of the NMR experiment, before applying the pulse, the magnetization in the $xy$-plane is zero. When transitions between the two states are induced by the radiofrequency pulse, the spins will start to precess in phase. This phase coherence comes from the common orientation of the spins when the radiofrequency is applied, and from their precession with the same Larmor frequency. Their in-phase precession leads to a net magnetization in the $xy$-plane that rotates with the Larmor frequency. Overall, the excitation pulse reduces the magnetization in the $z$-direction, and introduces a magnetization in the $xy$-plane. Vector addition of these components gives the overall magnetization $M$ that can be regarded as the magnetization in the $z$-direction that has been tilted away from the $z$-axis by an angle $\alpha$ (Figure 20.11). A pulse that causes equal populations of both energy levels will convert the initial macroscopic magnetization in the $z$-direction into magnetization in the $y$-direction. Such a pulse thus tilts the overall magnetization by 90°, and is therefore called a 90° pulse. A 180° pulse leads to population inversion and converts the initial $z$-magnetization into negative magnetization ($-z$). In general, the angle $\alpha$ is related to the duration $\Delta t$ of the pulse by

$$\alpha = \gamma\, B_1\, \Delta t$$

eq. 20.12

where $B_1$ is the magnetic field induced by the electromagnetic radiation of the pulse, and $\gamma$ is the magnetogyric ratio of the nucleus. When the excitation pulse stops, the system will return to equilibrium by *relaxation* (Figure 20.11). Two processes contribute to this relaxation. On the one hand, the population excess for the parallel orientation is restored by spin transitions, leading to an increase of the $z$-magnetization to its original value. This longitudinal relaxation (along the axis defined by the external field) occurs exponentially with the *longitudinal relaxation time* $T_1$, the *spin-lattice relaxation time*. On the other hand, the loss of phase coherence between the precessing spins reduces the $xy$-magnetization to zero with the *transverse relaxation time* $T_2$, the *spin-spin relaxation time*. $T_1$ and $T_2$ relaxation times can be measured by defined pulse sequences (Box 20.4).

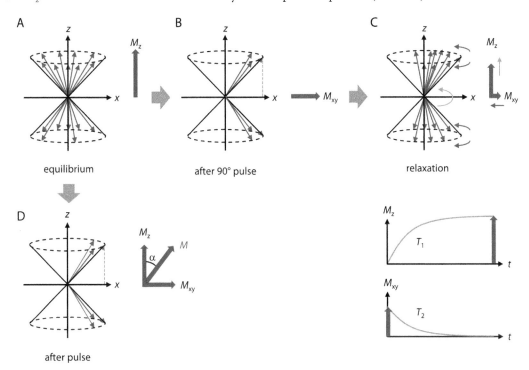

**FIGURE 20.11:    Magnetization, magnetization flipping, and relaxation.** A: In equilibrium, more nuclei have their magnetic dipole oriented parallel to the external magnetic field in the $z$-direction than antiparallel ($\gamma > 0$). The excess of spins with parallel orientation causes a net magnetization in the $z$-direction, $M_z$. B: Excitation leads to the inversion of spins into the antiparallel orientation of the magnetic moments, and a decrease of $M_z$. When both orientations are equally populated, $M_z$ is zero. At the same time, the precession of spins around the $z$-axis is synchronized: they precess in phase. The correlation of spin precession leads to a net magnetization in the $xy$-plane, $M_{xy}$. C: Relaxation to equilibrium occurs by two distinct processes. The inversion of spin orientations (orange arrow) restores the equilibrium population and the original $M_z$ magnetization with a time constant $T_1$, the longitudinal or spin-lattice relaxation time. At the same time, the coherence in precession around the $z$-axis is slowly lost (blue arrows), such that $M_{xy}$ decreases with the time constant $T_2$, the transverse or spin-spin relaxation time. D: In general, a pulse induces transitions between the two spin states, leading to a reduced population difference and phase coherence. The magnetic moment $M_z$ is reduced, and a magnetic moment appears in the $y$-direction. The overall magnetization $M$ is thus tilted by an angle $\alpha$.

We can measure our NMR signal after pulsed excitation by detecting the magnetization in the $xy$-plane as a function of time with a detection coil. The rotating magnetization will induce a sinusoidal current in the coil that oscillates with the Larmor frequency of the nucleus. This sinusoidal current decays to zero with the relaxation time $T_2$. The exponentially decaying sinusoidal oscillation is called the *free induction decay* (FID).

---

**BOX 20.4: MEASUREMENT OF SPIN-LATTICE ($T_1$)
AND SPIN-SPIN ($T_2$) RELAXATION TIMES.**

$T_1$ relaxation times can be measured in *inversion-recovery* experiments (Figure 20.12). First, the magnetization $M_z$ is inverted to $-M_z$ by a 180° pulse. After time $t$, the magnetization is reduced because of relaxation by spin inversion. To measure the size of the magnetization, a 90° pulse is applied that brings the magnetization into the $xy$-plane where it

*(Continued)*

can be detected. A series of measurements with different values for $t$ gives a set of NMR spectra from which the peak height for each resonance can be determined as a function of $t$. From the time constant $\tau$ derived from the exponential dependence of the peak height on $t$, the individual $T_1$ relaxation time for each resonance can be determined. $T_2$ relaxation times are typically measured in spin-echo experiments. The $M_z$ magnetization is brought into the xy plane by a 90° pulse. Transverse relaxation and the loss of phase coherence contribute to relaxation during the time $t$. Phase coherence is lost because of inhomogeneities of the external magnetic field, and chemically identical nuclei precess at slightly different Larmor frequencies. As a consequence, the magnetization vectors for individual spins spread out, with more slowly precessing vectors at the end, more rapidly precessing ones at the front of the set of vectors. After time $t$, a 180° pulse along the $y$-axis inverts the spins. Now their order is reversed: the slower ones are at the front, the faster ones at the end. Therefore, the faster precessing vectors can "catch up" with the others, leading to a refocusing of the $M_{xy}$ magnetization after the time $2\tau$. This effect is called the *spin echo*. The spin echo cannot compensate for the transverse relaxation that has taken place, and the spin echo will be smaller than the original magnetization. The decay of the spin echo with time provides the relaxation time $T_2$. Relaxation times are measured to obtain information on the dynamics of molecules.

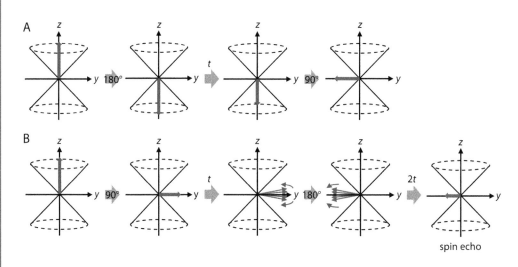

**FIGURE 20.12: Measurement of relaxation times**. A: Measurement of $T_1$ by inversion-recovery. B: Measurement of $T_2$ by spin-echo experiments.

For one type of nucleus with one Larmor frequency, the FID only contains one component. For several nuclei with different Larmor frequencies, however, the FID is the sum of all individual FIDs for each nucleus with its particular Larmor frequency and transverse relaxation time $T_2$. We can extract the different resonances and their Larmor frequencies encoded in the FID by the mathematical operation called *Fourier transformation* that converts the measured signal as function of time into a signal as a function of inverse time = frequency (see also Fourier transformations in IR-spectroscopy, Section 19.4.3, and in X-ray crystallography, Chapter 22). The Fourier transform of the measured magnetization as a function of time is the intensity as a function of frequency, which is our Fourier transform NMR (FT-NMR) spectrum (Figure 20.13).

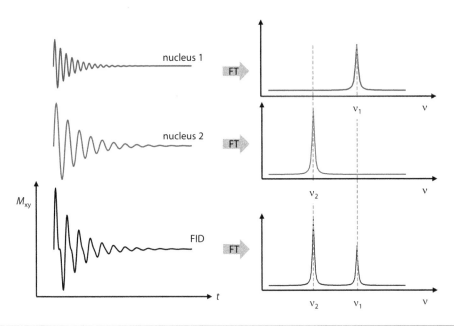

**FIGURE 20.13:** **The FID and Larmor frequencies of nuclei.** The nuclei 1 and 2 with their specific Larmor frequencies $\nu_1$ and $\nu_2$ cause a sinusoidal oscillation with the Larmor frequency that decays exponentially with the transverse relaxation time $T_2$ (red, blue). Fourier transformation (FT) of the exponentially decaying sinusoidal oscillation gives a Lorentzian line at the oscillation frequency $\nu_1$ or $\nu_2$. The measured FID ($M_{xy}$ as a function of time, black) is the sum of the individual FIDs for the two nuclei. Fourier transformation gives the two Lorentzian lines at the Larmor frequencies $\nu_1$ and $\nu_2$. The amplitude of the two individual FIDs determines the intensity of the lines.

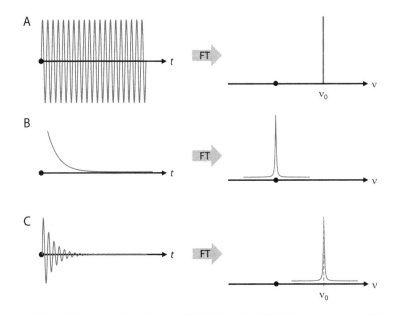

**FIGURE 20.14:** **Fourier transformation of functions describing the FID.** A: The Fourier transform of an infinite sinusoidal oscillation is a $\delta$-function at the frequency $\nu_0$. B: Fourier transformation of an exponential function gives a Lorentzian line centered at zero (marked by the black circle on the axis). C: The Fourier transform of an exponentially decaying sinusoidal oscillation, such as the FID, is a Lorentzian line centered at the frequency $\nu_0$.

One important feature we have not yet discussed is the consequence of using a short excitation pulse. By Fourier transformation of this pulse (intensity as a function of time), we can also analyze the excitation pulse for its frequency components (intensity as a function of frequency). The excitation pulse consists of an infinite sine wave, multiplied by a square function. The Fourier transform (FT) of the infinite sine wave is a δ-function (Figure 20.14) that has a value of 1 at the frequency $f$ of the wave, but is zero at all other $x$-values. The Fourier transform of a square function is a $(\sin x)/x$ function symmetric to the $y$-axis. The Fourier transform of the product of the two functions is the convolution (see Box 19.14, Section 29.15) of the two Fourier transformed functions, i.e. a $(\sin x)/x$ function (Figure 20.15), shifted along the $x$-axis such that it is centered at the frequency ν. Thus, the excitation with a short pulse provides different frequencies simultaneously. The shorter the pulse, the broader the spectrum of frequencies it contains. With sufficiently short excitation pulses, we can therefore induce transitions of all nuclei of interest simultaneously. NMR using short excitation pulses and measuring the FID is called *Fourier transform-NMR* (FT-NMR).

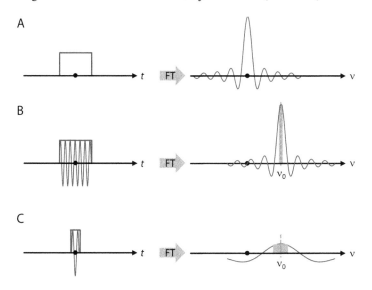

**FIGURE 20.15:    Fourier transformation of functions describing the excitation pulse.** A: The Fourier transformation of a rectangular pulse is a $(\sin x)/x$ function, centered at zero. The shorter the pulse, the broader the $(\sin x)/x$ function. B: Fourier transformation of a short pulse of a sinusoidally oscillating intensity gives a $(\sin x)/x$ function, centered at the frequency $\nu_0$ of the sine function. C: Shorter pulses correspond to a wider $(\sin x)/x$ function, and can provide a wider range of frequencies around $\nu_0$ (gray)

### 20.1.5 Two-Dimensional FT-NMR: COSY and NOESY

NMR spectra of larger molecules with increasing number of nuclei become more and more complex. It is therefore increasingly difficult to assign resonances to a particular nucleus, and to directly determine the structure of the molecule of interest. At some point, the spectrum will be so busy that resonances overlap and cannot be distinguished anymore. *Two-dimensional (2D) NMR* introduces a second frequency axis, and thereby enables separation of overlapping resonances along both axes. A conceptual analogy is the separation of proteins by gel electrophoresis in two dimensions. In the first dimension the proteins are separated according to their molecular mass. However, proteins with similar molecular mass will not be separated, and their bands will overlap. We now perform a separation in a second dimension perpendicular to the first. This time we separate not by the molecular mass but by a different parameter, such as the isoelectric point. If the proteins with similar molecular masses have different isoelectric points, they will now be separated in the second dimension into non-overlapping spots. We will now first introduce the general concept of 2D-NMR, and then see how 2D-NMR can be employed to determine high-resolution structures of proteins and nucleic acids.

### 20.1.5.1  Principle of a 2D-FT-NMR Experiment

Imagine all spins in our sample are in equilibrium, and we start our FT-NMR experiment by applying a 90° pulse, rotating the $z$-magnetization onto the $y$-axis. For a certain time interval $t_1$, we now let the system evolve. During this evolution time, the spins will precess with their Larmor frequencies around the $z$-axis. As a consequence, the macroscopic magnetization will rotate in the $xy$-plane with the Larmor frequency, leading to a decrease in $y$- and a concomitant increase in $x$-magnetization. At the same time, transitions of spins between the two states will increase (and eventually restore) the $z$-magnetization. After the time $t_1$, we apply a second 90° pulse that rotates the $y$-component of the magnetization into the $–z$-direction, and then measure the FID of the $x$-magnetization (Figure 20.16). After all spins have relaxed to equilibrium, we repeat this experiment with a different evolution time $t_1$. The series of FIDs with different $t_1$-times can then be converted into frequency space by Fourier transformation. Thus, in 2D-NMR, we perform Fourier transformations in two dimensions, one dimension with the evolution time $t_1$ and in the second dimension with the measurement time $t_2$ of the FIDs.

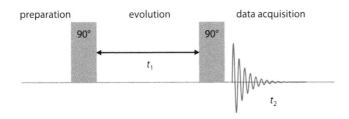

**FIGURE 20.16:**    **Principle of a 2D-NMR experiment.** During the preparation phase, the equilibrium between ground state (spin parallel to the external field) and excited state (antiparallel) is established, leading to a net magnetization $M_z$. The 90° pulse rotates the magnetization onto the $y$-axis. During the evolution time $t_1$, phase coherence can be transferred between nuclei. Longitudinal relaxation restores the $z$-magnetization, and transverse relaxation reduces the magnetization in the $xy$-plane. After $t_1$, a second 90° pulse is applied, and the FID is measured as a function of the time $t_2$.

After the first 90° pulse the in-phase precession of individual spins in the same chemical environment gives rise to the magnetization rotating in the $xy$-plane. The phase coherence of these spins can be transferred to a second group of spins. Phase coherence transfer is in principle possible between nuclei that interact, either by scalar $J$-coupling through bonds or by dipolar coupling through space by the nuclear Overhauser effect (NOE, Section 20.1.3). Phase coherence transfer is detected through a process called frequency labeling. For a group of spins in the same chemical environment, the entire magnetization will be in the $y$-direction after the first 90° pulse at $t_1 = 0$, with no component along the $x$-axis. Due to the precession of the spins and the rotation of the magnetization in the $xy$-plane with the Larmor frequency $ν$, the component of the magnetization along the $y$-axis will then decrease, and the component along the $x$-axis will increase. We can calculate the magnetization in the $x$- and $y$-directions, $M_x$ and $M_y$, at $t_1$ as

$$M_x = M_{xy}\sin\left(2\pi ν t_1\right)$$

eq. 20.13

and

$$M_y = M_{xy}\cos\left(2\pi ν t_1\right)$$

eq. 20.14

When we apply the second 90° pulse at $t_1$, the $y$-component of the magnetization ends up in the $–z$-direction. This makes the $y$-component undetectable, but the $x$-component is not affected by the second 90° pulse and can be detected as the measured signal. This magnetization depends on the Larmor frequency $ν$ and the time $t_1$ according to eq. 20.14. The measured signal oscillates sinusoidally, "labeled" with the frequency $ν$ (Figure 20.17).

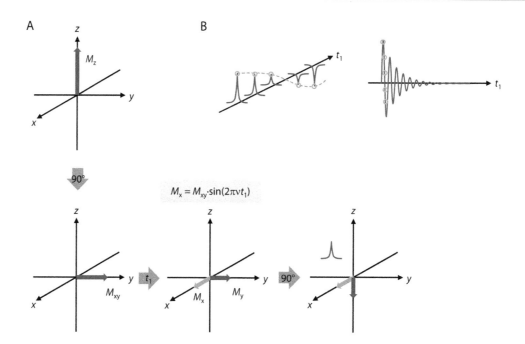

**FIGURE 20.17:    Magnetization and measured signal during a 2D-NMR experiment.** A: $z$-magnetization in equilibrium is turned onto the $y$-axis by a 90° pulse. After the evolution time $t_1$, the $y$-component is reduced to $M_{xy}$ cos$2\pi v t_1$, the $x$-component $M_x$ is $M_{xy}$ sin $2\pi v t_1$. A second 90° pulse converts the $y$-magnetization into the $-z$-direction. The $x$-magnetization is not affected by the detection pulse and is detected. B: The measured signal oscillates sinusoidally as a function of $t_1$ and decays with the transverse relaxation time $T_2$. Thus, the signal as a function of $t_1$ also looks like an FID.

In 2D-NMR spectroscopy, we thus measure a series of FIDs as a function of $t_2$ with different evolution times $t_1$. The signal along both time axes oscillates sinusoidally. Fourier transformation along $t_2$ generates a series of 1D spectra with resonances at the corresponding Larmor frequencies v. The intensity of each signal in these spectra is modulated sinusoidally and carries the label of the frequency v. Fourier transformation along $t_1$ also gives a 1D spectrum with a signal at the frequency v. Two-dimensional Fourier transformation along $t_1$ and $t_2$ generates the 2D-NMR spectrum. For one type of nucleus, the resulting signal is one single resonance at the coordinates (v,v). If we now manage to transfer phase coherence from one group of nuclei that precesses with the Larmor frequency $v_1$ during $t_1$ to a second group of nuclei with the Larmor frequency $v_2$ that we detect during our acquisition time $t_2$, the measured signal at the frequency $v_2$ is labeled with the frequency $v_1$. The resulting peak in the 2D-NMR spectrum obtained by two-dimensional Fourier transformation is located at the coordinates ($v_1$,$v_2$). This *cross-peak* at ($v_1$,$v_2$) tells us which groups of nuclei are coupled. Coupling can be observed between nuclei of the same element (*homonuclear coupling*) or between nuclei from different elements (*heteronuclear coupling*).

### 20.1.5.2  Correlated Spectroscopy

We now consider a two-spin system with two spins X and Y that show scalar coupling. The system has four possible energy states, depending on the state of the individual spins (Figure 20.18). In a one-dimensional NMR spectrum, we observe two doublets of resonances, one for spin X, one for spin Y. In *two-dimensional <u>co</u>rrelated <u>s</u>pectroscopy* (COSY), we apply an initial pulse that excites all resonances, in this case X and Y. A second pulse applied after the evolution time $t_1$ then leads to sharing of phase coherence between coupled spins. We detect the FID with the measurement time $t_2$, and repeat the experiment with a different evolution time $t_1$. The pulse sequence for a COSY experiment is 90°—$t_1$—90°—$t_2$.

By two-dimensional FT, we obtain the 2D-COSY spectrum that contains 16 peaks: for precession with the Larmor frequency of $x_1$ during $t_1$ and of $y_1$ during $t_2$, of $x_2$ during $t_1$ and of $y_1$ during $t_2$, of $y_1$ during $t_1$ and of $x_1$ during $t_2$, of $y_1$ during $t_1$ and of $y_2$ during $t_2$, and so on (Figure 20.18). From the

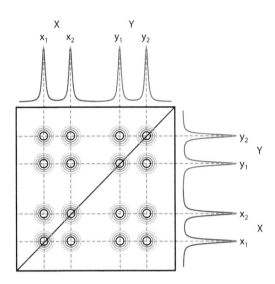

**FIGURE 20.18:** **COSY spectrum of two coupled nuclei.** Scalar coupling between the nuclei X and Y leads to splitting of their resonances into doublets (red, $x_1$ and $x_2$; blue, $y_1$ and $y_2$). In a two-dimensional spectrum, scalar coupling between nuclei gives rise to cross-peaks (off-diagonal concentric circles). The resonances on the diagonal correspond to the 1D spectrum.

presence of a cross-peak in a COSY spectrum at $(v_1, v_2)$, we can thus conclude that there are nuclei with resonance frequencies $v_1$ and $v_2$ whose spins interact by scalar coupling.

In the absence of coupling, we would only observe four peaks, namely $x_1/x_1$, $x_2/x_2$, $y_1/y_1$ and $y_2/y_2$. These are the peaks on the diagonal of the 2D spectrum. The diagonal of the 2D spectrum corresponds to the spectrum we would measure by one-dimensional NMR.

### 20.1.5.3 Nuclear Overhauser Enhancement Spectroscopy

We can also use 2D-NMR to detect coupling of spins through space, making use of the nuclear Overhauser effect (Section 20.1.3). In a COSY experiment, we apply the second pulse to transfer phase coherence. This pulse brings the magnetization in the $y$-direction into the $-z$-direction, such that it cannot be detected. In *Nuclear Overhauser Enhancement Spectroscopy* (NOESY), we apply a third pulse after the *mixing time* $t_m$. This pulse brings the magnetization from the $-z$-direction back into the $xy$-plane, and we can detect it. The pulse sequence for a NOESY experiment is $90° - t_1 - 90° - t_m - 90° - t_2$.

We now consider two nuclei close in space whose spins interact *via* dipolar coupling during the mixing time $t_m$. Because of cross-relaxation (see Section 20.1.3) during the mixing time $t_m$, spin X will be detected at the Larmor frequency of spin Y. After Fourier transformation, this coupling gives rise to a cross-peak at $v_1/v_2$ in the 2D NOESY spectrum.

NOESY spectra provide extremely useful information for structure determination because they originate from the strongly distance-dependent dipolar interaction (Section 15.3 and Figure 15.7). NOEs can be grouped into strong, intermediate and weak, corresponding to distances of the interacting nuclei of 0.18–0.25 nm, 0.18–0.35 nm, and 0.18–0.5 nm. The same lower boundary takes into account that the absence of an NOE cannot necessarily be interpreted as a large distance between nuclei.

### 20.1.5.4 Spin Systems and Sequential Assignment of Protein NMR Spectra

In the following, the general procedure for protein structure determination by homonuclear 2D-FT-NMR will be outlined, starting from a 2D-COSY and a 2D-NOESY $^1$H spectrum of the protein of interest. These spectra show numerous peaks along the diagonal, corresponding to the 1D spectrum, plus a number of cross-peaks outside the diagonal. The cross-peaks in the COSY spectrum provide information on scalar coupling, i.e. on protons that are connected through chemical bonds. The first step is to assign these resonances to individual protons. In light of the number of protons this task seems rather demanding, if not impossible. The assignment is simplified, however, by the limited number of building blocks in proteins. The important point to note is that detection of scalar coupling is limited to nuclei separated by up to three bonds. Scalar coupling does not extend over the peptide bond, which means that there is no scalar coupling between the $C_\alpha$ proton of an amino acid and the N-H proton of the next amino acid. Scalar coupling is therefore limited to protons of the same amino acid. These protons constitute a *spin system*. Each type of amino acid gives a characteristic coupling pattern that is determined by its chemical structure. Gly has the simplest spin system with only one proton resonance from its $C_\alpha$ proton. Ala gives rise to two resonances, one for the $C_\alpha$ proton and one for the three $C_\beta$ protons. The $C_\alpha$ proton and the $C_\beta$ protons are separated by three bonds and show scalar coupling, leading to a cross-peak in the COSY spectrum (Figure 20.19). Each Ala in the protein will give rise to a set of these resonances, shifted slightly relative to each other because of different chemical environments. Similarly, each Val, Leu, etc. constitutes a spin system and gives rise to a characteristic peak pattern (Figure 20.19).

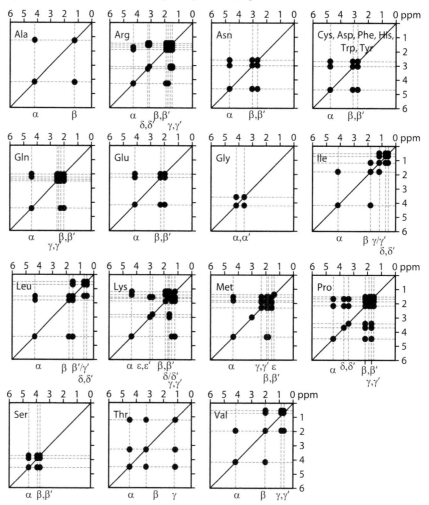

**FIGURE 20.19: Spin systems of the 20 proteinogenic amino acids.** The chemical structure of amino acid side chains determines the scalar coupling pattern of their protons. The expected COSY cross-peak pattern for the 20 proteinogenic amino acids is shown. Note that Cys, Asp, Phe, His, Tyr, and Trp show the same cross-peak pattern, but the individual values for the chemical shift of their protons are slightly different.

First, we have to identify the individual spin systems and assign their resonances to the individual amino acids in our protein. In a second step, we have to find out the exact positions in the protein sequence to which the spin systems belong. This information is not encoded in the COSY spectrum because scalar coupling does not cross the peptide bond. To identify amino acids that are sequential neighbors, we have to resort to NOESY spectra for what is called the *sequential assignment* (Figure 20.20). Cross-peaks from the NOESY spectrum give information on which protons are closer than 0.5 nm. The most important cross-peak for sequential assignment is the NOE cross-peak between the $C_\alpha$ of one amino acid and the N-H proton of the subsequent amino acid. Using this resonance, we can cover the gap from one amino acid to the next. The area of the NOESY spectrum where these cross-peaks are found is called the fingerprint region. We start with the $C_\alpha$ proton resonance for one amino acid (identified in the COSY spectrum) and look for cross-peaks to N-H protons in the NOESY spectrum. Information on the protein sequence is useful for this "chain tracing" because it tells us which spin systems should be connected. If we have an Ala-Val pair in the sequence, we have to look for $C_\alpha$ protons from an Ala spin system coupled to the N-H resonance of a Val spin system. Should we come to the conclusion that the resonance we have assigned to the $C_\alpha$ of an Ala is coupled to the N-H resonance of a Tyr, but we have no Ala-Tyr pair in our protein of interest, we know that we better check our assignment again. Prolines cannot be assigned sequentially this way because they lack a $C_\alpha$ proton. Hence, the sequential assignment (Figure 20.20) breaks off at each proline in the sequence, and we have to restart tracing at the amide proton of the amino acid that follows the Pro in the sequence.

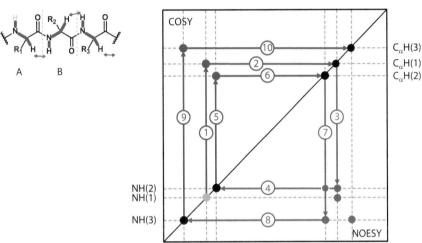

FIGURE 20.20:    **Chain tracing.** Starting at the amide proton resonance for one amino acid (A, orange), the corresponding cross-peak with the $C_\alpha$ proton of this amino acid is identified in the COSY spectrum (1), and traced to the diagonal (2). In the NOESY spectrum (bottom half), the cross-peak between this $C_\alpha$-proton resonance and the resonance for the N-H proton of the subsequent amino acid B is identified (3) and traced to the diagonal (4). By alternating between COSY and NOESY spectra, the tracing continues (5–10).

At the end of the chain tracing procedure, we have assigned groups of peaks to a spin system pertaining to a specific type of amino acid, resonances within each spin system to individual protons of this amino acid, and each spin system to a particular amino acid in the protein sequence. The remaining NOESY cross-peaks can now directly be assigned to coupling of individual protons from their position in the spectrum (resonance frequencies). NOEs between protons from amino acids distant in sequence provide us with important information on spatial vicinity of these protons in three dimensions. These NOEs are classified as weak, intermediate, or strong NOEs, and provide distance information that is used as conformational constraints for the calculation of the protein structure.

Determination of nucleic acid structures by NMR is more difficult because there are fewer types of building blocks. The nucleobases adenine, cytosine, guanine and thymine constitute four spin systems. Thymine can be identified by its signature COSY cross-peak between the methyl protons and the $C_6$ proton. In RNA, the number of different spin systems is reduced because C and U show identical coupling. However, both can be separated from purines by their coupling of $C_5$ and $C_6$ protons that is absent in the purine spin systems. The ribose/deoxyribose sugars constitute the fourth/fifth spin system in RNA/DNA. Protons from bases and sugars do not show scalar coupling, but NOESY cross-peaks are found between a base and its own sugar, and to the sugar at the adjacent nucleotide in the 5′-direction. Two principal chain-tracing procedures, based on these NOESY cross-peaks, are possible (Box 20.5). Isotope labeling (deuteration, $^{15}$N- or $^{13}$C-isotope labeling) can be used to simplify the interpretation of nucleic acid NMR spectra. An additional advantage lies with the presence of the NMR-active natural isotope $^{31}$P, which is directly amenable to NMR studies. In general, the number of long-range distance constraints per building block is smaller than for proteins, and the determination of nucleic acid structures therefore relies more on the correct local geometry. From

NOESY spectra, A- and B-DNA (Section 17.5.1) can be distinguished. A-DNA spectra show a high intensity cross-peak for coupling of the C6- (pyrimidine) or C8 (purine) proton and the C2′ proton. The cross-peak with the C2″ protons is smaller. In B-DNA, these intensities are inverted because of the different sugar pucker (Section 17.2.1). In addition, A-DNA has a higher helical pitch, and a larger inter-base distance (see Table 17.1). Cross-peaks between the C6/C8 protons of one base and the neighboring base are therefore stronger in B-DNA than in A-DNA.

---

**BOX 20.5: SEQUENTIAL ASSIGNMENT IN ¹H NMR OF NUCLEIC ACIDS.**

The proton resonances in nucleic acids can be assigned sequentially using NOESY cross-peaks (Figure 20.21). One route uses the cross-peaks between the C6/C8 proton of the base and the H1 of the ribose within the same nucleotide, and between the H1′ and the C6/C8 proton of the next base in the 3′-direction. The second route alternates between cross-peaks from the C6/C8 proton of the base to the H2′/2″ of the ribose in the same nucleotide and from the H2′/2″ to the C6/C8 proton of the next base in 3′-direction.

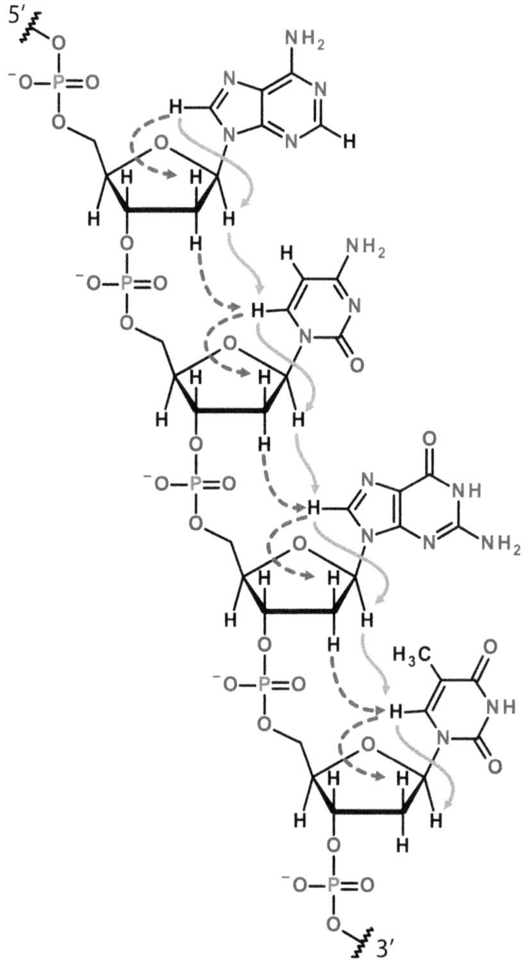

FIGURE 20.21:    **Sequential assignment of nucleic acid resonances.** Orange: sequential assignment by following the cross-peaks from the C6/C8 proton of the base to the C1′ proton of the ribose in the same nucleotide, and from the C1′ proton of the ribose to the C6/C8 proton of the base in the next nucleotide in the 3′-direction. Magenta: sequential assignment using cross-peaks from the C6/C8 proton of the base to the C2′/2″ protons of the ribose in the same nucleotide, and from the C2′/2″ protons to the C6/C8 proton of the base in the next nucleotide in the 3′-direction.

### 20.1.5.5 Structure Calculation

Finally, we have to convert our NMR restraints into a structural model by expressing all boundary conditions, not only the ones derived from NMR data, in terms of energy. Energy terms express the contributions from covalent and non-covalent interactions (Sections 15.2 and 15.3). Covalent interactions are specified by bond length, bond angle, and torsion angles for neighboring bonds. The energy of a chemical bond is expressed as a harmonic oscillator around the mean bond length $r_0$:

$$E = \frac{1}{2} k_{\text{bond}} \left( r - r_0 \right)^2 \qquad \text{eq. 20.15}$$

with the actual bond length $r$ and the bond stiffness $k_{\text{bond}}$. Energies associated with bond angles $\theta$ of two bonds from the same atom are expressed as

$$E = \frac{1}{2} k_\theta \left( \theta - \theta_0 \right)^2 \qquad \text{eq. 20.16}$$

with the mean angle $\theta_0$ and the force constant $k_\theta$. Energies from torsion angles (dihedral angles) for bonds emanating from two atoms bound to each other are expressed as

$$E = \frac{1}{2} k_\phi \left( 1 + \cos \left( n\phi - \delta \right) \right) \qquad \text{eq. 20.17}$$

with the dihedral angle $\phi$, the force constant $k_\phi$, and the empirical parameters $n$ and $\delta$ (see Section 15.2.2).

Non-covalent interactions include electrostatic and van der Waals interactions and hydrogen bonds. The energy of electrostatic interactions depends on the charges $q_1$ and $q_2$ of the two partners, their distance $r$ and the (dielectric) constant $D$:

$$E_{el} = \frac{q_1 q_2}{Dr^2} \qquad \text{eq. 20.18}$$

The energy of van der Waals interactions is described by a Lennard-Jones potential (Figure 15.14):

$$E_{\text{vdw}} = \frac{A}{r^6} - \frac{B}{r^{12}} \qquad \text{eq. 20.19}$$

with the constants $A$ and $B$, and the distance $r$. The energy of hydrogen bonds can be described by a function of similar form

$$E_{\text{Hbond}} = \frac{A'}{r^x} - \frac{B'}{r^y} f\left( \alpha, \beta \right) \qquad \text{eq. 20.20}$$

where $A'$ and $B'$ are constants, $r$ is the distance, $x$ and $y$ are empirical parameters, and $f(\alpha, \beta)$ is a function of the angles $\alpha$ and $\beta$ of the H-atom to donor and acceptor atoms.

In addition to these geometrically defined interaction energies, energy terms reflecting the experimental restraints from NMR such as $J$-coupling or intramolecular distances derived from NOEs are added. The combination of geometric, pre-defined and experimental restraints is the force field (Section 18.2.1), which allows one to calculate the energy of any given structure. Starting from a random coil conformation calculated from the protein sequence, NMR structure calculation is aimed at folding the extended chain into a compact form of minimal energy. Simulated annealing (Section 18.2.4.3) allows large conformational changes of the starting structure, followed by energy minimization. If an intermediate structure violates the NMR restraints, the corresponding energy terms of the force field apply energetic penalties, and the total energy of the respective structure is high. In this case, additional cycles of minimization are required. The whole process is repeated multiple times until a set of low-energy structures is obtained that fulfill the NMR restraints.

### 20.1.6 EXTENDING NMR TO STRUCTURE DETERMINATION OF LARGE MOLECULES

NMR of larger proteins becomes difficult because of the increasing number of resonances and the resulting overly crowded spectra. In principle, we can introduce a third (or more) dimension into NMR spectra by introducing a third time period in our measurements. A simple three-dimensional NMR experiment (3D-NMR) then consists of a 90° pulse, followed by an evolution time $t_1$, a second 90° pulse and a second evolution time $t_2$, and a third pulse, followed by the measurement time $t_3$. The NMR spectrum is then obtained by FT in three dimensions, leading to a cubic spectrum that can be regarded as layers of 2D-spectra (Figure 20.22). As a result, resonances that overlap or superimpose in a 2D spectrum can now be spread out on the third axis and are separated. Again, 3D-NMR can be performed by transferring magnetization from one type of nucleus to the same type during the evolution times (homonuclear 3D-NMR) or by following the interaction of spins that belong to different types of nuclei (heteronuclear NMR, e.g. $^1$H-$^{15}$N-$^{13}$C 3D-NMR). In principle, we can introduce a fourth or fifth dimension into NMR by extending this approach to more evolution times. The number of dimensions in NMR is limited by the relaxation of spins to the ground states and the concomitant loss in net magnetization with time, however.

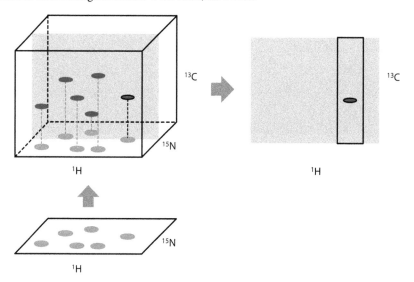

**FIGURE 20.22:** **Principle of 3D NMR.** Special pulse sequences allow detection of coupling between $^1$H, $^{15}$N, and $^{13}$C. Cross-peaks in a $^1$H/$^{15}$N spectrum (orange) are spread (gray dotted line) along the $^{13}$C axis according to the coupling with $^{13}$C. The 3D-NMR spectrum can be represented as a cube that is formed by a stack of 2D spectra. Coupling between $^1$H and $^{13}$C can be extracted by looking at the layer for a constant $^{15}$N chemical shift (gray plane). Often, information from 3D spectra is represented in form of strips (black rectangle).

We can also limit the number of resonances by *isotope labeling* (Box 20.6). If one type of amino acid is incorporated in isotopically labeled form into our protein of interest, resonances from these amino acids are selectively observed. Alternatively, a segment of the protein can be labeled, and resonances from the labeled area within the protein are selectively observed.

---

**BOX 20.6: ISOTOPE LABELING OF PROTEINS.**

Entire proteins can be labeled with $^{15}$N and $^{13}$C by growing bacteria that overproduce the protein of interest on minimal media supplemented with $^{15}$NH$_4$Cl and $^{13}$C glucose. Alternatively, only a region of interest, such as a specific domain, can be isotopically labeled. This segmental labeling exploits expressed protein ligation (EPL, Section 19.5.4). One part of the protein is produced in an isotopically labeled form as a C-terminal fusion with an intein. The second part of the protein is produced as an N-terminal intein fusion

*(Continued)*

without isotope labeling. Cleavage of this fusion protein generates this part of the protein with a cysteine at the N-terminus. The cysteine acts as a nucleophile and cleaves the isotopically labeled intein fusion, yielding the protein of interest, segmentally labeled in the N-terminal part. More refined tailored labeling of only one type of amino acids or specific groups in one amino acid are is also possible. In ILV labeling, isoleucine, leucine, and valine are specifically labeled by growing overproducing bacteria on minimal media, providing isotope-labeled forms of the amino acid precursors α-ketobutyrate and α-ketoisovalerate. Methionine methyl labeling results in proteins that are specifically $^{13}$C-labeled at methyl groups in methionines. Several other specific labeling schemes have been developed.

In addition to the mere number of resonances causing assignment problems, *line-broadening* is a problem for larger molecules. The line width of NMR resonances is related to the size of the molecule, or more exactly to its rotational correlation time. A small, rapidly rotating molecule gives rise to narrow lines, whereas a larger molecule with a large rotational correlation time shows reduced $T_2$ relaxation times. Line-broadening increases with the size of the protein until ultimately the resonances disappear. Two processes are dominant in relaxation, the dipole-dipole coupling and the *chemical shift anisotropy*. While relaxation through chemical shift anisotropy increases with the strength of the magnetic field, relaxation by dipole-dipole coupling is independent of the magnetic field. A special pulse sequence, called <u>t</u>ransverse <u>r</u>elaxation <u>o</u>ptimized <u>s</u>pectroscop<u>y</u> (TROSY) has been introduced to achieve cancellation of these two relaxation processes with different signs. Cancellation of these effects can only be achieved at high magnetic field strengths. In TROSY experiments, the $T_2$ relaxation time is increased, and resolution and sensitivity are increased (Figure 20.23). TROSY may optimally be applied to proteins of >100 kDa molecular mass.

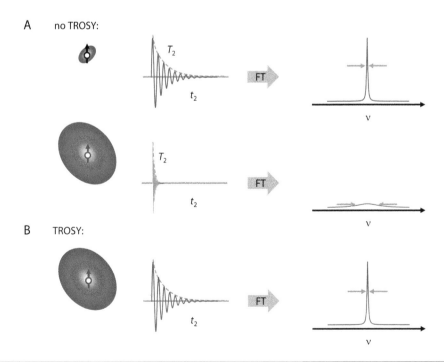

**FIGURE 20.23:** **TROSY.** A: The line width of NMR resonances depends on the transverse relaxation time $T_2$ and on the size of the molecule and its rotational correlation time. Spectra for small molecules consist of narrow lines, whereas the resonances are broadening for larger molecules that tumble more slowly. B: TROSY is a specific pulse sequence that leads to cancellation of two relaxation pathways. As a consequence, $T_2$ is increased, and the line width is reduced.

## 20.1.7 NMR AND DYNAMICS

In addition to structural information, NMR can provide information on the mobility of macromolecules on a large range of time scales from several minutes to microseconds. Protons of the macromolecule are constantly exchanged with solvent protons. The rate of this exchange depends on the exposure of the individual proton, and on the pH. Protons buried in the three-dimensional structure of a protein are protected from solvent exchange, and show low exchange rates. Readily accessible protons, on the other hand, exhibit high exchange rates. Engagement of protons in a hydrogen bond also slows down exchange by up to five orders of magnitude. *Amide proton exchange* rates are a measure of the accessibility of the protein backbone. To determine exchange rates of amide protons, proteins are freeze-dried from $H_2O$ solutions. Upon dissolution in $D_2O$, the protons exchange with deuterons, leading to a decrease of proton resonances over time at the rate of proton/deuteron exchange. Alternatively, deuterated proteins can be dissolved in $H_2O$, and the increasing signal from rapidly exchanging deuterons that are being replaced by protons can be observed. This approach offers the advantage of observing the resonances in the absence of other proton resonances. Amide proton exchange experiments allow the measurement of exchange rates on the time scale of seconds or slower. Amide proton exchange has been used extensively to monitor secondary and tertiary structure formation during protein folding, and to delineate protein folding pathways (see Section 16.3.5). Proton exchange can also be employed to probe flexibility and structure of nucleic acids, where hydrogen bonding also leads to protection from exchange.

The line width of individual NMR resonances depends on the transverse relaxation time $T_2$, which is related to molecular motions. Line-shape analysis can be used to obtain information on mobility. Line-broadening of small molecule resonances can also be used as a signal for macromolecule binding. Alternatively, insight into local dynamics can be gained from direct measurement of relaxation times (Box 20.4). Chemical exchange on the second to millisecond time scale between two species with different chemical shifts for a particular nucleus can be followed by line-shape analysis (Box 20.7). These two species can be two conformers that interconvert, or the free and bound state of a ligand. If the interconversion between the two forms is slow on the NMR time scale ($k \ll \omega_A - \omega_B$), the two species cause two different resonances at their characteristic chemical shift, with intensities that reflect their populations. If the exchange between the two forms is rapid ($k \gg \omega_A - \omega_B$), only one resonance is measured. This resonance is at a chemical shift that corresponds to the population-weighted average of the chemical shifts of the nucleus in the individual species. In an intermediate exchange regime, in-between these limiting cases ($k \approx \omega_A - \omega_B$), the two resonances will coalesce. From the line shape, the exchange rate can be determined (Box 20.7).

---

**BOX 20.7: CHEMICAL EXCHANGE AND LINE-SHAPE ANALYSIS.**

Two species A and B with resonance frequencies $\omega_A$ and $\omega_B$ that are interconverting slowly ($k \ll \omega_A - \omega_B$) are observed as separate lines, with amplitudes that reflect the relative population of A and B (Figure 20.24). The line width $\Gamma$ is affected by the rate of the chemical exchange between A and B, and corresponds to

$$\Gamma = \frac{1}{T_2} + k \qquad \text{eq. 20.21}$$

Slow exchange is the most frequent case in optical spectroscopy. In NMR spectroscopy, chemical exchange is often in the intermediate or rapid exchange regime. For intermediate exchange rates ($k \approx \omega_A - \omega_B$), the two resonances are broadened and coalesce. If

*(Continued)*

the two forms interconvert rapidly ($k \gg \omega_A - \omega_B$), a single resonance line is observed. Its position corresponds to the population-weighted average of the resonance frequencies of the two interconverting species. The width $\Gamma$ of this line is

$$\Gamma = \frac{1}{T_2} + \frac{\Delta\omega^2}{8k}$$

eq. 20.22

NMR *line-shape analysis* can thus provide information on the kinetics of binding and dissociation or on conformational changes on the millisecond to second time scale.

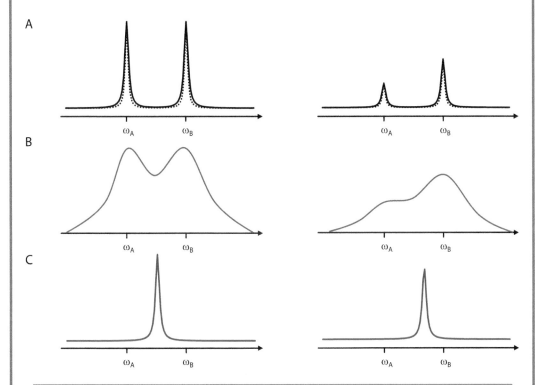

**FIGURE 20.24:    Line-shape analysis.** A: Slow exchange ($k \ll \omega_A - \omega_B$). From the line width, the rate constant of the chemical exchange $k$ can be determined (eq. 20.21). The natural line width in the absence of chemical exchange is indicated as a dotted line. B: At intermediate exchange ($k \approx \omega_A - \omega_B$), the two lines for the resonance in A and B coalesce. C: Rapid exchange ($k \gg \omega_A - \omega_B$) leads to the detection of one resonance line for the nucleus at a position that corresponds to the population-weighted resonance frequencies. The line width is related to the rate constant of chemical exchange (eq. 20.22). The left panel shows the spectra for equal populations of A and B, i.e. for $k_{A \to B} = k_{B \to A}$. The right panels show the spectra for a 1:2 ratio of A and B, corresponding to $k_{A \to B} = 2 \cdot k_{B \to A}$.

Line shapes are also affected by paramagnetic species, which can be exploited to map ion binding sites by *paramagnetic relaxation enhancement* (Box 20.8). Relaxation times provide information on dynamics on the microsecond to nanosecond time scale. Transverse relaxation is enhanced in the vicinity of paramagnetic species, and paramagnetic relaxation enhancement can be used to probe distances and dynamics (Box 20.8).

---

**BOX 20.8: PARAMAGNETIC RELAXATION ENHANCEMENT.**

Paramagnetic species enhance relaxation and therefore reduce $T_2$ times of nearby nuclei, leading to line-broadening and eventually disappearance of their NMR resonances. Paramagnetic relaxation enhancement (PRE) can be used to locate ion binding sites in proteins. For example, $Mg^{2+}$ binding sites can be determined by substituting with the paramagnetic $Mn^{2+}$ ions and identifying the signals that are reduced upon binding. PRE can also be used to study Fe-S clusters in proteins. The introduction of spin labels (see Section 20.2) at cysteine residues can be used to probe protein-protein interfaces. Spin-labeled fatty acids and lipids enable the investigation of membrane dynamics and for the identification of protein regions that insert into or span the membrane. Due to the distance dependence of the interaction between paramagnetic species and the nucleus, PRE experiments provide distance information in the 10–20 Å range and therefore complement NOE measurements (Section 20.1.3).

---

### 20.1.8 SOLID STATE NMR AND BIOLOGY

So far, we have restricted our considerations to NMR on molecules in solution. NMR spectra from molecules in solution show narrow lines in comparison to the very broad lines in spectra of solids. Line-broadening in solid-state spectra is caused by dipolar interactions between spins and by the chemical shift anisotropy. Dipolar coupling depends on the distance $r_{AB}$ between the two nuclei and the angle $\theta$ of the joining vector $\boldsymbol{r}_{AB}$ with respect to the magnetic field. Dipolar coupling of two spins A and B ($\mu_A$, $\mu_B$) leads to a shift $\Delta\nu$ in their Larmor frequency by

$$\Delta\nu \propto \frac{\mu_A\mu_B}{r_{AB}^3}\left(3\cos^2\theta - 1\right)$$

eq. 20.23

Because of the strong distance dependence, dipolar coupling is only strong between nuclei that are directly bonded or immediate neighbors. In solution, the orientation of the two spins and thus the angle $\theta$ is averaged by rapid tumbling of the molecules. The average of $\cos^2\theta$ over time in this case is ⅓, and the term in brackets in eq. 20.23 is zero. Therefore, the entire dipolar coupling term is zero and not relevant for solution NMR. In contrast, molecules in the solid state have a fixed orientation with respect to each other and to the external magnetic field, and the spin-spin interactions cannot be neglected. As a consequence, NMR lines of solid compounds are very broad. Nevertheless, solid-state NMR can provide important information on biomolecular structure and dynamics. Because of the dipolar coupling, the chemical shift and the shape of lines depend on the orientation of the sample, which is the chemical shift anisotropy. Experiments on oriented samples and measurements at different angles of the orientation axis with respect to the external magnetic field can provide information on the orientation of helices or entire proteins within membranes. *Magic angle spinning* (MAS) experiments achieve cancellation of dipolar coupling effects by rotating the sample around its axis at a 54.7° angle with respect to the external magnetic field. The cosine squared for the magic angle is ⅓, and the time-averaged term ($3\cos^2\theta-1$) is zero. As a result, the resonances are again measured as sharp lines. High-resolution structural information on membrane proteins can thereby be obtained. Solid state NMR experiments can also provide information on the backbone and side-chain dynamics of membrane proteins. Solid state NMR is particularly useful to study membrane proteins in their natural environment. Solid state methods have also been applied to fibril-forming proteins such as α-synuclein to study intrinsically disordered proteins and freeze-trapped protein folding intermediates.

## 20.1.9 NMR and Imaging

NMR can also be used as a probe for imaging purposes. In *magnetic resonance imaging* (MRI), the spatial information for each spin is encoded in its resonance frequency by employing a position-dependent external B-field. In a magnetic field gradient along the $x$-, $y$-, or $z$-axis, each nucleus experiences a magnetic field that now depends on its $x$-, $y$-, or $z$-coordinates. The Zeeman effect and the Larmor frequency therefore also depend on the location of the respective nucleus in this field. Thus, recording the NMR signal from each volume element of a three-dimensional specimen provides a signal whose frequency depends on the position, and whose intensity is determined by the number of nuclei it contains. By measuring the NMR signal as a function of resonance frequencies in a magnetic field gradient in the $x$-, $y$-, and $z$-directions, a map for the distribution of protons in a cell, an organ, or an entire organism can be re-constructed (Figure 20.25). Contrast can be achieved by adjusting the experimental conditions such that the measured intensity depends on the $T_1$- or $T_2$-relaxation times. For example, the relaxation times of protons in $H_2O$ depend on the cellular environment and cover a range from 100 ms to several seconds. Freely diffusing water can thereby be clearly distinguished from water within a membrane that shows little mobility. The chemical shift can also be used as the imaging parameter.

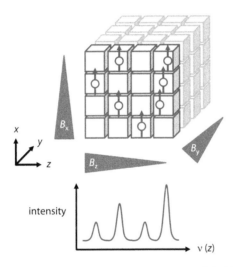

**FIGURE 20.25:    Magnetic resonance imaging (MRI).** The three-dimensional imaging object (cube array) is placed in a magnetic field gradient. As a consequence of the field gradient in $x$-, $y$-, or $z$-direction (gray arrows), the resonance frequency of each spin (blue) depends on its coordinates along the axis of the gradient. The signal intensity at each frequency and thus at each position is proportional to the number of spins. Scanning along the $z$-axis provides signals at each frequency that are proportional to the projection of the spins in the section marked in red onto the $z$-axis. From measurements in all three dimensions, the spin density at each point of the image in three dimensions can be reconstructed.

MRI is widely used in medical diagnostics. *Functional MRI* follows changes in the brain to external stimuli and is an important tool in neuroscience and neurosurgery. Brain activity can be followed by MRI probing the oxygenation state of hemoglobin: deoxygenated hemoglobin is more paramagnetic than oxygenated hemoglobin. Deoxygenated hemoglobin therefore causes paramagnetic relaxation, and thereby alters proton resonances (paramagnetic relaxation enhancement, see Box 20.8). Active brain regions with high blood flow contain a higher level of oxygenated hemoglobin, whereas inactive regions contain more deoxygenated hemoglobin. Imaging proton resonances therefore serves as a probe for brain activity.

## 20.2 ELECTRON PARAMAGNETIC RESONANCE

### 20.2.1 PRINCIPLE OF ELECTRON PARAMAGNETIC RESONANCE

Electron paramagnetic resonance (EPR) or electron spin resonance (ESR) is based on the same principles as NMR spectroscopy. Instead of nuclear spins, EPR probes the energy levels of *electron spins* and their transitions between different states. Similar to NMR-active nuclei, unpaired electrons of radicals and triplet states have a magnetic moment. Analogous to nuclear spins, any spatial orientation is possible for free electrons with spin ½ in the absence of an external magnetic field, and these states have the same energy. When an external magnetic field is present, only two orientations with respect to the field are possible, −½ and +½. (Figure 20.26). These states are of different energies. Their energy $E$ is the scalar product of the magnetic moment of the electron and the external magnetic field $B_0$. The magnetic moment is the product of the g-factor of the electron $g_e$, the Bohr magneton $\mu_\beta$, and the spin orientation $m_s$ (½, −½) of the electron.

$$E = g_e\mu_\beta B_0 m_s \qquad \text{eq. 20.24}$$

The factor $g_e$ is dimensionless and has a value of ~2 for a free electron. For electrons within molecules, it depends on the local magnetic field $B_{local}$ and thus on the chemical context. The unit of the magnetic moment is provided by the Bohr magneton $\mu_\beta$ ($\mu_\beta = 9.27 \cdot 10^{-24}$ J T$^{-1}$). The resulting energy difference between the two spin states of unpaired electrons is

$$\Delta E = g_e\mu_\beta B_0 \qquad \text{eq. 20.25}$$

Typical magnetic fields in EPR spectrometers are 0.1–5 T (see Table 20.2). According to eq. 20.25, transitions between the two states are induced by irradiation with electromagnetic radiation of wavelengths in the centimeter range (microwave radiation).

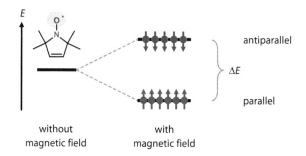

**FIGURE 20.26: Energy levels and transitions in EPR.** An unpaired electron, such as in a nitroxide radical, has a spin of ½. In the presence of a magnetic field, the spins align either in a parallel or an antiparallel orientation with respect to the external field. The energies of these two spin states are different, and transitions between them can be induced by microwave radiation.

EPR spectra can be measured at a fixed, constant magnetic field while scanning the frequency of the microwave radiation. Measurements can also be done in the opposite way by keeping the microwave radiation at a constant frequency and varying the strength of the magnetic field $B_0$, which is the typical configuration in modern EPR spectrometers. Spectrometers are named according to the wavelength used (Table 20.2). The majority of commercially available instruments are X-band spectrometers.

**TABLE 20.2**
**Frequency bands of EPR spectrometers.**

| band | typical frequency (range) (GHz) | typical wavelength (cm) | typical $B$-field (T) |
|---|---|---|---|
| L | 1.5 (1–2) | 20 | 0.0054 |
| S | 3.0 (2–4) | 10 | 0.11 |
| X | 9.5 (8–12) | 3 | 0.34 |
| Ku | 17 (12–18) | 1.7 | 0.6 |
| Q | 36 (30–50) | 0.8 | 1.28 |
| V | 70 (50–75) | 0.4 | 2.5 |
| W | 95 (75–110) | 0.3 | 3.39 |
| D | 149 (110–170) | 0.2 | 5 |

*Note:* According to the Radio Society of Great Britain.

Due to the small energy difference between the two states, their population is almost equal, with just a tiny excess in the lower energy parallel spin orientation. The absorption signal in EPR experiments is therefore small, and EPR is rather insensitive. To achieve higher sensitivity, instruments typically measure the change in signal upon modulation of the sweeping B-field instead of the signal itself, and EPR spectra are recorded as the first derivative of the absorption (Figure 20.27). This derivative signal is detected at an external magnetic field $B_0$ where the energy difference between the spin states equals the energy of the constant microwave radiation that is applied.

Similar to nuclear spins, relaxation of excited electron spins occurs by spin-lattice or spin-spin relaxation, giving rise to the *relaxation times* $T_1$ and $T_2$, respectively. As in NMR, the $T_2$ relaxation time determines the line width of the EPR resonance.

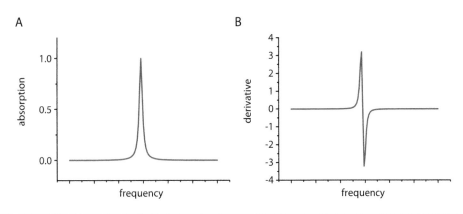

**FIGURE 20.27:    EPR spectrum.** A: Resonance line in the absorbance spectrum. B: First derivative.

In *pulsed EPR* (Section 20.2.7), electron spins are excited by short microwave pulses, similar to FT-NMR (Section 20.1.4).

## 20.2.2 SPIN-SPIN INTERACTIONS: HYPERFINE COUPLING OF UNPAIRED ELECTRONS WITH NUCLEI

The *hyperfine coupling* is caused by the interaction of an electron spin S with the spin I of a neighboring nucleus. Both the electron and the nuclear spin experience the Zeeman effect in the presence of an external magnetic field. The spin-spin interaction leads to mutual effects on the local magnetic field experienced by nucleus and electron: they sense the orientation of the other spin. The important implication for EPR spectroscopy is the splitting of an EPR signal into $2I + 1$ signals by a nucleus of spin $I$ (Figure 20.28). A nucleus with $I = \frac{1}{2}$ such as $^1$H splits the EPR resonance into two lines, a nucleus with $I = 1$ such as $^{14}$N splits it into three lines, analogous to the scalar coupling between several nuclei in NMR (Section 20.1.2 and Figure 20.4). Nitroxide spin labels that are frequently used in EPR (see Section 20.2.3) show a characteristic three-line resonance because of hyperfine coupling of the electron spin with the nuclear spin of the nitrogen. The hyperfine coupling constant is a measure of the extent of coupling and splitting, and is analogous to the coupling constant $J$ in NMR.

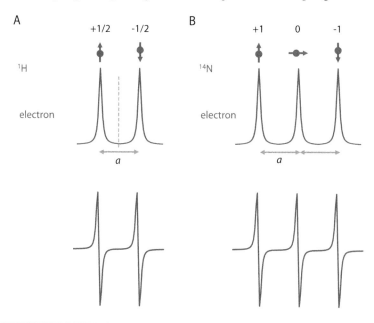

**FIGURE 20.28:** **Hyperfine coupling.** A: Splitting of an EPR resonance into two lines by an $I = \frac{1}{2}$ nucleus such as $^1H$. Top: absorption lines, bottom: first derivative. B: Splitting of an EPR resonance into three lines by an $I = 1$ nucleus such as $^{14}N$. $a$ is the hyperfine coupling constant.

The interaction of electron spins with nuclear spins can be exploited to determine the *hyperfine coupling constant* and to identify coupled nuclei. In *electron nuclear double resonance* (ENDOR) experiments, the electron spin is employed as a probe to detect coupled nuclei. A radio frequency pulse induces transitions of the nuclear spins, and the effect of these transitions on the intensity of the electron spin resonance is measured. The spectrum obtained corresponds to an NMR spectrum of the coupled nuclei. By ENDOR, the hyperfine coupling constant $a$ and the Larmor frequencies (Section 20.1.2) of the coupled nuclei can be determined. ENDOR enables the direct detection of hydrogen bonds to the radical and has been used to probe the hydrogen bond network around the active-site tyrosyl radical in ribonucleotide reductase.

### 20.2.3 EPR Probes and Spin Labeling

Most biomolecules do not contain unpaired electrons and are EPR-silent. Radicals occurring in biological reactions are only transiently populated and short-lived. Exceptions are metal ion cofactors in metalloenzymes that are stable for extended periods of time and can be probed by EPR. Most of the time, however, EPR is applied to molecules that have been modified with *spin labels* or *spin traps*. Spin labels are stable radicals, typically nitroxide radicals. A commonly used spin label is the thiol-reactive methanethiol sulfonate (MTSSL, SL for spin label), which forms disulfide bonds with Cys residues in proteins (Figure 20.29). Maleimide- and iodacetamide derivatives of spin labels are also available. Similar to fluorescent labeling (Section 19.5.4), alternative approaches include the incorporation of spin-labeled unnatural amino acids and ligation techniques (see Section 19.5.4). For nucleic acid labeling, spin labels are introduced directly during solid-phase synthesis. Spin labels can be attached to the ribose, the phosphates or the nucleobase (Figure 20.29). A number of flexibly attached and rigid spin labels for nucleic acids have been described. For the study of membranes, spin-labeled lipids are available that contain a nitroxide moiety either attached to their hydrophobic tail or to their polar head group.

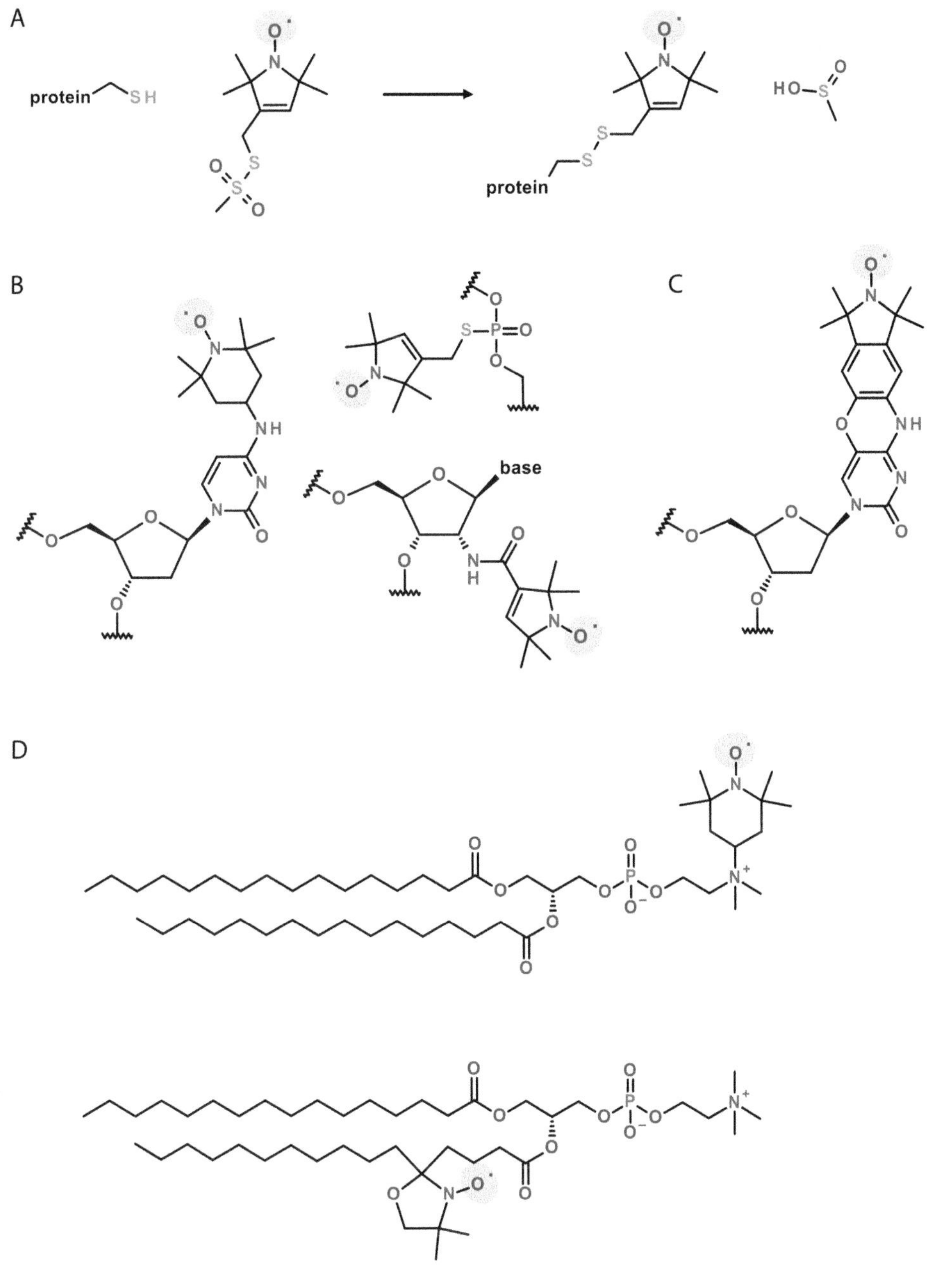

**FIGURE 20.29:    Spin labeling.** A: Incorporation of nitroxide spin labels into proteins by disulfide exchange with methanethiolsulfonate. B: Spin-labeled nucleic acids modified at the base, the ribose and the (thio)phosphate. C: A rigid spin label for nucleic acids. D: Spin-labeling of lipids at the polar head group or within the hydrophobic tail.

Spin labels are often attached to the molecule of interest by flexible linkers. Flexible attachment provides segmental flexibility and leads to sharp lines in the EPR spectrum. Rigid spin labels show broad EPR resonances, but are useful probes for the overall mobility of the molecule of interest (Section 20.2.4) and to determine the relative orientations of the two spin centers (Section 20.2.7).

In contrast to spin labels, spin traps either convert unstable radicals into a stable radical that is EPR-active or allow recombination of an unstable radical with an EPR-active radical to an EPR-silent moiety (Figure 20.29). Spin traps are typically used to monitor radical formation. Examples for spin traps are 2,2,6,6-tetramethylpiperidine (TMP) and 2,2,6,6-tetramethyl-1-piperidinyloxy (TEMPO, Figure 20.30).

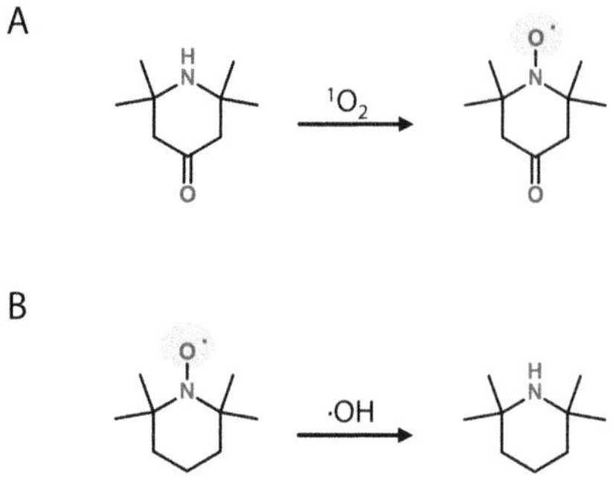

A

$^1O_2$

B

·OH

**FIGURE 20.30:** **Spin traps.** A: The EPR-silent 2,2,6,6-tetramethylpiperidin-4-one (TMP) is oxidized by singlet oxygen to the stable nitroxide 2,2,6,6-tetramethyl-1-piperidinyloxy-4-one. B: The nitroxide 2,2,6,6-tetramethyl-1-piperidinyloxy (TEMPO) reacts with hydroxide radicals to the EPR-silent 2,2,6,6-tetramethylpiperidine.

By the same principle, i.e. the generation or removal of a stable spin center by a transient radical, redox potentials of EPR-active cofactors can be determined in redox-titrations. The EPR signal of the cofactor is measured as a function of the buffered electrode potential of the solution during titration with oxidant or reductant, and the resulting curve is analyzed with the Nernst equation (see Section 5.4; eq. 5.15).

### 20.2.4 EPR AS A PROBE FOR MOBILITY AND DYNAMICS

Both *g*-factor and hyperfine coupling depend on the orientation of the spin label with respect to the external magnetic field. The extent of these orientation effects and the shape of the EPR signal depend on the mobility of the spin label. The spin label thus reports on the local mobility of the molecule of interest (Figure 20.31). For highly mobile spin labels, the orientation-dependent contributions are averaged out completely, leading to a nitroxide signal with three sharp lines. Conversely, immobile spin labels that do not move during the transverse relaxation time $T_2$ give rise to a broadened *rigid-limit spectrum*. In-between these extremes, intermediate shapes of the spectrum are observed.

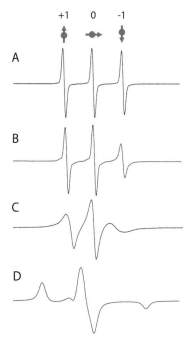

**FIGURE 20.31:** **Effect of local mobility on EPR signals.** A: EPR spectrum of the MTSSL spin label in solution. B: Spin label attached to a 15 amino acid peptide. C: Spin label attached to the same 15 amino acid peptide under conditions where it forms an α-helix. D: EPR spectrum of the frozen spin-labeled peptide. Modified from Klug, C. S. & Feix, J. B. (2008) Methods and applications of site-directed spin labeling EPR spectroscopy. *Methods Cell. Biol.* 84: 617–658, reprinted with permission.

From the shape of the EPR resonance, the rotational correlation time $\tau$ (Section 19.5.9) of the spin label can be determined as a measure of the local mobility. EPR can therefore be used to monitor local mobility and dynamics of biomolecules in solution. EPR measurements have been employed to probe membrane fluidity and dynamics, as well as protein and nucleic acid structure and dynamics. Plotting mobility data along the sequence of the protein of interest yields characteristic patterns for secondary structure elements. The dependence of the EPR signal on mobility can also be used as a probe for binding (Box 20.9) and local conformational changes. By recording EPR signals at constant magnetic field as a function of time at ambient temperature (*time-resolved EPR*), conformational changes of spin-labeled proteins can be probed at millisecond resolution. Time-resolved EPR has been used to monitor local conformational changes of spin-labeled bacteriorhodopsin in purple membranes during the photocycle in real time.

---

**BOX 20.9: EPR AS A PROBE FOR BINDING.**

The shape of an EPR spectrum depends on mobility and local dynamics. Binding of ligands to spin-labeled molecules therefore often leads to a change in the EPR spectrum. Provided that the two spectra are measurably different, EPR can be used as a probe for binding in titration experiments. Before saturation is reached, the measured spectrum is a linear combination of the reference spectra of the free and bound forms. By describing the measured spectra as a weighted sum of the reference spectra, the fractions of free and bound spin-labeled molecules can be determined. Binding curves can then be analyzed to determine the $K_d$ value (see Section 19.5.5.1). This method is also useful to determine partitioning constants of peptides and proteins into lipid bilayers.

---

The effect of spin label mobility on the shape of the measured EPR resonance is always dependent on the time scale of the experiment, i.e. on the frequency range used for excitation. The overall tumbling of spin-labeled proteins is a slow process with respect to the inverse spectral width in high frequency EPR (250 GHz) and does not contribute to the line shape. At these frequencies, more rapid motions such as backbone fluctuations of proteins become experimentally accessible and can be studied. The frequency of the microwave radiation therefore determines the time scale of motions that fall into the slow, intermediate, and rapid exchange regime (see Section 20.1.7), and that can be detected. Multi-frequency EPR takes advantage of these different time windows to study a wide range of processes occurring on different time scales.

## 20.2.5 EPR AS A PROBE FOR ACCESSIBILITY

EPR lines are broadened in the presence of paramagnetic compounds such as oxygen or paramagnetic ions, due to a decrease in relaxation time. Paramagnetic compounds can therefore be used to probe the accessibility of spin labels in an approach similar to fluorescence quenching (Section 19.5.7). This method has been employed to map the orientation of membrane proteins within the lipid bilayer. For such studies, oxygen and $Ni^{2+}$ containing compounds are typically used as paramagnetic reagents. Because of its hydrophobicity, oxygen partitions to the center of the membrane and into hydrophobic pockets of proteins, whereas its concentration in the aqueous phase is low. In contrast, $Ni^{2+}$ compounds such as nickel (II) ethylenediamine acetate are hydrophilic and do not partition much into lipid bilayers. The effect of these reagents on the relaxation time of the spin label and the line width of its EPR signal allows distinction between sites within the membrane, in the hydrophobic core of the protein, or on the solvent-accessible surface. The intrinsic relaxation time of the spin label in the particular position is measured in a reference measurement in the presence of nitrogen.

The different partitioning of oxygen and hydrophilic paramagnetic compounds into the membrane leads to inverse concentration gradients of these reagents within the lipid bilayer, an effect that can be used to determine the immersion depth of the spin label within the lipid bilayer. In a calibration experiment, the response of spin-labeled lipids with the spin labels attached to different positions and with known immersion depths in membranes to both paramagnetic reagents is determined. The response is either measured as a change in line width, or the collision frequencies of the spin label and the paramagnetic reagent are determined directly from relaxation times and the saturation behavior. The dependence of the response on the depth provides a calibration curve that can be used to determine the depth of a spin label attached to a membrane protein. Depths determined for different positions of the spin label then provide information on the orientation and positioning of integral membrane proteins or membrane-associated proteins (Figure 20.32).

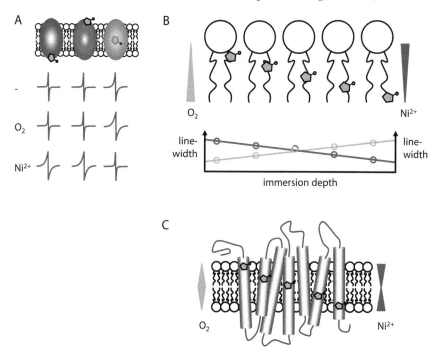

**FIGURE 20.32:    Relaxation enhancement by paramagnetic compounds and positioning of proteins within a membrane**. A: EPR lines for a spin-labeled membrane protein with the spin label attached to the surface (left, middle) or with the spin label inserted into the lipid bilayer (green). Resonances for solvent-exposed spin labels are not affected by relaxation enhancement and line-broadening in the presence of oxygen, but line-broadening by nickel compounds is observed. Resonances for spin labels deeply immersed into the membrane are broadened due to paramagnetic relaxation enhancement in the presence of oxygen, but not in the presence of nickel compounds. Note that for simplicity only the central resonance line is shown. B: Calibration experiment for the determination of the immersion depth of spin labels within the lipid bilayer. Using membranes that contain spin-labeled lipids with known immersion depths, the response to oxygen (blue) and nickel compounds (red) can be determined as a function of the immersion depth in EPR experiments. The concentration gradients of oxygen and nickel compounds are indicated by the colored triangles. C: From the calibration experiment, responses of resonances for spin-labeled proteins to the presence of paramagnetic compounds can be converted into the immersion depth to determine the orientation and position of the protein within the lipid bilayer.

## 20.2.6 MEASURING SPIN-SPIN DISTANCES

The shape of the EPR signal also depends on the concentration of spins, which can be exploited to measure spin-spin distances. Inter-spin distances can be determined by EPR using three possible approaches: from the line width of cw-EPR spectra of molecules in solution, from relaxation measurements, or from oscillating patterns of magnetization detected in pulsed EPR experiments of frozen samples (Section 20.2.7).

The magnetic dipoles of two spin labels can interact through space, analogous to the dipolar interaction of nuclear spins (Section 20.1.8; eq. 20.23). The *dipolar coupling* $\omega_{dd}$ between the two electron spins depends on their distance $r_{AB}$ and their relative orientation, described by the angle $\theta$ of the distance vector between spins A and B with the external magnetic field $B_0$, according to

$$\omega_{dd} \propto \frac{1}{r_{AB}^3}\left(1 - 3\cos^2\theta\right)$$

eq. 20.26

If the two spins are close-by (below 2 nm), the dipolar coupling is large, and line-splitting can be measured by cw-EPR. For larger distances, the line-splitting caused by dipolar coupling is very small and will be hidden within the resonance line. Dipolar coupling also leads to line-broadening. The line width is related to the $T_2$ relaxation time, and relaxation time measurements therefore also provide information on inter-spin distance.

### 20.2.7 DISTANCE DETERMINATION BY PULSED EPR: PELDOR/DEER

Larger inter-spin distances can be determined from pulsed EPR experiments that specifically select for dipolar coupling between electron spins. Pulsed EPR experiments to measure distances *via* dipolar coupling are called _double electron-electron resonance_ (DEER) or _pulsed electron-electron double resonance_ (PELDOR) experiments. The sample is excited by a series of nanosecond microwave pulses at a constant external field $B_0$. Such an experiment is analogous to FT-NMR with pulsed excitation. As discussed before for pulsed NMR experiments (Section 20.1.4), a 90° pulse induces transitions of spin A, and tilts the net magnetization of the spin ensemble A from the $z$-axis into the $xy$-plane. The $xy$ magnetization then decays due to the loss of phase correlation between the individual precessing spins. In *Hahn echo experiments*, the magnetization is inverted from the $+y$- to the $-y$-direction by a 180° pulse at time $t$. As a consequence, the previously diverging vectors for each precessing spin will now converge, leading to refocusing of the signal at the time $2t$ (*spin echo* or *Hahn echo*, Figure 20.33).

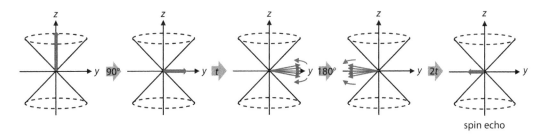

**FIGURE 20.33:** **Hahn echo for refocusing.** The net magnetization in the $z$-direction is rotated to the $y$-direction by a 90° pulse. After the time $t$, a 180° pulse inverts the magnetization to -$y$, leading to a refocusing of the magnetization vector at the time $2t$, which generates the spin echo.

In the simplest implementation of a PELDOR experiment, a three-pulse experiment, a pump pulse at a second frequency (for spin B) will be applied at a certain time $t$ during the experiment to invert the B spins. Because of this pump pulse and the transitions of B, the spins at the probe frequency will experience a different magnetic environment before and after the pump pulse. The size of the echo will therefore depend on the time at which the pump pulse was applied. Three-pulse PELDOR is limited by a dead time because the 90° pulse for spins A and the 180° pulse that inverts the B spins cannot be applied at the same time but have to have a minimal delay. The signal at short times $t$ that is missed is very important to determine short inter-spin distances. Four-pulse PELDOR avoids this dead time. Here, the 180° pulse to invert spins B is applied at different times after the spin echo, and therefore temporally separated from other pulses. A subsequent additional 180° pulse at the resonance frequency of spins A then introduces a second inversion of the magnetization, and generates a re-focused echo that is measured as a function of the delay time $T$ (Figure 20.34).

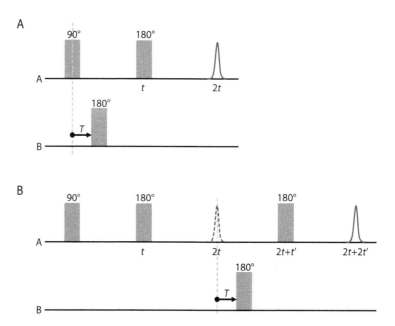

**FIGURE 20.34:   Three-pulse and four-pulse PELDOR.** A: Three-pulse PELDOR. At the beginning of the experiments, a 90° pulse is applied at the resonance frequency of spin A. At different times *T*, the magnetization due to spins B is inverted by a 180° pulse. The Hahn echo of resonance A at time 2*t* (blue) is measured as a function of the time *T*. B: Four-pulse PELDOR. A 90° pulse is applied at the resonance frequency of spin A. A 180° pulse at time *t* generates a spin echo at 2*t* (dotted blue line). At different times *T* after this spin echo, the magnetization due to spins B is inverted by a 180° pulse. A 180° pulse at 2*t* + *t'* generates a refocused spin echo for resonance A at time 2*t* + 2*t'* (blue) that is measured as a function of the time *T*.

Dipolar coupling between spins A and B leads to an oscillation of the (refocused) spin-echo intensity as a function of the time when the pump pulse was applied. The characteristic coupling frequency $\omega_{dd}$ depends on the coupling strength and thus on the inter-spin distance $r_{AB}$ (eq. 20.26). Thus, the distance between the spins can be calculated from the measured oscillation frequency of the spin echo. The measured (refocused) spin-echo intensity $V(T)$ can be described by

$$V(T) = \int_0^\infty P(r)\,dr \int_0^1 \left(1 - p\left(1 - \cos\left(\omega_{dd}T\right)\right)\right) d\theta \cos\theta$$   eq. 20.27

$P(r)$ is the distribution of inter-spin distances, $p$ denotes the fraction of spins B that are inverted by the probe pulse, and $\omega_{dd}$ is the dipolar coupling between these spins that depends on the angle $\theta$ of the distance vector between the spins with the external magnetic field $B_0$ (eq. 20.26). The signal $V(T)$ contains contributions from intramolecular coupling of spins on the same molecule and intermolecular coupling of spins in different molecules. For randomly distributed spins, the contribution from intermolecular coupling can be described by a single exponential function that depends on the spin concentration. Division of the measured signal by this function yields the contribution of intramolecular dipolar spin coupling to the measured signal.

The time-dependent signal due to intramolecular spin-coupling is either Fourier-transformed into frequency space or analyzed directly. The signal is integrated over all molecules with different orientations relative to the external field $B_0$, and thus different values for $\theta$. As a consequence, Fourier transformation does not yield a single line at a defined frequency, but a characteristic frequency distribution, the *Pake pattern* (Figure 20.35). The peaks of this pattern correspond to molecules with $\theta = 90°$, the edges represent molecules with $\theta = 0°$. The splitting of the doublet is $\omega_{dd}$, from which $r_{AB}$ can be calculated. Alternatively, the measured time-dependent signal can be directly analyzed with suitable programs that use the *Tikhonov regularization* to calculate a distance distribution. The distance distribution $P(r)$ is encoded in the dampening of the modulation of the measured signal. If the spins are fixed, the relative orientation of the two spin centers, i.e. the angle $\theta$ between the B-field and the distance vector joining the spin labels, can be extracted using more involved approaches. PELDOR can also be applied to systems with more than two spin systems. Multiple distances between spin labels have been determined in PELDOR experiments of tetrameric, heptameric, and octameric membrane proteins. The modulation depth of the measured signal depends on the number of interacting spins. Determination of the number of spin centers can be used to obtain information on complex stoichiometry.

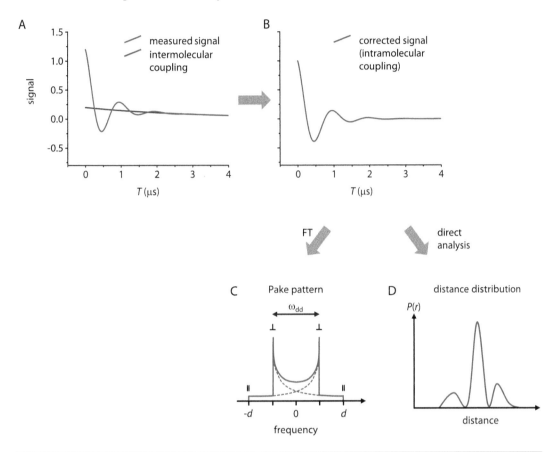

FIGURE 20.35: **Distance determination from DEER/PELDOR experiments**. A: Measured Hahn-echo intensity as a function of the pump pulse time $T$. B: Correction for contributions from intermolecular spin-spin interactions. The dampening of the modulation depends on the distribution of inter-spin distances, which can be extracted by directly analyzing the corrected signal. C: Conversion of the signal due to intramolecular spin-spin interactions into frequency space by Fourier transformation gives a Pake pattern. The peaks correspond to spin centers with vertical and parallel orientations to the external magnetic field. The inter-spin distance is calculated from the splitting of the doublet $\omega_{DD}$. D: Direct analysis of the corrected modulated signal by Tikhonov regularization gives a distance distribution $P(r)$.

DEER/PELDOR experiments require spin label concentrations in the range of 100 µM. They are performed on frozen samples, typically at ~50–80 K. Although this precludes dynamic measurements, distance distributions determined from these experiments reflect the distribution of inter-spin distances populated in solution at the time of freezing. Time-resolved studies by pulsed EPR rely on flash-freeze methods that capture reaction intermediates or conformations at different time points. With rapid-freeze approaches relying on microfluidic mixing, reactions down to the millisecond time scale have become amenable to EPR studies.

DEER/PELDOR experiments need to measure a full oscillation of the spin-echo intensity for precise determination of distances, limiting the range of measureable distances to 1–8 nm. The need for $T$ being long enough to cover a full oscillation requires sufficiently large relaxation times, otherwise the signal will have decayed before one oscillation is completed. The current distance record for distance determination by pulsed EPR is 10 nm (with deuterated proteins to increase the relaxation time).

Distances from EPR experiments have been used to probe conformational changes of proteins, and to validate structural models. Multiple distances, determined from molecules that are labeled in different positions, can be used to determine structures *ab initio* and to model global conformations by triangulation (Box 20.10). A recent study has demonstrated that PELDOR can also be used for distance measurements *in vivo*.

---

**BOX 20.10: DISTANCE RESTRAINTS FROM EPR AND
MODELING OF GLOBAL CONFORMATIONS.**

Inter-spin distances from PELDOR measurements can be used to probe structures of large biomolecules and biomolecular complexes in solution. A combined EPR- and NMR study has used long-distance restraints from EPR and short distance restraints from NMR experiments to model the structure of the RsmZ-RsmE ribonucleoprotein complex. This 70 kDa complex consists of the 72 nucleotide RsmE RNA and three homodimers of the RsmE protein and is involved in the regulation of bacterial translation initiation. The two spin labels were introduced into the RsmZ RNA by splint ligation of two RNA oligonucleotides spin-labeled at a 4-thiouridine. The ribonucleoprotein complex was formed by adding an excess of deuterated RsmE dimers to the spin-labeled RNA. Distances of > 8 nm were determined in high power Q-band four-pulse PELDOR experiments. From the calculated distance distributions, the center of the probability distribution for each radical was determined. The distance between the two centers was used as a single restraint, and allowed to vary within a range that includes 70% probability around the maximum of the experimentally determined distance during the calculation of the structural model. The structure was modeled using both NMR short-distance restraints to define local geometry and 21 long-distance restraints from EPR to define the global conformation, starting from random coil structures. The calculated structure was validated by back-calculations of distance distributions and comparison with the experimental data. Structure calculations gave comparable results when individual restraints were omitted, which demonstrates that the structure is overdetermined by the experimental parameters. Duss *et al.* (2014) *Nat. Commun.* 5: 3669.

---

## QUESTIONS

20.1 What is the magnetic field $B_0$ in a 600 MHz spectrometer and a 1 GHz spectrometer? The magnetogyric ratio of $^1$H is $\gamma = 268 \cdot 10^6$ rad s$^{-1}$ T$^{-1}$. What is the Larmor frequency of $^{13}$C and $^{15}$N in these spectrometers? $\gamma$ ($^{13}$C) = $67 \cdot 10^6$ rad s$^{-1}$ T$^{-1}$, $\gamma$ ($^{15}$N) = $-27 \cdot 10^6$ rad s$^{-1}$ T$^{-1}$. What is the difference in resonance frequency for protons with a chemical shift of 2.5 ppm and 6 ppm?

20.2 What are the possible alignments for the magnetic moments of a $^1$H and a $^{15}$N nucleus with scalar heteronuclear coupling in an external magnetic field in $z$-direction, and what are the associated relative energies?

20.3 Explain the expected features of a 1D-NMR spectrum of ethanol.

20.4 A 1D-NMR spectrum contains four doublets. How can you tell which nuclei show scalar coupling?

20.5 Show that $\Gamma$ is the full width at half maximum of an NMR line.

20.6 The dispersion of an NMR spectrum is defined as the range of chemical shifts covered by the observed resonances. Why is a large dispersion an indication for folding? What does the NMR spectrum of an intrinsically unfolded protein look like?

20.7    What NOE pattern between $C_\alpha$-protons and amide protons would you expect within an $\alpha$-helix and within a $\beta$-sheet?

20.8    In an NMR experiment at $10°C$, a protein-ligand binding reaction is in the slow exchange regime with $k = 0.01 \cdot \Delta\omega$, and two resonance lines are observed for a proton in the ligand binding site. Can you bring the reaction into the fast exchange regime ($k = 100 \cdot \Delta\omega$) to observe these resonances as a single line?

20.9    A protein P is titrated with a ligand L, and ligand binding is observed as a change in the chemical shift of a proton in the binding site (fast exchange regime). Derive an equation for the measured chemical shift $\delta_{obs}$ as a function of total protein and ligand concentrations, $P_0$ and $L_0$.

20.10   RNA helicases consist of two domains that are connected by a flexible linker. ATP binds at the interface between the domains. In the presence of RNA, ATP binding leads to closure of the cleft between the domains, and the domains tightly interact around the bound ligands. What result would you expect in a chemical shift perturbation experiment?

20.11   You want to determine the structure of a small globular domain that is part of a 400 kDa protein. Name three possible NMR approaches and state their limitations.

20.12   Calculate the energy difference and difference in the population of the two spin states of unpaired electrons for X-band (0.34 T), Q-band (1.28 T), and D-band EPR spectrometers (5 T). $g_e = 2.00$, $\mu_\beta = 9.274 \cdot 10^{-24}$ J T$^{-1}$, $T = 298.15$ K. What is the wavelength of the radiation required to excite transitions? How much energy is absorbed in 500 $\mu$L of a 1 mM aqueous solution of a spin label upon saturation of the transition at each condition (in the absence of relaxation), and what is the resulting change in temperature? $C_p$ (H$_2$O) = 4.18 kJ kg$^{-1}$ K$^{-1}$.

20.13   Why do paramagnetic compounds decrease the relaxation times of spin labels?

---

## REFERENCES

Rattle, H. (1995) An NMR primer for life scientists. UK, Partnership Press.
· concise introduction into the principles of NMR with a focus on structure determination of biomolecules

Derome, A. E. (1987) Modern NMR techniques for chemistry research. Oxford, Pergamon Press.
· thorough treatment of Fourier transform NMR

Wishart, D. S., Sykes, B. D. and Richards, F. M. (1991) Relationship between nuclear magnetic resonance chemical shift and protein secondary structure. *J Mol Biol* **222**(2): 311–333.
· discovery of the sensitivity of chemical shifts to secondary structure

Wishart, D. S., Sykes, B. D. and Richards, F. M. (1992) The chemical shift index: a fast and simple method for the assignment of protein secondary structure through NMR spectroscopy. *Biochemistry* **31**(6): 1647–1651.
· assignment of secondary structure elements in proteins from $C_\alpha$-proton chemical shift data

Zuiderweg, E. R. (2002) Mapping protein-protein interactions in solution by NMR spectroscopy. *Biochemistry* **41**(1): 1–7.
· review article about NMR approaches to investigate protein-protein interactions, including chemical shift perturbation experiments

Tugarinov, V. and Kay, L. E. (2003) Ile, Leu, and Val methyl assignments of the 723-residue malate synthase G using a new labeling strategy and novel NMR methods. *J Am Chem Soc* **125**(45): 13868–13878.
· specific isotope labeling of isoleucine, leucine, and valine in proteins and NMR experiments for assignment of $^{13}$C and $^1$H resonances from these residues in large proteins

Fischer, M., Kloiber, K., Hausler, J., Ledolter, K., Konrat, R. and Schmid, W. (2007) Synthesis of a $^{13}$C-methyl-group-labeled methionine precursor as a useful tool for simplifying protein structural analysis by NMR spectroscopy. *ChemBioChem* **8**(6): 610–612.
· synthesis of a $^{13}$C-labeled methionine precursor for isotope labeling of methionine methyl groups

Ruschak, A. M. and Kay, L. E. (2010) Methyl groups as probes of supra-molecular structure, dynamics and function. *J Biomol NMR* **46**(1): 75–87.
· review article about using methionine with labeled methyl groups for structure determination of large proteins

Cowburn, D. and Muir, T. W. (2001) Segmental isotopic labeling using expressed protein ligation. *Methods Enzymol* **339**: 41–54.
· review article about segmental isotope labeling for NMR using expressed protein ligation

Freiburger, L., Sonntag, M., Hennig, J., Li, J., Zou, P. and Sattler, M. (2015) Efficient segmental isotope labeling of multi-domain proteins using Sortase A. *J Biomol NMR* **63**(1): 1–8.
· sortase-mediated segmental isotope labeling

Gelis, I., Bonvin, A. M., Keramisanou, D., Koukaki, M., Gouridis, G., Karamanou, S., Economou, A. and Kalodimos, C. G. (2007) Structural basis for signal-sequence recognition by the translocase motor SecA as determined by NMR. *Cell* **131**(4): 756–769.
· application of methionine methyl labeling to the investigation of signal peptide binding to the SecA subunit of Sec translocase

Pervushin, K., Riek, R., Wider, G. and Wuthrich, K. (1997) Attenuated $T_2$ relaxation by mutual cancellation of dipole-dipole coupling and chemical shift anisotropy indicates an avenue to NMR structures of very large biological macromolecules in solution. *Proc Natl Acad Sci U S A* **94**(23): 12366–12371.
· first description of transverse relaxation-optimized spectroscopy (TROSY) and its potential for structure determination of larger biological macromolecules

Kleckner, I. R. and Foster, M. P. (2011) An introduction to NMR-based approaches for measuring protein dynamics. *Biochim Biophys Acta* **1814**(8): 942–968.
· review article about dynamic information from NMR

Palmer, A. G., 3rd, Kroenke, C. D. and Loria, J. P. (2001) Nuclear magnetic resonance methods for quantifying microsecond-to-millisecond motions in biological macromolecules. *Methods Enzymol* **339**: 204–238.
· in-depth quantitative treatment of NMR chemical exchange to monitor conformational changes on the millisecond to second time scale

Baker, L. A. and Baldus, M. (2014) Characterization of membrane protein function by solid-state NMR spectroscopy. *Curr Opin Struct Biol* **27**: 48–55.
· review on applications of solid-state NMR to membrane proteins

Comellas, G. and Rienstra, C. M. (2013) Protein structure determination by magic-angle spinning solid-state NMR, and insights into the formation, structure, and stability of amyloid fibrils. *Annu Rev Biophys* **42**: 515–536.
· review article on magic-angle spinning NMR and its applications to studies of amyloids

Felli, I. C. and Pierattelli, R. (2012) Recent progress in NMR spectroscopy: toward the study of intrinsically disordered proteins of increasing size and complexity. *IUBMB Life* **64**(6): 473–481.
· review on applications of solid-state NMR to intrinsically disordered proteins

Hu, K. N. and Tycko, R. (2010) What can solid state NMR contribute to our understanding of protein folding? *Biophys Chem* **151**(1–2): 10–21.
· review on applications of solid-state NMR in protein folding and on freeze-trapped folding intermediates

Viegas, A., Macedo, A. L. and Cabrita, E. J. (2009) Ligand-based nuclear magnetic resonance screening techniques. *Methods Mol Biol* **572**: 81–100.
· review on NMR methods for ligand screening with a focus on practical apsects

Viegas, A., Manso, J., Corvo, M. C., Marques, M. M. and Cabrita, E. J. (2011) Binding of ibuprofen, ketorolac, and diclofenac to COX-1 and COX-2 studied by saturation transfer difference NMR. *J Med Chem* **54**(24): 8555–8562.
· saturation transfer difference experiments to characterize the interaction of ibuprofen and diclofenac with COX-I and II

Angulo, J. and Nieto, P. M. (2011) STD-NMR: application to transient interactions between biomolecules-a quantitative approach. *Eur Biophys J* **40**(12): 1357–1369.
· review of theory of saturation transfer difference-NMR and quantitative applications

Argirevic, T., Riplinger, C., Stubbe, J., Neese, F. and Bennati, M. (2012) ENDOR spectroscopy and DFT calculations: evidence for the hydrogen-bond network within α2 in the PCET of *E. coli* ribonucleotide reductase. *J Am Chem Soc* **134**(42): 17661–17670.

Nick, T. U., Lee, W., Kossmann, S., Neese, F., Stubbe, J. and Bennati, M. (2015) Hydrogen bond network between amino acid radical intermediates on the proton-coupled electron transfer pathway of *E. coli* α2 ribonucleotide reductase. *J Am Chem Soc* **137**(1): 289–298.
· EPR studies on the hydrogen network around the tyrosyl radical in the catalytic site of ribonucleotide reductase

Bennati, M., Lendzian, F., Schmittel, M. and Zipse, H. (2005) Spectroscopic and theoretical approaches for studying radical reactions in class I ribonucleotide reductase. *Biol Chem* **386**(10): 1007–1022.
· review on insights into the ribonucleotide reductase mechanism from EPR studies

Altenbach, C., Flitsch, S. L., Khorana, H. G. and Hubbell, W. L. (1989) Structural studies on transmembrane proteins. 2. Spin labeling of bacteriorhodopsin mutants at unique cysteines. *Biochemistry* **28**(19): 7806–7812.
· site-specific introduction of spin labels into bacteriorhodopsin, reconstitution in lipid vesicles, and relaxation time measurements to map local dynamics

Schmidt, M. J., Fedoseev, A., Summerer, D. and Drescher, M. (2015) Genetically encoded spin labels for *in vitro* and in-cell EPR studies of native proteins. *Methods Enzymol* **563**: 483–502.
· spin labeling *via* incorporation of unnatural amino acids

Becker, C. F., Lausecker, K., Balog, M., Kalai, T., Hideg, K., Steinhoff, H. J. and Engelhard, M. (2005) Incorporation of spin-labelled amino acids into proteins. *Magn Reson Chem* **43**: S34–39.
· expressed protein ligation for site-specific spin labeling

Romainczyk, O., Elduque, X. and Engels, J. W. (2012) Attachment of nitroxide spin labels to nucleic acids for EPR. *Curr Protoc Nucleic Acid Chem* **7**: Unit 7.17.
· direct introduction of spin labels into nucleic acids during solid-phase synthesis

Cekan, P., Smith, A. L., Barhate, N., Robinson, B. H. and Sigurdsson, S. T. (2008) Rigid spin-labeled nucleoside C: a nonperturbing EPR probe of nucleic acid conformation. *Nucleic Acids Res* **36**(18): 5946–5954.
· use of a deoxycytidine analog as a rigid spin label to probe local mobility in nucleic acids

Hagedoorn, P. L., van der Weel, L. and Hagen, W. R. (2014) EPR monitored redox titration of the cofactors of Saccharomyces cerevisiae Nar1. *Journal of Visualized Experiments: JoVE*(93): e51611.
· redox titrations to determine the redox potential of EPR-active cofactors

Klug, C. S. and Feix, J. B. (2008) Methods and applications of site-directed spin labeling EPR spectroscopy. *Methods Cell Biol* **84**: 617–658.
· review article about spin label EPR and its applications to identify secondary structure elements, to study protein-protein- and protein-membrane interactions, folding/unfolding, and to probe protein conformation and dynamics

Rink, T., Riesle, J., Oesterhelt, D., Gerwert, K. and Steinhoff, H. J. (1997) Spin-labeling studies of the conformational changes in the vicinity of D36, D38, T46, and E161 of bacteriorhodopsin during the photocycle. *Biophys J* **73**(2): 983–993.
· relaxation time measurements of spin-labeled bacteriorhodopsin to study local protein mobility during the photocycle

Steinhoff, H. J., Mollaaghababa, R., Altenbach, C., Hideg, K., Krebs, M., Khorana, H. G. and Hubbell, W. L. (1994) Time-resolved detection of structural changes during the photocycle of spin-labeled bacteriorhodopsin. *Science* **266**(5182): 105–107.

Mollaaghababa, R., Steinhoff, H. J., Hubbell, W. L. and Khorana, H. G. (2000) Time-resolved site-directed spin-labeling studies of bacteriorhodopsin: loop-specific conformational changes in M. *Biochemistry* **39**(5): 1120–1127.
· time-resolved EPR studies on conformational changes in bacteriorhodopsin during the photocycle

Marsh, D., Kurad, D. and Livshits, V. A. (2005) High-field spin label EPR of lipid membranes. *Magn Reson Chem* **43**: S20–25.
· review article on EPR and membrane dynamics

Klare, J. P. (2013) Site-directed spin labeling EPR spectroscopy in protein research. *Biol Chem* **394**(10): 1281–1300.
· review article on information on protein structure and dynamics from EPR experiments

Krstic, I., Endeward, B., Margraf, D., Marko, A. and Prisner, T. F. (2012) Structure and dynamics of nucleic acids. *Top Curr Chem* **321**: 159–198.
· review article on applications of EPR experiments in structural and dynamic studies of nucleic acids

Bhargava, K. and Feix, J. B. (2004) Membrane binding, structure, and localization of cecropin-mellitin hybrid peptides: a site-directed spin-labeling study. *Biophys J* **86**(1 Pt 1): 329–336.
· EPR to monitor partitioning of peptides into lipid bilayers

Altenbach, C., Greenhalgh, D. A., Khorana, H. G. and Hubbell, W. L. (1994) A collision gradient method to determine the immersion depth of nitroxides in lipid bilayers: application to spin-labeled mutants of bacteriorhodopsin. *Proc Natl Acad Sci U S A* **91**(5): 1667–1671.
· collision gradient method to determine the immersion depth of spin labeled bacteriorhodopsin within the lipid bilayer by use of hydrophobic and hydrophilic paramagnetic compounds

Steinhoff, H. J. (2002) Methods for study of protein dynamics and protein-protein interaction in protein-ubiquitination by electron paramagnetic resonance spectroscopy. *Front Biosci* **7**: c97–110.
· review article on inter-spin distance determination by spectral moment analysis and possible applications to study protein ubiquitination

Pannier, M., Veit, S., Godt, A., Jeschke, G. and Spiess, H. W. (2000) Dead-time free measurement of dipole-dipole interactions between electron spins. *J Magn Reson* **142**(2): 331–340.
· first description of dead-time free four-pulse PELDOR and description of the Tikhonov regularization

Marko, A., Margraf, D., Yu, H., Mu, Y., Stock, G. and Prisner, T. (2009) Molecular orientation studies by pulsed electron-electron double resonance experiments. *J Chem Phys* **130**(6): 064102.
· determination of the relative orientation of the two spin centers by PELDOR experiments

Schiemann, O. and Prisner, T. F. (2007) Long-range distance determinations in biomacromolecules by EPR spectroscopy. *Q Rev Biophys* **40**(1): 1–53.
· review article about the principle of EPR and applications to structure probing and distance determinations in proteins and nucleic acids

Borbat, P. P. and Freed, J. H. (2007) Measuring distances by pulsed dipolar ESR spectroscopy: spin-labeled histidine kinases. *Methods Enzymol* **423**: 52–116.
· in-depth discussion of inter-spin distance determination by EPR and applications to the signaling complex of the histidine kinase CheA and CheW

Hagelueken, G., Ingledew, W. J., Huang, H., Petrovic-Stojanovska, B., Whitfield, C., ElMkami, H., Schiemann, O. and Naismith, J. H. (2009) PELDOR spectroscopy distance fingerprinting of the octameric outer-membrane protein Wza from *Escherichia coli*. *Angew Chem Int Ed* **48**(16): 2904–2906.
· determination of distances between multiple-spins in the octameric membrane protein Wza

Endeward, B., Butterwick, J. A., MacKinnon, R. and Prisner, T. F. (2009) Pulsed electron-electron double-resonance determination of spin-label distances and orientations on the tetrameric potassium ion channel KcsA. *J Am Chem Soc* **131**(42): 15246–15250.
· determination of inter-spin distances and orientation of multiple spin centers in the tetrameric ion channel KscA

Valera, S., Ackermann, K., Pliotas, C., Huang, H., Naismith, J. H. and Bode, B. E. (2016) Accurate Extraction of Nanometer Distances in Multimers by Pulse EPR. *Chem Eur J* **22**(14): 4700–4703.
· review on distance determinations for multiple spin centers in oligomers

Bode, B. E., Margraf, D., Plackmeyer, J., Durner, G., Prisner, T. F. and Schiemann, O. (2007) Counting the monomers in nanometer-sized oligomers by pulsed electron-electron double resonance. *J Am Chem Soc* **129**(21): 6736–6745.
· determination of complex stoichiometry from the presence of multiple spin centers

Kaufmann, R., Yadid, I. and Goldfarb, D. (2013) A novel microfluidic rapid freeze-quench device for trapping reactions intermediates for high field EPR analysis. *J Magn Reson* **230**: 220–226.
· rapid-mixing device to stop reactions by freezing at different time points

El Mkami, H. and Norman, D. G. (2015) EPR Distance Measurements in Deuterated Proteins. *Methods Enzymol* **564**: 125–152.
· methodology review on the determination of inter-spin distances of up to 10 nm by pulsed EPR

Ward, R., Bowman, A., Sozudogru, E., El-Mkami, H., Owen-Hughes, T. and Norman, D. G. (2010) EPR distance measurements in deuterated proteins. *J Magn Reson* **207**(1): 164–167.
· deuteration of proteins to reduce relaxation times and to determine larger inter-spin distances

Alexander, N., Bortolus, M., Al-Mestarihi, A., McHaourab, H. and Meiler, J. (2008) De novo high-resolution protein structure determination from sparse spin-labeling EPR data. *Structure* **16**(2): 181–195.
· *ab initio* structure determination with distance restraints from PELDOR experiments

Krstic, I., Hansel, R., Romainczyk, O., Engels, J. W., Dotsch, V. and Prisner, T. F. (2011) Long-range distance measurements on nucleic acids in cells by pulsed EPR spectroscopy. *Angew Chem Int Ed* **50**(22): 5070–5074.
· EPR distance measurements *in vivo*

Duss, O., Yulikov, M., Jeschke, G. and Allain, F. H. (2014) EPR-aided approach for solution structure determination of large RNAs or protein-RNA complexes. *Nat Commun* **5**: 3669.
· EPR-aided structure calculation of the RsmZ-RsmE ribonucleoprotein complex

Duss, O., Michel, E., Yulikov, M., Schubert, M., Jeschke, G. and Allain, F. H. (2014) Structural basis of the noncoding RNA RsmZ acting as a protein sponge. *Nature* **509**(7502): 588–592.
· structure of the RsmZ-RsmE ribonucleoprotein complex and functional implications

# Solution Scattering

The techniques described in Chapters 19 and 20 are based on absorption of electromagnetic radiation by molecules. Molecules that do not absorb may interact with electromagnetic radiation and alter its propagation direction, a phenomenon termed *scattering*. Scattering depends on molecular size and shape. Depending on the size of the molecule with respect to the wavelength, different information is gained. Scattering of visible light (Section 21.1) yields molecular masses or diffusion coefficients, while scattering of X-rays and neutrons (Section 21.2) provides additional information on the shape of molecules.

## 21.1 LIGHT SCATTERING

Molecules that scatter light alter its propagation direction, and sometimes also its wavelength and/or phase. Light scattering can be *elastic* or *inelastic*, and *coherent* or *incoherent*. Elastic light scattering, also termed *Thomson scattering*, does not change the energy (wavelength) of the scattered light. If the scattering goes along with absorption, the energy of the scattered light is changed and we have inelastic, or *Compton scattering*. As we will see shortly, the interaction of light with matter induces a dipole that acts as a scattering center. Multiple dipoles scattering in phase are coherent scatterers, which is always the case for light that has large wavelengths compared to the size of the scattering object. If, however, the wavelength is small compared to the scattering object, the scattering dipoles may emit light of different relative phases, and incoherent scattering is observed. Thus, in most cases inelastic scattering is also incoherent. The different light scattering experiments that can be performed by adjusting the wavelength relative to the size of the object provide different information on molecules in solution, ranging from molecular mass determinations to defining their dimensions, shapes, and sometimes their internal structure.

### 21.1.1 STATIC LIGHT SCATTERING

First, we consider the interaction of a single molecule with an electromagnetic wave. We assume that the molecule is located at the origin of a Cartesian coordinate system at (0,0,0), and exposed to linearly polarized light of a certain wavelength $\lambda$ propagating along the $x$-axis, with the electric field vector oscillating in direction of the $z$-axis. The magnetic field vector oscillates perpendicular to the electric field vector in the $y$-direction (Figure 21.1).

We will see that the size of the particle relative to the wavelength of the incident light is of importance for the scattering properties. For now, we assume that the molecule is small compared to the wavelength of the light. The electric field vector $E$ (Section 19.1.1) depends on the $x$-coordinate and the time according to

$$E(t,x) = E_0 \cos\left( 2\pi\left( \nu t - \frac{x}{\lambda} \right) \right)$$

eq. 19.1

$E_0$ is the amplitude of the electric field vector, $\lambda$ is the wavelength, and $\nu$ the frequency of the incident light. At the origin, where the molecule is located, the oscillation of the electric field vector over time follows eq. 21.1:

$$E(t) = E_0 \cos(2\pi\nu t)$$

eq. 21.1

An oscillating electric field $E(t)$ induces oscillations of the electrons of the molecule. The molecule therefore acts as an oscillating dipole $\mu(t)$ that depends on the polarizability $\alpha$ of the molecule (eq. 21.2)

$$\mu(t) = \alpha E(t) = \alpha E_0 \cos(2\pi\nu t)$$

eq. 21.2

Eq. 21.2 is identical to eq. 19.9 where we have used it to describe the induced dipole in a chromophore by the electric field of the incident light. Because of its oscillating dipole moment, the molecule will function as an antenna and emit radiation. For an isotropic molecule with a centrosymmetric electron distribution, the direction of the induced dipole corresponds to the direction of the E-field vector of the incident light, in this case the $z$-direction. The spatial distribution of the electric field emanating from this oscillating dipole can be described as a function of the distance $r$ from the dipole and the angle $\phi$ with respect to the polarization direction of the incident light:

$$E(r,\phi) = \frac{4\pi^2 \sin\phi}{r\lambda^2} \alpha E_0 \cos\left( 2\pi\left( \nu t - \frac{r}{\lambda} \right) \right)$$

eq. 21.3

The electric field thus depends on the sine of the angle $\phi$ to the $z$-axis. The intensity of light depends on the square of its amplitude. The intensity $I_s$ of the scattered light is thus the square of the prefactor in eq. 21.3

$$I_s = \left( \frac{4\pi^2 \sin\phi}{r\lambda^2} \alpha E_0 \right)^2$$

eq. 21.4

and the intensity of the incident light $I_0$ is

$$I_0 = E_0^2$$

eq. 21.5

From eq. 21.4 and eq. 21.5, we can calculate the ratio of the intensities of scattered and incident light as

$$\frac{I_s}{I_0} = \left( \frac{4\pi^2 \sin\phi}{r\lambda^2} \alpha \right)^2 = \frac{16\pi^4 \sin^2\phi}{r^2\lambda^4} \alpha^2$$

eq. 21.6

The relative intensity of the scattered light thus decreases with $r^2$, which is an inherent property of radiation from a point source. It also depends on the angle $\phi$. The sine of $0°$ is zero, meaning there is no scattering intensity in the direction of the induced dipole ($\phi = 0°$). A large part of the scattered radiation is thus emitted in directions other than the direction of the incident light (Figure 21.1).

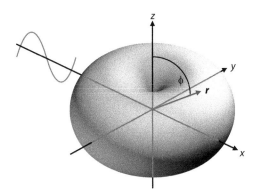

FIGURE 21.1:   **Scattering directions for polarized light.** Incident light polarized in the *z*-direction excites oscillation of a dipole along *z*, leading to scattering into a "doughnut shape" volume. The distance from the origin to the surface equals the scattered light intensity in the direction ***r***.

We have derived the angular and distance dependence of scattered light (eq. 21.6) for polarized incident light. To generalize this equation to scattering of unpolarized light, we consider unpolarized light as a superposition of wave trains with random polarization in the *yz*-plane (Figure 21.2). Each of these individual waves elicits an electric field described by eq. 21.3, and leads to scattering intensities that are described by eq. 21.6 (Figure 21.1). The overall scattering intensity in all three directions is obtained by averaging eq. 21.3 for all directions in the *yz*-plane, which gives

$$\frac{I_s}{I_0} = N \cdot \frac{8\pi^4}{r^2 \lambda^4} \cdot \alpha^2 \cdot \left(1 + \cos^2 \theta\right)$$

eq. 21.7

The angle $\theta$ is the observation angle with respect to the *x*-axis. The scattering intensity according to eq. 21.7 corresponds to a dumbbell-like shape with lowest intensity in the *yz*-plane (Figure 21.2). This shape is the result of the superposition of all doughnut-like scattering surfaces that are elicited by each wave train (Figure 21.1), each rotated by a different angle around the *x*-axis because of the different polarization directions.

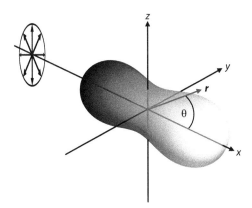

FIGURE 21.2:   **Scattering directions for unpolarized light.** Each wave train of unpolarized light (shown as different arrows) excites the dipole in the direction of its *E*-vector. The scattering of light in this direction is minimal. The sum of all scattered light has the shape of a dumbbell.

Eq. 21.7 describes the scattering intensity for a single molecule. In an ideal solution with $N$ particles distributed over a volume $V$, each solute molecule contributes equally to the scattering intensity, with the same distance dependence and angular dependence. For ideal, infinitely diluted solutions without any interactions between the scatterers, the scattering intensity of an ensemble of molecules therefore scales with $N$:

$$\frac{I_s}{I_0} = N \cdot \frac{8\pi^4}{r^2 \lambda^4} \cdot \alpha^2 \cdot \left(1 + \cos^2\theta\right)$$

eq. 21.8

If the scattering particles are small compared to the wavelength, this scattering process is called *Rayleigh scattering*. The *polarization factor* $1 + \cos^2\theta$ in the equation for Rayleigh scattering shows that scattering is present in all directions, but is twice as strong in the direction of the incident beam (forward and backward) as perpendicular to it. Scattering depends on the square of the molecular polarizability ($\alpha^2$), which increases with molecular size. Importantly, scattering has a massive wavelength dependence ($\lambda^4$), which is the reason for the colors of the sky we know: when viewed at large angles $\theta$ to the sun (at noon), the blue light is scattered more strongly out of the direct solar beam by particles in the atmosphere than light of longer wavelengths, reaching our eyes from all directions, so the whole hemisphere looks blue. Increased path length through the atmosphere leads to increased scattering. For this reason, the horizon appears a little whiter (less blue) than the zenith. During sunrise and sunset, the sky looks red in the direction of the sun (small $\theta$) due to efficient removal of blue light from the direct beam of the sun by scattering.

From Rayleigh scattering experiments (eq. 21.7), the Rayleigh ratio $R_\theta$ can be measured:

$$R_\theta = \frac{I_s}{I_0} \cdot \frac{r^2}{V \cdot \left(1 + \cos^2\theta\right)} = N' \cdot \frac{8\pi^4 \alpha^2}{\lambda^4}$$

eq. 21.9

$N'$ is the number density of particles and equals $N/V$. The Rayleigh ratio is the relative intensity of the scattered light and is related to the molecular mass of the scattering particle. In order to measure the Rayleigh ratio in practice, the polarizability $\alpha$ is expressed in terms of a measurable macroscopic quantity, the refractive index $n$, which in turn is proportional to the concentration of the scatterers and, thus, to their molecular mass $M_M$. The number of scatterers $N$ in the volume $V$ is related to the concentration $c$ by the molecular mass $M_M$ and Avogadro's number $N_A$

$$c = \frac{N \cdot M_M}{V \cdot N_A}$$

eq. 21.10

The polarizability can then be expressed as

$$\alpha = \frac{n_0}{2\pi} \cdot \frac{\partial n}{\partial c} \cdot \frac{M_M}{N_A}$$

eq. 21.11

$n_0$ is the refractive index of the pure solvent and $\partial n/\partial c$ is the *refractive index increment*, both of which can be measured. Substituting the above two equations into eq. 21.9 yields

$$R_\theta = \frac{2\pi^2 n_0^2}{N_A \cdot \lambda^4} \cdot \left(\frac{\partial n}{\partial c}\right)^2 \cdot c \cdot M_M = K \cdot c \cdot M_M$$

eq. 21.12

This result shows that the Rayleigh ratio linearly depends on the molecular mass. The equation has applications in size exclusion chromatography-multi-angle laser light scattering (SEC-MALLS), which provides inline monitoring of the molecular masses of the eluted proteins.

What happens if the scattering molecules are large compared to the wavelength of the incident light? In this case, dipole oscillations are induced in several parts of the molecule that are at an appreciable distance apart from each other. Each dipole will act as an antenna, and the scattered light from each dipole will come from a different origin and with a different phase. The individual scattering centers are fixed within the molecule, so their motion is correlated by the tumbling of the molecule, and hence their scattering cannot be considered independent anymore. Light that is scattered from different parts of the molecule will travel different path lengths until it reaches the detector. Therefore, the scattered light from individual scattering centers within one molecule gives an overall interference pattern for each molecule, determined by its *form factor* (see Section 21.2.4). Because of the random orientation of scattering molecules in solution, the overall observed scattering is the scattering from individual molecules averaged over all directions. The information content of the scattering experiments now extends to molecular shapes and dimensions and even the internal structure of the molecule (see Section 21.2.4.4). If the scattering is performed on a sample with ordered molecules, i.e. a crystal, the scattering patterns of each molecule interfere with each other, yielding an overall scattering that is termed the *diffraction pattern*. Such patterns are no longer determined by the form factors of the individual molecules but the *structure factor* of the crystal lattice (see Chapter 22).

## 21.1.2 DYNAMIC LIGHT SCATTERING

Instead of measuring the angular dependence of the scattering intensity, we can also follow the scattering intensity from a small volume element of our macromolecule in solution over time at a fixed angle. In this case, the scattering intensity will fluctuate because of the fluctuations in local concentration of the molecule due to Brownian motion. While on average there are N scattering molecules within the observed volume element, random thermal motion leads to fluctuations in their number, associated with an increase or decrease of the scattering intensity. The fluctuations in scattering intensity thus depend on the diffusion velocity of the molecules, which in turn is related to their molecular mass. Measuring the fluctuations of scattered light at one fixed angle $\theta$ permits the determination of the size of the molecule. If fluctuations are followed at multiple angles (multi-angle light scattering, MALS), information on the size distribution in mixtures of particles can be extracted.

The (translational) diffusion coefficient $D$ is defined as

$$D = \frac{RT}{N_A f} = \frac{k_B T}{f}$$

eq. 21.13

$R$ is the general gas constant, $k_B$ the Boltzmann constant, $T$ the absolute temperature, $N_A$ is the Avogadro constant, and $f$ the friction coefficient. The friction coefficient $f$ depends on the size, shape, and hydration of the molecule. For spherical molecules, $f$ depends on the radius $r$ as

$$f = 6\pi\eta r$$

eq. 21.14

For molecules of a certain molecular mass, $f$ is smallest when the molecule is spherical: for non-spherical molecules, the friction coefficient will be larger (discussed further for analytical ultra-centrifugation; see Section 26.2). The units of the diffusion coefficient $D$ are $cm^2\ s^{-1}$ (see Section 4.1). The diffusion coefficient can be extracted from the temporal fluctuations of the scattering intensity by calculating the autocorrelation function. The fluctuation of the scattering intensity $I$ at time $t$, $\Delta I(t)$, is defined as the difference between the current intensity $I(t)$ and the average scattering intensity $\bar{I}$ :

$$\Delta I(t) = I(t) - \bar{I}$$

eq. 21.15

At a point later in time, at $t + \tau$, the fluctuation is

$$\Delta I(t+\tau) = I(t+\tau) - \bar{I}$$

eq. 21.16

The autocorrelation function relates the fluctuation at time $t$ with the fluctuations at different time points $t + \tau$ as

$$A(\tau) = \overline{\Delta I(t) \cdot \Delta I(t + \tau)}$$

eq. 21.17

The bar indicates that the average over the time $t$ is taken for each fixed value of $\tau$. For small values of $\tau$, the fluctuations at $t$ and $t + \tau$ will be of equal sign more often than of opposite sign, and $A(\tau)$ will be positive. In contrast, for large $t$, the probability for $\Delta I(t)$ and $\Delta I(t + \tau)$ to be of opposite sign is very high, and $A(\tau)$ will become smaller and smaller (Figure 21.3).

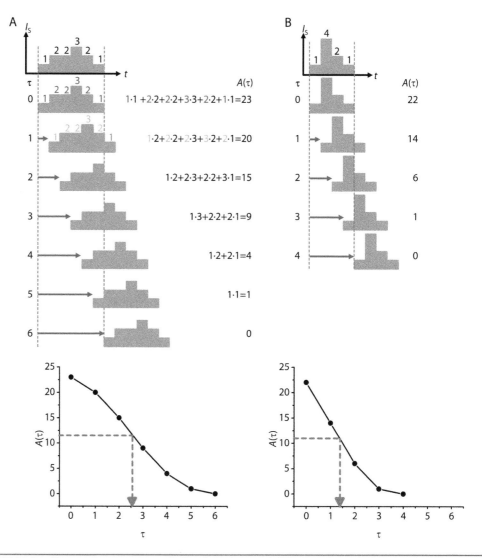

**FIGURE 21.3:** **Fluctuations, autocorrelation function, and diffusion times.** A: A molecule diffuses through the observation volume and causes a scattering signal (gray bars). After six time units, it leaves the observation volume, and the scattering signal ceases. The autocorrelation function $A(\tau)$ for this scattering pulse can be calculated by shifting the function relative to itself by 0, 1, 2, 3, 4, 5, and 6 time units, by multiplying the signal of the corresponding time points for the shifted and un-shifted function, and summing up the products. The inflection point of the autocorrelation curve (bottom, 2.6 time units) corresponds to the mean time the molecule spends in the observation volume. B: A second molecule diffuses more quickly through the confocal volume and gives rise to a shorter scattering pulse. The un-normalized autocorrelation function $A(\tau)$, calculated by shifting the function relative to itself by 0, 1, 2, 3, and 4 time units, by multiplying the signal of the corresponding time points and summing up the products decays to zero more rapidly. The inflection point is now at 1.4 time units, the second molecule diffuses almost twice as rapidly through the observation volume.

For a single type of scattering molecule, the autocorrelation function $A(\tau)$ at a fixed observation angle $\theta$ decays exponentially with $\tau$ (eq. 21.18):

$$A(\tau) = A_0 \exp\left(-D\frac{8\pi^2 n^2}{\lambda^2}\sin^2\frac{\theta}{2}\right)\tau$$

eq. 21.18

The diffusion coefficient D can be determined from the exponential function. Typical values for diffusion coefficients are summarized in Table 21.1. Often, the autocorrelation is plotted with a logarithmic time $\tau$ axis. The time $\tau$ at which the autocorrelation has been reduced to $A_0/e$ corresponds to the mean diffusion time $\tau_{diff}$ of molecules through the observation volume (Figure 21.3). The diffusion coefficient $D$ can also be calculated from $\tau_{diff}$. We will return to the autocorrelation function in the context of fluorescence correlation spectroscopy (Section 23.1.6).

**TABLE 21.1**
**Typical diffusion coefficients *D* for biological processes.**

| diffusion process | D (cm² s⁻¹) |
|---|---|
| small molecule in aqueous solution | $10^{-5}$ |
| protein in aqueous solution | $10^{-6}$ |
| phospholipid in membrane | $10^{-8}$ |
| protein in lipid membrane | $10^{-10}$ |

### 21.1.3 RAMAN SCATTERING

So far, we have considered elastic scattering, in which the incident and scattered light have the same wavelength. In other words, we have assumed that no energy is exchanged when the light interacts with the molecule. In inelastic scattering, energy is exchanged, and the wavelength of the incident and the scattered light are different. This can be caused by changes in the polarizability due to internal vibrations of the molecule. For example, the overall polarizability of diatomic molecules may change with the vibration of the connecting bond. If this is the case, the polarizability $\alpha$ can be expressed as a function of bond vibration with the frequency $\nu_{bond}$ as

$$\alpha(\nu_{bond}) = \alpha + \beta\cos\left(2\pi\nu_{bond}t\right)$$

eq. 21.19

The induced dipole then depends on $\alpha(\nu_{bond})$:

$$\mu = \alpha E_0 \cos\left(2\pi\nu t\right) + \beta E_0 \cos\left(2\pi\nu t\right)\cos\left(2\pi\nu_{bond}t\right)$$

eq. 21.20

The first term in eq. 21.20 gives rise to elastic scattering at the same frequency $\nu$ as the incident light, and thus to Rayleigh scattering. The second term corresponds to scattering at wavelengths distinct from the wavelength of the excitation light. We can rearrange the product of the cosine terms according to eq. 21.21:

$$\cos a \cdot \cos b = \frac{1}{2}\left(\cos(a+b) + \cos(a-b)\right)$$

eq. 21.21

and obtain

$$\mu = \alpha E_0 \cos\left(2\pi\nu t\right) + \frac{\beta E_0}{2}\left(\cos\left(2\pi(\nu+\nu_{bond})t\right) + \cos\left(2\pi(\nu-\nu_{bond})t\right)\right)$$

eq. 21.22

Eq. 21.22 thus reveals that, in addition to elastic Rayleigh scattering (first term), scattering occurs at higher and lower frequencies $\nu + \nu_{bond}$ (second term) and $\nu - \nu_{bond}$ (third term; Figure 21.4). This phenomenon is called Raman scattering. The Raman bands at lower frequency than the excitation light occur because a vibration is excited in a molecule, and are called Stokes bands. The bands at higher frequency than the excitation light occur when a molecule in a higher vibronic state is irradiated and returns to a lower vibrational state. They are called *anti*-Stokes bands. The energy of the vibration can thus be calculated from the spectral difference between Raman bands and excitation light. To measure a Raman spectrum, the sample is illuminated with monochromatic, plane-polarized light. The scattered light is measured at a 90° angle.

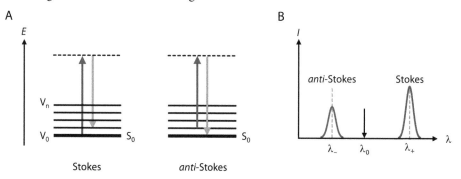

**FIGURE 21.4:** **Raman scattering**. A: Raman spectroscopy is based on inelastic scattering, caused by a change in polarizability of molecules with bond vibrations. The scattered light is of lower energy/lower frequency when a bond vibration is excited (left), giving rise to the Stokes band. If a vibronic state relaxes, the scattered light is of higher energy/higher frequency (right), giving rise to the *anti*-Stokes band. B: *anti*-Stokes and Stokes bands at $\lambda_-$ and $\lambda_+$ after irradiation at $\lambda_0$.

Raman bands are of much lower intensity than the Rayleigh band, and the coefficient $\beta$ is typically less than $10^{-4}$ of $\alpha$. Nevertheless, Raman spectroscopy can be very valuable for the study of biomolecules. Raman bands are observed when the polarizability of the molecule changes with the vibration. Vibrations can also give rise to absorption of infrared light (Section 19.4). Infrared absorption occurs when the vibration leads to a change in charge distribution, i.e. requires a non-zero transition dipole. Because of these different requirements, Raman spectroscopy is often complementary to infrared spectroscopy. It should be noted that Raman bands from water vibrations distort fluorescence spectra and need to be corrected for (see Section 19.5.6). On the scale of Raman bands from biomolecules, however, Raman bands from water are small, and water absorption does not interfere with Raman studies of biomolecules in aqueous solution. In Raman microscopy, Raman spectroscopy is used as a probe in imaging.

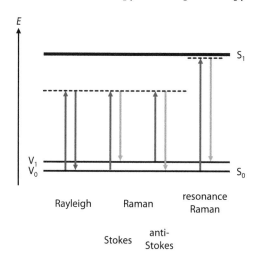

**FIGURE 21.5:** **Rayleigh, Raman, and resonance Raman scattering**. Rayleigh scattering is elastic, and the incident and scattered light are of the same wavelength. Raman scattering is inelastic, and the scattered light is either of higher (Stokes) or of lower wavelength (*anti*-Stokes) than the incident light. The energy difference excites a higher vibrational state (Stokes) or stems from the relaxation of a higher vibrational state (*anti*-Stokes). In resonance Raman scattering, the wavelength of the incident light lies within an absorption band. Here, the transition for the Stokes band is shown.

Raman spectroscopy encounters two major problems that limit its application to biomolecules. First, the intensity of Raman scattering is typically very low. Second, Raman spectroscopy reports on vibrational transitions, and the number of bands in a Raman spectrum increases with the number of degrees of freedom of the molecule (3*N*-6 for a nonlinear molecule of *N* atoms, see Section 19.4.2). Such complex Raman spectra provide a signature for molecules, and can be used, for example, as fingerprints of pharmaceutical compounds. G-quadruplex structures (see Section 17.5.2.2) show different Raman spectra depending on whether they are parallel or antiparallel.

The intensity of Raman bands can be enhanced by resonance and surface effects. Resonance Raman spectroscopy exploits the enhancement of Raman signals near electronic transitions, i.e. near an absorption band (Figure 21.5). This enhancement is caused by the large change of polarizability associated with absorption. In resonance Raman spectroscopy, vibrations of a chromophore can therefore be observed not only with high sensitivity, but also selectively within a larger molecule or molecular assembly. Resonance Raman spectroscopy has been used to study heme or flavin groups, carotenes, iron-sulfur clusters, and the retinal chromophore in rhodopsin.

Raman scattering is also intensified near surfaces of noble metals such as gold, an effect that is exploited in surface-enhanced Raman spectroscopy (SERS). When SERS is combined with the resonance Raman effect (surface enhanced resonance Raman spectroscopy, SERRS), single-molecule sensitivity is reached. SERS fingerprinting of molecules, cells and tissues is used in analytical applications. SERS has been used to study membrane proteins such as bacteriorhodopsin, rhodopsin, and photosynthetic complexes in model membranes or their natural membrane environment.

## 21.2 SMALL ANGLE SCATTERING

In Section 21.1.1 we discussed how light scattering from molecules that are small compared to the wavelength of the incident light is different from scattering by large molecules. X-rays and particle beams have wavelengths on the order of bond lengths, much smaller than the average molecule diameter, so in principle can provide higher resolution information than light scattering. In the following, we discuss the scattering of X-rays and neutrons by molecules. The intensity of scattered radiation decreases strongly at larger angles with respect to the incident light. Scattering intensity can therefore only be measured at small angles, hence the collective term small angle scattering (SAS).

### 21.2.1 SCATTERING OF X-RAYS AND NEUTRONS

X-rays are scattered by electrons, and neutrons are scattered by nuclei, but the physical principles of the two phenomena are the same. Both scattering processes can be elastic or inelastic. We will focus on the elastic scattering of X-rays, i.e. no energy is transferred from the incident light onto the sample. The incident radiation induces oscillations in electrons, which then emit secondary X-rays in all directions.

In the introduction to light scattering, we derived the equation for the overall scattering intensity as a function of scattering angle and direction (eq. 21.7). The polarizability $\alpha$ of electrons is a function of natural constants (electron charge $e$, electron mass $m_e$, and vacuum permittivity $\varepsilon_0$) and the frequency $\omega$ of the electron:

$$\alpha = \frac{e^2}{4\pi \cdot \varepsilon_0 \cdot m_e \cdot \omega^2}$$

eq. 21.23

The frequency $\omega$ can be expressed in terms of the speed of light $c$ and the wavelength $\lambda$ as $\omega = 2\pi c/\lambda$, and we can eliminate the polarization and wavelength from eq. 21.7 to arrive at the *Thomson equation*. The Thomson equation describes the intensity of the light scattered by a single electron $I_s$ relative to the absolute intensity of the incident beam $I_0$:

$$\frac{I_s}{I_0} = \frac{e^4}{16\pi^2 \cdot \varepsilon_0^2 \cdot m_e^2 \cdot c^4} \cdot \frac{\left(1+\cos^2\theta\right)}{2r^2} = 7.94 \cdot 10^{-30} m^2 \cdot \frac{\left(1+\cos^2\theta\right)}{2r^2}$$

eq. 21.24

The constant factor in the Thomson equation, $C_{Th} = 7.94 \cdot 10^{-30}$ m², is a very small number, reflecting the minute scattering intensity from a single electron. However, the sum of all electrons present in macromolecular solutions and crystals will scatter light at detectable intensities. The unit m² of $C_{Th}$ indicates a *cross-section*, a measure of the likelihood of a collision event between particles (Section 26.1.5.5) or photons. The constant is proportional to the scattered light intensity and is related to the extinction coefficient in absorption. Its square root amounts to 2.82 fm, the *Thomson scattering length* $b_e$. $b_e$ signifies the amplitude of the scattered radiation at $\theta = 0$ per unit solid angle (in steradian). In other words, it is an absolute measure for the scattering power of a single electron.

$$b_e = \frac{e^2}{4\pi \cdot \varepsilon_0 \cdot m_e \cdot c^2} = \frac{\mu_0 e^2}{4\pi m_e} = 2.82 \text{ fm}$$

eq. 21.25

In eq. 21.25, $\mu_0$ is the vacuum permeability, which depends on the vacuum permittivity and the speed of light as $1/(\varepsilon_0 c^2)$. $b_e$ is also known as the *classical electron radius*. Although electrons are typically considered as point charges with no spatial extent, $b_e$ can be viewed as the radius of a sphere of charge $e$ whose potential energy equals the potential energy of a particle with the electron rest mass $m_e$. The scattering length $b_{atom}$ of an atom with atomic number $Z$ normally scales with the number of electrons:

$$b_{atom} = Z \cdot b_e$$

eq. 21.26

Table 21.2 lists the X-ray and neutron scattering lengths for a few biologically interesting isotopes. While the linearity of $b_{atom}$ for X-rays is evident, neutron scattering lengths do not scale with $Z$. Neutron scattering lengths describe the more complicated interaction between a neutron beam and atomic nuclei. For example, the magnitudes of the neutron scattering lengths of hydrogen ($^1$H) and deuterium (D) are comparable to those of much heavier elements, such as sulfur and uranium (Table 21.2). Hydrogen atoms are therefore detectable by neutron scattering. Note also the negative value for the $^1$H neutron scattering length compared to D. This sign difference allows discrimination between $^1$H and D by neutron scattering. We will come back to this point in the description of small angle neutron scattering (SANS; Section 21.2.5).

| TABLE 21.2 | | | |
|:---|:---:|:---:|:---:|
| **Coherent scattering lengths for X-rays and neutrons.** | | | |
| | | **scattering length (in fm)** | |
| **isotope** | **Z** | **X-rays** | **neutrons** |
| $^1$H | 1 | 2.9 | −3.7 |
| $^2$H (D) | 1 | 2.9 | 6.7 |
| $^3$H (T) | 1 | 2.9 | 4.8 |
| $^{12}$C | 6 | 17.1 | 6.7 |
| $^{14}$N | 7 | 20.0 | 9.4 |
| $^{15}$N | 7 | 20.0 | 6.4 |
| $^{16}$O | 8 | 22.8 | 5.8 |
| $^{31}$P | 15 | 42.8 | 5.1 |
| $^{32}$S | 16 | 45.6 | 2.8 |
| $^{238}$U | 92 | 262 | 8.4 |

The relative scattering power of an atom is expressed in terms of the ratio of the scattering lengths of the atom and the electron (for X-rays; eq. 21.27). This ratio is the *atomic scattering factor f*, also known as the *atomic form factor*. In other words, $f$ is the ratio of the amplitudes of the scattered waves of the atom and a single electron:

$$f = \frac{\text{amplitude of wave scattered by the atom}}{\text{amplitude of wave scattered by a single electron}} = \frac{b_{\text{atom}}}{b_e} \qquad \text{eq. 21.27}$$

Mathematically, the atomic scattering factor is the Fourier transform (Section 29.14) of the electron density of a spherical atom. It is a function of the scattering angle θ. At θ = 0 and in the absence of absorption of the incident X-rays or other side effects, $f_0$ is equal to the number of electrons in the scattering particle (Z for atoms). The magnitude of $f$ decreases with larger scattering angles following a function of the form $\exp(-\sin^2\theta)$ because at larger scattering angles the waves emitted by each of the Z electrons in an atom will have larger phase shifts relative to each other, and thus show increasingly destructive interference (Figure 21.6). Because of this strong angular dependence, scattering intensity can only be measured at small angles (small angle scattering, SAS).

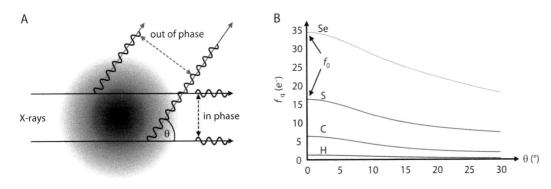

FIGURE 21.6:    **Atomic scattering factors.** A: Elastic scattering from two regions of the electron cloud of an atom introduces phase shifts depending on the scattering angle. In the forward direction (black, θ = 0), there is no phase shift because the path difference between the diffracted beams is zero. The path difference, and hence the phase shift, increases with increasing θ. B: Resulting angle dependence of the atomic scattering factor for a few elements. At θ = 0 the atomic scattering factor equals the number of electrons (H: 1, C: 6, S: 16, Se:34).

The scattering length $b_{\text{molecule}}$ of a molecule consisting of N atoms can be calculated as the sum of all atomic scattering lengths:

$$b_{\text{molecule}} = \sum_{i=1}^{N} b_{\text{atom},i} \qquad \text{eq. 21.28}$$

In analogy to the atomic form factor, the *molecular form factor* describes the overall relative scattering amplitude of a particle (molecule) in ideal solution relative to an electron. Rays scattered by different atoms in the same molecule will interfere with each other (Figure 21.7) and provide information on the extent of the molecule (its form). Eq. 21.28 only holds for ideal solutions, where the scattered radiation interferes only within the same molecule. In non-ideal solutions the molecules interact with each other, and scattered light from different molecules shows additional interference. This scattering is not described by a form factor any more, but by the *structure factor*. The structure factor is particularly important for ordered molecular arrays such as crystals, where it describes the expected intensity distribution of light diffracted by the array. Because scattering of solutions and diffraction by crystals are based on the same physical process, they share many aspects, which we will point out throughout this chapter and in the chapter on X-ray crystallography (Chapter 22).

For the diluted and practically ideal solutions used in small angle scattering (SAS) studies, the scattering length per volume is an informative parameter. The volume of a molecule can be calculated from the density ρ, the molecular mass $M_M$, and the Avogadro constant $N_A$. Division of the scattering length of a molecule by its volume yields its *scattering length density* SLD, sometimes also referred to as *scattering density*:

$$\text{SLD} = \frac{b_{\text{molecule}}}{V_{\text{molecule}}} = \frac{\rho \cdot N_A \cdot b_{\text{molecule}}}{M_M}$$

eq. 21.29

Scattering densities are useful to judge the amount of expected scattering from a particular sample. Carbohydrates, lipids, water, proteins, and nucleic acids all have characteristic X-ray and neutron scattering densities (Table 21.3). Nucleic acids have the highest scattering densities and thus give stronger signals for SAS. The scattering densities for X-rays and neutrons in Table 21.3 follow similar trends as the atomic form factors given in Table 21.2: substances with higher electron density scatter X-rays more strongly, and the neutron scattering densities for the hydrogen isotopes have the opposite sign.

**TABLE 21.3**
**Scattering densities and electron densities for biological substances.**

| substance | X-rays | | neutrons | |
|---|---|---|---|---|
| | SLD ($10^{-4}$ nm$^{-2}$) | $\rho_e$ (e nm$^{-3}$) | SLD ($10^{-4}$ nm$^{-2}$) | % D$_2$O |
| H$_2$O | 9.4 | 334 | −0.56 | 0 |
| D$_2$O | 9.4 | 334 | 6.4 | 100 |
| proteins | 11.5–12.6 | 410–450 | 0.02–0.03 | 40–45 |
| DNA | 16.5 | 590 | 0.04 | 65 |
| RNA | 16.7 | 600 | 0.05 | 72 |
| carbohydrates | 13.8 | 490 | 0.03 | 47 |
| lipids | 8.7–9.6 | 310–340 | 0.01–0.04 | 10–14 |

*Source:* Data from Perkins, S. J. (1988). *Modern Physical Methods in Biochemistry, Part B* (A. Neuberger & L. L. M. van Deenen, Eds.), with permission.

*Note:* For neutron scattering it is more convenient not to state the nuclear density but the percentage of D$_2$O that is needed to experimentally match the density of the given substance (the *neutron match point*).

The differences in X-ray and neutron scattering densities between water (H$_2$O and D$_2$O) and biological substances indicate that both are essentially difference measurements: the scattering *contrast* between solvent and solute is measured (Figure 21.7). The contrast depends on the difference in the scattering densities of the macromolecule and the solvent (Table 21.3). Given all other parameters identical, subtracting the X-ray scattering data of a buffer control from the scattering data of a sample leaves the scattering of the macromolecule, or more precisely, of the macromolecule plus its sphere of hydration (see radius of hydration; Box 21.1). For SANS, the change in nuclear density between solvent and solute is negligible in water (Table 21.3), but changes dramatically in the presence of D$_2$O. A technique called *density matching* or *contrast variation* is used in SANS to determine the shape of macromolecules (see Section 21.2.5.2).

## 21.2.2 SAS INTENSITY DISTRIBUTION

Atomic scattering factors decrease strongly as a function of scattering angle due to increased negative interference of the beams scattered at different parts of the atom (Figure 21.6). The same is true for scattering by molecules. The scattering into the same direction from two different points of a molecule will result in two beams of different path length and, hence, different phase. These beams interfere with each other and the detector only registers the *interference pattern* or *scattering pattern* (Figure 21.7).

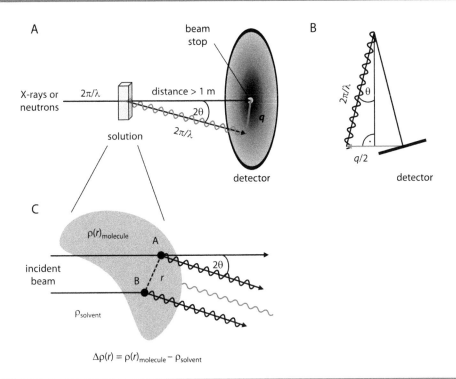

**FIGURE 21.7:    Schematic SAS experiment.** A: The incident (direct) beam of wavenumber $2\pi/\lambda$ illuminates a sample in solution. The detector is at several meters distance from the sample and is protected from the intense direct beam by a beam stop. The angle between the direct and the scattered beams is defined as $2\theta$. The scattering pattern on the detector is radially symmetric because the molecules in solution adopt all possible orientations. Darker shading towards the beam stop represents higher scattering intensity. B: The scattering vector $\boldsymbol{q}$ is related to the scattering angle and the wavenumber $2\pi/\lambda$ by trigonometry. C: Scattering from two different points A and B in a molecule (gray shape) leads to negative interference for $\theta > 0$, indicated by the blue sine wave of lower amplitude.

Within molecules, both the scattering angle $\theta$ and the distance $r$ between the scattering points influence the path difference. For $\theta$ approaching zero, the path difference is also zero, and hence interference is absent with no loss of scattering (intensity $I_0$) in the forward direction. At angles $\theta > 0$, the scattered intensity is conveniently described by the *scattering vector* $\boldsymbol{q}$ (Figure 21.7), which is the result of the interference of all beams scattered into certain directions by all points of the molecule.

The incident beam can be represented by the *spatial frequency* $2\pi/\lambda$, which represents the number of complete waves ($2\pi$) per unit distance. Together with the scattering angle $\theta$, the spatial frequency defines the direction and amplitude of the scattering vector $\boldsymbol{q}$. The magnitude $q$ of the scattering vector is related to the scattering angle $\theta$ by trigonometry (Figure 21.7).

$$q = \frac{4\pi}{\lambda} \sin \theta \qquad\qquad \text{eq. 21.30}$$

The parameter $q$, also known as the *momentum transfer*, is both a measure of resolution and of the change in directional momentum of the scattered photons: the larger the change in directional momentum, the higher resolution information can be gathered from the experiment (Figure 21.8). The unit of $q$ is inverse length ($\text{Å}^{-1}$ or $\text{nm}^{-1}$). We will see later that $q$ is related to the Bragg resolution $d$ in crystallography (see eq. 22.6) by $q = 2\pi/d$. The inverse length of the momentum transfer highlights that scattering happens in *reciprocal space*, in contrast to the *real space* of molecules. The mathematical connection between the two spaces is the Fourier transformation (see Section 29.14).

Because the molecules in solution have random orientations, $I(q)$ is proportional to the scattering of a single particle averaged over all of these orientations. The scattering from molecules in random orientations produces an isotropic pattern, which means it has radial symmetry about the incident X-ray beam (Figure 21.7). Different molecular shapes lead to different $I(q)$ distributions (Figure 21.8).

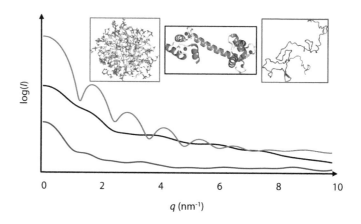

**FIGURE 21.8:** **Examples of scattering curves from molecules with different shapes.** Scattering curves for a small spherical micelle (blue), the elongated, helical $Ca^{2+}$-binding protein calmodulin (black), and an extended random coil (red). The more isometric the molecule, the more features appear in the intensity distribution at high $q$ values. Since the scattering intensity decreases strongly with increasing $q$ values, $\log(I)$ is plotted. Calculated from coordinates using the program CRYSOL at www.embl-hamburg.de/biosaxs/atsas-online.

For a monodisperse and ideal solution, i.e. a highly diluted solution without interactions between freely diffusing solute molecules, the scattering intensity as a function of small $q$ values is given by the *Guinier approximation* (Figure 21.9).

$$I(q) = I_0 \cdot e^{-\frac{R_g^2 q^2}{3}}$$

eq. 21.31

$R_g$ is the *radius of gyration* and $I_0$ is the total scattering intensity at $\theta = 0$. Mathematically, $R_g$ is the quadratic mean of the distances $r_i$ between atoms and the center of mass, weighted by the contrast $m_i$ of the atoms:

$$R_g = \sqrt{\frac{\sum m_i r_i^2}{\sum m_i}}$$

eq. 21.32

$R_g$ is a useful measure for the degree of particle size and anisometry, i.e. its degree of asymmetry. For example, spheres of radius $R$ follow the linear dependence $R_g = \sqrt{0.4} \cdot R$, while cylindrical rods of radius $R$ and length $L$ (nucleic acids) follow $R_g^2 = R^2/4 + L^2/12$. $R_g$ is closely related to another quantity, the *radius of hydration* $R_h$. Methods such as SAS are sensitive to the layer of water bound to the surface of macromolecules and measure $R_h$ rather than $R_g$ (Box 21.1).

**BOX 21.1: $R_G$ VERSUS $R_H$.**

Dissolved macromolecules bind a layer of water on their surface. This sphere of hydration corresponds to ca. 0.3 g $H_2O$ per gram protein and is about 15% more densely packed than bulk solvent (see Figure 16.35). The added solvent changes $R_g$ to $R_h$, the *hydrodynamic radius* or *radius of hydration*. $R_h$ equals the radius of an equivalent hard sphere that diffuses with the same velocity as the molecule under observation (including its hydration sphere). $R_h$ can be larger or smaller than $R_g$, depending on the shape and flexibility of the particle. For spherical particles, the hydration shell increases $R_g$ slightly. In theory, a single layer of water molecules on the surface of a spherical molecule will increase its radius by ca. 0.34 nm, but in practice several layers are present, and the two radii are then roughly related by $R_h \approx 1.3\, R_g$. $R_h$ is measured by methods that are sensitive to diffusion, such as light scattering, analytical ultracentrifugation (see Section 26.2), gel permeation chromatography, electrophoresis, and also SAXS.

The Guinier approximation is valid only for $q$ values satisfying the relation $R_g{\cdot}q < 1.3$. Thus, the larger the molecule, the larger $R_g$, and the smaller the $q$ values where $I(q)$ follows the Guinier approximation. Eq. 21.31 provides a convenient method to calculate $R_g$ from the measured scattering curves. Taking the logarithm of eq. 21.31 yields the *Guinier equation*.

$$\ln\big(I(q)\big) = I_0 - \frac{R_g^2}{3}\cdot q^2$$

eq. 21.33

The plot of $\ln(I(q))$ as a function of $q^2$ is known as a *Guinier plot* (Figure 21.9). A fit of the Guinier equation to the linearized scattering data at small $q$ values yields $R_g$ of the particle from the slope.

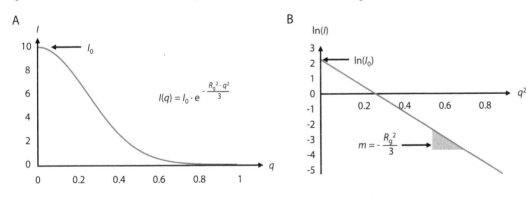

**FIGURE 21.9:** **Theoretical solution scattering curve.** A: Exponential intensity decay as a function of scattering angle. An arbitrary $I_0$ of 10 and $R_g$ of 5 nm were used for this calculation. B: Linearization of the intensity in a Guinier plot allows retrieval of $R_g$ and $I_0$ by a linear fit. In practice, the plot will only be linear for $q$ values satisfying $R_g{\cdot}q < 1.3$.

The second parameter obtained from the Guinier plot is $I_0$, the total forward scattering intensity as predicted from Rayleigh scattering (Section 21.1). $I_0$ is an important parameter in SAXS because (after buffer subtraction) it is proportional to the square of the effective number of electrons in the sample volume that was illuminated by the incident beam. Thus, $I_0$ is also proportional to the mass concentration of the solutes, which in principle allows computing molecular masses (Section 21.2.4.2), independent of the particle shape. Unfortunately, $I_0$ cannot be measured directly because this forward scattered beam lies in the direction of the incident beam. The beam stop not only blocks the direct beam, but also inevitably truncates the measured scattering curves at small $\theta$-angles (Figure 21.7). Therefore, the lowest resolution data have to be measured carefully to obtain an accurate value of $I_0$ at the intercept $\theta = 0$ ($q = 0$) from the linear fit.

### 21.2.3 DISTANCE DISTRIBUTION FUNCTION

Since the Guinier approximation to calculate $R_g$ and $I_0$ is limited to small $q$ values where only few data are available, it is preferable to include all of the scattering data in their calculation. The *pair distance distribution function P(r)*, or short *distribution function*, provides such an approach. $P(r)$ describes the electron distribution of a particle in real space, and the corresponding intensity distribution $I(q)$ in reciprocal space is given by a Fourier transform of $P(r)$:

$$I(q) = 4\pi \int_0^{D_{max}} P(r) \cdot \frac{\sin(q \cdot r)}{q \cdot r} \cdot dr \qquad \text{eq. 21.34}$$

$D_{max}$ is the maximum distance $r$ between two electrons in the molecule. Since $I(q)$ is measured, $P(r)$ can be calculated from the measured $I(q)$ data by a numerical *Fourier inversion*, also known as the indirect transformation procedure of Glatter. This operation converts the discrete number of measurement points of the scattering curve into the continuous $P(r)$ function.

$$P(r) = \frac{r}{2\pi^2} \cdot \int_0^\infty q \cdot I(q) \cdot \sin(q \cdot r) \cdot dq \qquad \text{eq. 21.35}$$

The beauty of the $I(q) \leftrightarrow P(r)$ Fourier pair is that both $I_0$ and $R_g$ can now be calculated using all of the available measurement data according to

$$I_0 = 4\pi \int_0^{D_{max}} P(r) \cdot dr \qquad \text{eq. 21.36}$$

and

$$R_g{}^2 = \frac{\int_0^{D_{max}} r^2 \cdot P(r) \cdot dr}{\int_0^{D_{max}} P(r) \cdot dr} \qquad \text{eq. 21.37}$$

Ideally, the $I_0$ and $R_g$ values calculated from the distribution function should coincide with the values from the Guinier plot: their comparison provides a useful consistency check for the data. In any case, $I_0$ and $R_g$ values calculated from $P(r)$ are more reliable since they use all available SAXS/SANS data.

The different scattering curves for molecules with different shapes (Figure 21.8) are caused by different distance distributions between the electrons in these molecules. $P(r)$ is an autocorrelation function of the electron density, or in other words, it represents an account of how often which distance between two electrons is present in the molecule. It now becomes clear why the point $P(r) = 0$ represents the maximum particle dimension $D_{max}$: it is impossible for two electrons within a molecule to have a distance exceeding $D_{max}$ (Figure 21.10).

The shape of the $P(r)$ function contains information on the shape of the molecule. Globular macromolecules have unimodal distance distributions, while multi-subunit proteins may display several distance distribution maxima. The first maximum corresponds to a set of short distances within a domain while other maxima represent sets of distances between the domains. Since there are, on average, fewer large distances within a molecule than shorter distances, $P(r)$ tails off towards $D_{max}$. Flexible proteins with many extended conformations have larger tails in their $P(r)$ functions, corresponding to a higher number of large distances.

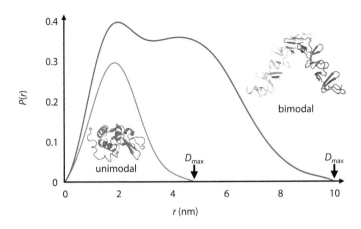

**FIGURE 21.10:**  *P(r)* **electron pair distribution functions.** Globular, single-domain proteins such as lyso-zyme have unimodal distance distributions (blue curve). The dimeric lytic amidase (LytA; red curve) shows a bimodal distance distribution. For the dimeric LytA, the large maximum at shorter distances represents the distances within one monomer. The set of larger distances between the monomers (shown as red and gray ribbons) is the second maximum on the right. Data are from the sasbdb.org database (accession IDs SASDA96 and SASDBJ4). $D_{max}$ is the maximum distance between electrons in the particle.

$P(r)$ is not only useful for comparing different proteins, but also to detect conformational changes within a molecule. These include large structural changes upon ligand binding, protein and nucleic acid folding, or oligomerization. For example, a change in relative domain positions in an oligomer affects many distance pairs between the domains, drastically changing the shape of the $P(r)$ distribution.

## 21.2.4  SMALL ANGLE X-RAY SCATTERING

Small angle X-ray scattering (SAXS) and X-ray diffraction experiments use the same X-ray sources. The generation of X-rays is described in Section 22.1. SAXS is one of the more powerful methods to assess molecular flexibility, without the size limitations of NMR and EM.

### 21.2.4.1  SAXS Experiment

Unless the aim is to analyze a mixture of macromolecules, for instance the dissociation of an oligomer, SAXS samples should be > 90% homogeneous. SAXS measurements are optimally performed at several concentrations, which have to be known accurately with < 10% error because they are used to normalize the scattering data. If the scattering curves collected at different concentrations superimpose after the scaling by concentration, the molecule does not change its aggregation state over the concentration range, and the sample behaves like an ideal solution. A typical sample for SAXS is about 15 µL in a quartz capillary at a concentration of 1–10 mg/mL for proteins. Nucleic acid samples can be less concentrated because of their higher electron density. To maximize the difference in electron density between solvent and solute, the contrast, a buffer of low ionic strength (< 0.5 M) is used with as few additives (< 0.5 mM) or co-solvents (< 5% glycerol) as possible. X-ray-induced radiation damage of the sample can be a problem in SAXS and X-ray crystallography. Radiation damage is assessed by repeated measurement of scattering from the same sample and comparison of the scattering curves, which should be identical. Co-solvents such as glycerol, formate, and thiols act as radical scavengers and can reduce radiation damage. If radiation damage poses a problem despite the presence of radical scavengers, a flow cell may be used that continuously offers fresh material to the X-ray beam at the expense of more sample. In case detergents have to be used to stabilize proteins for SAXS measurements, their concentration needs to be below the CMC (Section 16.2.8.3) to avoid contributions to X-ray scattering from micelles.

An exposure at a modern synchrotron beamline lasts about 1 s, depending on the wavelength, the beam size, and the molecular mass and concentration of the macromolecule. Scattering intensities are corrected for buffer contributions by subtracting the buffer scattering measured in a separate experiment. This allows measurement of SAXS data in the range of 0.1 nm$^{-1}$ < $q$ < 6 nm$^{-1}$, equivalent to scattering intensities over three orders of magnitude. Small $q$ values are required for accurate $R_g$ and $D_{max}$ determinations, while larger $q$ values (higher resolution data) are needed for reliable structure modeling (Section 21.2.4.4).

Two of the first parameters that are accessible from SAXS data are $I_0$ and $R_g$ from a Guinier plot (Figure 21.9). An accurate estimate for $R_g$ is obtained by limiting the data analysis to the *Guinier region* ($R_g \cdot q$ < 1.3). Since $R_g$ is not known initially, several rounds of fitting may be required to identify the best data region for $R_g$ estimation. For aggregated samples, the Guinier region is not linear, so $R_g$ and $I_0$ will be different compared to their values obtained from the $P(r)$ function. In this case, different buffer conditions or purification methods (filtering, centrifugation) are required to establish a monodisperse sample. Other parameters that can be calculated from the scattering data are the excluded volume, the molecular mass $M_M$, and a Kratky plot.

### 21.2.4.2  Excluded Volume and Molecular Mass

The excluded volume $V$ of a molecule is the volume occupied by the molecule plus its sphere of hydration that cannot be penetrated by bulk solvent (Box 21.1). The volumes of large (> 70 kDa) globular macromolecules can be approximately calculated from the low resolution scattering data using the *Porod equation* (eq. 21.38).

$$V = 2 \cdot \pi^2 \cdot \frac{I_0}{\int_0^\infty q^2 \cdot I(q) \cdot \mathrm{d}q} \qquad \text{eq. 21.38}$$

The integral in the denominator of eq. 21.38 is the *Porod invariant*. Since the integration is over all $q$ values, the Porod invariant is proportional to the total scattering intensity $I_0$, and the volume calculation does not require data normalization by the solute concentration. $V$ is also obtained from *ab initio* shape modeling (Section 21.2.4.4), and both methods should yield consistent values. Based on the fairly uniform density of macromolecules, the excluded volume in nm$^3$ is converted into a molecular mass in kDa using the rule-of-thumb relation $M_M \approx 0.6 \cdot V$. For non-spherical, asymmetric, and smaller macromolecules, the Porod volume calculation is inaccurate. In this case, the molecular mass is determined from the total forward scattering intensity $I_0$ of a solution with known concentration $c$:

$$\frac{I_0}{c} = \frac{N_A \cdot M_M}{\mu^2} \cdot \left(1 - \frac{\rho_0}{\rho_{sample}}\right)^2 \qquad \text{eq. 21.39}$$

The parameter $\mu$ is the ratio of the molecular mass to the number of electrons of the molecule, typically 1.87 for proteins. $\rho_0$ and $\rho_{sample}$ are the average electron densities of the solvent and the sample (see Table 21.3). If the sample is not monodisperse, the molecular mass calculated with eq. 21.39 will be a mass-average of all the constituents of the sample:

$$\overline{M_M} = \frac{\sum_i N_i \cdot M_i^2}{\sum_i N_i \cdot M_i} \qquad \text{eq. 21.40}$$

where $N_i$ is the number of molecules of molecular mass $M_i$.

### 21.2.4.3  Kratky Plot

The Kratky plot visualizes the compactness of a macromolecule and can be directly calculated from the scattering curve (Figure 21.11). It is simply a different way to plot the intensity data as a function of the scattering vector $q$, namely as the product $q^2 \cdot I(q)$ *versus* $q$. The Kratky plot for a globular, folded, and compact protein is bell-shaped. Unfolded and extended molecules, on the other hand, show a plateau in the Kratky plot that does not decay towards the baseline at high $q$ values.

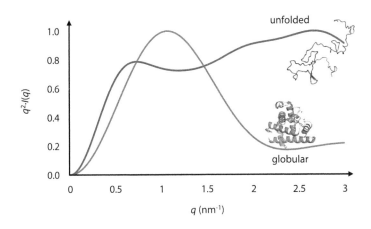

**FIGURE 21.11:** **Kratky plot.** $q^2 \cdot I(q)$ is plotted as a function of $q$. Globular proteins (blue curve) exhibit a peak in this plot, while unfolded (extended) proteins give rise to a plateau.

By way of the Kratky plot, SAXS provides similar information on protein folding and unfolding as far-UV CD spectroscopy (Section 19.3).

### 21.2.4.4  Modeling of Scattering Curves

The modeling of the entire scattering curve by a molecular envelope, termed *shape reconstruction*, extends the scattering analysis beyond the mere extraction of molecular parameters. Due to the limited data-to-parameter ratio inherent in SAXS data, care must be taken not to over-interpret the data. Often, modeling of molecular shapes from scattering curves uses crystal or NMR structures of macromolecules as rigid bodies. Flexible macromolecules pose a special problem for SAXS modeling because multiple molecular envelopes can be found that fit the data equally well and therefore undermine the reliability of the shape reconstruction. If experimental data from other methods are available, hybrid approaches (Box 18.1) can improve the shape reconstruction.

If no structural information is available, *ab initio* modeling of low resolution particle shapes can be performed without other information, directly from the scattering pattern. The starting model is either a random coil polypeptide, or an array of several thousand beads that are much smaller than the resolution of the SAXS data, and in total approximate the macromolecule as a *dummy atom model*. If the resolution of the SAXS data is better than 2 nm, the number of beads may be increased to match the number of amino acids or nucleotides (see Section 18.2.4.4). By using coarse-grained modeling and simulated annealing procedures (Sections 18.2.4.3 and 18.2.4.4), a compact arrangement of the beads that best fits the measured data is searched for. The quality of the fit can be assessed by the *normalized discrepancy function*, which is a measure of the differences between the measured data $I_{obs}$ and the calculated scattering data $I_{calc}$ from the final structural model:

$$\chi^2 = \frac{1}{N-1} \sum_{i=1}^{N} \left[ \frac{I_{obs}(q_i) - c \cdot I_{calc}(q_i)}{\sigma(q_i)} \right]^2 \qquad \text{eq. 21.41}$$

Minimizing $\chi^2$ during the molecular dynamics simulation strongly weights the low resolution (~2 nm) part of the SAXS data, and explicitly takes the errors of the data into account. $N$ is the number of data points $I_{obs}$ with their associated errors $\sigma_i$, and $c$ is a scaling constant between the measured and calculated intensities. The experimental errors are unknown and have to be correctly estimated for the $\chi^2$ statistic to work. This necessity is avoided by using a *correlation map*, where the correlation between observed and calculated intensity distributions is based only on intensity values. As a result, a set of different models is obtained that all describe the scattering data equally well. The smaller the differences between the individual models, the more reliable is the fitting result.

SAXS shape reconstructions are compared using the *normalized spatial discrepancy* (NSD). NSD calculations can be done in the absence of coordinates, only based on object shape. For each point in structure S1, all distances to all points in structure S2 are calculated, and the minimum distances are noted. The calculation is repeated for all points in structure 2, then all distances are added and normalized by the average distances between neighboring points in the structures:

$$NSD_{S1,S2} = \sqrt{\frac{1}{2}\left[\frac{1}{N_1 \cdot \overline{d_2}^2}\sum_{i=1}^{N_1}\delta_{S1i,S2}^2 + \frac{1}{N_2 \cdot \overline{d_1}^2}\sum_{i=1}^{N_2}\delta_{S2i,S1}^2\right]}$$

eq. 21.42

Identical structures would have NSD = 0. A low NSD value allows calculation of an average structure and to extract the most probable structure from the ensemble. For NSD > 1, the structures diverge significantly and in a systematic manner, indicating that there might be structures of very different shapes in the sample. For instance, a dynamic monomer-oligomer system will scatter as the sum of monomer and oligomer, weighted by the mole fractions. SAXS data for pure monomer at low concentration and oligomer at high concentration can be collected, and the mole fractions of monomer and oligomer at intermediate concentrations can be extracted by a weighted fit using the references. The shape reconstruction will then give envelopes for both the monomer and oligomer. For flexible molecules that adopt a plethora of conformations in solution, the shape reconstruction is more difficult. Sometimes, the calculated shapes can be grouped into clusters that may represent subpopulations of conformers in solution. The molecular shapes calculated from SAXS data contain an ambiguity. Both enantiomorphs (mirror images) of the shape envelope will describe the measured data equally well, and additional information from other techniques is required to decide on which enantiomorph is correct.

### 21.2.5 SMALL ANGLE NEUTRON SCATTERING

SAXS and small angle neutron scattering (SANS) are in many aspects complementary techniques, and all SAXS data evaluation methods also work for SANS data. In addition, SANS offers the method of *contrast variation*, which is based on the opposite sign of the $^1$H and D neutron scattering lengths. Another advantage of SANS compared to SAXS is the virtual absence of radiation damage because neutrons generate much less radicals than X-rays do, which enables long measurement times. However, while SAXS beam time is widely accessible at synchrotrons, SANS experiments are limited to a few large neutron production facilities.

#### 21.2.5.1 Generation of Neutrons

Neutrons, like any elementary particles, can be viewed both as particles or waves. The wave-particle dualism of quantum mechanics posits that a *de Broglie wavelength* can be assigned to a neutron of mass $m_n$ traveling at velocity $v$:

$$\lambda = \frac{h}{m_n \cdot v}$$

eq. 21.43

where $h$ is the Planck constant. For SANS measurements, slow or *thermal neutrons* are required, which travel with an average velocity of a few km s$^{-1}$. These neutrons are not easily available, they are mainly produced in two types of large facilities known as *fission reactors* and *spallation sources*. In a fission reactor, neutrons are generated in a controlled chain reaction where thermal neutrons hit a suitable fuel, for instance enriched $^{235}$U:

$$^{235}U + n \longrightarrow {}^{236}U \longrightarrow {}^{89}Kr + {}^{144}Ba + 3\,n$$

scheme 21.1

The fission of a uranium nucleus yields fast neutrons, which are decelerated by a moderator, often $H_2O$ or $D_2O$. The neutrons leave the reactor at a continuous flux of up to $8 \cdot 10^{14}$ neutrons per $cm^2$ and second.

The other major method to produce neutrons is spallation. A suitable target, e.g. lead (Pb) or Bismuth (Bi) is bombarded by protons that were accelerated to relativistic speeds using a synchrotron. The impact on the target atom produces 20–30 neutrons per absorbed proton. The Swiss SINQ source produces a flux of $10^{14}$ neutrons per $cm^2$ and second.

The average energy of neutrons from a water moderator at ambient temperature is about 25 meV. Most experiments require a monochromatic neutron beam, which is generated either by velocity selection (time of flight) or by Bragg reflection (diffraction) of the neutrons (see eq. 22.6) from a single crystal. For velocity selection, a beam cutter rotating at a certain speed selects neutrons for both a specific velocity and direction. The result is a parallel, or *collimated*, beam of a specified wavelength. In Bragg reflection, a monochromator passes only neutrons of a specific wavelength on to the experiment.

### 21.2.5.2 Contrast Variation

The information gained from a neutron scattering experiment of a macromolecule in solution depends on the difference density between the solute and the solvent. In SANS the nuclear difference density can be adjusted by a method termed *contrast variation* or *solvent matching*. We have introduced the neutron scattering length as a measure of the strength of the interaction between a neutron and a nucleus (Section 21.2.1). Recall that the scattering lengths of $^1H$ and D differ not only in their magnitude but also in their sign (Table 21.2). $^1H$ has a large negative scattering length, which means that the neutron beam scattered from $^1H$ is phase-shifted by 180° relative to beams scattered by any other atom type in the molecule. As usual, the total scattered intensity is the result of interference of all beams scattered by all atoms. The phase shift by $^1H$ therefore has a major effect on the total scattering. By contrast, deuterium has a large positive neutron scattering length, and therefore its contribution to the total scattering is almost opposite to that of $^1H$. If, for example, we prepare a deuterated macromolecule and dissolve it in $H_2O$, the scattering of the solvent will be very different from the scattering of the solute, greatly increasing the contrast between the two. The reverse experiment, dissolving a solute containing $^1H$ in $D_2O$, is also possible. By adjusting the $H_2O/D_2O$ ratio to match the solvent scattering density with that of the macromolecule, the macromolecule can effectively be made to disappear in SANS. The *neutron match point* is the percentage $D_2O$ in the solvent where scattering from macromolecule and solvent are equal. Table 21.3 lists the neutron match points of a few biological substances.

Contrast variation can be used to study macromolecular complexes. The trick is to find a scattering density ($H_2O/D_2O$ ratio) for the solvent that matches some part of a macromolecular complex and thus makes this part invisible to SANS so that the scattering from the complex corresponds only to the visible parts that do not match the solvent. For example, the layer of detergents that is used to solubilize a membrane protein can be made invisible to SANS by contrast variation. The SANS data then report only on the membrane protein itself, without the contribution of the detergent. Contrast matching is also applied in the distinction between proteins and nucleic acids, which have different neutron match points (Table 21.3). By matching the $H_2O/D_2O$ ratio to either protein or nucleic acid, the relative location of the two components can be addressed. An example is the nucleosome: originally, it was unknown whether in nucleosomes (Figure 17.29) the DNA was wrapped around protein or whether protein would enclose the DNA. By matching the solvent selectively to either the protein or the DNA and comparing the $R_g$ values from different SANS experiments, it was clear that the DNA was wrapped around the histone proteins (Figure 21.12).

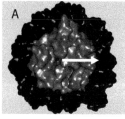
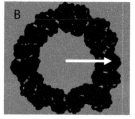
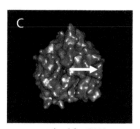

unmatched                matched for protein                matched for DNA

**FIGURE 21.12:   Principle of contrast variation in SANS.** The nucleosome serves as an example to distinguish whether DNA is wrapped around proteins or *vice versa*. From left to right the scattering density of the solvent is increased, indicated by the light gray, dark gray, and black backgrounds. $R_g$, indicated by a white arrow, is the measured variable. The structures themselves are shown as a reference but are not obtained from SANS. A: The solvent is 100% $H_2O$ with low scattering density, as is typically used in SAXS. SAXS data are therefore similar to SANS data collected in 0% $D_2O$, and both the DNA and the protein contribute to $R_g$. B: The solvent scattering density matches that of the protein (40–45% $D_2O$). SANS is dominated by the ring-shaped DNA, which leads to a larger $R_g$ than the whole complex in (A). C: When the solvent scattering density matches that of DNA (65% $D_2O$) only the protein remains visible with a small $R_g$.

Contrast variation in SANS is also useful to gather distance information on macromolecular complexes. In the 1970s, before the crystal structure of the ribosome (see Figure 17.37) was known, the location of the constituent ribosomal proteins was established by SANS: deuterated ribosomes were isolated from bacteria grown in deuterated media. Individual ribosomal proteins with [1]H instead of D were incorporated singly or in pairs into the deuterated ribosomes (a rather complicated assembly process on its own), and SANS data were collected in $D_2O$. $R_g$ and $D_{max}$ for the [1]H substructures (the two proteins within the ribosome) were extracted from the distance distribution functions. Three SANS experiments on the singly and doubly modified ribosomes resulted in the $R_g$ values for the individual proteins and the substructure, from which the distance between the two components was determined. Repeated SANS experiment with many different pairs of subunits yielded a triangulation map of the ribosomal proteins whose only uncertainty was its chirality, as in the *ab initio* SAXS calculations.

## QUESTIONS

21.1   The Rayleigh ratio is given by

$$R_\theta = \frac{I_s}{I_0} \cdot \frac{r^2}{V \cdot \left(1 + \cos^2 \theta\right)}$$

The absorption is defined as the logarithm of the ratio of the incident and transmitted light intensities:

$$A = \log \frac{I_0}{I}$$

Thus, the apparent absorption due to scattering is

$$A_s = \log \frac{I_0}{I_0 - I_s}$$

Calculate $A_s$ as a function of $R_\theta$. Choose a suitable volume $V$ and integrate over all angles $\theta$ to obtain the total scattering.

21.2  Raman scattering leads to two bands, the Stokes band and the *anti*-Stokes band. In fluorescence spectroscopy, the position of the Raman peak as a function of the excitation wavelength $\lambda_{ex}$ is given by:

$$\lambda_{\text{Raman}} = \frac{\lambda_{ex}}{1 - 36 \cdot 10^{-5} \cdot \lambda_{ex}}$$

To which of the two Stokes bands does the Raman peak correspond, and why? You want to titrate a 1 µM solution of a protein with a high-affinity ligand and use Trp fluorescence as the spectroscopic probe. The protein solution absorbs maximally at 287 nm and emits fluorescence maximally at 320 nm. Would you excite at the absorption maximum? Which excitation wavelength would place the Raman peak 10 nm red-shifted from the fluorescence maximum? Are such considerations important when using a 10-fold higher concentration of protein?

21.3  At a given scattering angle $\theta$, what is the relative scattering intensity of argon atoms Ar, chloride ions $Cl^-$, and calcium ions $Ca^{2+}$? What about $H_2O$, $F^-$, and $Mg^{2+}$?

21.4  Calculate the average electron density of water.

21.5  Derive a formula that relates the molecular mass $M_M$ of a spherical protein (in Da) to the radius of gyration $R_g$ (in nm). The partial specific volume of proteins is 0.73 cm$^3$ g$^{-1}$.

21.6  The figure shows the Kratky plots of an RNA molecule, the regulatory domain of the Lys riboswitch, both in the absence and in the presence of 5 mM $MgCl_2$ (all other buffer conditions are identical). What is the explanation for the different curve shapes?

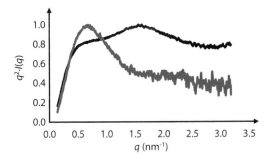

21.7  What are the frequency, temperature, velocity, and wavelength of neutrons with energy of 25 meV? One eV equals $1.602 \cdot 10^{-19}$ J, the Planck constant is $h = 6.62607 \cdot 10^{-34}$ J s, and the neutron mass is $m_n = 1.67493 \cdot 10^{-27}$ kg.

## REFERENCES

Tuma, R. (2005) Raman spectroscopy of protein: from peptides to large assemblies. *J Raman Spectrosc* **36**: 307–319.
· review article with a focus on assignment of Raman resonances and applications to proteins

Talari, A. C. S., Evans, C.A., Holen, I., Coleman, R.E., Rehman, I.U. (2015) Raman spectroscopic analysis differentiates between breast cancer cell lines. *J Raman Spectrosc* **46**: 421–427.
· original article demonstrating differences in lipid, nucleic acid, and protein composition of normal and malignant cell lines leading to different Raman fingerprints

Miura, T., Benevides, J. M. and Thomas, G. J., Jr. (1995) A phase diagram for sodium and potassium ion control of polymorphism in telomeric DNA. *J Mol Biol* **248**(2): 233–238.
· use of different Raman spectral signatures for parallel and antiparallel G-quadruplex structures to analyze the populations in telomeric DNA in equilibrium

Palacky, J., Vorlickova, M., Kejnovska, I. and Mojzes, P. (2013) Polymorphism of human telomeric quadruplex structure controlled by DNA concentration: a Raman study. *Nucleic Acids Res* **41**(2): 1005–1016.
· characterization of the concentration dependence of G-quadruplex polymorphism by Raman spectroscopy

Spiro, T. G. (1974) Resonance Raman Spectroscopy: a new structure probe for biological chromophores. *Acc Chem Res* **7**: 339–344.
· review article highlighting the principle of resonance Raman spectroscopy and its applications to biomolecules

Lewis, A., Fager, R.S., Abrahamson, E.W. (1973) Tunable laser resonance Raman spectroscopy of the visual process. *J Raman Spectr* **1**: 465–470.
· resonance Raman spectroscopy of rhodopsin

Weidinger, I. M. (2015) Analysis of structure-function relationships in cytochrome c oxidase and its biomimetic analogs via resonance Raman and surface enhanced resonance Raman spectroscopies. *Biochim Biophys Acta* **1847**(1): 119–125.
· review article on the application of resonance Raman spectroscopy to cytochrome c oxidase and other heme-copper oxidases

Cotton, T. M., Kim, J.H., Chumanov, G.D. (1991) Applications of surface-enhanced Raman spectroscopy to biological systems. *J Raman Spectr* **22**: 729–742.
· review article that summarizes the application of Raman spectroscopy to small molecules, DNA, soluble and membrane proteins, and membranes

Cotton, T. M., Schultz, S. G. and Van Duyne, R. P. (1980) Surface-enhanced resonance Raman scattering from cytochrome c and myoglobin adsorbed on a silver electrode. *J Am Chem Soc* **102**(27): 7960–7962.
· surface-enhanced resonance Raman scattering

Aydin, O., Altas, M., Kahraman, M., Bayrak, O. F. and Culha, M. (2009) Differentiation of healthy brain tissue and tumors using surface-enhanced Raman scattering. *Appl Spectrosc* **63**(10): 1095–1100.
· use of surface-enhanced Raman scattering to distinguish between healthy and tumor cells

Hudson, S. D. and Chumanov, G. (2009) Bioanalytical applications of SERS (surface-enhanced Raman spectroscopy). *Anal Bioanal Chem* **394**(3): 679–686.
· review on bioanalytical applications of surface-enhanced Raman spectroscopy

Hu, J. and Zhang, C. Y. (2012) Single base extension reaction-based surface enhanced Raman spectroscopy for DNA methylation assay. *Biosens Bioelectron* **31**(1): 451–457.
· surface-enhanced Raman spectroscopy to determine the degree of DNA methylation

Perkins, S. J. (1988) Modern Physical Methods in Biochemistry, Part B, Elsevier.
· scattering densities and electron densities for biological substances

Putnam, C. D., Hammel, M., Hura, G. L. and Tainer, J. A. (2007) X-ray solution scattering (SAXS) combined with crystallography and computation: defining accurate macromolecular structures, conformations and assemblies in solution. *Q Rev Biophys* **40**(3): 191–285.
· a very informative review on the combination of SAXS and X-ray crystallography that highlights the close physical relationship between the two methods

Kozin, M. B. and Svergun, D. I. (2001) Automated matching of high- and low-resolution structural models. *J Appl Cryst* **34**(1): 33–41.
· introduces the normalized spatial discrepancy as a means to compare structures not represented by atomic coordinates but by an array of points

Franke, D., Jeffries, C. M. and Svergun, D. I. (2015) Correlation Map, a goodness-of-fit test for one-dimensional X-ray scattering spectra. *Nat Methods* **12**(5): 419–422.
· use of correlation maps to judge the quality of models in *ab initio* modeling of SAXS curves

Svergun, D. I., Richard, S., Koch, M. H., Sayers, Z., Kuprin, S. and Zaccai, G. (1998) Protein hydration in solution: experimental observation by X-ray and neutron scattering. *Proc Natl Acad Sci U S A* **95**(5): 2267–2272.
· solvation of proteins can be detected by both SAXS and SANS, with the first hydration layer being about 10% more dense than bulk water

Tuukkanen, A. T. and Svergun, D. I. (2014) Weak protein-ligand interactions studied by small-angle X-ray scattering. *FEBS J* **281**(8): 1974–1987.
· SAXS can be used to analyze the change in protein oligomerization state upon ligand binding, and low-affinity ligands amounting to 10–15% of the protein mass can be tracked directly by SAXS

Jeffries, C. M. and Svergun, D. I. (2015) High-throughput studies of protein shapes and interactions by synchrotron small-angle X-ray scattering. *Methods Mol Biol* **1261**: 277–301.
· synchrotron small-angle X-ray scattering

# X-ray Crystallography

In crystallography, diffraction is the scattering of electromagnetic radiation by ordered arrays of molecules. X-ray crystallography measures diffraction patterns of (macromolecular) crystals to obtain structural information. The maximum resolution attainable is determined by the wavelength of the radiation, and is $> \lambda/2$ (Section 23.1.1). The wavelength of X-rays is on the scale of atomic distances (100–250 pm), and X-ray diffraction thus provides structural information with atomic resolution.

X-ray crystallography is by far the most successful structure determination method: $> 90\%$ of all macromolecular structures are determined with this method. In contrast to other techniques that deliver atomic models (NMR, electron microscopy), X-ray crystallographic objects can be of any size, from single atoms to many MDa objects such as ribosomes (see Figure 17.37) and viruses (see Figure 16.33). The quality and resolution of X-ray crystal structures (or short: crystal structures) is often high enough to infer enzymatic mechanisms, to reveal binding interfaces in large macromolecular complexes, and to significantly speed up structure-based drug design in search for new medications.

X-ray crystal structure determination is a multi-step process involving sample preparation, crystallization, data collection, phasing, model building into electron density, followed by refinement, validation, and interpretation of the model (Figure 22.1). After an outline describing how X-rays are generated, the remainder of the chapter will follow these steps.

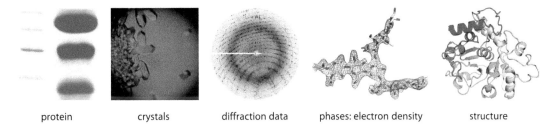

protein          crystals          diffraction data          phases: electron density          structure

FIGURE 22.1:   **Steps in X-ray crystal structure determination.** A macromolecule or a complex is purified at high concentration and crystallized. Diffraction data are collected, but during the process the phases of the diffracted light beams are lost. Phases are required to calculate an electron density map. The map is interpreted in terms of an atomic model. The model is often simplified as a ribbon diagram.

## 22.1 GENERATION OF X-RAYS

There are two widely used methods to produce high-brilliance X-rays, (1) emission during electron transfer from a high-energy state to a low-energy state between the inner shells of atoms and (2) emission from particles accelerated to relativistic velocities. X-rays in the laboratory are generated using the first method in *sealed tubes* and *rotating anodes*. Orders of magnitude higher brilliances are obtained in *synchrotrons* that use the second method.

The intensity and quality of an X-ray beam is described by several parameters, among them the number of photons in the beam, the beam angular divergence, the cross-section of the beam, and its monochromaticity. These parameters are important for crystallography. The higher the intensity the smaller the data acquisition time. A small cross-section limits background scattering and allows the study of tiny samples. Small angular divergence improves the signal strength from the sample, and a precisely defined wavelength is essential for several phasing methods.

The *flux* of an X-ray beam is the number of photons per second per 0.1% bandwidth, where 0.1% bandwidth at 0.1 nm would include photons of wavelengths between 0.0999–0.1001 nm. The *brightness* of a beam is the flux per $mrad^2$. The angular increment in millirad (mrad) is a measure of how quickly the beam diverges with distance from the source. Highly collimated beams where the photons are almost parallel to each other have high brightness. Finally, the *brilliance* is the brightness per $mm^2$, where $mm^2$ refers to the cross-sectional area of the source.

$$brilliance = \frac{\# \, photons}{s \cdot mm^2 \cdot mrad^2 \cdot 0.1\% \, bandwidth}$$

eq. 22.1

In other words, brilliance is the photon flux passing through 1 $mm^2$ area with a divergence given in millirad. High brilliance X-ray beams can be shaped with respect to cross-section, angular divergence, and monochromaticity while maintaining a high enough intensity for the experiment.

When Röntgen discovered X-rays in 1895, they were generated using the first method by electrons passing through a partially evacuated glass tube. The deflection, slowing, or sudden stopping of electrons by collisions with gas molecules in the glass tube produced a continuous X-ray spectrum, termed the *Bremsstrahlung* or *braking radiation*. Most applications require X-rays of defined wavelength, which are produced in the laboratory by shooting electrons from an *electron gun* onto a metal target (Figure 22.2). The electrons are evaporated from a cathode, a coiled metal filament heated to red-hot temperatures by an electric current of ~4 A. A magnetic field of 20–60 kV accelerates these primary electrons toward the metal-coated anode, where they eject other electrons from the inner K-shell of the anode atoms (Figure 22.2). The resulting "hole" is filled by an electron from an outer shell of higher energy, usually from the next higher

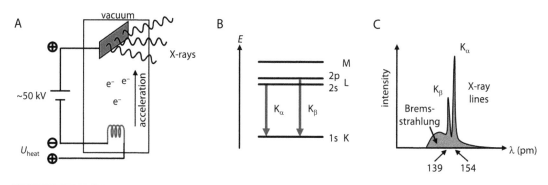

**FIGURE 22.2:** **X-ray emission by intra-atomic electron transitions.** A: Schematic representation of how electrons are evaporated from a cathode and accelerated towards the anode target in a sealed tube or rotating anode. The cathode, often a tungsten wire, is heated to 2000–3000°C by an electric current (red coil). The high voltage electric field accelerates the electrons towards the anode. B: Energy states in an anode atom. The transition for lighter metals (Table 22.1) is usually between K- and L-shells. Depending on the starting orbital, s or p, the emitted X-ray photon is termed α- or β-radiation. C: The X-ray spectrum resulting from the impact of cathode electrons on the anode is a superposition of the continuous spectrum from scattering and the line spectra from ionization of the anode material (here copper).

energy L-shell. The energy difference between the electronic states is emitted as an X-ray photon. Each element has its *characteristic X-ray line spectrum* that superimposes on the continuous Bremsstrahlung (Figure 22.2). The choice of anode material determines the wavelength of the X-rays generated.

The first X-ray sources, used in Röntgen's time, were *sealed tubes*, evacuated glass tubes that hold the anode and cathode. Higher X-ray intensities, used from the 1960s until the present, were achieved by increasing the electron flux onto the anode. The brilliance of the beams was improved by focusing the electrons on a smaller spot on the anode. As most of the energy of the impacting electrons is converted into heat, increase of the electron flux requires active cooling of the anode or else the impacting electrons will melt a hole in the anode. As the name implies, in a *rotating anode* generator, the anode is a rotating wheel, so the electrons hit the anode always at a different spot. A water circuit inside the anode ensures additional cooling of the anode material, but is a considerable source of machine failure. Due to the many moving parts in rotating anodes many laboratories now choose sealed tubes. Micro-focus sealed tubes achieve X-ray brilliances ($2 \cdot 10^9$ photons s$^{-1}$ mm$^{-2}$ mrad$^{-2}$) comparable to those of micro-focus rotating anodes ($6 \cdot 10^9$–$26 \cdot 10^9$ photons s$^{-1}$ mm$^{-2}$ mrad$^{-2}$) and require less maintenance. The X-rays emitted from the anode are focused by total reflection at confocal multi-layered mirrors.

| TABLE 22.1 X-ray lines of selected elements. | | |
|---|---|---|
| **target** | **$K_\alpha$ (pm)** | **$K_\beta$ (pm)** |
| Cu | 154 | 139 |
| Cr | 229 | 209 |
| Fe | 194 | 176 |
| Co | 179 | 162 |
| Ni | 166 | 150 |
| Mo | 63 | 71 |

The second widely used method to generate X-rays is by light emission from charged particles accelerated to relativistic velocities. This principle is the basis for the generation of X-rays in synchrotrons. Synchrotrons are large circular facilities, some with a diameter of several hundred meters, which produce highly focused light over a wide spectral range from infrared microwaves to hard X-rays. The X-rays produced by synchrotrons reach brilliances of > $10^{18}$ photons s$^{-1}$ mm$^{-2}$ mrad$^{-2}$. Light emission in synchrotrons is a relativistic effect. If a charged particle traveling at almost the speed of light is further accelerated, it emits part of its energy as light. Forcing a charged particle on an orbit by a magnetic field means continuous centripetal acceleration towards the center of the orbit, which results in continuous light emission. From each point on the orbit, the light is emitted tangentially as a tightly focused cone (Figure 22.3). The light is linearly polarized and has a continuous (white) spectrum.

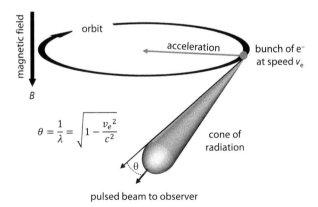

**FIGURE 22.3:** **Light emission at relativistic speed.** Charged particles traveling on a circular path experience continuous centripetal acceleration (arrow). Bunches of electrons orbit in synchrotrons at velocity $v_e$ approaching the speed of light. Energy is radiated in a focused cone tangentially to the orbit at all points where electrons are. The angle $\theta$ of the cone depends on the velocity of the particle. The light seen by the observer is pulsed with a frequency corresponding to the orbiting frequency of the electrons.

An important characteristic of a synchrotron is γ, the ratio of the total particle energy and its rest mass energy. For electrons, γ is

$$\gamma = \frac{E_0}{m_e \cdot c^2}$$

eq. 22.2

The reciprocal value of γ is also known as the *natural emission angle* of synchrotron radiation. This value is equivalent to half the opening angle of the radiation cone (Figure 22.3) and serves as a measure of how focused the light beam is. The larger γ, the better-collimated is the radiation (Figure 22.3). For an observer the synchrotron light appears pulsed with a frequency equivalent to the orbiting frequency of the electrons.

In synchrotrons, an electron bunch from an electron gun, also known as a LINAC (linear accelerator) is injected into the evacuated tube of a small booster synchrotron. Once the electrons have reached their final velocity, the bunch is injected into the larger *storage ring*, where it is kept on a stable orbit by the strong magnetic fields of superconducting *bending magnets*. (Figure 22.4). The synchrotron light emitted by the electrons is used at dozens of beamlines constructed tangentially to the storage ring. Flux, brightness, and wavelength of the X-rays are adjusted by filters, optics, and diffraction gratings located between the storage ring and the end station. A typical X-ray wavelength range for a synchrotron is 0.05–0.4 nm (0.5–4 Å).

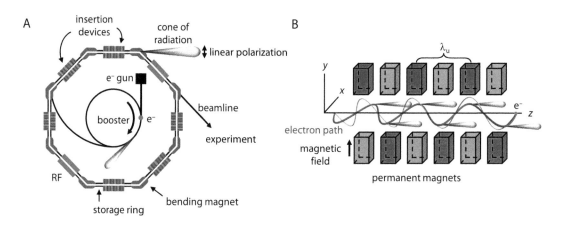

**FIGURE 22.4:** **Synchrotron and insertion devices.** A: Electrons from an electron gun are accelerated by a magnetic field and injected into a booster ring, where they are accelerated further by magnetic fields. The electrons are then injected into the storage ring. Bending magnets (magenta) keep the electrons in orbit. A continuous spectrum of linearly polarized synchrotron light is emitted in the forward direction, i.e. tangential to the orbiting electrons. Beamlines run tangential to the storage ring, and optics focus X-rays and select a certain wavelength for the experiment. Linear sections of the storage ring host insertion devices, termed wigglers and undulators, that deflect the electrons perpendicularly from their path in the plane of the storage ring, which intensifies the synchrotron light. The radiated energy that is lost by the electrons is replenished by radiofrequency coils (RF). B: Schematic view of wigglers and undulators. Permanent magnets are shown with north and south poles as red and green, spaced at a period of $\lambda_u$. The magnets construct an oscillating magnetic field (gray line) that sinusoidally deflects the orbiting electrons. As this deflection is equivalent to a net acceleration, the intensity of the emitted synchrotron light is boosted. In an undulator, the polarized light emitted at each peak is in phase with the other peaks. The resulting constructive interference further boosts the brilliance.

A storage ring is not circular but a polygon with dozens of linear sections connected by the same number of short circular arcs. Bending magnets sitting at these junctions change the direction of the orbiting electrons from one linear segment to another. While the main task of a bending magnet is to keep the electrons on their closed-path orbit in the storage ring, the resulting burst of electromagnetic radiation induced by a bending magnet has a huge brilliance of about $10^{13}$ photons·s$^{-1}$·mm$^{-2}$·mrad$^{-2}$. Even higher brilliances and much more focused light is produced by *insertion devices* known as *wigglers* and *undulators*, which increase the brilliance of the synchrotron light to about $10^{16}$ and $10^{19}$ photons·s$^{-1}$·mm$^{-2}$·mrad$^{-2}$, respectively. Both types of insertion devices are placed (inserted) at linear sections of the synchrotron. Wigglers and undulators are *multipole magnets*, i.e. linear arrays of magnets with alternating polarity (Figure 22.4). The name wiggler comes from the periodic deflection of the electron trajectory by the Lorentz force when they pass this device (they "wiggle"). Since the deflection amounts to a net acceleration of the electrons within the wiggler, the electrons emit light of high intensity at each magnetic pole. The total light intensity is therefore proportional to the number $N$ of magnets (a few to more than 100). Compared to bending magnets, wigglers produce about 1000-fold more brilliant light. Undulators, on the other hand, achieve yet another 1000-fold increase in brilliance over wigglers because the light intensity emitted by them scales with $N^2$. The difference between a wiggler and an undulator lies in the interference of light emitted at the magnetic poles. While no significant interference is present in wigglers, undulators are constructed such that light emitted at one magnetic pole constructively interferes with the light emitted from another magnetic pole. The light emitted by an electron at one magnetic pole travels faster than the electron. If the time difference between the electron and the emitted ray to reach the next magnetic pole corresponds to traveling one wavelength, constructive interference occurs. The interference condition for the light rays depends on the spacing of the magnets, $\lambda_u$, the energy of the electron beam, $\gamma m_e c^2$ (eq. 22.2), and the strength of the magnetic field. The magnetic field can be adjusted by changing the distance between north and south poles of the magnets (the *undulator gap*). Due to the interference, undulators produce a beam consisting of a fundamental frequency and several higher-frequency harmonics, one of which is chosen for the experiment by a monochromator. Undulators are now routinely used in synchrotron beamlines for SAXS and X-ray crystallography. The brilliance of light produced by undulators is ten orders of magnitude higher than cathode/anode devices, and only surpassed by the recently developed *free electron lasers* (FELs; Box 22.1).

---

### BOX 22.1: X-RAY FREE ELECTRON LASERS (XFELS).

When an electron bunch enters an undulator, its constituent electrons have a uniform spatial distribution, so the radiation they emit is incoherent, i.e. there is no phase relation between the X-rays. Because the emitted X-rays travel faster than the electrons, they overtake the electrons ahead in the bunch and interact with them on the way. The interaction accelerates some electrons and slows down others, depending on the phase relation between the electron and the X-rays. As a result, the electrons gradually drift into a series of thin disks, termed *micro-bunches*. The micro-bunches are separated by a distance equal to one wavelength, and now emit coherent (same phase) X-rays that reinforce themselves by constructive interference. The result is a series of extremely short (fs) and intense X-ray pulses with the laser-like properties of high spatial and temporal coherence. The separation of the electrons into micro-bunches in the undulator takes some time, so XFELs require very long undulators of ca. 100 m. XFELs can be easily tuned by changing either the electron beam energy of the storage ring, or the magnetic field strength of the undulator. The light has the high spatial and temporal coherence that is characteristic of lasers. The spatial coherence allows strong focusing of the laser light into a tiny spot, while the temporal coherence provides a sequence of pulses whose timing can be changed.

Data collection at synchrotrons requires crystals of at least a few µm in each dimension. XFELs have the advantage that diffraction from tiny micro-crystals of < 1 µm edges can be measured. Crystals exposed to an XFEL inevitably disintegrate in a matter of about 50 fs, so only a single exposure is recorded by ultrafast direct photon detectors (Section 22.6), and a complete dataset requires many thousand crystals. A stream of crystals is injected from a capillary into the XFEL beam for this purpose. Micro-crystals are

*(Continued)*

often easier to produce than larger crystals, especially for membrane proteins. Sometimes they even grow *in vivo* during protein production! A key advantage of the high time-resolution of XFELs is that the structure of reaction intermediates may be determined without worrying about radiation damage (data are collected before damage sets in). For example, metalloproteins can be studied in their oxidized form. The cleavage of the Fe-CO bond in carbonmonoxy-myoglobin after laser-photolysis or the double bond isomerization in the chromophore of photoactive yellow protein can be monitored. The next step would be to determine structures of single molecules by XFELs. This would require recording of the scattering patterns of a series of single molecules in random orientations, aligning the 2D diffraction patterns, assembling them into a three-dimensional intensity distribution, and phasing the structure. So far, this has not been achieved, and crystals are still required for structure determination.

## 22.2  PHASE PROBLEM AND REQUIREMENT FOR CRYSTALS

Unlike electrons, the problem with neutrons, X-rays, and light of very short wavelengths in general is that no lenses exist to focus such radiation. All that can be done is to collect the scattered light intensity as a function of scattering angle, but in doing so, the relative phase information between the scattered beams is lost. This is acceptable if only low resolution information on molecules is required, as in SAS. For high-resolution information on atomic positions, however, the phases are absolutely required. This issue is known as the *phase problem*. In crystallography, the image of the molecule is reconstructed from the intensity measurements using *phasing methods* (Section 22.7) and mathematical transformations that include the Fourier transformation (Section 29.14).

Another severe problem for high-resolution scattering studies is signal strength. The scattered light intensity given off by a single macromolecule is so tiny that currently not even the most intense X-ray sources, the free electron lasers (Box 22.1), are strong enough to retrieve structural information from single molecules. This problem can be solved by arranging molecules into crystals. A crystal acts as an amplifier for the scattering signal from a single molecule because all molecules have defined relative orientations, and thus their scattering follows defined phase relationships. The scattering from a crystal, termed *diffraction* (Section 22.5), is very different from that of a molecule in solution: the scattering intensity from crystals is proportional to the number of molecules and limited to certain diffraction angles, so the total diffraction intensity concentrates in certain areas and can be measured to high resolution. An important step in the determination of a *crystal structure* is therefore the generation of well-ordered crystals.

## 22.3  CRYSTALLIZATION OF MACROMOLECULES

A crystal is a regular arrangement of objects (Figure 22.5). Macromolecular crystals are held together by weak non-covalent interactions between surface atoms of neighboring molecules, particularly H-bonds, electrostatic interactions, and van der Waals interactions (see Section 15.3). The percentage of surface areas that provide these *crystal contacts* becomes smaller and smaller for larger molecules: if we consider a sphere of radius $r$, the volume scales with $r^3$, but the surface scales only with $r^2$. Consequently, larger macromolecules can form fewer crystal contacts than smaller macromolecules and tend to be more difficult to crystallize. The forces acting between the molecules are actually smaller than the forces exerted during conformational changes. For instance, ligand binding to a crystalline protein frequently causes the crystals to crack if the associated conformational changes are large.

As a result of the few crystal contacts, large solvent-filled voids are present between the macromolecules that form a network of *solvent channels* running through the crystal (Figure 22.5). The fractional

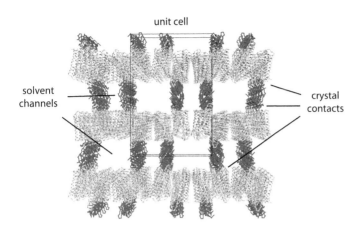

**FIGURE 22.5:**    **Macromolecular crystal.** A protein-protein complex between a nanobody (blue) and a GPCR (gray) forms a regular arrangement. The nanobody serves as a spacer between the GPCR molecules. Large solvent channels are visible between the proteins.

volume not occupied by macromolecules is the *solvent content* of a crystal. On average, the solvent content is 50% of the total crystal volume, but crystals with solvent contents from 25–90% have been characterized. Macromolecular crystals therefore resemble an ordered gel with large pores, quite unlike the hard small molecule crystals of sugar and salt, which are densely packed and contain no solvent. Together with the few and weak interactions present in macromolecular crystals, their high amount of water renders them quite fragile: they break easily during handling, change of temperature, or slight dehydration. Absorption of X-rays by the solvent in macromolecular crystals also induces *radiation damage*, a severe problem for X-ray crystallography that can be overcome by cryo-cooling (see Box 22.3).

The crystallization process can be conceptually divided into a kinetic and a thermodynamic part. A crystal will only form if the overall $\Delta G$ of the assembly is negative, which is equivalent to stating that molecules in a solution of certain composition (at supersaturation) have a higher chemical potential (Section 2.6.1) than the crystal. Whether the crystal forms on a reasonable time scale depends on the activation energy that is associated with forming a critical nucleus (Figure 22.6), a tiny ordered assembly capable of growing into a macroscopic crystal. Nucleus formation is reversible, and the smaller the nucleus, the more likely it is to dissolve again. A competing pathway to crystallization is aggregation, which is usually irreversible. Both aggregation and nucleus formation are favored by high concentrations of macromolecules and thus can occur at the same time under the same conditions. Successful crystallization of macromolecules is therefore a delicate balance between these competing pathways.

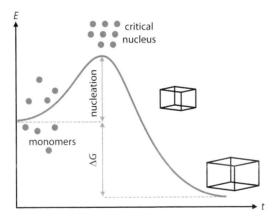

**FIGURE 22.6:**    **Energy diagram of crystallization.** Under conditions of supersaturation, molecules in solution represent a higher energy state than the crystal. Both states are separated by an activation energy barrier (nucleation). The ordered nucleus grows into a crystal until the solid phase is in equilibrium with the solution phase.

Unfortunately, it is impossible to predict the crystallization conditions from general properties such as sequence, pI, or molecular mass. Crystallization conditions may radically change even for proteins and nucleic acids that differ only by a single residue or that have been purified by different methods. Consequently, crystallization is a process of trial and error, and a lot of creativity goes into the design of crystallizing samples and the development of crystallization methods. Typically, several hundred to a few thousand crystallization trials are required until ordered crystals are found or it is judged that the macromolecule to be crystallized needs to be modified. The whole process can last from days to years and consume just 1 mg or several grams of sample.

The techniques used for crystallization include *dialysis*, *free interface diffusion*, *micro-batch*, *vapor diffusion*, and *seeding* (Figure 22.7). All crystallization methods use the same strategy. The macromolecular solution at 0.5–50 mg/mL concentration is mixed with a *precipitant* to bring the solution into a supersaturated state from which crystals might grow. Useful precipitants are salts (0.5–4M), organic solvents (ethanol, propanol, MPD or 2-methyl-2,4-pentanediol at 5–25%), and polyethylene glycol (PEG; 5–50%) of average molecular mass between 200 Da and 20 kDa. Other parameters that can be adjusted are pH (usually between 3 and 10) and temperature (4°C–40°C).

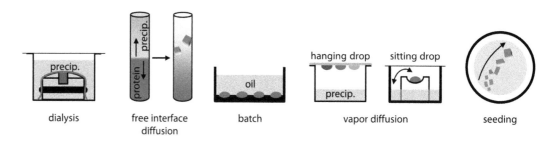

**FIGURE 22.7: Crystallization methods.** The methods are shown from left to right according to decreasing amount of sample that is usually needed. Protein solution is shown in blue color, and the precipitant (*reservoir*) is shown in gray. In dialysis, the protein is separated from the precipitant by a membrane, while free interface diffusion does not need a membrane. In batch crystallization the protein is mixed with precipitant and placed under oil. Vapor diffusion has two variants, hanging and sitting drop. Several hanging drops, e.g. different protein concentrations or ligand complexes, can be placed over the same reservoir. Seeding can be applied to any of the techniques. The direction of the arrow marks the trajectory of a hair dipped into a seed solution and dragged through a sitting drop. The seeds are diluted along the trajectory, leading to fewer crystals at its end. Batch and sitting drop vapor diffusion have been miniaturized into 96-well formats.

For dialysis, a protein solution is placed in a *dialysis button* of 5–50 µL volume, closed off by a semipermeable membrane, and equilibrated against a reservoir of precipitant. The rates of water and ion fluxes through the membrane are determined by the pore size of the membrane and the concentration gradients across it. Free interface diffusion is carried out in narrow capillaries where mass transport occurs predominantly by diffusion but not convection. The macromolecule and precipitant solutions are brought into direct contact (free interface). Diffusion of precipitant into the macromolecule solution and *vice versa* creates concentration gradients of opposite directions. At some point along the capillary, the system may reach supersaturation and form crystals. In micro-batch crystallization, supersaturation is achieved quasi instantaneously by mixing ca. 0.5–2 µL of a macromolecular solution with a precipitant in a 1:3 to 3:1 ratio, depending on the solubility of the macromolecule. The droplet is then sealed with a layer of oil. Probably the most popular crystallization method is vapor diffusion. Droplets of 0.02–1 µL volume are prepared by mixing macromolecule and precipitant solutions in various ratios (1:2, 1:1, 2:1, etc.), and equilibrating them against a 0.05–1 mL *reservoir* of precipitant. Only volatile substances such as water and alcohols can be exchanged between the droplet and the reservoir. The droplet will evaporate until equilibrium with the reservoir is reached. In the hanging drop method, droplets on a cover slip are suspended over the reservoir while in the sitting drop method, the droplet rests on a post above the reservoir solution. Usually carried out in 96-well plates by a robot, the sitting drop method is ideal for large-scale screening of hundreds of different crystallization conditions.

The different crystallization methods achieve supersaturation in different ways, and they follow different paths in the phase diagram towards the equilibrium (Figure 22.8). A crystallization trial is basically a search for the region in the phase diagram where the crystalline phase is stable.

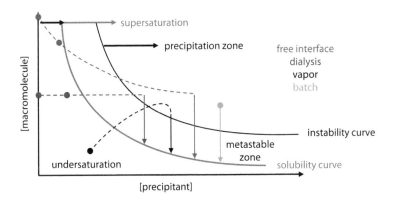

FIGURE 22.8:    **Schematic crystallization phase diagram.** The solubility curve (blue line) separates the undersaturation and supersaturation zones. The supersaturation zone is subdivided into a metastable and a precipitation zone. Crystals form in the precipitation zone and continue to grow in the metastable zone. Four crystallization methods and the routes they take to arrive *via* the nucleation zone at the solubility curve are highlighted in color. Dots represent possible starting concentrations of macromolecule and precipitant. Free interface diffusion and dialysis may start with premixed precipitant and protein. In free interface diffusion, every part of the capillary follows a different path. Figure modified after Chayen, N. E. (1998). *Acta Cryst. D* 54(Pt1): 8–15 and reproduced with permission of the IUCR (journals.iucr.org).

The *solubility curve* in the phase diagram represents the solubility of a macromolecule as a function of precipitant concentration (Figure 22.8). At lower precipitant concentrations is the stable zone of undersaturation, where the maximum protein concentration is not reached at the given precipitant concentration. Any crystals that might be present in this zone would dissolve. The supersaturation zone to the right of the solubility curve is conceptually divided into a *metastable zone* and a *precipitation zone* by the *instability curve*. The likelihood of nucleation is thought to increase from the solubility curve toward the instability curve. Nucleation can occur in both the metastable and precipitation zone. The metastable zone will at some point separate into a protein-rich phase (aggregate, droplets, or crystals), while the system right of the instability curve spontaneously decomposes into precipitate (aggregate, droplets, or crystals) and saturated macromolecule solution. Once nuclei have formed, the protein concentration diminishes and the system linearly traverses the metastable zone towards the solubility curve.

The formation of a nucleus in the supersaturation zone can be the rate-limiting step for crystallization (Figure 22.6). This step can be skipped by introducing seeds that function as external nuclei. Seeds are usually made by smashing unusable crystals, either of the same or another protein. It is not uncommon for seeds to work in very different conditions compared to where they originated from (*cross-seeding*). Alternatively, small mineral crystallites may be used, often serendipitously introduced in the form of dust particles. Seeds may be introduced by pipetting or with a smooth hair (historically a cat whisker or horse hair) dipped into a seed suspension and dragged through the crystallization drop. Using a hair dilutes the seeds attached to it and often leads to a shower of crystals along the trajectory with some larger crystals at the end (Figure 22.7).

Due to the many parameters that can be adjusted, searching the crystallization phase diagram for the crystallization zone is a multi-dimensional problem. In addition, the likelihood of crystallization also depends on properties of the sample: its purity, entropy, and surface hydrophobicity. In general, a pure sample with rigid molecules and polar surfaces is less likely to aggregate and will more likely crystallize than an impure sample with flexible and hydrophobic molecules. Impurities can bind to growing crystals and poison their growth by interrupting the regular pattern of the molecules. High entropy of the molecules in the sample translates into conformational heterogeneity, which reduces the likelihood of incorporation into the crystal. Very hydrophobic molecules also tend to aggregate faster than they form crystals. Two general strategies to increase the likelihood of crystallization are therefore reduction of the entropy and reduction of the amount of solvent-exposed hydrophobic surfaces.

Large multi-domain proteins with flexible linkers have high entropy and thus do not crystallize easily. By contrast, their individual domains, often identified by limited proteolysis, tend to crystallize more easily (*divide et impera* approach). Limited proteolysis of termini and loops is a general method to reduce the surface entropy of a protein by removing flexible parts. Amino acids with long side chains (Lys, Arg, and Glu) contributing to high surface entropy can also be mutated to smaller amino acids. This *surface entropy reduction* (SER) requires some prior knowledge on which side chains are likely to be on the surface. By contrast, proteins from thermophiles naturally have reduced surface entropy compared to their mesophilic and psychrophilic counterparts: on average, thermophilic proteins have shorter loops and less flexible side chains at their surface, and the reduced entropy contributes to their increased thermostability. Structures of orthologs are similar at the core but different at the surface where the non-conserved residues cluster, so choosing a thermophilic homolog is a natural way to exploit SER for crystallization.

Finally, if variation of the protein itself is unsuccessful, *crystallization helpers* may be used (Figure 22.9). Crystallization helpers are very stable and readily crystallizing proteins that bind with high affinity to hydrophobic areas or flexible loop regions, thus reducing the hydrophobicity and/or entropy of the sample and increasing the likelihood of crystallization. This is especially useful for the crystallization of membrane proteins (Box 22.2) and flexible RNA (Figure 22.9). Crystallization helpers include antibody Fabs, nanobodies, DARPins (designed ankyrin repeat proteins), and the U1A protein.

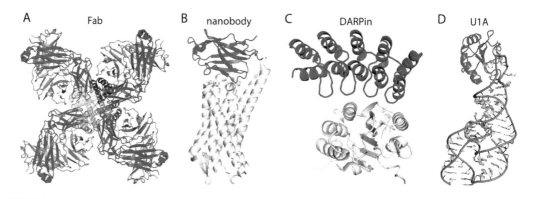

**FIGURE 22.9:** **Helper proteins for crystallization.** Helper proteins (blue) are often responsible for the majority of intermolecular contacts in a crystal. A: Antibodies have been raised against the voltage-gated K$^+$ channel KcsA and the antigen-binding fragments (Fab; light and dark blue) were used to generate a crystallizing complex (PDB-ID 4uuj). As the K$^+$ channel is a homotetramer, there are four Fabs bound in this complex. B: The μ-opioid receptor is a GPCR that in this example has a llama nanobody bound (blue; PDB-ID 5c1m). C: DARPins have an array of loops that can be engineered to bind to any epitope, as in this example of caspase-7 (PDB-ID 4jb8). D: The U1A protein binds to a conserved RNA tetraloop that was engineered into the hepatitis δ virus ribozyme (PDB-ID 1drz), a very successful method to crystallize RNA molecules in general.

Antibodies can be raised against both proteins and nucleic acids, and limited proteolysis is used to generate the Fab. A smaller version of Fab is nanobodies, which contain only a single Ig domain and are derived from camelid or shark antibodies. Like antibodies, DARPins have loop regions of variable sequence that can be generated to bind to virtually any protein or nucleic acid.

## BOX 22.2: CRYSTALLIZATION OF MEMBRANE PROTEINS.

Membrane proteins are more difficult to crystallize than soluble proteins. Once removed from their lipid environment, membrane proteins are unstable and tend to aggregate (Section 16.2.8). Much effort goes into the stabilization of solubilized membrane proteins for crystallization. Random mutagenesis and test for thermostability (see Box 16.7) is often successful but it may take a long time until a construct is obtained that can be crystallized. Binding of the crystallization helper proteins that mask hydrophobic surface areas and/or reduce aggregation is also possible (Figure 22.9). If these attempts fail, membrane proteins can be crystallized from a *lipidic cubic phase* (LCP, *in meso* crystallization) that mimics a membrane-like environment (Figure 22.10). LCP are bi-continuous crystalline (i.e. repeating) phases not unlike a sponge that form spontaneously upon mixing of suitable lipids with water. Examples of such lipids are the mono-glycerides monoolein and monopalmitolein, and the di-terpene alcohol phytantriol. The *mesophases* thus obtained are mixed with the detergent-solubilized membrane protein and serve as crystallization matrices within which the proteins can diffuse laterally and assemble into a crystal. The crystals obtained from membrane proteins, particularly those of higher eukaryotes, are usually tiny, extremely fragile, and sensitive to any change in buffer composition.

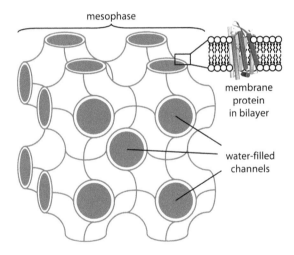

FIGURE 22.10: **Lipidic cubic phase.**

Crystallization of nucleic acids has the general problem of electrostatic repulsion between the phosphate backbones of molecules, so counter ions are essential for charge neutralization. Most of the contacts in nucleic acid crystals are made *via* the phosphate backbone and metal ions, but specific base-mediated contacts such as kissing loops (Section 17.6.1) can also be essential for crystallization. Kissing loops may be engineered into RNA to promote crystal formation. Alternatively, a 10 base pair stem-loop structure that is recognized by the 11 kDa RNA-binding domain of the spliceosomal U1A protein can be included in RNA. The biophysical properties of protein surfaces are more variable than those of nucleic acids, so the inclusion of U1A has become a popular technique in RNA crystallization (Figure 22.9). Because the isoelectric point of U1A is > 10, electrostatic interactions in addition to protein-RNA interactions can promote crystallization of RNA/U1A complexes.

**BOX 22.3: RADIATION DAMAGE AND CRYO-PROTECTION.**

In crystallography, the term radiation damage describes the detrimental effects of absorbed X-rays on a crystal. The first step in radiation damage is photoelectric absorption whereby an X-ray photon transfers its energy to an electron, ejecting it from its atomic orbital (see Figure 22.2). At room temperature this *photoelectron* can ionize hundreds of other atoms, producing a few hundred more electrons in the process. The atom that is left behind without the electron may return to its ground state by emission of an X-ray photon (see Figure 22.2), or it may emit yet another electron, the *Auger electron*. All these electrons traveling through the crystal will produce radical species, mostly from water (OH· and H·). Radicals and electrons induce bond breakage by chemical reactions either directly with the surrounding macromolecules, or, after diffusion through the crystal, with distant macromolecules. The result of radiation damage is a lower resolution crystal structure (*global damage*) with less information content than it could potentially have (*specific damage*). While global damage is a general reduction in diffraction intensity, specific damage includes reduction of disulfide bonds to Cys and of metal ions in metalloproteins ($Fe^{3+}$ to $Fe^{2+}$, for example), decarboxylation of acidic side chains, de-hydroxylation of Tyr, and destruction of aromatic rings, particularly in nucleic acid crystals.

It has been found that the extent of radiation damage is reduced if data collection is done at low temperature where the photoelectrons and radicals generated by X-rays diffuse more slowly. This leads to a dramatic increase of crystal life time in the X-ray beam. The first structures using cooling were determined at 4°C, well above the freezing point of the crystallization mother liquor. Most crystal structures are nowadays determined at cryogenic temperatures by placing the crystal in a stream of gaseous nitrogen at its boiling point (about 100 K). At this temperature, the crystallization mother liquor forms a solid, effectively abolishing diffusion within the crystal. The problem of this *cryo-cooling* is that water in the crystal will expand and eventually freeze into ice crystals, destroying the macromolecular crystal. *Cryo-protection* of macromolecular crystals seeks to avoid water ice formation and to reduce the expansion of the mother liquor upon cryo-cooling as much as possible. The standard way to inhibit ice formation is to disrupt the H-bonded water clusters using cryo-protectants that compete with water for H-bonds. These include polyethylene glycol (PEG), low molecular mass alcohols (ethylene glycol, glycerol, 2,3-butanediol, 2-methyl-2,4-pentanediol, 1,6-hexanediol, glucose, xylitol, saccharose) or cryosalts ($\geq$ 1 M $Li^+$ salts or the organic salts malonate, formate, citrate, tartrate, and acetate).

## 22.4 SYMMETRY AND SPACE GROUPS

A crystal is a regular arrangement of objects that has two main characteristics: periodicity and symmetry. The periodicity of crystals is described by their *lattice*, and their symmetry is determined by the *space group*. In practice, the lattice and space group are determined from the measured diffraction data.

If we cut a crystal in two pieces, we obtain two smaller crystals of the same periodicity and symmetry. We can repeat this exercise until we arrive at the smallest conceivable crystal, the *unit cell*. The unit cell is a parallelepiped, defined by three *crystal axes* a, b, and c that enclose three angles of magnitude α, β, and γ. The crystal axes form a right-handed coordinate system represented by three unit vectors $\boldsymbol{a}$, $\boldsymbol{b}$, and $\boldsymbol{c}$ (Figure 22.11). The magnitudes of the unit vectors are the *unit cell constants*, typically between 3 nm and 300 nm for macromolecular crystals.

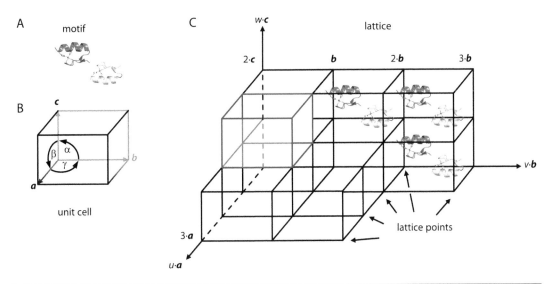

**FIGURE 22.11:** **Unit cell and crystal lattice.** A: The motif or asymmetric unit that builds up the unit cell by application of symmetry operators. In this case, there are two asymmetric units per unit cell. B: The unit cell is defined by three axes lengths and three angles. The *a*- and *b*-axes enclose the γ-angle, the *b*- and *c*-axes enclose α, and the *c*- and *a*-axes enclose β. Mathematically, the unit cell is described by three basis vectors **a**, **b**, and **c**. C: The macroscopic crystal is a result of three-dimensional translation of the unit cell. Three of the unit cells are filled with their asymmetric units. The lattice points (arrows) are located at the intersections of the lines. Unit cells that have lattice points only at their corners as in this example build up a primitive lattice.

Translation of the unit cell along its edges in all three dimensions builds up the macroscopic crystal. The crystal lattice is a mathematical concept that evenly divides space by sets of parallel and equidistant planes. The lattice is infinite in all three dimensions and covers all available space without gaps. The requirement to fill space completely means that only lattices with 1-fold, 2-fold 3-fold, 4-fold, and 6-fold rotational symmetries are possible. For example, 6-fold rotational symmetry means that after rotation of the lattice by 360°/6 = 60° the same 3D pattern of lattice points is achieved. The different rotational symmetries give rise to the seven *crystal systems* triclinic, monoclinic, orthorhombic, tetragonal, trigonal, hexagonal, and cubic (Figure 22.12). These crystal systems not only differ among themselves by their internal symmetry but also by the requirements for their unit cell constants, i.e. the relative length of the axes and the angles enclosed by the unit cell vectors (Table 22.2). For example, the cell dimensions of a triclinic unit cell have no restrictions, and it has 1-fold symmetry (i.e. none): only a full rotation by 360° will superimpose the lattice onto itself. By contrast, cubic cells must have all axes of the same length and all angles must be 90°, and there are four 3-fold axes along the body diagonals of the cube (Table 22.2). It is important to note that the variable parameters, indicated by a "≠" in Table 22.2, may be different but do not necessarily <u>have to</u> be different from the fixed parameters. It is entirely possible that by diffraction measurement a triclinic lattice is found with cell dimensions that resemble that of a cube.

At the intersections of the lattice planes lie the *lattice points*. All lattice points are equivalent because they have the same environment. The different lattice symmetries in three-dimensional crystals are abbreviated P, C, I, F, and R. Primitive (P) lattices have lattice points only at the corners of the unit cells (Figure 22.11). The lattice points are shared by eight adjoining unit cells, so there is a total of one lattice point per unit cell. Lattices other than primitive have additional lattice points. Single face-centered, or C-centered lattices have unit cells with an additional lattice point at the center of one pair of opposing faces. In face-centered, or F-centered lattices, all three pairs of opposing faces have a lattice point at the center of each face. I-centered (body-centered) lattices have an additional lattice point at the intersection of their body diagonals. Lastly, the rhombohedral R-lattice is an alternative description of a hexagonal lattice. Not all combinations of lattice types are possible with all crystal systems. For example, the triclinic crystal system is only compatible with a primitive lattice. In total, there are fourteen *Bravais lattice symmetries* (Figure 22.12).

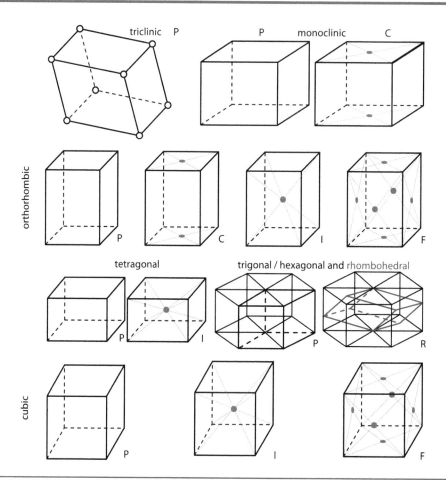

**FIGURE 22.12:** **The fourteen Bravais lattices and seven crystal systems.** Each parallelepiped is the building block of a Bravais lattice. The crystal systems are trigonal, monoclinic, orthorhombic, tetragonal, trigonal, hexagonal, and cubic. All of these can have a primitive (P) lattice. C, I, and F are the C-centered, face-centered, and body-centered lattice types, which are not possible for all crystal systems. The additional lattice points in C, I, and F lattices are marked by blue dots. C and I-lattices have one additional lattice point, and F-lattices have three additional lattice points. A rhombohedral cell (R-type lattice) is a cube stretched or compressed along one of its body diagonals. Rhombohedral cells can be described by a hexagonal cell three times as large, so the rhombohedral system is part of the hexagonal system. You can construct your own crystal systems by folding the two-dimensional cutouts provided at periodni.com/download/models_of_crystal_systems.pdf.

If we further disassemble the unit cell, we arrive at the motif or *asymmetric unit*. The asymmetric unit is the smallest entity from which the unit cell can be built by application of *symmetry operators*, which are mathematical operations that describe translations, rotations, mirror and inversion symmetry, and combinations thereof. An example of a symmetry operator is $(-x,-y,z + ½)$, a shorthand notation for a two-fold rotation of a point $(x,y,z)$ about the $z$-axis, followed by a translation of ½ its length along the $z$-axis:

$$\left(-x,-y,z+\frac{1}{2}\right) = \boldsymbol{R} \cdot \boldsymbol{x} + \boldsymbol{T} = \begin{pmatrix} -1 & 0 & 0 \\ 0 & -1 & 0 \\ 0 & 0 & 1 \end{pmatrix} \begin{pmatrix} x \\ y \\ z \end{pmatrix} + \begin{pmatrix} 0 \\ 0 \\ 0.5 \end{pmatrix} \qquad \text{eq. 22.3}$$

$\boldsymbol{R}$ is the *rotation matrix* operating on the vector $\boldsymbol{x} = (x,y,z)$, and $\boldsymbol{T}$ is the *translation vector*. The combination of this rotation and translation results in a screw axis, in this case a $2_1$-axis (Figure 22.13). In general, an $a_b$ screw axis rotates a point by $360°/a$ about an axis, and then applies a translation of $b/a$ along that axis. A $2_1$-axis is therefore a 180° rotation, followed by a shift of ½ along the axis.

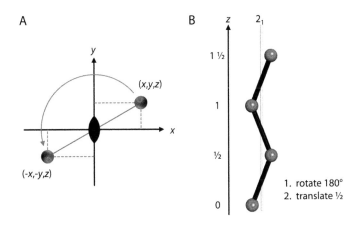

**FIGURE 22.13:** **Rotational symmetry.** A: View along the *z*-axis. The two-fold rotation (indicated by the lens-shape) about the *z*-axis converts coordinates (*x*,*y*,*z*) into (−*x*,−*y*,*z*), which is equivalent to the rotation matrix in eq. 22.3. B: Side-view from (A). The translation by ½ of the *z*-axis length along the *z*-axis is encoded in the translation vector **T** = (0,0, ½) in eq. 22.3.

In general, there are only 230 different ways to arrange objects in three dimensions so that the same pattern is repeated infinitely in all directions. These *space groups* combine the information of a Bravais lattice (P, C, I, F, R) with the internal symmetry of the unit cell. Each space group is a complete description of the symmetry of the entire crystal by its characteristic set of symmetry operators. A limitation comes into play with the chirality of biological (macro)molecules. In the vast majority of cases, the biologically active isomer is a single enantiomer. Omitting all space groups containing symmetry operators that change the handedness of a molecule (mirror and inversion symmetry) limits the number of space groups accessible for chiral molecules to the 65 *chiral space groups* listed in Table 22.2.

## TABLE 22.2
### Crystal systems, Bravais lattices, and chiral space groups.

| system | Bravais type | minimum internal symmetry | unit cell properties | chiral space groups |
|---|---|---|---|---|
| triclinic | P | none ("1-fold axis") | $a \neq b \neq c, \alpha \neq \beta \neq \gamma$ | P1 |
| monoclinic | P, C | 2-fold axis $\parallel$ **b** | $a \neq b \neq c, 90, \beta \neq 90, 90$ | $P2_{(1)}$, C2 |
| orthorhombic | P, C, I, F | three 2-fold axes, mutually perpendicular | $a \neq b \neq c, 90, 90, 90$ | $P222_{(1)}$, $P222_1$, $P2_12_12$, $P2_12_12_1$; $C222_{(1)}$; F222; $I2_{(1)}2_{(1)}2_{(1)}$ |
| tetragonal | P, I | one 4-fold axis $\parallel$ **c** | $a = b \neq c, 90, 90, 90$ | $P4_{(1,3)}$, $P4_{(2)}$, $P4_{(2)}22$, $P4_{(2)}2_{(1)}2$, $P4_{(1,3)}22$, $P4_{(1,3)}2_12$; $I4_{(1)}$, $I4_{(1)}22$ |
| trigonal | P | one 3-fold axis $\parallel$ **c** | $a = b \neq c, 90, 90, 120$ | P3, $P3_{(1,2)}$, $P3_{(1,2)}21$, $P3_{(1,2)}12$ |
| hexagonal | P | one 6-fold axis $\parallel$ **c** | $a = b \neq c, 90, 90, 120$ | $P6_{(1,3)}$, $P6_{(1,5)}$, $P6_{(2,4)}$, $P6_{(1,3)}22$, $P6_522$, $P6_{(2,4)}22$ |
| rhombohedral | R | one 3-fold axis | $a = b = c, \alpha = \beta = \gamma \neq 90$ | R3, R32 |
| cubic | P, F, I | four 3-fold axes along body diagonals | $a = b = c, 90, 90, 90$ | $P2_{(1)}3$, $P4_{(1,3)}32$, $P4_{(2)}32$; $I2_{(1)}3$, $I4_{(1)}32$; F23, $F4_{(1)}32$ |

*Note:* $\parallel$ indicates parallel. The "$\neq$" sign means "not necessarily". Non-subscripted numbers signify *n*-fold rotation axes. Subscripts denote screw axes. For example, $P3_{(1,2)}$ is short for the two space groups $P3_1$ and $P3_2$.

In contrast to the space group symmetries, which are limited to a certain set of rotations and translations, the asymmetric unit of a crystal can be anything from a single atom to a huge complex, and may display any kind of symmetry. The asymmetric unit can be a truly asymmetric macromolecule, it can be an oligomer of *n*-fold symmetry, or it can be a subset of a symmetric oligomer (Figure 22.14). The term "asymmetric unit" therefore does not mean that it must be asymmetric, but that if it displays symmetry, this *local* or *non-crystallographic symmetry (NCS)* is restricted to the asymmetric unit and different from the global crystallographic symmetry of the space group. NCS is observed in about ⅓ of all macromolecular crystals. For example, the asymmetric unit of the GroEL structure contains two seven-membered rings. The NCS is 7-fold, something that is impossible for crystallographic symmetry (Figure 22.14). On the other hand, the asymmetric unit may also contain only a part of a symmetric oligomer if a symmetry axis of the oligomer coincides with a space group axis. An example is HIV-1 capsid protein, where the biologically relevant unit is a hexamer but the asymmetric unit contains only a monomer. This protein crystallizes in space group P6, and the hexamer is generated by the crystallographic 6-fold symmetry. (Figure 22.14). Thus, the content of the asymmetric unit has no relation to the biologically active particle, and it is sometimes difficult to decipher the biological unit from the crystal structure. Methods analyzing the molecules in solution are then required.

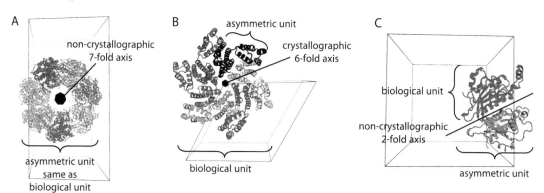

**FIGURE 22.14:** **Asymmetric unit and biological unit.** A: In the GroEL structure (PDB-ID 1sx3), the homoheptamer is both, the asymmetric and the biological unit. B: A single monomer is present in the asymmetric unit of the HIV-1 capsid structure in space group P6 (PDB-ID 4xfx). The biologically relevant hexamer is generated by the 6-fold crystallographic axis of this space group. C: The asymmetric unit may also contain more than one biological unit, which can be related by NCS. While the small GTPase Cdc42 forms a dimer in the crystal, it is a monomer in solution (PDB-ID 1a4r).

Once a possible space group and cell dimensions have been determined from the diffraction data, it is useful to estimate the likely content of the asymmetric unit, i.e. how many molecules there might be. If this number is less than unity, the macromolecule does not fit into the asymmetric unit and the space group may be wrong. If there are several molecules in the asymmetric unit, this information can be used for electron density improvement by NCS averaging (Section 22.8). The number of molecules is estimated by calculating the *Matthews parameter* $V_M$, a reciprocal crystal packing density given in units of Å³/Da (in crystallography, the Ångstrøm is still the preferred unit for length; it equals to 0.1 nm). $V_M$ is calculated by dividing the volume of the asymmetric unit by the molecular mass $M_M$ of the crystallized macromolecule. The volume of the asymmetric unit is the volume of the unit cell, $V_{cell}$, divided by the number of symmetry operators (symops) of the space group.

$$V_M = \frac{V_{cell}}{\#_{symops} \cdot M_M}$$

eq. 22.4

$V_{cell}$ is calculated as the volume of a general parallelepiped:

$$V_{cell} = abc \cdot \sqrt{1 + 2\cos\alpha \cdot \cos\beta \cdot \cos\gamma - \cos^2\alpha - \cos^2\beta - \cos^2\gamma}$$

eq. 22.5

In the case of all angles equal to 90°, eq. 22.5 reduces to $V_{cell} = a \cdot b \cdot c$. The molecular mass $M_M$ is chosen as integer multiples of the crystallized molecule to obtain candidate $V_M$ values for one, two, three, etc. molecules per asymmetric unit. $V_M$ has a range of 2.2–4 Å³/Da for most macromolecular crystals.

In summary, the entire macroscopic crystal is fully described by the asymmetric unit and the space group. This means that the crystal structure is determined once the structure of the asymmetric unit is known. All structure determination efforts aim at elucidation of the content of the asymmetric unit.

## 22.5 X-RAY DIFFRACTION FROM CRYSTALS

Because all molecules in a crystal are in defined relative orientations, the X-rays scattered from different molecules have a fixed phase relationship. This phase relationship leads to strong interference of the scattered beams with the result that crystals diffract X-rays only in certain directions. Geometrically, the diffraction of X-rays by crystals can be described as if the light were reflected at sets of equidistant parallel planes that run through the crystal and are oriented at a certain angle θ relative to the incident beam (Figure 22.15). Using this concept of reflection at planes, we can calculate under which conditions the reflected X-rays are in phase, interfere constructively, and can thus be measured as a *reflection* (a spot on a detector corresponding to a certain number of photons).

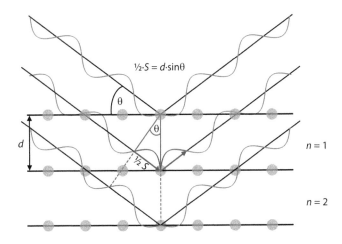

**FIGURE 22.15:** **Bragg's law.** Three planes of distance *d* are shown side-on, passing through atoms (gray dots). X-rays (blue waves) are reflected from these planes. The total reflection angle is 2θ. Constructive interference is only possible if the path difference *S* is an integer multiple of the wavelength: 1·λ between the first and second plane (*n* = 1), 2·λ between the first and third plane (*n* = 2), etc.

Reflection of X-rays at the first plane follows Snellius' law (Section 23.1.4; eq. 23.18): the angle of incidence is identical to the emergent angle (Figure 22.15). Thus, the total reflection angle is 2θ. X-rays reflected by the second plane travel a path difference *S* with respect to the first X-rays. In order for the two X-ray beams to remain in phase, *S* must be an integer number of the wavelength: $S = n \cdot \lambda$. By trigonometry, we obtain $\frac{1}{2} \cdot S = d \cdot \sin\theta$, where *d* is the interplanar spacing. Combining these two expressions, we obtain Bragg's law:

$$n \cdot \lambda = 2 \cdot d \cdot \sin\theta \qquad \text{eq. 22.6}$$

Bragg's law states that positive interference at an angle θ is observed as a *reflection*, if the path difference of X-rays reflected from planes of spacing *d* is an integer multiple of the wavelength λ. In other words, only if the path difference of the X-rays is an integer number of wavelengths will the reflected beams interfere constructively. If the path difference of X-rays reflected by the second plane is but only slightly different from an integer multiple of the wavelength compared to the first plane, negative interference will occur. Because there are so many planes in a crystal, there will always be a plane "further down" in the crystal where the phase difference is exactly 180° and the reflected X-rays will cancel. While in solution the light is scattered in all directions, it is diffracted only in certain directions by a crystal. In the absence of absorption, the total light intensity remains unchanged, so the few X-rays that interfere constructively are enormously enhanced compared to solution scattering, and the reflection intensities are therefore readily measurable.

The interplanar distance $d$ corresponds to the resolution of the diffraction experiment: the smaller $d$, the higher the resolution. The resolution is limited to $\lambda/2$ by the fact that $\sin\theta_{max} = 1$ for $\theta = 90°$, in accordance with the limiting resolution in optical microscopy (see Section 23.1.1.3).

For $n = 1$, Braggs law states that each *set of planes* with interplanar distance $d$ gives rise to a single reflection at angle $\theta$. The set of planes is indexed by three integer numbers, termed the *Miller indices* $h$, $k$, and $l$ (Figure 22.16). Many sets of equidistant planes can be drawn through a crystal lattice, but all planes within a set must intersect the unit cell axes rationally, or diffraction from different unit cells will lead to negative interference. The indices $h,k,l$ count the number of times that the set of planes cuts the unit cell axes. The same indices are also used to refer to the X-ray wave reflected from the set of planes, the reflection ($hkl$) with its intensity $I_{hkl}$. It is important to note that the larger the indices, the closer together lie the planes. Since closer planes mean more structural detail, reflections with larger indices have a higher resolution.

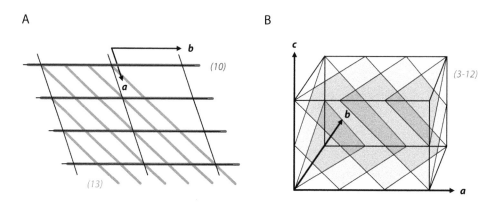

**FIGURE 22.16:   Lattice planes and Miller indices.** A: A three-dimensional lattice (black lines) is projected along the $c$-axis so it appears two-dimensional for simplicity. Planes parallel to the $c$-axis then show as lines (green and red). The unit cell constants $a$ and $b$ are shown as arrows. The green and red planes are two of the many possible sets of lattice planes that rationally intersect the $a$- and $b$-axes. Successive planes of the green set intersect the lattice at spacings of $a$ in the $a$-direction and at spacings of $\frac{1}{3}\cdot b$ in the $b$-direction. Thus, there is one plane every distance $a$, and three planes every distance $b$, hence the set of planes is named (*13*) in two dimensions. Since the set of red planes is parallel to the $b$-axis, it will never intersect it, so the index along this direction is zero. The red set also has a spacing of $a$ along the $a$-direction, so this set is indexed as (*10*). In three dimensions the green and red sets have indices (*130*) and (*100*). B: The three-dimensional example shows a (*3-12*) set of planes. The origin of the unit cell is at the bottom left. The negative index (*–1*) for the $b$-intersection comes from the fact that the planes intersect the $b$-axis in the negative direction.

Diffraction of X-rays into defined directions does not depend on the content of the unit cell, only on the symmetry of the crystal. However, the unit cell contents determine the reflection intensity $I_{hkl}$. In other words, the distribution of the reflections is determined by the space group and cell dimensions, and the reflection intensities are determined by the contents of the asymmetric unit. Mathematically, the observed diffraction pattern (see Figure 22.19) is a result of the convolution of the crystal lattice with the continuous molecular transform.

We mentioned before that the scattering process occurs in reciprocal space (Section 21.2.2), with the amplitude of the scattering vector in units of reciprocal length ($nm^{-1}$). The same real/reciprocal space relation holds for diffraction, and again the mathematical connection between the two spaces is the Fourier transformation. The set of planes with interplanar spacing $d$ can be represented by a single vector that is perpendicular to the planes and has the magnitude $1/d$. Doubling, tripling, etc. of $d$ will lead to the reciprocal magnitudes $1/(2d)$, $1/(3d)$, etc. The end points of the perpendicular vectors span a new lattice, known as the *reciprocal lattice* (Figure 22.17). The (*hkl*) triplets characterize equally the planes in real space and their associated planes in reciprocal space. The concept of a reciprocal lattice may be difficult to grasp, but it has the advantage that a whole set of planes in a crystal is described by a single reciprocal lattice point at which the reflection intensity is measured, and that it highlights the reciprocal relationship between the diffraction pattern and the real space crystal: a real space (direct) lattice with closely spaced points corresponding to small unit cell dimensions will give a diffraction pattern with widely-spaced reflections corresponding to large reciprocal cell dimensions. It is a general property of the Fourier transformation to invert the relation between parameters.

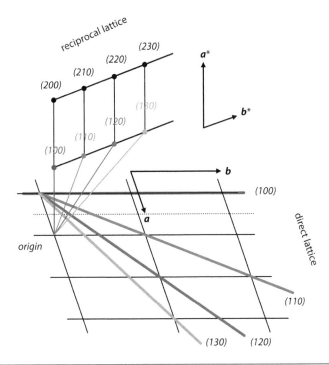

**FIGURE 22.17:**   **Real space and reciprocal space.** A three-dimensional lattice is shown projected along the *c*-axis, so it appears two-dimensional. The colored lines are lattice planes parallel to the *c*-axis. Each line represents the whole set of planes (*hkl*). A vector perpendicular to the planes with magnitude *1/d* marks a reciprocal lattice point (a colored dot). The largest distance *d* characterizes the red set of planes, hence the reciprocal vector (*100*) is the shortest. Doubling the magnitude of the red vector (end point then at *200*) means halving *d* for the real space planes (the dotted line). Repeating this process for all possible sets of real space planes constructs the reciprocal lattice. The origin of the real and reciprocal lattices is the same, and the real space unit cell dimensions (*a, b*) are reciprocal to the reciprocal unit cell dimensions (*a\*, b\**).

In solution scattering, the molecular form factor describes the combined scattering from all atoms within a molecule. Due to the ordering of molecules in crystals, the scattering by crystals is influenced not only by interference of waves within one molecule, but also by interference of scattered waves from different molecules. The analogy to the form factor in SAS is the *structure factor* in crystallography, which is given by

$$F_{hkl} = \left| F_{hkl} \right| \cdot e^{i\phi_{hkl}} = \sum_{n=1}^{N} f_{n,\theta_{hkl}} \cdot e^{2\pi i \cdot (hx_n + ky_n + lz_n)} \qquad \text{eq. 22.7}$$

$F_{hkl}$ is the diffracted beam of all the $N$ atoms in the unit cell in the direction given by the plane of indices *hkl*. Each atom $n$ contributes to the scattering according to its resolution-dependent atomic scattering factor $f_{n,\theta hkl}$. We recall from Section 21.2.1 that the magnitude of the scattering factor depends on the scattering angle θ. For diffraction, this means that the structure factor depends on the angle $\theta_{hkl}$ of the incident X-rays to the plane characterized by indices $hkl$ (see Figure 22.15), as we have seen with Bragg's law. An important notion from eq. 22.7 is that the phase of the structure factor is only determined by the atomic positions $(x_n, y_n, z_n)$, where $x_n$, $y_n$, and $z_n$ are the fractional coordinates of the $n^{th}$ atom in the unit cell. If we know the atomic positions $(x_n, y_n, z_n)$, we can calculate the structure factor. However, in crystallography, the problem is the reverse: we would like to know the atomic positions based on the knowledge of the structure factor. Unfortunately, measured diffraction data contain only the intensities $I_{hkl}$, which are related to the square of the structure factor amplitude $|F_{hkl}|$, but they do not contain the relative phase information $\phi_{hkl}$ for the reflections. This phase problem has to be solved in order to calculate the electron distribution in the unit cell of the crystal, which then allows us to place atoms $(x_n, y_n, z_n)$.

Until now, we have not used the complex notation for an electromagnetic wave because we were only dealing with amplitudes (proportional to intensities), not the phases. In crystallography, the phases are central to calculating the electron density, so we need to use the general formulation for electromagnetic waves. The structure factor $\boldsymbol{F}_{hkl}$ is an electromagnetic wave characterized by its amplitude $|F_{hkl}|$ and its phase $\phi_{hkl}$, which can be visualized in an Argand diagram (Section 29.10; Figure 22.18).

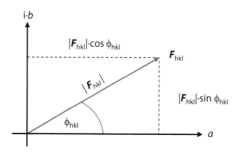

**FIGURE 22.18:** **Structure factor representation.** The two properties amplitude and phase of the structure factor are conveniently represented in an Argand diagram.

The atomic scattering factor (see Figure 21.6) is the Fourier transform of the electron density of a spherical atom. Since X-rays are scattered by electrons, the total diffraction of X-rays by a crystal is the sum of the contributions by all electrons in the unit cell, and consequently the Fourier transform of the total diffraction returns the electron density distribution of the unit cell. Re-formulation of eq. 22.7 replaces the atomic scattering factors $f_{n,\theta\,hkl}$ with the electron density $\rho_{xyz}$ at a point $(x,y,z)$.

$$\mathbf{F}_{hkl} = \left|F_{hkl}\right| \cdot e^{i\phi_{hkl}} = V \cdot \sum_{xyz} \rho_{xyz} \cdot e^{2\pi i \cdot (hx+ky+lz)} \qquad \text{eq. 22.8}$$

The electron density in the unit cell is then the Fourier transform of the structure factor:

$$\rho_{xyz} = \frac{1}{V} \cdot \sum_{hkl} \left|F_{hkl}\right| \cdot e^{i\phi_{hkl}} \cdot e^{-2\pi i \cdot (hx+ky+lz)} \qquad \text{eq. 22.9}$$

In contrast to the structure factor, which is a complex number, the electron density is a real number. If we evaluate eq. 22.9 at every point $(x,y,z)$ in the unit cell, we have defined the electron distribution of the entire crystal (by simple translation). In practice, it is sufficient to know the electron density distribution (representing the atomic structure) of the asymmetric unit, because the unit cell is built from the asymmetric unit by the symmetry operators of the space group. There are two phase factors in eq. 22.9, one is the phase $\phi_{hkl}$ for the diffracted X-ray, the other is a standing wave ($h$x + $k$y + $l$z) repeating each unit cell ($2\pi$) with a wavelength corresponding to integral fractions of the unit cell dimensions. It is the first phase, $\phi_{hkl}$, that is missing from the measured data. The next two sections will discuss what information can be gleaned from the intensity data without phases before continuing to discuss methods to retrieve the phases.

## 22.6  DIFFRACTION DATA COLLECTION AND ANALYSIS

The general setup for diffraction data collection is the same as that for SAS (Figure 21.7): the sample is placed in the X-ray beam and the scattered/diffracted light intensity is measured as a function of θ. While molecules in solution have random orientations, their relative orientations are fixed in crystals. Therefore, the difference to solution scattering is that the crystal has to be rotated in order to collect all the diffracted light. For this purpose, the crystal is mounted on a *goniometer* (Figure 22.19) that enables rotation of the crystal about one or more axes relative to the incident X-ray beam. For room temperature data collection (rarely done nowadays), the crystal is transferred into an X-ray transparent capillary and padded with mother liquor to avoid drying out. For the standard cryogenic data collection (Box 22.3), crystals in cryo-protectant are scooped up in a plastic loop at the end of a magnetic pin. The loop is rapidly cooled in liquid $N_2$, and placed on a magnetic base of the goniometer (Figure 22.19). The crystal is kept at 100 K in a cryo-stream of nitrogen gas and centered in the X-ray beam. A few hundred diffraction images are collected, which constitute a *dataset*. During each image collection, the crystal is rotated by a small angular increment of 0.05–1°, termed the *oscillation range*. The total oscillation range required for a dataset depends on the symmetry of the space group. The more symmetric the space group, the less angular space has to be covered for a dataset. For a highly symmetric cubic lattice, about 22° total rotation can be sufficient, but for the least symmetric triclinic lattice at least 180° are necessary.

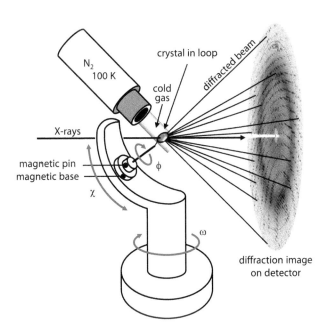

**FIGURE 22.19:  Diffractometer and cryo-cooling.** The crystal is in a loop at the end of a magnetic pin that fits onto the goniometer. A laminar $N_2$ gas stream keeps the crystal at 100 K for the duration of the experiment. While some goniometers have only a single axis of rotation (ω), three-circle goniometers can rotate the crystal about three axes, termed ω, φ, and χ. The diffraction image on the right-hand side is not to scale.

Apart from the content of the crystal, the intensity of the reflections, $I_{hkl}$, is proportional to the quality of the crystal, the intensity of the primary beam, the exposure time, and the time the crystal is kept in reflection condition. Bragg's law states that the reflection condition for an interplanar spacing is fulfilled at one exact angle θ. When continuously rotating a perfectly ordered crystal, this condition would only be fulfilled for an infinitely small time, and the intensity collected would be zero. However, macroscopic crystals are imperfect: they are a mosaic of perfect micro-crystals that are rotated slightly with respect to each other in all three dimensions. The mean angular rotation of the mosaic regions is termed the *mosaicity*, which is responsible for the fact that the crystal remains in reflection condition for about 0.1–2° angular rotation. This is large enough to collect the reflection intensities during continuous rotation with oscillation speeds between 0.01–1° s⁻¹. A typical diffraction image on a CCD (charge-coupled device; see below) detector is collected with an oscillation range of 0.5° and exposure times of 0.5–2 s. In contrast, detection with modern pixel detectors enables a complete data set to be collected in thin slices of 0.1–0.25° per image with an angular velocity of 1° s⁻¹ in a matter of minutes (3 min for 180°).

The first X-ray detectors used photographic film, which required tedious development and digitization of the reflection intensities. *Image plates* store the X-ray photon energy in pixels of a *phosphor* (crystalline BaFBr doped with a lanthanide) that luminesces after excitation with a suitable He-Ne laser. The phosphorescent light is amplified by a photomultiplier and counted. The detector is then erased by irradiation at high intensity to prepare it for the next exposure. The time for a readout/erasure cycle is about 1 min, which makes image plate detectors too slow for data collection at synchrotrons, but they are still used in some X-ray laboratories. CCDs follow an analogous strategy to image plates in that they store the X-ray photon energy in a phosphor: an X-ray photon liberates a fixed number of electrons in the respective pixel, which are attracted towards an anode, with the result of stable charge separation until readout. Readout of the CCD is initiated by alternating the polarity of a series of electrodes across detector lines (or columns, depending on the setup). The electron clouds travel along the lines, and at the end this current is converted into a voltage and recorded. CCD readouts are much faster (ca. 1 s) than those from image plates. The area covered by CCDs is smaller than that of image plates. If larger detector areas are required, several CCDs can be assembled into a 2 × 2 or 3 × 3 grid. Until the mid-2000s, such composite CCD detectors were the standard at synchrotron beamlines. The latest developments are single photon counting, or *direct pixel detectors*, which directly absorb and count X-rays in a silicon diode layer. Since every diode is connected to its own readout electronics (further discussed in Section 23.2.4), the readout dead time is on the order of a few μs and whole datasets can be collected in seconds. Pixel detectors have a dynamic range more than ten times that of CCDs, no electronic noise, and can be operated in continuous readout mode without closing the shutter between successive images, which facilitates time-resolved diffraction experiments (Box 22.4). These detectors are now the standard at synchrotrons.

### BOX 22.4: KINETIC CRYSTALLOGRAPHY.

There are two approaches to obtain time-resolved high-resolution structural information using crystallography: fast data collection of an ongoing reaction (continuous) and stopping a synchronized reaction *in cristallo* at different time points followed by offline analysis (discontinuous). In the discontinuous approach, a slow chemical reaction is started throughout an enzyme crystal by soaking a substrate into a number of crystals. The reaction is stopped at different time points by cryo-cooling a single crystal, and its structure is determined. The crystal structure will report on the progress of the reaction by way of accumulation of intermediates and products. An alternative way of synchronizing the

*(Continued)*

chemical reaction in crystals is laser flash photolysis (Section 25.3). *Cis/trans* isomerization of double bonds and CO-cleavage from heme groups can be initiated within the cryo-cooled crystal by a UV-laser pulse. Other reactions may be triggered by laser flash photolysis of an inactive substrate precursor, a *caged compound* (see Figures 25.6 and 25.7). Until fast pixel detectors were developed, reactions triggered by laser pulses were usually followed by *Laue diffraction*. In Laue diffraction, a spectrum of wavelengths is used to collect the diffraction data. The resulting diffraction images contain many more reflections than a single-wavelength exposure because the Bragg condition is fulfilled for several wavelengths at the same time, and a dataset can be collected much more quickly compared to a single-wavelength experiment. The time resolution for Laue diffraction is in the millisecond to nanosecond range. Even higher time resolutions in the femtosecond range may be possible with XFELs (see Box 22.1): synchronization of the reaction in the crystals is achieved by passing a slurry of micro-crystals, all executing the chemical reaction to be studied, at constant velocity perpendicularly across an XFEL beam. The time point collected by the diffraction experiment is the time the slurry needs from the start of the reaction to reach the XFEL beam. Images at different time points are collected by changing the distance between start and contact with the XFEL.

Once the first few images have been collected, the positions of the reflections in the diffraction patterns can be *indexed*, i.e. ascribed their *(hkl)* values using the best fitting Bravais lattice, crystal system, and unit cell constants. At this stage, it is decided whether the lattice is likely primitive, C-centered, etc. and whether the crystal system might be triclinic, monoclinic, etc. (Table 22.2). Using this information, a *data collection strategy* can be calculated that tells how to rotate the crystal with respect to the incident X-rays in order to rapidly collect a complete dataset with minimal X-ray exposure of the crystal for minimal radiation damage. Data processing involves integration of the reflections on all diffraction images and merging them into a list of intensity measurements with their associated errors. The errors are the mean standard deviations for *symmetry-related reflections*, which are reflections that should have identical intensities due to crystal symmetry and that have been measured more than once, which is almost always the case. A very useful piece of information that can be calculated using the measured intensities alone is the *Patterson function*. We recall that the electron density is the Fourier transform of the structure factor:

$$\rho_{xyz} = \frac{1}{V} \cdot \sum_{hkl} |F_{hkl}| \cdot e^{i\Phi_{hkl}} \cdot e^{-2\pi i \cdot (hx+ky+lz)} \qquad \text{eq. 22.9}$$

In the absence of phases, we can calculate the Patterson function by using only the measured intensities $I_{hkl} \sim |F_{hkl}|^2$ as the *Fourier coefficients*:

$$P_{xyz} = \frac{1}{V} \cdot \sum_{hkl} |F_{hkl}|^2 \cdot e^{-2\pi i \cdot (hx+ky+lz)} \qquad \text{eq. 22.10}$$

Mathematically, the Patterson function is the convolution of the electron density with itself. This convolution is positive at points that correspond to interatomic vectors and zero elsewhere. Thus, the Patterson function corresponds to a set of vectors between atoms in a given volume $V$ (a single molecule, an asymmetric unit, or a unit cell). A plot of this function, the *native Patterson map* (Figure 22.20), is a vector map where each vector magnitude corresponds to the distance between two atoms. The end points of the vectors tell about the relative positions of the atoms. The peak heights of the Patterson function are proportional to the product of the number of electrons of the atoms involved. As a consequence, very heavy atoms are more easily identified in the map than light atoms (see heavy atom derivatives; Section 22.7.1).

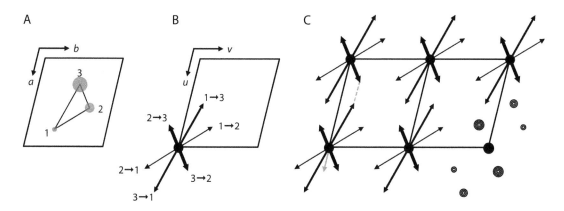

**FIGURE 22.20:** **Patterson map.** A: A hypothetical triangular molecule of three atoms with different mass is shown as blue dots of different radii in a two-dimensional unit cell. Atom 3 is the heaviest atom (highest electron density). B: Interatomic vectors are generated and translated to the origin. The unit cell has the same dimensions, but since the vectors denote relative, not absolute coordinates, the coordinates of their end points are denoted by $(u,v,w)$ rather than $(x,y,z)$. Since the vector $1{\rightarrow}2$ has the opposite direction to $2{\rightarrow}1$, the Patterson map has a center of inversion symmetry at its origin (black dot): it is centrosymmetric. The peak height at the tip of the vector between atoms 2 and 3 is the largest off-origin peak, indicated by the thickness of the arrow. C: Translation of the unit cell to generate the lattice shows that there are also vectors between atoms located in adjoining unit cells (one example is drawn in green). The vectors in the lower right-hand unit cell are shown as contour lines, an equivalent way of plotting the Patterson map.

The procedure to draw a Patterson map starting from a given molecular structure is conceptually simple (Figure 22.20). Each vector between pairs of atoms is moved such that it starts at the origin. For $N$ atoms, this will result in $N^2$ vectors, arranged centrosymmetrically about the origin. The largest peak is at the origin, because it collects all the $N$ *self-vectors*, the vectors of zero length from one atom to itself. If the Patterson map is calculated over a larger volume than a single molecule, additional *intermolecular* vectors between atoms from adjacent unit cells are present in the map (Figure 22.20), which are important in molecular replacement phasing (Section 22.7.3). Although the Patterson map does not contain phase information, it does in principal contain all the information necessary to determine the crystal structure. Working back from a Patterson map to the structure is feasible for small molecules that have only a few atoms. With increasing number of atoms, the number of peaks in the Patterson map grows quickly, and they start to overlap, rendering assignment and direct structure determination impossible. We will see in the next sections how the Patterson function is applied for phasing of macromolecular structures.

## 22.7 PHASING METHODS

Several methods have been developed to retrieve the missing phases of the measured reflections, which both are required for calculating the electron density. Three methods will be described, *isomorphous replacement*, *anomalous scattering*, and *molecular replacement*, all of which make use of the Patterson function. Isomorphous replacement and anomalous scattering simplify the problem of locating many atoms in the asymmetric unit by first locating only a small *substructure* of heavier atoms. The phases for the substructure then serve as starting phases for the macromolecule. In molecular replacement, the phases are calculated from a closely related structure that has been positioned in the unit cell of the unknown structure.

### 22.7.1 ISOMORPHOUS REPLACEMENT

The first step in isomorphous replacement is to obtain a heavy atom derivative of the crystal, where each macromolecule has bound one or just a few heavy atoms (Box 22.5). A dataset of the heavy atom derivative (*PH*) is collected and compared to a *native* dataset (*P*) that has no heavy atoms bound.

Equivalent reflection intensities with the same (*hkl*) indices are subtracted from one another, leaving only the contributions of the heavy atoms. For the subtraction to give useful results, the heavy atom derivative and the native crystal must be *isomorphous*: their space group and cell dimensions must be very similar, ideally identical, and the structure of the macromolecules in the derivative must not have changed too much by heavy atom binding. Using the *isomorphous differences* as the Fourier coefficients, a *difference Patterson map* is calculated (eq. 22.10). The difference Patterson map will have only a few prominent peaks that correspond to the distances between the heavy atoms, and is therefore much easier to interpret than the native Patterson map of a whole macromolecule. The positions of the heavy atoms, termed the substructure, are calculated by computers directly from the difference Patterson map. From these positions, the complete structure factors (amplitude and phase) for the heavy atom substructure can be calculated. The phases derived from the substructure allow calculation of starting phases for the macromolecule. In effect, the heavy atoms "replace" the large problem of diffraction by the macromolecule by the smaller problem of finding the substructure. This principle for phasing is also used in variations of isomorphous replacement and in *anomalous diffraction* (Section 22.7.2).

---

### BOX 22.5: SOAKING OF MACROMOLECULAR CRYSTALS.

The large solvent channels in macromolecular crystals allow molecules of considerable size to diffuse through them. *Soaking* of crystals is an important technique for heavy atom phasing and to determine protein-ligand complexes. To obtain a *heavy atom derivative* of a crystal, the crystals are soaked in heavy atom solutions for a few seconds to many days to allow either covalent or non-covalent binding to the macromolecule. For example, $Hg^{2+}$ and $[PtCl_4]^{2-}$ covalently bind to Cys, His, and nucleobases, whereas $I^-$ (iodide) non-covalently binds to positive charges (Lys, Arg) or hydrophobic sites. An alternative to soaking is to co-crystallize the heavy atom compound with the macromolecule, but soaking has the advantage that many trials with different heavy atom solutions can be carried out in parallel. It is impossible to know whether soaking or co-crystallization have produced a useful derivative without collecting diffraction data. Hence, it is sometimes better to chemically incorporate heavy atoms prior to crystallization (Box 22.6).

Soaking is also used to introduce small molecules such as substrates, products, or inhibitors into crystals. The binding site for such a ligand must not be occluded by crystal contacts, and no large conformational changes must occur upon binding or the order of the crystal might be destroyed. If a protein-ligand complex can be formed *in situ*, its structure can be determined. This is the preferred method in structure-based drug design where hundreds of protein-inhibitor complexes are determined starting from "empty" crystals. Some enzymes that do not undergo large conformational changes during catalysis not only bind their ligand but also retain their catalytic activity in the crystalline form. By soaking an enzyme crystal with a substrate and stopping the reaction at different time points by cryo-cooling, a series of structures representing the substrate, intermediate, and product complexes may be determined (Box 22.4). Ligand binding and enzymatic activity *in cristallo* are also good indications that the molecular conformation trapped in the crystal is of biological relevance, a question that arises for every crystal structure and normally needs to be answered by independent experiments in solution.

---

The contribution of the heavy atoms to diffraction and the determination of the phases for the reflections can be visualized by an Argand diagram of the structure factors involved (Figure 22.21). For each reflection (*hkl*) of the native dataset, a structure factor amplitude $|F_P|$ is calculated from the reflection intensity. The phase of this reflection is unknown, so the vector $F_P$ can have any direction

(Figure 22.21). A diffraction experiment with the derivative crystal gives the amplitude $|F_{PH}|$, also with unknown phase for the vector $\boldsymbol{F}_{PH}$. The connection between these measurements is the vector $\boldsymbol{F}_{H}$, the structure factor of the heavy atom:

$$\boldsymbol{F}_{PH} = \boldsymbol{F}_P + \boldsymbol{F}_H \qquad \text{eq. 22.11}$$

We know the magnitude and the phase of $\boldsymbol{F}_{H}$ from the difference Patterson map. A *Harker construction* geometrically combines the known magnitudes $|F_P|$, $|F_H|$, and $|F_{PH}|$, and the phase for $\boldsymbol{F}_H$, to estimate the phases of $\boldsymbol{F}_P$ and $\boldsymbol{F}_{PH}$ (Figure 22.21).

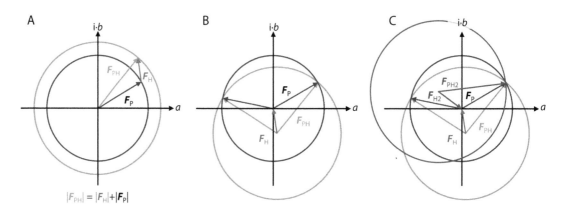

**FIGURE 22.21:** **Harker construction.** A: The measured structure factor amplitudes of the protein $|\boldsymbol{F_P}|$ and the derivative $|\boldsymbol{F_{PH}}|$ determine the radii of circles (black and green). The directions, or phases, of $\boldsymbol{F}_p$ and $\boldsymbol{F}_{PH}$ are unknown at first, so these vectors can point to any point on the circumference of their circle. By contrast, both the amplitude and the phase of the structure factor $\boldsymbol{F_H}$ may be calculated from a difference Patterson map (blue arrow), which will now determine the two possible directions of $\boldsymbol{F}_p$ and $\boldsymbol{F}_{PH}$. B: To find these directions, the Harker construction places the vector $\boldsymbol{F_H}$ with its tip at the origin. The circle for $\boldsymbol{F_P}$ (unknown phase; black) has its center at the end of $\boldsymbol{F_H}$, just as in panel (A). The center of the circle for $\boldsymbol{F_{PH}}$ (unknown phase; green) is moved to the beginning of $\boldsymbol{F_H}$. The two circles now intersect at two points, which define the two possible choices of phases for $\boldsymbol{F}_p$ and $\boldsymbol{F}_{PH}$. C: A second derivative (magenta with known $\boldsymbol{F_{H2}}$ in red) resolves the phase ambiguity. This Harker construction is repeated for every reflection (*hkl*).

The Harker construction shows that for each reflection there are two possible phases for $\boldsymbol{F_P}$ (and also for $\boldsymbol{F_{PH}}$). This phase ambiguity is characteristic for single isomorphous replacement (SIR), where we have only a single derivative. The ambiguity can be resolved by repeating the experiment with a second derivative (multiple isomorphous replacement; MIR) where the heavy atom must have bound to a different location on the macromolecule. The structure factor from this derivative (Figure 22.21) can be plotted as a third circle in the Harker construction, and will intersect the other two circles in only a single point, thus resolving the phase ambiguity for $\boldsymbol{F_P}$ (and $\boldsymbol{F_{PH}}$).

### 22.7.2 ANOMALOUS DIFFRACTION

Bragg's law relates the angle of incidence θ to the spacing of a set of planes (see Figure 22.15). The same relation holds for an angle of incidence −θ at the same set of planes, but which are now indexed as (−*h*–*k*–*l*). Reflection of X-rays from the "underside" of the (*hkl*) planes gives rise to the same reflection intensity, but with opposite phase. This relationship is known as *Friedel's law* (Figure 22.22).

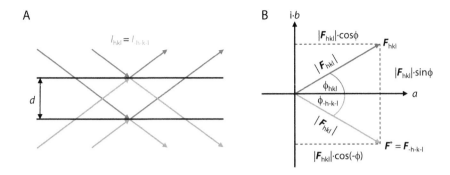

**FIGURE 22.22: Friedel's law.** A: A set of planes is equally described by the indices (*hkl*) and (–*h*–*k*–*l*). Diffraction from this set of planes is possible from angles θ and –θ. Friedel's law states, that the reflection intensities $I_{hkl}$ and $I_{-h-k-l}$ are identical. B: Argand diagram of the structure factors of a Friedel pair, showing that their magnitudes are identical, but the phases have opposite signs.

The reflections $I_{hkl}$ and $I_{-h-k-l}$ are termed *Friedel mates*, and their corresponding structure factors $F_{hkl}$ and $F_{-h-k-l}$ are complex conjugates (denoted by a star; Section 29.10): they have the same amplitude, but opposite phase.

$$F_{hkl} = F_{hkl}^{*} \qquad \text{eq. 22.12}$$

This is equivalent to writing:

$$\left| F_{hkl} \right| = \left| F_{-h-k-l} \right|$$

$$\phi_{hkl} = -\phi_{-h-k-l} \qquad \text{eq. 22.13}$$

As a consequence of Friedel's law, the diffraction pattern is *centrosymmetric*, meaning that the distribution of reflections has a center of symmetry just like the Patterson function.

Friedel's law is true for most elements at most X-ray wavelengths. The presence of *anomalous scatterers* in a crystal leads to a breakdown of Friedel's law, so that the intensities of the Friedel mates are not identical any more. Anomalous scattering occurs when the wavelength of the incident X-rays is close to an electronic transition in an atom. In this case, a fraction of the incident photons is absorbed by the atom and re-emitted with a phase shift relative to the incident radiation. This inelastic scattering may reduce the total light intensity by 1–3%, which can be measured and used for phasing. In principle, any atom exhibits anomalous scattering if X-rays of a suitable wavelength are used. The spectral range accessible with synchrotrons (ca. 0.04–0.4 nm) limits the atom types from which useful anomalous signal can be collected to atomic numbers > 16 (sulfur). The anomalous signal becomes increasingly larger for more electron-dense (heavier) atoms.

The presence of anomalous scattering is described by a modified atomic scattering factor:

$$f_{\text{total}} = f_0 + f' + i \cdot f'' \qquad \text{eq. 22.14}$$

$f_0$ is the wavelength-independent component of the atomic scattering factor (see Section 21.2.1; eq. 21.27). It depends on the resolution (and thus on the diffraction angle θ), and is equal to the atomic number at θ = 0. $f'$ and $f''$ are the real and imaginary parts of the wavelength-dependent atomic scattering (Figure 22.23). $f'$ is a *dispersive* correction to the normal scattering at a given wavelength. It is usually negative and out of phase by 180° relative to the normal scattering dictated by $f_0$. The absorption $f''$ is strongly wavelength-dependent (Figure 22.24) and has a phase shift of 90° relative to $f_0$, which is signified by the imaginary number $i$ in eq. 22.14. The value of $f''$ corresponds to the number of electrons contributing to anomalous scattering.

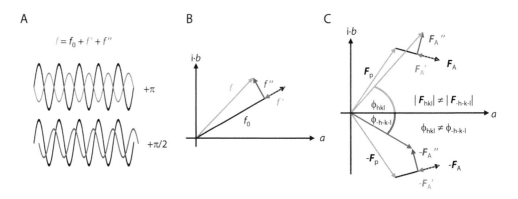

**FIGURE 22.23: Modified atomic scattering factor and breakdown of Friedel's law.** A: In the presence of anomalous scattering, the atomic scattering factor $f_0$ is modified by two components that are 90° phase shifted relative to each other. $f'$ is out of phase by 180° relative to $f_0$, and $f''$ has a phase advance of 90° relative to $f_0$. B: Vector representation of the scattering factor components. As $f'$ (blue) is normally negative, the magnitude of $f_0$ is reduced. Addition of the perpendicular component $f''$ (red) results in the new scattering factor $f$ (green). C: In the presence of an anomalous scatterer with structure factor $F_A$, the structure factors of the protein $F_P$ and its Friedel mate $-F_P$ (gray) are modified by the same scheme as shown in panel (B). Note that now both the magnitude and the phase of the former Friedel mates (green *versus* magenta) have changed.

To find out whether an anomalous scatterer is present in a crystal, an X-ray fluorescence emission spectrum is recorded, and the absorption and dispersion spectra are calculated (Figure 22.24). The wavelength of maximum anomalous absorption is termed the *absorption edge* or *peak*. The dispersion is maximal at the *inflection point*, which is often only $10^{-4}$ nm away from the absorption peak. Both $f'$ and $f''$ are characteristic for each element and the values are tabulated also for many ions and small molecules.

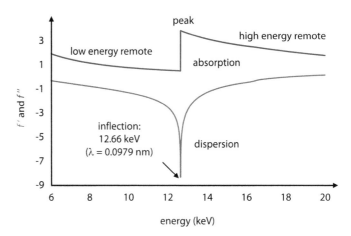

**FIGURE 22.24: X-ray absorption.** Absorption and dispersion spectra of Se as a function of X-ray energy. The magnitude of $f'$ and $f''$ depends on the energy (wavelength) of the X-rays. The maximum anomalous signal $f''$ for Se is 3.8 e⁻ at $\lambda = 0.0979$ nm (0.979 Å). Atomic scattering factors are tabulated for all elements and many ions.

A diffraction dataset is collected at a wavelength where anomalous diffraction occurs, and the *anomalous differences* of the Friedel mates are used to calculate the positions of the anomalous scatterers from an *anomalous difference Patterson map*. This procedure is similar to the difference Patterson map used to locate heavy atoms in isomorphous replacement, but in anomalous diffraction the differences are calculated using data within a single dataset, not from two different datasets, which eliminates the problem of non-isomorphism. A Harker construction then yields phase estimates for the protein structure factors. In *single wavelength anomalous dispersion* (SAD) the anomalous signal is collected at only one wavelength, often the peak of the elemental absorption with maximum $f''$ (Figure 22.24), such that the differences between $I_{hkl}$ and $I_{-h-k-l}$ are maximized. SAD has a phase ambiguity similar to SIR, but it can often be resolved by density modification (Section 22.8). Recent developments in detector sensitivity and speed of data acquisition enable the weak anomalous signal from the sulfur atoms of Cys and Met for SAD phasing of protein structures to be measured. Sulfur SAD phasing could become a standard technique for *de novo* protein crystal structure determination.

If SAD phasing fails, *multiple wavelength anomalous dispersion* (MAD) can be used. For MAD, diffraction at the *inflection* wavelength with maximum $f'$ is collected in addition to a dataset collected at the peak wavelength. Additional datasets below (*low energy remote*) and above (*high energy remote*) the absorption edge are also often collected and are treated as separate "derivatives" similar to MIR phasing, which usually gives improved phases over SAD, but data collection and all calculations take longer. Care must be taken to avoid radiation damage, which easily changes the reflection intensities far more than the magnitude of the anomalous signal. Anomalous phase information can be readily combined with phases from single and multiple isomorphous replacement, termed *SIRAS* and *MIRAS*. The result of a phasing experiment that includes anomalous signal can be an electron density map of stunning quality (see Figure 22.27).

---

### BOX 22.6: COVALENT INTRODUCTION OF HEAVY ATOMS AND ANOMALOUS SCATTERERS INTO MACROMOLECULES.

Many heavy atoms that are soaked into crystals for MIR phasing also have anomalous signal at the wavelengths to which synchrotrons can be tuned. However, often the soaked heavy atoms fail to bind to the protein, which is only found out after lengthy data collection and evaluation. An alternative is to covalently bind the anomalous scatterer to the macromolecule (Figure 22.25). 2'-Bromo and 2'-seleno nucleotides can be included in DNA and RNA during synthesis, provided there is enough space at the 2'-deoxy position (DNA) or the 2'-OH group is not involved in tertiary interactions (RNA). U and T have also been replaced by 5-bromouracil.

The most popular method for proteins is *Se-Met modification*. An *E. coli* culture in minimal medium without Met but including Se-Met will incorporate Se-Met into proteins. Alternatively, the culture can be grown in rich medium, and Met biosynthesis may be inhibited prior to induction of the culture by addition of Ile/Leu/Phe/Lys/Thr at high concentrations. The remaining Met is quickly depleted by protein biosynthesis, followed by incorporation of the added Se-Met. This approach is also used for introducing isotopes into proteins (see Box 20.6). The content of the incorporated Se-Met may be analyzed by mass spectrometry. It is useful to know the percentage of Se-incorporation before the time-consuming process of crystallization and data collection. A slight disadvantage of Se-Met is its hydrophobicity and slightly larger volume compared to Met. Se-Met proteins therefore tend to be less soluble and are sometimes less stable than their natural Met counterparts.

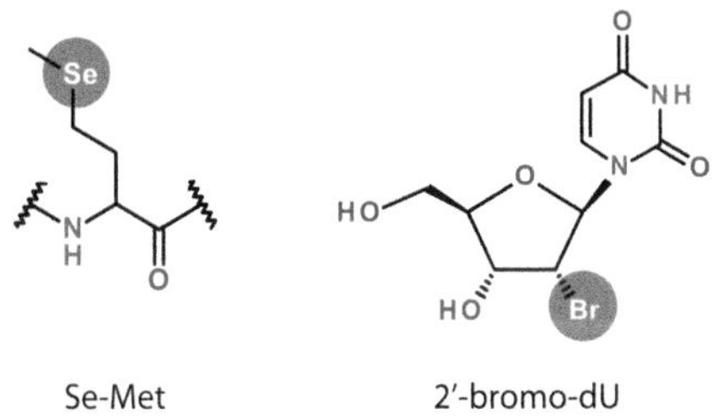

Se-Met    2'-bromo-dU

FIGURE 22.25:   **Seleno-methionine (Se-Met) and 2'-bromo-uridine (2'-bromo-dU).**

### 22.7.3 MOLECULAR REPLACEMENT

From a conceptional point of view, molecular replacement is the simplest way of obtaining phases. It is also the most widely used method for phasing: over 70% of all structures in the PDB have been determined using molecular replacement. The phases are calculated from a similar structure, the *search model*, after placement of that model in the unit cell of the observed diffraction data. "Placement" means that the search model needs to be rotated and translated to become superimposed on the unknown structure. Computationally, this is a six-dimensional problem because a superposition of two molecules is described by three rotation angles about the axes of the coordinate system, and three translations along the axes. The position $A'$ of the atoms in the target molecule is given by their position $A$ multiplied by a rotation matrix $R$ plus a translation vector $T$:

$$A' = A \cdot R + T$$

eq. 22.15

The problem of molecular replacement is solved when $R$ and $T$ have been found. Using the properties of the Patterson function (Section 22.6), the molecular replacement calculation can be split into separate rotation and translation steps (Figure 22.26).

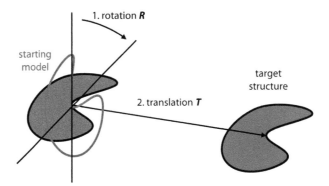

**FIGURE 22.26:** **Molecular replacement.** The starting model is first rotated by Patterson function comparison (rotation function). The translation function places the rotated model at the correct location in the target unit cell.

The rotation search makes use of the Patterson functions of the search model, $P_{calc}$, and the diffraction data, $P_{obs}$. We recall that the Patterson function can be calculated in the absence of phases and that it represents a set of distance vectors $u$ between atoms in a molecule. If we rotate one Patterson function in three dimensions with respect to the other, we should be able to find a relative orientation where the sets of vectors match closely. This calculation is done by a *cross-rotation function* $RF_R$, which multiplies the two different Patterson functions after one of them ($P_{calc}$) has been rotated by an angular increment with respect to the other ($P_{obs}$).

$$RF_R = \int P_{obs}(u) \cdot P_{calc}(R \cdot u) \cdot du$$

eq. 22.16

The angular increment is encoded in the matrix $R$. If the agreement between the Patterson functions after rotation is high, $RF_R$ has a maximum and the rotation of the search model with respect to the target structure is found. The maximum vector magnitude $u$ for the integration should correspond to the diameter of the search model in order to limit the calculation to predominantly intramolecular distances. This value can be estimated from the sequence or molecular mass of the target molecule, assuming that it is roughly spherical. If many intermolecular vectors enter the calculation by choosing a too large value for $u$, the cross-rotation function may get too noisy to find the correct rotation solution.

The next step is to find the position of the correctly oriented model in the unit cell by use of a *translation function* $TF_t$. Like the rotation function $RF_R$, $TF_t$ is the product of the two Patterson functions $P_{obs}$ and $P_{calc}$. Now, $P_{calc}$ is the Patterson function calculated from the search model after the rotation matrix found in the rotation step has been applied. $P_{calc}$ is then translated on a three-dimensional grid throughout the entire unit cell by a vector $\boldsymbol{T}$.

$$\mathrm{TF}_t = \int P_{obs}\left(\boldsymbol{u}\right) \cdot P_{calc}\left(\boldsymbol{u}+\boldsymbol{T}\right) \cdot \mathrm{d}u \qquad \text{eq. 22.17}$$

Upon translation of $P_{calc}$, the intramolecular vectors stay the same while the intermolecular vectors change. Modern implementations of the translation function subtract the set of intramolecular vectors to remove noise from the calculation and use only the intermolecular vectors to decide on the correct position of the search model. A translation vector $\boldsymbol{T}$ that leads to superimposition of $P_{obs}$ and $P_{calc}$ produces a maximum in $\mathrm{TF}_t$. Many solutions are usually found, and the next step is to distinguish the correct solution from the best wrong solution. This is done by *packing analysis*: the symmetry operators of the space group are applied to the rotated and translated search model to fill the unit cell, and the number of overlaps between symmetry-related molecules is counted. The best solution should produce a minimum number of overlaps, maybe between loop regions that differ between the search model and the correct structure. This is checked by calculating an electron density map using the phases from the molecular replacement solution and the amplitudes $|F_{obs}|$ from the measured reflections. If the electron density contains information that was not present in the search model, such as a different loop conformation or a bound ligand, the molecular replacement solution is very likely correct.

The search models used for molecular replacement might differ by just a few atoms from the true structure, for instance in variants of a few side chains or when a different ligand is bound. Such structure determinations are usually straightforward. Sometimes, however, the closest known search model is so different from the true structure that molecular replacement fails. The minimum structural similarity between a search model and the true structure is preferably less than 0.15 nm rmsd (see eq. 29.2), but this is not known *a priori*. The sequence identity is therefore used as a guide to judge the likelihood of success for a given search model. More than 40% identity should give a clear solution, unless there are large conformational differences between search model and the unknown structure. In this case sub-domains of the search model can be used one after the other. The minimum sequence identity between the search model and the target structure is about 20%, but low resolution of the data and a large number of molecules that have to be searched for may require better search models. Search models can be improved *in silico* by removing non-conserved parts identified from multiple sequence alignments (see Section 18.1.2). If search models of < 30% sequence identity are successful in determining the structure, it is because the three-dimensional fold is more conserved than expected. The conservation of three-dimensional folds despite large differences in sequence is the basis of novel developments in molecular replacement, which promise to determine many new crystal structures without the need for isomorphous and anomalous phasing (Box 22.7).

---

## BOX 22.7: MOLECULAR REPLACEMENT AS THE ULTIMATE PHASING TOOL?

Although the number of crystal structures in the PDB keeps growing at an almost exponential rate, fewer and fewer novel folds are added to the database (see Section 16.2.4). It is likely that the number of different folds in nature is limited, and that each fold is already represented by a crystal structure in the PDB. This raises the possibility that any new structure could be phased by molecular replacement using search models cleverly pieced together from known structures. Two promising approaches have been developed, one that uses homology models and another that uses secondary structure elements as search models. Both can be applied to either proteins or RNA (DNA is usually unproblematic due to its conserved structure). In the

*(Continued)*

long run, the techniques promise to determine any crystal structure by molecular replacement using only the information we already have from previous structures.

ROSETTA (rosettacommons.org), is a set of programs that uses multiple sequence alignment to identify parts of the sequence for which a similar structure is available. A homology model is made from this sequence using the known structure as a template and used as a search model. The electron density maps calculated after molecular replacement are interpreted automatically and the structures refined, so many trials can be done until hopefully a solution is found. The key to success of this method is to integrate structure prediction with automated electron density interpretation and structure refinement.

ARCIMBOLDO and BORGES (chango.ibmb.csic.es/ARCIMBOLDO) are two programs that do not use homology models for the search but thousands of secondary and super-secondary structure elements derived from the PDB. The required accuracy for a successful search model is below 0.06 nm rmsd, explaining the huge number of searches that have to be conducted. The approach currently works best with $\alpha$-helices as these show the least structural variation and have high electron density due to the helical arrangement of the residues. $\beta$-sheets and loops are more extended and have more variable structures, making their identification more difficult. A few dozen structures, predominantly $\alpha$-helical, have been determined so far. The diffraction data should extend to at least 0.25 nm resolution, and the asymmetric unit should have less than 400 residues.

## 22.8 ELECTRON DENSITY AND MODEL BUILDING

After approximate phases have been obtained by any of the methods described in the previous sections a first electron density map is calculated (see eq. 22.9) using the phases and the measured structure factor amplitudes $|F_{obs}|$. The next step on the way to the crystal structure is interpretation of the electron density in terms of atoms by threading a polypeptide or oligonucleotide chain through the density. However, in many cases the quality of the initial electron density map is too poor for interpretation. *Density modification* procedures improve the phases, and thus the electron density maps, by including additional information, as described in the following.

*Solvent flattening* relies on the fact that the solvent between the macromolecules has a different (usually lower) electron density than the macromolecules (see Table 21.3). An envelope around a volume corresponding to the macromolecule is identified, and the electron density outside the envelope is set to zero or to a fixed value (e.g. 330 e/nm³ for water). From these modified maps, new phases are calculated by inverse Fourier transformation. With the improved phases and the experimental structure factor amplitudes $|F_{obs}|$ a new electron density map is calculated that is often easier to interpret.

*Histogram matching* is based on a feature used in image processing and photography that assumes a rather constant or only gradually changing value across an object (color in images, electron density in molecules). The electron density should be rather constant within molecules, and if the initial electron density map has large discontinuities, these may be smoothed. For this purpose, the electron density map is binned into histograms and compared to the expected *electron density histogram* of proteins and nucleic acids. The reference histograms of proteins and nucleic acids are remarkably similar and rather independent of sequence or structure. Adjusting the density distribution of an initial map to match the expected values yields new and improved phases.

*Non-crystallographic symmetry (NCS) averaging* can be applied when at least two independent copies of the same molecule are present in the asymmetric unit. These molecules should have equal electron densities at corresponding sequence positions. The method requires an NCS operator for each pair of molecules for which the electron density should be averaged. NCS averaging can spectacularly improve the quality of the electron density maps, particularly when many NCS copies are present, such as in highly symmetric virus structures. In essence, the more NCS copies, the more restrained are the phases due to averaging.

If the resulting electron density map is of sufficiently high quality, the outline of the macromolecules are visible (also termed a *solvent boundary*), and secondary structure elements can be recognized (Figure 22.27). In favorable cases, the main chain can be traced and many side chains can be assigned just on the basis of electron density (Figure 22.27). Usually, however, knowledge of the sequence is required to assign all side chains. In general, the resolution of the electron density map is determined by the measured diffraction data and determines the level of detail (Figure 22.27): α-helices are the most compact secondary structure elements and are visible as rods in electron density maps down to 0.5 nm resolution. β-sheets become discernible at around 0.4 nm resolution. Tracing of the main chain and some flexible loops requires around 0.3 nm resolution. Side chains, especially the large aromatic Trp, Phe, Tyr, and His, also become well defined at this resolution. At 0.2 nm, most side chains and *cis*-peptides are clearly visible. Finally, atomic resolution structures (≤ 0.1 nm) allow visualization of individual atoms (Figure 22.27). Since hydrogen atoms contain only a single electron, they are normally not visible in electron density maps. Neutron crystallography (Box 22.8) is the preferred method to study hydrogens by diffraction methods.

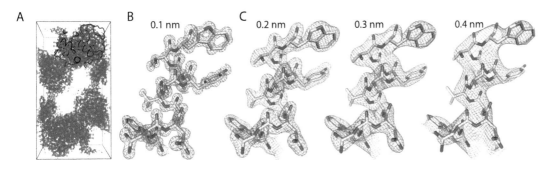

**FIGURE 22.27:    Electron density after density modification and effect of resolution.** A: The clear solvent boundary indicates accurate phases. One molecule of the unit cell is outlined by a black line. B: Magnification of a part of the density in (A) showing an α-helix with well resolved side chains at 0.1 nm resolution. C: The electron density map in (B) was re-calculated to lower resolutions of 0.2 nm, 0.3 nm, and 0.4 nm. Note the decreasing quality of side-chain density. The main-chain trace of the α-helix is still visible at 0.3 nm resolution. At 0.4 nm resolution, the α-helix resembles a rod. The information content of electron density strongly depends on the quality of the phases. In the examples shown, the phase error is small (~30°).

## BOX 22.8: NEUTRON CRYSTALLOGRAPHY.

If neutrons instead of X-rays are used in a diffraction experiment, a neutron crystal structure can be determined in the same way as an X-ray crystal structure. While X-rays are scattered at electrons, neutrons are scattered at nuclei. Currently available neutron beams are several orders of magnitude less intense than X-ray beams, and neutron detectors are not as sensitive as X-ray detectors. Therefore, the major requirement for a neutron crystal structure is large crystals of at least 1 mm in each direction. However, the advantage is that with neutron crystallography hydrogen atoms can be visualized, something that is difficult to achieve with X-rays. Neutron crystallography is not used for *de novo* structure determination (this is done with X-rays), but is used to study hydrogen atoms.

While the result of X-ray crystallography is an electron density map, neutron crystallography returns a *nuclear density map*, i.e. the distribution of atomic nuclei within a molecule. Hydrogen isotopes are not normally visible in electron density maps because they contain only a single electron. In contrast, hydrogen and deuterium show up very prominently in nuclear density maps as negative and positive peaks, respectively. This arises

*(Continued)*

since the scattering lengths of hydrogen (−3.7 e) and deuterium (+6.7 e) are comparable in magnitude to the scattering lengths of other atoms (see Table 21.2), but hydrogen is unique in having a negative sign, making it easy to distinguish from other nuclei.

This property of hydrogen and deuterium is used to study the protonation states of proteins: crystals are soaked in buffer containing $H_2O$, $D_2O$, or a mixture, then flash-cooled so that no further H/D exchange can take place, and a series of neutron crystal structures is determined by molecular replacement using the X-ray crystal structure as a search model. Enzyme reactions with shifts in hydrogen atoms are particularly suited for neutron crystallography. For example, the location of the proton in the catalytic Ser-His-Asp triad of serine proteases was detected by neutron crystallography to be at the His, rather than at the Asp. In D-xylose isomerase, a metalloenzyme catalyzing the isomerization of aldoses to ketoses, four reaction intermediates have been detected using neutron crystallography.

A particularly useful kind of map for model building is the *difference map*. It is calculated with Fourier coefficients $|F_{obs}|-|F_{calc}|$, where $|F_{obs}|$ comes from the measured data and $|F_{calc}|$ and $\phi_{hkl}$ are calculated from the model:

$$\Delta\rho_{xyz} = \frac{1}{V} \cdot \sum_{hkl} \left\| F_{obs} \right| - \left| F_{calc} \right\| \cdot e^{i\phi_{hkl}} \cdot e^{-2\pi i \cdot (hx+ky+lz)}$$

eq. 22.18

Since difference maps require $|F_{calc}|$ and $\phi_{hkl}$ from a model, they are usually calculated after model refinement (Section 22.9). Difference maps are positive in regions where atoms are missing, and negative in regions where too many electrons (atoms) are present in the model. Hence, a difference map provides important hints where and in which direction the model needs to be corrected to better describe the electron density (Figure 22.28). A second difference map using $2|F_{obs}|-|F_{calc}|$ as the Fourier coefficients is also calculated. This map weights $F_{obs}$ twice as strongly as $F_{calc}$ to emphasize the measured diffraction data (Figure 22.28) and represents the overall electron density of the molecule.

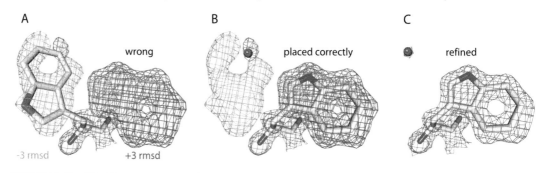

**FIGURE 22.28:** **Difference electron density.** A: The Trp side chain is in the wrong orientation as shown by the negative difference density (green). The correct placement would be in the positive difference density (red), which also shows a water molecule (red sphere) that occupies some of the space taken by the wrong Trp side chain. B: Corrected model before refinement. Another rotamer of Trp was chosen, and a water molecule was assigned to the isolated difference density on the top left (its binding partners are not shown). C: Maps after refinement. There is no difference density any more, just $2|F_{obs}|-|F_{calc}|$ density (blue).

The electron density map of a unit cell represents the average electron density of all unit cells that have contributed to the diffraction data. Thus, the electron density contains information on a huge number of molecules, which may differ slightly in their conformations. Sometimes, and especially at high resolutions of < 0.2 nm, the electron density maps show more than one conformation for part of a molecule. The most frequently encountered case is side chains in alternate conformations, which can be modeled by assigning different *occupancies* to the conformations (Figure 22.29). The occupancy is the fraction of each conformation throughout the crystal. For covalently bonded atoms, the sum of all occupancies must be unity, while non-covalently bound molecules, such as water, metal ions, cofactors, and ligands may have occupancies < 1 because they are not necessarily present in all unit cells.

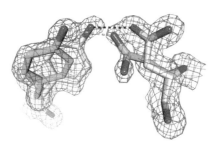

FIGURE 22.29:    **Alternate conformations.** These Tyr and Asp side chains have alternate conformations that depend on each other. The conformations in gray color form an H-bond (dashed black line), as does the pair shown in green. While the main-chain atoms are almost unchanged and have occupancies of unity, the side-chain atoms were duplicated and assigned occupancies of 0.7 (gray) and 0.3 (green).

The first electron density map does not usually allow building of a complete and error-free model. The initial model is refined against the diffraction data (Section 22.9) to obtain improved phases that are used to calculate new difference maps, which are then used in the next round of model building. This iterative cycling between model building and refinement is performed until the resulting electron density maps do not show novel features that should be included in the model.

## 22.9  MODEL REFINEMENT AND VALIDATION

Refinement modifies the coordinates and positional uncertainties (*B*-values; see later in this section) of the atoms in the model to match the diffraction data as closely as possible while maintaining a plausible geometry that obeys the covalent and non-covalent interactions discussed in Sections 15.2 and 15.3. These two aspects of refinement can be expressed by an energy function that is a weighted ($w_{xray}$) sum of geometric ($E_{geometry}$) and X-ray diffraction ($E_{xray}$) energy terms:

$$E_{tot} = E_{geometry} + w_{xray} \cdot E_{xray}$$

eq. 22.19

During the refinement $E_{tot}$ is minimized. The geometric energy includes the *restraints* on the structure: energy terms for bonds, angles, torsions, planes, chiralities, as well as ionic, dipolar, and van der Waals interactions (see eq. 18.2). The energy function can also include any other prior information that is independent of the diffraction data, such as non-crystallographic symmetry, if present. The purpose of including restraints is to decrease the number of parameters that are fitted during refinement, i.e. to increase the data-to-parameter ratio. For example, in the absence of geometric restraints, a typical crystal with 43% solvent content that diffracts to a resolution of 0.25 nm corresponds to a data-to-parameter ratio of unity, which is too small for finding the global minimum of the energy function. For high-resolution diffraction data with many reflections, the restraints are less important, and the X-ray energy term $E_{xray}$ may be weighted higher by increasing $w_{xray}$.

Traditionally, $E_{xray}$ was of the form

$$E_{xray} = \sum_{hkl} \left( |F_{obs}| - k \cdot |F_{calc}| \right)^2$$

eq. 22.20

where $|F_{obs}|$ and $|F_{calc}|$ are the structure factor amplitudes of the reflection (*hkl*), and *k* is a relative scale factor between $F_{obs}$ and $F_{calc}$. This form of $E_{xray}$ is the *least squares target*, which in older refinement algorithms was minimized using the conjugate gradient minimization procedure

(Section 18.2.2). The currently most successful formulation for the X-ray energy term is the *maximum likelihood* (ML) target,

$$E_{\mathrm{xray}} = \sum_{hkl} \frac{1}{\sigma_{ML}^2} \left( |F_{\mathrm{obs}}| - \langle |F_{\mathrm{obs}}| \rangle \right)^2 \qquad \text{eq. 22.21}$$

where $<|F_{\mathrm{obs}}|>$ is the expected value for $|F_{\mathrm{obs}}|$, and $\sigma_{\mathrm{ML}}^2$ is the variance. ML methods determine the probability of making a certain set of measurements (the diffraction dataset) given a molecular model, the diffraction data, and error estimates for the model and the data. The main problem is to estimate the unknown errors of the (incomplete) model. It can be shown that the probability distribution of a structure factor from the model, $F_{\mathrm{calc}}$, depends on the parameter $\sigma_A$, which can be estimated from the experimental data. $\sigma_A$ essentially represents the fraction of the structure factor that is expected to be correct, and its value ranges from zero (completely incorrect) to unity. Together with the known $|F_{\mathrm{obs}}|$ and the $|F_{\mathrm{calc}}|$ from the model, $\sigma_A$ allows one to calculate the expected values for $<|F_{\mathrm{obs}}|>$ and the corresponding variance $\sigma_{\mathrm{ML}}^2$ in eq. 22.21. The ML target has proven so successful that many crystallographic programs have it implemented.

As a result of the energy minimization during refinement, the difference between $F_{\mathrm{obs}}$ and $F_{\mathrm{calc}}$ is minimized. The crystallographic $R$-factor $R_{\mathrm{cryst}}$ is a measure of the residual magnitude of this difference:

$$R_{\mathrm{cryst}} = \frac{\sum_{hkl} \left| |F_{\mathrm{obs}}| - |F_{\mathrm{calc}}| \right|}{\sum_{hkl} |F_{\mathrm{obs}}|} \qquad \text{eq. 22.22}$$

The sum in the denominator of eq. 22.22 normalizes the sum of differences in the numerator, so $R_{\mathrm{cryst}}$ adopts values < 1. A perfect match between model ($F_{\mathrm{calc}}$) and data ($F_{\mathrm{obs}}$) would have $R_{\mathrm{cryst}}$ = 0. At the outset of refinement, $R_{\mathrm{cryst}}$ values of 0.5 (50%) are not unusual, but will eventually drop to values of < 25%, maybe < 15% for a well-refined structure. The final value of $R_{\mathrm{cryst}}$ depends on the resolution and quality of the data, and the stereochemical quality of the model built, but $R$-factors for macromolecular structures almost never approach zero because of measurement errors in the diffraction data and sub-optimal energy functions.

Sometimes, $R_{\mathrm{cryst}}$ is artificially reduced by refinement programs without any actual improvement of the molecular model. This corresponds to a decrease in the energy function by accumulation of systematic errors in the model, known as *overfitting*. The *free R-factor* $R_{\mathrm{free}}$ avoids this problem: $R_{\mathrm{free}}$ is calculated in the same manner as $R_{\mathrm{cryst}}$ (eq. 22.22), but from a test set of ideally > 1000 reflections that are never used in refinement. This test set is "free" from bias of the model for the diffraction data, which is introduced by any refinement procedure. $R_{\mathrm{free}}$ therefore represents a much more reliable quantity to judge the overall progress of refinement than $R_{\mathrm{cryst}}$, but at the cost of not using all available data. As a rule of thumb, $R_{\mathrm{free}}$ should roughly correspond to the resolution of the diffraction data in nm, e.g. a structure determined to 0.25 nm resolution should have $R_{\mathrm{free}} \leq 25\%$. Since a certain amount of overfitting is introduced by any refinement procedure, $R_{\mathrm{cryst}}$ will always be lower than $R_{\mathrm{free}}$ but should not differ by more than a few (3–8) percentage points.

At the beginning of the refinement the model is likely very incomplete and contains many errors, requiring methods such as rigid body refinement and simulated annealing (Section 18.2.4.3) that apply large changes to the coordinates. While simulated annealing usually moves secondary structure elements, loops, and side chains, rigid body refinement is a six-dimensional search ($3 \times$ rotation, $3 \times$ translation) for the best fit of whole domains to the diffraction data. Later refinements with a more correct model require smaller adjustments of atomic coordinates and *B-values*. As the electron density is a time and space average, the crystal structure also represents a time and space average. The *B*-value is used to approximate the observed "blurring" of the electron density due to the combined effect of atomic thermal vibrations (dynamic disorder) and small differences in the atomic positions between crystallographically "equivalent" molecules (static disorder). Both dynamic and static disorder are expressed together as the isotropic (spherical movement assumed) uncertainty $u$ of the center of gravity of an atom in any particular direction $x$:

$$B_{\mathrm{iso}} = 8\pi^2 \left\langle u_x^2 \right\rangle \qquad \text{eq. 22.23}$$

The radial displacement $u_r$, i.e. the total distance from the mean position, is calculated from the mean square radial displacement $\langle u_r^2 \rangle$, which is three times the mean square displacement $\langle u_x^2 \rangle$. Hence,

$$B_{\mathrm{iso}} = \frac{8}{3}\pi^2 \left\langle u_r^2 \right\rangle \qquad \text{eq. 22.24}$$

Regardless of which view is taken, displacement in a particular direction or radial, the larger the $B$-value, the less well-defined is the atom position. When the unit of $u$ is given in Å, the unit of the $B$-value is Å$^2$. The atomic resolution crystal structures of small molecules usually have tiny $B$-values around 1–2 Å$^2$, which reflect purely dynamic disorder, i.e. represent the uncertainty of the atom position as a harmonic oscillation. Macromolecules in crystals display much more static and dynamic disorder because they are packed less tightly than small molecules. The corresponding crystal structures have mean $B$-values of 20–80 Å$^2$. According to eq. 22.23, a $B$-value of 79 Å$^2$, which is not unusual for macromolecular structures, corresponds to an uncertainty in atom position of 1 Å (100 pm) in each direction. Because this uncertainty is on the order of the C-C bond length (154 pm), it is clear that such $B$-values must describe to a large extent static disorder.

Instrument parameters such as X-ray beam divergence, detector noise fluctuations, and vibrations of the crystal during data collection can also influence the $B$-value. Therefore, it is difficult to compare $B$-values across different crystal structures, but within a single crystal structure, the $B$-value is useful to compare different parts of the same molecule: a stretch of elevated $B$-values relative to the average could indicate high mobility of this region (Figure 22.30). Crystallographic B-values are not the best method to assess flexibility, but if several structures of the same molecule in different packing environments show the same trend, it may be assumed with some confidence that these regions are also flexible in solution.

If the resolution of the diffraction data is ≤ 0.15 nm, the isotropic $B$-value $B_{\mathrm{iso}}$ can replaced by an *anisotropic B-value* $B_{\mathrm{aniso}}$. Instead of describing the atomic uncertainty by a single parameter as identical in all directions, $B_{\mathrm{aniso}}$ uses a set of six parameters for the orientation and extent of an ellipsoid in which the atom may be found (Figure 22.30). Anisotropic refinement is also possible for lower resolution structures by assigning anisotropic properties to whole domains or secondary structure elements instead of individual atoms. This refinement technique is known as TLS (Translation, Libration, Screw) refinement.

When after a few cycles of building and refinement certain quality criteria such as reasonable R-factors and geometry are met, the iterative process may be aborted and the model considered final and ready for biological analysis. There are basically two classes of quality checks, those based purely on coordinates (does the model make sense?), and those that refer to diffraction data (how well does the model correspond with the diffraction data?). The parameters expressing the model quality can be global or local. Among the most important ways to test the quality of a crystal structure are $R_{\mathrm{free}}$, the Ramachandran plot, and electron density.

$R_{\mathrm{free}}$ is one of the most frequently used global quality indicators that refer to diffraction data. Other global quality indicators, such as the rmsd values for bonds, 1,3-angles, and torsion angles are not good descriptors for structure quality since they are used as targets in refinement: even a wrong crystal structure can have quite acceptable geometry. Provided that the main-chain torsion angles have not been restrained to the Ramachandran values during refinement, the Ramachandran plot (see Figure 16.16) is an invaluable tool to pinpoint residues in high-energy conformations, often located at enzyme active sites, metal binding sites, or as *cis*-peptides. If the electron density agrees with the peculiar geometry, the outlier is a genuine feature, not an error.

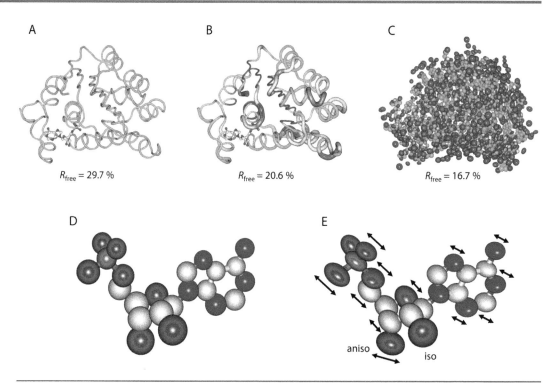

**FIGURE 22.30:** *B*-value refinement. The effect of isotropic and anisotropic *B*-value refinement is shown for a 0.12 nm resolution structure with the ligand AMP bound (PDB-ID 2gxq). A: No *B*-values are refined, every atom has the same *B*-value of 15.2 Å², and $R_{free}$ is almost 30%. B: Isotropic *B*-value refinement reduces $R_{free}$ to 21% and shows surface areas of potential flexibility (red and thick tubes). Coloring is from blue (low *B*-values) to red (highest *B*-values). C: Anisotropic *B*-value refinement reduces $R_{free}$ to about 17%. The atoms are shown as thermal ellipsoids. D: Isotropic *B*-values for the AMP ligand from (*B*) show higher apparent flexibility of the ribose part (larger spheres) than the adenine nucleobase. E: Anisotropic *B*-values for AMP from (C) reveals that the phosphate and adenine groups seem to oscillate as individual groups. The adenine atoms oscillate perpendicular to the plane of the nucleobase. Some atoms stay isotropic, for example the 2′-hydroxyl group (labeled *iso*), indicating that it has no preferred oscillation direction.

The best local quality indicator for a crystal structure is visual inspection of the electron density map. Deposition of diffraction data in the PDB is mandatory for new structures since 2008, so maps can be calculated for newer crystal structures and displayed in programs such as Coot (www2.mrc-lmb.cam.ac.uk/personal/pemsley/coot), CCP4mg (ccp4.ac.uk), VMD (www.ks.uiuc.edu/Research/vmd), and Pymol (pymol.org). Some programs, Coot for example, allow input of a PDB-ID and directly contact the PDB or the automated structure re-refinement database PDB_REDO (www.cmbi.ru.nl/pdb_redo), so electron density can be judged with minimal effort. The advantage of inspecting the electron density is that most researchers are interested in local aspects of a crystal structure (an active site, a ligand pose, an interaction surface), and this is directly represented by electron density. If, for example, the electron density for a ligand is weak compared to the electron density of the surrounding protein atoms, the ligand may have high mobility when bound to the protein, or an occupancy less than unity, or the whole ligand may simply be a modeling error.

## QUESTIONS

22.1 Moseley's law is an empirical law that relates the energy of the electronic transitions to the atomic number *Z*: for the $K_\alpha$ transitions, Moseley's law is

$$\lambda = \frac{121568.5}{\left(Z-1\right)^2} \, pm$$

Calculate the expected wavelengths for Cu and Ni.

22.2    The $K_\beta$ transitions for Cu and Ni are at 139 pm and 150 pm, respectively. Why can a thin Ni-foil be used to remove the $K_\beta$ radiation from the spectrum generated by a Cu anode?

22.3    Calculate the number of unit cells in a cube-shaped crystal of 0.1 mm edges for unit cell dimensions $a = b = c = 10$ nm. For some X-FEL experiments, crystals of only 100 unit cells along each edge are sufficient. What edge would a cube-shaped crystal with this number of unit cells have?

22.4    A protein of molecular mass 21 kDa has been crystallized in tetragonal space group $P4_12_12$ (8 symmetry operators) with cell dimensions $a = b = 98.5$ Å, $c = 104.2$ Å, $\alpha = \beta = \gamma = 90°$. How many molecules are possibly in the asymmetric unit and what is the solvent content in the most likely case? The solvent content is related to the Matthews parameter $V_M$ by the following rule of thumb:

$$\text{Solvent} = 1 - \frac{1.23}{V_M}$$

22.5    A protein of 89 amino acid residues has been crystallized in space group $P2_12_12_1$ (4 symmetry operators) with cell dimensions of $a = 28.4$ Å, $b = 36.3$ Å, and $c = 65.7$ Å. What is the likely number of molecules per asymmetric unit and the solvent content of this crystal? The average molecular mass for an amino acid residue is 112.5 Da.

## REFERENCES

Rupp, B. (2010) Biomolecular crystallography : principles, practice, and application to structural biology. New York, Garland Science.
· a comprehensive and up-to-date treatment of crystallography for the biological sciences

Derewenda, Z. S. (2010) Application of protein engineering to enhance crystallizability and improve crystal properties. *Acta Crystallographica Section D: Biological Crystallography* **66**(Pt 5): 604–615.
· overview of protein-engineering methods designed to enhance crystallizability

Holton, J. M. (2009) A beginner's guide to radiation damage. *J Synch Radiat* **16**(Pt 2): 133–142.
· review on radiation damage

Warkentin, M. and Thorne, R. E. (2007) A general method for hyperquenching protein crystals. *J Struct Funct Genomics* 8(4): 141–144.
· to reduce the concentration of cryo-protectant, a simple technique termed *hyperquenching* may be used

Matthews, B. W. (1968) Solvent content of protein crystals. *J Mol Biol* **33**: 491–497.
· introduction of the concept that $V_M$ and solvent content are distributed over a rather narrow range in protein crystals

Redecke, L., Nass, K., DePonte, D. P., White, T. A., Rehders, D., Barty, A., Stellato, F., Liang, M., Barends, T. R., Boutet, S., Williams, G. J., Messerschmidt, M., Seibert, M. M., Aquila, A., Arnlund, D., Bajt, S., Barth, T., Bogan, M. J., Caleman, C., Chao, T. C., Doak, R. B., Fleckenstein, H., Frank, M., Fromme, R., Galli, L., Grotjohann, I., Hunter, M. S., Johansson, L. C., Kassemeyer, S., Katona, G., Kirian, R. A., Koopmann, R., Kupitz, C., Lomb, L., Martin, A. V., Mogk, S., Neutze, R., Shoeman, R. L., Steinbrener, J., Timneanu, N., Wang, D., Weierstall, U., Zatsepin, N. A., Spence, J. C., Fromme, P., Schlichting, I., Duszenko, M., Betzel, C. and Chapman, H. N. (2013) Natively inhibited *Trypanosoma brucei* brucei cathepsin B structure determined by using an X-ray laser. *Science* **339**(6116): 227–230.
· the structure of the *Trypanosoma brucei* cysteine protease cathepsin B was determined using an XFEL from micro-crystals that spontaneously grew *in vivo*

Neutze, R., Wouts, R., van der Spoel, D., Weckert, E. and Hajdu, J. (2000) Potential for biomolecular imaging with femtosecond X-ray pulses. *Nature* **406**(6797): 752–757.
· suggests that single-molecule "crystallography" is in principle possible

Kern, J., Yachandra, V. K. and Yano, J. (2015) Metalloprotein structures at ambient conditions and in real-time: biological crystallography and spectroscopy using X-ray free electron lasers. *Curr Opin Struct Biol* **34**: 87–98.
· XFELs can help avoid the radiation damage typically observed at the metal centers of metalloproteins

Brünger, A. T. (1992) Free R value: a novel statistical quantity for assessing the accuracy of crystal structures. *Nature* **355**: 472–475.
· introduces the free R-factor for global model validation

Tickle, I. J., Laskowski, R. A. and Moss, D. S. (1998) $R_{free}$ and the $R_{free}$ ratio. I. Derivation of expected values of cross-validation residuals used in macromolecular least-squares refinement. *Acta Cryst D* **54**: 547–557.

Tickle, I. J., Laskowski, R. A. and Moss, D. S. (2000) $R_{free}$ and the $R_{free}$ ratio. II. Calculation Of the expected values and variances of cross-validation statistics in macromolecular least-squares refinement. *Acta Cryst D* **56**(Pt 4): 442–450.
· describes the relation of the crystallographic and free R-factors

Read, R. J. (1986) Improved fourier coefficients for maps using phases from partial structures with errors. *Acta Cryst A* **42**: 140–149.

Pannu, N. S. and Read, R. J. (1996) Improved structure refinement through maximum likelihood. *Acta Cryst A* **52**: 659–668.

Murshudov, G. N., Vagin, A. A. and Dodson, E. J. (1997) Refinement of macromolecular structures by the maximum-likelihood method. *Acta Cryst D* **53**(Pt 3): 240–255.
· the application of the maximum likelihood target for crystallographic refinement in modern software is described

Blakeley, M. P., Hasnain, S. S. and Antonyuk, S. V. (2015) Sub-atomic resolution X-ray crystallography and neutron crystallography: promise, challenges and potential. *IUCrJ* **2**(Pt 4): 464–474.
· a review on neutron crystallography

Glusker, J. P., Carrell, H. L., Kovalevsky, A. Y., Hanson, L., Fisher, S. Z., Mustyakimov, M., Mason, S., Forsyth, T. and Langan, P. (2010) Using neutron protein crystallography to understand enzyme mechanisms. *Acta Cryst D* **66**(Pt 11): 1257–1261.
· neutron crystallography reveals intermediates for aldose/ketose isomerization

Bourgeois, D. and Royant, A. (2005) Advances in kinetic protein crystallography. *Curr Opin Struct Biol* **15**(5): 538–547.
· review on the Laue method for kinetic crystallography and its complementation with other spectroscopic techniques

Winn, M. D., Isupov, M. N. and Murshudov, G. N. (2001) Use of TLS parameters to model anisotropic displacements in macromolecular refinement. *Acta Cryst D* **57**(Part 1): 122–133.
· TLS refinement for anisotropic B-value modeling at low resolution

## ONLINE RESOURCES

kinemage.biochem.duke.edu/teaching/BCH681/2013BCH681/elasticScattering
· Elastic scattering explained with relation to X-ray crystallography.

www.xtal.iqfr.csic.es/Cristalografia/index-en.html
· Comprehensive overview of crystallography with many figures, animations, and historic links.

www.doitpoms.ac.uk/tlplib/index.php
· The subsections on "Crystallography", "Reciprocal Lattice", and "X-ray Diffraction Techniques" contain many instructive animations.

www.ysbl.york.ac.uk/~cowtan
· Kevin Cowtan's tutorials on structure factors and crystallographic Fourier transforms.

# Imaging and Microscopy

I n the previous chapters, we have discussed methods that allow us to measure physical parameters for ensembles of isolated molecules (Chapter 19) or methods that provide structural information on biomolecules (Chapters 20–22). Microscopes are devices that generate magnified images of small objects such as cells and sub-cellular structures by using optical principles (light microscopy), streams of charged particles (particle microscopy), or scanning probes (scanning probe microscopy).

## 23.1 FLUORESCENCE MICROSCOPY

In fluorescence microscopy, a type of *light microscopy*, fluorescence emitted by molecules in the sample is used to generate images of tissues, cells, and sub-cellular structures. By providing spatial resolution, fluorescence microscopy allows localization of fluorescent(ly labeled) biomolecules within the cell. The cellular localization of biomolecules provides hints as to in which processes they are involved and what their particular function might be. Fluorescence microscopy can also be used to determine fluorescence parameters for single molecules, either in solution or immobilized on surfaces, which reveals the distribution of the respective parameter over all molecules (Section 23.1.7).

In a fluorescence microscope, the sample is illuminated by a light source, and the fluorescence from the object of interest is detected as a function of position. Depending on the illumination geometry, a small spot within the sample is illuminated while the rest remains in the dark, or a larger area is illuminated with uniform light intensity. The fluorescence emitted from the illuminated spot or area is collected by a microscope objective. Excitation light is rejected by a filter, and the emitted fluorescence light is passed on to a point or area detector. Typical point detectors are sensitive photomultipliers or avalanche photodiodes. To generate an image of the sample, the fluorescence of the sample is measured as a function of position, point by point. The intensity is detected by a point detector, one pixel after the other. Alternatively, and more commonly, the fluorescence intensity from an area is imaged onto an array detector, such as a CCD camera.

### 23.1.1 OPTICAL PRINCIPLES OF MICROSCOPY

Before we compare different types of fluorescence microscopy in more detail, we will first review the very basic principles of optics. Microscope objectives are nothing other than optical lenses, and to appreciate the differences between microscope types, we need to understand how these lenses focus and collect light.

#### 23.1.1.1 Focusing and Collecting Light by Optical Lenses

Light is collimated or focused by collecting lenses with convex surfaces. When illuminated by a parallel light bundle, the light is refracted at the air-glass interface when it enters the lens, and at the glass-air interface when it exits the lens. This refraction at the entry and exit interface leads to *focusing* of the light bundle into a small spot in the focal plane of the lens (Figure 23.1). The distance between the lens and the focal spot is called the *focal length f*. Focusing lenses are used to focus excitation light into small spots within the microscope sample.

Optical light paths are reversible, meaning that light emitted from a point source in the focal plane of a focusing lens will end up as a parallel light bundle on the opposite side of the lens, which is called *collimation*. This is the arrangement that allows us to use lenses to collect fluorescence light emitted from a certain spot within our microscope sample. The *light collection efficiency* of a lens is described by the *numerical aperture* NA (Figure 23.1). The numerical aperture is defined as

$$NA = n\sin\theta$$

eq. 23.1

$n$ is the refractive index of the medium between the point source and the lens, and $\theta$ is the half-angle of the outer rays of the collimated light with the optical axis. The larger $\theta$, the more light is collected by the lens. The angle $\theta$ depends on the focal length $f$ of the lens and its radius $r$ according to

$$\sin\theta = \frac{r}{f}$$

eq. 23.2

A higher collection efficiency, i.e. higher NA, is achieved either by using a lens with a shorter focal length $f$ and/or by a larger radius (eq. 23.2), or by using an immersion liquid with a refractive index $n$ higher than air, such as water or oil (eq. 23.1).

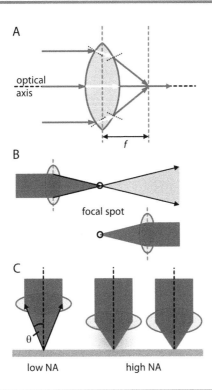

**FIGURE 23.1:    Focusing lenses and principles of optics**. A: A focusing lens. Only the outer and central rays of the parallel light bundle are depicted (red arrows). At the air-glass interface, the rays are refracted towards the optical axis (black dotted lines). The light path within the lens is indicated by white arrows. At the distal side of the lens, the exiting rays are again refracted towards the optical axis. As a consequence, all rays meet in the same spot on the optical axis in the focal plane. The distance between the center of the lens and the focal plane is called the focal distance $f$. B: Focusing and collimation of light. Top: Parallel light (red) is focused into a focal spot by a focusing lens. After the focal spot, the light bundle diverges again (light red). Bottom: Light from a point source in the focal plane of a collimating lens is collected by the lens and converted into a parallel light bundle. C: Numerical aperture. The numerical aperture NA depends on the highest angle $\theta$ of rays that are collected by a lens. The NA can be increased by placing a lens with a shorter focal length closer to the light source (right), or by using an immersion oil (yellow) with a higher refractive index than air (center).

Lenses can be combined to manipulate light beams in various ways. One example is the combination of two collecting lenses in a *telescopic lens system* to widen or narrow a light beam (Figure 23.2). A parallel light bundle of radius $r_1$ is focused by a first lens with a focal length $f_1$. A second lens is placed such that its distance from the focus equals its focal length $f_2$. The light diverges as it propagates from the focal spot, and is collimated and converted into a parallel light bundle after passing the second lens. The radius $r_2$ of the light bundle is determined by the ratio of the focal lengths:

$$\frac{r_1}{r_2} = \frac{f_1}{f_2} \qquad\qquad \text{eq. 23.3}$$

According to eq. 23.3, the beam is widened for $f_2 > f_1$, and the beam is compressed for $f_1 > f_2$.

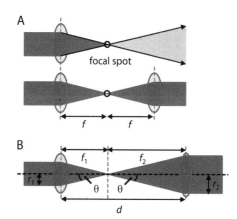

**FIGURE 23.2:** **Shaping beams by telescopic lens systems.** A: A parallel light bundle is focused by a lens and diverges after the focal spot (top). A second lens placed at its focal distance *f* from the focal point collects the light from the focal point and converts it back into a parallel light bundle (collimation, bottom). B: Two lenses with different focal lengths $f_1$ and $f_2$ are placed at a distance $d = f_1 + f_2$, such that their focal points coincide. Such a telescopic lens system increases or decreases the radius of the parallel beam. The ratio of the radii $r_1/r_2$ depends on the ratio of the focal lengths $f_1/f_2$ (eq. 23.3).

Special lens shapes create special effects. We have seen before that a convex lens focuses light into a focal spot. A *cylindrical lens*, on the other hand, is only curved in one dimension (Figure 23.3). Therefore, it focuses light only in or from the plane perpendicular to the cylinder axis. As a result, it generates a *focal line* instead of a focal point (Figure 23.3). We will see later that this property can be exploited to illuminate samples within defined areas (see Box 23.2).

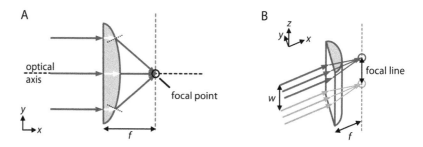

**FIGURE 23.3:** **A cylindrical lens focuses light only in one dimension.** A: Top view: A cylindrical lens is illuminated with a parallel light bundle that impinges on the planar side at a right angle and is not refracted. Light exiting the lens on the convex side is refracted and focused into a focal point. B: Light hitting the lens in different positions along the *z*-axis (red, orange) is focused into different points in the focal plane. A light bundle of the width *w* in the *z*-direction leads to a focal line extending from the focal point of one limiting beam to the focal point of the second limiting beam (red and orange circles).

### 23.1.1.2 Microscopes: How to Achieve Magnification with Optical Lenses

The size an object appears depends on the distance *d* of the eye from the object and the angle ε at which the outer rays reach the eye (Figure 23.4). A *magnification* of 1 is defined by the angle $ε_0$ at which we see an object that is located at a distance $d = 25$ cm from the eye. At larger distances ($ε < ε_0$), the same object appears smaller, at shorter distances ($ε > ε_0$), it appears larger. The magnification *m* is defined by the ratio of the corresponding angles

$$m = \frac{ε}{ε_0}$$

eq. 23.4

In principle, we can achieve increasing magnifications by moving our eyes closer and closer to the object. However, there is a limit to this approach. The eye cannot accomodate distances below ca. 10 cm. To reach higher magnifications, we therefore need optical instruments such as a magnifying glass or a microscope. A minimal *microscope* consists of two focusing lenses (Figure 23.4): the first lens, the objective, collects light from the object O and generates an intermediate image I that is observed through a second lens, the ocular.

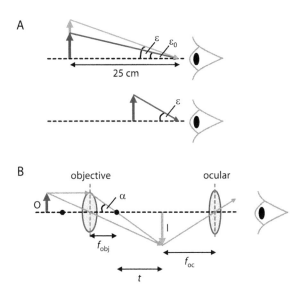

**FIGURE 23.4:** **Microscopes and magnification**. A: The size an object appears depends on the angle $\varepsilon$. A magnification of 1 is defined to correspond to an object at a distance of 25 cm from the eye, giving the angle $\varepsilon_0$ (top, red). Larger objects lead to larger angles $\varepsilon$ (orange), smaller objects lead to smaller angles. An object of the same size, at a shorter distance to the eye (bottom), is perceived larger. The magnification $m$ is defined by the ratio $\varepsilon/\varepsilon_0$. B: A microscope consists of an objective lens and an ocular lens. The objective generates an image I of the object that is viewed through the ocular. $f_{oc}$, $f_{obj}$: focal lengths of ocular and objective, $t$: tube length.

The ratio of the sizes of object O and intermediate image I defines the magnification $m_{obj}$ of the microscope objective:

$$m_{obj} = \frac{I}{O}$$

eq. 23.5

According to Figure 23.4, we can relate this ratio to the focal length $f_{obj}$ of the objective and the distance between the focus of the objective and the objective image, the tube length $t$. From the geometric relationship

$$\tan\alpha = \frac{O}{f_{obj}} = \frac{I}{t}$$

eq. 23.6

we obtain

$$m_{obj} = \frac{I}{O} = \frac{t}{f_{obj}}$$

eq. 23.7

The ocular also contributes to the magnification. The magnification $m_{oc}$ is given by the ratio of the intermediate image I and the image I' perceived by the eye:

$$m_{oc} = \frac{I'}{I}$$

eq. 23.8

The size of the perceived image I' depends on the distance $d$ of the eye from the ocular and on the angle $\alpha$ according to

$$\tan\alpha = \frac{I'}{d} = \frac{I}{f_{oc}}$$

eq. 23.9

By rearranging eq. 23.9, we obtain the magnification $m_{oc}$ as

$$m_{oc} = \frac{d}{f_{oc}}$$

eq. 23.10

The overall magnification $m$ of a microscope is the product of objective and ocular contributions:

$$m = m_{oc} \cdot m_{obj} = \frac{d}{f_{oc}} \cdot \frac{t}{f_{obj}}$$

eq. 23.11

In advanced microscopes, the image is not observed by eye, but is imaged onto a detector that determines the light intensity at each spot of the image.

### 23.1.1.3 The Diffraction Limit of Optical Resolution

Strictly speaking, a lens can only focus light into a focal point in the absence of diffraction. However, when a lens is illuminated, light is always diffracted. The diffraction widens the focal spot to an intensity distribution in the focal plane that is centrosymmetric around the optical axis, the *Fraunhofer diffraction pattern*. The Fraunhofer patterns of a uniformly illuminated centric aperture such as a lens consists of a bright circular region in the center of the focus, surrounded by a set of concentric rings of lower intensity (Figure 23.5). Such a pattern is also called a *diffraction-limited focus*. The diameter of the central bright area, the *Airy disc*, is defined by the position of the first minimum of the diffraction pattern. The angle $\theta_{min}$ at which the first minimum is observed depends on the wavelength $\lambda$ and the diameter $d$ of the lens:

$$\sin\theta_{min} = 1.22\frac{\lambda}{d}$$

eq. 23.12

The number 1.22 takes into account the circular shape of the lens aperture and the diffraction pattern. For small angles, we can approximate

$$\theta_{min} \approx 1.22\frac{\lambda}{d}$$

eq. 23.13

The diffraction-limited focus is the smallest possible focus we can achieve when focusing light by optical lenses. Conversely, when we collect light from a microscope sample and focus it onto a detector, a point source of light within the sample is detected as the same centrosymmetric intensity distribution, the *point-spread function* (PSF). The PSF is the intensity pattern obtained by imaging a point source of light through a microscope. Typically, the rings surrounding the Airy disc are not detected because of their low intensity. Hence, a point source of light within the sample is detected as a finite circular spot. The intensity distribution as a function of the $x$- and $y$-coordinates can be described by Gaussian distributions. This detected spot is called the *molecule detection function* (MDF; Figure 23.5).

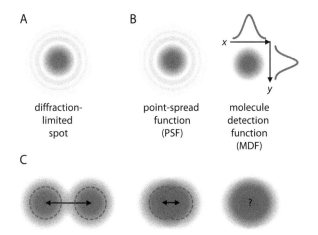

**FIGURE 23.5:    Diffraction, point-spread function, molecule detection function and optical resolution.** A: Due to diffraction, light cannot be focused into an infinitely small spot. The intensity distribution of a diffraction-limited spot is centrosymmetric around the optical axis. The central disc of high intensity, the Airy disc, is surrounded by concentric rings of decreasing intensity. B: A point source of light is imaged by a lens according to the point-spread function (PSF), with the same intensity pattern as a diffraction-limited spot (left). The intensity of the surrounding rings is often too low to be detected. The diffraction-limited image from a point source of light is therefore a centrosymmetric intensity distribution corresponding to the Airy disc (right). This intensity distribution is called the molecule detection function (MDF). C: The size of the Airy disc determines the resolution of a microscope. Two point sources of light can be detected as separate if the distance between their centers is larger than the radius of the Airy disc.

The *resolution* of a microscope is a measure of the minimal distance $d$ between two points that are still detected as distinct. Ernst Abbe found the relationship

$$d = \frac{\lambda}{2n\sin\theta} = \frac{\lambda}{2NA} \qquad \text{eq. 23.14}$$

for the resolution of light microscopes. For air objectives ($n = 1$), we obtain the well-known limiting resolution

$$d < \frac{\lambda}{2} \qquad \text{eq. 23.15}$$

Microscopy using light in the visible range (400–800 nm) therefore cannot achieve resolutions below 200–400 nm.

For microscopy of light-emitting samples, the resolution is determined by the size of the Airy disc of the PSF or MDF: two point sources that are far apart in the object are detected as two individual diffraction patterns (Figure 23.5). When the points come closer, the MDFs will start to overlap. The shortest distance at which the points can still be distinguished is reached when the first minimum of the pattern from one point source coincides with the central maximum from the other point source. In other words, the distance between the centers of the Airy discs must be larger than the disc diameter. This condition is called the *Rayleigh criterion*. The diameter $d$ of the Airy disc can be approximated as

$$d \approx 0.61\frac{\lambda}{n\sin\theta} = 0.61\frac{\lambda}{NA} \qquad \text{eq. 23.16}$$

For light in the visible range (400–800 nm) and standard oil immersion objectives with numerical apertures of 1.4, the diameter of the Airy disc and thus the diffraction-limited resolution is ca. 200–400 nm (in the xy plane).

Diffraction only becomes relevant when light propagates over long distances. Near-field methods avoid blurring because of diffraction by obtaining the signal near the emitting molecule. In *near-field scanning optical microscopy*, a resolution of 20–50 nm is achieved by scanning a sharp tip across the surface to probe the sample. However, because of their restriction near to surfaces, these methods are of limited value for high-resolution microscopy on biological specimen. We will see later that far-field microscopy can also be taken to higher resolutions beyond the diffraction limit by introducing modifications that compensate for diffraction effects (see Section 23.1.8).

## 23.1.2 WIDE-FIELD FLUORESCENCE MICROSCOPY

Fluorescence microscopy uses fluorescence excitation and detects fluorescence emitted from the sample as a function of position. In epi-fluorescence microscopy, the objective is used to illuminate the sample with light and to collect the fluorescence emission. Standard fluorescence microscopes illuminate the sample from above, whereas *inverted microscopes* illuminate the sample from below, and allow easy access for additional manipulation from above. In *wide-field microscopy*, a parallel light bundle is focused into the back-focal plane of the objective (Figure 23.6). As a result, the light is again parallel on the distal side of the objective lens, allowing homogenous illumination of a defined area on the microscope slide. Fluorophores within this illuminated area are excited and emit fluorescence in all three directions of space. The fraction emitted towards the objective is collected by the objective lens, passed through a filter (Box 23.1) to reject any scattered excitation light, and imaged onto a detector.

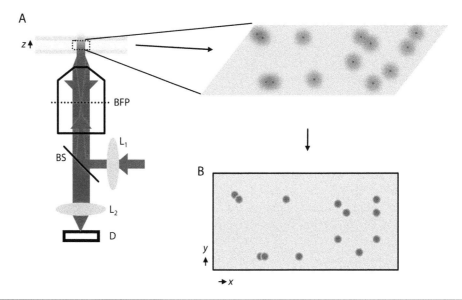

FIGURE 23.6: **Wide-field microscopy.** A: Excitation light (blue) is focused into the back-focal plane (BFP) of the objective by lens L1. After passing the objective lens, a parallel light bundle is generated that illuminates an area of the sample (inset). Fluorescent light emitted from the sample (red) is collimated by the objective, passes the beam splitter (BS) and is focused onto a detector D by a second lens (L₂). B: On the detector, each fluorescent emitter in the sample generates a point-spread function. In a wide-field microscope, all fluorophores in the light path along the z-axis are illuminated and detected. The image is a projection of all these emitters into one plane.

**BOX 23.1: OPTICAL FILTERS.**

Optical filters enable the selection of defined spectral regions of light, and are used to separate excitation light from red-shifted emission, or to separate fluorescence from different emitters with different fluorescence wavelengths (Figure 23.7). A longpass (LP) filter transmits light above a cut-off wavelength and rejects all wavelengths below. These filters are named LP, followed by a number that corresponds to the wavelength at which the light is transmitted (cut-off wavelength). Long-pass filters with a very sharp transition between transparency and absorption are called edge filters. A short-pass (SP) filter is transparent for light below a threshold wavelength, but rejects light above this wavelength. A bandpass (BP) filter is transparent for a certain range of wavelengths, but rejects all wavelengths above and below. These filters are named BP, followed by the lower and higher cut-off wavelength. A beam splitter or dichroic mirror (DM) is an optical element that is transparent for light above a certain wavelength. Light of lower wavelengths is reflected.

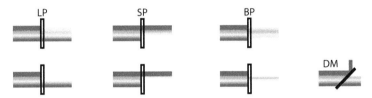

**FIGURE 23.7:** Optical filters. LP: longpass filter, SP: short path filter, BP: bandpass filter, DM: dichroic mirror. The top row shows filters with a gradual cut-off, the bottom row shows edge filters with sharp cut-offs.

Wide-field microscopy does not allow selection of a certain plane within the sample: the parallel light used for excitation in wide-field fluorescence microscopy penetrates through all layers of the sample. All molecules in all planes of the sample are excited, and their emission is focused by the objective and imaged onto the detector. The image on the detector is therefore a superposition of images of all individual planes in the sample: the position of the molecule in the $z$-direction that causes one particular spot is not encoded in the image.

### 23.1.3 CONFOCAL SCANNING MICROSCOPY

By *confocal microscopy*, optical selection of a specific layer of the sample becomes possible, and images of individual planes of the sample can be generated. In confocal microscopy, a parallel bundle of light is focused into the sample by an objective (Figure 23.8). The diffraction-limited focus in three dimensions is an ellipsoid, with its long axis along the beam in the $z$-direction, and a shorter axis in the $xy$-plane. Within this ellipsoid, the light intensity $I(x,y,z)$ decays exponentially in the $x$- and $y$-directions and in the $z$-direction from its center according to eq. 23.17:

$$I(x,y,z) = I_0 \exp\left( -\frac{2(x^2 + y^2)}{\omega_1^2} - \frac{2z^2}{\omega_2^2} \right)$$

eq. 23.17

$I_0$ is the light intensity in the center, $\omega_1$ is the distance from the center where the intensity has decayed to $I_0/e$ in the $x$- and $y$-directions, and $\omega_2$ is the corresponding distance from the center in the $z$-direction. The volume defined by the surface where the intensity has decayed to $I_0/e$ is called the *confocal volume.* Molecules that reside in the confocal volume experience high light intensity, are excited, and emit fluorescence. Fluorescence from the confocal volume is collected by the objective, focused into a small centric aperture, a *pinhole* (Figure 23.8), and eventually detected by a point detector such as an avalanche photodiode. The pinhole is the central element for the selection of a defined plane within the sample. Light emitted from the plane of interest is focused into the plane of the pinhole, and can therefore pass the pinhole completely. Light emitted from above or below is focused into different planes, is rejected by the pinhole and does not reach the detector (Figure 23.8).

To generate an image of the entire plane, we now have to repeat the detection process at a different spot of the sample, piecing together the fluorescence intensity over the area of interest in the sample point by point and line by line. This technique is called *scanning confocal microscopy* (SCM). Either the confocal volume is scanned across the sample (beam scanning), or the sample is scanned relative to the fixed objective and confocal volume (sample scanning).

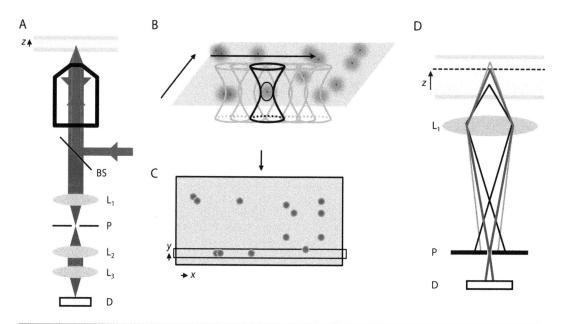

**FIGURE 23.8:** **(Scanning) confocal microscopy.** A: In a confocal microscope, parallel excitation light (blue) is coupled into a high numerical aperture objective and focused into a small volume, the *confocal volume.* Fluorophores within the volume are excited, and their fluorescence (red) is collected by the objective. The fluorescence is focused by lens $L_1$. A pinhole (P, a small circular aperture) is centered in the focal plane of $L_1$ and of the collecting lens $L_2$. The pinhole is the central element for optical selection of a specific plane within the sample. $L_3$ focuses the fluorescence light onto the point detector (D). B: To construct the image, fluorescence from each point has to be measured, one after the other. To achieve this, the focus (double cone for the beam waist, the confocal volume is indicated by the ellipse in the center) is scanned over the sample point by point and line by line. C: The scanned image shows an Airy disc for each emitter within the sample. D: optical selection for a plane within the sample. Light emitted from the plane of interest (black broken line; red) is focused into the aperture of the pinhole P by $L_1$. Light from planes above (gray) or below (black) is focused into different planes, and will be rejected by the pinhole. Only light from the plane of interest reaches the detector D.

Scanning confocal microscopy can also be used to generate three-dimensional images of specimen. To achieve this, imaging is performed layer by layer, and the optically selected focal plane is changed incrementally between two subsequent images. The three-dimensional image is then reconstructed from the set of two-dimensional images from different focal planes.

## 23.1.4 TOTAL INTERNAL REFLECTION MICROSCOPY

Wide-field microscopy allows simultaneous imaging of the entire surface area, but does not select a specific plane in the $z$-direction. Using confocal microscopy, we can selectively image a layer of the sample, but we need to generate images point by point, which is time-consuming. *Total internal reflection fluorescence* (TIRF) *microscopy* combines the advantage of simultaneously detecting fluorescence from a larger area with selection of a certain layer of the sample (Figure 23.9). In TIRF microscopy, the incident laser beam is coupled into the objective off the optical axis, such that it will impinge at a defined angle on the interface between the slide surface and the sample (Figure 23.9). The refractive index $n_2$ of the sample (aqueous solution) is smaller than the refractive index $n_1$ of the glass or quartz slide. The excitation light therefore will be refracted away from the optical axis according to *Snellius' law* of refraction:

$$n_1 \sin\theta_1 = n_2 \sin\theta_2 \qquad \text{eq. 23.18}$$

$n_1$ is the refractive index of the microscope slide, $n_2$ the refractive index of the sample, $\theta_1$ and $\theta_2$ are the incident and exiting angles of the beam with the optical axis. Above a certain angle $\theta$, the *critical angle* $\theta_c$, total internal reflection will occur. From Snellius' law, we can calculate the critical angle $\theta_c$ as

$$\theta_c = \sin^{-1}\left(\frac{n_2}{n_1}\right) \qquad \text{eq. 23.19}$$

The beam will then be reflected at the glass-water interface, and does not penetrate into the sample (Figure 23.9). Under these conditions, an *evanescent field* is generated on the distal side of the interface, within the sample. The intensity $I(z)$ of this evanescent field decays exponentially with increasing distance from the interface (see SPR in Section 26.3.1)

$$I(z) = I(0) \cdot e^{-z/d} \qquad \text{eq. 23.20}$$

where $I(0)$ is the intensity at the interface, $z$ is the distance from the interface, and $d$ is the decay parameter that depends on the wavelength $\lambda$ of the excitation light, the refractive indices $n_1$ and $n_2$ of the slide and the sample, and on the angle $\theta$ at which the excitation light impinges on the glass-water interface:

$$d = \frac{\lambda_0}{4\pi} \cdot \frac{1}{\sqrt{n_1^2 \sin^2\theta - n_2^2}} \qquad \text{eq. 23.21}$$

Fluorophores within a few 100 nm from the interface are within the range of the evanescent field and are excited, whereas fluorophores further away from the interface are not excited. TIRF microscopy therefore allows selective excitation of molecules on cell surfaces, or of molecules immobilized on the slide surface.

Instead of using an objective (objective-type TIRF microscopy), the conditions for total internal reflection can also be generated using prisms (prism-type TIRF microscopy; Figure 23.9). In both cases, the fluorescence emission is collected by an objective and imaged onto a CCD camera.

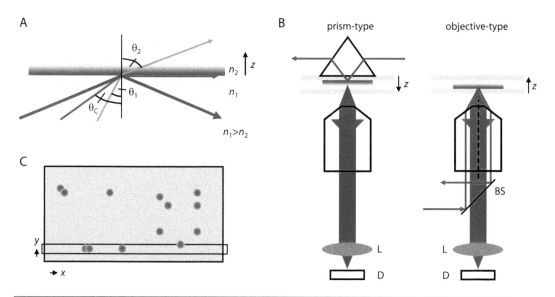

**FIGURE 23.9:** **TIRF microscopy.** A: Total internal reflection. When light passes the interface from an optically dense medium (glass/quartz) with a refractive index $n_1$ to a medium with lower refractive index ($n_2 < n_1$; water), it is refracted away from the optical axis ($\theta_2 > \theta_1$) according to Snellius' law (eq. 23.18; red). At incident angles above a critical angle $\theta_c$, the light will not enter the second medium, but is reflected at the interface (blue). At the critical angle $\theta_c$, the light beam travels along the interface (gray). Total internal reflection leads to an evanescent field on the distal side of the interface (blue gradient), and to light intensity close to the surface that decays exponentially along the $z$-axis. As a consequence, fluorophores at the surface will be excited and emit light. B: Prism-type and objective-type TIRF microscopes. In the prism-type TIRF microscope, the critical angle is achieved by passing the excitation light through a prism. The evanescent field is generated at the top interface (blue gradient). In objective-type TIRF microscopy, it is achieved by coupling the excitation light into the objective off the optical axis. In this case, the evanescent field is generated at the bottom interface (blue gradient). Fluorescence from molecules at the surface is collected by an objective and imaged onto an area detector D (BS: beam splitter, L: lens). C: The image shows an Airy disc for each emitter on the surface.

Sectioning of the sample can also be achieved by *light-sheet microscopy*, where a section of the sample, perpendicular to the direction of observation or inclined with respect to the optical axis, is selectively illuminated (Box 23.2). Light-sheet microscopy is not limited to selective excitation near surfaces and avoids artifacts from surface effects.

---

### BOX 23.2: OPTICAL SECTIONING OF SAMPLES BY LIGHT-SHEET ILLUMINATION.

In light-sheet microscopy, the excitation light is passed through a cylindrical lens. This lens only focuses the light in one direction (see Section 23.1.1.1), leading to the generation of a thin light sheet that illuminates a thin area within the sample (Figure 23.10). In *highly-inclined and laminated optical sheet microscopy* (HILO; Figure 23.10), a variation of TIRF microscopy, the excitation light is coupled into the objective off the optical axis, but such that it impinges on the interface between the glass slide and the sample at an angle below the critical angle for total internal reflection. The light is refracted away from the optical axis and passes through the sample at an acute angle with respect to the surface. Light-sheet microscopy does not reach the signal-to-noise ratio of TIRF microscopy, but offers the advantage of selectively observing molecules in a narrow plane away from the microscope slide surface. The illumination of a limited area within the sample reduces photodamage. Tokunaga *et al.* (2008) *Nat. Methods* 5(2): 159–161.

*(Continued)*

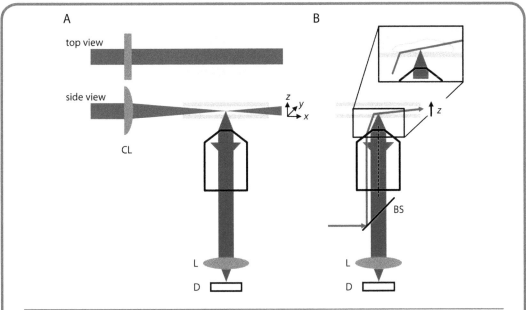

**FIGURE 23.10:** **Light-sheet microscopy.** A: Light is focused in one dimension only by a cylindrical lens (CL), leading to illumination of the sample by a sheet of light in the *xy*-plane. Fluorescent molecules within this plane are excited and their fluorescence is imaged onto a detector (D) by a lens (L). B: In highly-inclined and laminated optical sheet microscopy (HILO), the light sheet is generated at an acute angle with the surface by illuminating the glass-water interface at an angle larger than the critical angle for total internal reflection. Only an inclined plane within the specimen (inset, gray) is observed. BS: beam splitter, L: lens, D: detector.

Wide-field, confocal scanning, and TIRF microscopes can be equipped with multiple excitation lasers to excite different fluorophores and to image two or more differentially labeled molecules simultaneously. Images obtained from the same specimen using different fluorophores can then be superimposed. Appearance of fluorescence from different molecules in the same position within a cell is often used as evidence for colocalization and a possible functional interaction. In *FRET microscopy* (see Section 19.5.10), the donor fluorophore is excited, and the emission from donor and acceptor fluorophore is separated using a dichroic mirror. Focusing the donor and acceptor fluorescence separately, but next to each other on the CCD chip of the camera provides one image in terms of donor fluorescence, and one in terms of acceptor fluorescence. The FRET efficiency can be calculated pixel by pixel as a function of position, providing information on colocalization and interaction between two fluorescently labeled entities. We will come back to FRET microscopy in Section 23.1.7.6.

### 23.1.5 Fluorescence Lifetime Imaging Microscopy

Wide-field, scanning confocal and TIRF microscopy yield images of the sample, i.e. fluorescence intensity as a function of position. Sometimes it is important to detect and distinguish two different fluorophores, or to detect different environments of the same fluorophore. If a pulsed excitation source is used, the fluorescence lifetime at each spot of the sample can be measured (see Section 19.5.9). Different fluorophores can then be distinguished by their lifetimes. *Fluorescence lifetime imaging microscopy* (FLIM) requires spatial resolution to generate the image, combined with temporal resolution on the nanosecond time scale at each spot to measure the fluorescence decay. Such measurements can be achieved by using a *streak camera* as the detector. Just like any conventional photodetector, a streak camera converts incoming photons into secondary (photo)electrons. Streak cameras can imprint information on the arrival time of the photon on the photoelectron. This is achieved by passing the photoelectrons through an electric field whose intensity varies with time (Figure 23.11). Photoelectrons are deflected when they pass this electric field, and the extent of the deflection depends on the magnitude of the electric field at this time point. If the magnitude of the electric field increases with time, electrons that arrive earlier will then be deflected less than electrons that pass the electric field later. In other words, electrons are deflected proportionally to their arrival time. The time-dependent deflection in one direction thus introduces a second axis that reflects the arrival time of the photon. The electrons then pass a microchannel plate (Section 19.5.9; Figure 26.11 in Section 26.1.3) for amplification of the signal, and are detected by a phosphor screen. The light from the phosphor screen is imaged onto a CCD camera. One round of measurements thus yields one line of the image. Line by line, the complete image is then build up, yielding a three-dimensional dataset with $x$- and $y$-axes and an additional t-axis with the fluorescence decay at each $x,y$-coordinate. The fluorescence lifetime can be calculated from the fluorescence decay along the t-axis. The whole dataset is represented as a color-coded image in which the color corresponds to the lifetime at a particular spot (Figure 23.11).

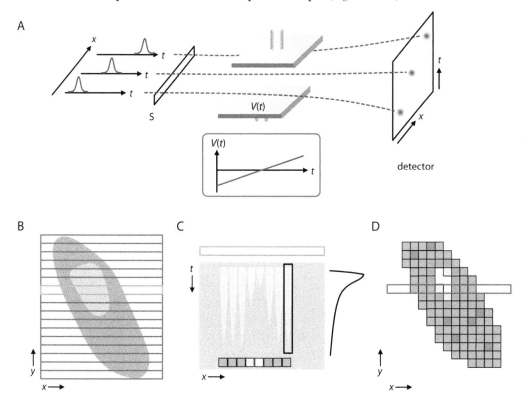

**FIGURE 23.11:   Streak camera and fluorescence lifetime imaging.** A: Principle of a streak camera. Photons arriving at different positions along the $x$-axis and at different time points $t$ pass the entry slit of the streak camera and generate photoelectrons that are accelerated towards the detector. Along the way, they pass a set of deflection plates. A time-dependent voltage leads to a time-dependent deflection of the electrons. The electrons arrive at the detector at their $x$-position and are spread along a second axis according to their arrival times $t$. B: A cell (nucleus in light gray, cytoplasm in dark gray) is imaged line by line. C: Imaging of the highlighted line gives a streak image with photon arrival times along the t-axis and their position along the $x$-axis. From the time-dependent photon count, fluorescence decays and fluorescence lifetimes are calculated at each spot in the line. The lifetime is color-coded (green-orange-yellow-white from large to small lifetimes). D: The lifetime image is constructed from the lifetimes measured at each spot per line for all lines.

Since fluorescence lifetimes are sensitive to pH, to the local oxygen concentration, or to the presence of fluorescence acceptors, FLIM can provide spatially resolved information on these parameters within cells.

### 23.1.6 Fluorescence (Cross-)Correlation Spectroscopy

We have seen in Section 23.1.3 that confocal microscopy can be used to construct fluorescence images by scanning the confocal volume across the sample. Instead of measuring fluorescence intensities as a function of position, we can also measure fluorescence from the confocal volume as a function of time. The measured fluorescence will fluctuate around its mean value because of diffusion (see Section 4.1), and from the fluctuation speed, we can gain information on diffusion velocities.

### 23.1.6.1 Fluorescence Correlation Spectroscopy

The confocal volume is tiny (on the order of femtoliter), and at any given time we will have a limited number of fluorescent molecules within the focus. At sufficiently low concentrations, the average number of fluorescent molecules is on the order of ten. Because of Brownian motion (see Section 4.1), fluorescent molecules constantly enter and leave the confocal volume, leading to fluctuations in the average number of molecules within the confocal volume, and a fluctuation in the fluorescence intensity. At moderate concentrations of the fluorescent molecule, the fluorescence will therefore measurably fluctuate over time around the mean intensity. The fluctuation $\delta I(t)$ is the difference between the fluorescence intensity $I(t)$ and the mean fluorescence intensity $\langle I \rangle$ (averaged over time):

$$\delta I(t) = I(t) - \langle I \rangle \qquad \text{eq. 23.22}$$

The *autocorrelation function* $G(\tau)$ (see Section 21.1.2) relates the fluorescence intensity at time $t$, $I(t)$, to the fluorescence at a later time $t+\tau$, $I(t+\tau)$:

$$G(\tau) = \frac{\langle I(t) \cdot I(t+\tau) \rangle}{\langle I \rangle^2} \qquad \text{eq. 23.23}$$

The brackets indicate that we calculate the mean of the respective parameters over time. We express the intensity at each point in time as the mean intensity $\langle I(t) \rangle$ plus the fluctuation $\delta I(t)$, and obtain

$$G(\tau) = \frac{\langle (\langle I \rangle + \delta I(t)) \cdot (\langle I \rangle + \delta I(t+\tau)) \rangle}{\langle I \rangle^2} \qquad \text{eq. 23.24}$$

We can factorize the numerator to

$$G(\tau) = \frac{\langle \langle I \rangle \langle I \rangle + \langle I \rangle \cdot \langle \delta I(t) \rangle + \langle I \rangle \cdot \langle \delta I(t+\tau) \rangle + \langle \delta I(t) \cdot \delta I(t+\tau) \rangle \rangle}{\langle I \rangle^2} \qquad \text{eq. 23.25}$$

The means of the fluctuations, $\langle \delta I(t) \rangle$ and $\langle \delta I(t+\tau) \rangle$, over time are zero, and their products with $\langle I \rangle$ are also zero. The autocorrelation function $G(\tau)$ thus becomes

$$G(\tau) = 1 + \frac{\langle \delta I(t) \cdot \delta I(t+\tau) \rangle}{\langle I \rangle^2} \qquad \text{eq. 23.26}$$

The autocorrelation curve decays from its initial value to 1. It is often plotted with a logarithmic $\tau$-axis, resulting in a sigmoidal shape (Figure 23.12). The inflection point of the curve marks the average diffusion time $\tau_D$ of the molecule through the confocal volume. The amplitude of the autocorrelation curve is inversely proportional to the average number $N$ of molecules in the confocal volume.

The second term in eq. 23.26 summarizes the time-dependent contributions to the autocorrelation function. This time-dependent part decays from the starting value to zero. The magnitude of the autocorrelation at time $\tau$ reflects the decreasing probability of detecting a photon at increasing times time $\tau$ after detection of a photon at $\tau = 0$.

The time the molecule spends within the confocal volume depends on its translational diffusion coefficient $D_t$ (Section 4.1) and on the geometry of the confocal volume. We have seen before that the light intensity can be described according to eq. 23.17, and the confocal volume is an ellipsoid with the short half axis $\omega_1$ in the $x$- and $y$-directions, and long half axis $\omega_2$ along the $z$-direction (Figure 23.12):

$$I(x,y,z) = I_0 \exp\left(-\frac{2(x^2+y^2)}{\omega_1^2} - \frac{2z^2}{\omega_2^2}\right) \qquad \text{eq. 23.17}$$

Taking into account diffusion as the (only) source of the measured fluctuations, the autocorrelation curve can be described by

$$G(\tau) = 1 + \frac{1}{N}\left(\frac{1}{1+4D_t\tau/\omega_1^2}\right)\left(\frac{1}{1+4D_t\tau/\omega_2^2}\right)^{0.5}$$

eq. 23.27

$N$ is the average number of molecules in the confocal volume, $D_t$ is the diffusion coefficient, and $\omega_1$ and $\omega_2$ are the distances from the optical axes in the $x/y$- and $z$-directions at which the light intensity has decayed by $1/e$ from the maximum. Note that the $\tau$ remaining in eq. 23.27 is the autocorrelation time, not the diffusion time! Thus, from autocorrelation curves we obtain diffusion times (inflection point) and the number of molecules within the confocal volume (inverse of the amplitude). We can convert these values to diffusion coefficients and concentrations only if we know the dimensions of the confocal volume. To determine diffusion coefficients and concentrations from autocorrelation curves, we describe the experimentally obtained curves by eq. 23.27, with $D_t$ and $N$ as fitting parameters and $\omega_1$ and $\omega_2$ as fixed parameters that define the dimensions of the confocal volume.

Diffusion of molecules into and out of the confocal volume is not the only process that leads to fluctuations of the fluorescence intensity, though. Fluctuations can also be caused by the photophysical properties of the fluorophore, such as transitions to the triplet state, or by chemical reactions and binding events. If polarized light is used for excitation, rotational motion of the fluorophores also causes fluorescence fluctuations. Any process that causes fluctuations of the fluorescence intensity can be studied by *fluorescence correlation spectroscopy* (FCS) (Figure 23.12).

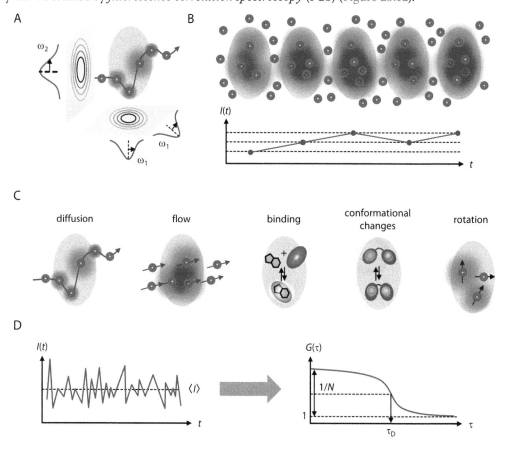

FIGURE 23.12:   **Fluorescence correlation spectroscopy (FCS).** A: Dimensions of the confocal volume. The ellipsoidal volume can be described in terms of the half-axes $\omega_1$ (in the $x$- and $y$-directions) and $\omega_2$ (in the $z$-direction). B: Principle of FCS: diffusion of molecules into and out of the confocal volume leads to fluctuations of the measured fluorescence. C: Processes that lead to fluctuation: diffusion and flow alter the number of molecules within the confocal volume and cause thereby fluorescence fluctuations. Binding events or conformational changes can lead to changes in fluorophore brightness and cause fluorescence fluctuations. If polarized light is used for excitation, the rotation of molecules within the confocal volume also leads to fluorescence fluctuations. D: Measurement (left) and autocorrelation curve (right). The amplitude of the sigmoidal autocorrelation curve $G(\tau)$ is proportional to the inverse average number of molecules ($1/N$) within the confocal volume. The inflection point corresponds to the diffusion time $\tau_D$.

In principle, FCS measurements require only one detector that measures fluorescence from the confocal volume as a function of time. However, any noise from this detector will also correlate with itself and contribute to the autocorrelation curve. To avoid the correlation of noise, FCS measurements are often performed by splitting the fluorescence signal and focusing each half onto a separate detector. The correlation curve obtained from the two halves of the signal then only contains information from correlated fluctuations, whereas the noise from the two detectors is uncorrelated and cancels.

### 23.1.6.2 FCS to Monitor Binding Events

FCS can be used to monitor (high affinity) ligand binding by changes in diffusion time. Typically, increasing concentrations of the higher molecular mass binding partner will be added to a fluorescently-labeled smaller ligand. The short diffusion time of the free ligand increases when it binds to the macromolecule. This increase in diffusion time serves as a probe for binding. In FCS, typical ligand concentrations are in the low nanomolar range, and $K_d$ values in the high picomolar to low nanomolar range can be determined reliably. There are only few other techniques that can quantify such high-affinity interactions.

When we measure binding by FCS, two processes occur during our measurement: diffusion leads to fluorescence fluctuations, and binding and dissociation alter the diffusion coefficient. Binding and dissociation can be fast or slow on the time scale of the diffusion. If binding and dissociation are slower than diffusion ($\tau_{\mathrm{binding}} \gg \tau_{\mathrm{D}}$), no binding or dissociation occurs during the passage of the molecules through the confocal volume: molecules will diffuse in and out in their free form, with the diffusion coefficient $D_{\mathrm{free}}$, or they enter and exit in the bound state and diffuse more slowly with $D_{\mathrm{bound}}$. The autocorrelation curve therefore shows two inflection points, one corresponds to the diffusion of the free ligand, the second corresponds to the slower diffusion of the bound ligand (Figure 23.13). The autocorrelation can be described in terms of the two diffusion coefficients $D_t^{\mathrm{free}}$ and $D_t^{\mathrm{bound}}$ and the fractions of free and bound ligand, $\alpha$ and 1-$\alpha$, respectively:

$$G(\tau) = 1 + \frac{1}{N}\left(\alpha X_{free} + (1-\alpha) X_{bound}\right) \qquad \text{eq. 23.28}$$

with

$$X_{free} = \left(\frac{1}{1+4D_t^{free}\tau/\omega_1^2}\right)\left(\frac{1}{1+4D_t^{free}\tau/\omega_2^2}\right)^{0.5} \qquad \text{eq. 23.29}$$

and

$$X_{bound} = \left(\frac{1}{1+4D_t^{bound}\tau/\omega_1^2}\right)\left(\frac{1}{1+4D_t^{bound}\tau/\omega_2^2}\right)^{0.5} \qquad \text{eq. 23.30}$$

The diffusion coefficients for the free and bound ligand can be determined in reference measurements. The fractions bound and free for each curve can then be determined from the amplitudes of the corresponding component of the autocorrelation curve (Figure 23.13). Plotting the fraction bound as a function of the concentration of the added macromolecule yields a binding curve that can then be analyzed for the $K_d$ value (eq. 19.70 in Section 19.5.5.1).

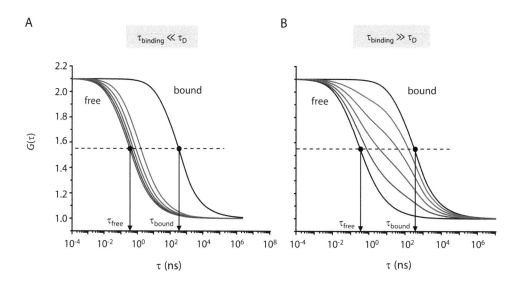

**FIGURE 23.13: FCS to monitor binding and to determine $K_d$ values.** A: Autocorrelation curves for a rapid equilibrium ($\tau_{binding} \ll \tau_D$) from 0% binding (free ligand, black) over 20% (dark red), 40% (red), 60% (violet), 80% (blue) to 100% bound ligand (black). From the autocorrelation curves, a mean diffusion time $\tau_D$ can be determined. This mean diffusion time represents the population-weighted average of the diffusion times $\tau_{free}$ and $\tau_{bound}$ for free and bound ligand. From the mean diffusion time, the fraction free and bound ligand can be determined. B: Autocorrelation curves for a slow equilibrium ($\tau_{binding} \gg \tau_D$). The autocorrelation curves are population-weighted linear combinations of the curves for free and bound ligand. From the amplitudes of the two components of the measured curve, the fraction free and bound ligand can be determined.

If the binding reaction is fast ($\tau_{binding} \ll \tau_D$), binding and dissociation are in dynamic equilibrium while the molecules reside in the confocal volume. In this case, the set of autocorrelation curves of a titration experiment show a single inflection point that continuously shifts towards longer times for increasing concentrations of macromolecule. (Figure 23.13). This *apparent diffusion coefficient* $\langle D_t \rangle$ from each autocorrelation curve is the population-weighted average of the diffusion times for the free and bound ligand.

$$G(\tau) = 1 + \frac{1}{N}\left(\frac{1}{1 + 4\langle D_t \rangle \tau/\omega_1^2}\right)\left(\frac{1}{1 + 4\langle D_t \rangle \tau/\omega_2^2}\right)^{0.5} \qquad \text{eq. 23.31}$$

To obtain a binding curve, the apparent diffusion coefficient is plotted as a function of the concentration of the macromolecule, and the $K_d$ value can be determined by data fitting (see eq. 19.70).

For the quantitative analysis of FCS experiments to determine diffusion coefficients or numbers of molecules, the size and shape of the confocal volume have to be known, which is often difficult to achieve and experimentally difficult to control. These problems for quantitative FCS can be circumvented by using *dual-focus FCS* (2f-FCS). Here, two overlapping foci are generated that are laterally shifted by a known distance. The distance between the centers of the foci provides an external length ruler for the exact determination of diffusion coefficients (Box 23.3).

---

**BOX 23.3: DUAL-FOCUS FCS (2F-FCS) AND ABSOLUTE DIFFUSION COEFFICIENTS.**

2f-FCS is based on two confocal volumes that overlap, but are laterally shifted by a known distance. To generate the two foci, two excitation lasers with orthogonal polarization are used. Their beams are passed through a Nomarski prism that generates a lateral shift between the two components because of their difference in polarization. An objective

*(Continued)*

then focuses the two beams in separate spots (Figure 23.14). The distance between the two foci can be determined in reference measurements with compounds of known diffusion coefficients, or by imaging the two foci with fluorescent beads. The inter-focal distance depends only on the type of prism and on the objective used. From the auto-correlation curves of fluorescence from focus 1 or from focus 2, as well as the cross-correlation between fluorescence from both foci (see Section 23.1.6.3, Figure 23.14), the diffusion coefficient can be determined by global analysis. 2f-FCS allows the determination of diffusion coefficients *in vivo*. 2f-FCS has been used to measure lipid diffusion in black lipid membranes. In combination with line scanning, ligand affinities of membrane receptors have been determined within the membrane of living cells. Weiss & Enderlein (2012) *Chem. Phys. Chem.* 13(4): 990–1000. Dorlich *et al.* (2015) *Sci. Rep.* 5: 10149.

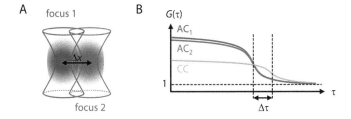

**FIGURE 23.14:**    **2f-FCS**. A: Principle. Two overlapping, laterally shifted foci serve as external ruler. B: Autocorrelation curves from each focus ($AC_1$, $AC_2$) and cross-correlation curve (CC) from both foci. The cross-correlation curve is shifted to longer times $\tau$ because of the distance between the foci. Global analysis of auto- and cross-correlation curves permits determination of absolute diffusion coefficients.

The size of the confocal volume depends on the wavelength of the light used, the width of the incoming beam, and the numerical aperture of the objective. In confocal scanning microscopy, higher spatial resolution is achieved with smaller foci. Although the size of the confocal volumes can be reduced somewhat by widening the incoming excitation beam or by increasing the numerical aperture of the objective, a much larger reduction in size is achieved by using *two-photon excitation* (Box 23.4).

### BOX 23.4: FCS WITH TWO-PHOTON EXCITATION.

Instead of excitation of a fluorophore by illumination with light corresponding to the energy difference between $S_0$ and $S_1$ states, excitation can in principle also be achieved by absorption of two photons of twice the wavelength that each provide half of the energy for the transition (Figure 23.15). The simultaneous absorption of two photons is an unlikely process, and high light intensities are required to achieve two-photon excitation. As a consequence, two-photon excitation can only occur close to the center of the confocal volume. The effective confocal volume for two-photon excitation is therefore much smaller. A smaller confocal volume provides increased resolution for confocal imaging. Excitation with infrared light instead of visible light is advantageous for imaging of cells: the background fluorescence is lower, and photodamage to the cell is reduced.

*(Continued)*

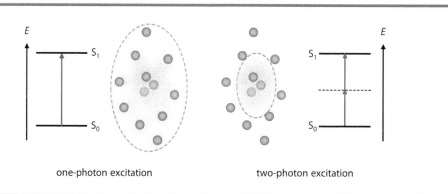

one-photon excitation                    two-photon excitation

**FIGURE 23.15:   One- and two-photon excitation.** Left: A green fluorophore can be excited by illumination with blue light whose energy corresponds to the energy difference between the $S_0$ and $S_1$ states. The excitation is most efficient in the center of the confocal volume and decays exponentially in the $x$-, $y$-, and $z$-directions. Right: Alternatively, a green fluorophore can be excited by absorption of two red photons of half the energy. The dotted line indicates a virtual intermediate state between $S_0$ and $S_1$. The absorption of two photons is unlikely and requires higher light intensities than excitation by one photon. As a consequence, the fluorophore is only excited efficiently close to the center of the confocal volume, but not at its rim. The effective confocal volume (broken gray line) is significantly reduced compared to one-photon excitation.

While information on rates of diffusion, photophysical processes, and chemical reactions is determined from autocorrelation curves, information on the number of molecules and their brightness is typically obtained from *photon counting histograms* (PCH) that count how often a certain intensity has been measured (Figure 23.16). The PCH reveals the presence of different species with different brightness and their relative populations, and thereby allows detection of dimer formation or oligomerization.

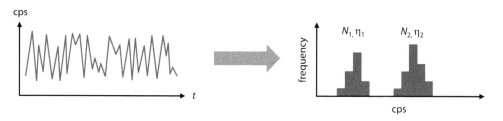

**FIGURE 23.16:   Photon counting histograms, number of molecules, and molecular brightness.** The fluctuating fluorescence signal from the confocal volume (left) is re-plotted in terms of the frequency with which a certain count rate (cps, counts per second) has been detected (photon counting histogram, PCH; right). The PCH shows two species, one with a low brightness $\eta_1$ ($N_1$ particles), and a second with a higher brightness $\eta_2$ ($N_2$ particles). These species could be a monomer and a dimer. Note that photon counting statistics leads to more complex shapes and more complex analyses of photon counting histograms.

### 23.1.6.3 Fluorescence Cross-Correlation Spectroscopy

Instead of correlating fluctuations of a fluorescence signal with itself, it is also possible to correlate the signal from two fluorophores with each other. In *fluorescence cross-correlation spectroscopy* (FCCS), two dyes are used. The fluorescence emitted from molecules in the confocal volume is split into the spectral regions of the two dyes for quantification on two separate detectors. The detected signals $I_1(t)$ and $I_2(t)$ are then cross-correlated to calculate the *cross-correlation function* $G(\tau)$:

$$G(\tau) = 1 + \frac{\langle \delta I_1(t) \cdot \delta I_2(t+\tau) \rangle}{\langle I_1 \rangle \cdot \langle I_2 \rangle}$$

eq. 23.32

The shape of the cross-correlation function is identical to the autocorrelation function, but the meaning is somewhat different. The amplitude is now proportional to the number of species in the confocal volume that carry both fluorophores, the inflection point corresponds to the diffusion time of the species that carries both fluorophores. The cross-correlation curve describes the probability to detect a photon in channel 2 at time point $\tau$ after detection of a photon in channel 1 at $\tau = 0$ (Figure 23.17).

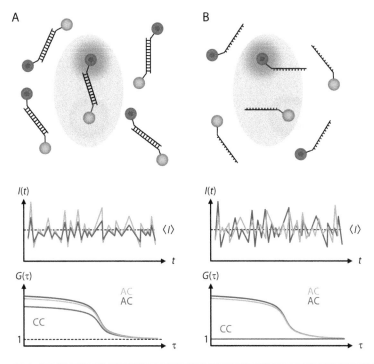

**FIGURE 23.17:** **Fluorescence cross-correlation spectroscopy (FCCS).** A: When two fluorophores diffuse together, their fluorescence fluctuations are correlated. Whenever green fluorescence is decreasing, red fluorescence also decreases, and *vice versa*. The cross-correlation therefore has a measureable amplitude. The inflection points of the cross-correlation function as well as of the autocorrelation functions for green and red fluorescence are identical and correspond to the diffusion time of the entity that carries both fluorophores. B: If the fluorophores belong to two non-interacting molecules, the green and red fluorescence fluctuations are independent of each other and uncorrelated. The amplitude of the cross-correlation function is zero. The diffusion time of each molecule can be extracted from the inflection point of the green and red autocorrelation functions.

FCCS can be applied to study interactions of molecules both *in vitro* and *in vivo* (Box 23.5). Because of the high sensitivity of confocal fluorescence microscopy, high-affinity interactions can be studied.

---

### BOX 23.5: EXAMPLES OF FC(C)S STUDIES *IN VIVO*.

FCCS has been used to study the interaction between fluorescently labeled transcription factors jun and fos *in vivo*. Jun and fos were fused to the fluorescent proteins EGFP and mRFP, respectively, and overproduced in living cells. Cells producing an EGFP-mRFP fusion protein served as a positive control, cells producing EGFP and mRFP individually were used as a negative control. Cross-correlation curves with significant amplitudes were obtained from cells containing jun-EGFP and fos-RFP, demonstrating their heterodimerization *in vivo*. When the dimerization domains of both proteins were deleted, the cross-correlation amplitude was lost, demonstrating the loss of dimerization. Baudendistel *et al.* (2005) *Chem. Phys. Chem.* 6(5): 984–990.

An *in vivo* FCS/FCCS study of the RNA-induced silencing complex (RISC), a protein-RNA complex involved in RNA interference, has shown that this complex exists in a cytoplasmic and a nuclear form with different diffusion coefficients. Nuclear RISC (ca. 160 kDa) diffuses much more rapidly than the ca. 3 MDa cytoplasmic RISC. Together with biochemical data, this study provided evidence for a mechanism in which the Argonaute protein Ago2

*(Continued)*

and a small regulatory RNA form a RISC complex in the cytoplasm that is then shuttled to the nucleus where it interacts with its target mRNA and mediates RNA-dependent regulation of gene expression. Ohrt *et al.* (2008) *Nucleic Acids Res.* 36(20): 6439–6449.

FCS (and FCCS) can also be used in high-throughput format to study the dynamics of large numbers of proteins in living cells. Using automated screening and time-lapse acquisition of FCS data at specific subcellular locations, and automated data analysis, the dynamics of more than fifty nuclear proteins has been studied by high-throughput FCS. Wachsmuth *et al.* (2015) *Nat. Biotechnol.* 33(4): 384–389.

## 23.1.7 SINGLE-MOLECULE FLUORESCENCE MICROSCOPY

### 23.1.7.1 Principles of Single-Molecule Microscopy

In Sections 23.1.2–23.1.5 we have treated different types of fluorescence microscopy to generate images of tissues, cells, and sub-cellular structures. Fluorescence microscopy with laser excitation and ultrasensitive detectors provides the sensitivity to detect fluorescence from single molecules. Wide-field, confocal scanning, and TIRF microscopy can be used to obtain images of single molecules immobilized on surfaces (Box 23.6, see also Section 26.3.4). Each emitting molecule gives rise to a bright spot in the image of the surface. Integration over the entire spot yields the fluorescence intensity of the particular molecule. In contrast, FCS-type confocal microscopy allows measurement of fluorescence from single molecules in solution as a function of time while they are traversing the confocal volume. Each molecule gives rise to a burst of fluorescence photons that is detected by an avalanche photodiode.

### BOX 23.6: SURFACE IMMOBILIZATION OF SINGLE MOLECULES FOR TIRF MICROSCOPY.

Microscope slides for surface immobilization of single molecules need to be passivated to minimize unspecific interactions of the molecule of interest with the surface. Passivation is achieved with the protein bovine serum albumin (BSA) that interacts non-specifically with the glass surface, with polyethylene glycol (PEG) that forms polymer brushes on the glass slide, or by depositing a lipid bilayer on the glass surface. Surface immobilization for single-molecule fluorescence studies is often performed using the biotin-streptavidin interaction (Figure 23.18). A streptavidin functionalization of the surface is achieved by doping the passivation layer on the microscope surface with biotinylated BSA, PEG, or lipids to which streptavidin is coupled. Proteins are biotinylated either *in vivo* or *in vitro*, and immobilized on the streptavidin-functionalized surface. Biotinylated nucleic acids are either chemically synthesized (DNA, RNA), biotinylated by adding a biotinylated starter nucleotide to *in vitro* transcription reactions (RNA), or by ligation or hybridization of a biotinylated oligonucleotide (DNA, RNA). Alternatively, proteins and nucleic acids can be incorporated into lipid vesicles that are surface-immobilized. This method minimizes surface effects because the molecule of interest is in solution within the vesicle, yet spatially confined for long-time observation. Vesicles containing protein pores allow addition of small molecules to the molecule of interest, but the addition of larger interaction partners is not possible.

*(Continued)*

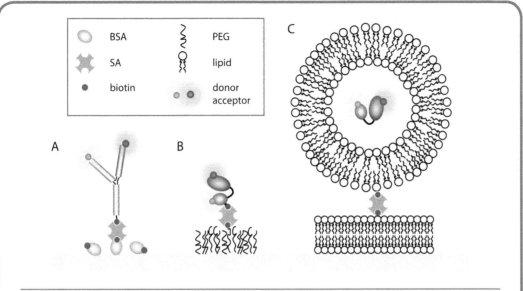

FIGURE 23.18:    **Immobilization schemes for proteins and nucleic acids.** A: Immobilization of biotinylated nucleic acids (cylinders) on glass surfaces functionalized and passivated with biotinylated bovine serum albumin (BSA). B: Immobilization of biotinylated proteins (ovals) on polyethylene glycol-passivated glass surfaces doped with biotin. C: Incorporation of proteins (or nucleic acids) into lipid vesicles and immobilization of lipid vesicles on surfaces with biotin-doped lipid bilayers.

The observation time for each single molecule in solution by FCS-type confocal microscopy is limited to milliseconds, the diffusion time through the confocal volume. For surface-immobilized molecules the observation time is in principle unlimited: we can take a new image and determine the fluorescence of each molecule at any time. In practice, the observation time will be limited by the photostability of the dye(s) used. Fluorophores undergo *photobleaching* and *blinking*, and transiently or permanently stop emitting fluorescence. Fluorescence emission stops when the fluorophore undergoes a transition to a non-fluorescent dark state, typically a triplet state (see Figure 19.34; Figure 23.19). While in the triplet state, the fluorophore does not contribute to fluorescence. Only after its return to the ground state is the fluorophore available for further excitation-emission cycles and continues to emit fluorescence. The transient loss of fluorescence due to reversible transitions into "dark states" is called blinking. Blinking can be suppressed by addition of *triplet state quenchers*, such as the vitamin E analog Trolox (Box 23.7) that mediates rapid relaxation of triplet states. Molecular oxygen is also a triplet state quencher. Quenching of triplet states by molecular oxygen generates reactive singlet oxygen that causes photodestruction of the fluorophore and/or covalent modifications, accompanied by a permanent loss of fluorescence. This irreversible process is called photobleaching. Photobleaching can be minimized by using *oxygen scavengers* (Box 23.7), which inevitably leads to more pronounced blinking. The combined use of oxygen scavengers and triplet state quenchers affords typical observation times of tens of seconds up to several minutes.

## BOX 23.7: MINIMIZING PHOTOBLEACHING AND BLINKING.

The observation time of a fluorophore is limited by the occurrence of intermittent dark phases (blinking) and irreversible loss of fluorescence due to photobleaching. Oxygen scavenging systems remove molecular oxygen from solutions, and suppress the generation of reactive oxygen species that cause irreversible photodestruction of fluorophores and photobleaching. Two examples for frequently used oxygen scavenging systems employ protocatechuic acid in combination with the enzyme protocatechuate-3,4-dioxygenase, or glucose with glucose oxidase and catalase for oxygen removal (Figure 23.19). Molecular oxygen acts as a triplet state quencher. Its removal not only suppresses photobleaching, but at the same time increases the triplet state lifetime, and blinking becomes more pronounced. Blinking has therefore to be suppressed by adding other triplet state quenchers such as β-mercaptoethanol or Trolox, a vitamin E derivative (Figure 23.19). However, high concentrations of β-mercaptoethanol can lead to the population of long-lived dark states.

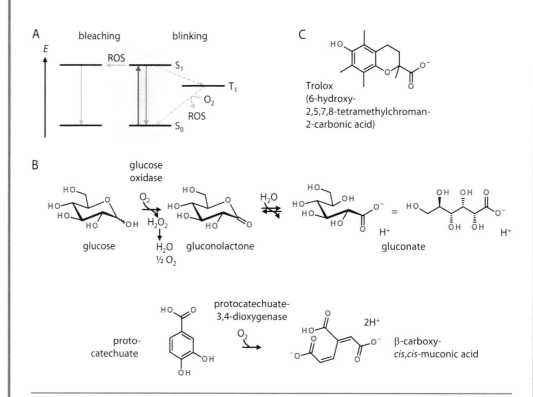

**FIGURE 23.19: Photobleaching and blinking limit observation times for single molecules.** A: Photobleaching and blinking convert fluorophores into permanent (bleaching) or transient (blinking) dark states. ROS: reactive oxygen species. B: Oxygen scavenging systems suppress photobleaching, but promote blinking. The glucose/glucose oxidase/catalase system removes oxygen by oxidizing glucose to gluconate. Catalase converts the resulting $H_2O_2$ to $H_2O$ and $O_2$. Mismatched activities of glucose oxidase and catalase may lead to accumulation of $H_2O_2$, which can create more reactive oxygen species. The protocatechuate/protocatechuate dioxygenase system removes oxygen by oxidizing protocatechuic acid to β-carboxy-*cis, cis*-muconic acid. C: Triplet state quenchers such as Trolox (6-hydroxy-2,5,7,8-tetramethylchroman-2-carbonic acid) suppress blinking by accelerating the relaxation of triplet states.

### 23.1.7.2 Why Single Molecules?

When we measure fluorescence parameters such as intensity, anisotropy, lifetime, or FRET efficiency from solutions over large ensembles of molecules, we obtain values that represent the average over all species that are present in the sample. The average is a meaningful number if the sample is homogeneous, i.e. if individual molecules in the ensemble exhibit identical values for the measured parameter or at least similar values narrowly distributed around a mean value (Figure 23.20). The average value does not provide any information on the width of the distribution of the measured parameter or on symmetry or asymmetry of the distribution. The measured value is not meaningful for heterogeneous samples with different species that exhibit different fluorescence parameters. In this case, the measured value is population-averaged over all molecules in the sample. The average value has very limited meaning on a molecular scale: most likely, this particular value is not represented by a single molecule in our sample. We have discussed this limitation in the context of the FRET efficiency (Section 19.5.10). This limitation is overcome if we determine fluorescence parameters for single molecules individually, without averaging over a large and possibly heterogeneous population. Each single molecule will have one defined value for the respective parameter. By measuring the fluorescence parameter for many single molecules, either simultaneously by imaging a surface on which many molecules are immobilized or one after the other by watching single molecules in solution pass the confocal volume, we directly obtain the distribution of the measured parameter over many molecules in the sample (Figure 23.20). Plots of the frequency with which a particular value is observed are called histograms. We have encountered histograms before: photon arrival time histograms to determine fluorescence lifetimes (Figure 19.64 in Section 19.5.9) and photon counting histograms to identify species with different brightness (Figure 23.16 in Section 23.1.6.2).

Single-molecule measurements also eliminate the disadvantage of averaging out temporal changes in the measured parameter. If the molecule of interest can switch between two states with different values for the measured parameter, the transitions of individual molecules in an ensemble from one state to the other are unsynchronized and do not occur in a coordinated manner. The random fluorescence changes are averaged at each time point, and the fluorescence intensity from this ensemble of molecules will be constant over time. The only possibility to follow the transition from one state to the other is to synchronize the molecules and to make sure that they all start the transition from one state, and at the same time. We will see how we can measure the velocity of transitions in synchronized ensembles in Chapter 25. Often, it is difficult or impossible to trigger a certain transition or reaction, and the molecules cannot be synchronized. If we focus on a single molecule, however, we do not need to synchronize it with others to observe the transition: a single molecule can only be in one of the two possible states at any given time. Monitoring the fluorescence intensity of a single molecule over time will therefore provide a fluorescence trace with jumps between the two fluorescence levels. From the fluorescence traces, we obtain information about how long the molecule remains in each state, the *dwell time* (Figure 23.20). Again, we can histogram the individual dwell times of the molecule in one state, and directly see the *dwell time distribution*. From dwell-time histograms, rate constants for the underlying transition can be determined. We will revisit these types of information when we discuss single molecule FRET (Section 23.1.7.6).

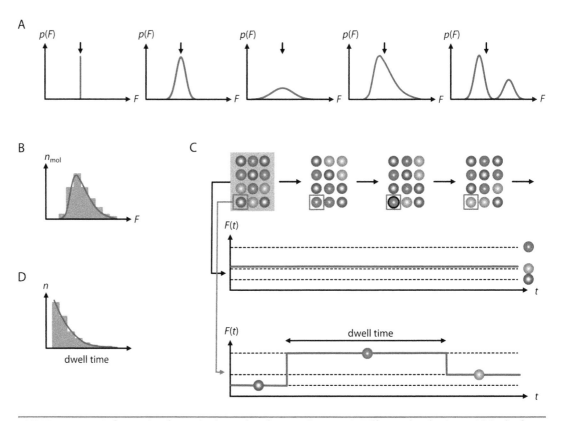

**FIGURE 23.20: Information from single-molecule experiments.** A: Different distributions $p(F)$ (red) of the observable parameter $F$ and the average value (arrow) obtained from ensemble experiments. B: Single-molecule experiments avoid averaging over the entire population and determine the parameter F individually for a large number of single molecules. The distribution is obtained by plotting histograms. C: Single-molecule experiments avoid temporal averaging and reveal time-dependent changes of an observable for a single molecule over time (bottom). This information is hidden in ensemble experiments (top). From time traces of the observed parameter for a single molecule, dwell times in individual states can be determined. D: Dwell time histograms provide kinetic information on interconversion of different states (see Section 23.1.7.4).

### 23.1.7.3 Localization and Tracking of Single Molecules

We have discussed before (Section 23.1.1.3) that a single emitting molecule on the surface appears as a bright spot with a radially symmetric intensity distribution in the fluorescence image, the molecule detection function that corresponds to the Airy disc. The intensity is highest in the center (i.e. where the fluorescent molecule is really located), and decays with increasing distance from the center. This intensity distribution is spread over a limited number of pixels of the area detector. Each pixel typically corresponds to a square on the surface of about 0.1 μm by 0.1 μm. Thus, by selecting the highest-intensity pixel, we can only locate the center of the distribution and the position of the emitter with an accuracy in the 0.1 μm range. However, we can exploit the fact that the intensity profile is described by a Gaussian distribution. By describing the intensities measured pixel-wise in the $x$- and $y$-directions, we can determine the center of the Gaussian distribution to much higher accuracy, down to 1 nm. This procedure for the localization of single fluorophores is called *fluorescence imaging with one nanometer accuracy* (FIONA, Figure 23.21). The principle of FIONA is also exploited in super-resolution microscopy (Section 23.1.8).

Single-molecule microscopy can also be used to track the movement of single molecules over time. When a single fluorophore is imaged at different time points, its $x,y$-coordinates can be determined as a function of time. From this information, *trajectories* of its path can be constructed. This approach has been used to study virus entry into cells, and to investigate lateral diffusion of membrane proteins.

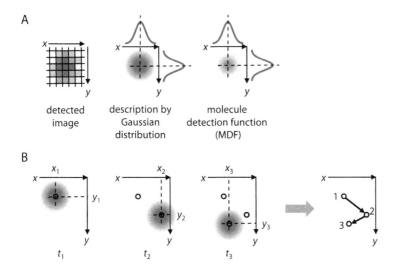

**FIGURE 23.21:** **Localization of single molecules by FIONA and single-molecule tracking.** A: Fluorescence imaging with one nanometer accuracy (FIONA). The continuous intensity distribution from a single emitter (point-spread function, right) is detected by a limited number of pixels on the detector (left). Description of the measured intensity profile by a Gaussian intensity distribution in the $x$- and $y$-directions allows determination of the center of the distribution with nanometer accuracy (center). B: From the position of a single molecule at different time points $t_1$, $t_2$, $t_3$ its trajectory can be constructed (right).

### 23.1.7.4 Kinetic Information from Single-Molecule Microscopy

The fluorescence intensity of a single fluorophore can be measured as a function of time, for example to follow enzymatic reactions in real time. For example, cholesterol oxidase, a bacterial flavoenzyme involved in steroid oxidation and isomerization, uses FAD to oxidize cholesterol to cholest-4-en-3-one. FAD is fluorescent, whereas the reduced form $FADH_2$ is non-fluorescent (Section 19.5.4). Enzymatic turnover can therefore be followed by fluorescence. The time the molecule spends in the on- and off-states, the dwell times, can be measured and compiled into a histogram (Figure 23.22). The mean time

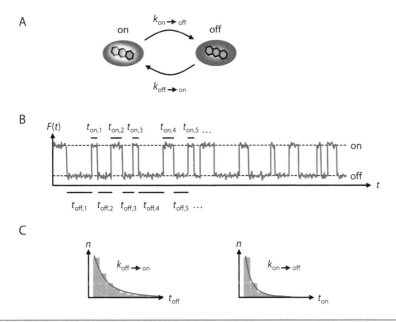

**FIGURE 23.22:** **Dwell times, dwell-time histograms, and rate constants.** A: A single molecule switches between two states with low and high fluorescence. From the fluorescence measured over time, the dwell times in the low and high fluorescence states can be determined. B: Dwell-time histograms yield lifetimes of the low and high fluorescence states of the molecule. The lifetimes are the inverse of the rate constants associated with the process away from the respective state.

a molecule spends in a particular state is its lifetime, which is related to the rate constant of the reaction away from this state (see Chapter 7). From the distribution of dwell times (dwell-time histograms) in the on- and off-states, rate constants for the two processes can therefore be obtained.

In *protein-induced fluorescence enhancement* (PIFE) experiments, the increase in fluorescence quantum yield upon binding of proteins in the vicinity of a fluorescent dye is used as a probe for protein binding and can be used to monitor movement of proteins along DNA (Box 23.8). PIFE is sensitive in the distance range of 1–4 nm. Conversely, static quenching of a reporter fluorophore by certain groups can be a useful probe for binding and movement (Box 23.8).

---

**BOX 23.8: PROTEIN-INDUCED FLUORESCENCE ENHANCEMENT (PIFE) AND STATIC QUENCHING AS PROBES FOR BINDING AND MOVEMENT.**

Binding of proteins near fluorophores can either lead to an increase in fluorescence (protein-induced fluorescence enhancement, PIFE) or to fluorescence quenching. In both cases, the change in fluorescence is a sensitive probe for the distance between the protein and the respective fluorescent group. Changes in fluorescence intensity report on binding of the protein, and can also report on movement of the protein of interest towards or away from the fluorescent dye (Figure 23.23). PIFE: Hwang & Myong (2014) *Chem. Soc. Rev.* 43(4): 1221–1229. Quenching: Honda *et al.* (2009) *Mol. Cell* 35(5): 694–703.

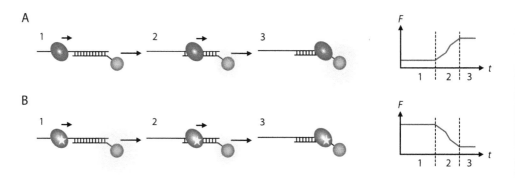

**FIGURE 23.23: Fluorescence enhancement and quenching as a probe for binding and movement.** A: Binding of a protein close to a fluorescent dye on the DNA may lead to an enhancement of fluorescence. Protein-induced fluorescence enhancement (PIFE) has been used to monitor movement of helicases along DNA. B: Quenching of fluorescence can also serve as a probe for protein movement along DNA. Quenching of fluorescence by the FeS cluster of the protein XPD has been exploited to measure translocation of XPD along the DNA.

---

### 23.1.7.5 Colocalization of Molecules

Multiple color single-molecule fluorescence microscopy can be used to detect individual subunits in multi-component complexes. In <u>co</u>localization <u>single-molecule spectroscopy</u> (CoSMoS), one compound of the complex of interest is immobilized on a surface, and its position is determined by fluorescence microscopy. A second component, labeled with a second dye whose fluorescence is spectrally separated from the fluorescence of the first, is then added, and its position is determined. Superposition of the two images reveals if the two fluorophores colocalize, pointing towards interaction of the partners. In principle, this procedure can be extended to three and more dyes (Figure 23.24). The number of fluorophores that can be used simultaneously is limited by the requirement for spectral separation. The necessity to install separate lasers for excitation of each fluorophore makes multi-color applications rather expensive. CoSMoS provides information on the composition of complexes in equilibrium, and can be used to study complex assembly and disassembly pathways and velocities.

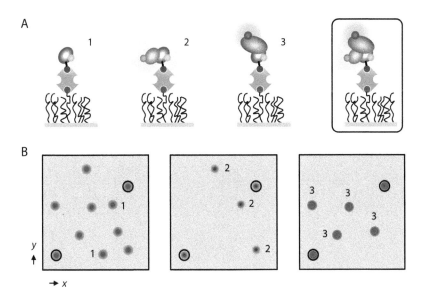

**FIGURE 23.24:   Colocalization single-molecule spectroscopy (CoSMoS).** A: One component of a complex is fluorescently labeled (blue), biotinylated (brown sphere), and immobilized on streptavidin-functionalized microscope slide. A second and third compound are fluorescently labeled with different dyes (green, red). B: The surface is imaged in all three colors, and each labeled component can be localized in the corresponding image. By superimposing the three images, the composition of each complex can be determined. From the ratio of incomplete complexes 1, 2, and 3, and the fully assembled complex (black rim), complex stabilities can be determined. From images at different time points, association and dissociation events can be followed as a function of time.

CoSMoS has been used to study the dynamics of transcription and splicing complexes. The application of CoSMoS is limited to systems that can be assembled *in vitro* from individually purified or synthesized components. Alternatively, the molecule or complex of interest can be pulled down from cell extracts using specific antibodies, and is then surface-immobilized through the antibody for analysis on the single molecule level. This procedure is called *single-molecule pull-down* (SiMPull; Box 23.9).

---

**BOX 23.9: SINGLE-MOLECULE PULL-DOWN (SiMPull).**

Single-molecule pull-down combines the principle of pull-down experiments with single-molecule analysis. Proteins or protein complexes can be directly pulled down onto a microscope slide functionalized with an antibody against the protein of interest (Figure 23.25). An extract from cells expressing the protein of interest is incubated with the slide surface, and the protein of interest, either alone or in complex with interaction partners, is immobilized on the surface. Unbound proteins are removed by washing. Alternatively, the antibody-bound complex of interest can be immobilized on a protein A-functionalized microscope slide. Protein A recognizes the constant region of immunoglobulins. If the protein has been produced in a fluorescently tagged version, e.g. as a fusion with GFP, the surface can be imaged directly. Otherwise, immunolabeling procedures can be employed. Here, a primary antibody against (one of) the immobilized protein(s) is added, and stained

*(Continued)*

by a secondary antibody that is fluorescently labeled. With SiMPull functional complexes from cell extracts may be captured that cannot be reconstituted from individual components *in vitro*. SiMPull of fluorescently labeled proteins and protein complexes enables the determination of complex composition and stoichiometry by analyzing the number of photobleaching steps. Jain *et al.* (2011) *Nature* 473(7348): 484–488.

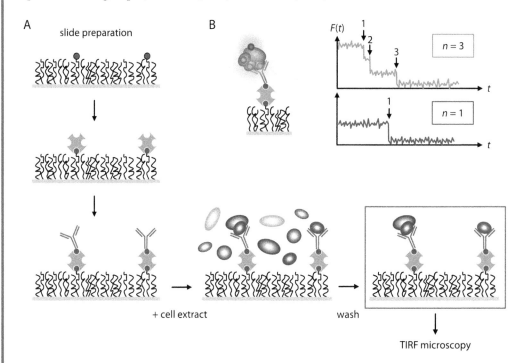

**FIGURE 23.25: Single-molecule pull-down (SiMPull).** A: Microscope slides are passivated with polyethylene glycol. A small fraction of the polyethylene glycol molecules carries a biotin group (brown sphere) that binds streptavidin (orange). Biotinylated antibodies are immobilized on the streptavidin-functionalized surface. Cell extract is added, and unbound proteins are removed by washing. The antibody captures the protein of interest (blue) and possible interaction partners (brown). The protein (complexes) captured on the surface can be imaged directly if one or both components have been labeled with genetically encoded dyes, such as fluorescent proteins. Alternatively, the surface is immunostained by adding a primary antibody against the interaction partner and a secondary antibody that is fluorescently labeled. B: With SiMPull the composition of isolated complexes can be determined in multi-color experiments. The green fluorescence shows three photobleaching steps, the complex thus contains three of these subunits. The red fluorescence photobleaches in a single step, indicating that only one subunit is part of the complex.

### 23.1.7.6 Single-Molecule FRET

The presence of two fluorophores enables the measurement of FRET efficiencies and changes of FRET efficiencies due to conformational changes (see Section 19.5.10). Single-molecule FRET microscopy offers the advantage that donor and acceptor emission is measured for single molecules individually instead of averaging intensities over an ensemble. The FRET efficiency from a single molecule therefore directly reports on the inter-dye distance. Histograms of FRET efficiencies for a set of single molecules yield the distribution of FRET efficiencies in the sample, without a prior assumption of the shape of the distribution. When FRET is measured as a function of time, conformational changes can be followed in real time. Single-molecule FRET microscopy can be performed on donor/acceptor-labeled molecules in solution, using confocal microscopy, or on surface-immobilized molecules using confocal scanning, wide-field, or TIRF microscopy. The emission light is separated into donor and acceptor emission by dichroic mirrors (Box 23.1) and detected by separate avalanche photodiodes or imaged onto adjacent areas of a CCD camera. For single-molecule FRET microscopy on freely diffusing molecules, the raw data is donor and acceptor fluorescence as a function of time. Spikes of fluorescence stem from a single molecule that passes the confocal volume. Using threshold criteria, single-molecule events can be identified, and the FRET efficiency for each single molecule can be calculated from the donor and acceptor intensities integrated over the duration of the fluorescence burst (Figure 23.26). FRET histograms reveal the distribution of FRET efficiencies over all molecules observed one after the other, and allow the identification of different conformational states. Single-molecule FRET experiments on surface-immobilized molecules yield pairs of images with diffraction-limited spots for each single molecule: one image reflects the donor fluorescence for each of these molecules, the second the acceptor fluorescence. Integration over the spots allows calculation of the FRET efficiencies for all observed molecules, and for the construction of FRET histograms. From a series of images taken at different time points, the donor acceptor fluorescence and the FRET efficiency can be determined as a function of time. FRET time traces reveal the sequence of different conformational states and the dwell times of the molecule in each state. From dwell-time histograms, rate constants of conformational changes can be determined (Section 23.1.7.4).

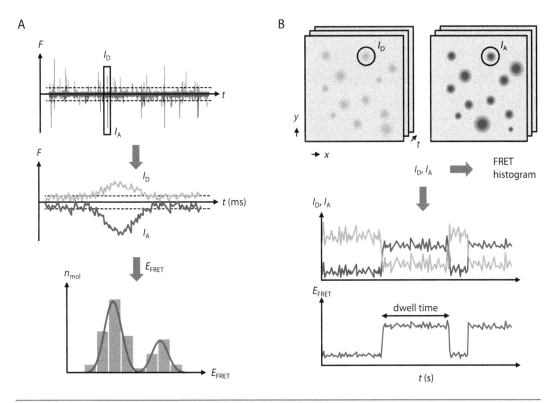

**FIGURE 23.26: Single-molecule FRET on molecules in solution or on surfaces.** A: Single-molecule FRET in solution by confocal microscopy. Top: raw data: donor (green) and acceptor fluorescence (red, inverted) as a function of time. Each spike of fluorescence is caused by a single molecule diffusing through the confocal volume. The broken lines mark the threshold: bursts with higher intensity are counted as single-molecule events. Middle: Close-up of a single fluorescence burst. The donor and acceptor fluorescence intensities $I_D$ and $I_A$ are obtained by integrating the areas under the curves. From $I_D$ and $I_A$, the FRET efficiency ($E_{FRET}$) for the single molecule is calculated. Bottom: FRET efficiencies for all observed molecules are plotted as a FRET histogram. B: Single-molecule FRET of molecules immobilized on a surface. Top: Images of donor (green) and acceptor fluorescence (red) are acquired at different time points. Each molecule appears as a bright spot in the images. The donor and acceptor fluorescence intensities $I_D$ and $I_A$ are determined by integrating the intensities in one spot (black circle). From $I_D$ and $I_A$, the FRET efficiencies for each molecule can be obtained and plotted as histograms. Middle, bottom: More importantly, the donor and acceptor intensities and thus the FRET efficiency for each single molecule can be calculated as a function of time, revealing conformational changes of the molecule in real time. FRET time traces permit the determination of dwell times in each conformational state. Dwell-time histograms yield rate constants for the transitions between these states.

Single-molecule FRET has been widely applied to study protein and nucleic acid conformational changes. The conformational cycle of the translation initiation factor eIF4A, an RNA helicase involved in ATP-dependent unwinding of secondary structures in the 5′-untranslated region of mRNAs during translation initiation, has been investigated by single-molecule FRET confocal and TIRF microscopy. eIF4A adopts three different conformations, termed open, half-open, and closed states. Other translation initiation factors regulate eIF4A activity by modulating the population of these states and the kinetics of their interconversion. Single-molecule FRET can also delineate a series of sequential conformational changes in molecular machines. By introducing donor and acceptor fluorophores in different positions into the DNA topoisomerase gyrase, a cascade of DNA- and nucleotide-induced conformational changes at the beginning of the catalytic cycle of DNA supercoiling has been observed. While most studies detect FRET between a pair of dyes and monitor changes of a single distance, single-molecule FRET experiments can be extended to three colors to monitor two (or more) different distances simultaneously (Box 23.10).

---

### BOX 23.10: THREE-COLOR FRET.

Three-color FRET has been used to investigate the conformational changes in Holliday junctions, four-way DNA helical junctions that are intermediates in homologous recombination (Section 17.5.3.1). The donor fluorophore Cy3 and two acceptor fluorophores, Cy5 and Cy5.5, were attached at the end of three arms of the four-way helical junction, and the Holliday junction was surface-immobilized *via* biotin attached to its fourth arm for single-molecule FRET experiments by TIRF microscopy (Figure 23.27). The Holliday junction can form two X-shaped forms in which two helical arms are co-axially stacked. Two different stacking patterns are possible, with different pairwise stacking of the four helices. Each of these stacking conformers can adopt a parallel or antiparallel form. The stability of stacking conformers is determined by the sequence in the junction region. The measured fluorescence intensities of all three fluorophores were consistent with completely stacked antiparallel Holliday junctions. No intermediates with incomplete stacking were detected. The interconversion of the two antiparallel conformers leads to a decrease in FRET efficiency from donor to one acceptor, and an increase in FRET efficiency between donor and the other acceptor (Figure 23.27). Fluorescence time traces revealed a correlation of the fluorescence increase of one and decrease of the other acceptor (on a time scale of 20 ms), indicating that interconversion of the two antiparallel stacking conformers occurs by a simultaneous and coordinated unstacking of one helix pair and restacking of the alternative pair. Hohng *et al.* (2004) *Biophys. J.* 87(2): 1328–1337.

*(Continued)*

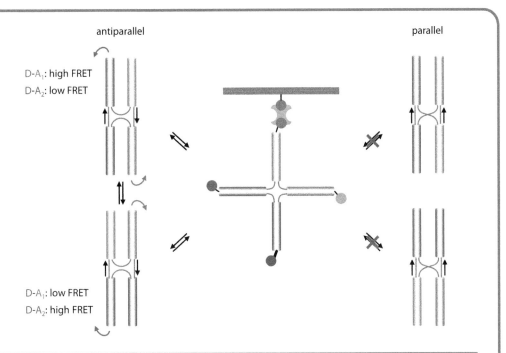

antiparallel                                                                parallel

D-A₁: high FRET
D-A₂: low FRET

D-A₁: low FRET
D-A₂: high FRET

**FIGURE 23.27:   Three-color single-molecule FRET to study the dynamics of Holliday junctions.**
A Holliday junction was labeled with donor (blue), acceptor 1 (green) and acceptor 2 (red) on the ends
of three helical arms. The fourth helix carried a biotin for surface immobilization. In single-molecule
FRET experiments only antiparallel stacking isoforms were detected. During interconversion of these
two isoforms by unstacking and restacking of the four helices (colored arrows), the FRET efficiencies
between the donor to each of the two acceptors change in opposite directions. The fluorescence
intensity of one acceptor increases, the intensity of the second acceptor decreases. In FRET time traces,
changes in fluorescence of the two acceptors occurred within a time window of 20 ms, in agreement
with a simultaneous movement of the two helices (red, green) during rearrangement.

Distances determined from FRET efficiencies of donor/acceptor-labeled single molecules have
a molecular meaning. Provided that the structures of the individual domains or components are
known, FRET distances can therefore be used as distance restraints to map the global conformation of multi-domain molecules or to gain insight into the architecture of larger complexes. To map
global conformations, donor and acceptor fluorophores are introduced pairwise at various positions
in the two parts whose relative orientation is unknown, and the corresponding FRET efficiency is
determined in single-molecule experiments (Figure 23.28). To obtain correct FRET efficiencies and
correct inter-dye distances, the measured donor and acceptor fluorescence intensities have to be
corrected for background fluorescence, for cross-talk of donor fluorescence into the acceptor channel ($\alpha$) and cross-talk of acceptor fluorescence into the donor channel ($\beta$), for different quantum
yields of donor and acceptor and different sensitivities of the detectors at donor and acceptor emission wavelengths ($\gamma$), and for direct excitation of the acceptor at the excitation wavelength of the
donor ($\delta$). The corrected FRET efficiency devoid of these effects is

$$E_{FRET} = \frac{(1+\beta\gamma\delta)\cdot\left(I_A - \dfrac{\alpha+\gamma\delta}{1+\beta\gamma\delta}\cdot I_D\right)}{(1+\beta\gamma\delta)\cdot\left(I_A - \dfrac{\alpha+\gamma\delta}{1+\beta\gamma\delta}\cdot I_D\right)+(\gamma+\gamma\delta)\cdot(I_D-\beta I_A)}$$

eq. 23.33

The corrected FRET efficiencies are then converted into distances (eq. 19.131). Conversion of FRET
efficiencies into distances requires the knowledge of the Förster distance for the donor-acceptor

pair under the experimental conditions, i.e. in the complex studied. Once a set of distances has been obtained, the two domains of the molecule of interest or the components of a multi-subunit complex are then placed such that the distance restraints from the FRET experiments are fulfilled. FRET-restrained structural modeling takes into account the length of the linker with which the dyes are attached to the molecule of interest and the resulting uncertainties of their positions, and calculates the most probable global conformation from FRET data. Although the procedure is somewhat tedious and requires a large set of donor/acceptor-labeled constructs and calculation effort, nano-positioning by FRET can provide valuable structural information on large complexes whose structures are inherently difficult to determine by high-resolution methods.

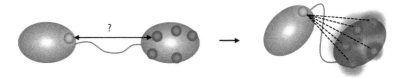

**FIGURE 23.28: Principle of FRET-restrained modeling.** A two-domain protein of unknown relative domain orientation (left) is labeled with a donor in one domain and an acceptor in different places within the second domain. From the measured pairwise FRET efficiencies, inter-dye distances can be determined (broken lines). The two domains are then placed relative to each other such that the distance constraints are fulfilled.

In standard single-molecule FRET experiments, the donor fluorescence is excited, and donor and acceptor emission are detected. Molecules that do not show acceptor fluorescence might reflect low FRET states, might carry an acceptor fluorophore in a dark state because of blinking or photobleaching, or might not carry an acceptor at all. Integration of a second laser into the microscope enables intermittent direct excitation of the acceptor to probe if the acceptor is still present (Figure 23.29). For excitation with continuous wave lasers, this technique is called alternating laser excitation (ALEX). For pulsed lasers, it is called pulsed interleaved excitation (PIE). In ALEX or PIE experiments, the FRET efficiency E is determined from the donor and acceptor fluorescence intensities upon excitation of the donor:

$$E = \frac{F_{Dex}^{Aem}}{F_{Dex}^{Dem} + F_{Dex}^{Aem}}$$

eq. 23.34

Dex denotes donor excitation, Dem and Aem denote donor and acceptor emission, respectively. Note that eq. 23.34 does not take into account corrections for instrument non-idealities (see eq. 23.33).

In addition to the FRET efficiency, the stoichiometry $S$ of the fluorescent dyes can be determined by relating the sum of donor and acceptor fluorescence intensities upon donor excitation and acceptor excitation:

$$S = \frac{F_{Dex}^{D+A}}{F_{Dex}^{D+A} + F_{Aex}^{D+A}}$$

eq. 23.35

For a molecule that carries donor and acceptor, the sums of donor and acceptor fluorescence ($F^{D+A}$) will be equal for donor and acceptor excitation, and a value of $S = 0.5$ is obtained. For molecules that carry only a donor, $F_{Aex}$ is small, and $S \rightarrow 1$, whereas for molecules that carry only an acceptor, $F_{Dex}$ is much smaller than $F_{Aex}$, and $S \rightarrow 0$. FRET efficiency $E$ and stoichiometry $S$ can be plotted in two-dimensional histograms (Figure 23.29). From these histograms, it is evident that molecules with the same FRET efficiency that carry different labels can now clearly be distinguished. ALEX and PIE afford spectral sorting of species that carry both the donor and the acceptor from other (unwanted) species that contain only one of the dyes. Donor-only and acceptor-only peaks in FRET histograms can be suppressed, and the subsequent data analysis is restricted to donor/acceptor-labeled molecules. This approach is particularly important for molecules where complete labeling cannot be achieved.

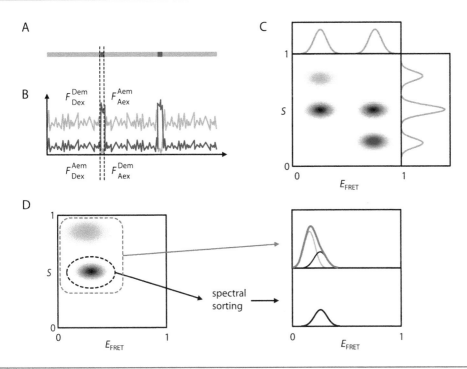

**FIGURE 23.29:   ALEX/PIE and spectral sorting.** A: Illumination scheme. In ALEX/PIE, the sample is illuminated with light to excite donor fluorescence (green) and intermittently with light of a different wavelength to excite the acceptor fluorescence directly (red). B: While the donor fluorescence is excited, the donor and acceptor fluorescence intensities, $F_{Dex}^{Dem}$ and $F_{Dex}^{Aem}$ report on FRET. In the intervals when the acceptor fluorescence is excited, the donor fluorescence $F_{Aex}^{Dem}$ drops to background level, and $F_{Aex}^{Aem}$ reports on the presence of the acceptor. From the individual intensities, FRET efficiency $E$ (eq. 23.34) and stoichiometry $S$ (eq. 23.35) can be calculated. C: Two-dimensional histogram of FRET efficiency $E$ and stoichiometry $S$ for a donor/acceptor-labeled molecule. 25% of the molecules carry only a donor (green), 50% carry both donor and acceptor and occur in a high- and low-FRET state (black), and 25% carry only the acceptor (red). The individual FRET and $S$ histograms are shown on top and on the right, respectively. D: Spectral sorting. FRET-$S$-histogram for a sample with a low FRET donor/acceptor-labeled molecule (black) that contains a large fraction of molecules labeled only with the donor (green). Without spectral sorting, the FRET histogram (top right, gray) is dominated by the FRET = 0 peak from the donor only molecules. The donor/acceptor-labeled molecules cause a shoulder, and their FRET efficiency distribution (black) cannot be extracted reliably. The FRET histogram constructed from only those molecules that carry both dyes and give rise to FRET (bottom right) reveals a clear FRET distribution with a well-defined maximum.

### 23.1.8 SUPER-RESOLUTION MICROSCOPY

We have seen in Section 23.1.1.3 that focusing excitation light by a lens does not lead to a tiny, infinitesimally small focus, but to a spot of finite dimensions because of diffraction effects. The diffraction-limited resolution of light microscopy is a physical law and cannot be overcome. Nevertheless, developments have been made to circumvent these limitations, and to obtain higher resolution.

Moderate increases in optical resolution are achieved by confocal microscopy using two-photon excitation (Box 23.4), by *structured illumination microscopy* where the sample is illuminated in periodic patterns, or by *4Pi microscopy* that uses two opposing objectives for excitation and emission to increase the apparent numerical aperture of the lens system. More drastic reductions of the point-spread function and increases in resolution can be achieved by introducing sub-diffraction-limit features into the excitation light. For example, the size of the point-spread function can be reduced by taking advantage of stimulated emission (Figure 23.30). In a first step, the excitation light is focused into a diffraction-limited spot, leading to excitation of molecules within this area. In a second step, an annular (ring-shaped) laser beam is used to illuminate the rim of the excitation spot. In the ring illuminated by this second laser, stimulated emission occurs, i.e. the laser light induces transitions of the fluorophores from the excited state back to the ground state. This causes a depletion of molecules in the excited state at the rim of the spot (*stimulated emission depletion, STED*). As a result, a central spot of excited molecules is left that is now smaller than the diffraction limit, effectively reducing the point-spread function. The development of STED microscopy has been recognized by the Nobel Prize in Chemistry in 2014 to Stefan Hell.

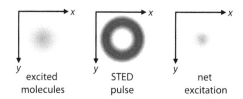

**FIGURE 23.30:** **Stimulated emission depletion (STED) microscopy.** In STED microscopy, a diffraction-limited spot of the sample is illuminated and molecules within this spot are excited (left, orange). With a second high-power laser (STED laser) with an annular intensity profile (red, center), molecules at the rim of the focal spot are illuminated and undergo stimulated emission from the excited to the ground state. As a result, a spot of excited molecules remains that is smaller than the diffraction-limited spot (orange, right). With this smaller focal spot, higher resolution imaging is achieved.

Because of the high laser powers needed for stimulated emission depletion, the application of STED is limited to highly photostable fluorophores. In super-resolution microscopy based on the *reversible saturable optically linear fluorescence transitions* (RESOLFT) technique, the diameter of the diffraction-limited spot is reduced by depletion of fluorophores that can be switched reversibly between on- and off-states. This switching can be achieved with lower laser power than stimulated emission depletion, and at the same time expands the range of suitable fluorophores. With STED and RESOLFT microscopy, lateral resolutions (in the xy plane) of 30–50 nm are achieved (Figure 23.31).

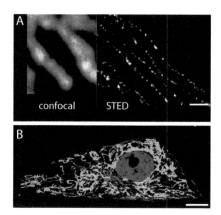

**FIGURE 23.31:** **STED microscopy.** A: Comparison of STED and confocal images of mitochondria. The TOM complex was immunostained using fluorescently labeled antibodies. The white bar corresponds to 500 nm. B: STED image of the mitochondrial network of a Ptk2 kangaroo rat cell. The TOM complex (green) and the microtubules (red) were immunostained using fluorescently labeled antibodies, the DNA in the nucleus (blue) is stained with DAPI. The white bar corresponds to 10 μm. Adapted from Schmidt, R., Wurm, C.A., Punge, A., Egner, A., Jacobs, S., Hell, S.W. (2009) Mitochondrial christae revealed with focused light, *Nano Lett.* 9(6) 2508–2510 with permission. Copyright 2009, American Chemical Society.

A conceptually different set of high-resolution imaging approaches exploits the defined shape of the point-spread and molecule detection functions for a single emitter and locates the emitter with nanometer accuracy from the center of the Gaussian intensity profile (FIONA, Section 23.1.7.3). The location of a single fluorescent molecule can be determined more accurately than the Abbe limit predicts. However, imaging of cells or sub-cellular structures requires the accurate localization of a large number of fluorophores. Each of these fluorophores causes a fluorescence spot on the detector.

If these fluorescence spots overlap, localization of individual molecules becomes difficult or impossible. Under certain circumstances, the FIONA localization principle can also be applied for imaging of such large ensembles of fluorophores. All we need to do is to selectively activate a subset of the fluorophores present that are separated by a distance exceeding the diffraction limit. We can then determine the position of these fluorophores accurately. In the next round, we activate a different subset of fluorophores, and determine their positions. After several rounds, we will statistically have determined the position of each fluorophore at least once. A complete image is constructed from all partial images (Figure 23.32). This principle is the basis of *(fluorescence) photoactivated localization microscopy* (PALM or FPALM). While the fluorophores are switched on and off by illumination in (F)PALM, a related type of microscopy, *stochastic optical reconstruction microscopy* (STORM; Figure 23.33), relies on stochastic switching between on and off states of the fluorophore to limit the number of fluorophores in each partial image.

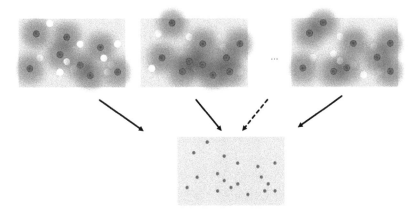

**FIGURE 23.32:   Principle of PALM and STORM.** A fluorescent image is the sum of a large number of single emitters. Each emitter gives rise to an Airy disc in the image, but their Airy discs overlap and lead to a blurred image. PALM and STORM use fluorescent dyes that can be switched on and off by light (PALM) or that stochastically switch between dark and emitting states (STORM). In both cases, only a small subset of emitters is "on" at any given point in time. The Airy discs from these few emitters are now clearly separated in the image (top), and their centers can be determined with nanometer accuracy (FIONA, see Figure 23.21). From a series of images with only a subset of dyes activated, the positions of all emitters can be determined, and the entire image can be constructed (bottom).

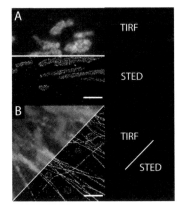

**FIGURE 23.33:   STORM microscopy.** A: Comparison of STORM and TIRF images of African green monkey kidney cells. Green: outer mitochondrial membrane protein Tom20, red: inner mitochondrial membrane protein ATP synthase. Proteins were immunostained using fluorescently labeled secondary antibodies. B: STORM and TIRF image of Tom20 (red) and microtubules (green). White bars correspond to 1 μm. Adapted with permission from Tam, J. & Merino, D. (2015) Stochastic optical reconstruction microscopy (STORM) in comparison with stimulated emission depletion (STED) and other imaging methods. *J. Neurochem.* 135(4), 643–658. Copyright 2015, John Wiley and Sons.

## 23.2 ELECTRON MICROSCOPY

### 23.2.1 PRINCIPLE OF ELECTRON MICROSCOPY

Electron microscopy (EM) is an example of charged particle microscopy. Instead of a light source, a stream of accelerated electrons is used for imaging. These electrons are generated by an electron gun, in which a tungsten filament or a field emission cathode is heated to induce thermoionic emission of electrons. The electrons are pulled away and accelerated in a strong electric field (Figure 23.34) towards the anode. A hole in the anode allows passage of the electron beam towards the sample. The sample is placed on metal grids with 30–100 μm spacing. Instead of optical lenses, electrostatic or electromagnetic lenses are used to focus the electron beam. The focal length of an electromagnetic lens depends on the current running through the magnetic coil. The entire electron path in an electron microscope is under vacuum (Figure 23.34) to avoid collision and dispersion of electrons with gas molecules. To prevent drying of biological samples in the vacuum, they are frozen, embedded, or fixed (Section 23.2.2).

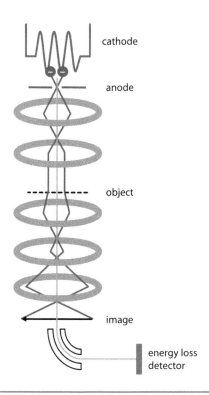

**FIGURE 23.34:  Schematic depiction of an electron microscope.** The electron beam is generated by thermoemission from a cathode (red). Electrons are accelerated in an electric field towards the anode (dark blue). The beam is focused by magnetic lenses (gray) onto the sample. Transmitted electrons are detected by a detector in the image plane (black arrow). Electrons can also be analyzed for their energy loss due to inelastic scattering by the object.

Electrons interact with the sample and undergo elastic or inelastic scattering. The strong interaction of electrons with matter enables imaging of single molecules. Electron microscopy can be performed in different modes, depending on the signals measured. In *transmission electron microscopy* (TEM), the sample is illuminated by the electron beam, and electrons transmitted by the sample are used for imaging, similar to light microscopy. The transmitted beam is imaged onto a high-resolution phosphor screen that is coupled to a CCD sensor. TEM requires thin samples with a thickness < 100 nm, and is useful to image proteins and protein complexes or biological membranes, or ultrathin sections of thicker samples. By sectioning a thick specimen and imaging the different layers separately, a three-dimensional image can be constructed. In *scanning electron microscopy* (SEM), the electron beam is focused into a nanometer spot that is scanned over the sample. Interaction with the sample leads to scattering of electrons in all directions (see Section 23.2.3). Back-scattered electrons are detected to obtain topological information on the surface of the sample at (sub-)nanometer resolution. Although SEM yields images at lower resolution than TEM (Figure 23.35), it offers the advantage of imaging the surface of large, thick specimen. SEM is frequently used in material science, but small organisms and cells can also be imaged. For thin samples, SEM can be used in transmission mode, termed <u>*scanning transmission electron microscopy*</u> (STEM), to achieve sub-nanometer resolution.

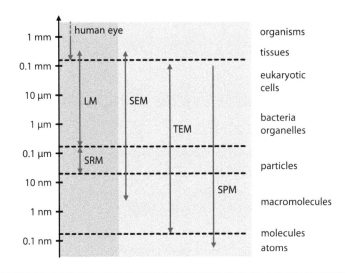

**FIGURE 23.35:**    **Resolution of electron microscopy**. The human eye can resolve objects down to 0.2 mm. LM: light microscopy, SRM: super-resolution microscopy (Section 23.1.8), SEM: scanning electron microscopy, TEM: transmission electron microscopy, SPM: scanning probe microscopy.

By *scanning probe microscopy* (SPM) imaging of surfaces becomes possible down to atomic resolution. *Scanning tunneling microscopy* (STM) is based on the tunnel effect and measures the resulting current when the sample surface is scanned by a small probe. *Scanning force* and *atomic force microscopy* (SFM, AFM) measure forces with which a probe is scanned across a surface and provide surface images. In single particle cryo-EM, particles are spread on a surface, and imaged by electron microscopy. The particles are deposited on the surface in random orientations, and the images contain the projections in these different orientations. From the different projection, a three-dimensional image of the particle can be reconstructed, revealing its dimensions and shape (Section 23.2.4).

### 23.2.2 SAMPLE PREPARATION

To enhance structural features of the sample and to increase the contrast of electron microscopy, samples are often coated with metals and heavy atoms that show strong scattering. Negative staining of proteins, protein complexes, and biological membranes is achieved by applying heavy atom salts that envelop the structure and provide more contrast. In metal shadowing, metal vapor is applied to the sample surface at an angle. After removal of the original sample, a replica is obtained and imaged. To prevent dehydration in the evacuated chamber of the electron microscope, samples are dehydrated either by freeze-drying, by replacing water with organic solvents followed by drying, or by embedding the sample in resins. Resin-embedded cells and tissues can be sectioned into layers with < 100 nm thickness for transmission applications after polymerization of the resin. In cryo-EM, objects or molecules in aqueous solutions are flash-frozen and imaged in vitreous ice. Additional fixation can be achieved by chemical crosslinking: protein structures are stabilized by crosslinking with glutaraldehyde, lipids can be stabilized with osmium tetroxide. Imaging can be improved by studying a large number of molecules that are ordered and aligned in two-dimensional crystals. Membrane proteins can be forced to associate regularly into two-dimensional crystals in lipid bilayers. Two-dimensional crystals of soluble proteins are obtained by ordered assembly on a surface, such as mica or a lipid monolayer.

### 23.2.3 IMAGE GENERATION AND ANALYSIS

According to the wave-particle dualism, particles such as electrons can be described as electromagnetic waves. Their wavelength $\lambda$ depends on the impulse $p$, the product of the particle mass $m$ and its velocity $v$, and the Planck constant $h$ according to the de Broglie relationship:

$$\lambda = \frac{h}{p} = \frac{h}{mv}$$

eq. 21.43

In electron microscopy, electrons are accelerated to velocities approaching the speed of light $c$. Under these conditions, the wavelength $\lambda_e$ depends on the kinetic energy $E$ of the electrons according to

$$\lambda_e = \frac{h}{\sqrt{2m_0 E\left(1 + \dfrac{E}{2m_0 c^2}\right)}}$$

eq. 23.36

where $m_0$ is the resting mass of the electron. The product of $m_0$ and $c^2$ is the resting energy $E_0$ of the electron. By substitution and rearrangement we obtain

$$\lambda_e = \frac{hc}{\sqrt{2E_0 E + E^2}}$$

eq. 23.37

The kinetic energy $E$ of the electrons is the product of the accelerating voltage $U_{acc}$ and the electron charge $e$. Typical voltages of $U_{acc} = 10^5$ V (100 kV) lead to a kinetic energy of $1.6 \cdot 10^{-19}$ J (100 keV). These electrons correspond to radiation with wavelengths $\ll 1$ nm, much shorter than that of light. Because of this, electron microscopy achieves much higher resolution (< 0.1 nm, eq. 23.16) than light microscopy (> 200 nm, see Section 23.1.1.3). In high voltage electron microscopy (> $10^6$ V), electrons reach wavelengths below 1 pm, and individual atoms can be resolved.

The electrons of the electron beam interact with the object and are scattered. Elastic and inelastic scattering provide information for imaging (Figure 23.36). Elastic scattering occurs when the electrons are deflected by the nuclei of the object. This deflection is higher the closer the electron passes to the nucleus, the higher the charge of the nucleus, and the slower the electron. In addition to the deflection, the electron experiences a phase shift. Interference with the non-obstructed beam can then lead to a phase contrast. Inelastic scattering occurs when the electron interacts with electrons from the object, and energy is transferred from the electron to the object. The loss of energy corresponds to an increase in wavelength of the scattered electron. Instead of the uniform wavelength before the object, scattered electrons will show a spectrum of wavelengths after the sample. The stronger the interaction, the more energy is transferred. The loss of energy therefore depends on the elemental composition and can provide information on electron distribution with spatial resolution.

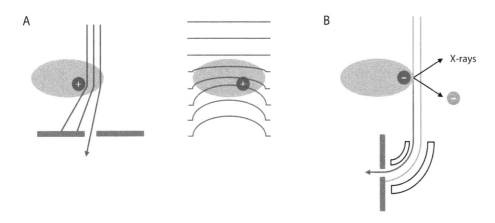

FIGURE 23.36:   **Interactions of electrons with the object**. A: Elastic scattering of electrons by nuclei. The electrostatic interaction causes a deflection of the electrons (left). In TEM, an aperture (dark gray) rejects the highly deflected rays. The slightly deflected rays are phase-shifted relative to the incident beam and give rise to interference contrast upon interference with the unobstructed beam (right). In STEM, an annular detector instead of the aperture detects the highly deflected electron, which provide information on molecular mass. B: Inelastic scattering of electrons by surface electrons leads to a reduction in electron energy and a change in wavelength (indicated by the color change from yellow to red). In a magnetic field, the electrons are deflected depending on their wavelength, and an aperture selects for the wavelength that is detected. Upon interaction of electrons with surface electrons, X-rays can be emitted, or secondary electrons can be generated.

The electron microscopy image obtained is a convolution of the actual structure of the object and the *contrast transfer function* of the microscope. This relationship is analogous to the description of measured fluorescence decays as convolutions of the instrument response and the actual fluorescence decay (see Section 19.5.9). The contrast transfer function is analogous to the instrument response of the fluorescence lifetime spectrometers, and can be measured in a reference image from carbon foil. The contrast transfer function is a pattern of concentric rings whose dimensions depend on the position of the focus. Dark rings correspond to gaps in which no information can be obtained. These detection gaps in the image can be circumvented by adjusting the focus such that the central region is large enough to image the area of interest evenly. Alternatively, the differences in contrast can be corrected by combining images with different focal settings.

In STEM, thin objects are scanned with a focused electron beam. An annular detector permits detection of electrons that are strongly deflected. These electrons contain information on the mass of the object. When the elemental composition is known (e.g. for proteins), the molecular mass can be determined. The calculation is typically performed over several hundreds of molecules to obtain more precise masses, and information on mass distributions (oligomers, fibrils).

### 23.2.4 THREE-DIMENSIONAL ELECTRON MICROSCOPY: CRYO-ELECTRON TOMOGRAPHY AND SINGLE PARTICLE CRYO-EM

Electron microscopy images of three-dimensional objects are projections of the object along the direction of the incident electron beam. The goal of both electron tomography and single particle EM is to reconstruct a three-dimensional image of an object from a series of two-dimensional projections collected from the object in different orientations. In electron tomography, thin sections of a biological specimen kept at liquid nitrogen temperature (100 K) are imaged at a range of tilt angles up to $\pm 70°$. Thousands of different projections need to be carefully aligned and are then combined into a three-dimensional image of the object. Complete data would require tilt angles of 90°, but unfortunately these cannot be accomplished for technical reasons: at some angle, the sample support obstructs the incident electron beam. Another aspect is the increased path length of the electrons through the sample. It becomes so large that most electrons are either absorbed, or inelastically scattered, or suffer multiple events of elastic scattering. Absorption of electrons contributes to radiation damage, while inelastically and multiply scattered electrons blur the image. Inelastically scattered electrons have less energy than elastically scattered electrons and are focused in different planes, which explains the blurring effect (*chromatic aberration*). To eliminate inelastically scattered electrons, magnetic energy filters are used to divert them before they hit the detector. However, as a result of the incomplete tilting range, some EM data are always missing along the axis perpendicular to the two-dimensional sample (the *z*-axis). If several samples are imaged at different starting orientations, the missing data part can be reduced from a wedge to a cone, but in any case the three-dimensional reconstruction will be distorted (elongated) and be of lower resolution in the *z*-direction.

As the samples are not metal stained, the signal in electron tomography as well as in unstained cryo-EM, comes from the intrinsic scattering of the incident electrons by the proteins and nucleic acids. This signal is inherently weak. To improve the contrast in the three-dimensional tomogram reconstructions, it is sometimes possible to average the electron density in sub-volumes of the tomogram that have the same internal structure. Electron tomography has enabled studies of the interior structure of cells (organelle localization, cytoskeleton structure, nucleic acid organization), cellular fusion, virus entry into cells, and the workings of the flagellar motor. Imaging by tilting of the specimen is also applied to ordered single molecules and two-dimensional crystals (Box 23.11).

---

**BOX 23.11: ELECTRON CRYSTALLOGRAPHY.**

The low signal in EM images is one of the main reasons for the low resolution of EM structures. To improve the signal in cryo-EM, images are taken from a large number of molecules, aligned, and averaged. Alternatively, pre-aligned molecules can be studied by electron crystallography. Aligning and averaging images of single molecules from different orientations is conceptually identical to pre-ordering the molecules first in a crystal lattice and then imaging them. As crystals are arrays of very well-aligned molecules, the electron dose used in electron crystallography is smaller than that needed for single particle cryo-EM. The wavelength of the incident electrons is small compared to atomic distances, so electron crystallography has the potential to produce high-resolution information on atomic positions. Electron crystallography combines advantages from microscopy and crystallography by collecting both images and electron diffraction patterns from crystals. The phases of the object may be retrieved from the images (no phase problem exists in electron crystallography), and the signal in the diffraction pattern is amplified by the crystal. Due to the strong interaction of electrons with matter, the crystals used for electron crystallography need to be thin, ideally less than 100 nm. Two-dimensional sheets or helical crystals produced by fibers are commonly used, but three-dimensional micro-crystals also work. Membrane proteins, which are naturally dissolved in two-dimensional lipid bilayers *in vivo*, can sometimes be made to associate into two-dimensional crystals. Two-dimensional or thin three-dimensional crystals of soluble proteins may also be obtained by ordered assembly on a helper surface, such as mica or a lipid monolayer. The diffraction by these sheet-like crystals is continuous in the direction perpendicular to the crystal plane. The resulting lattice lines are sampled by collecting data at different tilt angles (±70°), just as with cryo-EM imaging and electron tomography, but there will almost always be a missing part of data that affects the quality of the resulting electron density map. In addition, the thin crystals are prone to bending during sample preparation, an effect that can be computationally corrected if it is small (*un-bending*).

*(Continued)*

The protein databank holds ~50 structures determined by electron crystallography, among them amyloid-forming peptides, soluble proteins (lysozyme, catalase), muscle and fiber proteins (actin, myosin, tubulin), and membrane proteins (bacteriorhodopsin, $K^+$ channel, $Ca^{2+}$ ATPase, prostaglandin E synthase). The first structure of bacteriorhodopsin had a resolution of only 0.35 nm, showing the orientation of α-helices. More details were visible in the structure of the membrane protein aquaporin AQP0, which has been determined from double-layered, two-dimensional crystals at a resolution of 0.19 nm, including the lipid bilayer surrounding the AQP0 tetramers and the water molecules within the pores (Figure 23.37).

FIGURE 23.37:    **Structure of aquaporin.** The membrane protein aquaporin forms homotetramers of four water-conducting pores. At the center of each channel there are three water molecules (red spheres) that are too far apart to form H-bonds. The lipids (yellow stick models) packing around the tetramer adopt preferred conformations. Gonen *et al.* (2005). *Nature* 438(7068): 633–638.

An alternative to tilting of the sample is to deposit single molecules in random orientations on a grid, such that all projections are obtained in a single image. The projections are sorted into classes, each class is averaged, and the averages are combined into a three-dimensional image. Averaging of $N$ images improves the signal-to-noise ratio by $N^{\frac{1}{2}}$, provided that the noise in the images is not correlated with the structure. The procedure of classification and averaging is at the heart of single particle cryo-EM, perhaps the most widely used type of electron microscopy. Cryo-EM enables determination of the three-dimensional structures of large and flexible macromolecular complexes that cannot be studied to comparable resolutions by other techniques.

A few decades ago, cryo-EM was limited to rather low resolution on the order of 0.7–3 nm and required large, preferably highly symmetric specimen of typically 1–10 MDa molecular mass. Large particles give high contrast in EM, which allows accurate determination of particle orientation. High symmetry enables averaging of the recorded images over the internal symmetry elements of the particles, strongly improving the signal-to-noise ratio. Consequently, the first EM structures were those of highly symmetric virus capsids of several MDa molecular mass that exhibit at least 60-fold symmetry. For smaller and less symmetric molecules, the electron density maps often did not contain much information on the internal structure of the molecules but, like SAXS, provided an envelope of the molecular shape. In the case of macromolecular complexes, these shapes are useful for docking of individual crystal structures and NMR ensembles to get an integrative view of the complex.

The images in EM contain the very phase information that is initially absent in crystallographic techniques. As a result, the electron density maps from an EM experiment are rarely improved by phase information from an atomic model (as is the case in X-ray crystallography). Other maps simply cannot be interpreted by an atomic model due to their limited resolution. It is therefore customary to deposit the electron density maps themselves in the *EM databank* (Box 23.12).

---

**BOX 23.12: EM DATABANK.**

Electron density maps from EM experiments are deposited in the EM databank (emdatabank.org), which in 2015 held 3367 maps and 983 models. Icosahedral viruses and ribosome structures make up almost half of all entries. The other half is dominated by molecules that display high internal symmetry, such as GroEL (see Figure 16.33) and the proteasome. About 30% of entries have resolutions of < 1 nm and 10% of the maps have a resolution of < 0.5 nm. $\alpha$-helices start to become visible in the maps at resolutions of 0.6–0.9 nm, and individual $\beta$-strands and large side chains may be discerned at resolutions beyond 0.45 nm. Hence, most EM structures do not contain detailed structural information, but they nevertheless answer many biological questions about stoichiometry and conformational plasticity of large complexes. The recent developments in detector and image processing allow more and more maps to be interpreted in terms of atomic models, and these models are also deposited in the PDB, so the two databases interface with each other.

---

Until recently, X-ray crystallography and NMR spectroscopy were far more successful than EM in determining high-resolution macromolecular structures. However, a critical disadvantage of crystallography is the requirement for the time-consuming production of crystals. For NMR, the problem lies in the size limitation to 40–50 kDa molecular mass that is amenable to study, and both techniques usually require large amounts of highly pure sample. All these limitations are in principle absent in single particle cryo-EM, which may require as little as 0.1–1 mg of sample and is less sensitive to impurities because their contributions may be eliminated during image selection. The remaining obstacle, limited resolution, has been addressed in recent years by two developments that radically changed the field: *single electron counting detectors* and new image processing procedures. Single particle cryo-EM structures begin to rival the resolution and quality of X-ray crystal structures, although the current lower limit of ca. 100 kDa molecular mass still remains. The development of cryo-EM is exemplified by the ribosome: EM structures of ribosomes (see Figure 17.37) had a resolution of 4 nm in the 1990s. A decade later, the resolution was improved to 0.7–0.9 nm, mainly due to improvements in sample preparation and electron beams. The current resolution of a cryo-EM ribosome structure is less than 0.3 nm, enabled by the recent improvements in detectors and image processing.

Inelastic scattering of electrons transfers energy onto the sample, liberates more electrons, and produces radicals. To limit the radiation damage caused by these effects, the dose is limited to ca. 2000–6000 electrons per $nm^2$. Despite limiting the dose, EM samples degrade quickly, severely limiting the overall signal that can be obtained from a single molecule. As a consequence, EM images are inherently noisy (Figure 23.38).

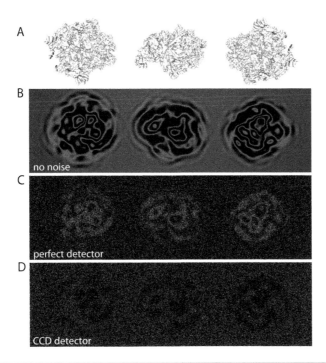

FIGURE 23.38:    **Noise in EM images.** The structure of the mitochondrial ribosome ("mitoribosome"; PDB-ID 5aj4), determined at a resolution of 0.38 nm, was used for these projection calculations. A: Three orientations of the ribosome. B: The projection of the orientations in the absence of any noise. C: EM image from a perfect detector with detective quantum efficiency of unity. D: CCD detector. In reality, the projections of the molecules are often barely visible by eye in the images.

The signal coming from the sample is further diminished by inefficient recording in the detector, described by a parameter termed *detective quantum efficiency* (DQE). A perfect detector would have a DQE of unity because it will not add additional (electronic) noise to the signal, and it will count each and every single electron. Photographic film and CCD cameras have approximate DQEs of 0.3 and 0.1, respectively. The relatively large DQE and the small pixel size of photographic film explain why the time-consuming film development was still used in cryo-EM when other techniques already adopted CCD detectors. Ultimately, more efficient and fast CCD detectors became the standard also in cryo-EM. During the last few years a new class of *direct electron detectors* has been developed. These detectors consist of an array of semi-conducting pixels, each coupled to its own readout electronics. When impinged by a single electron, the pixel either stores the energy from the collision or the electron is counted in the pixel. Data is read out in a matter of microseconds by dedicated CMOS (complementary metal-oxide semiconductor) electronics: each pixel has its own amplifier. This setup is similar to the pixel array detectors now routinely used in X-ray crystallography. A support matrix is needed behind the pixel array, but back-scatters some electrons towards the pixel array, introducing noise. Thus, the support matrix is kept as thin as possible (*back thinning*), and the overall thickness of the detector is only about 50 mm. Some detectors (e.g. the Falcon) integrate the energy deposited by the electrons, while others (e.g. the K2) count the number of electrons that hit the detector. The detectors are quickly saturated at higher electron fluxes, so they operate in *movie mode*: the energy of the recorded electrons is spread over several hundred images per second and later merged into a single exposure. Currently, the DQE of such detectors is about 0.5, but values > 0.8 should be possible.

Image processing procedures have improved in two areas, dealing with sample heterogeneity and correcting for electron beam-induced movement of the sample. Almost all macromolecules, but especially the large macromolecular complexes most amenable to cryo-EM, exhibit considerable conformational heterogeneity. The two-dimensional projections of these conformations must be grouped into separate classes of the same conformation because averaging over different conformations not only drastically decreases the resolution of the average, but the resulting three-dimensional reconstruction will be blurred as well. Distinguishing between projections from different directions and genuinely different conformations is difficult. Modern algorithms rely on maximum likelihood and Bayesian statistics to automatically sort the projections into classes representing a single conformation. The second improvement relates to correction of beam-induced movements. The particle nature of electrons endows them with an impulse $p = m_e v$. The energy deposited in the sample by the electron beam induces motion, both in the vitreous ice layer and the macromolecule. Movement of the sample on the time scale of the image recording leads to blurring of the resulting three-dimensional reconstruction. The movie mode of the new direct electron detectors is used to follow the sample movement, and then to back-calculate the position of the macromolecule for each image in the movie, essentially re-aligning the images in the movie so that averaging them improves the signal-to-noise ratio. The procedure is improved when several particles are in the field of view, so that their trajectories can be compared. There is synergy between the improvements in detectors and sorting algorithms: the detectors record higher resolution images that can more easily be assigned into classes and motion-corrected by the improved algorithms. The number of particles that have to be imaged and classified for the reconstructions is also reduced by 1–2 orders of magnitude: only 35000 particles were needed for a 0.4 nm resolution reconstruction of the *T. thermophilus* ribosome where many side chains were visible in the electron density map.

Because the phases are measured in EM, electron density maps calculated from EM data tend to show more details than do density maps from X-ray crystallography at the same nominal resolution. The same real-space refinement techniques and stereochemical restraints that are used for X-ray crystallography have been adapted for refinement of EM models. Even small molecules can now be visualized by EM methods, provided that they are bound to macromolecules large enough to allow reliable image processing of the two-dimensional projections. Examples are the hexameric 540 kDa ATPase p97, a cancer drug target, and the tetrameric 465 kDa β-galactosidase (Figure 23.39). Both structures were determined in complex with small molecule inhibitors and/or allosteric regulators at resolutions of 0.23 nm and 0.22 nm, respectively, in principle opening cryo-EM to the field of structure-based drug design.

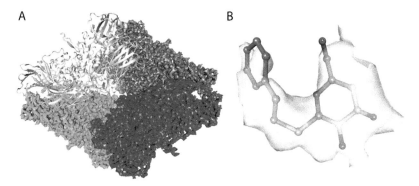

**FIGURE 23.39: High-resolution EM structure of β-galactosidase.** The electron density map for this tetrameric enzyme of 465 kDa is shown for three monomers in different colors. The fourth monomer is shown as a ribbon diagram. On the right is the electron density at 1.8 rmsd for the ligand (PDB-ID 5a1a).

The current developments in EM aim to determine structures of smaller and asymmetric macromolecules with molecular masses approaching 100 kDa or less. Until now, the success of high-resolution cryo-EM has relied on large (many MDa), highly symmetric (> 60-fold) molecules, such as icosahedral virus capsids. One of the smallest, asymmetric proteins that has been studied using cryo-EM is γ-secretase, a heterotetrameric membrane protein of 170 kDa molecular mass. However, the current resolution of 0.45 nm is not yet sufficient for *de novo* model building or visualization of small molecule ligands, showing the impact that internal symmetry of the sample has on resolution. Other developments in this field aim for reduction or even prevention of electron beam-induced sample movement by reducing the electron dose. Radiation damage due to charging of the sample by the electron beam can be reduced by suspending the sample on sheets of graphene.

### 23.2.5 SCANNING PROBE MICROSCOPY: SCANNING TUNNELING, SCANNING FORCE, AND ATOMIC FORCE MICROSCOPY

In scanning probe microscopy, surfaces are scanned by a small probe, and distance-dependent interactions of the probe with the surface are measured. *Scanning tunneling microscopy* (STM) measures the tunnel current (0.01–10 pA) between the probe and the surface. In *scanning force* or *atomic force microscopy* (SFM/AFM), a small tip is scanned across the surface, and the sum of all attractive and repulsive forces between the tip and the surface is measured (Figure 23.40). These forces are in the range of several pN to 100 nN. Biological samples are spread on mica, whereas hydrophobic organic molecules are analyzed on graphite surfaces.

AFM provides information on surface topology with nanometer resolution, depending on the dimensions of the tip. The pointed tip is attached to a cantilever. Attractive or repulsive forces between the tip and the sample surface lead to deflection of the cantilever which is measured as a function of position, yielding an image of the surface in terms of height (Figure 23.40). AFM measures the sum of several forces with different range, such as van der Waals forces, electrostatic forces, or capillary forces. By functionalization of the tip, specific forces can be measured, including conductivity, hydrophobicity, or binding forces between interacting molecules.

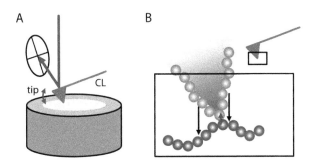

**FIGURE 23.40: Principle of an AFM.** A: A fine tip attached to a cantilever (CL) is scanned over the sample (light gray). Changes in sample height lead to a deflection of the cantilever (blue double-headed arrow). The deflection of the cantilever is detected by a laser (red) that is reflected from the cantilever to a position-sensitive detector (D). B: Schematic view of the tip (gray) and the surface (blue). The spheres show the tip and surface atoms. The force between tip and surface consists of attractive forces (black arrow) and repulsive forces (red arrow).

The extent of the deflection $\Delta x$ of the AFM tip associated with a certain force $F$ depends on the stiffness $k$ of the cantilever, and follows Hooke´s law (see Section 19.4.1):

$$F = -k\Delta x$$ eq. 23.38

The lower the stiffness $k$, the higher the deflection associated with a certain force. Thus, cantilevers with lower stiffness allow measurements of smaller surface features, but are also more susceptible to perturbations that generate noise.

AFM can be performed in different modes. In the *contact mode* (Figure 23.41), the AFM tip is pulled directly across the surface. The deflection is either measured directly, or the tip is kept at a constant distance by applying a feedback force. Such an experiment reveals the contours of the surface and the surface topology. For soft biological samples, the contact of the AFM tip with the surface can be destructive, and modes with intermittent or no contact between tip and sample are preferred. In the *tapping mode* (Figure 23.41), the AFM tip oscillates about 100–200 nm above the surface. Interactions with the surface lead to a dampening of these oscillations. A feedback force keeps the tip at constant distance to maintain the oscillation frequency. The image obtained in the tapping mode contains information on surface topology (height), but also on its elasticity and hardness. In *frequency-modulated AFM* (non-contact mode, Figure 23.41), the AFM tip undergoes a low amplitude frequency-modulated oscillation (less than in the tapping mode) during surface scanning. The modulation is altered by (van der Waals) interactions with the surface. The same feedback system is used as for the tapping-mode experiment.

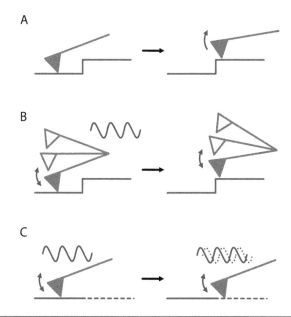

**FIGURE 23.41: AFM modes and examples for AFM images.** A: Contact mode. The tip (gray) is scanned over the surface (red) in direct contact. The topology of the surface leads to a deflection of the tip (right). B: Tapping mode and topology. The oscillating tip is moved across the surface, which minimizes the contact time. The step in the surface is detected as a deflection of the oscillating tip. C: Tapping mode and softness. The oscillating tip is scanned across the surface. The softer the surface, the larger the phase shift between the actual tip oscillation and the driving oscillation.

Recent developments in AFM methodology aim at imaging with increased, sub-second time-resolution (high-speed AFM) and at topological imaging in combination with fluorescence detection. In addition to imaging applications, AFM can also be used to measure inter- or intramolecular forces (see Section 24.1). Force spectroscopy by AFM is a powerful tool to analyze protein and nucleic acid folding and stability, forces exerted by enzymes on nucleic acids or filaments, and binding forces in general.

## QUESTIONS

23.1   You need to expand a laser beam from $d_1 = 0.2$ mm to $d_2 = 5$ cm to fully illuminate the back-focal plane of your objective. The microscope has two empty lens holders with a distance of 26 cm. What lenses can you use?

23.2   A rapidly diffusing molecule has a very short residence time in the confocal volume of an FCS instrument. How do you have to change the size of the confocal volume to be able to measure the diffusion time? Does this change require a widening or a focusing of the excitation beam?

23.3    In an FCS experiment, the following fluorescence intensities are measured as a function of time.

| *t* (ns) | intensity |
|---|---|
| 0.1 | 1 |
| 0.2 | 1 |
| 0.3 | 2 |
| 0.4 | 4 |
| 0.5 | 1 |
| 0.6 | 1 |
| 0.7 | 1 |
| 0.8 | 1 |
| 0.9 | 0 |
| 1 | 0 |

Calculate the autocorrelation curve for $\tau = 0$ ns to $\tau = 1$ ns (in 0.1 ns steps). What is the diffusion time of the molecule through the confocal volume?

23.4    By FCS, the diffusion time of a double-stranded DNA is measured as $\tau_{\text{diff,DNA}} = 3$ ns. In presence of saturating concentrations of a restriction enzyme, the diffusion time increases to $\tau_{\text{diff,complex}} = 10$ ns. At a concentration of $c_{\text{DNA}} = 100$ nM and an enzyme concentration $c_{\text{enzyme}} = 500$ nM, an apparent diffusion time of 8 ns is measured. What is the $K_d$ value of the enzyme-DNA complex?

23.5    How can you use fluorescence cross-correlation (FCCS) to test if a FRET sample contains species that carry both the donor and the acceptor fluorophore?

23.6    In a total internal reflection fluorescence (TIRF) microscope the evanescent field is created at the glass-water interface ($n_1 = n_{\text{glass}} = 1.51$ and $n_2 = n_{\text{water}} = 1.33$). Calculate the critical angle $\theta_c$ for total internal reflection. What is the penetration depth $d$ of the evanescent field into the sample (water) at $\theta_c$ for excitation light with $\lambda = 280$ nm, $\lambda = 495$ nm, and $\lambda = 800$ nm. What is $d$ if the beam hits the interface with $\theta = 70°$ and $\theta = 80°$?

23.7    Compare the information from dwell-time histograms obtained by single molecule microscopy with the information from a histogram of photon arrival times in time-correlated single photon counting.

23.8    Why do fluorescence bursts from single molecule FRET by confocal microscopy show a rather wide distribution of intensities and durations?

23.9    Derive a formula that relates the acceleration voltage in an electron microscope to the wavelength of electrons. How should this formula be modified to take relativistic effects into account? What are the wavelengths for acceleration voltages of 100 kV, 200 kV, and $10^6$ V?

## REFERENCES

Selvin, M., Ha, T. (2008) Single-molecule techniques – A laboratory manual. New York, Cold Spring Harbour Press.
· an overview about single-molecule techniques, including single-molecule fluorescence microscopy, FRET and FCS, with a focus on practical aspects and protocols.

Gell, C., Brockwell, D. and Smith, A. (2006) Handbook of single-molecule fluorescence spectroscopy. Oxford, Oxford University Press.
· single molecule microscopy

Okumus, B., Arslan, S., Fengler, S. M., Myong, S. and Ha, T. (2009) Single molecule nanocontainers made porous using a bacterial toxin. *J Am Chem Soc* **131**(41): 14844–14849.
· lipid vesicles containing the pore-forming toxin hemolysin as nano-containers for single-molecule experiments

Tokunaga, M., Imamoto, N. and Sakata-Sogawa, K. (2008) Highly inclined thin illumination enables clear single-molecule imaging in cells. *Nat Methods* **5**(2): 159–161.
· description of highly-inclined and laminated optical sheet microscopy (HILO) and its application to *in vivo* imaging

Dertinger, T., Pacheco, V., von der Hocht, I., Hartmann, R., Gregor, I. and Enderlein, J. (2007) Two-focus fluorescence correlation spectroscopy: a new tool for accurate and absolute diffusion measurements. *Chemphyschem* **8**(3): 433–443.
· accurate determination of diffusion coefficients by dual-focus FCS

Weiss, K. and Enderlein, J. (2012) Lipid diffusion within black lipid membranes measured with dual-focus fluorescence correlation spectroscopy. *Chemphyschem* **13**(4): 990–1000.
· application of dual-focus FCS to lipid diffusion in black lipid membranes

Dorlich, R. M., Chen, Q., Niklas Hedde, P., Schuster, V., Hippler, M., Wesslowski, J., Davidson, G. and Nienhaus, G. U. (2015) Dual-color dual-focus line-scanning FCS for quantitative analysis of receptor-ligand interactions in living specimens. *Scientific reports* **5**: 10149.
· determination of ligand affinities of membrane receptors within the membrane of living cells by dual-focus FCS and line-scanning

Baudendistel, N., Muller, G., Waldeck, W., Angel, P. and Langowski, J. (2005) Two-hybrid fluorescence cross-correlation spectroscopy detects protein-protein interactions *in vivo*. *Chemphyschem* **6**(5): 984–990.
· *in vivo* study of the interaction between fluorescently labeled transcription factors jun and fos by FCCS

Ohrt, T., Mutze, J., Staroske, W., Weinmann, L., Hock, J., Crell, K., Meister, G. and Schwille, P. (2008) Fluorescence correlation spectroscopy and fluorescence cross-correlation spectroscopy reveal the cytoplasmic origination of loaded nuclear RISC *in vivo* in human cells. *Nucleic Acids Res* **36**(20): 6439–6449.
· *in vivo* FCS/FCCS study on the RNA-induced silencing complex that identified cytoplasmic and nuclear forms with different diffusion coefficients

Wachsmuth, M., Conrad, C., Bulkescher, J., Koch, B., Mahen, R., Isokane, M., Pepperkok, R. and Ellenberg, J. (2015) High-throughput fluorescence correlation spectroscopy enables analysis of proteome dynamics in living cells. *Nat Biotechnol* **33**(4): 384–389.
· high-throughput FCS (and FCCS) format to study the dynamics of large numbers of proteins in living cells by time-lapse acquisition of FCS data at specific subcellular locations

Seisenberger, G., Ried, M. U., Endress, T., Buning, H., Hallek, M. and Brauchle, C. (2001) Real-time single-molecule imaging of the infection pathway of an adeno-associated virus. *Science* **294**(5548): 1929–1932.
· single-virus tracking to study virus entry pathways into cells

Lommerse, P. H., Snaar-Jagalska, B. E., Spaink, H. P. and Schmidt, T. (2005) Single-molecule diffusion measurements of H-Ras at the plasma membrane of live cells reveal microdomain localization upon activation. *J Cell Sci* **118**(Pt 9): 1799–1809.
· single-molecule tracking to investigate lateral diffusion of the membrane-associated protein Ras

Schaaf, M. J., Koopmans, W. J., Meckel, T., van Noort, J., Snaar-Jagalska, B. E., Schmidt, T. S. and Spaink, H. P. (2009) Single-molecule microscopy reveals membrane microdomain organization of cells in a living vertebrate. *Biophys J* **97**(4): 1206–1214.
· single-molecule tracking of Ras in membranes of living zebrafish embryos

Yildiz, A. and Selvin, P. R. (2005) Fluorescence imaging with one nanometer accuracy: application to molecular motors. *Acc Chem Res* **38**(7): 574–582.
· description of fluorescence imaging with one nanometer accuracy (FIONA)

Hwang, H. and Myong, S. (2014) Protein induced fluorescence enhancement (PIFE) for probing protein-nucleic acid interactions. *Chem Soc Rev* **43**(4): 1221–1229.
· review article on applications of protein-induced fluorescence enhancement (PIFE)

Honda, M., Park, J., Pugh, R. A., Ha, T. and Spies, M. (2009) Single-molecule analysis reveals differential effect of ssDNA-binding proteins on DNA translocation by XPD helicase. *Mol Cell* **35**(5): 694–703.
· measurement of translocation by the DNA helicase XPD along DNA by fluorescence quenching

Friedman, L. J., Chung, J. and Gelles, J. (2006) Viewing dynamic assembly of molecular complexes by multi-wavelength single-molecule fluorescence. *Biophys J* **91**(3): 1023–1031.
· principle of multi-color detection that is the basis for colocalization single-molecule spectroscopy (CoSMoS)

Jain, A., Liu, R., Ramani, B., Arauz, E., Ishitsuka, Y., Ragunathan, K., Park, J., Chen, J., Xiang, Y. K. and Ha, T. (2011) Probing cellular protein complexes using single-molecule pull-down. *Nature* **473**(7348): 484–488.
· original description of single-molecule pull-down experiments

Rasnik, I., McKinney, S. A. and Ha, T. (2006) Nonblinking and long-lasting single-molecule fluorescence imaging. *Nat Methods* **3**(11): 891–893.
· validation of the triplet state quencher Trolox to minimize blinking

Hohng, S., Joo, C. and Ha, T. (2004) Single-molecule three-color FRET. *Biophys J* **87**(2): 1328–1337.
· three-color FRET study that monitors two distance vectors in Holliday junctions simultaneously

Harms, U., Andreou, A. Z., Gubaev, A. and Klostermeier, D. (2014) eIF4B, eIF4G and RNA regulate eIF4A activity in translation initiation by modulating the eIF4A conformational cycle. *Nucleic Acids Res* **42**(12): 7911–7922.
· investigation of the kinetics of the conformational cycle of the translation initation factor eIF4A and its regulation by interaction partners by single-molecule FRET using TIRF microscopy

Andreou, A. Z. and Klostermeier, D. (2014) eIF4B and eIF4G jointly stimulate eIF4A ATPase and unwinding activities by modulation of the eIF4A conformational cycle. *J Mol Biol* **426**(1): 51–61.
· identification of three conformational states of the translation initation factor eIF4A by confocal microscopy

Gubaev, A. and Klostermeier, D. (2014) The mechanism of negative DNA supercoiling: a cascade of DNA-induced conformational changes prepares gyrase for strand passage. *DNA Repair (Amst)* **16**: 23–34.
· review article summarizing single-molecule FRET studies that identified a series of conformational changes in the catalytic cycle of DNA gyrase

Sustarsic, M. and Kapanidis, A. N. (2015) Taking the ruler to the jungle: single-molecule FRET for understanding biomolecular structure and dynamics in live cells. *Curr Opin Struct Biol* **34**: 52–59.
· review article highlighting advances in applications of single-molecule FRET *in vivo*

Huang, B., Bates, M. and Zhuang, X. (2009) Super-resolution fluorescence microscopy. *Annu Rev Biochem* **78**: 993–1016.
· review article comparing different super-resolution microscopy techniques

Egner, A. and Hell, S. W. (2005) Fluorescence microscopy with super-resolved optical sections. *Trends Cell Biol* **15**(4): 207–215.
· review article on 4Pi super-resolution microscopy

Hell, S. W., Kroug, M. (1995) Ground-state depletion fluorescence microscopy, a concept for breaking the diffraction resolution limit. *Appl Phys B* **60**: 495–497.
· original description of the concept of stimulated emission depletion (STED) microscopy

Klar, T. A., Jakobs, S., Dyba, M., Egner, A. and Hell, S. W. (2000) Fluorescence microscopy with diffraction resolution barrier broken by stimulated emission. *Proc Natl Acad Sci U S A* **97**(15): 8206–8210.
· first achievement of super-resolution by stimulated emission depletion (STED) microscopy

Schmidt, R., Wurm, C. A., Punge, A., Egner, A., Jakobs, S. and Hell, S. W. (2009) Mitochondrial cristae revealed with focused light. *Nano Lett* **9**(6): 2508–2510.
· imaging of mitochondrial fine-structure using stimulated emission depletion (STED) microscopy

Tam, J. and Merino, D. (2015) Stochastic optical reconstruction microscopy (STORM) in comparison with stimulated emission depletion (STED) and other imaging methods. *J Neurochem* **135**(4): 643–658.
· review article comparing different super-resolution and conventional microscopy techniques

Fischer, N., Neumann, P., Konevega, A. L., Bock, L. V., Ficner, R., Rodnina, M. V. and Stark, H. (2015) Structure of the *E. coli* ribosome-EF-Tu complex at < 3 Å resolution by Cs-corrected cryo-EM. *Nature* **520**(7548): 567–570.
· EM structure of the ribosome/EF-Tu complex

Bai, X. C., Fernandez, I. S., McMullan, G. and Scheres, S. H. (2013) Ribosome structures to near-atomic resolution from thirty thousand cryo-EM particles. *Elife* **2**: e00461.
· ribosome structures to near-atomic resolution by cryo-EM

Banerjee, S., Bartesaghi, A., Merk, A., Rao, P., Bulfer, S. L., Yan, Y., Green, N., Mroczkowski, B., Neitz, R. J., Wipf, P., Falconieri, V., Deshaies, R. J., Milne, J. L., Huryn, D., Arkin, M. and Subramaniam, S. (2016) 2.3 A resolution cryo-EM structure of human p97 and mechanism of allosteric inhibition. *Science* **351**(6275): 871–875.
· EM structure of the 540 kDa ATPase p97, an anti-cancer drug target

Bartesaghi, A., Merk, A., Banerjee, S., Matthies, D., Wu, X., Milne, J. L. and Subramaniam, S. (2015) 2.2 A resolution cryo-EM structure of β-galactosidase in complex with a cell-permeant inhibitor. *Science* **348**(6239): 1147–1151.
· cryo-EM structure of a comparatively small protein, the tetrameric 465 kDa β-galactosidase

Lu, P., Bai, X. C., Ma, D., Xie, T., Yan, C., Sun, L., Yang, G., Zhao, Y., Zhou, R., Scheres, S. H. and Shi, Y. (2014) Three-dimensional structure of human γ-secretase. *Nature* **512**(7513): 166–170.
· EM structure of human γ-secretase

Scheres, S. H. (2014) Beam-induced motion correction for sub-megadalton cryo-EM particles. *Elife* **3**: e03665.
· describes how motion correction due to the electron beam can be corrected to yield higher resolution data

Kalle, W. and Strappe, P. (2012) Atomic force microscopy on chromosomes, chromatin and DNA: a review. *Micron* **43**(12): 1224–1231.
· review article about applications of AFM to image DNA and DNA-protein complexes

Ando, T. (2014) High-speed AFM imaging. *Curr Opin Struct Biol* **28**: 63–68.
· review article about instrumentation and applications in high-speed AFM

Uchihashi, T., Iino, R., Ando, T. and Noji, H. (2011) High-speed atomic force microscopy reveals rotary catalysis of rotorless F(1)-ATPase. *Science* **333**(6043): 755–758.
· real-time imaging of the rotation of $F_1$ ATPase by high-speed AFM at sub-second resolution

Kodera, N., Yamamoto, D., Ishikawa, R. and Ando, T. (2010) Video imaging of walking myosin V by high-speed atomic force microscopy. *Nature* **468**(7320): 72–76.
· imaging of myosin V movement along actin by high-speed AFM

Hecht, E., Knittel, P., Felder, E., Dietl, P., Mizaikoff, B. and Kranz, C. (2012) Combining atomic force-fluorescence microscopy with a stretching device for analyzing mechanotransduction processes in living cells. *Analyst* **137**(22): 5208–5214.
· original work describing the combination of topological imaging by AFM with fluorescence imaging

# Force Measurements

ntra- and intermolecular forces hold together biomolecular structures (Chapter 15), and mediate molecular interactions. Forces are also central in any kind of motion, such as the movement of motor proteins along tracks in muscle contraction, chromosome segregation or organelle transport, or the translocation of proteins along nucleic acids. In force methods, as the name implies, calibrated forces may be applied to biomolecules or forces associated with biological processes may be measured. By applying forces and pulling on molecules their mechanical response is probed to analyze structure, stability, and folding pathways, or interactions with binding partners. In this type of experiment, the force is measured as a function of the extension of the molecule, either by atomic force microscopy (Section 24.1), *optical tweezers* (Section 24.2), or *magnetic tweezers* (Section 24.3). In force experiments, forces that biomolecules exert on one another can be measured, and the mechanochemical coupling of enzymes and motor proteins can be interrogated. For example, motor proteins generate forces when they move along their track, such as myosin that moves along actin filaments during muscle contraction, or kinesin that moves along tubulin filaments to transport its cargo through the cell. DNA and RNA polymerases exert forces on their DNA template during replication and transcription. Ribosomes exert forces on the translated mRNA. These forces can be measured in force experiments using optical tweezers (Section 24.2). Torsional stress in DNA leads to supercoiling (Section 17.5.3.2). Magnetic tweezers allow to apply torque to DNA and to study DNA twisting, untwisting, and supercoiling (Section 24.3).

Force experiments are performed on single molecules, and therefore avoid limitations due to ensemble averaging (Section 23.1.7.2). They measure the mechanical response of single molecules without the need to synchronize, and directly reveal the distribution of the observed parameter instead of the mean. Force experiments therefore reveal folding pathways, their probabilities, the existence of off-pathway and rare on-pathway intermediates, and kinetically trapped species. They detect movement and pausing of enzymes, their pausing probabilities, and pause length.

Force is a state variable that affects equilibrium constants and rate constants, much as the state variables pressure $p$ and temperature $T$. The free energy change $dG$ now not only depends on temperature $T$ and pressure $p$ (see eq. 2.113 in Section 2.5.2), but also on the force $F$ according to

$$dG = -SdT + Vdp - xdF \qquad \text{eq. 24.1}$$

At constant pressure and temperature, eq. 24.1 simplifies to

$$dG = -x dF$$

eq. 24.2

Integration from $F = 0$ (no force applied) to $F$ (*force $F$* applied)

$$\int_{F=0}^{F} dG = -\int_{F=0}^{F} x dF$$

eq. 24.3

yields

$$G(F) = G_0 - xF$$

eq. 24.4

$G_0$ is the free energy in the absence of force. The linear term $x \cdot F$ leads to a tilting of energy profiles (Figure 24.1).

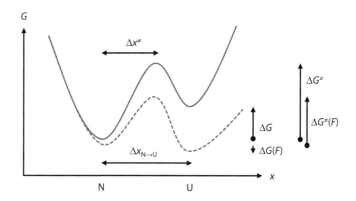

**FIGURE 24.1:   Tilting of energy landscapes by external forces**. The energy landscape for unfolding of a protein from its native state N to its unfolded state U is tilted in the presence of an external force. High forces favor states with larger extension $x$ (eq. 24.4). The external force $F$ therefore stabilizes the transition state of unfolding, and even more so the unfolded state with the highest extension. As a consequence, the energy barrier for unfolding, $\Delta G^{\neq}$, is decreased to $\Delta G^{\neq}(F)$, and unfolding is accelerated. The free energy change for unfolding is changed from $\Delta G$ to $\Delta G(F)$, and the equilibrium constant $K_{\text{unfold}}$ increases.

From eq. 24.4, we see that the application of force stabilizes more extended species, and therefore changes equilibrium constants between species with different dimensions. For the force dependence of changes in free energy $\Delta G(F)$, we obtain

$$\Delta G(F) = \Delta G_0 - F \Delta x$$

eq. 24.5

where $\Delta x$ is the change in extension from reactant to product. In accordance with the principle of Le Chatelier (Section 3.1.2), the application of force thus favors reactions with an increase in extension, i.e. unfolding reactions. In contrast to $p$ and $T$, force is a vector. The effect of force on the free energy $\Delta G$ thus depends on the direction of force application.

Forces also affect rate constants in reactions where the extension of the molecule is changed in the transition state, such as unfolding reactions:

$$k(F) = k_0 \exp\left( \frac{F dx^{\neq}}{k_B T} \right)$$

eq. 24.6

$k_B$ is the Boltzmann constant, $dx^{\neq}$ is the change in extension in the transition state, and $k_0$ is the unfolding rate constant at zero force. Note that unfolding pathways in mechanically induced unfolding transitions are most likely different from the pathways in thermal or solvent-induced unfolding.

The rate constant $k_0$ therefore may or may not be comparable to unfolding rates from ensemble experiments. Rate constants for folding and unfolding can be determined in force experiments at constant force, where the molecule dynamically changes between the folded state (compact, low extension) and the unfolded state (high extension). From the distribution of dwell times in the folded and unfolded state, rate constants of unfolding and folding can be determined (Section 23.1.7.4). Alternatively, unfolding can be performed at different pulling rates, and rate constants are determined from the distribution of unfolding forces.

The dynamic range of force spectroscopy covers extensions as small as 0.3 nm corresponding to movements on DNA by a single bp up to 100 μm when entire living cells are manipulated. Forces range from fractions of a pN applied in unfolding of RNA molecules up to the high pN to nN range, which is sufficient to break chemical bonds. In AFM-based force spectroscopy, forces range from a few pN up to nN and extensions between 0.5 and $10^4$ nm can be measured. With optical and magnetic tweezers lower forces can be measured or applied (0.1–100 pN for optical tweezers, 0.001–100 pN for magnetic tweezers). Magnetic tweezers reach about 5 nm resolution; with optical tweezers, distance changes down to 0.1 nm can be resolved.

## 24.1 FORCE SPECTROSCOPY BY AFM

We have discussed before how we can obtain images of a surface by atomic force microscopy (AFM), either in terms of height or elasticity/softness with sub-nanometer resolution (Section 23.2.5). For force spectroscopy applications of AFM, the cantilever is not scanned laterally over the sample as in the imaging mode, but moved in the vertical direction. The molecule of interest is attached to a surface, and its free end is coupled to the cantilever (Figure 24.2). It is often sufficient to attach the biomolecule of interest to the tip by non-specific adsorption. Alternatively, the molecule can be attached specifically to functionalized AFM tips that are commercially available. The cantilever functions as a handle: by moving the sample away from the cantilever, the surface-cantilever distance increases. The molecule is extended between AFM tip and surface as the pulling force increases. The force causes deflection of the cantilever, and the force magnitude can be calculated from the measured deflection.

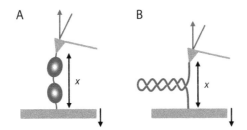

**FIGURE 24.2:  Force spectroscopy by AFM**. A: Unfolding of a two-domain protein. The protein is attached to the surface on one end, and to the AFM cantilever on the other end. By moving the stage downward (black arrow), the distance between surface and cantilever is increased, and the molecule is stretched. The extension *x* of the molecule is the distance between surface and cantilever, the force is determined from the deflection of the cantilever (measured by laser light reflection, red). B: Unfolding of a DNA or RNA hairpin.

Before forces can be measured, the AFM needs to be calibrated. For small displacements, the cantilever deflection follows Hooke's law (eq. 23.38), meaning the force and the resulting displacement from its equilibrium position are proportional. By applying known forces and measuring the corresponding change in position $\Delta x$, the stiffness $k$ of the cantilever can be determined. Once the stiffness is known, the deflection of the cantilever is recorded during force experiments along with the distance of the cantilever from the surface. The deflection is then converted to forces using Hooke's law. Plotting the force as a function of the extension of the molecule, i.e. the distance between surface and tip, yields a *force-extension curve* (Figure 24.5). The forces associated with the stretching and unfolding of biomolecules are typically in the piconewton range, and changes in extension are on the nanometer scale.

If the biopolymer is rigid, the force required for its extension is high and roughly proportional to the extension. For elastic molecules, larger extensions are achieved with relatively small forces. The response of elastic polymers to applied forces can be described by different polymer models, such as the *Gaussian chain* (GC), *freely jointed chain* (FJC), and *worm-like chain* (WLC) models (Box 24.1). These models provide relationships between force and extension for elastic polymers that can be used to describe experimentally obtained force-extension curves and to extract information on the mechanical properties of the molecule of interest.

---

**BOX 24.1: EXTENSION OF ELASTIC POLYMERS BY FORCE: GAUSSIAN CHAIN, FREELY JOINTED CHAIN, AND WORM-LIKE CHAIN.**

There are two limiting cases for the behavior of a linear polymer: it can behave as a rigid rod, or it can act as a completely flexible chain (Figure 24.3). The *Gaussian chain* model describes the conformation of a polymer by a Gaussian distribution of the end to end vector $x$. For small end-to-end distances $x$ compared to the contour length $L$, the force to stretch the polymer is a linear function of the stretching distance $x$:

$$F = 3\frac{k_B T}{L \cdot l_K} \cdot x \qquad \text{eq. 24.7}$$

$k_B$ is the Boltzmann constant, and $l_K$ is the Kuhn length (see below). Thus, a Gaussian chain behaves like an ideal spring and follows Hooke's law (eq. 23.38) for small forces. For larger forces and an increasing alignment and extension of the molecule, however, eq. 24.7 does not hold, and the polymer model needs to be refined. The *freely-jointed chain* (FJC) model describes the polymer as a chain of $N$ rigid elements of length $l_K$, the Kuhn length. The *contour length L* of the polymer, i.e. the length along its backbone, is $N$ times $l_K$. From the FJC model, the following relationship between force and extension of the polymer can be derived (eq. 24.8).

$$x = N \cdot l_K \left( \coth\left( \frac{F \cdot l_K}{k_B T} \right) - \frac{k_B T}{F \cdot l_K} \right) \qquad \text{eq. 24.8}$$

Eq. 24.8 can be used to describe measured force-extension curves to extract $l_K$ and $N$. For small forces ($F \ll k_B T/l_K$), eq. 24.8 can be approximated and rearranged to

$$F = 3\frac{k_B T}{N \cdot l_K^2} \cdot x \qquad \text{eq. 24.9}$$

which is identical to eq. 24.7 and corresponds to the behavior of a Gaussian chain following Hooke's law (eq. 23.38). For large forces ($F \gg kT/l_K$), eq. 24.8 gives

$$F = \frac{k_B T}{l_K} \cdot \frac{1}{1 - \dfrac{x}{L}} \qquad \text{eq. 24.10}$$

Eq. 24.10 diverges for $x \to L$, i.e. when the molecule approaches its contour length. The behavior of polysaccharides under force is often well-described by the FJC model.

The *worm-like chain* (WLC) model assumes that the polymer is flexible along the chain, with an irregular, position-dependent curvature. It covers all possible cases between rigid rods and fully flexible chains. The polymer is described by its contour length $L$ and the

*(Continued)*

*persistence length* $l_p$, a measure of the polymer stiffness. The persistence length describes over which distance the correlation in direction of the tangents to the curve is lost, or on which length scale the polymer appears linear. Below the persistence length, the polymer behaves like a rigid rod. Expressions for the force $F$ as a function of the extension $x$ have been derived from the WLC model using numerical procedures. The dependence of force on extension is described by

$$F\left(x\right) = \frac{k_B T}{l_p} \cdot \left( \frac{1}{4\left(1 - \dfrac{x}{L}\right)^2} + \frac{x}{L} - \frac{1}{4} \right) \qquad \text{eq. 24.11}$$

Eq. 24.11 can be used to describe the measured force-extension curves for biopolymers (Figure 24.3). The WLC model describes stretching of single and double-stranded DNA and RNA, and of polypeptides without structure. The persistence length of double-stranded DNA is 50 nm, for RNA it is 62 nm. Single-stranded nucleic acids have a much smaller persistence length of ~1 nm.

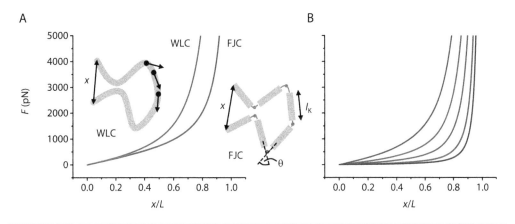

**FIGURE 24.3: Force-extension curves according to FJC and WLC models.** A: Comparison of force-extension curves according to the WLC and FJC models. Contour length $L = 3400$ nm (100 bp), persistence length $l_p = 50$ nm, T = 298 K, Kuhn length $l_k = 100$ nm. B: Force-extension curves according to the WLC model with different persistence lengths. Dark blue: $l_p = 50$ nm, blue: $l_p = 100$ nm, purple: $l_p = 200$ nm, red: $l_p = 500$ nm, brown: $l_p = 1000$ nm.

In the presence of structural changes, the force-extension curve is not continuous as described by the different polymer models (eq. 24.8, eq. 24.11), but contains jumps representing structural transitions. Single- and double-stranded DNA, for example, show different elasticities: double-stranded DNA is more rigid than single-stranded DNA. In other words, the persistence length for double-stranded DNA is larger than for single-stranded DNA. The persistence length of double-stranded DNA is 50 nm, whereas single-stranded DNA has a much smaller persistence length of less than 1 nm. The WLC model thus predicts different force-extension curves for stretching of single- and double-stranded DNA (Figure 24.4). Transitions between double- and single-stranded DNA, such as the dissociation of double-stranded DNA, appear as jumps in the force-extension curve, from one WLC curve to the other.

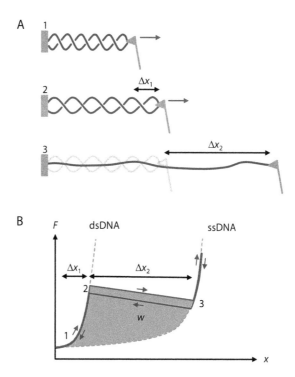

A

1

2

$\Delta x_1$

3

$\Delta x_2$

B

$F$    dsDNA              ssDNA

$\Delta x_1$          $\Delta x_2$

2

3

$w$

1

$x$

**FIGURE 24.4:** **Force-extension curve for double-stranded DNA.** A: To measure force-extension curves, double-stranded DNA is attached to a surface on one end. The other end is attached to the cantilever (1). When the cantilever is moved, the DNA is stretched (1→2), and its extension increases by $\Delta x_1$. At a certain force (2), the DNA strands dissociate (2→3), accompanied by a sudden increase in extension ($\Delta x_2$). B: Force-extension curve for stretching (red) and release (blue). During stretching of the double-stranded DNA (1→2), the force-extension curve can be described by the WLC model (eq. 24.11) with the persistence length and contour length of the double-stranded DNA as parameters. When the DNA strands dissociate (2→3), a sudden increase in extension is observed. After that (3), single-stranded DNA is extended, and the force-extension curve follows a WLC model with the persistence length and contour length of the single-stranded DNA. The area under the force-extension curve (orange) reflects the work $w$ performed during strand separation. For reversible transitions, it equals the free energy change $\Delta G$. Reversibility can be tested by measuring stretch-release curves. For reversible transitions, both curves superimpose. For irreversibility, a hysteresis is observed: unfolding upon stretching (red) occurs at higher forces than refolding during release (blue).

A force-extension curve directly provides information on the change in end-to-end distance associated with the underlying structural change, which is related to molecular dimensions. From the areas under the force-extension curves, the work $w$ performed during the structural transition can be calculated by integrating the force (in pN) over the distance (in nm):

$$\int F \mathrm{d}x = w$$

eq. 24.12

The values are typically given in pN·nm, which corresponds to $10^{-21}$ J or 1 zJ:

$$1\,\mathrm{pN} \cdot 1\,\mathrm{nm} = 10^{-12}\mathrm{N} \cdot 10^{-9}\mathrm{m} = 10^{-21}\,\mathrm{J} = 1\,\mathrm{zJ}$$

eq. 24.13

1 pN·nm is on the order of the thermal energy of molecules (0.24 $k_B T$ at 298 K).

For molecules that consist of multiple independently (un)folding domains, force-extension curves contain multiple transitions reflecting the unfolding of each domain (Figure 24.5). In these cases, AFM experiments provide information on the underlying change in extension and the work performed during each structural transition.

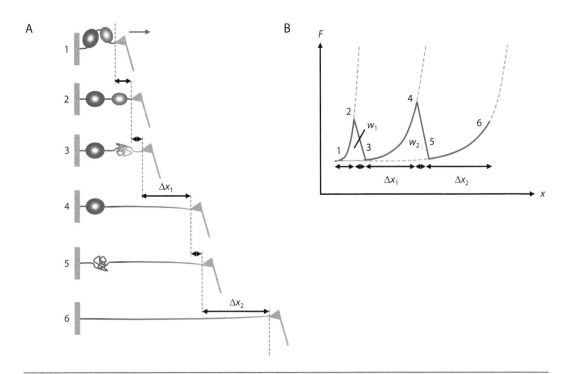

**FIGURE 24.5:** **Measuring force-extension curves by AFM.** A: Unfolding of a two-domain protein by AFM. The protein is immobilized on a surface (gray) with one end, and attached to a cantilever (gray triangle) with the other end. By moving the cantilever, the molecule is extended and unfolded by force. At the beginning (1→2), the coiled termini are stretched, and little force is required to extend the protein. After that, increasing force achieves little change in end-to-end distance. The energy is required to overcome interactions within the folded protein domains. The orange domain unfolds first. Once the interactions have been overcome (2), the force drops (2→3). While the polypeptide chain of the orange domain is extended (3→4), increasing the end-to-end distance requires little force. Further pulling then unfolds the blue domain (4→5). After that, small forces again achieve large changes in end-to-end distance until the polypeptide chain is stretched to complete extension (6). The area under the curve corresponds to the work performed during unfolding of each domain ($w_1$, $w_2$), and equals $\Delta G$ for reversible transitions.

For reversible transitions, this work is the associated change in free energy $\Delta G$. Reversibility can be tested by measuring *stretch-release curves* (Figure 24.4), where the molecule is unfolded by increasing force (stretch), and then allowed to refold while the force is decreasing (release). If both curves superimpose, the transition is reversible. A hysteresis indicates lack of reversibility.

AFM has been used to study the mechanical properties of nucleic acids, polypeptides, and polysaccharides, and to investigate the unfolding of proteins and DNA. AFM is also suitable to measure unfolding of membrane proteins in the context of a biological membrane (Figure 24.6).

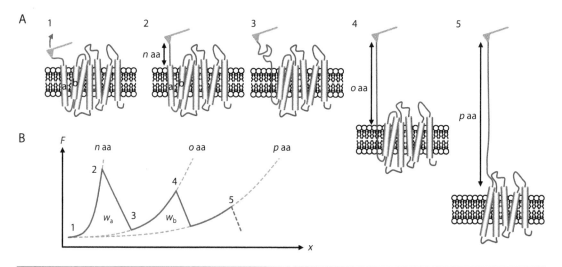

**FIGURE 24.6: Force-induced unfolding of membrane proteins**. A: Application of force to membrane proteins unfolds one structural element after the other (here: helices *a* and *b*). First, the terminus is extended (1→2), then helix *a* becomes unfolded (2→3) and fully extended (3→4), followed by unfolding of helix *b* and its full extension (4→5). B: Force-extension curve. The broken lines indicate the force-extension curves according to the WLC model. $w_a$, $w_b$: work associated with unfolding of helices *a* and *b*, respectively. *n*, *o*, and *p* are the numbers of amino acids (aa) of the unfolded chain.

The membrane is deposited on the surface, and the AFM tip is moved close to the surface and attaches to the protein of interest. Pulling the cantilever upwards then leads to unfolding of the attached protein until the polypeptide chain is removed completely from its membrane environment. During this process, force-extension curves for unfolding can be measured. AFM unfolding studies of bacteriorhodopsin and halorhodopsin, the water channel aquaporin, and a bacterial $Na^+/H^+$ antiporter have shown that membrane proteins fold and unfold differently from soluble proteins. While soluble proteins typically show cooperative transitions, membrane proteins can be unfolded one element after the other. By attaching the AFM tip to the terminus of the membrane protein, the first α-helix can be unfolded and extracted from the membrane, followed by the second, third, and so on. Reducing the applied force after unfolding of one or two helices allows reformation of the helices and reinsertion into the membrane: the molecule of interest refolds. The force-extension curve for a seven-helix transmembrane protein such as bacteriorhodopsin therefore consists of multiple transitions. From these curves, the energy involved in the underlying structural transition and the length change per transition can be determined. By comparing different constructs lacking individual structural elements of the molecule of interest, observed transitions can be assigned to unfolding of individual elements of the molecule.

Force spectroscopy by AFM also allows measuring rupture forces to determine binding energies and equilibrium dissociation constants (Figure 24.7). In such an experiment, a ligand is covalently coupled to a functionalized AFM tip, and brought close to its surface-attached binding partner to allow complex formation. By pulling the cantilever away from the surface (or strictly speaking by moving the surface away from the cantilever), the force on the complex increases until the external force exceeds the mechanical stability of the complex. At this *rupture force*, the ligand dissociates from its binding partner, and the force drops to zero. As the method is not an equilibrium approach, the work performed, i.e. the integral under the force-extension curve, does not correspond to $\Delta G$ of the dissociation reaction. However, $\Delta G$ can be calculated from the distribution of rupture forces and the distribution of work values determined over many experiments (eq. 24.14):

$$\Delta G = -k_B T \cdot \ln \left\langle \exp\left(-\frac{w}{k_B T}\right) \right\rangle \qquad \text{eq. 24.14}$$

$k_\text{B}$ is the Boltzmann constant.

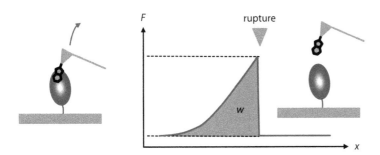

FIGURE 24.7:   **Measuring rupture forces by AFM.** A ligand is attached to the AFM tip, and the tip is moved close to the surface to allow binding of the ligand to its binding partner. Retraction of the cantilever leads to disruption of the interaction, associated with a certain rupture force and work. As the process is irreversible, the work $w$ performed is not equal to the free energy change $\Delta G$. The change in free energy and thus the stability of the interaction can be calculated from the distribution of the work for many rupture events (eq. 24.14).

Novel developments in AFM force spectroscopy aim at combining different types of information, such as AFM imaging with force mapping and force spectroscopy at individual points, or a combination of AFM with fluorescence detection on isolated molecules and in living cells.

## 24.2 OPTICAL TWEEZERS

In force experiments with optical tweezers, the molecule of interest is attached with one end to a polystyrene or silica bead that is trapped in a laser focus. The second end of the molecule is attached to a second bead that is held and moved to exert a force. Before we discuss how these experiments are performed and what we can learn from them, we will first introduce the concept of optical trapping. Optical tweezers are based on trapping of micrometer-sized dielectric objects in a laser focus. The trapping effect is caused by the gradient in light intensity, from high intensity in the focus to low and finally no intensity further away from the focus. The incident laser light is refracted by the object, and induces fluctuating dipoles in the object. These dipoles, as any dipole in an inhomogeneous electric field, experience a force in the direction of the field gradient. This *dipole force* is proportional to the polarizability of the object and to the intensity of the gradient. With tightly focused laser light, strong field gradients are generated, leading to large forces. If the object is moved out of the focus, this force acts as a three-dimensional restoring force, pulling the object back toward the focus (Figure 24.8). Scattering and reflection of light by the object, however, lead to forces that cause a movement of the object in the propagation direction of the light, and thus outside of the focus. If the gradient force due to refraction over-compensates the forces due to scattering, the object is stably trapped in the focus (Figure 24.8).

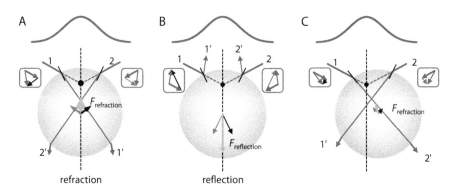

**FIGURE 24.8: Principle of optical trapping.** A polystyrene bead (blue) in the light path refracts and reflects the incident light. For clarity, only the two outer rays of the focused light bundle are depicted (1, 2). The red Gaussian distribution indicates the intensity gradient around the optical axis (black broken line). A: Refraction leads to a change in direction of the incident light (1, 2) after exiting the polystyrene bead (1´, 2´). The change in direction (black for beam 1, gray for beam 2, insets) and thus impulse of the light leads to an opposite impulse change of the bead, and to a resulting force (yellow) on the bead. The restoring force is always directed towards the light focus (black circle), and leads to a trapping of the bead in the focus. B: Reflection of the incident light at the surface of the bead also changes the propagation direction of the light (inset). The opposite force on the bead now pulls it out of the focus, and counteracts the trapping force. C: Displacement of the bead off the optical axis leads to a lateral force that traps the bead on the optical axis.

Typically, an infrared laser is used for trapping. The laser light is focused with a high numerical aperture objective. The rays at highest angles contribute most to the *trapping force*. The trapping force and the stability of the optical trap therefore increase with the numerical aperture of the objective (Figure 24.9). Trapping forces can be increased even further using dual-beam optical tweezers, where two opposing laser beams are focused in the same spot by two objectives (Figure 24.9). In this case, the scattering forces cancel, but the gradient forces are added, leading to more stable trapping. Optical trapping forces allow direct trapping of cells or organelles, as well as lipid vesicles, but are too small to trap biological molecules. Smaller molecules are attached to polystyrene or silica beads of about 1 µm in diameter, and these are trapped instead.

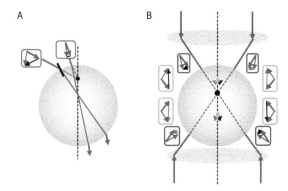

**FIGURE 24.9: Increasing the trapping force.** A: Outer rays contribute more to the trapping force than the inner rays of the light bundle. The higher the numerical aperture of the objective used, the higher the trapping force. The direction for an outer ray and an inner ray of the light bundle before and after the bead (red) and the change (black) are depicted in the inset. B: A dual-beam optical trap. If two opposite beams are focused into the same spot, the trapping forces acting on the bead add up, whereas the forces from reflection cancel. Both effects lead to more stable trapping. Orange inset: Changes in the propagation direction of the light due to reflection. Blue inset: Change in propagation direction due to refraction.

The optical trap holds the bead in its equilibrium position. However, external forces on the bead lead to small displacements $\Delta x$ of the center of the bead from this equilibrium (Figure 24.10). Within short distances from the focus, the optical trap follows Hooke's law (eq. 23.38): displacements of the bead out of the focus depend linearly on the associated force. The ratio $F/\Delta x$ is the stiffness $k$ of the trap. As discussed before for AFM measurements, the stiffness of optical tweezers has to be calibrated before experiments can be performed. This can be done through the hydrodynamic drag on the bead, by analyzing the thermal fluctuations of the bead, or by stretching a tethered DNA in a reference measurement. Once the stiffness is known, the displacement of the bead within the optical trap is measured and converted into the force using Hooke's law.

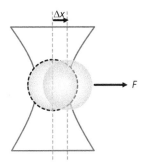

FIGURE 24.10:    **Response of a trapped bead to external forces**. An external force $F$ leads to a lateral displacement of the bead from its equilibrium position (black broken line) by $\Delta x$. When the stiffness $k$ of the optical trap is calibrated, the force $F$ acting on the bead can be calculated from the displacement $\Delta x$ according to Hooke's law.

To perform a stretching experiment in an optical tweezers setup, the biomolecule of interest is attached on one end to the bead. On the other end it is attached to a second bead that is held in place by suction with a micropipette or by a second optical trap (Figure 24.11). The attachment of the biomolecule to the bead must be strong enough to withstand the forces applied during the experiment, and high-affinity interactions are necessary. DNA and also proteins are often labeled with biotin and coupled to streptavidin-coated beads (see also SPR in Section 26.3.4.2 and TIRF in Section 23.1.7.5; Figure 23.25). Another popular attachment method for DNA uses the interaction between digoxygenin and the anti-digoxygenin antibody, and immobilizes digoxygenin-labeled DNA on beads functionalized with anti-digoxygenin antibodies. Proteins can also be immobilized *via* hexahistidine tags on $Ni^{2+}$-NTA-functionalized beads, although this interaction is less strong. Functionalized polystyrene beads are commercially available.

In optical tweezers experiments, the position of the bead relative to the trap center needs to be determined to calculate the force exerted on the bead. This is typically done by imaging the interference pattern formed by light from the trapping laser and the light scattered by the bead. As long as the bead is centered in the optical trap, the interference pattern is centrosymmetric. Small lateral changes in position lead to a displacement of the interference pattern and deviations from symmetry. The position is calibrated before the experiment by moving a bead on the surface of the sample cell by a known distance and measuring the positional signal. Interference pattern analysis allows detection of lateral movements of the bead by 0.1 nm on the millisecond time scale.

Optical tweezers can be operated in different modes. By applying *force ramps*, molecules attached to beads can be unfolded. In unfolding experiments, one end of the molecule is attached to a trapped bead, the other end to a second bead that is held in place with a micropipette or a second optical trap (Figure 24.11). While proteins and DNA are attached directly to the beads, RNA molecules are often attached through DNA/RNA handles. To stretch the molecule of interest, the micropipette or the second optical trap is moved away. The force exerted is then determined from the displacement of the trapped bead from its equilibrium position. From the distance of the bead centers (minus the radii of the beads), the extension is calculated. Variations in bead radius lead to uncertainties for absolute extensions, but extension changes can be measured down to < 1 nm. The force is continuously increased or decreased to induce unfolding of the biomolecule, and force-extension curves are measured. A structural transition of the molecule is associated with an abrupt change in force and extension: unfolding of a molecule leads to a sudden increase in its extension, with a concomitant drop in force, such that the trapped bead moves back towards the center of the optical trap. Conversely, a folding transition is measured as an abrupt decrease in extension and increase in force. For reversibly unfolding molecules, the work performed during the transition equals the underlying free energy change $\Delta G$.

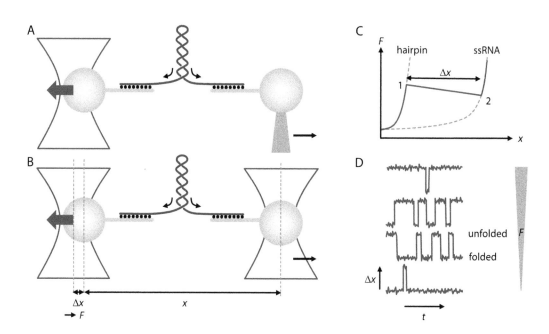

**FIGURE 24.11:** **Optical tweezers experiments to monitor unfolding.** The ends of the molecule of interest, here an RNA hairpin, are attached to polystyrene beads using DNA/RNA handles. One bead is in an optical trap, the second is either held by a micropipette (A) or by a second optical trap (B). Moving the micropipette or the second optical trap laterally (black arrow) exerts force on the molecule that pulls the trapped bead out of its equilibrium position. The force can be calculated from the displacement $\Delta x$. The extension $x$ of the molecule is determined from the distance between the centers of the two beads. When the hairpin unfolds, the extension increases, and the force decreases. The trapped bead therefore moves to the left (red arrow). C: In force-ramp experiments, force-extension curves for unfolding of the hairpin (1→2) are measured. D: In force-clamp experiments, the force is maintained constant, and the extension of the molecule is measured as a function of time. The molecule switches back and force between the folded state (low extension) and the unfolded state (high extension). Depending on the applied force, the molecule spends more time in the folded state (low force) or in the unfolded state (high force).

In the *force-clamp mode*, the position of the second bead is continuously adjusted during the experiment to maintain a constant force on the bead and the molecule of interest (force feedback). In this case, the displacement of the trap reflects changes in extension of the attached molecule. In force-clamp experiments, dynamic changes in the extension of the molecule attached to the beads can be monitored, e.g. due to dynamic transitions between folded (short) and unfolded (extended) states in equilibrium at constant force (Figure 24.11), due to shortening of single-stranded DNA that is converted into double-stranded DNA by polymerase activity, or DNA or RNA extension through action of a helicase. Using force-clamp experiments, the movement of RNA polymerase along DNA has been followed with single base pair resolution. Rate constants of structural transitions can be determined from the dwell time distribution of the molecule in each state at constant force. The equilibrium constant for folding is obtained from the ratio of the rate constants (see Section 8.1), and also allows the calculation of free energy changes. By performing force-clamp experiments at different forces, the force dependence of rate and equilibrium constants can be determined. Force-clamp experiments also allow determination of stalling forces, i.e. forces at which the net displacement is zero and translocation and activity cease. Stalling forces provide information on the force generated by these enzymes. The stalling force for RNA polymerase is ~25 pN, for DNA polymerase the stalling force is 34 pN. The stalling forces of the motor proteins myosin/actin and kinesin/microtubules are only around 5 pN.

Optical tweezers can also be operated in an *extension-clamp mode* where the extension is fixed by a feedback mechanism that constantly adjusts the position of the second bead. In this mode, structural transitions are measured as a change in force.

*Force-jump experiments* are used to measure unfolding and folding rate constants for the same molecule. First, the force is set to a high value at which the molecule unfolds. The time until unfolding occurs, i.e. the dwell time in the folded state, is measured. After that, the molecule is fully unfolded by further increasing the force. In the second part of the experiment, the force is then suddenly dropped to a value that allows refolding, and the dwell time of the unfolded state of the molecule is monitored. From dwell times, rate constants can be determined. In force-jump experiments, different structural transitions that occur at different forces can be studied separately. Each step is observed at a suitable force, as in force-clamp experiments, and rate constants are extracted from individual dwell times. These experiments allow to delineate multi-step folding pathways and the rate constants associated with each step, and to identify rate-limiting steps.

Optical tweezers have been used to study folding and unfolding of DNA and RNA. DNA and RNA hairpins unfold at forces of ~15 pN. For protein unfolding, AFM is often used because of the larger forces required in comparison to unfolding of nucleic acids.

Optical tweezers may also be used to measure forces on DNA or RNA exerted by processive enzymes such as polymerases, helicases, or ribosomes, or forces involved in movement of motors on their track, such as myosin moving along actin filaments, or kinesin moving along microtubules. Depending on the application and the type of information sought, different experimental configurations are used (Figure 24.12). In the *interaction assay*, one partner (for example myosin) is attached to a bead in an optical trap, and the other partner (for myosin, an actin filament) is immobilized on the surface of the sample cell. The displacement of the bead in the optical trap reports on the relative movement of the partners to each other, and on the forces involved. This configuration is particularly suited to study movement of motor proteins along their track. In the *tethered assay* (Figure 24.12), an enzyme (for example, a polymerase moving along DNA) is attached to the optically trapped bead. One end of the DNA is attached to the surface, or to a second bead held by a micropipette. When the enzyme starts to move along the DNA, it either reels the DNA in or lets it out, depending on the direction of translocation. Translocation therefore leads to a displacement of the bead. At constant force, this configuration allows one to follow enzyme movement over long

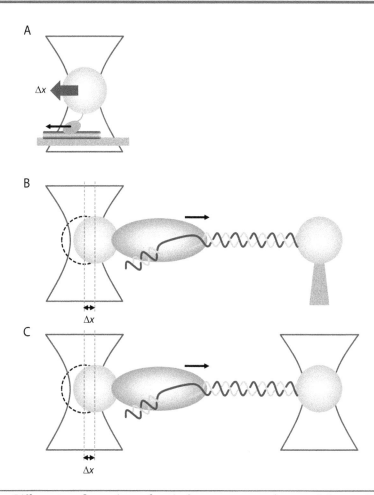

**FIGURE 24.12: Different configurations of optical tweezers experiments.** A: Interaction assay. A motor protein (orange) is immobilized on the polystyrene bead that is trapped. The motor protein moves along a track (blue) that is immobilized on the surface (gray). Movement of the motor along the track generates a force and causes a displacement of the trapped bead. B: Tethered assay. An enzyme (RNA polymerase, orange) is immobilized on the optically trapped bead. The DNA template (blue) is attached to a second bead that is held by a micropipette, the other end of the DNA is free. Movement of the RNA polymerase along the DNA leads to a displacement of the optically trapped bead. C: Dumbbell assay. In contrast to (B), the second bead is held by a second optical trap. Note that beads and molecules are not drawn to scale.

periods of activity. The position of the trapped bead directly reports on translocation of the enzyme. By measuring translocation rates at different forces, the mechanochemical coupling of the enzyme can be interrogated. The tethered assay is also used to study unfolding of nucleic acids. The *dumbbell configuration* (Figure 24.12) is a variation of the tethered assay in which the DNA end is attached to a second bead which is held in its own, independent optical trap. During the enzymatic action on the DNA, one of the traps is continuously moved to keep the force constant. In this configuration less drift of the bead is observed than in a tethered assay.

The dumbbell configuration offers the advantage of being able to move both optical traps within the sample chamber. Experiments can thereby be performed away from surfaces. Using microfluidic sample cells allows movement of the two optical traps and thus of the molecule of interest between different flow channels and for facile change of buffer conditions (Figure 24.13).

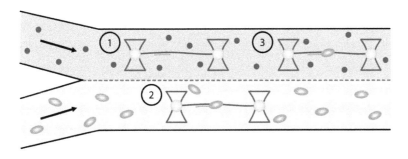

FIGURE 24.13: **Optical tweezers experiments and microfluidic sample cells**. A microfluidic sample cell with two flow channels contains deoxynucleotide triphosphates (blue spheres) in one channel (light red), and DNA polymerase (orange) in the other (light green). Beads (light blue) attached to the end of a single-stranded DNA (dark blue) with a hybridized primer (turquoise) are captured by optical traps (red, 1). By moving the optical traps, the DNA is moved into the second flow channel (light green, 2), where a polymerase molecule is allowed to bind to the DNA. The DNA-polymerase complex is then transferred back to the first flow channel where DNA synthesis starts (3).

Optical tweezers experiments have been used to study the effect of DNA or RNA binding proteins on DNA structure, such as wrapping of DNA by nucleosomes, to monitor movement of DNA and RNA polymerases, helicases, and ribosomes along DNA or RNA, and to investigate the mechano-chemical cycles of motor proteins such as actomyosin and kinesin moving along microtubules.

The application spectrum of optical tweezers has been expanded by introducing torque, and by combining optical tweezers experiments with fluorescence detection by TIRF, confocal or wide-field microscopy, or FRET. Optical trapping and force measurements are also possible *in vivo*. Multiplexing, i.e. observing multiple optical traps in parallel, allows manipulating more than one molecule at the same time (Box 24.2).

---

### BOX 24.2: MULTIPLEXING WITH OPTICAL TWEEZERS: SIMULTANEOUS MANIPULATION OF TWO DNA MOLECULES.

Many proteins bridge DNA molecules by interacting with different regions of double-stranded DNA. The DNA in the bacterial nucleoid is associated with numerous proteins that mediate the compact folding of the DNA. The histone-like nucleoid structuring protein (H-NS) is a major determinant for the compaction and the structural organization of DNA in the nucleoid, and has also been ascribed a regulatory role for gene expression. H-NS binds to two DNA molecules and forms a bridge between them. The energy landscape of the DNA-DNA interaction mediated by H-NS has been probed with four optical traps that are used pairwise to trap two molecules of DNA. The quadruple trap (Q-trap) allows the application of force on bridged DNA molecules, and to interrogate their mechanical stability (Figure 24.14). By moving the two bridged DNA molecules away from each other, the rupture of individual bridges was measured. Dame *et al.* (2006) *Nature* 444(7117): 387–390.

*(Continued)*

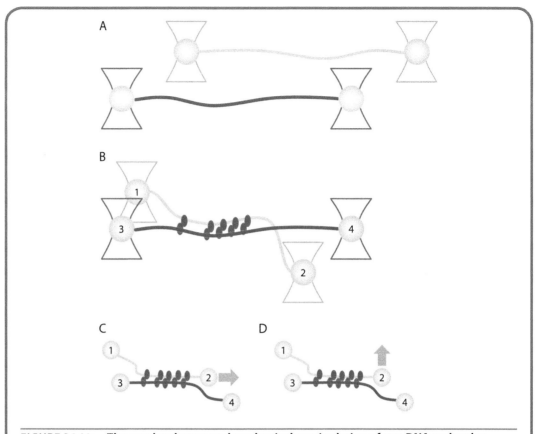

**FIGURE 24.14:   The quadruple trap and mechanical manipulation of two DNA molecules.**
A: With two pairwise optical traps (orange, red), two double-stranded DNA molecules (light blue, dark blue) can be manipulated. B: By moving the optical traps, the DNA molecules can be brought into close proximity, such that a bridging protein (red) can bind. C: By moving bead 2 to the right, shearing experiments can be performed. D: Moving bead 2 upwards allows unzipping experiments in which stepwise disruption of individual bridges is observed.

## 24.3 MAGNETIC TWEEZERS

Conceptually, magnetic tweezers work according to the same principle as optical tweezers, and also serve to apply forces to biomolecules, or to monitor forces generated by biomolecules. In magnetic tweezers, magnetic beads are held in place by a pair of magnets. The magnetic field induces a magnetic moment in the bead. As a result the bead experiences a force in the direction of the field gradient and proportional to the gradient. The field gradient is constant over the length scale of the movements of the magnetic bead, and no force feedback is required as in optical tweezers (see Section 24.2).

The molecule of interest is attached with one end to the magnetic bead, with the other end to the surface of the sample cell. Functionalized magnetic beads for the attachment of biomolecules are commercially available. By moving the magnets upwards, away from the bead, the bead is pulled upward, and the attached molecule of interest is stretched as in optical tweezers stretching experiments. This type of experiment yields force-extension curves.

The strength of magnetic tweezers experiments is the capability to rotate the magnetic bead by rotation of the magnets. The torsion of the bead introduces twist into the attached molecule. By turning the magnetic bead, DNA can be over- or undertwisted. At a certain degree of twisting, further twist is converted into writhe (Section 17.5.3.2) and leads to "buckling" of the DNA and plectoneme formation (Figure 24.15). Plectoneme formation leads to a strong decrease of the DNA extension. Twist and torque play an important role in all processes on DNA, including transcription, replication, and recombination.

The maximum force of a magnetic tweezers instrument is determined by the magnitude of the magnetic moment, which in turn depends on the size of the bead and is proportional to its volume.

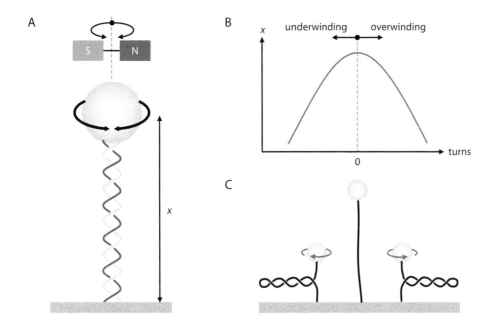

**FIGURE 24.15:** **Twisting DNA by magnetic tweezers.** A: A DNA molecule is attached with one end to the magnetic bead, with the other end to the surface. The magnetic bead is held by permanent magnets (red/green) at a height at which the DNA is extended. By turning the magnets, the DNA is under- or overwound. B: Extension of the DNA as a function of turns of the magnet. At low force, the DNA forms plectonemes when it is over- or underwound. C: Under torque, double-stranded DNA forms supercoiled regions, termed plectonemes.

The second factor is the minimal distance of the magnets from the bead. The typical force range is 10–100 pN. The magnetic bead is the means to introduce torsional stress and, as for optical tweezers, is the object that is followed during the experiment by detecting the interference pattern of light scattered by the bead with unscattered light. Vertical movement of the bead moves the bead out of focus and changes the intensity distribution of the interference pattern. From changes in the number of concentric rings of the interference pattern, the height of the bead along the $z$-axis can be determined with about 10 nm accuracy. The height is calibrated before the experiment.

Magnetic tweezers have been used to probe the mechanical properties of DNA and RNA. They are also very useful to investigate the mechanisms of enzymes that act on DNA or RNA, such as DNA or RNA polymerases and helicases, or DNA repair and replication proteins. Because of the facile introduction of twist and torque, magnetic tweezers are particularly suited to monitor the supercoiling or relaxation activity of DNA topoisomerases on DNA (Box 24.3), and to interrogate the mechanochemical coupling of topoisomerases. Magnetic tweezers have been used to study the mechanism of negative DNA supercoiling by gyrase and of positive supercoiling by reverse gyrase. They have also revealed the mechanism of topoisomerase I poisoning by the drug camptothecin.

---

**BOX 24.3: MAGNETIC TWEEZERS, DNA TOPOLOGY, AND DNA TOPOISOMERASES.**

Using magnetic tweezers, negative or positive supercoils can be introduced into DNA by simple turning of the magnetic bead and thus under- or overtwisting of the DNA (Figure 24.16). Relaxation of these supercoils by DNA topoisomerases can then be monitored as an extension of the DNA. Measuring the extension as a function of time provides information on the step size of relaxation, and thus on the change in linking number per step. The time passing between two relaxation events provides kinetic information on DNA relaxation. Instead of generating plectonemes, magnetic tweezers can also be used to introduce crossings into DNA by twisting two double strands around each other (Figure 24.16).

*(Continued)*

Using magnetic tweezers, it was shown that topoisomerase IV relaxes positive supercoils more rapidly than negative supercoils. Neuman *et al.* (2009) *Proc. Natl. Acad. Sci. U. S. A.* 106(17): 6986–6991.

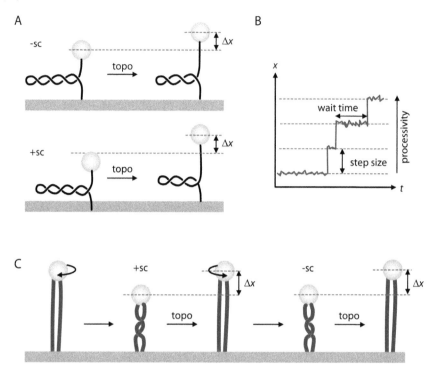

**FIGURE 24.16: Monitoring relaxation of positive and negative DNA supercoils by topoisomerases.** A: Under- and overtwisting of DNA generates plectonemes with negative and positive DNA supercoils. Relaxation of these supercoils leads to an increase in extension $\Delta x$ of the DNA. B: Time dependent measurement of DNA extension during DNA relaxation. B: Schematic time trace of DNA extension as a function of time. Time traces provide information on the step size, the pausing time before a relaxation event, and the processivity of the DNA topoisomerase. C: Negative and positive crossings can also be introduced into DNA by twisting two double strands (red, blue) around each other. Again, relaxation of these supercoils by DNA topoisomerases (topo) is measured as an increase in DNA extension.

Although magnetic tweezers allow twisting of DNA molecules, they only provide a read-out for the extension of the molecule, not for its rotation. In a *rotor-bead assay* (Box 24.4), angular movement of the molecule of interest is monitored simultaneously with changes in extension. This method has been applied to follow topological changes in DNA during negative supercoiling by DNA gyrase and to interrogate mechanochemical coupling (Box 24.4).

Recently developed magnetic tweezers make use of electromagnets instead of permanent magnets. The magnetic field of electromagnets can be changed by altering the current, which is much faster than moving conventional ferromagnets, and decreases the dead time. Recent developments also include multiplexing, and the combination of magnetic tweezers with fluorescence detection and FRET.

## BOX 24.4: ROTOR-BEAD ASSAY AND DNA SUPERCOILING.

A rotor-bead assay uses an additional bead that is attached to the DNA internally to provide a read-out on twisting of DNA in optical or magnetic tweezers. The DNA molecule contains a nick that acts as a swivel. The rotor bead is attached below the nick. The DNA is held between magnetic tweezers at sufficiently high force to prevent buckling and plectoneme formation. To investigate the mechanochemical coupling of DNA gyrase, a binding site for gyrase was introduced in the bottom part of the DNA. When gyrase binds to the DNA and changes the linking number, the topological change can manifest as twist or writhe (Section 17.5.3.2). In the presence of force, writhing associated with a decrease in extension is disfavored, and the change in linking number leads to a change in twist of the DNA. This twist leads to a torque on the rotor bead. The angular position of the bead is a direct readout for the twist of the DNA below the nick, and its angular velocity reports on the torque. A change in linking number by 1 would result in a complete turn (360°) of the rotor bead, and a change in linking number by 2 is detected as a rotation by 720°. Using this procedure, it was confirmed that gyrase leads to two complete turns of the rotor bead during one catalytic cycle, in agreement with the linking number change by two. Gyrase binds DNA and wraps it around itself at the beginning of the catalytic cycle. In the absence of ATP, the rotor-bead angle alternates between two values, corresponding to DNA wrapped around gyrase (1.3 rotations) and DNA that is not wrapped (no rotation). When the force is increased, the time gyrase spends in the wrapped state is decreased. At high ATP concentrations, the rotor bead undergoes two complete rotations during each catalytic cycle. When the ATP concentration is reduced, a pause at about one rotation is detected. Under these conditions, progression through the catalytic cycle is reduced, and unwrapping efficiently competes with ATP binding and DNA supercoiling (Figure 24.17). Gore *et al.* (2006) *Nature* 439(7072): 100–104.

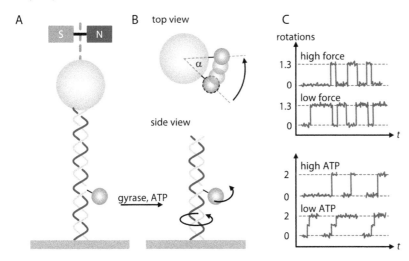

**FIGURE 24.17: A rotor-bead assay to measure twist and torque.** A: DNA with a nick is held under force in a magnetic tweezers setup. The DNA can swivel at the nick. A decrease in linking number of the DNA due to gyrase activity manifests itself as twist of the DNA below the nick, which exerts a torque on the rotor bead. B: Top view: The angle α reports on the twist, and the angular velocity of the rotor bead on the torque. The angle of the rotor bead is measured as a function of time. C: Schematic traces of the angle as a function of time to illustrate the behavior of DNA in the presence of gyrase at low and high force in the absence of ATP (top) and at low and high ATP concentrations (bottom). DNA wrapping and unwrapping in the absence of ATP is reflected in jumps between 0 and 1.3 rotations. At high ATP concentrations, each catalytic cycle leads to two complete turns of the rotor bead around the DNA. At low ATP concentration, pauses at one rotation are observed. Schematic traces for three different events are shown.

## QUESTIONS

24.1    An optical trap has a stiffness of $k = 0.05$ pN nm$^{-1}$. A polymerase pulls with $F = 22$ pN on the DNA attached to the bead in the optical trap. What is the displacement $\Delta x$? What would the displacement be in a magnetic tweezers set-up ($k = 0.5$ pN nm$^{-1}$) or in an AFM experiment ($k = 100$ pN nm$^{-1}$)?

## REFERENCES

Selvin, M., Ha, T. (2008) Single-molecule techniques –A laboratory manual. New York, Cold Spring Harbour Press.
· an overview about single-molecule techniques including force methods with a focus on practical aspects and protocols.

Neuman, K. C. and Nagy, A. (2008) Single-molecule force spectroscopy: optical tweezers, magnetic tweezers and atomic force microscopy. *Nat Methods* **5**(6): 491–505.
· review article that compares optical and magnetic tweezers and atomic force microscopy

Janshoff, A., Neitzert, M., Oberdorfer, Y. and Fuchs, H. (2000) Force spectroscopy of molecular systems-single molecule spectroscopy of polymers and biomolecules. *Angew Chem Int Ed* **39**(18): 3212–3237.
· review article on mechanochemical models for stretching of biopolymers, providing a theoretical treatment of the worm-like chain, freely-jointed chain, and Gaussian chain models

Bustamante, C., Marko, J. F., Siggia, E. D. and Smith, S. (1994) Entropic elasticity of lambda-phage DNA. *Science* **265**(5178): 1599–1600.
· mechanochemical study of double-stranded DNA and description of the force-extension curve according to the worm-like chain model

Abels, J. A., Moreno-Herrero, F., van der Heijden, T., Dekker, C. and Dekker, N. H. (2005) Single-molecule measurements of the persistence length of double-stranded RNA. *Biophys J* **88**(4): 2737–2744.
· magnetic tweezers and atomic force study of the mechanical properties of double-stranded RNA

Smith, S. B., Cui, Y. and Bustamante, C. (1996) Overstretching B-DNA: the elastic response of individual double-stranded and single-stranded DNA molecules. *Science* **271**(5250): 795–799.
· mechanochemical study of single- and double-stranded DNA

Marko, J. F. and Siggia, E. D. (1995) Stretching DNA. *Macromolecules* **28**(26): 8759–8770.
· statistical mechanics treatment of the worm-like chain model and its applicability to mechanical unfolding of DNA

Rief, M., Clausen-Schaumann, H. and Gaub, H. E. (1999) Sequence-dependent mechanics of single DNA molecules. *Nat Struct Biol* **6**(4): 346–349.
· study of double-stranded DNA stability as a function of sequence by atomic force microscopy

Francius, G., Alsteens, D., Dupres, V., Lebeer, S., De Keersmaecker, S., Vanderleyden, J., Gruber, H. J. and Dufrene, Y. F. (2009) Stretching polysaccharides on live cells using single molecule force spectroscopy. *Nat Protoc* **4**(6): 939–946.
· protocol for force studies of cell-surface polysaccharides by atomic force microscopy and description by the freely-jointed chain model

Rief, M., Gautel, M., Oesterhelt, F., Fernandez, J. M. and Gaub, H. E. (1997) Reversible unfolding of individual titin immunoglobulin domains by AFM. *Science* **276**(5315): 1109–1112.
· multi-step unfolding of a multi-domain protein by atomic force microscopy

Kedrov, A., Janovjak, H., Sapra, K. T. and Muller, D. J. (2007) Deciphering molecular interactions of native membrane proteins by single-molecule force spectroscopy. *Annu Rev Biophys Biomol Struct* **36**: 233–260.
· review article about force-induced unfolding of membrane proteins by atomic force microscopy

Sapra, K. T., Besir, H., Oesterhelt, D. and Muller, D. J. (2006) Characterizing molecular interactions in different bacteriorhodopsin assemblies by single-molecule force spectroscopy. *J Mol Biol* **355**(4): 640–650.
· atomic force microscopy study of the effect of dimer and trimer formation on the stability of bacteriorhodopsin

Janovjak, H., Muller, D. J. and Humphris, A. D. (2005) Molecular force modulation spectroscopy revealing the dynamic response of single bacteriorhodopsins. *Biophys J* **88**(2): 1423–1431.
· stepwise unfolding of secondary structure elements of bacteriorhodopsin in native purple membranes

Cisneros, D. A., Oesterhelt, D. and Muller, D. J. (2005) Probing origins of molecular interactions stabilizing the membrane proteins halorhodopsin and bacteriorhodopsin. *Structure* **13**(2): 235–242.
· comparative atomic force spectroscopy study of the mechanostability of halorhodopsin and bacteriorhodopsin

Moller, C., Fotiadis, D., Suda, K., Engel, A., Kessler, M. and Muller, D. J. (2003) Determining molecular forces that stabilize human aquaporin-1. *J Struct Biol* **142**(3): 369–378.
· investigation of the folding pathways of the water channel aquaporin by atomic force microscopy

Kedrov, A., Ziegler, C., Janovjak, H., Kuhlbrandt, W. and Muller, D. J. (2004) Controlled unfolding and refolding of a single sodium-proton antiporter using atomic force microscopy. *J Mol Biol* **340**(5): 1143–1152.
· mechanical unfolding study of the bacterial Na$^+$/H$^+$ antiporter NhaA

Hoffmann, T. and Dougan, L. (2012) Single molecule force spectroscopy using polyproteins. *Chem Soc Rev* **41**(14): 4781–4796.
· review article highlighting insights into protein folding by atomic force microscopy

Jarzynski, C. (1997) Nonequilibrium Equality for Free Energy Differences. *Phys Rev Lett* **78**(14): 2690–2693.
· derivation of the calculation of ΔG from the distribution of rupture forces and the distribution of work values determined over many experiments

Liphardt, J., Dumont, S., Smith, S. B., Tinoco, I., Jr. and Bustamante, C. (2002) Equilibrium information from nonequilibrium measurements in an experimental test of Jarzynski's equality. *Science* **296**(5574): 1832–1835.
· experimental proof of the Jarzynski equation that allows calculation of ΔG values from non-equilibrium unfolding

Muller, S. A., Muller, D. J. and Engel, A. (2011) Assessing the structure and function of single biomolecules with scanning transmission electron and atomic force microscopes. *Micron* **42**(2): 186–195.
· combination of AFM imaging with force mapping and force spectroscopy at individual points

Hecht, E., Knittel, P., Felder, E., Dietl, P., Mizaikoff, B. and Kranz, C. (2012) Combining atomic force-fluorescence microscopy with a stretching device for analyzing mechanotransduction processes in living cells. *Analyst* **137**(22): 5208–5214.
· combination of AFM with fluorescence detection on isolated molecules and in living cells

Ashkin, A. (1970) Acceleration and trapping of particles by radiation pressure. *Phys Rev Lett* **24**(4): 156–159.
· first description of the principle of optical trapping

Tinoco, I., Jr., Li, P. T. and Bustamante, C. (2006) Determination of thermodynamics and kinetics of RNA reactions by force. *Q Rev Biophys* **39**(4): 325–360.
· review article on the analysis of the thermodynamics and kinetics of RNA folding by optical tweezers, including a detailed treatment of calibration procedures

Dumont, S., Cheng, W., Serebrov, V., Beran, R. K., Tinoco, I., Jr., Pyle, A. M. and Bustamante, C. (2006) RNA translocation and unwinding mechanism of HCV NS3 helicase and its coordination by ATP. *Nature* **439**(7072): 105–108.
· analysis of translocation of hepatitis C NS3 helicase on RNA by optical tweezers with two base pair resolution

Abbondanzieri, E. A., Greenleaf, W. J., Shaevitz, J. W., Landick, R. and Block, S. M. (2005) Direct observation of base-pair stepping by RNA polymerase. *Nature* **438**(7067): 460–465.
· investigation of the movement of RNA polymerase along DNA by force-clamp optical tweezers experiments with single base pair resolution

Bustamante, C., Chemla, Y. R., Forde, N. R. and Izhaky, D. (2004) Mechanical processes in biochemistry. *Annu Rev Biochem* **73**: 705–748.
· review article on the effects of forces on folding, stability, and activity of proteins, highlighting motor proteins and enzymes as examples

Wang, M. D., Schnitzer, M. J., Yin, H., Landick, R., Gelles, J. and Block, S. M. (1998) Force and velocity measured for single molecules of RNA polymerase. *Science* **282**(5390): 902–907.
· mechanochemical study of the response of RNA polymerase translocation to forces using optical tweezers

Yin, H., Wang, M. D., Svoboda, K., Landick, R., Block, S. M. and Gelles, J. (1995) Transcription against an applied force. *Science* **270**(5242): 1653–1657.
· optical tweezers study of transcription by RNA polymerase under force

Wuite, G. J., Smith, S. B., Young, M., Keller, D. and Bustamante, C. (2000) Single-molecule studies of the effect of template tension on T7 DNA polymerase activity. *Nature* **404**(6773): 103–106.
· optical tweezers study of the effect of forces on DNA polymerase activity

Heller, I., Hoekstra, T. P., King, G. A., Peterman, E. J. and Wuite, G. J. (2014) Optical tweezers analysis of DNA-protein complexes. *Chem Rev* **114**(6): 3087–3119.
· review article with a focus on applications to DNA-protein interactions

Visscher, K., Schnitzer, M. J. and Block, S. M. (1999) Single kinesin molecules studied with a molecular force clamp. *Nature* **400**(6740): 184–189.
· force study of the movement of myosin along actin filaments

Svoboda, K. and Block, S. M. (1994) Force and velocity measured for single kinesin molecules. *Cell* **77**(5): 773–784.
· mechanochemical study of kinesin movement along microtubules

Li, P. T., Collin, D., Smith, S. B., Bustamante, C. and Tinoco, I., Jr. (2006) Probing the mechanical folding kinetics of TAR RNA by hopping, force-jump, and force-ramp methods. *Biophys J* **90**(1): 250–260.
· comparative study of force-induced unfolding of HIV TAR RNA using force hopping, force jump- and force-ramp optical tweezers experiments

Liphardt, J., Onoa, B., Smith, S. B., Tinoco, I., Jr. and Bustamante, C. (2001) Reversible unfolding of single RNA molecules by mechanical force. *Science* **292**(5517): 733–737.
· force-induced unfolding studies on the folding pathway of the P5abc region of the *Tetrahymena* ribozyme

Bockelmann, U., Thomen, P., Essevaz-Roulet, B., Viasnoff, V. and Heslot, F. (2002) Unzipping DNA with optical tweezers: high sequence sensitivity and force flips. *Biophys J* **82**(3): 1537–1553.
· study on the mechanical stability of DNA hairpins

Michaelis, J. and Treutlein, B. (2013) Single-molecule studies of RNA polymerases. *Chem Rev* **113**(11): 8377–8399.
· review article summarizing the insight into the mechanism of transcription by RNA polymerase obtained from a variety of single-molecule approaches

Qu, X., Wen, J. D., Lancaster, L., Noller, H. F., Bustamante, C. and Tinoco, I., Jr. (2011) The ribosome uses two active mechanisms to unwind messenger RNA during translation. *Nature* **475**(7354): 118–121.
· optical tweezers study of ribosome-mediated RNA unwinding

Molloy, J. E., Burns, J. E., Kendrick-Jones, J., Tregear, R. T. and White, D. C. (1995) Movement and force produced by a single myosin head. *Nature* **378**(6553): 209–212.
· optical tweezers study of the force and step size during myosin movement along actin filaments

Simmons, R. (1996) Molecular motors: single-molecule mechanics. *Curr Biol* **6**(4): 392–394.
· review article on optical tweezer studies on myosin, kinesin, and RNA polymerase

Arslan, S., Khafizov, R., Thomas, C. D., Chemla, Y. R. and Ha, T. (2015) Protein structure. Engineering of a superhelicase through conformational control. *Science* **348**(6232): 344–347.
· analysis of DNA unwinding processivity of the DNA helicase RepX using optical tweezers in combination with microfluidics

Herbert, K. M., La Porta, A., Wong, B. J., Mooney, R. A., Neuman, K. C., Landick, R. and Block, S. M. (2006) Sequence-resolved detection of pausing by single RNA polymerase molecules. *Cell* **125**(6): 1083–1094.
· optical tweezers study on sequence-dependent pausing of RNA polymerase

Veigel, C., Coluccio, L. M., Jontes, J. D., Sparrow, J. C., Milligan, R. A. and Molloy, J. E. (1999) The motor protein myosin-I produces its working stroke in two steps. *Nature* **398**(6727): 530–533.
· analysis of the mechanochemical cycle of actomyosin by optical tweezers

Ma, J., Bai, L. and Wang, M. D. (2013) Transcription under torsion. *Science* **340**(6140): 1580–1583.
· investigation of the effect of DNA supercoiling on transcription by RNA polymerases transcription with optical tweezers and torque

Moffitt, J. R., Chemla, Y. R., Smith, S. B. and Bustamante, C. (2008) Recent advances in optical tweezers. *Annu Rev Biochem* **77**: 205–228.
· review article about applications and limitations of optical tweezers

Comstock, M. J., Whitley, K. D., Jia, H., Sokoloski, J., Lohman, T. M., Ha, T. and Chemla, Y. R. (2015) Protein structure. Direct observation of structure-function relationship in a nucleic acid-processing enzyme. *Science* **348**(6232): 352–354.
· a combined optical tweezers/confocal fluorescence microscopy study on the DNA helicase UvrD

Comstock, M. J., Ha, T. and Chemla, Y. R. (2011) Ultrahigh-resolution optical trap with single-fluorophore sensitivity. *Nat Methods* **8**(4): 335–340.
· combination of optical tweezers with fluorescence detection at single-molecule sensitivity

Lang, M. J., Fordyce, P. M., Engh, A. M., Neuman, K. C. and Block, S. M. (2004) Simultaneous, coincident optical trapping and single-molecule fluorescence. *Nat Methods* **1**(2): 133–139.
· combination of optical tweezers with fluorescence detection and TIRF microscopy

Zhou, R., Schlierf, M. and Ha, T. (2010) Force-fluorescence spectroscopy at the single-molecule level. *Methods Enzymol* **475**: 405–426.
· combination of optical tweezers with confocal microscopy and fluorescence detection

Lee, S. and Hohng, S. (2013) An optical trap combined with three-color FRET. *J Am Chem Soc* **135**(49): 18260–18263.
· combination of optical tweezers with three-color fluorescence detection for three-color FRET

Blehm, B. H., Schroer, T. A., Trybus, K. M., Chemla, Y. R. and Selvin, P. R. (2013) *In vivo* optical trapping indicates kinesin's stall force is reduced by dynein during intracellular transport. *Proc Natl Acad Sci U S A* **110**(9): 3381–3386.
· *in vivo* optical trapping and force measurements on kinesin and dynein

Dame, R. T., Noom, M. C. and Wuite, G. J. (2006) Bacterial chromatin organization by H-NS protein unravelled using dual DNA manipulation. *Nature* **444**(7117): 387–390.
· analysis of the energy landscape of the DNA-DNA interaction mediated by H-NS by pairwise trapping of two molecules of DNA in a quadruple optical trap (Q-trap)

Zlatanova, J. and Leuba, S. H. (2003) Magnetic tweezers: a sensitive tool to study DNA and chromatin at the single-molecule level. *Biochem Cell Biol* **81**(3): 151–159.
· review article highlighting the applications of magnetic tweezers to DNA and DNA-dependent enzymes and its potential for studies on chromatin

Strick, T. R., Croquette, V. and Bensimon, D. (2000) Single-molecule analysis of DNA uncoiling by a type II topoisomerase. *Nature* **404**(6780): 901–904.
· magnetic tweezers study of DNA relaxation by DNA topoisomerase II

Neuman, K. C., Charvin, G., Bensimon, D. and Croquette, V. (2009) Mechanisms of chiral discrimination by topoisomerase IV. *Proc Natl Acad Sci U S A* **106**(17): 6986–6991.
· magnetic tweezers study on the relaxation of positive and negative DNA supercoils by topoisomerase IV

Nollmann, M., Stone, M. D., Bryant, Z., Gore, J., Crisona, N. J., Hong, S. C., Mitelheiser, S., Maxwell, A., Bustamante, C. and Cozzarelli, N. R. (2007) Multiple modes of *Escherichia coli* DNA gyrase activity revealed by force and torque. *Nat Struct Mol Biol* **14**(4): 264–271.
· study on the response of different DNA gyrase activities to external forces by magnetic tweezers

Ogawa, T., Yogo, K., Furuike, S., Sutoh, K., Kikuchi, A. and Kinosita, K., Jr. (2015) Direct observation of DNA overwinding by reverse gyrase. *Proc Natl Acad Sci U S A* **112**(24): 7495–7500.
· monitoring positive DNA supercoiling by reverse gyrase with magnetic tweezers

Koster, D. A., Palle, K., Bot, E. S., Bjornsti, M. A. and Dekker, N. H. (2007) Antitumour drugs impede DNA uncoiling by topoisomerase I. *Nature* **448**(7150): 213–217.
· probing the mechanism of topoisomerase I poisoning by the drug camptothecin using magnetic tweezers

Gore, J., Bryant, Z., Stone, M. D., Nollmann, M., Cozzarelli, N. R. and Bustamante, C. (2006) Mechanochemical analysis of DNA gyrase using rotor bead tracking. *Nature* **439**(7072): 100–104.
· investigation of topological changes in DNA and mechanochemical coupling during negative supercoiling by DNA gyrase using the magnetic tweezers rotor-bead assay

Bryant, Z., Stone, M. D., Gore, J., Smith, S. B., Cozzarelli, N. R. and Bustamante, C. (2003) Structural transitions and elasticity from torque measurements on DNA. *Nature* **424**(6946): 338–341.
· torque measurements on DNA using optical tweezers and a rotor-bead assay

Ribeck, N. and Saleh, O. A. (2008) Multiplexed single-molecule measurements with magnetic tweezers. *Rev Sci Instrum* **79**(9): 094301.
· multiplexing of magnetic tweezers experiments

Long, X., Parks, J. W., Bagshaw, C. R. and Stone, M. D. (2013) Mechanical unfolding of human telomere G-quadruplex DNA probed by integrated fluorescence and magnetic tweezers spectroscopy. *Nucleic Acids Res* **41**(4): 2746–2755.
· analysis of mechanical unfolding of telomeric G-quadruplex sequences by combining magnetic tweezers with FRET

Swoboda, M., Grieb, M. S., Hahn, S. and Schlierf, M. (2014) Measuring two at the same time: combining magnetic tweezers with single-molecule FRET. *EXS* **105**: 253–276.
· review article about combining magnetic tweezers with fluorescence detection and FRET

# Transient Kinetic Methods

In Part II, we derived rate laws for different types of reactions, and formally treated pre-steady state and steady-state kinetics. In kinetic experiments, a reaction is typically started by manual or automated mixing of the components. The disappearance of reactants or appearance of intermediates and products is followed either continuously by monitoring a spectroscopic signal that reports on reactant or product concentration in real time, or discontinuously by stopping the reaction at different time points, followed by quantification of reactant(s) and/or product(s) at each time point. This chapter focuses on experimental techniques to initiate reactions of different velocities. Manual mixing typically requires a few seconds before the first data point or sample can be taken, which is the *dead time* of the experiment. Slow reactions will not have progressed much during this dead time, but fast reactions might already have completed before the measurement has even started or the first sample has been taken. In simple words: for rapid reactions, rapid mixing is essential.

## 25.1 STOPPED FLOW

A *stopped flow* instrument (Figure 25.1) contains the reactants in two syringes that both connect to a mixing or reaction chamber. The reaction is started by movement of a plate, either driven by expansion of compressed gas or by a motor, which rapidly pushes the pistons of the syringes. The reactants flow into the mixing chamber where they displace the old solution from the previous run into a stop syringe. As soon as the piston of the stop syringe contacts a stop plate, the mixing process is finished, and the measurement starts. The overall dead times achieved by stopped flow instruments are about 1 ms. The reaction progress is typically followed *via* absorption, fluorescence intensity or anisotropy, or circular dichroism. Any spectroscopic signal that is a measure of concentration can be used to monitor the reaction, even NMR. From the time dependence of the spectroscopic signal, the rate constant for the reaction may be determined (see Chapters 6–8). By varying the concentrations of reactants, the underlying kinetic mechanism can be deduced. Rate constants can be determined as a function of temperature to determine activation energies (Chapter 13).

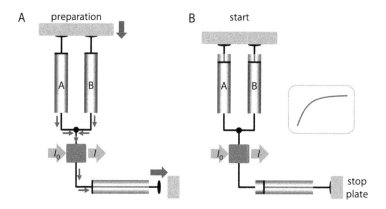

FIGURE 25.1: **Single-mixing stopped flow instrument.** A: The top plate pushes the syringes containing the reactants A (blue) and B (red), leading to rapid mixing in the mixing chamber (black circle). B: The mixing process ends when the solution pushed into the mixing and observation chamber (gray) reaches the stop syringe (purple), and the plunger hits the stop plate (right). At this point, the measurement of a spectroscopic signal (here: absorption) for the disappearance of reactants or appearance of products is started.

*Single-mixing* stopped flow instruments have two syringes that are moved simultaneously to mix two reactants. *Double-mixing* stopped flow instruments (Figure 25.2) can realize more complex mixing schemes, such as pre-formation of a complex, followed by mixing of this complex with a third compound. This way, transient intermediates can be populated and their reactions with other compounds can be studied.

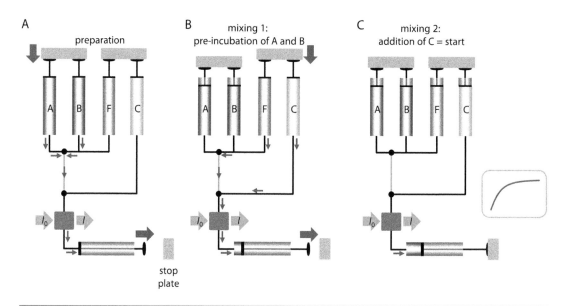

FIGURE 25.2: **Double-mixing stopped flow instrument.** A: A double-mixing stopped flow typically consists of four syringes, and has two plates to move the syringes independently in pairs. B: In the first mixing event, plate 1 pushes the two syringes containing reactants A and B, leading to mixing of A and B, and allowing for a reaction, such as AB complex formation, in the pre-incubation loop (orange). C: After the pre-incubation time, plate 2 pushes syringes F (for flush) and C, leading to mixing of C and AB. Syringe F is filled with buffer and pushes the aged mixture of A and B forward. The measurement starts when the plunger of the stop syringe hits the stop plate at the end of the second mixing event.

Double-mixing experiments are often used in protein folding studies, where the protein of interest is first unfolded for a certain time, and then brought back into native conditions to initiate refolding. Alternatively, the unfolded protein is refolded for a certain time, and then again transferred to denaturing conditions. From the number of phases in the refolding/unfolding reactions and their amplitudes, information on folding intermediates and folding pathways can be deduced (Box 25.1).

## BOX 25.1: DOUBLE-MIXING STOPPED FLOW EXPERIMENTS TO ANALYZE PATHWAYS OF PROTEIN FOLDING.

Ribonuclease $T_1$ has been used as a model system to study the role of proline isomerization in protein folding. In the native state of RNase $T_1$ the peptide bonds preceding Pro39 and Pro55 are in the *cis*-configuration. Equilibrium unfolding experiments of RNase $T_1$ are consistent with the two-state model (Section 16.3.4), but the folding and unfolding reactions are more complex. After loss of the native structure, the *cis*-prolyl peptide bonds isomerize to the more stable trans configuration. As a result, the unfolded state of RNase $T_1$ is an equilibrium mixture of four species with different configurations of the prolyl peptide bonds: 39*cis*/55*cis*, 39*cis*/55*trans*, 39*trans*/55*cis* and 39*trans*/55*trans* (Figure 25.3). 39*cis*/55*cis* folds directly into the native state, whereas all other species rapidly form partially folded intermediates that reach the native state through slow *trans-cis* isomerization of one or two prolines. Double-mixing stopped flow experiments were used to first populate the unfolded state 39*cis*/55*cis* that is only present to 2–4% in equilibrium by short unfolding for a time $\Delta t_u$, and to then monitor its refolding by an increase in fluorescence. Refolding curves show double-exponential behavior, with a fast phase ($k = 7.6$ s$^{-1}$) and a slow phase ($k = 1$ s$^{-1}$). The amplitudes of the two phases depend on the unfolding time $\Delta t_u$ in the first step which determines the populations of the different prolyl isomers. Both amplitudes increase with increasing unfolding time, reach a maximum, and then decrease to a small final value (Figure 25.3). This behaviour is characteristic of consecutive reactions (see Section 8.3), and the curves provide the rate constants of production and decay for the species that refolds in each phase (Figure 25.3). The rapid refolding phase was assigned to the refolding of the unfolded RNase $T_1$ with 39*cis*/55*cis* configuration, generated by unfolding of N and decaying to other species by prolyl isomerization. The second phase was assigned to one or several species that are formed from 39*cis*/55*cis* by prolyl isomerization and converted into other species also by prolyl isomerization. In conjunction with folding data of variants with substitutions at Pro39 and Pro55, the rate constants for prolyl isomerization of the peptide bonds preceding Pro39 and Pro55 in the unfolded state were extracted by global analysis of the refolding reactions. Mayr *et al.* (1996) *Biochemistry* 35(17): 5550–5561.

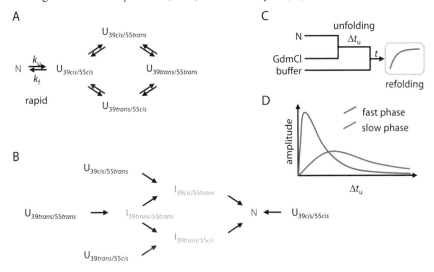

**FIGURE 25.3:    Analysis of RNase $T_1$ folding by double-mixing stopped flow experiments.** A: Equilibrium unfolding scheme. B: Folding kinetics. U: unfolded, I: intermediate, N: native state. C: Schematic depiction of double-mixing experiments. Native RNase $T_1$ is unfolded for the time $\Delta t_u$ by dilution into buffer with 6 M GdmCl, pH 1.6, and refolded by dilution to 1 M GdmCl, pH 4.6. Folding is monitored by fluorescence. D: Amplitude of fast and slow phases of refolding as a function of the unfolding time $\Delta t_u$.

## 25.2  QUENCH FLOW

A quench flow instrument (Figure 25.4) also contains the reactants in two syringes. In addition to the stopped flow instrument, it contains a third syringe with a quenching solution that stops the reaction of interest at a certain time point. Quench flow is thus a discontinuous method to monitor the progress of a reaction. To initiate the reaction, the motor plate pushes all three syringes, such that the reactants are mixed and passed through a reaction loop. The quenching solution from the third syringe is made to travel a longer path. The quench solution is combined with the reaction mixture only after the reaction mixture has passed the reaction loop, and the reaction time is thus determined by the volume of the reaction loop. Using reaction loops of different volumes, the reaction can be stopped at different time points. The individual samples are expelled from the instrument into a collection vial and subjected to analysis by any suitable method such as thin layer chromatography, HPLC, quantitative mass spectrometry, etc. The dead time, i.e. the shortest reaction time that can be realized in a quench flow instrument, is about 1–2 ms. Longer reaction times above the travel time through the ageing loop can be realized by a push-wait-push sequence: a first push of the mixing plate mixes A and B, the reaction can proceed during the waiting time, and the second push adds the quench solution and expels the sample from the instrument. On the one hand, quench flow is more versatile than stopped flow because no spectroscopic signal is needed. On the other hand, each sample has to be analyzed separately, which is cumbersome and yields fewer data points. Again, more complex implementations of double mixing followed by quenching allow the realization of sequential reactions.

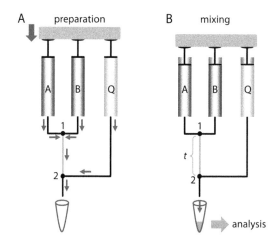

**FIGURE 25.4:  Rapid mixing by quench flow methods.** A: A quench flow system consists of three syringes, filled with the reactants A and B and the quencher compound Q that stops the reaction. A moving plate pushes all three syringes simultaneously, leading to mixing of A and B in the first mixing chamber (black circle, 1). The quencher is added later in the path in mixer 2 (black circle, 2) by a second push after a delay time. B: During the time the solution passes the reaction loop (orange), the reaction between A and B occurs. When the reaction mixture reaches mixer 2, addition of the quench solution stops the reaction. The length of the reaction loop thus determines the reaction time, and quench flow instruments contain several reaction loops of different lengths to stop the reaction at different time points. The sample is expelled from the instrument, collected and subjected to analysis for reactant and/or product content.

Quench flow experiments are used for reactions that cannot be followed by a spectroscopic signal in real time. Quench flow analysis has been used to study the kinetic mechanism of primer/ template binding and the incorporation of nucleotides by HIV reverse transcriptase, and the stimulation of ATP hydrolysis by the chaperone DnaK in presence of its co-chaperone DnaJ (Box 25.2).

**BOX 25.2: QUENCH FLOW EXPERIMENTS TO DETERMINE
THE STIMULATION OF ATP HYDROLYSIS BY CHAPERONES
IN THE PRESENCE OF CO-CHAPERONES.**

The chaperone DnaK assists protein folding by interacting with exposed hydrophobic regions of unfolded proteins, preventing their aggregation. DnaK is an ATP-binding protein. In the ATP state, binding and dissociation of peptides is rapid, but the peptide affinity is low. ATP hydrolysis converts DnaK into the ADP state that shows slow peptide binding and dissociation, but has a high peptide affinity. Dissociation of ADP and re-binding of ATP starts a new catalytic cycle. The cycle of ATP hydrolysis is the timer for binding and release of peptides, and central for the chaperone function of DnaK. The rate of ATP hydrolysis is increased in the presence of the co-chaperone DnaJ, the rate of nucleotide exchange is stimulated by the co-chaperone GrpE. The stimulation of ATP hydrolysis by DnaJ was determined in quench flow experiments. The DnaK-ATP complex was preformed during 10 s, and subsequently mixed with DnaJ in a quench flow apparatus (Figure 25.5). After different reaction times $t$, the reaction was stopped by addition of perchloric acid, ATP and phosphate were separated by thin layer chromatography, and the amount of ATP was quantified. The rate constant of ATP hydrolysis in the presence of DnaJ was $k = 0.14\ \mathrm{s}^{-1}$, which is 100-fold higher than the rate constant for ATP hydrolysis by DnaK alone ($k = 1.5 \cdot 10^{-3}\ \mathrm{s}^{-1}$). The maximal stimulation was determined by performing the same reaction at different concentrations of DnaJ. The maximal rate constant was $k = 0.79\ \mathrm{s}^{-1}$, corresponding to a stimulation by 530-fold. From the plot of the rate constant as a function of the DnaJ concentration, a $K_d$ value of 16 µM for dissociation of DnaJ from the DnaK-ATP complex was obtained. Laufen *et al.* (1999) *Proc. Natl. Acad. Sci. U. S. A.* 96(10): 5452–5457.

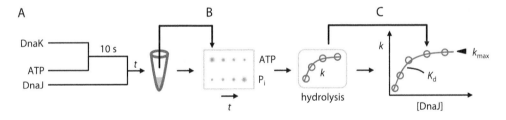

**FIGURE 25.5:   Stimulation of ATP hydrolysis by DnaK in the presence of DnaJ analyzed by quench flow.** A: Schematic depiction of a double-mixing experiment to preincubate DnaK with ATP, with subsequent mixing of the DnaK-ATP complex with DnaJ. B: The reaction is stopped at times $t$, and the ATP and inorganic phosphate ($P_i$) content is analyzed by thin layer chromatography to obtain the rate constant of hydrolysis. C: The rate constant as a function of DnaJ concentration provides the rate constant at maximal stimulation, $k_{max}$, and the $K_d$ value for dissociation of DnaJ from the DnaK-ATP complex.

## 25.3 LASER FLASH PHOTOLYSIS

For very rapid reactions that occur on the millisecond time scale or faster, starting the reaction by mechanical mixing is not fast enough. Light-sensitive processes, such as the *cis/trans* isomerization of retinal in rhodopsins or the reactions of other photoreceptors, can be triggered on the microsecond time scale by illumination. Certain processes can be initiated rapidly by illumination although they are not naturally triggered by light, such as the dissociation of CO from heme. Many processes can be rendered light-sensitive using *caged compounds*. These reagents are inactivated by photolabile protective groups that can be removed by illumination with (typically) UV light. Removal of the protective group liberates the active component and starts the reaction *in situ*. The reaction can then be followed continuously using spectroscopic probes. The dead time of these procedures is on the order of 1 μs.

Examples for caged compounds include caged ATP (Figure 25.6) for the studies of ATPases such as the muscle protein myosin, and caged GTP to investigate GTPases, such as G-proteins. Caged $Ca^{2+}$ allows the rapid release of $Ca^{2+}$ to trigger signaling cascades. Nitrophenyl-EGTA (Figure 25.7) cages $Ca^{2+}$ by high-affinity binding. Photolysis cleaves the chelator in two, leading to a drop in $Ca^{2+}$ affinity and its release. Caged glutamate (4-methoxy-7-nitro-indolinyl-Glu; Figure 25.7) does not bind to postsynaptic glutamate-gated ion channels. Illumination at 350 nm ($\varepsilon$ = 4300 $M^{-1}$ $cm^{-1}$), liberates glutamate that acts as a neurotransmitter and activates ion channels.

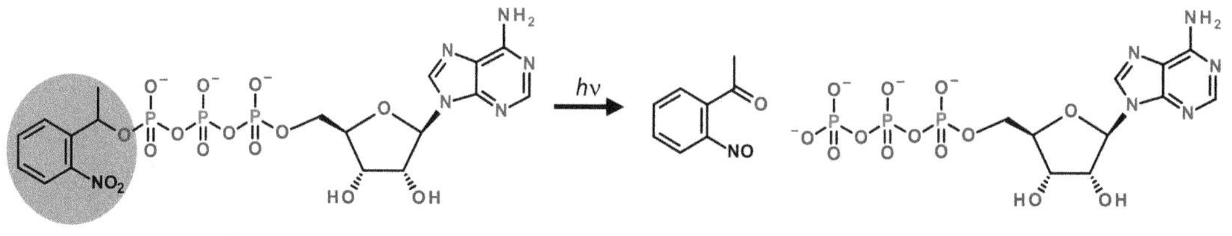

**FIGURE 25.6: Activation of caged compounds.** In caged ATP (1-(2-nitrophenylethyl)-ATP, NPE-ATP), the γ-phosphate is protected by a 2-nitrophenylethyl group (orange). In this state, ATP is not hydrolyzed by ATPases. The NPE group has an extinction coefficient of 660 $M^{-1}$ $cm^{-1}$ at 347 nm, and the caged compound is activated by photo-induced cleavage of the group (*hv*) to rapidly liberate ATP.

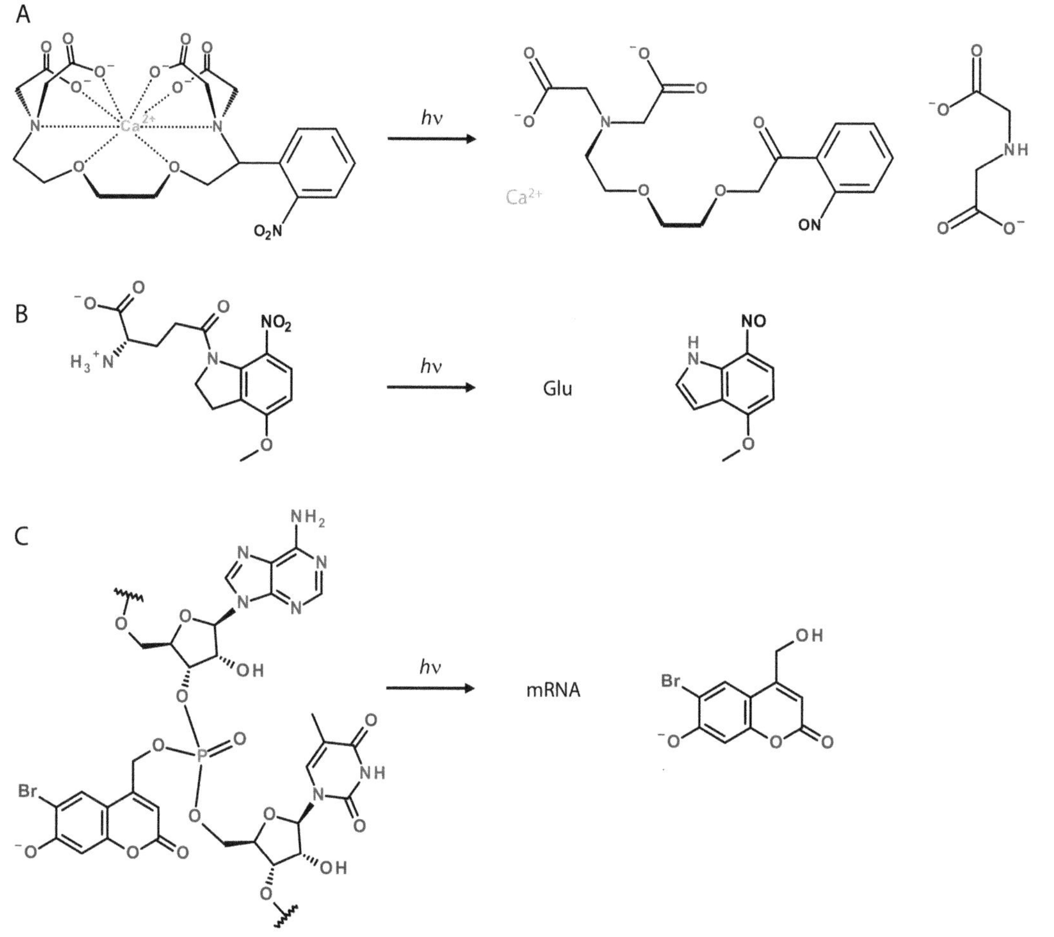

**FIGURE 25.7: Examples of caged compounds.** A: Caged $Ca^{2+}$, a complex of nitrophenyl-EGTA and $Ca^{2+}$. B: Caged glutamate (4-methoxy-7-nitro-indolinyl-glutamate). C: Caged mRNA, protected by the bromo-hydroxycoumarin group at the phosphate groups.

Larger molecules can also be inactivated by caging. In caged mRNA (Figure 25.7), the backbone phosphates are protected by the bromohydroxycoumarin group that is introduced during chemical synthesis. Caged mRNA is activated by photocleavage of the bromohydroxycoumarin group and then serves as a template for protein biosynthesis. It is also possible to cage individual groups in proteins. Using chemically synthesized aminoacyl-tRNAs that carry photocleavable o-nitrophenyl-modified phosphoserine, phosphothreonine, or phosphotyrosine, caged phosphoproteins can be generated by *in vitro* translation and suppression of stop codons. Using suppression of stop codons, caged cysteine and caged tyrosine (modified with the photocleavable o-nitrobenzyl group) have been incorporated into the nicotinic acetylcholine receptor *in vivo*. By photocleavage of the o-nitrobenzyl groups, active receptor molecules were recovered with a time constant of 500 ms, which is slower than the photoreaction (1 ms) and most likely caused by a conformational change of the receptor.

## 25.4 RELAXATION KINETICS: PRESSURE- AND TEMPERATURE-JUMP

*Relaxation kinetics* is based on a small, but rapid, disturbance of a reaction in equilibrium, either by applying a jump in temperature or in pressure. Due to the temperature and pressure dependence of equilibrium constants (see Section 3.1), the position of the equilibrium will be different after the disturbance. The system therefore relaxes to the new equilibrium in a process that depends on the rate constants for the forward and backward reaction. The relaxation to the new equilibrium is followed continuously using a spectroscopic signal. For conversions of two species, such as folding and unfolding

$$A \underset{k_{-1}}{\overset{k_1}{\rightleftharpoons}} B$$    scheme 25.1

the observed relaxation time $\tau$, i.e. the time constant for reaching the new equilibrium position, is

$$\tau = \frac{1}{k_1 + k_{-1}}$$    eq. 25.1

The constants $k_1$ and $k_{-1}$ are the rate constants at the temperature or pressure after the jump. For a simple bimolecular reaction of

$$A + B \underset{k_{-1}}{\overset{k_1}{\rightleftharpoons}} AB$$    scheme 25.2

the observed relaxation time $\tau$ for the approach of the new equilibrium is

$$\tau = \frac{1}{k_1([A]+[B])+k_{-1}}$$    eq. 25.2

Thus, from the relaxation times measured at different concentrations of A and B, the forward (association) rate constant $k_1$ and the reverse (dissociation) rate constant $k_{-1}$ can be determined. *Temperature jumps* are usually achieved by illumination with infrared laser light. A temperature jump by 5–10°C can easily be completed within 0.1–1 µs. *Pressure jumps* in a pressure chamber can be achieved by a quick release valve, or by bursting of a membrane. While these approaches are limited to pressure drops, modern instruments achieve pressure changes in both directions by using piezoelectric actuators. This approach offers the additional advantage of performing a large number of repetitions of the experiment. Accumulation of the measured transients significantly improves the signal-to-noise ratio. The dead time of such a pressure jump instrument is on the order of 50–100 µs (Figure 25.8).

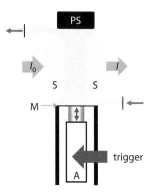

FIGURE 25.8: **Principle of a pressure-jump instrument.** An electric trigger pulse to the piezo actuator (A) causes its expansion and moves the piston against the membrane (M) that delimits the sample chamber to the bottom (voltage increase/pressure increase) or away from it (voltage decrease/pressure decrease; double-headed blue arrow). A pressure sensor (PS) is attached to the top of the sample chamber. Valves at the inlet and outlet (red arrows) enable changing of the sample. The change in absorption ($I_0$, $I$, orange) during relaxation to the new equilibrium after the pressure jump can be monitored through sapphire windows (S). Fluorescence can also be used as a probe in pressure jump experiments.

*T*-jump and pressure-jump relaxation kinetics have been applied to protein folding down to the microsecond time scale (Box 25.3). Relaxation methods can be combined with NMR, SAXS, FT-IR, and CD/ORD.

---

**BOX 25.3: PRESSURE JUMP AND PROTEIN FOLDING.**

A number of small proteins fold into their native states on the time scale of milliseconds and faster. The folding of these proteins is complete within the dead time of mixing of stopped flow devices, and therefore cannot be studied with this technique. One example of a rapidly folding protein is the cold-shock protein CspB. Folding and unfolding cycles of CspB were investigated during cycles of pressure release and increase in a pressure-jump instrument. Folding/unfolding was monitored on the time scale from 50 μs to 70 s by the associated increase/decrease in fluorescence, and rate constants for folding and unfolding were determined over a wide range of temperatures. As the folding/unfolding traces provide the sum of the folding and unfolding rates, the equilibrium constant of folding was determined independently in thermal unfolding experiments. From the temperature dependence of rate constants and equilibrium constants, the enthalpies, entropies, and heat capacities of activation, and the corresponding thermodynamic parameters for folding and unfolding were determined. From the pressure dependence of the equilibrium constant of folding, the activation volume for folding and unfolding can be calculated. The activation volume was positive for both reactions: the transition state of folding/unfolding has a higher volume than the unfolded or native states. By comparing the thermodynamic parameters of folding and unfolding of wild-type CspB with variants in which exposed phenylalanine residues were replaced by alanines, the effect of these residues on folding and stability was interrogated. Jacob *et al.* (1999) *Biochemistry* 38(10): 2882–2891.

---

## QUESTIONS

25.1 You want to measure GTP hydrolysis by a small G-protein in complex with a regulatory protein. The regulatory protein stimulates the GTPase activity, and the reaction becomes too fast for hand-mixing. How can you do that (1) using stopped flow or (2) using quench flow?

25.2 Why is pressure jump useful to study protein folding?

# REFERENCES

Mayr, L. M., Odefey, C., Schutkowski, M. and Schmid, F. X. (1996) Kinetic analysis of the unfolding and refolding of ribonuclease T1 by a stopped-flow double-mixing technique. *Biochemistry* **35**(17): 5550–5561.
· double-jump stopped flow study on prolyl isomerization during folding and unfolding of RNase T1

Laufen, T., Mayer, M. P., Beisel, C., Klostermeier, D., Mogk, A., Reinstein, J. and Bukau, B. (1999) Mechanism of regulation of Hsp70 chaperones by DnaJ cochaperones. *Proc Natl Acad Sci U S A* **96**(10): 5452–5457.
· quench flow experiments to determine the stimulation of ATP hydrolysis by the chaperone DnaK in the presence of its co-chaperone DnaJ

Wohrl, B. M., Krebs, R., Goody, R. S. and Restle, T. (1999) Refined model for primer/template binding by HIV-1 reverse transcriptase: pre-steady-state kinetic analyses of primer/template binding and nucleotide incorporation events distinguish between different binding modes depending on the nature of the nucleic acid substrate. *J Mol Biol* **292**(2): 333–344.
· analysis of the kinetic mechanism of primer/template binding and nucleotide incorporation by HIV reverse transcriptase

Schlichting, I. and Goody, R. S. (1997) Triggering methods in crystallographic enzyme kinetics. *Meth Enzymol* **277**: 467–490.
· review on methods to trigger enzymatic reactions in crystals

Barends, T. R., Foucar, L., Ardevol, A., Nass, K., Aquila, A., Botha, S., Doak, R. B., Falahati, K., Hartmann, E., Hilpert, M., Heinz, M., Hoffmann, M. C., Kofinger, J., Koglin, J. E., Kovacsova, G., Liang, M., Milathianaki, D., Lemke, H. T., Reinstein, J., Roome, C. M., Shoeman, R. L., Williams, G. J., Burghardt, I., Hummer, G., Boutet, S. and Schlichting, I. (2015) Direct observation of ultrafast collective motions in CO myoglobin upon ligand dissociation. *Science* **350**(6259): 445–450.
· flash photolysis of the Fe-CO bond in myoglobin and structural studies on ultrafast protein conformational changes

Rapp, G., Poole, K. J., Maeda, Y., Guth, K., Hendrix, J. and Goody, R. S. (1986) Time-resolved structural studies on insect flight muscle after photolysis of caged-ATP. *Biophys J* **50**(5): 993–997.
· time-resolved structural study of insect flight muscle after photolysis of caged ATP

Ellis-Davies, G. C. (2007) Caged compounds: photorelease technology for control of cellular chemistry and physiology. *Nat Methods* **4**(8): 619–628.
· review about photocaging and triggering of biochemical reactions by flash photolysis

Schlichting, I., Rapp, G., John, J., Wittinghofer, A., Pai, E. F. and Goody, R. S. (1989) Biochemical and crystallographic characterization of a complex of c-Ha-ras p21 and caged GTP with flash photolysis. *Proc Natl Acad Sci U S A* **86**(20): 7687–7690.
· time-resolved crystallography study of GTP hydrolysis by H-Ras in crystals using caged GTP and flash photolysis

Marx, A., Jagla, A. and Mandelkow, E. (1990) Microtubule assembly and oscillations induced by flash photolysis of caged-GTP. *Eur Biophys J* **19**(1): 1–9.
· study of tubulin assembly using caged GTP, flash photolysis, and time-resolved X-ray scattering

Corrie, J. E., DeSantis, A., Katayama, Y., Khodakhah, K., Messenger, J. B., Ogden, D. C. and Trentham, D. R. (1993) Postsynaptic activation at the squid giant synapse by photolytic release of L-glutamate from a 'caged' L-glutamate. *J Physiol* **465**: 1–8.
· postsynaptic activation of squid giant synapse using flash photolysis of caged glutamate.

Ando, H., Furuta, T., Tsien, R. Y. and Okamoto, H. (2001) Photo-mediated gene activation using caged RNA/DNA in zebrafish embryos. *Nat Genet* **28**(4): 317–325.
· triggering gene expression by flash photolysis of caged mRNA in zebrafish embryos

Rothman, D. M., Petersson, E. J., Vazquez, M. E., Brandt, G. S., Dougherty, D. A. and Imperiali, B. (2005) Caged phosphoproteins. *J Am Chem Soc* **127**(3): 846–847.
· chemical synthesis of photocaged phosphorylated amino acids and incorporation into nicotinic acetyl choline receptor and the vasodilator-stimulated phosphoprotein VASP by amber stop codon suppression and *in vitro* translation

Philipson, K. D., Gallivan, J. P., Brandt, G. S., Dougherty, D. A. and Lester, H. A. (2001) Incorporation of caged cysteine and caged tyrosine into a transmembrane segment of the nicotinic ACh receptor. *Am J Physiol Cell Physiol* **281**(1): C195–206.
· incorporation of caged cysteine and caged tyrosine into the nicotinic acetylcholine receptor *in vivo* by stop codon suppression and recovery of active receptors by flash photolysis

Wirth, A. J., Liu, Y., Prigozhin, M. B., Schulten, K. and Gruebele, M. (2015) Comparing fast pressure jump and temperature jump protein folding experiments and simulations. *J Am Chem Soc* **137**(22): 7152–7159.
· a comparative study of rapid folding of β-sheets using pressure jump, temperature jump, and computational analyses

Dumont, C., Emilsson, T. and Gruebele, M. (2009) Reaching the protein folding speed limit with large, submicrosecond pressure jumps. *Nat Methods* **6**(7): 515–519.
· fluorescence-detected refolding study of λ repressor applying microsecond pressure drops reveals folding without activation barrier

Pearson, D. S., Holtermann, G., Ellison, P., Cremo, C. and Geeves, M. A. (2002) A novel pressure-jump apparatus for the microvolume analysis of protein-ligand and protein-protein interactions: its application to nucleotide binding to skeletal-muscle and smooth-muscle myosin subfragment-1. *Biochem J* **366**(Pt 2): 643–651.
· report of a pressure jump instrument that achieves pressure increase and decrease by using piezoelectric actuators

Matsumoto, T., Nakagawa, T. and Kuwata, K. (2009) Cold destabilization and temperature jump of the murine prion protein mPrP(23-231). *Biochim Biophys Acta* **1794**(4): 669–673.
· temperature jump study on stability and cold denaturation of prions

Herberhold, H. and Winter, R. (2002) Temperature- and pressure-induced unfolding and refolding of ubiquitin: a static and kinetic Fourier transform infrared spectroscopy study. *Biochemistry* **41**(7): 2396–2401.
· temperature-dependent folding study of ubiquitin using pressure-jump and Fourier transform infrared spectroscopy on the time scale of seconds to minutes

Kremer, W., Arnold, M., Munte, C. E., Hartl, R., Erlach, M. B., Koehler, J., Meier, A. and Kalbitzer, H. R. (2011) Pulsed pressure perturbations, an extra dimension in NMR spectroscopy of proteins. *J Am Chem Soc* **133**(34): 13646–13651.
· combination of millisecond pressure-jump with NMR spectroscopy

Woenckhaus, J., Kohling, R., Thiyagarajan, P., Littrell, K. C., Seifert, S., Royer, C. A. and Winter, R. (2001) Pressure-jump small-angle X-ray scattering detected kinetics of staphylococcal nuclease folding. *Biophys J* **80**(3): 1518–1523.
· combined pressure-jump and small-angle X-ray scattering study on folding of staphylococcal nuclease on the seconds to minutes time scale

Panick, G. and Winter, R. (2000) Pressure-induced unfolding/refolding of ribonuclease A: static and kinetic Fourier transform infrared spectroscopy study. *Biochemistry* **39**(7): 1862–1869.
· pressure-induced unfolding of RNase A studied by Fourier transform infrared spectroscopy on the time scale of minutes

Gruenewald, B. and Knoche, W. (1978) Pressure jump method with detection of optical rotation and circular dichroism. *Rev Sci Instrum* **49**(6): 797.
· pressure-jump instrument with circular dichroism/optical rotation dispersion detection

Jacob, M., Holtermann, G., Perl, D., Reinstein, J., Schindler, T., Geeves, M. A. and Schmid, F. X. (1999) Microsecond folding of the cold shock protein measured by a pressure-jump technique. *Biochemistry* **38**(10): 2882–2891.
· pressure-jump analysis of folding of the protein CspB on the microsecond time scale

# Molecular Mass, Size, and Shape

Accurate mass determination is the key to identify macromolecules and their complexes. Mass spectrometry (MS, Section 26.1) is the only method that determines masses directly. MS relies on the ionization of molecules and the separation of these ions according to their mass in an electric field. Analytical ultracentrifugation (AUC, Section 26.2) measures the sedimentation behavior of macromolecules in a gravitational field, which not only provides information on their mass, but also on their size and shape. In surface plasmon resonance (SPR, Section 26.3), the change in refractive index reports on mass changes due to interactions of binding partners with surface-immobilized molecules. All of these methods are suitable to characterize macromolecular interactions.

## 26.1 MASS SPECTROMETRY

Mass spectrometry (MS) starts with the ionization of molecular entities in vacuum, followed by the separation of these ions according to molecular mass, and then their detection. A mass spectrum measures the abundance of species with a certain mass to charge ratio, $m/z$. The determination of molecular masses is the most important application of MS (Section 26.1.5.1). MS can identify non-covalently bound adducts and covalent modifications, and can be used to identify and sequence proteins and nucleic acids (Section 26.1.5.3). Native MS (Section 26.1.5.4) uses gentle ionization protocols and sophisticated ion manipulation methods to determine the composition, stoichiometry, and conformational heterogeneity of macromolecular complexes, giving insight into the biological pathway of complex assembly. Kinetic information down to the millisecond time scale can be gained by measuring the time-dependent exchange of deuterium for hydrogen atoms (Section 26.1.5.2) or measuring exchange of whole protein subunits in multi-protein complexes (Section 26.1.5.4). Finally, MS is often combined with chromatographic techniques, such as reversed phase (RP)-HPLC or size exclusion chromatography, to analyze the components of a mixture one after the other.

## 26.1.1  Ionization

The first step in MS is ionization of the molecule(s) of interest. Ions are manipulated and analyzed by applying electric fields. There are two main methods for non-destructive ionization of a molecule, *electrospray ionization* (ESI) and *matrix-assisted laser desorption ionization* (MALDI). In MALDI the ionization is done out of the solid state under denaturing conditions whereas ESI can be performed under both denaturing and physiological conditions.

### 26.1.1.1  Matrix-Assisted Laser Desorption Ionization

MALDI vaporizes macromolecules that are embedded in a *matrix*, which is often an organic acid of low molecular mass (Figure 26.1). Sinapinic acid, or *trans*-3,5-dimethoxy-4-hydroxycinnamic acid, is efficient for ionization of proteins, peptides, and lipids. 2,5-dihydroxybenzoic acid and its aceto-phenone are good options for (glyco)peptides, oligosaccharides, nucleotides, and oligonucleotides. Nucleic acids are also efficiently ionized from 3-hydroxy picolinic acid. For sample preparation, the macromolecule (analyte) in a low ionic strength buffer is mixed with a $10^3$-fold excess of the matrix dissolved in a volatile solvent such as acetonitrile/water. Addition of 0.1% trifluoroacetic acid to the solvents promotes protonation of the macromolecule. The mixture is dried and placed under vacuum in the mass spectrometer.

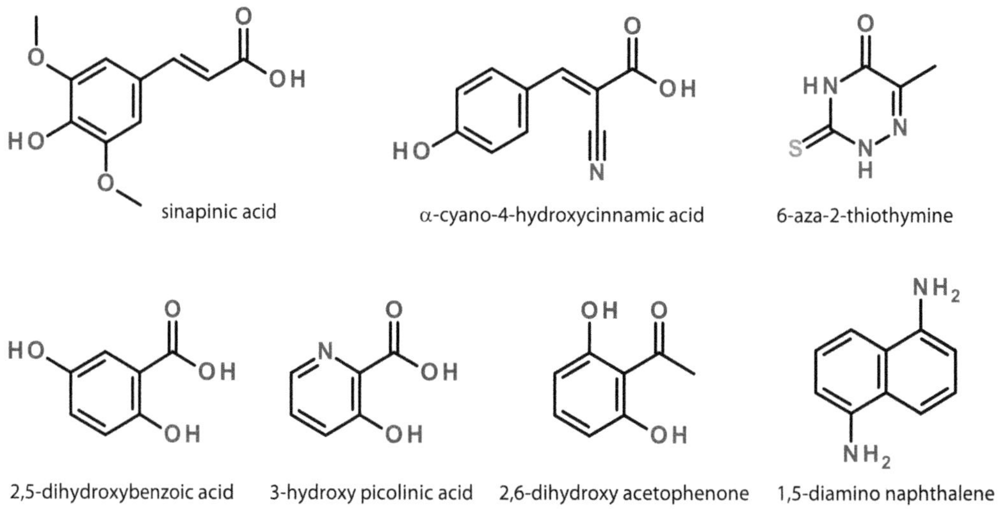

sinapinic acid          α-cyano-4-hydroxycinnamic acid          6-aza-2-thiothymine

2,5-dihydroxybenzoic acid     3-hydroxy picolinic acid     2,6-dihydroxy acetophenone     1,5-diamino naphthalene

**FIGURE 26.1:   Examples of matrices used in MALDI ionization.** The aromatic systems are excited at wavelengths of 266, 337 (nitrogen laser), or 355 nm. Water, acetonitrile, acetone, and chloroform plus 0.1% trifluoroacetic acid are popular solvents to dissolve the matrix. Many similar matrices to the ones shown exist that differ only by a few substituents. 1,5-diaminonaphthalene is not an acid but forms radicals upon laser irradiation that reduce disulfide bonds and cleave peptide bonds in proteins.

A pulsed UV laser, typically a nitrogen laser operating at 337 nm, excites and vaporizes the matrix. The matrix will transfer energy onto the macromolecule, thereby lifting it out of the solid state into the vacuum. Since the matrix is a weak acid it also helps to protonate the macromolecule, generating a positive ion. The number of protons transferred from the matrix onto the macromolecules is limited, and MALDI ionization yields ions with few charges. Peptide ions often carry only a single charge, whereas proteins are converted into a set of ions with different charges. In contrast, nucleic acids may retain their overall negative charge. Depending on the charge, the subsequent manipulation needs electric fields of different polarities. The different modes are termed *positive* and *negative ion modes* (Figure 26.2). The throughput of a MALDI measurement can be increased by using a sample plate holding 96, 384, or 1536 matrix-embedded samples that are ionized in sequence.

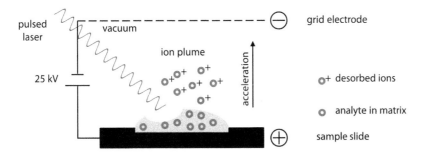

FIGURE 26.2:   **Sample preparation and ionization in MALDI.** The macromolecule is embedded into a suitable matrix and placed under vacuum in the mass spectrometer. A laser pulse is applied to the matrix, which vaporizes both the matrix and the analyte into a plume. An electric field accelerates the ions according to their mass-to-charge ratio *m/z*. The depicted polarity corresponds to the positive ion mode.

### 26.1.1.2  Electrospray Ionization

In ESI the dissolved analyte is directly ionized from its solvent (Figure 26.3). The analyte is passed at a low flow rate of ~100 nL min$^{-1}$ through a metal-coated glass capillary of ca. 100 μm diameter under atmospheric pressure. Special micro- and nano-ESI instruments use capillaries of down to 3–25 μm in diameter. The metal serves as the anode for a high voltage electric field of a few kV against a cathode in the direction of the detector. Charged droplets are ejected from the capillary by pressure and drawn into the spectrometer that is kept under vacuum. The droplets evaporate during this process. This is promoted by the vacuum, a carrier gas (usually nitrogen), and sometimes heating. With decreasing drop volume the charge density increases, up to a point where Coulomb repulsion between the ions in the droplets exceeds the surface tension of water that keeps the droplet intact, and the droplets separate into smaller and smaller droplets. At some point the solvent completely evaporates, leaving a varying number of charges on the surface of the ionized analyte (Figure 26.3). Proteins and peptides usually form positive ions, and similar to MALDI the protonation efficiency of the solvent can be increased by a trace of formic acid. Carbohydrates and nucleic acids form negative ions, and deprotonation can be promoted by addition of volatile amines or ammonia to the solvent. Like MALDI, ESI can be performed in positive and negative ion modes. In contrast to MALDI, more charges are transferred to the analyte, and the *m/z* distribution of the generated ions is also larger. Peptides usually carry a single charge after MALDI and 1–4 charges after ESI. For larger proteins, the charge differences are larger: 1–3 charges for MALDI and 20 to >100 charges for ESI. As a rule of thumb, ESI of proteins produces ions that typically carry about one positive charge per five peptide bonds plus the charges on the basic amino acids Lys and Arg.

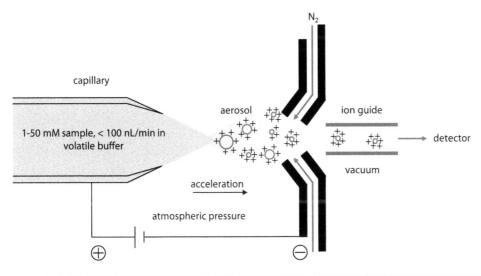

FIGURE 26.3:   **Principle of electrospray ionization.** A sample in a volatile buffer is sprayed into an ionization chamber at atmospheric pressure where the solvent evaporates, aided by a stream of nitrogen gas. An electric field between the metal-coated quartz capillary and a cathode accelerates the ions, which are then guided into the evacuated mass spectrometer.

## 26.1.2 ION STORAGE AND MANIPULATION

The next step after ionization is acceleration of the ions in an electric field to select and separate them according to their *m/z*. Such devices are collectively called *mass analyzers*.

### 26.1.2.1 Time of Flight Analysis

One way to determine *m/z* is by measuring the time of flight (TOF) that an ion needs to travel through the electric field until it arrives at the detector. The simplest arrangement for TOF is an ionization chamber, an accelerating electric field, a field-free *drift tube* through which ions with small *m/z* pass more quickly than ions with large *m/z*, and a detector that measures the time-dependent impact of the ions. For this linear arrangement, the kinetic energy $E_{kin}$ of the molecule after passing the electric field is given by

$$E_{kin} = z \cdot e \cdot U = \frac{1}{2} \cdot m \cdot v^2 \qquad \text{eq. 26.1}$$

where *z* is the number of elementary charges *e*, *U* is the voltage, *m* the mass of the ion, and *v* its velocity. We can rearrange eq. 26.1 to obtain *m/z* as

$$\frac{m}{z} = \frac{2 \cdot e \cdot U}{v^2} \qquad \text{eq. 26.2}$$

The velocity is the distance *L* the ion travels through the drift tube, divided by the time *t*, the TOF measured by the detector. Substitution of *v* in eq. 26.1 by *L/t* gives an expression that allows calculating *m/z* from the measured TOF:

$$\frac{m}{z} = \frac{2 \cdot e \cdot U}{L^2} \cdot t^2 \qquad \text{eq. 26.3}$$

According to eq. 26.3, *m/z* depends only on the TOF, and on instrument parameters that are constant and known in principle. Nevertheless, TOF mass analyzers are always calibrated to achieve higher reproducibility, either by running a sample of known molecular mass or by internal calibration, i.e. mixing analyte and reference with the matrix (for MALDI) or in the same buffer (for ESI).

The laser pulse in MALDI instruments generates a plume with a small spatial and time distribution of ions (Figure 26.2). Therefore, the kinetic energy after acceleration and hence the TOF along the drift tube vary for ions of the same *m/z*, and the spread of TOFs leads to lower resolution of the mass spectrum. The starting time of the acceleration can be synchronized when ion generation and acceleration are separated by switching off the accelerating field for a brief time interval after the laser pulse (*delayed extraction*). Ions leave the matrix with different kinetic energies. During the time between their generation and the acceleration, they now collide with molecules from the carrier gas, transferring some of their kinetic energy to the gas (Figure 26.3). Faster ions that have emerged with higher kinetic energy collide more often with gas molecules and lose more energy than slower ions, leading to a narrowing of the distribution of kinetic energies $E_{kin}$. In addition, faster ions have traveled further away from the anode toward the cathode by the time the accelerating field is switched on. As a consequence, they have a shorter distance to travel through the electric field, and experience less acceleration than slower ions, which further narrows the distribution of kinetic energies.

The spread in $E_{kin}$ can be further reduced by increasing the drift length for high-energy ions and reducing that for low-energy ions using a *reflectron* (Figure 26.4). A reflectron is an electric field of opposite polarity to the acceleration field placed at the end of the drift tube, serving as an ion mirror. Faster ions will continue to travel to the reflecting electrode for a longer time than slower ions with the same *m/z*. After reaching the reversal point, the faster ions are accelerated in the opposite direction for a longer time than the slower ions, entering a second drift tube. The net result is a focusing of the ions at a certain point where the detector is placed. The total TOF is then a composite of the two drift times and the acceleration time in the reflectron.

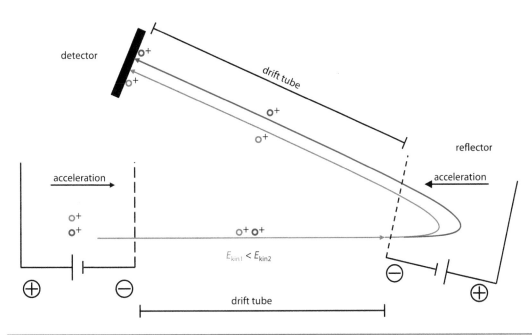

FIGURE 26.4:   **Harmonizing the TOF of ions with the same *m/z* by a reflectron.** An ion with higher kinetic energy (red) travels deeper against the electric field into the reflectron than an ion with lower kinetic energy (blue). As a result, the ion with higher kinetic energy travels a longer distance (red trajectory) than the slower ion (blue trajectory). The overall effect is a harmonization of the TOFs for ions with different kinetic energies but equal *m/z*.

TOF analysis can be combined equally well with MALDI and ESI. Compared to MALDI, where a pulsed laser is used, and the ions are analyzed in the forward direction, the ionization in ESI is continuous, which means that ESI-TOF measurements need to be done in a direction perpendicular to the stream of droplets, not parallel to it. Other methods to analyze ions in the forward direction include quadrupole magnets and ion traps. Both of these can therefore be applied to MALDI and ESI.

### 26.1.2.2 Quadrupoles and Ion Traps

Quadrupoles and ion traps are used to select for ions with *m/z* ratios in a defined range. In a quadrupole, four rod-shaped metal electrodes are placed parallel to each other under superimposed (AC) alternate and constant (DC) voltages. Opposing pairs of rods have the same polarity and voltage. The constant voltage $U_{DC}$ of a few kV serves to accelerate the ions between the electrodes: positively charged ions will oscillate between the positive electrodes and be attracted to the negative electrodes. The alternating voltage component $U_{AC}$ of radio frequency ω in the MHz range creates an alternating electric field that induces alternating deflections from the straight path along the quadrupole in the *x*- and *y*-directions. The amplitude of these deflections depends on *m/z*, the frequency ω, and the magnitudes of $U_{AC}$ and $U_{DC}$. The combination of these parameters determines whether an ion may pass the quadrupole or has an unstable trajectory, either leaving the quadrupole or colliding with the electrodes (Figure 26.5). The quadrupole thus acts as a selective *mass analyzer* by only letting certain *m/z* ions pass. This may be understood by looking at the electrode pairs separately: positive ions with large $m/z$ (heavy ions) are sent back and forth between the DC field of the positive rods but are not very much affected by the positive AC field because of their inertia. Thus, heavier ions pass the positive voltage filter of the quadrupole. Ions with too small $m/z$ (light ions) are sensitive to the positive AC field: they are deflected too much and will crash into the electrodes or be ejected from the quadrupole, which puts a lower limit on the $m/z$ range. A reverse argument holds for the negative voltage pair of electrodes: ions with large $m/z$ are attracted by the negative electrodes and will eventually crash into them or leave the quadrupole, unless they are "rescued" by a tuned negative AC field. Only lighter ions are sensitive enough to the radiofrequency field so that their trajectory is stabilized throughout the quadrupole, putting an upper limit on the $m/z$ range. Together, the positive and negative electrode pairs in the quadrupole create an $m/z$ window that is adjusted by tuning the voltages and $\omega$.

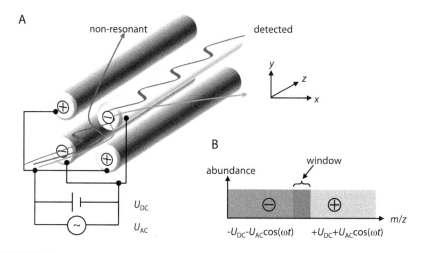

**FIGURE 26.5:   Schematic view of a quadrupole.** A: One of the electrodes is rendered transparent to provide a view onto the path of the filtered $m/z$ ions that are passed on to the detector. The negative electrodes are connected and have the same voltage. The same is true for the positive electrodes. Only ions with a certain $m/z$ undergo stable oscillation while traveling through the linear quadrupole and can pass to the detector (blue trajectory). Non-resonant ions (magenta and green trajectories) will leave the quadrupole or crash into the electrodes. B: Only low $m/z$ ions pass between the negative rods, and only high $m/z$ ions pass between the positive rods. Those ions that meet both conditions form a window of detection, which is scanned by changing the frequency $\omega$ and the AC and DC voltages.

Quadrupoles act on moving ion beams, as do reflectrons in TOF analysis. Ion traps, on the other hand, not only select ions with certain $m/z$, but in addition have cleverly shaped electric fields that can decelerate and store ions for subsequent analysis. An electrostatic ion gate controls entry and exit of ions, and thus the number of ions stored in the trap. The trap contains a small amount of a collision gas (ca. 0.1 Pa of He). Upon entry into the trap, the velocity of ions is reduced by collisions with the gas molecules. The ions are then focused by an electromagnetic lensing system, a three-dimensional circular electric field that is created by a ring-shaped electrode operating with voltage $U_1$ and a constant radiofrequency $\omega$ of typically 1.1 MHz (Figure 26.6).

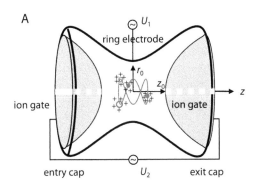

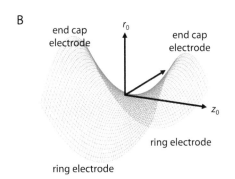

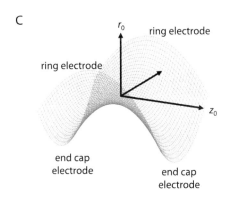

**FIGURE 26.6: Schematic of an ion trap.** A: Overview of the trap. Ions of a certain *m/z* range, defined by the amplitude and frequency of the alternating voltage at the ring and cap electrodes, $U_1$ and $U_2$, are kept on stable orbits in the trap. Varying the parameters of $U_2$ permits fragmentation and/or ejection of ions with a defined *m/z* for further analysis. The holes in the entry and exit caps are opened and closed by an electrostatic ion gate. The trap geometry is defined by the curvatures of the ring electrode and the caps, and by the radii $r_0$ and $z_0$. B: The shape of the electromagnetic field in the trap is shown by a blue mesh. Ions are focused toward the saddle point in the *z*-direction but may diffuse perpendicular to the *rz*-plane. C: The field spanned by the ring and cap electrodes is inverted to re-focus the ions toward the saddle point in the *rz*-plane. Oscillation of the field forces the ions into a stable orbit.

The electric field $U$ generated at position $(r, z)$ by the ring electrode can be described by eq. 26.4.

$$U(r,z) = A \cdot \left( z^2 - \frac{r^2}{2} \right)$$

eq. 26.4

$A$ is a constant related to trap geometry. The electric field has the shape of a saddle with a minimum at the center of the ion trap. The saddle focuses the ions in direction $r$ but they can continue to move in the $z$-direction. This defocusing is counterbalanced by switching the polarity of the ring and end cap electrodes, which flips the radiofrequency field: the ions now focus back in the $z$-direction, but can defocus in $r$. The balance between these field directions keeps ions of a certain $m/z$ orbiting in the trap. At the same time, the ions are bunched up in the center of the trap, which is useful for pulsed ejection into a mass analyzer such as the orbitrap (Section 26.1.2.3). The balance of the fields is key for the performance of the ion trap, which is why the number of ions within the trap is controlled: too many ions result in *space charge* effects that distort the electric fields of the trap, which in turn reduces its performance.

The selectivity of the ion trap, i.e. the $m/z$ of ions that have stable orbits, is determined by the geometry of the trap, the magnitude of $U_1$, and the radiofrequency $\omega$. While a quadrupole is usually tuned to allow passage of a narrow $m/z$ range, an ion trap can hold a wider range of $m/z$.

The ejection of ions from the trap in a controlled manner is achieved by applying an oscillating electric field $U_2$ to the end caps of the trap. Depending on the magnitude of $U_2$, the orbit of ions with a narrow range of $m/z$ becomes unstable. In this *mass instability mode*, the ions are in resonance with the oscillating electric field $U_2$ and ejected from the trap along the $z$-axis. Quadrupoles and ion traps can be placed in sequence for finer sorting and more accurate manipulation of ions, an arrangement referred to as tandem-MS, or MS/MS.

### 26.1.2.3 Orbitraps

The orbitrap is a type of mass analyzer that is frequently used in modern MS equipment for proteomics and native mass spectrometry. Like ion traps, orbitraps serve to select, store, and manipulate ions of a certain $m/z$ range. Orbitraps can hold many ions at once and of very different $m/z$ (Figure 26.7). In contrast to ion traps, orbitraps combine mass analysis with ion detection and $m/z$ determination.

The orbitrap consists of an inner spindle-shaped electrode that is co-axially surrounded by a cylindrical outer electrode, and two end cap electrodes. A static DC voltage is applied between the inner and outer electrodes, and an alternating AC voltage between the end caps. The DC voltage produces a radial logarithmic potential while the AC voltage produces an axial quadrupole potential, which together establish a "quadro-logarithmic" potential (eq. 26.5):

$$U(r,z) = A \cdot \left( z^2 - \frac{r^2}{2} + B \cdot \ln r \right)$$

eq. 26.5

The constants $A$ and $B$ in eq. 26.5 are specific to the geometry and voltages of the orbitrap, and determine the shape of the electric field. Ion bunches are injected tangentially to the inner electrode and immediately begin to orbit the electrode due to the Lorentz force (Figure 26.7). The centrifugal force of the ions counteracts the Lorentz force until an equilibrium radius is reached. Ions of different $m/z$ have different velocities and therefore different rotational frequencies and orbiting radii. The radii can be reduced to a desired minimum value by increasing the voltage to 3.5–5 kV, which in turn increases the detectable mass range because ions with larger $m/z$ can now be stored in the orbitrap. At the final voltage of the orbitrap, those ions with the same $m/z$ organize into rings.

Importantly, by injecting the ions away from the center (the thickest part) of the inner electrode at $z \neq 0$, they experience an additional lateral (axial) movement along the $z$-direction because of the quadrupole field. The special shape of the orbitrap electrodes ensures that the axial oscillation along the $z$-direction is harmonic, meaning that it is independent of the perpendicular radial oscillation around $z$. The angular frequency $\omega$ of this axial oscillation along $z$ is

$$\omega = \sqrt{\frac{z}{m} \cdot k}$$

eq. 26.6

where $k$ is the force constant of the potential, similar to a spring constant (Sections 15.2.2 and 19.4.1). In fact, the axial oscillation solely depends on the $m/z$ of the ions, and it is this axial oscillation that is measured in an orbitrap.

Once the voltage of the orbitrap has reached its final value, the electromagnetic field becomes static and detection can start. The trapped ion rings induce a current in the outer electrode, the *image current* (Figure 26.7). The current is measured over a few milliseconds to several seconds and Fourier transformed (Section 29.14) from the time domain into the frequency domain to give the frequencies ω of the ions. The *m/z* ratios for all ions are then calculated using eq. 26.6. Thus, all ions are detected simultaneously over a given time. The resolution of the orbitrap is increased both by increased observation time and field strength.

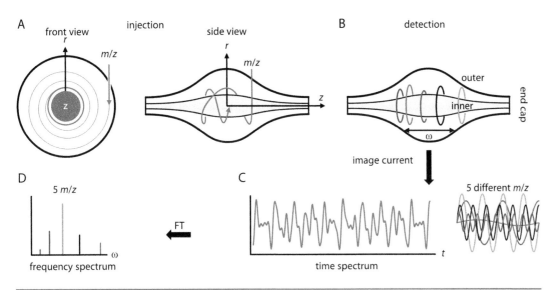

**FIGURE 26.7:    Ions in an orbitrap.** A: Orbitraps have specially shaped inner (gray fill) and outer electrodes to which a constant DC voltage of 3.5–5 kV is applied. The radial distance *r* between the electrodes varies as a function of the axial coordinate *z*. Ions are injected in bunches from another ion source (blue arrow in the front view), and oscillate radially around the cathode until centrifugal and electrostatic forces are in equilibrium. By injecting the ions away from the center of the inner electrode (side view, *z* ≠ 0), an oscillation in the *z*-direction is started that is independent of the oscillation in the r-direction. The side view shows an example of a stable ion trajectory, which resembles a helix. B: The ions on stable orbits organize into rings that oscillate laterally with frequency ω. C: For analysis, the image current at constant voltage is observed over a period of milliseconds to several seconds. Here, the image current is the sum of five sine waves of different amplitudes and frequencies shown on the right-hand side. D: Fourier transformation (FT) of the time spectrum into the frequency spectrum yields the *m/z* values of all ions simultaneously according to eq. 26.6.

### 26.1.2.4 Ion Fragmentation and Sequencing

In the analysis of macromolecular ions it is often desirable to fragment them into smaller ions with *m/z* ratios that can be handled with greater accuracy and measured to higher precision. Sequencing of proteins and in principle also nucleic acids by mass spectrometry relies on formation of a series of ions that differ by one amino acid or one nucleotide. The general strategy for sequencing is to first collect a complete (and very crowded) mass spectrum of a sample that has been fragmented and then use mass analyzers to isolate a certain fragment from the sample by its *m/z* that is then analyzed in a second MS run.

Ion fragmentation involves breaking of chemical bonds (Figure 26.8). Conditions for fragmentation need to be controlled to ensure reproducible fragmentation and to minimize side-reactions to different ions that complicate the spectra obtained. Fragmentation of macromolecular ions is achieved by electron bombardment (not selective, but useful for small molecules), light irradiation (in MALDI), and collision of ions with other ions and with carrier gas molecules in drift tubes and ion traps.

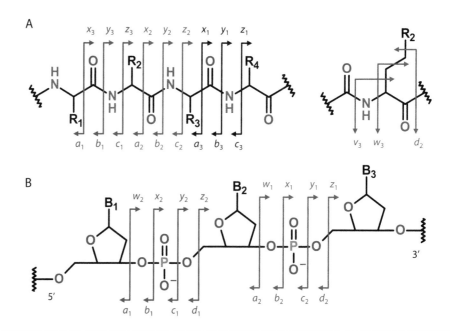

**FIGURE 26.8:** **Backbone fragmentation of peptides and nucleic acids.** A: Bond breakage in the peptide backbone yields the N-terminal *a*, *b*, and *c* series and the C-terminal *x*, *y*, and *z* series of fragments. Breaking of the peptide bond leads to the *b/y* fragments (Figure 26.9). Certain side chains can also break during fragmentation (top right), leading to the *d*, *v*, and *w* series. B: In nucleic acids, the bonds around the phosphodiester preferentially break, giving the 5′-series *a*, *b*, *c*, *d* and the 3′-series *w*, *x*, *y*, *z*. The subscripts refer to the position of the amino acid or nucleotide. Usually one fragment is neutral and cannot be detected. If two or more bond cleavage events occur within a single peptide, internal fragments are produced.

The first possible source of ion fragments is MALDI when the laser light is of such high intensity to not only ionize the macromolecule to a *parent ion*, but to immediately produce *daughter ions* by direct light absorption or temperature effects that lead to bond fission. Further ions are generated in MALDI by collisions between ions within the desorption plume shortly (ns–μs) after excitation of the sample by the laser pulse, while the ions are accelerated by the electric field (Figure 26.2). This mode of ion fragmentation is termed *in-source activation* or *in-source decay* (ISD). Some ions produced by MALDI are metastable and spontaneously decay while traveling along the drift tube, which is termed *post-source decay* (PSD). A third method often used in modern mass spectrometers to produce ion fragments is *collision-induced decay* (CID). Drift tubes or ion traps are filled with a *carrier gas* or *dampening gas*. Collisions between ions and gas molecules lead to ion fragmentation. For ion traps, the procedure is as follows: in a first MS run, a subset of ions with the same $m/z$ is selected by discarding all other ions. Next, the selected ions are fragmented directly within the ion trap. Instead of ejecting an ion by application of a large alternating voltage $U_2$ (Figure 26.6), a slightly smaller voltage, the *tickle voltage*, is applied to the end caps. The ions are accelerated and collide more and more frequently with the carrier gas. When the kinetic energy of the ions is sufficient, collisions lead to bond fission and generation of daughter ions. Only then are the daughter ions ejected from the ion trap and directed towards the detector in a second MS. Such concatenation of several mass analyzing steps is known as *tandem-MS* or MS/MS.

Peptides fragment mainly at the peptide bond (Figure 26.9). The other two backbone bonds in the polypeptide may also be broken, but these processes are less efficient than breakage of an amide because they involve the more stable C–C bonds.

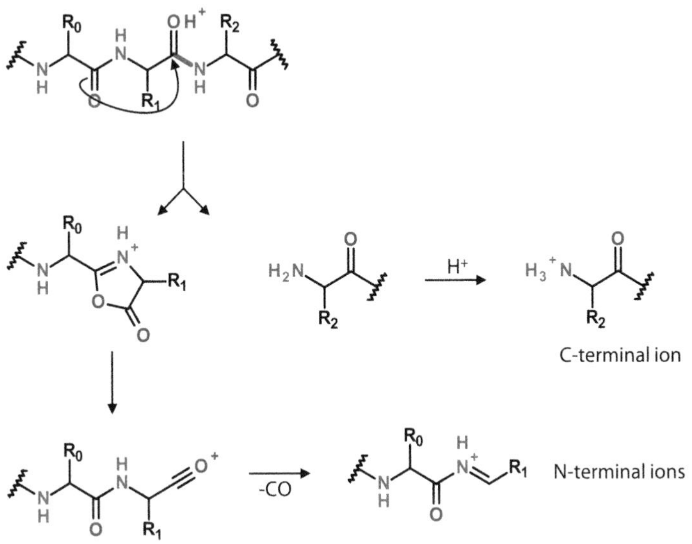

FIGURE 26.9: **Main products from cleavage of peptide ions.** Protonated carbonyl groups are electrophilic, and can be attacked by the carbonyl group of a preceding amino acid. The amino group of the C-terminal part attracts a proton to form the *C-terminal ion* (*y*-ion). The N-terminal intermediate ring can open to form an oxonium ion, which may or may not eliminate carbon monoxide before detection as an *N-terminal ion* (*b*-ion). Usually, one part of the peptide remains neutral and is not detected.

For a singly charged parent ion, the localization of the positive charge determines whether the N- or C-terminal daughter ion is detectable; the neutral fragment is not detected. If the parent ion has two or more charges, both daughter ions may be charged and are detected. The fragmentation reaction thus produces a series of daughter ions that differ by a single amino acid residue. The identity of this amino acid is inferred from the mass difference between two successive daughter ions. From the series of daughter ions derived from a peptide fragmented at different peptide bonds, the peptide amino acid sequence is deduced (*de novo* sequencing). Although normally each peptide only undergoes a single fragmentation reaction, sometimes fragmentation occurs at two sites, generating internal ions and more complex mass spectra. Ions that undergo fragmentation through reactions at their side chains (Figure 26.8) can also complicate the interpretation of mass spectra.

## 26.1.3 DETECTION

We have now treated the generation of ions and their analysis and separation according to the *m/z* ratio. All of these ions eventually need to be detected to generate mass spectra. Detector types include ion-to-photon detectors, electron multipliers, and microchannel plates. In ion-to-photon detectors, ion collisions with a scintillating compound generate photons that are then detected as a photocurrent by a photomultiplier tube. In *electron multipliers* (Figure 26.10), the ions impinge on an electrode, the *conversion dynode*, and liberate primary electrons. This initial signal is too small to be detected accurately, and is therefore amplified by acceleration of the primary electrons in an electric field towards another dynode, where more electrons are liberated. A cascade of dynodes leads to an avalanche of electrons that are detected at the last electrode, the *detection anode*.

FIGURE 26.10:   **Principle of a discrete electron multiplier.** An ion collides with the conversion dynode, liberating a few secondary electrons. Attraction of these electrons by a sequence of dynodes produces more and more electrons until a gain of typically $10^6$ is reached.

*Microchannel plate detectors* (MCPs) are particularly sensitive and versatile arrays of dynodes (Figure 26.11). About ten million glass tubes of ~6–10 µm diameters and 1 mm length, coated with a semi-conductor on the inside, are densely packed into a plate. The tubes or channels are parallel to each other, but often inclined by ca. 8° from the plate normal to ensure that any electron (from a conversion dynode) entering from the cathode side collides with the inside wall of the channel. The collision with the semi-conducting walls liberates more electrons, which continue to collide with the walls and generate more secondary electrons while passing along the channel. At the anode side of the MCP, a cloud of several thousand electrons exits over a narrow area, and is detected as a signal current. The large area of MCPs offers high spatial resolution. Since only a few channels out of several millions are used for the detection of a certain $m/z$, many different ions hitting the MCP at different spots can be detected simultaneously. MCPs have fast time responses of a few nanoseconds to below one nanosecond, which makes them especially useful for TOF measurements because they do not spread the TOF signal appreciably.

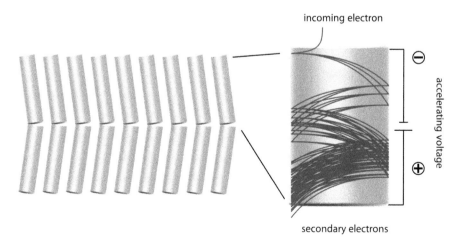

FIGURE 26.11:   **Schematic principle of an MCP.** Two plates are stacked on top of each other to limit ion feedback: any signal going into the opposite direction is blocked at the interface of the two plates. The individual micro channels have a slight inclination from the normal. The right-hand figure visualizes how electrons are multiplied in a single tube for detection upon exit at the bottom.

## 26.1.4 Mass Spectra

The collection of peaks corresponding to ions with different $m/z$ ratios that is detected after ionization, manipulation of ions, and mass analysis, is the mass spectrum. MALDI generates only a few ions from the molecule analyzed, with a single or a few charges. ESI generates a series of ions with multiple charges and hence different $m/z$ ratios from each molecule in the sample analyzed, giving rise to a series of peaks. As a consequence, MALDI spectra with only a few peaks can be analyzed directly, whereas ESI spectra are more complex (Figure 26.12).

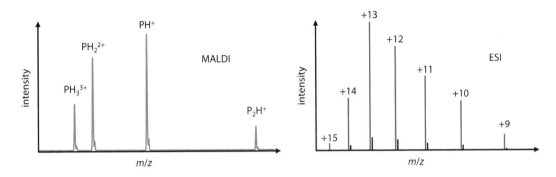

**FIGURE 26.12:** **Schematic MALDI and ESI spectra of a peptide.** For MALDI the most prominent ion is the singly charged molecular ion. Ions with two or three charges are less abundant. A small population of ions $P_2H^+$ with protein dimer is also sometimes detectable. Ions from ESI contain many more charges per molecule than from MALDI. A smaller series of adduct peaks is visible to the right of each $m/z$ peak.

The $m/z$ ratio for each peak in a mass spectrum follows the relation

$$\frac{m}{z} = \frac{M_M + n\left(H^+\right)}{z}$$

eq. 26.7

where $M_M$ is the molecular mass, $n(H^+)$ the number of protons attached, and $z$ the number of positive charges. The smaller number of peaks in a MALDI spectrum can often be interpreted directly. In addition to molecule ions with multiple charges, non-covalent adducts of ions such as sodium, potassium, matrix ions, and phosphate may be observed. The identity of these adducts can be determined from the mass difference to the unmodified ion. For instance, in sodium adducts, $Na^+$ (atomic mass 23) replaces a proton without changing $z$, which results in a +22 Da peak (Figure 26.12). Similarly, matrix adducts are distinguished from the parent ion by tabulated molecular mass shifts.

The more complex ESI spectra typically require deconvolution to obtain the molecular mass of the detected ions. Initially, the absolute number of protons $n$ (eq. 26.7) is not known, but must be an integer. To determine the molecular mass of the molecule that gives rise to a series of peaks, $n$ is guessed for one of the detected peaks, and $M$ is calculated according to eq. 26.7, taking into account that ions in neighboring peaks must differ by one charge. The value for $n$ which gives identical masses $M$ for all peaks of the series is the correct number of charges, and $M$ is the molecular mass of the molecule analyzed. Ion adducts give rise to smaller peaks on the right-hand side of the main $m/z$ peaks in an ESI spectrum. They occur at the same $m/z$ difference to each main peak, and can be identified from this mass difference.

## 26.1.5 Applications

### 26.1.5.1 Mass Analysis for the Identification of Molecules

Determination of the molecular mass of the molecule analyzed is the main purpose of MS experiments. Mass determination on its own allows the nature of a compound, its isotope distribution, and its purity to be established. Mass determination by MS achieves accuracies of below 0.01%. In other words the calculated and the measured molecular masses for a macromolecule of 10 kDa can be compared to an accuracy of 1 Da. Therefore it is possible to distinguish reduced from oxidized forms of cysteines (2 Da difference), which enables the number of disulfide bonds in small proteins to be counted. Preparations of macromolecules are often heterogeneous due to chemical modifications. If a mass difference between expected and measured molecular masses is found, databases can be searched to tentatively assign the discrepancy in mass to a type of chemical modification. Post-translational modifications of proteins (Section 16.1.4) can also be identified by MS. While defined modifications such as phosphorylation or methylation are readily identified, other modifications may be heterogeneous and the resulting complex mass spectra may be difficult to assign. For example, proteins may be glycosylated at Asn, Ser, or Thr residues. Depending on the source of the protein, the glycosylation trees can be heterogeneous. In such cases, identification is not possible without fragmentation and analysis of the glycosylated fragments. MS is also routinely used to identify products in crosslinking experiments of macromolecular complexes from their mass, and to map binding interfaces by identification of the crosslinking sites (Section 26.1.5.6).

### 26.1.5.2 Isotope Distribution and Isotope Exchange

The resolution of mass spectrometry is sufficient to detect differences in isotopic composition, for example in the content of $^{14}N$ and $^{15}N$, or $^{12}C$ and $^{13}C$. Isotope analysis is helpful to test the percentage incorporation of $^{13}C$ or $^{15}N$ of pure NMR samples. The natural abundance of $^{13}C$ is about 1.1% (the rest is mainly $^{12}C$, and traces of radioactive $^{14}C$ (see Box 7.2). The probability to contain more than one $^{13}C$ increases with the number of carbon atoms in the molecule. For organic molecules or small peptides, the likelihood of incorporation of more than one $^{13}C$ is small. For proteins, the likelihood of incorporation of several $^{13}C$ nuclei increases with size. The distribution of peak intensities within a series reflects the probability for the particular isotope content (Figure 26.13). The shape of the peak distribution is a skewed Gaussian distribution, and the relative peak intensities can be calculated from the isotope abundances using the binomial theorem (Section 29.3).

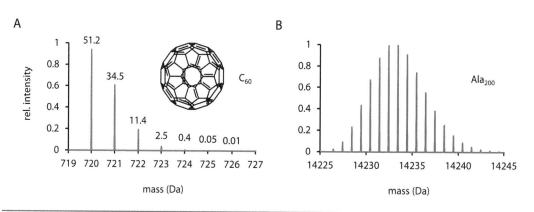

**FIGURE 26.13:** **Skewed Gaussian peak distribution in small molecule and protein ESI spectra.** All peaks have $z = 1$. A: Small organic molecules such as the fullerene $C_{60}$ and short peptides have a quasi-exponential distribution of intensities for their $m/z$ peaks. B: Isotope distribution of the hypothetical protein poly-Ala$_{200}$ of $M_M = 14233.6$ Da shows a skewed Gaussian distribution of peak intensities.

For mixtures of macromolecules, the natural isotope distributions can lead to very complex mass spectra. For example, a 100 kDa protein contains ca. 50 $^{13}C$ atoms, leading to at least 50 different peaks in the mass spectrum for the same molecule. Two macromolecules of similar masses will have overlapping mass spectra. Some of the peaks cannot be resolved, which limits the resolution attainable from mixtures.

Isotope exchange experiments, particularly hydrogen-deuterium (H/D) exchange, are frequently used to probe the accessibility of hydrogens in proteins or other macromolecules. Deuterium ($^2H$ or D) has an extra neutron and thus twice the mass of $^1H$. Under mild alkaline conditions, hydrogens in peptide bonds and any side chain that carries a NH or OH group can exchange with deuterium from the solvent. When a protein is placed in an aqueous buffer based on $D_2O$ instead of $H_2O$, the velocity of H/D exchange depends on the hydrogen accessibility: in solvent-exposed regions, hydrogens exchange quickly, while hydrogens in buried parts exchange slowly or not at all. After stopping the reaction by lowering the pH, the protein is fragmented, and the peptides are analyzed for deuterium content (and sequence) by MS. Peptides with high deuterium content reflect parts of the protein that are solvent-exposed. H/D exchange experiments can also be performed in a time-dependent manner by stopping the exchange reactions at different time intervals. Time-resolved H/D exchange is frequently employed to study protein (un)folding mechanisms (Section 16.3.5.3). Also, binding partners protect their binding sites from hydrogen exchange because they exclude access of solvent. H/D exchange monitored by MS can thus be used to follow complex formation. Changes in protection patterns can also give insight into conformational changes.

### 26.1.5.3 Protein Identification from One- and Two-Dimensional Gels

MS is routinely used to identify individual proteins separated by one- or two-dimensional gel electrophoresis. A band or spot containing the protein of interest is excised from the gel, and the protein is hydrolyzed *in situ* by a specific protease (*in-gel digest*), e.g. trypsin. The resulting mixture of peptides is first separated by liquid chromatography (LC) before determining their masses by MS. LC-MS couples the chromatographic separation directly to the ionization chamber of an ESI mass spectrometer. If the peptides are too large, further fragmentation of the proteolytic peptides in the mass spectrometer is done by CID or PSD (Section 26.1.2.4). The peptide masses from the unknown protein are compared to calculated masses (using the same protease *in silico*) of all proteins in a sequence database. In favorable cases, the unknown protein can be identified based solely on the masses of its proteolytic peptides, without any sequence information. At least two unique peptide masses are required to identify the unknown protein. For a peptide mass hit to be significant, i.e. to not occur by chance, a low *Expect value (E-value)* is required. The *E*-value describes the number of hits for a certain peptide mass we expect to get purely by chance when searching a database of a particular size for that mass: the *E*-value represents the random noise of the database. The larger the peptide, the more significant will a hit from this peptide be in the database, because the likelihood of finding the same peptide mass twice or more times in the database is smaller for longer sequences.

Protein identification by MS becomes more difficult when post-translational modifications such as glycosylation and disulfide bonds are present, or when side-chain fragmentation increases the complexity of the mass spectrum. However, the bioinformatics tools to calculate expected fragmentation patterns from the sequence databases are being constantly expanded to include masses for such unusual peptides. Hence, the identification of proteins from peptide masses becomes more and more facile. However, if the mass-based approach does not work, some representative peptides are sequenced by MS, and run against the sequence database using the additional sequence information.

### 26.1.5.4 Native Mass Spectrometry

The crystalline matrices, acids, and organic solvents normally used in MALDI or ESI will dissociate non-covalent macromolecular complexes and unfold proteins. While MALDI is usually incompatible with maintaining the native state, ESI can be performed on native proteins and macromolecular complexes by replacing the organic solvent with a volatile buffer such as ammonium acetate, which in vacuum decomposes into $NH_3$ and acetic acid, leaving the molecules in their native state. Despite the more complex spectra obtained by ESI (more peaks than MALDI) and the associated difficulties in their interpretation (overlap of peaks), ESI is the preferred method of ionization for native MS.

Proteins are stabilized to a large extent by the hydrophobic effect (Section 16.3.1). The hydrophobic effect vanishes in the vacuum of the mass spectrometer, where the dielectric constant $\varepsilon_r$ is unity. It was initially believed that proteins would instantaneously unfold and complexes held together by hydrophobic interactions would immediately dissociate *in vacuo*.

Nevertheless, it is now firmly established that native protein conformations can prevail in a mass spectrometer, at least on the millisecond time scale of data acquisition. There are several reasons why proteins and their complexes remain folded and intact *in vacuo*: (1) attractive van der Waals interactions, (2) increased H-bonding of surface residues with each other when bulk water is absent, and (3) stronger charge-charge interactions between ionized residues in complex interfaces. These arguments also hold for membrane proteins in the absence of detergent molecules. It is now possible to ionize membrane proteins and strip them of detergents without triggering their unfolding.

Advances in instrumentation and ionization procedures have made large molecular complexes amenable to native MS studies. In general, the larger the molecular mass, the more adducts ($H^+$ and other cations) are formed, and also the broader the isotope distributions. The higher complexity of ESI spectra limits the resolution with which masses can be determined from native samples of high molecular mass. However, complexes isolated in their native form from cell extracts or from membrane preparations are analyzed by native MS for their conformational distributions (by ion mobility; Section 26.1.5.5), their stoichiometry (by quantitative MS; Section 26.1.5.8), their subunit interactions, and their assembly pathway, which is not possible with denaturing ESI.

Subunit interactions within a large complex are studied by generating sub-complexes with an overlapping set of subunits in solution. The decomposition of these protein sub-complexes *in vacuo* follows a series of well-defined steps: first, water molecules and counter ions are removed by collision with gas molecules followed by structural changes of the protein complex. The next steps are unfolding and dissociation of single protein subunits from the complex, one after the other. The two masses of the liberated monomer and the remaining complex are detected separately, establishing, which subunit has dissociated from the complex. In case different decomposition pathways occur at the same time, more than one pair of liberated monomer and remaining complex is produced. At the end, a complete assembly chart for the complex may be constructed.

In analogy to H/D exchange (Section 26.1.5.2), whole subunits can be exchanged in a multi-protein complex, and the exchange kinetics can be monitored. For these experiments, the mass of the subunits may either be known or they need some sort of mass tag, such as an isotopic label or a chemical modification. Exchange experiments with several subunits of the complex can thus reveal the assembly pathway. Protein-nucleic acid complexes and complexes of membrane proteins with soluble proteins can also be studied by native MS. Examples for membrane proteins studied are the rotary V-ATPase and the P-glycoprotein. Large multi-protein/nucleic acid complexes that have been studied by native MS are the ribosome, the TRAP (tryptophan regulator attenuation protein), and the spliceosome.

### 26.1.5.5 Ion Mobility and Molecular Shape

In contrast to reflectrons, quadrupoles, and ion traps, which all select ions based on $m/z$, ion mobility measurements separate ions that have the same $m/z$ ratio but differ in shape. Different mobilities of ions of the same $m/z$ are caused by different conformations of the macromolecule. Ions with extended conformations have large *cross-sections* while more compact ions have smaller cross-sections. To resolve ions by shape, they are passed through a drift tube where they are slowed down by collisions with a carrier gas. Ions with large cross-sections experience more collisions with gas molecules compared to compact ions. The resulting distribution of velocities and the resulting distribution of drift times, measured by TOF, allow conclusions to be made on the conformational space that is sampled by the protein. Ligand-induced conformational changes can also be detected by ion mobility. The drift times can be converted to orientationally averaged collision cross-sections, which represent the cross-sections of all possible orientations of a particular macromolecule or complex. If structural information such as a crystal structure or an envelope from EM or SAXS is available, the cross-sections calculated from the structures can be compared with that from MS to find which conformation is present *in vacuo*.

### 26.1.5.6 Identifying Protein-RNA Interaction Sites after Photo-Crosslinking

RNA-binding proteins (RNPs) in the cell are involved in vital processes such as transcription, splicing, and translation. To identify the interaction sites between proteins and RNA molecules in the complex, they can be subjected to photo-crosslinking, followed by MS analysis. Photo-crosslinking of RNA-protein complexes can be done in the cell or in solution by irradiation with UV light at 260 nm, where the nucleobases absorb (Section 19.2.6) and crosslink to nearby proteins. Alternatively, 4-thiouridine can be incorporated into the RNA, and crosslinking is performed by irradiation at 365 nm. The complexes are isolated from the cells and enriched by chromatography. Instead of analyzing the entire protein-RNA complex by MS, the analysis is simplified by trimming the complex with RNase and proteases to generate fragments of only 1–3 nucleotides that are cross-linked to a short peptide. The masses of the thousands of peptides and crosslinked RNA-protein adducts are then determined by LC-MS. To find the masses corresponding only to the adducts, an identical control experiment with a non-crosslinked sample is run in parallel. The identified RNA-protein species are then sequenced by MS/MS to identify both the RNA and protein sequences that are in contact in the complex.

In an experiment that crosslinked proteins and RNA in yeast ribosomes, from almost 10000 determined masses only 184 (ca. 2%) remained as candidates for cross-linked species. Their identification provided the crosslinking sites and thus information on the RNA-protein interfaces within the ribosome. The analysis also showed that proteins can crosslink to RNA through almost all amino acids except D, N, E, and Q, whereas RNA forms crosslinks almost exclusively through uridine or 4-thiouridine (if it was included during RNA synthesis). While most of the identified RNPs were ribosomal proteins, some unexpected enzymes including enolase and inorganic pyrophosphatase were identified as RNA-binding proteins. These proteins might have ribosome-related functions that were previously unknown.

### 26.1.5.7 Secondary Ion Mass Spectrometry

Secondary ion mass spectrometry (SIMS) is a MS-based method for label-free two- and three-dimensional imaging of biochemical surfaces such as cell membranes. The surface is bombarded with heavy *primary* ions with energies of tenths of a keV that are focused on sub-μm spots. The primary ions generate *secondary* ions of 1 kDa mass or less from lipids and amino acids, reporting on the distribution of membranes and proteins, respectively. The secondary ions are detected in parallel, yielding a full mass spectrum for each spot of the surface interrogated. From the mass distributions and abundances of the secondary ions, the chemical composition of the surface is inferred as a function of position in two dimensions.

The technique was applied to monitor the change in lipid composition of the *Tetrahymena* cell membrane during mating. *Tetrahymena* is a common freshwater protozoon that forms curved fusion pores at the conjugation site during mating. For SIMS analysis, the cells before and during mating were freeze-fractured and the lipid composition was determined using an $In^+$ primary ion beam. Analysis of the secondary ions by MS revealed a reduction of the phosphocholine signal at $m/z = 184$, matched by an increased signal for the head group of 2-aminoethylphosphonolipids at $m/z = 126$, suggesting a dramatic change in lipid composition during mating.

Three-dimensional information with sub-cellular resolution can be gained by a series of two-dimensional scans that are stacked on top of each other. However, because bombardment of surfaces with primary ions partly destroys sub-surface structures, a fresh layer of intact material needs to be exposed by a gentle ablation step prior to the next two-dimensional scan. The three-dimensional distribution of triacylglycerides in *Xenopus* oocytes at ambient temperature was analyzed this way by cycles of two-dimensional imaging of the surface by SIMS using a heavy metal ion beam, followed by gentle ablation of the imaged surface layer using $C_{60}^+$ ions. From the series of two-dimensional images, the three-dimensional image of triacyl glyceride distribution ($m/z$ 815–960) was reconstructed with a volume of 1200 × 1200 × 75

$\mu m^3$. The lateral resolution in this experiment was lower than in the *Tetrahymena* example because the cells were not frozen and therefore exhibit some flexibility when bombarded. Nevertheless, it shows the applicability of mass spectrometry to whole cells in almost physiological environments.

### 26.1.5.8 Quantitative Mass Spectrometry

Mass spectrometry is not an inherently quantitative method because the ionization efficiency is different for different molecules. Consequently, the intensity of peaks in a mass spectrum is not a good measure for the abundance of the respective species. Hence, quantitative interpretation of mass spectra requires an internal reference. The term *quantitative mass spectrometry* refers to several proteomic techniques that quantify proteins from cells of different tissues or in different physiological states relative to one another or, in favorable cases, on an absolute scale.

In the simplest case we know the nature of the peptide that we wish to quantify and its sequence and mass. An isotopically labeled peptide of the same sequence, added to the sample at a known concentration, can then serve as an *internal standard*. To quantify a peptide with the natural isotopes $^{12}C$ and $^{14}N$, a reference peptide containing the isotopes $^{13}C$ and $^{15}N$ is synthesized. The two peptides are called *isotopologs* –physico-chemically identical molecules that differ only in their isotope composition. Target and reference peptide co-elute during LC pre-fractionation of the sample, and are detected simultaneously as two peaks with different *m/z* by MS. Their ionization properties are also the same, and the peak intensities for the two peptides in the mass spectrum are therefore equal to their relative abundances. Since the concentration of the reference peptide is known, the absolute quantity of the target peptide in the sample can be calculated from the relative peak intensities.

In SILAC, *stable isotope labeling by amino acids in cell culture*, a cell culture to be analyzed is fed with modified amino acids, e.g. $^{13}C$-Lys, to incorporate the isotope $^{13}C$ into proteins, a process termed *metabolic labeling*. A control cell culture is grown under identical conditions, but provided with the natural amino acid, e.g. $^{12}C$-Lys. Both cells therefore produce isotopologous proteins. Equal amounts of protein extracts from labeled and non-labeled cell cultures are then mixed and analyzed by tandem LC-MS. The ratio of the peak intensities of the isotopologs equals the relative abundance of the proteins. As a control, cell cultures that are grown under identical conditions must have a ratio of unity for the isotopologs. By comparing the ratios of isotopologs between cell cultures grown under different conditions, changes in the abundance of proteins and the protein composition of the cell (the proteome) due to the different conditions can be detected. SILAC allows the effect of stress conditions, chemical treatment, genetic manipulation, or mutations in a cancer cell line to be analyzed on protein levels. SILAC depends on the isotope-labeled reference and is therefore limited to cell cultures, whose growth is time-consuming and costly. Samples taken directly from tissues cannot be analyzed by SILAC because they lack the isotope-labeled reference.

Mass spectrometry using *isobaric tags for relative and absolute quantification* (iTRAQ) introduces the internal standard by chemically modifying the sample with an iTRAQ reagent, which circumvents the requirement for metabolic labeling. The iTRAQ reagent consists of three parts: a charged reporter group, a neutral balance group, and a peptide-reactive group (Figure 26.14). Proteins are extracted from the samples that are to be compared, the total protein concentration is determined using a standard method (Section 19.2.7), and peptides are generated by proteolysis of equal amounts of each protein extract. The peptide mixtures from each sample are then reacted with an iTRAQ reagent that modifies the primary amino groups of N-termini and Lys side chains (see Section 26.3.4.1), leaving only the reporter and balance groups on the protein (Figure 26.14). A different iTRAQ reagent with a different mass distribution between reporter and balance groups is used for each peptide mixture.

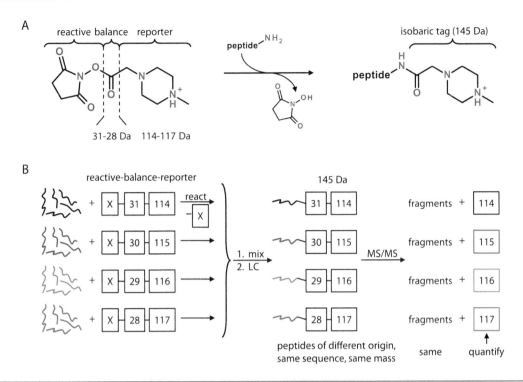

**FIGURE 26.14:  iTRAQ principle.** A: The reagent reacts with primary amino groups *via* its activated N-hydroxysuccinimidyl ester, which transfers the reporter group, here a charged N-methylpiperazine, and the balance group, here a carbonyl group, to the peptides of the sample. By using groups with different isotope compositions, *m/z* of the reporter group is varied between 114 Da and 117 Da. The balance accordingly must range in mass between 31 Da to 28 Da, such that the total mass of the tag is constant (145 Da). B: Peptide mixtures from four sources are labeled with four different iTRAQ reagents, then mixed, and the peptides separated by LC. Peptides of the same mass but with different balance-reporter groups are analyzed by MS/MS. Their fragmentation gives the same peptide fragments, but different reporter ions, which are quantified to reflect the relative abundances of the proteins in the original cell extracts.

The iTRAQ reagents are designed such that the sum of the masses of the reporter and balance group transferred to the peptide is the same in all reagents (*isobaric tag*), but their isotope composition is different. The labeled peptides from all three samples are pooled and jointly analyzed by tandem LC-MS. Identical peptides from the different samples elute from the LC at the same time and are therefore ionized together and with the same efficiencies. Peptide sequences are determined by fragmentation to identify the protein they originated from (Section 26.1.2.4). Also, the isobaric tag decomposes in the mass spectrometer into the balance and reporter groups. The reporter groups are now not isobaric any more but have different masses for the different samples. The peak intensities of the reporter groups are measured and reflect the relative abundances of the *same* peptide from *different* samples. If the instrument is calibrated with a known concentration of the reporter groups, the peptides in all samples can be quantified on an absolute scale.

With improvements of chromatographic techniques (nano-HPLC for exact retention times), higher reproducibility from run to run in MS analysis (no drift in *m/z*), and more sophisticated data analysis software, MS moves towards a quantitative method without the requirement of internal standards, and thus to label-free quantitative MS. Label-free experiments require less amount of sample and less preparation time. The data are therefore often acquired more quickly and interpreted more easily. However, label-free MS requires more careful control of the experimental parameters than quantification based on internal standards. In label-free MS, quantification is based on chromatographic peak areas, which requires highly reproducible peptide separation between samples prior to MS analysis. By calibrating the spectrometer using increasing concentrations of a known protein, its unknown concentration in a complex sample can be quantified. This method relies on the observation that the signal intensity from ESI correlates with ion concentration. Relative abundances of unknown proteins from different samples can be detected by comparing results from MS runs performed under identical conditions, which is not always possible. A more reliable method for label-free quantification is *spectral counting*, which relates the number of mass spectra obtained for a specific protein in many tandem MS/MS experiments to its abundance. This works because an increase of protein abundance will also increase the number of unique proteolytic peptides, which in turn leads to an increase in spectral count. Comparison of the spectral counts gives the relative protein abundances from different samples.

## 26.2 ANALYTICAL ULTRACENTRIFUGATION

Centrifuges have been used for hundreds of years. Here we will concentrate on the analytical uses of centrifugation by observation of the sedimentation behavior of particles, from small molecules of only 100 Da to cellular organelles reaching 10 GDa molecular mass. In analytical ultracentrifugation (AUC), a solution is subjected to a strong gravitational field. The solution components will sediment and establish a concentration gradient that is opposed by diffusion. The resulting time-dependent shape of the concentration gradient (Figure 26.15) is detected and analyzed to determine the hydro-dynamic properties of a macromolecule, its molecular mass, and interactions between components of the solution. The analysis of concentration gradients is based on the forces that are experienced by the molecules, and requires no additional assumptions.

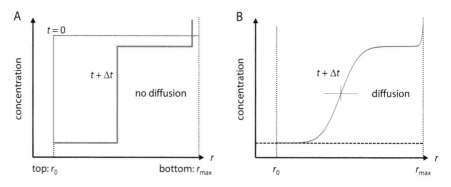

**FIGURE 26.15:** **Change of concentration in a sample upon centrifugation.** A: A sample once uniformly distributed (gray line) from the top ($r_0$) to the bottom ($r_{max}$) of the centrifuge tube is centrifuged, and the concentration is measured at a time point $t + \Delta t$ as a function of $r$. In the absence of diffusion, molecules sediment with a defined boundary towards the bottom of the cell, leading to a step gradient. B: Diffusion counteracts sedimentation and spreads the boundary region.

AUC experiments can be performed in two modes, termed *sedimentation velocity* and *sedimentation equilibrium*. In sedimentation velocity experiments, the molecules are sedimented at high rotor speeds, and the rate of the movement of the *boundary* is measured. From this rate, the *sedimentation coefficient* is calculated, which allows conclusions about purity, size, shape, disper-sity, and molecular mass of particles. In sedimentation equilibrium experiments, rotor speeds are chosen such that sedimentation and diffusion balance each other over a time frame of hours to days. This establishes a stable concentration profile that can be analyzed with respect to molecular mass. For interacting particles, the association constant and complex stoichiometry can be determined. Before we discuss the analysis of sedimentation velocity and sedimentation equilibrium experi-ments (Section 26.2.3 and 26.2.4), we will begin by introducing the instrumentation and the dif-ferent options to detect the concentration gradients in AUC, followed by a general treatment of the forces that are exerted on molecules in solution in a gravitational field (Section 26.2.2).

### 26.2.1 INSTRUMENTATION AND DETECTION SYSTEMS

In an AUC experiment, a rotor carrying a sample operates at an angular velocity ω. For the analytical purposes in AUC, a temperature-controllable fixed angle rotor is used (Figure 26.16). The high rotor speeds require evacuation of the rotor chamber in order to avoid friction between the rotor and air, which would heat the sample. Most importantly, the rotor must allow passage of light through the sample for measuring the concentration of the solute as a function of distance from the center of the rotor.

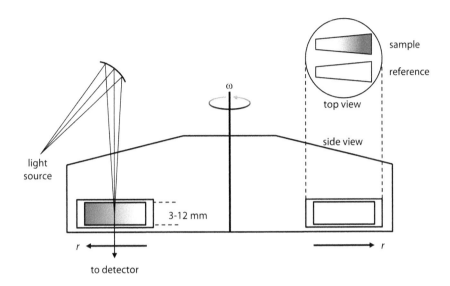

**FIGURE 26.16:  Schematic view of a fixed angle AUC rotor.** The angle is defined by the angle between the rotation axis and the longitudinal axis of the cell, in this case 90°. Rotors may have up to eight bores, each of which can hold a sector-shaped sample/reference cell. For absorbance measurements as a function of the cell radius $r$, light passes the sample cell from top to bottom. The standard path length of the cell is 3–12 mm.

The gravitational force generated by the rotation of the rotor leads to sedimentation of molecules in sample and reference cells (Figure 26.17) along radial paths. In parallel cells, molecules would therefore accumulate on the cell walls because of increased friction, and the resulting concentration gradient from the wall toward the center of the cell would lead to convection. The sector shape of the cell prevents accumulation of molecules on the cell walls.

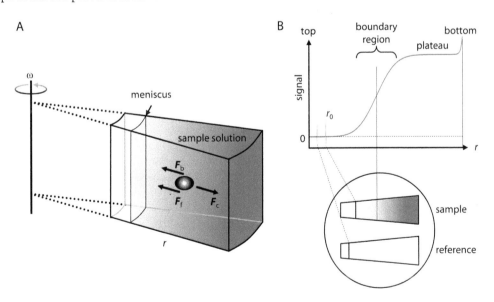

**FIGURE 26.17:  Principle of an AUC experiment.** A: The rotor operates at an angular velocity ω, spinning a cell that holds the sample (blue). The sector shape of the cell prevents accumulation of molecules at the cell walls upon sedimentation. B: Top view of a double sector cell containing a reference (buffer) and the sample. The reference cell is filled slightly higher to distinguish one meniscus from the other (shown as disturbances in the signal above). Upon centrifugation the molecules accumulate towards the bottom of the cell, leading to a gradient that is characterized by a boundary and a plateau region.

In principle, any kind of signal from the sample analyzed is suitable to follow the sedimentation process. AUC instruments are available that measure absorption (Section 19.2), fluorescence (Section 19.5), changes in the refractive index, or interference.

For absorption and fluorescence measurements, a suitable wavelength from a xenon lamp is selected by a monochromator. The light is directed toward the cell from the top by a grating, and transmitted light is detected by a detector below the cells. Absorption, the most frequently used analytical property in AUC, has the advantage of being directly proportional to concentration according to the Lambert-Beer law (Section 19.2.3). The presence of chromophores in proteins and nucleic acids make absorption a sensitive and selective spectroscopic probe for AUC. Given their very high extinction coefficients, nucleic acids and their complexes with proteins can be detected at micromolar concentrations.

Fluorescence typically provides higher detection sensitivity than absorbance, down to the sub-nanomolar range. Higher sensitivity enables the analysis of higher affinity interactions by AUC. Trp and Tyr in proteins are fluorescent, but have low quantum yields, so extrinsic fluorophores are preferred. Nucleic acids are generally non-fluorescent, and labeling with extrinsic fluorophores is required. When fluorescence is used as a probe, it must be established that the fluorescence signal is proportional to the concentration of the solute, and quenching (Section 19.5.7) or inner filter effects (Section 19.2.8; Figure 19.21) at higher solute concentrations should be excluded.

A series of absorption or fluorescence scans of sample cells at different time points is shown in Figure 26.18.

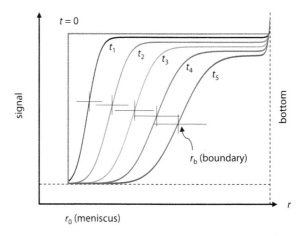

**FIGURE 26.18:   Spectroscopic signal as a function of radial distance is a measure of concentration.** At $t = 0$ the whole cell has uniform concentration. The concentration continuously decreases close to the meniscus and increases at the bottom during centrifugation. Schematic concentration distributions are shown at constant time intervals at times $t_1 - t_5$ after start of the centrifugation. Due to diffusion the concentration gradient covers a wider range of distances at later time points (crosses). The plateau decreases because of sedimented material. The radial position $r_b$ is close to the inflection point of the distribution and characterizes the position of the boundary.

Solutions of polysaccharides or lipid vesicles do not contain chromophores or fluorophores for absorption or fluorescence detection. In this case, refractometric methods such as Schlieren optics or Rayleigh interference can be used. These methods measure differences in the refractive index, which are proportional to changes in concentrations.

### 26.2.2 BEHAVIOR OF A MOLECULE IN A GRAVITATIONAL FIELD

The exact treatment of molecular transport processes requires thermodynamic approaches that explicitly treat Brownian motion of molecules (Section 4.1). A more intuitive understanding of the behavior of molecules in a gravitational field is obtained from a mechanical view of the forces involved. A particle sedimenting in a gravitational field experiences three forces, the *gravitational* or *sedimenting* force $F_c$ towards the bottom of the cell, and the *buoyancy* $F_b$ and the *friction* $F_f$ towards the top of the cell (Figure 26.19).

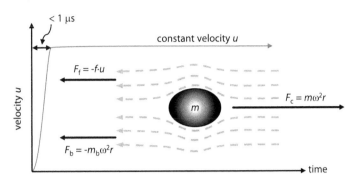

FIGURE 26.19:    **Forces acting on a particle in a gravitational field.** A particle sedimenting in a gravitational field experiences the gravitational force $F_c$ towards the bottom of the cell (positive sign), and the buoyancy $F_b$ and the friction $F_f$ towards the top of the cell (negative sign). The particle reaches a constant velocity $u$ when these forces are in equilibrium, less than a microsecond after the start of the centrifugation. The broken arrows indicate the solvent resistance the particle experiences while moving to the right.

The gravitational (or centrifugal) force $F_c$ generated by the centrifugal field is defined by the product of acceleration $a = \omega^2 r$ and the mass $m$ of the molecule.

$$F_c = m \cdot a = m \cdot \omega^2 r = m \cdot (2\pi v)^2 \cdot r \qquad \text{eq. 26.8}$$

The angular velocity $\omega$ is the frequency $v$ of rotation in revolutions per second times $2\pi$. Division of the centrifugal acceleration by the earth gravitational acceleration $g = 9.81$ m·s$^{-2}$ results in the relative centrifugal force (RCF), a dimensionless number that expresses the acceleration as multiples of $g$.

$$\text{RCF} = \frac{\omega^2 r}{g} \qquad \text{eq. 26.9}$$

The RCF describes centrifugal forces independent of the rotational radius and hence independent of the particular rotor and centrifuge used, which facilitates comparison. For the sedimentation of biological molecules, RCFs of several hundred thousand are required.

The second force we have to take into account is the buoyancy $F_b$ toward the top of the cell. According to Archimedes' principle, $F_b$ is equal to the mass of solvent $m_0$ displaced by the submerged molecule multiplied by the gravitational acceleration, $\omega^2 r$:

$$F_b = -m_0 \cdot \omega^2 r \qquad \text{eq. 26.10}$$

The negative sign indicates that this force acts opposite to $F_c$. Any submerged body displaces a volume of solvent equal to its own volume. The displaced mass of solvent $m_0$ hence equals the product of the volume of the particle in solution and the density $\rho$ of the solvent. The volume of the particle can be expressed as the product of the partial specific volume of the molecule $\bar{v}$ and its mass $m$:

$$m_0 = V \cdot \rho = m \cdot \bar{v} \cdot \rho \qquad \text{eq. 26.11}$$

The *partial specific volume* is the occupied volume of the molecule in solution per mass unit (Table 26.1; see also Section 2.6.4). Its unit is that of a reciprocal density so that the factor $\bar{v}\rho$ is dimensionless. As a result of the buoyancy, the mass of the molecule is reduced to an *effective mass* $m_{\text{eff}}$ given by

$$m_{\text{eff}} = m \cdot \left(1 - \bar{v} \cdot \rho\right) \qquad \text{eq. 26.12}$$

From eq. 26.12, we can calculate the effective mass $m_{\text{eff}}$ for proteins ($\bar{v}$ of $0.70 - 0.75$ L kg$^{-1}$, Table 26.1) in aqueous solution ($\rho = 1$ kg L$^{-1}$) as 25–30% of their mass $m$. The effective mass of nucleic acids, in contrast, is higher, about 45% of their mass $m$, because of their smaller partial specific volume. We will see later that the relation in eq. 26.12 can be used to separate molecules according to their density (Section 26.2.5).

**TABLE 26.1**
**Partial specific volumes of macromolecules.**

| molecule | value range (L kg$^{-1}$) |
|---|---|
| proteins | 0.70–0.75 |
| polysaccharides | 0.59–0.65 |
| lipids | 0.95–1.00 |
| DNA | 0.55–0.59 |
| RNA | 0.47–0.55 |

*Note:* Partial specific volumes are temperature-dependent and sensitive to solution conditions (see Section 2.6.4), and only ranges can be given.

The third and last force acting on the molecule is the frictional force $F_f$. Friction counteracts gravitational forces. In the absence of friction, the increase of $F_c$ with increasing distance from the rotor axis (eq. 26.8) would continuously accelerate the molecule to higher and higher velocities. However, this is not observed because the gravitational effect is counteracted by friction (Figure 26.19), which depends on the velocity $u$:

$$F_f = -f \cdot u \qquad \text{eq. 26.13}$$

The *translational frictional coefficient f* depends on both the size and shape of the molecule. Spherical molecules have the lowest frictional coefficients, whereas the friction coefficient is large for elongated, rod-shaped molecules. The frictional coefficient is a measure of how difficult it is for the solvent to flow around the molecule, or how difficult it is for the molecule to diffuse within the solvent. The Stokes-Einstein equation (eq. 26.14) relates the frictional coefficient $f$ to the thermal energy of the molecule (Section 4.1) and the diffusion coefficient $D$:

$$f = \frac{k_B T}{D} = \frac{RT}{N_A D} \qquad \text{eq. 26.14}$$

The reciprocal relationship shows that rod-shaped molecules with large frictional coefficients diffuse more slowly than spherical molecules with lower frictional coefficients.

The frictional coefficient $f_0$ of a sphere depends on the sphere radius $R_e$ and the solvent viscosity $\eta$. For movements of particles through continuous media (e.g. a protein dissolved in buffer) the relation between the frictional coefficient and the sphere radius is known as *Stokes' Law*:

$$f_0 = 6\pi\eta R_e \qquad \text{eq. 26.15}$$

More complicated shapes are spheroids and rods, which can serve as approximations for non-spherical proteins and elongated nucleic acids, respectively. A spheroid is an ellipse rotated about one of its axes, and can be described by the axial and equatorial half-axes $a$ and $b$. Rotation of an ellipse about its major axis generates a prolate or cigar-shaped spheroid. Rotation about the minor axis leads to an oblate or lentil-shaped spheroid. A rod is a cylinder of length $a$ and radius $b$. The frictional coefficients of spheroids and rods depend on their half-axes $a$ and $b$ (Table 26.2). It is convenient to normalize frictional coefficients with that of a sphere of equal volume (radius $R_e$). As spherical molecules have the lowest friction coefficients, frictional ratios $f/f_0$ are always larger than unity (Figure 26.20). However, for an axial ratio <2, oblate and prolate spheroids have similar frictional ratios that are not too different from unity. Globular proteins have frictional ratios of $f/f_0 \approx 1.2$.

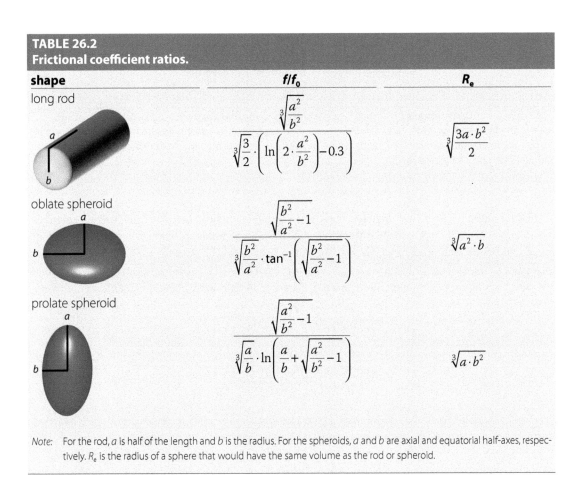

**TABLE 26.2**
**Frictional coefficient ratios.**

| shape | $f/f_0$ | $R_e$ |
|---|---|---|
| long rod | $\dfrac{\sqrt[3]{\dfrac{a^2}{b^2}}}{\sqrt[3]{\dfrac{3}{2}} \cdot \left( \ln\left(2 \cdot \dfrac{a^2}{b^2}\right) - 0.3 \right)}$ | $\sqrt[3]{\dfrac{3a \cdot b^2}{2}}$ |
| oblate spheroid | $\dfrac{\sqrt{\dfrac{b^2}{a^2} - 1}}{\sqrt[3]{\dfrac{b^2}{a^2}} \cdot \tan^{-1}\left( \sqrt{\dfrac{b^2}{a^2} - 1} \right)}$ | $\sqrt[3]{a^2 \cdot b}$ |
| prolate spheroid | $\dfrac{\sqrt{\dfrac{a^2}{b^2} - 1}}{\sqrt[3]{\dfrac{a}{b}} \cdot \ln\left( \dfrac{a}{b} + \sqrt{\dfrac{a^2}{b^2} - 1} \right)}$ | $\sqrt[3]{a \cdot b^2}$ |

*Note:* For the rod, $a$ is half of the length and $b$ is the radius. For the spheroids, $a$ and $b$ are axial and equatorial half-axes, respectively. $R_e$ is the radius of a sphere that would have the same volume as the rod or spheroid.

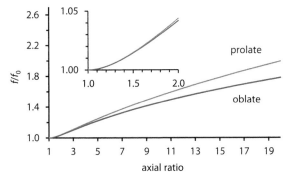

**FIGURE 26.20: The frictional coefficient of spheroids depends on the shape of the particle.** For particles with an axial ratio up to 2, oblate and prolate spheroids behave very similarly to each other (inset), and their frictional ratio $f/f_0$ is not too different from that of a sphere of equal volume (deviations <5%).

Frictional coefficients for irregular shapes can be approximated by applying *hydrodynamic bead modeling*, which divides a trial shape into segments and assigns a (usually spherical) bead to each segment (Figure 26.21).

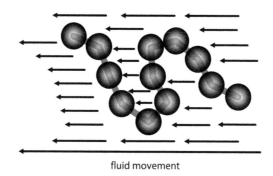

fluid movement

**FIGURE 26.21: Calculation of frictional coefficients for irregular shapes using beads.** A macromolecule is divided into segments, each of which is assigned a spherical bead with known radius. Any shape can thus be approximated as a string of beads. Some beads influence the movement of others by blocking their paths (indicated by smaller arrows).

For $N$ segments we have $N$ spheres, each of frictional coefficient $f_1 = 6\pi\eta R_1$. If the movement of one sphere (bead) did not influence the movement of nearby beads, the total frictional coefficient would just be the sum of the frictional coefficients of all spheres, $f_N = N{\cdot}f_1$. In reality each segment somewhat decreases the fluid movement in its vicinity. The frictional force and thus the frictional coefficient therefore become smaller than $N{\cdot}f_1$. For a molecule with $N$ identical subunits, either a homo-oligomer or a polymer with $N$ segments approximated by $N$ beads, the total frictional coefficient is given by the *Kirkwood approximation* (eq. 26.16).

$$f_N = Nf_1 \left(1 + \frac{f_1}{6\pi\eta N} \sum_{i=1}^{N} \sum_{j\neq i}^{N} \frac{1}{R_{ij}}\right)^{-1} \qquad \text{eq. 26.16}$$

$R_{ij}$ is the distance between the subunit pair $ij$. The summation is over all pairs with $i \neq j$. The Kirkwood approximation is especially useful for symmetric homo-oligomeric complexes (see Section 26.2.3). In principle, it allows the calculation of frictional coefficients for any shape by modeling the shape with a large number of spheres. However, the Kirkwood approximation is inexact for highly asymmetric particles. In such cases, it is better to determine the frictional coefficient experimentally. This can be done by measuring the sedimentation coefficient in sedimentation velocity experiments, which we will discuss in the next section, and where the equations above are used to describe the properties of the molecules in the centrifuge.

### 26.2.3 SEDIMENTATION VELOCITY

Newton's first law of motion states that a particle rests or moves uniformly with a constant velocity in a straight direction when no external forces act upon it. We find this scenario in the ultracentrifuge after the particle has been accelerated to a constant velocity (Figure 26.19). The sum of the forces must then be zero:

$$F_c + F_b + F_f = 0 = m \cdot \omega^2 r - m \cdot \bar{v} \cdot \rho \cdot \omega^2 r - f \cdot u \qquad \text{eq. 26.17}$$

By expressing the mass of the molecule in eq. 26.17 as the ratio of the molar mass $M$ and Avogadro's constant $N_A$ and rearranging we obtain

$$\frac{M}{N_A} \cdot (1 - \bar{v}\rho) \cdot \omega^2 r - f \cdot u = 0 \qquad \text{eq. 26.18}$$

and

$$\frac{M(1-\bar{v}\rho)}{N_A \cdot f} = \frac{u}{\omega^2 r} \equiv s \qquad \text{eq. 26.19}$$

We define the velocity $u$ of the particle divided by the gravitational acceleration $\omega^2 r$ as the *sedimentation coefficient s*. Sedimentation coefficients are measured in *Svedberg* (S), which is a time unit with $1\ S = 10^{-13}$ s, or 100 fs, named after Theodor Svedberg who built the first analytical ultracentrifuge in the mid-1920s. By AUC, Svedberg clearly established that proteins were indeed macromolecules of defined molecular masses in times where the concept of macromolecules did not yet exist, which earned him the Nobel Prize for Chemistry in 1926. Svedberg units are frequently used in biochemistry to describe large molecular entities (Table 26.3).

According to the left-hand side of eq. 26.19, the sedimentation coefficient only depends on the properties of the particle. It is inversely proportional to the frictional coefficient and directly proportional to the *buoyant* molecular mass $m_{eff}$ (eq. 26.12). Molecules with different shapes and sizes will therefore have different sedimentation coefficients, and there is no strict correlation between $s$ values and molecular mass. $s$ values are not additive because of differences in the molecular shapes of complexes compared to their constituents.

**TABLE 26.3**
**Sedimentation coefficients of some biological macromolecules.**

| particle | s value (S) | $M_M$ (MDa) |
|---|---|---|
| prokaryotic ribosome, large subunit, small subunit | 70, 50, 30 | ca. 2.0 / 1.3 / 0.7 |
| eukaryotic ribosome, large subunit, small subunit | 80, 60, 40 | ca. 3.2 / 2.0 / 1.2 |
| prokaryotic ribosomal RNA (*E. coli*) | 5, 16, 23 | 0.041 / 0.52 / 0.99 |
| eukaryotic ribosomal RNA | 18, 28 | 0.64 / 1.7 |
| eukaryotic proteasome | 20 | 0.75 |
| spliceosome | 50–60 | ca. 4.8 |

By expressing the frictional coefficient in eq. 26.19 in terms of the diffusion coefficient (eq. 26.14), the *Svedberg equation* is obtained.

$$\frac{s}{D} = \frac{M_M \cdot (1-\bar{v}\cdot\rho)}{RT} \qquad \text{eq. 26.20}$$

The Svedberg equation is used for molecular mass determination after measurement of $s$ from sedimentation velocity and $D$ from sedimentation equilibrium centrifugation experiments, which are described in the next sections.

### 26.2.3.1 Determination of Sedimentation Coefficients

Sedimentation coefficients can be determined from the velocity $u$ at the known acceleration $\omega^2 r$. The velocity $u$ is measured by following the position of the boundary as a function of time. Since the sedimentation force $F_c$ increases with radius $r$, so does the velocity $u$. We can re-write eq. 26.19 to take the dependency of $u(r)$ into account and express $u$ as $dr/dt$:

$$s = \frac{u(r)}{\omega^2 r} = \frac{\frac{dr}{dt}}{\omega^2 r} \qquad \text{eq. 26.21}$$

Separation of the variables in this differential equation prepares it for integration:

$$s\omega^2 dt = \frac{dr}{r} \qquad \text{eq. 26.22}$$

Integration from the meniscus $r_0$ to the radial distance of the boundary $r_b$ yields

$$s\omega^2 t = \int_{r_0}^{r_b} \frac{dr}{r} = \ln\left(\frac{r_b}{r_0}\right)$$

eq. 26.23

or

$$s\omega^2 t + \ln(r_0) = \ln(r_b)$$

eq. 26.24

A plot of $\ln(r_b)$ *versus* time gives a straight line with slope $s\omega^2$ from which the sedimentation coefficient of the solute can be derived.

Instead of confining data analysis to the movement of the boundary, sedimentation coefficients can also be measured by following the whole concentration distribution over time, i.e. the concentration of solute as a function of both time and radial distance. The diffusion coefficient $D$ and the sedimentation coefficient $s$ are obtained by globally analyzing the set of concentration distributions at different times with the *Lamm equation* (eq. 26.25):

$$\frac{\partial c}{\partial t} = \frac{1}{r} \cdot \frac{\partial}{\partial r}\left(r \cdot D \cdot \frac{\partial c}{\partial r} - s\omega^2 r^2 c\right)$$

eq. 26.25

The Lamm equation is a differential equation which results from a combination of Fick's law of diffusion (Section 4.1) and the Svedberg equation, and provides a general description of any AUC experiment, both sedimentation velocity and sedimentation equilibrium. It takes into account *radial dilution* due to the sector-shape of the cells as well as end effects due to accumulation of material at the bottom. If we execute the $\partial/\partial r$ differential in eq. 26.25 we arrive at a form of the Lamm equation that assumes that $s$ and $D$ do not depend on the concentration and that nicely shows the separate diffusion and sedimentation terms:

$$\frac{\partial c}{\partial t} = D \cdot \left(\frac{\partial c}{r \cdot \partial r} + \frac{\partial^2 c}{\partial r^2}\right) - s\omega^2 \cdot \left(2c + r \cdot \frac{\partial c}{\partial r}\right)$$

eq. 26.26

The Lamm equation has no analytical solution, but various approximations to the Lamm equation have been used in AUC data analysis. Modern AUC sedimentation velocity analysis software such as Sedfit numerically solves the Lamm equation to extract $s$ and $D$. To deduce the possible shape of the solute molecule, the derived values for $s$ and $D$ can be compared to those calculated for various trial shapes by hydrodynamic bead modeling (Figure 26.21 and eq. 26.16) using programs such as Hydropro.

### 26.2.3.2 Solvent and Concentration Dependence of the Sedimentation Coefficient

The value of $s$ is not only characteristic of the molecular shape, but also depends on the viscosity and density of the solvent, and the temperature at which the experiment was conducted. Measured $s_{obs}$ values in a certain solvent and at a certain temperature are therefore often converted into a standard $s_{20,w}$ value, which corresponds to the sedimentation coefficient at 20°C in water as reference solvent according to eq. 26.27:

$$s_{20,w} = s_{obs} \cdot \frac{1 - \overline{v} \cdot \rho_{20,w}}{1 - \overline{v} \cdot \rho_{obs}} \cdot \frac{\eta_{obs}}{\eta_{20,w}}$$

eq. 26.27

$\rho_{20,w}$ is the density of water at 20°C and $\rho_{obs}$ that of the solvent at the temperature of the measurement. Likewise, $\eta_{20,w}$ is the viscosity of water at 20°C and $\eta_{obs}$ the viscosity of the solvent at the temperature of the measurement.

The value of $s$ also depends on the solute concentration, and $s$ becomes smaller with higher concentration. One reason for this decrease in $s$ is the slowdown of sedimentation due to the increased viscosity of the concentrated solution. In addition, sedimenting particles must continuously displace solvent during sedimentation. At high concentrations, they also have to displace other solute macromolecules (molecular crowding), which are more difficult to displace than the solvent. This effect is larger for extended molecules with high frictional coefficients such as nucleic acids, unfolded proteins, or glycosylated proteins. The concentration dependence of the sedimentation coefficient follows a hyperbolic relation (eq. 26.28):

$$ s_{20,w}(c) = \frac{s_{20,w}(c=0)}{(1+k_s \cdot c)} \qquad \text{eq. 26.28} $$

The constant $k_s$ has units of inverse concentration and depends on the shape of the molecule. It is usually very small for globular proteins (ca. 0.005 mL·mg$^{-1}$) but larger for extended molecules (Figure 26.22). Extrapolation of $s_{20,w}$ to zero concentration is useful for comparing $s$ values from experiments performed at different concentrations.

The concentration dependence of $s$ may lead to sharpening of the boundary region. Molecules preceding the boundary $r_b$ are slowed down because of the higher concentrations whereas molecules in the trailing edge of the boundary can move faster. The net effect is a partial counteraction of diffusion, leading to a sharper boundary region.

A change in sedimentation coefficient with solute concentration may also reflect self-association. Such a concentration dependence of $s$ values reflects the higher molecular mass and the changed shape of the complex (Figure 26.22), and can be used to determine the equilibrium association constant of the complex (Section 3.2).

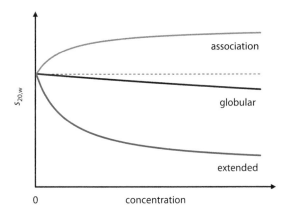

**FIGURE 26.22:** **Concentration dependence of s.** The $s_{20,w}$ value at zero concentration ($c = 0$) changes in a different manner with increasing concentration depending on the shape and association state of the molecule.

### 26.2.3.3 Measuring Polydispersity and Association

If there are several components in a sample, the concentration gradients measured in sedimentation velocity experiments are a superposition of the individual traces for each component. The presence of more than one component might not be apparent from the shape of the traces when the molecular masses are similar or the concentration of a component is very low. Two solutes with sufficiently different molecular masses that are both present in detectable amounts will give rise to double sigmoidal traces (Figure 26.23). In this case, the polydispersity, i.e. the number of non-interacting species and their abundance, can be assessed.

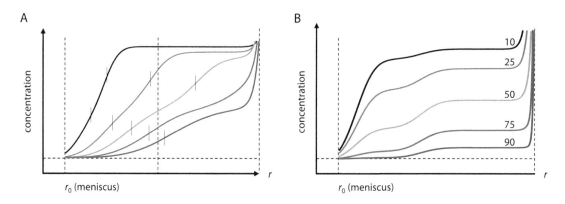

**FIGURE 26.23: Simulated sedimentation velocity experiments for mixtures of two species.**
A: Equimolar mixture of two non-interacting molecules with molecular masses of 20 kDa and 80 kDa. Traces are calculated for different time points 5 min apart. B: Traces at one particular time point for different ratios of the two molecules. The numbers indicate the percentage of the 80 kDa species.

AUC data for mixtures can be evaluated using a linear combination of Lamm equations, one for each component, weighted by the component concentration. The analysis provides a distribution of $s$ values from which the relative abundance and the $s$ value of each species can be extracted (Figure 26.24).

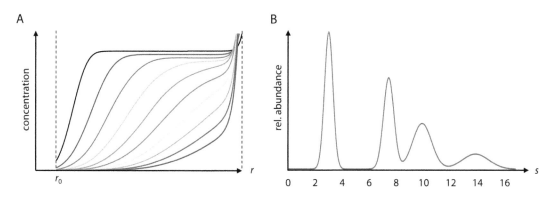

**FIGURE 26.24: Sedimentation of a polydisperse mixture.** A: Sedimentation data of a mixture of four components with $s$ values 3 S, 7.5 S, 10 S, and 14 S. B: Distribution of $s$ values obtained from analysis of the sedimentation data with a concentration-weighted sum of Lamm equations. The width of the distributions increases with increasing $s$ values.

If the molecules in the solution interact, sedimentation velocity experiments provide information on complex formation. Depending on the time scale of association and dissociation reactions with respect to the time scale of the AUC experiment, each trace may contain information about any species present in the equilibrium at any time point and at any radial position. The traces obtained are therefore more complex than those from mixtures of non-interacting molecules. The simplest case is the monomer-dimer equilibrium:

$$A + A \underset{k_{off}}{\overset{k_{on}}{\rightleftharpoons}} A_2 \qquad \text{scheme 26.1}$$

If binding and dissociation is slow compared to the time scale of the AUC experiment, the mixture will behave like a polydisperse solution with two boundaries per trace. In this case, analysis of the traces with the Lamm equation yields the distribution of $s$, the $s$ values for monomer and dimer, and their relative abundance in equilibrium. Otherwise, the traces show a single asymmetric boundary, and analysis of the velocity of its movement gives a weight-averaged sedimentation coefficient. By measuring this weight-averaged coefficient as a function of monomer concentration, the equilibrium association constant can be determined.

## 26.2.4 SEDIMENTATION EQUILIBRIUM

In AUC experiments with lower rotor speeds, on the order of 10000 *g* depending on the size of the molecules, diffusion plays a major role in shaping the concentration profile of the sedimenting molecules. The rate of diffusion is proportional to the concentration gradient (Section 4.1; eq. 4.1). Therefore, diffusion is higher near the bottom of the cell where the concentration is higher, and net diffusion is thus directed toward the top of the cell. Because of the radial dependence of the centrifugation force, sedimentation is faster near the bottom of the cell, and it is directed toward the bottom of the cell (Figure 26.25). After many hours, these opposing effects establish a stable concentration gradient following an exponential function, where at each point of the profile sedimentation and diffusion are balanced.

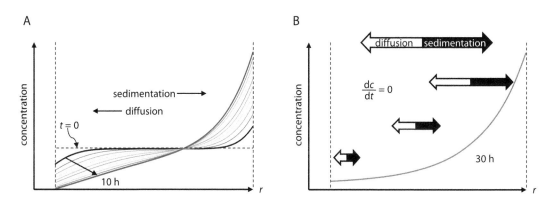

**FIGURE 26.25:** **Approach to sedimentation equilibrium.** A: Simulated traces for a 50 kDa protein centrifuged at 10000 *g*. At the outset of the experiment, the concentration of solute is constant throughout the cell (dashed horizontal line). After 1 h (black), the meniscus region has lower concentration than the bottom of the cell. The sedimentation force increases with radial distance, leading to a higher concentration at larger *r*. Diffusion opposes sedimentation because it redistributes solute to regions of smaller concentration (i.e. smaller *r*), leading to establishment of a stable concentration gradient over the experiment. B: Invariant solute distribution is reached after 30 h of centrifugation. The solute profile follows an exponential function. The size of the arrows shows that the magnitude of the opposing forces increase at higher concentrations.

Both sedimentation and diffusion depend on the diffusion coefficient $D$ or the frictional coefficient $f$ (eq. 26.14). When diffusion and sedimentation have reached equilibrium, the data obtained are independent of $D$ and $f$, and thus are independent of the shape of the molecule.

### 26.2.4.1 Determination of Molecular Mass Using Sedimentation Equilibrium

Analysis of the equilibrium concentration distribution as a function of radial position directly yields the molecular mass, which can be compared to the results from mass spectrometry (Section 26.1) or SAXS (Section 21.2.4). To derive an expression that allows determination of the molecular mass from the radial concentration dependence c(r) in equilibrium, we begin by recalling the definition of the chemical potential (eq. 2.156):

$$\mu = \mu^0 + RT \ln \frac{c}{c^0} = \mu^0 + RT \ln c(r) \qquad \text{eq. 2.156}$$

$c^0$ is the concentration at standard conditions (1 M). $\mu^0$ is the chemical potential at standard conditions. The solute concentration $c$ is a function of the radial position $r$. The total potential $\mu_{tot}$ of the solution in the centrifuge is the sum of the chemical potential μ and a centrifugal potential $U(r)$ that is also a function of the radial position:

$$\mu_{tot} = \mu + U(r) = \mu^0 + RT \ln c(r) + U(r) \qquad \text{eq. 26.29}$$

At equilibrium, the potential $\mu_{tot}$ is at a minimum, hence its derivative is zero. We can calculate the derivative of the sum as the sum of the individual derivatives, and obtain

$$\frac{d\mu_{tot}}{dr} = \frac{d\mu^0}{dr} + RT \frac{d \ln c(r)}{dr} + \frac{dU(r)}{dr} = 0 \qquad \text{eq. 26.30}$$

Since $\mu^0$ is a constant, its derivative is zero, and we have the following condition for equilibrium:

$$RT \frac{d \ln c(r)}{dr} = -\frac{dU(r)}{dr} \qquad \text{eq. 26.31}$$

The term $-dU/dr$ is the negative *gradient* of the centrifugal potential, which is equivalent to the centrifugal force $F_c$ that we have defined in the preceding chapter as $F_c = m\omega^2 r$ (eq. 26.8). In general, a force is the negative gradient of a potential function. Substituting the effective mass (eq. 26.12) into this equation, we obtain:

$$\frac{dU(r)}{dr} = -F_c = -m \cdot \omega^2 r = -m(1 - \bar{v}\rho) \cdot \omega^2 r \qquad \text{eq. 26.32}$$

The chemical potential in eq. 2.156 uses molar concentrations, and in order to reflect molar energy changes of the centrifugal field, we have to multiply $F_c$ with $N_A$, which gives

$$-\frac{dU(r)}{dr} = F_{c,\,molar} = M(1 - \bar{v}\rho) \cdot \omega^2 r \qquad \text{eq. 26.33}$$

We can then combine eq. 26.31 and eq. 26.33 to

$$RT \frac{d \ln c(r)}{dr} = M(1 - \bar{v}\rho) \cdot \omega^2 r \qquad \text{eq. 26.34}$$

separate the variables

$$RT \, d \ln c(r) = M(1 - \bar{v}\rho) \cdot \omega^2 r \cdot dr \qquad \text{eq. 26.35}$$

and integrate from a reference concentration $c_0$ at radial position $r_0$ to any concentration $c$ at its respective radial position $r$:

$$\int_{c_0}^{c} RT \, d \ln c(r) = \int_{r_0}^{r} M(1 - \bar{v}\rho) \cdot \omega^2 r \cdot dr \qquad \text{eq. 26.36}$$

Evaluation of the integral gives

$$RT \left[ \ln c(r) \right]_{c_0}^{c} = M(1 - \bar{v}\rho) \cdot \omega^2 \cdot \left[ \frac{r^2}{2} \right]_{r_0}^{r} \qquad \text{eq. 26.37}$$

and

$$\ln \frac{c(r)}{c_0} = \frac{M(1 - \bar{v}\rho) \omega^2 (r^2 - r_0^2)}{2RT} \qquad \text{eq. 26.38}$$

According to eq. 26.38, a plot of $\ln(c/c_0)$ as a function of $r^2$ is linear, and from its slope, the molar mass $M$ can be calculated. To determine the molar mass $M$ from such a plot, the reference concentration $c_0$, for instance at the meniscus of the cell $r_0$, has to be known. With modern fitting programs, linearization is not necessary but the data can be described directly by the exponential function we obtain by solving eq. 26.38 for $c(r)$:

$$c(r) = c_0 \cdot e^{\frac{M(1-\bar{v}\rho)\omega^2\left(r^2-r_0^2\right)}{2RT}}$$

eq. 26.39

In cases where several components of different molecular masses are present in the sample, the observed concentration distribution will be a linear combination of several instances of eq. 26.39. The dependence of $\ln(c/c_0)$ on $r^2$ (eq. 26.38) will then not be linear anymore, which serves as a diagnostic for polydispersity.

### 26.2.4.2 Association in Sedimentation Equilibrium

In a mixture of interacting components an equilibrium between all possible species is established at every point in the centrifuge. For the simplest case, a monomer-dimer equilibrium (Scheme 26.1), the equilibrium association constant is

$$K = \frac{c_2}{c_1^{\,2}}$$

eq. 26.40

$c_2$ and $c_1$ are the concentrations of the dimer and monomer, respectively. Eq. 26.40 must be fulfilled at each and every point within the sample cell. We can formulate the radial dependence of the concentration (eq. 26.39) for dimer and monomer, and include the fact that the mass of the dimer is twice that of the monomer. Division of the concentration profiles gives:

$$\frac{c_2(r)}{c_1(r)} = \frac{c_2(r_0)e^{\frac{2M(1-\bar{v}\rho)\omega^2\left(r^2-r_0^2\right)}{2RT}}}{c_1(r_0)e^{\frac{M(1-\bar{v}\rho)\omega^2\left(r^2-r_0^2\right)}{2RT}}} = \frac{c_2(r_0)}{c_1(r_0)}e^{\frac{M(1-\bar{v}\rho)\omega^2\left(r^2-r_0^2\right)}{2RT}}$$

eq. 26.41

The exponential term on the right-hand side is just $c_1(r)$ divided by $c_1(r_0)$ (eq. 26.39). Substituting the exponential term in eq. 26.41 with $c_1(r)/c_1(r_0)$ yields

$$\frac{c_2(r)}{c_1(r)} = \frac{c_2(r_0)}{c_1(r_0)} \cdot \frac{c_1(r)}{c_1(r_0)}$$

eq. 26.42

or

$$\frac{c_2(r)}{c_1^{\,2}(r)} = \frac{c_2(r_0)}{c_1^{\,2}(r_0)} = K$$

eq. 26.43

This result tells us that the equilibrium constant is the same at every position of the cell, as it should be. Furthermore, eq. 26.43 is also independent of the rotor speed $\omega$, and the equilibrium distribution of $c_1$ and $c_2$ will be the same at any position at any rotor speed.

In the centrifuge, monomer and dimer are not observed independently. Only the total signal corresponding to a weighted average of the contributing species is observed (Figure 26.26).

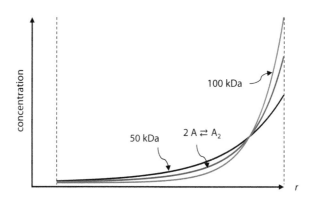

FIGURE 26.26: **Concentration profile for the sedimentation equilibrium of a monomer that self-associates to a dimer.** The black and blue traces are the expected sedimentation profiles of a 50 kDa monomer and a 100 kDa dimer, respectively. The red trace is the observed concentration profile for a monomer-dimer equilibrium. The equilibrium association constant of the complex can be derived by fitting a combination of the expected traces for the pure monomer and dimer to the observed data.

The dimer contributes twice to the detected trace compared to the monomer. The total absorption signal is then a weighted sum of monomer and dimer contributions according to the Lambert-Beer law:

$$A_{tot}(r) = \varepsilon_1 c_1(r) d + \varepsilon_2 c_2(r) d = \varepsilon_1 c_1(r) d + 2\varepsilon_1 \cdot K c_1(r)^2 \qquad \text{eq. 26.44}$$

Here, $d$ is the path length of the cell and $\varepsilon_1 = 0.5\varepsilon_2$ represents the assumption that the extinction coefficient of the dimer is twice that of the monomer. The right-hand side of eq. 26.44 just expresses $c_2$ by $c_1$ using eq. 26.40. By expressing $c_1(r)$ and $c_2(r)$ according to eq. 26.39, the total signal $A_{tot}(r)$ can be written as

$$A_{tot}(r) = \varepsilon_1 c_{1,r_0} d \cdot e^{\frac{M(1-\bar{v}\rho)\omega^2 \left(r^2 - r_0^2\right)}{2RT}} + 2\varepsilon_1 K c_{1,r_0}^2 d \cdot e^{\frac{2M(1-\bar{v}\rho)\omega^2 \left(r^2 - r_0^2\right)}{2RT}} \qquad \text{eq. 26.45}$$

Using eq. 26.45, we can calculate the equilibrium association constant from the measured absorption profile, provided that the concentration of the monomer at a reference point $r_0$ and the molecular mass of the monomer are known. We see that the derived expression is rather complex even for the comparatively simple monomer-dimer equilibrium. Equations for equilibria between more than two components and for complex stoichiometries different from unity have been derived and implemented in AUC fitting programs such as Sedfit. Sedimentation equilibrium experiments are suitable for measuring a broad range of equilibrium association constants ($10^4$–$10^8$ M$^{-1}$).

### 26.2.5 ZONAL, BAND, OR ISOPYCNIC CENTRIFUGATION

Concentration gradients such as those established in sedimentation equilibrium experiments can also be used for preparative purposes to separate molecules according to their density. When the densities of the solute $(1/\bar{v})$ and the solvent $(\rho)$ are equal (*isopycnic*), the factor $\bar{v}\rho$ in eq. 26.12 is unity and the effective mass $m_{eff}$ of the particle is zero. In this situation, no net force acts on the molecule, there is no tendency to move, and the sample is concentrated in a *zone* or *band* at the corresponding position of the density gradient.

Density gradients can be prepared in continuous and discontinuous forms. For *step gradients*, a highly concentrated solution of sucrose or the polysaccharide ficoll is overlaid with layers of less and less concentrated solutions (Figure 26.27). In contrast, gradual mixing of glucose, sucrose, or glycerol solutions of different density delivers a *continuous gradient* of desired steepness (Figure 26.27). Such gradients are then used to fractionate by centrifugation components in complex mixtures according to their density. Both types of gradient are stable for several hours. The centrifugation is usually performed in swing-out rotors to minimize convection. In step gradients, the particles accumulate on top of the first layer with a higher density than the density of the particle. In continuous gradients the sedimenting particles form bands at positions where their density equals that of the gradient. The fractions are recovered by piercing the tubes with a syringe at the position of the band, or by drop-wise collection (Figure 26.27). Gradient centrifugation allows fractionation of different blood cell types from blood plasma, and cellular organelles or large macromolecules such as ribosomes and ribosomal subunits from cell extracts.

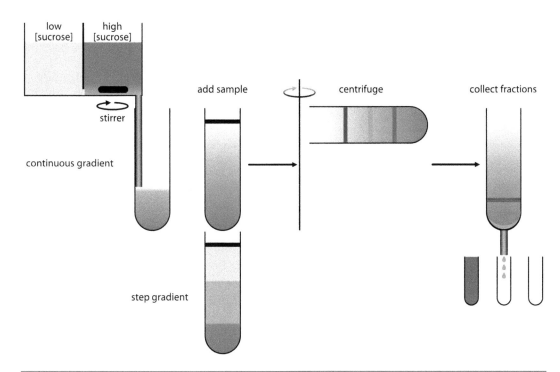

FIGURE 26.27:   **Principle of sucrose gradient centrifugation.** A continuous sucrose gradient is prepared in a gradient mixer by mixing of two sucrose solutions of different concentrations. Alternatively, a step gradient can be generated by overlaying sucrose solutions of different densities layer by layer. The sample, of lower density than the top of the gradient, is applied to the gradient surface. The gradient is centrifuged in a swing-out rotor to separate the components of the sample into distinct bands. Fractions are recovered by piercing the tube to extract the band of interest, or by collecting fractions from bottom to top.

Instead of manual mixing of gradients, sedimentation equilibrium can be used to generate concentration gradients *in situ* during the separation (see Figure 26.25). In this type of experiment, the sample is mixed with a salt solution, typically CsCl or $Cs_2SO_4$, and centrifuged for several hours. During the centrifugation, the density gradient is established, and the particles move to their isopycnic positions. The concentration distribution of a pure solute centrifuged in a density gradient follows a symmetric Gaussian function of the radial position (Figure 26.28):

$$c_{solute}(r) = c_{solute}(r_{eq})e^{-\frac{(r-r_{eq})^2}{2\sigma^2}}$$

eq. 26.46

The gradient has a concentrating effect because of the balancing of diffusion and sedimentation forces, which have different signs on either side of the equilibrium position $r_{eq}$ (Figure 26.28). The standard deviation $\sigma$ of the Gaussian (at 68% of the peak area) is inversely related to the molar mass of the solute:

$$\sigma^2 = \frac{RT}{M\bar{v}\left(\dfrac{d\rho}{dr}\right)_{r_{eq}}\omega^2 r_{eq}}$$

eq. 26.47

Thus, measuring the width $\sigma$ and position $r_{eq}$ of the band allows calculation of the molar mass of the solute, provided that the steepness of the gradient is known.

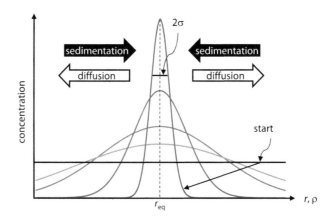

**FIGURE 26.28:** **Band formation in linear gradients.** At the outset, the cell contains a homogeneous mixture of solute (black line). During centrifugation the solute molecules are concentrated at a radial position $r_{eq}$ that equals their buoyant density (dashed line). The opposite directions of movement due to sedimentation and diffusion lead to a sharpening effect over the time of the experiment (green, blue, magenta, and finally red curve).

Isopycnic centrifugation in self-establishing gradients is particularly successful for the analysis of nucleic acids that form highly focused bands because of their high frictional coefficients and hence very slow diffusion. CsCl density gradients allow the separation of AT-rich and GC-rich DNA, or of supercoiled plasmids from contaminating RNA. Historically, the most prominent contribution of CsCl gradient centrifugation was the proof that DNA replication is semi-conservative (Box 26.1).

---

**BOX 26.1: SEMI-CONSERVATIVE DNA REPLICATION –
THE MESELSON-STAHL EXPERIMENT.**

When in 1953 James Watson and Francis Crick, based on the X-ray diffraction data of Rosalind Franklin, published the structure of the DNA double helix, a replication model for DNA was immediately apparent. Each single strand of DNA can serve as a template to reproduce the other strand. This model predicts that the two new DNA double helices each contain one "old" and one "new" strand. Using CsCl gradient centrifugation, Matthew Meselson and Frank Stahl proved this in 1958 (Figure 26.29). *E. coli* was grown on medium with $^{15}NH_4Cl$ as the only nitrogen source for several generations to produce a heavy version of its DNA. The bacteria were then quickly shifted to a medium containing the usual $^{14}N$ isotope. DNA samples were analyzed by centrifugation in a time-dependent manner. After one generation of bacterial growth (ca. 20 min), only a single band with a density exactly between that of the $^{15}N/^{15}N$ (both strands $^{15}N$-labeled) and $^{14}N/^{14}N$ double-stranded DNA was detected, which corresponded to an $^{14}N/^{15}N$ hybrid DNA. No parent DNA ($^{15}N/^{15}N$) was detected anymore. This proved that all daughter DNA inherited exactly half of the parent DNA. After two generations, two bands of equal concentration were observed, corresponding to $^{14}N/^{15}N$ and $^{14}N/^{14}N$, which is also in agreement with semi-conservative replication.

*(Continued)*

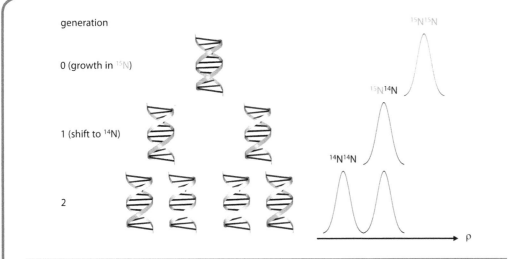

generation

0 (growth in ¹⁵N)

1 (shift to ¹⁴N)

2

¹⁵N¹⁵N

¹⁵N¹⁴N

¹⁴N¹⁴N

ρ

**FIGURE 26.29:** **The Meselson-Stahl experiment.** Bacteria are grown in medium with ¹⁵NH₄Cl to label both strands of their DNA with ¹⁵N (green; generation 0). In generation 1, grown on medium with ¹⁴N, the ¹⁵N/¹⁵N DNA is converted into a DNA with a density between ¹⁵N/¹⁵N and ¹⁴N/¹⁴N (detected by density gradient centrifugation, right), consistent with the formation of a ¹⁴N/¹⁵N hybrid (green/gray). In generation 2, equal amounts of ¹⁴N/¹⁵N and ¹⁴N/¹⁴N DNA are observed, in agreement with predictions from the model of semi-conservative DNA replication.

The density of proteins in CsCl solutions is observed to change with pH. The increased number of negative charges at higher pH leads to binding of Cs⁺ ions near carboxylates. Deprotonation of His and Lys, on the other hand, will release less dense, hydrated anions. Both effects increase the density of the protein. The extent of salt binding to proteins may also change depending on their final position in the gradient. In general, due to interaction with the gradient material, which is a situation far away from an ideal solution in a thermodynamic sense, no precise molecular masses can be determined. A very useful application for this equilibrium density gradient centrifugation is the detection of *changes* in densities due to protein-nucleic acid or protein-lipid complex formation. The partial specific volumes of these components are quite different (Table 26.1), leading to measurable changes in density although the molecular masses may not change dramatically.

## 26.3 SURFACE PLASMON RESONANCE

Surface plasmon resonance (SPR) has widespread applications in the biophysical characterization of molecular interactions. The principle of the method is rather simple. A *receptor* molecule is fixed to a sensor surface and brought into contact with a binding partner, termed the *analyte*. The refractive index near the surface changes upon binding, and this change is measured in a time-dependent fashion. The SPR signal contains kinetic and thermodynamic information: association and dissociation rates, affinities, and stoichiometries of binding can be obtained. SPR measurements are fast and can be automated, which makes them particularly useful for screening of small molecules for binding to pharmaceutical drug targets.

### 26.3.1 PHYSICAL BACKGROUND OF SPR

The SPR phenomenon can be understood starting from total internal reflection (TIR) at an interface of two non-absorbing materials (see Figure 23.9 in Section 23.1.4). Light propagating from a medium of high refractive index $n_1$ to a medium of low refractive index $n_2$ will be either refracted or reflected, depending on the incidence angle θ relative to the optical axis (Figure 26.30). At small angles θ the light is refracted away from the optical axis. Once the incidence angle θ matches the *critical angle*

$\theta_c$, the light will be refracted perpendicular to the optical axis and then propagates parallel to the interface. At higher incident angles $\theta$, total internal reflection occurs. The fully reflected beam leaks electric field intensity, called an *evanescent wave*, into the medium of lower refractive index. The evanescent wave amplitude decreases exponentially with increasing distance from the interface (see eq. 23.20, Section 23.1.4).

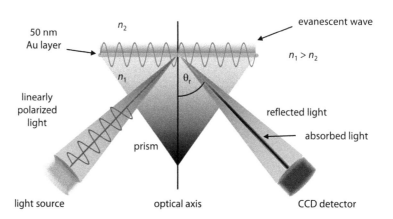

**FIGURE 26.30:**  **Principle of surface plasmon generation.** Linearly polarized light is coupled into a glass prism, such that it impinges on the glass-sample interface at different angles $\theta$ with respect to the optical axis. At a certain angle $\theta$, the light is totally reflected at the interface. If a gold layer is placed at the $n_1/n_2$ interface, the evanescent wave excites a surface plasmon (blue sine wave) in this layer, which leads to a slight decrease of the light intensity in the reflected beam.

In SPR, the evanescent wave generated by total internal reflection is allowed to interact with an electrically conducting material, a thin metal layer that is placed at the glass-sample interface. Gold is frequently used because it is chemically almost inert, but silver and aluminium also work well. The incident light has vector components perpendicular and parallel to the plane of the metal layer, and causes displacement of the metal electrons in both directions. The lateral movement of electrons due to displacement in the plane of the metal layer is observed as electrical conductivity. In contrast, displacement perpendicular to the plane of the metal layer increases the distance between electrons and nuclei (that are fixed in the plane), leading to charge separation. The vector component of the electromagnetic field perpendicular to the metal layer oscillates, and the electrons therefore also oscillate, giving rise to a *charge density wave* of the same wavelength as the incident light (Figure 26.30). The energy quantum of this charge density wave is termed a surface *plasmon*, in analogy to the energy quantum *photon* for a free electromagnetic wave. The excitation of the surface plasmon leads to a reduction of the light intensity in the reflected beam, which is measured in an SPR experiment using a CCD detector (Figure 26.30). The energy of the surface plasmon itself is eventually dissipated into heat.

The intensity of the surface plasmon at the metal-glass prism interface in an SPR experiment depends on the magnitude of the vector component of the incident light that is perpendicular to the metal layer. This component is maximized by using laser light polarized in the *plane of incidence*, which is the plane that is defined by the optical axis and the propagation direction of the laser beam. In addition, the angle of incidence $\theta$ needs to be optimized by rotating the prism with respect to the laser beam (Figure 26.31). The *resonance angle* $\theta_r$ with maximal intensity of the surface plasmon depends on the laser wavelength and the refractive indices $n_1$ and $n_2$. At this angle, the minimal light intensity of the reflected beam is detected.

The amplitude of the surface plasmon decreases exponentially with distance from the glass-sample interface. The plasmon extends only about 1 $\lambda$ (a few hundred nm) away from the gold layer into the medium (the sample) of refractive index $n_2$. Within the narrow layer near the surface that is probed by the surface plasmon, SPR is therefore sensitive to changes in the refractive index $n_2$ of the sample. This property makes surface plasmons useful as optical probes to follow binding events on surfaces. Binding of a molecule to a partner on the surface leads to an increase in mass and to a change in the refractive index $n_2$ near the surface. This change in refractive index in turn causes a change in the resonance angle $\theta_r$ at which the surface plasmon intensity is maximal and the intensity of the reflected beam has a minimum (Figure 26.31). During an SPR experiment, the change in resonance angle is measured as a *response unit* (*RU*). SPR experiments can detect changes in SPR angle by 0.001°, which corresponds to changes in $n_2$ by as little as $10^{-5}$. Because of the local confinement of the surface plasmon, SPR is insensitive to properties further away from the surface, so that samples may be colored, opaque, or turbid.

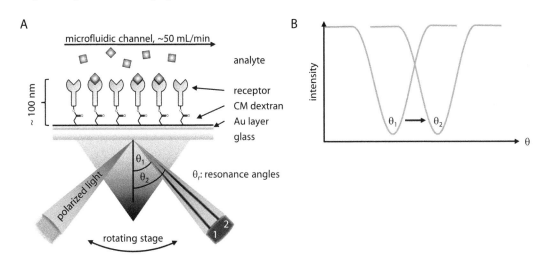

FIGURE 26.31: **Principle of an SPR experiment.** A: The gold layer is functionalized with the first binding partner (receptor). A solution containing the second binding partner (analyte) is passed over the sensor chip. Binding causes a change in mass near the gold layer and changes the refractive index, which in turn alters the resonance condition θ for the surface plasmon. B: Intensity of the reflected light as a function of θ. $\theta_1$ is the resonance angle in the absence of the binding partner, $\theta_2$ is the resonance angle after binding. The change can either be monitored by changing the angle of incidence or by moving the prism on a rotating stage.

### 26.3.2 PRINCIPLE AND INFORMATION CONTENT OF AN SPR EXPERIMENT

In an SPR experiment, the first binding partner (receptor), typically the macromolecule, is covalently or non-covalently immobilized on the metal layer of the sensor surface of the SPR chip (see Section 26.3.4). The receptor can be a small molecule, lipid, carbohydrate, protein, nucleic acid, or a large particle such as a virus or cell. The receptor is also often called "ligand" in SPR literature, which may be confusing to biochemists because ligand usually stands for a small molecule or a mobile binding partner. In the following, we will use the term receptor for the molecule that is attached to the sensor chip, and *analyte* for the second binding partner in solution.

The SPR measurement is carried out in a microfluidic device that allows rapid switching between solutions that are in contact with the sensor carrying the immobilized receptor (Figure 26.32). First, a baseline of the signal (the *response*) is established by washing the sensor surface with buffer. In a *contact* or *association phase*, the sensor surface is then brought into contact with the solution that contains the analyte, and binding is followed as an increase in the SPR signal as a function of time. Finally, the *dissociation phase* is started by flowing buffer over the sensor chip again. During the dissociation phase, bound analyte dissociates, which is recorded as a reduction in SPR signal (Figure 26.32). After dissociation has completed, the sensor chip is in principle ready for another binding reaction. Regeneration of the chip by an extra washing step is usually required to remove unspecifically bound analyte. The SPR response as a function of time is termed the *sensorgram*. The sensorgram is corrected for signals unrelated to the actual binding event, such as differences in the refractive indices of running buffer and analyte solution, or unspecific binding of analyte to the chip. These effects are measured in a control experiment performed without receptor, and subtracted.

Differences of a few millimolar in salt concentrations between buffers or just 1% organic solvent from an analyte stock solution give rise to a significant SPR signal.

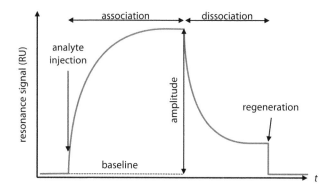

**FIGURE 26.32:** **SPR sensorgram.** The sensorgram is the change in SPR angle in resonance units (RU) as a function of time (blue curve). A buffer control is subtracted from the binding curve for analysis of association and dissociation phases. At the end of the experiment, unspecifically bound material is washed away (regeneration).

The association and dissociation phases of the sensorgram can be analyzed separately. From the association phase, two important values can be extracted: the amplitude, i.e. the total signal change relative to the baseline, and an observed association rate constant $k_{obs}$ that is calculated by fitting a single exponential function to the data. Both amplitude and observed rate constant depend on the analyte concentration. For a reversible one-step binding reaction, the microscopic association rate constants $k_{on}$ and $k_{off}$ can be obtained from the linear dependence of $k_{obs}$ on the analyte concentration (see Chapter 10). The dissociation rate constant $k_{off}$ is directly obtained from the dissociation phase of the sensorgram. The dissociation of the complex is concentration-independent, and sensorgrams at different analyte concentrations should therefore yield the same $k_{off}$ values. The ratio of $k_{off}$ and $k_{on}$ then yields a kinetically determined equilibrium dissociation constant $K_d$ (eq. 26.48).

$$K_d = \frac{k_{off}}{k_{on}}$$

eq. 26.48

In order for $K_d$ to represent the binding equilibrium in solution, the surface-immobilized receptor molecules must behave the same way as in solution. Although $K_d$ values determined by SPR are often comparable to the values determined in solution, fixing of the receptor to the sensor surface may change the binding entropy $\Delta S$. In these cases, the equilibrium constant from SPR will be different from the $K_d$ value measured in solution, and has to be considered as an apparent $K_d$ value. These apparent $K_d$ values still have qualitative value and can be compared within a series of experiments performed under identical conditions.

An alternative method estimates $K_d$ from the amplitudes of sensorgrams measured at different analyte concentrations (Figure 26.33). Under non-saturating conditions, the sensor signal approaches a plateau termed the *equilibrium response*, where analyte binding and dissociation are in equilibrium. The maximum or *saturation response* of the binding equilibrium is obtained under saturating conditions, i.e. with the analyte concentration much larger than the dissociation constant. Under these conditions, any receptor-analyte complex that dissociates will immediately be re-established by the sheer amount of analyte in the running buffer. The response units as a function of analyte concentration therefore provide a binding curve from which $K_d$ can be determined. Global fitting procedures simultaneously evaluate a whole set of sensorgrams to derive $k_{on}$, $k_{off}$, and $K_d$ from all curves.

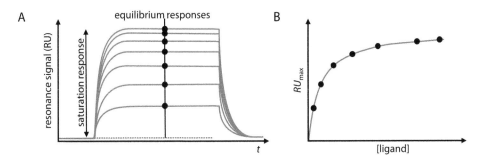

**FIGURE 26.33:** **Binding constant from a set of sensorgrams.** A: Several sensorgrams on top of each other with fast association and dissociation phases. B: Plot of amplitudes as a function of analyte concentration to yield $K_d$. This $K_d$ value should be the same as the $K_d$ derived from the kinetic constants.

SPR is suitable for the determination of association rate constants $<10^5$ $M^{-1}$ $s^{-1}$, and dissociation rate constants $>10^{-5}$ $s^{-1}$. If $k_{on}$ exceeds $10^5$ $M^{-1}$ $s^{-1}$, which is often the case for protein-protein interactions, there may be too few data points in the association phase for a reliable fit. Furthermore, when analyte binding is rapid, the analyte is depleted near the binding surface and its local concentration is lower than the concentration in the bulk solution, leading to errors in the analysis (Section 26.3.3). For slow dissociation reactions with $k_{off}$ values below $10^{-5}$ $s^{-1}$ (equivalent to a half-life of ca. 19 h), the dissociation phase cannot be recorded fully within a sensible time frame, and $k_{off}$ has to be estimated from the initial slope of the dissociation curve and the expected amplitude. These limitations make SPR suitable for the kinetic determination of $K_d$ values in the nanomolar range.

In principle, SPR experiments provide a set of thermodynamic values for binding. $\Delta G^0$ can be calculated from the determined $K_d$. The enthalpy change $\Delta H$ can be obtained from the temperature dependence of $K_d$ and a van't Hoff plot (Section 3.1.1), and the entropy can be calculated from the Gibbs-Helmholtz equation (see Section 2.5; eq. 2.105). However, the effect of receptor immobilization on thermodynamic parameters should always be tested.

### 26.3.3 MASS TRANSPORT LIMITATION

When an analyte binds to its immobilized receptor during the association phase of the SPR experiment, the liquid layer close to the sensor chip will be depleted of analyte. Normally, convection and diffusion quickly replenish the depleted layer with analyte. If, however, the receptor density on the chip is very high and/or binding is very rapid, neither of these mechanisms can keep up with analyte depletion. The actual analyte concentration at the sensor chip will consistently be smaller than that in the flowing solution, a situation termed *mass transport limitation*. As a result, the observed association rate constants $k_{on}$ will also be smaller than the true value. A similar effect is observed during the dissociation phase. Any analyte dissociating from a complex can either be washed away or re-binds to the receptor. When receptor densities are high and/or the flow rate is low, re-binding is favored, and the observed dissociation rate constant is smaller than the true dissociation rate constant. One way to reduce the effects of mass transport is to increase the flow rate (from typically 10 μL/min to 100 μL/min), which will deliver analyte faster during binding or wash it away instantaneously during dissociation. A second way to reduce or sometimes entirely avoid mass transport limitation is to reduce the receptor density on the sensor chip by changing the immobilization conditions. This includes varying the receptor concentration, pH, ionic strength, and the activation and contact times.

### 26.3.4 RECEPTOR IMMOBILIZATION ON THE SENSOR SURFACE

In order to couple the receptor to the sensor surface, the gold layer first needs to be functionalized. This is achieved by coupling of a matrix that carries functional groups to the gold layer. Popular matrices include cellulose, agarose, alginate, polycarboxylate hydrogel, and in particular carboxymethyl-functionalized (CM-)dextran. By using dextran with carboxymethyl groups in various densities, different receptor densities can be achieved. The carboxymethyl groups are then activated and used for introduction of other functionalities to which the receptor is either bound covalently or non-covalently.

### 26.3.4.1  Covalent Receptor Immobilization

*Amine coupling* is the most frequently used technique for direct immobilization of receptors. The carboxyl groups of the CM-dextran matrix are activated by a carbodiimide (Figure 26.34) and converted into N-hydroxysuccinimide (NHS) esters. These esters then react with amine nucleophiles of the receptor such as the abundant Lys side chains or the N-terminus. The result is a stable amide bond of the receptor with the sensor surface. Unreacted functional groups are blocked with an excess of ethanolamine.

Due to the high abundance of lysine in proteins and the large number of amino groups, amine coupling may lead to a wide spectrum of different receptor orientations relative to the surface, and not all of the receptors may be capable of binding to their binding partner.

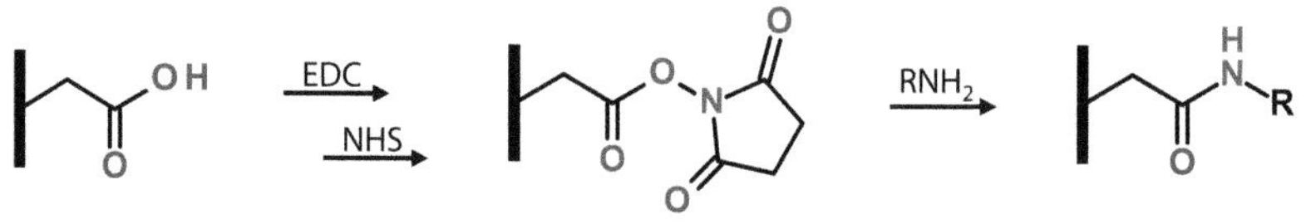

**FIGURE 26.34:    Amine coupling.** Carboxylate groups are activated for instance by the soluble 1-ethyl-3-(3-dimethylaminopropyl)carbodiimide (EDC). Substitution with N-hydroxysuccinimide (NHS) gives a reactive ester and the urea derivative of EDC. Substitution of the NHS-ester by an amine-containing receptor couples the receptor to the surface *via* an amide bond. Ethanolamine is then used in excess to block unreacted esters.

*Thiol coupling* uses disulfide exchange reactions to attach a receptor containing free thiol groups to the sensor chip *via* a disulfide bond (Figure 26.35). A disulfide group in the matrix is generated by activating the carboxylate groups of the dextran with EDC and NHS just as in the amine coupling procedure, and the resulting NHS ester is reacted with a disulfide-containing amine that contains a good leaving group. The receptor is then coupled to the surface by disulfide exchange.

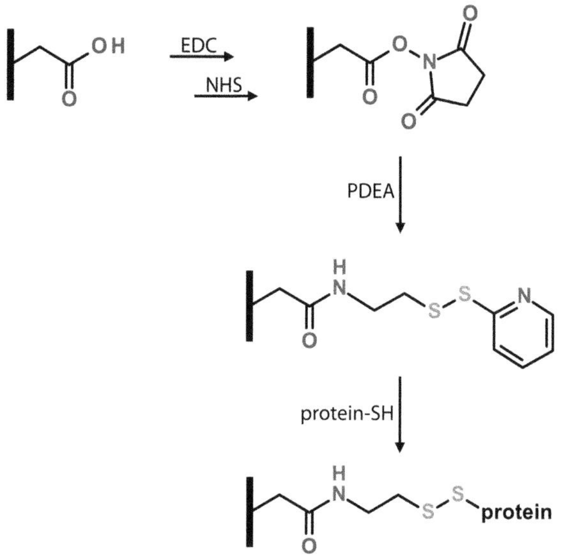

**FIGURE 26.35:    Thiol coupling.** 2-(2-pyridinyldithio)ethaneamine (PDEA) is amine-coupled to the dextran matrix, followed by disulfide exchange with the receptor. 2-thiopyridine is a good leaving group. Excess free Cys is used to block unreacted PDEA groups.

Cysteines in the receptor can be used for thiol-coupling. Because of the lower abundance of cysteines compared to lysines, thiol coupling leads to more uniform orientations of the receptor on the surface than amine coupling. If the receptor lacks cysteines, they can be introduced by site-directed mutagenesis.

*Maleimide coupling* is also based on thiol chemistry. Here, Cys residues in proteins react with a matrix derivatized with maleimide groups (Figure 26.36). The resulting esters are stable in basic conditions and insensitive to reducing agents that may be required in the SPR buffer.

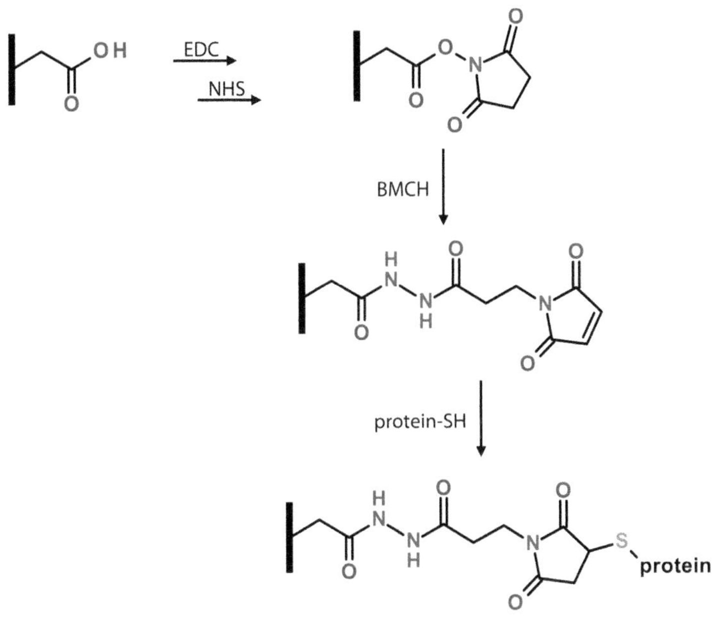

FIGURE 26.36: **Maleimide coupling.** Any reactive maleimide can be used to substitute the NHS-ester, for instance BMCH (N-[β-maleimidopropionic acid]-hydrazide), or EMCH (N-[ε-maleimidocaprocic acid]-hydrazide), which has a longer alkyl linker.

*Aldehyde coupling* (Figure 26.37) is useful for immobilizing glycoproteins *via* their carbohydrate moieties on hydrazide-functionalized surfaces. If the carbohydrate possesses a reducing end, this aldehyde group can be directly used for coupling. If not, a *cis*-diol in the carbohydrate is oxidized to form aldehyde groups. The aldehyde is then condensed with the hydrazide on the surface to form a hydrazone, which is further reduced to a stable amine. As carbohydrates occur at specific sites on the protein surface (Section 16.1.4.1), carbohydrate coupling leads to homogenously immobilized receptors on the sensor surface.

Apart from these organic chemical coupling reactions, biochemical coupling of modified peptides to the receptor using intein-mediated protein ligation or the transpeptidase *sortase* is also possible (see Section 19.5.4).

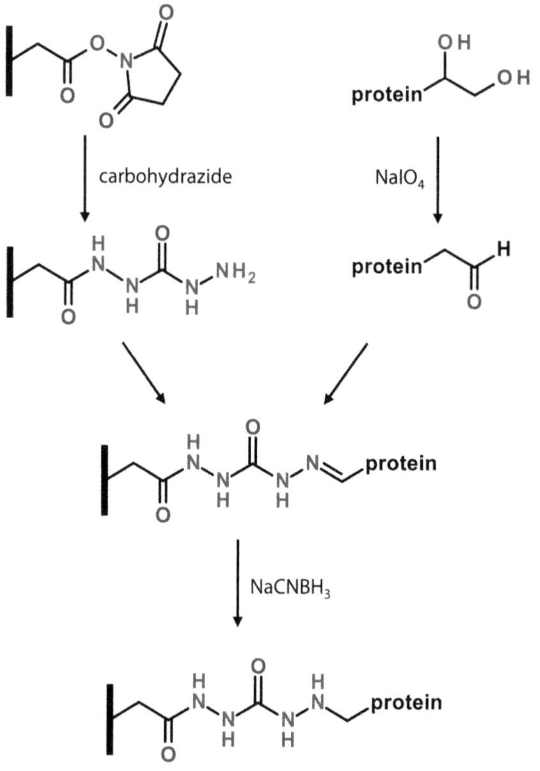

**FIGURE 26.37:   Aldehyde coupling.** The matrix is functionalized by reaction of the NHS ester with carbohydrazide. The receptor aldehyde groups, generated by oxidation of *cis*-diols if necessary, are condensed with the hydrazide to a hydrazone that is reduced to a stable amine by NaCNBH₃.

### 26.3.4.2 Non-Covalent Receptor Immobilization

Non-covalent immobilization uses the interaction of surface-coupled binding modules such as monoclonal antibodies, metal chelators, or streptavidin with affinity tags on the receptor. The binding module is covalently coupled to the surface by the same techniques used for covalent receptor coupling (Section 26.3.4.1). The receptor is produced with an affinity tag that is recognized by the binding module, such as the epitope for an antibody, a His-tag, or a biotin group, and coupled to the surface. Non-covalent coupling allows complete regeneration of the sensor chip, and re-use with the same or a different receptor, which is impossible with covalently attached receptors.

Non-covalent immobilization uses the interaction with a unique moiety on the receptor, and usually leads to a rather homogeneous orientation of the receptor on the sensor. The affinity of the interaction between the affinity tag on the receptor and the binding module on the surface should be high enough to prevent receptor dissociation during the experiment, which leads to a baseline drift.

The *biotin-streptavidin* interaction is among the strongest known in nature ($K_d$ ca. $10^{-15}$ M). Proteins can be chemically biotinylated at Lys residues using biotinyl-N-hydroxysuccinimide esters. Site-specific biotinylation can be achieved by expressed protein ligation of the receptor with a biotinylated peptide, by introduction of a biotinylation sequence and *in vivo* or *in vitro* biotinylation at a specific lysine in this sequence with biotin ligase, or by incorporation of a biotinylated amino acid through *in vitro* translation of mRNA and stop-codon suppression (Box 16.2; see also Section 19.5.4.2). Biotinylated proteins are then immobilized on streptavidin-functionalized surfaces. The dissociation rate of the streptavidin-biotin complex is low ($k_{off}$ ca. $10^{-6}$ s$^{-1}$), leading to negligible baseline drifts during experiments that may take several hours to complete.

*Monoclonal antibodies* can be covalently coupled to the sensor chip. The epitope recognized by the antibody, such as glutathione S transferase (GST), the His-tag, or the FLAG tag (see also Section 19.5.4.2), is fused to the receptor, which is then directly coupled to the surface by antibody-antigen interactions.

*Metal chelation* is a standard affinity chromatographic step during protein purification. The same principle can be applied for coupling of receptors to SPR sensor chips. The dextran matrix of the sensor chip is coupled to nitrilotriacetic acid (NTA) that chelates divalent cations (mostly $Ni^{2+}$, but $Zn^{2+}$ and $Co^{2+}$ are also used). The protein of interest carries an N- or C-terminal His-tag, which binds to the NTA-metal complex.

### 26.3.5 STOICHIOMETRY OF BINDING IN AN SPR EXPERIMENT

SPR experiments can be used to determine binding stoichiometries. The maximum amount of analyte that can be bound depends on the receptor density $RU_{receptor}$ (see Section 26.3.3) and on the fraction $P$ of the receptor that is active, i.e. which is immobilized in an orientation that does not occlude the binding site. With random chemical coupling, the fraction $P$ of active receptor can be as low as 0.2. $P$ can be determined by measuring the maximal SPR response $RU_{known\ analyte,max}$ by an analyte that binds to the receptor with known stoichiometry $N_{known}$:

$$P = \frac{1}{N_{known}} \cdot \frac{RU_{known\ analyte,\ max}}{RU_{receptor}} \cdot \frac{M_{M,\ receptor}}{M_{M,\ known\ analyte}} \qquad \text{eq. 26.49}$$

Eq. 26.49 takes into account that the measured response only depends on the molecular masses $M_M$ of analyte and receptor. Once $P$ is known for the reference system, the stoichiometry for an unknown analyte binding to the same sensor chip coupled with this receptor can be calculated from the maximum signal $RU_{unknown\ analyte,max}$ obtained by the unknown analyte.

$$N = \frac{RU_{unknown\ analyte,max}}{P \cdot RU_{receptor}} \cdot \frac{M_{M,receptor}}{M_{M,unknown\ analyte}} \qquad \text{eq. 26.50}$$

The stoichiometry of binding for an unknown analyte can thus be obtained in comparison to a reference measurement. Stoichiometries deviating from the expected number of binding sites may point to unspecific binding of the analyte to the sensor surface. Stoichiometries larger than the number of binding sites (super-stoichiometric binding) are often observed in the analysis of weak interactions, where the analyte concentrations have to be very high (Box 26.2).

---

**BOX 26.2: USING SPR IN SMALL MOLECULE SCREENING.**

SPR can be automated to test binding of several thousand compounds to a receptor within a matter of days or a few weeks. In SPR-based screening of small molecule binding to macromolecular drug targets, the drug target is coupled to the surface as the receptor, and a small molecule library is tested for binding at a single analyte concentration. The SPR response scales with the molecular mass of the compound, and testing the binding of small molecules ($M_M$ 100–250 Da) to a macromolecule brings SPR near its current

*(Continued)*

---

detection limit (ca. 50 Da or a few *RU*). To maximize the measured signal, high densities of receptor and high fractions of active receptors are needed. The resulting mass transport limitation is acceptable as long as no kinetic information is desired. High receptor densities are affordable because the overall receptor consumption in SPR is small, typically less than 100 μg per screen. Small molecules usually bind with low affinity to the receptor, with $K_d$ values in the micromolar to millimolar range, and SPR experiments therefore require high concentrations of analyte (10–200 μM). Unspecific binding that leads to false positive hits therefore needs to be accounted for by suitable controls (see Section 26.3.6). Molecules that elicit an SPR response (*positive hits*) are validated in a *dose-response* or titration experiment to obtain binding curves and to possibly gain insight into the binding kinetics. Further validation of binding uses an orthogonal technique such as NMR or structure determination by X-ray crystallography. Using SPR-based small molecule screening, starting molecules for drug design have been identified for a number of pharmaceutical targets, including HIV-1 protease and HIV-1 reverse transcriptase, but also for the G-protein coupled receptors $A_{2A}$ and CCR5.

### 26.3.6 SPECIFICITY OF BINDING IN AN SPR EXPERIMENT

A response signal in SPR experiment reports on any mass deposition on the sensor surface, and may either originate from specific binding of the analyte to the receptor binding site or from unspecific binding to the surface of the receptor or directly to the sensor. Specific and unspecific binding can be distinguished in control experiments. To monitor unspecific binding to the sensor surface, a reference channel is used that has no receptor immobilized. This type of control experiment can be used to minimize unspecific binding by optimization of buffer composition, temperature, and flow rate. Unspecific binding to the immobilized receptor outside the binding site can be estimated in a control experiment with an inactive receptor. This inactive receptor may have an amino acid substitution in the binding site that interferes with binding of the analyte, or the binding site may be blocked by modification by a covalent inhibitor. If an SPR response is obtained with such a blocked receptor, the signal is likely due to unspecific binding. Careful controls for unspecific binding are particularly important in the validation of binding events during small molecule screening with SPR (Box 26.2).

SPR is now a standard technique to characterize molecular interactions. SPR is particularly well-suited to study ligand binding to membrane receptors, because the immobilization of membrane proteins on the surface emulates their two-dimensional arrangement in a lipid bilayer. A recent development is to solubilize membrane proteins in *nanodiscs*, which are small patches of lipid bilayers where the hydrophobic rim is shielded from solvent by the *Membrane Scaffold Protein* (MSP). MSP is a derivative of the human apolipoprotein 1A, which forms ring-shaped dimers that encircle the nanodisc. The whole nanodisc, including the dissolved membrane protein and MSP, is then attached to a sensor chip. SPR has also been used to detect conformational changes in proteins by way of changes in mass distribution of surface-immobilized proteins.

## QUESTIONS

26.1   A rotor spins at 60000 rpm. What is the radial position at which a centrifugal field of 250000 $g$ is generated? What would be the apparent mass of an imbalance of 100 mg at this position?

26.2    You want to study the assembly of a nucleotide-binding protein by AUC. In order for the protein to form a complex, high concentrations of nucleotide are required. What detection methods are best suited? Another protein binds very tightly to its nucleotide, and upon binding of the nucleotide it forms oligomers at very low (nanomolar) concentrations. What probe located at which binding partner would you choose to detect such an assembly?

26.3    Using l'Hôpital's rule, show that for an oblate spheroid:

$$\lim_{a \to b} \frac{f}{f_0} = \lim_{a \to b} \frac{\sqrt{\dfrac{b^2}{a^2} - 1}}{\sqrt[3]{\dfrac{b^2}{a^2}} \cdot \tan^{-1}\left(\sqrt{\dfrac{b^2}{a^2} - 1}\right)} = 1$$

The derivative of $\tan^{-1}(x)$ is:

$$\frac{\mathrm{d}}{\mathrm{d}x} \tan^{-1}(x) = \frac{1}{x^2 + 1}$$

26.4    Using l'Hôpital's rule, show that for a prolate spheroid:

$$\lim_{a \to b} \frac{f}{f_0} = \lim_{a \to b} \frac{\sqrt{\dfrac{a^2}{b^2} - 1}}{\sqrt[3]{\dfrac{a}{b}} \cdot \ln\left(\dfrac{a}{b} + \sqrt{\dfrac{a^2}{b^2} - 1}\right)} = 1$$

26.5    Given is the Kirkwood approximation,

$$f_N = N f_1 \left(1 + \frac{f_1}{6\pi\eta N} \sum_{i=1}^{N} \sum_{j \neq i}^{N} \frac{1}{R_{ij}}\right)^{-1} \tag{1}$$

and the relation of molar mass and friction coefficient to the sedimentation coefficient,

$$s = \frac{M(1 - \bar{v}\rho)}{N_A \cdot f} \tag{2}$$

and Stokes' law:

$$f_1 = 6\pi\eta R_e \tag{3}$$

Show that for a homo-oligomer the following relation holds:

$$\frac{s_N}{s_1} = 1 + \frac{R_e}{N} \sum_{i=1}^{N} \sum_{j \neq i}^{N} \frac{1}{R_{ij}} \tag{4}$$

where $R_e$ is the Stokes radius of the monomer with friction coefficient $f_1$, and $s_N$ and $s_1$ are the sedimentation coefficients of the oligomer and the monomer, respectively.

26.6    Using the result from the previous question, show that for a dimer ($N = 2$) and trimer ($N = 3$) of spheres, the expected ratios $s_N/s_1$ are 1.25 and 1.5, respectively.

26.7    A tetramer sediments with a coefficient of $s_4 = 33.5$ S. The monomer sediments with $s_1 = 20$ S. Is the tetramer square planar or tetrahedral?

26.8    Starting from this form of the Lamm equation:

$$\frac{\partial c}{\partial t} = \frac{1}{r} \cdot \frac{\partial}{\partial r}\left(r \cdot D \cdot \frac{\partial c}{\partial r} - s\omega^2 r^2 c\right)$$

derive the Lamm equation with the separate contributions of diffusion and sedimentation:

$$\frac{\partial c}{\partial t} = D \cdot \left(\frac{\partial c}{r \cdot \partial r} + \frac{\partial^2 c}{\partial r^2}\right) - s\omega^2 \cdot \left(2c + r \cdot \frac{\partial c}{\partial r}\right)$$

26.9    How much (in %) does the observed $s$ value for a globular protein at 10 mg mL$^{-1}$ differ from that extrapolated to zero concentration ($k_s$ = 0.005 mL·mg$^{-1}$)? What concentration would be needed to have only a 0.1% change of $s$?

26.10    A 50 kDa protein is sedimented under equilibrium conditions at 300 K in a buffer of density 1 kg/L. Calculate the rotation speed required to produce a concentration ratio of 100 between the meniscus ($r_0$ = 6.2 cm) and a radial position 5 mm downwards of the meniscus. The partial specific volume can be assumed as $\bar{v}$ = 0.73 L/kg. The concentration ratio is given by

$$\ln\frac{c(r)}{c_0} = \frac{M(1-\bar{v}\rho)\omega^2\left(r^2 - r_0^2\right)}{2RT}$$

26.11    For $r_0$ = 1 cm, $z_0$ = 0.783 cm, $U_1$ = 1000 V, and $\omega$ = 1.1 MHz, what is the minimum $m/z$ that can be stored in the ion trap of a mass spectrometer? The Avogadro constant is 6.022·10$^{23}$ mol$^{-1}$ and 1 V = 1 kg m$^2$ s$^{-2}$ C$^{-1}$.

26.12    In a mass spectrometer we would like to measure the current of an ion with a single positive charge for as long as it takes to reach a signal-to-noise ratio of ten. The noise level of the detector is 1 fA and we assume a counting efficiency of 100%. The ion current is 300 Hz. How long do we have to measure? Hint: For counting processes (Poisson statistics), the signal-to-noise ratio increases with the square root of the integration time:

$$\frac{S}{N} \sim \sqrt{t}$$

26.13    To circumvent the long integration time from the previous question, we use a photomultiplier with a gain of 10$^5$ to boost the signal. How long would we need to measure for $S/N$ = 10, assuming a counting efficiency of 100%?

26.14    The balance group in the iTRAQ reagent has only two atoms, carbon and oxygen. Which isotope combinations are used to achieve the four different masses of 28–31 Da?

26.15    The time of flight of an ion after MALDI is 0.2 ms. The acceleration voltage is 15 kV, and the length of the drift tube is 2 m. Only a single ion type is observed. What is the molecular mass of the ion in Da? Is it a peptide? The elemental charge is 1.602·18·10$^{-19}$ C and 1 Da equates to 1.6605·10$^{-27}$ kg. What would be the time of flight for a peptide of molecular mass 1000 Da under these conditions? Would the same peptide with two charges take half the time?

26.16    Se-Met modified proteins are used for phasing in X-ray crystallography, and $^{15}$N-labeled proteins are used for 2D NMR spectroscopy. How can the percentage of incorporation of these modifications be measured by mass spectrometry?

26.17    What is the expected relative mass peak distribution for ethane? What are the first three relative peak intensities for fullerene C$_{60}$? The natural abundances of $^{12}$C and $^{13}$C are 98.9% and 1.1%, respectively.

26.18   A protein is analyzed by native and denaturing ESI. The following spectra are obtained:

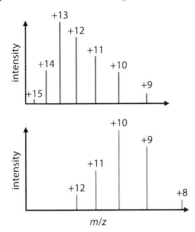

Which of the spectra was recorded for the native protein and why is the ionization pattern different?

26.19   Lysozyme from chicken has a calculated molecular mass of 16238.6 Da. The sequence contains eight cysteine residues. ESI mass analysis gives the following $m/z$ data:

| m/z |
| --- |
| 2706.08 |
| 2319.68 |
| 2029.81 |
| 1804.40 |
| 1624.04 |
| 1476.53 |
| 1353.57 |

Assign the correct charges to the $m/z$ data and calculate the mean of the mass. What could explain the mass difference to the calculated value?

26.20   Why can samples be colored, turbid, or even opaque, or why does absorption or scattering not pose a problem for SPR?

26.21   To reduce mass transport during the dissociation phase in an SPR experiment, someone suggests injecting an inhibitor for either the receptor or the analyte in order to avoid rebinding. What conditions must such an inhibitor meet?

26.22   Why is Tris buffer not suitable for amine coupling?

26.23   Thiol coupling is often used for acidic proteins with a pI of <3.5. Why could that be?

26.24   In some SPR instruments, one *response unit* or *resonance unit* (*RU*) corresponds to 0.0001° shift in SPR angle, which is equivalent to a refractive index change of $10^{-6}$ and a mass change on the sensor surface of about 1 pg mm$^{-2}$. A receptor of molecular mass 50 kDa (for example an antibody Fab) is immobilized on an SPR chip, which leads to an increase in the response by 1000 *RU*. What are the changes in resonance angle and refractive index? What are the weight and molar receptor densities? A ligand of molecular mass 500 Da that is known to bind in a 1:1 stoichiometry to the Fab is passed over the chip and elicits an additional response signal of 5 *RU*. What are the expected signal and, consequently, the fraction of active receptor?

## REFERENCES

Meselson, M. and Stahl, F. W. (1958) The Replication of DNA in *Escherichia coli*. *Proc Natl Acad Sci U S A* **44**(7): 671–682.
· original description of the Meselson-Stahl experiment

Glish, G. L. and Vachet, R. W. (2003) The basics of mass spectrometry in the twenty-first century. *Nat Rev Drug Discov* **2**(2): 140–150.
· review of mass spectrometry

Kramer, K., Sachsenberg, T., Beckmann, B. M., Qamar, S., Boon, K. L., Hentze, M. W., Kohlbacher, O. and Urlaub, H. (2014) Photo-cross-linking and high-resolution mass spectrometry for assignment of RNA-binding sites in RNA-binding proteins. *Nat Methods* **11**(10): 1064–1070.
· identifying protein-RNA interaction sites by photo-crosslinking and mass-spectrometric analysis

Fletcher, J. S., Vickerman, J. C. and Winograd, N. (2011) Label free biochemical 2D and 3D imaging using secondary ion mass spectrometry. *Curr Opin Chem Biol* **15**(5): 733–740.
· description of the SIMS method

Ostrowski, S. G., Van Bell, C. T., Winograd, N. and Ewing, A. G. (2004) Mass spectrometric imaging of highly curved membranes during *Tetrahymena* mating. *Science* **305**(5680): 71–73.
· dramatic change in lipid composition during mating of *Tetrahymena* detected by MS

Fletcher, J. S., Lockyer, N. P., Vaidyanathan, S. and Vickerman, J. C. (2007) TOF-SIMS 3D biomolecular imaging of *Xenopus laevis* oocytes using buckminsterfullerene (C60) primary ions. *Anal Chem* **79**(6): 2199–2206.
· $C_{60}^+$ ions were used to gently etch the surface of *Xenopus* oocytes at ambient temperature and reconstruct a 3D image of triacyl glyceride distribution

Huber, W. and Mueller, F. (2006) Biomolecular interaction analysis in drug discovery using surface plasmon resonance technology. *Curr Pharm Des* **12**(31): 3999–4021.

Navratilova, I. and Hopkins, A. L. (2011) Emerging role of surface plasmon resonance in fragment-based drug discovery. *Future Med Chem* **3**(14): 1809–1820.
· reviews of the physics and applications of SPR with an emphasis on drug design

Hall, D. R., Cann, J. R. and Winzor, D. J. (1996) Demonstration of an upper limit to the range of association rate constants amenable to study by biosensor technology based on surface plasmon resonance. *Anal Biochem* **235**(2): 175–184.
· discusses how mass transport limitation puts an upper limit to the association rate constants measurable by SPR

Lue, R. Y., Chen, G. Y., Hu, Y., Zhu, Q. and Yao, S. Q. (2004) Versatile protein biotinylation strategies for potential high-throughput proteomics. *J Am Chem Soc* **126**(4): 1055–1062.
· intein-mediated biotinylation of proteins

Tsao, K. L., DeBarbieri, B., Michel, H. and Waugh, D. S. (1996) A versatile plasmid expression vector for the production of biotinylated proteins by site-specific, enzymatic modification in *Escherichia coli*. *Gene* **169**(1): 59–64.

Klatt, S., Hartl, D., Fauler, B., Gagoski, D., Castro-Obregon, S. and Konthur, Z. (2013) Generation and characterization of a *Leishmania tarentolae* strain for site-directed *in vivo* biotinylation of recombinant proteins. *J Proteome Res* **12**(12): 5512–5519.
· use of biotin ligase for biotinylation

Navratilova, I., Papalia, G. A., Rich, R. L., Bedinger, D., Brophy, S., Condon, B., Deng, T., Emerick, A. W., Guan, H. W., Hayden, T., Heutmekers, T., Hoorelbeke, B., McCroskey, M. C., Murphy, M. M., Nakagawa, T., Parmeggiani, F., Qin, X., Rebe, S., Tomasevic, N., Tsang, T., Waddell, M. B., Zhang, F. F., Leavitt, S. and Myszka, D. G. (2007) Thermodynamic benchmark study using Biacore technology. *Anal Biochem* **364**(1): 67–77.
· the thermodynamic parameters $K_d$, $\Delta H$, and $\Delta S$ of carbonic anhydrase II measured by SPR and ITC were in agreement, indicating that fixing of the enzyme on the sensor surface does not significantly alter the entropy

Levary, D. A., Parthasarathy, R., Boder, E. T. and Ackerman, M. E. (2011) Protein-protein fusion catalyzed by sortase A. *PLoS One* **6**(4): e18342.
· protein-protein coupling using sortase

Clow, F., Fraser, J. D. and Proft, T. (2008) Immobilization of proteins to biacore sensor chips using *Staphylococcus aureus* sortase A. *Biotechnol Lett* **30**(9): 1603–1607.
· protein coupling to surfaces using sortase

Perspicace, S., Banner, D., Benz, J., Muller, F., Schlatter, D. and Huber, W. (2009) Fragment-based screening using surface plasmon resonance technology. *J Biomol Screen* **14**(4): 337–349.
· small molecule screening of a library identifies inhibitors for the serine protease chymase

Bocquet, N., Kohler, J., Hug, M. N., Kusznir, E. A., Rufer, A. C., Dawson, R. J., Hennig, M., Ruf, A., Huber, W. and Huber, S. (2015) Real-time monitoring of binding events on a thermostabilized human A2A receptor embedded in a lipid bilayer by surface plasmon resonance. *Biochim Biophys Acta* **1848**(5): 1224–1233.
· ligand binding to a GPCR embedded in nanodiscs is studied by SPR

Sota, H., Hasegawa, Y. and Iwakura, M. (1998) Detection of conformational changes in an immobilized protein using surface plasmon resonance. *Anal Chem* **70**(10): 2019–2024.

Boussaad, S., Pean, J. and Tao, N. J. (2000) High-resolution multiwavelength surface plasmon resonance spectroscopy for probing conformational and electronic changes in redox proteins. *Anal Chem* **72**(1): 222–226.
· detection of conformational changes in proteins by SPR

## ONLINE RESOURCES

www.analyticalultracentrifugation.com
· description of the Sedfit program

leonardo.inf.um.es/macromol/programs/programs.htm
· description of the Hydropro program

masspec.scripps.edu/book_toc.php
· Scripps Center for Metabolomics

www.astbury.leeds.ac.uk/facil/MStut/mstutorial.htm
· MS tutorials

www.sprpages.nl
· background information on SPR

# Calorimetry

Calorimetric methods measure the heat released (exothermic process) or taken up (endothermic process) during a reaction and are among the few methods that directly yield the associated changes in thermodynamic functions including $\Delta H$, $\Delta G$, $\Delta S$, and $\Delta C_p$. There are two implementations of calorimetry, isothermal titration calorimetry (ITC) and differential scanning calorimetry (DSC). ITC (Section 27.1) provides thermodynamic information on binding events and chemical reactions, whereas DSC (Section 27.2) finds applications in the thermodynamic characterization of phase transitions.

## 27.1 ISOTHERMAL TITRATION CALORIMETRY

### 27.1.1 GENERAL PRINCIPLE

In isothermal titration calorimetry (ITC), heat is used as a direct probe to follow binding reactions and to obtain binding curves. An isothermal titration calorimeter (Figure 27.1) consists of two cells, one for the sample and one for the reference buffer. The cells are made of a non-corroding metal alloy and are thermally insulated from the surroundings by an *adiabatic jacket*. The sample cell contains one of the binding partners in a suitable buffer. A syringe that holds the second binding partner in the same buffer is inserted into the sample cell. During an isothermal titration, this binding partner is injected in small steps into the sample cell. The syringe rotates throughout the experiment, and a paddle at its end acts as a stirrer to ensure rapid mixing. The reference cell contains the buffer in which the binding partners are dissolved.

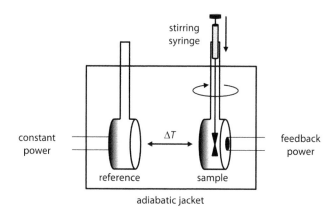

**FIGURE 27.1:   Principle of an ITC.** In ITC, the temperature of the reference cell is held constant (at the chosen temperature of the experiment), and the feedback power is used to compensate for any change in heat released or taken up by the reaction in the measurement cell. The adiabatic jacket keeps the whole setup at the temperature of the experiment. The paddle at the end of the syringe holding the second binding partner serves as a stirrer. Injection of a certain volume from the syringe (usually 3–15 μL at a rate of 0.5 μL s⁻¹) displaces the same volume from the measurement cell. This loss of material has to be taken into account during data analysis.

The primary readout for heat in modern ITC (and DSC) instruments is electric power. Electrical energy (in form of a heating current) is used to keep the reference cell and the sample cell at a constant temperature difference $\Delta T$. The sample cell is kept at a slightly higher temperature than the reference cell by a positive *feedback power*. Maintaining a constant temperature difference between the cells, instead of keeping both cells at the same temperature, offers the advantage that the feedback system can immediately respond to a temperature increase and to a temperature decrease in the reaction cell, simply by adjusting the feedback power. In exothermic reactions, heat is released in the sample cell in each titration step, leading to an increase in temperature. To maintain the constant temperature difference to the reference cell, the feedback power to the sample cell is reduced. For endothermic reactions, on the other hand, the feedback power needs to be increased to compensate for heat taken up. In either case the feedback power remains positive. It is important to equilibrate the entire system at the beginning, and to not start the first injection before a stable baseline of feedback power is recorded. Similarly, after each titration step the baseline has to be reached again before the next injection can follow. In principle, the sample and reference cells could also be maintained at the same temperature throughout the experiment. In this case, a temperature decrease in the sample cell during endothermic reactions would have to be compensated by cooling of the reference cell instead of heating of the sample cell. However, cooling is much slower than the instantaneous reduction in feedback power. Accordingly, the response time of a calorimeter operated in this *T-difference mode* is smaller (a few seconds) than the response time of a calorimeter operating at constant temperature. ITC experiments in the T-difference mode even allow measurement of slow enzymatic reactions (Section 27.1.6).

The raw data from an isothermal titration is a time trace of the feedback power, with a negative peak for each titration step for exothermic reactions (less feedback power required; Figure 27.2), and a positive peak for each step for endothermic reactions (more feedback power required; Figure 27.2 would appear flipped about the horizontal axis). Power equals energy (or work) per time, and has the unit Watt, with 1 W = 1 J s⁻¹. Integration of the electric power over time therefore directly provides the heat exchanged. Before the peaks can be integrated, the raw titration data need to be corrected for a general drift of the instrument baseline. Since introduction of the first titration calorimeter in 1989, ITC instruments have become more and more sensitive, and modern instruments measure heat changes in the μJ range.

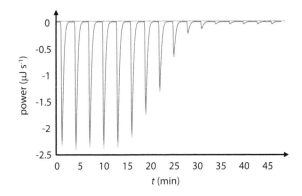

**FIGURE 27.2:    Schematic raw ITC data for an exothermic reaction.** After correction for the instrument baseline drift, the baseline for the titration is horizontal. Heat released in each titration step is measured as a reduced feedback power. At the end of the experiment, only heat of dilution is measured. An endothermic reaction would give a signal flipped along the horizontal axis. The number of injections (10–50) and the equilibration time per injection (1–5 min) determine the total experiment time. For data analysis, each peak is integrated.

During data analysis, the heat exchange in each titration step is calculated by integration of the power curve over time for each injection. From the known concentrations of the binding partners in the sample cell and the syringe, the cell volume, and the volume of the injection, the concentration of both binding partners at each titration step is calculated. Plotting the heat per injection as a function of the molar ratio of the binding partners yields a sigmoidal binding curve (Figure 27.3). This curve can then be analyzed to obtain the binding enthalpy, $\Delta H$, the equilibrium association constant $K_a$ (typically, ITC data are evaluated as $K_a = 1/K_d$), and the binding stoichiometry $n$.

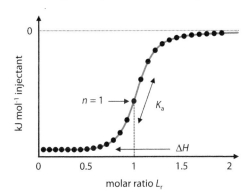

**FIGURE 27.3:    A binding curve resulting from integration of the raw ITC data.** The shape of the curve allows conclusions on the thermodynamic parameters. The inflection point corresponds to the stoichiometry of binding $n$. The constant heat at the beginning of the titration corresponds to $\Delta H$ of the reaction. The steepness of the curve is related to the association constant $K_a$, the affinity.

## 27.1.2  ITC DATA ANALYSIS

For a simple 1:1 binding event of ligand L and macromolecule M, the association constant $K_a$ is

$$K_a = \frac{[ML]}{[M] \cdot [L]} \qquad \text{eq. 27.1}$$

$[L]$, $[M]$, and $[ML]$ are the equilibrium concentrations of free ligand, free macromolecule, and complex, respectively. The starting concentration of ligand is $L_0$, and the starting concentration of the macromolecule is $M_0$. Substituting the mass conservation $[M] = M_0 - [ML]$ and $[L] = L_0 - [ML]$ into eq. 27.1 followed by rearrangement yields

$$[ML]^2 - \left(L_0 + M_0 + \frac{1}{K_a}\right) \cdot [ML] + 4M_0L_0 = 0 \qquad \text{eq. 27.2}$$

This quadratic equation has only one physically sensible solution (the other would yield ML concentrations exceeding $M_0$, which is impossible; see Section 3.2):

$$[ML] = \frac{1}{2} \cdot \left(L_0 + M_0 + \frac{1}{K_a} - \sqrt{\left(L_0 + M_0 + \frac{1}{K_a}\right)^2 - 4M_0L_0}\right) \qquad \text{eq. 27.3}$$

The heat we determine for each titration step of an ITC experiment is related to the additional formation of ML at the ligand concentration $L$ in this step. We therefore differentiate eq. 27.3 with respect to the total ligand concentration $L_0$ and simplify to

$$\frac{d[ML]}{dL_0} = \frac{1}{2} \cdot \left(1 - \frac{L_0 - M_0 + \frac{1}{K_a}}{\sqrt{\left(L_0 + M_0 + \frac{1}{K_a}\right)^2 - 4M_0L_0}}\right) \qquad \text{eq. 27.4}$$

This equation can be further simplified by defining the molar ratio $L_r$ of the binding partners

$$L_r = \frac{L_0}{M_0} \qquad \text{eq. 27.5}$$

and

$$r = \frac{1}{C} = \frac{1}{K_a M_0} \qquad \text{eq. 27.6}$$

$C$ is the *Wiseman constant*. We will see later in this chapter that the Wiseman constant helps us in designing ITC experiments. In order to substitute these helper variables into eq. 27.4 we multiply both numerator and denominator with $1/M_0$ and rearrange to

$$\frac{d[ML]}{dL_0} = \frac{1}{2} \cdot \left(1 - \frac{L_r - 1 + r}{\sqrt{L_r^2 - 2L_r(1-r) + (1+r)^2}}\right) \qquad \text{eq. 27.7}$$

We now have to relate the change in concentration of the ML complex in each titration step, d[ML], to the measured heat d$q$. The heat d$q$ required for compensating the release or take-up of heat during the reaction depends on the amount of complex d(ML) that is formed, the reaction enthalpy $\Delta H$, and the volume of the sample cell $V_0$:

$$dq = d[ML] \cdot \Delta H \cdot V_0 \qquad \text{eq. 27.8}$$

Upon combining eq. 27.7 and eq. 27.8 we arrive at the expression that describes our ITC binding curve (Figure 27.3):

$$\frac{dq}{dL_0} = \frac{\Delta H \cdot V_0}{2} \cdot \left(1 - \frac{L_r - 1 + r}{\sqrt{L_r^2 - 2L_r(1-r) + (1+r)^2}}\right) \qquad \text{eq. 27.9}$$

Non-linear least square fitting of eq. 27.9 to the binding curve then yields $\Delta H$, the molar ratio of the binding partners $L_r$, and the association constant $K_a$ (through the parameter $r$, see eq. 27.6). In practice, a correction is needed to account for the volume that is displaced from the sample cell by injection of ligand from the syringe (Figure 27.1). To keep this correction small, it is necessary to limit injections to small volumes relative to the sample cell volume $V_0$. For injections of 3–15 μL into a 1.4 mL cell, the effect of displacement of solution from the sample cell can be neglected. Injection volumes can be kept small by placing the more soluble binding partner into the syringe at a high concentration.

Eq. 27.9 is valid for 1:1 binding of ligand to the macromolecule. For $n$ independent binding sites, eq. 27.9 needs to be modified by multiplication of $M_0$ by $n$. The parameter $n$ can be varied in the fitting procedure to determine the stoichiometry of the reaction. However, it is important to note that any error in the concentrations of the binding partners will affect $n$, and experimental values for $n$ typically differ from integer numbers. If $n$ is 0.9 or 1.1, for example, we may look at a 10% error in concentration determination for one binding partner while the concentration for the other is accurate. If both concentrations are inaccurate, $n$ may not differ much from integer (if the errors compensate each other) or it may differ even more (if the errors add up). Accurate concentration determination of the binding partners is therefore essential for a reliable determination of binding stoichiometries by ITC.

The free energy of binding $\Delta G^0$ can be calculated from $K_a$, and the change in entropy, $\Delta S$, can be calculated according to the Gibbs-Helmholtz-equation from $\Delta G$ and $\Delta H$ (see Section 2.5; eq. 2.105). With these parameters, an almost complete set of thermodynamic parameters for the binding reaction is obtained from a single experiment. The last parameter missing in the thermodynamic profile is the heat capacity $\Delta C_p$. The heat capacity is defined by the temperature dependence of $\Delta H$ (Section 2.3; eq. 2.62). $\Delta C_p$ can be obtained from a set of ITC experiments that determine $\Delta H$ at different temperatures (Figure 27.4).

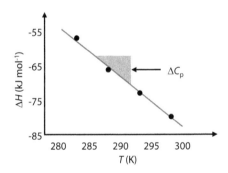

**FIGURE 27.4:    Heat capacity as the temperature dependence of ΔH.** Data are for thymidine binding to thymidine kinase (www.bindingdb.org).

A substantial change in heat capacity upon ligand binding can convert the reaction from enthalpy-driven to entropy-driven as a function of temperature (see Section 2.1.5). For instance, if $\Delta C_p$ amounts to –500 J mol$^{-1}$ K$^{-1}$, $\Delta H$ can change by –20 kJ mol$^{-1}$ over a temperature range from 5–25°C, which for many binding reactions is on the same scale as $\Delta H$ itself. An endothermic reaction at low temperature may become exothermic at higher temperatures. It also shows that a single $\Delta H$ value measured at a certain temperature is of little value unless $\Delta C_p$ is known or unless it is compared with other experiments that have been carried out at the same temperature.

### 27.1.3 ORIGIN OF ENTHALPIC CHANGES

$\Delta H$ values for the binding of a ligand to a single site on a macromolecule are distributed over a wide range from +50 kJ mol$^{-1}$ to –140 kJ mol$^{-1}$, although most values are in the range between –10 kJ mol$^{-1}$ and –20 kJ mol$^{-1}$ (data from www.bindingdb.org). The enthalpy measured in an ITC experiment is the sum of all heat changes that occur. Thus, $\Delta H$ is a composite of all enthalpic changes during each injection, including the breaking and formation of (hydrogen) bonds, van der Waals, ionic, and other polar and non-polar interactions (see Section 15.3). Decomposing the measured $\Delta H$ into individual molecular contributions is almost impossible. Apart from the binding event itself, protonation and/or deprotonation of the binding partners may occur. Heat changes associated with protonation events can be determined from a set of ITC experiments performed in different buffers at the same temperature and pH. If $n$ protons are exchanged upon ligand binding, the total enthalpy $\Delta H$ is the enthalpy due to ligand binding $\Delta H_{ligand}$ plus $n$ times the ionization enthalpy $\Delta H_{prot}$ of the buffer:

$$\Delta H = \Delta H_{ligand} + n \cdot \Delta H_{prot} \qquad \text{eq. 27.10}$$

To determine the number of protons that are exchanged, the observed enthalpy changes $\Delta H$ are plotted as a function of the tabulated buffer ionization enthalpies $\Delta H_{prot}$ (Table 27.1). Linear regression will give $n$ as the slope and $\Delta H_{ligand}$ (the sum of all remaining contributions to $\Delta H$) as the intercept with the $y$-axis.

**TABLE 27.1**
**$\Delta H$ and $\Delta C_p$ changes for the dissociation of protonated buffers in 0.1 M KCl at 25°C.**

| buffer | p$K_A$ | $\Delta H_{prot}$ (kJ mol$^{-1}$) | $\Delta C_p$ (J mol$^{-1}$ K$^{-1}$) |
|---|---|---|---|
| acetate | 4.62 | 0.49 ± 0.02 | 2128 ± 2 |
| MES | 6.07 | 15.53 ± 0.03 | 16 ± 2 |
| cacodylate | 6.14 | 21.96 ± 0.02 | 278 ± 2 |
| glycerol 2-phosphate | 6.26 | 20.72 ± 0.02 | 2179 ± 2 |
| PIPES | 6.71 | 11.45 ± 0.04 | 19 ± 4 |
| ACES | 6.75 | 31.41 ± 0.05 | 227 ± 4 |
| phosphate | 6.81 | 5.12 ± 0.03 | 2187 ± 3 |
| BES | 7.06 | 25.17 ± 0.07 | 2 ± 5 |
| MOPS | 7.09 | 21.82 ± 0.03 | 39 ± 3 |
| imidazole | 7.09 | 36.59 ± 0.06 | 216 ± 5 |
| TES | 7.42 | 32.74 ± 0.03 | 233 ± 3 |
| HEPES | 7.45 | 21.01 ± 0.07 | 49 ± 5 |
| EPPS | 7.87 | 21.55 ± 0.05 | 56 ± 4 |
| triethanolamine | 7.88 | 33.59 ± 0.04 | 48 ± 3 |
| tricine | 8.00 | 31.97 ± 0.05 | 245 ± 4 |
| bicine | 8.22 | 27.05 ± 0.05 | 2 ± 4 |
| TAPS | 8.38 | 41.49 ± 0.06 | 23 ± 5 |
| CAPS | 10.39 | 48.54 ± 0.07 | 29 ± 6 |

*Source:* Reproduced from Fukada & Takahasi (1998) *Proteins* 33(2):159–166, with permission.

Because many ligand binding reactions involve exchange of protons, the measured $\Delta H$ can be changed strongly by as much as 20 kJ mol$^{-1}$ (compare PIPES and ACES in Table 27.1) simply by changing the identity of the buffer at the same pH. Sometimes, the overall (apparent) $\Delta H$ value can be further decomposed into contributions from metal binding or conformational changes of the protein.

In principle, the enthalpy change $\Delta H$ can also be determined from a series of ITC experiments at different temperatures. The van't Hoff equation (eq. 3.11) relates the temperature dependence of $K_a$ to the enthalpy change $\Delta H$, and $\Delta H$ can be determined from the slope of a van't Hoff plot ($K_a$ versus $1/T$, see Figure 3.1). Van't Hoff plots are linear only if $\Delta C_p$ is constant over the respective temperature range (see Section 3.1.1). A difference between the van't Hoff enthalpy $\Delta H_{vH}$ and the enthalpy $\Delta H_{cal}$ determined directly from analysis of the calorimetric titration is an indication that other processes in addition to binding contribute to $\Delta H_{cal}$. In the absence of such complications the two enthalpies should be identical.

## 27.1.4 Practical Considerations

Since ITC directly measures heat, it is important that any change in heat that does not come from the binding event is corrected for. Common sources of unspecific heat generation are frictional heat, heats of dilution from unmatched buffers of the binding partners, and heats from side-reactions such as aggregation or unfolding.

Aggregates are a common source for frictional heat, and should be removed from all samples by centrifugation or at least by filtering prior to use. Aggregates can also form during the ITC experiment. Protein aggregation can often be prevented by a change in buffer composition, temperature, or pH. Some proteins aggregate when the solution is avidly stirred. In this case, the stirring speed of the syringe can be reduced. To ensure that mixing is complete before the next injection, the time interval between injections might have to be increased. On the other hand, the resulting longer total experiment times may not be tolerated by aggregation-prone or unstable proteins. Sometimes, mild detergents or higher salt concentrations can help avoid aggregation by reducing hydrophobic and ionic interactions, respectively. The frictional heat from aggregates decreases the signal-to-noise ratio of the baseline, which makes them easy to detect (although the experiment has failed at this stage). Baseline correction is an essential step in data processing, and data that suffer from the presence of aggregates are difficult if not impossible to integrate. If protein stability continues to be a problem, the single injection method might be a more suitable approach that allows the determination of thermodynamic parameters within minutes (Box 27.1).

---

### BOX 27.1: SINGLE INJECTION METHOD.

A typical ITC experiment of 1 h duration can be shortened to about 20 min by the single injection method. Instead of using several discrete injections and waiting until the baseline is reached after each step, the binding partner is added in a single, slow and continuous injection (Figure 27.5). This approach is only applicable if the association rate of the binding partners is sufficiently high, such that complex formation is complete within the 1–2 s response time of the instrument. The injection velocity in such an experiment is drastically decreased to 0.1 $\mu Ls^{-1}$ in order to ensure instant mixing of all reactant delivered by the syringe. The recorded curve is a binding isotherm in real time. The constant heat released at the beginning of the injection is caused by instantaneous binding of all ligand that is injected. At higher ligand concentrations, later in the injection, saturation is reached. The heat release decreases and finally returns to the baseline when the binding partner in the cell is saturated with ligand. For data analysis, the time scale is converted to ligand concentration, and the $K_a$ value can be determined from the binding curve using eq. 27.9. The overall saving in experimental time needs to be balanced with the usually less accurate data from single injection experiments compared to the multiple injection method.

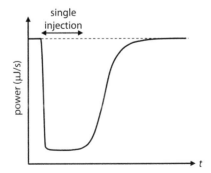

**FIGURE 27.5:**   **Single injection method.**

Towards the end of the ITC experiment, when 100% complex has been formed, additional injections should lead to no or only very small heat changes. Any heat change that is observed after saturation has been reached can typically be attributed to heats of dilution. Sizeable heats of dilution arise when the buffers of the binding partners do not match exactly with respect to concentration, ionic strength, or pH, and such differences should hence be avoided. To ensure identical buffer conditions for both binding partners, macromolecular binding partners should be dialyzed into the same buffer. Small molecule ligands that cannot be dialyzed should be as pure as possible and be dissolved in the same buffer as the macromolecule. It is important to monitor the pH during dilution of small molecules. Concentrated nucleotide solutions, for example, can easily change the pH of buffers with low capacity. If one of the two binding partners is not very soluble, it should be placed in the cell where a lower concentration is sufficient. Some small molecule ligands are only soluble in organic solvents, often DMSO, at the concentrations required for ITC. Organic solvents, particularly DMSO, have large heats of dilution. In this case, matching the buffers by including similar concentrations of organic solvents in the syringe and sample cell is necessary. If buffer matching is not possible, the heats of dilutions can be estimated in separate titrations of the binding partners into buffer and subtracting the two reference datasets from the binding data.

Side-reactions can also affect the measured heat. Unfolding of a protein during the ITC experiment adds the $\Delta H$ of unfolding to the measured heat and severely distorts the signal. Although ITC cells are made of inert materials, unstable or hydrophobic proteins may unfold and attach to the walls of the cell. This effect can be exacerbated by the presence of high detergent concentrations that are necessary to stabilize membrane proteins in aqueous buffers. Unfolding usually leads to a smooth baseline drift. As unfolding changes the total concentration of the active molecule in the cell, the determined values for $\Delta H$, $K_a$, and the stoichiometry $n$ are incorrect. Chemical reactions in the sample cell also pose problems. For example, a contaminating protease may generate heat by catalyzing hydrolysis of the binding partners.

In principle, equilibrium dissociation constants from the millimolar to the nanomolar range can be measured by ITC, but the accuracy of these constants depends on the concentrations of the binding partners used in the experiments. As a rule of thumb, the binding partner in the cell should be 10-fold more concentrated than the expected $K_d$, and the concentration of the molecule in the syringe should be 10-fold larger than that of the molecule in the cell. Under these conditions, the volume change in the cell due to repeated injections is kept small, and the calculation of $\Delta H$, $K_a$, and $n$ from the shape of the binding curve is straightforward. A more quantitative measure to design an ITC experiment is the Wiseman constant $C$, the product of the association constant $K_a$ and the concentration of the molecule in the cell $M_0$. The Wiseman constant influences the shape of the binding curve and hence its information content (Figure 27.6).

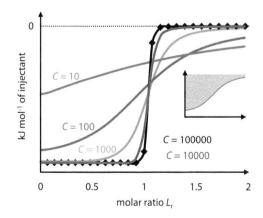

**FIGURE 27.6:    Wiseman constant and the shape of ITC curves.** Simulated ITC data for 1:1 stoichiometry, $M_0 = 0.1$ mM, and $K_a$ varying by a factor of ten from $10^9$ M$^{-1}$ (black curve with diamonds; $C = 10000$) to $10^5$ M$^{-1}$ (flat, blue curve; $C = 10$). The high-affinity measurement has no data points in the transition region. The flat low-affinity data does not allow accurate estimation of $n$, or $K_a$. The green curve with $C = 1000$ represents the best combination of $K_a$ and $M_0$. From the curves with $C = 10$ or $C = 100$, $\Delta H$ cannot be obtained from the $y$-axis intercept, but has to be calculated from the total area under the curve (inset).

When the Wiseman constant $C$ is between 100 and 1000, binding curves show a pronounced sigmoidal shape and are moderately steep. Binding curves that are too steep will not have sufficient data points (if any) in the transition region, making determination of $K_a$ difficult. Curves that are too flat, on the other hand, will result in large errors on $n$, $\Delta H$, and $K_a$. For a binding curve with a Wiseman constant of $C = 100$, the enthalpy change $\Delta H$ cannot be derived from the intercept of the binding curve with the $y$-axis. In these cases, $\Delta H$ can still be obtained from the total area under the binding curve (Figure 27.6). $K_a$ and $n$, on the other hand, cannot be determined accurately.

### 27.1.5 MEASURING HIGH AFFINITIES WITH ITC BY COMPETITION

The higher the affinity (the larger $K_a$), the lower is the required concentration of the binding partner in the sample cell to reach a Wiseman constant between 100 and 1000. With decreasing concentrations, the exchanged heat becomes smaller and smaller, and the measured signal might become too small to be detected reliably. In this case, displacement of a weakly bound ligand by a high-affinity ligand is an alternative to still determine the $K_a$ value. For this type of experiment, a low-affinity ligand B with known $K_{a,low}$ and $\Delta H_{low}$ is added to the macromolecule in the measurement cell, and the resulting low-affinity complex is then titrated with the high-affinity ligand. The heat is calculated from the integrated peaks to generate the displacement titration curve that can be analyzed for the true binding constant of the high-affinity ligand. The known $K_{a,low}$ value for the low-affinity ligand is fixed during data analysis. The concentration of the high-affinity complex is described by the solution of a cubic equation (see Section 19.5.5.1; Box 19.10). The apparent high-affinity binding constant $K_{app}$ is given by

$$K_{app} = \frac{K_{a,high}}{1 + K_{a,low} \cdot [B]}$$

eq. 27.11

The apparent binding constant $K_{app}$ is smaller than $K_{a,high}$ for the high-affinity interaction and can therefore be measured with greater accuracy. By adjusting the concentration of the low-affinity ligand $[B]$ in the sample cell, the apparent binding constant $K_{app}$ can be modulated.

The apparent enthalpy change is approximated by

$$\Delta H_{app} = \Delta H_{high} - \Delta H_{low} \cdot \frac{K_{a,low} \cdot [B]}{1 + K_{a,low} \cdot [B]}$$

eq. 27.12

If the low-affinity ligand B is present in large excess over its binding partner, the free concentration [B] in eq. 27.11 and eq. 27.12 can be replaced by the known total concentration of $B_0$.

### 27.1.6 MEASURING MICHAELIS-MENTEN ENZYME KINETICS WITH ITC

ITC can also be used to measure the steady-state enzyme kinetic parameters $k_{cat}$ and $K_M$ for enzymatic (Sections 11.1 and 11.2). The enzymatic reaction must be associated with a heat change and must be slow on the time scale of the instrument response. Typically, the enzyme is placed in the measurement cell, and the substrate is injected from the syringe. The heat is released over a larger time-window compared to a titration experiment, while the enzyme turns over its substrate. Once all substrate has been converted to products, the signal returns to the baseline (Figure 27.7).

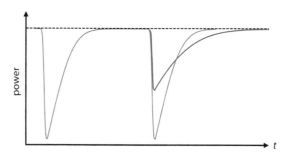

FIGURE 27.7: **Enzyme reactions in ITC.** Schematic raw ITC data upon injection of a small amount of substrate into a reaction cell containing an enzyme. The peak areas are larger than for a single injection in a binding equilibrium, and the signal takes much longer to return to the baseline. A smaller signal in a second, identical injection indicates whether inhibition of the enzyme by the product generated during the first injection occurs. The red curve would indicate such product inhibition.

The experiment is useful to derive $\Delta H$ of the reaction and to test for product inhibition by a second injection of substrate. If the second heat change is identical to the first, no product inhibition has occurred. This aspect is further discussed later in this chapter. To evaluate the data we differentiate eq. 27.8 by time (product formation now substitutes for the complex ML) and see how the feedback power depends on the formation of product P in the measuring cell:

$$power = \frac{dq}{dt} = \frac{d[P]}{dt} \cdot \Delta H \cdot V_0$$

eq. 27.13

Integration of the power signal $dq/dt$ over the time the system needs to return to the baseline yields the total heat $\Delta H$ evolved during the enzymatic reaction:

$$\Delta H = \frac{1}{V_0 \cdot [P]} \cdot \int_{t=0}^{t=\infty} \frac{dq}{dt} dt$$

eq. 27.14

The determination of $\Delta H$ is possible even if the substrate concentrations are not saturating. In order to obtain Michaelis-Menten parameters, the substrate concentration in the reaction cell has to be increased such that the enzyme is saturated. Subsequent injections are performed well before all substrate has been turned over. Upon each further injection, the baseline is then shifted in a stepwise manner (Figure 27.8).

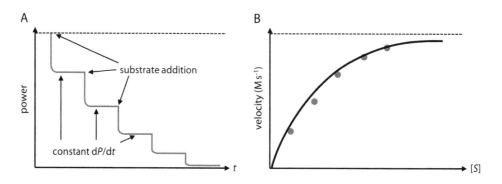

FIGURE 27.8: **Schematic ITC data upon injection of an excess of ligand into a reaction cell containing an enzyme.** A: The continuous reaction under steady-state conditions shifts the baseline after each substrate injection. B: The rates calculated from the baseline shifts are plotted as a function of substrate concentration [S]. Note that $\Delta H$ cannot be derived from this type of experiment because the substrate is not completely converted prior to starting the next injection. $\Delta H$ needs to be estimated from an experiment that reaches the baseline (Figure 27.7).

The shifts of the baseline can be converted into reaction velocities according to eq. 27.15, which we obtain by re-arrangement of eq. 27.13 and equating to the Michaelis-Menten equation. Under substrate saturation the reaction velocity $v$ is constant (Figure 27.8).

$$v = \frac{d[P]}{dt} = \frac{1}{\Delta H \cdot V_0} \frac{dq}{dt} = \frac{k_{cat} \cdot E_0 \cdot S}{S + K_M} \qquad \text{eq. 27.15}$$

The reaction velocities are plotted as a function of substrate concentration $[S]$, and the curve obtained is described by eq. 27.15 to determine $k_{cat}$ and $K_M$. For such experiments, the data need to be corrected for dilution of the total enzyme concentration $E_0$ by the displacement of solution from the cell in every injection.

A continuous enzyme assay is also possible with ITC. In this type of experiment, a large amount ($[S] \gg K_M$) of substrate is injected, such that product is formed at the maximum rate $v_{max}$. The signal is recorded until substrate depletion, where the power signal returns to the baseline (Figure 27.9). Under these conditions, inhibition of the enzyme can be investigated by performing the reaction at different concentrations of inhibitor. The areas under curves with or without inhibitor are identical, yielding $\Delta H$ of the reaction (eq. 27.14). In addition, the inhibition constant $K_i$ can be calculated from the reaction velocities (see Section 11.6).

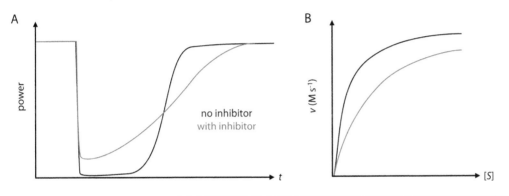

**FIGURE 27.9:**   **Substrate turnover and enzyme inhibition.** A: Power signal for substrate turnover by an enzyme without (black) and with (blue) inhibitor. B: Conversion of the heat signal to reaction velocities as a function of substrate concentration $[S]$.

The actual substrate concentration $[S](t)$ at any time $t$ along the curve can be calculated by subtraction of the actual product concentration $[P](t)$ from the total substrate concentration $S_0$ that was injected. $P$ can be calculated from eq. 27.14, which gives

$$[S](t) = S_0 - [P](t) = S_0 - \frac{\int_{t=0}^{t} \frac{dq}{dt} dt}{\Delta H \cdot V_0} \qquad \text{eq. 27.16}$$

The resulting hyperbolic curves can be described by the Michaelis-Menten equation (right-hand side of eq. 27.15) to obtain $k_{cat}$ and $K_M$ in the absence and presence of inhibitor.

Kinetic information on binding and dissociation reactions can also be obtained from ITC experiments by analysis of the shape of the individual injection profiles (Box 27.2).

## BOX 27.2: PRE-STEADY-STATE KINETIC INFORMATION FROM ITC – kinITC.

The analysis of ITC curves for thermodynamic parameters described above only uses the information of the integrated heat, but does not take into account the shape of the curve, i.e. the velocity with which the signal returns to the baseline. The shape of this signal is determined by the instrument response, by the finite duration of injection and mixing, and by the rate constants of the reaction. If the response time of the instrument as well as the duration of the injection and mixing are shorter than the time scale of the association reaction, the rate constants of the binding equilibrium can be determined by analyzing the return of the signal to the baseline. The fixed instrument response therefore puts an upper limit on the rate constants that can be determined. To extract the rate constants for binding and dissociation, a global analysis of the return to the baseline from all peaks can be performed in terms of a kinetic 1:1 binding model. This approach is termed kinetic ITC or kinITC, a label-free method that has the advantage of not requiring a spectroscopic signal (Figure 27.10). Burnouf *et al.* (2012) *J. Am. Chem. Soc.* 134(1): 559–565.

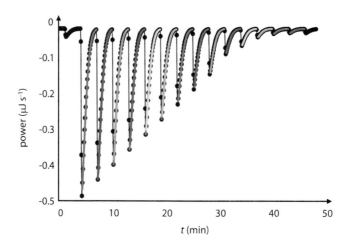

**FIGURE 27.10:**   **kinITC.** Only the exponential parts of the power signal are evaluated by exponential fits.

## 27.2 DIFFERENTIAL SCANNING CALORIMETRY

We have seen in the previous section how ITC is used to measure the heat associated with binding or catalytic events. In contrast, *differential scanning calorimetry* (DSC) measures the heat released or taken up during structural or phase transitions. DSC is most often used to study temperature-induced protein and nucleic acid folding and unfolding. DSC directly measures thermodynamic quantities of structural transitions, and does not rely on a spectroscopic probe. Measurements can be performed under pressure, which expands the experimentally accessible temperature range to $>100°C$.

### 27.2.1 GENERAL PRINCIPLE

Similar to ITC instruments, a DSC instrument consists of a reference cell, filled with buffer, and a sample cell, filled with the molecule of interest in buffer. Both cells are enclosed by an adiabatic jacket. During the measurement, both cells are heated, and held at a constant temperature difference $\Delta T$ by an electrical feedback system. When a structural transition such as a protein unfolding event takes place in the sample cell, heat is absorbed (protein unfolding is endothermic). To maintain a constant temperature difference $\Delta T$ between the reference and sample cells, the heating power to the sample cell is adjusted; in this case it is increased. The heating power is the primary signal and is proportional to the overall (excess) heat capacity $C_p$, which represents the temperature dependence of the enthalpy $H$. The raw data of a DSC measurement is a curve of $C_p$ as a function of $T$, the *thermogram*. After correction for the instrument baseline (caused by the heat capacity of the instrument itself), this thermogram displays a linear baseline before and after the structural transition, and a bell-shaped peak for the transition (Figure 27.11). The instrument baseline is recorded in a separate experiment with buffer in the sample cell and subtracted from the raw DSC data.

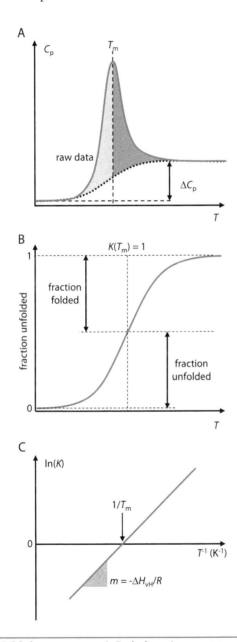

**FIGURE 27.11:   Schematic DSC thermograms.** A: Endothermic two-state protein unfolding transition corrected for the instrument baseline. The total area under the curve yields the enthalpy of the reaction (after normalization with the protein concentration). If the offsets of the two baselines before and after the experiment are different, there is an overall change in heat capacity ($\Delta C_p$). Should the slopes of the pre- or post-transition baselines not be zero, then $\Delta C_p$ itself is temperature-dependent. The areas shaded in gray correspond to the fractions of folded and unfolded protein at a given temperature. B: At any given temperature, the ratio of the areas in (A) is the equilibrium constant. Plotting the unfolded fractions as a function of temperature results in an unfolding curve. C: The van't Hoff plot resulting from the equilibrium constants $K(T)$ in panel (B). The enthalpies from the thermogram and the van't Hoff plot should be identical.

$T_m$, the "melting" temperature of the transition, is accurately obtained from the peak of the thermogram. For single-domain proteins, $T_m$ is the temperature at which half of the molecules are unfolded. $T_m$ correlates with the overall protein stability. It should be similar to $T_m$ values from temperature-induced unfolding experiments followed by spectroscopic methods such as CD, UV, and fluorescence. Further analysis of thermograms to obtain thermodynamic parameters is only valid if the measured transitions are reversible. To test for reversibility, a second thermogram is measured after the first, where the temperature is gradually decreased from the highest to the lowest temperature. For reversible transitions both thermograms are identical, whereas differences are indicative of irreversible transitions. In this case, the information content of the thermogram is limited to the apparent melting temperature $T_m$, the maximum of the peak.

The total change in heat capacity $\Delta C_p$ during the transition is obtained from the difference between the baselines. A positive value for $\Delta C_p$ (Figure 27.11) indicates solvation of non-polar surfaces and is a hallmark for protein unfolding. Indeed, $\Delta C_p$ seems to be the thermodynamic parameter dominating the hydrophobic effect, which drives protein folding. For some proteins, a correlation between $\Delta C_p$ and the amount of (non-polar) surface area that is buried upon complex formation or protein folding has been inferred, but overall it is difficult to relate surface properties to heat capacity. On the other hand, negative values for $\Delta C_p$ signify solvation of polar surfaces, typically seen with sequence-specific binding of proteins to DNA. For example, binding of the *trp* repressor to the *trp* operator DNA has a large negative $\Delta C_p$ of almost $-4$ kJ mol$^{-1}$ K$^{-1}$. By contrast, non-specific DNA binding by proteins does not change $\Delta C_p$ very much.

For reversible transitions, the thermodynamic parameters for the structural transition can be determined from the thermogram. The change in heat capacity $\Delta C_p$ during the transition is obtained from the difference between the baselines before and after the transition. The overall change in enthalpy $\Delta H$ is obtained from the peak area by integration (Figure 27.11) and normalization for protein concentration.

$$\Delta H(T) = \int_{T_0}^{T} C_p(T)\,\mathrm{d}T \qquad \text{eq. 27.17}$$

Because the baseline after the transition often does not return to the value before the transition, the shape of the baseline during the transition needs to be extrapolated before the area under the peak can be integrated to determine $\Delta H$. The most basic model assumes a jump of the baseline from the value before the transition to the final value at the melting temperature $T_M$, whereas more sophisticated models involve a linear or quadratic interpolation. However, for every data point in the transition region, a mixture of two states is present, and the baseline is perhaps best assumed as a sigmoidal curve (Figure 27.11) that reflect the progress of the protein unfolding transition (see Figure 16.47). The choice of baseline type affects the area of the transition, and hence the value of $\Delta H$, but these errors are often smaller than other uncertainties such as inaccurate determination of protein concentrations.

The equilibrium constant $K(T)$ for the structural transition can be calculated for any given temperature in the experiment within the transition region (Figure 27.11). For each temperature $T$ the two peak areas below and above $T$ are integrated separately. The area above $T$ (right-hand side) reflects the relative amount of folded protein, the area below $T$ (left-hand side) represents the fraction that has unfolded at this temperature. The ratio of these areas corresponds to $K(T)$. As usual, $\Delta G(T)$ can be calculated from $K(T)$, and the extrapolation to the standard temperature yields $\Delta G^0$. Similar to ITC, $\Delta S^0$ is then calculated according to the Gibbs-Helmholtz equation from $\Delta G^0$ and $\Delta H^0$. The complete set of thermodynamic parameters for the structural transition is obtained from one experiment, this time including $\Delta C_p$.

Sometimes multiphasic transitions are observed (Figure 27.12). In this case, the thermogram needs to be described by two or more peaks, and individual $\Delta C_p$, $T_m$, and $\Delta H$ values, as well as $K(T)$, $G(T)$ and $S(T)$ for each transition are determined.

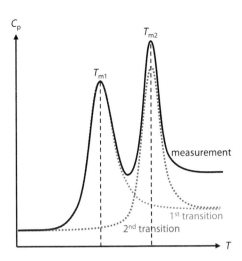

FIGURE 27.12:   **Multiple transitions.** The separate unfolding of two domains is shown. When the $T_m$ values are very different, the sequential unfolding of a two-domain protein can be deconvoluted into single transitions (blue and red dashes).

### 27.2.2 Two-State Unfolding of Macromolecules

DSC can measure the cooperative unfolding of any macromolecule. We would now like to derive a fitting function to describe a DSC thermogram. In the following we consider a two-state protein unfolding of the form:

$$ N \underset{}{\overset{K_{unfold}}{\rightleftharpoons}} U $$

scheme 27.1

N and U are the folded and unfolded states, respectively. The equilibrium constant for unfolding can either be expressed in terms of the concentrations of the folded and unfolded states, or by the fraction of unfolded protein $f_U$:

$$ K_{unfold} = K_U = \frac{[U]}{[N]} = \frac{f_u}{1 - f_u} $$

eq. 27.18

Solving eq. 27.18 for $f_U$ yields

$$ f_u = \frac{K_u}{1 + K_u} $$

eq. 27.19

The fraction of unfolded protein helps us in the next step to describe the total enthalpy of the system. The Hess law states that the enthalpies of a system are additive (Box 2.5). Thus, the total molar enthalpy $H$ in the DSC cell is

$$ H(T) = H_N(T) + f_u \cdot \Delta H_U(T) $$

eq. 27.20

The total enthalpy is the enthalpy of the folded state, $H_N(T)$, plus the enthalpy change of unfolding, $\Delta H_U(T)$, weighted by the fraction of unfolded protein $f_U$. If there are several unfolding transitions, an equivalent number of terms $f_{U,i} \cdot \Delta H_i(T)$ has to be added to eq. 27.20. The total molar heat capacity at constant pressure, $C_p$, is the temperature derivative of eq. 27.20:

$$ C_p(T) = C_{pN}(T) + f_u \cdot \Delta C_{pU}(T) + \Delta H_U(T) \cdot \left( \frac{\partial f_u}{\partial T} \right) $$

eq. 27.21

$\Delta C_{pN}(T)$ and $\Delta C_{pU}(T)$ are the heat capacities of the folded and unfolded states, respectively. In this equation, the differential $\partial f_U/\partial T$ can be calculated by differentiating eq. 27.19:

$$\frac{\partial f_U}{\partial T} = \frac{1}{\left(1+K_U\right)^2} \cdot \frac{\partial K_U}{\partial T} \qquad \text{eq. 27.22}$$

Realizing that $K$ is an exponential function of the temperature, we can re-write the differential $\partial K_U/\partial T$:

$$\frac{\partial K_U}{\partial T} = K_U \cdot \frac{\partial \ln K_U}{\partial T} \qquad \text{eq. 27.23}$$

The van't Hoff enthalpy is given by the temperature dependence of the equilibrium constant:

$$\frac{\partial \ln K_U}{\partial T} = \frac{\Delta H_{vH}}{RT^2} \qquad \text{eq. 27.24}$$

So we can re-write eq. 27.21

$$C_p(T) = C_{pN}(T) + \frac{K_U}{1+K_U} \cdot \Delta C_{pU}(T) + \Delta H_U(T) \cdot \frac{K_U}{\left(1+K_U\right)^2} \cdot \frac{\Delta H_{vH}}{RT^2} \qquad \text{eq. 27.25}$$

This equation is the general formulation of the heat capacity of a two-state system at a specific temperature. More transitions require extra terms for each $K_U$, so eq. 27.25 can be adapted for more complicated unfolding scenarios. If we assume that the van't Hoff enthalpy equals the calorimetric enthalpy of unfolding $\Delta H_U(T)$, eq. 27.25 simplifies to

$$C_p(T) = C_{pN}(T) + \frac{K_U(T) \cdot \Delta C_{pU}(T)}{1+K_U(T)} + \frac{K_U \cdot \left[\Delta H_U(T)\right]^2}{\left(1+K_U(T)\right)^2 \cdot RT^2} \qquad \text{eq. 27.26}$$

All that we have to do now is to express $\Delta H_U(T)$ and $K_U(T)$ in terms of the heat capacity and some easily accessible parameters. We choose as parameters the enthalpy of unfolding $\Delta H_U(T_m)$ at the transition midpoint temperature $T_m$ (Figure 27.11). We have derived expressions for $\Delta H_U(T)$ and $K_U(T)$ before in this book. $\Delta H_U(T)$ is accessible from integration of Kirchhoff's law (eq. 2.69):

$$\int_{T_m}^{T} d(\Delta H_U) = \int_{T_m}^{T} \Delta C_{pU} \cdot dT \qquad \text{eq. 27.27}$$

This yields

$$\Delta H_U(T) = \Delta H_U(T_m) + \Delta C_{pU} \cdot (T - T_m) \qquad \text{eq. 27.28}$$

$\Delta H_U(T_m) = \Delta H_m$ is the enthalpy of unfolding at the melting temperature $T_m$. From eq. 16.13 (see Section 16.3.4), we know an expression for $\ln K_U(T)$:

$$-RT \ln K_U(T) = \Delta H_m \cdot \left(1 - \frac{T}{T_m}\right) + \Delta C_{pU} \cdot \left(T - T_m - T \cdot \ln \frac{T}{T_m}\right) \qquad \text{eq. 27.29}$$

$K_U(T)$ is readily obtained from eq. 27.29 by rearrangement:

$$K_U = e^{-\frac{\Delta H_m}{RT}\left(1-\frac{T}{T_m}\right) - \frac{\Delta C_{pU}}{RT}\left(T - T_m - T \cdot \ln \frac{T}{T_m}\right)} \qquad \text{eq. 27.30}$$

Substituting eq. 27.29 and eq. 27.30 into eq. 27.26 and expressing the temperature dependence of $\Delta C_{\mathrm{pN}}(T)$ by a linear function, the baseline, as $\Delta C^0_{\mathrm{pN}} + m_{\mathrm{N}} \cdot T$, yields quite a complex function that we will refrain from reproducing here. It allows fitting of the measured DSC data with the five fit parameters $\Delta C^0_{\mathrm{pN}}$, $m_{\mathrm{N}}$, $T_m$, $\Delta H_m$, and $\Delta C_{\mathrm{pU}}$. If necessary, a second baseline for the unfolded state can be added to the fitting equation. Baseline correction can also be done before fitting, so the values of $C_{\mathrm{pN}}(T)$ and $\Delta C_{\mathrm{pU}}$ are effectively set to zero, and the $C_{\mathrm{p}}(T)$ equation (eq. 27.25) simplifies to

$$C_p(T) = \frac{K_U \cdot \left[\Delta H_U(T)\right]^2}{\left(1 + K_U(T)\right)^2 \cdot RT^2} \qquad \text{eq. 27.31}$$

with

$$\Delta H_U(T) = \Delta H_U(T_m) = \Delta H_m \qquad \text{eq. 27.32}$$

and

$$K_U = e^{-\frac{\Delta H_m}{RT}\left(1 - \frac{T}{T_m}\right)} \qquad \text{eq. 27.33}$$

Substituting eq. 27.32 and eq. 27.33 into eq. 27.31, we arrive at the more manageable expression for $C_{\mathrm{p}}(T)$ that has only $T_{\mathrm{m}}$ and $\Delta H_{\mathrm{m}}$ as fit parameters:

$$C_p(T) = \frac{e^{-\frac{\Delta H_m}{RT}\left(1 - \frac{T}{T_m}\right)} \cdot \left[\Delta H_m\right]^2}{\left(1 + e^{-\frac{\Delta H_m}{RT}\left(1 - \frac{T}{T_m}\right)}\right)^2 \cdot RT^2} \qquad \text{eq. 27.34}$$

### 27.2.3 TWO-STATE UNFOLDING WITH SUBUNIT DISSOCIATION

A frequently observed transition that is amenable to DSC is the coupled dissociation of oligomers and unfolding of the constituent monomers. In contrast to the separate transitions that are possible with multi-domain proteins (Figure 27.12), the DSC thermogram in this case shows only a single transition. A slight variation of the expression of $K(T)$ accommodates the oligomer-monomer equilibrium. We assume an oligomer of $n$ subunits $\mathrm{N}_n$ that simultaneously dissociates and unfolds into $n$ unfolded monomers U.

$$\mathrm{N}_n \underset{}{\overset{K_{\text{unfold}}}{\rightleftharpoons}} n\,\mathrm{U} \qquad \text{scheme 27.2}$$

The overall equilibrium constant is then

$$K = \frac{[U]^n}{[N_n]} = \frac{f_U}{1 - f_U} \qquad \text{eq. 27.35}$$

The total molar concentration in terms of oligomers is

$$c_{\text{tot}} = [N_n] + \frac{[U]}{n} \qquad \text{eq. 27.36}$$

and the fraction $f_{\mathrm{U}}$ of unfolded monomer is:

$$f_U = \frac{[U]}{n \cdot c_{\text{tot}}} \qquad \text{eq. 27.37}$$

Combining eq. 27.36 and eq. 27.37 to express $[N_n]$ and $[U]$ in terms of $f_U$ yields the following expression for $K$:

$$K = \frac{f_U^{\,n}}{1 - f_U} \cdot n^n \cdot c_{tot}^{\,n-1}$$

eq. 27.38

We know that at the melting temperature $T_m$, the fraction of unfolded monomers is 0.5:

$$K(T_m) = 0.5^{n-1} \cdot n^n \cdot c_{tot}^{\,n-1}$$

eq. 27.39

Unlike in the previous two-state unfolding transition (eq. 27.29), $K(T_m)$ is now different from unity. In a dimer-monomer dissociation/unfolding equilibrium with $n = 2$, $K(T_m)$ equals $2 \cdot c_{tot}$. Thus, $K(T)$ from eq. 27.29 modifies to

$$K(T) = 0.5^{n-1} \cdot n^n \cdot c_{tot}^{\,n-1} \cdot e^{-\frac{\Delta H_m}{RT}\left(1 - \frac{T}{T_m}\right) - \frac{\Delta C_{pU}}{RT}\left(T - T_m - T \cdot \ln\frac{T}{T_m}\right)}$$

eq. 27.40

$\Delta C_{pU}$ in the equation above represents the change in heat capacity that accompanies both dissociation of the oligomer, and coupled unfolding of the monomers. The same holds true for $\Delta H_m$. Once $K(T)$ is known from eq. 27.40, numerical evaluation is required to solve eq. 27.38 for $f_U(T)$. The temperature-dependent change $\partial f_U / \partial T$ in eq. 27.21 is then calculated and used to describe $C_p(T)$ of the DSC experiment. The same simplifications with respect to baseline subtraction and temperature-independence of $\Delta C_{pU}$ as discussed above can be applied to this case.

## QUESTIONS

27.1    An ITC with a cell volume of 1.4 mL is filled with 10 µM protein solution. The injection volumes are 5 µL of a 50 µM ligand solution every 2 min. After the titration is finished and evaluated, a $K_a$ of $10^9$ M$^{-1}$ and $\Delta H$ of −40 kJ mol$^{-1}$ is found. What is the heat evolved during the first injection and what average power is recorded by the instrument?

27.2    For the competitive titration of a low-affinity protein-ligand complex $MB$ characterized by $K_{a,low}$ with a high-affinity ligand $A$, the apparent equilibrium constant is

$$K_{app} = \frac{K_{a,high}}{1 + K_{a,low} \cdot [B]}$$

Using the laws of mass action for the two equilibria, derive the formula above.

27.3    The data given in the table below are from thymidine binding to thymidine kinase. Estimate the heat capacity from these data.

| $T$ (°C) | $\Delta H$ (kJ mol$^{-1}$) |
|---|---|
| 25 | −79.91 |
| 20 | −73.08 |
| 15 | −66.36 |
| 10 | −57.12 |

---

## REFERENCES

Velazquez-Campoy, A., Ohtaka, H., Nezami, A., Muzammil, S. and Freire, E. (2004) Isothermal titration calorimetry. *Curr Protoc Cell Biol* **17**: Unit 17.8.
· practical review on ITC

Wiseman, T., Williston, S., Brandts, J. F. and Lin, L. N. (1989) Rapid measurement of binding constants and heats of binding using a new titration calorimeter. *Anal Biochem* **179**(1): 131–137.
· introduction of the first microcalorimeter

Fukada, H. and Takahashi, K. (1998) Enthalpy and heat capacity changes for the proton dissociation of various buffer components in 0.1 M potassium chloride. *Proteins* **33**(2): 159–166.
· treatment of changes in enthalpy and heat capacity for common buffers

Armstrong, K. M. and Baker, B. M. (2007) A comprehensive calorimetric investigation of an entropically driven T cell receptor-peptide/major histocompatibility complex interaction. *Biophys J* **93**(2): 597–609.
· binding of a T-cell receptor to a peptide-MHC complex is analyzed by ITC - the enthalpy is a composite of metal binding and/or conformational changes

Horn, J. R., Russell, D., Lewis, E. A. and Murphy, K. P. (2001) Van't Hoff and calorimetric enthalpies from isothermal titration calorimetry: are there significant discrepancies? *Biochemistry* **40**(6): 1774–1778.
· shows that the van't Hoff enthalpy and $\Delta H$ from ITC should not differ

Burnouf, D., Ennifar, E., Guedich, S., Puffer, B., Hoffmann, G., Bec, G., Disdier, F., Baltzinger, M. and Dumas, P. (2012) kinITC: a new method for obtaining joint thermodynamic and kinetic data by isothermal titration calorimetry. *J Am Chem Soc* **134**(1): 559–565.
· introduces the idea of obtaining kinetic information from ITC data

Prabhu, N. V. and Sharp, K. A. (2005) Heat capacity in proteins. *Annu Rev Phys Chem* **56**: 521–548.
· review on heat capacity with special attention to protein folding

Ladbury, J. E., Wright, J. G., Sturtevant, J. M. and Sigler, P. B. (1994) A thermodynamic study of the *trp* repressor-operator interaction. *J Mol Biol* **238**(5): 669–681.
· binding of the *trp* repressor to its operator DNA sequence has a large negative heat capacity

# APPENDIX

# Prefixes, Units, Constants

## 28.1 PREFIXES

The prefixes listed in Table 28.1 span approximately the orders of magnitudes one might encounter in biological systems. For instance, a single molecule in a liter of water formally has a concentration of 1.7 yM (yoctomolar) while the mass of a proton is about 1.7 yg (yoctogram). At the other end of the spectrum we arrive at the yotta prefix when we consider the Avogadro constant of $6.022 \cdot 10^{23}$ particles per mol, which for example equals the number of water molecules in 18 mL.

| TABLE 28.1<br>Prefixes. | | | | | |
|---|---|---|---|---|---|
| name | prefix | scale | name | prefix | scale |
| yocto | y | $10^{-24}$ | yotta | Y | $10^{24}$ |
| zepto | z | $10^{-21}$ | zetta | Z | $10^{21}$ |
| atto | a | $10^{-18}$ | exa | E | $10^{18}$ |
| femto | f | $10^{-15}$ | peta | P | $10^{15}$ |
| pico | p | $10^{-12}$ | tera | T | $10^{12}$ |
| nano | n | $10^{-9}$ | giga | G | $10^{9}$ |
| micro | μ | $10^{-6}$ | mega | M | $10^{6}$ |
| milli | m | $10^{-3}$ | kilo | k | $10^{3}$ |
| centi | c | $10^{-2}$ | hecto | h | $10^{2}$ |
| deci | d | $10^{-1}$ | deka | da | $10^{1}$ |

## 28.2 SI (SYSTÈME INTERNATIONAL) OR BASE UNITS

**TABLE 28.2**
**SI units.**

| name | symbol | definition |
|---|---|---|
| length | 1 m | meter |
| mass | 1 kg | kilogram |
| time | 1 s | second |
| current | 1 A | Ampère |
| temperature | 1 K | Kelvin |
| amount | 1 mol | mole |
| luminous intensity | 1 cd | Candela |

There are only seven SI or base units (Table 28.2). Note that the kilogram is the only SI unit that is defined with a prefix (kilo). Only the base units kilogram, second, and Kelvin are defined independently. The other four are derived from these independent units: the meter is related to the second by the speed of light, the mol is related to the kilogram, and the definition of both the Ampère and the Candela involves a combination of meter, kilogram, and second.

## 28.3 DERIVED UNITS USED IN THIS BOOK

**TABLE 28.3**
**Derived units.**

| name | symbol | description | unit |
|---|---|---|---|
| °Celsius | °C | temperature | T (in K) – 273.15 = T (in °C) |
| Coulomb | C | charge | $A \cdot s$ |
| Farad | F | electric capacitance | $A^2 \cdot s^4 \cdot kg^{-1} \cdot m^{-2}$ |
| Hertz | Hz | frequency | $s^{-1}$ |
| Joule | 1 J = 1 Nm | energy, work, heat | $kg \cdot m^2 \cdot s^{-2}$ |
| Joule per mole | 1 J mol$^{-1}$ | molar energy | $kg \cdot m^2 \cdot s^{-2} \cdot mol^{-1}$ |
| Newton | 1 N | force | $kg \cdot m \cdot s^{-2}$ |
| Pascal | 1 Pa = 1 N m$^{-2}$ | pressure | $kg \cdot m^{-1} \cdot s^{-2}$ |
| Radian | rad | angle | $m \cdot m^{-1}$ |
| Tesla | T | magnetic field strength | $kg \cdot s^{-2} \cdot A^{-1}$ |
| Volt | 1 V = 1 W A$^{-1}$ | voltage | $kg \cdot m^2 \cdot s^{-3} \cdot A^{-1}$ |
| Watt | W = 1 J s$^{-1}$ | power | $kg \cdot m^2 \cdot s^{-3}$ |

The voltage denominates an electrical potential difference. The Coulomb describes either the electric unit charge or a quantity of electricity. Some units are not based on the SI system but enjoy common use (Table 28.4).

**TABLE 28.4**
**Non-SI derived units.**

| name | symbol | description | definition |
|---|---|---|---|
| Ångstrøm | Å | length | $10^{-10}$ m |
| atmosphere | atm | pressure | 101325 Pa |
| bar | bar | pressure | $10^5$ Pa |
| calorie | cal | energy | 4.184 J |
| Dalton | Da | molecular mass | $1.6605 \cdot 10^{-27}$ kg |
| Debye | D | dipole moment | $3.33564 \cdot 10^{-30}$ C·m |
| electron volt | eV | energy | $1.602 \cdot 10^{-19}$ J |
| inch | in | length | 0.0254 m |
| liter | L | volume | $10^{-3}$ m³ |
| mile | mi | length | 1.609.344 m |
| molar | M | concentration | mol·L⁻¹ |

*Note:*   While these units are not recommended, some of them are useful because they result in numbers that are easy to calculate with (compare for instance standard atmospheric pressure of 1 bar with the SI-based $10^5$ Pa). The Da, L, and M are also used in this book. The Dalton (1 Da) is 1/12$^{th}$ of the mass of carbon $^{12}$C.

## 28.4  NATURAL CONSTANTS USED IN THIS BOOK

**TABLE 28.5**
**Natural constants.**

| name | symbol | value |
|---|---|---|
| **universal** | | |
| gravitational acceleration | $g$ | 9.80665 m·s⁻² ≈ 9.81 m s⁻² |
| speed of light in vacuum | $c$ | 299792458 m·s⁻¹ ≈ 3 · 10⁸ m s⁻¹ |
| Planck constant | $h$ | $6.62607 \cdot 10^{-34}$ J·s |
| reduced Planck constant | $\hbar = h/(2\pi)$ | $1.05457 \cdot 10^{-34}$ J·s |
| **electromagnetic** | | |
| vacuum permeability | $\mu_0$ | $4\pi \cdot 10^{-7}$ N·A⁻² ≈ 1.2566 μN·A⁻² |
| vacuum permittivity | $\varepsilon_0 = 1/(\mu_0 c^2)$ | $8.85419 \cdot 10^{-12}$ F·m⁻¹ ≈ 8.854 pF·m⁻¹ |
| elementary charge | $e$ | $1.60218 \cdot 10^{-19}$ C |
| Bohr magneton | $\mu_B = e\,\hbar/(2m_e)$ | $9.27401 \cdot 10^{-24}$ J·T⁻¹ |
| nuclear magneton | $\mu_N = e\,\hbar/(2m_p)$ | $5.05078 \cdot 10^{-27}$ J·T⁻¹ |
| **atomic and nuclear** | | |
| electron mass | $m_e$ | $9.10938 \cdot 10^{-31}$ kg |
| proton mass | $m_p$ | $1.67262 \cdot 10^{-27}$ kg |
| neutron mass | $m_n$ | $1.67493 \cdot 10^{-27}$ kg |
| **physico-chemical** | | |
| **atomic mass** | $m_u$ | $1.66054 \cdot 10^{-27}$ kg |
| Avogadro's number | $N_A$ | $6.02214 \cdot 10^{23}$ mol⁻¹ |
| Boltzmann constant | $k_B$ | $1.38065 \cdot 10^{-23}$ J·K⁻¹ |
| Faraday constant | $F = N_A \cdot e$ | 96485.3365 C·mol⁻¹ |
| general gas constant | $R = k_B \cdot N_A$ | 8.31446 J·mol⁻¹·K⁻¹ |

*Note:*   The molar volume of an ideal gas at $T$ = 273.15 K and $p$ = 101325 Pa (1 atm pressure) is 22.413968 L· mol⁻¹.

# Mathematical Concepts Used in This Book

The mathematics included in this chapter is treated only to the depth necessary to follow the transformations described in the book and to solve the questions. It is incomplete, and many of the topics will be familiar to the reader. They are merely summarized here for quick reference.

## 29.1 SUMS AND PRODUCTS

Infinite sums and products can be represented in short forms:

sum:
$$\sum_{i=1}^{n} x_i = x_1 + x_1 + \ldots + x_n$$
eq. 29.1

For example, the root mean square distance between two superimposed molecules is defined as the following sum:

$$\text{rmsd} = \sqrt{\frac{1}{N} \cdot \sum_{i=1}^{N} \delta_i^2}$$
eq. 29.2

where $\delta_i$ is the distance between the $i^{\text{th}}$ atom pair and $N$ is the total number of atom pairs chosen for the calculation, e.g. equivalent $C_\alpha$ atoms or backbone atoms in proteins and phosphorous or ribose atoms in nucleic acids.

product:
$$\prod_{i=1}^{n} x_i = x_1 \cdot x_1 \cdot \ldots \cdot x_n$$
eq. 29.3

## 29.2  QUADRATIC EQUATION

The two solutions of a quadratic equation of the form

$$a \cdot x^2 + b \cdot x + c = 0 \qquad \text{eq. 29.4}$$

are given by

$$x_{1/2} = \frac{-b \pm \sqrt{b^2 - 4ac}}{2a} \qquad \text{eq. 29.5}$$

Usually only one of the solutions makes physical sense. A cubic equation of the form

$$a \cdot x^3 + b \cdot x^2 + c \cdot x + d = 0 \qquad \text{eq. 29.6}$$

can also be solved analytically, but needs distinguishing of several cases. These are described in Box 19.10.

## 29.3  BINOMIAL COEFFICIENTS

Polynomial expansion of the binomial power

$$P(x) = (1 - x)^n \qquad \text{eq. 29.7}$$

is often useful in calculus and combinatorics. The coefficient for the $k^{\text{th}}$ term is calculated using the "$n$ choose $k$" formula:

$$\binom{n}{k} = \frac{n!}{k!(n-k)!} \qquad \text{eq. 29.8}$$

Arrangement of the coefficients in $n$ rows of $k$ colums with $k$ ranging from zero to $n$ gives Pascal's triangle. The coefficients of the row $k + 1$ are the sums of the coefficients to the left and right of row $k$ (Figure 29.1).

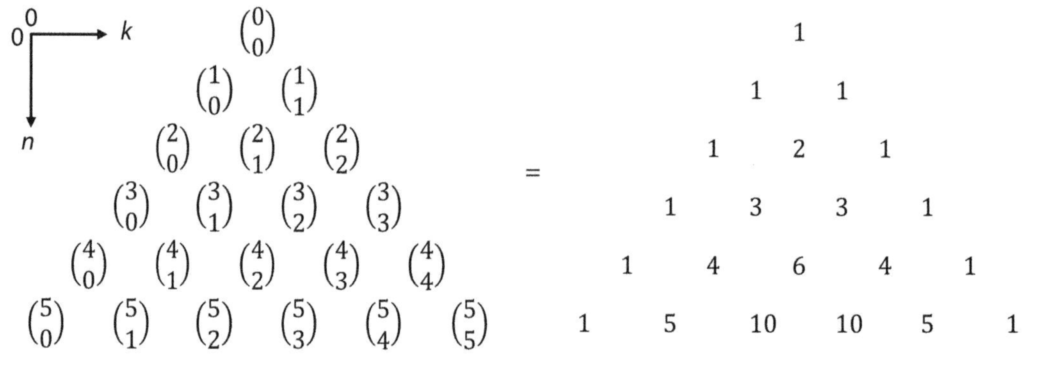

**FIGURE 29.1:**  **Pascal's triangle.** Triangular array of binomial coefficients.

The polynomial expansion is therefore

$$(a+b)^n = \sum_{k=0}^{n} \binom{n}{k} \cdot a^k \cdot b^{n-k} \qquad \text{eq. 29.9}$$

The relative intensity distributions of isotopologs in mass spectrometry (Section 26.1.5.8), the number of entropic microstates (Section 2.4.5), and the number of species in enzymes with several interacting active sites (Section 11.4) follow the binomial theorem.

## 29.4 TRIGONOMETRY

In a rectangular triangle (Figure 29.2), the following rules apply:

$$\sin\beta = \frac{b}{c} \qquad \cos\beta = \frac{a}{c} \qquad \tan\beta = \frac{b}{a} = \frac{\sin\beta}{\cos\beta}$$

eq. 29.10

For non-rectangular triangles the sine and cosine rule apply:

sine rule:
$$\frac{a}{\sin\alpha} = \frac{b}{\sin\beta} = \frac{c}{\sin\gamma}$$

eq. 29.11

cosine rule:
$$a^2 = b^2 + c^2 - 2bc\cos\gamma$$
$$b^2 = a^2 + c^2 - 2ac\cos\beta$$
$$c^2 = a^2 + b^2 - 2ab\cos\alpha$$

eq. 29.12

Thus, if $\gamma$ is 90° we obtain the Pythagorean theorem $a^2 + b^2 = c^2$ as a special case of the cosine rule. The angles in a triangle add to 180°, and for the two unique angles these conversions can be useful:

$$\sin\alpha \pm \sin\beta = 2\sin\frac{1}{2}(\alpha\pm\beta)\cos\frac{1}{2}(\alpha\mp\beta)$$

eq. 29.13

$$\cos\alpha + \cos\beta = 2\cos\frac{1}{2}(\alpha+\beta)\cos\frac{1}{2}(\alpha-\beta)$$

eq. 29.14

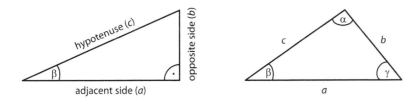

**FIGURE 29.2:   Rectangular and irregular triangle.** Angles are often named arbitrarily, but if the sides of the triangle have letters, the angle opposite to the side is named with the Greek symbol of the side. Thus, the angle opposite the $b$-side is $\beta$.

## 29.5 LOGARITHMS AND EXPONENTIALS

Exponentials and their inverse functions, the logarithms, are ubiquitous in nature. Logarithms are particularly useful to linearize and thus compress large value ranges. The pH and $pK_A$ notation (see Section 3.3) is based on the decadic logarithm, while many thermodynamic functions contain the natural logarithm.

Some useful operations on exponential terms include:

$$e^{\left(a^b\right)} \neq e^{\left(a\cdot b\right)} = \left(e^a\right)^b$$

eq. 29.15

$$e^a \cdot e^b = e^{\left(a+b\right)}$$

eq. 29.16

$$\frac{e^a}{e^b} = e^a \cdot \left(e^b\right)^{-1} = e^a \cdot e^{-b} = e^{(a-b)}$$

eq. 29.17

The operation given in eq. 29.17 follows directly from a combination of eq. 29.15 and eq. 29.16. The base of the exponential function must be the same for these operations to work. The two most frequently encountered bases for exponentials in nature are Euler's number $e$ and ten. In the rare case one encounters exponentials of different bases but with the same exponent, the following relation holds:

$$a^x \cdot b^x = \left(a \cdot b\right)^x$$

eq. 29.18

Logarithms and exponential functions are inverse functions to one another. The graphs of inverse functions are related to one another by mirror symmetry along the first bisectrix $y = x$. Thus, the natural logarithm $\ln(x)$ is the inverse function of $e^x$ (Figure 29.3). Analogously, the decadic logarithm $\log(x)$, sometimes written $\lg(x)$, is the inverse function of $10^x$.

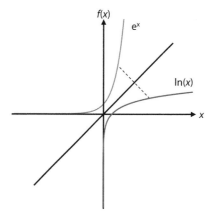

**FIGURE 29.3:** **Inverse functions.** A plot of the exponential function $e^x$ and the natural logarithm $\ln(x)$ shows that these functions are related by a mirror operation along the bisecting line $y = x$.

Since exponentials can have any base, so can logarithms. The base is usually just not written, because from the sign *ln* or *log* the base is clear ($e$ and 10). For other bases, the logarithm is written as a subscript to *log*, for instance

$$\log_e a = \ln a$$

eq. 29.19

or

$$\log_{10} a = \log a$$

eq. 29.20

The inverse operation of logarithms and exponentials can be used to convert the base of any logarithm into another base:

$$\log_b a = \frac{\log_b a}{\log_b b} = \frac{\ln a}{\ln b} = \frac{\log a}{\log b}$$

eq. 29.21

The term $\log_b b$ equals unity, so if we choose a new base $b$, we can switch between logarithms, for instance

$$\log_{10} a = \log a = \frac{\ln a}{\ln 10}$$

eq. 29.22

A simple factor of $\ln 10 = 2.303$ relates the natural and decadic logarithm. This factor is frequently observed in thermodynamic calculations.

Operations for logarithms are similar to those for exponentials.

1$^{st}$ law:
$$\ln a^c = \ln\left(a^c\right) = \ln\left(a\right)^c = c \cdot \ln a \neq \left(\ln a\right)^c$$
eq. 29.23

It is important that if there are parentheses they enclose the argument of the logarithm. Potentiation has priority, so if the parentheses enclose the logarithm itself the exponent cannot be taken as a factor in front of the logarithm.

2$^{nd}$ law:
$$\ln a + \ln b = \ln\left(a \cdot b\right)$$
eq. 29.24

3$^{rd}$ law:
$$\ln a - \ln b = \ln\left(\frac{a}{b}\right)$$
eq. 29.25

The third law is a consequence of the first two:

$$-\ln b = \ln\left(b^{-1}\right) = \ln\frac{1}{b}$$
eq. 29.26

As a consequence of eq. 29.25 and eq. 29.26 we have

$$\ln\left(\frac{a}{b}\right) = -\ln\left(\frac{b}{a}\right)$$
eq. 29.27

and

$$\sum_i \ln\left(x_i\right) = \ln\prod_i x_i$$
eq. 29.28

## 29.6 DIFFERENTIATION AND INTEGRATION

The first derivative of a function $f(x)$ of a single variable $x$ is written as $df(x)/dx$ or $f'(x)$, the second derivative is written as $d^2f(x)/dx^2$, etc. The first derivative defines the slope of f(x) at the point $(x|f(x))$. Setting the first derivative to zero allows calculation of the *maxima* ($d^2f(x)/dx^2 < 0$) and *minima* ($d^2f(x)/dx^2 > 0$) for the function. If the second derivative $d^2f(x)/dx^2 = 0$ and $d^3f(x)/dx^3 \neq 0$ the function has an *inflection point* at $(x|f(x))$. A few helpful rules are reiterated in the following.

Differentiation of the product of two functions:

product:
$$\frac{d\left(u\left(x\right) \cdot v\left(x\right)\right)}{dx} = u'\left(x\right) \cdot v\left(x\right) + u\left(x\right) \cdot v'\left(x\right)$$
eq. 29.29

Differentiation of the ratio of two functions:

ratio:
$$\frac{d\left(\dfrac{u\left(x\right)}{v\left(x\right)}\right)}{dx} = \frac{u'\left(x\right) \cdot v\left(x\right) - u\left(x\right) \cdot v'\left(x\right)}{\left(v\left(x\right)\right)^2}$$
eq. 29.30

A composite of two functions, denoted $u(x) \circ v(x)$, is differentiated using the chain rule:

composition:
$$\frac{d\big(u(x) \circ v(x)\big)}{dx} = u'\big(v(x)\big) \cdot v'(x)$$
eq. 29.31

The rule is used as many times as necessary with functions that have more than two composites. As an example, the function

$$f(x) = e^{\cos x^2}$$
eq. 29.32

is a triple composite of $e^u$, $u = \cos v$, and $v = x^2$ and therefore differentiates to

$$f'(x) = \frac{d}{du} e^u \cdot \frac{d}{dv} \cos v \cdot \frac{d}{dx}\big(x^2\big) = e^{\cos x^2} \cdot \big(-\sin x^2\big) \cdot (2x)$$
eq. 29.33

Functions with two or more variables, i.e. $y = f(x_1, x_2)$ can have partial or total differentials. The partial differential $\partial/\partial_i$ for variable $x_i$ is obtained when differentiating the function for $x_i$ while treating the others as constants:

$$\left( \frac{\partial f(x_1, x_2)}{\partial x_1} \right)_{x_2}$$
eq. 29.34

The total differential is the sum of all partial differentials.

$$\frac{dy}{dx_i} = \left( \frac{\partial f(x_1, x_2)}{\partial x_1} \right)_{x_2} + \left( \frac{\partial f(x_1, x_2)}{\partial x_2} \right)_{x_1}$$
eq. 29.35

Note that the letter d signifies an infinitesimal change, while the letter $\partial$ indicates the partial differential. A further source of confusion is the Greek letter $\delta$, which signifies a state function (Section 2.1).

Differential equations contain both a function and one or more of its derivatives. The highest derivative determines the order of the differential equation. A second-order differential equation therefore has the form $y = f(x) = k_1 \cdot d/dx + k_2 \cdot d^2/dx^2$. Kinetic rate laws (Chapters 6 and 7) are derived by establishing the differential equations for the change of concentrations of the reactants. Simpler equations can be solved by separating the variables and integration.

Integration of a function yields the area under the curve. General rules for integration include:

factoring out constants:
$$\int c \cdot f(x) dx = c \cdot \int f(x) dx$$
eq. 29.36

rule of sums:
$$\int \big(u(x) + v(x)\big) dx = \int u(x) dx + \int v(x) dx$$
eq. 29.37

product rule:
$$\int \big(u'(x) \cdot v(x)\big) dx = u(x) \cdot v(x) - \int u(x) \cdot v'(x) \cdot dx$$
eq. 29.38

substitution rule:
$$\int u\big(v(x)\big) dx = \int u(w) \frac{dw}{w'}$$
eq. 29.39

where $v(x) = w$ is a clever choice of substitution that results in a product that can be integrated. The integration variable changes from d$x$ to d$w$ according to $w' = \mathrm{d}w/\mathrm{d}x$. For example the following function can be integrated by substituting $w = (ax + b)$ and d$x = \mathrm{d}w/a$:

$$\int f(x)\,\mathrm{d}x = \int \frac{1}{ax+b}\,\mathrm{d}x = \int \frac{1}{w}\frac{\mathrm{d}w}{a} = \frac{1}{a}\int \frac{\mathrm{d}w}{w} = \frac{1}{a}\cdot \ln w = \frac{\ln(ax+b)}{a} \qquad \text{eq. 29.40}$$

In the last step the substitution is reverted.

Some useful integrals and derivatives are shown in Table 29.1. The integration constants have been omitted for clarity.

**TABLE 29.1**
**Useful integrals and differentials.**

| $\dfrac{\mathrm{d}f(x)}{\mathrm{d}x}$ | $f(x)$ | $F(x)=\int f(x)\,\mathrm{d}x$ |
|:---:|:---:|:---:|
| $0$ | $n \in \mathbb{R}$ | $n \cdot x$ |
| $n \cdot x^{n-1}$ | $x^n$ | $\dfrac{1}{n+1}\cdot x^{n+1}$ |
| $n \cdot e^{nx}$ | $e^{nx}$ | $\dfrac{1}{n}e^{nx}$ |
| $\ln n \cdot n^x$ | $n^x\,(n>0)$ | $\dfrac{n^x}{\ln n}$ |
| $\dfrac{n}{x}$ | $n \cdot \ln x$ | $n \cdot x \cdot \ln x - n \cdot x$ |
| $x^{n-1}\cdot(n\cdot\ln x+1)$ | $x^n \cdot \ln x\,(n\geq 0)$ | $\dfrac{x^{n+1}}{n+1}\cdot\left(\ln x - \dfrac{1}{n+1}\right)$ |
| $n \cdot \cos(nx)$ | $\sin(nx)$ | $-\dfrac{\cos(nx)}{n}$ |
| $\cosh x$ | $\sinh x$ | $\cosh x$ |
| $2\cdot\sin x\cdot\cos x$ | $\sin^2 x$ | $\dfrac{1}{2}\cdot(x-\sin x\cdot\cos x)$ |
| $-2\cdot\sin x\cdot\cos x$ | $\cos^2 x$ | $\dfrac{1}{2}\cdot(x+\sin x\cdot\cos x)$ |
| $1+\tan^2 x$ | $\tan x$ | $-\ln|\cos x|$ |
| $n \cdot \cos(2nx)$ | $\sin nx \cdot \cos nx$ | $-\dfrac{\cos^2(nx)}{2n}$ |
| $\dfrac{u''(x)\cdot u(x)-\left(u'(x)\right)^2}{\left(u(x)\right)^2}$ | $\dfrac{u'(x)}{u(x)}$ | $\ln|u(x)|$ |
| $\left(u(x)\right)^n\cdot\left(u''(x)+n\cdot\dfrac{\left(u'(x)\right)^2}{u(x)}\right)$ | $u'(x)\cdot\left(u(x)\right)^n$ | $\dfrac{1}{n+1}\cdot\left(u(x)\right)^{n+1}$ |

## 29.7  PARTIAL FRACTIONS

Expansion into partial fractions is useful to simplify a rational function before integrating it. The method is based on the fact that any rational function can be expressed as the sum of a polynomial function $P(x)$ and another rational function $Q(x)$:

$$P(x) + Q(x) = \sum_{i=1}^{n} a_i \cdot x^i + \frac{Z(x)}{N(x)}$$

eq. 29.41

where $Z(x)$ is of lower degree than $N(x)$. Arriving at this form may require polynomial long division. $Q(x)$ is of the form

$$Q(x) = \frac{Z(x)}{N(x)} = \sum_{j=1}^{m} \frac{c_j}{\left(x - x_j\right)^k}$$

eq. 29.42

The numbers $x_j$ in the rational function are the poles, where $Q(x)$ reaches infinity. They are also the zeros of multiplicity $k$ of $N(x)$. If the zeros of $N(x)$ are known, expansion into partial fractions reduces to finding the coefficients $c_j$. An example that is used in the kinetics part of this book is the function:

$$\frac{1}{\left(x^2 - 2xA_0B_0 + A_0^2B_0^2\right)} = \frac{1}{\left(A_0 - x\right)\left(B_0 - x\right)} = \frac{c_1}{\left(A_0 - x\right)} + \frac{c_2}{\left(B_0 - x\right)}$$

eq. 29.43

The poles are $A_0$ and $B_0$ and $Z(x) = 1$. $Q(x)$ should be written as a sum of ratios with the same denominators in the form shown to the right of the eq. 29.43, which can then be integrated without much difficulty. Multiplying eq. 29.43 with the denominator $(A_0 - x)(B_0 - x)$, and rearrangement yields

$$1 = c_1 \cdot \left(B_0 - x\right) + c_2 \cdot \left(A_0 - x\right) = -\left(c_1 + c_2\right) \cdot x + c_1 B_0 + c_2 A_0$$

eq. 29.44

Equating the coefficients in eq. 29.44 requires that

$$1 = c_1 B_0 + c_2 A_0$$

eq. 29.45

and

$$0 = -\left(c_1 + c_2\right) \cdot x$$

eq. 29.46

This system of linear equations leads to $c_1 = -c_2$ and

$$c_1 = -c_2 = -\frac{1}{A_0 - B_0}$$

eq. 29.47

Going back to eq. 29.43 then yields the end result:

$$\frac{1}{\left(A_0 - x\right)\left(B_0 - x\right)} = \frac{1}{\left(A_0 - B_0\right)} \cdot \left(\frac{1}{\left(B_0 - x\right)} - \frac{1}{\left(A_0 - x\right)}\right)$$

eq. 29.48

## 29.8  L'HÔPITAL'S RULE

This rule is useful for finding the limes or limiting value of a function which is the ratio of two other functions and where both of these functions approach zero or diverge to infinity. An example is the function

$$f(x) = \frac{g(x)}{h(x)} = \frac{\sin(x)}{x} \qquad \text{eq. 29.49}$$

The function $f(x)$ is undefined for $x = 0$, but the limes of $f(x)$ approaches unity for $x \rightarrow 0$. Simply using the limes for $g(x)$ and $h(x)$, as is usually done, is not helpful because both are zero.

$$\lim_{x \to 0} f(x) = \frac{\lim_{x \to 0} g(x)}{\lim_{x \to 0} h(x)} = \frac{0}{0} \qquad \text{eq. 29.50}$$

L'Hôpital's rule states that the limes of the function $f(x)$ is the same as the ratio of the limes of the derivatives of the component functions:

$$\lim_{x \to x_0} f(x) = \frac{\lim_{x \to x_0} \dfrac{d}{dx} g(x)}{\lim_{x \to x_0} \dfrac{d}{dx} h(x)} \qquad \text{eq. 29.51}$$

In the $\sin(x)/x$ example above this leads to

$$\lim_{x \to x_0} \frac{\sin(x)}{x} = \frac{\lim_{x \to x_0} \dfrac{d}{dx} \sin(x)}{\lim_{x \to x_0} \dfrac{d}{dx}(x)} = \frac{\lim_{x \to x_0} \cos(x)}{\lim_{x \to x_0} 1} \qquad \text{eq. 29.52}$$

The cosine function approaches unity for $x \rightarrow 0$.

## 29.9  VECTORS

A vector is a quantity with a defined magnitude and direction, for example velocity. In the three-dimensional Euclidian space that we live in, three cartesian coordinates describe a point. The coordinates can be viewed as the tip of a vector with a magnitude that is the shortest distance between the point $(x,y,z)$ and the origin of the coordinate system. The vector $\boldsymbol{u}$ then points from the origin $(0, 0, 0)$ to $(x,y,z)$. Using velocity as an example, the three coordinates are the component movements along the $x$-, $y$-, and $z$-directions that contribute to the overall movement. In molecular dynamics calculations, a vector sum (the resulting force) is calculated for all individual forces acting simultaneously on an atom due to its interactions with other atoms. Vectors are usually written in boldface to distinguish them from numbers.

$$\boldsymbol{u} = (x, y, z) = \begin{pmatrix} x \\ y \\ z \end{pmatrix} = \begin{pmatrix} u_1 \\ u_2 \\ u_3 \end{pmatrix} = u_1 \boldsymbol{i} + u_2 \boldsymbol{j} + u_3 \boldsymbol{k} \qquad \text{eq. 29.53}$$

The components $u_i$ of the vector can be multiplied by the orthogonal unit vectors $\boldsymbol{i}$, $\boldsymbol{j}$, and $\boldsymbol{k}$ along the x, y, and z axes to write the vector $\boldsymbol{u}$ as a *vector sum*. It does not matter if we denote the coordinates of the vector horizontally or vertically, or with subscripts, as long as we use a consistent notation. Vectors can be translated without changing their properties, which is useful to illustrate vector addition and subtraction (and is used in the Patterson map; Figure 22.20). Vectors are added and subtracted from one another by operating on the individual elements:

$$\boldsymbol{u} - \boldsymbol{v} = \begin{pmatrix} u_1 \\ u_2 \\ u_3 \end{pmatrix} - \begin{pmatrix} v_1 \\ v_2 \\ v_3 \end{pmatrix} = \begin{pmatrix} u_1 - v_1 \\ u_2 - v_2 \\ u_3 - v_3 \end{pmatrix}$$ 

eq. 29.54

In geometric terms, vector addition results in another vector (Figure 29.4).

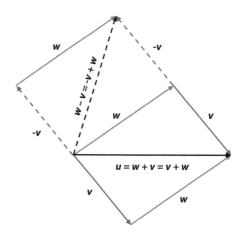

**FIGURE 29.4: Vector addition and subtraction.** Vectors in opposite directions are drawn as dashed lines. The operations are commutative, i.e. the order of the vectors for addition and subtraction is irrelevant.

The magnitude of a vector is given by the Pythagoras formula

$$|\boldsymbol{u}| = \|\boldsymbol{u}\| = \sqrt{u_1^2 + u_2^2 + u_3^2}$$ 

eq. 29.55

The magnitude of vectors is sometimes stated between double vertical bars to distinguish it from an absolute value of a number.

### 29.9.1 DOT PRODUCT

Angles between two vectors, e.g. $\boldsymbol{u}$ and $\boldsymbol{r}$, are calculated using the dot product ($\boldsymbol{u} \cdot \boldsymbol{r}$), which is also known as the *scalar product* because the result of the operation is a scalar. The dot product is defined as the product of the vector magnitudes and the cosine of their angle:

$$(\boldsymbol{u} \cdot \boldsymbol{v}) = u_1 v_1 + u_2 v_2 + u_3 v_3 = |\boldsymbol{u}||\boldsymbol{v}| \cos\theta$$ 

eq. 29.56

### 29.9.2 CROSS PRODUCT

The result of the cross product or *vector product* is another vector. This vector ($\boldsymbol{u} \times \boldsymbol{v}$) is perpendicular to the plane that contains the original vectors $\boldsymbol{u}$ and $\boldsymbol{v}$, and its magnitude is the area of the parallelogram that is spanned by $\boldsymbol{u}$ and $\boldsymbol{v}$:

$$(\boldsymbol{u} \times \boldsymbol{v}) = (u_2 v_3 - v_2 u_3)\boldsymbol{i} + (u_1 v_3 - v_1 u_3)\boldsymbol{j} + (u_1 v_2 - v_1 u_2)\boldsymbol{k}$$ 

eq. 29.57

The three vectors form a right-handed system, which means that the cross product is not commutative, or ($\boldsymbol{u} \times \boldsymbol{v}$) ≠ ($\boldsymbol{v} \times \boldsymbol{u}$). Swapping the order of the vectors in a cross product results in a vector of opposite direction.

A *scalar triple product* $\boldsymbol{u} \cdot (\boldsymbol{v} \cdot \boldsymbol{w})$ is the scalar product of $\boldsymbol{u}$ with the vector product of $\boldsymbol{v}$ and $\boldsymbol{w}$. It is the volume of the parallelepiped spanned by the three vectors, e.g. the unit cell volume of a crystal.

## 29.10  COMPLEX NUMBERS

A complex number has two parts, termed *real* (*a*) and *imaginary* (*i · b*) (eq. 29.58). These terms are just naming conventions for two properties of a quantity that need to be separated.

$$z = a + i \cdot b$$

$$i = \sqrt{-1}$$

eq. 29.58

$$i^2 = -1$$

The separator is *i*, the square root of –1. Complex numbers allow equations to be solved that have no real solutions, e.g. $z^2 - 2z + 2$, which has the complex solutions $z_{1/2} = 1 \pm i$. A graphical representation of complex numbers is the *Argand diagram* or *complex plane* (Figure 29.5).

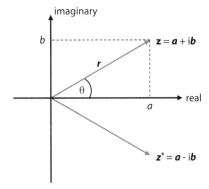

**FIGURE 29.5:    Argand diagram.** Two equivalent representations of *z* are possible, one using Cartesian notation *a* and *b*, the other polar notation *r* and θ, where *r* is the magnitude and θ is the phase angle. If θ is a function of time *t*, the arrow will rotate at a frequency ω*t* in the *ab*-plane. *z* and *z*\* are complex conjugates.

The components of the point *z* can also be expressed by polar coordinates using the magnitude *r* and the angle θ:

$$a = r \cdot \cos \theta$$
$$b = r \cdot \sin \theta$$

eq. 29.59

This leads to

$$z = a + i \cdot b = r \cdot (\cos \theta + i \cdot \sin \theta)$$

eq. 29.60

With the help of the Euler formula eq. 29.61, complex numbers in the polar coordinate notation can be written in a more compact form that is also mathematically simpler to process:

$$e^{i\theta} = \cos \theta + i \cdot \sin \theta$$

eq. 29.61

Substituting –θ for θ in eq. 29.61 gives an equivalent Euler formula:

$$e^{-i\theta} = \cos \theta - i \cdot \sin \theta$$

eq. 29.62

For the special case of $\theta = 2\pi$, the complex number becomes real:

$$e^{\pm i \cdot 2\pi} = \cos 2\pi \pm i \cdot \sin 2\pi = 1 \qquad \text{eq. 29.63}$$

Similarly:

$$e^{\pm i \cdot \pi} = -1$$

$$e^{\pm i \cdot \frac{\pi}{2}} = \pm i \qquad \text{eq. 29.64}$$

Adding and subtracting equations eq. 29.61 and eq. 29.62 re-defines the sine and cosine functions as complex exponentials:

$$\cos\theta = \frac{e^{i\theta} + e^{-i\theta}}{2}$$

$$\sin\theta = \frac{e^{i\theta} - e^{-i\theta}}{2i} \qquad \text{eq. 29.65}$$

Two complex numbers that just differ by the sign of the complex part are *complex conjugates*, termed $z^*$ (Figure 29.5). In other words, the complex conjugate can be found by replacing all instances of $i$ with $-i$:

$$z^* = (a + i \cdot b)^* = (a - i \cdot b)$$

$$z^* = r \cdot (\cos\theta - i \cdot \sin\theta) \qquad \text{eq. 29.66}$$

$$z^* = r \cdot e^{-i\theta}$$

The square root of the product of a complex number with its conjugate is the *magnitude, absolute value,* or *modulus* of the complex number:

$$|z| = \sqrt{z \cdot z^*}$$

$$|z| = \sqrt{(a + i \cdot b) \cdot (a - i \cdot b)} = \sqrt{a^2 + b^2} = r \qquad \text{eq. 29.67}$$

The same result of $|z| = r$ is obtained for the other two expressions of the complex conjugate in eq. 29.66. The magnitude of a complex number is equivalent to $r$ in the Argand diagram (Figure 29.5).

Arithmetic operations are carried out similar to vectors. For two complex numbers $x = a+ib$ and $y = c+id$

$$x + y = (a + c) + i \cdot (b + d)$$

$$x - y = (a - c) + i \cdot (b - d)$$

$$x \cdot y = (ac - bd) + i \cdot (ad + bc) \qquad \text{eq. 29.68}$$

$$\frac{x}{y} = \frac{a + ib}{c + id} = \frac{a + ib}{c + id} \cdot \frac{c - id}{c - id} = \frac{(ac - bd)}{c^2 + d^2} + i \cdot \frac{(ad + bc)}{c^2 + d^2}$$

When dividing complex numbers, the complex conjugate is used to make the denominator real. Multiplication and division of complex numbers is easier in their exponential form because the standard rules described for exponentials in Section 29.5 apply:

$$r_1 \cdot e^{i\theta_1} \cdot r_2 \cdot e^{i\theta_2} = r_1 \cdot r_2 \cdot e^{i(\theta_1 + \theta_2)}$$

$$\frac{r_1 \cdot e^{i\theta_1}}{r_2 \cdot e^{i\theta_2}} = \frac{r_1}{r_2} \cdot e^{i(\theta_1 - \theta_2)}$$

eq. 29.69

The Euler formula describes a periodic function with a repeat of $i \cdot 2\pi$. This can be understood by taking $\theta_2 = 2\pi$, as in eq. 29.63 above, and adding this angle to $\theta_1$:

$$e^{i\theta_1} \cdot e^{i \cdot 2\pi} = e^{i(\theta_1 + 2\pi)} = e^{i\theta_1}$$

eq. 29.70

The ease of angle addition, which equals multiplication of complex exponentials, is one of the reasons why harmonic wave functions are often expressed as complex exponentials (see Sections 22.5 and 22.7) rather than sine and cosine waves. Both the real ($r \cdot \cos\theta$) and imaginary ($r \cdot \sin\theta$) parts can be chosen to describe a harmonic wave function. In practice, calculations are done with the complex form, and later the real part is extracted from the solution to represent the actual wave. In addition, the observable quantity e.g. in scattering and diffraction is the intensity, which is proportional to the square of the amplitude (modulus or absolute value) of the light wave, which is always a real number.

## 29.11 BASIC ELEMENTS OF STATISTICS

Statistics provides the tools to analyze numerical data. One of them is the *arithmetic mean* $\langle x \rangle$ of a set of $N$ data values $x_i$, which is defined as the sum of all values divided by the total number of values:

$$\langle x \rangle = \bar{x} = \frac{1}{N} \cdot \sum_{i=1}^{N} x_i$$

eq. 29.71

The arithmetic mean can also be written with an overstrike as $\bar{x}$. A sometimes more robust estimate for the center of the dataset is the *median* value, which is the number half way along a sorted list of data points. The median is less influenced by extreme outliers than the mean. For instance, the set (1,2,3,4,100) has a mean of 22, which is larger than most data points, and a median of 3, which is closer to most data points. In this case, an outlier rejection would be appropriate. A measure of how closely the data points cluster around the mean is the *variance* $\sigma^2$, which is the average of the squared deviations from the mean:

$$\sigma^2 = \frac{1}{N} \cdot \sum_{i=1}^{N} (x_i - \bar{x})^2$$

eq. 29.72

The positive square root of the variance is the *standard deviation* $\sigma$:

$$\sigma = \sqrt{\frac{1}{N} \cdot \sum_{i=1}^{N} (x_i - \bar{x})^2}$$

eq. 29.73

Note that $N$ in eq. 29.73 changes to $(N - 1)$ if the set of data values is only a fraction of the whole dataset and thus serves to *estimate* the standard deviation for the whole dataset. The data values frequently have a *normal distribution* or *Gaussian distribution* (Figure 29.6), which follows the general formula:

$$f(x) = a \cdot e^{-\frac{x^2}{b}}$$

eq. 29.74

Specifically, the probability density distribution is defined as

$$f(x) = \frac{1}{\sigma\sqrt{2\pi}} \cdot e^{-\frac{(x-\bar{x})^2}{2\sigma^2}}$$

eq. 29.75

When integrated from $-\infty$ to $+\infty$ this function has an area of 1. The shape of the distribution is strongly influenced by the variance. Large variances (values scattered widely around the mean) give a broad distribution while small variances give a narrow distribution.

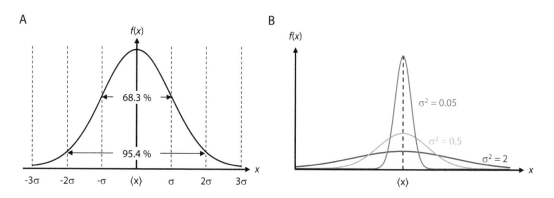

**FIGURE 29.6: Normal distribution.** A: 68.3% of all data points are within $\langle x \rangle \pm \sigma$ and 95.4% are within $\langle x \rangle \pm 2\sigma$. B: The functions have the same mean and the same area (they are normalized) but different variances. Large values of $\sigma^2$ give a broader distribution.

For an array of data, the *correlation coefficient CC* can establish if a correlation exists, and if so, how good it is. A two-dimensional array of $(x,y)$ number pairs has a *CC* of

$$CC = \frac{\sum_{i=1}^{N}(x_i - \bar{x})(y_i - \bar{y})}{\sqrt{\sum_{i=1}^{N}(x_i - \bar{x})^2 \cdot \sum_{i=1}^{N}(y_i - \bar{y})^2}}$$

eq. 29.76

*CC* values range from $-1 \leq CC \leq 1$. While for $CC = 0$ there is no correlation, positive and negative values for *CC* mean correlation and anti-correlation, respectively, between the values in the pairs $(x,y)$.

## 29.12 ERROR PROPAGATION

For a function $x = f(a, b, c,...)$ that depends on the variables $a, b, c...$ with the associated errors $\sigma_a$, $\sigma_b$, $\sigma_c$..., the overall error $\sigma_x$ is defined as

$$\sigma_x = \sqrt{\sigma_a^2 \cdot \left(\frac{\partial x}{\partial a}\right)^2 + \sigma_b^2 \cdot \left(\frac{\partial x}{\partial b}\right)^2 + \sigma_c^2 \cdot \left(\frac{\partial x}{\partial c}\right)^2 + \cdots}$$

eq. 29.77

For example, if we have a series of intensity measurements $I$ with associated errors $\sigma_I$ and would like to convert the $I$ into $\ln(I)$, the error is propagated to

$$\sigma_{I,\text{new}} = \sqrt{\sigma_I^2 \cdot \left(\frac{\partial \ln(I)}{\partial I}\right)^2} = \sqrt{\sigma_I^2 \cdot \left(\frac{1}{I}\right)^2} = \frac{\sigma_I}{I}$$

eq. 29.78

## 29.13 SERIES EXPANSION

### 29.13.1 TAYLOR SERIES

*Taylor series* expansion is used to approximate a continuous function $f(x)$ about a point $(x_0 \lfloor f(x_0))$. The function must be infinitely differentiable and single-valued.

$$f(x) = \sum_{n=0}^{\infty} \frac{f^n(x_0)}{n!} \cdot (x - x_0)^n$$

$$= f(x_0) + f'(x_0) \cdot (x - x_0) + \frac{f''(x_0)}{2!} \cdot (x - x_0)^2 + \frac{f'''(x_0)}{3!} \cdot (x - x_0)^3 + \cdots$$

eq. 29.79

If we stop the series expansion after $n = 1$ we obtain a linear function, which is the tangent to the function $f(x)$ at the point $(x_0 \lfloor f(x_0))$. While the tangent is not a good approximation for the function, the parabolic function obtained after the third term is often already good enough and is used in some energy minimization algorithms. The more terms we include in the expansion, the better will the power series describe the function $f(x)$ also far away from $x_0$. For $x_0 = 0$ the Taylor series is known as a *Maclaurin series*. For instance, the Maclaurin series for the exponential function $f(x) = e^x$ is

$$e^x = \sum_{n=0}^{\infty} \frac{x^n}{n!} = 1 + x + \frac{x^2}{2} + \frac{x^3}{3!} + \frac{x^4}{4!} + \cdots$$

eq. 29.80

and the sine function expands to

$$\sin x = \sum_{n=0}^{\infty} \frac{(-1)^n \cdot x^{2n+1}}{(2n+1)!} = x - \frac{x^3}{3!} + \frac{x^5}{5!} - \cdots$$

eq. 29.81

In order for the series to converge, $|x| < 1$. A number of trigonometric relations can be verified using the Maclaurin series:

$$\sin^2 x + \cos^2 x = 1$$

$$\cos(x_1 + x_2) = \cos x_1 \cdot \cos x_2 - \sin x_1 \cdot \sin x_2$$

eq. 29.82

$$\cos x_1 + \cos x_2 = 2 \cdot \cos \frac{x_1 + x_2}{2} \cdot \cos \frac{x_1 - x_2}{2}$$

The Euler formula relates the exponential function to the sine and cosine functions, as can be shown by Maclaurin expansion. This form of the exponential function is used to evaluate Fourier transformations for electron density calculations (Section 22.5):

$$e^{ix} = \cos x + i \cdot \sin x$$

$$e^{-ix} = \cos x - i \cdot \sin x$$

eq. 29.83

### 29.13.2 FOURIER SERIES

In 1807 the French physicist Joseph Fourier introduced the concept that every function $f(x)$ of periodicity $2L$ (frequency $2\pi \nu n_0 = \omega_0 = \pi/L$) can be expanded into an infinite series of sine and cosine waves of frequency $\omega = n\pi/L$, now called its *Fourier series*:

$$f(x) = \frac{1}{2}a_o + \sum_{n=1}^{\infty} a_n \cos\frac{n\pi x}{L} + \sum_{n=1}^{\infty} b_n \sin\frac{n\pi x}{L} \qquad \text{eq. 29.84}$$

The coefficients of the series are

$$a_0 = \frac{1}{L}\int_0^{2L} f(x)\,\mathrm{d}x$$

$$a_n = \frac{1}{L}\int_0^{2L} f(x)\cos\frac{n\pi x}{L}\,\mathrm{d}x \qquad \text{eq. 29.85}$$

$$b_n = \frac{1}{L}\int_0^{2L} f(x)\sin\frac{n\pi x}{L}\,\mathrm{d}x$$

Using the Euler formula, the Fourier series can also be written in complex form

$$f(x) = \sum_{n=0}^{\infty} c_n e^{in\omega_0 x}\mathrm{d}x \qquad \text{eq. 29.86}$$

with the Fourier coefficients $c_n$:

$$c_n = \frac{1}{L}\int_0^{2L} f(x)e^{in\omega_0 x}\mathrm{d}x \qquad \text{eq. 29.87}$$

To illustrate the application of the Fourier series, a *square wave* of length $2L$ in the time domain that we would like to approximate is plotted in Figure 29.7. Its Fourier series expansion has only sine terms:

$$\frac{4}{\pi}\cdot\sum_{n=1,3,5\ldots}^{\infty}\frac{1}{n}\cdot\sin\frac{n\pi t}{L} \qquad \text{eq. 29.88}$$

Since the number $n\pi/L$ corresponds to the frequency of the sine term $n$, later terms in the series correspond to higher frequencies and thus higher resolutions.

From Figure 29.7, we see that more terms better describe the function. In practice, as many terms as are needed or can be computed in a certain amount of time are used for such calculations. It can also be seen that near the corners of the square function a "ringing" or spike remains in the Fourier series that will not go away with larger numbers of terms. This *Gibbs phenomenon* is an overshoot of the Fourier series typical near points of discontinuity and can lead to artifacts in electron density maps. More worrysome, however, are errors in phases (see Section 22.8). Reversal of the phase in just a single term results in strong changes of the whole Fourier series. As expected, phase errors in early terms ($n = 3$ or $n = 4$ in Figure 29.7) have a stronger effect than phase errors for higher $n$. Still, phase reversal of a later term ($n = 14$ out of 16 total terms in Figure 29.7) has a noticeable effect on the shape of the series.

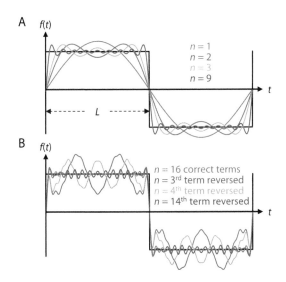

**FIGURE 29.7:    Fourier series.** The square wave is shown in black. A: The first approximation is a sine wave (blue) corresponding to only one term ($n = 1$). More sine wave terms are summed in the other functions, and in the progress ever better approximate the square function. B: Effect of wrong phase on the Fourier series. The blue curve is the Fourier series with 16 correct terms. Phase reversal of a single term, from $\sin(nx)$ to $\sin(-nx) = -\sin(nx)$, strongly affects the Fourier series.

## 29.14  FOURIER TRANSFORMATION

The Fourier transform (FT) is an extension of the Fourier series concept to include non-periodic functions (Figure 29.8). As the period of the function $f(t)$ increases, an increasing number of frequency terms is needed to describe it. For periods approaching infinity, where the function is essentially non-periodic, the frequency spectrum is continuous, i.e. has an infinite number of frequency terms.

The relationship between a time-domain function $f(t)$ and its frequency domain function $F(v)$ is

$$f(t) = \int_{-\infty}^{\infty} F(v)e^{2\pi i v t}\,dv \qquad\qquad \text{eq. 29.89}$$

and the inverse Fourier transform is

$$F(v) = \int_{-\infty}^{\infty} f(t)e^{-2\pi i v t}\,dt \qquad\qquad \text{eq. 29.90}$$

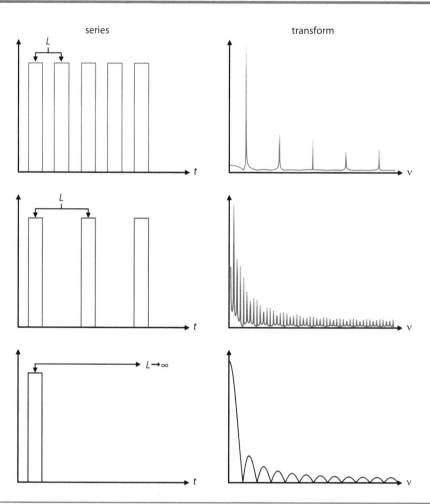

**FIGURE 29.8:**   **Fourier series and Fourier transforms.** Periodic functions (left) and their Fourier transforms (right). The smaller the period, the less peaks there are in the FT and the farther apart they are spaced. For infinite periods, the FT spectrum is continuous.

Since $f(t)$ is usually not continuous but a list of $M$ measurement values taken at small regular time intervals $\delta t$, a *discrete Fourier transform* is needed that uses summation rather than integration.

$$F\left(k\delta\nu\right) = \delta t \sum_{n=0}^{M-1} f\left(n\delta t\right) e^{-\frac{2\pi i n k}{M}}$$

eq. 29.91

The function $f(n\delta t)$ expresses the experimental values and the result is a set of $M$ frequency values $F(k\delta\nu)$ or *Fourier coefficients*. These coefficients are separated by $\delta\nu = 1/M\delta t$, showing the reciprocal relationship established by the FT. To calculate each $F(k\delta\nu)$, eq. 29.91 has to be evaluated for each value of $k$, so the computational time needed for FT scales with $M^2$, a severe limitation for large datasets (diffraction data, Sections 22.2, 22.5; NMR FIDs, Section 20.1.4). The introduction of the fast Fourier transform (FFT) algorithm, which scales with $M\cdot\ln M$, has therefore had a huge impact on scientific computation. The inverse operation from frequencies to time data using discrete FT is analogous:

$$f\left(n\delta t\right) = \frac{1}{M} \sum_{k=0}^{M-1} F\left(k\delta\nu\right) e^{\frac{2\pi i n k}{M}}$$

eq. 29.92

Figure 29.9 shows two classical Fourier transforms that are often encountered in biophysics. The FT of a square function (just one period of a square wave) is of the form $\sin(\nu)/\nu$. Square functions are, for example, light pulses (Figure 20.15) or diffraction slits. The measured responses oscillate with a frequency that is inversely correlated to the pulse length or the slit width. Another example is the exponentially dampened oscillation such as the FID in NMR spectroscopy (Figure 20.16). The FT of this function is a *Lorentzian function* that has both real and imaginary components. The real part is the *absorption spectrum* and the imaginary part is the *dispersion spectrum*. The line width of the Lorentzian function is proportional to the dampening constant of the exponential decay, i.e. the faster the decay of the signal, the broader the absorption and dispersion spectra.

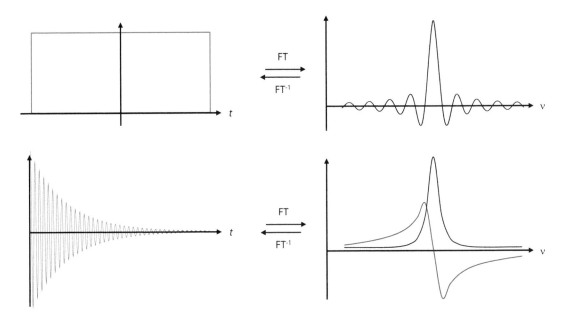

**FIGURE 29.9: Example Fourier transforms.** The top is the FT of a square function (pulse) in the time domain that transforms into frequency. For light diffraction at a slit, the slit width $x$ will transform into a frequency spectrum with the distance between the diffraction maxima proportional to $1/x$. On the bottom, the FT of an exponentially dampend oscillation yields a Lorentzian function with real (black) and imaginary (red) spectral components.

## 29.15 CONVOLUTION

A convolution (sometimes called German: Faltung) is the integral that reports the amount of overlap of two functions as one function, $g(t)$, is shifted across the other function, $f(t)$:

$$f(t) \otimes g(t) = \int_{-\infty}^{\infty} f(\tau) \cdot g(t-\tau) d\tau = \int_{-\infty}^{\infty} g(\tau) \cdot f(t-\tau) d\tau \qquad \text{eq. 29.93}$$

The convolution therefore blends one function with another function. Its operator symbol is the tensor product $\otimes$. The connection of convolution to Fourier transformation is the *convolution theorem*, which states that the FT of a convolution of functions $f(t)$ and $g(t)$ is the product of the individual FTs $F(v)$ and $G(v)$, and *vice versa*, that the FT of the product of two functions is the convolution of their FTs:

$$FT\big(f(t) \otimes g(t)\big) = F(v) \cdot G(v)$$

$$FT\big(f(t) \cdot g(t)\big) = F(v) \otimes G(v) \qquad \text{eq. 29.94}$$

Thus, convolution in the time domain, for example in the measurement of fluorescence lifetimes (Section 19.5.9.1), is equivalent to pointwise multiplication in the frequency domain. An important property of the convolution is that the convolution of two Gaussians is another Gaussian. Convolution is also a central aspect in crystallography. Mathematically speaking, a crystal is the convolution of an object (the content of the unit cell) with a crystal lattice (a symmetry function). The diffraction pattern of a crystal is therefore the product of the diffraction patterns of the lattice (defining the h,k,l triplets) and the unit cell (defining the intensities). The *deconvolution* of the observed diffraction pattern separates the symmetry aspect from the unit cell content.

# Index